Pearson New International Edition

Essentials of Genetics

Klug Cummings Spencer Palladino
Eighth Edition

PEARSON®

Pearson Education Limited
Edinburgh Gate
Harlow
Essex CM20 2JE
England and Associated Companies throughout the world

Visit us on the World Wide Web at: www.pearsoned.co.uk

 ISBN 10: 1-292-03922-1
ISBN 13: 978-1-292-03922-0

British Library Cataloguing-in-Publication Data
A catalogue record for this book is available from the British Library

Printed in the United States of America

Table of Contents

Glossary
William S. Klug/Michael R. Cummings/Charlotte A. Spencer/Michael A. Palladino **1**

1. Introduction to Genetics
William S. Klug/Michael R. Cummings/Charlotte A. Spencer/Michael A. Palladino **15**

2. Mitosis and Meiosis
William S. Klug/Michael R. Cummings/Charlotte A. Spencer/Michael A. Palladino **29**

3. Mendelian Genetics
William S. Klug/Michael R. Cummings/Charlotte A. Spencer/Michael A. Palladino **51**

4. Modification of Mendelian Ratios
William S. Klug/Michael R. Cummings/Charlotte A. Spencer/Michael A. Palladino **79**

5. Sex Determination and Sex Chromosomes
William S. Klug/Michael R. Cummings/Charlotte A. Spencer/Michael A. Palladino **115**

6. Chromosome Mutations: Variation in Number and Arrangement
William S. Klug/Michael R. Cummings/Charlotte A. Spencer/Michael A. Palladino **135**

7. Linkage and Chromosome Mapping in Eukaryotes
William S. Klug/Michael R. Cummings/Charlotte A. Spencer/Michael A. Palladino **159**

8. Genetic Analysis and Mapping in Bacteria and Bacteriophages
William S. Klug/Michael R. Cummings/Charlotte A. Spencer/Michael A. Palladino **185**

9. DNA Structure and Analysis
William S. Klug/Michael R. Cummings/Charlotte A. Spencer/Michael A. Palladino **207**

10. DNA Replication and Recombination
William S. Klug/Michael R. Cummings/Charlotte A. Spencer/Michael A. Palladino **231**

11. Chromosome Structure and DNA Sequence Organization
William S. Klug/Michael R. Cummings/Charlotte A. Spencer/Michael A. Palladino **253**

12. The Genetic Code and Transcription
William S. Klug/Michael R. Cummings/Charlotte A. Spencer/Michael A. Palladino **271**

13. Translation and Proteins
William S. Klug/Michael R. Cummings/Charlotte A. Spencer/Michael A. Palladino   **297**

14. Gene Mutation, DNA Repair, and Transposition
William S. Klug/Michael R. Cummings/Charlotte A. Spencer/Michael A. Palladino   **319**

15. Regulation of Gene Expression
William S. Klug/Michael R. Cummings/Charlotte A. Spencer/Michael A. Palladino   **343**

16. The Genetics of Cancer
William S. Klug/Michael R. Cummings/Charlotte A. Spencer/Michael A. Palladino   **373**

17. Recombinant DNA Technology
William S. Klug/Michael R. Cummings/Charlotte A. Spencer/Michael A. Palladino   **391**

18. Genomics, Bioinformatics, and Proteomics
William S. Klug/Michael R. Cummings/Charlotte A. Spencer/Michael A. Palladino   **415**

19. Applications and Ethics of Genetic Engineering and Biotechnology
William S. Klug/Michael R. Cummings/Charlotte A. Spencer/Michael A. Palladino   **449**

20. Developmental Genetics
William S. Klug/Michael R. Cummings/Charlotte A. Spencer/Michael A. Palladino   **479**

21. Quantitative Genetics and Multifactorial Traits
William S. Klug/Michael R. Cummings/Charlotte A. Spencer/Michael A. Palladino   **499**

22. Population and Evolutionary Genetics
William S. Klug/Michael R. Cummings/Charlotte A. Spencer/Michael A. Palladino   **521**

23. Special Topics in Modern Genetics: Epigenetics
William S. Klug/Michael R. Cummings/Charlotte A. Spencer/Michael A. Palladino   **547**

24. Special Topics in Modern Genetics: DNA Forensics
William S. Klug/Michael R. Cummings/Charlotte A. Spencer/Michael A. Palladino   **557**

25. Special Topics in Modern Genetics: Genomics and Personalized Medicine
William S. Klug/Michael R. Cummings/Charlotte A. Spencer/Michael A. Palladino   **569**

Index   **581**

Glossary

abortive transduction An event in which transducing DNA fails to be incorporated into the recipient chromosome.

accession number An identifying number or code assigned to a nucleotide or amino acid sequence for entry and cataloging in a database.

acentric chromosome Chromosome or chromosome fragment with no centromere.

acridine dyes A class of organic compounds that bind to DNA and intercalate into the double-stranded structure, producing local disruptions of base pairing. These disruptions result in nucleotide additions or deletions in the next round of replication.

acrocentric chromosome Chromosome with the centromere located very close to one end. Human chromosomes 13, 14, 15, 21, and 22 are acrocentric.

additive variance Genetic variance attributed to the substitution of one allele for another at a given locus. This variance can be used to predict the rate of response to phenotypic selection in quantitative traits.

allele One of the possible alternative forms of a gene, often distinguished from other alleles by phenotypic effects.

allele-specific oligonucleotide (ASO) Synthetic nucleotides, usually 15–20 bp in length, that under carefully controlled conditions will hybridize only to a perfectly matching complementary sequence.

allopatric speciation Process of speciation associated with geographic isolation.

allopolyploid Polyploid condition formed by the union of two or more distinct chromosome sets with a subsequent doubling of chromosome number.

allotetraploid An allopolyploid containing two genomes derived from different species.

allozyme An allelic form of a protein that can be distinguished from other forms by electrophoresis.

alternative splicing Generation of different protein molecules from the same pre-mRNA by incorporation of a different set and order of exons into the mRNA product.

Alu sequence A DNA sequence of approximately 300 bp found interspersed within the genomes of primates that is cleaved by the restriction enzyme *Alu* I. In humans, 300,000–600,000 copies are dispersed throughout the genome and constitute some 3–6 percent of the genome. See *short interspersed elements*.

Ames test A bacterial assay developed by Bruce Ames to detect mutagenic compounds; it assesses reversion to histidine independence in the bacterium *Salmonella typhimurium*.

aminoacyl tRNA A covalently linked combination of an amino acid and a tRNA molecule. Also referred to as a charged tRNA.

amniocentesis A procedure in which fluid and fetal cells are withdrawn from the amniotic layer surrounding the fetus; used for genetic testing of the fetus.

aneuploidy A condition in which the chromosome number is not an exact multiple of the haploid set.

annotation Analysis of genomic nucleotide sequence data to identify the protein-coding genes, the nonprotein-coding genes, and the regulatory sequences and function(s) of each gene.

anticodon In a tRNA molecule, the nucleotide triplet that binds to its complementary codon triplet in an mRNA molecule.

antiparallel A term describing molecules in parallel alignment but running in opposite directions. Most commonly used to describe the opposite orientations of the two strands of a DNA molecule.

antisense oligonucleotide A short, single-stranded DNA or RNA molecule complementary to a specific sequence.

antisense RNA An RNA molecule (synthesized *in vivo* or *in vitro*) with a ribonucleotide sequence that is complementary to part of an mRNA molecule.

apoptosis A genetically controlled program of cell death, activated as part of normal development or as a result of cell damage.

artificial selection See *selection*.

ascospore A meiotic spore produced in certain fungi.

ascus In fungi, the sac enclosing the four or eight ascospores.

attached-X chromosome Two conjoined X chromosomes that share a single centromere and thus migrate together during cell division.

attenuator A nucleotide sequence between the promoter and the structural gene of some bacterial operons that regulates the transit of RNA polymerase, reducing transcription of the neighboring structural gene.

autogamy A process of self-fertilization resulting in homozygosis.

autonomously replicating sequences (ARS) Origins of replication, about 100 nucleotides in length, found in yeast chromosomes.

autopolyploidy Polyploid condition resulting from the duplication of one diploid set of chromosomes.

autoradiography Production of a photographic image by radioactive decay. Used to localize radioactively labeled compounds within cells and tissues or to identify radioactive probes in various blotting techniques. See *Southern blotting*.

autosomes Chromosomes other than the sex chromosomes. In humans, there are 22 pairs of autosomes.

autotetraploid An autopolyploid condition composed of four copies of the same genome.

auxotroph A mutant microorganism or cell line that requires the addition of a nutritional substance for growth. Wild-type strains can synthesize this substance, and do not require it added for growth.

backcross A cross between an F_1 heterozygote and one of the P_1 parents (or an organism with a genotype identical to one of the parents).

bacteriophage A virus that infects bacteria, using it as the host for reproduction (also, *phage*).

balanced lethals Recessive, nonallelic lethal genes, each carried on different homologous chromosomes. When organisms carrying balanced lethal genes are interbred, only organisms with genotypes identical to the parents (heterozygotes) survive.

balanced translocation carrier An individual with a chromosomal translocation in which there has been an exchange of genetic information with no associated extra or missing genetic material.

balancer chromosome A chromosome containing one or more inversions that suppress crossing over with its homolog and which carries a dominant marker that is usually lethal when homozygous.

Barr body Densely staining DNA-positive mass seen in the somatic nuclei of mammalian females. Discovered by Murray

Barr, this body represents an inactivated X chromosome.

base analog A purine or pyrimidine base that differs structurally from one normally used in biological systems but whose chemical behavior is the same.

base substitution A single base change in a DNA molecule that produces a mutation.

bidirectional replication A mechanism of DNA replication in which two replication forks move in opposite directions from a common origin.

bioinformatics A field that focuses on the design and use of software and computational methods for the storage, analysis, and management of biological information such as nucleotide or amino acid sequences.

biometry The application of statistics and statistical methods to biological problems.

biotechnology Commercial and/or industrial processes that utilize biological organisms or products.

bivalents Synapsed homologous chromosomes in the first prophase of meiosis.

broad heritability That proportion of total phenotypic variance in a population that can be attributed to genotypic variance.

CAAT box A highly conserved DNA sequence found in the untranslated promoter region of eukaryotic genes. This sequence is recognized by transcription factors.

cancer stem cells Tumor-forming cells in a cancer that can give rise to all the cell types in a particular form of cancer. These cells have the properties of normal stem cells: self-renewal and ability to differentiate into multiple cell types.

capillary electrophoresis A collection of analytical methods that separates large and small charged molecules in a capillary tube by their size to charge ratio. Analysis of separated components takes place in the capillary usually by use of a UV detector.

carrier An individual heterozygous for a recessive trait.

cDNA (complementary DNA) DNA synthesized from an RNA template by the enzyme reverse transcriptase.

cell cycle The sequence of growth phases of an individual cell; divided into G1 (gap 1), S (DNA synthesis), G2 (gap 2), and M (mitosis). Cells that temporarily or permanently withdraw from the cell cycle are said to enter the G0 stage.

CEN The DNA region of centromeres critical to their function. In yeasts, fragments of chromosomal DNA, about 120 bp in length, that when inserted into plasmids confer the ability to segregate during mitosis.

centimorgan (cM) A unit of distance between genes on chromosomes representing 1 percent crossing over between two genes. Equivalent to 1 map unit (m.u.).

central dogma The concept that genetic information flow progresses from DNA to RNA to proteins. Although exceptions are known, this idea is central to an understanding of gene function.

centriole A cytoplasmic organelle composed of nine groups of microtubules, generally arranged in triplets. Centrioles function in the generation of cilia and flagella and serve as foci for the spindles in cell division.

centromere The specialized heterochromatic chromosomal region at which sister chromatids remain attached after replication, and the site to which spindle fibers attach to the chromosome during cell division. Location of the centromere determines the shape of the chromosome during the anaphase portion of cell division. Also known as the primary constriction.

centrosome Region of the cytoplasm containing a pair of centrioles.

chaperone A protein that regulates the folding of a polypeptide into a functional three-dimensional shape.

chiasma (pl., chiasmata) The crossed strands of nonsister chromatids seen in diplotene of the first meiotic division. Regarded as the cytological evidence for exchange of chromosomal material, or crossing over.

chi-square (χ^2) analysis Statistical test to determine whether or not an observed set of data is equivalent to a theoretical expectation.

chorionic villus sampling (CVS) A technique of prenatal diagnosis in which chorionic fetal cells are retrieved intravaginally or transabdominally and used to detect cytogenetic and biochemical defects in the embryo.

chromatid One of the longitudinal subunits of a replicated chromosome.

chromatin The complex of DNA, RNA, histones, and nonhistone proteins that make up uncoiled chromosomes, characteristic of the eukaryotic interphase nucleus.

chromatin immunoprecipitation (ChIP) An analytical method used to identify DNA-binding proteins that bind to DNA sequences of interest. In ChIP, antibodies to specific proteins are used to isolate DNA sequences that bind these proteins.

chromatin remodeling A process in which the structure of chromatin is chemically altered by a protein complex, resulting in changes in the transcriptional state of genes within the altered region.

chromomere A coiled, beadlike region of a chromosome, most easily visualized during cell division. The aligned chromomeres of polytene chromosomes are responsible for their distinctive banding pattern.

chromosomal aberration Any duplication, deletion, or rearrangement of the otherwise diploid chromosomal content of an organism. Sometimes referred to as a chromosomal mutation.

chromosome In prokaryotes, a DNA molecule containing the organism's genome; in eukaryotes, a DNA molecule complexed with proteins and RNA to form a threadlike structure containing genetic information; a structure that is visible during mitosis and meiosis.

chromosome banding Technique for the differential staining of mitotic or meiotic chromosomes to produce a characteristic banding pattern; or selective staining of certain chromosomal regions such as centromeres, the nucleolus organizer regions, and GC- or AT-rich regions. Not to be confused with the banding pattern present in polytene chromosomes, which is produced by the alignment of chromomeres.

chromosome map A diagram showing the location of genes on chromosomes.

chromosome puff A localized uncoiling and swelling in a polytene chromosome, usually regarded as a sign of active transcription.

chromosome theory of inheritance The idea put forward independently by Walter Sutton and Theodore Boveri that chromosomes are the carriers of genes and the basis for the Mendelian mechanisms of segregation and independent assortment.

chromosome walking A method for analyzing long stretches of DNA. The end of a cloned segment of DNA is subcloned and used as a probe to identify other clones that overlap the first clone.

***cis*-acting sequence** A DNA sequence that regulates the expression of a gene located on the same chromosome. This contrasts with a trans-acting element where regulation is under the control of a sequence on the homologous chromosome.

cis configuration The arrangement of two genes (or two mutant sites within a gene) on the same homolog, such as

$$\frac{a^1 \qquad a^2}{+ \qquad +}$$

cis–trans test A genetic test to determine whether two mutations are located within the same cistron (or gene).

cline A gradient of genotype or phenotype distributed over a geographic range.

clone Identical molecules, cells, or organisms derived from a single ancestor by asexual or parasexual methods; for example, a DNA segment that has been inserted into a plasmid or chromosome of a phage or a bacterium and replicated to produce many copies, or an organism with a genetic composition identical to that used in its production.

codominance Condition in which the phenotypic effects of a gene's alleles are fully and simultaneously expressed in the heterozygote.

codon A triplet of nucleotides that specifies a particular amino acid or a start or stop signal in the genetic code. Sixty-one codons specify the amino acids used in proteins, and three codons, called stop codons, signal termination of growth of the polypeptide chain. One codon acts as a start codon in addition to specifying an amino acid.

coefficient of coincidence A ratio of the observed number of double crossovers divided by the expected number of such crossovers.

coefficient of inbreeding The probability that two alleles present in a zygote are descended from a common ancestor.

coefficient of selection (s) A measurement of the reproductive disadvantage of a given genotype in a population. For example, for genotype *aa* if only 99 of 100 individuals reproduce, then the selection coefficient is 0.01.

cohesin A protein complex that holds sister chromatids together during mitosis and meiosis and facilitates attachments of spindle fibers to kinetochores.

colchicine An alkaloid compound that inhibits spindle formation during cell division. In the preparation of karyotypes, it is used for collecting a large population of cells inhibited at the metaphase stage of mitosis.

colinearity The linear relationship between the nucleotide sequence in a gene (or the RNA transcribed from it) and the order of amino acids in the polypeptide chain specified by the gene.

Combined DNA Index System (CODIS) A standardized set of 13 short tandem repeat (STR) DNA sequences used by law enforcement and government agencies in preparing DNA profiles.

competence In bacteria, the transient state or condition during which the cell can bind and internalize exogenous DNA molecules, making transformation possible.

complementarity Chemical affinity between nitrogenous bases of nucleic acid strands as a result of hydrogen bonding. Responsible for the base pairing between the strands of the DNA double helix and between DNA and RNA strands during genetic expression in cells and during the use of molecular hybridization techniques.

complementation test A genetic test to determine whether two mutations occur within the same gene (or cistron). If two mutations are present in a cell at the same time and produce a wild-type phenotype (i.e., they complement each other), they are often nonallelic. If a mutant phenotype is produced, the mutations are noncomplementing and are often allelic.

complete linkage A condition in which two genes are located so close to each other that no recombination occurs between them.

complex trait A trait whose phenotype is determined by the interaction of multiple genes and environmental factors.

concordance Pairs or groups of individuals with identical phenotypes. In twin studies, a condition in which both twins exhibit or fail to exhibit a trait under investigation.

conditional mutation A mutation expressed only under a certain condition; that is, a wild-type phenotype is expressed under certain (permissive) conditions and a mutant phenotype under other (restrictive) conditions.

conjugation Temporary fusion of two single-celled organisms for the sexual transfer of genetic material.

consanguineous Related by a common ancestor within the previous few generations.

consensus sequence The sequence of nucleotides in DNA or amino acids in proteins most often present in a particular gene or protein under study in a group of organisms.

contig A continuous DNA sequence reconstructed from overlapping DNA sequences derived by cloning or sequence analysis.

continuous variation Phenotype variation in which quantitative traits range from one phenotypic extreme to another in an overlapping or continuous fashion.

copy number variation (CNV) DNA segments larger than 1 kb that are repeated a variable number of times in the genome.

cosmid A vector designed to allow cloning of large segments of foreign DNA. Cosmids are composed of the *cos* sites of phage λ inserted into a plasmid. In cloning, the recombinant DNA molecules are packaged into phage protein coats, and after infection of bacterial cells, the recombinant molecule replicates and can be maintained as a plasmid.

CpG island A short region of regulatory DNA found upstream of genes that contain unmethylated stretches of sequence with a high frequency of C and G nucleotides.

crossing over The exchange of chromosomal material (parts of chromosomal arms) between homologous chromosomes by breakage and reunion. The exchange of material between nonsister chromatids during meiosis is the basis of genetic recombination.

C value The haploid amount of DNA present in a genome.

C value paradox The apparent paradox that there is no relationship between the size of the genome and the evolutionary complexity of species. For example, the C value (haploid genome size) of amphibians varies by a factor of 100.

cytogenetics A branch of biology in which the techniques of both cytology and genetics are used in genetic investigations.

cytokinesis The division or separation of the cytoplasm during mitosis or meiosis.

cytoplasmic inheritance Non-Mendelian form of inheritance in which genetic information is transmitted through the cytoplasm rather than the nucleus, usually by DNA-containing, self-replicating cytoplasmic organelles such as mitochondria and chloroplasts.

dalton (Da) A unit of mass equal to that of the hydrogen atom, which is 1.67×10^{-24} gram. A unit used in designating molecular weights.

degenerate code The representation of a given amino acid by more than one codon.

deletion A chromosomal mutation, also referred to as a deficiency, involving the loss of chromosomal material.

deme A local interbreeding population.

de novo Newly arising; synthesized from less complex precursors rather than being produced by modification of an existing molecule.

density gradient centrifugation A method of separating macromolecular mixtures by the use of centrifugal force and solutions of varying density.

deoxyribonuclease (DNAse) A class of enzymes that breaks down DNA into oligonucleotide fragments by introducing single-stranded or double-stranded breaks into the double helix.

deoxyribonucleic acid (DNA) A macromolecule usually consisting of nucleotide polymers comprising antiparallel chains in which the sugar residues are deoxyribose and which are held together by hydrogen bonds. The primary carrier of genetic information.

determination Establishment of a specific pattern of gene activity and developmental fate for a given cell, usually prior to any manifestation of the cell's future phenotype.

dicentric chromosome A chromosome having two centromeres, which can be pulled in opposite directions during anaphase of cell division.

dideoxynucleotide A nucleotide containing a deoxyribose sugar lacking a 3′ hydroxyl group. It stops further chain elongation when incorporated into a growing polynucleotide and is used in the Sanger method of DNA sequencing.

differentiation The complex process of change by which cells and tissues attain their adult structure and functional capacity.

dihybrid cross A genetic cross involving two characters in which the parents possess different forms of each character (e.g., yellow, round × green, wrinkled peas).

diploid (2n) A condition in which each chromosome exists in pairs; having two of each chromosome.

directed mutagenesis See *gene targeting*.

directional selection A selective force that changes the frequency of an allele in a given direction, either toward fixation (frequency of 100%) or toward elimination (frequency of 0%).

discontinuous replication of DNA The synthesis of DNA in discontinuous fragments on the lagging strand during replication. The fragments, known as Okazaki fragments, are subsequently joined by DNA ligase to form a continuous strand.

discontinuous variation Pattern of variation for a trait whose phenotypes fall into two or more distinct classes.

discordance In twin studies, a situation where one twin expresses a trait but the other does not.

disjunction The separation of chromosomes during the anaphase stage of cell division.

disruptive selection Simultaneous selection for phenotypic extremes in a population, usually resulting in the production of two phenotypically discontinuous strains.

dizygotic twins Twins produced from separate fertilization events; two ova fertilized independently. Also known as fraternal twins.

DNA fingerprinting A molecular method for identifying an individual member of a population or species. A unique pattern of DNA fragments is obtained by restriction enzyme digestion followed by Southern blot hybridization using minisatellite probes.

DNA footprinting A technique for identifying a DNA sequence that binds to a particular protein, based on the idea that the phosphodiester bonds in the region covered by the protein are protected from digestion by deoxyribonucleases.

DNA microarray An ordered arrangement of DNA sequences or oligonucleotides on a substrate (often glass). Microarrays are used in quantitative assays of DNA–DNA or DNA–RNA binding to measure profiles of gene expression (for example, during development or to compare the differences in gene expression between normal and cancer cells).

DNA profiling A method for identification of individuals that uses variations in the length of short tandem repeating DNA sequences (STRs) that are widely distributed in the genome.

dominant negative mutation A mutation whose gene product acts in opposition to the normal gene product, usually by binding to it to form dimers.

dosage compensation A genetic mechanism that equalizes the levels of expression of genes at loci on the X chromosome. In mammals, this is accomplished by random inactivation of one X chromosome, leading to Barr body formation.

double crossover Two separate events of chromosome breakage and exchange occurring within the same tetrad during meiosis.

double helix The model for DNA structure proposed by James Watson and Francis Crick, in which two antiparallel hydrogen-bonded polynucleotide chains are wound into a right-handed helical configuration 2 nm in diameter, with 10 base pairs per full turn.

driver mutation A mutation in a cancer cell that contributes to tumor progression.

duplication A chromosomal aberration in which a segment of the chromosome is repeated.

dyad The products of tetrad separation or disjunction at meiotic prophase I. Each dyad consists of two sister chromatids joined at the centromere.

effective population size In a population, the number of individuals with an equal probability of contributing gametes to the next generation.

electrophoresis A technique that separates a mixture of molecules by their differential migration through a stationary medium (such as a gel) under the influence of an electrical field.

electroporation A technique that uses an electric pulse to move polar molecules across the plasma membrane into the cell.

ELSI (Ethical, Legal, Social Implications) A program established by the National Human Genome Research Institute in 1990 as part of the Human Genome Project to sponsor research on the ethical, legal, and social implications of genomic research and its impact on individuals and social institutions.

embryonic stem cells (ESC) Cells derived from the inner cell mass of early blastocyst mammalian embryos. These cells are pluripotent, meaning they can differentiate into any of the embryonic or adult cell types characteristic of the organism.

endomitosis An increased DNA content in multiples of the haploid amount occurring in the absence of nuclear or cytoplasmic division. Polytene chromosomes are formed by endomitosis.

enhancer A DNA sequence that enhances transcription and the expression of structural genes. Enhancers can act over a distance of thousands of base pairs and can be located upstream, downstream, or internal to the gene they affect, differentiating them from promoters.

epigenesis The idea that an organism or organ arises through the sequential appearance and development of new structures, in contrast to preformationism, which holds that development is the result of the assembly of structures already present in the egg.

epigenetics The study of modifications in an organism's pattern of gene expression or phenotypic expression that are not attributable to alterations in the nucleotide sequence (mutations) of the organism's DNA.

epistasis Nonreciprocal interaction between nonallelic genes such that one gene influences or interferes with the expression of another gene, leading to a specific phenotype.

equational division A division stage where the number of centromeres is not reduced by half.

euchromatin Chromatin or chromosomal regions that are lightly staining and are relatively uncoiled during the interphase portion of the cell cycle. Euchromatic regions contain most of the structural genes.

eugenics A movement advocating the improvement of the human species by selective breeding. Positive eugenics refers to the promotion of breeding between people thought to possess favorable genes, and negative eugenics refers to the discouragement of breeding among those thought to have undesirable traits.

euphenics Medical or genetic intervention to reduce the impact of defective genotypes.

euploid Polyploid with a chromosome number that is an exact multiple of a basic chromosome set.

evolution Descent with modification. The emergence of new kinds of plants and animals from preexisting types.

excision repair Removal of damaged DNA segments followed by repair synthesis with the correct nucleotide sequence. Excision can include the removal of individual bases (base repair) or of a stretch of damaged nucleotides (nucleotide repair).

exon The DNA segments of a gene that contain the sequences that, through transcription and translation, are eventually represented in the amino acid sequence of the final polypeptide product.

expressed sequence tag (EST) All or part of the nucleotide sequence of a cDNA clone. ESTs are used as markers in the construction of genetic maps.

expression vector Plasmids or phages carrying promoter regions designed to cause expression of inserted DNA sequences.

expressivity The degree to which a phenotype for a given trait is expressed.

extranuclear inheritance Transmission of traits by genetic information contained in cytoplasmic organelles such as mitochondria and chloroplasts. Sometimes called *extrachromosomal inheritance*.

F⁻ cell A bacterial cell that does not contain a fertility factor and that acts as a recipient in bacterial conjugation.

F⁺ cell A bacterial cell that contains a fertility factor and that acts as a donor in bacterial conjugation.

F factor An episomal plasmid in bacterial cells that confers the ability to act as a donor in conjugation.

F′ factor A fertility factor that contains a portion of the bacterial chromosome.

F₁ generation First filial generation; the progeny resulting from the first cross in a series.

F₂ generation Second filial generation; the progeny resulting from a cross of the F₁ generation.

F pilus On bacterial cells possessing an F factor, a filament-like projection that plays a role in conjugation.

familial trait A trait transmitted through and expressed by members of a family. Usually used to describe a trait that runs in families, but whose precise mode of inheritance is not clear.

fate map A diagram of an embryo showing the location of cells whose developmental fate is known.

fetal cell sorting A noninvasive method of prenatal diagnosis that recovers and tests fetal cells from the maternal circulation.

filial generations See F_1, F_2 *generations*.

fitness A measure of the relative survival and reproductive success of a given individual or genotype.

fluctuation test A statistical test developed by Salvadore Luria and Max Delbrück demonstrating that bacterial mutations arise spontaneously, in contrast to being induced by selective agents.

fluorescence *in situ* hybridization (FISH) A method of *in situ* hybridization that utilizes probes labeled with a fluorescent tag, causing the site of hybridization to fluoresce when viewed using ultraviolet light.

flush–crash cycle A period of rapid population growth followed by a drastic reduction in population size.

folded-fiber model A model of eukaryotic chromosome organization in which each sister chromatid consists of a single chromatin fiber composed of double-stranded DNA and proteins wound like a tightly coiled skein of yarn.

forensic science The use of laboratory scientific methods to obtain data used in criminal and civil law cases.

forward genetics The classical approach used to identify a gene controlling a phenotypic trait in the absence of knowledge of the gene's location in the genome or its DNA sequence. Accomplished by isolating mutant alleles and mapping the gene's location, most traditionally using recombination analysis. Once mapped, the gene may be cloned and further studied at the molecular level. An approach contrasted with *reverse genetics*.

founder effect A form of genetic drift. The establishment of a population by a small number of individuals whose genotypes carry only a fraction of the alleles in the parental population.

fragile site A heritable gap, or nonstaining region, of a chromosome that can be induced to generate chromosome breaks.

frameshift mutation A mutational event leading to the insertion of one or more base pairs in a gene, shifting the codon reading frame in all codons that follow the mutational site.

G1 checkpoint A point in the G1 phase of the cell cycle when a cell either becomes committed to initiating DNA synthesis and continuing the cycle or withdraws into the G0 resting stage.

G0 A nondividing but metabolically active state that cells may enter from the G1 phase of the cell cycle.

gain-of-function mutation A mutation that produces a phenotype different from that of the normal allele and from any loss-of-function alleles.

gamete A specialized reproductive cell with a haploid number of chromosomes.

gap genes Genes expressed in contiguous domains along the anterior–posterior axis of the *Drosophila* embryo that regulate the process of segmentation in each domain.

GC box In eukaryotes, a region in a promoter containing a 5′-GGGCGG-3′ sequence, which is a binding site for transcriptional regulatory proteins.

gene The fundamental physical unit of heredity, whose existence can be confirmed by allelic variants and which occupies a specific chromosomal locus. A DNA sequence coding for a single polypeptide.

gene amplification The process by which gene sequences are selected and differentially replicated either extrachromosomally or intrachromosomally.

gene conversion The process of nonreciprocal recombination by which one allele in a heterozygote is converted into the corresponding allele.

gene duplication An event leading to the production of a tandem repeat of a gene sequence during replication.

gene family A number of closely related genes derived from a common ancestral gene by duplication and sequence divergence over evolutionary time.

gene flow The gradual exchange of genes between two populations; brought about by the dispersal of gametes or the migration of individuals.

gene interaction Production of novel phenotypes by the interaction of alleles of different genes.

gene knockout The introduction of a *null mutation* into a gene that is subsequently introduced into an organism using transgenic techniques, whereby the organism loses the function of the gene.

gene pool The total of all alleles possessed by the reproductive members of a population.

gene targeting A transgenic technique used to create and introduce a specifically altered gene into an organism. Gene targeting often involves the induction of a specific mutation in a cloned gene that is then introduced into the genome of a gamete involved in fertilization. The organism produced is bred to produce adults homozygous for the mutation, for example, the creation of a *gene knockout*.

genetically modified organism (GMO) A plant or animal whose genome carries a gene transferred from another species by recombinant DNA technology that is expressed to produce a gene product.

genetic anticipation The phenomenon in which the severity of symptoms in genetic disorders increases from generation to generation and the age of onset decreases from generation to generation. It is caused by the expansion of trinucleotide repeats within or near a gene and was first observed in myotonic dystrophy.

genetic background The impact of the collective genome of an organism on the expression of a gene under investigation.

genetic code The deoxynucleotide triplets that encode the 20 amino acids or specify termination of translation.

genetic drift Random variation in allele frequency from generation to generation, most often observed in small populations.

genetic engineering The technique of altering the genetic constitution of cells or individuals by the selective removal, insertion, or modification of individual genes or gene sets.

genetic equilibrium A condition in which allele frequencies in a population are neither increasing nor decreasing.

genetic erosion The loss of genetic diversity from a population or a species.

genetic fine structure analysis Intragenic recombinational analysis that provides intragenic mapping information at the level of individual nucleotides.

genetic load Average number of recessive lethal genes carried in the heterozygous condition by an individual in a population.

genetic polymorphism The stable coexistence of two or more distinct genotypes for a given trait in a population. When the frequencies of two alleles for such a trait are in equilibrium, the condition is called a balanced polymorphism.

Genetic Testing Registry (GTR) A centralized, online database, sponsored by the U.S. government to provide information about the scientific basis and availability of genetic tests.

genetic variability A measure of the tendency of the genotypes of individuals in a population to vary from one another. Genetic variability can be measured by determining the rate of mutation of specific genes.

genetics The branch of biology concerned with study of inherited variation. More specifically, the study of the origin, transmission, expression, and evolution of genetic information.

genome The set of hereditary information encoded in the DNA of an organism, including both the protein-coding and non–protein-coding sequences.

genome-wide association studies (GWAS) Analysis of genetic variation across an entire genome, searching for linkage (associations) between variations in DNA sequences and a genome region encoding a specific phenotype.

genomic imprinting The process by which the expression of an allele depends on whether it has been inherited from a male or a female parent. Also referred to as parental imprinting.

genomic library A collection of clones that contains all the DNA sequences of an organism's genome.

genomics A subdiscipline of the field of genetics generated by the union of classical and molecular biology with the goal of sequencing and understanding genes, gene interaction, genetic elements, as well as the structure and evolution of genomes.

genotype The allelic or genetic constitution of an organism; often, the allelic composition of one or a limited number of genes under investigation.

germ line An embryonic cell lineage that forms the reproductive cells (eggs and sperm).

germ plasm Hereditary material transmitted from generation to generation.

Goldberg–Hogness box A short nucleotide sequence 20–30 bp upstream from the initiation site of eukaryotic genes to which RNA polymerase II binds. The consensus sequence is TATAAAA. Also known as a TATA box.

gynandromorphy An individual composed of cells with both male and female genotypes.

haploid (*n*) A cell or an organism having one member of each pair of homologous chromosomes. Also referred to as the gametic chromosome number.

haploinsufficiency In a diploid organism, a condition in which an individual possesses only one functional copy of a gene with the other inactivated by mutation. The amount of protein produced by the single copy is insufficient to produce a normal phenotype, thus leading to an abnormal phenotype. In humans, this condition is present in many autosomal dominant disorders.

haplotype A set of alleles from closely linked loci carried by an individual inherited as a unit.

HapMap Project An international effort by geneticists to identify haplotypes (closely linked genetic markers on a single chromosome) shared by certain individuals as a way of facilitating efforts to identify, map, and isolate genes associated with disease or disease susceptibility.

Hardy–Weinberg law The principle that genotype frequencies will remain in equilibrium in an infinitely large, randomly mating population in the absence of mutation, migration, and selection.

heat shock A transient genetic response following exposure of cells or organisms to elevated temperatures. The response includes activation of a small number of loci, inactivation of some previously active loci, and selective translation of heat-shock mRNA.

helix–turn–helix (HTH) motif In DNA-binding proteins, the structure of a region in which a turn of four amino acids holds two α helices at right angles to each other.

hemizygous Having a gene present in a single dose in an otherwise diploid cell. Usually applied to genes on the X chromosome in heterogametic males.

heritability A relative measure of the degree to which observed phenotypic differences for a trait are genetic.

heterochromatin The heavily staining, late-replicating regions of chromosomes that are prematurely condensed in interphase.

heteroduplex A double-stranded nucleic acid molecule in which each polynucleotide chain has a different origin. It may be produced as an intermediate in a recombinational event or by the *in vitro* reannealing of single-stranded, complementary molecules.

heterogametic sex The sex that produces gametes containing unlike sex chromosomes. In mammals, the male is the heterogametic sex.

heterokaryon A somatic cell containing nuclei from two different sources.

heterozygote An individual with different alleles at one or more loci. Such individuals will produce unlike gametes and therefore will not breed true.

Hfr Strains of bacteria exhibiting a high frequency of recombination. These strains have a chromosomally integrated F factor that is able to mobilize and transfer part of the chromosome to a recipient F^- cell.

high-throughput DNA sequencing A collection of DNA sequencing methods that outperform the standard (Sanger) method of DNA sequencing by a factor of 100–1000 and reduce sequencing costs by more than 99 percent. Also called *next generation sequencing.*

Holliday structure In DNA recombination, an intermediate seen in transmission electron microscope images as an X-shaped structure showing four single-stranded DNA regions.

homeobox A sequence of about 180 nucleotides that encodes a sequence of 60 amino acids called a *homeodomain,* which is part of a DNA-binding protein that acts as a transcription factor.

homeotic mutation A mutation that causes a tissue normally determined to form a specific organ or body part to alter its pathway of differentiation and form another structure.

homogametic sex The sex that produces gametes that do not differ with respect to sex-chromosome content; in mammals, the female is homogametic.

homologous chromosomes Chromosomes that synapse or pair during meiosis and that are identical with respect to their genetic loci and centromere placement.

homozygote An individual with identical alleles for a gene or genes of interest. These individuals will produce identical gametes (with respect to the gene or genes in question) and will therefore breed true.

horizontal gene transfer The nonreproductive transfer of genetic information from an organism to another, across species and higher taxa (even domains). This mode is contrasted with vertical gene transfer, which is the transfer of genetic information from parent to offspring. In some species of bacteria and archaea, up to 5 percent of the genome may have originally been acquired through horizontal gene transfer.

hot spots Genome regions where mutations are observed with a high frequency. These include a predisposition toward single-nucleotide substitutions or unequal crossing over.

human immunodeficiency virus (HIV) An RNA-containing retrovirus associated with the onset and progression of acquired immunodeficiency syndrome (AIDS).

hybrid An individual produced by crossing parents from two different genetic strains.

hybrid vigor The general superiority of a hybrid over a purebred.

inborn error of metabolism A genetically controlled biochemical disorder; usually an enzyme defect that produces a clinical syndrome.

inbreeding depression A decrease in viability, vigor, or growth in progeny after several generations of inbreeding.

incomplete dominance Expressing a heterozygous phenotype that is distinct from the phenotype of either homozygous parent. Also called *partial dominance.*

independent assortment The independent behavior of each pair of homologous chromosomes during their segregation in meiosis I. The random distribution of maternal and paternal homologs into gametes.

inducible enzyme system An enzyme system under the control of an inducer, a regulatory molecule that acts to block a repressor and allow transcription.

initiation codon The nucleotide triplet AUG that in an mRNA molecule codes for incorporation of the amino acid methionine as the first amino acid in a polypeptide chain.

interference (I) A measure of the degree to which one crossover affects the incidence of another crossover in an adjacent region of the same chromatid. Negative interference increases the chance of another crossover; positive interference reduces the probability of a second crossover event.

interphase In the cell cycle, the interval between divisions.

intron Any segment of DNA that lies between coding regions in a gene. Introns are transcribed but are spliced out of the RNA product and are not represented in the polypeptide encoded by the gene. Also known as an intervening sequence.

inversion A chromosomal aberration in which a chromosomal segment has been reversed.

in vitro Literally, *in glass;* outside the living organism; occurring in an artificial environment.

in vivo Literally, *in the living;* occurring within the living body of an organism.

isotopes Alternate forms of atoms with identical chemical properties that have the same atomic number but differ in the number of neutrons (and thus their mass) contained in the nucleus.

isozyme Any of two or more distinct forms of an enzyme with identical or nearly identical chemical properties but differ in some property such as net electrical charge, pH optima, number and type of subunits, or substrate concentration.

κ particles DNA-containing cytoplasmic particles found in certain strains of *Paramecium aurelia* capable of releasing a toxin, paramecin, that kills other, sensitive strains.

karyokinesis The process of nuclear division.

karyotype The chromosome complement of a cell or an individual. Often used to refer to the arrangement of metaphase chromosomes in a sequence according to length and centromere position.

kinetochore A fibrous structure with a size of about 400 nm, located within the centromere. It is the site of microtubule attachment during cell division.

knockout mice Mice created by a process in which a normal gene is cloned, inactivated by the insertion of a marker (such as an antibiotic resistance gene), and transferred to embryonic stem cells, where the altered gene will replace the normal gene (in some cells).

Kozak sequence A short nucleotide sequence adjacent to the initiation codon that is recognized as the translational start site in eukaryotic mRNA.

lagging strand During DNA replication, the strand synthesized in a discontinuous fashion, in the direction opposite of the replication fork.

lampbrush chromosomes Meiotic chromosomes characterized by extended lateral loops that reach maximum extension during diplotene. Although most intensively studied in amphibians, these structures occur in meiotic cells of organisms ranging from insects to humans.

lariat structure A structure formed during pre-mRNA processing by formation of a 5' to 3' bond in an introns, leading to removal of that intron from an mRNA molecule.

leader sequence That portion of an mRNA molecule from the 5' end to the initiating codon, often containing regulatory or ribosome binding sites.

leading strand During DNA replication, the strand synthesized continuously in the direction of the replication fork.

lethal gene A gene whose expression results in premature death of the organism at some stage of its life cycle.

leucine zipper In DNA-binding proteins, a structural motif characterized by a stretch in which every seventh amino acid residue is leucine, with adjacent regions containing positively charged amino acids. Leucine zippers on two polypeptides may interact to form a dimer that binds to DNA.

linking number The number of times that two strands of a closed, circular DNA duplex cross over each other.

locus (pl., loci) The site or place on a chromosome where a particular gene is located.

long interspersed elements (LINEs) Long, repetitive sequences found interspersed in the genomes of higher organisms.

long terminal repeat (LTR) A sequence of several hundred base pairs found at both ends of a retroviral DNA.

loss-of-function mutation Mutations that produce alleles that encode proteins with reduced or no function.

Lyon hypothesis The proposal describing the random inactivation of the maternal or paternal X chromosome in somatic cells of mammalian females early in development.

lysis The disintegration of a cell brought about by the rupture of its membrane.

lysogenic bacterium A bacterial cell carrying the DNA of a temperate bacteriophage integrated into its chromosome.

lysogeny The process by which the DNA of an infecting phage becomes repressed and integrated into the chromosome of the bacterial cell it infects.

map unit A measure of the genetic distance between two genes, corresponding to a recombination frequency of 1 percent.

maternal effect Phenotypic effects in offspring attributable to genetic information transmitted through the oocyte derived from the maternal genome.

maternal inheritance The transmission of traits strictly through the maternal parent, usually due to DNA found in the cytoplasmic organelles, the mitochondria, or chloroplasts.

meiosis The process of cell division in gametogenesis or sporogenesis during which the diploid number of chromosomes is reduced to the haploid number.

melting profile (T_m) The temperature at which a population of double-stranded nucleic acid molecules is half-dissociated into single strands.

merozygote A partially diploid bacterial cell containing, in addition to its own chromosome, a chromosome fragment introduced into the cell by transformation, transduction, or conjugation.

messenger RNA (mRNA) An RNA molecule transcribed from DNA and translated into the amino acid sequence of a polypeptide.

metacentric chromosome A chromosome that has a centrally located centromere and therefore chromosome arms of equal lengths.

metafemale In *Drosophila*, a poorly developed female of low viability with a ratio of X chromosomes to sets of autosomes that exceeds 1.0.

metagenomics The study of DNA recovered from organisms collected from the environment as opposed to those grown as laboratory cultures. Often used for estimating the diversity of organisms in an environmental sample.

metamale In *Drosophila*, a poorly developed male of low viability with a ratio of X chromosomes to sets of autosomes that is below 0.5.

metastasis The process by which cancer cells spread from the primary tumor and establish malignant tumors in other parts of the body.

methylation Enzymatic transfer of methyl groups from S-adenosylmethionine to biological molecules, including phospholipids, proteins, RNA, and DNA. Methylation of DNA is associated with the regulation of gene expression and with epigenetic phenomena such as imprinting.

microRNA Single-stranded RNA molecules approximately 20–23 nucleotides in length that regulate gene expression by participating in the degradation of mRNA.

microsatellite A short, highly polymorphic DNA sequence of 1–4 base pairs, widely distributed in the genome, that is used as a molecular marker in a variety of methods. Also called a *simple sequence repeat (SSR)*.

migration coefficient A measure of the proportion of migrant genes entering the population per generation.

minimal medium A medium containing only the essential nutrients needed to support the growth and reproduction of wild-type strains of an organism. Usually comprised of inorganic components that include a carbon and nitrogen source.

minisatellite Series of short tandem repeat sequences (STRs) 10–100 nucleotides in length that occur frequently throughout the genome of eukaryotes. Because the number of repeats at each locus is variable, the loci are known as variable number tandem repeats (VNTRs). Used in DNA fingerprinting and DNA profiles.

mismatch repair A form of excision repair of DNA in which the repair mechanism is able to distinguish between the strand with the error and the strand that is correct.

missense mutation A mutation that changes a codon to that of another amino acid and thus results in an amino acid substitution in the translated protein. Such changes can make the protein nonfunctional.

mitosis A form of cell division producing two progeny cells identical genetically to the parental cell—that is, the production of two cells from one, each having the same chromosome complement as the parent cell.

model genetic organism An experimental organism conducive to efficiently conducted research whose genetics is intensively studied on the premise that the findings can be applied to other organisms; for example, the fruit fly (*Drosophila melanogaster*) and the mouse (*Mus musculus*) are model organisms used to study the causes and development of human genetic diseases.

molecular clock In evolutionary studies, a method that counts the number of differences in DNA or protein sequences as a way of measuring the time elapsed since two species diverged from a common ancestor.

monohybrid cross A genetic cross involving only one character (e.g., $AA \times aa$).

monophyletic group A taxon (group of organisms) consisting of an ancestor and all its descendants.

monosomic An aneuploid condition in which one member of a chromosome pair is missing; having a chromosome number of $2n - 1$.

monozygotic twins Twins produced from a single fertilization event; the first division of the zygote produces two cells, each of which develops into an embryo. Also known as *identical twins*.

multigene family A set of genes descended from a common ancestral gene usually by duplication and subsequent sequence divergence. The globin genes are an example of a multigene family.

multiple alleles The presence of three or more alleles of the same gene in a population of organisms.

mutagen Any agent that causes an increase in the spontaneous rate of mutation.

mutation The process that produces an alteration in DNA or chromosome structure; in genes, the source of new alleles.

mutation rate The frequency with which mutations take place at a given locus or in a population.

natural selection Differential reproduction among members of a species owing to variable fitness conferred by genotypic differences.

neutral mutation A mutation with no immediate adaptive significance or phenotypic effect.

noncrossover gamete A gamete whose chromosomes have undergone no genetic recombination.

nondisjunction A cell division error in which homologous chromosomes or the sister chromatids fail to separate and migrate to opposite poles; responsible for defects such as monosomy and trisomy.

noninvasive prenatal genetic diagnosis (NIPGD) A noninvasive method of fetal genotyping that uses a maternal blood sample to analyze thousands of fetal loci using fetal DNA fragments present in the maternal blood.

nonsense codons The nucleotide triplets (UGA, UAG, and UAA) in an mRNA molecule that signal the termination of translation.

nonsense mutation A mutation that changes a codon specifying an amino acid into a termination codon, leading to premature termination during translation of mRNA.

Northern blot An analytic technique in which RNA molecules are separated by electrophoresis and transferred by capillary action to a nylon or nitrocellulose membrane. Specific RNA molecules can then be identified by hybridization to a labeled nucleic acid probe.

nuclease An enzyme that breaks bonds in nucleic acid molecules.

nucleoid The DNA-containing region within the cytoplasm in prokaryotic cells.

nucleolar organizer region (NOR) A chromosomal region containing the genes for rRNA; most often found in physical association with the *nucleolus*.

nucleolus The nuclear site of ribosome biosynthesis and assembly; usually associated with or formed in association with the DNA comprising the *nucleolar organizer region*.

nucleoside In nucleic acid chemical nomenclature, a purine or pyrimidine base covalently linked to a ribose or deoxyribose sugar molecule.

nucleosome In eukaryotes, a nuclear complex consisting of four pairs of histone molecules wrapped by two turns of a DNA molecule. The major structure associated with the organization of chromatin in the nucleus.

nucleotide In nucleic acid chemical nomenclature, a nucleoside covalently linked to one or more phosphate groups. Nucleotides containing a single phosphate linked to the 5′ carbon of the ribose or deoxyribose are the building blocks of nucleic acids.

nucleus The membrane-bound cytoplasmic organelle of eukaryotic cells that contains the chromosomes and nucleolus.

null allele A mutant allele that produces no functional gene product. Usually inherited as a recessive trait.

null hypothesis (H$_0$) Used in statistical tests, the hypothesis that there is no real difference between the observed and expected datasets. Statistical methods such as chi-square analysis are used to test the probability associated with this hypothesis.

Okazaki fragment The short, discontinuous strands of DNA produced on the lagging strand during DNA synthesis.

oligonucleotide A linear sequence of about 10–20 nucleotides connected by 5′-3′ phosphodiester bonds.

oncogene A gene whose activity promotes uncontrolled proliferation in eukaryotic cells. Usually a mutant gene derived from a *proto-oncogene*.

Online Mendelian Inheritance in Man (OMIM) A database listing all known genetic disorders and disorders with genetic components. It also contains a listing of all known human genes and links genes to genetic disorders.

open reading frame (ORF) A nucleotide sequence organized as triplets that encodes the amino acid sequence of a polypeptide, including an initiation codon and a termination codon.

operator region In bacterial DNA, a region that interacts with a specific repressor protein to regulate the expression of an adjacent gene or gene set.

operon A genetic unit consisting of one or more structural genes encoding polypeptides, and an adjacent operator gene that regulates the transcriptional activity of the structural gene or genes.

outbreeding depression Reduction in fitness in the offspring produced by mating genetically diverse parents. It is thought to result from a lowered adaptation to local environmental conditions.

overlapping code A hypothetical genetic code in which any given triplet is shared by more than one adjacent codon.

pair-rule genes Genes expressed as stripes around the blastoderm embryo during development of the *Drosophila* embryo.

palindrome In genetics, a double stranded DNA segment where each strand's base sequence is identical when read 5′ to 3′. For example:

$$5'\text{-GAATTC-}3'$$
$$3'\text{-CTTAAG-}5'$$

Palindromic sequences are noteworthy as recognition and cleavage sites for restriction endonucleases.

paracentric inversion A chromosomal inversion that does not include the region containing the centromere.

pedigree In human genetics, a diagram showing the ancestral relationships and transmission of genetic traits over several generations in a family.

P element In *Drosophila*, a transposable DNA element responsible for hybrid dysgenesis.

penetrance The frequency, expressed as a percentage, with which individuals of a given genotype manifest at least some degree of a specific mutant phenotype associated with a trait.

pericentric inversion A chromosomal inversion that involves both arms of the chromosome and thus the centromere.

pharmacogenomics The study of how genetic variation influences the action of pharmaceutical drugs in individuals.

PharmGKB database A repository for genetic, molecular, cellular, and genomic information about individuals who have participated in clinical pharmacogenomic research studies.

phenotype The overt appearance of a genetically controlled trait.

Philadelphia chromosome The product of a reciprocal translocation in humans that contains the short arm of chromosome 9, carrying the *C-ABL* oncogene, and the long arm of chromosome 22, carrying the *BCR* gene.

phosphodiester bond In nucleic acids, the system of covalent bonds by which a phosphate group links adjacent nucleotides, extending from the 5′ carbon of one pentose sugar (ribose or deoxyribose) to the 3′ carbon of the pentose sugar in the neighboring nucleotide. Phosphodiester bonds create the backbone of nucleic acid molecules.

photoreactivation repair Light-induced repair of damage caused by exposure to ultraviolet light. Associated with an intracellular enzyme system.

phyletic evolution The gradual transformation of one species into another over time; so-called vertical evolution.

pilus A filamentlike projection from the surface of a bacterial cell. Often associated with cells possessing F factors.

plaque On an otherwise opaque bacterial lawn, a clear area caused by the growth and reproduction of a single bacteriophage.

plasmid An extrachromosomal, circular DNA molecule that replicates independently of the host chromosome.

pleiotropy Condition in which a single mutation causes multiple phenotypic effects.

ploidy A term referring to the basic chromosome set or to multiples of that set.

point mutation A mutation that can be mapped to a single locus. At the molecular level, a mutation that results in the substitution of one nucleotide for another. Also called a *gene mutation*.

polar body Produced in females at either the first or second meiotic division of gametogenesis, a discarded cell that contains one of the nuclei of the division process, but almost no cytoplasm as a result of an unequal cytokinesis.

polycistronic mRNA A messenger RNA molecule that encodes the amino acid sequence of two or more polypeptide chains in adjacent structural genes.

polygenic inheritance The transmission of a phenotypic trait whose expression depends on the additive effect of a number of genes.

polylinker A segment of DNA that has been engineered to contain multiple sites for restriction enzyme digestion. Polylinkers are usually found in engineered vectors such as plasmids.

polymerase chain reaction (PCR) A method for amplifying DNA segments that depends on repeated cycles of denaturation, primer annealing, and DNA polymerase–directed DNA synthesis.

polymerases Enzymes that catalyze the formation of DNA and RNA from deoxynucleotides and ribonucleotides, respectively.

polymorphism The existence of two or more discontinuous, segregating phenotypes in a population.

polynucleotide A linear sequence of 20 or more nucleotides, joined by 5'-3' phosphodiester bonds.

polypeptide A molecule composed of amino acids linked together by covalent peptide bonds. This term is used to denote the amino acid chain before it assumes its functional three-dimensional configuration.

polyploid A cell or individual having more than two haploid sets of chromosomes.

polysome A structure composed of two or more ribosomes associated with an mRNA and associated tRNAs engaged in translation. Also called a *polyribosome*.

polytene chromosome Literally, a many-stranded chromosome; one that has undergone numerous rounds of DNA replication without separation of the replicated strands, which remain in exact parallel register. The result is a giant chromosome with aligned chromomeres displaying a characteristic banding pattern, most often studied in *Drosophila* larval salivary gland cells.

population A local group of actually or potentially interbreeding individuals belonging to the same species.

population bottleneck A drastic reduction in population size and consequent loss of genetic diversity, followed by an increase in population size. The rebuilt population has a gene pool with reduced diversity caused by genetic drift.

positional cloning The identification and subsequent cloning of a gene in the absence of knowledge of its polypeptide product or function. The process uses cosegregation of mutant phenotypes with DNA markers to identify the chromosome containing the gene; the position of the gene is identified establishing linkage with additional markers.

position effect Change in expression of a gene associated with a change in the gene's location within the genome.

posttranslational modification The processing or modification of the translated polypeptide chain by enzymatic cleavage, or the addition of phosphate groups, carbohydrate chains, or lipids.

posttranscriptional modification Changes made to pre-mRNA molecules during conversion to mature mRNA. These include the addition of a methylated cap at the 5' end and a poly-A tail at the 3' end, excision of introns, and exon splicing.

postzygotic isolation mechanism A barrier that prevents or reduces inbreeding by acting after fertilization to produce nonviable, sterile hybrids or hybrids of lowered fitness.

preadaptive mutation A mutational event that later becomes of adaptive significance.

preimplantation genetic diagnosis (PGD) The removal and genetic analysis of unfertilized oocytes, polar bodies, or single cells from an early embryo (3–5 days old).

prezygotic isolation mechanism A barrier that reduces inbreeding by preventing courtship, mating, or fertilization.

Pribnow box In prokaryotic genes, a 6-bp sequence to which the sigma (σ) subunit of RNA polymerase binds, upstream from the beginning of transcription. The consensus sequence for this box is TATAAT.

primary sex ratio Ratio of males to females at fertilization, often expressed in decimal form (e.g., 1.06).

primer In nucleic acids, a short length of RNA or single-stranded DNA required for initiating synthesis directed by polymerases.

prion An infectious pathogenic agent devoid of nucleic acid and composed of a protein, PrP, with a molecular weight of 27,000–30,000 Da. Prions are known to cause scrapie, a degenerative neurological disease in sheep; bovine spongiform encephalopathy (BSE, or mad cow disease) in cattle; and similar diseases in humans, including kuru and Creutzfeldt–Jakob disease.

proband An individual who is the focus of a genetic study leading to the construction of a pedigree tracking the inheritance of a genetically determined trait of interest. Formerly known as a *propositus*.

probe A macromolecule such as DNA or RNA that has been labeled and can be detected by an assay such as autoradiography or fluorescence microscopy. Probes are used to identify target molecules, genes, or gene products.

promoter element An upstream regulatory region of a gene to which RNA polymerase binds prior to the initiation of transcription.

proofreading A molecular mechanism for scanning and correcting errors in replication, transcription, or translation.

prophage A bacteriophage genome integrated into a bacterial chromosome that is replicated along with the bacterial chromosome. Bacterial cells carrying prophages are said to be *lysogenic* and to be capable of entering the *lytic cycle*, whereby the phage is replicated.

propositus (female, proposita) See *proband*.

protein domain Amino acid sequences with specific conformations and functions that are structurally and functionally distinct from other regions on the same protein.

proteome The entire set of proteins expressed by a cell, tissue, or organism at a given time. The study of the proteome is referred to as proteomics.

proto-oncogene A gene that functions to initiate, facilitate, or maintain cell growth and division. Proto-oncogenes can be converted to *oncogenes* by mutation.

protoplast A bacterial or plant cell with the cell wall removed. Sometimes called a *spheroplast*.

prototroph A strain (usually of a microorganism) that is capable of growth on a defined, minimal medium. Wild-type strains are usually regarded as prototrophs and contrasted with *auxotrophs*.

pseudoalleles Genes that behave as alleles to one another by complementation but can be separated from one another by recombination.

pseudoautosomal region A region on the human Y chromosome that is also represented on the X chromosome. Genes found in this region of the Y chromosome have a pattern of inheritance that is indistinguishable from genes on autosomes.

pseudodominance The expression of a recessive allele on one homolog owing to the deletion of the dominant allele on the other homolog.

pseudogene A nonfunctional gene with sequence homology to a known structural gene present elsewhere in the genome. It differs from the functional version by insertions or deletions and by the presence of flanking direct-repeat sequences of 10–20 nucleotides.

punctuated equilibrium A pattern in the fossil record of long periods of species stability, punctuated with brief periods of species divergence.

pyrosequencing A high-throughput method of DNA sequencing that determines the sequence of a single-stranded DNA molecule by synthesis of a complementary strand. During synthesis, the sequence is determined by the chemiluminescent detection of pyrophosphate release that accompanies nucleotide incorporation into a newly synthesized strand of DNA.

quantitative real-time PCR (qPCR) A variation of PCR (polymerase chain reaction) that uses fluorescent probes to quantitate the amount of DNA or RNA product present after each round of amplification.

quantitative trait loci (QTLs) Two or more genes that act on a single polygenic trait in a quantitative way.

quantum speciation Formation of a new species within a single or a few generations by a combination of selection and drift.

rad A unit of absorbed dose of radiation with an energy equal to 100 ergs per gram of irradiated tissue.

radioactive isotope An unstable isotope with an altered number of neutrons that emits ionizing radiation during decay as it is transformed to a stable atomic configuration.

random amplified polymorphic DNA (RAPD) A PCR method that uses random primers about 10 nucleotides in length to amplify unknown DNA sequences.

reading frame A linear sequence of codons in a nucleic acid.

reannealing Formation of double-stranded DNA molecules from denatured single strands.

recessive An allele whose potential genetic expression is overridden in the heterozygous condition by a dominant allele.

reciprocal translocation A chromosomal aberration in which nonhomologous chromosomes exchange parts.

recombinant DNA technology A collection of methods used to create DNA molecules by *in vitro* ligation of DNA from two different organisms, and the replication and recovery of such recombinant DNA molecules.

recombination The process that leads to the formation of new allele combinations on chromosomes.

reductional division The chromosome division that halves the number of centromeres and thus reduces the chromosome number by half in the daughter cells. The first division of meiosis is a reductional division

rem Radiation equivalent in humans; the dosage of radiation that will cause the same biological effect as one roentgen of X rays.

renaturation The process by which a denatured protein or nucleic acid returns to its normal three-dimensional structure.

repetitive DNA sequence A DNA sequence present in many copies in the haploid genome.

replication fork The Y-shaped region of a chromosome associated with the site of DNA replication.

replicon The unit of DNA replication, beginning with DNA sequences necessary for the initiation of DNA replication. In bacteria, the entire chromosome is a replicon.

replisome The complex of proteins, including DNA polymerase, that assembles at the bacterial replication fork to synthesize DNA.

repressible enzyme system An enzyme or group of enzymes whose synthesis is regulated by the intracellular concentration of certain metabolites.

repressor A protein that binds to a regulatory sequence adjacent to a gene and blocks transcription of the gene.

reproductive isolation Absence of interbreeding between populations, subspecies, or species.

resistance transfer factor (RTF) A component of R plasmids that confers the ability to transfer the R plasmid between bacterial cells by conjugation.

restriction endonuclease A bacterial nuclease that recognizes specific nucleotide sequences in a DNA molecule, often a *palindrome*, and cleaves or nicks the DNA at those sites.

restriction fragment length polymorphism (RFLP) Variation in the length of DNA fragments generated by restriction endonucleases. These variations are caused by mutations that create or abolish cutting sites for restriction enzymes. RFLPs are inherited in a codominant fashion and are extremely useful as genetic markers.

restriction site A DNA sequence, often palindromic, recognized by a restriction endonuclease. The enzyme binds to the restriction site and cleaves the DNA at that site.

retrotransposon Mobile genetic elements that are major components of many eukaryotic genomes; these elements are copied by means of an RNA intermediate and can be inserted at a distant chromosomal site.

retrovirus A type of virus that uses RNA as its genetic material and employs the enzyme reverse transcriptase during its life cycle.

reverse genetics An experimental approach used to discover gene function after the gene has been identified, cloned, and sequenced.

reversion A mutation that restores the wild-type phenotype.

R factor (R plasmid) A bacterial plasmid that carries antibiotic resistance genes. Most R plasmids have two components: an r-determinant which carries the antibiotic resistance genes, and the resistance transfer factor (RTF).

Rh factor An antigenic system first described in the rhesus monkey.

11

ribonucleic acid (RNA) A nucleic acid similar to DNA but characterized by the pentose sugar ribose, the pyrimidine uracil, and the single-stranded nature of the polynucleotide chain. Several forms are recognized, including ribosomal RNA, messenger RNA, transfer RNA, and a variety of small regulatory RNA molecules.

ribose The five-carbon sugar associated with ribonucleosides and ribonucleotides associated with RNA.

ribosomal RNA (rRNA) The RNA molecules that are the structural components of the ribosomal subunits. In prokaryotes, these are the $16S$, $23S$, and $5S$ molecules; in eukaryotes, they are the $18S$, $28S$, and $5S$ molecules.

ribosome A ribonucleoprotein organelle consisting of two subunits, each containing RNA and protein molecules. Ribosomes are the site of translation of mRNA codons into the amino acid sequence of a polypeptide chain.

RNA interference (RNAi) Inhibition of gene expression in which a protein complex (RNA-induced silencing complex, or RISC), containing a partially complementary RNA strand binds to an mRNA, leading to degradation or reduced translation of the mRNA.

Robertsonian translocation A chromosomal aberration created by breaks in the short arms of two acrocentric chromosomes followed by fusion of the long arms of these chromosomes at the centromere. Also called *centric fusion*.

roentgen (R) A unit of measure of radiation emission, corresponding to the amount that generates 2.083×10^9 ion pairs in one cubic centimeter of air at 0°C and an atmospheric pressure of 760 mm of mercury.

Sanger sequencing DNA sequencing by synthesis of DNA chains that are randomly terminated by incorporation of a nucleotide analog (dideoxynucleotides) followed by sequence determination by analysis of resulting fragment lengths in each reaction.

satellite DNA DNA that forms a minor band when genomic DNA is centrifuged in a cesium salt gradient. This DNA usually consists of short repetitive sequences.

secondary sex ratio The ratio of males to females at birth, usually expressed in decimal form (e.g., 1.05).

segment polarity genes Genes that regulate the spatial pattern of differentiation within each segment of the developing *Drosophila* embryo.

segregation The separation of maternal and paternal homologs of each homologous chromosome pair into gametes during meiosis.

selection Changes in the frequency of alleles and genotypes in populations as a result of differential reproduction.

selection coefficient (*s*) A quantitative measure of the relative fitness of one genotype compared with another. Same as *coefficient of selection*.

semiconservative replication A mode of DNA replication in which a double-stranded molecule replicates in such a way that the daughter molecules are each composed of one parental (old) and one newly synthesized strand.

semisterility A condition in which a percentage of all zygotes are inviable.

sex chromosome A chromosome, such as the X or Y in humans, which is involved in sex determination.

sexduction Transmission of chromosomal genes from a donor bacterium to a recipient cell by means of the F factor.

sex-influenced inheritance Phenotypic expression conditioned by the sex of the individual. A heterozygote may express one phenotype in one sex and an alternate phenotype in the other sex (e.g., pattern baldness in humans).

sex-limited inheritance A trait that is expressed in only one sex even though the trait may not be X-linked.

Shine–Dalgarno sequence The nucleotides AGGAGG that serve as a ribosome-binding site in the leader sequence of prokaryotic genes. The $16S$ RNA of the small ribosomal subunit contains a complementary sequence to which the mRNA binds.

short interspersed elements (SINEs) Repetitive sequences found in the genomes of higher organisms. The 300-bp *Alu* sequence is a SINE element.

shotgun cloning The cloning of random fragments of genomic DNA into a vector (a plasmid or phage), usually to produce a library from which clones can be selected for use, as in sequencing.

sibling species Species that are morphologically similar but reproductively isolated from one another.

sigma (σ) factor In RNA polymerase, a polypeptide subunit that recognizes the DNA binding site for the initiation of transcription.

single molecule sequencing A high-throughput sequencing by synthesis involving direct sequencing of single DNA or RNA molecules without the need for amplification by PCR before sequencing.

single-nucleotide polymorphism (SNP) A variation in a single nucleotide pair in DNA, usually detected during genomic analysis. Present in at least 1 percent of a population, a SNP is useful as a genetic marker.

single-stranded binding proteins (SSBs) In DNA replication, proteins that bind to and stabilize the single-stranded regions of DNA that result from the action of unwinding proteins.

sister chromatid exchange (SCE) A crossing over event in meiotic or mitotic cells involving the reciprocal exchange of chromosomal material between sister chromatids joined by a common centromere. Such exchanges can be detected cytologically after BrdU incorporation into the replicating chromosomes.

site-directed mutagenesis A process that uses a synthetic oligonucleotide containing a mutant base or sequence as a primer for inducing a mutation at a specific site in a cloned gene.

small nuclear RNA (snRNA) Abundant species of small RNA molecules ranging in size from 90 to 400 nucleotides that in association with proteins form RNP particles known as snRNPs or *snurps*. Located in the nucleoplasm, snRNAs have been implicated in the processing of pre-mRNA and may have a range of cleavage and ligation functions.

solenoid structure A feature of eukaryotic chromatin conformation generated by nucleosome supercoiling.

somatic mutation A nonheritable mutation occurring in a somatic cell.

SOS response The induction of enzymes for repairing damaged DNA in *Escherichia coli*. The response involves activation of an enzyme that cleaves a repressor, activating a series of genes involved in DNA repair.

Southern blotting A technique developed by Edwin Southern in which DNA fragments produced by restriction enzyme digestion are separated by electrophoresis and transferred by capillary action to a nylon or nitrocellulose membrane. Specific DNA fragments can be identified by hybridization to a complementary radioactively labeled nucleic acid probe using the technique of *autoradiography*.

spacer DNA DNA sequences found between genes. Usually, these are repetitive DNA segments.

species A group of actually or potentially interbreeding individuals that is reproductively isolated from other such groups.

spectral karyotype A display of all the chromosomes in an organism as a karyotype with each chromosome stained in a different color.

spliceosome The nuclear macromolecule complex within which splicing reactions occur to remove introns from pre-mRNAs.

spontaneous mutation A mutation that arises in the absence of an external force; one that is not induced.

spore A unicellular body or cell encased in a protective coat. Produced by some bacteria, plants, and invertebrates, spores are capable of surviving in unfavorable environmental conditions and give rise to a new individual upon germination. In plants, spores are the haploid products of meiosis.

SRY The sex-determining region of the Y chromosome, found near the chromosome's pseudoautosomal boundary. Accumulated evidence indicates that this gene's product is the testis-determining factor (TDF).

stabilizing selection Preferential reproduction of individuals with genotypes close to the mean for the population. A selective elimination of genotypes at both extremes.

standard deviation (*s*) A quantitative measure of the amount of variation in a sample of measurements from a population calculated as the square root of the variance.

strain A group of organisms with common ancestry with physiological or morphological characteristics of interest for genetic analysis or domestication.

STR sequences Short tandem repeats 2–9 base pairs long that are found within minisatellites. These sequences are used to prepare DNA profiles in forensics, paternity identification, and other applications.

structural gene A gene that encodes the amino acid sequence of a polypeptide chain.

sublethal gene A mutation causing lowered viability, with death before maturity in less than 50 percent of the individuals carrying the gene.

submetacentric chromosome A chromosome with the centromere placed so that one arm of the chromosome is slightly longer than the other.

sum law The law that holds that the probability of one of two mutually exclusive outcomes occurring, where that outcome can be achieved by two or more events, is equal to the sum of their individual probabilities.

suppressor mutation A mutation that acts to completely or partially restore the function lost by a mutation at another site.

sympatric speciation Speciation occurring in populations that inhabit, at least in part, the same geographic range.

synapsis The pairing of homologous chromosomes at meiosis.

synaptonemal complex (SC) A sub-microscopic structure consisting of a tripartite nucleoprotein ribbon that forms between the paired homologous chromosomes in the pachytene stage of the first meiotic division.

syndrome A group of characteristics or symptoms associated with a disease or an abnormality. An affected individual may express a number of these characteristics but not necessarily all of them.

synthetic biology A field of research that combines science and engineering to construct novel biological-based systems and/or organisms that do not exist in nature.

synthetic genome A genome assembled from chemically synthesized DNA fragments that is transferred to a host cell without a genome.

systems biology A field that identifies and analyzes gene and protein networks to gain an understanding of intracellular regulation of metabolism, intra- and intercellular communication, and complex interactions within, between, and among cells.

TATA box See *Goldberg–Hogness box*.

tautomeric shift A reversible isomerization in a molecule, brought about by a shift in the location of a hydrogen atom. In nucleic acids, tautomeric shifts in the bases of nucleotides can cause changes in other bases during replication and can act as a source of mutations.

telocentric chromosome A chromosome in which the centromere is located at its very end.

telomerase The enzyme that adds short, tandemly repeated DNA sequences to the ends of eukaryotic chromosomes.

telomere The heterochromatic terminal region of a chromosome.

telomere repeat-containing RNA (TERRA) Large noncoding RNA molecules transcribed from telomere repeats that are an integral part of telomeric heterochromatin.

temperate phage A bacteriophage that can become a prophage, integrating its DNA into the chromosome of the host bacterial cell and making the latter lysogenic.

temperature-sensitive mutation A conditional mutation that produces a mutant phenotype at one temperature range and a wild-type phenotype at another.

template The single-stranded DNA or RNA molecule that specifies the sequence of a complementary nucleotide strand synthesized by DNA or RNA polymerase.

testcross A cross between an individual whose genotype at one or more loci may be unknown and an individual who is homozygous recessive for the gene or genes in question.

tetrad The four chromatids that make up paired homologs in the prophase of the first meiotic division. In eukaryotes with a predominant haploid stage (some algae and fungi), a tetrad also denotes the four haploid cells produced by a single meiotic division.

tetrad analysis A method that analyzes gene linkage and recombination in organisms with a predominant haploid phase in their life cycle.

tetranucleotide hypothesis An early theory of DNA structure proposing that the molecule was composed of repeating units, each consisting of the four nucleotides represented by adenine, thymine, cytosine, and guanine.

thymine dimer In a polynucleotide strand, a lesion consisting of two adjacent thymine bases that become joined by a covalent bond. Usually caused by exposure to ultraviolet light, this lesion inhibits DNA replication.

totipotent The capacity of a cell or an embryo part to differentiate into all cell types characteristic of an adult. This capacity is usually progressively restricted during development. Used interchangeably with *pluripotent*.

trait Any detectable phenotypic variation of a particular inherited character.

***trans*-acting element** A gene product (usually a diffusible protein or an RNA molecule) that acts to regulate the expression of a target gene.

***trans* configuration** An arrangement in which two mutant alleles are on opposite homologs, such as

$$\frac{a^1 \quad +}{+ \quad a^2}$$

transcription Transfer of genetic information from DNA by the synthesis of a complementary RNA molecule using a DNA template.

transcriptome The set of mRNA molecules present in a cell at any given time.

transdetermination Change in developmental fate of a cell or group of cells.

transduction Virally mediated bacterial recombination. Also used to describe the transfer of eukaryotic genes mediated by a retrovirus.

transfer RNA (tRNA) A small ribonucleic acid molecule that "adapts" a triplet codon to its corresponding amino acid during translation.

transformation Heritable change in a cell or an organism brought about by exogenous DNA. Known to occur naturally and also used in *recombinant DNA* studies.

transgenic organism An organism whose genome has been modified by the introduction of external DNA sequences into the germ line.

transition mutation A mutational event in which one purine is replaced by another or one pyrimidine is replaced by another.

translation The derivation of the amino acid sequence of a polypeptide from the base sequence of an mRNA molecule in association with a ribosome and tRNAs.

translocation A chromosomal mutation associated with the reciprocal or nonreciprocal transfer of a chromosomal segment from one chromosome to another. Also denotes the movement of mRNA through the ribosome during translation.

transmission genetics The field of genetics concerned with heredity and the mechanisms by which genes are transferred from parent to offspring.

transposable element A DNA segment that moves to other sites in the genome, essentially independent of sequence homology. Usually, such elements are flanked at each end by short inverted repeats of 20–40 base pairs.

transversion mutation A mutational event in which a purine is replaced by a pyrimidine or a pyrimidine is replaced by a purine.

trinucleotide repeat A tandemly repeated cluster of three nucleotides (such as CTG) within or near a gene. Certain diseases (myotonic dystrophy, Huntington disease) are caused by expansion in copy number of such repeats.

triploidy The condition in which a cell or an organism possesses three haploid sets of chromosomes.

trisomy The condition in which a cell or an organism possesses two copies of each chromosome except for one, which is present in three copies (designated $2n + 1$).

tumor-suppressor gene A gene that encodes a product that normally functions to suppress cell division. Mutations in tumor-suppressor genes activate cell division and cause tumor formation.

unequal crossing over A crossover between two improperly aligned homologs, producing one homolog with three copies of a region and the other with one copy of that region.

variable number tandem repeats (VNTRs) Short, repeated DNA sequences (of 2–20 nucleotides) present as tandem repeats between two restriction enzyme sites. Variation in the number of repeats creates DNA fragments of differing lengths following restriction enzyme digestion. Used in early versions of *DNA fingerprinting*.

variance (s^2) A statistical measure of the variation of values from a central value, calculated as the square of the standard deviation.

variegation Patches of differing phenotypes, such as color, in a tissue.

vector In recombinant DNA, an agent such as a phage or plasmid into which a foreign DNA segment will be inserted and used to transform host cells.

vertical gene transfer The transfer of genetic information from parents to offspring generation after generation.

virulent phage A bacteriophage that infects, replicates within, and lyses bacterial cells, releasing new phage particles.

Western blot An analytical technique in which proteins are separated by gel electrophoresis and transferred by capillary action to a nylon membrane or nitrocellulose sheet. A specific protein can be identified through hybridization to a labeled antibody.

wild type The most commonly observed phenotype or genotype, designated as the norm or standard.

wobble hypothesis An idea proposed by Francis Crick, stating that the third base in an anticodon can align in several ways to allow it to recognize more than one base in the codons of mRNA.

W, Z chromosomes The sex chromosomes in species where the female is the heterogametic sex (WZ).

X chromosome The sex chromosome present in species where females are the homogametic sex (XX).

X chromosome inactivation In mammalian females, the random cessation of transcriptional activity of the maternally or paternally derived X chromosome. This event, which occurs early in development, is a mechanism of dosage compensation, and all progeny cells inactivate the same X chromosome.

XIST A locus in the X-chromosome inactivation center that controls inactivation of the X chromosome in mammalian females.

X-linkage The pattern of inheritance resulting from genes located on the X chromosome.

X-ray crystallography A technique for determining the three-dimensional structure of molecules by analyzing X-ray diffraction patterns produced by bombarding crystals of the molecule under study with X-rays.

YAC Yeast artificial chromosome. A cloning vector constructed using chromosomal components including telomeres (from a ciliate), and centromeres, origin of replication, and marker genes from yeast. YACs are used to clone long stretches of eukaryotic DNA.

Y chromosome The sex chromosome in species where the male is heterogametic (XY).

Z-DNA An alternative "zig-zag" structure of DNA in which the two antiparallel polynucleotide chains form a left-handed double helix. Implicated in regulation of gene expression.

zinc finger A class of DNA-binding domains seen in proteins. They have a characteristic pattern of cysteine and histidine residues that complex with zinc ions, throwing intermediate amino acid residues into a series of loops or fingers.

zygote The diploid cell produced by the fusion of haploid gametic nuclei.

Newer model organisms in genetics include the roundworm *C. elegans*, the zebrafish, *D. rerio*, and the mustard plant *A. thaliana*.

Introduction to Genetics

CHAPTER CONCEPTS

- Genetics in the twenty-first century is built on a rich tradition of discovery and experimentation stretching from the ancient world through the nineteenth century to the present day.

- Transmission genetics is the general process by which traits controlled by genes are transmitted through gametes from generation to generation.

- Mutant strains can be used in genetic crosses to map the location and distance between genes on chromosomes.

- The Watson-Crick model of DNA structure explains how genetic information is stored and expressed. This discovery is the foundation of molecular genetics.

- Recombinant DNA technology revolutionized genetics, was the foundation for the Human Genome Project, and has generated new fields that combine genetics with information technology.

- Biotechnology provides genetically modified organisms and their products that are used across a wide range of fields including agriculture, medicine, and industry.

- Model organisms used in genetics research are now utilized in combination with recombinant DNA technology and genomics to study human diseases.

- Genetic technology is developing faster than the policies, laws, and conventions that govern its use.

In December 1998, the Icelandic Parliament passed a law granting a biotechnology company, deCODE Genetics, a license to create and operate a database drawn from medical records of all of Iceland's 270,000 residents. The records in this Icelandic Health Sector Database (or HSD) were encoded to ensure anonymity. The new law also allowed deCODE Genetics to cross-reference the medical information from the HSD with a comprehensive genealogical database from the National Archives and to correlate information in these two databases with results from the analysis of deoxyribonucleic acid (DNA) samples collected from Icelandic donors. This combination of medical, genealogical, and genetic information forms a powerful resource available exclusively to deCODE Genetics for marketing to researchers and companies for a period of 12 years, ending in 2012.

This is not a science fiction scenario from a movie such as *Gattaca* but a real example of the increasingly complex interaction of genetics and society in the first decades of the twenty-first century. The development and use of these databases in Iceland have generated similar large-scale projects in Great Britain, Estonia, Latvia, the Kingdom of Tonga, and other countries. In the United States, smaller-scale programs, involving tens of thousands of individuals, are underway. All these databases will be used to search for susceptibility genes that control complex human diseases.

From Chapter 1 of *Essentials of Genetics, Eighth edition*, William S. Klug, Michael R. Cummings, Charlotte A. Spencer, Michael A. Palladino. © 2013 by Pearson Education, Inc. All rights reserved.

15

deCODE Genetics selected Iceland for this unprecedented project because the people of Iceland have a high level of genetic relatedness that derives from the founding of Iceland about 1000 years ago by a small population drawn mainly from Scandinavian and Celtic sources. Several population reductions by disease and natural disasters further reduced genetic diversity, and until the last few decades, few immigrants brought new genes into the population. Moreover, because Iceland's health-care system is state-supported, medical records for all residents go back as far as the early 1900s. Genealogical information is available in the National Archives and church records for almost every current resident and for more than 500,000 of the estimated 750,000 individuals who have ever lived in Iceland. For all these reasons, the Icelandic datasets are a tremendous asset for geneticists searching for genes that control complex disorders. The project already has a number of successes, including the identification and isolation of genes associated with more than a dozen common diseases including asthma, heart disease, stroke, and osteoporosis.

On the flip side are questions of privacy, consent, and commercialization—issues at the heart of many controversies arising from the applications of genetic technology. Scientists and nonscientists alike are debating the role of law, the individual, and society in decisions about how and when genetic technology is used. For example, how will knowledge of the complete nucleotide sequence of the human genome be used? Will disclosure of someone's genetic information lead to discrimination in jobs or insurance? Should genetic technology such as prenatal diagnosis or gene therapy be available to all, regardless of ability to pay? More than at any other time in the history of science, addressing the ethical questions surrounding an emerging technology is as important as the information gained from that technology.

This chapter provides an overview of genetics and a survey of the high points in its history and gives a preliminary description of its central principles and emerging developments. This text will enable you to achieve a thorough understanding of modern-day genetics and its underlying principles. Along the way, enjoy your studies, but take your responsibilities as a novice geneticist very seriously.

1   Genetics Has a Rich and Interesting History

We don't know when people first recognized the hereditary nature of certain traits, but archaeological evidence (e.g., primitive art, preserved bones and skulls, and dried seeds)

documents the successful domestication of animals and the cultivation of plants thousands of years ago by artificial selection of genetic variants from wild populations. Between 8000 and 1000 B.C., horses, camels, oxen, and wolves were domesticated, and selective breeding of these species soon followed. Cultivation of many plants, including maize, wheat, rice, and the date palm, began around 5000 B.C. Such evidence documents our ancestors' successful attempts to manipulate the genetic composition of species.

During the Golden Age of Greek culture, the writings of the Hippocratic School of Medicine (500–400 B.C.) and of the philosopher and naturalist Aristotle (384–322 B.C.) discussed heredity as it relates to humans. The Hippocratic treatise *On the Seed* argued that active "humors" in various parts of the body served as the bearers of hereditary traits. Drawn from various parts of the male body to the semen and passed on to offspring, these humors could be healthy or diseased, with the diseased humors accounting for the appearance of newborns with congenital disorders or deformities. It was also believed that these humors could be altered in individuals before they were passed on to offspring, explaining how newborns could "inherit" traits that their parents had "acquired" in response to their environment.

Aristotle extended Hippocrates' thinking and proposed that the male semen contained a "vital heat" with the capacity to produce offspring of the same "form" (i.e., basic structure and capacities) as the parent. Aristotle believed that this heat cooked and shaped the menstrual blood produced by the female, which was the "physical substance" that gave rise to an offspring. The embryo developed not because it already contained the parts of an adult in miniature form (as some Hippocratics had thought) but because of the shaping power of the vital heat. Although the ideas of Hippocrates and Aristotle sound primitive and naive today, we should recall that prior to the 1800s neither sperm nor eggs had been observed in mammals.

1600–1850: The Dawn of Modern Biology

Between about 300 B.C. to 1600 A.D., there were few significant new ideas about genetics. However, between 1600 and 1850, major strides provided insight into the biological basis of life. In the 1600s, William Harvey studied reproduction and development and proposed the theory of epigenesis, which states that an organism develops from the fertilized egg by a succession of developmental events that eventually transform the egg into an adult. The theory of epigenesis directly conflicted with the theory of **preformation,** which stated that the fertilized egg contains a complete miniature adult, called a **homunculus (Figure 1).** Around 1830, Matthias Schleiden and Theodor Schwann proposed the **cell theory,** stating that all organisms are composed of basic units

FIGURE 1 Depiction of the *homunculus*, a sperm containing a miniature adult, perfect in proportion and fully formed.

(Hartsoeker, N. Essay de dioptrique Paris, 1694, p. 230. National Library of Medicine)

called cells, which are derived from similar preexisting structures. The idea of **spontaneous generation,** the creation of living organisms from nonliving components, was disproved by Louis Pasteur later in the century, and living organisms were considered to be derived from preexisting organisms and to consist of cells.

In the mid-1800s the revolutionary work and principles presented by Charles Darwin and Gregor Mendel set the stage for the rapid development of genetics in the twentieth and twenty-first centuries.

Charles Darwin and Evolution

With this background, we turn to a brief discussion of the work of Charles Darwin, who published *The Origin of Species*, in 1859, describing his ideas about evolution. Darwin's geological, geographical, and biological observations convinced him that existing species arose by descent with modification from ancestral species. Greatly influenced by his voyage on the HMS *Beagle* (1831–1836), Darwin's thinking culminated in his formulation of the theory of **natural selection,** which presented an explanation of the mechanism of evolutionary change. Formulated and proposed independently by Alfred Russel Wallace, natural selection is based on the observation that populations tend to contain more offspring than the environment can support, leading to a struggle for survival among individuals. Those individuals with heritable traits that allow them to adapt to their environment are better able to survive and reproduce than those with less adaptive traits. Over a long period of time, advantageous variations, even very slight ones, will accumulate. If a population carrying these inherited variations becomes reproductively isolated, a new species may result.

Darwin, however, lacked an understanding of the genetic basis of variation and inheritance, a gap that left his theory open to reasonable criticism well into the twentieth century. Shortly after Darwin published his book, Gregor Johann Mendel published a paper in 1866 showing how traits were passed from generation to generation in pea plants, offering a general model of how traits are inherited. His research was little known until it was partially duplicated and brought to light by Karl Correns, Hugo de Vries, and Erich Tschermak around 1900.

By the early part of the twentieth century, it became clear that heredity and development were dependent on genetic information residing in genes contained in chromosomes, which were then contributed to each individual by gametes—the so-called **chromosomal theory of inheritance.** The gap in Darwin's theory was closed, and Mendel's research has continued to serve as the foundation of genetics.

2 Genetics Progressed from Mendel to DNA in Less Than a Century

Because genetic processes are fundamental to life itself, the science of genetics unifies biology and serves as its core. The starting point for this branch of science was a monastery garden in central Europe in the late 1850s.

Mendel's Work on Transmission of Traits

Gregor Mendel, an Augustinian monk, conducted a decade-long series of experiments using pea plants. He applied quantitative data analysis to his results and showed that traits are passed from parents to offspring in predictable ways. He further concluded that each trait in the plant is controlled by a pair of genes and that during gamete formation (the formation of egg cells and sperm), members of a gene pair separate from each other. His work was published in 1866 but was largely unknown until it was partially duplicated and cited in papers published by others around 1900. Once confirmed, Mendel's findings became recognized as explaining the transmission of traits in pea plants and all other higher organisms. His work forms the foundation for **genetics,** which is defined as the branch of biology concerned with the study of heredity and variation.

The Chromosome Theory of Inheritance: Uniting Mendel and Meiosis

Mendel did his experiments before the structure and role of chromosomes were known. About 20 years after his work was published, advances in microscopy allowed researchers to identify chromosomes (**Figure 2**) and establish that, in most eukaryotes, members of each species have a characteristic number of chromosomes called the **diploid number** (**2n**) in most of their cells. For example, humans have a diploid number of 46 (**Figure 3**). Chromosomes in diploid cells exist in pairs, called **homologous chromosomes.**

Researchers in the last decades of the nineteenth century also described chromosome behavior during two forms of cell division, **mitosis** and **meiosis**. In mitosis (**Figure 4**), chromosomes are copied and distributed so that each daughter cell receives a set of chromosomes identical to those in the parental cell. Meiosis is associated with gamete formation. Cells produced by meiosis receive only one chromosome from each chromosome pair, and the resulting number of chromosomes is called the **haploid (n) number.** This reduction in chromosome number is essential if the offspring arising from the fusion of egg and sperm are to maintain the constant number of chromosomes characteristic of their parents and other members of their species.

Early in the twentieth century, Walter Sutton and Theodor Boveri independently noted that the behavior of chromosomes during meiosis is identical to the behavior of genes during gamete formation described by Mendel. For example, genes and chromosomes exist in pairs, and members of a

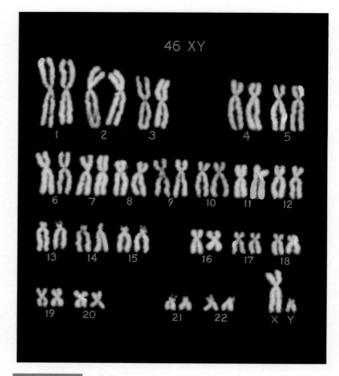

FIGURE 3   A colorized image of the human male chromosome set. Arranged in this way, the set is called a karyotype.

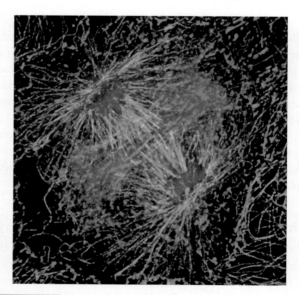

FIGURE 4   A late stage in mitosis after the chromosomes (stained blue) have separated.

gene pair and members of a chromosome pair separate from each other during gamete formation. Based on these parallels, Sutton and Boveri each proposed that genes are carried on chromosomes (**Figure 5**). They independently formulated the **chromosome theory of inheritance,** which states that inherited traits are controlled by genes residing on chromosomes faithfully transmitted through gametes, maintaining genetic continuity from generation to generation.

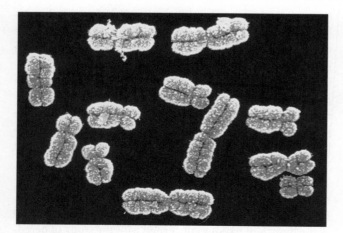

FIGURE 2   A colorized image of human chromosomes that have duplicated in preparation for cell division, as visualized under the scanning electron microscope.

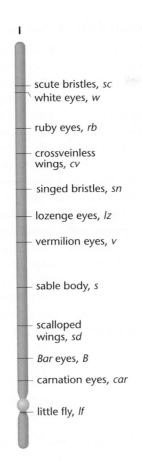

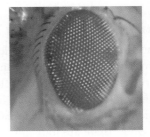

FIGURE 6 The normal red-eye color in *D. melanogaster* (bottom) and the white-eyed mutant (top).

FIGURE 5 A drawing of chromosome I (the X chromosome, one of the sex-determining chromosomes) of *D. melanogaster,* showing the location of several genes. Chromosomes can contain hundreds of genes.

ESSENTIAL POINT ■   ■   ■

The chromosome theory of inheritance explains how genetic information is transmitted from generation to generation.

Genetic Variation

About the same time that the chromosome theory of inheritance was proposed, scientists began studying the inheritance of traits in the fruit fly, *Drosophila melanogaster.* Early in this work, a white-eyed fly **(Figure 6)** was discovered among normal (wild-type) red-eyed flies. This variation was produced by a **mutation** in one of the genes controlling eye color. Mutations are defined as any heritable change in the DNA sequence and are the source of all genetic variation.

The white-eye color variant of the eye color gene discovered in *Drosophila* is an **allele** of a gene controlling eye color. Alleles are defined as alternative forms of a gene. Different alleles may produce differences in the observable features, or **phenotype,** of an organism. The set of alleles for a given trait carried by an organism is called the **genotype.** Using mutant

genes as markers, geneticists can map the location of genes on chromosomes (Figure 5).

The Search for the Chemical Nature of Genes: DNA or Protein?

Work on white-eyed *Drosophila* showed that the mutant trait could be traced to a single chromosome, confirming the idea that genes are carried on chromosomes. Once this relationship was established, investigators turned their attention to identifying which chemical component of chromosomes carried genetic information. By the 1920s, scientists knew that proteins and DNA were the major chemical components of chromosomes. There are a large number of different proteins, and because of their universal distribution in the nucleus and cytoplasm, many researchers thought proteins were the carriers of genetic information.

In 1944, Oswald Avery, Colin MacLeod, and Maclyn McCarty, researchers at the Rockefeller Institute in New York, published experiments showing that DNA was the carrier of genetic information in bacteria. This evidence, though clear-cut, failed to convince many influential scientists. Additional evidence for the role of DNA as a carrier of genetic information came from other researchers who worked with viruses. This evidence that DNA carries genetic information, along with other research over the next few years, provided solid proof that DNA, not protein, is the genetic material, setting the stage for work to establish the structure of DNA.

3 Discovery of the Double Helix Launched the Era of Molecular Genetics

Once it was accepted that DNA carries genetic information, efforts were focused on deciphering the structure of the DNA molecule and the mechanism by which information stored in it produces a phenotype.

The Structure of DNA and RNA

One of the great discoveries of the twentieth century was made in 1953 by James Watson and Francis Crick, who described the structure of DNA. DNA is a long, ladder-like macromolecule that twists to form a double helix (Figure 7). Each linear strand of the helix is made up of subunits called **nucleotides.** In DNA, there are four different nucleotides, each of which contains a nitrogenous base, abbreviated A (adenine), G (guanine), T (thymine), or C (cytosine). These four bases, in various sequence combinations, ultimately encode genetic information. The two strands of DNA are exact complements of one another, so that the rungs of the ladder in the double helix always consist of A=T and G≡C base pairs. Along with Maurice Wilkins, Watson and Crick were awarded a Nobel Prize in 1962 for their work on the structure of DNA.

Another nucleic acid, RNA, is chemically similar to DNA but contains a different sugar (ribose rather than deoxyribose) in its nucleotides and contains the nitrogenous base uracil in place of thymine. RNA, however, is generally a single-stranded molecule.

Gene Expression: From DNA to Phenotype

The genetic information encoded in the order of nucleotides in DNA is expressed in a series of steps that results in the expression of a phenotype. In eukaryotic cells, this process begins in the nucleus with **transcription,** in which the nucleotide sequence in one strand of DNA is used to construct a complementary RNA sequence (top part of **Figure 8**). Once an RNA molecule is produced, it moves to the cytoplasm, where the RNA—called **messenger RNA,** or **mRNA** for short—binds to a **ribosome.** The synthesis of proteins under the direction of mRNA is called **translation** (center part of Figure 8). The information encoded in mRNA (called the **genetic code**) consists of a linear series of nucleotide triplets. Each triplet, called a **codon,** is complementary to the information stored in DNA and specifies the insertion of a specific amino acid into a protein. Proteins (lower part of Figure 8) are polymers made up of amino acid monomers. There are 20 different amino acids commonly found in proteins.

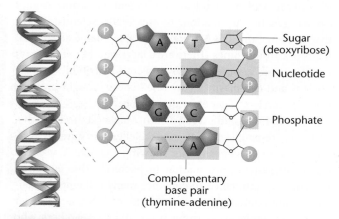

FIGURE 7 Summary of the structure of DNA, illustrating the arrangement of the double helix (on the left) and the chemical components making up each strand (on the right). The dotted lines on the right represent weak chemical bonds, called hydrogen bonds, which hold together the two strands of the DNA helix.

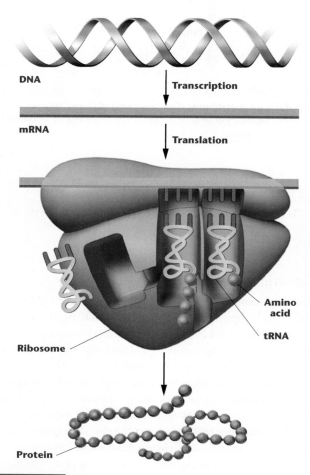

FIGURE 8 Gene expression consists of transcription of DNA into mRNA (top) and the translation (center) of mRNA (with the help of a ribosome) into a protein (bottom).

Protein assembly is accomplished with the aid of adapter molecules called **transfer RNA (tRNA).** Within the ribosome, tRNAs recognize the information encoded in the mRNA codons and carry the proper amino acids for construction of the protein during translation.

Proteins and Biological Function

Proteins are the end products of gene expression. The diversity of proteins and the biological functions they perform—the diversity of life itself—arises from the fact that proteins are made from combinations of 20 different amino acids. Consider that a protein chain containing 100 amino acids can have at each position any one of 20 amino acids; the number of possible different 100 amino acid proteins, each with a unique sequence, is therefore equal to

$$20^{100}$$

Obviously, proteins are molecules with the potential for enormous structural diversity and serve as the mainstay of biological systems.

Enzymes form the largest category of proteins. These molecules serve as biological catalysts, lowering the energy of activation in reactions, and allowing cellular metabolism to proceed at body temperature.

Proteins other than enzymes are critical components of cells and organisms. These include hemoglobin, the oxygen-binding molecule in red blood cells; insulin, a pancreatic hormone; collagen, a connective tissue molecule; and actin and myosin, the contractile muscle proteins. A protein's shape and chemical behavior are determined by its linear sequence of amino acids, which in turn are dictated by the stored information in the DNA of a gene that is transferred to RNA, which then directs the protein's synthesis. To repeat, DNA encodes RNA, which then directs the synthesis of protein.

Linking Genotype to Phenotype: Sickle-Cell Anemia

Once a protein is made, its biochemical or structural properties play a role in producing a phenotype. When mutation alters a gene, it may modify or even eliminate the encoded protein's usual function and cause an altered phenotype. To trace this chain of events, we will examine sickle-cell anemia, a human genetic disorder.

Sickle-cell anemia is caused by a mutant form of hemoglobin, the protein that transports oxygen from the lungs to cells in the body. Hemoglobin is a composite molecule made up of two different proteins, α-globin and β-globin, each encoded by a different gene. In sickle-cell anemia, a mutation in the gene encoding β-globin causes an amino acid substitution in 1 of the 146 amino acids in the protein. **Figure 9** shows part of the DNA sequence, and the corresponding mRNA codons and amino acid sequence, for the normal and

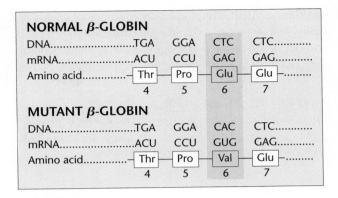

FIGURE 9 A single-nucleotide change in the DNA encoding β-globin (CTC→CAC) leads to an altered mRNA codon (GAG→GUG) and the insertion of a different amino acid (Glu→Val), producing the altered version of the β-globin protein that is responsible for sickle-cell anemia.

mutant forms of β-globin. Notice that the mutation in sickle-cell anemia consists of a change in one DNA nucleotide, which leads to a change in codon 6 in mRNA from GAG to GUG, which in turn changes amino acid number 6 in β-globin from glutamic acid to valine. The other 145 amino acids in the protein are not changed by this mutation.

Individuals with two mutant copies of the β-globin gene have sickle-cell anemia. Their mutant β-globin proteins cause hemoglobin molecules in red blood cells to polymerize when the blood's oxygen concentration is low, forming long chains of hemoglobin that distort the shape of red blood cells **(Figure 10).** The deformed cells are fragile and break easily, reducing the number of red blood cells in circulation (anemia is an insufficiency of red blood cells). Sickle-shaped blood cells block blood flow in capillaries and small blood vessels, causing severe pain and damage to the heart, brain, muscles,

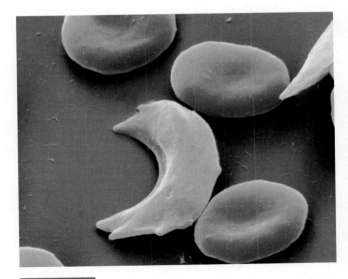

FIGURE 10 Normal red blood cells (round) and sickled red blood cells. The sickled cells block capillaries and small blood vessels.

21

and kidneys. All the symptoms of this disorder are caused by a change in a single nucleotide in a gene that changes one amino acid out of 146 in the β-globin molecule, demonstrating the close relationship between genotype and phenotype.

ESSENTIAL POINT ■ ■ ■

The central dogma of molecular biology—that DNA is a template for making RNA, which in turn directs the synthesis of proteins—explains how genes control phenotypes.

4 Development of Recombinant DNA Technology Began the Era of DNA Cloning

The era of recombinant DNA began in the early 1970s, when researchers discovered that **restriction enzymes,** used by bacteria to cut the DNA of invading viruses could be used to cut any organism's DNA at specific nucleotide sequences, producing a reproducible set of fragments.

Soon after, researchers discovered ways to insert the DNA fragments from restriction enzymes into carrier DNA molecules called **vectors** to form **recombinant DNA** molecules. When transferred into bacterial cells, thousands of copies, or **clones,** of the combined vector and DNA fragments are produced during bacterial reproduction. Large amounts of cloned DNA fragments can be isolated from these bacterial host cells. These DNA fragments can be used to isolate genes, to study their organization and expression, and to study their nucleotide sequence and evolution.

Collections of clones that represent an organism's **genome,** defined as the complete haploid DNA content of a specific organism, are called genomic libraries. Genomic libraries are now available for hundreds of species.

Recombinant DNA technology has not only accelerated the pace of research but also gave rise to the biotechnology industry, which has grown to become a major contributor to the U.S. economy.

5 The Impact of Biotechnology Is Continually Expanding

The use of recombinant DNA technology and other molecular techniques to make products is called **biotechnology.** In the United States, biotechnology has quietly revolutionized many aspects of everyday life; products made by biotechnology are now found in the supermarket, in health care, in agriculture, and in the court system.

Plants, Animals, and the Food Supply

The use of recombinant DNA technology to genetically modify crop plants has revolutionized agriculture. Genes for traits such as resistance to herbicides, insects, and nutritional enhancement have been introduced into crop plants (Table 1). The transfer of heritable traits across species using recombinant DNA technology creates **transgenic** organisms. Currently, more than a dozen different transgenic strains of crop plants are grown in the United States, with over 75 more being tested in field trials. Herbicide-resistant corn and soybeans were first planted in the mid-1990s, and transgenic strains now represent about 85 percent of the U.S. corn crop and 95 percent of the U.S. soybean crop. It is estimated that more than 75 percent of the processed food in the United States contains ingredients from transgenic crops.

Critics of this agricultural transformation are concerned that the use of herbicide-resistant crop plants may eventually result in the emergence of herbicide-resistant weeds and that traits in genetically engineered crops could be transferred to wild plants and will lead to irreversible changes in the ecosystem.

New methods of cloning livestock such as sheep and cattle have also changed the way we use these animals. In 1996, Dolly the sheep (**Figure 11**) was cloned by nuclear transfer, a method in which the nucleus of an adult cell is transferred into an egg that has had its nucleus removed. This method now makes it possible to produce dozens or hundreds of offspring with desirable traits. Cloning by nuclear transfer has many applications in agriculture, sports, and medicine.

Biotechnology has also changed the way human proteins for medical use are produced. Through use of gene transfer, transgenic animals now synthesize these therapeutic proteins. In 2009, an anticlotting protein derived from the milk of transgenic goats was approved by the U.S. Food and Drug Adminstration for use in the United States. Other

TABLE 1	Some Genetically Altered Traits in Crop Plants
Herbicide Resistance	
Corn, soybeans, rice, cotton, sugarbeets, canola	
Insect Resistance	
Corn, cotton, potato	
Virus Resistance	
Potato, yellow squash, papaya	
Nutritional Enhancement	
Golden rice	
Altered Oil Content	
Soybeans, canola	
Delayed Ripening	
Tomato	

human proteins from transgenic animals are now being used in clinical trials to treat several diseases, including emphysema. The biotechnology revolution will continue to expand as new methods are developed to make an increasing array of products.

Biotechnology in Genetics and Medicine

More than 10 million children or adults in the United States suffer from some form of genetic disorder, and every child-bearing couple faces an approximately 3 percent risk of having a child with some form of genetic anomaly. The molecular basis for hundreds of genetic disorders is now known (Figure 12), and many of these genes have been isolated and cloned. Biotechnology-derived genetic testing is now available for prenatal diagnosis of heritable disorders, and to test parents for their status as "carriers" of more than 100 inherited disorders. Newer methods now under development offer the possibility of scanning an entire genome to

FIGURE 11 Dolly, a Finn Dorset sheep cloned from the genetic material of an adult mammary cell, shown next to her first-born lamb, Bonnie.

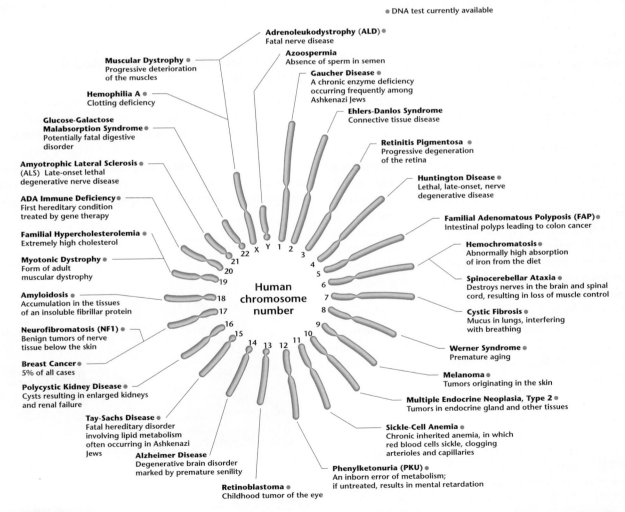

FIGURE 12 Diagram of the human chromosome set, showing the location of some genes whose mutant forms cause hereditary diseases. Conditions that can be diagnosed using genetic testing are indicated by a red dot.

establish an individual's risk of developing a genetic disorder or having an affected child. The use of genetic testing and related technologies raises ethical concerns that have yet to be resolved.

6 Genomics, Proteomics, and Bioinformatics Are New and Expanding Fields

The use of recombinant DNA technology to create genomic libraries prompted scientists to consider sequencing all the clones in a library to derive the nucleotide sequence of an organism's genome. This sequence information would be used to identify each gene in the genome and establish its function.

One such project, the Human Genome Project, began in 1990 as an international effort to sequence the human genome. By 2003, the publicly funded Human Genome Project and a private, industry-funded genome project completed sequencing of the gene-containing portion of the genome.

As more genome sequences were acquired, several new biological disciplines arose. One, called **genomics** (the study of genomes), studies the structure, function, and evolution of genes and genomes. A second field, **proteomics,** identifies the set of proteins present in a cell under a given set of conditions, and studies their functions and interactions. To store, retrieve, and analyze the massive amount of data generated by genomics and proteomics, a specialized subfield of information technology called **bioinformatics** was created to develop hardware and software for processing nucleotide and protein data.

Geneticists and other biologists now use information in databases containing nucleic acid sequences, protein sequences, and gene-interaction networks to answer experimental questions in a matter of minutes instead of months and years.

7 Genetic Studies Rely on the Use of Model Organisms

After the rediscovery of Mendel's work in 1900, work on a wide range of organisms confirmed that the principles of inheritance he described were of universal significance among plants and animals. Geneticists gradually came to focus attention on a small number of organisms, including the fruit fly (*Drosophila melanogaster*) and the mouse (*Mus musculus*) (Figure 13). This trend developed for two main reasons: first, it was clear that genetic mechanisms were the same in most organisms, and second, these organisms had characteristics that made them especially suitable for genetic research. They were easy to grow, had relatively short life cycles, produced many offspring, and their genetic analysis was fairly straightforward. Over time, researchers created a large catalog of mutant strains for these species, and the mutations were carefully studied, characterized, and mapped. Because of their well-characterized genetics, these species became **model organisms,** defined as organisms used for the study of basic biological processes.

The Modern Set of Genetic Model Organisms

Gradually, geneticists added other species to their collection of model organisms: viruses (such as the T phages and lambda phage) and microorganisms (the bacterium *Escherichia coli* and the yeast *Saccharomyces cerevisiae*) (Figure 14).

More recently, additional species have been developed as model organisms, three of which are shown in the chapter opening photograph. Each species was chosen to allow

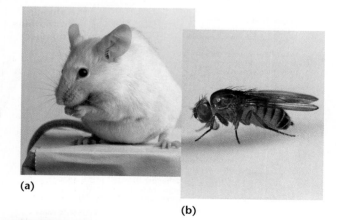

(a)

(b)

FIGURE 13 The first generation of model organisms in genetic analysis included (a) the mouse, *Mus musculus* and (b) the fruit fly, *Drosophila melanogaster*.

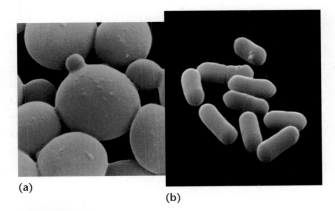

(a)

(b)

FIGURE 14 Microbes that have become model organisms for genetic studies include (a) the yeast *Saccharomyces cerevisiae* and (b) the bacterium *Eschericia coli*.

TABLE 2	Model Organisms Used to Study Some Human Diseases
Organism	**Human Diseases**
E. coli	Colon cancer and other cancers
S. cerevisiae	Cancer, Werner syndrome
D. melanogaster	Disorders of the nervous system, cancer
C. elegans	Diabetes
D. rerio	Cardiovascular disease
M. musculus	Lesch–Nyhan disease, cystic fibrosis, fragile-X syndrome, and many other diseases

study of some aspect of embryonic development. The nematode *Caenorhabditis elegans* was chosen as a model system to study the development and function of the nervous system because its nervous system contains only a few hundred cells and the developmental fate of these and all other cells in the body has been mapped out. *Arabidopsis thaliana*, a small plant with a short life cycle, has become a model organism for the study of many aspects of plant biology. The zebrafish, *Danio rerio*, is used to study vertebrate development: it is small, it reproduces rapidly, and its egg, embryo, and larvae are all transparent.

Model Organisms and Human Diseases

The development of recombinant DNA technology and the results of genome sequencing have confirmed that all life has a common origin. Because of this, genes with similar functions in different organisms tend to be similar or identical in structure and nucleotide sequence. Much of what scientists learn by studying the genetics of model organisms can therefore be applied to humans to serve as the basis for understanding and treating human diseases. In addition, the ability to create transgenic organisms by transferring genes between species has enabled scientists to develop models of human diseases in organisms ranging from bacteria to fungi, plants, and animals (Table 2).

The idea of studying a human disease such as colon cancer by using *E. coli* may strike you as strange, but the basic steps of DNA repair (a process that is defective in some forms of colon cancer) are the same in both organisms, and a gene involved (*mutL* in *E. coli* and *MLH1* in humans) is found in both organisms. More importantly, *E. coli* has the advantage of being easier to grow (the cells divide every 20 minutes), so that researchers can easily create and study new mutations in the bacterial *mutL* gene in order to figure out how it works. This knowledge may eventually lead to the development of drugs and other therapies to treat colon cancer in humans.

The fruit fly, *D. melanogaster*, is also being used to study specific human diseases. Mutant genes have been identified in *D. melanogaster* that produce phenotypes with abnormalities of the nervous system, including abnormalities of brain structure, and adult-onset degeneration of the nervous system. The information from genome-sequencing projects indicates that almost all these genes have human counterparts. For example, genes involved in a complex human disease of the retina called retinitis pigmentosa are identical to *Drosophila* genes involved in retinal degeneration. Study of these mutations in *Drosophila* is helping to dissect this complex disease and identify the function of the genes involved.

Another approach to studying diseases of the human nervous system is to transfer human disease genes into *Drosophila* by means of recombinant DNA technology. The transgenic flies are then used for studying the mutant human genes themselves, genes that affect the expression of the human disease genes, and the effects of therapeutic drugs on the action of those genes—all studies that are difficult or impossible to perform in humans. This gene-transfer approach is being used to study almost a dozen human neurodegenerative disorders, including Huntington disease, Machado–Joseph disease, myotonic dystrophy, and Alzheimer's disease.

As you read this, text you will encounter these model organisms again and again. Remember that each time you meet them; they not only have a rich history in basic genetics research but are also at the forefront in the study of human genetic disorders and infectious diseases. As discussed in the next section, however, we have yet to reach a consensus on how and when some of this technology will be determined to be safe and ethically acceptable.

ESSENTIAL POINT

The study of model organisms for understanding human health and disease is one of many ways genetics and biotechnology are rapidly changing everyday life.

8   We Live in the Age of Genetics

Mendel described his decade-long project on inheritance in pea plants in an 1865 paper presented at a meeting of the Natural History Society of Brünn in Moravia. Less than 100 years later, the 1962 Nobel Prize was awarded to James Watson, Francis Crick, and Maurice Wilkins for their work on the structure of DNA. This time span encompassed the years leading up to the acceptance of Mendel's work, the discovery that genes are on chromosomes, the experiments that proved DNA encodes genetic information, and the elucidation of the molecular basis for DNA replication. The rapid development of genetics from Mendel's monastery garden to the Human Genome Project and beyond is summarized in a timeline in **Figure 15**.

The Nobel Prize and Genetics

No other scientific discipline has experienced the explosion of information and the level of excitement generated by the discoveries in genetics. This impact is especially apparent in the list of Nobel Prizes related to genetics, beginning with those awarded in the early and mid-twentieth century and continuing into the present (see inside front cover). Nobel Prizes in Medicine or Physiology and Chemistry have been consistently awarded for work in genetics and related fields. The first Nobel Prize awarded for such work was given to Thomas Morgan in 1933 for his research on the chromosome theory of inheritance. That award was followed by many others, including prizes for the discovery of genetic recombination, the relationship between genes and proteins, the structure of DNA, and the genetic code. In this century, geneticists continue to be recognized for their impact on biology in the current millennium, including Nobel Prizes awarded in 2002, 2006, and 2007. In 2009, the Nobel Prize for Medicine or Physiology was awarded to Elizabeth Blackburn, Carol Greider, and Jack Szostak for their discovery of the structure and function of telomeres at the ends of eukaryotic chromosomes. The 2010 prize was given to Robert Edwards for the development of *in vitro* fertilization, and the 2010 Nobel Prize in Chemistry was awarded to Venkatraman Ramakrishnan, Thomas Steitz, and Ada Yonath for their work on the structure and function of the ribosome.

Genetics and Society

Just as there has never been a more exciting time to study genetics, the impact of this discipline on society has never been more profound. Genetics and its applications in biotechnology are developing much faster than the social conventions, public policies, and laws required to regulate their use. As a society, we are grappling with a host of sensitive genetics-related issues, including concerns about prenatal testing, genetic discrimination, ownership of genes, access to and safety of gene therapy, and genetic privacy. By the time you finish this course, you will have seen more than enough evidence to convince yourself that the present is the Age of Genetics, and you will understand the need to think about and become a participant in the dialogue concerning genetic science and its use.

ESSENTIAL POINT   ■   ■   ■

Genetic technology is having a profound effect on society, but policies and legislation governing its use are lagging behind the resulting innovations.

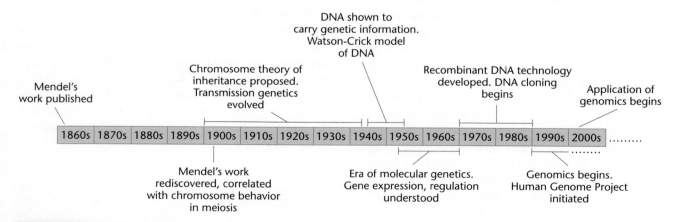

FIGURE 15   A timeline showing the development of genetics from Gregor Mendel's work on pea plants to the current era of genomics and its many applications in research, medicine, and society. Having a sense of the history of discovery in genetics should provide you with a useful framework as you proceed through this textbook.

GENETICS, TECHNOLOGY, AND SOCIETY

The Scientific and Ethical Implications of Modern Genetics

Today, genetics touches all aspects of life, bringing rapid changes in medicine, agriculture, law, biotechnology, and the pharmaceutical industry. Physicians use hundreds of genetic tests to diagnose and predict the course of disease and to detect genetic defects *in utero*. Scientists employ DNA-based methods to trace the path of evolution taken by many species, including our own. Farmers grow disease-resistant and drought-resistant crops, and raise more productive farm animals, created by gene transfer techniques. Law enforcement agencies apply DNA profiling methods to paternity, rape, and murder investigations. The biotechnology industry itself generates over 700,000 jobs and $50 billion in revenue each year and doubles in size every decade.

Along with these rapidly changing gene-based technologies comes a challenging array of ethical dilemmas. Who owns and controls genetic information? Are gene-enhanced agricultural plants and animals safe for humans and the environment? How can we ensure that genomic technologies will be available to all and not just to the wealthy? What are the likely social consequences of the new reproductive technologies? It is a time when everyone needs to understand genetics in order to make complex personal and societal decisions.

PROBLEMS AND DISCUSSION QUESTIONS

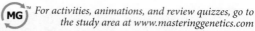

For activities, animations, and review quizzes, go to the study area at www.masteringgenetics.com

1. Describe Mendel's conclusions about how traits are passed from generation to generation.
2. What is the chromosome theory of inheritance, and how is it related to Mendel's findings?
3. Define genotype and phenotype, and describe how they are related.
4. What are alleles? Is it possible for more than two alleles of a gene to exist?
5. Given the state of knowledge at the time of the Avery, MacLeod, and McCarty experiment, why was it difficult for some scientists to accept that DNA is the carrier of genetic information?
6. Contrast chromosomes and genes.
7. How is genetic information encoded in a DNA molecule?
8. Describe the central dogma of molecular genetics and how it serves as the basis of modern genetics.
9. How many different proteins, each with a unique amino acid sequence, can be constructed with a length of five amino acids?
10. Outline the roles played by restriction enzymes and vectors in cloning DNA.
11. What are some of the impacts of biotechnology on crop plants in the United States?
12. Summarize the arguments for and against patenting genetically modified organisms.
13. We all carry about 20,000 genes in our genome. So far, patents have been issued for more than 6000 of these genes. Do you think that companies or individuals should be able to patent human genes? Why or why not?
14. How has the use of model organisms advanced our knowledge of the genes that control human diseases?
15. If you knew that a devastating late-onset inherited disease runs in your family (in other words, a disease that does not appear until later in life) and you could be tested for it at the age of 20, would you want to know whether you are a carrier? Would your answer be likely to change when you reach age 40?
16. Why do you think discoveries in genetics have been recognized with so many Nobel Prizes?

SOLUTIONS TO SELECTED PROBLEMS AND DISCUSSION QUESTIONS

2. Based on the parallels between Mendel's model of heredity and the behavior of chromosomes, the chromosome theory of inheritance emerged. It states that inherited traits are controlled by genes residing on chromosomes that are transmitted by gametes.

4. A gene variant is called an allele. There can be many such variants in a population, but for a diploid organism, only two such alleles can exist in any given individual.

6. *Genes*, the linear sequence of nucleotides, usually exert their influence by producing polypeptides through the process of transcription and translation. Genes are the functional units of heredity. They associate, sometimes with proteins, to form *chromosomes*. During the cell cycle, chromosomes, and therefore genes, are duplicated by a variety of enzymes so that daughter cells inherit copies of the parental hereditary information.

12. Unique transgenic plants and animals can be patented, as ruled by the United States Supreme Court in 1980. Supporters of organismic patenting argue that it is needed to encourage innovation and allow the costs of discovery to be recovered. Capital investors assume that there is a likely chance that their investments will yield positive returns. Others argue that natural substances should not be privately owned and that once they are owned by a small number of companies, free enterprise will be stifled.

16. For approximately 60 years discoveries in genetics have guided our understanding of living systems, aided rational drug design, and dominated many social discussions. Genetics provides the framework for universal biological processes and helps explain species stability and diversity. Given the central focus of genetics in so many of life's processes, it is understandable why so many genetic scientists have been awarded the Nobel Prize.

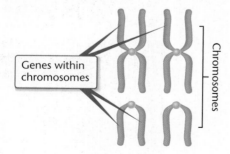

Genes within chromosomes

Chromosomes

CREDITS

Credits are listed in order of appearance.

Photo

CO,(a) Sinclair Stammers/Photo Researchers, Inc., (b) Mark Smith/ Photo Researchers, Inc., (c) Albert Salguero; F-1, Hartsoeker, N. Essay de dioptrique, Paris, 1694, p. 230. National Library of Medicine; F-2, Biophoto Associates/Science Source/Photo Researchers, Inc.; F-3, Medical-on-Line/Alamy; F-4, Dr. Alexey

Khodjakov/ Photo Researchers, Inc.; F-6, Science Source/Photo Researchers, Inc.; F-10, Oliver Meckes & Nicole Ottawa/Photo Researchers, Inc.; F-11, Roslin Institute; F-13 (a) John Paul Endress, (b) Hermann Eisenbeiss/Photo Researchers, Inc.; F-14 (a) Jeremy Burgess/Photo Researchers, Inc., (b) David McCarthy/Photo Researchers, Inc.

Chromosomes in the prometaphase stage of mitosis, from a cell in the flower of *Haemanthus*

Mitosis and Meiosis

CHAPTER CONCEPTS

- Genetic continuity between generations of cells and between generations of sexually reproducing organisms is maintained through the processes of mitosis and meiosis, respectively.

- Diploid eukaryotic cells contain their genetic information in pairs of homologous chromosomes, with one member of each pair being derived from the maternal parent and one from the paternal parent.

- Mitosis provides a mechanism to distribute chromosomes that have duplicated into progeny cells during cell reproduction.

- Mitosis converts a diploid cell into two diploid daughter cells.

- The process of meiosis distributes one member of each homologous pair of chromosomes into each gamete or spore, thus reducing the diploid chromosome number to the haploid chromosome number.

- Meiosis generates genetic variability by distributing various combinations of maternal and paternal members of each homologous pair of chromosomes into gametes or spores.

- It is during the stages of mitosis and meiosis that the genetic material has been condensed into discrete structures called chromosomes.

In every living thing, there exists a substance referred the genetic material. Except in certain viruses, this material is composed of the nucleic acid, DNA. A molecule of DNA is organized into units called genes, the products of which direct the metabolic activities of cells. DNA, with its array of genes, is organized into structures called chromosomes, which serve as vehicles for transmitting genetic information. The manner in which chromosomes are transmitted from one generation of cells to the next and from organisms to their descendants must be exceedingly precise. In this chapter, we consider exactly how genetic continuity is maintained between cells and organisms.

Two major processes are involved in eukaryotes: **mitosis** and **meiosis.** Although the mechanisms of the two processes are similar in many ways, the outcomes are quite different. On one hand, mitosis leads to the production of two cells, each with the same number of chromosomes as the parent cell. Meiosis, on the other hand, reduces the genetic content and the number of chromosomes to precisely half. This reduction is essential if sexual reproduction is to occur without doubling the amount of genetic material at each generation. Strictly speaking, mitosis is that portion of the cell cycle during which the hereditary components are equally divided into daughter cells. Meiosis is part of a special type of cell division that leads to the production of sex cells: **gametes** or **spores.** This process is an essential step in the transmission of genetic information from an organism to its offspring.

From Chapter 2 of *Essentials of Genetics, Eighth edition,* William S. Klug, Michael R. Cummings, Charlotte A. Spencer, Michael A. Palladino. © 2013 by Pearson Education, Inc. All rights reserved.

Normally, chromosomes are ~going division, the ge-
and meiosis. When cells are not~ only during mitosis
netic material making up ch~nes unfolds and uncoils
into a diffuse network with~cleus, generally referred to
as chromatin. Before d~~ ~nitosis and meiosis, we will
briefly review the st~~~ce to genetic function. We will
that are of particu~~ells, emphasizing components
also compare ~en differences between the prokaryotic (nonn~ ~bacteria and the eukaryotic cells of
higher o~ ~devote the remainder of the chapter to ~~nromosomes during cell division.

Structure Is Closely Tied ~Genetic Function

~40, our knowledge of cell structure was limited to
~could see with the light microscope. Around 1940,
~nsmission electron microscope was in its early stages of

development, and by 1960 many details of cell ultrastructure
were emerging. Under the electron microscope, cells were
seen as highly organized, precise structures. A new world
of whorled membranes, organelles, microtubules, granules,
and filaments was revealed. These discoveries revolutionized
thinking in the entire field of biology, but we will be concerned only with those aspects of cell structure that relate to
genetic study. The typical animal cell shown in **Figure 1** illustrates most of the structures we will discuss.

All cells are surrounded by a **plasma membrane,** an
outer covering that defines the cell boundary and delimits
the cell from its immediate external environment. This membrane is not passive but instead actively controls the movement of materials into and out of the cell. In addition to this
membrane, plant cells have an outer covering called the **cell
wall** whose major component is a polysaccharide called cellulose.

Many, if not most, animal cells have a covering over
the plasma membrane, referred to as the **glycocalyx,** or **cell
coat.** Consisting of glycoproteins and polysaccharides, this

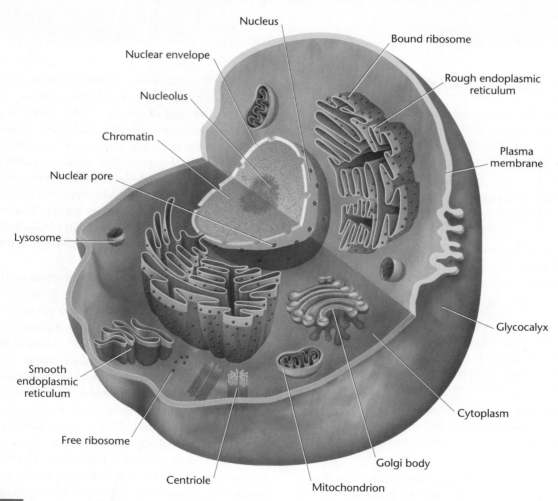

FIGURE 1 A generalized animal cell. The cellular components discussed in the text are emphasized here.

Labels: Nucleus, Nuclear envelope, Nucleolus, Chromatin, Nuclear pore, Lysosome, Smooth endoplasmic reticulum, Free ribosome, Centriole, Mitochondrion, Golgi body, Cytoplasm, Glycocalyx, Plasma membrane, Rough endoplasmic reticulum, Bound ribosome covering has a chemical composition that differs from comparable structures in either plants or bacteria. The glycocalyx, among other functions, provides biochemical identity at the surface of cells, and the components of the coat that establish cellular identity are under genetic control. For example, various cell surface markers that you may have heard of—the ABO, AB, Rh, and MN antigens—are found on the surface of red blood cells, among other cell types. On the surface of other cells, histocompatibility antigens, which elicit an immune response during tissue and organ transplants, are present. Various **receptor molecules** are also found on the surfaces of cells. These molecules act as recognition sites that transfer specific chemical signals across the cell membrane into the cell.

Living organisms are categorized into two major groups depending on whether or not their cells contain a nucleus. The presence of a nucleus and other membranous organelles characterizes **eukaryotic cells.** The **nucleus** houses the genetic material, DNA, which is complexed with an array of acidic and basic proteins into thin fibers. During nondivisional phases of the cell cycle, these fibers are uncoiled and dispersed into **chromatin.** During mitosis and meiosis, chromatin fibers coil and condense into structures called **chromosomes.** Also present in the nucleus is the **nucleolus,** an amorphous component where ribosomal RNA (rRNA) is synthesized and where the initial stages of ribosomal assembly occur. The areas of DNA that encode rRNA are collectively referred to as the **nucleolus organizer region,** or the **NOR.**

Prokaryotic cells lack a nuclear envelope and membranous organelles. In many bacteria such as *Escherichia coli,* the genetic material is present as a long, circular DNA molecule compacted into the **nucleoid** area. Part of the DNA may be attached to the cell membrane, but in general the nucleoid constitutes a large area throughout the cell. Although the DNA is compacted, it does not undergo the extensive coiling characteristic of the stages of mitosis where, in eukaryotes, chromosomes become visible. Nor is the DNA in these organisms associated as extensively with proteins as is eukaryotic DNA. **Figure 2,** in which two bacteria are forming during cell division, illustrates the nucleoid regions that house the bacterial chromosome. Prokaryotic cells do not have a distinct nucleolus, but they do contain genes that specify rRNA molecules.

The remainder of the eukaryotic cell enclosed by the plasma membrane, excluding the nucleus, is composed of **cytoplasm** and all associated cellular organelles. Cytoplasm is a nonparticulate, colloidal material referred to as the cytosol, which surrounds and encompasses the cellular organelles. Beyond these components, an extensive system of tubules and filaments comprising the cytoskeleton provides a lattice of support structures within the cytoplasm. Consisting primarily of tubulin-derived microtubules and actin-derived microfilaments, this structural framework maintains cell shape, facilitates cell mobility, and anchors the various organelles.

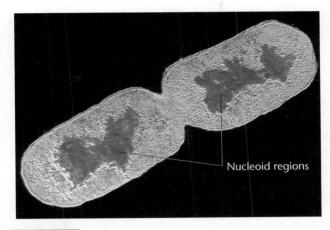

FIGURE 2   Color-enhanced electron micrograph of *E. coli* undergoing cell division. Particularly prominent are the two chromosomal areas (shown in red), called nucleoids, that have been partitioned into the daughter cells.

One such organelle, the membranous **endoplasmic reticulum (ER),** compartmentalizes the cytoplasm, greatly increasing the surface area available for biochemical synthesis. The ER may appear smooth, in which case it serves as the site for synthesizing fatty acids and phospholipids, or it may appear rough because it is studded with ribosomes. Ribosomes serve as sites where genetic information contained in messenger RNA (mRNA) is translated into proteins.

Three other cytoplasmic structures are very important in the eukaryotic cell's activities: mitochondria, chloroplasts, and centrioles. **Mitochondria** are found in both animal and plant cells and are the sites of the oxidative phases of cell respiration. These chemical reactions generate large amounts of adenosine triphosphate (ATP), an energy-rich molecule. **Chloroplasts** are found in plants, algae, and some protozoans. These organelles are associated with photosynthesis, the major energy-trapping process on Earth. Both mitochondria and chloroplasts contain a type of DNA distinct from that found in the nucleus. Furthermore, these organelles can duplicate themselves and transcribe and translate their genetic information. It is interesting to note that the genetic machinery of mitochondria and chloroplasts closely resembles that of prokaryotic cells. This and other observations have led to the proposal that these organelles were once primitive free-living organisms that established a symbiotic relationship with a primitive eukaryotic cell. This theory, which describes the evolutionary origin of these organelles, is called the **endosymbiotic hypothesis.**

Animal cells and some plant cells also contain a pair of complex structures called the **centrioles.** These cytoplasmic bodies, located in a specialized region called the centrosome, are associated with the organization of spindle fibers that function in mitosis and meiosis. In some organisms, the centriole is derived from another structure, the basal body, which is associated with the formation of cilia and flagella.

Over the years, many reports have suggested that centrioles and basal bodies contain DNA, which could be involved in the replication of these structures. Currently, this is thought not to be the case.

The organization of **spindle fibers** by the centrioles occurs during the early phases of mitosis and meiosis. These fibers play an important role in the movement of chromosomes as they separate during cell division. They are composed of arrays of microtubules consisting of polymers of polypeptide subunits of the protein tubulin.

ESSENTIAL POINT ■ ■ ■

Most components of cells are involved directly or indirectly with genetic processes.

---

2 Chromosomes Exist in Homologous Pairs in Diploid Organisms

As we discuss the processes of mitosis and meiosis, it is important that you understand the concept of homologous chromosomes. Such an understanding will also be of critical importance in our future discussions of Mendelian genetics. Chromosomes are most easily visualized during mitosis. When they are examined carefully, distinctive lengths and shapes are apparent. Each chromosome contains a condensed or constricted region called the **centromere,** which establishes the general appearance of each chromosome. **Figure 3** shows chromosomes with centromere placements at different points along their lengths. Extending from either side of the centromere are the arms of the chromosome. Depending on the position of the centromere, different arm ratios are produced. As Figure 3 illustrates, chromosomes are classified as **metacentric, submetacentric, acrocentric,** or **telocentric** on the basis of the centromere location. The shorter arm, by convention, is shown above the centromere and is called the **p arm** (p, for "petite"). The longer arm is shown below the centromere and is called the **q arm** (because q is the next letter in the alphabet).

When studying mitosis, several observations are of particular relevance. First, all somatic cells derived from members of the same species contain an identical number of chromosomes. In most cases, this represents the **diploid number (2n).** When the lengths and centromere placements of the chromosomes are examined, a second general feature is apparent. Nearly all chromosomes exist in pairs with regard to these two criteria, and the members of each pair are called **homologous chromosomes.** So for each chromosome exhibiting a specific length and centromere placement, another exists with identical features.

There are exceptions to this rule. Most bacteria and viruses have only one chromosome, and organisms such as yeasts and molds and certain plants such as bryophytes (mosses) spend the predominant phase of the life cycle in the haploid

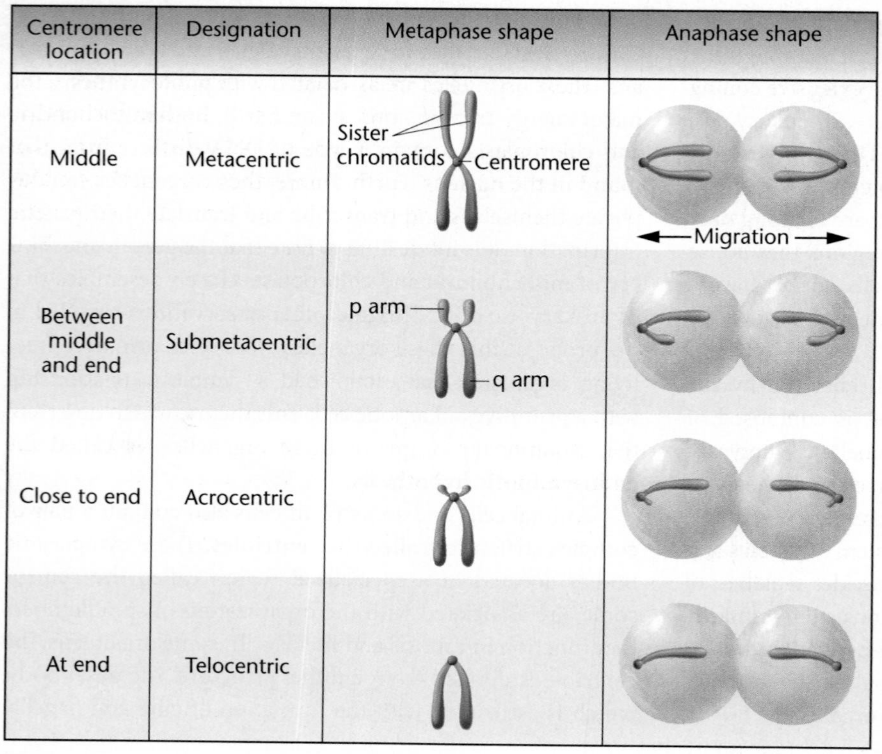

Centromere location	Designation	Metaphase shape	Anaphase shape
Middle	Metacentric	Sister chromatids / Centromere	Migration
Between middle and end	Submetacentric	p arm / q arm	
Close to end	Acrocentric		
At end	Telocentric		

FIGURE 3 Centromere locations and designations of chromosomes based on centromere location. Note that the shape of the chromosome during anaphase is determined by the position of the centromere.

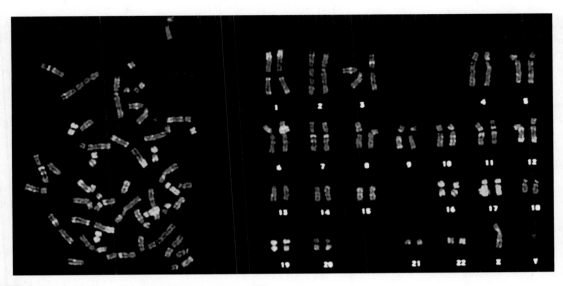

FIGURE 4   A metaphase preparation of chromosomes derived from a dividing cell of a human male (left), and the karyotype derived from the metaphase preparation (right). All but the X and Y chromosomes are present in homologous pairs. Each chromosome is clearly a double structure, constituting a pair of sister chromatids joined by a common centromere.

stage. That is, they contain only one member of each homologous pair of chromosomes during most of their lives.

Figure 4 illustrates the physical appearance of different pairs of homologous chromosomes. There, the human mitotic chromosomes have been photographed, cut out of the print, and matched up, creating a **karyotype.** As you can see, humans have a 2*n* number of 46, which on close examination exhibit a diversity of sizes and centromere placements. Note also that each of the 46 chromosomes is clearly a double structure consisting of two parallel **sister chromatids** connected by a common centromere. Had these chromosomes been allowed to continue dividing, the sister chromatids, which are replicas of one another, would have separated into two new cells as division continued.

The **haploid number (*n*)** of chromosomes is equal to one-half the diploid number. Collectively, the genetic information contained in a haploid set of chromosomes constitutes the **genome** of the species. This, of course, includes copies of all genes as well as a large amount of noncoding DNA. The examples listed in **Table 1** demonstrate the wide range of *n* values found in plants and animals.

Homologous pairs of chromosomes have important genetic similarities. They contain identical gene sites along their lengths, each called a **locus** (pl., loci). Thus, they are identical in their genetic potential. In sexually reproducing organisms, one member of each pair is derived from the maternal parent (through the ovum), and one is derived from the paternal parent (through the sperm). Therefore, each diploid organism contains two copies of each gene as a consequence of **biparental inheritance.** As we shall see in the following chapters on transmission genetics, the members of each pair of genes, while influencing the same characteristic or trait, need not be identical. In a population of members of the same species, many different alternative forms of the same gene, called **alleles,** can exist. The concepts of haploid number, diploid number, and homologous chromosomes are important in understanding the process of meiosis. During the formation of gametes or spores, meiosis converts the diploid number of chromosomes to the haploid number. As a result, haploid gametes or spores contain precisely one member of each homologous pair of chromosomes—that is, one complete haploid set. Following fusion of two gametes in fertilization, the diploid number is reestablished; that is, the zygote contains two complete sets of haploid chromosomes, one set from each parent. The constancy of genetic material is thus maintained from generation to generation.

There is one important exception to the concept of homologous pairs of chromosomes. In many species, one pair, the **sex-determining chromosomes,** is often not homologous in size, centromere placement, arm ratio, or genetic content. For example, in humans, while females carry two homologous X chromosomes, males carry one Y chromosome and one X chromosome (Figure 4). These X and Y chromosomes are not strictly homologous. The Y is considerably smaller and lacks most of the loci contained on the X. Nevertheless, they contain homologous regions and behave as homologs in meiosis so that gametes produced by males receive either one X or one Y chromosome.

ESSENTIAL POINT   ■   ■   ■

In diploid organisms, chromosomes exist in homologous pairs, where each member is identical in size, centromere placement, and gene sites. One member of each pair is derived from the maternal parent, and one is derived from the paternal parent.

TABLE 1   The Haploid Number of Chromosomes for a Variety of Organisms

Common Name	Scientific Name	Haploid Number
Black bread mold	Aspergillus nidulans	8
Broad bean	Vicia faba	6
Cat	Felis domesticus	19
Cattle	Bos taurus	30
Chicken	Gallus domesticus	39
Chimpanzee	Pan troglodytes	24
Corn	Zea mays	10
Cotton	Gossypium hirsutum	26
Dog	Canis familiaris	39
Evening primrose	Oenothera biennis	7
Frog	Rana pipiens	13
Fruit fly	Drosophila melanogaster	4
Garden onion	Allium cepa	8
Garden pea	Pisum sativum	7
Grasshopper	Melanoplus differentialis	12
Green alga	Chlamydomonas reinhardii	18
Horse	Equus caballus	32
House fly	Musca domestica	6
House mouse	Mus musculus	20
Human	Homo sapiens	23
Jimson weed	Datura stramonium	12
Mosquito	Culex pipiens	3
Mustard plant	Arabidopsis thaliana	5
Pink bread mold	Neurospora crassa	7
Potato	Solanum tuberosum	24
Rhesus monkey	Macaca mulatta	21
Roundworm	Caenorhabditis elegans	6
Silkworm	Bombyx mori	28
Slime mold	Dictyostelium discoidium	7
Snapdragon	Antirrhinum majus	8
Tobacco	Nicotiana tabacum	24
Tomato	Lycopersicon esculentum	12
Water fly	Nymphaea alba	80
Wheat	Triticum aestivum	21
Yeast	Saccharomyces cerevisiae	16
Zebrafish	Danio rerio	25

3   Mitosis Partitions Chromosomes into Dividing Cells

The process of mitosis is critical to all eukaryotic organisms. In some single-celled organisms, such as protozoans and some fungi and algae, mitosis (as a part of cell division) provides the basis for asexual reproduction. Multicellular diploid organisms begin life as single-celled fertilized eggs called **zygotes.** The mitotic activity of the zygote and the subsequent daughter cells is the foundation for the development and growth of the organism. In adult organisms, mitotic activity is prominent in wound healing and other forms of cell replacement in certain tissues. For example, the epidermal skin cells of humans are continuously sloughed off and replaced. Cell division also results in the continuous production of reticulocytes (immature red blood cells) that eventually shed their nuclei and replenish the supply of red blood cells in vertebrates. In abnormal situations, somatic cells may lose control of cell division and form a tumor.

The genetic material is partitioned into daughter cells during nuclear division or **karyokinesis.** This process is quite complex and requires great precision. The chromosomes must first be exactly replicated and then accurately partitioned. The end result is the production of two daughter nuclei, each with a chromosome composition identical to that of the parent cell.

Karyokinesis is followed by cytoplasmic division, or **cytokinesis.** The less complex division of the cytoplasm requires a mechanism that partitions the volume into two parts, then encloses both new cells within a distinct plasma membrane. Cytoplasmic organelles either replicate themselves, arise from existing membrane structures, or are synthesized *de novo* (anew) in each cell. The subsequent proliferation of these structures is a reasonable and adequate mechanism for reconstituting the cytoplasm in daughter cells.

Following cell division, the initial size of each new daughter cell is approximately one-half the size of the parent cell. However, the nucleus of each new cell is not appreciably smaller than the nucleus of the original cell. Quantitative measurements of DNA confirm that there is an amount of genetic material in the daughter nuclei equivalent to that in the parent cell.

Interphase and the Cell Cycle

Many cells undergo a continuous alternation between division and nondivision. The events that occur from the completion of one division until the beginning of the next division constitute the **cell cycle (Figure 5)**. We will consider the initial **interphase** stage of the cycle as the interval between divisions. It was once thought that the biochemical activity during interphase was devoted solely to the cell's growth and its normal function. However, we now know that another biochemical step critical to the ensuing mitosis occurs during interphase: *replication of the DNA of each chromosome.* This period during which DNA is synthesized occurs before the cell enters mitosis and is called the **S phase.** The initiation and completion of DNA synthesis can be detected by monitoring the incorporation of radioactive precursors into DNA.

Investigations of this nature show two periods during interphase when no DNA synthesis occurs, one before and one after S phase. These are designated **G1 (gap I) and G2 (gap II),** respectively. During both of these intervals, as well as during S, intensive metabolic activity, cell growth, and cell differentiation occur. By the end of G2, the volume of the cell has roughly doubled, DNA has been replicated, and mitosis (M) is initiated. Following mitosis, continuously dividing

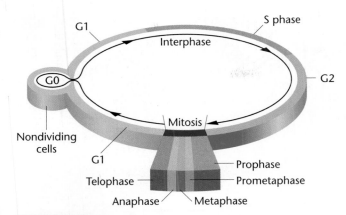

The intervals comprising an arbitrary cell cycle. Following mitosis, cells enter the G1 stage of interphase, initiating a new cycle. Cells may become nondividing (G0) or continue through G1, where they become committed to begin DNA synthesis (S) and complete the cycle (G2 and mitosis). Following mitosis, two daughter cells are produced and the cycle begins anew for both cells.

cells then repeat this cycle (G1, S, G2, M) over and over, as shown in Figure 5.

Much is known about the cell cycle based on *in vitro* (in glass) studies. When grown in culture, many cell types in different organisms traverse the complete cycle in about 16 hours. The actual process of mitosis occupies only a small part of the overall cycle, often less than an hour. The lengths of the S and G2 phases of interphase are fairly consistent among different cell types. Most variation is seen in the length of time spent in the G1 stage. **Figure 6** shows the length of these intervals in a typical cell.

G1 is of great interest in the study of cell proliferation and its control. At a point late in G1, all cells follow one of two paths. They either withdraw from the cycle, become quiescent, and enter the **G0 stage** (see Figure 5), or they become committed to initiating DNA synthesis and completing the cycle. Cells that enter G0 remain viable and metabolically active but are nonproliferative. Cancer cells apparently avoid entering G0 or

Interphase			Mitosis
G1	S	G2	M
5	7	3	1

Hours

Pro	Met	Ana	Tel
36	3	3	18

Minutes

The time spent in each phase of one complete cell cycle of a human cell in culture. Times vary according to cell types and conditions.

pass through it very quickly. Other cells enter G0 and never reenter the cell cycle. Still others remain in G0, but they can be stimulated to return to G1 and thereby reenter the cycle.

Cytologically, interphase is characterized by the absence of visible chromosomes. Instead, the nucleus is filled with chromatin fibers that are formed as the chromosomes are uncoiled and dispersed after the previous mitosis [**Figure 7(a)**,]. Once G1, S, and G2 are completed, mitosis is initiated. Mitosis is a dynamic period of vigorous and continual activity. For discussion purposes, the entire process is subdivided into discrete stages, and specific events are assigned to each one. These stages, in order of occurrence, are prophase, prometaphase, metaphase, anaphase, and telophase. They are diagrammed in Figure 7 along with photomicrographs of each stage.

Prophase

Often, over half of mitosis is spent in **prophase** [**Figure 7(b)**], a stage characterized by several significant activities. One of the early events in prophase of all animal cells involves the migration of two pairs of centrioles to opposite ends of the cell. These structures are found just outside the nuclear envelope in an area of differentiated cytoplasm called the **centrosome.** It is believed that each pair of centrioles consists of one mature unit and a smaller, newly formed centriole.

The centrioles migrate to establish poles at opposite ends of the cell. After migration, the centrosomes, in which the centrioles are localized, are responsible for organizing cytoplasmic microtubules into the spindle fibers that run between these poles, creating an axis along which chromosomal separation occurs. Interestingly, the cells of most plants (there are a few exceptions), fungi, and certain algae seem to lack centrioles. Spindle fibers are nevertheless apparent during mitosis.

As the centrioles migrate, the nuclear envelope begins to break down and gradually disappears. In a similar fashion, the nucleolus disintegrates within the nucleus. While these events are taking place, the diffuse chromatin fibers have begun to condense, until distinct threadlike structures, the chromosomes, become visible. It becomes apparent near the end of prophase that each chromosome is actually a double structure split longitudinally except at a single point of constriction, the centromere. The two parts of each chromosome are called **sister chromatids** because the DNA contained in each of them is genetically identical, having formed from a single replicative event. Sister chromatids are held together by a multi-subunit protein complex called **cohesin.** This molecular complex is originally formed between them during the S phase of the cell cycle when the DNA of each chromosome is replicated. Thus, even though we cannot see chromatids in interphase because the chromatin is uncoiled and dispersed in the nucleus, the chromosomes are already double struc-

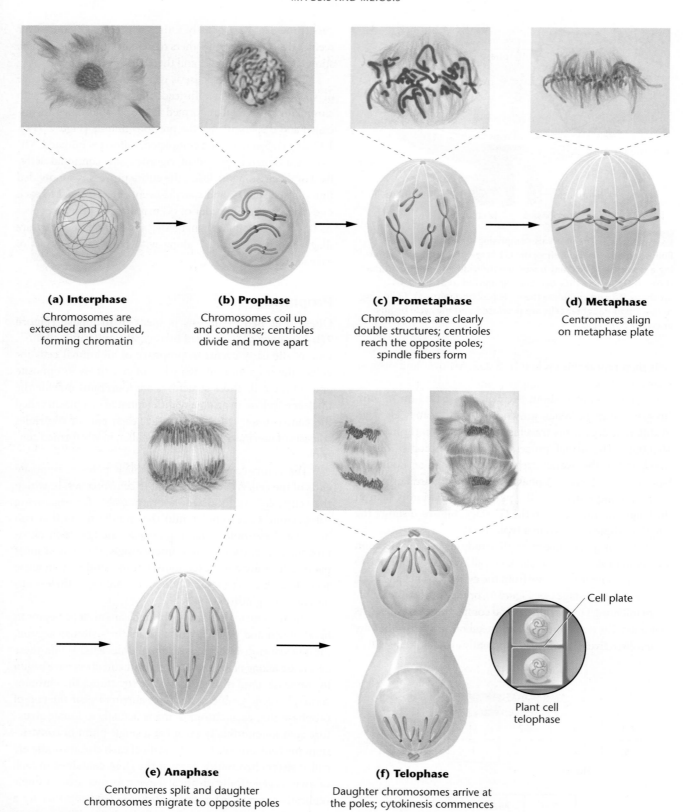

(a) Interphase
Chromosomes are extended and uncoiled, forming chromatin

(b) Prophase
Chromosomes coil up and condense; centrioles divide and move apart

(c) Prometaphase
Chromosomes are clearly double structures; centrioles reach the opposite poles; spindle fibers form

(d) Metaphase
Centromeres align on metaphase plate

(e) Anaphase
Centromeres split and daughter chromosomes migrate to opposite poles

(f) Telophase
Daughter chromosomes arrive at the poles; cytokinesis commences

Cell plate

Plant cell telophase

FIGURE 7   Drawings depicting mitosis in an animal cell with a diploid number of 4. The events occurring in each stage are described in the text. Of the two homologous pairs of chromosomes, one contains longer, metacentric members and the other shorter, submetacentric members. The maternal chromosome and the paternal chromosome of each pair are shown in different colors. In (f), the late telophase stage in a plant cell illustrates the formation of the cell plate and lack of centrioles. The cells shown in light micrographs came from the flower of *Haemanthus,* a plant that has a diploid number of 8.

tures, which becomes apparent in late prophase. In humans, with a diploid number of 46, a cytological preparation of late prophase reveals 46 chromosomes randomly distributed in the area formerly occupied by the nucleus.

Prometaphase and Metaphase

The distinguishing event of the two ensuing stages is the migration of every chromosome, led by its centromeric region, to the equatorial plane. The equatorial plane, also referred to as the *metaphase plate,* is the midline region of the cell, a plane that lies perpendicular to the axis established by the spindle fibers. In some descriptions, the term **prometaphase** refers to the period of chromosome movement **[Figure 7(c)]**, and the term **metaphase** is applied strictly to the chromosome configuration following migration.

Migration is made possible by the binding of spindle fibers to the chromosome's **kinetochore,** an assembly of multilayered plates of proteins associated with the centromere. This structure forms on opposite sides of each paired centromere, in intimate association with the two sister chromatids. Once properly attached to the spindle fibers, cohesin is degraded by an enzyme, appropriately named *separase,* and the sister chromatid arms disjoin, except at the centromere region. A unique protein family called **shugoshin** (from the Japanese meaning guardian spirit) protects cohesin from being degraded by separase at the centromeric regions. The involvement of the cohesin and shugoshin complexes with a pair of sister chromatids during mitosis is depicted in **Figure 8.**

At the completion of metaphase, each centromere is aligned at the metaphase plate with the chromosome arms extending outward in a random array. This configuration is shown in **Figure 7(d).**

Anaphase

Events critical to chromosome distribution during mitosis occur during **anaphase,** the shortest stage of mitosis. During this phase, sister chromatids of each chromosome *disjoin* (separate) from each other and migrate to opposite ends of the cell. For complete disjunction to occur, each centromeric region must be split in two. This event signals the initiation of anaphase. Once it occurs, each chromatid is referred to as a **daughter chromosome.**

Movement of daughter chromosomes to the opposite poles of the cell is dependent on the kinetochore–spindle fiber attachment. Recent investigations reveal that chromosome migration results from the activity of a series of specific proteins, generally called motor proteins. These proteins use the energy generated by the hydrolysis of ATP, and their activity is said to constitute **molecular motors** in the cell. These motors act at several positions within the dividing cell, but all of them are involved in the activity of microtubules and ultimately serve to propel the chromosomes to opposite ends of the cell. The centromeres of each chromosome *appear* to lead the way during migration, with the chromosome arms trailing behind. The location of the centromere determines the shape of the chromosome during separation, as you saw in Figure 3.

The steps that occur during anaphase are critical in providing each subsequent daughter cell with an identical set of chromosomes. In human cells, there would now be 46 chromosomes at each pole, one from each original sister pair. **Figure 7(e)** shows anaphase just prior to its completion.

Telophase

Telophase is the final stage of mitosis and is depicted in **Figure 7(f).** At its beginning, there are two complete sets of chromosomes, one set at each pole. The most significant event

> ### NOW SOLVE THIS
>
> **1** If an organism has a diploid number of 16, how many chromatids are visible at the end of mitotic prophase? How many chromosomes are moving to each pole during anaphase of mitosis?
>
> ■ **Hint** *This problem involves an understanding of what happens to each pair of homologous chromosomes during mitosis, asking you to extrapolate your understanding to chromosome behavior in an organism with a diploid number of 16. The major insight needed to solve this problem is to understand that, throughout mitosis, members of each homologous pair do not pair up but instead behave individually.*

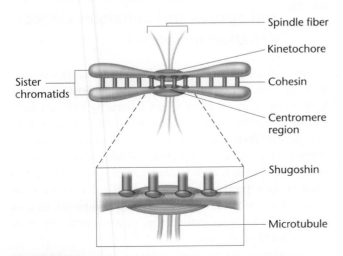

Spindle fiber

Kinetochore

Cohesin

Sister chromatids

Centromere region

Shugoshin

Microtubule

FIGURE 8 The depiction of the alignment, pairing, and disjunction of sister chromatids during mitosis, involving the molecular complexes cohesin and shugoshin and the enzyme separase.

is **cytokinesis,** the division or partitioning of the cytoplasm. Cytokinesis is essential if two new cells are to be produced from one cell. The mechanism differs greatly in plant and animal cells. In plant cells, a **cell plate** is synthesized and laid down across the region of the metaphase plate. Animal cells, however, undergo a constriction of the cytoplasm in the same way a loop of string might be tightened around the middle of a balloon. The end result is the same: Two distinct cells are formed.

It is not surprising that the process of cytokinesis varies among cells of different organisms. Plant cells, which are more regularly shaped and are structurally rigid, require a mechanism for depositing new cell wall material around the plasma membrane. The cell plate, laid down during telophase, becomes the **middle lamella.** Subsequently, the primary and secondary layers of the cell wall are deposited between the cell membrane and middle lamella on both sides of the boundary between the two daughter cells. In animals, complete constriction of the cell membrane produces a **cell furrow** characteristic of newly divided cells.

Other events necessary for the transition from mitosis to interphase are initiated during late telophase. They are generally a reversal of the events that occurred during prophase. In each new cell, the chromosomes begin to uncoil and become diffuse chromatin once again while the nuclear envelope re-forms around them. The spindle fibers disappear, and the nucleolus gradually re-forms and becomes visible in the nucleus during early interphase. At the completion of telophase, the cell enters interphase.

Cell-Cycle Regulation and Checkpoints

The cell cycle (Figure 5), culminating in mitosis, is fundamentally the same in all eukaryotic organisms. This similarity in many diverse organisms suggests that the cell cycle is governed by a genetically regulated program that has been conserved throughout evolution. Because disruption of this regulation may underlie the uncontrolled cell division characterizing malignancy, interest in how genes regulate the cell cycle is particularly strong.

A mammoth research effort over the past 20 years has paid high dividends, and we now have knowledge of many genes involved in the control of the cell cycle. This work was recognized by the awarding of the 2001 Nobel Prize in Medicine or Physiology to Lee Hartwell, Paul Nurse, and Tim Hunt. As with other studies of genetic control over essential biological processes, investigation has focused on the discovery of mutations that interrupt the cell cycle and on the effects of those mutations.

Many mutations are now known that exert an effect at one or another stage of the cell cycle. First discovered in yeast, but now evident in all organisms, including humans, such mutations were originally designated as *cell division cycle (cdc)* **mutations.** The normal products of many of the mutated genes are enzymes called **kinases** that can add

phosphates to other proteins. They serve as "master control" molecules functioning in conjunction with proteins called **cyclins.** Cyclins bind to these kinases, activating them at appropriate times during the cell cycle. Activated kinases then phosphorylate other target proteins that regulate the progress of the cell cycle.

The study of *cdc* mutations has established that the cell cycle contains at least three major **checkpoints,** when the processes culminating in normal mitosis are monitored, or "checked," by these master control molecules before the next stage of the cycle commences. The importance of cell-cycle control and these checkpoints is illustrated by considering what happens when this regulatory system is impaired. Let's assume, for example, that the DNA of a cell has incurred damage leading to one or more mutations impairing cell-cycle control. If allowed to proceed through the cell cycle as one of the population of dividing cells, this genetically altered cell would divide uncontrollably—precisely the definition of a tumor cell. If instead the cell cycle is arrested at one of the checkpoints, the cell may effectively be removed from the population of dividing cells, preventing its potential malignancy.

> ## ESSENTIAL POINT ■ ■ ■
>
> Mitosis is subdivided into discrete stages that initially depict the condensation of chromatin into the diploid number of chromosomes, each of which is initially a double structure, each composed of a pair of sister chromatids. During mitosis, sister chromatids are pulled apart and directed toward opposite poles, after which cytoplasmic division creates two new cells with identical genetic information.

4 Meiosis Creates Haploid Gametes and Spores and Enhances Genetic Variation in Species

Whereas in diploid organisms, mitosis produces two daughter cells with full diploid complements, **meiosis** produces gametes or spores that are characterized by only one haploid set of chromosomes. During sexual reproduction, haploid gametes then combine at fertilization to reconstitute the diploid complement found in parental cells. Meiosis must be highly specific since, by definition, haploid gametes or spores must contain precisely one member of each homologous pair of chromosomes. When successfully completed, meiosis provides the basis for maintaining genetic continuity from generation to generation.

Another major accomplishment of meiosis is to ensure that during sexual reproduction an enormous amount of genetic variation is produced among members of a species. Such variation occurs in two forms. First, meiosis produces gametes with many unique combinations of maternally and paternally derived chromosomes among the haploid complement, thus assuring that following fertilization, a large number of unique chromosome combinations are possible. process is the underlying basis of Mendel's principles of segregation and independent assortment. The second source of variation is created by the meiotic event referred to as **crossing over,** which results in genetic exchange between members of each homologous pair of chromosomes prior to one or the other finding its way into a haploid gamete or spore. This creates intact chromosomes that are mosaics of the maternal and paternal homologs from which they arise, further enhancing genetic variation. Sexual reproduction therefore significantly reshuffles the genetic material, producing highly diverse offspring.

Meiosis: Prophase I As in mitosis, the process in meiosis begins with a diploid cell duplicating its genetic material in the interphase stage preceding chromosome division. To achieve haploidy, two divisions are thus required. The meiotic achievements, as described above, are largely dependent on the behavior of chromosomes during the initial stage of the first division, called **prophase I.** Recall that in mitosis the paternally and maternally derived members of each homologous pair of chromosomes behave autonomously during division. Each chromosome is duplicated, creating genetically identical **sister chromatids,** and subsequently, one chromatid of each pair is distributed to each new cell. The major difference in meiosis is that once the chromatin characterizing interphase has condensed into visible structures, the homologous chromosomes are not autonomous, but are instead seen to be paired up, having undergone the process called **synapsis. Figure 9** illustrates this process as well as the ensuing events of prophase I. Each synapsed pair of homologs is initially called a **bivalent,** and the number of bivalents is equal to the haploid number. In Figure 9, we have depicted two homologous pairs of chromosomes and thus two bivalents. As the homologs condense and shorten, each bivalent gives rise to a unit called a **tetrad,** consisting of two pairs of sister chromatids, each of which is joined at a common centromere. Remember that one pair of sister chromatids is maternally derived, and the other pair paternally derived. The presence of tetrads is visible evidence that *both* homologs have, in fact, duplicated. As prophase progresses within each tetrad, each pair of sister chromatids is seen to pull apart. However, one or more areas remain in contact where chromatids are intertwined. Each such area, called a **chiasma** (pl., chiasmata), is thought to represent a point where **nonsister chromatids** (one paternal and one maternal chromatid) have undergone genetic exchange through the process of crossing over. Since crossing over is thought to occur one or more times in each tetrad, mosaic chromosomes are routinely created during every meiotic event. During the final period of prophase I, the nucleolus and nuclear envelope break down, and the two centromeres of each tetrad attach to the recently formed spindle fibers.

Meiotic prophase I

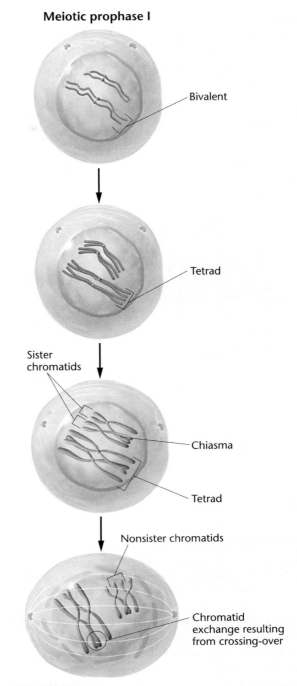

Bivalent

Tetrad

Sister chromatids

Chiasma

Tetrad

Nonsister chromatids

Chromatid exchange resulting from crossing-over

FIGURE 9 The events characterizing meiotic prophase I for the chromosomes depicted in Figure 7.

Metaphase I, Anaphase I, and Telophase I

The remainder of the meiotic process is depicted in **Figure 10.** After meiotic prophase I, steps similar to those of mitosis occur. In the first division, **metaphase I,** the chromosomes have maximally shortened and thickened. The terminal chiasmata of each tetrad are visible and appear to be the only factor holding the nonsister chromatids together. Each tetrad interacts with spindle fibers, facilitating movement to the metaphase plate. The alignment of each tetrad prior to the first anaphase is

Metaphase I   Anaphase I   Telophase I

Prophase II

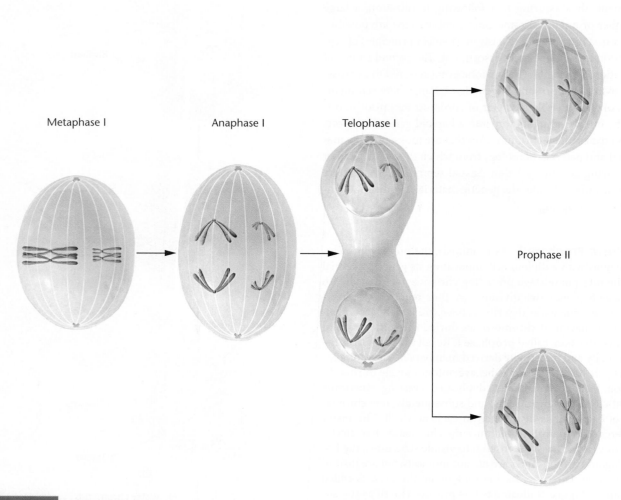

FIGURE 10   The major events in meiosis in an animal cell with a diploid number of 4, beginning with metaphase I. Note that the combination of chromosomes in the cells produced following telophase II is dependent on the random alignment of each tetrad and dyad on the equatorial plate during metaphase I and metaphase II. Several other combinations, which are not shown, can also be formed. The events depicted here are described in the text.

random. Half of each tetrad is pulled randomly to one or the other pole, and the other half then moves to the opposite pole.

During the stages of meiosis I, a single centromere holds each pair of sister chromatids together. It does *not* divide. At **anaphase I,** one-half of each tetrad (the dyad) is pulled toward each pole of the dividing cell. This separation process is the physical basis of **disjunction,** the separation of chromosomes from one another. Occasionally, errors in meiosis occur and separation is not achieved. The term **nondisjunction** describes such an error. At the completion of a normal anaphase I, a series of dyads equal to the haploid number is present at each pole.

If crossing over had not occurred in the first meiotic prophase, each dyad at each pole would consist solely of either paternal or maternal chromatids. However, the exchanges produced by crossing over create mosaic chromatids of paternal and maternal origin.

In many organisms, **telophase I** reveals a nuclear membrane forming around the dyads. Next, the nucleus enters into a short interphase period. If interphase occurs, the chromosomes do not replicate since they already consist of two chromatids. In other organisms, the cells go directly from anaphase I to meiosis II. In general, meiotic telophase is much shorter than the corresponding stage in mitosis.

The Second Meiotic Division

A second division, **meiosis II,** is essential if each gamete or spore is to receive only one chromatid from each original tetrad. The stages characterizing meiosis II are shown in the right half of Figure 10. During **prophase II,** each dyad is composed of one pair of sister chromatids attached by a common centromere. During **metaphase II,** the centromeres are positioned on the metaphase plate. When they divide, **anaphase II** is

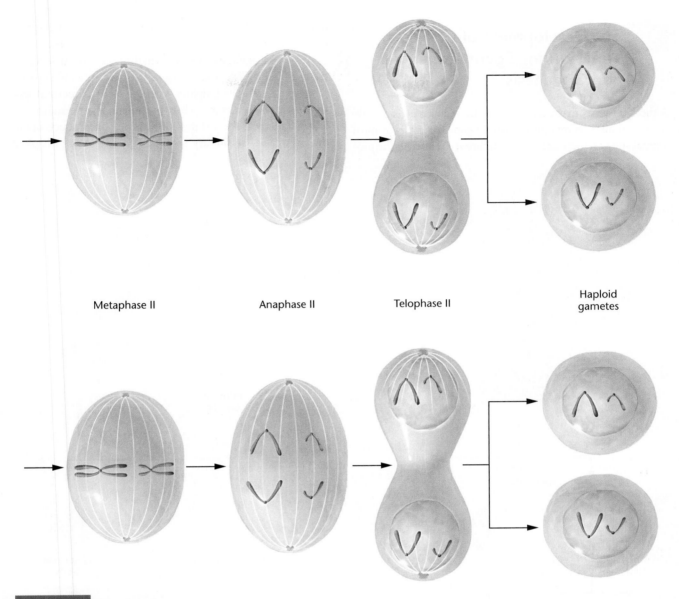

| Metaphase II | Anaphase II | Telophase II | Haploid gametes |

FIGURE 10 (Continued)

initiated, and the sister chromatids of each dyad are pulled to opposite poles. Because the number of dyads is equal to the haploid number, **telophase II** reveals one member of each pair of homologous chromosomes at each pole. Each chromosome is now a monad. Following cytokinesis in telophase II, four haploid gametes may result from a single meiotic event. At the conclusion of meiosis II, not only has the haploid state been achieved, but if crossing over has occurred, each monad is also a combination of maternal and paternal genetic information. As a result, the offspring produced by any gamete receives a mixture of genetic information originally present in his or her grandparents. Meiosis thus significantly increases the level of genetic variation in each ensuing generation.

NOW SOLVE THIS

2 An organism has a diploid number of 16 in a primary oocyte. (a) How many tetrads are present in prophase I? (b) How many dyads are present in prophase II? (c) How many monads migrate to each pole during anaphase II?

■ **Hint** *This problem involves an understanding of what happens to the maternal and paternal members of each pair of homologous chromosomes during meiosis, asking you to extrapolate your understanding to chromosome behavior in an organism with a diploid number of 16. The major insight needed to solve this problem is to understand that maternal and paternal homologs synapse during meiosis. Once it is evident that each chromatid has duplicated, creating a tetrad in the early phases of meiosis, each original pair behaves as a unit and leads to two dyads during anaphase I.*

ESSENTIAL POINT ■ ■ ■

Meiosis converts a diploid cell into a haploid gamete or spore, making sexual reproduction possible. As a result of chromosome duplication and two subsequent meiotic divisions, each haploid cell receives one member of each homologous pair of chromosomes.

5 The Development of Gametes Varies during Spermatogenesis and Oogenesis

Although events that occur during the meiotic divisions are similar in all cells that participate in gametogenesis in most animal species, there are certain differences between the pro-

duction of a male gamete (spermatogenesis) and a female gamete (oogenesis). **Figure 11** summarizes these processes.

Spermatogenesis takes place in the testes, the male reproductive organs. The process begins with the expanded growth of an undifferentiated diploid germ cell called a **spermatogonium**. This cell enlarges to become a **primary spermatocyte,** which undergoes the first meiotic division. The products of this division, called **secondary spermatocytes,** contain a haploid

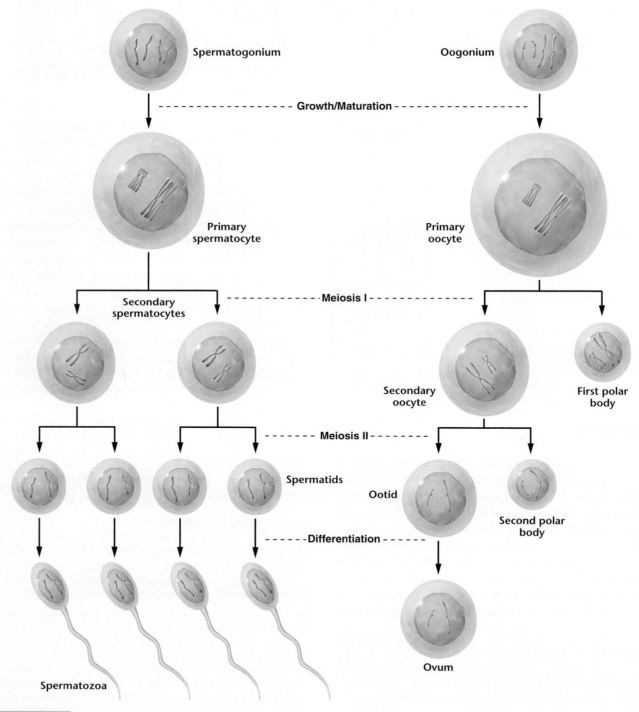

FIGURE 11 Spermatogenesis and oogenesis in animal cells.

number of dyads. The secondary spermatocytes then undergo meiosis II, and each of these cells produces two haploid **spermatids.** Spermatids go through a series of developmental changes, **spermiogenesis,** and become highly specialized, motile **spermatozoa** or **sperm.** All sperm cells produced during spermatogenesis contain the haploid number of chromosomes and equal amounts of cytoplasm.

Spermatogenesis may be continuous or may occur periodically in mature male animals; its onset is determined by the species' reproductive cycle. Animals that reproduce year-round produce sperm continuously, whereas those whose breeding period is confined to a particular season produce sperm only during that time.

In animal **oogenesis,** the formation of **ova** (sing., ovum), or eggs, takes place in the ovaries, the female reproductive organs. The daughter cells resulting from the two meiotic divisions receive equal amounts of genetic material, but they do *not* receive equal amounts of cytoplasm. Instead, during each division almost all the cytoplasm of the **primary oocyte,** which is derived from the **oogonium,** is concentrated in one of the two daughter cells. This concentration of cytoplasm is necessary because a major function of the mature ovum is to nourish the developing embryo after fertilization.

During anaphase I in oogenesis, the tetrads of the primary oocyte separate, and the dyads move toward opposite poles. During telophase I, the dyads at one pole are pinched off with very little surrounding cytoplasm to form the **first polar body.** The first polar body may or may not divide again to produce two small haploid cells. The other daughter cell produced by this first meiotic division contains most of the cytoplasm and is called the **secondary oocyte.** The mature ovum will be produced from the secondary oocyte during the second meiotic division. During this division, the cytoplasm of the secondary oocyte again divides unequally, producing an **ootid** and a **second polar body.** The ootid then differentiates into the mature ovum.

Unlike the divisions of spermatogenesis, the two meiotic divisions of oogenesis may not be continuous. In some animal species, the two divisions may directly follow each other. In others, including humans, the first division of all oocytes begins in the embryonic ovary but arrests in prophase I. Many years later, meiosis resumes in each oocyte just prior to its ovulation. The second division is completed only after fertilization.

ESSENTIAL POINT ■ ■ ■

There is a major difference between meiosis in males and in females. On the one hand, spermatogenesis partitions the cytoplasmic volume equally and produces four haploid sperm cells. Oogenesis, on the other hand, collects the bulk of cytoplasm in one egg cell and reduces the other haploid products to polar bodies. The extra cytoplasm in the egg contributes to zygote development following fertilization.

NOW SOLVE THIS

3 Examine Figure 11, which shows oogenesis in animal cells. Will the genotype of the second polar body (derived from meiosis II) always be identical to that of the ootid? Why or why not?

■ **Hint** *This problem involves an understanding of meiosis during oogenesis, asking you to demonstrate your knowledge of polar bodies. To answer this question, you must take into account that crossing over occurred during meiosis I between each pair of homologs.*

---

6 Meiosis Is Critical to the Sexual Reproduction Cycle of All Diploid Organisms

The process of meiosis is critical to the successful sexual reproduction of all diploid organisms. It is the mechanism by which the diploid amount of genetic information is reduced to the haploid amount. In animals, meiosis leads to the formation of gametes, whereas in plants, haploid spores are produced, which in turn leads to the formation of haploid gametes.

Each diploid organism contains its genetic information in the form of homologous pairs of chromosomes. Each pair consists of one member derived from the maternal parent and one from the paternal parent. Following meiosis, haploid cells potentially contain either the paternal or maternal representative of each homologous pair of chromosomes. However, the process of crossing over, which occurs in the first meiotic prophase, reshuffles alleles between the maternal and paternal members of each homologous pair, which then segregate and assort independently into gametes. This results in the great amounts of genetic variation in gametes.

It is important to touch briefly on the significant role that meiosis plays in the life cycles of fungi and plants. In many fungi, the predominant stage of the life cycle consists of haploid vegetative cells. They arise through meiosis and proliferate by mitotic cell division. In multicellular plants, the life cycle alternates between the diploid **sporophyte stage** and the haploid **gametophyte stage (Figure 12).** While one or the other predominates in different plant groups during this "alternation of generations," the processes of meiosis and fertilization constitute the "bridges" between the sporophyte and gametophyte stages. Therefore, meiosis is an essential component of the life cycle of plants.

ESSENTIAL POINT ■ ■ ■

Meiosis results in extensive genetic variation by virtue of the exchange during crossing over between maternal and paternal chromatids and their random segregation into gametes. In addition, meiosis plays an important role in the life cycles of fungi and plants, serving as the bridge between alternating generations.

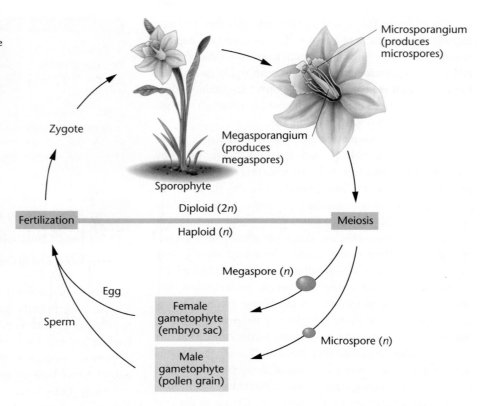

FIGURE 12 Alternation of generations between the diploid sporophyte (2*n*) and the haploid gametophyte (*n*) in a multicellular plant. The processes of meiosis and fertilization bridge the two phases of the life cycle. This is an angiosperm, where the sporophyte stage is the predominant phase.

7 Electron Microscopy Has Revealed the Cytological Nature of Mitotic and Meiotic Chromosomes

Thus far in this chapter, we have focused on mitotic and meiotic chromosomes, emphasizing their behavior during cell division and gamete formation. An interesting question is why chromosomes are invisible during interphase but visible during the various stages of mitosis and meiosis. Studies using electron microscopy clearly show why this is the case.

Recall that during interphase, only dispersed chromatin fibers are present in the nucleus [**Figure 13(a)**]. Once mitosis begins, however, the fibers coil and fold, condensing into typical metaphase chromosomes, as shown in the scanning electron micrograph in **Figure 13(b)**. At the periphery of the chromosome, individual fibers appear similarly to those seen in interphase chromatin. Individual fibers always seem to loop back into the interior where they are twisted and coiled around one another, forming the regular pattern of the mitotic chromosome. Starting in late telophase of mitosis and continuing during G1 of interphase, the process is reversed and

chromosomes unwind to form the long fibers characteristic of chromatin. It is in this physical arrangement that DNA can most efficiently function during transcription and replication.

Electron microscopic observations of metaphase chromosomes in varying states of coiling led Ernest DuPraw to postulate the **folded-fiber model** shown in **Figure 13(c)**. During metaphase, each chromosome consists of two sister chromatids joined at the centromeric region. Each arm of the chromatid appears to be a single fiber wound much like a skein of yarn. An orderly coiling–twisting–condensing process appears to be involved in the transition of the interphase chromatin to the more condensed, mitotic chromosomes. Geneticists believe that during the transition from interphase to prophase, a 5000-fold compaction occurs in the length of DNA within the chromatin fiber! This process must be extremely precise given the highly ordered and consistent appearance of mitotic chromosomes in all eukaryotes.

ESSENTIAL POINT ■ ■ ■

Mitotic chromosomes are produced as a result of the coiling and condensation of chromatin fibers characteristic of interphase and are thus visible only during cell division.

FIGURE 13 Comparison of (a) the chromatin fibers characteristic of the interphase nucleus with (b) and (c) a metaphase chromosome that was derived from chromatin during mitosis. Part (c) diagrams the mitotic chromosome and its various components, showing how chromatin is condensed into it. Part (a) is a transmission electron micrograph, while part (b) is a scanning electron micrograph.

EXPLORING GENOMICS

PubMed: Exploring and Retrieving Biomedical Literature

MG Study Area: *Exploring Genomics*

In this era of rapidly expanding information on genomics and the biomedical sciences, scientists must be conversant in the use of multiple online databases. These resources provide access to DNA and protein sequences, genomic data, chromosome maps, microarray gene-expression networks, and molecular structures, as well as to the bioinformatics tools necessary for data manipulation. Perhaps the most central database resource is PubMed, an online tool for conducting literature searches and accessing biomedical publications.

PubMed is an Internet-based search system developed by the National Center of Biotechnology Information (NCBI) at the National Library of Medicine. Using PubMed, one can access over 15 million articles in over 4600 biomedical journals. The full text of many of the journals can be obtained electronically through college or university libraries, and some journals (such as *Proceedings of the National Academy of Sciences USA; Genome Biology;* and *Science*) provide free public access to articles within certain time frames.

In this exercise, we will explore PubMed to answer questions about relationships between tubulin, human cancers, and cancer therapies, as well as the genetics of spermatogenesis.

■ Exercise I – Tubulin, Cancer, and Mitosis

In this **text** you were introduced to tubulin and the dynamic behavior of microtubules during the cell cycle. Cancer cells are characterized by continuous and uncontrolled mitotic divisions.

Is it possible that tubulin and microtubules contribute to the development of cancer? Could these important structures be targets for cancer therapies?

1. To begin your search for the answers, access the PubMed site at www.ncbi.nlm.nih.gov/sites/entrez?db=pubmed.
2. In the SEARCH box, type "tubulin cancer" and then select the "Go" button to perform the search.
3. Select several research papers and read the abstracts.

To answer the question about tubulin's association with cancer, you may want to limit your search to fewer papers, perhaps those that are review articles. To do this:

1. Select the "Limits" tab near the top of the page.
2. Scroll down the page and select "Review" in the "Type of Article" list.
3. Select "Go" to perform the search.

Explore some of the articles, as abstracts or as full text, available in your library or by free public access. Prepare a brief report or verbally share your experiences with your class. Describe two of the most important things you learned during your exploration and identify the information sources you encountered during the search.

■ Exercise II – Human Disorders of Spermatogenesis

Using the methods described in Exercise I, identify some human disorders associated with defective spermatogenesis. Which human genes are involved in spermatogenesis? How do defects in these genes result in fertility disorders? Prepare a brief written or verbal report on what you have learned and what sources you used to acquire your information.

CASE STUDY » Timing is everything

A man in his early 20s received chemotherapy and radiotherapy as treatment every 60 days for Hodgkin's disease. After unsuccessful attempts to have children, he had his sperm examined at a fertility clinic, upon which multiple chromosomal irregularities were discovered. When examined within 5 days of a treatment, extra chromosomes were often present or one or more chromosomes were completely absent. However, such irregularities were not observed at day 38 after treatment nor anytime after day 38.

1. How might a geneticist explain the time-related differences in chromosomal irregularities?
2. Do you think that exposure to chemotherapy and radiotherapy of a spermatogonium would cause more problems than exposure to a secondary spermatocyte?
3. What advice would you provide regarding fertility while the man remained under treatment?

INSIGHTS AND SOLUTIONS

1. In an organism with a diploid number of $2n = 6$, how many individual chromosomal structures will align on the metaphase plate during (a) mitosis, (b) meiosis I, and (c) meiosis II? Describe each configuration.

Solution:

(a) In mitosis, where homologous chromosomes do not synapse, there will be six double structures, each consisting of a pair of sister chromatids. The number of structures is equivalent to the diploid number.

(b) In meiosis I, the homologs have synapsed, reducing the number of structures to three. Each is a tetrad and consists of two pairs of sister chromatids.

(c) In meiosis II, the same number of structures exist (three), but in this case they are dyads. Each dyad is a pair of sister chromatids. When crossing over has occurred, each chromatid may contain parts of one of its nonsister chromatids obtained during exchange in prophase I.

2. Disregarding crossing over, draw all possible alignment configurations that can occur during metaphase I for the chromosomes shown in Figure 10.

Solution: As shown at the bottom of the page, four cases are possible when $n = 2$.

3. For the chromosomes in the previous problem, assume one gene is present on both of the larger chromosomes with two alleles, A and a,

as shown. Also assume a second gene with two alleles (B, b) is present on the smaller chromosomes. Calculate the probability of generating each gene combination (AB, Ab, aB, ab) following meiosis I.

Solution: As shown at the bottom of the page:

Case I	AB and ab	
Case II	Ab and aB	
Case III	aB and Ab	
Case IV	ab and AB	
Total:	$AB = 2$	$(p = 1/4)$
	$Ab = 2$	$(p = 1/4)$
	$aB = 2$	$(p = 1/4)$
	$ab = 2$	$(p = 1/4)$

4. Describe the composition of a meiotic tetrad as it exists during prophase I, assuming no crossover event has occurred. What impact would a single crossover event have on this structure?

Solution: Such a tetrad contains four chromatids, existing as two pairs. Members of each pair are sister chromatids. They are held together by a common centromere. Members of one pair are maternally derived, whereas members of the other are paternally derived. Maternal and paternal members are nonsister chromatids. A single crossover event has the effect of exchanging a portion of a maternal *and* a paternal chromatid, leading to a chiasma, where the two chromatids overlap physically in the tetrad.

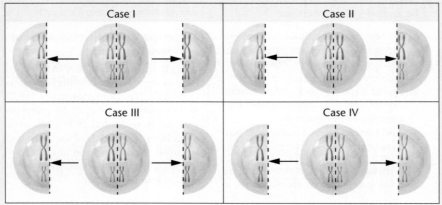

Solution for #2

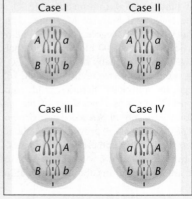

Solution for #3

PROBLEMS AND DISCUSSION QUESTIONS

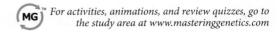

For activities, animations, and review quizzes, go to the study area at www.masteringgenetics.com

1. In this chapter, we have focused on how chromosomes are distributed during cell division, both in dividing somatic cells (mitosis) and in gamete- and spore-forming cells (meiosis). We found many opportunities to consider the methods and reasoning by which much of this information was acquired. From the explanations given in the chapter, answer the following fundamental questions:

   (a) How do we know that chromosomes exist in homologous pairs?

   (b) How do we know that DNA replication occurs during interphase, and not early in mitosis?

   (c) How do we know that mitotic chromosomes are derived from chromatin?

2. What role do the following cellular components play in the storage, expression, or transmission of genetic information: (a) chromatin, (b) nucleolus, (c) ribosome, (d) mitochondrion, (e) centriole, (f) centromere?

3. Discuss the concepts of homologous chromosomes, diploidy, and haploidy. What characteristics are shared between two homologous chromosomes?

4. If two chromosomes of a species are the same length and have similar centromere placements, yet are not homologous, what is different about them?

5. Describe the events that characterize each stage of mitosis.

6. How are chromosomes named on the basis of their centromere placement?

7. Contrast telophase in plant and animal mitosis.

8. Describe the phases of the cell cycle and the events that characterize each phase.

9. Contrast the end results of meiosis with those of mitosis.

10. Define and discuss these terms: (a) synapsis, (b) bivalent, (c) chiasmata, (d) crossing over, (e) sister chromatids, (f) tetrad, (g) dyad, (h) monad.

11. Contrast the genetic content and the origin of sister versus nonsister chromatids during their earliest appearance in prophase I of meiosis. How might the genetic content of these change by the time tetrads have aligned at the metaphase plate during metaphase?

12. Given the end results of the two types of division, why is it necessary for homologs to pair during meiosis and not desirable for them to pair during mitosis?

13. Contrast spermatogenesis and oogenesis. What is the significance of the formation of polar bodies?

14. Explain why meiosis leads to significant genetic variation while mitosis does not.

15. A diploid cell contains three pairs of homologous chromosomes designated C1 and C2, M1 and M2, and S1 and S2; no crossing over occurs. What possible combinations of chromosomes will be present in (a) daughter cells following mitosis, (b) the first meiotic metaphase, and (c) haploid cells following both divisions of meiosis?

16. Considering the preceding problem, predict the number of different haploid cells that will occur if a fourth chromosome pair (W1 and W2) is added.

17. If one follows 50 primary oocytes in an animal through their various stages of oogenesis, how many secondary oocytes would be formed? How many first polar bodies would be formed? How many ootids would be formed? If one follows 50 primary spermatocytes in an animal through their various stages of spermatogenesis, how many secondary spermatocytes would be formed? How many spermatids would be formed?

18. What is the probability that, in an organism with a haploid number of 10, a sperm will be formed that contains all 10 chromosomes whose centromeres were derived from maternal homologs?

19. Describe the genetic events that characterize meiotic prophase I.

20. Describe the role of meiosis in the life cycle of a plant.

21. Contrast the chromatin fiber with the mitotic chromosome. How are the two structures related?

22. Describe the folded-fiber model of the mitotic chromosome.

23. You are given a metaphase chromosome preparation (a slide) from an unknown organism that contains 12 chromosomes. Two are clearly smaller than the rest, appearing identical in length and centromere placement. Describe all that you can about these two chromosomes.

For Problems 24–29, consider a diploid cell that contains three pairs of chromosomes designated AA, BB, and CC. Each pair contains a maternal and a paternal member (e.g., A^m and A^p, etc.). Using these designations, demonstrate your understanding of mitosis and meiosis by drawing chromatid combinations as requested. Be sure to indicate when chromatids are paired as a result of replication and/or synapsis. You may wish to use a large piece of brown manila wrapping paper or a large cut-up paper bag and work with another student as you deal with these problems. Such cooperative learning may be a useful approach as you solve problems throughout the text.

24. In mitosis, what chromatid combination(s) will be present during metaphase? What combination(s) will be present at each pole at the completion of anaphase?

25. During meiosis, assuming no crossing over, what chromatid combination(s) will be present at the completion of prophase I? Draw all possible alignments of chromatids as migration begins during early anaphase I.

26. Are any possible combinations present during prophase II of meiosis other than those you drew in Problem 25? If so, draw them.

27. Draw all possible combinations of chromatids during anaphase II in meiosis.

28. Assume that during meiosis I, none of the C chromosomes disjoin at metaphase, but they separate into dyads (instead of monads) during meiosis II. How would this change the alignments that you constructed during the anaphase stages in meiosis I and II? Draw them.

29. Assume that each resultant gamete (Problem 28) participated in fertilization with a normal haploid gamete. What combinations will result? What percentage of zygotes will be diploid, containing one paternal and one maternal member of each chromosome pair?

30. The nuclear DNA content of a single sperm cell in *Drosophila melanogaster* is approximately 0.18 pg. What would be the expected nuclear DNA content of a primary spermatocyte in *Drosophila*? What would be the expected nuclear DNA content of a somatic cell (non-sex cell) in the G1 phase? What would be the expected nuclear DNA content of a somatic cell at metaphase?

SOLUTIONS TO SELECTED PROBLEMS AND DISCUSSION QUESTIONS

Answers to Now Solve This

1. 32 chromatids, 16 chromosomes moving to each pole
2. **(a)** eight tetrads **(b)** eight dyads **(c)** eight monads
3. Not necessarily. If crossing over occurred in meiosis I, then the chromatids in the secondary oocyte are not identical. Once they separate during meiosis II, dissimilar chromatids reside in the ootid and the second polar body.

Solutions to Problems and Discussion Questions

4. Genetic content in nonhomologous chromosomes is expected to be quite different. Other factors, including banding pattern and time of replication during S phase, would also be expected to vary among nonhomologous chromosomes.

12. During meiosis I, chromosome number is reduced to haploid complements. This is achieved by synapsis of homologous chromosomes and their subsequent separation. It would seem to be more mechanically difficult for genetically identical daughters to form from mitosis if homologous chromosomes paired. By having chromosomes unpaired at metaphase of mitosis, only centromere division is required for daughter cells to eventually receive identical chromosomal complements.

14. First, through independent assortment of chromosomes at anaphase I of meiosis, daughter cells (secondary spermatocytes and secondary oocytes) may contain different sets of maternally and paternally derived chromosomes. Second, crossing over, which happens at a much higher frequency in meiotic cells as compared to mitotic cells, allows maternally and paternally derived chromosomes to exchange segments, thereby increasing the likelihood that daughter cells (that is, secondary spermatocytes and secondary oocytes) are genetically unique.

16. There would be 16 combinations with the addition of another chromosome pair.

18. One-half of each tetrad will have a maternal homolog: $(1/2)^{10}$.

24. Duplicated chromosomes A^m, A^p, B^m, B^p, C^m, and C^p will align at metaphase, with the centromeres dividing and sister chromatids going to opposite poles at anaphase.

26. As long as you have accounted for eight possible combinations in the previous problem, there would be no new ones added in this problem.

28. See the products of nondisjunction of chromosome C at the end of meiosis I as follows.

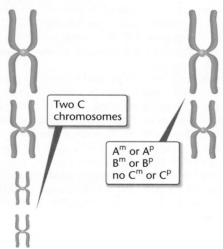

At the end of meiosis II, assuming that, as the problem states, the C chromosomes separate as dyads instead of monads during meiosis II, you would have monads for the A and B chromosomes and dyads (from the cell on the left) for both C chromosomes as one possibility.

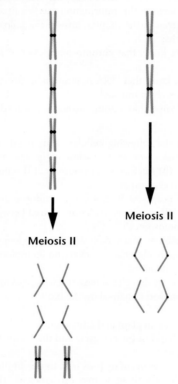

30. 0.72 picograms; 0.36 picograms; 0.72 picograms

CREDITS

Credits are listed in order of appearance.

Photo

CO, Dr. Andrew S. Bajer, University of Oregon; F-2, CNRI/Science Photo Library/Photo Researchers, Inc.; F-4, Dr. David Ward, Yale University; F-13 (a) Biophoto Associates/Photo Researchers, Inc., (b) Andrew Syred/Photo Researchers, Inc., (c) Biophoto Associates/Science Source/Photo Researchers, Inc.

Mendelian Genetics

From Chapter 3 of *Essentials of Genetics,, Eighth edition,* William S. Klug, Michael R. Cummings, Charlotte A. Spencer, Michael A. Palladino. ©2013 by Pearson Education, Inc. All rights reserved.

Gregor Johann Mendel, who in 1866 put forward the major postulates of transmission genetics as a result of experiments with the garden pea.

Mendelian Genetics

CHAPTER CONCEPTS

- Inheritance is governed by information stored in discrete factors called genes.

- Genes are transmitted from generation to generation on vehicles called chromosomes.

- Chromosomes, which exist in pairs, provide the basis of biparental inheritance.

- During gamete formation, chromosomes are distributed according to postulates first described by Gregor Mendel, based on his nineteenth-century research with the garden pea.

- Mendelian postulates prescribe that homologous chromosomes segregate from one another and assort independently with other segregating homologs during gamete formation.

- Genetic ratios, expressed as probabilities, are subject to chance deviation and may be evaluated statistically.

Although inheritance of biological traits has been recognized for thousands of years, the first significant insights into the mechanisms involved occurred about 140 years ago. In 1866, Gregor Johann Mendel published the results of a series of experiments that would lay the foundation for the formal discipline of genetics. Mendel's work went largely unnoticed until the turn of the century, but in the ensuing years the concept of the gene as a distinct hereditary unit was established. The ways in which genes, as members of chromosomes, are transmitted to offspring and control traits were clarified. Research has continued unabated throughout the twentieth century and into the present century—indeed, studies in genetics, most recently at the molecular level, have remained continually at the forefront of biological research since the early 1900s.

When Mendel began his studies of inheritance using *Pisum sativum*, the garden pea, chromosomes and the role and mechanism of meiosis were totally unknown. Nevertheless, he determined that discrete *units of inheritance* exist and predicted their behavior during the formation of gametes. Subsequent investigators, with access to cytological data, were able to relate their observations of chromosome behavior during meiosis to Mendel's principles of inheritance. Once this correlation was made, Mendel's postulates were accepted as the basis for the study of what is known as **transmission genetics.**

1 Mendel Used a Model Experimental Approach to Study Patterns of Inheritance

Johann Mendel was born in 1822 to a peasant family in the central European village of Heinzendorf. An excellent student in high school, he studied philosophy for several years afterward, and in 1843 he was admitted to the Augustinian Monastery of St. Thomas in Brno, now part of the Czech Republic, taking the name Gregor. In 1849, he was relieved of pastoral duties and accepted a teaching appointment that lasted several years. From 1851 to 1853, he attended the University of Vienna, where he studied physics and botany. He returned to Brno in 1854, where he taught physics and natural science for the next 16 years. Mendel received support from the monastery for his studies and research throughout his life.

In 1856, Mendel performed his first set of hybridization experiments with the garden pea. The research phase of his career lasted until 1868, when he was elected abbot of the monastery. Although he retained his interest in genetics, his new responsibilities demanded most of his time. In 1884, Mendel died of a kidney disorder. The local newspaper paid him the following tribute:

> "His death deprives the poor of a benefactor, and mankind at large of a man of the noblest character, one who was a warm friend, a promoter of the natural sciences, and an exemplary priest."

Mendel first reported the results of some simple genetic crosses between certain strains of the garden pea in 1865. Although his was not the first attempt to provide experimental evidence pertaining to inheritance, Mendel's success where others failed can be attributed, at least in part, to his elegant model of experimental design and analysis.

Mendel showed remarkable insight into the methodology necessary for good experimental biology. He chose an organism that is easy to grow and hybridize artificially. The pea plant is self-fertilizing in nature but is easy to cross-breed experimentally. It reproduces well and grows to maturity in a single season. Mendel followed seven visible features (unit characters), each represented by two contrasting forms, or **traits (Figure 1)**. For the character stem height, for example, he experimented with the traits *tall* and *dwarf*. He selected six other visibly contrasting pairs of traits involving

Character	Contrasting traits		F$_1$ results	F$_2$ results	F$_2$ ratio
Seed shape	round/wrinkled		all round	5474 round 1850 wrinkled	2.96:1
Seed color	yellow/green		all yellow	6022 yellow 2001 green	3.01:1
Pod shape	full/constricted		all full	882 full 299 constricted	2.95:1
Pod color	green/yellow		all green	428 green 152 yellow	2.82:1
Flower color	violet/white	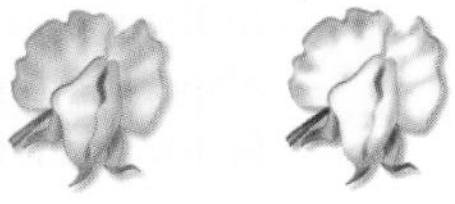	all violet	705 violet 224 white	3.15:1
Flower position	axial/terminal	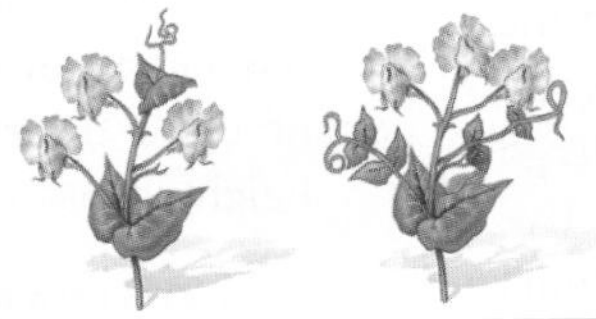	all axial	651 axial 207 terminal	3.14:1
Stem height	tall/dwarf		all tall	787 tall 277 dwarf	2.84:1

FIGURE 1 Seven pairs of contrasting traits and the results of Mendel's seven monohybrid crosses of the garden pea (*Pisum sativum*). In each case, pollen derived from plants exhibiting one trait was used to fertilize the ova of plants exhibiting the other trait. In the F$_1$ generation, one of the two traits was exhibited by all plants. The contrasting trait reappeared in approximately 1/4 of the F$_2$ plants.

seed shape and color, pod shape and color, and pod and flower arrangement. From local seed merchants, Mendel obtained true-breeding strains—those in which each trait appeared unchanged generation after generation in self-fertilizing plants.

There were several reasons for Mendel's success. In addition to his choice of a suitable organism, he restricted his examination to one or very few pairs of contrasting traits in each experiment. He also kept accurate quantitative records, a necessity in genetic experiments. From the analysis of his data, Mendel derived certain postulates that became principles of transmission genetics.

The results of Mendel's experiments were unappreciated until the turn of the century, well after his death. However, once Mendel's publications were rediscovered by geneticists investigating the function and behavior of chromosomes, the implications of his postulates were immediately apparent. He had discovered the basis for the transmission of hereditary traits!

2   The Monohybrid Cross Reveals How One Trait Is Transmitted from Generation to Generation

Mendel's simplest crosses involved only one pair of contrasting traits. Each such experiment is a **monohybrid cross,** which is made by mating true-breeding individuals from two parent strains, each exhibiting one of the two contrasting forms of the character under study. Initially, we examine the first generation of offspring of such a cross, and then we consider the results of **selfing,** the process by which self-fertilizing individuals from this first generation produce offspring. The original parents constitute the P_1 or **parental generation,** their offspring are the F_1 or **first filial generation,** and the individuals resulting from the selfed F_1 generation are the F_2 or **second filial generation.** We can continue to follow subsequent generations.

The cross between true-breeding pea plants with tall stems and dwarf stems is representative of Mendel's monohybrid crosses. *Tall* and *dwarf* are contrasting traits of the character of stem height. Unless tall or dwarf plants are crossed together or with another strain, they will undergo self-fertilization and breed true, producing their respective traits generation after generation. However, when Mendel crossed tall plants with dwarf plants, the resulting F_1 generation consisted only of tall plants. When members of the F_1 generation were selfed, Mendel observed that 787 of 1064 F_2 plants were tall, while the remaining 277 were dwarf. Note that in this cross (Figure 1) the dwarf trait disappears in the F_1 only to reappear in the F_2 generation.

Genetic data are usually expressed and analyzed as ratios. In this particular example, many identical P_1 crosses were made, and many F_1 plants—all tall—were produced. As

noted, of the 1064 F_2 offspring, 787 were tall and 277 were dwarf—a ratio of 2.84:1.0, or about 3:1.

Mendel made similar crosses between pea plants exhibiting other pairs of contrasting traits; the results of these crosses are shown in Figure 1. In every case, the outcome was similar to the tall/dwarf cross just described. All F_1 offspring were identical to one of the parents, but in the F_2 offspring, an approximate ratio of 3:1 was obtained. That is, three-fourths looked like the F_1 plants, while one-fourth exhibited the contrasting trait, which had disappeared in the F_1 generation.

We will point out one further aspect of Mendel's monohybrid crosses. In each cross, the F_1 and F_2 patterns of inheritance were similar regardless of which P_1 plant served as the source of pollen (sperm) and which served as the source of the ovum (egg). The crosses could be made either way—pollination of dwarf plants by tall plants or vice versa. These are called **reciprocal crosses.** Therefore, the results of Mendel's monohybrid crosses were not sex-dependent.

To explain these results, Mendel proposed the existence of particulate unit factors for each trait. He suggested that these factors serve as the basic units of heredity and are passed unchanged from generation to generation, determining the various traits expressed by each individual plant. Using these general ideas, Mendel proceeded to hypothesize precisely how unit factors could account for the results of the monohybrid crosses.

Mendel's First Three Postulates

Using the consistent pattern of results in the monohybrid crosses, Mendel derived the following three postulates or principles of inheritance.

1. UNIT FACTORS IN PAIRS
Genetic characters are controlled by unit factors that exist in pairs in individual organisms.

In the monohybrid cross involving tall and dwarf stems, a specific **unit factor** exists for each trait. Because the factors occur in pairs, three combinations are possible: two factors for tallness, two factors for dwarfness, or one factor for each trait. Every individual contains one of these three combinations, which determines stem height.

2. DOMINANCE/RECESSIVENESS
When two unlike unit factors responsible for a single character are present in a single individual, one unit factor is dominant to the other, which is said to be recessive.
In each monohybrid cross, the trait expressed in the F_1 generation is controlled by the **dominant** unit factor. The trait not expressed is controlled by the **recessive** unit factor. Note that this dominance/recessiveness relationship pertains only when unlike unit factors are present in pairs. The terms *dominant* and *recessive* are also used

to designate traits. In this case, tall stems are said to be dominant over the recessive dwarf stems.

3. SEGREGATION
During the formation of gametes, the paired unit factors separate or segregate randomly so that each gamete receives one or the other with equal likelihood.

If an individual contains a pair of like unit factors (e.g., both specific for tall), then all gametes receive one tall unit factor. If an individual contains unlike unit factors (e.g., one for tall and one for dwarf), then each gamete has a 50 percent probability of receiving either the tall or the dwarf unit factor.

These postulates provide a suitable explanation for the results of the monohybrid crosses. Let's use the tall/dwarf cross to illustrate. Mendel reasoned that P_1 tall plants contain identical paired unit factors, as do the P_1 dwarf plants. The gametes of tall plants all receive one tall unit factor as a result of **segregation.** Similarly, the gametes of dwarf plants all receive one dwarf unit factor. Following fertilization, all F_1 plants receive one unit factor from each parent: a tall factor from one and a dwarf factor from the other, reestablishing the paired relationship—but because tall is dominant to dwarf, all F_1 plants are tall.

When F_1 plants form gametes, the postulate of segregation demands that each gamete randomly receives either the tall or the dwarf unit factor. Following random fertilization events during F_1 selfing, four F_2 combinations result in equal frequency

1. tall/tall

2. tall/dwarf

3. dwarf/tall

4. dwarf/dwarf

Combinations (1) and (4) result in tall and dwarf plants, respectively. According to the postulate of dominance/recessiveness, combinations (2) and (3) both yield tall plants. Therefore, the F_2 is predicted to consist of 3/4 tall and 1/4 dwarf, or a ratio of 3:1. This is approximately what Mendel observed in the cross between tall and dwarf plants. A similar pattern was observed in each of the other monohybrid crosses (see Figure 1).

ESSENTIAL POINT ■ ■ ■

Mendel's postulates help describe the basis for the inheritance of phenotypic traits. He hypothesized that unit factors exist in pairs and exhibit a dominant/recessive relationship in determining the expression of traits. He further postulated that unit factors segregate during gamete formation, such that each gamete receives one or the other factor, with equal probability.

Modern Genetic Terminology

To analyze the monohybrid cross and Mendel's first three postulates, we must first introduce several new terms as well as a symbol convention for the unit factors.

Traits such as tall or dwarf are visible expressions of the information contained in unit factors. The physical appearance of a trait is the **phenotype** of the individual. Mendel's unit factors represent units of inheritance called **genes** by modern geneticists. For any given character, such as plant height, the phenotype is determined by alternative forms of a single gene called **alleles.** For example, the unit factors representing tall and dwarf are alleles determining the height of the pea plant.

Geneticists have several different systems for using symbols to represent genes. We will adopt one of these systems to use consistently throughout this chapter. According to this convention, the first letter of the recessive trait symbolizes the character in question; in lowercase italic, it designates the allele for the recessive trait, and in uppercase italic, it designates the allele for the dominant trait. Thus for Mendel's pea plants, we use *d* for the *d*warf allele and *D* for the tall allele. When alleles are written in pairs to represent the two unit factors present in any individual (*DD, Dd,* or *dd*), the resulting symbol is called the **genotype.** This term reflects the genetic makeup of an individual, whether it is haploid or diploid. By reading the genotype, we know the phenotype of the individual: *DD* and *Dd* are tall, and *dd* is dwarf. When both alleles are the same (*DD* or *dd*), the individual is **homozygous** or a **homozygote;** when the alleles are different (*Dd*), we use the term **heterozygous** or a **heterozygote.** These symbols and terms are used in **Figure 2** to illustrate the monohybrid cross.

Because he operated without the hindsight that modern geneticists enjoy, Mendel's analytical reasoning must be considered a truly outstanding scientific achievement. On the basis of rather simple but precisely executed breeding experiments, he not only proposed that discrete particulate units of heredity exist, he also explained how they are transmitted from one generation to the next.

Punnett Squares

The genotypes and phenotypes resulting from combining gametes during fertilization can be easily visualized by constructing a **Punnett square,** named after Reginald C. Punnett, who first devised this approach. **Figure 3** demonstrates this method of analysis for our $F_1 \times F_1$ monohybrid cross. Each of the possible gametes is assigned a column or a row; the vertical columns represent those of the female parent, and the horizontal rows represent those of the male parent. After assigning the gametes to the rows and columns, we predict the new generation by entering the male and female gametic information into each box, thus producing every possible resulting genotype. By filling out the Punnett square,

NOW SOLVE THIS

1 Pigeons exhibit a checkered or plain feather pattern. In a series of controlled matings, the following data were obtained

	F₁ Progeny	
P₁ Cross	**Checkered**	**Plain**
(a) checkered × checkered	36	0
(b) checkered × plain	38	0
(c) plain × plain	0	35

Then F₁ offspring were selectively mated with the following results. (The P₁ cross giving rise to each F₁ pigeon is indicated in parentheses.)

	F₂ Progeny	
F₁ × F₁ Crosses	**Checkered**	**Plain**
(d) checkered (a) × plain (c)	34	0
(e) checkered (b) × plain (c)	17	14
(f) checkered (b) × checkered (b)	28	9
(g) checkered (a) × checkered (b)	39	0

How are the checkered and plain patterns inherited? Select and assign symbols for the genes involved, and determine the genotypes of the parents and offspring in each cross.

■ **Hint** *This problem asks you to analyze the data produced from several crosses involving pigeons and to determine the mode of inheritance and the genotypes of the parents and offspring in a number of instances. The key to its solution is to first determine whether or not this is a monohybrid cross. To do so, convert the data to ratios that are characteristic of Mendelian crosses. In the case of this problem, ask first whether any of the F₂ ratios match Mendel's 3:1 monohybrid ratio. If so, the second step is to determine which trait is dominant and which is recessive.*

we are listing all possible random fertilization events. The genotypes and phenotypes of all potential offspring are ascertained by reading the combinations in the boxes.

The Punnett square method is particularly useful when you are first learning about genetics and how to solve problems. Note the ease with which the 3:1 phenotypic ratio and the 1:2:1 genotypic ratio is derived in the F₂ generation in Figure 3.

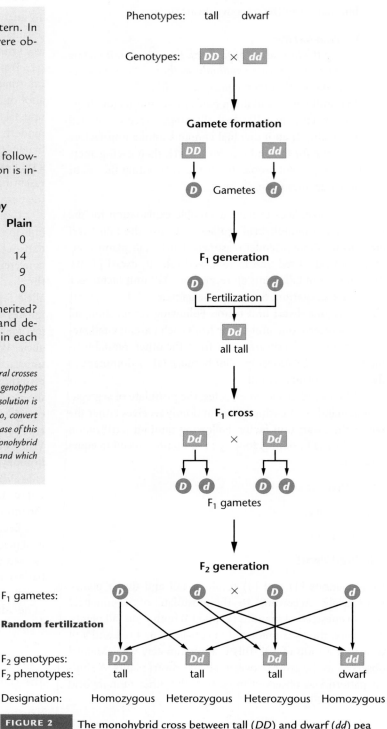

FIGURE 2 The monohybrid cross between tall (*DD*) and dwarf (*dd*) pea plants. Individuals are shown in rectangles, and gametes in circles.

The Testcross: One Character

Tall plants produced in the F₂ generation are predicted to be either the *DD* or *Dd* genotype. You might ask if there is a way to distinguish the genotype. Mendel devised a rather simple method that is still used today in breeding plants and animals: the **testcross**. The organism express-ing the dominant phenotype, but of unknown genotype, is crossed to a known *homozygous recessive individual*. For example, as shown in **Figure 4(a)**, if a tall plant of genotype *DD* is testcrossed to a dwarf plant, which must have the *dd* genotype, all offspring will be tall phenotypically and *Dd* genotypically. However, as shown in **Figure 4(b)**, if a tall

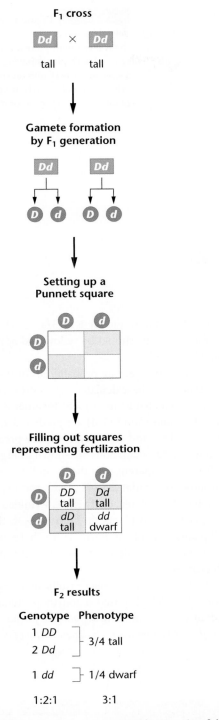

F₁ cross

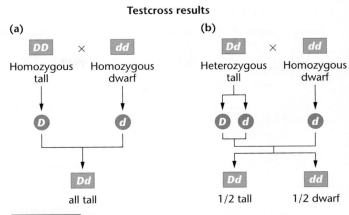

Testcross results

FIGURE 4   Testcross of a single character. In (a), the tall parent is homozygous, but in (b), the tall parent is heterozygous. The genotype of each tall P₁ plant can be determined by examining the offspring when each is crossed to a homozygous recessive dwarf plant.

FIGURE 3   A Punnett square generating the F₂ ratio of the F₁ × F₁ cross shown in Figure 2.

3 Mendel's Dihybrid Cross Generated a Unique F₂ Ratio

As a natural extension of the monohybrid cross, Mendel also designed experiments in which he examined two characters simultaneously. Such a cross, involving two pairs of contrasting traits, is a **dihybrid cross,** or *two-factor cross*. For example, if pea plants having yellow seeds that are round are bred with those having green seeds that are wrinkled, the results shown in **Figure 5** will occur: the F₁ offspring will all be yellow and round. It is therefore apparent that yellow is dominant to green and that round is dominant to wrinkled. When the F₁ individuals are selfed, approximately 9/16 of the F₂ plants express yellow and round, 3/16 express yellow and wrinkled, 3/16 express green and round, and 1/16 express green and wrinkled.

A variation of this cross is also shown in Figure 5. Instead of crossing one P₁ parent with both dominant traits (yellow, round) and one with both recessive traits (green, wrinkled), plants with yellow, wrinkled seeds are crossed with plants with green, round seeds. In spite of the change in the P₁ phenotypes, both the F₁ and F₂ results remain unchanged. It will become clear in the next section why this is so.

Mendel's Fourth Postulate: Independent Assortment

We can most easily understand the results of a dihybrid cross if we consider it theoretically as consisting of two monohybrid crosses conducted separately. Think of the two sets of traits as inherited independently of each other; that is, the chance of any plant having yellow or green seeds is not at all influenced by the chance that this plant will have round or

plant is *Dd* and it is crossed to a dwarf plant (*dd*), then one-half of the offspring will be tall (*Dd*) and the other half will be dwarf (*dd*). Therefore, a 1:1 tall/dwarf ratio demonstrates the heterozygous nature of the tall plant of unknown genotype. The testcross reinforced Mendel's conclusion that separate unit factors control traits.

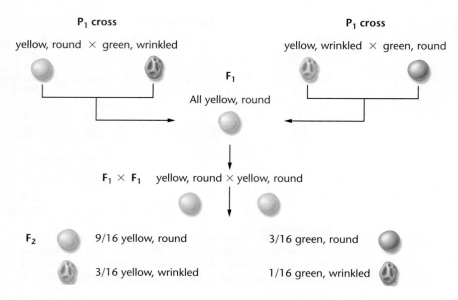

P₁ cross

yellow, round × green, wrinkled

P₁ cross

yellow, wrinkled × green, round

F₁

All yellow, round

F₁ × F₁   yellow, round × yellow, round

F₂   9/16 yellow, round   3/16 green, round

3/16 yellow, wrinkled   1/16 green, wrinkled

FIGURE 5   F₁ and F₂ results of Mendel's dihybrid crosses, where the plants on the top left with yellow, round seeds are crossed with plants having green, wrinkled seeds, and the plants on the top right with yellow, wrinkled seeds are crossed with plants having green, round seeds.

wrinkled seeds. Thus, because yellow is dominant to green, all F₁ plants in the first theoretical cross would have yellow seeds. In the second theoretical cross, all F₁ plants would have round seeds because round is dominant to wrinkled. When Mendel examined the F₁ plants of the dihybrid cross, all were yellow and round, as we just predicted.

The predicted F₂ results of the first cross are 3/4 yellow and 1/4 green. Similarly, the second cross would yield 3/4 round and 1/4 wrinkled. Figure 5 shows that in the dihybrid cross, 12/16 F₂ plants are yellow while 4/16 are green, exhibiting the expected 3:1 (3/4 : 1/4) ratio. Similarly, 12/16 F₂ plants have round seeds while 4/16 have wrinkled seeds, again revealing the 3:1 ratio.

These numbers demonstrate that the two pairs of contrasting traits are inherited independently, so we can predict the frequencies of all possible F₂ phenotypes by applying the **product law** of probabilities: *When two independent events occur simultaneously, the combined probability of the two outcomes is equal to the product of their individual probabilities of occurrence.* For example, the probability of an F₂ plant

having yellow *and* round seeds is (3/4) (3/4), or 9/16 because 3/4 of all F₂ plants should be yellow and 3/4 of all F₂ plants should be round.

In a like manner, the probabilities of the other three F₂ phenotypes can be calculated: yellow (3/4) and wrinkled (1/4) are predicted to be present together 3/16 of the time; green (1/4) and round (1/4) are predicted 1/16 of the time; and green (1/4) and wrinkled (1/4) are predicted (1/16) of the time. These calculations are shown in **Figure 6**.

It is now apparent why the F₁ and F₂ results are identical whether the initial cross is yellow, round plants bred with green, wrinkled plants, or if yellow, wrinkled plants are bred with green, round plants. In both crosses, the F₁ genotype of all plants is identical. Each plant is heterozygous for both gene pairs. As a result, the F₂ generation is also identical in both crosses.

On the basis of similar results in numerous dihybrid crosses, Mendel proposed a fourth postulate called **independent assortment:** *During gamete formation, segregating pairs of unit factors assort independently of each other.*

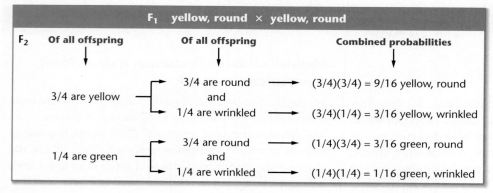

F₁	yellow, round × yellow, round		
F₂	**Of all offspring**	**Of all offspring**	**Combined probabilities**
	↓	↓	↓
	3/4 are yellow	3/4 are round and	(3/4)(3/4) = 9/16 yellow, round
		1/4 are wrinkled	(3/4)(1/4) = 3/16 yellow, wrinkled
	1/4 are green	3/4 are round and	(1/4)(3/4) = 3/16 green, round
		1/4 are wrinkled	(1/4)(1/4) = 1/16 green, wrinkled

FIGURE 6   Computation of the combined probabilities of each F₂ phenotype for two independently inherited characters. The probability of each plant's seeds being yellow or green is independent of the probability of its seeds being round or wrinkled.

This postulate stipulates that segregation of any pair of unit factors occurs independently of all others. As a result of random segregation, each gamete receives one member of every pair of unit factors. For one pair, whichever unit factor is received does not influence the outcome of segregation of any other pair. Thus, according to the postulate of independent assortment, all possible combinations of gametes are formed in equal frequency.

The Punnett square in **Figure 7** shows how independent assortment works in the formation of the F_2 generation. Examine the formation of gametes by the F_1 plants; segregation prescribes that every gamete receives either a *G* or *g* allele and a *W* or *w* allele. Independent assortment stipulates that all four combinations (*GW, Gw, gW,* and *gw*) will be formed with equal probabilities.

In every $F_1 \times F_1$ fertilization event, each zygote has an equal probability of receiving one of the four combinations from each parent. If many offspring are produced, 9/16 have yellow, round seeds, 3/16 have yellow, wrinkled seeds, 3/16 have green, round seeds, and 1/16 have green, wrinkled seeds, yielding what is designated as **Mendel's 9:3:3:1 dihybrid ratio.** This is an ideal ratio based on probability events involving segregation, independent assortment, and random fertilization. Because of deviation due strictly to chance,

particularly if small numbers of offspring are produced, actual results are highly unlikely to match the ideal ratio.

ESSENTIAL POINT ■ ■ ■

Mendel's postulate of independent assortment states that each pair of unit factors segregates independently of other such pairs. As a result, all possible combinations of gametes are formed with equal probability.

The Testcross: Two Characters

The testcross can also be applied to individuals that express two dominant traits but whose genotypes are unknown. For example, the expression of the yellow, round seed phenotype in the F_2 generation just described may result from the *GGWW, GGWw, GgWW,* and *GgWw* genotypes. If an F_2 yellow, round plant is crossed with a homozygous recessive green, wrinkled plant (*ggww*), analysis of the offspring will indicate the actual genotype of that yellow, round plant. Each of the above genotypes results in a different set of gametes, and in a testcross, a different set of phenotypes in the resulting offspring. You should work out the results of each of these four crosses to be sure you understand this concept.

How Mendel's Peas Become Wrinkled: A Molecular Explanation

Only recently, well over a hundred years after Mendel used wrinkled peas in his groundbreaking hybridization experiments, have we come to find out how the *wrinkled* gene makes peas wrinkled. The wild-type allele of the gene encodes a protein called **starch-branching enzyme (SBEI).** This enzyme catalyzes the formation of highly branched starch molecules as the seed matures.

Wrinkled peas, which result from the homozygous presence of the mutant form of the gene, lack the activity of this enzyme. The production of branch points is inhibited during the synthesis of starch within the seed, which in turn leads to the accumulation of more sucrose and a higher water content while the seed develops. Osmotic pressure inside rises, causing the seed to lose water internally and ultimately results in the wrinkled appearance of the seed at maturation. In contrast, developing seeds that bear at least one copy of the normal gene (being either homozygous or heterozygous for the dominant allele) synthesize starch and reach an osmotic balance that minimizes the loss of water. The end result is a smooth-textured outer coat.

The *SBEI* gene has been cloned and analyzed, providing greater insight into the relationship between genotypes and phenotypes. Interestingly, the mutant gene contains a foreign sequence of some 800 base pairs that disrupts the normal coding region. This foreign segment closely resembles other such units, called **transposable elements.** These elements have the ability to move from place to place in the genome of organisms. Transposable elements have been found in maize (corn), parsley, snapdragons, fruit flies, and humans, among many other organisms.

A wrinkled and round garden pea, the phenotypic traits in one of Mendel's monohybrid crosses.

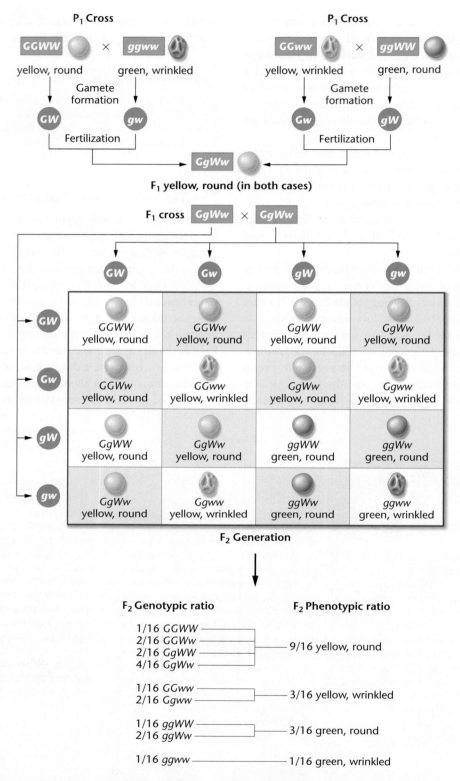

FIGURE 7 Analysis of the dihybrid crosses shown in Figure 5. The F_1 heterozygous plants are self-fertilized to produce an F_2 generation, which is computed using a Punnett square. Both the phenotypic and genotypic F_2 ratios are shown.

2 Determine the genotypes of the parental plants by analyzing the phenotypes of the offspring from the following crosses

Parental Plants	Offspring
(a) round, yellow × round, yellow	3/4 round, yellow 1/4 wrinkled, yellow
(b) round, yellow × wrinkled, yellow	6/16 wrinkled, yellow 2/16 wrinkled, green 6/16 round, yellow 2/16 round, green
(c) round, yellow × wrinkled, green	1/4 round, yellow 1/4 round, green 1/4 wrinkled, yellow 1/4 wrinkled, green

■ **Hint** *This problem involves a series of Mendelian dihybrid crosses where you are asked to determine the genotypes of the parents in a number of instances. The key to its solution is to write down everything that you know for certain. This reduces the problem to its bare essentials, clarifying what you need to determine. For example, the wrinkled, yellow plant in case (b) must be homozygous for the recessive wrinkled alleles and bear at least one dominant allele for the yellow trait. Having established this, you need only determine the remaining allele for cotyledon color.*

4   The Trihybrid Cross Demonstrates That Mendel's Principles Apply to Inheritance of Multiple Traits

Thus far, we have considered inheritance by individuals of up to two pairs of contrasting traits. Mendel demonstrated that the identical processes of segregation and independent assortment apply to three pairs of contrasting traits in what is called a **trihybrid cross,** or *three-factor cross.*

Although a trihybrid cross is somewhat more complex than a dihybrid cross, its results are easily calculated if the principles of segregation and independent assortment are followed. For example, consider the cross shown in **Figure 8**, where the gene pairs of theoretical contrasting traits are represented by the symbols *A, a, B, b, C,* and *c.* In the cross between *AABBCC* and *aabbcc* individuals, all F_1 individuals are heterozygous for all three gene pairs. Their genotype, *AaBbCc,* results in the phenotypic expression of the dominant *A, B,* and *C* traits. When F_1 individuals serve as parents, each produces eight different gametes in equal frequencies. At this point, we could construct a Punnett square with 64 separate boxes and read out the phenotypes—but such a method is cumbersome in a cross involving so many factors. Therefore, another method has been devised to calculate the predicted ratio.

Trihybrid gamete formation

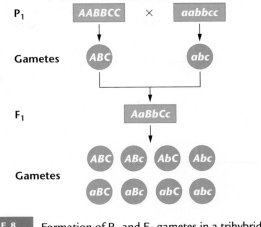

FIGURE 8   Formation of P_1 and F_1 gametes in a trihybrid cross.

The Forked-Line Method

It is much less difficult to consider each contrasting pair of traits separately and then to combine these results by using the **forked-line method,** first shown in Figure 6. This method, also called a **branch diagram,** relies on the simple application of the laws of probability established for the dihybrid cross. Each gene pair is assumed to behave independently during gamete formation.

When the monohybrid cross *AA* × *aa* is made, we know that:

1. All F_1 individuals have the genotype *Aa* and express the phenotype represented by the *A* allele, which is called the *A* phenotype in the following discussion.

2. The F_2 generation consists of individuals with either the *A* phenotype or the *a* phenotype in the ratio of 3:1.

The same generalizations can be made for the *BB* × *bb* and *CC* × *cc* crosses. Thus, in the F_2 generation, 3/4 of all organisms express phenotype *A*, 3/4 express *B*, and 3/4 express *C*. Similarly, 1/4 of all organisms express phenotype *a*, 1/4 express *b*, and 1/4 express *c*. The proportions of organisms that express each phenotypic combination can be predicted by assuming that fertilization, following the independent assortment of these three gene pairs during gamete formation, is a random process—we simply apply the product law of probabilities once again. **Figure 9,** uses the forked-line method to calculate the phenotypic proportions of the F_2 generation. They fall into the trihybrid ratio of 27:9:9:9:3:3:3:1. The same method can be used to solve crosses involving any number of gene pairs, *provided that all gene pairs assort independently of each other.* We shall see later that this is not always the case. However, it appeared to be true for all of Mendel's characters.

Generation of F₂ trihybrid phenotypes

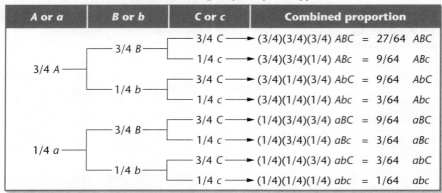

FIGURE 9 Generation of the F₂ trihybrid phenotypic ratio using the forked-line method. This method is based on the expected probability of occurrence of each phenotype.

ESSENTIAL POINT ■ ■ ■

The forked-line method is less complex than, but just as accurate as, the Punnett square in predicting the probabilities of phenotypes or genotypes from crosses involving two or more gene pairs.

NOW SOLVE THIS

3 Using the forked-line method, determine the genotypic and phenotypic ratios of these trihybrid crosses:

    a. *AaBbCc × AaBBCC*
    b. *AaBBCc × aaBBCc*
    c. *AaBbCc × AaBbCc*

■ **Hint** *This problem asks you to use the forked-line method to determine the outcome of a number of trihybrid crosses. The key to its solution is to realize that in using the forked-line method, you must consider each gene pair separately. For example, in this problem, first predict the outcome of each cross for the A/a genes, then for the B/b genes, and finally, for the C/c genes. Then you are prepared to pursue the outcome of each cross using the forked-line method.*

---

5   Mendel's Work Was Rediscovered in the Early Twentieth Century

Mendel initiated his work in 1856, presented it to the Brünn Society of Natural Science in 1865, and published it the following year. Although his findings were often cited and discussed, their significance went unappreciated for about 35 years. Many explanations have been proposed for this delay.

First, Mendel's adherence to mathematical analysis of probability events was an unusual approach in those days for biological studies. Perhaps his approach seemed foreign to his contemporaries. More important, his conclusions did not fit well with existing theories on the cause of variation among organisms. The source of natural variation intrigued students of evolutionary theory. These individuals, stimulated by the proposal developed by Charles Darwin and Alfred Russel Wallace, believed in **continuous variation,** which held that offspring were a *blend* of their parents' phenotypes. As we mentioned earlier, Mendel theorized that variation was due to discrete or particulate units, resulting in **discontinuous variation.** For example, Mendel proposed that the F₂ offspring of a dihybrid cross are expressing traits produced by new combinations of previously existing unit factors. As a result, Mendel's theories did not fit well with the evolutionists' preconceptions about causes of variation.

It is also likely that Mendel's contemporaries failed to realize that Mendel's postulates explained *how* variation was transmitted to offspring. Instead, they may have attempted to interpret his work in a way that addressed the issue of *why* certain phenotypes survive preferentially. It was this latter question that had been addressed in the theory of natural selection, but it was not addressed by Mendel. The collective vision of Mendel's scientific colleagues may have been obscured by the impact of Darwin's extraordinary theory of organic evolution.

The Chromosomal Theory of Inheritance

In the latter part of the nineteenth century, a remarkable observation set the scene for the rebirth of Mendel's work: Walter Flemming's discovery of chromosomes in the nuclei of salamander cells. In 1879, Flemming described the behavior of these threadlike structures during cell division. As a result of his findings and the work of many other cytologists, the presence of discrete units within the nucleus soon became an integral part of ideas about inheritance. It was this mindset that prompted scientists to reexamine Mendel's findings.

In the early twentieth century, research led to renewed interest in Mendel's work. Hybridization experiments similar to Mendel's were performed independently by three botanists: Hugo de Vries, Karl Correns, and Erich Tschermak. De Vries's work demonstrated the principle of segregation in his experiments with several plant species. Apparently, he searched the existing literature and found that Mendel's work anticipated his own conclusions! Correns and Tschermak also reached conclusions similar to those of Mendel.

About the same time, two cytologists, Walter Sutton and Theodor Boveri, independently published papers linking their discoveries of the behavior of chromosomes during meiosis to the Mendelian principles of segregation and independent assortment. They pointed out that the separation of chromosomes during meiosis could serve as the cytological basis of these two postulates. Although they thought Mendel's unit factors were probably chromosomes rather than genes on chromosomes, their findings also reestablished the importance of Mendel's work, which became the basis of ensuing genetic investigations. Sutton and Boveri are credited with initiating the **chromosomal theory of inheritance**—the idea that the genetic material in living organisms is contained within chromosomes, which are transmitted from generation to generation. This fundamental theory continued to gain support during the next two decades.

ESSENTIAL POINT ■ ■ ■

The discovery of chromosomes in the late 1800s, along with subsequent studies of their behavior during meiosis, led to the rebirth of Mendel's work, linking the behavior of his unit factors to that of chromosomes during meiosis.

The Correlation between Mendel's Unit Factors and the Behavior of Chromosomes during Meiosis

Because the correlation between Sutton's and Boveri's observations and Mendelian principles serves as the foundation for the modern interpretation of transmission genetics, we will examine this relationship in some depth before moving on to other topics.

As we know, each species possesses a specific number of chromosomes in each somatic cell nucleus (except in gametes). For diploid organisms, this number is called the **diploid number (2n)** and is characteristic of that species. During the formation of gametes, this number is precisely halved (n), and when two gametes combine during fertilization, the diploid number is reestablished. During meiosis, however, the chromosome number is not reduced in a random manner. It was apparent to early cytologists that the diploid number of chromosomes is composed of homologous pairs identifiable by their morphological appearance and behavior. The gametes contain one member of each pair—thus the chromosome complement of a gamete is quite specific, and

the number of chromosomes in each gamete is equal to the haploid number.

With this basic information, we can see the correlation between the behavior of unit factors and chromosomes and genes. **Figure 10** shows three of Mendel's postulates and the chromosomal explanation of each. Unit factors are really genes located on homologous pairs of chromosomes [**Figure 10(a)**]. Members of each pair of homologs separate, or segregate, during gamete formation [**Figure 10(b)**]. Two different alignments are possible, both of which are shown.

To illustrate the principle of independent assortment, we must distinguish between members of any given homologous pair of chromosomes. One member of each pair comes from the **maternal parent,** while the other member comes from the **paternal parent.** (We represent the different parental origins with different colors.) As shown in **Figure 10(c)**, following independent segregation of each pair of homologs, each gamete receives one member from each pair of chromosomes. All possible combinations are formed with equal probability. If we add the symbols used in Mendel's dihybrid cross (G, g and W, w) to the diagram, we can see why equal numbers of the four types of gametes are formed. The independent behavior of Mendel's pairs of unit factors (G and W in this example) is due to their presence on separate pairs of homologous chromosomes.

Observations of the phenotypic diversity of living organisms make it logical to assume that there are many more genes than chromosomes. Therefore, each homolog must carry genetic information for more than one trait. The currently accepted concept is that a chromosome is composed of a large number of linearly ordered, information-containing *genes*. Mendel's unit factors (which determine tall or dwarf stems, for example) actually constitute a pair of genes located on one pair of homologous chromosomes. The location on a given chromosome where any particular gene occurs is called its **locus** (pl., loci). The different forms taken by a given gene, called *alleles* (G or g), contain slightly different genetic information that determines the same character (seed color in this case). Although we have examined only genes with two alternative alleles, most genes have more than two allelic forms. We conclude this section by reviewing the criteria necessary to classify two chromosomes as a homologous pair:

1. During mitosis and meiosis, when chromosomes are visible as distinct figures, both members of a homologous pair are the same size and exhibit identical centromere locations. The sex chromosomes are an exception.

2. During early stages of meiosis, homologous chromosomes form pairs, or synapse.

3. Although not generally microscopically visible, homologs contain identical, linearly ordered gene loci.

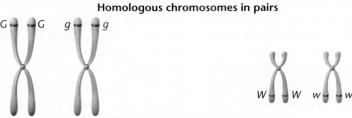

(a) Unit factors in pairs (first meiotic prophase)

Homologous chromosomes in pairs

Genes are part of chromosomes

(b) Segregation of unit factors during gamete formation (first meiotic anaphase)

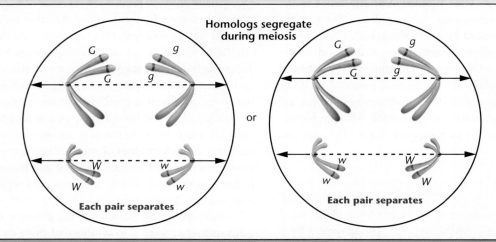

Homologs segregate
during meiosis

or

Each pair separates   Each pair separates

(c) Independent assortment of segregating unit factors (following many meiotic events)

Nonhomologous chromosomes assort independently

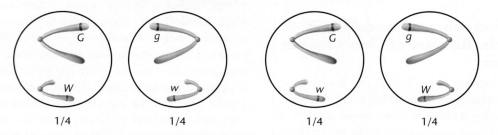

1/4   1/4   1/4   1/4

All possible gametic combinations are formed with equal probability

FIGURE 10 Illustrated correlation between the Mendelian postulates of (a) unit factors in pairs, (b) segregation, and (c) independent assortment, showing the presence of genes located on homologous chromosomes and their behavior during meiosis.

6 Independent Assortment Leads to Extensive Genetic Variation

One major consequence of independent assortment is the production by an individual of genetically dissimilar gametes. Genetic variation results because the two members of any homologous pair of chromosomes are rarely, if ever, genetically identical. As the maternal and paternal members of all pairs are distributed to gametes through independent assortment, all possible chromosome combinations are produced, leading to extensive genetic diversity.

We have seen that the number of possible gametes, each with different chromosome compositions, is 2^n, where n equals the haploid number. Thus, if a species has a haploid number of $n = 4$, then $2^4 = 16$ different gamete combinations can be formed as a result of independent assortment.

Tay–Sachs Disease: The Molecular Basis of a Recessive Disorder in Humans

An interesting question involving Mendelian traits centers around how mutant genes result in mutant phenotypes. Insights are gained by considering a modern explanation of the gene that causes **Tay–Sachs disease** (TSD), a devastating inherited recessive disorder involving unalterable destruction of the central nervous system. Infants with TSD are unaffected at birth and appear to develop normally until they are about six months old. Then, a progressive loss of mental and physical abilities occurs. Afflicted infants eventually become blind, deaf, mentally retarded, and paralyzed, often within only a year or two, seldom living beyond age 5. Typical of rare autosomal recessive disorders, two unaffected heterozygous parents, who most often have no immediate family history of the disorder, have a probability of one in four of having a Tay–Sachs child.

We know that proteins are the end products of the expression of most genes. The protein product involved in TSD has been identified, and we now have a clear understanding of the underlying molecular basis of the disorder. TSD results from the loss of activity of a single enzyme **hexosaminidase A (Hex-A)**. Hex-A, normally found in lysosomes within cells, is needed to break down the ganglioside GM2, a lipid component of nerve cell membranes. Without functional Hex-A, ganglio-sides accumulate within neurons in the brain and cause deterioration of the nervous system. Heterozygous carriers of TSD with one normal copy of the gene produce only about 50 percent of the normal amount of Hex-A, but they show no symptoms of the disorder. The observation that the activity of only one gene (one wild-type allele) is sufficient for the normal development and function of the nervous system explains and illustrates the molecular basis of recessive mutations. Only when both genes are disrupted by mutation is the mutant phenotype evident. The responsible gene is located on chromosome 15 and codes for the alpha subunit of the Hex-A enzyme. More than 50 different mutations within the gene have been identified that lead to TSD phenotypes.

Although this number is not high, consider the human species, where $n = 23$. If 2^{23} is calculated, we find that in excess of 8×10^6, or over 8 million, different types of gametes are represented. Because fertilization represents an event involving only one of approximately 8×10^6 possible gametes from each of two parents, each offspring represents only one of $(8 \times 10^6)^2$, or one of only 6.4×10^{13} potential genetic combinations! No wonder that, except for identical twins, each member of the human species demonstrates a distinctive appearance and individuality—this number of combinations is far greater than the number of humans who have ever lived on Earth! Genetic variation resulting from independent assortment has been extremely important to the process of evolution in all sexually reproducing organisms.

---

7   Laws of Probability Help to Explain Genetic Events

Recall that genetic ratios are expressed as probabilities—for example, 3/4 tall:1/4 dwarf. These values predict the outcome of each fertilization event, such that the probability of each zygote having the genetic potential for becoming tall is 3/4 while the potential for becoming dwarf is 1/4. Probabilities range from 0.0, when an event is *certain not to occur,* to 1.0, when an event is *certain to occur.* When two or more events occur independently but at the same time, we can calculate the probability of possible outcomes when they occur together.

This is accomplished by applying the **product law**—the probability of two or more events occurring simultaneously is equal to the product of their individual probabilities. Two or more events are independent of one another if the outcome of each one does not affect the outcome of any of the others under consideration.

To illustrate the product law, consider the possible results if you toss a penny (P) and a nickel (N) at the same time and examine all combinations of heads (H) and tails (T) that can occur. There are four possible outcomes

$$(P_H : N_H) = (1/2)\,(1/2) = 1/4$$
$$(P_T : N_H) = (1/2)\,(1/2) = 1/4$$
$$(P_H : N_T) = (1/2)\,(1/2) = 1/4$$
$$(P_T : N_T) = (1/2)\,(1/2) = 1/4$$

The probability of obtaining a head or a tail in the toss of either coin is 1/2 and is unrelated to the outcome of the toss of the other coin. Thus, all four possible combinations are predicted to occur with equal probability.

If we want to calculate the probability where the possible outcomes of two events are independent of one another but can be accomplished in more than one way, we apply the **sum law.** For example, what is the probability of tossing our penny and nickel and obtaining one head and one tail? In such a case, we do not care whether it is the penny or the nickel that comes up heads, provided the other coin has the alternative outcome. As we saw above, there are two ways in which the desired outcome can be accomplished, each with a probability of 1/4. Thus, according to the sum law, the overall probability is equal to

$$(1/4) + (1/4) = 1/2$$

One-half of all coin tosses are predicted to yield the desired outcome.

These simple probability laws will be useful throughout our discussions of transmission genetics and for solving genetics problems. In fact, we already applied the product law when we used the forked-line method to calculate the phenotypic results of Mendel's dihybrid and trihybrid crosses. When we wish to know the results of a cross, we need only calculate the probability of each possible outcome. The results of this calculation then allow us to predict the proportion of offspring expressing each phenotype or each genotype.

An important point to remember when you deal with probability is that predictions of possible outcomes are based on large sample sizes. If we predict that 9/16 of the offspring of a dihybrid cross will express both dominant traits, it is very unlikely that, in a small sample, exactly 9 of every 16 offspring will express this phenotype. Instead, our prediction is that, of a large number of offspring, approximately 9/16 of them will do so. The deviation from the predicted ratio in smaller sample sizes is attributed to chance, a subject we examine in our discussion of statistics in the next section. As you shall see, the impact of deviation due strictly to chance diminishes as the sample size increases.

8 | Chi-Square Analysis Evaluates the Influence of Chance on Genetic Data

Mendel's 3:1 monohybrid and 9:3:3:1 dihybrid ratios are hypothetical predictions based on the following assumptions: (1) Each allele is dominant or recessive; (2) segregation is operative; (3) independent assortment occurs; and (4) fertilization is random. The final two assumptions are influenced by chance events and are therefore subject to random fluctuation. This concept of **chance deviation** is most easily illustrated by tossing a single coin numerous times and recording the number of heads and tails observed. In each toss, there is a probability of 1/2 that a head will occur and a probability of 1/2 that a tail will occur. Therefore, the expected ratio

of many tosses is 1:1. If a coin is tossed 1000 times, usually *about* 500 heads and 500 tails will be observed. Any reasonable fluctuation from this hypothetical ratio (e.g., 486 heads and 514 tails) is attributed to chance.

As the total number of tosses is reduced, the impact of chance deviation increases. For example, if a coin is tossed only four times, you would not be too surprised if all four tosses result in only heads or only tails. For 1000 tosses, however, 1000 heads or 1000 tails would be most unexpected. In fact, you might believe that such a result would be impossible. Actually, all heads or all tails in 1000 tosses can be predicted to occur with a probability of $(1/2)^{1000}$. Since $(1/2)^{20}$ is equivalent to less than one in a million times, an event occurring with a probability of $(1/2)^{1000}$ is virtually impossible. Two major points should be noted before we consider *chi-square analysis*:

1. The outcomes of independent assortment and fertilization, like coin tossing, are subject to random fluctuations from their predicted occurrences as a result of chance deviation.

2. As the sample size increases, the average deviation from the expected results decreases. Therefore, a larger sample size diminishes the impact of chance deviation on the final outcome.

Chi-Square Calculations and the Null Hypothesis

In genetics, the ability to evaluate observed deviation is a crucial skill. When we assume that data will fit a given ratio such as 1:1, 3:1, or 9:3:3:1, we establish what is called the **null hypothesis (H_0)**. It is so named because the hypothesis assumes that *no real difference* exists between *measured values* (or ratio) and *predicted values* (or ratio). The apparent difference can be attributed purely to chance. The null hypothesis is evaluated using statistical analysis. On this basis, the null hypothesis may either (1) be rejected or (2) fail to be rejected. If it is rejected, the observed deviation from the expected result is not attributed to chance alone. The null hypothesis and the underlying assumptions leading to it must be reexamined. If the null hypothesis fails to be rejected, any observed deviations are attributed to chance.

One of the simplest statistical tests devised to assess the null hypothesis is **chi-square (χ^2) analysis.** This test takes into account the observed deviation in each component of an expected ratio as well as the sample size and reduces them to a single numerical value. The value for χ^2 is then used to estimate how frequently the observed deviation can be expected to occur strictly as a result of chance. The formula for chi-square analysis is

$$\chi^2 = \sum \frac{(o - e)^2}{e}$$

TABLE 1	Chi-Square Analysis				
(a) Monohybrid Cross Expected Ratio	**Observed (o)**	**Expected (e)**	**Deviation ($o - e$)**	**Deviation (d^2)**	**d^2/e**
3/4	740	3/4(1000) = 750	740 − 750 = −10	$(-10)^2 = 100$	100/750 = 0.13
1/4	260	1/4(1000) = 250	260 − 250 = +10	$(+10)^2 = 100$	100/250 = 0.40
	Total = 1000				$\chi^2 = 0.53$
					$p = 0.48$
(b) Dihybrid Cross Expected Ratio	**o**	**e**	**$(o - e)$**	**d^2**	**d^2/e**
9/16	587	567	+20	400	0.71
3/16	197	189	+8	64	0.34
3/16	168	189	−21	441	2.33
1/16	56	63	−7	49	0.78
	Total = 1008				$\chi^2 = 4.16$
					$p = 0.26$

where o is the observed value for a given category, e is the expected value for that category, and Σ (the Greek letter sigma) represents the sum of the calculated values for each category of the ratio. Because $(o - e)$ is the deviation (d) in each case, the equation reduces to

$$\chi^2 = \Sigma \frac{d^2}{e}$$

Table 1(a) shows a χ^2 calculation for the F_2 results of a hypothetical monohybrid cross. To analyze these data, you work from left to right, calculating and entering the appropriate numbers in each column. Regardless of whether the deviation d is positive or negative, d^2 always becomes positive after the number is squared. In Table 1(b) the F_2 results of a hypothetical dihybrid cross are analyzed. Be sure that you understand how each number was calculated in the dihybrid example.

The final step in chi-square analysis is to interpret the (χ^2) value. To do so, you must initially determine the value of the **degrees of freedom (df)**, which is equal to $n - 1$ where n is the number of different categories into which each datum point may fall. For the 3:1 ratio, $n = 2$ so $df = 2 - 1 = 1$. For the 9:3:3:1 ratio, $n = 4$ and $df = 3$. Degrees of freedom must be taken into account because the greater the number of categories, the more deviation is expected as a result of chance.

Once you have determined the degrees of freedom, we can interpret the χ^2 value in terms of a corresponding **probability value (p).** Since this calculation is complex, we usually take the p value from a standard table or graph. **Figure 11** shows a wide range of χ^2 and p values for various degrees of freedom in both a graph and a table. Let's use the graph to determine the p value. The caption for **Figure 11(b)** explains how to use the table.

To determine p, execute the following steps:

1. Locate the χ^2 value on the abscissa (the horizontal or x-axis).

2. Draw a vertical line from this point up to the angled line on the graph representing the appropriate df.

3. Extend a horizontal line from this point to the left until it intersects the ordinate (the vertical or y-axis).

4. Estimate, by interpolation, the corresponding p value.

We used these steps for the monohybrid cross in Table 1(a) to estimate the p value of 0.48 shown in **Figure 11(a)**. For the dihybrid cross, try this method to see if you can determine the p value. Since the χ^2 value is 4.16 and $df = 3$, an approximate p value is 0.26. Checking this result in the table confirms that p values for both the monohybrid and dihybrid crosses are between 0.20 and 0.50.

Interpreting Probability Values

So far, we have been concerned with calculating χ^2 values and determining the corresponding p values. The most important aspect of chi-square analysis is understanding the meaning of the p value. Let's use the example of the dihybrid cross in Table 1(b) ($p = 0.26$) In these discussions, it is simplest to think of the p value as a percentage (e.g., $0.26 = 26\%$). In our example, the p value indicates that if we repeat the same experiment many times, 26 percent of the trials would be expected to exhibit chance deviation as great as or greater than that seen in the initial trial. Conversely, 74 percent of the trials would show less deviation than initially observed as a result of chance. Thus, the p value reveals that a hypothesis (the 9:3:3:1 ratio in this case) is never proved or disproved absolutely. Instead, a relative standard is set that enables us to either *reject* or *fail to reject* the null hypothesis—this standard is most often a p value

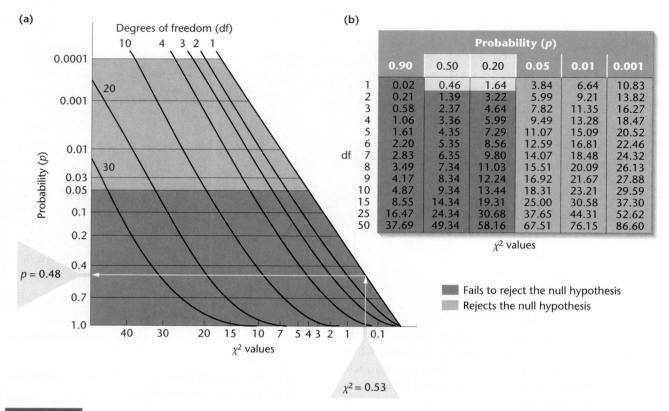

(a)

Degrees of freedom (df)

(b)

	Probability (p)					
	0.90	0.50	0.20	0.05	0.01	0.001
1	0.02	0.46	1.64	3.84	6.64	10.83
2	0.21	1.39	3.22	5.99	9.21	13.82
3	0.58	2.37	4.64	7.82	11.35	16.27
4	1.06	3.36	5.99	9.49	13.28	18.47
5	1.61	4.35	7.29	11.07	15.09	20.52
6	2.20	5.35	8.56	12.59	16.81	22.46
df 7	2.83	6.35	9.80	14.07	18.48	24.32
8	3.49	7.34	11.03	15.51	20.09	26.13
9	4.17	8.34	12.24	16.92	21.67	27.88
10	4.87	9.34	13.44	18.31	23.21	29.59
15	8.55	14.34	19.31	25.00	30.58	37.30
25	16.47	24.34	30.68	37.65	44.31	52.62
50	37.69	49.34	58.16	67.51	76.15	86.60

χ^2 values

■ Fails to reject the null hypothesis
■ Rejects the null hypothesis

FIGURE 11 (a) Graph for converting χ^2 values to p values. (b) Table of χ^2 values for selected values of df and p. χ^2 values that lead to a p value of 0.05 or greater (darker blue areas) justify failure to reject the null hypothesis. Values leading to a p value of less than 0.05 (lighter blue areas) justify rejecting the null hypothesis. For example, using the table in part (b), where $\chi^2 = 0.053$ for 1 degree of freedom, the corresponding p value is between 0.20 and 0.50. The graph in (a) gives a more precise p value of 0.48 by interpolation. Thus, we fail to reject the null hypothesis.

of 0.05. When applied to chi-square analysis, a p value less than 0.05 means that the observed deviation in the set of results will be obtained by chance alone less than 5 percent of the time. Such a p value indicates that the difference between the observed and predicted results is significant and thus enables us to reject the null hypothesis.

On the other hand, p values of 0.05 or greater (0.05 to 1.0) indicate that the observed deviation will be obtained by chance alone 5 percent or more of the time. The conclusion is not to reject the null hypothesis. Thus, for the p value of 0.26, assessing the hypothesis that independent assortment accounts for the results fails to be rejected. Therefore, the observed deviation can be reasonably attributed to chance.

A final note is relevant here for the case where the null hypothesis is rejected, that is, where $p \leq 0.05$. Suppose we had tested a data set to assess a possible 9:3:3:1 ratio, as in Table 1(b), but we rejected the null hypothesis based on our χ^2 calculation. What are alternative interpretations of the data? Researchers will reassess the assumptions that underlie the null hypothesis. In our example, we assumed that segregation operates faithfully for both gene pairs. We also assumed that fertilization is random and that the viability of all gametes is equal regardless of genotype—that is, that all gametes

are equally likely to participate in fertilization. Finally, we assumed that, following fertilization, all preadult stages and adult offspring are equally viable, regardless of their genotype. If any of these assumptions is incorrect, the original hypothesis is not necessarily invalid.

An example will clarify this issue. Suppose our null hypothesis is that a dihybrid cross between fruit flies will result in 3/16 mutant wingless fly zygotes. However, not as many of the mutant embryos may survive their preadult development or as young adults, compared to flies whose genotype gives rise to wings. As a result, when the data are gathered, there are fewer than 3/16 wingless flies. Rejection of the null hypothesis is not in itself cause for us to reject the validity of the postulates of segregation and independent assortment, because other factors we are unaware of may also be affecting the outcome.

ESSENTIAL POINT ■ ■ ■

Chi-square analysis allows us to assess the null hypothesis, which states that there is no real difference between the expected and observed values. As such, it tests the probability of whether observed variations can be attributed to chance deviation.

4 In one of Mendel's dihybrid crosses, he observed 315 round, yellow; 108 round, green; 101 wrinkled, yellow; and 32 wrinkled, green F_2 plants. Analyze these data using chi-square analysis to see whether (a) they fit a 9:3:3:1 ratio; (b) the round, wrinkled traits fit a 3:1 ratio; and (c) the yellow, green traits fit a 3:1 ratio.

■ **Hint** *This problem asks you to apply χ^2 analysis to a set of data and to determine whether those data fit any of several ratios. The key to its solution is to first calculate χ^2 by initially determining the expected outcomes using the predicted ratios. Then follow a stepwise approach, determining the deviation in each case, and calculating d^2/e for each category. Once you have determined the χ^2 value, you must then determine and interpret the p value for each ratio.*

9 Pedigrees Reveal Patterns of Inheritance of Human Traits

We now explore how to determine the mode of inheritance of phenotypes in humans, where designed crosses are not possible and where relatively few offspring are available for study. The traditional way to study inheritance has been to construct a family tree, indicating the presence or absence of the trait in question for each member of each generation. Such a family tree is called a **pedigree.** By analyzing a pedigree, we may be able to predict how the trait under study is inherited—for example, is it due to a dominant or recessive allele? When many pedigrees for the same trait are studied, we can often ascertain the mode of inheritance.

Pedigree Conventions

Figure 12 illustrates a number of conventions geneticists follow in constructing pedigrees. Circles represent females and squares designate males. Parents are connected by a single horizontal line, and vertical lines lead to their offspring. If the parents are related—that is, they are **consanguineous,** such as first cousins—they are connected by a double line. Offspring are called **sibs** (short for **siblings**) and are connected by a horizontal **sibship line.** Sibs are placed from left to right according to birth order and are labeled with Arabic numerals. Each generation is indicated by a Roman numeral. If the sex of an individual is unknown, a diamond is used. When a pedigree traces only a single trait, the circles, squares, and diamonds are shaded if the phenotype being considered is expressed and unshaded if not. In some pedigrees, those individuals that fail to express a recessive trait, but are known with certainty to be a heterozygous carrier, have a shaded dot within their unshaded circle or square. If an individual is deceased and the phenotype is unknown, a diagonal line is placed over the circle or square.

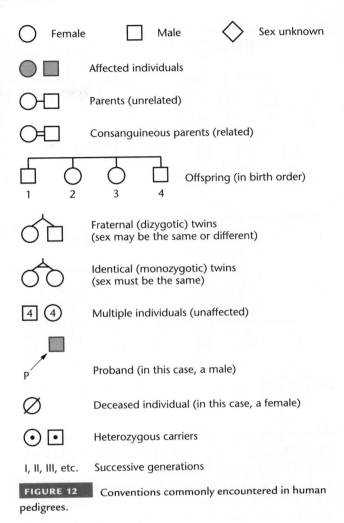

FIGURE 12 Conventions commonly encountered in human pedigrees.

Twins are indicated by diagonal lines stemming from a vertical line connected to the sibship line. For **identical (or monozygotic) twins,** the diagonal lines are linked by a horizontal line. **Fraternal (or dizygotic) twins** lack this connecting line. A number within one of the symbols represents numerous sibs of the same or unknown phenotypes. The individual whose phenotype first brought attention to the investigation and construction of the pedigree is called the **proband** and is indicated by an arrow connected to the designation **p.** This term applies to either a male or a female.

Pedigree Analysis

In **Figure 13**, two pedigrees are shown. The first illustrates a representative pedigree for a trait that demonstrates autosomal recessive inheritance, such as **albinism.** The male parent of the first generation (I-1) is affected. Characteristic of a rare recessive trait with an affected parent, the trait "disappears" in the offspring of the next generation. Assuming recessiveness, we might predict that the unaffected female parent (I-2) is a homozygous normal individual because none of the offspring show the disorder. Had she been heterozygous, one-half

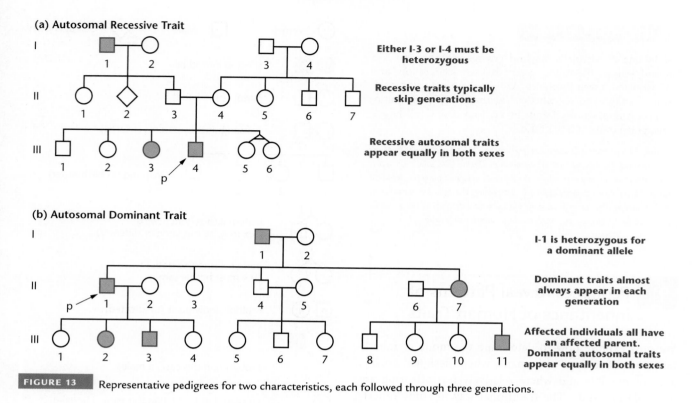

FIGURE 13 Representative pedigrees for two characteristics, each followed through three generations.

of the offspring would be expected to exhibit albinism, but none do. However, such a small sample (three offspring) prevents us from knowing for certain.

Further evidence supports the prediction of a recessive trait. If albinism were inherited as a dominant trait, individual II-3 would have to express the disorder in order to pass it to his offspring (III-3 and III-4), but he does not. Inspection of the offspring constituting the third generation (row III) provides still further support for the hypothesis that albinism is a recessive trait. If it is, parents II-3 and II-4 are both heterozygous, and approximately one-fourth of their offspring should be affected. Two of the six offspring do show albinism. This deviation from the expected ratio is not unexpected in crosses with few offspring. Once we are confident that albinism is inherited as an autosomal recessive trait, we could portray the II-3 and II-4 individuals with a shaded dot within their larger square and circle. Finally, we can note that, characteristic of pedigrees for autosomal traits, both males and females are affected with equal probability. There are certain limitations imposed on the transmission of X-linked traits, such as that these traits are more prevalent in male offspring and are never passed from affected fathers to their sons.

The second pedigree illustrates the pattern of inheritance for a trait such as Huntington disease, which is caused by an autosomal dominant allele. The key to identifying such a pedigree that reflects a dominant trait is that all affected off-spring will have a parent that also expresses the trait. It is also possible, by chance, that none of the offspring will inherit the dominant allele. If so, the trait will cease to exist in future generations. Like recessive traits, provided that the gene is autosomal, both males and females are equally affected.

When autosomal dominant diseases are rare within the population, and most are, then it is highly unlikely that affected individuals will inherit a copy of the mutant gene from both parents. Therefore, in most cases, affected individuals are heterozygous for the dominant allele. As a result, approximately one-half of the offspring inherit it. This is borne out in the second pedigree in Figure 13. Furthermore, if a mutation is dominant, and a single copy is sufficient to produce a mutant phenotype, homozygotes are likely to be even more severely affected, perhaps even failing to survive. An illustration of this is the dominant gene for **familial hypercholesterolemia.** Heterozygotes display a defect in their receptors for low-density lipoproteins, the so-called LDLs. As a result, too little cholesterol is taken up by cells from the blood, and elevated plasma levels of LDLs result. Such heterozygous individuals have heart attacks during the fourth decade of their life, or before. While heterozygotes have LDL levels about double that of a normal individual, rare homozygotes have been detected. They lack LDL receptors altogether and have LDL levels nearly ten times above the normal range. They are likely to have a heart attack very early in life, even before age 5, and almost inevitably before they reach the age of 20.

Pedigree analysis of many traits has historically been an extremely valuable research technique in human genetic

TABLE 2 Representative Recessive and Dominant Human Traits

Recessive Traits	Dominant Traits
Albinism	Achondroplasia
Alkaptonuria	Brachydactyly
Ataxia telangiectasia	Congenital stationary night blindness
Color blindness	Ehler–Danlos syndrome
Cystic fibrosis	Hypotrichosis
Duchenne muscular dystrophy	Huntington disease
Galactosemia	Hypercholesterolemia
Hemophilia	Marfan syndrome
Lesch–Nyhan syndrome	Neurofibromatosis
Phenylketonuria	Phenylthiocarbamide (PTC) tasting
Sickle-cell anemia	Porphyria
Tay-Sachs disease	Widow's peak

studies. However, the approach does not usually provide the certainty in drawing conclusions afforded by designed crosses yielding large numbers of offspring. Nevertheless, when many independent pedigrees of the same trait or disorder are analyzed, consistent conclusions can often be drawn. **Table 2** lists numerous human traits and classifies them according to their recessive or dominant expression.

ESSENTIAL POINT ■ ■ ■

Pedigree analysis is a method for studying the inheritance pattern of human traits over several generations, providing the basis for predicting the mode of inheritance of characteristics and disorders in the absence of extensive genetic crossing and large numbers of offspring.

NNANACTGACNCAC TA TAGGGCGAA TTCGAGCTCGG TACCCGGNGGA TCCTC AGAG CGACC GCAGGCA GCAAGC GAGA
10   20   30   40   50   60   70   80

EXPLORING GENOMICS

Online Mendelian Inheritance in Man

 Study Area: *Exploring Genomics*

The Online Mendelian Inheritance in Man (OMIM) database is a catalog of human genes and human genetic disorders that are inherited in a Mendelian manner. Genetic disorders that arise from major chromosomal aberrations, such as monosomy or trisomy (the loss of a chromosome or the presence of a superfluous chromosome, respectively), are not included. The OMIM database is a daily-updated version of the book *Mendelian Inheritance in Man,* edited by Dr. Victor McKusick of Johns Hopkins University. Scientists use OMIM as an important information source to accompany the sequence data generated by the Human Genome Project.

The OMIM entries will give you links to a wealth of information, including DNA and protein sequences, chromosomal maps, disease descriptions, and relevant scientific publications. In this exercise, you will explore OMIM to answer questions about the recessive human disease sickle-cell anemia and other Mendelian inherited disorders.

■ Exercise I – Sickle-cell Anemia

In this chapter, you were introduced to recessive and dominant human traits. You will now discover more about sickle-cell anemia as an autosomal recessive disease by exploring the OMIM database.

1. To begin the search, access the OMIM site at: **www.omim.org**.
2. In the "SEARCH" box, type "sickle-cell anemia" and click on the "Go" button to perform the search.
3. Select the first entry (#603903).
4. Examine the list of subject headings in the left-hand column and read some of the information about sickle-cell anemia.
5. Select one or two references at the bottom of the page and follow them to their abstracts in PubMed.
6. Using the information in this entry, answer the following questions:
   a. Which gene is mutated in individuals with sickle-cell anemia?
   b. What are the major symptoms of this disorder?

   c. What was the first published scientific description of sickle-cell anemia?
   d. Describe two other features of this disorder that you learned from the OMIM database and state where in the database you found this information.

■ Exercise II – Other Recessive or Dominant Disorders

Select another human disorder that is inherited as either a dominant or recessive trait and investigate its features, following the general procedure presented above. Follow links from OMIM to other databases if you choose.

Describe several interesting pieces of information you acquired during your exploration and cite the information sources you encountered during the search.

CASE STUDY » To test or not to test

Thomas first discovered a potentially devastating piece of family history when he learned the medical diagnosis for his brother's increasing dementia, muscular rigidity, and frequency of seizures. His brother, at age 49, was diagnosed with Huntington disease (HD), a dominantly inherited condition that typically begins with such symptoms around the age of 45 and leads to death in one's early 60s. As depressing as the news was to Thomas, it helped explain his father's suicide. Thomas, 38, now wonders what are his chances of carrying the gene for HD, and he and his wife have discussed the pros and cons of him getting tested. Thomas and his wife have two teenaged children, a boy and a girl.

1. What role might a genetic counselor play in this real-life scenario?
2. How might the preparation and analysis of a pedigree help explain the dilemma facing Thomas and his family?
3. If Thomas decides to go ahead with the genetic test, what should be the role of the health insurance industry in such cases?
4. If Thomas tests positive for HD, and you were one of his children, would you want to be tested?

INSIGHTS AND SOLUTIONS

As a student, you will be asked to demonstrate your knowledge of transmission genetics by solving genetics problems. Success at this task represents not only comprehension of theory but its application to more practical genetic situations. Most students find problem solving in genetics to be challenging and rewarding. This section will provide you with basic insights into the reasoning essential to this process.

Genetics problems are in many ways similar to word problems in algebra. The approach to solving them is identical: (1) Analyze the problem carefully; (2) translate words into symbols, first defining each one; and (3) choose and apply a specific technique to solve the problem. The first two steps are critical. The third step is largely mechanical.

The simplest problems state all necessary information about the P_1 generation and ask you to find the expected ratios of the F_1 and F_2 genotypes and/or phenotypes. Always follow these steps when you encounter this type of problem:

(a) Determine insofar as possible the genotypes of the individuals in the P_1 generation.

(b) Determine what gametes may be formed by the P_1 parents.

(c) Recombine gametes by the Punnett square or the forked-line methods, or if the situation is very simple, by inspection. Read the F_1 phenotypes.

(d) Repeat the process to obtain information about the F_2 generation.

Determining the genotypes from the given information requires that you understand the basic theory of transmission genetics. Consider this problem: *A recessive mutant allele, black, causes a very dark body in* Drosophila *(a fruit fly) when homozygous. The wild-type (normal) color is gray. What F_1 phenotypic ratio is predicted when a black female is crossed with a gray male whose father was black?*

To work out this problem, you must understand dominance and recessiveness, as well as the principle of segregation. Furthermore, you must use the information about the male parent's father. Here is one way to solve this problem:

(a) The female parent is black, so she must be homozygous for the mutant allele (*bb*).

(b) The male parent is gray; therefore, he must have at least one dominant allele (*B*). His father was black (*bb*), and he received one of the chromosomes bearing these alleles, so the male parent must be heterozygous (*Bb*).

With this information, the problem is simple:

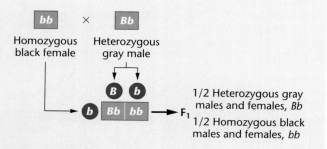

Homozygous black female *bb* × Heterozygous gray male *Bb*

F_1
1/2 Heterozygous gray males and females, *Bb*
1/2 Homozygous black males and females, *bb*

Apply this approach to the following problems.

1. Mendel found that full pods are dominant over constricted pods, while round seeds are dominant over wrinkled seeds. One of his crosses was between full, round plants and constricted, wrinkled plants. From this cross, he obtained an F_1 generation that was all full and round. In the F_2 generation, Mendel obtained his classic 9:3:3:1 ratio. Using this information, determine the expected F_1 and F_2 results of a cross between homozygous constricted, round plants and full, wrinkled plants.

Solution: Define gene symbols for each pair of contrasting traits. Use the lowercase first letter of the recessive traits to designate those phenotypes and the uppercase first letter to designate the dominant traits. Thus, *C* and *c* indicate full and constricted, and *W* and *w* indicate round and wrinkled phenotypes, respectively.

Determine the genotypes of the P_1 generation, form the gametes, reconstitute the F_1 generation, and read off the phenotype(s):

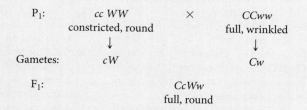

P_1:	*cc WW* constricted, round	×	*CCww* full, wrinkled
Gametes:	*cW*		*Cw*
F_1:		*CcWw* full, round	

You can immediately see that the F_1 generation expresses both dominant phenotypes and is heterozygous for both gene pairs. Thus, you expect that the F_2 generation will yield the classic Mendelian ratio of 9:3:3:1. Let's work it out anyway just to confirm this expectaton, using the forked-line method. Both gene pairs are heterozygous and can be expected to assort independently, so we can predict the outcomes from each gene pair separately and then proceed with the forked-line method.

Every F_2 offspring is subject to the following probabilities:

$$Cc \times Cc \qquad\qquad Ww \times Ww$$
$$\downarrow \qquad\qquad\qquad \downarrow$$

$$\left.\begin{array}{l} CC \\ Cc \\ cC \end{array}\right\} \text{full} \qquad \left.\begin{array}{l} WW \\ Ww \\ wW \end{array}\right\} \text{round}$$
$$cc \quad \text{constricted} \qquad ww \quad \text{wrinkled}$$

The forked-line method then confirms the 9:3:3:1 phenotypic ratio. Remember that this represents proportions of 9/16:3/16:3/16:1/16. Note that we are applying the product law as we compute the final probabilities:

3/4 full
— 3/4 round $\xrightarrow{(3/4)(3/4)}$ 9/16 full, round
— 1/4 wrinkled $\xrightarrow{(3/4)(1/4)}$ 3/16 full, wrinkled

1/4 constricted
— 3/4 round $\xrightarrow{(1/4)(3/4)}$ 3/16 constricted, round
— 1/4 wrinkled $\xrightarrow{(1/4)(1/4)}$ 1/16 constricted, wrinkled

2. In another cross involving parent plants of unknown genotype and phenotype, the following offspring were obtained.

F_1:
3/8	full, round
3/8	full, wrinkled
1/8	constricted, round
1/8	constricted, wrinkled

Determine the genotypes and phenotypes of the parents.

Solution: This problem is more difficult and requires keener insight because you must work backward. The best approach is to consider the outcomes of pod shape separately from those of seed texture.

Of all the plants, $3/8 + 3/8 = 3/4$ are full and $1/8 + 1/8 = 1/4$ are constricted. Of the various genotypic combinations that can serve as parents, which combination will give rise to a ratio of 3/4:1/4? This ratio is identical to Mendel's monohybrid F_2 results, and we can propose that both unknown parents share the same genetic characteristic as the monohybrid F_1 parents; they must both be heterozygous for the genes controlling pod shape and thus are Cc.

Before we accept this hypothesis, let's consider the possible genotypic combinations that control seed texture. If we consider this characteristic alone, we see that the traits are expressed in a ratio of $3/8 + 1/8 = 1/2$ round: $3/8 + 1/8 = 1/2$ wrinkled. To generate such a ratio, the parents cannot both be heterozygous, or their offspring would yield a 3/4:1/4 phenotypic ratio. They cannot both be homozygous, or all of their offspring would express a single phenotype. Thus, we are left with testing the hypothesis that one parent is homozygous and one is heterozygous for the alleles controlling texture. The potential case of $WW \times Ww$ does not work, since it yields only a single phenotype. This leaves us with the potential case of $Ww \times ww$. Offspring in such a mating will yield 1/2 Ww (round): 1/2 ww (wrinkled), exactly the outcome we are seeking.

Now, let's combine the hypotheses and predict the outcome of the cross. In our solution, we use a dash (–) to indicate that the second allele may be either dominant or recessive, since we are only predicting phenotypes.

3/4 $C-$
— 1/2 $Ww \rightarrow$ 3/8 $C-Ww$ full, round
— 1/2 $ww \rightarrow$ 3/8 $C-ww$ full, wrinkled

1/4 cc
— 1/2 $Ww \rightarrow$ 1/8 $ccWw$ constricted, round
— 1/2 $ww \rightarrow$ 1/8 $ccww$ constricted, wrinkled

As you can see, this cross produces offspring according to our initial information, and we have solved the problem. Note that in this solution, we used genotypes in the forked-line method, in contrast to the use of phenotypes in the earlier solution.

3. Determine the probability that a plant of genotype $CcWw$ will be produced from parental plants with the genotypes $CcWw$ and $Ccww$.

Solution: The two gene pairs demonstrate straightforward dominance and recessiveness and assort independently during gamete formation. We need only calculate the individual probabilities of obtaining the two separate outcomes (Cc and Ww) and apply the product law to calculate the final probability:

$$Cc \times Cc \rightarrow \quad 1/4\ CC : 1/2\ Cc : 1/4\ cc$$
$$Ww \times ww \rightarrow 1/2\ Ww : 1/2\ ww$$
$$p = (1/2\ Cc)\,(1/2\ Ww) = 1/4\ CcWw$$

4. In the laboratory, a genetics student crossed flies that had normal, long wings with flies expressing the *dumpy* mutation (truncated wings), which she believed was a recessive trait. In the F_1 generation, all flies had long wings. The following results were obtained in the F_2 generation:

792 long-winged flies

208 dumpy-winged flies

The student tested the hypothesis that the dumpy wing is inherited as a recessive trait, using chi-square analysis of the F_2 data.

(a) What ratio was hypothesized?

(b) Did the analysis support the hypothesis?

(c) What do the data suggest about the *dumpy* mutation?

Solution:

(a) The student hypothesized that the F_2 data (792:208) fit Mendel's 3:1 monohybrid ratio for recessive genes.

(b) The initial step in χ^2 analysis is to calculate the expected results (e) if the ratio is 3:1. Then we can compute deviation $o - e$ (d) and the remaining numbers.

Ratio	o	e	d	d^2	d^2/e
3/4	792	750	42	1764	2.35
1/4	208	250	−42	1764	7.06

Total = 1000

$$\chi^2 = \sum d^2/e$$
$$= 2.35 + 7.06$$
$$= 9.41$$

We consult Figure 11 to determine the probability (p) and determine whether the deviations can be attributed to chance. There are two possible outcomes (*n*), so the degrees of freedom (*df*) = *n* − 1 or 1. The table in Figure 11(b) shows that *p* is a value between 0.01 and 0.001; the graph in Figure 11(a) gives an estimate of about 0.001. Since *p* < 0.05. we reject the null hypothesis. The data do not fit a 3:1 ratio.

(c) When we accepted Mendel's 3:1 ratio as a valid expression of the monohybrid cross, numerous assumptions were made. Examining our underlying assumptions may explain why the null hypothesis

was rejected. We assumed that all genotypes are equally viable— that genotypes yielding long wings are equally likely to survive from fertilization through adulthood as the genotype yielding dumpy wings. Further study may reveal that dumpy-winged flies are somewhat less viable than normal flies. As a result, we would expect less than 1/4 of the total offspring to express dumpy wings. This observation is borne out in the data, although we have not proven that this is true.

PROBLEMS AND DISCUSSION QUESTIONS

 For activities, animations, and review quizzes, go to the study area at www.masteringgenetics.com

When working out genetics problems in this chapters, always assume that members of the P₁ generation are homozygous, unless the information given, or the data, indicates otherwise.

HOW DO WE KNOW?

1. In this chapter, we focused on the Mendelian postulates, probability, and pedigree analysis. We also considered some of the methods and reasoning by which these ideas, concepts, and techniques were developed. On the basis of these discussions, what answers would you propose to the following questions:

   (a) How was Mendel able to derive postulates concerning the behavior of "unit factors" during gamete formation, when he could not directly observe them?

   (b) How do we know whether an organism expressing a dominant trait is homozygous or heterozygous?

   (c) In analyzing genetic data, how do we know whether deviation from the expected ratio is due to chance rather than to another, independent factor?

   (d) Since experimental crosses are not performed in humans, how do we know how traits are inherited?

2. In a cross between a black and a white guinea pig, all members of the F₁ generation are black. The F₂ generation is made up of approximately 3/4 black and 1/4 white guinea pigs. Diagram this cross, and show the genotypes and phenotypes.

3. Albinism in humans is inherited as a simple recessive trait. Determine the genotypes of the parents and offspring for the following families. When two alternative genotypes are possible, list both.

   (a) Two nonalbino (normal) parents have five children, four normal and one albino.

   (b) A normal male and an albino female have six children, all normal.

4. In a problem involving albinism (see Problem 3), which of Mendel's postulates are demonstrated?

5. Why was the garden pea a good choice as an experimental organism in Mendel's work?

6. Mendel crossed peas having round seeds and yellow cotyledons with peas having wrinkled seeds and green cotyledons. All the F₁ plants had round seeds with yellow cotyledons. Diagram this cross through the F₂ generation, using both the Punnett square and forked-line methods.

7. Refer to the Now Solve This Problem 2. Are any of the crosses in this problem testcrosses? If so, which one(s)?

8. Which of Mendel's postulates can be demonstrated in the Now Solve This Problem 2 but not in Problem 2 above? Define this postulate.

9. Correlate Mendel's four postulates with what is now known about homologous chromosomes, genes, alleles, and the process of meiosis.

10. What is the basis for homology among chromosomes?

11. Distinguish between homozygosity and heterozygosity.

12. In *Drosophila*, gray body color is dominant over ebony body color, while long wings are dominant over vestigial wings. Work the following crosses through the F₂ generation, and determine the genotypic and phenotypic ratios for each generation. Assume that the P₁ individuals are homozygous: (a) gray, long × ebony, vestigial, and (b) gray, vestigial × ebony, long, and (c) gray, long × gray, vestigial.

13. How many different types of gametes can be formed by individuals of the following genotypes? What are they in each case? (a) *AaBb*, (b) *AaBB*, (c) *AaBbCc*, (d) *AaBBcc*, (e) *AaBbcc*, and (f) *AaBbCcDdEe*?

14. Mendel crossed peas with round, green seeds with peas having wrinkled, yellow seeds. All F₁ plants had seeds that were round and yellow. Predict the results of testcrossing these F₁ plants.

15. Shown are F₂ results of two of Mendel's monohybrid crosses. State a null hypothesis that you will test using chi-square analysis. Calculate the χ^2 value and determine the *p* value for both crosses, then interpret the *p* values. Which cross shows a greater amount of deviation?

(a) Full pods	882
Constricted pods	299
(b) Violet flowers	705
White flowers	224

16. A geneticist, in assessing data that fell into two phenotypic classes, observed values of 250:150. He decided to perform chi-square analysis using two different null hypotheses: (a) The data fit a 3:1 ratio; and (b) the data fit a 1:1 ratio. Calculate the χ^2 values for each hypothesis. What can you conclude about each hypothesis?

17. The basis for rejecting any null hypothesis is arbitrary. The researcher can set more or less stringent standards by deciding to raise or lower the critical *p* value. Would the use of a standard of *p* = 0.10 be more or less stringent in failing to reject the null hypothesis? Explain.

18. Consider three independently assorting gene pairs, *A/a*, *B/b*, and *C/c* where each demonstrates typical dominance (*A−*, *B−*, *C−*) and recessiveness (*aa*, *bb*, *cc*). What is the probability of obtaining an offspring that is *AABbCc* from parents that are *AaBbCC* and *AABbCc*?

19. What is the probability of obtaining a triply recessive individual from the parents shown in Problem 18?

20. Of all offspring of the parents in Problem 18, what proportion will express all three dominant traits?

21. For the following pedigree, predict the mode of inheritance and the resulting genotypes of each individual. Assume that the alleles A and a control the expression of the trait.

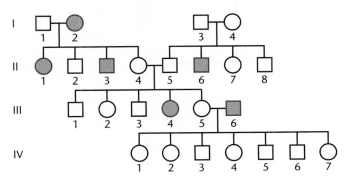

22. Which of Mendel's postulates are demonstrated by the pedigree in Problem 21? List and define these postulates.

23. The following pedigree follows the inheritance of myopia (near-sightedness) in humans. Predict whether the disorder is inherited as a dominant or a recessive trait. Based on your prediction, indicate the most probable genotype for each individual.

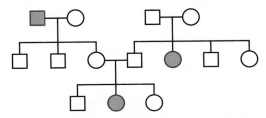

24. Draw all possible conclusions concerning the mode of inheritance of the trait expressed in each of the following limited pedigrees. (Each case is based on a different trait.)

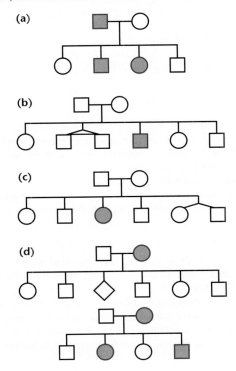

(a)

(b)

(c)

(d)

25. Two true-breeding pea plants are crossed. One parent is round, terminal, violet, constricted, while the other expresses the contrasting phenotypes of wrinkled, axial, white, full. The four pairs of contrasting traits are controlled by four genes, each located on a separate chromosome. In the F_1 generation, only round, axial, violet, and full are expressed. In the F_2 generation, all possible combinations of these traits are expressed in ratios consistent with Mendelian inheritance.

(a) What conclusion can you draw about the inheritance of these traits based on the F_1 results?
(b) Which phenotype appears most frequently in the F_2 results? Write a mathematical expression that predicts the frequency of occurrence of this phenotype.
(c) Which F_2 phenotype is expected to occur least frequently? Write a mathematical expression that predicts this frequency.
(d) How often is either P_1 phenotype likely to occur in the F_2 generation?
(e) If the F_1 plant is testcrossed, how many different phenotypes will be produced?

26. The wild-type (normal) fruit fly, *Drosophila melanogaster,* has straight wings and long bristles. Mutant strains have been isolated with either curled wings or short bristles. The genes representing these two mutant traits are located on separate chromosomes. Carefully examine the data from the five crosses below.
(a) For each mutation, determine whether it is dominant or recessive. In each case, identify which crosses support your answer; and (b) Define gene symbols, and determine the genotypes of the parents for each cross.

	Number of Progeny			
Cross	straight wings, long bristles	straight wings, short bristles	curled wings, long bristles	curled wings, short bristles
1. straight, short × straight, short	30	90	10	30
2. straight, long × straight, long	120	0	40	0
3. curled, long × straight, short	40	40	40	40
4. straight, short × straight, short	40	120	0	0
5. curled, short × straight, short	20	60	20	60

27. To assess Mendel's law of segregation using tomatoes, a true-breeding tall variety (SS) is crossed with a true-breeding short variety (ss). The heterozygous tall plants (Ss) were crossed to produce the two sets of F_2 data as follows:

Set I	Set II
30 tall	300 tall
5 short	50 short

(a) Using chi-square analysis, analyze the results for both data sets. Calculate χ^2 values, and estimate the p values in both cases.
(b) From the analysis in part (a), what can you conclude about the importance of generating large data sets in experimental settings?

SOLUTIONS TO SELECTED PROBLEMS AND DISCUSSION QUESTIONS

Answers to Now Solve This

1. P = checkered; p = plain.

Checkered is tentatively assigned the dominant function because in a casual examination of the data, especially cross (b), we see that checkered types are more likely to be produced than plain types.

P_1 Cross	Progeny	
	Checkered	Plain
(a) $PP \times PP$	PP	
(b) $PP \times pp$	Pp	
(c) $pp \times pp$		pp
(d) $PP \times pp$	Pp	
(e) $Pp \times pp$	Pp	pp
(f) $Pp \times Pp$	PP, Pp	pp
(g) $PP \times Pp$	PP, Pp	

2. Symbolism as before:

w = wrinkled seeds g = green cotyledons
W = round seeds G = yellow cotyledons

Examine each characteristic (seed shape vs. cotyledon color) separately.

(a) Notice a 3:1 ratio for seed shape; therefore, $Ww \times Ww$, and no green cotyledons; therefore, $GG \times GG$ or $GG \times Gg$. Putting the two characteristics together gives

$WwGG \times WwGG$ or $WwGG \times WwGg$.

(b) $WwGg \times wwGg$.

(c) $WwGg \times wwgg$.

3. (a)

Genotypes	Ratio	Phenotypes
AABBCC	(1/16)	
AABBCc	(1/16)	
AABbCC	(1/16)	
AABbCc	(1/16)	
AaBBCC	(2/16)	$A\_B\_C\_ = 12/16$
AaBBCc	(2/16)	
AaBbCC	(2/16)	
AaBbCc	(2/16)	
aaBBCC	(1/16)	
aaBBCc	(1/16)	
aaBbCC	(1/16)	$aaB\_C\_ = 4/16$
aaBbCc	(1/16)	

(b)

Genotypes	Ratio	Phenotypes
AaBBCC	1/8	$A\_BBC\_ = 3/8$
AaBBCc	2/8	
AaBBcc	1/8	$A\_BBcc = 1/8$
aaBBCC	1/8	$aaBBC\_ = 3/8$
aaBBCc	2/8	
aaBBcc	1/8	$aaBBcc = 1/8$

(c) There will be eight (2^n) different kinds of gametes from each of the parents and therefore a 64-box Punnett square. Doing this problem by the forked-line method helps considerably.

Simply multiply through each component to arrive at the final genotypic frequencies.

For the phenotypic frequencies, set up the problem in the following manner

4. (a) $\chi^2 = 0.47$

The χ^2 value is associated with a probability greater than 0.90 for 3 degrees of freedom (because there are now four classes in the χ^2 test). The observed and expected values do not deviate significantly.

To deal with parts (b) and (c), it is easier to see the observed values for the monohybrid ratios if the phenotypes are listed:

smooth, yellow	315	smooth, green	108
wrinkled, yellow	101	wrinkled, green	32

For the smooth: wrinkled *monohybrid component*, the smooth types total 423 (315 + 108), while the wrinkled types total 133 (101 + 32).

The χ^2 value is 0.35, and in examining the text for 1 degree of freedom, the p value is greater than 0.50 and less than 0.90. We fail to reject the null hypothesis and conclude that the observed values do not differ significantly from the expected values.

(c) For the yellow:green portion of the problem, see that there are 416 yellow plants (315 + 101) and 140 (108 + 32) green plants.

The χ^2 value is 0.01, and in examining the text for 1 degree of freedom, the p value is greater than 0.90. We fail to reject the null hypothesis and conclude that the observed values do not differ significantly from the expected values.

Solutions to Problems and Discussion Questions

2. $Ww \times Ww$ $\frac{1}{2} = Ww$ black $\frac{1}{4} = WW$ black $\frac{1}{4} = ww$ white

6. Symbolism:

w = wrinkled seeds g = green cotyledons
W = round seeds G = yellow cotyledons

P_1: $WWGG \times wwgg$

F_1: $WwGg$

$F_1 \times F_1$: $WwGg \times WwGg$

Collecting the phenotypes according to the dominance scheme presented above gives the following:

9/16 $W\_G\_$ round seeds, yellow cotyledons
3/16 W_gg round seeds, green cotyledons
3/16 $wwG\_$ wrinkled seeds, yellow cotyledons
1/16 $wwgg$ wrinkled seeds, green cotyledons

Notice that an underscore ($\_$) is used where, because of dominance, it makes no difference as to the dominant/recessive status of the allele.

8. Because independent assortment may be defined as one gene pair segregating independently of another gene pair, one would need at least two gene pairs in order to demonstrate independent assortment. The problem satisfies criteria for Mendel's postulate of independent assortment.

10. Briefly, the factors that specify chromosomal homology are the following:

Overall length
Position of the centromere
Banding patterns
Type and location of genes
Autoradiographic pattern
Function

12. First, assign meaningful gene symbols.

E = gray body color V = long wings
e = ebony body color v = vestigial wings

(a) P_1: $EEVV \times eevv$
F_1: $EeVv$ (gray, long)
F_2: This will be the result of a Punnett square with 16 boxes with the following outcome

Phenotypes	Ratio	Genotypes	Ratio
gray, long	9/16	$EEVV$	1/16
		$EEVv$	2/16
		$EeVV$	2/16
		$EeVv$	4/16
gray, vestigial	3/16	$EEvv$	1/16
		$Eevv$	2/16
ebony, long	3/16	$eeVV$	1/16
		$eeVv$	2/16
ebony, vestigial	1/16	$eevv$	1/16

(b) P_1: $EEvv \times eeVV$
F_1: The F_2 ratio will be the same as in (a) also.

(c) P_1: $EEVV \times EEvv$
F_1: $EEVv$ (gray, long)
F_2:

Phenotypes	Ratio	Genotypes	Ratio
gray, long	3/4	$EEVV$	1/4
		$EEVv$	2/4
gray, vestigial	1/4	$EEvv$	1/4

14. Symbols:

Seed shape **Seed color**
W = round G = yellow
w = wrinkled g = green

F_1: $WwGg$ cross to $wwgg$
1/4 $WwGg$ (round, yellow)
1/4 $Wwgg$ (round, green)
1/4 $wwGg$ (wrinkled, yellow)
1/4 $wwgg$ (wrinkled, green)

Seed shape	Cotyledon		Phenotypes	Color
3/4 round	3/4 yellow	→	(9/16) round	yellow
	1/4 green	→	(3/16) round	green
1/4 wrinkled	3/4 yellow	→	(3/16) wrinkled	yellow
	1/4 green	→	(1/16) wrinkled	green

16. **(a, b)** For the test of a 3:1 ratio, the χ^2 value is 33.3, with an associated p value of less than 0.01 for 1 degree of freedom. For the test of a 1:1 ratio, the χ^2 value is 25.0, again with an associated p value of less than 0.01 for 1 degree of freedom. Based on these probability values, both null hypotheses should be rejected.

18. 1/8

20. 3/4

22. *Unit factors in pairs, Dominance and recessiveness, Segregation*

24. **(a)** There are two possibilities. Either the trait is dominant, in which case I-1 is heterozygous, as are II-2 and II-3, or the trait is recessive and I-1 is homozygous and I-2 is heterozygous. Under the condition of recessiveness, both II-1 and II-4 would be heterozygous; II-2 and II-3 would be homozygous.

   (b) Recessive: Parents *Aa, Aa*

   (c) Recessive: Parents *Aa, Aa*

   (d) Recessive or dominant: if recessive, parents *AA* (probably), *aa*. Second pedigree: recessive or dominant, not sex-linked, if recessive, parents *Aa, aa*

26. **(a)** Notice in cross #1 that the ratio of straight wings to curled wings is 3:1 and the ratio of short bristles to long bristles is also 3:1. This would indicate that straight is dominant to curled and short is dominant to long. Possible symbols would be (using standard *Drosophila* symbolism):

   straight wings $= w^+$ curled wings $= w$
   short bristles $= b^+$ long bristles $= b$

   (b) Cross #1: $w^+/w \; ; b^+/b \;\; \times \; w^+/w \; ; b^+/b$
   Cross #2: $w^+/w \; ; b\, /b \;\; \times \; w^+/w \; ; b/b$
   Cross #3: $w\, /w \; ; b\, /b \;\; \times \; w^+/w \; ; b^+/b$
   Cross #4: $w^+/w^+ \; ; b^+/b \; \times \; w^+/w^+ \; ; b^+/b$
   (one parent could be w^+/w)
   Cross #5: $w\, /w \; ; b^+/b \times w^+/w \; ; b^+/b$

CREDITS

Credits are listed in order of appearance.

Photo

CO, Archiv/Photo Researchers, Inc.; UNF4-1, Martin Shields/Alamy

Modification of Mendelian Ratios

From Chapter 4 of *Essentials of genetics, Eighth edition,* William S. Klug, Michael R. Cummings, Charlotte A. Spencer, Michael A. Palladino. ©2013 by Pearson Eduction, Inc. All rights reserved.

Labrador retriever puppies, which may display brown (also called chocolate), golden (also called yellow), or black coats.

Modification of Mendelian Ratios

CHAPTER CONCEPTS

- While alleles are transmitted from parent to offspring according to Mendelian principles, they sometimes fail to display the clear-cut dominant/recessive relationship observed by Mendel.

- In many cases, in contrast to Mendelian genetics, two or more genes are known to influence the phenotype of a single characteristic.

- Still another exception to Mendelian inheritance is the presence of genes on sex chromosomes, whereby one of the sexes contains only a single member of that chromosome.

- Phenotypes are often the combined result of both genetics and the environment within which genes are expressed.

- The result of the various exceptions to Mendelian principles is the occurrence of phenotypic ratios that differ from those resulting from standard monohybrid, dihybrid, and trihybrid crosses.

- Extranuclear inheritance, resulting from the expression of genes present in the DNA found in mitochondria and chloroplasts, modifies Mendelian inheritance patterns. Such genes are most often transmitted through the female gamete.

Genes are present on homologous chromosomes and that these chromosomes segregate from each other and assort independently with other segregating chromosomes during gamete formation. These two postulates are the basic principles of gene transmission from parent to offspring. However, when gene expression does not adhere to a simple dominant/recessive mode or when more than one pair of genes influences the expression of a single character, the classic 3:1 and 9:3:3:1 ratios are usually modified. In this chapters, we consider more complex modes of inheritance. In spite of the greater complexity of these situations, the fundamental principles set down by Mendel still hold.

In this chapter, we restrict our initial discussion to the inheritance of traits controlled by only one set of genes. In diploid organisms, which have homologous pairs of chromosomes, two copies of each gene influence such traits. The copies need not be identical because alternative forms of genes (alleles) occur within populations. How alleles influence phenotypes is our primary focus. We will then consider gene interaction, a situation in which a single phenotype is affected by more than one set of genes. Numerous examples will be presented to illustrate a variety of heritable patterns observed in such situations.

Thus far, we have restricted our discussion to chromosomes other than the X and Y pair. By examining cases where genes are present on the X chromosome, illustrating X-linkage, we will see yet another modification of Mendelian ratios. Our discussion of modified ratios also includes the consideration of sex-limited and sex-influenced inheritance, cases where the sex of the individual, but not necessarily genes on the X chromosome, influences the phenotype. We will also consider how a given phenotype often varies depending on the overall environment in which a gene, a cell, or an organism finds itself. This discussion points out that phenotypic expression depends on more than just the genotype of an organism. Finally, we conclude with a discussion of extranuclear inheritance, cases where DNA within organelles influences an organism's phenotype.

1 Alleles Alter Phenotypes in Different Ways

After Mendel's work was rediscovered in the early 1900s, researchers focused on the many ways in which genes influence an individual's phenotype. Each type of inheritance was more thoroughly investigated when observations of genetic data did not conform precisely to the expected Mendelian ratios, and hypotheses that modified and extended the Mendelian principles were proposed and tested with specifically designed crosses. The explanations were in accord with the principle that a phenotype is under the control of one or more genes located at specific loci on one or more pairs of homologous chromosomes.

To understand the various modes of inheritance, we must first examine the potential function of alleles. Alleles are alternative forms of the same gene. The allele that occurs most frequently in a population, the one that we arbitrarily designate as normal, is called the **wild-type allele.** This is often, but not always, dominant. Wild-type alleles are responsible for the corresponding wild-type phenotype and are the standards against which all other mutations occurring at a particular locus are compared.

A mutant allele contains modified genetic information and often specifies an altered gene product. For example, in human populations, there are many known alleles of the gene that encodes the β chain of human hemoglobin. All such alleles store information necessary for the synthesis of the β-chain polypeptide, but each allele specifies a slightly different form of the same molecule. Once the allele's product has been manufactured, the function of the product may or may not be altered.

The process of mutation is the source of alleles. For a new allele to be recognized when observing an organism, it must cause a change in the phenotype. A new phenotype results from a change in functional activity of the cellular product specified by that gene. Often, the mutation causes the diminution or the loss of the specific wild-type function. For example, if a gene is responsible for the synthesis of a specific enzyme, a mutation in that gene may ultimately change the conformation of this enzyme and reduce or eliminate its affinity for the substrate. Such a case is designated as a **loss-of-function mutation.** If the loss is complete, the mutation has resulted in what is called a **null allele.**

Conversely, other mutations may enhance the function of the wild-type product. Most often when this occurs, it is the result of increasing the quantity of the gene product. In such cases, the mutation may be affecting the regulation of transcription of the gene under consideration. Such cases are designated **gain-of-function mutations,** which generally result in dominant alleles since one copy in a diploid organism is sufficient to alter the normal phenotype. Examples of gain-of-function mutations include the genetic conversion of proto-oncogenes, which regulate the cell cycle, to oncogenes, where regulation is overridden by excess gene product. The result is the creation of a cancerous cell.

Having introduced the concept of gain- or loss-of-function mutations, it is important to note the possibility that a mutation will create an allele where no change in function can be detected. In this case, the mutation would not be immediately apparent since no phenotypic variation would be evident. However, such a mutation could be detected if the DNA sequence of the gene was examined directly. These are sometimes referred to as **neutral mutations** because the gene product presents no change to either the phenotype or the evolutionary fitness of the organism.

Finally, we note here that while a phenotypic trait may be affected by a single mutation in one gene, traits are often influenced by more than one gene. For example, enzymatic reactions are most often part of complex metabolic pathways leading to the synthesis of an end product, such as an amino acid. Mutations in any of the various reactions have a common effect—the failure to synthesize the end product. Therefore, phenotypic traits related to the end product are often influenced by more than one gene.

In many crosses, only one or a few gene pairs are involved. Keep in mind that in each cross discussed, all genes that are not under consideration are assumed to have no effect on the inheritance patterns described.

2 Geneticists Use a Variety of Symbols for Alleles

There is a standard convention that is used to symbolize alleles for very simple Mendelian traits. The initial letter of the name of a recessive trait, lowercased and italicized, denotes the recessive allele, and the same letter in uppercase refers to the dominant allele. Thus, in the case of *tall* and *dwarf,* where *dwarf* is recessive, *D* and *d* represent the alleles responsible for these respective traits. Mendel used upper- and lowercase letters such as these to symbolize his unit factors.

Another useful system was developed in genetic studies of the fruit fly *Drosophila melanogaster* to discriminate between wild-type and mutant traits. This system uses the initial letter, or a combination of two or three letters, of the name of the mutant trait. If the trait is recessive, lowercase is used; if it is dominant, uppercase is used. The contrasting wild-type trait is denoted by the same letter, but with a superscript +. For example, *ebony* is a recessive body color mutation in *Drosophila.* The normal wild-type body color is gray. Using this system, we denote *ebony* by the symbol *e,* and we denote gray by e^+. The responsible locus may be occupied by either the wild-type allele (e^+) or the mutant allele (e). A diploid fly may thus exhibit one of three possible genotypes:

e^+/e^+ gray homozygote (wild type)

e^+/e gray heterozygote (wild type)

$e\ /e$ ebony homozygote (mutant)

The slash between the letters indicates that the two allele designations represent the same locus on two homologous chromosomes. If we instead consider a dominant wing mutation such as *Wrinkled* (*Wr*) wing in *Drosophila,* the three possible designations are Wr^+/Wr^+, Wr^+/Wr, and Wr/Wr. The latter two genotypes express the wrinkled-wing phenotype.

One advantage of this system is that further abbreviation can be used when convenient: the wild-type allele may simply be denoted by the + symbol. With *ebony* as an example, the designations of the three possible genotypes become

+/+   gray homozygote (wild type)

+/e   gray heterozygote (wild type)

e/e   ebony homozygote (mutant)

Another variation is utilized when no dominance exists between alleles. We simply use uppercase italic letters and superscripts to denote alternative alleles (e.g., R^1 and R^2, L^M and L^N, I^A and I^B). Their use will become apparent later in this chapter.

Many diverse systems of genetic nomenclature are used to identify genes in various organisms. Usually, the symbol selected reflects the function of the gene or even a disorder caused by a mutant gene. For example, the yeast *cdk* is the abbreviation for the *c*yclin *d*ependent *k*inase gene, whose product is involved in cell-cycle regulation. In bacteria, leu^- refers to a mutation that interrupts the biosynthesis of the amino acid leucine, where the wild-type gene is designated leu^+. The symbol *dnaA* represents a bacterial gene involved in DNA replication (and DnaA is the protein made by that gene). In humans, capital letters are used to name genes: *BRCA*1 represents the first gene associated with susceptibility to *br*east *ca*ncer. Although these different systems may seem complex, they are useful ways to symbolize genes.

3   Neither Allele Is Dominant in Incomplete, or Partial, Dominance

A cross between parents with contrasting traits may generate offspring with an intermediate phenotype. For example, if plants such as four-o'clocks or snapdragons with red flowers are crossed with white-flowered plants, the offspring have pink flowers. Some red pigment is produced in the F_1 intermediate pink-colored flowers. Therefore, neither red nor white flower color is dominant. This situation is known as **incomplete,** or **partial, dominance.**

If this phenotype is under the control of a single gene and two alleles where neither is dominant, the results of the F_1 (pink) $\times$ F_1 (pink) cross can be predicted. The resulting F_2 generation shown in **Figure 1** confirms the hypothesis that only one pair of alleles determines these phenotypes. The *genotypic ratio* (1:2:1) of the F_2 generation is identical to that of Mendel's monohybrid cross. However, because neither allele is dominant, the *phenotypic ratio* is identical to the *genotypic ratio.* Note that because neither allele is recessive, we have chosen not to use upper- and lowercase letters as symbols. Instead, we denoted the red and white alleles as R^1 and R^2, respectively. We could have used W^1 and W^2 or still other designations such as C^w and C^R, where *C* indicates "color" and the *W* and *R* superscripts indicate white and red.

Clear-cut cases of incomplete dominance, which result in intermediate expression of the overt phenotype, are relatively rare. However, even when complete dominance seems apparent, careful examination of the gene product, rather than the phenotype, often reveals an intermediate level of gene expression. An example is the human biochemical disorder **Tay–Sachs disease,** in which homozygous recessive individuals are severely affected with a fatal lipid storage disorder. There is almost no activity of the enzyme **hexosaminidase** in afflicted individuals. Heterozygotes, with only a single copy of the mutant gene, are phenotypically normal but express only about 50 percent of the enzyme activity found in homozygous normal individuals. Fortunately, this level of enzyme activity is adequate to achieve normal biochemical function—a situation not uncommon in enzyme disorders.

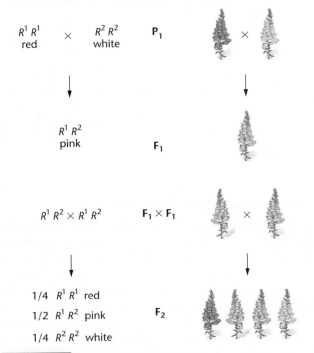

$R^1 R^1$ $\times$ $R^2 R^2$ **P₁**
red   white

$R^1 R^2$
pink   **F₁**

$R^1 R^2 \times R^1 R^2$ **F₁ × F₁**

1/4 $R^1 R^1$ red
1/2 $R^1 R^2$ pink   **F₂**
1/4 $R^2 R^2$ white

FIGURE 1   Incomplete dominance shown in the flower color of snapdragons.

4   In Codominance, the Influence of Both Alleles in a Heterozygote Is Clearly Evident

If two alleles of a single gene are responsible for producing two distinct, detectable gene products, a situation different from incomplete dominance or dominance/recessiveness arises. In this case, *the joint expression of both alleles in a heterozygote* is called **codominance**. The **MN blood group** in humans illustrates this phenomenon and is characterized by an antigen called a glycoprotein, found on the surface of red blood cells. In the human population, two forms of this glycoprotein exist, designated M and N; an individual may exhibit either one or both of them.

The MN system is under the control of an autosomal locus found on chromosome 4 and two alleles designated

L^M and L^N. Humans are diploid, so three combinations are possible, each resulting in a distinct blood type:

Genotype	Phenotype
$L^M L^M$	M
$L^M L^N$	MN
$L^N L^N$	N

As predicted, a mating between two heterozygous MN parents may produce children of all three blood types, as follows:

$$L^M L^N \times L^M L^N$$
$$\downarrow$$
$$1/4 \; L^M L^M$$
$$1/2 \; L^M L^N$$
$$1/4 \; L^N L^N$$

Once again the genotypic ratio, 1:2:1, is upheld.

Codominant inheritance is characterized by *distinct expression of the gene products of both alleles*. This characteristic distinguishes it from incomplete dominance, where heterozygotes express an intermediate, blended phenotype. We shall see another example of codominance when we examine the ABO blood-type system in the following section.

5   Multiple Alleles of a Gene May Exist in a Population

The information stored in any gene is extensive, and mutations can modify this information in many ways. Each change produces a different allele. Therefore, for any specific gene, the number of alleles within members of a population need not be restricted to two. When three or more alleles of the same gene are found, **multiple alleles** are present that create a unique mode of inheritance. It is important to realize that *multiple alleles can be studied only in populations*. An individual diploid organism has, at most, two homologous gene loci that may be occupied by different alleles of the same gene. However, among many members of a species, numerous alternative forms of the same gene can exist.

The ABO Blood Group

The simplest case of multiple alleles is that in which three alternative alleles of one gene exist. This situation is illustrated by the **ABO blood group** in humans, discovered by Karl Landsteiner in the early 1900s. The ABO system, like the MN blood group, is characterized by the presence of antigens on the surface of red blood cells. The A and B antigens are distinct from MN antigens and are under the control of a different gene, located on chromosome 9. As in the MN system, one combination of alleles in the ABO system exhibits a codominant mode of inheritance.

When individuals are tested using antisera that contain antibodies against the A or B antigen, four phenotypes are revealed. Each individual has either the A antigen (A phenotype), the B antigen (B phenotype), the A and B antigens (AB phenotype), or neither antigen (O phenotype). In 1924, it was hypothesized that these phenotypes were inherited as the result of three alleles of a single gene. This hypothesis was based on studies of the blood types of many different families.

Although different designations can be used, we use the symbols I^A, I^B, and I to distinguish these three alleles; the I designation stands for *isoagglutinogen*, another term for antigen. If we assume that the I^A and I^B alleles are responsible for the production of their respective A and B antigens and that I is an allele that does not produce any detectable A or B antigens, we can list the various genotypic possibilities and assign the appropriate phenotype to each:

Genotype	Antigen	Phenotype
$I^A I^A$	A	
$I^A I$	A	A
$I^B I^B$	B	
$I^B I$	B	B
$I^A I^B$	A, B	AB
$I I$	Neither	O

In these assignments the I^A and I^B alleles are dominant to the I allele but are codominant to each other. Our knowledge of human blood types has several practical applications, the most important of which are compatible blood transfusions and organ transplantations.

The Bombay Phenotype

The biochemical basis of the ABO blood-type system has been carefully worked out. The A and B antigens are actually carbohydrate groups (sugars) that are bound to lipid molecules (fatty acids) protruding from the membrane of the red blood cell. The specificity of the A and B antigens is based on the terminal sugar of the carbohydrate group. Both the A and B antigens are derived from a precursor molecule called the **H substance,** to which one or two terminal sugars are added.

In extremely rare instances, first recognized in a woman in Bombay in 1952, the H substance is incompletely formed. As a result, it is an inadequate substrate for the enzyme that normally adds the terminal sugar. This condition results in the expression of blood type O and is called the **Bombay phenotype.** Research has revealed that this condition is due to a rare recessive mutation at a locus separate from that controlling the A and B antigens. The gene is now designated *FUT1* (encoding an enzyme, fucosyl transferase), and individuals that are homozygous for the mutation cannot synthesize the complete H substance. Thus, even though they may

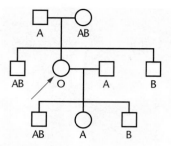

FIGURE 2   A partial pedigree of a woman with the Bombay phenotype. Functionally, her ABO blood group behaves as type O. Genetically, she is type B.

have the I^A and/or I^B alleles, neither the A nor B antigen can be added to the cell surface. This information explains why the woman in Bombay expressed blood type O, even though one of her parents was type AB (thus she should not have been type O), and why she was able to pass the I^B allele to her children (**Figure 2**).

The *white* Locus in *Drosophila*

Many other phenotypes in plants and animals are known to be controlled by multiple allelic inheritance. In *Drosophila*, many alleles are known at practically every locus. The recessive mutation that causes white eyes, discovered by Thomas H. Morgan and Calvin Bridges in 1912, is one of over 100 alleles that can occupy this locus. In this allelic series, eye colors range from complete absence of pigment in the *white* allele to deep ruby in the *white-satsuma* allele, to orange in the *white-apricot* allele, to a buff color in the *white-buff* allele. These alleles are designated w, w^{sat}, w^a, and w^{bf}, respectively. In each case, the total amount of pigment in these mutant eyes is reduced to less than 20 percent of that found in the brick-red, wild-type eye. **Table 1** lists these and other *white* alleles and their color phenotypes.

TABLE 1   Alleles of the *white* Locus of *Drosophila melanogaster* and Their Phenotype

Allele	Name	Eye Color
w	white	pure white
w^a	white-apricot	yellowish orange
w^{bf}	white-buff	light buff
w^{bl}	white-blood	yellowish ruby
w^{cf}	white-coffee	deep ruby
w^e	white-eosin	yellowish pink
w^{mo}	white-mottled orange	light mottled orange
w^{sat}	white-satsuma	deep ruby
w^{sp}	white-spotted	fine grain, yellow mottling
w^t	white-tinged	light pink

1 In the guinea pig, one locus involved in the control of coat color may be occupied by any of four alleles: C (full color), c^k (sepia), c^d (cream), or c^a (albino), with an order of dominance of: $C > c^k > c^d > c^a$. (C is dominant to all others, c^k is dominant to c^d and c^a, but not C, etc.) In the following crosses, determine the parental genotypes and predict the phenotypic ratios that would result:

(a) sepia × cream, where both guinea pigs had an albino parent

(b) sepia × cream, where the sepia guinea pig had an albino parent and the cream guinea pig had two sepia parents

(c) sepia × cream, where the sepia guinea pig had two full-color parents and the cream guinea pig had two sepia parents

(d) sepia × cream, where the sepia guinea pig had a full-color parent and an albino parent and the cream guinea pig had two full-color parents

■ **Hint** *This problem involves an understanding of multiple alleles. The key to its solution is to note particularly the hierarchy of dominance of the various alleles. Remember also that even though there can be more than two alleles in a population, an individual can have at most two of these. Thus, the allelic distribution into gametes adheres to the principle of segregation.*

6 Lethal Alleles Represent Essential Genes

Many gene products are essential to an organism's survival. Mutations resulting in the synthesis of a gene product that is nonfunctional can often be tolerated in the heterozygous state; that is, one wild-type allele may be sufficient to produce enough of the essential product to allow survival. However, such a mutation behaves as a *recessive lethal allele,* and homozygous recessive individuals will not survive. The time of death will depend on when the product is essential. In mammals, for example, this might occur during development, early childhood, or even adulthood.

In some cases, the allele responsible for a lethal effect when homozygous may also result in a distinctive mutant phenotype when present heterozygously. It is behaving as a recessive lethal allele but is dominant with respect to the phenotype. For example, a mutation that causes a yellow coat in mice was discovered in the early part of this century. The yellow coat varies from the normal agouti (wild-type) coat phenotype, as shown in **Figure 3**. Crosses between the

FIGURE 3 Inheritance patterns in three crosses involving the normal wild-type *agouti* allele (A) and the mutant *yellow* allele (A^Y) in the mouse. Note that the mutant allele behaves dominantly to the normal allele in controlling coat color, but it also behaves as a homozygous recessive lethal allele. The genotype $A^Y A^Y$ mice do not survive.

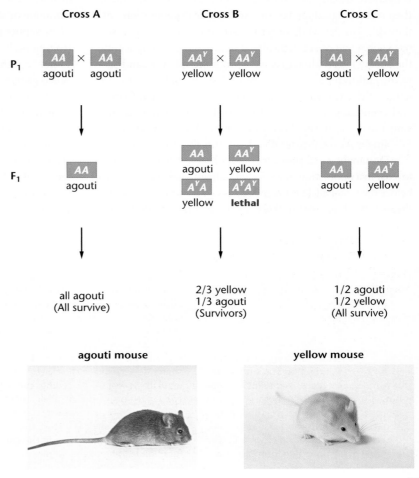

various combinations of the two strains yield unusual results:

Crosses			
(A) agouti	×	agouti	⟶ all agouti
(B) yellow	×	yellow	⟶ 2/3 yellow:
			1/3 agouti
(C) agouti	×	yellow	⟶ 1/2 yellow:
			1/2 agouti

These results are explained on the basis of a single pair of alleles. With regard to coat color, the mutant *yellow* allele (A^Y) is dominant to the wild-type *agouti* allele (*A*), so heterozygous mice will have yellow coats. However, the *yellow* allele is also a homozygous recessive lethal. When present in two copies, the mice die before birth. Thus, there are no homozygous yellow mice. The genetic basis for these three crosses is shown in Figure 3.

In other cases, a mutation may behave as a *dominant lethal allele*. In such cases, the presence of just one copy of the allele results in the death of the individual. In humans, a disorder called **Huntington disease** (previously referred to as Huntington's chorea) is due to a dominant autosomal allele *H*, where the onset of the disease in heterozygotes (*Hh*) is delayed, usually well into adulthood. Affected individuals then undergo gradual nervous and motor degeneration until they die. This lethal disorder is particularly tragic because it has such a late onset, typically at about age 40. By that time, the affected individual may have produced a family, and each of the children has a 50 percent probability of inheriting the lethal allele, transmitting the allele to his or her offspring, and eventually developing the disorder. The American folk singer and composer Woody Guthrie (father of modern-day folk singer Arlo Guthrie) died from this disease at age 39.

Dominant lethal alleles are rarely observed. For these alleles to exist in a population, the affected individuals must reproduce before the lethal allele is expressed, as can occur in Huntington disease. If all affected individuals die before reaching reproductive age, the mutant gene will not be passed to future generations, and the mutation will disappear from the population unless it arises again as a result of a new mutation.

ESSENTIAL POINT ■ ■ ■

Since Mendel's work was rediscovered, transmission genetics has been expanded to include many alternative modes of inheritance, including the study of incomplete dominance, codominance, multiple alleles, and lethal alleles.

---

7 Combinations of Two Gene Pairs with Two Modes of Inheritance Modify the 9:3:3:1 Ratio

Each example discussed so far modifies Mendel's 3:1 F_2 monohybrid ratio. Therefore, combining any two of these modes of inheritance in a dihybrid cross will likewise modify the classical 9:3:3:1 ratio. Having established the foundation for the modes of inheritance of incomplete dominance, codominance, multiple alleles, and lethal alleles, we can now deal with the situation of two modes of inheritance occurring simultaneously. Mendel's principle of independent assortment applies to these situations, provided that the genes controlling each character are not linked on the same chromosome—in other words, that they do not demonstrate what is called *genetic linkage*.

Consider, for example, a mating that occurs between two humans who are both heterozygous for the autosomal recessive gene that causes albinism and who are both of blood type AB. What is the probability of a particular phenotypic combination occurring in each of their children? Albinism is inherited in the simple Mendelian fashion, and the blood types are determined by the series of three multiple alleles, I^A, I^B, and I. The solution to this problem is diagrammed in **Figure 4**, using the

---

The Molecular Basis of Dominance and Recessiveness: The Agouti Gene

Molecular analysis of the gene resulting in agouti and yellow mice has provided insight into how a mutation can be both dominant for one phenotypic effect (hair color) and recessive for another (embryonic development). The A^Y allele is a classic example of a gain-of-function mutation.

Animals homozygous for the wild-type *A* allele have yellow pigment deposited as a band on the otherwise black hair shaft, resulting in the agouti phenotype (see Figure 3). Heterozygotes deposit yellow pigment along the entire length of hair shafts as a result of the deletion of the regulatory region preceding the DNA coding region of the A^Y allele. Without any means to regulate its expression, one copy of the A^Y allele is always turned on in heterozygotes, resulting in the gain of function leading to the dominant effect.

The homozygous lethal effect has also been explained by molecular analysis of the mutant gene. The extensive deletion of genetic material that produced the A^Y allele actually extends into the coding region of an adjacent gene (*Merc*), rendering it nonfunctional. It is this gene that is critical to embryonic development, and the loss of its function in A^Y/A^Y homozygotes is what causes lethality. Heterozygotes exceed the threshold level of the wild-type *Merc* gene product and thus survive.

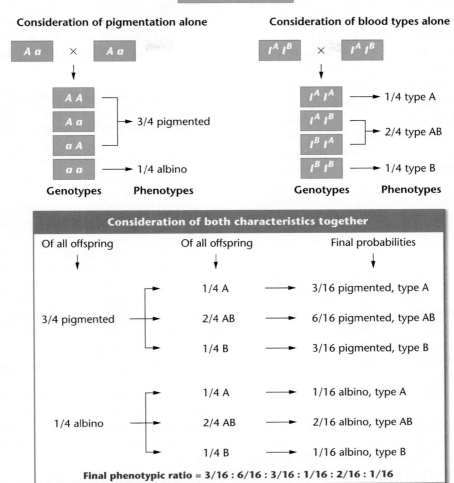

FIGURE 4 Calculation of the mating probabilities involving the ABO blood type and albinism in humans, using the forked-line method.

forked-line method. This dihybrid cross does not yield the classical four phenotypes in a 9:3:3:1 ratio. Instead, six phenotypes occur in a 3:6:3:1:2:1 ratio, establishing the expected probability for each phenotype. This is just one of the many variants of modified ratios that are possible when different modes of inheritance are combined.

8 Phenotypes Are Often Affected by More Than One Gene

Soon after Mendel's work was rediscovered, experimentation revealed that individual characteristics displaying discrete phenotypes are often under the control of more than one gene. This was a significant discovery because it revealed that genetic influence on the phenotype is often much more complex than Mendel had envisioned. Instead of single genes controlling the development of individual parts of the plant or animal body, it soon became clear that phenotypic characters can be influenced by the interactions of many different genes and their products.

The term **gene interaction** is often used to describe the idea that several genes influence a particular characteristic. This does not mean, however, that two or more genes, or their products, necessarily interact directly with one another to influence a particular phenotype. Rather, the cellular function of numerous gene products contributes to the development of a common phenotype. For example, the development of an organ such as the compound eye of an insect is exceedingly complex and leads to a structure with multiple phenotypic manifestations—such as specific size, shape, texture, and color. The formation of the eye results from a complex cascade of events during its development. This process exemplifies the developmental concept of **epigenesis**, whereby each step of development increases the complexity of this sensory organ and is under the control and influence of one or more genes.

An enlightening example of *epigenesis* and *multiple gene interaction* involves the formation of the inner ear in mammals. The inner ear consists of distinctive anatomical features to capture, funnel, and transmit external sound waves and to convert them into nerve impulses. During the formation of the ear, a cascade of intricate developmental events occur, influenced by many genes. Mutations that interrupt many of the steps of ear development lead to a common phenotype: **hereditary deafness.** In a sense, these many genes "interact" to produce a common phenotype. In such situations, the mutant phenotype is described as a **heterogeneous trait,** reflecting the many genes involved. In humans, while a few common alleles are responsible for the vast majority of cases of hereditary deafness, over 50 genes are involved in development of the ability to discern sound.

Epistasis

Some of the best examples of gene interaction are those that reveal the phenomenon of **epistasis** where the expression of one gene or gene pair masks or modifies the expression of another gene or gene pair. Sometimes the genes involved control the expression of the same general phenotypic characteristic in an antagonistic manner, as when masking occurs. In other cases, however, the genes involved exert their influence on one another in a complementary, or cooperative, fashion.

For example, the homozygous presence of a recessive allele prevents or overrides the expression of other alleles at a second locus (or several other loci). In this case, the alleles at the first locus are said to be **epistatic** to those at the second locus, and the alleles at the second locus are **hypostatic** to those at the first locus. In another example, a single dominant allele at the first locus influences the expression of the alleles at a second gene locus. In a third example, two gene pairs are said *to complement one another* such that at least one dominant allele at each locus is required to express a particular phenotype.

The Bombay phenotype discussed earlier is an example of the homozygous recessive condition at one locus masking the expression of a second locus. There, we established that the homozygous presence of the mutant form of the *FUT1* gene masks the expression of the I^A and I^B alleles. Only individuals containing at least one wild-type *FUT1* allele can form the A or B antigen. As a result, individuals whose genotypes include the I^A or I^B allele and who lack a wild-type allele are of the type O phenotype, regardless of their potential to make either antigen. An example of the outcome of matings between individuals heterozygous at both loci is illustrated in **Figure 5.** If many such individuals have children, the phenotypic ratio of 3 A: 6 AB: 3 B: 4 O is expected in their offspring.

It is important to note the following points when examining this cross and the predicted phenotypic ratio:

1. A key distinction exists in this cross compared to the modified dihybrid cross shown in Figure 4: *only one characteristic—blood type—is being followed.* In the modified dihybrid cross of Figure 4, blood type *and* skin pigmentation are followed as separate phenotypic characteristics.

2. Even though only a single character was followed, the phenotypic ratio is expressed in sixteenths. If we knew nothing about the H substance and the genes controlling it, we could still be confident that a second gene pair, other than that controlling the A and B antigens, is involved in the phenotypic expression. *When studying a single character, a ratio that is expressed in 16 parts (e.g., 3:6:3:4) suggests that two gene pairs are "interacting" during the expression of the phenotype under consideration.*

The study of gene interaction reveals inheritance patterns that modify the classical Mendelian dihybrid F_2 ratio (9:3:3:1) in other ways as well. In these examples, epistasis combines one or more of the four phenotypic categories in various ways. The generation of these four groups is reviewed in **Figure 6**, along with several modified ratios.

As we discuss these and other examples, we will make several assumptions and adopt certain conventions:

1. In each case, distinct phenotypic classes are produced, each clearly discernible from all others. Such traits illustrate *discontinuous variation,* where phenotypic categories are discrete and qualitatively different from one another.

2. The genes considered in each cross are not linked and therefore assort independently of one another during gamete formation. To allow you to easily compare the results of different crosses, we designated alleles as *A, a* and *B, b* in each case.

3. When we assume that complete dominance exists between the alleles of any gene pair, such that *AA* and *Aa* or *BB* and *Bb* are equivalent in their genetic effects, we use the designations *A*– or *B*– for both combinations, where the dash (–) indicates that either allele may be present, without consequence to the phenotype.

4. All P_1 crosses involve homozygous individuals (e.g., *AABB* × *aabb, AAbb* × *aaBB,* or *aaBB* × *AAbb*). Therefore, each F_1 generation consists of only heterozygotes of genotype *AaBb.*

5. In each example, the F_2 generation produced from these heterozygous parents is our main focus of analysis. When two genes are involved (as in Figure 6), the F_2 genotypes fall into four categories: 9/16 *A–B–*, 3/16 *A–bb*, 3/16 *aaB–*, and 1/6 *aabb*. Because of dominance, all genotypes in each category have an equivalent effect on the phenotype.

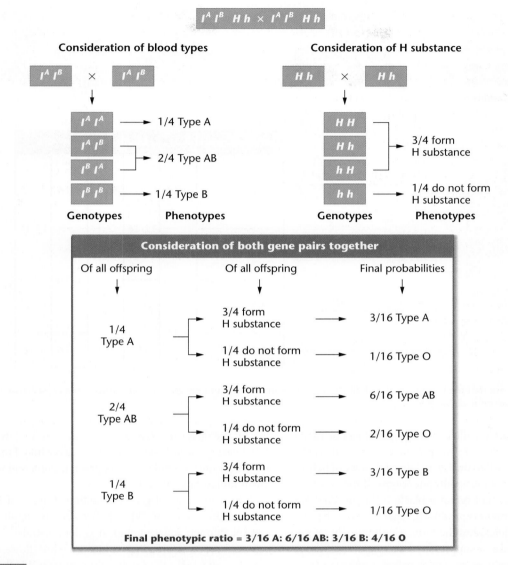

FIGURE 5 The outcome of a mating between individuals who are heterozygous at two genes determining their ABO blood type. Final phenotypes are calculated by considering both genes separately and then combining the results using the forked-line method.

Case 1 is the inheritance of coat color in mice (**Figure 7**). Normal wild-type coat color is agouti, a grayish pattern formed by alternating bands of pigment on each hair. Agouti is dominant to black (non-agouti) hair, which is caused by a recessive mutation, *a*. Thus, *A*– results in agouti, while *aa* yields black coat color. When it is homozygous, a recessive mutation, *b*, at a separate locus, eliminates pigmentation altogether, yielding albino mice (*bb*), regardless of the genotype at the other locus. The presence of at least one *B* allele allows pigmentation to occur in much the same way that the *H* allele in humans allows the expression of the ABO blood types. In a cross between agouti (*AABB*) and albino (*aabb*), members of the F_1 are all *AaBb* and have agouti coat color. In the F_2 progeny of a cross between two F_1 heterozygotes, the following genotypes and phenotypes are observed:

F_1: *AaBb* × *AaBb*

↓

F_2 Ratio	Genotype	Phenotype	Final Phenotypic Ratio
9/16	*A–B–*	agouti	9/16 agouti
3/16	*A–bb*	albino	3/16 black
3/16	*aa B–*	black	4/16 albino
1/16	*aa bb*	albino	

We can envision gene interaction yielding the observed 9:3:4 F_2 ratio as a two-step process:

	Gene B		Gene A	
Precursor	↓	Black	↓	Agouti
Molecule	⟶	Pigment	⟶	Pattern
(colorless)	*B–*		*A–*	

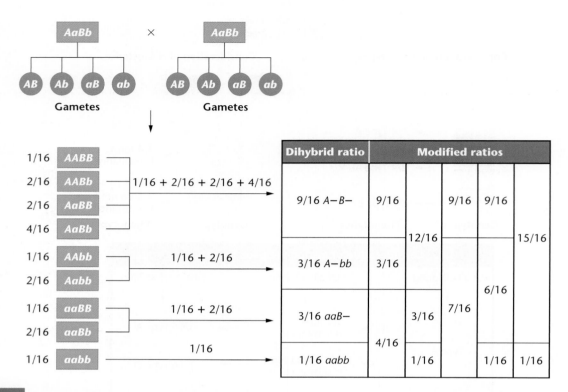

FIGURE 6 Generation of the various modified dihybrid ratios from the nine unique genotypes produced in a cross between individuals who are heterozygous at two genes.

In the presence of a *B* allele, black pigment can be made from a colorless substance. In the presence of an *A* allele, the black pigment is deposited during the development of hair in a pattern that produces the agouti phenotype. If the *aa* genotype occurs, all of the hair remains black. If the *bb* genotype occurs, no black pigment is produced, regardless of the presence of the *A* or *a* alleles, and the mouse is albino. Therefore, the *bb* genotype masks or suppresses the expression of the *A* gene. As a result, this is referred to as *recessive epistasis*.

A second type of epistasis, called *dominant epistasis,* occurs when a dominant allele at one genetic locus masks the expression of the alleles at a second locus. For instance, Case 2 of Figure 7 deals with the inheritance of fruit color in summer squash. Here, the dominant allele *A* results in white fruit color regardless of the genotype at a second locus, *B*. In the absence of the dominant *A* allele (the *aa* genotype), *BB* or *Bb* results in yellow color, while *bb* results in green color. Therefore, if two white-colored double heterozygotes (*AaBb*) are crossed, this type of epistasis generates an interesting phenotypic ratio:

$$F_1: AaBb \times AaBb$$
$$\downarrow$$

F$_2$ Ratio	Genotype	Phenotype	Final Phenotypic Ratio
9/16	*A–B–*	white	12/16 white
3/16	*A–bb*	white	3/16 yellow
3/16	*aaB–*	yellow	1/16 green
1/16	*aabb*	green	

Of the offspring, 9/16 are *A–B–* and are thus white. The 3/16 bearing the genotypes *A–bb* are also white. Finally, 3/16 are yellow (*aaB–*) while 1/16 are green (*aabb*); and we obtain the modified ratio of 12:3:1.

Our third type of gene interaction (Case 3 of Figure 7) was first discovered by William Bateson and Reginald Punnett (of Punnett square fame). It is demonstrated in a cross between two true-breeding strains of white-flowered sweet peas. Unexpectedly, the results of this cross yield all purple F$_1$ plants, and the F$_2$ plants occur in a ratio of 9/16 purple to 7/16 white. The proposed explanation suggests that the presence of at least one dominant allele of each of two gene pairs is essential for flowers to be purple. Thus, this cross represents a case of *complementary gene interaction*. All other genotype combinations yield white flowers because the homozygous condition of *either* recessive allele masks the expression of the dominant allele at the other locus. The cross is shown as follows:

$$P_1: AAbb \times aaBB$$
white   white
$$\downarrow$$
$$F_1: \text{All } AaBb \text{ purple}$$
$$\downarrow$$

F$_2$ Ratio	Genotype	Phenotype	Final Phenotypic Ratio
9/16	*A–B–*	purple	9/16 purple
3/16	*A–bb*	white	7/16 white
3/16	*aa B–*	white	
1/16	*aabb*	white	

Case	Organism	Character	F₂ Phenotypes				Modified ratio
			9/16	3/16	3/16	1/16	
1	Mouse	Coat color	agouti	albino	black	albino	9:3:4
2	Squash	Color	white		yellow	green	12:3:1
3	Pea	Flower color	purple	white			9:7
4	Squash	Fruit shape	disc	sphere		long	9:6:1
5	Chicken	Color	white		colored	white	13:3
6	Mouse	Color	white-spotted	white	colored	white-spotted	10:3:3
7	Shepherd's purse	Seed capsule	triangular			ovoid	15:1
8	Flour beetle	Color	6/16 sooty	3/16 red	black	jet	black	6:3:3:4

FIGURE 7 The basis of modified dihybrid F₂ phenotypic ratios, resulting from crosses between doubly heterozygous F₁ individuals. The four groupings of the F₂ genotypes shown in Figure 6 and across the top of this figure are combined in various ways to produce these ratios.

We can now see how two gene pairs might yield such results:

Gene A

Gene B

| Precursor Substance (colorless) | → A— | Intermediate Product (colorless) | → B— | Final Product (purple) |

At least one dominant allele from each pair of genes is necessary to ensure both biochemical conversions to the final product, yielding purple flowers. In our cross, this will occur in 9/16 of the F₂ offspring. All other plants (7/16) have flowers that remain white.

The preceding examples illustrate how the products of two genes "interact" to influence the development of a common phenotype. In other instances, more than two genes and their products are involved in controlling phenotypic expression.

Novel Phenotypes

Other cases of gene interaction yield novel, or new, phenotypes in the F₂ generation, in addition to producing modified dihybrid ratios. Case 4 in Figure 7 depicts the inheritance of fruit shape in the summer squash *Cucurbita pepo*. When plants with disc-shaped fruit (*AABB*) are crossed to plants with long fruit (*aabb*), the F₁ generation all have disc fruit. However, in the F₂ progeny, fruit with a novel shape—

sphere—appear, along with fruit exhibiting the parental phenotypes. A variety of fruit shapes are shown in **Figure 8**.

The F₂ generation, with a modified 9:6:1 ratio, is generated as follows:

F₁: *AaBb* × *AaBb*

disc   disc

↓

F₂ Ratio	Genotype	Phenotype	Final Phenotypic Ratio
9/16	*A—B—*	disc	9/16 disc
3/16	*A—bb*	sphere	6/16 sphere
3/16	*aaB—*	sphere	1/16 long
1/16	*aabb*	long	

FIGURE 8 Summer squash exhibiting the fruit-shape phenotypes disc, long, and sphere.

In this example of gene interaction, both gene pairs influence fruit shape equally. A dominant allele at either locus ensures a sphere-shaped fruit. In the absence of dominant alleles, the fruit is long. However, if both dominant alleles (*A* and *B*) are present, the fruit displays a flattened, disc shape.

2 In some plants a red flower pigment, cyanidin, is synthesized from a colorless precursor. The addition of a hydroxyl group (OH⁻) to the cyanidin molecule causes it to become purple. In a cross between two randomly selected purple varieties, the following results were obtained:

94 purple

31 red

43 white

How many genes are involved in the determination of these flower colors? Which genotypic combinations produce which phenotypes? Diagram the purple × purple cross.

■ **Hint** *This problem describes a plant in which flower color, a single characteristic, can take on one of three variations. The key to its solution is to first analyze the raw data and convert the numbers to a meaningful ratio. This will guide you in determining how many gene pairs are involved. Then you can group the genotypes in a way that corresponds to the phenotypic ratio.*

Other Modified Dihybrid Ratios

The remaining cases (5–8) in Figure 7 show additional modifications of the dihybrid ratio and provide still other examples of gene interactions. However, all eight cases have two things in common. First, we have not violated the principles of segregation and independent assortment to explain the inheritance pattern of each case. Therefore, the added complexity of inheritance in these examples does not detract from the validity of Mendel's conclusions. Second, the F_2 phenotypic ratio in each example has been expressed in sixteenths. When similar observations are made in crosses where the inheritance pattern is unknown, it suggests to geneticists that two gene pairs are controlling the observed phenotypes. You should make the same inference in your analysis of genetics problems.

ESSENTIAL POINT   ■   ■   ■

Mendel's classic F_2 ratio is often modified in instances when gene interaction controls phenotypic variation. Such instances can be identified when the final ratio is divided into eighths or sixteenths.

9 Complementation Analysis Can Determine If Two Mutations Causing a Similar Phenotype Are Alleles of the Same Gene

An interesting situation arises when two mutations, both of which produce a similar phenotype, are isolated independently. Suppose that two investigators independently isolate and establish a true-breeding strain of wingless *Drosophila* and demonstrate that each mutant phenotype is due to a recessive mutation. We might assume that both strains contain mutations in the same gene. However, since we know that many genes are involved in the formation of wings, mutations in any one of them might inhibit wing formation during development. The experimental approach called **complementation analysis** allows us to determine whether two such mutations are in the same gene—that is, whether they are alleles of the same gene or whether they represent mutations in separate genes.

To repeat, our analysis seeks to answer this simple question: *Are two mutations that yield similar phenotypes present in the same gene or in two different genes?* To find the answer, we cross the two mutant strains and analyze the F_1 generation. Two alternative outcomes and interpretations of this cross are shown in **Figure 9**. We discuss both cases, using the designations m^a for one of the mutations and m^b for the other one. Now we will determine experimentally whether or not m^a and m^b are alleles of the same gene.

Case 1. *All offspring develop normal wings.*
Interpretation: The two recessive mutations are in separate genes and are not alleles of one another. Following the cross, all F_1 flies are heterozygous for both genes. *Complementation* is said to occur. Since each mutation is in a separate gene and each F_1 fly is heterozygous at both loci, the normal products of both genes are produced (by the one normal copy of each gene), and wings develop.

Case 2. *All offspring fail to develop wings.*
Interpretation: The two mutations affect the same gene and are alleles of one another. Complementation does *not* occur. Since the two mutations affect the same gene, the F_1 flies are homozygous for the two mutant alleles (the m^a allele and the m^b allele). No normal product of the gene is produced, and in the absence of this essential product, wings do not form.

Complementation analysis, as originally devised by the Nobel Prize-winning *Drosophila* geneticist Edward B. Lewis, may be used to screen any number of individual mutations that result in the same phenotype. Such an analysis may reveal that only a single gene is involved or that two or more

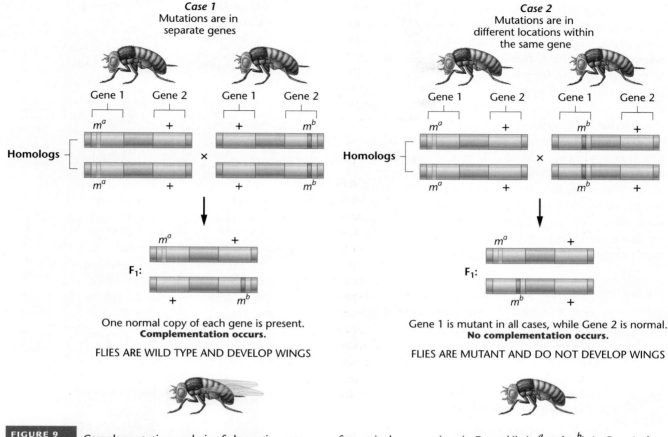

Case 1
Mutations are in
separate genes

Case 2
Mutations are in
different locations within
the same gene

One normal copy of each gene is present.
Complementation occurs.

FLIES ARE WILD TYPE AND DEVELOP WINGS

Gene 1 is mutant in all cases, while Gene 2 is normal.
No complementation occurs.

FLIES ARE MUTANT AND DO NOT DEVELOP WINGS

FIGURE 9 Complementation analysis of alternative outcomes of two wingless mutations in *Drosophila* (m^a and m^b). In Case 1, the mutations are not alleles of the same gene, whereas in Case 2, the mutations are alleles of the same gene.

genes are involved. All mutations determined to be present in any single gene are said to fall into the same **complementation group,** and they will complement mutations in all other groups. When large numbers of mutations affecting the same trait are available and studied using complementation analysis, it is possible to predict the total number of genes involved in the determination of that trait.

ESSENTIAL POINT ■ ■ ■

Complementation analysis determines whether independently isolated mutations producing similar phenotypes are alleles of one another or whether they represent separate genes.

---

10 Expression of a Single Gene May Have Multiple Effects

While the previous sections have focused on the effects of two or more genes on a single characteristic, the converse situation, where expression of a single gene has multiple

phenotypic effects, is also quite common. This phenomenon, which often becomes apparent when phenotypes are examined carefully, is referred to as **pleiotropy.** We will review two such cases involving human genetic disorders to illustrate this point.

Marfan syndrome is a human malady resulting from an autosomal dominant mutation in the gene encoding the connective tissue protein fibrillin. Because this protein is widespread in many tissues in the body, one would expect multiple effects of such a defect. In fact, fibrillin is important to the structural integrity of the lens of the eye, to the lining of vessels such as the aorta, and to bones, among other tissues. As a result, the phenotype associated with Marfan syndrome includes lens dislocation, increased risk of aortic aneurysm, and lengthened long bones in limbs. This disorder is of historical interest in that speculation abounds that Abraham Lincoln was afflicted.

Our second example involves another human autosomal dominant disorder, **porphyria variegata.** Afflicted individuals cannot adequately metabolize the porphyrin component of hemoglobin when this respiratory pigment is broken down as red blood cells are replaced. The accumulation of excess porphyrins is immediately evident in the urine, which takes on a deep red color. The severe features of the disorder are due to the toxicity of the buildup of porphyrins in the body, particularly in the brain. Complete phenotypic characterization includes abdominal pain, muscular weakness, fever, a racing pulse, insomnia, headaches, vision problems (that can lead to blindness), delirium, and ultimately convulsions. As you can see, deciding which phenotypic trait best characterizes the disorder is impossible.

Like Marfan syndrome, porphyria variegata is also of historical significance. George III, king of England during the American Revolution, is believed to have suffered from episodes involving all of the above symptoms. He ultimately became blind and senile prior to his death. We could cite many other examples to illustrate pleiotropy, but suffice it to say that if one looks carefully, most mutations display more than a single manifestation when expressed.

> ### ESSENTIAL POINT ■ ■ ■
> Pleiotropy refers to multiple phenotypic effects caused by a single mutation.

11 X-Linkage Describes Genes on the X Chromosome

In many animal and some plant species, one of the sexes contains a pair of unlike chromosomes that are involved in sex determination. In many cases, these are designated as the X and Y. For example, in both *Drosophila* and humans, males contain an X and a Y chromosome, whereas females contain two X chromosomes. While the Y chromosome must contain a region of pairing homology with the X chromosome if the two are to synapse and segregate during meiosis, much of the remainder of the Y chromosome in humans and other species is considered to be relatively inert genetically. Thus, it lacks most genes that are present on the X chromosome. As a result, genes present on the X chromosome exhibit unique patterns of inheritance in comparison with autosomal genes. The term **X-linkage** is used to describe these situations.

In the following discussion, we will focus on inheritance patterns resulting from genes present on the X but absent from the Y chromosome. This situation results in a modification of Mendelian ratios, the central theme of this chapter.

X-Linkage in *Drosophila*

One of the first cases of X-linkage was documented by Thomas H. Morgan around 1920 during his studies of the *white* mutation in the eyes of *Drosophila* **(Figure 10)**. The normal wild-type red eye color is dominant to white. We will use this case to illustrate X-linkage.

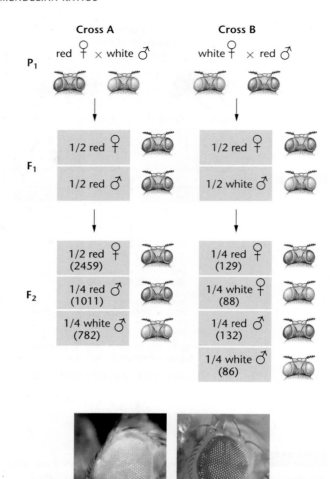

FIGURE 10 The F_1 and F_2 results of T. H. Morgan's reciprocal crosses involving the X-linked *white* mutation in *Drosophila melanogaster*. The actual F_2 data are shown in parentheses. The photographs show a white eye and a brick-red eye.

Morgan's work established that the inheritance pattern of the white-eye trait is clearly related to the sex of the parent carrying the mutant allele. Unlike the outcome of the typical monohybrid cross, reciprocal crosses between white- and red-eyed flies did not yield identical results. In contrast, in all of Mendel's monohybrid crosses, F_1 and F_2 data were similar regardless of which P_1 parent exhibited the recessive mutant trait. Morgan's analysis led to the conclusion that the *white* locus is present on the X chromosome rather than on one of the autosomes. As such, both the gene and the trait are said to be X-linked.

Results of reciprocal crosses between white-eyed and red-eyed flies are shown in Figure 10. The obvious differences in phenotypic ratios in both the F_1 and F_2 generations are dependent on whether or not the P_1 white-eyed parent was male or female.

Morgan was able to correlate these observations with the difference found in the sex-chromosome composition

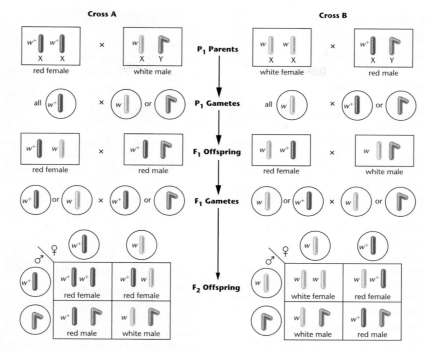

FIGURE 11 The chromosomal explanation of the results of the X-linked crosses shown in Figure 10.

between male and female *Drosophila*. He hypothesized that the recessive allele for white eyes is found on the X chromosome, but its corresponding locus is absent from the Y chromosome. Females thus have two available gene sites, one on each X chromosome, whereas males have only one available gene site on their single X chromosome.

Morgan's interpretation of X-linked inheritance, shown in **Figure 11**, provides a suitable theoretical explanation for his results. Since the Y chromosome lacks homology with most genes on the X chromosome, whatever alleles are present on the X chromosome of the males will be expressed directly in their phenotype. Males cannot be homozygous or heterozygous for X-linked genes, and this condition is referred to as being **hemizygous.**

One result of X-linkage is the **crisscross pattern of inheritance,** whereby phenotypic traits controlled by recessive X-linked genes are passed from homozygous mothers to all sons. This pattern occurs because females exhibiting a recessive trait carry the mutant allele on both X chromosomes. Because male offspring receive one of their mother's two X chromosomes and are hemizygous for all alleles present on

that X, all sons will express the same recessive X-linked traits as their mother.

Morgan's work has taken on great historical significance. By 1910, the correlation between Mendel's work and the behavior of chromosomes during meiosis had provided the basis for the **chromosome theory of inheritance.** Work involving the X chromosome around 1920 is considered to be the first solid experimental evidence in support of this theory. In the ensuing two decades, these findings inspired further research, which provided indisputable evidence in support of this theory.

X-Linkage in Humans

In humans, many genes and the respective traits they control are recognized as being linked to the X chromosome (see **Table 2**). These X-linked traits can be easily identified in a pedigree because of the crisscross pattern of inheritance. A pedigree for one form of human **color blindness** is shown in **Figure 12**. The mother in generation I passes the trait to all her sons but to none of her daughters. If the

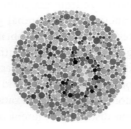

FIGURE 12 A human pedigree of the X-linked color-blindness trait. The photograph is of an Ishihara color-blindness chart. Those with normal vision will see the number 15, while those with red-green color blindness will see the number 17.

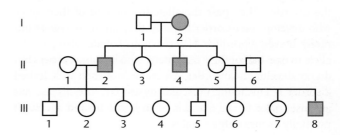

TABLE 2   Human X-Linked Traits

Condition	Characteristics
Color blindness, deutan type	Insensitivity to green light.
Color blindness, protan type	Insensitivity to red light.
Fabry disease	Deficiency of galactosidase A; heart and kidney defects, early death.
G-6-PD deficiency	Deficiency of glucose-6-phosphate dehydrogenase; severe anemic reaction following intake of primaquines in drugs and certain foods, including fava beans.
Hemophilia A	Classical form of clotting deficiency; absence of clotting factor VIII.
Hemophilia B	Christmas disease; absence of clotting factor IX.
Hunter syndrome	Mucopolysaccharide storage disease resulting from iduronate sulfatase enzyme deficiency; short stature, clawlike fingers, coarse facial features, slow mental deterioration, and deafness.
Ichthyosis	Deficiency of steroid sulfatase enzyme; scaly dry skin, particularly on extremities.
Lesch–Nyhan syndrome	Deficiency of hypoxanthine-guanine phosphoribosyl transferase enzyme (HGPRT) leading to motor and mental retardation, self-mutilation, and early death.
Duchenne muscular dystrophy	Progressive, life-shortening disorder characterized by muscle degeneration and weakness; sometimes associated with mental retardation; absence of the protein dystrophin.

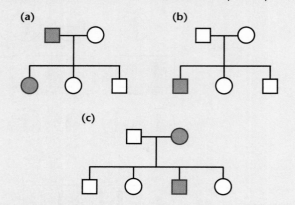
offspring in generation II marry normal individuals, the color-blind sons will produce all normal male and female offspring (III-1, 2, and 3); the normal-visioned daughters will produce normal-visioned female offspring (III-4, 6, and 7), as well as color-blind (III-8) and normal-visioned (III-5) male offspring.

The way in which X-linked genes are transmitted causes unusual circumstances associated with recessive X-linked disorders, in comparison to recessive autosomal disorders. For example, if an X-linked disorder debilitates or is lethal to the affected individual prior to reproductive maturation, the disorder occurs exclusively in males. This is so because the only sources of the lethal allele in the population are in heterozygous females who are "carriers" and do not express the disorder. They pass the allele to one-half of their sons, who develop the disorder because they are hemizygous but rarely, if ever, reproduce. Heterozygous females also pass the allele to one-half of their daughters, who become carriers but do not develop the disorder. An example of such an X-linked disorder is Duchenne muscular dystrophy. The disease has an onset prior to age 6 and is often lethal around age 20. It normally occurs only in males.

12   In Sex-Limited and Sex-Influenced Inheritance, an Individual's Sex Influences the Phenotype

In contrast to X-linked inheritance, patterns of gene expression may be affected by the sex of an individual even when the genes are not on the X chromosome. In numerous examples in different organisms, the sex of the individual plays a determining role in the expression of certain phenotypes. In some cases, the expression of a specific phenotype is absolutely limited to one sex; in others, the sex of an individual influences the expression of a phenotype that is not limited to one sex or the other. This distinction differentiates sex-limited inheritance from sex-influenced inheritance. In both types of inheritance, autosomal genes are responsible for the existence of contrasting phenotypes, but the expression of these genes is dependent on the hormone constitution of the individual. Thus, the heterozygous genotype may exhibit one phenotype in males and the contrasting one in females. In domestic fowl, for example, tail and neck plumage is often distinctly different in males and females **(Figure 13)**,

FIGURE 13   Hen feathering (left) and cock feathering (right) in domestic fowl. Note that the hen's feathers are shorter and less curved.

FIGURE 14   Pattern baldness, a sex-influenced autosomal trait in humans.

demonstrating **sex-limited inheritance.** Cock feathering is longer, more curved, and pointed, whereas hen feathering is shorter and less curved. Inheritance of these feather phenotypes is controlled by a single pair of autosomal alleles whose expression is modified by the individual's sex hormones.

As shown in the following chart, hen feathering is due to a dominant allele, *H*, but regardless of the homozygous presence of the recessive *h* allele, all females remain hen-feathered. Only in males does the *hh* genotype result in cock feathering.

Genotype	Phenotype	
	Females	Males
HH	Hen-feathered	Hen-feathered
Hh	Hen-feathered	Hen-feathered
hh	Hen-feathered	Cock-feathered

In certain breeds of fowl, the hen-feathering or cock-feathering allele has become fixed in the population. In the Leghorn breed, all individuals are of the *hh* genotype; as a result, males always differ from females in their plumage. Sebright bantams are all *HH*, resulting in no sexual distinction in feathering phenotypes.

Another example of sex-limited inheritance involves the autosomal genes responsible for milk yield in dairy cattle. Regardless of the overall genotype that influences the quantity of milk production, those genes are obviously expressed only in females.

Cases of **sex-influenced inheritance** include pattern baldness in humans, horn formation in certain breeds of sheep (e.g., Dorset Horn sheep), and certain coat-color patterns in cattle. In such cases, autosomal genes are responsible for the contrasting phenotypes displayed by both males and females, but the expression of these genes is dependent on the hormonal constitution of the individual. Thus, the heterozygous genotype exhibits one phenotype in one sex and the contrasting one in the other. For example, **pattern baldness** in humans, where the hair is very thin on the top of the head (**Figure 14**), is inherited in this way:

Genotype	Phenotype	
	Females	Males
BB	Bald	Bald
Bb	Not bald	Bald
bb	Not bald	Not bald

Females can display pattern baldness, but this phenotype is much more prevalent in males. When females do inherit the *BB* genotype, the phenotype is less pronounced than in males and is expressed later in life.

ESSENTIAL POINT   ■   ■   ■

Sex-limited and sex-influenced inheritance occurs when the sex of the organism affects the phenotype controlled by a gene located on an autosome.

13   Genetic Background and the Environment Affect Phenotypic Expression

We now focus on **phenotypic expression.** In previous discussions, we assumed that the genotype of an organism is always directly expressed in its phenotype. For example, pea plants homozygous for the recessive *d* allele (*dd*) will always be dwarf. We discussed gene expression as though the genes operate in a closed system in which the presence or absence of functional products directly determines the collective phenotype of an individual. The situation is actually much more complex. Most gene products function within the internal milieu of the cell, and cells interact with one another in various ways. Furthermore, the organism exists under diverse environmental influences. Thus, gene expression and the resultant phenotype are often modified through the interaction between an individual's particular genotype and the external environment. Here, we deal with several important variables that are known to modify gene expression.

Penetrance and Expressivity

Some mutant genotypes are always expressed as a distinct phenotype, whereas others produce a proportion of individuals whose phenotypes cannot be distinguished from normal (wild type). The degree of expression of a particular trait can be studied quantitatively by determining the *penetrance* and *expressivity* of the genotype under investigation. The percentage of individuals who show at least some degree of expression of a mutant genotype defines the **penetrance** of the mutation. For example, the phenotypic expression of many mutant alleles in *Drosophila* can overlap with wild type. If 15 percent of mutant flies show the wild-type appearance, the mutant gene is said to have a penetrance of 85 percent.

By contrast, **expressivity** reflects the *range of expression* of the mutant genotype. Flies homozygous for the recessive mutant *eyeless* gene yield phenotypes that range from the presence of normal eyes to a partial reduction in size to the complete absence of one or both eyes (**Figure 15**). Although the average reduction of eye size is one-fourth to one-half, expressivity ranges from complete loss of both eyes to completely normal eyes.

Examples such as the expression of the *eyeless* gene provide the basis for experiments to determine the causes of phenotypic variation. If, on one hand, a laboratory environment is held constant and extensive phenotypic variation is still observed, other genes may be influencing or modifying the *eyeless* phenotype. On the other hand, if the genetic background is not the cause of the phenotypic variation, environmental factors such as temperature, humidity, and nutrition may be involved. In the case of the *eyeless* phenotype, experiments have shown that both genetic background and environmental factors influence its expression.

Genetic Background: Position Effects

Although it is difficult to assess the specific effect of the **genetic background** and the expression of a gene responsible for determining a potential phenotype, one effect of genetic background has been well characterized, the **position effect**. In such instances, the physical location of a gene in relation to other genetic material may influence its expres-

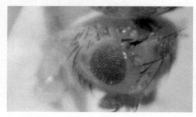
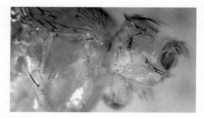

FIGURE 15 Variable expressivity, as shown in flies homozygous for the *eyeless* mutation in *Drosophila*. Gradations in phenotype range from wild type to partial reduction to eyeless.

sion. For example, if a region of a chromosome is relocated or rearranged (called a translocation or an inversion event), normal expression of genes in that chromosomal region may be modified. This is particularly true if the gene is relocated to or near certain areas of the chromosome that are prematurely condensed and genetically inert, referred to as **heterochromatin**. An example of a position effect involves female *Drosophila* heterozygous for the X-linked recessive eye color mutant *white* (*w*). The w^+/w genotype normally results in a wild-type brick-red eye color. However, if the region of the X chromosome containing the wild-type w^+ allele is translocated so that it is close to a heterochromatic region, expression of the w^+ allele is modified. Instead of having a red color, the eyes are variegated, or mottled with red and white patches. Apparently, heterochromatic regions inhibit the expression of adjacent genes.

Temperature Effects—An Introduction to Conditional Mutations

Chemical activity depends on the kinetic energy of the reacting substances, which in turn depends on the surrounding temperature. We can thus expect temperature to influence phenotypes. An example is seen in the evening primrose, which produces red flowers when grown at 23°C and white flowers when grown at 18°C. An even more striking example is seen in Siamese cats and Himalayan rabbits, which exhibit

(a)

(b)

FIGURE 16 (a) A Himalayan rabbit. (b) A Siamese cat. Both species show dark fur color on the snout, ears, and paws. The patches are due to the temperature-sensitive allele responsible for pigment production.

dark fur in certain regions where their body temperature is slightly cooler, particularly the nose, ears, and paws **(Figure 16)**. In these cases, it appears that the enzyme normally responsible for pigment production is functional only at the lower temperatures present in the extremities, but it loses its catalytic function at the slightly higher temperatures found throughout the rest of the body.

Mutations whose expression is affected by temperature, called **temperature-sensitive mutations,** are examples of **conditional mutations,** whereby phenotypic expression is determined by environmental conditions. Examples of temperature-sensitive mutations are known in viruses and a variety of organisms, including bacteria, fungi, and *Drosophila*. In extreme cases, an organism carrying a mutant allele may express a mutant phenotype when grown at one temperature but express the wild-type phenotype when reared at another temperature. This type of temperature effect is useful in studying mutations that interrupt essential processes during development and are thus normally detrimental or lethal. For example, if bacterial viruses are cultured under *permissive conditions* of 25°C, the mutant gene product is functional, infection proceeds normally, and new viruses are produced and can be studied. However, if bacterial viruses carrying temperature-sensitive mutations infect bacteria cultured at 42°C—the *restrictive condition*—infection progresses up to the point where the essential gene product is required (e.g., for viral assembly) and then arrests. Temperature-sensitive mutations are easily induced and isolated in viruses, and have added immensely to the study of viral genetics.

Onset of Genetic Expression

Not all genetic traits become apparent at the same time during an organism's life span. In most cases, the age at which a mutant gene exerts a noticeable phenotype depends on events during the normal sequence of growth and development.

In humans, the prenatal, infant, preadult, and adult phases require different genetic information. As a result, many severe inherited disorders are not manifested until after birth. For example, **Tay–Sachs disease,** inherited as an autosomal recessive, is a lethal lipid metabolism disease involving an abnormal enzyme, hexosaminidase A. Newborns appear to be phenotypically normal for the first few months. Then developmental retardation, paralysis, and blindness ensue, and most affected children die around the age of 3.

The **Lesch–Nyhan syndrome,** inherited as an X-linked recessive disease, is characterized by abnormal nucleic acid metabolism (biochemical salvage of nitrogenous purine bases), leading to the accumulation of uric acid in blood and tissues, mental retardation, palsy, and self-mutilation of the lips and fingers. The disorder is due to a mutation in the gene encoding hypoxanthine-guanine phosphoribosyl transferase (HGPRT). Newborns are normal for six to eight months prior to the onset of the first symptoms.

Still another example involves **Duchenne muscular dystrophy (DMD),** an X-linked recessive disorder associated with progressive muscular wasting. It is not usually diagnosed until the child is 3 to 5 years old. Even with modern medical intervention, the disease is often fatal in the early 20s.

Perhaps the most age-variable of all inherited human disorders is **Huntington disease.** Inherited as an autosomal dominant, Huntington disease affects the frontal lobes of the cerebral cortex, where progressive cell death occurs over a period of more than a decade. Brain deterioration is accompanied by spastic uncontrolled movements, intellectual and emotional deterioration, and ultimately death. Onset of this disease has been reported at all ages, but it most frequently occurs between ages 30 and 50, with a mean onset age of 38 years.

These examples support the concept that the critical expression of genes varies throughout the life cycle of all organisms, including humans. Gene products may play more essential roles at certain life stages, and it is likely that the internal physiological environment of an organism changes with age.

Genetic Anticipation

Interest in studying the genetic onset of phenotypic expression has intensified with the discovery of heritable disorders that exhibit a progressively earlier age of onset and an increased severity of the disorder in each successive generation. This phenomenon is called **genetic anticipation.**

Myotonic dystrophy (DM), the most common type of adult muscular dystrophy, clearly illustrates genetic anticipation. Individuals afflicted with this autosomal dominant disorder exhibit extreme variation in the severity of symptoms. Mildly affected individuals develop cataracts as adults but have little or no muscular weakness. Severely affected individuals demonstrate more extensive myopathy and may be mentally retarded. In its most extreme form, the disease is fatal just after birth. In 1989, C. J. Howeler and colleagues confirmed the correlation of increased severity and earlier onset with successive generations. They studied 61 parent–child pairs, and in 60 cases, age of onset was earlier and more severe in the child than in his or her affected parent.

In 1992, an explanation was put forward for the molecular cause of the mutation responsible for DM, as well as the basis of genetic anticipation. A particular region of the DM gene—a short trinucleotide DNA sequence—is repeated a variable number of times and is unstable. Normal individuals average about five copies of this region; minimally affected individuals have about 50 copies; and severely affected individuals have over 1000 copies. The most remarkable observation was that in successive generations, the size of the repeated segment increases. Although it is not yet clear how this expansion in size affects onset and phenotypic expression, the correlation is extremely strong. Several other inherited human disorders, including the fragile-X syndrome, Kennedy disease, and Huntington disease, also reveal an association between the size of specific regions of the responsible gene and disease severity.

14 Genomic (Parental) Imprinting and Gene Silencing

A final example involving genetic background involves what is called **genomic,** or **parental, imprinting,** whereby the process of selective *gene silencing* occurs during early development, impacting subsequent phenotypic expression. Examples involve cases where genes or regions of a chromosome are imprinted on one homolog but not the other. The impact of silencing depends on the parental origin of the genes or regions that are involved. Such silencing leads to the direct phenotypic expression of the allele(s) on the homolog that is not silenced. Thus, the imprinting step, the critical issue in understanding this phenomenon, is thought to occur before or during gamete formation, leading to differentially marked genes (or chromosome regions) in sperm-forming versus egg-forming tissues.

The first example of genomic imprinting was discovered in 1991, in three specific mouse genes. One is the gene encoding insulin-like growth factor II (*Igf2*). A mouse that carries two normal alleles of this gene is normal in size, whereas a mouse that carries two mutant alleles lacks the growth factor and is dwarf. The size of a heterozygous mouse—one allele normal and one mutant—depends on the parental origin of the wild-type allele. The mouse is normal in size if the normal allele comes from the father, but it is dwarf if the normal allele came from the mother. From this, we can deduce that the normal *Igf2* gene is imprinted and thus silenced during egg production, but it functions normally when it has passed through sperm-producing tissue in males. The imprint is inherited in the sense that the *Igf2* gene in all progeny cells formed during development remain silenced. Imprinting in the next generation then depends on whether the gene passes through sperm-producing or egg-forming tissue.

An example in humans involves two distinct genetic disorders thought to be caused by differential imprinting of the same region of the long arm of chromosome 15 (15q1). In both cases, the disorders are due to an identical deletion of this region in one member of the chromosome 15 pair. The first disorder, **Prader–Willi syndrome (PWS),** results when the paternal segment is deleted and an undeleted maternal chromosome remains. If the maternal segment is deleted and an undeleted paternal chromosome remains, an entirely different disorder, **Angelman syndrome (AS),** results.

These two conditions exhibit different phenotypes. PWS entails mental retardation, a severe eating disorder marked by an uncontrollable appetite, obesity, diabetes, and growth retardation. Angelman syndrome also involves mental retardation, but involuntary muscle contractions (chorea) and seizures characterize the disorder. We can conclude that the involved region of chromosome 15 is imprinted differently in male and female gametes and that both an undeleted maternal and a paternal region are required for normal development.

Although numerous questions remain unanswered regarding genomic imprinting, it is now clear that many genes are subject to this process. More than 50 have been identified in mammals thus far. It appears that regions of chromosomes rather than specific genes are imprinted. This phenomenon is an example of the more general topic of **epigenetics,** where genetic expression is *not* the direct result of the information stored in the nucleotide sequence of DNA. Instead, the DNA is altered in a way that affects its expression. These changes are stable in the sense that they are transmitted during cell division to progeny cells, and often through gametes to future generations.

The precise molecular mechanism of imprinting and other epigenetic events is still a matter for conjecture, but it seems certain that **DNA methylation** is involved. In most eukaryotes, methyl groups can be added to the carbon atom at position 5 in cytosine as a result of the activity of the enzyme DNA methyltransferase. Methyl groups are added when the dinucleotide CpG or groups of CpG units (called CpG islands) are present along a DNA chain.

DNA methylation is a reasonable mechanism for establishing a molecular imprint, since there is evidence that a high level of methylation can inhibit gene activity and that active genes (or their regulatory sequences) are often undermethylated. We will encounter other examples throughout the text.

ESSENTIAL POINT ■   ■   ■

Phenotypic expression is not always the direct reflection of the genotype. A percentage of organisms may not express the expected phenotype at all, the basis of the penetrance of a mutant gene. In addition, the phenotype can be modified by genetic background, temperature, and nutrition. The onset of expression of a gene may vary during the lifetime of an organism, and it may even be imprinted so that it is expressed differently depending on parental origin.

---

15 Extranuclear Inheritance Modifies Mendelian Patterns

Throughout the history of genetics, occasional reports have challenged the basic tenet of Mendelian transmission genetics—that the phenotype is determined solely by nuclear genes located on the chromosomes of both parents. In this final section of the chapter, we consider several examples of inheritance patterns that vary from those predicted by the traditional biparental inheritance of nuclear genes, phenomena that are designated as **extranuclear inheritance.** In the following cases, we will focus on two broad categories. In the first, an organism's phenotype is affected by the expression of genes contained in the DNA of mitochondria or chloroplasts rather than the nucleus, generally referred to as organelle heredity. In the second category, referred to as a maternal effect, an organism's phenotype is determined by genetic information expressed in the gamete of the mother—such that, following fertilization, the developing zygote's phenotype is influenced not by the individual's genotype, but by gene products directed by the genotype of the mother.

Initially, such observations met with skepticism. However, with increasing knowledge of molecular genetics and the discovery of DNA in mitochondria and chloroplasts, the phenomenon of extranuclear inheritance came to be recognized as an important aspect of genetics.

Organelle Heredity: DNA in Chloroplasts and Mitochondria

We begin by examining examples of inheritance patterns related to chloroplast and mitochondrial function. Before DNA was discovered in these organelles, the exact mechanism of transmission of the traits was not clear, except that their inheritance appeared to be linked to something in the cytoplasm rather than to genes in the nucleus. Furthermore, transmission was most often from the maternal parent through the ooplasm, causing the results of reciprocal crosses to vary. Such an extranuclear pattern of inheritance is now appropriately called **organelle heredity.**

Analysis of the inheritance patterns resulting from mutant alleles in chloroplasts and mitochondria has been difficult for two reasons. First, the function of these organelles is dependent on gene products from both nuclear and organelle DNA, making the discovery of the genetic origin of mutations affecting organelle function difficult. Second, many mitochondria and chloroplasts are contributed to each progeny. Thus, if only one or a few of the organelles contain a mutant gene in a cell among a population of mostly normal mitochondria, the corresponding mutant phenotype may not be revealed. This condition, referred to as **heteroplasmy,** may lead to normal cells since the organelles lacking the mutation provide the basis of wild-type function. Analysis is therefore much more complex than for Mendelian characters.

Chloroplasts: Variegation in Four-o'clock Plants

In 1908, Karl Correns (one of the rediscoverers of Mendel's work) provided the earliest example of inheritance linked to chloroplast transmission. Correns discovered a variant of the four-o'clock plant, *Mirabilis jalapa*, that had branches with either white, green, or variegated white-and-green leaves. The white areas in variegated leaves and in the completely white leaves lack chlorophyll that provides the green color to normal leaves. Chlorophyll is the light-absorbing pigment made within chloroplasts.

Correns was curious about how inheritance of this phenotypic trait occurred. As shown in **Figure 17**, inheritance in all possible combinations of crosses is strictly determined by the phenotype of the ovule source. For example, if the seeds (representing the progeny) were derived from ovules on branches with green leaves, all progeny plants bore only green leaves, regardless of the phenotype of the source of pollen. Correns concluded that inheritance was transmitted through the cytoplasm of the maternal parent because the pollen, which contributes little or no cytoplasm to the zygote, had no apparent influence on the progeny phenotypes.

Since leaf coloration is related to the chloroplast, genetic information contained either in that organelle or somehow present in the cytoplasm and influencing the chloroplast

	Location of Ovule		
Source of Pollen	White branch	Green branch	Variegated branch
White branch	White	Green	White, green, or variegated
Green branch	White	Green	White, green, or variegated
Variegated branch	White	Green	White, green, or variegated

FIGURE 17 Offspring from crosses between flowers from various branches of four-o'clock plants. The photograph illustrates a mixture of white, green, and variegated leaves.

must be responsible for the inheritance pattern. It now seems certain that the genetic "defect" that eliminates the green chlorophyll in the white patches on leaves is a mutation in the DNA housed in the chloroplast.

Mitochondrial Mutations: *poky* in *Neurospora* and *petite* in *Saccharomyces*

Mutations affecting mitochondrial function have been discovered and studied, revealing that they too contain a distinctive genetic system. As with chloroplasts, mitochondrial mutations are transmitted through the cytoplasm. In our current discussion, we will emphasize the link between mitochondrial mutations and the resultant extranuclear inheritance patterns.

In 1952, Mary B. Mitchell and Hershel K. Mitchell studied the bread mold *Neurospora crassa*. They discovered a slow-growing mutant strain and named it *poky*. Slow growth is associated with impaired mitochondrial function, specifically in relation to certain cytochromes essential for electron transport. Results of genetic crosses between wild-type and *poky* strains suggest that *poky* is an extranuclear trait inherited through the cytoplasm. If one mating type is *poky* and the other is wild type, all progeny colonies are *poky*. The reciprocal cross, where *poky* is transmitted by the other mating type, produces normal wild-type colonies.

Another extensive study of mitochondrial mutations has been performed with the yeast *Saccharomyces cerevisiae*. The first such mutation, described by Boris Ephrussi and his co-workers in 1956, was named *petite* because of the small size of the yeast colonies (**Figure 18**). Many independent *petite*

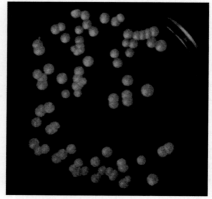

Normal colonies

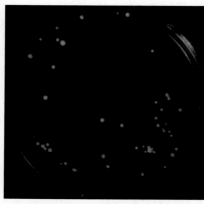

Petite colonies

FIGURE 18 Photos comparing normal versus petite colonies of the yeast *Saccharomyces cerevisiae*.

mutations have since been discovered and studied, and all have a common characteristic—a deficiency in cellular respiration involving abnormal electron transport. The majority of them demonstrate cytoplasmic transmission, indicating mutations in the DNA of the mitochondria. This organism is a facultative anaerobe and can grow by fermenting glucose through glycolysis; thus, it may survive the loss of mitochondrial function by generating energy anaerobically.

The complex genetics of *petite* mutations has revealed that a small proportion are the result of nuclear DNA changes. They exhibit Mendelian inheritance and illustrate that mitochondria function depends on both nuclear and organellar gene products.

Mitochondrial Mutations: Human Genetic Disorders

Our knowledge of the genetics of mitochondria has now greatly expanded. The DNA found in human mitochondria has been completely sequenced and contains 16,569 base pairs. Mitochondrial gene products have been identified and include the following:

13 proteins, required for aerobic cellular respiration

22 transfer RNAs (tRNAs), required for translation

2 ribosomal RNAs (rRNAs), required for translation

Because a cell's energy supply is largely dependent on aerobic cellular respiration, disruption of any mitochondrial gene by mutation may potentially have a severe impact on that organism, such as we saw in our previous discussion of the *petite* mutation in yeast. In fact, mtDNA is particularly vulnerable to mutations for two possible reasons. First, the ability to repair mtDNA damage does not appear to be equivalent to that of nuclear DNA. Second, the concentration of highly mutagenic free radicals generated by cell respiration that accumulate in such a confined space very likely raises the mutation rate in mtDNA.

Fortunately, a zygote receives a large number of organelles through the egg, so if only one organelle or a few of them contain a mutation (an illustration of *heteroplasmy*), the impact is greatly diluted by the many mitochondria that lack the mutation and function normally. If a deleterious mutation arises or is present in the initial population of organelles, adults will have cells with a variable mixture of both normal and abnormal organelles. From a genetic standpoint, this condition of heteroplasmy makes analysis quite difficult.

Several disorders in humans are known to be due to mutations in mitochondrial genes. For example, **myoclonic epilepsy and ragged-red fiber disease (MERRF)** demonstrates a pattern of inheritance consistent with maternal transmission. Only the offspring of affected mothers inherit this disorder, while the offspring of affected fathers are normal. Individuals with this rare disorder express ataxia (lack of muscular coordination), deafness, dementia, and epileptic seizures. The disease is named for the presence of "ragged-red" skeletal-muscle fibers that exhibit blotchy red patches resulting from the proliferation of aberrant mitochondria. Brain function, which has a high energy demand, is also affected in this disorder, leading to the neurological symptoms described above.

The mutation that causes MERRF has now been identified and is in a mitochondrial gene whose altered product interferes with the capacity for translation of proteins within the organelle. This, in turn, leads to the various manifestations of the disorder. The cells of MERRF individuals exhibit heteroplasmy, containing a mixture of normal and abnormal mitochondria. Different patients display different proportions of the two, and even different cells from the same patient exhibit various levels of abnormal mitochondria. Were it not for heteroplasmy, the mutation would very likely be lethal, testifying to the essential nature of mitochondrial function and its reliance on the genes encoded by DNA within the organelle.

A second human disorder, **Leber's hereditary optic neuropathy (LHON),** also exhibits maternal inheritance as well as mitochondrial DNA lesions. The disease is characterized by sudden bilateral blindness. The average age of vision loss is 27, but onset is quite variable. Four mutations have been identified, all of which disrupt normal oxidative phosphorylation. Over 50 percent of cases are due to a mutation at a specific position in the mitochondrial gene encoding a subunit of NADH dehydrogenase, an enzyme essential to cell respiration. It is interesting to note that in many instances of LHON, there is no family history; a significant number of cases are "sporadic," resulting from newly arisen mutations.

Maternal Effect

In **maternal effect,** also referred to as maternal influence, an offspring's phenotype for a particular trait is under the control of the mother's *nuclear gene products* present in the egg. This is in contrast to biparental inheritance, where both parents transmit information on genes in the nucleus that determines the offspring's phenotype. In cases of maternal effect, the nuclear genes of the female gamete are transcribed, and the genetic products (either proteins or yet untranslated mRNAs) accumulate in the egg ooplasm. After fertilization, these products are distributed among newly formed cells and influence the patterns or traits established during early development. Two examples will illustrate such an influence of the maternal genome on particular traits.

Ephestia Pigmentation

A straightforward illustration of a maternal effect is seen in the Mediterranean meal moth *Ephestia kuehniella*. The wild-type larva of this moth has a pigmented skin and brown eyes as a

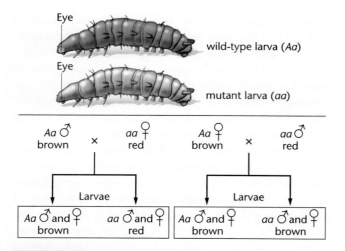

Eye

wild-type larva (*Aa*)

Eye

mutant larva (*aa*)

Aa ♂ × *aa* ♀   *Aa* ♀ × *aa* ♂
brown   red   brown   red

Larvae   Larvae

| *Aa* ♂ and ♀ | *aa* ♂ and ♀ | *Aa* ♂ and ♀ | *aa* ♂ and ♀ |
| brown | red | brown | brown |

FIGURE 19   Maternal influence in the inheritance of eye pigment in the meal moth *Ephestia kuehniella*.

result of the dominant gene *A*. The pigment is derived from a precursor molecule, kynurenine, which is in turn a derivative of the amino acid tryptophan. A mutation, *a*, interrupts the synthesis of kynurenine and, when homozygous, may result in red eyes and little pigmentation in larvae. However, as illustrated in **Figure 19**, results of the cross *Aa* × *aa* depend on which parent carries the dominant gene. When the male is the heterozygous parent, a 1:1 brown- to red-eyed ratio is observed in larvae, as predicted by Mendelian segregation. When the female is heterozygous for the *A* gene, however, all larvae are pigmented and have brown eyes, in spite of half of them being *aa*. As these larvae develop into adults, one-half of them gradually develop red eyes, reestablishing the 1:1 ratio.

One explanation for these results is that the *Aa* oocytes synthesize kynurenine or an enzyme necessary for its synthesis and accumulate it in the ooplasm prior to the completion of meiosis. Even in *aa* progeny, if the mothers were *Aa*, this pigment is distributed in the cytoplasm of the cells of the developing larvae; thus, they develop pigmentation and brown eyes. In these progeny, however, the pigment is eventually diluted among many cells and depleted, resulting in the conversion to red eyes as adults. The *Ephestia* example demonstrates the

maternal effect in which a cytoplasmically stored nuclear gene product influences the larval phenotype and, at least temporarily, overrides the genotype of the progeny.

Embryonic Development in *Drosophila*

A more recently documented example of maternal effect involves various genes that control embryonic development in *Drosophila melanogaster*. The genetic control of embryonic development in *Drosophila* is a fascinating story. The protein products of the maternal-effect genes function to activate other genes, which may in turn activate still other genes. This cascade of gene activity leads to a normal embryo whose subsequent development yields a normal adult fly. The extensive work by Edward B. Lewis, Christiane Nüsslein-Volhard, and Eric Wieschaus (who shared the 1995 Nobel Prize for Physiology or Medicine for their findings) has clarified how these and other genes function. Genes that illustrate maternal effect have products that are synthesized by the developing egg and stored in the oocyte prior to fertilization. Following fertilization, these products specify molecular gradients that determine spatial organization as development proceeds.

For example, the gene *bicoid* (*bcd*) plays an important role in specifying the development of the anterior portion of the fly. Embryos derived from mothers who are homozygous for this mutation (bcd^-/bcd^-) fail to develop anterior areas that normally give rise to the head and thorax of the adult fly. Embryos whose mothers contain at least one wild-type allele (bcd^+) develop normally, even if the genotype of the embryo is homozygous for the mutation. Consistent with the concept of maternal effect, the *genotype of the female parent*, not the *genotype of the embryo*, determines the phenotype of the offspring.

ESSENTIAL POINT ■   ■   ■

When patterns of inheritance vary from that expected due to biparental transmission of nuclear genes, phenotypes are often found to be under the control of DNA present in mitochondria or chloroplasts, or are influenced during development by the expression of the maternal genotype in the egg.

GENETICS, TECHNOLOGY, AND SOCIETY

Improving the Genetic Fate of Purebred Dogs

For dog lovers, nothing is quite so heartbreaking as watching a dog slowly go blind, struggling to adapt to a life of perpetual darkness. That's what happens in progressive retinal at-

rophy (PRA), a group of inherited disorders first described in Gordon setters in 1909. Since then, PRA has been detected in more than 100 other breeds of dogs, including Irish setters, border collies,

Norwegian elkhounds, toy poodles, miniature schnauzers, cocker spaniels, and Siberian huskies.

The products of many genes are required for the development and maintenance of healthy retinas, and a defect in any one of these genes may cause retinal dysfunction. Decades of research have led to the identification of five such genes (PDE6A, PDE6B, PRCD, rhodopsin, and PRGR), and more may be discovered. Different mutant alleles are present in different breeds, and each allele is associated with a different form of PRA that varies slightly in its clinical symptoms and rate of progression. Mutations of PDE6A, PDE6B, and PRCD genes are inherited in a recessive pattern, mutations of the rhodopsin gene (such as those found in Mastiffs) are dominant, and PRGR mutations (in Siberian huskies and Samoyeds) are X-linked.

PRA is almost ten times more common in certain purebred dogs than in mixed breeds. The development of distinct breeds of dogs has involved intensive selection for desirable attributes, such as a particular size, shape, color, or behavior. Many desired characteristics are determined by recessive alleles. The fastest way to increase the homozygosity of these alleles is to mate close relatives, which are likely to carry the same alleles. For example, dogs may be mated to a cousin or a grandparent. Some breeders, in an attempt to profit from impressive pedigrees, also produce hundreds of offspring from individual dogs that have won major prizes at dog shows. This "popular sire effect," as it has been termed, further increases the homozygosity of alleles in purebred dogs.

Unfortunately, the generations of inbreeding that have established favorable characteristics in purebreeds have also increased the homozygosity of certain harmful recessive alleles, resulting in a high incidence of inherited diseases.

More than 300 genetic diseases have been characterized in purebred dogs, and many breeds have a predisposition to more than 20 of them. According to researchers at Cornell University, purebred dogs suffer the highest incidence of inherited disease of any animal: 25 percent of the 20 million purebred dogs in America are affected with one genetic ailment or another.

Fortunately, advances in canine genetics are beginning to provide new tools to increase the health of purebred dogs. As of 2007, genetic tests are available to detect 30 different retinal diseases in dogs. Tests for PRA are now being used to identify heterozygous carriers of PRCD mutations. These carriers show no symptoms of PRA but, if mated with other carriers, pass the trait on to about 25 percent of their offspring. Eliminating PRA carriers from breeding programs has almost eradicated this condition from Portuguese water dogs and has greatly reduced its prevalence in other breeds.

Scientists will be able to identify more genes underlying canine inherited diseases thanks to the completion of the Dog Genome Project in 2005. In addition, new therapies that correct gene-based defects will emerge.

The Dog Genome Project may have benefits for humans beyond the reduction of disease in their canine companions. Eighty-five percent of the genes in the dog genome have equivalents in humans, and over 300 diseases affecting dogs also affect humans, including heart disease, epilepsy, allergies, and cancer. The identification of a disease-causing gene in dogs can be a shortcut to the isolation of the corresponding gene in humans. By contributing to the cure of human diseases, dogs may prove to be "man's best friend" in an entirely new way.

Your Turn

Take time, individually or in groups, to answer the following questions. Investigate the references and links, to help you understand some of the issues surrounding the genetics of purebred dogs.

1. What are some of the limitations of genetic tests, especially as they apply to purebred dog genetic diseases?

   This topic is discussed on the OptiGen website (http://www.optigen.com). OptiGen is a company that offers gene tests for all known forms of PRA in dogs and is developing tests for other inherited disorders. From their TESTS list, select prcd-PRA, and visit the link "Benefits and Limitations of All Genetic Tests."

2. Which human disease is similar to PRA in the Siberian husky?

   To learn more about these genes and diseases, visit the "Inherited Diseases in Dogs" database (http://www.vet.cam.ac.uk/idid/) and search the database for Progressive Retinal Atrophy in the Siberian husky. Once there, follow the OMIM reference link to learn about the human version of PRA in the Siberian husky.

3. Recently, commercial laboratories have cloned dogs for research purposes and for people who want their beloved pet to return. Do you approve of cloning pet dogs? Why or why not? Do you think that a cloned dog would be identical to the original dog?

   To learn about a recent pet dog cloning, read a Manchester Guardian article entitled "Pet cloning service bears five baby Boogers." (http://www.guardian.co.uk/science/2008/aug/05/-genetics.korea)

CASE STUDY » But he isn't deaf

Researching their family histories, a deaf couple learns that each of them has relatives through several generations who are deaf. They also learn that one form of deafness can be inherited as an autosomal recessive trait. They plan to have children, and based on the above information, they assume that all of their children may be deaf. To their surprise, their first child has normal hearing. The couple turns to you as a geneticist to help explain this situation.

1. Is it likely that these parents inherited their deafness as an autosomal recessive trait?
2. If two deaf parents have a hearing child, what conclusions can be drawn about the genetic control of deafness?
3. Is it likely that a future child will be deaf?

INSIGHTS AND SOLUTIONS

Genetic problems take on added complexity if they involve two independent characters and multiple alleles, incomplete dominance, or epistasis. The most difficult types of problems are those that pioneering geneticists faced during laboratory or field studies. They had to determine the mode of inheritance by working backward from the observations of offspring to parents of unknown genotype.

1. Consider the problem of comb-shape inheritance in chickens, where walnut, pea, rose, and single are the observed distinct phenotypes (see the photographs below). How is comb shape inherited, and what are the genotypes of the P_1 generation of each cross? Use the following data:

Cross 1 :	single	×	single	⟶ all single
Cross 2 :	walnut	×	walnut	⟶ all walnut
Cross 3 :	rose	×	pea	⟶ all walnut
Cross 4 :	$F_1 \times F_1$ of cross 3			
	walnut	×	walnut	⟶ 93 walnut
				28 rose
				32 pea
				10 single

Walnut

Pea

Rose

Single

Solution: At first glance, this problem appears quite difficult. However, applying a systematic approach and breaking the analysis into steps usually simplifies it. Our approach involves two steps. First, analyze the data carefully for any useful information. Then, once you identify something that is clearly helpful, follow an empirical approach—that is, formulate a hypothesis and, in a sense, test it against the given data. Look for a pattern of inheritance that is consistent with all cases.

This problem gives two immediately useful facts. First, in cross 1, P_1 singles breed true. Second, while P_1 walnuts breed true in cross 2, a walnut phenotype is also produced in cross 3 between rose and pea. When these F_1 walnuts are crossed in cross 4, all four comb shapes are produced in a ratio that approximates

9:3:3:1. This observation immediately suggests a cross involving two gene pairs, because the resulting data display the same ratio as in Mendel's dihybrid crosses. Since only one trait is involved (comb shape), epistasis may be occurring. This could serve as your working hypothesis, and you must now propose how the two gene pairs "interact" to produce each phenotype.

If you call the allele pairs *A*, *a* and *B*, *b*, you can predict that because walnut represents 9/16 in cross 4, *A–B–* will produce walnut. You might also hypothesize that in cross 2, the genotypes are *AABB* × *AABB*, where walnut bred true. (Recall that *A–* and *B–* mean *AA* or *Aa* and *BB* or *Bb*, respectively.)

The phenotype representing 1/16 of the offspring of cross 4 is single; therefore, you could predict that this phenotype is the result of the *aabb* genotype. This is consistent with cross 1.

Now you have only to determine the genotypes for rose and pea. The most logical prediction is that at least one dominant *A* or *B* allele combined with the double recessive condition of the other allele pair accounts for these phenotypes. For example,

$$A\text{–}bb \quad \longrightarrow \quad \text{rose}$$
$$aaB\text{–} \quad \longrightarrow \quad \text{pea}$$

If *AAbb* (rose) is crossed with *aaBB* (pea) in cross 3, all offspring will be *AaBb* (walnut). This is consistent with the data, and you must now look at only cross 4. We predict these walnut genotypes to be *AaBb* (as above), and from the cross *AaBb* (walnut) × *AaBb* (walnut) we expect

9/16	*A–B–*	(walnut)
3/16	*A–bb*	(rose)
3/16	*aaB–*	(pea)
1/16	*aabb*	(single)

Our prediction is consistent with the information given. The initial hypothesis of the epistatic interaction of two gene pairs proves consistent throughout, and the problem is solved.

This problem demonstrates the need for a basic theoretical knowledge of transmission genetics. Then, you can search for appropriate clues that will enable you to proceed in a stepwise fashion toward a solution. Mastering problem solving requires practice but can give you a great deal of satisfaction. Apply this general approach to the following problems.

2. In radishes, flower color may be red, purple, or white. The edible portion of the radish may be long or oval. When only flower color is studied, no dominance is evident, and red × white crosses yield all purple. If these F_1 purples are interbred, the F_2 generation consists of 1/4 red: 1/2 purple: 1/4 white. Regarding radish shape, long is dominant to oval in a normal Mendelian fashion.

(a) Determine the F_1 and F_2 phenotypes from a cross between a true-breeding red, long radish and one that is white and oval. Be sure to define all gene symbols initially.

Solution: This is a modified dihybrid cross in which the gene pair controlling color exhibits incomplete dominance. Shape is controlled conventionally. First, establish gene symbols:

$$RR = \text{red} \qquad O- = \text{long}$$
$$Rr = \text{purple} \qquad oo = \text{oval}$$
$$rr = \text{white}$$

P₁: *RROO* × *rroo*
(red long)   (white oval)

F₁: all *RrOo* (purple long)

F₁ × F₁: *RrOo* × *RrOo*

F₂:
$\begin{cases} 1/4\ RR \end{cases}$

1/4 RR	3/4O— 3/16 RRO—	red long
	1/4oo 1/16 RRoo	red oval
2/4 Rr	3/4O— 6/16 RrO—	purple long
	1/4oo 2/16 Rroo	purple oval
1/4 rr	3/4O— 3/16 rrO—	white long
	1/4oo 1/16 rroo	white oval

Note that to generate the F₂ results, we have used the forked-line method. First, we consider the outcome of crossing F₁ parents for the color genes (*Rr* × *Rr*). Then the outcome of shape is considered (*Oo* × *Oo*).

3. In humans, red–green color blindness is inherited as an X-linked recessive trait. A woman with normal vision whose father is color blind marries a male who has normal vision. Predict the color vision of their male and female offspring.

Solution: The female is heterozygous since she inherited an X chromosome with the mutant allele from her father. Her husband is normal. Therefore, the parental genotypes are

$$Cc \times C \uparrow \ (\ \uparrow \text{ represents the Y chromosome})$$

All female offspring are normal (*CC* or *Cc*). One-half of the male children will be color blind (*c* ↑), and the other half will have normal vision (*C* ↑).

4. Consider the two very limited unrelated pedigrees shown below. Of the four combinations of X-linked recessive, X-linked dominant, autosomal recessive, and autosomal dominant, which modes of inheritance can be absolutely ruled out in each case?

Solution: For both pedigrees, X-linked recessive and autosomal recessive remain possible, provided that the maternal parent is heterozygous in pedigree (b). At first glance autosomal dominance seems unlikely in pedigree (a), since at least half of the offspring should express a dominant trait expressed by one of their parents. However, while it is true that if the affected parent carries an autosomal dominant gene heterozygously, each offspring has a 50 percent chance of inheriting and expressing the mutant gene, the sample size of four offspring is too small to rule out this possibility. In pedigree (b), autosomal dominance is clearly possible. In both cases, one can rule out X-linked dominance because the female offspring would inherit and express the dominant allele, and they do not express the trait in either pedigree.

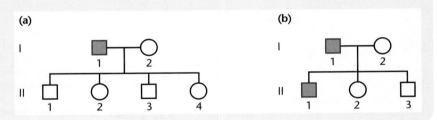

PROBLEMS AND DISCUSSION QUESTIONS

MG™ *For activities, animations, and review quizzes, go to the study area at www.masteringgenetics.com*

HOW DO WE KNOW?

1. In this chapter, we focused on many extensions and modifications of Mendelian principles and ratios. In the process, we encountered many opportunities to consider how this information was acquired. Answer the following fundamental questions:

   (a) How were early geneticists able to ascertain inheritance patterns that did not fit typical Mendelian ratios?

   (b) How did geneticists determine that inheritance of some phenotypic characteristics involves the interactions of two or more gene pairs? How were they able to determine how many gene pairs were involved?

   (c) How do we know that specific genes are located on the sex-determining chromosomes rather than on autosomes?

   (d) For genes whose expression seems to be tied to the sex of individuals, how do we know whether a gene is X-linked in contrast to exhibiting sex-limited or sex-influenced inheritance?

   (e) How was extranuclear inheritance discovered?

2. In Shorthorn cattle, coat color may be red, white, or roan. Roan is an intermediate phenotype expressed as a mixture of red and white hairs. The following data are obtained from various crosses:

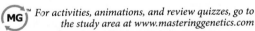

red	×	red	→	all red
white	×	white	→	all white
red	×	white	→	all roan
roan	×	roan	→	1/4 red: 1/2 roan: 1/4 white

How is coat color inherited? What are the genotypes of parents and offspring for each cross?

3. Contrast incomplete dominance and codominance.

4. With regard to the ABO blood types in humans, determine the genotypes of the male parent and female parent:

Male parent: blood type B whose mother was type O

Female parent: blood type A whose father was type B

Predict the blood types of the offspring that this couple may have and the expected ratio of each.

5. In foxes, two alleles of a single gene, *P* and *p*, may result in lethality (*PP*), platinum coat (*Pp*), or silver coat (*pp*). What ratio is obtained when platinum foxes are interbred? Is the *P* allele behaving dominantly or recessively in causing (a) lethality; (b) platinum coat color?

6. Three gene pairs located on separate autosomes determine flower color and shape as well as plant height. The first pair exhibits incomplete dominance, where color can be red, pink (the heterozygote), or white. The second pair leads to the dominant personate or recessive peloric flower shape, while the third gene pair produces either the dominant tall trait or the recessive dwarf trait. Homozygous plants that are red, personate, and tall are crossed with those that are white, peloric, and dwarf. Determine the F_1 genotype(s) and phenotype(s). If the F_1 plants are interbred, what proportion of the offspring will exhibit the same phenotype as the F_1 plants?

7. As in the plants of Problem 6, color may be red, white, or pink; and flower shape may be personate or peloric. Determine the P_1 and F_1 genotypes for the following crosses:

(a) red, peloric × white, personate
  └──────────→ F_1: all pink, personate

(b) red, personate × white, peloric
  └──────────→ F_1: all pink, personate

(c) pink, personate × red, peloric
  └──────────→ F_1:
  { 1/4 red, personate
    1/4 red, peloric
    1/4 pink, personate
    1/4 pink, peloric

(d) pink, personate × white, peloric
  └──────────→ F_1:
  { 1/4 white, personate
    1/4 white, peloric
    1/4 pink, personate
    1/4 pink, peloric

(e) What phenotypic ratios woud result from crossing the F_1 of (a) to the F_1 of (b)?

8. The following genotypes of two independently assorting autosomal genes determine coat color in rats:

A–B–(gray); *A–bb* (yellow); *aaB–*(black); *aabb* (cream)

A third gene pair on a separate autosome determines whether any color will be produced. The *CC* and *Cc* genotypes allow color according to the expression of the *A* and *B* alleles. However, the *cc* genotype results in albino rats regardless of the *A* and *B* alleles present. Determine the F_1 phenotypic ratio of the following crosses: (a) *AAbbCC* × *aaBBcc*; (b) *AaBBcc* × *AABbcc*; (c) *AaBbCc* × *AaBbcc*.

9. Given the inheritance pattern of coat color in rats described in Problem 8, predict the genotype and phenotype of the parents that produced the following F_1 offspring: (a) 9/16 gray: 3/16 yellow: 3/16 black: 1/16 cream; (b) 9/16 gray: 3/16 yellow: 4/16 albino; (c) 27/64 gray: 16/64 albino: 9/64 yellow: 9/64 black: 3/64 cream.

10. A husband and wife have normal vision, although both of their fathers are red–green color-blind, inherited as an X-linked recessive condition. What is the probability that their first child will

be (a) a normal son, (b) a normal daughter, (c) a color-blind son, (d) a color-blind daughter?

11. In humans, the ABO blood type is under the control of autosomal multiple alleles. Red–green color blindness is a recessive X-linked trait. If two parents who are both type A and have normal vision produce a son who is color-blind and type O, what is the probability that their next child will be a female who has normal vision and is type O?

12. In goats, development of the beard is due to a recessive gene. The following cross involving true-breeding goats was made and carried to the F_2 generation:

P_1: bearded female × beardless male
↓
F_1: all bearded males and beardless females

$F_1 \times F_1 \rightarrow$
{ 1/8 beardless males
  3/8 bearded males
  3/8 beardless females
  1/8 bearded females

Offer an explanation for the inheritance and expression of this trait, diagramming the cross. Propose one or more crosses to test your hypothesis.

13. In cats, orange coat color is determined by the *b* allele, and black coat color is determined by the *B* allele. The heterozygous condition results in a coat pattern known as tortoiseshell. These genes are X-linked. What kinds of offspring would be expected from a cross of a black male and a tortoiseshell female? What are the chances of getting a tortoiseshell male?

14. In *Drosophila*, an X-linked recessive mutation, *scalloped* (*sd*), causes irregular wing margins. Diagram the F_1 and F_2 results if (a) A scalloped female is crossed with a normal male. (b) A scalloped male is crossed with a normal female.

Compare these results to those that would be obtained if the *scalloped* gene were autosomal.

15. Another recessive mutation in *Drosophila*, *ebony* (*e*), is on an autosome (chromosome 3) and causes darkening of the body compared with wild-type flies. What phenotypic F_1 and F_2 male and female ratios will result if a scalloped-winged female with normal body color is crossed with a normal-winged ebony male? Work this problem by both the Punnett square method and the forked-line method.

16. While *vermilion* is X-linked in *Drosophila* and causes eye color to be bright red, *brown* is an autosomal recessive mutation that causes the eye to be brown. Flies carrying both mutations lose all pigmentation and are white-eyed. Predict the F_1 and F_2 results of the following crosses:

(a) vermilion females × brown males
(b) brown females × vermilion males
(c) white females × wild males

17. In pigs, coat color may be sandy, red, or white. A geneticist spent several years mating true-breeding pigs of all different color combinations, even obtaining true-breeding lines from different parts of the country. For crosses 1 and 4 in the following table, she encountered a major problem: her computer crashed and she lost the F_2 data. She nevertheless persevered and, using the limited data shown here, was able to predict the mode of inheritance and the number of genes involved, as well as to assign genotypes to each coat color. Attempt to duplicate her analysis, based on the available data generated from the crosses shown.

Cross	P₁	F₁	F₂
1	sandy × sandy	All red	Data lost
2	red × sandy	All red	3/4 red: 1/4 sandy
3	sandy × white	All sandy	3/4 sandy: 1/4 white
4	white × red	All red	Data lost

When you have formulated a hypothesis to explain the mode of inheritance and assigned genotypes to the respective coat colors, predict the outcomes of the F_2 generations where the data were lost.

18. A geneticist from an alien planet that prohibits genetic research brought with him two true-breeding lines of frogs. One frog line croaks by *uttering* "rib-it rib-it" and has purple eyes. The other frog line croaks by *muttering* "knee-deep knee-deep" and has green eyes. He mated the two frog lines, producing F_1 frogs that were all utterers with blue eyes. A large F_2 generation then yielded the following ratios:

27/64 blue, utterer
12/64 green, utterer
9/64 blue, mutterer
9/64 purple, utterer
4/64 green, mutterer
3/64 purple, mutterer

(a) How many total gene pairs are involved in the inheritance of both eye color and croaking?

(b) Of these, how many control eye color, and how many control croaking?

(c) Assign gene symbols for all phenotypes, and indicate the genotypes of the P_1, F_1, and F_2 frogs.

(d) After many years, the frog geneticist isolated true-breeding lines of all six F_2 phenotypes. Indicate the F_1 and F_2 phenotypic ratios of a cross between a blue, mutterer and a purple, utterer.

19. In another cross, the frog geneticist from Problem 18 mated two purple, utterers with the results shown here. What were the genotypes of the parents?

9/16 purple, utterers
3/16 purple, mutterers
3/16 green, utterers
1/16 green, mutterers

20. In cattle, coats may be solid white, solid black, or black-and-white spotted. When true-breeding solid whites are mated with true-breeding solid blacks, the F_1 generation consists of all solid white individuals. After many $F_1 \times F_1$ matings, the following ratio was observed in the F_2 generation:

12/16 solid white
3/16 black-and-white spotted
1/16 solid black

Explain the mode of inheritance governing coat color by determining how many gene pairs are involved and which genotypes yield which phenotypes. Is it possible to isolate a true-breeding strain of black-and-white spotted cattle? If so, what genotype would they have? If not, explain why not.

21. Consider the following three pedigrees, all involving the same human trait:

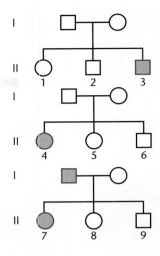

(a) Which sets of conditions, if any, can be excluded?

dominant and X-linked
dominant and autosomal
recessive and X-linked
recessive and autosomal

(b) For any set of conditions that you excluded, indicate the *single individual* in generation II (1–9) that was instrumental in your decision to exclude that condition. If none were excluded, answer "none apply."

(c) Given your conclusions in parts (a) and (b), indicate the *genotype* of individuals II-1, II-6, and II-9. If more than one possibility applies, list all possibilities. Use the symbols A and a for the genotypes.

22. Labrador retrievers may be black, brown, or golden in color (see the chapter opening photograph). Although each color may breed true, many different outcomes occur if numerous litters are examined from a variety of matings, where the parents are not necessarily true-breeding. The following results show some of the possibilities. Propose a mode of inheritance that is consistent with these data, and indicate the corresponding genotypes of the parents in each mating. Indicate as well the genotypes of dogs that breed true for each color.

(a) black	×	brown	⟶ all black
(b) black	×	brown	⟶ 1/2 black
			1/2 brown
(c) black	×	brown	⟶ 3/4 black
			1/4 golden
(d) black	×	golden	⟶ all black
(e) black	×	golden	⟶ 4/8 golden
			3/8 black
			1/8 brown
(f) black	×	golden	⟶ 2/4 golden
			1/4 black
			1/4 brown
(g) brown	×	brown	⟶ 3/4 brown
			1/4 golden
(h) black	×	black	⟶ 9/16 black
			4/16 golden
			3/16 brown

23. Three autosomal recessive mutations in *Drosophila*, all with tan eye color (*r1*, *r2*, and *r3*), are independently isolated and

subjected to complementation analysis. Of the results shown below, which, if any, are alleles of one another? Predict the results of the cross that is not shown—that is, $r2 \times r3$.

Cross 1: $r1 \times r2 \longrightarrow$ F$_1$: all wild-type eyes
Cross 2: $r1 \times r3 \longrightarrow$ F$_1$: all tan eyes

24. Horses can be cremello (a light cream color), chestnut (a reddish brown color), or palomino (a golden color with white in the horse's tail and mane).

Chestnut

Palomino

Cremello

Of these phenotypes, only palominos never breed true. The following results have been observed:

cremello × palomino $\longrightarrow$ 1/2 cremello
1/2 palomino

chestnut × palomino $\longrightarrow$ 1/2 chestnut
1/2 palomino

palomino × palomino $\longrightarrow$ 1/4 chestnut
1/2 palomino
1/4 cremello

(a) From these results, determine the mode of inheritance by assigning gene symbols and indicating which genotypes yield which phenotypes.
(b) Predict the F$_1$ and F$_2$ results of many initial matings between cremello and chestnut horses.

25. Pigment in the mouse is produced only when the C allele is present. Individuals of the cc genotype have no color. If color is present, it may be determined by the A and a alleles. AA or Aa results in agouti color, whereas aa results in black coats.
(a) What F$_1$ and F$_2$ genotypic and phenotypic ratios are obtained from a cross between $AACC$ and $aacc$ mice?
(b) In the three crosses shown here between agouti females whose genotypes were unknown and males of the $aacc$ genotype, what are the genotypes of the female parents for each of the following phenotypic ratios?

(1) 8 agouti	(2) 9 agouti	(3) 4 agouti
8 colorless	10 black	5 black
		10 colorless

26. Five human matings numbered 1–5 are shown in the following table. Included are both maternal and paternal phenotypes for ABO and MN blood-group antigen status.

	Parental Phenotypes				Offspring		
(1)	A,	M	×	A,	N	(a)	A,	N
(2)	B,	M	×	B,	M	(b)	O,	N
(3)	O,	N	×	B,	N	(c)	O,	MN
(4)	AB,	M	×	O,	N	(d)	B,	M
(5)	AB,	MN	×	AB,	MN	(e)	B,	MN

Each mating resulted in one of the five offspring shown to the right (a–e). Match each offspring with one correct set of parents, using each parental set only once. Is there more than one set of correct answers?

27. Two mothers give birth to sons at the same time at a busy urban hospital. The son of mother 1 is afflicted with hemophilia, a disease caused by an X-linked recessive allele. Neither parent has the disease. Mother 2 has a normal son, despite the fact that the father has hemophilia. Several years later, couple 1 sues the hospital, claiming that these two newborns were swapped in the nursery following their birth. As a genetic counselor, you are called to testify. What information can you provide the jury concerning the allegation?

28. In Dexter and Kerry cattle, animals may be polled (hornless) or horned. The Dexter animals have short legs, whereas the Kerry animals have long legs. When many offspring were obtained from matings between polled Kerrys and horned Dexters, half were found to be polled Dexters and half polled Kerrys. When these two types of F$_1$ cattle were mated to one another, the following F$_2$ data were obtained:

3/8 polled Dexters   3/8 polled Kerrys

1/8 horned Dexters   1/8 horned Kerrys

A geneticist was puzzled by these data and interviewed farmers who had bred these cattle for decades. She learned that Kerrys were true-breeding. Dexters, on the other hand, were not true-breeding and never produced as many offspring as Kerrys. Provide a genetic explanation for these observations.

29. What genetic criteria distinguish a case of extranuclear inheritance from (a) a case of Mendelian autosomal inheritance; (b) a case of X-linked inheritance?

30. The specification of the anterior–posterior axis in *Drosophila* embryos is initially controlled by various gene products that are synthesized and stored in the mature egg following oogenesis. Mutations in these genes result in abnormalities of the axis during embryogenesis, illustrating maternal effect. How do such mutations vary from those involved in organelle heredity that illustrate extranuclear inheritance? Devise a set of parallel crosses and expected outcomes involving mutant genes that contrast maternal effect and organelle heredity.

31. The maternal-effect mutation *bicoid* (*bcd*) is recessive. In the absence of the bicoid protein product, embryogenesis is not completed. Consider a cross between a female heterozygous for the bicoid mutation (bcd^+/bcd^-) and a homozygous male (bcd^-/bcd^-)
(a) How is it possible for a male homozygous for the mutation to exist?
(b) Predict the outcome (normal vs. failed embryogenesis) in the F$_1$ and F$_2$ generations of the cross described.

32. *Chlamydomonas*, a eukaryotic green alga, is sensitive to the antibiotic erythromycin, which inhibits protein synthesis in prokaryotes. There are two mating types in this alga, mt^+ and mt^-. If an mt^+ cell sensitive to the antibiotic is crossed with an mt^- cell that is resistant, all progeny cells are sensitive. The reciprocal cross (mt^+ resistant and mt^- sensitive) yields all resistant progeny cells. Assuming that the mutation for resistance is in

the chloroplast DNA, what can you conclude from the results of these crosses?

33. Below is a partial pedigree of hemophilia in the British Royal Family descended from Queen Victoria, who is believed to be the original "carrier" in this pedigree. Analyze the pedigree and indicate which females are also certain to be carriers. What is the probability that Princess Irene is a carrier?

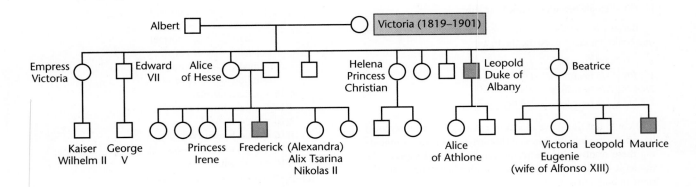

SOLUTIONS TO SELECTED PROBLEMS AND DISCUSSION QUESTIONS

Answers to Now Solve This

1. (a) $c^k c^a \times c^d c^a \Longrightarrow$ 2/4 sepia; 1/4 cream; 1/4 albino
   (b) ($c^k c^d \times c^k c^d$ or $c^k c^d \times c^k c^a$)
   The cream parent could be $c^d c^d$ or $c^d c^a$.
   $c^k c^a \times c^d c^d \Longrightarrow$ 1/2 sepia; 1/2 cream*
   *(If the cream parent is assumed to be homozygous)
   or $c^k c^a \times c^d c^a \Longrightarrow$ 1/2 sepia; 1/4 cream; 1/4 albino
   (c) Crosses possible:
   $c^k c^k \times c^d c^d \Longrightarrow$ all sepia
   $c^k c^k \times c^d c^a \Longrightarrow$ all sepia
   $c^k c^d \times c^d c^d \Longrightarrow$ 1/2 sepia; 1/2 cream
   $c^k c^d \times c^d c^a \Longrightarrow$ 1/2 sepia; 1/2 cream
   $c^k c^a \times c^d c^d \Longrightarrow$ 1/2 sepia; 1/2 cream
   $c^k c^a \times c^d c^a \Longrightarrow$ 1/2 sepia; 1/4 cream; 1/4 albino
   (d) Crosses possible:
   $c^k c^a \times c^d c^d \Longrightarrow$ 1/2 sepia; 1/2 cream
   $c^k c^a \times c^d c^a \Longrightarrow$ 1/2 sepia; 1/4 cream; 1/4 albino

2. A = pigment; a = pigmentless (colorless) B = purple; b = red
   $AaBb \times AaBb$
   $\Downarrow$
   $A\_B\_$ = purple
   A_bb = red
   $aaB\_$ = colorless
   $aabb$ = colorless

3. For all three pedigrees, let a represent the mutant gene and A represent its normal allele.
   (a) This pedigree is consistent with an X-linked recessive trait because the male would contribute an X chromosome carrying the a mutation to the aa daughter. The mother would have to be heterozygous Aa.
   (b) This pedigree is consistent with an X-linked recessive trait because the mother could be Aa and transmit her a allele to her one son (a/Y) and her A allele to her other son.

(c) This pedigree is not consistent with an X-linked mode of inheritance because the aa mother has an A/Y son.

Solutions to Problems and Discussion Questions

2. Looking at the data given, notice that a cross of the "extremes" (red × white) gives roan, suggesting its heterozygous nature and the homozygous nature of the parents. Seeing the 1:2:1 ratio in the offspring of

   roan × roan

   confirms the hypothesis of incomplete dominance as the mode of inheritance.
   Symbolism:
   AA = red aa = white Aa = roan
   Crosses:
   $AA \times AA \Longrightarrow AA$
   $aa \times aa \Longrightarrow aa$
   $AA \times aa \Longrightarrow Aa$
   $Aa \times Aa \Longrightarrow$ 1/4 AA; 2/4 Aa; 1/4 aa

4. Parents: $I^A i^O \times I^B i^O$

	I^B	i^O
I^A	$I^A I^B$ (AB)	$I^A i^O$ (A)
i^O	$I^B i^O$ (B)	$i^O i^O$ (O)

   The ratio would be 1(A):1(B):1(AB):1(O).

6. Flower color: RR = red; Rr = pink; rr = white
   Flower shape: P = personate; p = peloric
   Plant height: D = tall; d = dwarf
   (a) $RRPPDD \times rrppdd \Longrightarrow RrPpDd$ (pink, personate, tall)
   (b) 2/4 pink × 3/4 personate × 3/4 tall = 18/64

8. (a) $AaBbCc \Longrightarrow$ gray (C allows pigment)
   (b) $A\_B\_cc \Longrightarrow$ albino
   (c) 16/32 albino   9/32 gray   3/32 black
       3/32 yellow   1/32 cream

111

10. (a) 1/4 (b) 1/2 (c) 1/4 (d) zero

12. This situation is similar to sex-influenced pattern baldness in humans. Consider two alleles that are autosomal and let

BB = beardless in both sexes
Bb = beardless in females
Bb = bearded in males
bb = bearded in both sexes

P$_1$: female: bb (bearded) × male: BB (beardless)
F$_1$: Bb = female beardless; male bearded

Because half of the offspring are males and half are females, one could, for clarity, rewrite the F$_2$ as:

	1/2 females	**1/2 males**
1/4 BB	1/8 beardless	1/8 beardless
2/4 Bb	2/8 beardless	2/8 bearded
1/4 bb	1/8 bearded	1/8 bearded

One could test the above model by crossing F$_1$ (heterozygous) beardless females with bearded (homozygous) males. Comparing these results with the reciprocal cross would support the model if the distributions of sexes with phenotypes were the same in both crosses.

14. Symbolism: Normal wing margins = sd^+; scalloped = sd

(a) P$_1$: $X^{sd}X^{sd} \times X^+/Y$

F$_1$: 1/2 X^+X^{sd} (female, normal)
1/2 X^{sd}/Y (male, scalloped)

F$_2$: 1/4 X^+X^{sd} (female, normal)
1/4 $X^{sd}X^{sd}$ (female, scalloped)
1/4 X^+/Y (male, normal)
1/4 X^{sd}/Y (male, scalloped)

(b) P$_1$: $X^+/X^+ \times X^{sd}/Y$

F$_1$: 1/2 X^+X^{sd} (female, normal)
1/2 X^+/Y (male, normal)

F$_2$: 1/4 X^+X^+ (female, normal)
1/4 X^+X^{sd} (female, normal)
1/4 X^+/Y (male, normal)
1/4 X^{sd}/Y (male, scalloped)

If the *scalloped* gene were not X-linked, then all of the F$_1$ offspring would be wild-type (phenotypically) and a 3:1 ratio of normal to scalloped would occur in the F$_2$.

16. (a) P$_1$: X^vX^v; $+/+ \times X^+/Y$; b^r/b^r

F$_1$: 1/2 X^+X^v; $+/b^r$ (female, normal)
1/2 X^v/Y; $+/b^r$ (male, vermilion)

F$_2$: **Eye color (X)**   **Eye color (autosomal)**

1/4 female, ⟨ 3/4 normal 3/16
normal   1/4 brown 1/16
1/4 female, ⟨ 3/4 normal 3/16
vermilion   1/4 brown 1/16
1/4 male, ⟨ 3/4 normal 3/16
normal   1/4 brown 1/16
1/4 male, ⟨ 3/4 normal 3/16
vermilion   1/4 brown 1/16

3/16 = female, normal
1/16 = female, brown eyes
3/16 = female, vermilion eyes
1/16 = female, white eyes

3/16 = male, normal
1/16 = male, brown eyes
3/16 = male, vermilion eyes
1/16 = male, white eyes

(b) P$_1$: X^+X^+; $b^r/b^r \times X^v/Y$; $+/+$
F$_1$: 1/2 X^+X^v; $+/b^r$ (female, normal)
1/2 X^+/Y; $+/b^r$ (male, normal)

F$_2$: **Eye color (X)**   **Eye color (autosomal)**

2/4 female, ⟨ 3/4 normal
normal   1/4 brown
1/4 male, ⟨ 3/4 normal
normal   1/4 brown
1/4 male, ⟨ 3/4 normal
vermilion   1/4 brown

6/16 = female, normal
2/16 = female, brown eyes
3/16 = male, normal
1/16 = male, brown eyes
3/16 = male, vermilion eyes
1/16 = male, white eyes

(c) P$_1$: X^vX^v; $b^r/b^r \times X^+/Y$; $+/+$
F$_1$: 1/2 X^+X^v; $+/b^r$ (female, normal)
1/2 X^v/Y; $+/b^r$ (male, vermilion)

F$_2$: **Eye color (X)**   **Eye color (autosomal)**

1/4 female, ⟨ 3/4 normal
normal   1/4 brown
1/4 female, ⟨ 3/4 normal
vermilion   1/4 brown
1/4 male, ⟨ 3/4 normal
normal   1/4 brown
1/4 male, ⟨ 3/4 normal
vermilion   1/4 brown

3/16 = female, normal
1/16 = female, brown eyes
3/16 = female, vermilion eyes
1/16 = female, white eyes
3/16 = male, normal
1/16 = male, brown eyes
3/16 = male, vermilion eyes
1/16 = male, white eyes

18. (a) Because the denominator in the ratios is 64, one would begin to consider that three independently assorting gene pairs were operating in this problem. Because there are only two characteristics (eye color and croaking), however, one might hypothesize that two gene pairs are involved in the inheritance of one trait while one gene pair is involved in the other.

(b) Notice that there is a 48:16 (or 3:1) ratio of rib-it to knee-deep and a 36:16:12 (or 9:4:3) ratio of blue to green to purple eye color. Because of these relationships, one would conclude that croaking is due to one (dominant/recessive) gene pair while eye color is due to two gene pairs. Because there is a 9:4:3 ratio regarding eye color, some gene interaction (epistasis) is indicated.

(c) Symbolism: Croaking: $R\_$ = utterer; rr = mutterer. Eye color: Since the most frequent phenotype is blue eye, let $A\_B\_$ represent the genotypes. For the purple class, a 3/16 group uses the A_bb genotypes. The 4/16 class (green) would be the $aaB\_$ and the $aabb$ groups.

(d) The cross involving a blue-eyed, mutterer frog and a purple-eyed, utterer frog would have the genotypes

$AABBrr \times AAbbRR$

which would produce an F_1 of *AABbRr*, which would be blue-eyed and utterer. The F_2 will follow a pattern of a 9:3:3:1 ratio because of homozygosity for the *A* locus and heterozygosity for both the *B* and *R* loci.

9/16 *AAB_R_* = blue-eyed, utterer
3/16 *AAB_rr* = blue-eyed, mutterer
3/16 *AAbbR_* = purple-eyed, utterer
1/16 *AAbbrr* = purple-eyed, mutterer

20. A 12:3:1 ratio is obtained, which is a clear sign that epistasis has modified a typical 9:3:3:1 ratio. In this case, cattle in one of the 3/16 classes has the same phenotype as cattle in the 9/16 class. Since the 9/16 class typically takes the genotype of *A_B_*, it seems reasonable to think of the following genotypic classifications:

*A_B_* = solid white (9/16)
*aaB_* = solid white (3/16)
A_bb = black and white spotted (3/16)
aabb = solid black (1/16)

The selection of *bb* as giving the spotted phenotype is arbitrary. One could obtain *AAbb* true-breeding black and white spotted cattle.

22. Symbolism:

*A_B_* = black   *A_bb* = golden
aabb = golden   *aaB_* = brown

The combination of *bb* is epistatic to the *A* locus.

(a) *AAB_* × *aaBB* (other configurations are possible, but each must give all offspring with *A* and *B* dominant alleles)

(b) *AaB_* × *aaBB* (other configurations are possible, but both parents cannot be *Bb*)

(c) *AABb* × *aaBb*

(d) *AABB* × *aabb*

(e) *AaBb* × *Aabb*

(f) *AaBb* × *aabb*

(g) *aaBb* × *aaBb*

(h) *AaBb* × *AaBb*

Those genotypes that will breed true will be as follows:

black = *AABB*
golden = all genotypes that are *bb*
brown = *aaBB*

24. (a) $C^{ch}C^{ch}$ = chestnut C^cC^c = cremello $C^{ch}C^c$ = palomino

(b) The F_1 resulting from matings between cremello and chestnut horses would be expected to be all palomino. The F_2 would be expected to fall in a 1:2:1 ratio as in the third cross in part (a) above.

26. Cross #1 = (c)   Cross #4 = (e)
Cross #2 = (d)   Cross #5 = (a)
Cross #3 = (b)

Given that each parental/offspring grouping can only be used once, there are no other combinations.

28. The homozygous dominant type is lethal. Polled is caused by an independently assorting dominant allele, while horned is caused by the recessive allele to polled.

30. Maternal effect genes produce products that are not carried over for more than one generation, as is the case with organelle and infectious heredity. Crosses that illustrate the transient nature of a maternal effect could include the following

Female *Aa* × male *aa* ⟶ all offspring of the A phenotype.

Take a female A phenotype from the above cross and conduct the following mating: *aa* × male *Aa*. All offspring may be of the *a* phenotype because all of the offspring will reflect the *genotype* of the mother, not her *phenotype*. This cross illustrates that maternal effects last only one generation. However, depending on particular biochemical/developmental parameters, all crosses may not give these types of patterns.

32. The mt^+ strain is the donor of the cpDNA since the inheritance of resistance or sensitivity is dependent on the status of the mt^+ gene.

CREDITS

Credits are listed in order of appearance.

Photo

CO, Juniors Bildarchiv GmbH/Alamy; F-1, John Kaprielian/Photo Researchers, Inc.; F-3 (a, b) The Jackson Laboratory; F8, Rosemary Buffoni/iStockphoto.com; F-10, Science Source/Photo Researchers, Inc.; F-12, Prisma Bildagentur AG/Alamy; F-13, Hans Reinhard/Photoshot Holdings, Ltd.; F-14, Hans Reinhard/Photoshot Holdings, Ltd.; F-15 (a, b, c) Tanya Wolff; F-16 (a) Jane Burton/Photoshot Holdings, Ltd. (b) Dr. William S. Klug; F-17, Ryushi Abura; F18 (a, b) Dr. Ronald A. Butow, Department of Molecular Biology and Oncology, University of Texas Southwestern Medical Center; Dr. Ralph Somes, University of Wisconsin, Animal Sciences Department, Dr. Ralph Somes, University of Wisconsin, Animal Sciences Department

Text

Source: Sargan, D.R. IDID: inherited diseases in dogs: web-based information for canine inherited disease genetics. Mamm Genome. 2004 Jun;15(6):503-6. Source: Guardian UK

Sex Determination
and Sex Chromosomes

From Chapter 5 of *Essentials of Genetics, Eighth edition,* William S. Klug, Michael R. Cummings, Charlote A. Spencer, Michael A. Palladino. ©2013 by Pearson Education, Inc. All rights reserved.

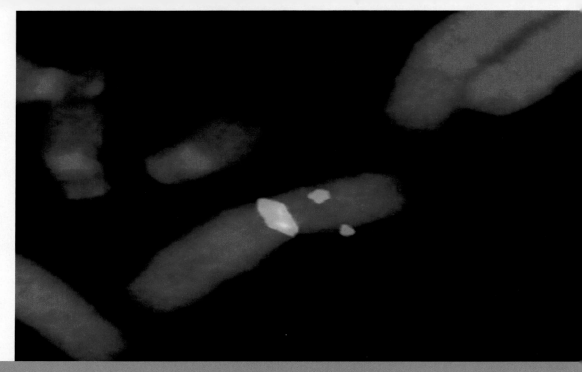

A human X chromosome highlighted using fluorescence *in situ* hybridization (FISH), a method in which specific probes bind to specific sequences of DNA. The probe producing green fluorescence binds to DNA at the centromere of X chromosomes. The probe producing red fluorescence binds to the DNA sequence of the X-linked Duchenne muscular dystrophy (DMD) gene.

Sex Determination and Sex Chromosomes

CHAPTER CONCEPTS

- Sexual reproduction, which greatly enhances genetic variation within species, requires mechanisms that result in sexual differentiation.

- A wide variety of genetic mechanisms lead to sexual dimorphism.

- Often, specific genes, usually on a single chromosome, cause maleness or femaleness during development.

- In humans, the presence of extra X or Y chromosomes beyond the diploid number may be tolerated but often leads to syndromes demonstrating distinctive phenotypes.

- While segregation of sex-determining chromosomes should theoretically lead to a one-to-one sex ratio of males to females, in humans the actual ratio greatly favors males at conception.

- In mammals, females inherit two X chromosomes compared to one in males, but the extra genetic information in females is compensated for by random inactivation of one of the X chromosomes early in development.

- In some reptilian species, temperature during incubation of eggs determines the sex of offspring.

In the biological world, a wide range of reproductive modes and life cycles are observed. Some organisms are entirely asexual, displaying no evidence of sexual reproduction. Other organisms alternate between short periods of sexual reproduction and prolonged periods of asexual reproduction. In most diploid eukaryotes, however, sexual reproduction is the only natural mechanism for producing new members of the species. Orderly transmission of genetic material from parents to offspring, and the resultant phenotypic variability, rely on the processes of segregation and independent assortment that occur during meiosis. Meiosis produces haploid gametes so that, following fertilization, the resulting offspring maintain the diploid number of chromosomes characteristic of their kind. Thus, meiosis ensures genetic constancy within members of the same species.

These events, seen in the perpetuation of all sexually reproducing organisms, depend ultimately on an efficient union of gametes during fertilization. In turn, successful fertilization depends on some form of sexual differentiation in the reproductive organisms. Even though it is not overtly evident, this differentiation occurs in organisms as low on the evolutionary scale as bacteria and single-celled eukaryotic algae. In more complex forms of life, the differentiation of the sexes is more evident as phenotypic dimorphism of males and females. The ancient symbol for iron and for Mars, depicting a shield and spear (♂), and the ancient symbol for copper and for Venus, depicting a mirror (♀), have also come to symbolize maleness and femaleness, respectively.

Dissimilar, or **heteromorphic, chromosomes,** such as the XY pair in mammals, characterize one sex or the other in a wide range of species, resulting in their label as **sex chromosomes.** Nevertheless, it is genes, rather than chromosomes, that ultimately serve as the underlying basis of **sex determination.** As we will see, some of these genes are present on sex chromosomes, but others are autosomal. Extensive investigation has revealed a wide variation in sex-chromosome systems—even in closely related organisms—suggesting that mechanisms controlling sex determination have undergone rapid evolution many times in the history of life.

In this chapter, we review some representative modes of sexual differentiation by examining the life cycles of two model organisms often studied in genetics: the maize plant (corn), *Zea mays*; and the nematode (roundworm), *Caenorhabditis elegans*. These organisms contrast the different roles that sexual differentiation plays in the lives of diverse organisms. Then, we delve more deeply into what is known about the genetic basis for the determination of sexual differences, with a particular emphasis on two organisms: our own species, representative of mammals; and *Drosophila*, on which pioneering sex-determining studies were performed.

1 Life Cycles Depend on Sexual Differentiation

In describing sexual dimorphism (differences between males and females) in multicellular animals, biologists distinguish between **primary sexual differentiation,** which involves only the gonads, where gametes are produced, and **secondary sexual differentiation,** which involves the overall appearance of the organism, including clear differences in such organs as mammary glands and external genitalia. In plants and animals, the terms **unisexual, dioecious,** and **gonochoric** are equivalent; they all refer to an individual containing only male *or* only female reproductive organs. Conversely, the terms **bisexual, monoecious,** and **hermaphroditic** refer to individuals containing both male *and* female reproductive organs, a common occurrence in both the plant and animal kingdoms. These organisms can produce both eggs and sperm. The term **intersex** is usually reserved for individuals of an intermediate sexual condition; these are most often sterile.

Next, we discuss the role of sexual differentiation during the life cycle of a representative plant and invertebrate animal.

Zea mays

The life cycles of many plants alternate between the haploid gametophyte stage and the diploid sporophyte stage (see Figure 12). The processes of meiosis and fertilization link the two phases during the life cycle. The relative amount of time spent in the two phases varies between the major plant groups. In some nonseed plants, such as mosses, the haploid gametophyte phase and the morphological structures representing this stage predominate. The reverse is true in seed plants.

Maize (*Zea mays*), familiar to you as corn, exemplifies a monoecious seed plant in which the sporophyte phase and the morphological structures representing that phase predominate during the life cycle. Both male and female structures are present on the adult plant. Thus, sex determination occurs differently in different tissues of the same organism, as shown in the life cycle of this plant (Figure 1). The stamens, which collectively constitute the tassel, produce diploid microspore mother cells, each of which undergoes meiosis and gives rise to four haploid microspores. Each haploid microspore in turn develops into a mature male microgametophyte—the pollen grain—which contains two haploid sperm nuclei.

Equivalent female diploid cells, known as megaspore mother cells, are produced in the pistil of the sporophyte. Following meiosis, only one of the four haploid megaspores survives. It usually divides mitotically three times, producing a total of eight haploid nuclei enclosed in the embryo sac. Two of these nuclei unite near the center of the embryo sac, becoming the endosperm nuclei. At the micropyle end of the sac, where the sperm enters, sit three other nuclei: the oocyte nucleus and two synergids. The remaining three, antipodal nuclei, cluster at the opposite end of the embryo sac.

Pollination occurs when pollen grains make contact with the silks (or stigma) of the pistil and develop long pollen tubes that grow toward the embryo sac. When contact is made at the micropyle, the two sperm nuclei enter the embryo sac. One sperm nucleus unites with the haploid oocyte nucleus, and the other sperm nucleus unites with two endosperm nuclei. This process, known as *double fertilization*, results in the diploid zygote nucleus and the triploid endosperm nucleus, respectively. Each ear of corn can contain as many as 1000 of these structures, each of which develops into a single kernel. Each kernel, if allowed to germinate, gives rise to a new plant, the *sporophyte*.

The mechanism of sex determination and differentiation in a monoecious plant such as *Zea mays*, where the tissues that form the male and female gametes have the same genetic constitution, was difficult to unravel at first. However, the discovery of a large number of mutant genes that disrupt normal tassel and pistil formation supports the concept that normal products of these genes play an important role in sex determination by affecting the differentiation of male or female tissue in several ways.

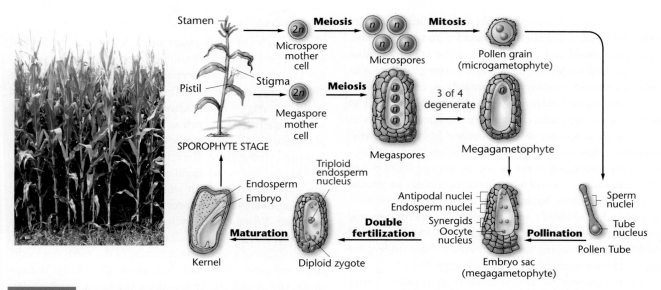

FIGURE 1 The life cycle of maize (*Zea mays*). The diploid sporophyte bears stamens and pistils that give rise to haploid microspores and megaspores, which develop into the pollen grain and the embryo sac that ultimately house the sperm and oocyte, respectively. Following fertilization, the embryo develops within the kernel and is nourished by the endosperm. Germination of the kernel gives rise to a new sporophyte (the mature corn plant), and the cycle repeats itself.

For example, mutant genes that cause sex reversal provide valuable information. When homozygous, all mutations classified as *tassel seed (ts)* interfere with tassel production and induce the formation of female structures instead. Thus, a single gene can cause a normally monoecious plant to become exclusively female. On the other hand, the recessive mutations *silkless (sk)* and *barren stalk (ba)* interfere with the development of the pistil, resulting in plants with only male-functioning reproductive organs.

Data gathered from studies of these and other mutants suggest that the products of many wild-type alleles of these genes interact in controlling sex determination. During development, certain cells are "directed" to become male or female structures. Following sexual differentiation into either male or female structures, male or female gametes are produced.

Caenorhabditis elegans

The nematode worm *Caenorhabditis elegans* [*C. elegans*; Figure 2(a)] has become a popular organism in genetic studies, particularly for investigating the genetic control of development. Its usefulness is based on the fact that adults consist of approximately 1000 cells, the precise lineage of which can be traced back to specific embryonic origins. There are two sexual phenotypes in these worms: males, which have only testes, and hermaphrodites, which contain both testes and ovaries. During larval development of hermaphrodites, testes form that produce sperm, which is stored. Ovaries are also produced, but oogenesis does not occur until the adult

(a)

(b)

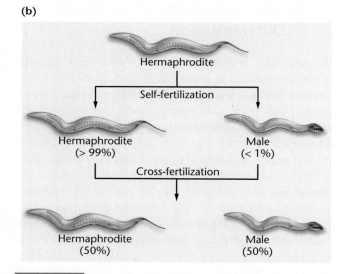

FIGURE 2 (a) Photomicrograph of a hermaphroditic nematode, *C. elegans*; (b) the outcomes of self-fertilization in a hermaphrodite, and a mating of a hermaphrodite and a male worm.

stage is reached several days later. The eggs that are produced are fertilized by the stored sperm in a process of self-fertilization.

The outcome of this process is quite interesting [**Figure 2(b)**]. The vast majority of organisms that result are hermaphrodites, like the parental worm; less than 1 percent of the offspring are males. As adults, males can mate with hermaphrodites, producing about half male and half hermaphrodite offspring.

The genetic signal that determines maleness in contrast to hermaphroditic development is provided by genes located on both the X chromosome and autosomes. *C. elegans* lacks a Y chromosome altogether—hermaphrodites have two X chromosomes, while males have only one X chromosome. It is believed that the ratio of X chromosomes to the number of sets of autosomes ultimately determines the sex of these worms. A ratio of 1.0 (two X chromosomes and two copies of each autosome) results in hermaphrodites, and a ratio of 0.5 results in males. The absence of a heteromorphic Y chromosome is not uncommon in organisms.

NOW SOLVE THIS

1 The marine echiurid worm *Bonellia viridis* is an extreme example of environmental influence on sex determination. Undifferentiated larvae either remain free-swimming and differentiate into females, or they settle on the proboscis of an adult female and become males. If larvae that have been on a female proboscis for a short period are removed and placed in seawater, they develop as intersexes. If larvae are forced to develop in an aquarium where pieces of proboscises have been placed, they develop into males. Contrast this mode of sexual differentiation with that of mammals. Suggest further experimentation to elucidate the mechanism of sex determination in *B. viridis*.

▪ **Hint** *This problem asks you to analyze experimental findings related to sex determination. The key to its solution is to devise further testing with the goal of isolating the unknown factor affecting sex determination and testing it experimentally.*

ESSENTIAL POINT ▪ ▪ ▪

Sexual reproduction depends on the differentiation of male and female structures responsible for the production of male and female gametes, which in turn is controlled by specific genes, most often housed on specific sex chromosomes.

---

2 X and Y Chromosomes Were First Linked to Sex Determination Early in the Twentieth Century

How sex is determined has long intrigued geneticists. In 1891, H. Henking identified a nuclear structure in the sperm of certain insects, which he labeled the X-body. Several years later, Clarence McClung showed that some of the sperm in grasshoppers contain an unusual genetic structure, which he called a *heterochromosome*, but the remainder of the sperm

lack such a structure. He mistakenly associated the presence of the heterochromosome with the production of male progeny. In 1906, Edmund B. Wilson clarified Henking and McClung's findings when he demonstrated that female somatic cells in the butterfly *Protenor* contain 14 chromosomes, including two X chromosomes. During oogenesis, an even reduction occurs, producing gametes with seven chromosomes, including one X chromosome. Male somatic cells, on the other hand, contain only 13 chromosomes, including one X chromosome. During spermatogenesis, gametes are produced containing either six chromosomes, without an X, or seven chromosomes, one of which is an X. Fertilization by X-bearing sperm results in female offspring, and fertilization by X-deficient sperm results in male offspring [**Figure 3(a)**].

The presence or absence of the X chromosome in male gametes provides an efficient mechanism for sex determination in this species and also produces a 1:1 sex ratio in the resulting offspring. This mechanism, now called the **XX/XO,** or *Protenor,* **mode of sex determination,** depends on the random distribution of the X chromosome into one-half of the male gametes during segregation. As we saw earlier, *C. elegans* exhibits this system of sex determination.

Wilson also experimented with the milkweed bug *Lygaeus turcicus,* in which both sexes have 14 chromosomes. Twelve of these are autosomes (A). In addition, the females have two X chromosomes, while the males have only a single X and a smaller heterochromosome labeled the **Y chromosome.**

(a) *Protenor* **mode**

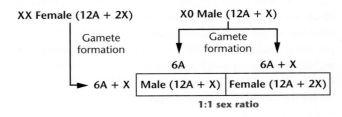

(b) *Lygaeus* **mode**

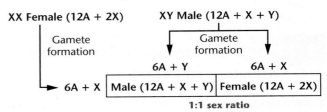

FIGURE 3 (a) The *Protenor* mode of sex determination where the heterogametic sex (the male in this example) is X0 and produces gametes with or without the X chromosome; (b) the *Lygaeus* mode of sex determination, where the heterogametic sex (again, the male in this example) is XY and produces gametes with either an X or a Y chromosome.

Females in this species produce only gametes of the (6A + X) constitution, but males produce two types of gametes in equal proportions, (6A + X) and (6A + Y). Therefore, following random fertilization, equal numbers of male and female progeny will be produced with distinct chromosome complements. This mode of sex determination is called the *Lygaeus,* or **XX/XY, mode of sex determination** [Figure 3(b)].

In *Protenor* and *Lygaeus* insects, males produce unlike gametes. As a result, they are described as the **heterogametic sex,** and in effect, their gametes ultimately determine the sex of the progeny in those species. In such cases, the female, who has like sex chromosomes, is the **homogametic sex,** producing uniform gametes with regard to chromosome numbers and types.

The male is not always the heterogametic sex. In some organisms, the female produces unlike gametes, exhibiting either the *Protenor* XX/XO or *Lygaeus* XX/XY mode of sex determination. Examples include certain moths and butterflies; most birds; some fish; reptiles; amphibians; and at least one species of plants (*Fragaria orientalis*). To immediately distinguish situations in which the female is the heterogametic sex, some geneticists use the notation **ZZ/ZW,** where ZW is the heterogamous female, instead of the XX/XY notation.

Geneticists' experience with fowl (chickens) illustrates the difficulty of establishing which sex is heterogametic and whether the *Protenor* or *Lygaeus* mode is operable. Although genetic evidence supported the hypothesis that the female is the heterogametic sex, the cytological identification of the sex chromosome was not accomplished until 1961, because of the large number of chromosomes (78) characteristic of chickens. When the sex chromosomes were finally identified, the female was shown to contain an unlike chromosome pair including a heteromorphic chromosome (the W chromosome). Thus, in fowl, the female is indeed heterogametic and is characterized by the *Lygaeus* type of sex determination.

ESSENTIAL POINT ■ ■ ■

Specific sex chromosomes contain genetic information that controls sex determination and sexual differentiation.

---

3 The Y Chromosome Determines Maleness in Humans

The first attempt to understand sex determination in our own species occurred almost 100 years ago and involved the visual examination of chromosomes in dividing cells. Efforts were made to accurately determine the diploid chromosome number of humans, but because of the relatively large number of chromosomes, this proved to be quite difficult. Then, in 1956, Joe Hin Tjio and Albert Levan discovered an effective way to prepare chromosomes for accurate viewing. This technique led to a strikingly clear demonstration of metaphase stages showing that 46 was indeed the human diploid number. Later that same year, C. E. Ford and John L. Hamerton, also working with testicular tissue, confirmed this finding. The familiar karyotypes of a human male (Figure 4) illustrate the difference in size between the human X and Y chromosomes.

Of the normal 23 pairs of human chromosomes, one pair was shown to vary in configuration in males and females. These two chromosomes were designated the X and Y sex chromosomes. The human female has two X chromosomes, and the human male has one X and one Y chromosome.

We might believe that this observation is sufficient to conclude that the Y chromosome determines maleness. However, several other interpretations are possible. The Y could play no role in sex determination; the presence of two X chromosomes could cause femaleness; or maleness could result from the lack of a second X chromosome. The evidence that clarified which explanation was correct came from study of the effects of human sex-chromosome variations, described in the following section. As such investigations revealed, the Y chromosome does indeed determine maleness in humans.

Klinefelter and Turner Syndromes

Around 1940, scientists identified two human abnormalities characterized by aberrant sexual development, **Klinefelter syndrome (47,XXY)** and **Turner syndrome (45,X).**[*] Individuals with Klinefelter syndrome are generally tall and have long arms and legs and large hands and feet. They usually have genitalia and internal ducts that are male, but their testes are rudimentary and fail to produce sperm. At the same time, feminine sexual development is not entirely suppressed. Slight enlargement of the breasts (gynecomastia) is common, and the hips are often rounded. This ambiguous sexual development, referred to as intersexuality, can lead to abnormal social development. Intelligence is often below the normal range as well.

In Turner syndrome, the affected individual has female external genitalia and internal ducts, but the ovaries are rudimentary. Other characteristic abnormalities include short stature (usually under 5 feet), skin flaps on the back of the neck, and underdeveloped breasts. A broad, shieldlike chest is sometimes noted. Intelligence is usually normal.

In 1959, the karyotypes of individuals with these syndromes were determined to be abnormal with respect to the sex chromosomes. Individuals with Klinefelter syndrome have more than one X chromosome. Most often they have an XXY complement in addition to 44 autosomes [Figure 4(a)], which is why people with this karyotype are designated 47,XXY. Individuals with Turner syndrome most of-

---

[*]Although the possessive form of the names of eponymous syndromes is sometimes used (e.g., Klinefelter's syndrome), the current preference is to use the nonpossessive form.

(a)   (b)

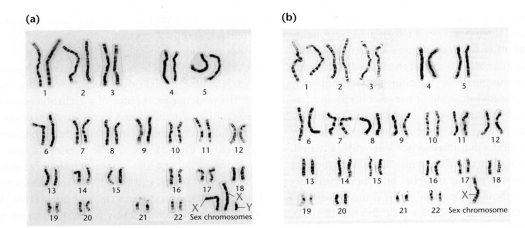

FIGURE 4   The karyotypes of individuals with (a) Klinefelter syndrome (47,XXY) and (b) Turner syndrome (45,X).

ten have only 45 chromosomes, including just a single X chromosome; thus, they are designated 45,X **[Figure 4(b)]**. Note the convention used in designating these chromosome compositions. The number states the total number of chromosomes present, and the information after the comma indicates the deviation from the normal diploid content. Both conditions result from **nondisjunction,** the failure of the sex chromosomes to segregate properly during meiosis.

These Klinefelter and Turner karyotypes and their corresponding sexual phenotypes led scientists to conclude that the Y chromosome determines maleness in humans. In its absence, the person's sex is female, even if only a single X chromosome is present. The presence of the Y chromosome in the individual with Klinefelter syndrome is sufficient to determine maleness, even though male development is not complete. Similarly, in the absence of a Y chromosome, as in the case of individuals with Turner syndrome, no masculinization occurs. Note that we cannot conclude anything regarding sex determination under circumstances where a Y chromosome is present without an X because Y-containing human embryos lacking an X chromosome (designated 45,Y) do not survive.

Klinefelter syndrome occurs in about 1 of every 660 male births. The karyotypes **48,XXXY, 48,XXYY, 49,XXXXY,** and **49,XXXYY** are similar phenotypically to 47,XXY, but manifestations are often more severe in individuals with a greater number of X chromosomes.

Turner syndrome can also result from karyotypes other than 45,X, including individuals called **mosaics,** whose somatic cells display two different genetic cell lines, each exhibiting a different karyotype. Such cell lines result from a mitotic error during early development, the most common chromosome combinations being **45,X/46,XY** and **45,X/46,XX.** Thus, an embryo that began life with a normal karyotype can give rise to an individual whose cells show a mixture of karyotypes and who exhibits varying aspects of this syndrome.

Turner syndrome is observed in about 1 in 2000 female births, a frequency much lower than that for Klinefelter syndrome. One explanation for this difference is the observation that a substantial majority of 45,X fetuses die *in utero* and are aborted spontaneously. Thus, a similar frequency of the two syndromes may occur at conception.

47,XXX Syndrome

The abnormal presence of three X chromosomes along with a normal set of autosomes (**47,XXX**) results in female differentiation. The highly variable syndrome that accompanies this genotype, often called **triplo-X,** occurs in about 1 of 1000 female births. Frequently, 47,XXX women are perfectly normal and may remain unaware of their abnormality in chromosome number unless a karyotype is done. In other cases, underdeveloped secondary sex characteristics, sterility, delayed development of language and motor skills, and mental retardation may occur. In rare instances, **48,XXXX** (tetra-X) and **49,XXXXX** (penta-X) karyotypes have been reported. The syndromes associated with these karyotypes are similar to but more pronounced than the 47,XXX syndrome. Thus, in many cases, the presence of additional X chromosomes appears to disrupt the delicate balance of genetic information essential to normal female development.

47,XYY Condition

Another human condition involving the sex chromosomes is **47,XYY.** Studies of this condition, in which the only deviation from diploidy is the presence of an additional Y chromosome in an otherwise normal male karyotype, were initiated in 1965 by Patricia Jacobs. She discovered that 9 of 315 males in a Scottish maximum security prison had the 47,XYY karyotype. These males were significantly above average in height and had been incarcerated as a result of dangerous, violent, or criminal propensities. Of the nine males studied, seven were of subnormal intelligence, and all suffered personality disorders. Several other studies produced similar findings.

The possible correlation between this chromosome composition and criminal behavior piqued considerable interest, and extensive investigation of the phenotype and frequency of the 47,XYY condition in both criminal and noncriminal populations ensued. Above-average height (usually over 6 feet) and subnormal intelligence were substantiated, and the frequency of males displaying this karyotype was indeed revealed to be higher in penal and mental institutions compared with unincarcerated populations (one study showed 29 XYY males when 28,366 were examined (0.10%). A particularly relevant question involves the characteristics displayed by the XYY males who are not incarcerated. The only nearly constant association is that such individuals are over 6 feet tall.

A study to further address this issue was initiated in 1974 to identify 47,XYY individuals at birth and to follow their behavioral patterns during preadult and adult development. While the study was considered unethical and soon abandoned, it has became clear that there are many XYY males present in the population who do not exhibit antisocial behavior and who lead normal lives. Therefore, we must conclude that there is a high, but not constant, correlation between the extra Y chromosome and the predisposition of these males to exhibit behavioral problems.

Sexual Differentiation in Humans

Once researchers had established that, in humans, it is the Y chromosome that houses genetic information necessary for maleness, they attempted to pinpoint a specific gene or genes capable of providing the "signal" responsible for sex determination. Before we delve into this topic, it is useful to consider how sexual differentiation occurs in order to better comprehend how humans develop into sexually dimorphic males and females. During early development, every human embryo undergoes a period when it is potentially hermaphroditic. By the fifth week of gestation, gonadal primordia (the tissues that will form the gonad) arise as a pair of **gonadal (genital) ridges** associated with each embryonic kidney. The embryo is potentially hermaphroditic because at this stage its gonadal phenotype is sexually indifferent—male or female reproductive structures cannot be distinguished, and the gonadal ridge tissue can develop to form male or female gonads. As development progresses, primordial germ cells migrate to these ridges, where an outer cortex and inner medulla form (*cortex* and *medulla* are the outer and inner tissues of an organ, respectively). The cortex is capable of developing into an ovary, while the medulla may develop into a testis. In addition, two sets of undifferentiated ducts called the Wolffian and Müllerian ducts exist in each embryo. Wolffian ducts differentiate into other organs of the male reproductive

tract, while Müllerian ducts differentiate into structures of the female reproductive tract.

Because gonadal ridges can form either ovaries or testes, they are commonly referred to as **bipotential gonads.** What switch triggers gonadal ridge development into testes or ovaries? The presence or absence of a Y chromosome is the key. If cells of the ridge have an XY constitution, development of the medulla into a testis is initiated around the seventh week. However, in the absence of the Y chromosome, no male development occurs, the cortex of the ridge subsequently forms ovarian tissue, and the Müllerian duct forms oviducts (Fallopian tubes), uterus, cervix, and portions of the vagina. Depending on which pathway is initiated, parallel development of the appropriate male or female duct system then occurs, and the other duct system degenerates. If testes differentiation is initiated, the embryonic testicular tissue secretes hormones that are essential for continued male sexual differentiation. As we will discuss in the next section, the presence of a Y chromosome and the development of the testes also inhibit formation of female reproductive organs.

In females, as the twelfth week of fetal development approaches, the oogonia within the ovaries begin meiosis, and primary oocytes can be detected. By the twenty-fifth week of gestation, all oocytes become arrested in meiosis and remain dormant until puberty is reached some 10 to 15 years later. In males, on the other hand, primary spermatocytes are not produced until puberty is reached (see Figure 11).

The Y Chromosome and Male Development

The human Y chromosome, unlike the X, was long thought to be mostly blank genetically. It is now known that this is not true, even though the Y chromosome contains far fewer genes than does the X. Data from the Human Genome Project indicate that the Y chromosome has at least 75 genes, compared to 900–1400 genes on the X. Current analysis of these genes and regions with potential genetic function reveals that some have homologous counterparts on the X chromosome and others do not. For example, present on both ends of the Y chromosome are so-called **pseudoautosomal regions (PARs)** that share homology with regions on the X chromosome and synapse and recombine with it during meiosis. The presence of such a pairing region is critical to segregation of the X and Y chromosomes during male gametogenesis. The remainder of the chromosome, about 95 percent of it, does not synapse or recombine with the X chromosome. As a result, it was originally referred to as the *nonrecombining region of the Y (NRY)*. More recently, researchers have designated this region as the **male-specific region of the Y (MSY).** Some portions of the MSY share homology with genes on the X chromosome, and others do not.

The human Y chromosome is diagrammed in **Figure 5.** The MSY is divided about equally between *euchromatic* re-

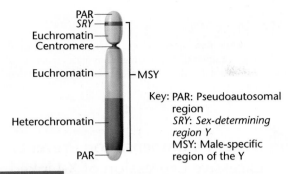

FIGURE 5 The regions of the human Y chromosome.

gions, containing functional genes, and *heterochromatic* regions, lacking genes. Within euchromatin, adjacent to the PAR of the short arm of the Y chromosome, is a critical gene that controls male sexual development, called the ***sex-determining region Y (SRY)***. In humans, the absence of a Y chromosome almost always leads to female development; thus, this gene is absent from the X chromosome. At six to eight weeks of development, the *SRY* gene becomes active in XY embryos. *SRY* encodes a protein that causes the undifferentiated gonadal tissue of the embryo to form testes. This protein is called the **testis-determining factor (TDF)**. *SRY* (or a closely related version) is present in all mammals thus far examined, indicative of its essential function throughout this diverse group of animals.[*]

Our ability to identify the presence or absence of DNA sequences in rare individuals whose sex-chromosome composition does not correspond to their sexual phenotype has provided evidence that *SRY* is the gene responsible for male sex determination. For example, there are human males who have two X and no Y chromosomes. Often, attached to one of their X chromosomes is the region of the Y that contains *SRY*. There are also females who have one X and one Y chromosome. Their Y is almost always missing the *SRY* gene. These observations argue strongly in favor of the role of *SRY* in providing the primary signal for male development.

Further support of this conclusion involves an experiment using **transgenic mice**. These animals are produced from fertilized eggs injected with foreign DNA that is subsequently incorporated into the genetic composition of the developing embryo. In normal mice, a chromosome region designated *Sry* has been identified that is comparable to *SRY* in humans. When mouse DNA containing *Sry* is injected into normal XX mouse eggs, most of the offspring develop into males.

The question of how the product of this gene triggers development of embryonic gonadal tissue into testes rather than ovaries is the key question under investigation. TDF is now believed to function as a *transcription factor*, a DNA-binding protein that interacts directly with the regulatory

sequences of other genes to stimulate their expression. Thus, TDF behaves as a master switch that controls other genes downstream in the process of sexual differentiation. Interestingly, many identified thus far reside on autosomes, including the human *SOX9* gene located on chromosome 17 and the subject of the next Now Solve This entry.

NOW SOLVE THIS

2 Campomelic dysplasia (CMD1) is a congenital human syndrome featuring malformation of bone and cartilage. It is caused by an autosomal dominant mutation of a gene located on chromosome 17. Consider the following observations in sequence, and in each case, draw whatever appropriate conclusions are warranted.

(a) Of those with the syndrome who are karyotypically 46,XY, approximately 75 percent are sex reversed, exhibiting a wide range of female characteristics.

(b) The nonmutant form of the gene, called *SOX9*, is expressed in the developing gonad of the XY male, but not the XX female.

(c) The *SOX9* gene shares 71 percent amino acid coding sequence homology with the Y-linked *SRY* gene.

(d) CMD1 patients who exhibit a 46,XX karyotype develop as females, with no gonadal abnormalities.

■ **Hint** *This problem asks you to apply the information presented in this chapter to a real-life example. The key to its solution is knowing that some genes are activated and produce their normal product as a result of expression of products of other genes found on different chromosomes.*

ESSENTIAL POINT ■ ■ ■

The presence or absence of a Y chromosome that contains an intact *SRY* gene is responsible for causing maleness in humans.

4 The Ratio of Males to Females in Humans Is Not 1.0

The presence of heteromorphic sex chromosomes in one sex of a species but not the other provides a potential mechanism for producing equal proportions of male and female offspring. This potential depends on the segregation of the X and Y (or Z and W) chromosomes during meiosis, such that half of the gametes of the heterogametic sex receive one of the chromosomes and half receive the other one. As we learned in the previous section, small pseudoautosomal regions of pairing homology do exist at both ends of the human X and Y chromosomes, suggesting that the X and Y chromosomes do synapse and then segregate into different gametes. Provided that both types of gametes are equally successful in fertilization and that the two sexes are equally

---

[*] It is interesting to note that in chickens, a similar gene has recently been identified. Called *DMRTI*, it is located on the Z chromosome. This gene is the subject of Problem 29 in the Problems section at the end of the chapter.

viable during development, a 1:1 ratio of male and female offspring should result.

The actual proportion of male to female offspring, referred to as the **sex ratio**, has been assessed in two ways. The **primary sex ratio** reflects the proportion of males to females conceived in a population. The **secondary sex ratio** reflects the proportion of each sex that is born. The secondary sex ratio is much easier to determine but has the disadvantage of not accounting for any disproportionate embryonic or fetal mortality.

When the secondary sex ratio in the human population was determined in 1969 by using worldwide census data, it did not equal 1.0. For example, in the Caucasian population in the United States, the secondary ratio was a little less than 1.06, indicating that about 106 males were born for each 100 females. (In 1995, this ratio dropped to slightly less than 1.05.) In the African-American population in the United States, the ratio was 1.025. In other countries the excess of male births is even greater than is reflected in these values. For example, in Korea, the secondary sex ratio was 1.15.

Despite these ratios, it is possible that the primary sex ratio is 1.0 and is later altered between conception and birth. For the secondary ratio to exceed 1.0, then, prenatal female mortality would have to be greater than prenatal male mortality. However, this hypothesis has been examined and shown to be false. In fact, just the opposite occurs. In a Carnegie Institute study, reported in 1948, the sex of approximately 6000 embryos and fetuses recovered from miscarriages and abortions was determined, and fetal mortality was actually higher in males. On the basis of the data derived from that study, the primary sex ratio in U.S. Caucasians was estimated to be 1.079. It is now believed that the figure is much higher—between 1.20 and 1.60—suggesting that many more males than females are conceived in the human population.

It is not clear why such a radical departure from the expected primary sex ratio of 1.0 occurs. To come up with a suitable explanation, researchers must examine the assumptions on which the theoretical ratio is based:

1. Because of segregation, males produce equal numbers of X- and Y-bearing sperm.

2. Each type of sperm has equivalent viability and motility in the female reproductive tract.

3. The egg surface is equally receptive to both X- and Y-bearing sperm.

No direct experimental evidence contradicts any of these assumptions; however, the human Y chromosome is smaller than the X chromosome and therefore of less mass. Thus, it has been speculated that Y-bearing sperm are more motile than X-bearing sperm. If this is true, then the probability of a fertilization event leading to a male zygote is increased, providing one possible explanation for the observed primary ratio.

ESSENTIAL POINT ■ ■ ■

In humans, while many more males than females are conceived, and although the mortality rate is higher in male embryos and fetuses than in females, there are nevertheless more males than females born.

5 Dosage Compensation Prevents Excessive Expression of X-Linked Genes in Humans and Other Mammals

The presence of two X chromosomes in normal human females and only one X in normal human males is unique compared with the equal numbers of autosomes present in the cells of both sexes. On theoretical grounds alone, it is possible to speculate that this disparity should create a "genetic dosage" difference between males and females, with attendant problems, for all X-linked genes. There is the potential for females to produce twice as much of each product of all X-linked genes. The additional X chromosomes in both males and females exhibiting the various syndromes discussed earlier in this chapter are thought to compound this dosage problem. Embryonic development depends on proper timing and precisely regulated levels of gene expression. Otherwise, disease phenotypes or embryonic lethality can occur. In this section, we will describe research findings regarding X-linked gene expression that demonstrate a genetic mechanism of **dosage compensation** that balances the dose of X chromosome gene expression in females and males.

Barr Bodies

Murray L. Barr and Ewart G. Bertram's experiments with female cats, as well as Keith Moore and Barr's subsequent study in humans, demonstrate a genetic mechanism in mammals that compensates for X chromosome dosage disparities. Barr and Bertram observed a darkly staining body in the interphase nerve cells of female cats that was absent in similar cells of males. In humans, this body can be easily demonstrated in female cells derived from the buccal mucosa (cheek cells) or in fibroblasts (undifferentiated connective tissue cells), but not in similar male cells (**Figure 6**). This highly condensed structure, about 1 μm in diameter, lies against the nuclear envelope of interphase cells, and it stains positively for a number of different DNA-binding dyes.

This chromosome structure, called a **sex chromatin body,** or simply a **Barr body,** is an inactivated X chromosome. Susumo Ohno was the first to suggest that the Barr body arises from one of the two X chromosomes. This hypothesis is attractive because it provides a possible mechanism for dosage compensation. If one of the two X chromosomes is inactive in

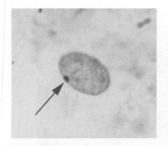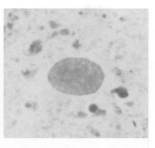

FIGURE 6 Photomicrographs comparing cheek epithelial cell nuclei from a male that fails to reveal Barr bodies (right) with a nucleus from a female that demonstrates a Barr body (indicated by the arrow in the left image). This structure, also called a sex chromatin body, represents an inactivated X chromosome.

the cells of females, the dosage of genetic information that can be expressed in males and females will be equivalent. Convincing, though indirect, evidence for this hypothesis comes from study of the sex-chromosome syndromes described earlier in this chapter. Regardless of how many X chromosomes a somatic cell possesses, all but one of them appear to be inactivated and can be seen as Barr bodies. For example, no Barr body is seen in the somatic cells of Turner 45,X females; one is seen in Klinefelter 47,XXY males; two in 47,XXX females; three in 48,XXXX females; and so on (**Figure 7**). Therefore, the number of Barr bodies follows an $N - 1$ rule, where N is the total number of X chromosomes present.

Although this apparent inactivation of all but one X chromosome increases our understanding of dosage compensation, it further complicates our perception of other matters. For example, because one of the two X chromosomes is in-

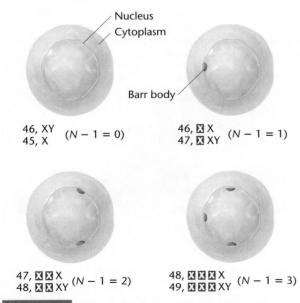

Nucleus
Cytoplasm

Barr body

46, XY
45, X $(N - 1 = 0)$

46, ☒ X
47, ☒ XY $(N - 1 = 1)$

47, ☒ ☒ X
48, ☒ ☒ XY $(N - 1 = 2)$

48, ☒ ☒ ☒ X
49, ☒ ☒ ☒ XY $(N - 1 = 3)$

FIGURE 7 Occurrence of Barr bodies in various human karyotypes, where all X chromosomes except one ($N - 1$) are inactivated.

activated in normal human females, why then is the Turner 45,X individual not entirely normal? Why aren't females with the triplo-X and tetra-X karyotypes (47,XXX and 48,XXXX) completely unaffected by the additional X chromosome? Furthermore, in Klinefelter syndrome (47,XXY), X chromosome inactivation effectively renders the person 46,XY. Why aren't these males unaffected by the extra X chromosome in their nuclei?

One possible explanation is that chromosome inactivation does not normally occur in the very early stages of development of those cells destined to form gonadal tissues. Another possible explanation is that not all genes on each X chromosome forming a Barr body are inactivated. Recent studies have indeed demonstrated that as many as 15 percent of the human X chromosomal genes actually escape inactivation. Clearly, then, not every gene on the X requires inactivation. In either case, excessive expression of certain X-linked genes might still occur at critical times during development despite apparent inactivation of superfluous X chromosomes.

The Lyon Hypothesis

In mammalian females, one X chromosome is of maternal origin, and the other is of paternal origin. Which one is inactivated? Is the inactivation random? Is the same chromosome inactive in all somatic cells? In 1961, Mary Lyon and Liane Russell independently proposed a hypothesis that answers these questions. They postulated that the inactivation of X chromosomes occurs randomly in somatic cells at a point early in embryonic development, most likely sometime during the blastocyst stage of development. Once inactivation has occurred, all descendant cells have the same X chromosome inactivated as their initial progenitor cell.

This explanation, which has come to be called the **Lyon hypothesis**, was initially based on observations of female mice heterozygous for X-linked coat-color genes. The pigmentation of these heterozygous females was mottled, with large patches expressing the color allele on one X and other patches expressing the allele on the other X. This is the phenotypic pattern that would be expected if different X chromosomes were inactive in adjacent patches of cells. Similar mosaic patterns occur in the black and yellow-orange patches of female tortoiseshell and calico cats (**Figure 8**). Such X-linked coat-color patterns do not occur in male cats because all their cells contain the single maternal X chromosome and are therefore hemizygous for only one X-linked coat-color allele.

The most direct evidence in support of the Lyon hypothesis comes from studies of gene expression in clones of human fibroblast cells. Individual cells are isolated following biopsy and cultured *in vitro*. A culture of cells derived from a single cell is called a **clone**. The synthesis of the enzyme glucose-6-phosphate dehydrogenase (G6PD) is controlled by an X-linked gene. Numerous mutant alleles of this gene

125

(a)

(b)

FIGURE 8 (a) The random distribution of orange and black patches in a calico cat illustrates the Lyon hypothesis. The white patches are due to another gene, distinguishing calico cats from tortoiseshell cats (b), which lack the white patches.

have been detected, and their gene products can be differentiated from the wild-type enzyme by their migration pattern in an electrophoretic field.

Fibroblasts have been taken from females heterozygous for different allelic forms of *G6PD* and studied. The Lyon hypothesis predicts that if inactivation of an X chromosome occurs randomly early in development, and thereafter all progeny cells have the same X chromosome inactivated as their progenitor, such a female should show two types of clones, each containing only one electrophoretic form of *G6PD*, in approximately equal proportions. This prediction has been confirmed experimentally, and studies involving modern techniques in molecular biology have clearly established that X chromosome inactivation occurs.

One ramification of X-inactivation is that mammalian females are mosaics for all heterozygous X-linked alleles—some areas of the body express only the maternally derived alleles, and others express only the paternally derived alleles. An especially interesting example involves red–green color blindness, an X-linked recessive disorder. In humans, hemizygous males are fully color-blind in all retinal cells. However, heterozygous females display mosaic retinas, with patches of defective color perception and surrounding areas with normal color perception. In this example, random inactivation of one or the other X chromosome early in the development of heterozygous females has led to these phenotypes.

The Mechanism of Inactivation

The least understood aspect of the Lyon hypothesis is the mechanism of X chromosome inactivation. Somehow, either DNA, the attached histone proteins, or both DNA and histone proteins, are chemically modified, silencing most genes that are part of that chromosome. Once silenced, a memory is created that keeps the same homolog inactivated following chromosome replications and cell divisions. Such a process,

whereby expression of genes on one homolog, but not the other, is affected, is referred to as **imprinting.** This term also applies to a number of other examples in which genetic information is modified and gene expression is repressed.

NOW SOLVE THIS

3 Carbon Copy (CC), the first cat produced from a clone, was created from an ovarian cell taken from her genetic donor, Rainbow, a calico cat. The diploid nucleus from the cell was extracted and then injected into an enucleated egg. The resulting zygote was then allowed to develop in a petri dish, and the cloned embryo was implanted in the uterus of a surrogate mother cat, who gave birth to CC. CC's surrogate mother was a tabby (see the photo below). Geneticists were very interested in the outcome of cloning a calico cat because they were not certain if the cloned cat would have patches of orange and black, just orange, or just black. Taking into account the Lyon hypothesis, explain the basis of the uncertainty. Would you expect CC to appear identical to Rainbow? Explain why or why not.

Carbon Copy with her surrogate mother.

■ **Hint** *This problem involves an understanding of the Lyon hypothesis. The key to its solution is to realize that the donor nucleus was from a differentiated ovarian cell of an adult female cat, which itself had inactivated one of its X chromosomes.*

Collectively, such events are part of the growing field of **epigenetics**.

Ongoing investigations are beginning to clarify the mechanism of inactivation. A region of the mammalian X chromosome is the major control unit. This region, located on the proximal end of the p arm in humans, is called the X inactivation center (**Xic**), and its genetic expression *occurs only on the X chromosome that is inactivated*. The *Xic* is about 1 Mb (10^6 base pairs) in length and is known to contain several putative regulatory units and four genes. One of these, *X-inactive specific transcript (XIST)*, is now known to be a critical gene for X-inactivation.

Several interesting observations have been made regarding the RNA that is transcribed from the *XIST* gene, many coming from experiments that used the equivalent gene in the mouse (*Xist*). First, the RNA product is quite large and does not encode a protein, and thus is not translated. The RNA products of *Xist* spread over and coat the X chromosome *bearing the gene that produced them*. Two other noncoding genes at the *Xic* locus, *Tsix* (an antisense partner of *Xist*) and *Xite*, are also believed to play important roles in X-inactivation.

A second observation is that transcription of *Xist* initially occurs at low levels on all X chromosomes. As the inactivation process begins, however, transcription continues, and is enhanced, only on the X chromosome that becomes inactivated. In 1996, a research group led by Neil Brockdorff and Graeme Penny provided convincing evidence that transcription of *Xist* is the critical event in chromosome inactivation. These researchers introduced a targeted deletion (7 kb) into this gene, disrupting its sequence. As a result, the chromosome bearing the deletion lost its ability to become inactivated.

ESSENTIAL POINT ■ ■ ■

In mammals, female somatic cells randomly inactivate one of two X chromosomes during early embryonic development, a process important for balancing the expression of X-chromosome linked genes in males and females.

6 | The Ratio of X Chromosomes to Sets of Autosomes Determines Sex in *Drosophila*

Because males and females in *Drosophila melanogaster* (and other *Drosophila* species) have the same general sex-chromosome composition as humans (males are XY and females are XX), we might assume that the Y chromosome also causes maleness in these flies. However, the elegant work of Calvin Bridges in 1916 showed this not to be true.

His studies of flies with quite varied chromosome compositions led him to the conclusion that the Y chromosome is not involved in sex determination in this organism. Instead, Bridges proposed that the X chromosomes and autosomes together play a critical role in sex determination. Recall that in the nematode *C. elegans*, which lacks a Y chromosome, the sex chromosomes and autosomes are also both critical to sex determination.

Bridges' work can be divided into two phases: (1) A study of offspring resulting from nondisjunction of the X chromosomes during meiosis in females and (2) subsequent work with progeny of females containing three copies of each chromosome, called triploid ($3n$) females. As we have seen previously in this chapter (and as you will see in Figure 1), nondisjunction is the failure of paired chromosomes to segregate or separate during the anaphase stage of the first or second meiotic divisions. The result is the production of two types of abnormal gametes, one of which contains an extra chromosome ($n + 1$) and the other of which lacks a chromosome ($n - 1$). Fertilization of such gametes with a haploid gamete produces ($2n + 1$) or ($2n - 1$) zygotes. As in humans, if nondisjunction involves the X chromosome, in addition to the normal complement of autosomes, both an XXY and an X0 sex-chromosome composition may result. (The "0" signifies that neither a second X nor a Y chromosome is present, as occurs in X0 genotypes of individuals with Turner syndrome.)

Contrary to what was later discovered in humans, Bridges found that the XXY flies were normal females and the X0 flies were sterile males. The presence of the Y chromosome in the XXY flies did not cause maleness, and its absence in the X0 flies did not produce femaleness. From these data, Bridges concluded that the Y chromosome in *Drosophila* lacks male-determining factors, but since the X0 males were sterile, it does contain genetic information essential to male fertility.

Bridges was able to clarify the mode of sex determination in *Drosophila* by studying the progeny of triploid females ($3n$), which have three copies each of the haploid complement of chromosomes. *Drosophila* has a haploid number of 4, thereby possessing three pairs of autosomes in addition to its pair of sex chromosomes. Triploid females apparently originate from rare diploid eggs fertilized by normal haploid sperm. Triploid females have heavy-set bodies, coarse bristles, and coarse eyes, and they may be fertile. Because of the odd number of each chromosome (3), during meiosis, a variety of different chromosome complements are distributed into gametes that give rise to offspring with a variety of abnormal chromosome constitutions. Correlations between the sexual morphology and chromosome composition, along with Bridges' interpretation, are shown in **Figure 9**.

Bridges realized that the critical factor in determining sex is the ratio of X chromosomes to the number of haploid sets

Normal diploid male

2 sets of autosomes
+
X   Y

Chromosome formulation	Ratio of X chromosomes to autosome sets	Sexual morphology
3X/2A	1.5	Metafemale
3X/3A	1.0	Female
2X/2A	1.0	Female
3X/4A	0.75	Intersex
2X/3A	0.67	Intersex
X/2A	0.50	Male
XY/2A	0.50	Male
XY/3A	0.33	Metamale

FIGURE 9 The ratios of X chromosomes to sets of autosomes and the resultant sexual morphology seen in *Drosophila melanogaster*.

of autosomes (A) present. Normal (2X:2A) and triploid (3X:3A) females each have a ratio equal to 1.0, and both are fertile. As the ratio exceeds unity (3X:2A, or 1.5, for example), what was once called a *superfemale* is produced. Because such females are most often inviable, they are now more appropriately called **metafemales.**

Normal (XY:2A) and sterile (X0:2A) males each have a ratio of 1:2, or 0.5. When the ratio decreases to 1:3, or 0.33, as in the case of an XY:3A male, infertile **metamales** result. Other flies recovered by Bridges in these studies had an (X:A) ratio intermediate between 0.5 and 1.0. These flies were generally larger, and they exhibited a variety of morphological abnormalities and rudimentary bisexual gonads and genitalia. They were invariably sterile and expressed both male and female morphology, thus being designated as **intersexes.**

Bridges' results indicate that in *Drosophila*, factors that cause a fly to develop into a male are not located on the sex chromosomes but are instead found on the autosomes. Some female-determining factors, however, are located on the X chromosomes. Thus, with respect to primary sex determination, male gametes containing one of each autosome plus a Y chromosome result in male offspring not because of the presence of the Y but because they fail to contribute an X chromosome. This mode of sex determination is explained by the **genic balance theory.** Bridges proposed that a threshold for maleness is reached when the X:A ratio is 1:2 (X:2A), but that the presence of an additional X (XX:2A) alters the balance and results in female differentiation.

Numerous genes involved in sex determination in *Drosophila* have been identified. The recessive autosomal gene *transformer* (*tra*), discovered over 50 years ago by Alfred H. Sturtevant, clearly demonstrated that a single autosomal gene could have a profound impact on sex determination. Females homozygous for *tra* are transformed into sterile males, but homozygous males are unaffected. More recently, another

gene, *Sex-lethal* (*Sxl*), has been shown to play a critical role, serving as a "master switch" in sex determination. Activation of the X-linked *Sxl* gene, which relies on a ratio of X chromosomes to sets of autosomes that equals 1.0, is essential to female development. In the absence of activation—as when, for example, the X:A ratio is 0.5—male development occurs.

Although it is not yet exactly clear how this ratio influences the *Sxl* locus, we do have some insights into the question. The *Sxl* locus is part of a hierarchy of gene expression and exerts control over other genes, including *tra* (discussed in the previous paragraph) and *dsx* (*doublesex*). The wild-type allele of *tra* is activated by the product of *Sxl* only in females and in turn influences the expression of *dsx*. Depending on how the initial RNA transcript of *dsx* is processed (spliced, as explained below), the resultant dsx protein activates either male- or female-specific genes required for sexual differentiation. Each step in this regulatory cascade requires a form of processing called **RNA splicing,** in which portions of the RNA are removed and the remaining fragments are "spliced" back together prior to translation into a protein. In the case of the *Sxl* gene, the RNA transcript may be spliced in different ways, a phenomenon called **alternative splicing.** A different RNA transcript is produced in females than in males. In potential females, the transcript is active and initiates a cascade of regulatory gene expression, ultimately leading to female differentiation. In potential males, the transcript is inactive, leading to a different pattern of gene activity, whereby male differentiation occurs.

7   Temperature Variation Controls Sex Determination in Reptiles

We conclude this chapter by discussing several cases involving reptiles, in which the environment—specifically temperature—has a profound influence on sex determination. In contrast to **chromosomal, or genotypic, sex determination (CSD or GSD),** in which sex is determined genetically (as is true of all examples thus far presented in the chapter), the cases that we will now discuss are categorized as **temperature-dependent sex determination (TSD).** As we shall see, the investigations leading to this information may well have come closer to revealing the true nature of the underlying basis of sex determination than any findings previously discussed.

In many species of reptiles, sex is predetermined at conception by sex-chromosome composition, as is the case in many organisms already considered in this chapter. For example, in many snakes, including vipers, a ZZ/ZW mode is in effect, in which the female is the heterogamous sex (ZW). However, in boas and pythons, it is impossible to distinguish one sex chromosome from the other in either sex. In many lizards, both the XX/XY and ZZ/ZW systems are found, depending on the species.

In still other reptilian species, however, TSD is the norm, including all crocodiles, most turtles, and some lizards, where sex determination is achieved according to the incubation temperature of eggs during a critical period of embryonic development. Three distinct patterns of TSD emerge (cases I–III in **Figure 10**). In case I, low temperatures yield 100 percent females, and high temperatures yield 100 percent males. Just the opposite occurs in case II. In case III, low *and* high temperatures yield 100 percent females, while intermediate temperatures yield various proportions of males. The third pattern is seen in various species of crocodiles, turtles, and lizards, although other members of these groups are known to exhibit the other patterns.

Two observations are noteworthy. First, in all three patterns, certain temperatures result in both male and female offspring; second, this pivotal temperature T_P range is fairly narrow, usually spanning less than 5°C, and sometimes only 1°C. The central question raised by these observations is: What are the metabolic or physiological parameters affected by temperature that lead to the differentiation of one sex or the other?

The answer is thought to involve steroids (mainly estrogens) and the enzymes involved in their synthesis. Studies clearly demonstrate that the effects of temperature on estrogens, androgens, and inhibitors of the enzymes controlling their synthesis are involved in the sexual differentiation of ovaries and testes. One enzyme in particular, **aromatase,** converts androgens (male hormones such as testosterone) to estrogens (female hormones such as estradiol). The activity of this enzyme is correlated with the pathway of reactions that occurs during gonadal differentiation activity and is high in developing ovaries and low in developing testes. Researchers in this field, including Claude Pieau and colleagues, have proposed that a thermosensitive factor mediates the transcription of the reptilian aromatase gene, leading to temperature-dependent sex determination. Several other genes are likely to be involved in this mediation.

The involvement of sex steroids in gonadal differentiation has also been documented in birds, fishes, and amphibians. Thus, sex-determining mechanisms involving estrogens seem to be characteristic of nonmammalian vertebrates. The regulation of such systems, while temperature-dependent in many reptiles, appears to be controlled by sex chromosomes (XX/XY or ZZ/ZW) in many of these other organisms. A final intriguing thought on this matter is that the product of *SRY*, a key component in mammalian sex determination, has been shown to bind *in vitro* to a regulatory portion of the aromatase gene, suggesting a mechanism whereby it could act as a repressor of ovarian development.

ESSENTIAL POINT ■ ■ ■

Many reptiles show temperature-dependent effects on sex determination. Although specific sex chromosomes determine genotypic sex in many reptiles, temperature effects on genes involved in sexual determination affect whether an embryo develops a male or female phenotype.

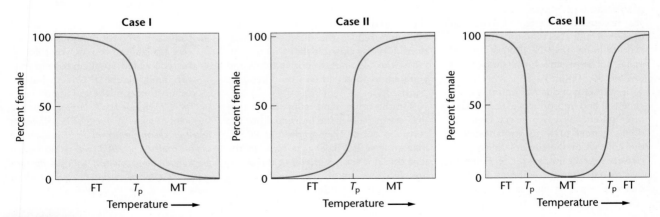

FIGURE 10　Three different patterns of temperature-dependent sex determination (TSD) in reptiles, as described in the text. The relative pivotal temperature T_p is crucial to sex determination during a critical point during embryonic development. FT = Female-determining temperature; MT = male-determining temperature

GENETICS, TECHNOLOGY, AND SOCIETY

A Question of Gender: Sex Selection in Humans

Throughout history, people have attempted to influence the gender of their unborn offspring by following varied and sometimes bizarre procedures. In medieval Europe, prospective parents would place a hammer under the bed to help them conceive a boy, or a pair of scissors to conceive a girl. Other practices were based on the ancient belief that semen from the right testicle created male offspring and that from the left testicle created females. As late as the eighteenth century, European men might tie off or remove their left testicle to increase the chances of getting a male heir. In some cultures, efforts to control the sex of offspring has had a darker side—female infanticide. In ancient Greece, the murder of female infants was so common that the male:female ratio in some areas approached 4:1. In some parts of rural India, hundreds of families admitted to female infanticide as late as the 1990s. In 1997, the World Health Organization reported population data showing that about 50 million women were "missing" in China, likely because of selective abortion of female fetuses and institutionalized neglect of female children.

In recent times, sex-specific abortion has replaced much of the traditional female infanticide. For a fee, some companies offer amniocentesis and ultrasound tests for prenatal sex determination. Studies in India estimate that hundreds of thousands of fetuses are aborted each year because they are female. As a result of sex-selective abortion, the female:male ratio in India was 927:1000 in 1991. In some northern states, the ratio was as low as 600:1000.

In Western industrial countries, new genetics and reproductive technologies offer parents ways to select their children's gender prior to implantation of the embryo in the uterus—called *preimplantation gender selection (PGS)*. Following *in vitro* fertilization, embryos are biopsied and assessed for gender. Only sex-selected embryos are then implanted.

The simplest method involves separating X and Y chromosome-bearing spermatozoa based on their DNA content. Because of the difference in size of the X and Y chromosomes, X-bearing sperm contain 2.8 to 3.0 percent more DNA than Y-bearing sperm. Sperm samples are treated with a fluorescent DNA stain, and then passed through a laser beam in a Fluorescence-Activated Cell Sorter machine that separates the sperm into two fractions based on the intensity of their DNA-fluorescence. The sorted sperm are then used for standard intrauterine insemination.

The emerging PGS methods raise a number of legal and ethical issues. Some people feel that prospective parents have the legal right to use sex-selection techniques as part of their fundamental procreative liberty. Proponents state that PGS will reduce the suffering of many families. For example, people at risk for transmitting X-linked diseases such as hemophilia or Duchenne muscular dystrophy would be able to enhance their chance of conceiving a female child, who will not express the disease.

The majority of people who undertake PGS, however, do so for nonmedical reasons—to "balance" their families. A possible argument in favor of this use is that the ability to intentionally select the sex of an offspring may reduce overpopulation and economic burdens for families who would repeatedly reproduce to get the desired gender. By the same token, PGS may reduce the number of abortions. It is also possible that PGS may increase the happiness of both parents and children, as the children would be more "wanted."

On the other hand, some argue that PGS serves neither the individual nor the common good. They argue that PGS is inherently sexist, having its basis in the idea that one sex is superior to the other, and leads to an increase in linking a child's worth to gender. Other critics fear that social approval of PGS will open the door to other genetic manipulations of children's characteristics. It is difficult to predict the full effects that PGS will bring to the world. But the gender-selection genie is now out of the bottle and is unwilling to return.

Your Turn

Take time, individually or in groups, to answer the following questions. Investigate the references and links to help you understand some of the issues that surround the topic of gender selection.

1. What do you think are valid arguments for and against the use of PGS?

2. A generally accepted moral and legal concept is that of reproductive autonomy—the freedom to make individual reproductive decisions without external interference. Are there circumstances under which reproductive autonomy should be restricted?

The above questions, and others, are explored in a series of articles in the American Journal of Bioethics, *Volume 1 (2001). See the article by J. A. Robertson on pages 2–9, for a summary of the moral and legal issues surrounding PGS.*

3. What do you think are the reasons that some societies practice female infanticide and prefer the birth of male children?

For a discussion of this topic, visit the "Gendercide Watch" Web site http://www.gendercide.org.

4. If safe and efficient methods of PGS were available to you, do you think that you would use them to help you with family planning? Under what circumstances might you use them?

The Genetics and IVF Institute (Fairfax, Virginia) is presently using PGS techniques based on sperm sorting, in an FDA-approved clinical trial. As of 2008, over 1000 human pregnancies have resulted, with an approximately 80 percent success rate. Read about these methods on their Web site: http://www.microsort.net.

CASE STUDY » Doggone it!

A dog breeder discovers one of her male puppies has abnormal genitalia. After a visit to the veterinary clinic at a nearby university, the breeder learns that the dog's karyotype lacks a Y chromosome, but instead has an XX chromosome pair, with one of the X chromosomes slightly larger than usual (being mammals, male dogs are normally XY and females are XX). The veterinarian tells her that in other breeds, some females display an XY chromosome pair, with the Y chromosome being slightly shorter than normal. These observations raise several interesting questions:

1. Can you offer a chromosomal explanation of these two cases?
2. How could such cases be used to locate the gene(s) responsible for maleness?
3. Suppose you discover a female dog with a normal-sized XY chromosome pair. What kind of mutation might be involved in this case?
4. Suppose you discover a female dog with only a single X chromosome. What predictions might you make about the sex organs and reproductive capacity of this dog?

INSIGHTS AND SOLUTIONS

1. In *Drosophila*, the X chromosomes may become attached to one another ($\hat{X}X$) such that they always segregate together. Some flies thus contain a set of attached X chromosomes plus a Y chromosome.

(a) What sex would such a fly be? Explain why this is so.

(b) Given the answer to part (a), predict the sex of the offspring that would occur in a cross between this fly and a normal one of the opposite sex.

(c) If the offspring described in part (b) are allowed to interbreed, what will be the outcome?

Solution:

(a) The fly will be a female. The ratio of X chromosomes to sets of autosomes—which determines sex in *Drosophila*—will be 1.0, leading to normal female development. The Y chromosome has no influence on sex determination in *Drosophila*.

(b) All progeny flies will have two sets of autosomes along with one of the following sex-chromosome compositions:

(1) $\hat{X}XX \rightarrow$ a metafemale with 3 X's (called a trisomic)

(2) $\hat{X}XY \rightarrow$ a female like her mother

(3) XY $\rightarrow$ a normal male

(4) YY $\rightarrow$ no development occurs

(c) A stock will be created that maintains attached-X females generation after generation.

2. The Xg cell-surface antigen is coded for by a gene located on the X chromosome. No equivalent gene exists on the Y chromosome. Two codominant alleles of this gene have been identified: *Xg1* and *Xg2*. A woman of genotype *Xg2/Xg2* bears children with a man of genotype *Xg1/Y*, and they produce a son with Klinefelter syndrome of genotype *Xg1/Xg2Y*. Using proper genetic terminology, briefly explain how this individual was generated. In which parent and in which meiotic division did the mistake occur?

Solution: Because the son with Klinefelter syndrome is *Xg1/Xg2Y*, he must have received both the *Xg1* allele and the Y chromosome from his father. Therefore, nondisjunction must have occurred during meiosis I in the father.

PROBLEMS AND DISCUSSION QUESTIONS

 For activities, animations, and review quizzes, go to the study area at www.masteringgenetics.com

HOW DO WE KNOW?

1. In this chapter, we will focus on sex differentiation, sex chromosomes, and genetic mechanisms involved in sex determination. At the same time, we will find many opportunities to consider the methods and reasoning by which much of this information was acquired. From the explanations given in the chapter, you should answer the following fundamental questions:

   (a) How do we know that in humans the X chromosomes play no role in sex determination, while the Y chromosome causes maleness and its absence causes femaleness?

   (b) How did we learn that, although the sex ratio at birth in humans favors males slightly, the sex ratio at conception favors them much more?

   (c) How do we know that X chromosomal inactivation of either the paternal or maternal homolog is a random event during early development in mammalian females?

   (d) How do we know that *Drosophila* utilizes a different sex-determination mechanism than mammals, even though it has the same sex-chromosome compositions in males and females?

2. As related to sex determination, what is meant by

   (a) homomorphic and heteromorphic chromosomes; and
   (b) isogamous and heterogamous organisms?

3. Contrast the life cycle of a plant such as *Zea mays* with an animal such as *C. elegans*.

4. Distinguish between the concepts of sexual differentiation and sex determination.

5. Contrast the *Protenor* and *Lygaeus* modes of sex determination.

6. Describe the major difference between sex determination in *Drosophila* and in humans.

7. How do mammals, including humans, solve the "dosage problem" caused by the presence of an X and Y chromosome in one sex and two X chromosomes in the other sex?

8. What specific observations (evidence) support the conclusions about sex determination in *Drosophila* and humans?

9. Describe how nondisjunction in human female gametes can give rise to Klinefelter and Turner syndrome offspring following fertilization by a normal male gamete.

10. An insect species is discovered in which the heterogametic sex is unknown. An X-linked recessive mutation for *reduced wing* (*rw*)

is discovered. Contrast the F_1 and F_2 generations from a cross between a female with reduced wings and a male with normal-sized wings when

(a) the female is the heterogametic sex; and

(b) the male is the heterogametic sex.

11. Given your answers to Problem 10, is it possible to distinguish between the *Protenor* and *Lygaeus* mode of sex determination based on the outcome of these crosses?

12. When cows have twin calves of unlike sex (fraternal twins), the female twin is usually sterile and has masculinized reproductive organs. This calf is referred to as a freemartin. In cows, twins may share a common placenta and thus fetal circulation. Predict why a freemartin develops.

13. An attached-X female fly, $\hat{X}XY$ (see the "Insights and Solutions" box), expresses the recessive X-linked *white*-eye phenotype. It is crossed to a male fly that expresses the X-linked recessive miniature wing phenotype. Determine the outcome of this cross in terms of sex, eye color, and wing size of the offspring.

14. Assume that on rare occasions the attached X chromosomes in female gametes become unattached. Based on the parental phenotypes in Problem 13, what outcomes in the F_1 generation would indicate that this has occurred during female meiosis?

15. It is believed that any male-determining genes contained on the Y chromosome in humans are not located in the limited region that synapses with the X chromosome during meiosis. What might be the outcome if such genes were located in this region?

16. What is a Barr body, and where is it found in a cell?

17. Indicate the expected number of Barr bodies in interphase cells of individuals with Klinefelter syndrome; Turner syndrome; and karyotypes 47,XYY, 47,XXX, and 48,XXXX.

18. Define the Lyon hypothesis.

19. Can the Lyon hypothesis be tested in a human female who is homozygous for one allele of the X-linked *G6PD* gene? Why, or why not?

20. Predict the potential effect of the Lyon hypothesis on the retina of a human female heterozygous for the X-linked red–green color-blindness trait.

21. Cat breeders are aware that kittens expressing the X-linked calico coat pattern and tortoiseshell pattern (Figure 8) are almost invariably females. Why?

22. What does the apparent need for dosage compensation mechanisms suggest about the expression of genetic information in normal diploid individuals?

23. How does X chromosome dosage compensation in *Drosophila* differ from that process in humans?

24. Devise as many hypotheses as you can that might explain why so many more human male conceptions than human female conceptions occur.

25. In mice, the *Sry* gene (see Section 3) is located on the Y chromosome very close to one of the pseudoautosomal regions that pairs with the X chromosome during male meiosis. Given this information, propose a model to explain the generation of unusual males who have two X chromosomes (with an *Sry*-containing piece of the Y chromosome attached to one X chromosome).

26. The genes encoding the red- and green-color-detecting proteins of the human eye are located next to one another on the X chromosome and probably evolved from a common ancestral pigment gene. The two proteins demonstrate 76 percent homology in their amino acid sequences. A normal-visioned woman with both genes on each of her two X chromosomes has a red-color-blind son who was shown to have one copy of the green-detecting gene and no copies of the red-detecting gene. Devise an explanation for these observations at the chromosomal level (involving meiosis).

27. In mice, the X-linked dominant mutation *Testicular feminization* (*Tfm*) eliminates the normal response to the testicular hormone testosterone during sexual differentiation. An XY mouse bearing the *Tfm* allele on the X chromosome develops testes, but no further male differentiation occurs—the external genitalia of such an animal are female. From this information, what might you conclude about the role of the *Tfm* gene product and the X and Y chromosomes in sex determination and sexual differentiation in mammals? Can you devise an experiment, assuming you can "genetically engineer" the chromosomes of mice, to test and confirm your explanation?

28. Shown here are graphs that plot the percentage of fertilized eggs containing males against the atmospheric temperature during early development in (a) snapping turtles and (b) most lizards. Interpret these data as they relate to the effect of temperature on sex determination.

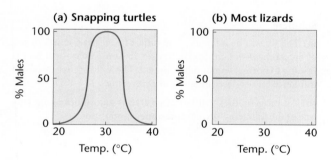

29. In chickens, a key gene involved in sex determination has recently been identified. Called *DMRT1*, it is located on the Z chromosome and is absent on the W chromosome. Like *SRY* in humans, it is male determining. Unlike *SRY* in humans, however, female chickens (ZW) have a single copy while males (ZZ) have two copies of the gene. Nevertheless, it is transcribed only in the developing testis. Working in the laboratory of Andrew Sinclair (a co-discoverer of the human *SRY* gene), Craig Smith and colleagues were able to "knock down" expression of *DMRT1* in ZZ embryos using RNA interference techniques. In such cases, the developing gonads look more like ovaries than testes [*Nature* 461: 267 (2009)]. What conclusions can you draw about the role that the *DMRT1* gene plays in chickens in contrast to the role the *SRY* gene plays in humans?

SOLUTIONS TO SELECTED PROBLEMS AND DISCUSSION QUESTIONS

Answers to Now Solve This

1. In mammals, the scheme of sex determination is dependent on the presence of a piece of the Y chromosome. If present, a male is produced. In *Bonellia viridis*, the female proboscis produces some substance that triggers a morphological, physiological, and behavioral developmental pattern that produces males. To elucidate the mechanism, one could attempt to isolate and characterize the active substance by testing different chemical fractions of the proboscis. Mutant analysis usually provides critical approaches to studying developmental processes. Depending on characteristics of the organism, one could attempt to isolate mutants that lead to changes in male or female development. Third, by using micro-tissue transplantations, one could attempt to determine which anatomical "centers" of the embryo respond to the chemical cues of the female.

2. (a) Something is missing from the male-determining system of sex determination at the level of the genes, gene products, or receptors, and so on.

   (b) The *SOX9* gene, or its product, is probably involved in male development. Perhaps it is activated by *SRY*.

   (c) There is probably some evolutionary relationship between the *SOX9* gene and *SRY*. There is considerable evidence that many other genes and pseudogenes are also homologous to *SRY*.

   (d) Normal female sexual development does not require the *SOX9* gene or gene product(s).

3. Because of X chromosome inactivation in mammals, scientists would be interested in determining whether the nucleus taken from Rainbow (donor) would continue to show such inactivation. Would the inactivated X chromosome retain the property of inactivation? Since X chromosome inactivation is random, CC would have a different patch pattern from her genetic mother based on the random X inactivation alone.

Solutions to Problems and Discussion Questions

6. In *Drosophila* it is the balance between the number of X chromosomes and the number of haploid sets of autosomes that determines sex. In humans there is a small region on the Y chromosome that determines maleness.

8. The Y chromosome is male determining in humans, and it is a particular region of the Y chromosome that causes maleness, the sex-determining region (SRY). SRY releases a product called the testis-determining factor (TDF), which causes the undifferentiated gonadal tissue to form testes. Individuals with the 47, XXY complement are males, while 45, XO produces females. In *Drosophila* it is the balance between the number of X chromosomes and the number of haploid sets of autosomes that determines sex. In contrast to humans, XO *Drosophila* are males and the XXY complement is female.

10. (a) female $X^{rw}Y$ × male X^+X^+

   F_1: females: X^+Y (normal)

           males: $X^{rw}X^+$ (normal)

   F_2: females: X^+Y (normal)

                   $X^{rw}Y$ (reduced wing)

         males: $X^{rw}X^+$ (normal)

                X^+X^+ (normal)

(b) female $X^{rw}X^{rw}$ × male X^+Y

   F_1: females: $X^{rw}X^+$ (normal)

           males: $X^{rw}Y$ (reduced wing)

   F_2: females: $X^{rw}X^+$ (*normal*)

                  $X^{rw}X^{rw}$ (reduced wing)

         males: X^+Y (normal)

                $X^{rw}Y$ (reduced wing)

12. Males and females share a common placenta and therefore hormonal factors carried in blood. Hormones and other molecular species (transcription factors, perhaps) triggered by the presence of a Y chromosome lead to a cascade of developmental events that both suppress female organ development and enhance masculinization. Other mammals also exhibit a variety of similar effects depending on the sex of their uterine neighbors during development.

14. If the male offspring had white eyes and the female offspring were wild-type, one might suspect that the attached-X had become unattached.

20. Females may display mosaic retinas with patches of defective color perception. Under these conditions, their color vision may be influenced.

22. Many organisms have evolved over millions of years under the fine balance of numerous gene products, usually occurring with two copies (identical or similar) of each gene. Many genes required for normal cellular and organismic function in *both* males and females are located on the X chromosome where only one copy occurs. These gene products have nothing to do with sex determination or sex differentiation, but their output must be balanced in some manner.

24. One could account for the significant departures from a 1:1 ratio of males to females by suggesting that at anaphase I of meiosis, the Y chromosome more often goes to the pole that produces the more viable sperm cells. One could also speculate that the Y-bearing sperm has a higher likelihood of surviving in the female reproductive tract, or that the egg surface is more receptive to Y-bearing sperm. At this time the mechanism is unclear.

26. Because of the homology between the *red* and *green* genes, there exists the possibility for an irregular synapsis (see the figure below) that, following crossing over, would give a chromosome with only one (*green*) of the duplicated genes. When this X chromosome combines with the normal Y chromosome, the son's phenotype can be explained.

Normal synapsis:

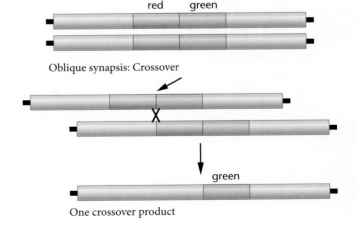

Oblique synapsis: Crossover

One crossover product

28. In snapping turtles, sex determination is strongly influenced by temperature such that males are favored in the 26–34°C range. Lizards, on the other hand, appear to have their sex determined by factors other than temperature in the 20–40°C range.

Basic set of nine unique chromosomes (*n*)

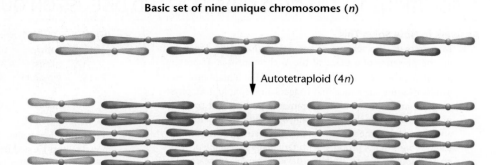

Autotetraploid (4*n*)

CREDITS

Credits are listed in order of appearance.

Photo

CO, Wessex Reg. Genetics Centre/Wellcome Images; F-1, Artem Solovev/Fotolia LLC; F-2 (a) Dr. Maria Gallegos, F-4 (a, b)

Catherine G. Palmer, F-6 (a, b) Michael Abbey/Photo Researchers, Inc.; F-8 (a) Sari O'Neal/Shutterstock.com; F-8 (b) Dr. William S. Klug; Texas A&M University UPI Photo Service/Newscom

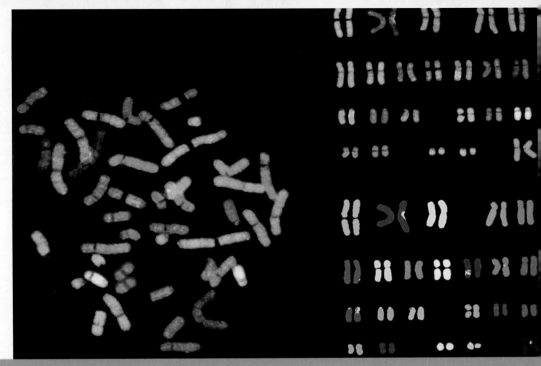

Spectral karyotyping of human chromosomes utilizing differentially labeled "painting" probes.

Chromosome Mutations: Variation in Number and Arrangement

CHAPTER CONCEPTS

- The failure of chromosomes to properly separate during meiosis results in variation in the chromosome content of gametes and subsequently in offspring arising from such gametes.

- Plants often tolerate an abnormal genetic content, but, as a result, they often manifest unique phenotypes. Such genetic variation has been an important factor in the evolution of plants.

- In animals, genetic information is in a delicate equilibrium whereby the gain or loss of a chromosome, or part of a chromosome, in an otherwise diploid organism often leads to lethality or to an abnormal phenotype.

- The rearrangement of genetic information within the genome of a diploid organism may be tolerated by that organism but may affect the viability of gametes and the phenotypes of organisms arising from those gametes.

- Chromosomes in humans contain fragile sites—regions susceptible to breakage, which lead to abnormal phenotypes.

Mutations and the resulting alleles affect an organism's phenotype and traits are passed from parents to offspring according to Mendelian principles. In this chapter, we look at phenotypic variation that results from more substantial changes than alterations of individual genes—modifications at the level of the chromosome.

Although most members of diploid species normally contain precisely two haploid chromosome sets, many known cases vary from this pattern. Modifications include a change in the total number of chromosomes, the deletion or duplication of genes or segments of a chromosome, and rearrangements of the genetic material either within or among chromosomes. Taken together, such changes are called **chromosome mutations** or **chromosome aberrations,** to distinguish them from gene mutations. Because the chromosome is the unit of genetic transmission, according to Mendelian laws, chromosome aberrations are passed to offspring in a predictable manner, resulting in many unique genetic outcomes.

Because the genetic component of an organism is delicately balanced, even minor alterations of either content or location of genetic information within the genome can result in some form of phenotypic variation. More substantial changes may be lethal, particularly in animals. Throughout this chapter, we consider many types of chromosomal aberrations, the phenotypic consequences for the organism that harbors an aberration, and the impact of the aberration on the offspring of an affected individual. We will also discuss the role of chromosome aberrations in the evolutionary process.

From Chapter 6 of *Essentials of Genetics, Eighth edition,* William S. Klug, Michael R. Cummings, Charlotte A. Spencer, Michael A. Palladino. ©2013 by Pearson Education, Inc. All rights reserved.

1 Variation in Chromosome Number: Terminology and Origin

Variation in chromosome number ranges from the addition or loss of one or more chromosomes to the addition of one or more haploid sets of chromosomes. Before we embark on our discussion, it is useful to clarify the terminology that describes such changes. In the general condition known as **aneuploidy,** an organism gains or loses one or more chromosomes but not a complete set. The loss of a single chromosome from an otherwise diploid genome is called *monosomy.* The gain of one chromosome results in *trisomy.* These changes are contrasted with the condition of **euploidy,** where complete haploid sets of chromosomes are present. If more than two sets are present, the term **polyploidy** applies. Organisms with three sets are specifically *triploid,* those with four sets are *tetraploid,* and so on. **Table 1** provides an organizational framework for you to follow as we discuss each of these categories of aneuploid and euploid variation and the subsets within them.

As we consider cases that include the gain or loss of chromosomes, it is useful to examine how such aberrations originate. For instance, how do the syndromes arise where the number of sex-determining chromosomes in humans is altered? As you may recall, the gain (47,XXY) or loss (45,X) of an X chromosome from an otherwise diploid genome affects the phenotype, resulting in **Klinefelter syndrome** or **Turner**

TABLE 1	Terminology for Variation in Chromosome Numbers
Term	**Explanation**
Aneuploidy	$2n \pm x$ chromosomes
Monosomy	$2n - 1$
Disomy	$2n$
Trisomy	$2n + 1$
Tetrasomy, pentasomy, etc.	$2n + 2$, $2n + 3$, etc.
Euploidy	Multiples of n
Diploidy	$2n$
Polyploidy	$3n$, $4n$, $5n$, . . .
Triploidy	$3n$
Tetraploidy, pentaploidy, etc.	$4n$, $5n$, etc.
Autopolyploidy	Multiples of the same genome
Allopolyploidy (amphidiploidy)	Multiples of closely related genomes

syndrome, respectively (See Figure 4). Human females may contain extra X chromosomes (e.g., 47,XXX, 48,XXXX), and some males contain an extra Y chromosome (47,XYY).

Such chromosomal variation originates as a random error during the production of gametes, a phenomenon referred to as **nondisjunction,** whereby paired homologs fail to disjoin during segregation. This process disrupts the normal distribution of chromosomes into gametes. The results of nondisjunction during meiosis I and meiosis II for a single chromosome of a diploid organism are shown in **Figure 1.** As you can see, abnormal gametes can form that contain

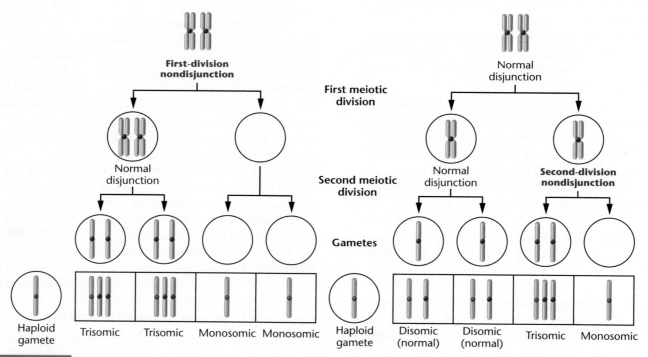

FIGURE 1   Nondisjunction during the first and second meiotic divisions. In both cases, some of the gametes that are formed either contain two members of a specific chromosome or lack that chromosome. After fertilization by a gamete with normal haploid content, monosomic, disomic (normal), or trisomic zygotes are produced.

either two members of the affected chromosome or none at all. Fertilizing these with a normal haploid gamete produces a zygote with either three members (trisomy) or only one member (monosomy) of this chromosome. Nondisjunction leads to a variety of aneuploid conditions in humans and other organisms.

ESSENTIAL POINT ■ ■ ■

Alterations of the precise diploid content of chromosomes are referred to as chromosomal aberrations or chromosomal mutations.

2 Monosomy and Trisomy Result in a Variety of Phenotypic Effects

We turn now to a consideration of variations in the number of autosomes and the genetic consequence of such changes. The most common examples of aneuploidy, where an organism has a chromosome number other than an exact multiple of the haploid set, are cases in which a single chromosome is either added to, or lost from, a normal diploid set.

Monosomy

The loss of one chromosome produces a $2n - 1$ complement called **monosomy.** Although monosomy for the X chromosome occurs in humans, as we have seen in 45,X Turner syndrome, monosomy for any of the autosomes is not usually tolerated in humans or other animals. In *Drosophila*, flies that are monosomic for the very small chromosome IV (containing less than 5 percent of the organism's genes) develop more slowly, exhibit reduced body size, and have impaired viability. Monosomy for the larger chromosomes II and III is apparently lethal because such flies have never been recovered.

The failure of monosomic individuals to survive is at first quite puzzling, since at least a single copy of every gene is present in the remaining homolog. However, one explanation is that if just one of those genes is represented by a lethal allele, monosomy unmasks the recessive lethal that is tolerated in heterozygotes carrying the corresponding wild-type allele, leading to the death of the organism. In other cases, a single copy of a recessive gene due to monosomy may be insufficient to provide life-sustaining function for the organism, a phenomenon called **haploinsufficiency.**

Aneuploidy is better tolerated in the plant kingdom. Monosomy for autosomal chromosomes has been observed in maize, tobacco, the evening primrose (*Oenothera*), and the jimson weed (*Datura*), among many other plants. Nevertheless, such monosomic plants are usually less viable than their diploid derivatives. Haploid pollen grains, which undergo extensive development before participating in fertilization, are particularly sensitive to the lack of one chromosome and are seldom viable.

Trisomy

In general, the effects of **trisomy** $(2n + 1)$ parallel those of monosomy. However, the addition of an extra chromosome produces somewhat more viable individuals in both animal and plant species than does the loss of a chromosome. In animals, this is often true, provided that the chromosome involved is relatively small. However, the addition of a large autosome to the diploid complement in both *Drosophila* and humans has severe effects and is usually lethal during development.

In plants, trisomic individuals are viable, but their phenotype may be altered. A classical example involves the jimson weed, *Datura,* whose diploid number is 24. Twelve primary trisomic conditions are possible, and examples of each one have been recovered. Each trisomy alters the phenotype of the plant's capsule sufficiently to produce a unique phenotype. These capsule phenotypes were first thought to be caused by mutations in one or more genes.

Still another example is seen in the rice plant (*Oryza sativa*), which has a haploid number of 12. Trisomic strains for each chromosome have been isolated and studied—the plants of 11 strains can be distinguished from one another and from wild-type plants. Trisomics for the longer chromosomes are the most distinctive, and the plants grow more slowly. This is in keeping with the belief that larger chromosomes cause greater genetic imbalance than smaller ones. Leaf structure, foliage, stems, grain morphology, and plant height also vary among the various trisomies.

Down Syndrome: Trisomy 21

The only human autosomal trisomy in which a significant number of individuals survive longer than a year past birth was discovered in 1866 by Langdon Down. The condition is now known to result from trisomy of chromosome 21, one of

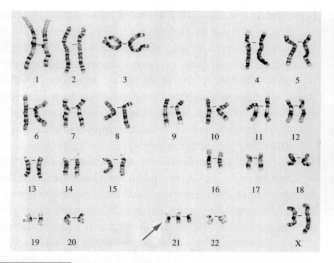

FIGURE 2   The karyotype and a photograph of a child with Down syndrome (hugging her unaffected sister on the right). In the karyotype, three members of the G-group chromosome 21 are present, creating the 47,21+ condition.

the G group* **(Figure 2)**, and is called **Down syndrome** or simply **trisomy 21** (designated 47,21+). This trisomy is found in approximately 1 infant in every 800 live births. While this might seem to be a rare, improbable event, there are approximately 4000–5000 such births annually in the United States, and there are currently over 250,000 individuals with Down syndrome.

Typical of other conditions classified as syndromes, many phenotypic characteristics *may* be present in trisomy 21, but any single affected individual usually exhibits only a subset of these. In the case of Down syndrome, there are 12 to 14 such characteristics, with each individual, on average, expressing 6 to 8 of these. Nevertheless, the outward appearance of these individuals is very similar, and they bear a striking resemblance to one another. This is, for the most part, due to a prominent epicanthic fold in each eye** and the typically flat face and round head. People with Down syndrome are also characteristically short and may have a protruding, furrowed tongue (which causes the mouth to remain partially open) and short, broad hands with characteristic palm and fingerprint patterns. Physical, psychomotor, and mental development are retarded, and poor muscle tone is characteristic. While life expectancy is shortened to an average of about 50 years, individuals are known to survive into their 60s.

Children afflicted with Down syndrome are prone to respiratory disease and heart malformations, and they show

an incidence of leukemia approximately 20 times higher than that of the normal population. However, careful medical scrutiny and treatment throughout their lives can extend their survival significantly. A striking observation is that death in older Down syndrome adults is frequently due to Alzheimer disease. The onset of this disease occurs at a much earlier age than in the normal population.

Because Down syndrome is common in our population, a comprehensive understanding of the underlying genetic basis has long been a research goal. Investigations have given rise to the idea that a critical region of chromosome 21 contains the genes that are dosage sensitive in this trisomy and responsible for the many phenotypes associated with the syndrome. This hypothetical portion of the chromosome has been called the **Down syndrome critical region (DSCR)**. A mouse model was created in 2004 that is trisomic for the DSCR, although some mice do not exhibit the characteristics of the syndrome. Nevertheless, this remains an important investigative approach.

Current studies of the DSCR region in both humans and mice have led to several interesting findings. We now believe that the three copies of the genes present in this region are necessary, but themselves not sufficient for the cognitive deficiencies characteristic of the syndrome. Another finding involves the important observation that Down syndrome individuals have a decreased risk of developing a number of cancers involving solid tumors, including lung cancer and melanoma. This health benefit has been correlated with the presence of an extra copy of the *DSCR1* gene, which encodes a protein that suppresses *vascular endothelial growth factor* (*VEGF*). This suppression, in turn, blocks the process of angiogenesis. As a result, the overexpression of this gene inhibits tumors from forming proper vascularization, diminishing their growth. A 14-year study published in 2002 involving 17,800 Down

---

* On the basis of size and centromere placement, human autosomal chromosomes are divided into seven groups: A (1–3), B (4–5), C (6–12), D (13–15), E (16–18), F (19–20), and G (21–22).

** The epicanthic fold, or epicanthus, is a skin fold of the upper eyelid, extending from the nose to the inner side of the eyebrow. It covers and appears to lower the inner corner of the eye, giving the eye a slanted, or almond-shaped, appearance. The epicanthus is a prominent normal component of the eyes in many Asian groups.

syndrome individuals revealed an approximate 10 percent reduction in cancer mortality in contrast to a control population. No doubt, further information will be forthcoming from the study of the DSCR region.

The Origin of the Extra 21st Chromosome in Down Syndrome

Most frequently, this trisomic condition occurs through nondisjunction of chromosome 21 during meiosis. Failure of paired homologs to disjoin during either anaphase I or II may lead to gametes with the $n + 1$ chromosome composition. About 75 percent of these errors leading to Down syndrome are attributed to nondisjunction during the first meiotic division. Subsequent fertilization with a normal gamete creates the trisomic condition.

Chromosome analysis has shown that, while the additional chromosome may be derived from either the mother or father, the ovum is the source in about 95 percent of 47,21+ trisomy cases. Before the development of techniques using polymorphic markers to distinguish paternal from maternal homologs, this conclusion was supported by the more indirect evidence derived from studies of the age of mothers giving birth to infants afflicted with Down syndrome. **Figure 3** shows the relationship between the incidence of Down syndrome births and maternal age, illustrating the dramatic increase as the age of the mother increases. While the frequency is about 1 in 1000 at maternal age 30, a tenfold increase to a frequency of 1 in 100 is noted at age 40. The frequency increases still further to about 1 in 30 at age 45. A very alarming statistic is that as the age of childbearing women exceeds 45, the probability of a Down syndrome

birth continues to increase substantially. In spite of this high probability, substantially more than half of Down syndrome births occur to women younger than 35 years, because the overwhelming proportion of pregnancies in the general population involve women under that age.

Although the nondisjunctional event that produces Down syndrome seems more likely to occur during oogenesis in women over the age of 35, we do not know with certainty why this is so. However, one observation may be relevant. Meiosis is initiated in all the eggs of a human female when she is still a fetus, until the point where the homologs synapse and recombination has begun. Then oocyte development is arrested in meiosis I. Thus, all primary oocytes have been formed by birth. When ovulation begins at puberty, meiosis is reinitiated in one egg during each ovulatory cycle and continues into meiosis II. The process is once again arrested after ovulation and is not completed unless fertilization occurs.

The end result of this progression is that each ovum that is released has been arrested in meiosis I for about a month longer than the one released during the preceding cycle. As a consequence, women 30 or 40 years old produce ova that are significantly older and that have been arrested longer than those they ovulated 10 or 20 years previously. However, no direct evidence proves that ovum age is the cause of the increased incidence of nondisjunction leading to Down syndrome.

These statistics obviously pose a serious problem for the woman who becomes pregnant late in her reproductive years. Genetic counseling early in such pregnancies is highly recommended. Counseling informs prospective parents about the probability that their child will be affected and educates them about Down syndrome. Although some individuals with Down syndrome must be institutionalized, others benefit greatly from special education programs and may be cared for at home. Down syndrome children in general are noted for their affectionate, loving nature. A genetic counselor may also recommend a prenatal diagnostic technique in which fetal cells are isolated and cultured.

In **amniocentesis** and **chorionic villus sampling (CVS)**, the two most familiar approaches, fetal cells are obtained from the amniotic fluid or the chorion of the placenta, respectively. In a newer approach, fetal cells and DNA are derived directly from the maternal circulation, a technique referred to as **noninvasive prenatal genetic diagnosis (NIPGD)**. Requiring only a 10 mL maternal blood sample, this procedure will become increasingly more common because it poses no risk to the fetus. After fetal cells are obtained and cultured, the karyotype can be determined by cytogenetic analysis. If the fetus is diagnosed as being affected, a therapeutic abortion is one option currently available to parents. Obviously, this is a difficult decision involving a number of religious and ethical issues.

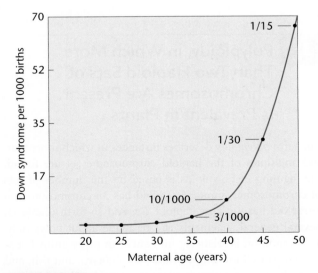

FIGURE 3 Incidence of Down syndrome births related to maternal age.

Since Down syndrome is caused by a random error—nondisjunction of chromosome 21 during maternal or paternal meiosis—the occurrence of the disorder is *not* expected to be inherited. Nevertheless, Down syndrome occasionally runs in families. These instances, referred to as familial Down syndrome, involve a translocation of chromosome 21, another type of chromosomal aberration, which we will discuss later in the chapter.

Human Aneuploidy

Besides Down syndrome, only two human trisomies, and no autosomal monosomies, survive to term: **Patau** and **Edwards syndromes** (47,13+ and 47,18+, respectively). Even so, these individuals manifest severe malformations and early lethality. **Figure 4** illustrates the abnormal karyotype and the many defects characterizing Patau infants.

The above observation leads us to ask whether many other aneuploid conditions arise but that the affected fetuses do not survive to term. That this is the case has been confirmed by karyotypic analysis of spontaneously aborted fetuses. These studies reveal two striking statistics: (1) Approximately 20 percent of all conceptions terminate in spontaneous abortion (some estimates are considerably higher); and (2) about 30 percent of all spontaneously aborted fetuses

demonstrate some form of chromosomal imbalance. This suggests that at least 6 percent (0.20×0.30) of conceptions contain an abnormal chromosome complement. A large percentage of fetuses demonstrating chromosomal abnormalities are aneuploids.

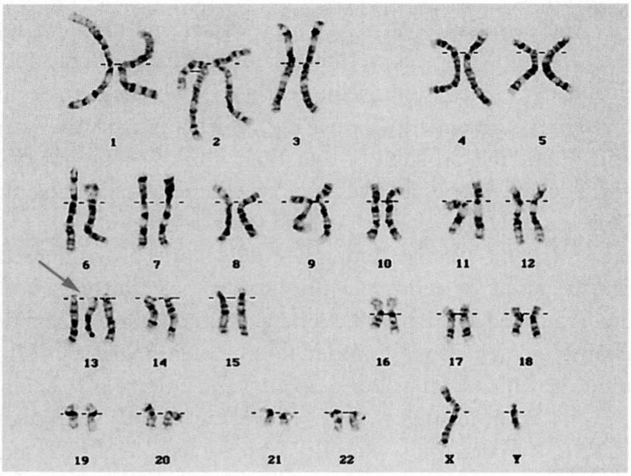

An extensive review of this subject by David H. Carr has revealed that a significant percentage of aborted fetuses are trisomic for one of the chromosome groups. Trisomies for every human chromosome have been recovered. Interestingly, the monosomy with the highest incidence among abortuses is the 45,X condition, which produces an infant with Turner syndrome if the fetus survives to term. Autosomal monosomies are seldom found, however, even though nondisjunction should produce $n - 1$ gametes with a frequency equal to $n + 1$ gametes. This finding suggests that gametes lacking a single chromosome are functionally impaired to a serious degree or that the embryo dies so early in its development that recovery occurs infrequently. We discussed the potential causes of monosomic lethality earlier in this chapter. Carr's study also found various forms of polyploidy and other miscellaneous chromosomal anomalies.

These observations support the hypothesis that normal embryonic development requires a precise diploid complement of chromosomes to maintain the delicate equilibrium in the expression of genetic information. The prenatal mortality of most aneuploids provides a barrier against the introduction of these genetic anomalies into the human population.

ESSENTIAL POINT ■ ■ ■

Studies of monosomic and trisomic disorders are increasing our understanding of the delicate genetic balance that is essential for normal development.

Mental retardation
Growth failure
Low-set, deformed ears
Deafness
Atrial septal defect
Ventricular septal defect
Abnormal polymorphonuclear granulocytes

Microcephaly
Cleft lip and palate
Polydactyly
Deformed finger nails
Kidney cysts
Double ureter
Umbilical hernia
Developmental uterine abnormalities
Cryptorchidism

FIGURE 4 The karyotype and phenotypic description of an infant with Patau syndrome, where three members of the D-group chromosome 13 are present, creating the 47,13+ condition.

3 | Polyploidy, in Which More Than Two Haploid Sets of Chromosomes Are Present, Is Prevalent in Plants

The term *polyploidy* describes instances in which more than two multiples of the haploid chromosome set are found. The naming of polyploids is based on the number of sets of chromosomes found: A triploid has $3n$ chromosomes; a tetraploid has $4n$; a pentaploid, $5n$; and so forth (Table 1). Several general statements can be made about polyploidy. This condition is relatively infrequent in many animal species but is well known in lizards, amphibians, and fish, and is much more common in plant species. Usually, odd numbers of chromosome sets are not reliably maintained from generation to generation because a polyploid organism with an uneven number of homologs often does not produce genetically balanced gametes. For this reason, triploids, pentaploids, and so on, are not usually found in plant species that depend solely on sexual reproduction for propagation.

Polyploidy originates in two ways: (1) The addition of one or more extra sets of chromosomes, identical to the normal haploid complement of the same species, resulting in **autopolyploidy;** or (2) the combination of chromosome sets from different species occurring as a consequence of hybridization, resulting in **allopolyploidy** (from the Greek word *allo,* meaning "other" or "different"). The distinction between auto- and allopolyploidy is based on the genetic origin of the extra chromosome sets, as shown in **Figure 5**.

In our discussion of polyploidy, we use the following symbols to clarify the origin of additional chromosome sets. For example, if A represents the haploid set of chromosomes of any organism, then

$$A = a_1 + a_2 + a_3 + a_4 + \cdots + a_n$$

where $a_1, a_2,$ and so on, are individual chromosomes and n is the haploid number. A normal diploid organism is represented simply as AA.

Autopolyploidy

In autopolyploidy, each additional set of chromosomes is identical to the parent species. Therefore, triploids are represented as AAA, tetraploids are $AAAA$, and so forth.

Autotriploids arise in several ways. A failure of all chromosomes to segregate during meiotic divisions can produce a diploid gamete. If such a gamete is fertilized by a haploid gamete, a zygote with three sets of chromosomes is produced. Or, rarely, two sperm may fertilize an ovum, resulting in a triploid zygote. Triploids are also produced under experimental conditions by crossing diploids with tetraploids. Diploid organisms produce gametes with n chromosomes, while tetraploids produce $2n$ gametes. Upon fertilization, the desired triploid is produced.

Because they have an even number of chromosomes, **autotetraploids** ($4n$) are theoretically more likely to be found in nature than are autotriploids. Unlike triploids, which often produce genetically unbalanced gametes with odd numbers of chromosomes, tetraploids are more likely to produce balanced gametes when involved in sexual reproduction.

How polyploidy arises naturally is of great interest to geneticists. In theory, if chromosomes have replicated, but the parent cell never divides and instead reenters interphase, the chromosome number will be doubled. That this very likely occurs is supported by the observation that tetraploid cells can be produced experimentally from diploid cells. This is accomplished by applying cold or heat shock to meiotic cells or by applying colchicine to somatic cells undergoing mitosis. Colchicine, an alkaloid derived from the autumn crocus, interferes with spindle formation, and thus replicated chromosomes cannot separate at anaphase and do not migrate to the poles. When colchicine is removed, the cell can reenter interphase. When the paired sister chromatids separate and uncoil, the nucleus contains twice the diploid number of chromosomes and is therefore $4n$. This process is shown in **Figure 6**.

In general, autopolyploids are larger than their diploid relatives. This increase seems to be due to larger cell size rather than greater cell number. Although autopolyploids do not contain new or unique information compared with their diploid relatives, the flower and fruit of plants are often increased in size, making such varieties of greater horticultural or commercial value. Economically important triploid plants include several potato species of the genus *Solanum,* Winesap apples, commercial bananas, seedless watermelons, and the cultivated tiger lily *Lilium tigrinum.* These plants are propagated asexually. Diploid bananas contain hard seeds, but the commercial, triploid, "seedless" variety has edible seeds. Tetraploid alfalfa, coffee, peanuts, and McIntosh apples are also of economic value because they are either larger or grow more vigorously than do their diploid or triploid counterparts. Many of the most popular varieties of hosta plant

FIGURE 5 Contrasting chromosome origins of an autopolyploid versus an allopolyploid karyotype.

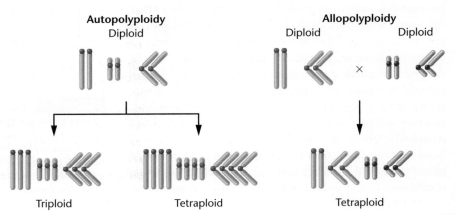

FIGURE 6 The potential involvement of colchicine in doubling the chromosome number. Two pairs of homologous chromosomes are shown. While each chromosome had replicated its DNA earlier during interphase, the chromosomes do not appear as double structures until late prophase. When anaphase fails to occur normally, the chromosome number doubles if the cell reenters interphase.

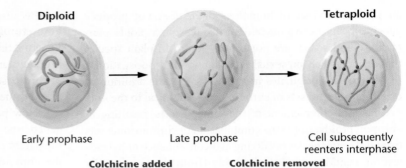

are tetraploid. In each case, leaves are thicker and larger, the foliage is more vivid, and the plant grows more vigorously. The commercial strawberry is an octoploid.

How cells with increased ploidy values express different phenotypes from their diploid counterparts has been investigated. Gerald Fink and his colleagues have been able to create strains of the yeast *Saccharomyces cerevisiae* with one, two, three, or four copies of the genome and then examined the expression levels of all genes during the cell cycle. Using the stringent standards of at least a tenfold increase or decrease of gene expression, Fink and coworkers identified numerous cases where, as ploidy increased, gene expression either increased or decreased at least tenfold. Among these cases are two genes that encode **G1 cyclins,** which are repressed when ploidy increases. G1 cyclins facilitate the cell's movement through G1 of the cell cycle, which is thus delayed when expression of these genes is repressed. The polyploid cell stays in the G1 phase longer and, on average, grows to a larger size before it moves beyond the G1 stage of the cell cycle, providing a clue as to how other polyploids demonstrate increased cell size.

Allopolyploidy

Polyploidy can also result from hybridizing two closely related species. If a haploid ovum from a species with chromosome sets AA is fertilized by sperm from a species with sets BB, the resulting hybrid is AB, where $A = a_1, a_2, a_3, \ldots a_n$ and $B = b_1, b_2, b_3, \ldots b_n$. The hybrid organism may be sterile because of its inability to produce viable gametes. Most often, this occurs when some or all of the a and b chromosomes are not homologous and therefore cannot synapse in meiosis. As a result, unbalanced genetic conditions result. If, however, the new AB genetic combination undergoes a natural or an induced chromosomal doubling, two copies of all a chromosomes and two copies of all b chromosomes will be present, and they will pair during meiosis. As a result, a fertile $AABB$ tetraploid is produced. These events are shown in **Figure 7**. Since this polyploid contains the equivalent of four haploid genomes derived from separate species, such an organism is called an **allotetraploid.** When both original species are

known, an equivalent term, **amphidiploid,** is preferred in describing the allotetraploid.

Amphidiploid plants are often found in nature. Their reproductive success is based on their potential for forming balanced gametes. Since two homologs of each specific chromosome are present, meiosis occurs normally (Figure 7) and fertilization successfully propagates the plant sexually. This discussion assumes the simplest situation, where none of the

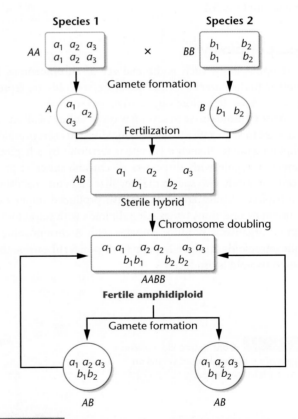

FIGURE 7 The origin and propagation of an amphidiploid. Species 1 contains genome A consisting of three distinct chromosomes, a_1, a_2, and a_3. Species 2 contains genome B consisting of two distinct chromosomes, b_1 and b_2. Following fertilization between members of the two species and chromosome doubling, a fertile amphidiploid containing two complete diploid genomes ($AABB$) is formed.

chromosomes in set *A* are homologous to those in set *B*. In amphidiploids formed from closely related species, some homology between *a* and *b* chromosomes is likely. Allopolyploids are rare in most animals because mating behavior is most often species-specific, and thus the initial step in hybridization is unlikely to occur.

A classical example of amphidiploidy in plants is the cultivated species of American cotton, *Gossypium* (**Figure 8**). This species has 26 pairs of chromosomes: 13 are large and 13 are much smaller. When it was discovered that Old World cotton had only 13 pairs of large chromosomes, allopolyploidy was suspected. After an examination of wild American cotton revealed 13 pairs of small chromosomes, this speculation was strengthened. J. O. Beasley reconstructed the origin of cultivated cotton experimentally by crossing the Old World strain with the wild American strain and then treating the hybrid with colchicine to double the chromosome number. The result of these treatments was a fertile amphidiploid variety of cotton. It contained 26 pairs of chromosomes as well as characteristics similar to the cultivated variety.

Amphidiploids often exhibit traits of both parental species. An interesting example involves the grasses wheat and rye. Wheat (genus *Triticum*) has a basic haploid genome of seven chromosomes. In addition to normal diploids ($2n = 14$), cultivated autopolyploids exist, including tetraploid ($4n = 28$) and hexaploid ($6n = 42$) species. Rye (genus *Secale*) also has a genome consisting of seven chromosomes. The only cultivated species is the diploid plant ($2n = 14$).

Using the technique outlined in Figure 7, geneticists have produced various hybrids. When tetraploid wheat is crossed with diploid rye and the F$_1$ plants are treated with colchicine, a hexaploid variety ($6n = 42$) is obtained; the hybrid, designated *Triticale,* represents a new genus. Other *Triticale* varieties have been created. These hybrid plants demonstrate characteristics of both wheat and rye. For example, they combine the high-protein content of wheat with rye's high

content of the amino acid lysine, which is low in wheat and thus is a limiting nutritional factor. Wheat is considered to be a high-yielding grain, whereas rye is noted for its versatility of growth in unfavorable environments. *Triticale* species that combine both traits have the potential of significantly increasing grain production. This and similar programs designed to improve crops through hybridization have long been under way in several developing countries.

2 When two plants belonging to the same genus but different species are crossed, the F$_1$ hybrid is viable and has more ornate flowers. Unfortunately, this hybrid is sterile and can only be propagated by vegetative cuttings. Explain the sterility of the hybrid and what would have to occur for the sterility of this hybrid to be reversed.

■ **Hint** *This problem involves an understanding of allopolyploid plants. The key to its solution is to focus on the origin and composition of the chromosomes in the F$_1$ and how they might be manipulated.*

ESSENTIAL POINT ■ ■ ■

When complete sets of chromosomes are added to the diploid genome, these sets can have an identical or a diverse genetic origin, creating either autopolyploidy or allopolyploidy, respectively.

4 Variation Occurs in the Composition and Arrangement of Chromosomes

The second general class of chromosome aberrations includes changes that delete, add, or rearrange substantial portions of one or more chromosomes. Included in this broad category are deletions and duplications of genes or part of a chromosome and rearrangements of genetic material in which a chromosome segment is inverted, exchanged with a segment of a nonhomologous chromosome, or merely transferred to another chromosome. Exchanges and transfers are called translocations, in which the locations of genes are altered within the genome. These types of chromosome alterations are illustrated in **Figure 9**.

In most instances, these structural changes are due to one or more breaks along the axis of a chromosome, followed by either the loss or rearrangement of genetic material. Chromosomes can break spontaneously, but the rate of breakage may increase in cells exposed to chemicals or radiation. The ends produced at points of breakage are "sticky" and can rejoin other broken ends. If breakage and rejoining do not reestablish the original relationship and if the alteration

FIGURE 8 The pods of the amphidiploid form of *Gossypium,* the cultivated American cotton plant.

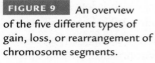

FIGURE 9 An overview of the five different types of gain, loss, or rearrangement of chromosome segments.

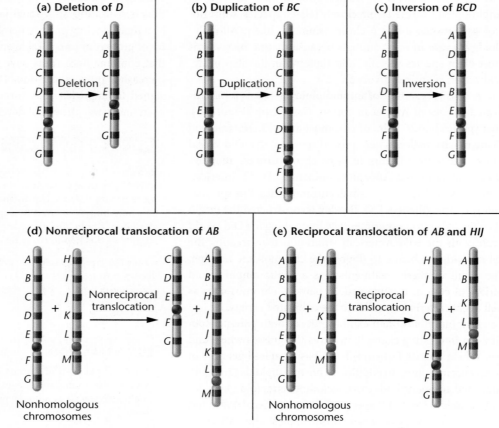

occurs in germ plasm, the gametes will contain the structural rearrangement, which is heritable.

If the aberration is found in one homolog, but not the other, the individual is said to be *heterozygous for the aberration*. In such cases, unusual but characteristic pairing configurations are formed during meiotic synapsis. These patterns are useful in identifying the type of change that has occurred. If no loss or gain of genetic material occurs, individuals bearing the aberration "heterozygously" are likely to be unaffected phenotypically. However, the unusual pairing arrangements often lead to gametes that are duplicated or deficient for some chromosomal regions. When this occurs, the offspring of "carriers" of certain aberrations have an increased probability of demonstrating phenotypic changes.

5   A Deletion Is a Missing Region of a Chromosome

When a chromosome breaks in one or more places and a portion of it is lost, the missing piece is called a **deletion** (or a **deficiency**). The deletion can occur either near one end or within the interior of the chromosome. These are **terminal** and **intercalary deletions,** respectively **[Figure 10(a) and**

(b)]. The portion of the chromosome that retains the centromere region is usually maintained when the cell divides, whereas the segment without the centromere is eventually lost in progeny cells following mitosis or meiosis. For synapsis to occur between a chromosome with a large intercalary deletion and a normal homolog, the unpaired region of the normal homolog must "buckle out" into a **deletion, or compensation, loop [Figure 10(c)]**.

If only a small part of a chromosome is deleted, the organism might survive. However, a deletion of a portion of a chromosome need not be very great before the effects become severe. We see an example of this in the following discussion of the cri du chat syndrome in humans. If even more genetic information is lost as a result of a deletion, the aberration is often lethal, in which case the chromosome mutation never becomes available for study.

Cri du Chat Syndrome in Humans

In humans, the **cri du chat syndrome** results from the deletion of a small terminal portion of chromosome 5. It might be considered a case of *partial monosomy,* but since the region that is missing is so small, it is better referred to as a **segmental deletion**. This syndrome was first reported by Jérôme LeJeune in 1963, when he described the clinical symptoms,

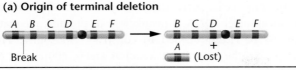

(a) Origin of terminal deletion

(b) Origin of intercalary deletion

(c) Formation of deletion loop

FIGURE 10   Origins of (a) a terminal and (b) an intercalary deletion. In (c), pairing occurs between a normal chromosome and one with an intercalary deletion by looping out the undeleted portion to form a deletion (or compensation) loop.

including an eerie cry similar to the meowing of a cat, after which the syndrome is named. This syndrome is associated with the loss of a small, variable part of the short arm of chromosome 5 **(Figure 11)**. Thus, the genetic constitution may be designated as 46,5p−, meaning that the individual has all 46 chromosomes but that some or all of the p arm (the petite, or short, arm) of one member of the chromosome 5 pair is missing.

Infants with this syndrome may exhibit anatomic malformations, including gastrointestinal and cardiac complications, and they are often mentally retarded. Abnormal development

of the glottis and larynx (leading to the characteristic cry) is typical of this syndrome.

Since 1963, hundreds of cases of cri du chat syndrome have been reported worldwide. An incidence of 1 in 25,000–50,000 live births has been estimated. Most often, the condition is not inherited but instead results from the sporadic loss of chromosomal material in gametes. The length of the short arm that is deleted varies somewhat; longer deletions appear to have a greater impact on the physical, psychomotor, and mental skill levels of those children who survive. Although the effects of the syndrome are severe, most individuals achieve motor and language skills and may be home-cared. In 2004, it was reported that the portion of the chromosome that is missing contains the *TERT* gene, which encodes telomerase reverse transcriptase, an enzyme essential for the maintenance of telomeres during DNA replication. Whether the absence of this gene on one homolog is related to the multiple phenotypes of cri du chat infants is still unknown.

6 A Duplication Is a Repeated Segment of a Chromosome

When any part of the genetic material—a single locus or a large piece of a chromosome—is present more than once in the genome, it is called a **duplication**. As in deletions, pairing in heterozygotes can produce a compensation loop. Duplications may arise as the result of unequal crossing over between synapsed chromosomes during meiosis **(Figure 12)** or through a replication error prior to meiosis. In the former case, both a duplication and a deletion are produced.

We consider three interesting aspects of duplications. First, they may result in gene redundancy. Second, as with deletions, duplications may produce phenotypic variation. Third, according to one convincing theory, duplications have also been an important source of genetic variability during evolution.

FIGURE 11   A representative karyotype and a photograph of a child exhibiting cri du chat syndrome (46,5p−). In the karyotype, the arrow identifies the absence of a small piece of the short arm of one member of the chromosome 5 homologs.

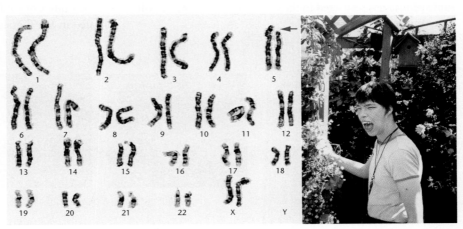

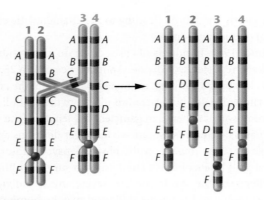

The origin of duplicated and deficient regions of chromosomes as a result of unequal crossing over. The tetrad on the left is mispaired during synapsis. A single crossover between chromatids 2 and 3 results in the deficient (chromosome 2) and duplicated (chromosome 3) chromosomal regions shown on the right. The two chromosomes uninvolved in the crossover event remain normal in gene sequence and content.

Gene Amplification—Ribosomal RNA Genes

Although many gene products are not needed in every cell of an organism, other gene products are known to be essential components of all cells. For example, ribosomal RNA must be present in abundance to support protein synthesis. The more metabolically active a cell is, the higher the demand for this molecule. We might hypothesize that a single copy of the gene encoding rRNA is inadequate in many cells. Studies using the technique of molecular hybridization, which enables us to determine the percentage of the genome that codes for specific RNA sequences, show that our hypothesis is correct. Indeed, multiple copies of genes code for rRNA. Such DNA is called **rDNA,** and the general phenomenon is referred to as **gene amplification.** For example, in the common intestinal bacterium *Escherichia coli* (*E. coli*), about 0.7 percent of the haploid genome consists of rDNA—the equivalent of seven copies of the gene. In *Drosophila melanogaster,* 0.3 percent of the haploid genome, equivalent to 130 gene copies, consists of rDNA. Although the presence of multiple copies of the same gene is not restricted to those coding for rRNA, we will focus on them in this section.

In some cells, particularly oocytes, even the normal amplification of rDNA is insufficient to provide adequate amounts of rRNA needed to construct ribosomes. For example, in the amphibian *Xenopus laevis*, 400 copies of rDNA are present per haploid genome. These genes are all found in a single area of the chromosome known as the **nucleolar organizer region (NOR).** In *Xenopus* oocytes, the NOR is selectively replicated to further increase rDNA copies, and each new set of genes is released from its template. Each set forms a small nucleolus, and as many as 1500 of these "micronucleoli" have been observed in a single oocyte. If we multiply the number of micronucleoli (1500) by the number of gene copies in each NOR (400), we see that amplification in *Xenopus* oocytes can result in over half a million gene copies! If each copy is transcribed only 20 times during the maturation of the oocyte, in theory, sufficient copies of rRNA are produced to result in well over 12 million ribosomes.

The *Bar* Mutation in *Drosophila*

Duplications can cause phenotypic variation that might at first appear to be caused by a simple gene mutation. The *Bar*-eye phenotype in *Drosophila* (**Figure 13**) is a classic example. Instead of the normal oval-eye shape, *Bar*-eyed flies have narrow, slitlike eyes. This phenotype is inherited in the same way as a dominant X-linked mutation.

In the early 1920s, Alfred H. Sturtevant and Thomas H. Morgan discovered and investigated this "mutation." Normal wild-type females (B^+/B^+) have about 800 facets in each eye. Heterozygous females (B/B^+) have about 350 facets, while homozygous females (B/B) average only about 70 facets. Females were occasionally recovered with even fewer facets and were designated as *double Bar* (B^D/B^+)

About 10 years later, Calvin Bridges and Herman J. Muller compared the polytene X chromosome banding pattern of the *Bar* fly with that of the wild-type fly. These chromosomes contain specific banding patterns that have been well categorized into regions. Their studies revealed that one copy of the region designated as 16A is present on both X-chromosomes of wild-type flies but that this region was duplicated in *Bar* flies and triplicated in *double Bar* flies. These

Bar-eye phenotypes in contrast to the wild-type eye in *Drosophila* (shown in the left panel).

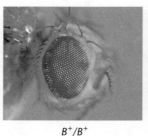

B^+/B^+

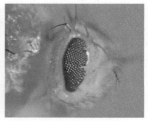

B/B^+

B/B

observations provided evidence that the *Bar* phenotype is not the result of a simple chemical change in the gene but is instead a duplication.

The Role of Gene Duplication in Evolution

During the study of evolution, it is intriguing to speculate on the possible mechanisms of genetic variation. The origin of unique gene products present in more recently evolved organisms but absent in ancestral forms is a topic of particular interest. In other words, how do "new" genes arise?

In 1970, Susumo Ohno published a provocative monograph, *Evolution by Gene Duplication,* in which he suggested that gene duplication is essential to the origin of new genes during evolution. Ohno's thesis is based on the supposition that the gene products of many genes, present as only a single copy in the genome, are indispensable to the survival of members of any species during evolution. Therefore, unique genes are not free to accumulate mutations sufficient to alter their primary function and give rise to new genes.

However, if an essential gene is duplicated in the germ line, major mutational changes in this extra copy will be tolerated in future generations because the original gene provides the genetic information for its essential function. The duplicated copy will be free to acquire many mutational changes over extended periods of time. Over short intervals, the new genetic information may be of no practical advantage. However, over long evolutionary periods, the duplicated gene may change sufficiently so that its product assumes a divergent role in the cell. The new function may impart an "adaptive" advantage to organisms, enhancing their fitness. Ohno has outlined a mechanism through which sustained genetic variability may have originated.

Ohno's thesis is supported by the discovery of genes that have a substantial amount of their organization and DNA sequence in common, but whose gene products are distinct. For example, trypsin and chymotrypsin fit this description, as do myoglobin and the various forms of hemoglobin. The DNA sequence is so similar (homologous) in each case that we may conclude that members of each pair of genes arose from a common ancestral gene through duplication. During

evolution, the related genes diverged sufficiently that their products became unique.

A new debate has begun concerning a second aspect of Ohno's thesis—that major evolutionary jumps, such as the transition from invertebrates to vertebrates, may have involved the duplication of entire genomes. Ohno has suggested that this might have occurred several times during the course of evolution. Although it seems clear that genes, and even segments of chromosomes, have been duplicated, there is not yet any compelling evidence to convince evolutionary biologists that genome duplication has been responsible.

ESSENTIAL POINT ■ ■ ■

Deletions or duplications of segments of a gene or a chromosome may be the source of mutant phenotypes such as cri du chat syndrome in humans and *Bar* eyes in *Drosophila,* while duplications can be particularly important as a source of redundant or new genes.

---

7 Inversions Rearrange the Linear Gene Sequence

The **inversion**, another class of structural variation, is a type of chromosomal aberration in which a segment of a chromosome is turned around 180 degrees within a chromosome. An inversion does not involve a loss of genetic information but simply rearranges the linear gene sequence. An inversion requires breaks at two points along the length of the chromosome and subsequent reinsertion of the inverted segment. **Figure 14** illustrates how an inversion might arise. By forming a chromosomal loop prior to breakage, the newly created "sticky ends" are brought close together and rejoined.

The inverted segment may be short or quite long and may or may not include the centromere. If the centromere is not part of the rearranged chromosome segment, it is a **paracentric inversion,** which is the type shown in Figure 14. If the centromere is part of the inverted segment, it is described as a **pericentric inversion.**

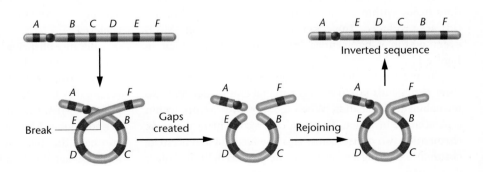

FIGURE 14 One possible origin of a paracentric inversion.

Copy Number Variants (CNVs)—Duplications and Deletions at the Molecular Level

Genomic investigations that focus on the DNA sequences in humans are providing new insights into our understanding of duplications and deletions. These variations, often involving thousands or even millions of base pairs, and even entire genes, may prove to play crucial roles in many of our individual attributes, such as sensitivity to drugs and susceptibility to disease. These differences, because they represent large DNA sequences, are termed **copy number variants (CNVs),** and are found in both coding and noncoding regions of the genome. Mutant CNVs show an increase or a decrease in copy number in comparison to a reference genome from a normal individual.

In 2004, two research groups independently described the presence of CNVs in the genomes of healthy individuals with no known genetic disorders. CNVs were defined as regions of DNA at least 1 kb in length (1000 base pairs) that display at least 90 percent sequence identity. This initial study revealed 50 loci consisting of CNVs, and in 2005, several other groups began sifting through the genome in search of CNVs, defining almost 300 additional sites. The current number of CNV sites has now risen to include approximately 12 percent of the human genome.

Current CNV studies have focused on finding associations with human diseases. CNVs appear to have both positive and negative associations with many diseases in which the genetic basis is not yet fully understood. For example, an association has been reported between CNVs and autism, a neurodevelopmental disorder that impairs communication, behavior, and social interaction. Interestingly, a mutant CNV site has been found to appear *de novo* (anew) in 10 percent of so-called sporadic cases of autism, where unaffected parents lack the CNV mutation. This is in contrast to only 2 percent of affected individuals where the disease appears to be familial (run in the family). Similarly, a higher than average copy number of the gene *CCL3L1* imparts an HIV-suppressive effect during viral infection, diminishing the progression to AIDS. Another research group has associated specific mutant CNV sites with certain subset populations of individuals with lung cancer—the greater number of copies of the *EGFR* (*Epidermal Growth Factor Receptor*) gene, the more responsive are patients with non–small-cell lung cancer to treatment. Finally, the greater the reduction in copy number of the gene designated *DEFB*, the greater the risk of developing Crohn's disease, a condition affecting the colon. Relevant to this chapter, these findings reveal that duplications and deletions are no longer restricted to textbook examples of these chromosomal mutations.

Although inversions appear to have a minimal impact on the individuals bearing them, their consequences are of great interest to geneticists. Organisms that are heterozygous for inversions may produce aberrant gametes that have a major impact on their offspring.

Consequences of Inversions during Gamete Formation

If only one member of a homologous pair of chromosomes has an inverted segment, normal *linear synapsis* during meiosis is not possible. Organisms with one inverted chromosome and one noninverted homolog are called **inversion heterozygotes.** Pairing between two such chromosomes in meiosis is accomplished only if they form an **inversion loop (Figure 15)**.

If crossing over does not occur within the inverted segment of the inversion loop, the homologs will segregate, which results in two normal and two inverted chromatids that are distributed into gametes. However, if crossing over does occur within the inversion loop, abnormal chromatids are produced. The effect of a single crossover (SCO) event within a paracentric inversion is diagrammed in Figure 15.

In any meiotic tetrad, a single crossover between non-sister chromatids produces two parental chromatids and two recombinant chromatids. When the crossover occurs within a paracentric inversion, however, one recombinant **dicentric chromatid** (two centromeres) and one recombinant **acentric chromatid** (lacking a centromere) are produced. Both contain duplications and deletions of chromosome segments as well.

During anaphase, an acentric chromatid moves randomly to one pole or the other or may be lost, while a dicentric chromatid is pulled in two directions. This polarized movement produces *dicentric bridges* that are cytologically recognizable. A dicentric chromatid usually breaks at some point so that part of the chromatid goes into one gamete and part into another gamete during the reduction divisions. Therefore, gametes containing either recombinant chromatid are deficient in genetic material. In animals, when such a gamete participates in fertilization, the zygote most often develops abnormally, inviable embryos are produced, and lethality is the final result. In plants, gametes receiving such aberrant chromatids fail to develop normally, leading to aborted pollen or ovules. Thus, fertilization is not achieved.

Because offspring bearing crossover gametes are inviable and not recovered, it *appears* as if the inversion suppresses crossing over. Actually, in inversion heterozygotes, the inversion has the effect of *suppressing the recovery of crossover products* when chromosome exchange occurs within the inverted region. Moreover, up to one-half of the viable gametes have the inverted chromosome, and the inversion will be perpetuated within the species. The cycle will be repeated continuously during meiosis in future generations.

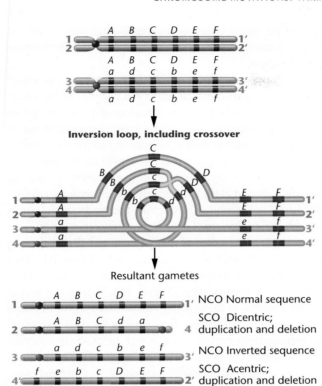

FIGURE 15 The effects of a single crossover (SCO) within an inversion loop in a paracentric inversion heterozygote, where two altered chromosomes are produced, one acentric and one dicentric. Both chromosomes also contain duplicated and deficient regions.

Evolutionary Advantages of Inversions

Because recovery of crossover products is suppressed in inversion heterozygotes, groups of specific alleles at adjacent loci within inversions may be preserved from generation to generation. If the alleles of the involved genes confer a survival advantage on the organisms maintaining them, the inversion is beneficial to the evolutionary survival of the species. For example, if a set of alleles *ABcDef* is more adaptive than sets *AbCdeF* or *abcdEF*, effective gametes will contain this favorable set of genes, undisrupted by crossing over.

In laboratory studies, the same principle is applied using **balancer chromosomes,** which contain inversions. When an organism is heterozygous for a balancer chromosome, desired sequences of alleles are preserved during experimental work.

NOW SOLVE THIS

3 What is the effect of a rare double crossover within a chromosome segment that is heterozygous for a paracentric inversion?

■ **Hint** *This problem involves an understanding of how homologs synapse in the presence of a heterozygous paracentric inversion. The key to its solution is to draw out the tetrad and follow the chromatids undergoing a double crossover.*

8 Translocations Alter the Location of Chromosomal Segments in the Genome

Translocation, as the name implies, is the movement of a chromosomal segment to a new location in the genome. Reciprocal translocation, for example, involves the exchange of segments between two nonhomologous chromosomes. The least complex way for this event to occur is for two nonhomologous chromosome arms to come close to each other so that an exchange is facilitated. **Figure 16(a)** shows a simple reciprocal translocation in which only two breaks are required. If the exchange includes internal chromosome segments, four breaks are required, two on each chromosome.

The genetic consequences of reciprocal translocations are, in several instances, similar to those of inversions. For example, genetic information is not lost or gained. Rather, there is only a rearrangement of genetic material. The presence of a translocation does not, therefore, directly alter the viability of individuals bearing it.

Homologs that are heterozygous for a reciprocal translocation undergo unorthodox synapsis during meiosis. As shown in **Figure 16(b),** pairing results in a crosslike configuration. As with inversions, genetically unbalanced gametes are also produced as a result of this unusual alignment during meiosis. In the case of translocations, however, aberrant gametes are not necessarily the result of crossing over. To see how unbalanced gametes are produced, focus on the homologous centromeres in Figure 16(b) and **Figure 16(c).** According to the principle of independent assortment, the chromosome containing centromere 1 migrates randomly toward one pole of the spindle during the first meiotic anaphase; it travels along with *either* the chromosome having centromere 3 *or* the chromosome having centromere 4. The chromosome with centromere 2 moves to the other pole along with the chromosome containing *either* centromere 3 *or* centromere 4. This results in four potential meiotic products. The 1,4 combination contains chromosomes that are not involved in the translocation. The 2,3 combination, however, contains translocated chromosomes. These contain a complete complement of genetic information and are balanced. The other two potential products, the 1,3 and 2,4 combinations, contain chromosomes displaying duplicated and deleted segments. To simplify matters, crossover exchanges are ignored here.

When incorporated into gametes, the resultant meiotic products are genetically unbalanced. If they participate in fertilization, lethality often results. As few as 50 percent of the progeny of parents that are heterozygous for a reciprocal translocation survive. This condition, called **semisterility,** has an impact on the reproductive fitness of organisms, thus playing a role in evolution. Furthermore, in humans, such

(a) Possible origin of a reciprocal translocation between two nonhomologous chromosomes

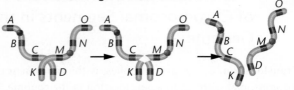

(b) Synapsis of translocation heterozygote

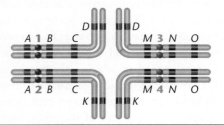

(c) Two possible segregation patterns leading to gamete formation

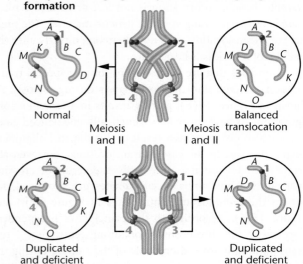

FIGURE 16 (a) Possible origin of a reciprocal translocation. (b) Synaptic configuration formed during meiosis in an individual that is heterozygous for the translocation. (c) Two possible segregation patterns, one of which leads to a normal and a balanced gamete (called alternate segregation) and one that leads to gametes containing duplications and deficiencies (called adjacent segregation).

an unbalanced condition results in partial monosomy or trisomy, leading to a variety of birth defects.

Translocations in Humans: Familial Down Syndrome

Research conducted since 1959 has revealed numerous translocations in members of the human population. One common type of translocation involves breaks at the extreme ends of the short arms of two nonhomologous acrocentric chromosomes. These small segments are lost, and the larger segments fuse at their centromeric region. This type of translocation produces a new, large submetacentric or metacentric chromosome, often called a **Robertsonian translocation.**

One such translocation accounts for cases in which Down syndrome is familial (inherited). Earlier in this chapter, we pointed out that most instances of Down syndrome are due to trisomy 21. This chromosome composition results from nondisjunction during meiosis in one parent. Trisomy accounts for over 95 percent of all cases of Down syndrome. In such instances, the chance of the same parents producing a second affected child is extremely low. However, in the remaining families with a Down child, the syndrome occurs in a much higher frequency over several generations.

Cytogenetic studies of the parents and their offspring from these unusual cases explain the cause of **familial Down syndrome.** Analysis reveals that one of the parents contains a 14/21, D/G translocation **(Figure 17)**. That is, one parent has the majority of the G-group chromosome 21 translocated to one end of the D-group chromosome 14. This individual is phenotypically normal, even though he or she has only 45 chromosomes. During meiosis, one-fourth of the individual's gametes have two copies of chromosome 21: a normal chromosome and a second copy translocated to chromosome 14. When such a gamete is fertilized by a standard haploid gamete, the resulting zygote has 46 chromosomes but three copies of chromosome 21. These individuals exhibit Down syndrome. Other potential surviving offspring contain either the standard diploid genome (without a translocation) or the balanced translocation like the parent. Both cases result in normal individuals. Although not illustrated in Figure 17, two other gametes may be formed, though rarely. Such gametes are unbalanced, and upon fertilization, lethality occurs.

Knowledge of translocations, as described above, has allowed geneticists to resolve the seeming paradox of an inherited trisomic phenotype in an individual with an apparent diploid number of chromosomes. It is also unique that the "carrier," who has 45 chromosomes and exhibits a normal phenotype, does not contain the *complete* diploid amount of genetic material. A small region is lost from both chromosomes 14 and 21 during the translocation event. This occurs because the ends of both chromosomes have broken off prior to their fusion. These specific regions are known to be two of many chromosomal locations housing multiple copies of the genes encoding rRNA, the major component of ribosomes. Despite the loss of up to 20 percent of these genes, the carrier is unaffected.

> **ESSENTIAL POINT** ■ ■ ■
>
> Inversions and translocations may initially cause little or no loss of genetic information or deleterious effects. However, heterozygous combinations of the involved chromosome segments may result in genetically abnormal gametes following meiosis, with lethality or inviability often ensuing.

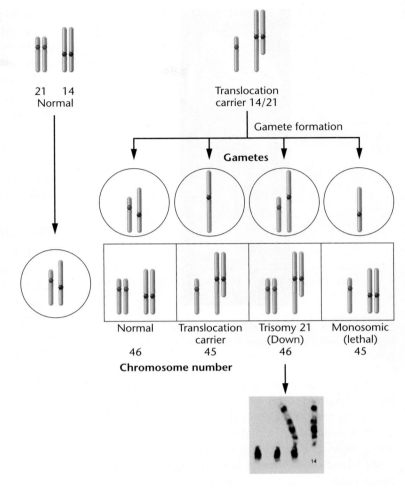

FIGURE 17 Chromosomal involvement and translocation in familial Down syndrome. The photograph shows the relevant chromosomes from a trisomy 21 offspring produced by a translocation carrier parent.

9 Fragile Sites in Human Chromosomes Are Susceptible to Breakage

We conclude this chapter with a brief discussion of the results of an intriguing discovery made around 1970 during observations of metaphase chromosomes prepared following human cell culture. In cells derived from certain individuals, a specific area along one of the chromosomes failed to stain, giving the appearance of a gap. In other individuals whose chromosomes displayed such morphology, the gaps appeared at other positions within the set of chromosomes. Such areas eventually became known as **fragile sites**, since they appeared to be susceptible to chromosome breakage when cultured in the absence of certain chemicals such as folic acid, which is normally present in the culture medium. Fragile sites were at first considered curiosities, until a strong association was subsequently shown to exist between one of the sites and a form of mental retardation.

The cause of the fragility at these sites is unknown. Because they represent points along the chromosome that are susceptible to breakage, these sites may indicate regions where the chromatin is not tightly coiled. Note that even though almost all studies of fragile sites have been carried out *in vitro* using mitotically dividing cells, clear associations have been established between several of these sites and the corresponding altered phenotype, including mental retardation and cancer.

Fragile-X Syndrome (Martin-Bell Syndrome)

Most fragile sites do not appear to be associated with any clinical syndrome. However, individuals bearing a folate-sensitive site on the X chromosome **(Figure 18)** exhibit the **fragile-X syndrome** (or **Martin-Bell syndrome**), the most common form of inherited mental retardation. This syndrome affects about 1 in 4000 males and 1 in 8000 females. Females carrying only one fragile X chromosome can be mentally retarded because it is a dominant trait. Fortunately, the trait is not fully expressed, as only about 30 percent of fragile X females are retarded, whereas about 80 percent of fragile X males are mentally retarded. In addition to mental retardation, affected males have characteristic long, narrow faces with protruding chins, enlarged ears, and increased testicular size.

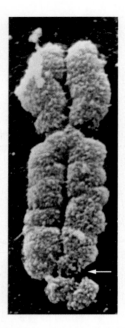

FIGURE 18 A fragile human X chromosome. The "gap" region, identified by the arrow, is associated with the fragile-X syndrome.

A gene that spans the fragile site may be responsible for this syndrome. This gene, known as *FMR-1,* is one of a growing number of genes that have been discovered in which a sequence of three nucleotides is repeated many times, expanding the size of the gene. This phenomenon, called **trinucleotide repeats,** is also recognized in other human disorders, including Huntington disease. In *FMR-1,* the trinucleotide sequence CGG is repeated in an untranslated area adjacent to the coding sequence of the gene (called the "upstream" region). The number of repeats varies immensely within the human population, and a high number correlates directly with expression of fragile-X syndrome. Normal individuals have between 6 and 54 repeats, whereas those with 55 to 230 repeats are considered "carriers" of the disorder. More than 230 repeats lead to expression of the syndrome.

It is thought that when the number of repeats reaches this level, the CGG regions of the gene become chemically modified so that the bases within and around the repeat are methylated, causing inactivation of the gene. The normal product of the gene is an RNA-binding protein, FMRP, known to be expressed in the brain. Evidence is now accumulating that directly links the absence of the protein with the cognitive defects associated with the syndrome.

From a genetic standpoint, a most interesting aspect of fragile-X syndrome is the instability of the CGG repeats. An individual with 6 to 54 repeats transmits a gene containing the same number to his or her offspring. However, those with 55 to 230 repeats, though not at risk to develop the syndrome, may transmit to their offspring a gene with an increased number of repeats. The number of repeats continues to increase in future generations, demonstrating the

phenomenon known as **genetic anticipation.** Once the threshold of 230 repeats is exceeded, expression of the malady becomes more severe in each successive generation as the number of trinucleotide repeats increases. Interestingly, expansion from the carrier status (55 to 230 repeats) to the syndrome status (over 230 repeats) occurs only during the transmission of the gene by the maternal parent, not by the paternal parent. Thus, a "carrier" male may transmit a stable chromosome to his daughter, who may subsequently transmit an unstable chromosome with an increased number of repeats to her offspring. Their grandfather was the source of the original chromosome.

The Link between Fragile Sites and Cancer

While the study of the fragile-X syndrome first brought unstable chromosome regions to the attention of geneticists, a link between an autosomal fragile site and lung cancer was reported in 1996 by Carlo Croce, Kay Huebner, and their colleagues. They have subsequently postulated that the defect is associated with the formation of a variety of different tumor types. Croce and Huebner first showed that the *FHIT* gene (standing for *f*ragile *h*istidine *t*riad), located within the well-defined fragile site designated as *FRA3B* on the p arm of chromosome 3, is often altered or missing in cells taken from tumors of individuals with lung cancer. More extensive studies have now revealed that the normal protein product of this gene is absent in cells of many other cancers, including those of the esophagus, breast, cervix, liver, kidney, pancreas, colon, and stomach. Genes such as *FHIT* that are located within fragile regions undoubtedly have an increased susceptibility to mutations and deletions.

Subsequently, Muller Fabbri and Kay Huebner, working with others in Croce's lab, identified and studied another fragile site, with most interesting results. Found within the *FRA16D* site on chromosome 16 is the *WWOX* gene. Like the *FHIT* gene, it has been implicated in a range of human cancers. In particular, like *FHIT,* it has been found to be either lost or genetically silenced in the large majority of lung tumors, as well as in cancer tissue of the breast, ovary, prostate, bladder, esophagus, and pancreas. When the gene is present but silent, its DNA is thought to be heavily methylated, rendering it inactive. Furthermore, the active gene is also thought to behave as a tumor suppressor, providing a surveillance function by recognizing cancer cells and inducing apoptosis, effectively eliminating them before malignant tumors can be initiated.

ESSENTIAL POINT

Fragile sites in human mitotic chromosomes have sparked research interest because one such site on the X chromosome is associated with the most common form of inherited mental retardation, while other autosomal sites have been linked to various forms of cancer.

GENETICS, TECHNOLOGY, AND SOCIETY

Down Syndrome and Prenatal Testing—The New Eugenics?

Down syndrome is the most common chromosomal abnormality seen in newborn babies. Each year, about 4000–5000 babies are born with Down syndrome in the United States. Prenatal diagnostic tests for Down syndrome have been available for decades, especially to older pregnant women who have an increased risk of bearing a Down syndrome child. Scientists estimate that there is an abortion rate of about 30 percent for fetuses that test positive for Down syndrome in the United States, and rates of up to 85 percent in other parts of the world, such as Taiwan and France. The reasons for aborting a Down syndrome fetus are varied but include the wish to have a normal healthy child and the desire to avoid the financial, physical, and emotional suffering that may occur in families that must raise these children.

In addition to Down syndrome testing, prenatal genetic tests for other chromosomal and gene-specific defects are entering medical practice. Over 1000 gene mutation tests are now offered to pregnant women. In addition, ultrasound and enzyme tests are standard procedures in obstetrical care. As new genomic methods gain prevalence, we will soon see inexpensive genetic screens for hundreds of genetic defects and gene variants.

Some prenatal tests have had a marked effect on infant mortality and disease rates. For example, in Taiwan, the genetic screening program for thalassemia mutations (resulting in severe anemia) has reduced the newborn disease rate from 5.6 per 100,000 births to 1.2. In some countries, infant mortality rates have declined by one-half, due to prenatal screening for congenital defects.

Many people agree that it is morally acceptable to prevent the birth of a genetically abnormal fetus. However, many others argue that prenatal genetic testing and the elimination of congenital disorders is unethical. These arguments often center around their views about abortion and self-determination. In addition, some people argue that prenatal genetic tests followed by selective abortion is eugenic.

These eugenics arguments are highly emotionally charged, and some people contend they are inappropriate arguments to use when discussing screening for genetic diseases. It is safe to say that most people on both sides of the eugenics debate are unclear about the definition of eugenics and even less clear about how it may relate to prenatal genetic testing.

So, what is eugenics? And how does eugenics apply, if at all, to screening for Down syndrome and other human genetic defects? The term *eugenics* was first defined by Francis Galton in 1883 as "the science which deals with all influences that improve the inborn qualities of a race; also with those that develop them to the utmost advantage." Galton believed that human traits such as intelligence and personality were hereditary and that humans could selectively mate with each other to create gifted groups of people—analogous to the creation of purebred dogs with specific traits. Galton did not propose coercion but thought that people would voluntarily enter matings for beneficial outcomes.

In the early to mid-twentieth century, countries throughout the world adopted eugenic policies with the aim of enhancing desirable human traits and eliminating undesirable ones. Two distinct approaches to eugenics developed: *positive eugenics,* in which those with desirable mental and physical traits were encouraged to reproduce, and *negative eugenics,* in which those with undesirable traits were discouraged or prevented from reproducing. Many countries, including Britain, Canada, and the United States, enacted compulsory sterilization programs for the "feebleminded," mentally ill, and criminals. The eugenic policies of Nazi Germany were particularly infamous, resulting in forced human genetic experimentation and the killing of tens of thousands of disabled people. The eugenics movement was discredited after World War II, and the evils perpetuated in its name have tainted the term *eugenics* ever since.

Given the history of the eugenics movement, is it fair to use the term *eugenics* when we speak about genetic testing for Down syndrome and other genetic disorders? Some people argue that it is not eugenic to select for healthy children because there is no coercion, the state is not involved, and the goal is the elimination of suffering. Others point out that such voluntary actions still constitute eugenics, since they involve a form of bioengineering for better human beings.

Now that we are entering an era of unprecedented knowledge about our genomes and our predisposition to genetic disorders, we must make decisions about whether our attempts to control or improve human genomes is ethical, and what limits we should place on these efforts. In theory, current and future genetic technologies may make possible a "new eugenics" with a scope and power that Francis Galton could not have imagined. The story of the eugenics movement provides us with a powerful cautionary tale about the potential misuses of genetic information.

Your Turn

Take time, individually or in groups, to answer the following questions. Investigate the references and links to help you discuss some of the issues surrounding genetic testing and eugenics.

1. Do you think that modern prenatal and preimplantation genetic testing followed by selective abortion is eugenic? Why or why not?

For background on these questions, see McCabe, L. and McCabe, E., 2011. Down syndrome: coercion and eugenics. *Genet. Med.* 13: 708–710. *Another useful discussion can be found in* Scully, J. L., 2008. Disability and genetics in the era of genomic medicine. *Nature Reviews Genetics* 9: 797–802.

2. Is it desirable, or even possible, for genetic counseling to be nondirective?

One of the arguments that genetic tests followed by selective abortion are eugenic is that there are strong pressures on pregnant women to produce "normal" children. Some of these pressures may come from genetic counselors. For a discussion of this topic, see: King, D. S., 1999. Preimplantation genetic diagnosis

and the "new" eugenics. *J. Med. Ethics* 25: 176–182.

3. If genetic technologies were more advanced than today, and you could choose the traits of your children, would you take advantage of that option? Which traits would you choose—height, weight, intellectual

abilities, athleticism, artistic talents? If so, would this be eugenic?

To read about similar questions answered by groups of Swiss law and medical students, read: Elger, B. and Harding, T., 2003. Huntington's disease: Do future physicians and lawyers think eugenically? *Clin. Genet.* 64: 327–338.

CASE STUDY » Fish tales

Aquatic vegetation overgrowth, usually controlled by dredging or herbicides, represents a significant issue in maintaining private and public waterways. In 1963, diploid grass carp were introduced in Arkansas to consume vegetation, but they reproduced prodigiously and spread to eventually become a hazard to aquatic ecosystems in 35 states. In the 1980s, many states adopted triploid grass carp as an alternative because of their high, but not absolute, sterility and their longevity of seven to ten years. Today, most states require permits for vegetation control by triploid carp, requiring their containment in the body of water to which they are introduced. Genetic modifications of organisms to achieve specific outcomes will certainly become more common in the future and raise several interesting questions.

1. Taking triploid carp as an example, what controversies may emerge as similar modified species become available for widespread use?
2. If you were a state employee in charge of a specific waterway, what questions would you ask before you approved the introduction of a laboratory-produced, polyploid species into your waterway?
3. Why would the creation and use of a tetraploid carp species be less desirable in the above situation?

INSIGHTS AND SOLUTIONS

1. In a cross using maize that involves three genes, *a, b,* and *c,* a heterozygote (*abc*/+++) is testcrossed to *abc/abc,* Even though the three genes are separated along the chromosome, thus predicting that crossover gametes and the resultant phenotypes should be observed, only two phenotypes are recovered: *abc* and +++. In addition, the cross produced significantly fewer viable plants than expected. Can you propose why no other phenotypes were recovered and why the viability was reduced?

Solution: One of the two chromosomes may contain an inversion that overlaps all three genes, effectively precluding the recovery of any "crossover" offspring. If this is a paracentric inversion and the genes are clearly separated (assuring that a significant number of crossovers occurs between them), then numerous acentric and dicentric chromosomes will form, resulting in the observed reduction in viability.

2. A male *Drosophila* from a wild-type stock is discovered to have only seven chromosomes, whereas normally 2*n* = 8. Close examination reveals that one member of chromosome IV (the smallest chromosome) is attached to (translocated to) the distal end of chromosome II and is missing its centromere, thus accounting for the reduction in chromosome number.

(a) Diagram all members of chromosomes II and IV during synapsis in meiosis I.

Solution:

(b) If this male mates with a female with a normal chromosome composition who is homozygous for the recessive chromosome IV mutation *eyeless* (*ey*), what chromosome compositions will occur in the offspring regarding chromosomes II and IV?

Solution:

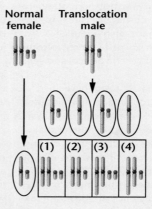

Normal female   Translocation male

(1) (2) (3) (4)

(c) Referring to the diagram in the solution to part (b), what phenotypic ratio will result regarding the presence of eyes, assuming all abnormal chromosome compositions survive?

Solution:

1. normal (heterozygous)
2. eyeless (monosomic, contains chromosome IV from mother)
3. normal (heterozygous; trisomic and may die)
4. normal (heterozygous; balanced translocation)

The final ratio is 3/4 normal: 1/4 eyeless.

PROBLEMS AND DISCUSSION QUESTIONS

 For activities, animations, and review quizzes, go to the study area at www.masteringgenetics.com

HOW DO WE KNOW ?

1. In this chapter, we have focused on chromosomal mutations resulting from a change in number or arrangement of chromosomes. In our discussions, we found many opportunities to consider the methods and reasoning by which much of this information was acquired. From the explanations given in the chapter, what answers would you propose to the following fundamental questions?

   (a) How do we know that the extra chromosome causing Down syndrome is usually maternal in origin?

   (b) How do we know that human aneuploidy for each of the 22 autosomes occurs at conception, even though most often human aneuploids do not survive embryonic or fetal development and thus are never observed at birth?

   (c) How do we know that specific mutant phenotypes are due to changes in chromosome number or structure?

   (d) How do we know that the mutant *Bar*-eye phenotype in *Drosophila* is due to a duplicated gene region rather than to a change in the nucleotide sequence of a gene?

2. For a species with a diploid number of 18, indicate how many chromosomes will be present in the somatic nuclei of individuals that are haploid, triploid, tetraploid, trisomic, and monosomic.

3. Define these pairs of terms, and distinguish between them.

   aneuploidy/euploidy
   monosomy/trisomy
   Patau syndrome/Edwards syndrome
   autopolyploidy/allopolyploidy
   autotetraploid/amphidiploid
   paracentric inversion/pericentric inversion

4. Contrast the relative survival times of individuals with Down, Patau, and Edwards syndromes. Speculate as to why such differences exist.

5. What explanation has been proposed to explain why Down syndrome is more often the result of nondisjunction during oogenesis rather than during spermatogenesis?

6. Contrast the fertility of an allotetraploid with an autotriploid and an autotetraploid.

7. Why do human monosomics most often fail to survive prenatal development?

8. Describe the origin of cultivated American cotton.

9. Predict how the synaptic configurations of homologous pairs of chromosomes might appear when one member is normal and the other member has sustained a deletion or duplication.

10. Inversions are said to "suppress crossing over." Is this terminology technically correct? If not, restate the description accurately.

11. Predict the genetic composition of gametes derived from tetrads of inversion heterozygotes where crossing over occurs within a pericentric inversion.

12. Human adult hemoglobin is a tetramer containing two alpha (α) and two beta (β) polypeptide chains. The α gene cluster on chromosome 16 and the β gene cluster on chromosome 11 share amino acid similarities such that 61 of the amino acids of the α-globin polypeptide (141 amino acids long) are shared in identical sequence with the β-globin polypeptide (146 amino acids long). How might one explain the existence of two polypeptides with partially shared function and structure on two different chromosomes? Include in your answer a link to Ohno's hypothesis regarding the origin of new genes during evolution.

13. The primrose, *Primula kewensis*, has 36 chromosomes that are similar in appearance to the chromosomes in two related species, *P. floribunda* ($2n = 18$) and *P. verticillata* ($2n = 18$). How could *P. kewensis* arise from these species? How would you describe *P. kewensis* in genetic terms?

14. Certain varieties of chrysanthemums contain 18, 36, 54, 72, and 90 chromosomes; all are multiples of a basic set of nine chromosomes. How would you describe these varieties genetically? What feature do the karyotypes of each variety share? A variety with 27 chromosomes has been discovered, but it is sterile. Why?

15. *Drosophila* may be monosomic for chromosome 4, yet remain fertile. Contrast the F_1 and F_2 results of the following crosses involving the recessive chromosome 4 trait, *bent* bristles: (a) monosomic IV, bent bristles × normal bristles; (b) monosomic IV, normal bristles × bent bristles.

16. Mendelian ratios are modified in crosses involving autotetraploids. Assume that one plant expresses the dominant trait green seeds and is homozygous (*WWWW*). This plant is crossed to one with white seeds that is also homozygous (*wwww*). If only one dominant allele is sufficient to produce green seeds, predict the F_1 and F_2 results of such a cross. Assume that synapsis between chromosome pairs is random during meiosis.

17. Having correctly established the F_2 ratio in Problem 16, predict the F_2 ratio of a "dihybrid" cross involving two independently assorting characteristics (e.g., $P_1 = WWWWAAAA \times wwwwaaaa$).

18. In a cross between two varieties of corn, $gl_1gl_1Ws_3Ws_3$ (egg parent) $\times Gl_1Gl_1ws_3ws_3$ (pollen parent), a triploid offspring was produced with the genetic constitution $Gl_1Gl_1gl_1Ws_3ws_3ws_3$. From which parent, egg or pollen, did the $2n$ gamete originate? Is another explanation possible? Explain.

19. A couple planning their family are aware that through the past three generations on the husband's side a substantial number of stillbirths have occurred and several malformed babies were born who died early in childhood. The wife has studied genetics and urges her husband to visit a genetic counseling clinic, where a complete karyotype-banding analysis is performed. Although the tests show that he has a normal complement of 46 chromosomes, banding analysis reveals that one member of the chromosome 1 pair (in group A) contains an inversion covering 70 percent of its length. The homolog of chromosome 1 and all other chromosomes show the normal banding sequence. (a) How would you explain the high incidence of past stillbirths? (b) What can you predict about the probability of abnormality/normality of their future children? (c) Would you advise the woman that she will have to bring each pregnancy to term to determine whether the fetus is normal? If not, what else can you suggest?

20. A woman who sought genetic counseling is found to be heterozygous for a chromosomal rearrangement between the second and third chromosomes. Her chromosomes, compared to those in a normal karyotype, are diagrammed here:

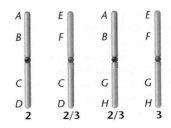

(a) What kind of chromosomal aberration is shown?

(b) Using a drawing, demonstrate how these chromosomes would pair during meiosis. Be sure to label the different segments of the chromosomes.

(c) This woman is phenotypically normal. Does this surprise you? Why or why not? Under what circumstances might you expect a phenotypic effect of such a rearrangement?

21. The woman in Problem 20 has had two miscarriages. She has come to you, an established genetic counselor, with these questions: (a) Is there a genetic explanation of her frequent miscarriages? (b) Should she abandon her attempts to have a child of her own? (c) If not, what is the chance that she could have a normal child? Provide an informed response to her concerns.

22. In a recent cytogenetic study on 1021 cases of Down syndrome, 46 were the result of translocations, the most frequent of which was symbolized as t(14;21). What does this designation represent, and how many chromosomes would you expect to be present in t(14;21) Down syndrome individuals?

23. A boy with Klinefelter syndrome (47,XXY) is born to a mother who is phenotypically normal and a father who has the X-linked skin condition called anhidrotic ectodermal dysplasia. The mother's skin is completely normal with no signs of the skin abnormality. In contrast, her son has patches of normal skin and patches of abnormal skin. (a) Which parent contributed the abnormal gamete? (b) Using the appropriate genetic terminology,

describe the meiotic mistake that occurred. Be sure to indicate in which division the mistake occurred. (c) Using the appropriate genetic terminology, explain the son's skin phenotype.

24. In a human genetic study, a family with five phenotypically normal children was investigated. Two children were "homozygous" for a Robertsonian translocation between chromosomes 19 and 20 (they contained two identical copies of the fused chromosome). They have only 44 chromosomes but a complete genetic complement. Three of the children were "heterozygous" for the translocation and contained 45 chromosomes, with one translocated chromosome plus a normal copy of both chromosomes 19 and 20. Two other pregnancies resulted in stillbirths. It was later discovered that the parents were first cousins. Based on this information, determine the chromosome compositions of the parents. What led to the stillbirths? Why was the discovery that the parents were first cousins a key piece of information in understanding the genetics of this family?

25. A 3-year-old child exhibited some early indication of Turner syndrome, which results from a 45,X chromosome composition. Karyotypic analysis demonstrated two cell types: 46,XX (normal) and 45,X. Propose a mechanism for the origin of this mosaicism.

26. A normal female is discovered with 45 chromosomes, one of which exhibits a Robertsonian translocation containing most of chromosomes 18 and 21. Discuss the possible outcomes in her offspring when her husband contains a normal karyotype.

SOLUTIONS TO SELECTED PROBLEMS AND DISCUSSION QUESTIONS

Answers to Now Solve This

1. If the father had hemophilia, it is likely that the Turner syndrome individual inherited the X chromosome from the father and no sex chromosome from the mother. If nondisjunction occurred in the mother, either during meiosis I or meiosis II, an egg with no X chromosome can be the result. See the text for a diagram of primary and secondary nondisjunction.

2. The sterility of interspecific hybrids is often caused from a high proportion of univalents in meiosis I. As such, viable gametes are rare and the likelihood of two such gametes "meeting" is remote. Even if partial homology of chromosomes allows some pairing, sterility is usually the rule. The horticulturist may attempt to reverse the sterility by treating the sterile hybrid with colchicine. Such a treatment, if successful, may double the chromosome number, so each chromosome would now have a homolog with which to pair during meiosis.

3. The rare double crossovers within the boundaries of a paracentric inversion heterozygote produce only minor departures from the standard chromosomal arrangement as long as the crossovers involve the same two chromatids. With two-strand double crossovers, the second crossover negates the first. However, three-strand and four-strand double crossovers have consequences that lead to anaphase bridges as well as a high degree of genetically unbalanced gametes.

Solutions to Problems and Discussion Questions

2. With a diploid chromosome number of 18 ($2n$), a haploid (n) would have nine chromosomes, a triploid ($3n$) would have 27 chromosomes, and a tetraploid ($4n$) would have 36 chromosomes. A trisomic would have one extra chromosome (19) and a monosomic one less than the diploid (17).

4. Comparing the different sizes of the involved chromosomes (21, 13, and 18, respectively), for example, suggests that the larger the chromosome, the lower the likelihood of lengthy

survival. In addition, it would be expected that certain chromosomes, because of their genetic content, may have different influences on development.

6. Because an allotetraploid has a possibility of producing bivalents at meiosis I, it would be considered the most fertile of the three. Having an even number of chromosomes to match up at the metaphase I plate, autotetraploids would be considered to be more fertile than autotriploids.

10. While there is the appearance that crossing over is suppressed in inversion heterozygotes, the phenomenon extends from the fact that the crossover chromatids end up being abnormal in genetic content. As such, they fail to produce viable (or competitive) gametes or lead to zygotic or embryonic death.

14. Given the basic chromosome set of nine unique chromosomes (a haploid complement), other forms with the "n multiples" are forms of autotetraploidy. In the illustration below the n basic set is multiplied to various levels as is the autotetraploid in the example.

Individual organisms with 27 chromosomes are triploids ($3n$) and are more likely to be sterile because there are trivalents at meiosis I that cause a relatively high number of unbalanced gametes to be formed.

16. The cross would be as follows:

$$WWWW \times wwww$$
(assuming that chromosomes pair at meiosis)

F_1: $WWww$

F_2: $1WW$ $4Ww$ $1ww$

$1WW$	
$4Ww$	$35W\_\_\_$ and $1wwww$
$1ww$	

18. Since two Gl_1 alleles and two ws_3 alleles are present in the triploid, they must have come from the pollen parent. By the wording of the problem, it is implied that the pollen parent contributed an unreduced ($2n$) gamete; however, another explanation, dispermic fertilization, is possible. In this case two Gl_1ws_3 gametes could have fertilized the ovule.

20. (a) reciprocal translocation

(b)

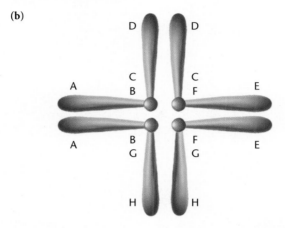

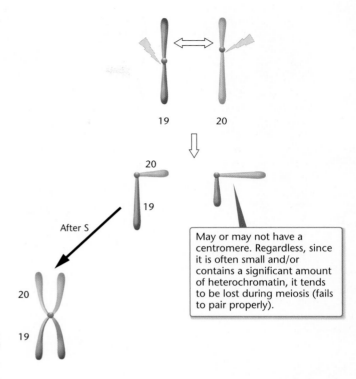

May or may not have a centromere. Regardless, since it is often small and/or contains a significant amount of heterochromatin, it tends to be lost during meiosis (fails to pair properly).

(c) Notice that all chromosomal segments are present and there is no apparent loss of chromosomal material. However, if the breakpoints for the translocation occurred within genes, then an abnormal phenotype may be the result. In addition, a gene's function is sometimes influenced by its position—its neighboring genes, in other words. If such "position effects" occur, then a different phenotype may result.

22. The symbol t(14;21) indicates that part of chromosome 21 is translocated to chromosome 14. When a gamete containing such a chromosome plus a normal chromosome 21 is fertilized by a standard haploid gamete, the individual has 46 chromosomes but effectively has three copies of chromosome 21.

24. Below is a description of breakage/reunion events that illustrate a Robertsonian translocation in the relatively small, similarly sized chromosomes 19 (metacentric) and 20 (metacentric/sub-metacentric). The case described here is shown occurring before S phase duplication. Since the likelihood of such a translocation is fairly small in a general population, inbreeding played a significant role in allowing the translocation to "meet itself."

26. This female will produce meiotic products of the following types:

normal: $18 + 21$ *translocated plus 21:* $18/21 + 21$
translocated: $18/21$ *deficient:* 18 only

Note: The $18/21 + 18$ gamete is not formed because it would require separation of primarily homologous chromosomes at anaphase I.

Fertilization with a normal $18 + 21$ sperm cell will produce the following offspring:

normal: 46 chromosomes
translocation carrier: 45 chromosomes $18/21 + 18 + 21$
trisomy 21: 46 chromosomes $18/21 + 21 + 21$
monosomic: 45 chromosomes $18 + 18 + 21$, lethal

CREDITS

Credits are listed in order of appearance.

Photo

CO, Evelin Schrock, Stan du Manoir and Tom Reid, NIH; F-2 (a) Courtesy of the Greenwood Genetic Center, Greenwood, SC; F-2 (b) Design Pics Inc./Alamy; F-4, David D. Weaver, M.D.; F-8, Courtesy of National Cotton Council of America; F-11 (a) Courtesy of University of Washington Medical Center Pathology, F-11 (b) Ray Clarke, Cri du chat Syndrome Support Group, UK; F6-13 (a, b, c) Mary Lilly/Carnegie Institution of Washington; F6-17 (b) Dr. Jorge Yunis. From Yunis and Chandler, 1979; F6-18, Custom Medical Stock Photo

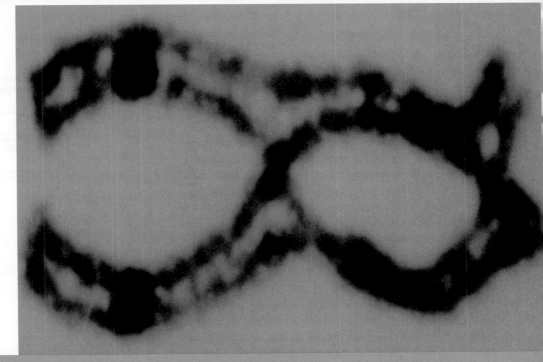

Chiasmata between synapsed homologs during the first meiotic prophase.

Linkage and Chromosome Mapping in Eukaryotes

CHAPTER CONCEPTS

- Chromosomes in eukaryotes contain many genes whose locations are fixed along the length of the chromosomes.

- Unless separated by crossing over, alleles present on a chromosome segregate as a unit during gamete formation.

- Crossing over between homologs is a process of genetic recombination during meiosis that creates gametes with new combinations of alleles that enhance genetic variation within species.

- Crossing over between homologs serves as the basis for the construction of chromosome maps.

- While exchange occurs between sister chromatids during mitosis, no new recombinant chromatids are created.

Walter Sutton, along with Theodor Boveri, was instrumental in uniting the fields of cytology and genetics. As early as 1903, Sutton pointed out the likelihood that there must be many more "unit factors" than chromosomes in most organisms. Soon thereafter, genetics investigations revealed that certain genes segregate as if they were somehow joined or linked together. Further investigations showed that such genes are part of the same chromosome and may indeed be transmitted as a single unit. We now know that most chromosomes contain a very large number of genes. Those that are part of the same chromosome are said to be *linked* and to demonstrate **linkage** in genetic crosses.

Because the chromosome, not the gene, is the unit of transmission during meiosis, linked genes are not free to undergo independent assortment. Instead, the alleles at all loci of one chromosome should, in theory, be transmitted as a unit during gamete formation. However, in many instances this does not occur. During the first meiotic prophase, when homologs are paired or synapsed, a reciprocal exchange of chromosome segments can take place. This crossing over results in the reshuffling, or **recombination,** of the alleles between homologs, and it always occurs during the tetrad stage.

The frequency of crossing over between any two loci on a single chromosome is proportional to the distance between them. Therefore, depending on which loci are being studied, the percentage of recombinant gametes varies. This correlation allows us to construct chromosome maps, which give the relative locations of genes on chromosomes.

In this chapter, we will discuss linkage, crossing over, and chromosome mapping in more detail. We will conclude by entertaining the rather intriguing question of why Mendel, who studied seven genes in an organism with seven chromosomes, did not encounter linkage. Or did he?

From Chapter 7 of *Essentials of Genetics, Eighth edition,* William S. Klug, Michael R. Cummings, Charlotte A. Spencer, Michael A. Palladino. ©2013 by Pearson Education, Inc. All rights reserved.

1 Genes Linked on the Same Chromosome Segregate Together

A simplified overview of the major theme of this chapter is given in **Figure 1**, which contrasts the meiotic consequences of (a) independent assortment, (b) linkage *without* crossing over, and (c) linkage *with* crossing over. In **Figure 1(a)**, we see the results of independent assortment of two pairs of chromosomes, each containing one heterozygous gene pair. No linkage is exhibited. When a large number of meiotic events are observed, four genetically different gametes are formed in equal proportions, and each contains a different combination of alleles of the two genes.

Now let's compare these results with what occurs if the same genes are linked on the same chromosome. If no crossing over occurs between the two genes [**Figure 1(b)**], only two genetically different gametes are formed. Each gamete receives the alleles present on one homolog or the other, which is transmitted intact as the result of segregation. This case demonstrates *complete linkage,* which produces only **parental** or **noncrossover gametes.** The two parental gametes are formed in equal proportions. Though complete linkage between two genes seldom occurs, it is useful to consider the theoretical consequences of this concept.

Figure 1(c) shows the results of crossing over between two linked genes. As you can see, this crossover involves only two nonsister chromatids of the four chromatids present in the tetrad. This exchange generates two new allele combinations, called **recombinant** or **crossover gametes.** The two chromatids not involved in the exchange result in noncrossover gametes, like those in Figure 1(b). The frequency with which crossing over occurs between any two linked genes is generally proportional to the distance separating the respective loci along the chromosome. In theory, two randomly selected genes can be so close to each other that crossover events are too infrequent to be detected easily. As shown in Figure 1(b), this complete linkage produces only parental gametes. On the other hand, if a small but distinct distance separates two genes, few recombinant and many parental gametes will be formed. As the distance between the two genes increases, the proportion of recombinant gametes increases and that of the parental gametes decreases.

FIGURE 1 Results of gamete formation when two heterozygous genes are (a) on two different pairs of chromosomes; (b) on the same pair of homologs, but with no exchange occurring between them; and (c) on the same pair of homologs, but with an exchange occurring between two nonsister chromatids. Note in this and the following figures that members of homologous pairs of chromosomes are shown in two different colors.

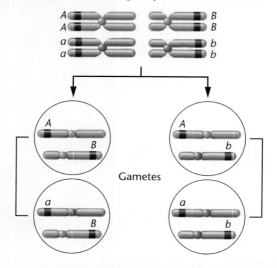

(a) Independent assortment: Two genes on two different homologous pairs of chromosomes

Gametes

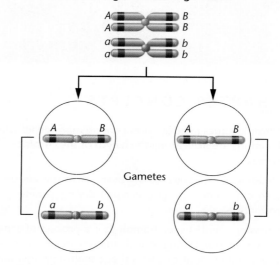

(b) Linkage: Two genes on a single pair of homologs; no exchange occurs

Gametes

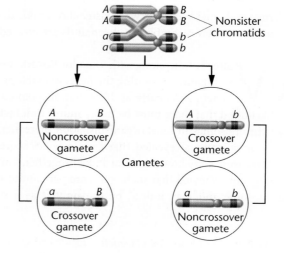

(c) Linkage: Two genes on a single pair of homologs; exchange occurs between two nonsister chromatids

Nonsister chromatids

Noncrossover gamete

Crossover gamete

Gametes

Crossover gamete

Noncrossover gamete

As we will discuss later in this chapter, when the loci of two linked genes are far apart, the number of recombinant gametes approaches, but does not exceed, 50 percent. If 50 percent recombinants occur, the result is a 1:1:1:1 ratio of the four types (two parental and two recombinant gametes). In this case, transmission of two linked genes is indistinguishable from that of two unlinked, independently assorting genes. That is, the proportion of the four possible genotypes is identical, as shown in Figure 1(a) and (c).

The Linkage Ratio

If complete linkage exists between two genes because of their close proximity, and organisms heterozygous at both loci are mated, a unique F_2 phenotypic ratio results, which we designate the **linkage ratio.** To illustrate this ratio, let's consider a cross involving the closely linked, recessive, mutant genes *heavy* wing vein (*hv*) and *brown eye* (*bw*) in *Drosophila melanogaster* (**Figure 2**). The normal, wild-type alleles hv^+ and bw^+ are both dominant and result in thin wing veins and red eyes, respectively.

In this cross, flies with normal thin wing veins and mutant brown eyes are mated to flies with mutant heavy wing veins and normal red eyes. In more concise terms, brown-eyed flies are crossed with heavy-veined flies. Linked genes are represented by placing their allele designations above and below a single or double horizontal line. Those above the line are located at loci on one homolog, and those below are located at the homologous loci on the other homolog. Thus, we represent the P_1 generation as follows:

$$P_1: \frac{hv^+ \; bw}{hv^+ \; bw} \times \frac{hv \; bw^+}{hv \; bw^+}$$

thin, brown   heavy, red

These genes are located on an autosome, so no distinction between males and females is necessary.

In the F_1 generation, each fly receives one chromosome of each pair from each parent. All flies are heterozygous for both gene pairs and exhibit the dominant traits of thin wing veins and red eyes:

$$F_1: \frac{hv^+ \; bw}{hv \; \; bw^+}$$

thin, red

As shown in **Figure 2(a)**, when the F_1 generation is interbred, each F_1 individual forms only parental gametes because of complete linkage. After fertilization, the F_2 generation is produced in a 1:2:1 phenotypic and genotypic ratio. One-fourth of this generation shows thin wing veins and brown eyes; one-half shows both wild-type traits, namely, thin wing veins and red eyes; and one-fourth shows heavy wing veins and red eyes. In more concise terms, the ratio is 1 heavy:2

wild:1 brown. Such a 1:2:1 ratio is characteristic of complete linkage. Complete linkage is usually observed only when genes are very close together and the number of progeny is relatively small.

Figure 2(b) also gives the results of a testcross with the F_1 flies. Such a cross produces a 1:1 ratio of thin, brown and heavy, red flies. Had the genes controlling these traits been incompletely linked or located on separate autosomes, the testcross would have produced four phenotypes rather than two.

When large numbers of mutant genes present in any given species are investigated, genes located on the same chromosome show evidence of linkage to one another. As a result, **linkage groups** can be established, one for each chromosome. In theory, the number of linkage groups should correspond to the haploid number of chromosomes. In diploid organisms in which large numbers of mutant genes are available for genetic study, this correlation has been confirmed.

NOW SOLVE THIS

1 Consider two hypothetical recessive autosomal genes *a* and *b,* where a heterozygote is testcrossed to a double-homozygous mutant. Predict the phenotypic ratios under the following conditions:

(a) *a* and *b* are located on separate autosomes.
(b) *a* and *b* are linked on the same autosome but are so far apart that a crossover always occurs between them.
(c) *a* and *b* are linked on the same autosome but are so close together that a crossover almost never occurs.

■ **Hint** *This problem involves an understanding of linkage, crossing over, and independent assortment. The key to its solution is to be aware that results are indistinguishable when two genes are unlinked compared to the case where they are linked but so far apart that crossing over always intervenes between them during meiosis.*

ESSENTIAL POINT   ■   ■   ■

Genes located on the same chromosome are said to be linked. Alleles of linked genes located close together on the same homolog are usually transmitted together during gamete formation.

2 Crossing Over Serves as the Basis of Determining the Distance between Genes during Mapping

It is highly improbable that two randomly selected genes linked on the same chromosome will be so close to one another along the chromosome that they demonstrate complete linkage. Instead, crosses involving two such genes almost always produce a percentage of offspring resulting

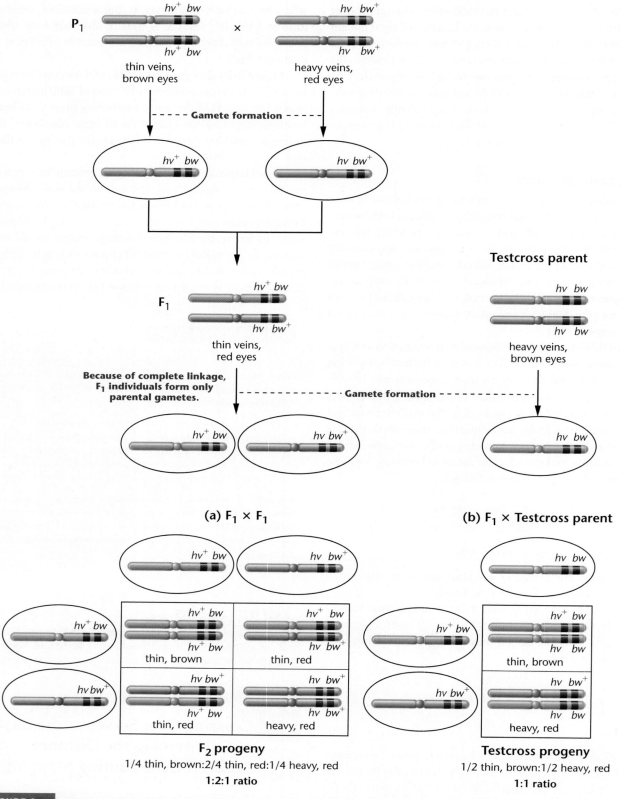

FIGURE 2 Results of a cross involving two genes located on the same chromosome and demonstrating complete linkage. (a) The F_2 results of the cross. (b) The results of a testcross involving the F_1 progeny.

from recombinant gametes. This percentage is variable and depends on the distance between the two genes along the chromosome. This phenomenon was first explained around 1910 by two *Drosophila* geneticists, Thomas H. Morgan and his undergraduate student, Alfred H. Sturtevant.

Morgan, Sturtevant, and Crossing Over

In his studies, Morgan investigated numerous *Drosophila* mutations located on the X chromosome. When he analyzed crosses involving only one trait, he deduced the mode of X-linked inheritance. However, when he made crosses involving two X-linked genes, his results were initially puzzling. For example, female flies expressing the mutant *yellow* body (*y*) and *white* eyes (*w*) alleles were crossed with wild-type males (gray bodies and red eyes). The F_1 females were wild type, while the F_1 males expressed both mutant traits. In the F_2 generation, the vast majority of the offspring showed the expected parental phenotypes—either yellow-bodied, white-eyed flies or wild-type flies (gray-bodied, red-eyed). However, the remaining flies, less than 1.0 percent, were either yellow-bodied with red eyes or gray-bodied with white eyes. It was as if the two mutant alleles had somehow separated from each other on the homolog during gamete formation in the F_1 female flies. This cross is illustrated in cross A of **Figure 3**, using data later compiled by Sturtevant.

When Morgan studied other X-linked genes, the same basic pattern was observed, but the proportion of the unexpected F_2 phenotypes differed. For example, in a cross involving the mutant *white* eye (*w*), *miniature* wing (*m*) alleles, the majority of the F_2 again showed the parental phenotypes, but a much higher proportion of the offspring appeared as if the mutant genes had separated during gamete formation. This is illustrated in cross B of Figure 3, again using data subsequently compiled by Sturtevant.

Morgan was faced with two questions: (1) What was the source of gene separation, and (2) why did the frequency of the apparent separation vary depending on the genes being studied? The answer he proposed for the first question was based on his knowledge of earlier cytological observations made by F.A. Janssens and others. Janssens had observed that synapsed homologous chromosomes in meiosis wrapped around each other, creating **chiasmata** (sing., *chiasma*) where points of overlap are evident. Morgan proposed that chiasmata could represent points of genetic exchange.

In the crosses shown in Figure 3, Morgan postulated that if an exchange occurs during gamete formation between the mutant genes on the two X chromosomes of the F_1 females, the unique phenotypes will occur. He suggested that such exchanges led to **recombinant gametes** in both the *yellow–white* cross and the *white–miniature* cross, in contrast to the **parental gametes** that have undergone no exchange. On the basis of this and other experiments, Morgan concluded that linked genes exist in a linear order along the chromosome and that a variable amount of exchange occurs between any two genes during gamete formation.

In answer to the second question, Morgan proposed that two genes located relatively close to each other along a chromosome are less likely to have a chiasma form between them than if the two genes are farther apart on the chromosome. Therefore, the closer two genes are, the less likely a genetic exchange will occur between them. Morgan was the first to propose the term **crossing over** to describe the physical exchange leading to recombination.

Sturtevant and Mapping

Morgan's student, Alfred H. Sturtevant, was the first to realize that his mentor's proposal could be used to map the sequence of linked genes. According to Sturtevant,

> In a conversation with Morgan . . . I suddenly realized that the variations in strength of linkage, already attributed by Morgan to differences in the spatial separation of the genes, offered the possibility of determining sequences in the linear dimension of a chromosome. I went home and spent most of the night (to the neglect of my undergraduate homework) in producing the first chromosomal map.

Sturtevant compiled data from numerous crosses made by Morgan and other geneticists involving recombination between the genes represented by the *yellow, white,* and *miniature* mutants. These data are shown in Figure 3. The following recombination between each pair of these three genes, published in Sturtevant's paper in 1913, is as follows:

(1)	*yellow–white*	0.5%
(2)	*white–miniature*	34.5%
(3)	*yellow–miniature*	35.4%

Because the sum of (1) and (2) approximately equals (3), Sturtevant suggested that the recombination frequencies between linked genes are additive. On this basis, he predicted that the order of the genes on the X chromosome is *yellow–white–miniature*. In arriving at this conclusion, he reasoned as follows: the *yellow* and *white* genes are apparently close to each other because the recombination frequency is low. However, both of these genes are much farther apart from *miniature* genes because the *white–miniature* and *yellow–miniature* combinations show larger recombination frequencies. Because *miniature* shows more recombination with *yellow* than with *white* (35.4 percent versus 34.5 percent), it follows that *white* is located between the other two genes, not outside of them.

Sturtevant knew from Morgan's work that the frequency of exchange could be used as an estimate of the distance

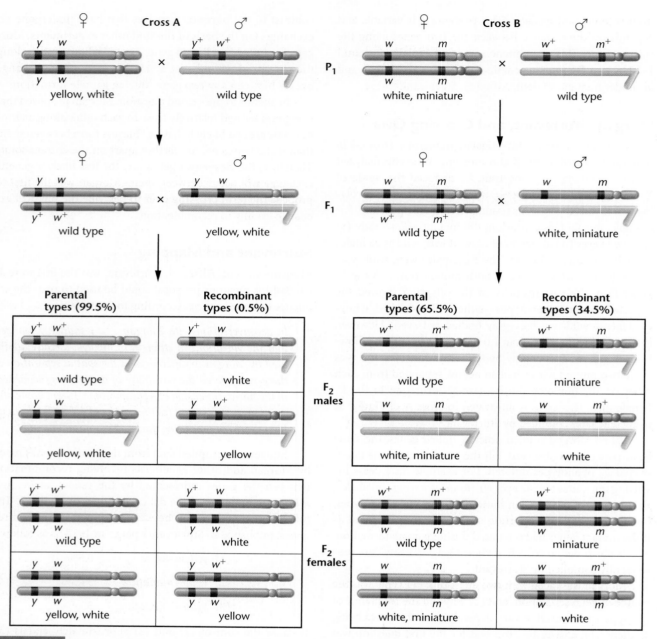

FIGURE 3 The F_1 and F_2 results of crosses involving the *yellow* (*y*), *white* (*w*) mutations (cross A), and the *white, miniature* (*m*) mutations (cross B), as compiled by Sturtevant. In cross A, 0.5 percent of the F_2 flies (males and females) demonstrate recombinant phenotypes, which express either *white* or *yellow*. In cross B, 34.5 percent of the F_2 flies (males and females) demonstrate recombinant phenotypes, which are either *miniature* or *white* mutants.

between two genes or loci along the chromosome. He constructed a **chromosome map** of the three genes on the X chromosome, setting 1 map unit (mu) equal to 1 percent recombination between two genes.* In the preceding example, the distance between *yellow* and *white* is thus 0.5 mu, and the distance between *yellow* and *miniature* is 35.4 mu. It follows that the distance between *white* and *miniature* should be 35.4 − 0.5 = 34.9 mu. This estimate is close to the

actual frequency of recombination between *white* and *miniature* (34.5 percent). The map for these three genes is shown in **Figure 4**. The fact that these do not add up perfectly is due to the imprecision of independently conducted mapping experiments.

In addition to these three genes, Sturtevant considered two other genes on the X chromosome and produced a more extensive map that included all five genes. He and a colleague, Calvin Bridges, soon began a search for autosomal linkage in *Drosophila*. By 1923, they had clearly shown

---

* In honor of Morgan's work, map units are often referred to as centi-Morgans (cM).

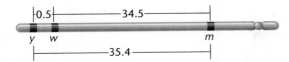

FIGURE 4 A map of the *yellow* (*y*), *white* (*w*), and *miniature* (*m*) genes on the X chromosome of *Drosophila melanogaster*. Each number represents the percentage of recombinant offspring produced in one of three crosses, each involving two different genes.

that linkage and crossing over are not restricted to X-linked genes but can also be demonstrated with autosomes. During this work, they made another interesting observation. Crossing over in *Drosophila* was shown to occur only in females. The fact that no crossing over occurs in males made genetic mapping much less complex to analyze in *Drosophila*. However, crossing over does occur in both sexes in most other organisms.

Although many refinements in chromosome mapping have been developed since Sturtevant's initial work, his basic principles are considered to be correct. These principles are used to produce detailed chromosome maps of organisms for which large numbers of linked mutant genes are known. Sturtevant's findings are also historically significant to the broader field of genetics. In 1910, the **chromosomal theory of inheritance** was still widely disputed—even Morgan was skeptical of this theory before he conducted his experiments. Research has now firmly established that chromosomes contain genes in a linear order and that these genes are the equivalent of Mendel's unit factors.

Single Crossovers

Why should the relative distance between two loci influence the amount of recombination and crossing over observed between them? During meiosis, a limited number of crossover events occur in each tetrad. These recombinant events occur randomly along the length of the tetrad. Therefore, the closer two loci reside along the axis of the chromosome, the less likely any single-crossover event will occur between them. The same reasoning suggests that the farther apart two linked loci are, the more likely a random crossover event will occur between them.

In **Figure 5(a)**, a single crossover occurs between two nonsister chromatids but not between the two loci; therefore, the crossover is not detected because no recombinant gametes are produced. In **Figure 5(b)**, where two loci are quite far apart, a crossover does occur between them, yielding gametes in which the traits of interest are recombined.

When a single crossover occurs between two nonsister chromatids, the other two chromatids of the tetrad are not involved in this exchange and enter the gamete unchanged. Even if a single crossover occurs 100 percent of the time between two linked genes, recombination is subsequently ob-

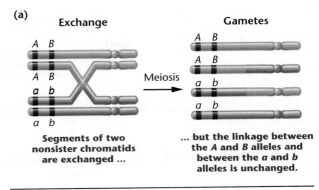

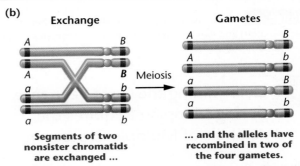

FIGURE 5 Two examples of a single crossover between two nonsister chromatids and the gametes subsequently produced. In (a) the exchange does not alter the linkage arrangement between the alleles of the two genes, only parental gametes are formed, and the exchange goes undetected. In (b) the exchange separates the alleles, resulting in recombinant gametes, which are detectable.

served in only 50 percent of the potential gametes formed. This concept is diagrammed in **Figure 6**. Theoretically, if we consider only single exchanges and observe 20 percent recombinant gametes, crossing over actually occurred in 40 percent of the tetrads. Under these conditions, the general rule is that the percentage of tetrads involved in an exchange between two genes is twice the percentage of recombinant gametes produced. Therefore, the theoretical limit of observed recombination due to crossing over is 50 percent.

When two linked genes are more than 50 mu apart, a crossover can theoretically be expected to occur between them in 100 percent of the tetrads. If this prediction were achieved, each tetrad would yield equal proportions of the four gametes shown in Figure 6, just as if the genes were on different chromosomes and assorting independently. However, this theoretical limit is seldom achieved.

> **ESSENTIAL POINT** ■ ■ ■
>
> Crossover frequency between linked genes during gamete formation is proportional to the distance between genes, providing the experimental basis for mapping the location of genes relative to one another along the chromosome.

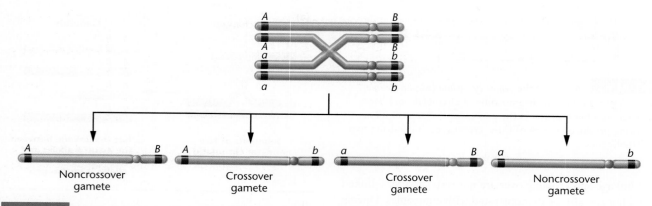

The consequences of a single exchange between two nonsister chromatids occurring in the tetrad stage. Two noncrossover (parental) and two crossover (recombinant) gametes are produced.

3 Determining the Gene Sequence during Mapping Requires the Analysis of Multiple Crossovers

The study of single crossovers between two linked genes provides the basis of determining the *distance* between them. However, when many linked genes are studied, their *sequence* along the chromosome is more difficult to determine. Fortunately, the discovery that multiple exchanges occur between the chromatids of a tetrad has facilitated the process of producing more extensive chromosome maps. As we shall see next, when three or more linked genes are investigated simultaneously, it is possible to determine first the sequence of the genes and then the distances between them.

Multiple Crossovers

It is possible that in a single tetrad, two, three, or more exchanges will occur between nonsister chromatids as a result of several crossover events. Double exchanges of genetic material result from **double crossovers (DCOs),** as shown in **Figure 7**. For a double exchange to be studied, three gene pairs must be investigated, each heterozygous for two alleles. Before we determine the frequency of recombination among all three loci, let's review some simple probability calculations.

As we have seen, the probability of a single exchange occurring between the *A* and *B* or the *B* and *C* genes relates directly to the distance between the respective loci. The closer *A* is to *B* and *B* is to *C,* the less likely a single exchange will occur between either of the two sets of loci. In the case of a double crossover, two separate and independent events or exchanges must occur simultaneously. The mathematical probability of two independent events occurring simultaneously is equal to the product of the individual probabilities (the **product law**).

Suppose that crossover gametes resulting from single exchanges are recovered 20 percent of the time ($p = 0.20$) between *A* and *B*, and 30 percent of the time ($p = 0.30$) between *B* and *C*. The probability of recovering a double-crossover gamete arising from two exchanges (between *A* and *B,* and between *B* and *C*) is predicted to be $(0.20)(0.30) = 0.06$, or 6 percent. It is apparent from this calculation that the frequency of double-crossover gametes is always expected to be much lower than that of either single-crossover class of gametes.

If three genes are relatively close together along one chromosome, the expected frequency of double-crossover gametes is extremely low. For example, suppose the *A–B* distance in Figure 7 is 3 mu and the *B–C* distance is 2 mu. The expected double-crossover frequency is $(0.03)(0.02) = 0.0006$, or 0.06 percent. This translates to only 6 events in 10,000. Thus, in a mapping experiment where closely linked genes are

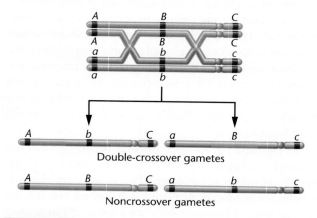

Consequences of a double exchange occurring between two nonsister chromatids. Because the exchanges involve only two chromatids, two noncrossover gametes and two double-crossover gametes are produced. The chapter opening photograph The opening paragraph of this chapter illustrates two chiasmata present in a tetrad isolated during the first meiotic prophase stage.

involved, very large numbers of offspring are required to detect double-crossover events. In this example, it is unlikely that a double crossover will be observed even if 1000 offspring are examined. Thus, it is evident that if four or five genes are being mapped, even fewer triple and quadruple crossovers can be expected to occur.

NOW SOLVE THIS

2 With two pairs of genes involved (P/p and Z/z), a testcross ($ppzz$) with an organism of unknown genotype indicated that the gametes produced were in the following proportions

PZ, 42.4%; Pz, 6.9%; pZ, 7.1%; pz, 43.6%

Draw all possible conclusions from these data.

■ **Hint** *This problem involves an understanding of the proportionality between crossover frequency and distance between genes. The key to its solution is to be aware that noncrossover and crossover gametes occur in reciprocal pairs of approximately equal proportions.*

ESSENTIAL POINT ■ ■ ■

Determining the sequence of genes in a three-point mapping experiment requires analysis of the double-crossover gametes, as reflected in the phenotype of the offspring receiving those gametes.

Three-Point Mapping in *Drosophila*

The information in the preceding section enables us to map three or more linked genes in a single cross. To illustrate the mapping process in its entirety, we examine two situations involving three linked genes in two quite different organisms.

To execute a successful mapping cross, three criteria must be met:

1. The genotype of the organism producing the crossover gametes must be heterozygous at all loci under consideration.

2. The cross must be constructed so that genotypes of all gametes can be determined accurately by observing the phenotypes of the resulting offspring. This is necessary because the gametes and their genotypes can never be observed directly. To overcome this problem, each phenotypic class must reflect the genotype of the gametes of the parents producing it.

3. A sufficient number of offspring must be produced in the mapping experiment to recover a representative sample of all crossover classes.

These criteria are met in the three-point mapping cross from *Drosophila* shown in **Figure 8**. In this cross, three X-linked recessive mutant genes—*yellow* body color (y),

white eye color (w), and *echinus* eye shape (ec)—are considered. To diagram the cross, *we must assume some theoretical sequence, even though we do not yet know if it is correct.* In Figure 8, we initially assume the sequence of the three genes to be *y–w–ec*. If this assumption is incorrect, our analysis will demonstrate this and reveal the correct sequence.

In the P_1 generation, males hemizygous for all three wild-type alleles are crossed to females that are homozygous for all three recessive mutant alleles. Therefore, the P_1 males are wild type with respect to body color, eye color, and eye shape. They are said to have a *wild-type phenotype*. The females, on the other hand, exhibit the three mutant traits—yellow body color, white eyes, and echinus eye shape.

This cross produces an F_1 generation consisting of females that are heterozygous at all three loci and males that, because of the Y chromosome, are hemizygous for the three mutant alleles. Phenotypically, all F_1 females are wild type, while all F_1 males are yellow, white, and echinus. The genotype of the F_1 females fulfills the first criterion for mapping; that is, it is heterozygous at the three loci and can serve as the source of recombinant gametes generated by crossing over. Note that because of the genotypes of the P_1 parents, all three mutant alleles in the F_1 female are on one homolog and all three wild-type alleles are on the other homolog. With other females, *other arrangements are possible that could produce a heterozygous genotype.* For example, a heterozygous female could have the *y* and *ec* mutant alleles on one homolog and the *w* allele on the other. This would occur if, in the P_1 cross, one parent was yellow, echinus and the other parent was white.

In our cross, the second criterion is met by virtue of the gametes formed by the F_1 males. Every gamete contains either an X chromosome bearing the three mutant alleles or a Y chromosome, which is genetically inert for the three loci being considered. Whichever type participates in fertilization, the genotype of the gamete produced by the F_1 female will be expressed phenotypically in the F_2 male and female offspring derived from it. Thus, all F_1 noncrossover and crossover gametes can be detected by observing the F_2 phenotypes.

With these two criteria met, we can now construct a chromosome map from the crosses shown in Figure 8. First, we determine which F_2 phenotypes correspond to the various noncrossover and crossover categories. To determine the noncrossover F_2 phenotypes, we must identify individuals derived from the parental gametes formed by the F_1 female. Each such gamete contains an X chromosome *unaffected by crossing over.* As a result of segregation, approximately equal proportions of the two types of gametes and, subsequently, the F_2 phenotypes, are produced. Because they derive from a heterozygote, the genotypes of the two parental gametes and the resultant F_2 phenotypes complement one another. For example, if one is wild type, the other is

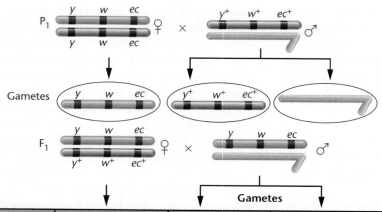

Origin of female gametes	Gametes			F₂ phenotype	Observed number	Category, total, and percentage
NCO	① y w ec	y w ec / y w ec	y w ec	y w ec	4685	Non-crossover
	② y⁺ w⁺ ec⁺	y⁺ w⁺ ec⁺ / y w ec	y⁺ w⁺ ec⁺	y⁺ w⁺ ec⁺	4759	9444 94.44%
SCO	③ y w⁺ ec⁺	y w⁺ ec⁺ / y w ec	y w⁺ ec⁺	y w⁺ ec⁺	80	Single crossover between y and w
	④ y⁺ w ec	y⁺ w ec / y w ec	y⁺ w ec	y⁺ w ec	70	150 1.50%
SCO	⑤ y w ec⁺	y w ec⁺ / y w ec	y w ec⁺	y w ec⁺	193	Single crossover between w and ec
	⑥ y⁺ w⁺ ec	y⁺ w⁺ ec / y w ec	y⁺ w⁺ ec	y⁺ w⁺ ec	207	400 4.00%
DCO	⑦ y w⁺ ec	y w⁺ ec / y w ec	y w⁺ ec	y w⁺ ec	3	Double crossover between y and w and between w and ec
	⑧ y⁺ w ec⁺	y⁺ w ec⁺ / y w ec	y⁺ w ec⁺	y⁺ w ec⁺	3	6 0.06%

Map of *y*, *w*, and *ec* loci

├─1.56─┼──4.06──┤

FIGURE 8   A three-point mapping cross involving the *yellow* (*y* or *y⁺*), *white* (*w* or *w⁺*), and *echinus* (*ec* or *ec⁺*) genes in *Drosophila melanogaster*. NCO, SCO, and DCO refer to noncrossover, single-crossover, and double-crossover groups, respectively. Centromeres are not drawn on the chromosomes, and only two nonsister chromatids are initially shown in the left-hand column.

completely mutant. This is the case in the cross being considered. In other situations, if one chromosome shows one mutant allele, the second chromosome shows the other two mutant alleles, and so on. They are therefore called **reciprocal classes** of gametes and phenotypes.

The two noncrossover phenotypes are most easily recognized because *they exist in the greatest proportion*. Figure 8 shows that gametes 1 and 2 are present in the greatest numbers. Therefore, flies that express yellow, white, and echinus phenotypes and flies that are normal (or wild type) for all three characters constitute the noncrossover category and represent 94.44 percent of the F_2 offspring.

The second category that can be easily detected is represented by the double-crossover phenotypes. Because of their low probability of occurrence, *they must be present in the least numbers*. Remember that this group represents two independent but simultaneous single-crossover events. Two reciprocal phenotypes can be identified: gamete 7, which shows the mutant traits yellow, echinus but normal eye color; and gamete 8, which shows the mutant trait white but normal body color and eye shape. Together these double-crossover phenotypes constitute only 0.06 percent of the F_2 offspring.

The remaining four phenotypic classes represent two categories resulting from single crossovers. Gametes 3 and 4, reciprocal phenotypes produced by single-crossover events occurring between the *yellow* and *white* loci, are equal to 1.50 percent of the F_2 offspring; gametes 5 and 6, constituting 4.00 percent of the F_2 offspring, represent the reciprocal phenotypes resulting from single-crossover events occurring between the *white* and *echinus* loci.

The map distances separating the three loci can now be calculated. The distance between y and w or between w and ec is equal to the percentage of all detectable exchanges occurring between them. For any two genes under consideration, this includes all appropriate single crossovers as well as all double crossovers. *The latter are included because they represent two simultaneous single crossovers.* For the y and w genes, this includes gametes 3, 4, 7, and 8, totaling 1.50% + 0.06%, or 1.56 mu. Similarly, the distance between w and ec is equal to the percentage of offspring resulting from an exchange between these two loci: gametes 5, 6, 7, and 8, totaling 4.00% + 0.06%, or 4.06 mu. The map of these three loci on the X chromosome is shown at the bottom of Figure 8.

Determining the Gene Sequence

In the preceding example, the sequence (or order) of the three genes along the chromosome was assumed to be y–w–ec. Our analysis shows this sequence to be consistent with the data. However, in most mapping experiments the gene sequence is not known, and this constitutes another variable in the analysis. In our example, had the gene sequence been unknown, it could have been determined using a straightforward method.

This method is based on the fact that there are only three possible arrangements, each containing one of the three genes between the other two:

$$(\text{I}) \quad w\text{–}y\text{–}ec \qquad (y \text{ in the middle})$$
$$(\text{II}) \quad y\text{–}ec\text{–}w \qquad (ec \text{ in the middle})$$
$$(\text{III}) \quad y\text{–}w\text{–}ec \qquad (w \text{ in the middle})$$

Use the following steps during your analysis to determine the gene order:

1. Assuming any one of the three orders, first determine the *arrangement of alleles* along each homolog of the heterozygous parent giving rise to noncrossover and crossover gametes (the F_1 female in our example).

2. Determine whether a double-crossover event occurring within that arrangement will produce the *observed double-crossover phenotypes*. Remember that these phenotypes occur least frequently and are easily identified.

3. If this order does not produce the predicted phenotypes, try each of the other two orders. One must work!

In **Figure 9**, the above steps are applied to each of the three possible arrangements (I, II, and III above). A full analysis can proceed as follows:

1. Assuming that y is between w and ec, arrangement I of alleles along the homologs of the F_1 heterozygote is

$$\frac{w \quad y \quad ec}{w^+ \quad y^+ \quad ec^+}$$

We know this because of the way in which the P_1 generation was crossed: The P_1 female contributes an X chromosome bearing the w, y, and ec alleles, while the P_1 male contributes an X chromosome bearing the w^+, y^+, and ec^+ alleles.

2. A double crossover within that arrangement yields the following gametes

$$\underline{w \quad y^+ \quad ec} \quad \text{and} \quad \underline{w^+ \quad y \quad ec^+}$$

Following fertilization, if y is in the middle, the F_2 double-crossover phenotypes will correspond to these gametic genotypes, yielding offspring that express the white, echinus phenotype and offspring that express the yellow phenotype. Instead, determination of the actual double-crossover phenotypes reveals them to be yellow, echinus flies and white flies. *Therefore, our assumed order is incorrect.*

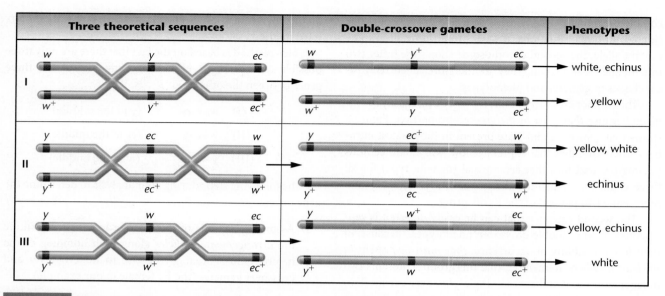

Three theoretical sequences	Double-crossover gametes	Phenotypes

FIGURE 9 The three possible sequences of the *white, yellow,* and *echinus* genes, the results of a double crossover in each case, and the resulting phenotypes produced in a testcross. For simplicity, the two noncrossover chromatids of each tetrad are omitted.

3. If we consider arrangement II with the ec/ec^+ alleles in the middle or arrangement III with the w/w^+ alleles in the middle

$$\text{(II)} \quad \frac{y \quad ec \quad w}{y^+ \quad ec^+ \quad w^+} \quad \text{or} \quad \text{(III)} \quad \frac{y \quad w \quad ec}{y^+ \quad w^+ \quad ec^+}$$

we see that arrangement II again provides *predicted* double-crossover phenotypes that *do not* correspond to the *actual* (observed) double-crossover phenotypes. The predicted phenotypes are yellow, white flies and echinus flies in the F_2 generation. *Therefore, this order is also incorrect.* However, arrangement III produces the observed phenotypes—yellow, echinus flies and white flies. *Therefore, this arrangement, with the w gene in the middle, is correct.*

To summarize, first determine the arrangement of alleles on the homologs of the heterozygote yielding the crossover gametes by locating the reciprocal noncrossover phenotypes. Then, test each of three possible orders to determine which yields the observed double-crossover phenotypes—*the one that does so represents the correct order.*

Solving an Autosomal Mapping Problem

Having established the basic principles of chromosome mapping, we will now consider a related problem in maize (corn). This analysis differs from the preceding example in two ways. First, the previous mapping cross involved X-linked genes. Here, we consider autosomal genes. Second, in the discussion of this cross we have

changed our use of symbols. Instead of using the gene symbols and superscripts (e.g., bm^+, v^+, and pr^+), we simply use $+$ to denote each wild-type allele. This system is easier to manipulate but requires a better understanding of mapping procedures.

When we look at three autosomally linked genes in maize, the experimental cross must still meet the same three criteria we established for the X-linked genes in *Drosophila*: (1) One parent must be heterozygous for all traits under consideration; (2) the gametic genotypes produced by the heterozygote must be apparent from observing the phenotypes of the offspring; and (3) a sufficient sample size must be available for complete analysis.

In maize, the recessive mutant genes *brown* midrib (*bm*), *virescent* seedling (*v*), and *purple* aleurone (*pr*) are linked on chromosome 5. Assume that a female plant is known to be heterozygous for all three traits, but we do not know (1) the arrangement of the mutant alleles on the maternal and paternal homologs of this heterozygote, (2) the sequence of genes, or (3) the map distances between the genes. What genotype must the male plant have to allow successful mapping? To meet the second criterion, the male must be homozygous for all three recessive mutant alleles. Otherwise, offspring of this cross showing a given phenotype might represent more than one genotype, making accurate mapping impossible.

Figure 10 diagrams this cross. As shown, we know neither the arrangement of alleles nor the sequence of loci in the heterozygous female. Several possibilities are shown, but we have yet to determine which is correct. We don't know the sequence in the testcross male parent either, and so we must

(a) Some possible allele arrangements and gene sequences in a heterozygous female

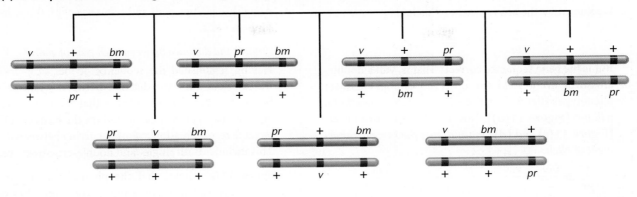

Which of the above is correct?

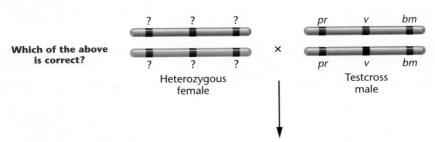

Heterozygous female   ×   Testcross male

(b) Actual results of mapping cross*

Phenotypes of offspring			Number	Total and percentage	Exchange classification
+	v	bm	230	467 42.1%	Noncrossover (NCO)
pr	+	+	237		
+	+	bm	82	161 14.5%	Single crossover (SCO)
pr	v	+	79		
+	v	+	200	395 35.6%	Single crossover (SCO)
pr	+	bm	195		
pr	v	bm	44	86 7.8%	Double crossover (DCO)
+	+	+	42		

*The sequence *pr – v – bm* may or may not be correct.

FIGURE 10 (a) Some possible allele arrangements and gene sequences in a heterozygous female. The data from a three-point mapping cross, depicted in (b), where the female is testcrossed, provide the basis for determining which combination of arrangement and sequence is correct. [See Figure 11(d)].

designate it randomly. Note that we have initially placed *v* in the middle. *This may or may not be correct.*

The offspring are arranged in groups of two for each pair of reciprocal phenotypic classes. The two members of each reciprocal class are derived from no crossing over (NCO), one of two possible single-crossover events (SCO), or a double crossover (DCO).

To solve this problem, refer to Figures 10 and 11 as you consider the following questions.

1. *What is the correct heterozygous arrangement of alleles in the female parent?*

Determine the two noncrossover classes, those that occur with the highest frequency. In this case, they are $+$ v bm and pr $+$ $+$. Therefore, the alleles on the homologs of the female parent must be arranged as shown in **Figure 11(a)**, . These homologs segregate into gametes, unaffected by any recombination event. Any other arrangement of alleles will not yield the observed noncrossover classes. (Remember that $+$ v bm is equivalent to pr^+ v bm, and that pr $+$ $+$ is equivalent to pr v^+ bm^+.)

2. *What is the correct sequence of genes?*

We know that the arrangement of alleles is

$$\frac{+ \quad v \quad bm}{pr \quad + \quad +}$$

But is the gene sequence correct? That is, will a double-crossover event yield the observed double-crossover phenotypes after fertilization? *Observation shows that it will not* [**Figure 11(b)**]. Now try the other two orders [**Figure 11(c)** and **(d)**] *maintaining the same arrangement of alleles*

$$\frac{+ \quad bm \quad v}{pr \quad + \quad +} \quad \text{or} \quad \frac{v \quad + \quad bm}{+ \quad pr \quad +}$$

Only the order on the right yields the observed double-crossover gametes [Figure 11(d)]. Therefore, the *pr* gene is in the middle. From this point on, work the problem using this arrangement and sequence, with the *pr* locus in the middle.

3. *What is the distance between each pair of genes?*

Having established the sequence of loci as *v–pr–bm*, we can determine the distance between *v* and *pr* and between *pr* and *bm*. Remember that the map distance between two genes is calculated on the basis of all detectable recombination events occurring between them. This includes both single- and double-crossover events.

Figure 11(e) shows that the phenotypes $\underline{v \quad pr \quad +}$ and $\underline{+ \quad + \quad bm}$ result from single crossovers between the *v* and *pr* loci, accounting for 14.5 percent of the offspring [according to data in Figure 10(b)]. By adding

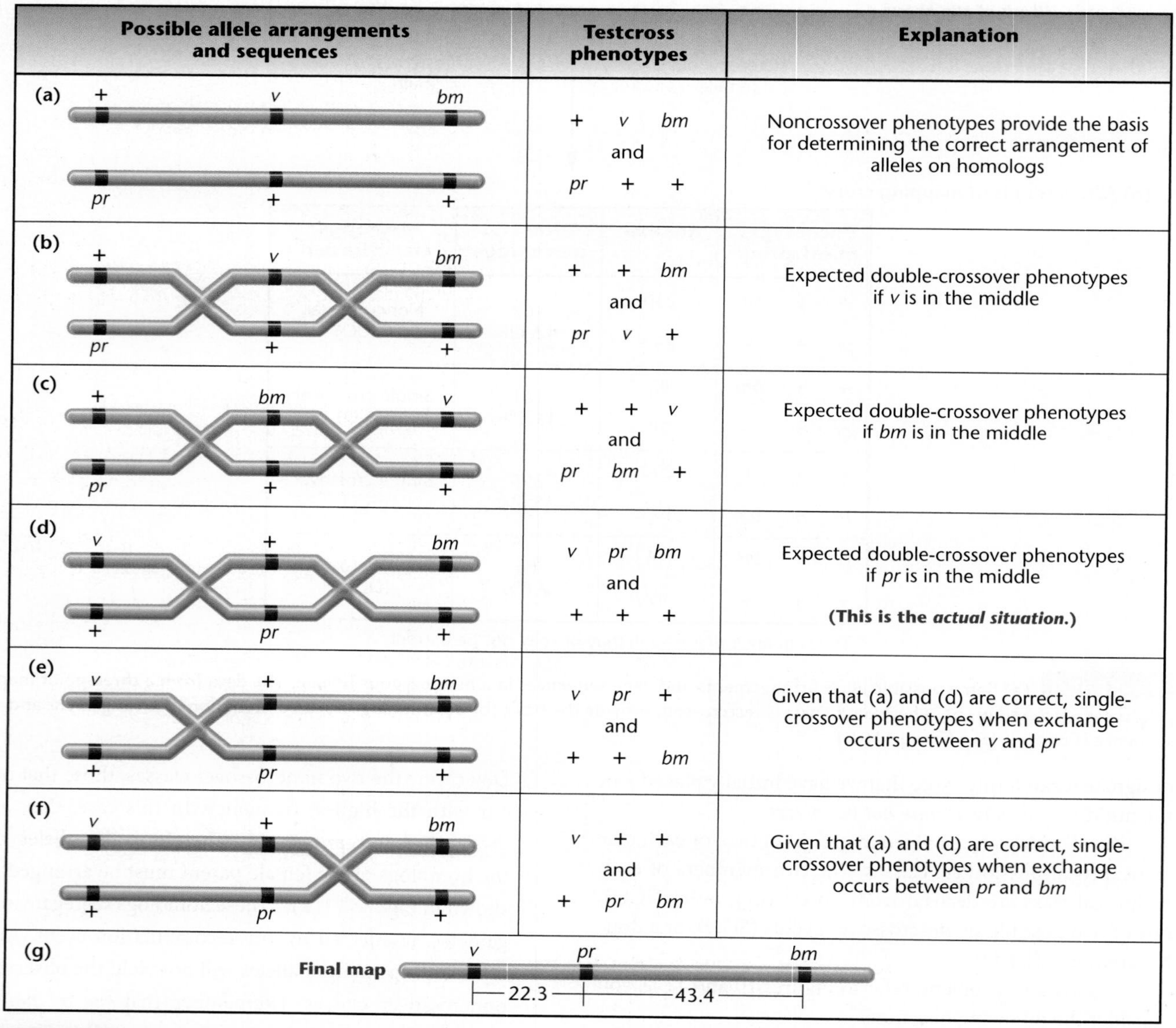

Possible allele arrangements and sequences	Testcross phenotypes	Explanation
(a) + v bm / pr + +	+ v bm and pr + +	Noncrossover phenotypes provide the basis for determining the correct arrangement of alleles on homologs
(b) + v bm / pr + +	+ + bm and pr v +	Expected double-crossover phenotypes if *v* is in the middle
(c) + bm v / pr + +	+ + v and pr bm +	Expected double-crossover phenotypes if *bm* is in the middle
(d) v + bm / + pr +	v pr bm and + + +	Expected double-crossover phenotypes if *pr* is in the middle **(This is the *actual situation*.)**
(e) v + bm / + pr +	v pr + and + + bm	Given that (a) and (d) are correct, single-crossover phenotypes when exchange occurs between *v* and *pr*
(f) v + bm / + pr +	v + + and + pr bm	Given that (a) and (d) are correct, single-crossover phenotypes when exchange occurs between *pr* and *bm*
(g) Final map	v ——22.3—— pr ———43.4——— bm	

FIGURE 11 Steps utilized in producing a map of the three genes in the cross in Figure 10, where neither the arrangement of alleles nor the sequence of genes in the heterozygous female parent is known.

the percentage of double crossovers (7.8 percent) to the number obtained for single crossovers, the total distance between the *v* and *pr* loci is calculated to be 22.3 mu.

Figure 11(f) shows that the phenotypes *v* + + and + *pr bm* result from single crossovers between the *pr* and *bm* loci, totaling 35.6 percent. Added to the double crossovers (7.8 percent), the distance between *pr* and *bm* is calculated to be 43.4 mu. The final map for all three genes in this example is shown in **Figure 11(g)**.

NOW SOLVE THIS

3 In *Drosophila*, a heterozygous female for the X-linked recessive traits *a*, *b*, and *c* was crossed to a male that phenotypically expressed *a*, *b*, and *c*. The offspring occurred in the following phenotypic ratios.

+	*b*	*c*	460
a	+	+	450
a	*b*	*c*	32
+	+	+	38
a	+	*c*	11
+	*b*	+	9

No other phenotypes were observed.

(a) What is the genotypic arrangement of the alleles of these genes on the X chromosome of the female?

(b) Determine the correct sequence and construct a map of these genes on the X chromosome.

(c) What progeny phenotypes are missing? Why?

■ **Hint** *This problem involves a three-point mapping experiment where only six phenotypic categories are observed, even though eight categories are typical of such a cross. The key to its solution is to be aware that if the distances between the loci are relatively small, the sample size may be too small for the predicted number of double crossovers to be recovered, even though reciprocal pairs of single crossovers are seen. You should write the missing gametes down as double crossovers and record zeros for their frequency of appearance.*

---

4 As the Distance between Two Genes Increases, Mapping Estimates Become More Inaccurate

So far, we have assumed that crossover frequencies are directly proportional to the distance between any two loci along the chromosome. However, it is not always possible to detect all crossover events. A case in point is a double exchange that occurs between the two loci in question. As shown in **Figure 12(a)**, if a double exchange occurs, the original arrangement of alleles on each nonsister homolog is recovered. Therefore, even though crossing over has occurred, it is impossible to detect. This

phenomenon is true for all even-numbered exchanges between two loci.

Furthermore, as a result of complications posed by **multiple-strand exchanges**, mapping determinations usually underestimate the actual distance between two genes. The farther apart two genes are, the greater the probability that undetected crossovers will occur. While the discrepancy is minimal for two genes relatively close together, the degree of inaccuracy increases as the distance increases, as shown in the graph of map distance versus recombination frequency in **Figure 12(b)**. There, the theoretical frequency where a direct correlation between recombination and map distance exists is contrasted with the actual frequency observed as the distance between two genes increases. The most accurate maps are constructed from experiments where genes are relatively close together.

Interference and the Coefficient of Coincidence

As shown in our maize example, we can predict the expected frequency of multiple exchanges, such as double crossovers, once the distance between genes is established. For example, in the maize cross, the distance between *v* and *pr* is 22.3 mu, and the distance between *pr* and *bm* is 43.4 mu. If the two single crossovers that make up a double crossover occur independently of one another, we can calculate the expected frequency of double crossovers (DCO_{exp}):

$$DCO_{exp} = (0.223) \times (0.434) = 0.097 = 9.7\%$$

Often in mapping experiments, the observed DCO frequency is less than the expected number of DCOs. In the maize cross, for example, only 7.8 percent DCOs are observed when 9.7 percent are expected. **Interference (*I*)**, the phenomenon through which a crossover event in one region of the chromosome inhibits a second event in nearby regions, causes this reduction.

To quantify the disparities that result from interference, we calculate the **coefficient of coincidence (*C*)**

$$C = \frac{\text{Observed DCO}}{\text{Expected DCO}}$$

In the maize cross, we have

$$C = \frac{0.078}{0.097} = 0.804$$

Once we have found *C*, we can quantify interference using the simple equation

$$I = 1 - C$$

In the maize cross, we have

$$I = 1.000 - 0.804 = 0.196$$

173

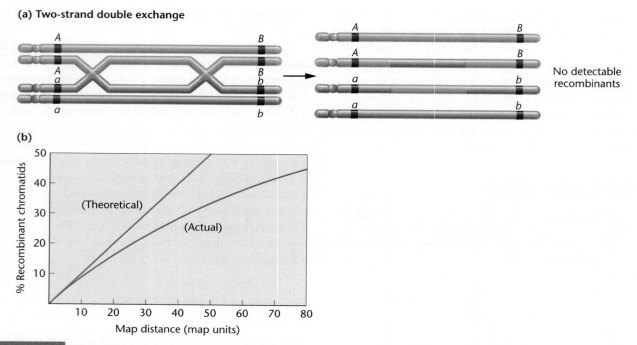

(a) Two-strand double exchange

No detectable recombinants

(b)

FIGURE 12 (a) A double crossover is undetected because no rearrangement of alleles occurs. (b) The theoretical and actual percentage of recombinant chromatids versus map distance. The straight line shows the theoretical relationship if a direct correlation between recombination and map distance exists. The curved line is the actual relationship derived from studies of *Drosophila*, *Neurospora*, and *Zea mays*.

If interference is complete and no double crossovers occur, then $I = 1.0$. If fewer DCOs than expected occur, I is a positive number and positive interference has occurred. If more DCOs than expected occur, I is a negative number and negative interference has occurred. In the maize example, I is a positive number (0.196), indicating that 19.6 percent fewer double crossovers occurred than expected.

Positive interference is most often the rule in eukaryotic systems. In general, the closer genes are to one another along the chromosome, the more positive interference occurs. In fact, interference in *Drosophila* is often complete within a distance of 10 mu, and no multiple crossovers are recovered. This observation suggests that physical constraints preventing the formation of closely aligned chiasmata contribute to interference. This interpretation is consistent with the finding that interference decreases as the genes in question are located farther apart. In the maize cross in Figures 10 and 11, the three genes are relatively far apart, and 80 percent of the expected double crossovers are observed.

ESSENTIAL POINT ■ ■ ■

Interference describes the extent to which a crossover in one region of a chromosome influences the occurrence of a crossover in an adjacent region of the chromosome and is quantified by calculating the coefficient of coincidence (*C*).

5 | **Chromosome Mapping Is Now Possible Using DNA Markers and Annotated Computer Databases**

Although traditional methods based on recombination analysis have produced detailed chromosomal maps in several organisms, such maps in other organisms (including humans) that do not lend themselves to such studies are greatly limited. Fortunately, the development of technology allowing direct analysis of DNA has greatly enhanced mapping in those organisms. We will address this topic using humans as an example.

Progress has initially relied on the discovery of DNA markers (mentioned earlier) that have been identified during recombinant DNA and genomic studies. These markers are short segments of DNA whose sequence and location are known, making them useful *landmarks* for mapping purposes. The analysis of human genes in relation to these markers has extended our knowledge of the location within the genome of countless genes, which is the ultimate goal of mapping.

The earliest examples are the DNA markers referred to as restriction fragment length polymorphisms (RFLPs) and microsatellites . RFLPS are polymorphic sites generated when specific DNA sequences are recognized and cut by restriction enzymes. Microsatellites are short repetitive sequences that are found throughout the genome, and they vary in the number of repeats at any given site. For example, the two-nucleotide sequence CA is repeated 5–50 times per site [$(CA)_n$] and appears throughout the genome approximately every 10,000 bases, on average. Microsatellites may be identified not only by the number of repeats but by the DNA sequences that flank them. More recently, variation in single nucleotides (called single-nucleotide polymorphisms or SNPs) has been utilized. Found throughout the genome, up to several million of these variations may be screened for an association with a disease or trait of interest, thus providing geneticists with a means to identify and locate related genes.

Cystic fibrosis offers an early example of a gene located by using DNA markers. It is a life-shortening autosomal recessive exocrine disorder resulting in excessive, thick mucus that impedes the function of organs such as the lung and pancreas. After scientists established that the gene causing this disorder is located on chromosome 7, they were then able to pinpoint its exact location on the long arm (the q arm) of that chromosome.

Several years ago (June 2007), using SNPs as DNA markers, associations between 24 genomic locations were established with seven common human diseases: *Type 1* (insulin dependent) and *Type 2 diabetes*, *Crohn's disease* (inflammatory bowel disease), *hypertension, coronary artery disease, bipolar* (manic-depressive) *disorder*, and *rheumatoid arthritis*. In each case, an inherited susceptibility effect was mapped to a specific location on a specific chromosome within the genome. In some cases, this either confirmed or led to the identification of a specific gene involved in the cause of the disease. In other cases, new genes will no doubt soon be discovered as a result of the identification of their location.

The many Human Genome Project databases that have been completed now make it possible to map genes along a human chromosome in base-pair distances rather than recombination frequency. This distinguishes what is referred to as a **physical map** of the genome from the *genetic maps* described above. Distances can then be determined relative to other genes and to features such as the DNA markers discussed. Through this approach, geneticists will soon be able to construct chromosome maps for individuals that designate specific allele combinations at each gene site.

> **ESSENTIAL POINT** ■ ■ ■
> Human linkage studies have been enhanced by the use of newly discovered molecular DNA markers.

6 Other Aspects of Genetic Exchange

Careful analysis of crossing over during gamete formation allows us to construct chromosome maps in many organisms. However, we should not lose sight of the real biological significance of crossing over, which is to generate genetic variation in gametes and, subsequently, in the offspring derived from the resultant eggs and sperm. Many unanswered questions remain, which we consider next.

Crossing Over—A Physical Exchange between Chromatids

Once genetic mapping was understood, it was of great interest to investigate the relationship between chiasmata observed in meiotic prophase I and crossing over. Are chiasmata visible manifestations of crossover events? If so, then crossing over in higher organisms appears to result from an actual physical exchange between homologous chromosomes. That this is the case was demonstrated independently in the 1930s by Harriet Creighton and Barbara McClintock in *Zea mays* and by Curt Stern in *Drosophila*.

Since the experiments are similar, we will consider only the work with maize. Creighton and McClintock studied two linked genes on chromosome 9. At one locus, the alleles *colorless* (*c*) and *colored* (*C*) control endosperm coloration. At the other locus, the alleles *starchy* (*Wx*) and *waxy* (*wx*) control the carbohydrate characteristics of the endosperm. The maize plant studied is heterozygous at both loci. The key to this experiment is that one of the homologs contains two unique cytological markers. The markers consist of a densely stained knob at one end of the chromosome and a translocated piece of another chromosome (8) at the other end. The arrangements of alleles and cytological markers can be detected cytologically and are shown in **Figure 13**.

Creighton and McClintock crossed this plant to one homozygous for the *colored* allele (*c*) and heterozygous for the endosperm alleles. They obtained a variety of different phenotypes in the offspring, but they were most interested in a crossover result involving the chromosome with the unique cytological markers. They examined the chromosomes of this plant with the colorless, waxy phenotype (Case I in Figure 13) for the presence of the cytological markers. If physical exchange between homologs accompanies genetic crossing over, the translocated chromosome will still be present, but the knob will not—this is exactly what happened. In a second plant (Case II), the phenotype colored, starchy should result from either nonrecombinant gametes or crossing over. Some of the plants then ought to contain chromosomes with the dense knob but not the translocated chromosome. This condition was also found, and the conclusion that a physical

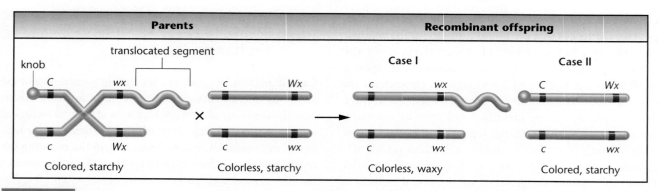

Parents		Recombinant offspring	
		Case I	Case II
knob, C, wx (translocated segment) / c, Wx — Colored, starchy	c, Wx / c, wx — Colorless, starchy	c, wx / c, wx — Colorless, waxy	C, Wx / c, wx — Colored, starchy

FIGURE 13 The phenotypes and chromosome compositions of parents and recombinant offspring in Creighton and McClintock's experiment in maize. The knob and translocated segment served as cytological markers, which established that crossing over involves an actual exchange of chromosome arms.

exchange takes place was again supported. Along with Stern's findings with *Drosophila,* this work clearly established that crossing over has a cytological basis.

> ### ESSENTIAL POINT
>
> Cytological investigations of both maize and *Drosophila* reveal that crossing over involves a physical exchange of segments between nonsister chromatids.

Sister Chromatid Exchanges between Mitotic Chromosomes

Considering that crossing over occurs between synapsed homologs in meiosis, we might ask whether a similar physical exchange occurs between homologs during mitosis. While homologous chromosomes do not usually pair up or synapse in somatic cells (*Drosophila* is an exception), each individual chromosome in prophase and metaphase of mitosis consists of two identical sister chromatids, joined at a common centromere. Surprisingly, several experimental approaches have demonstrated that reciprocal exchanges similar to crossing over occur between sister chromatids. These **sister chromatid exchanges (SCEs)** do not produce new allelic combinations, but evidence is accumulating that attaches significance to these events.

Identification and study of SCEs are facilitated by several modern staining techniques. In one technique, cells replicate for two generations in the presence of the thymidine analog **bromodeoxyuridine (BrdU).** Following two rounds of replication, each pair of sister chromatids has one member with one strand of DNA "labeled" with BrdU and the other member with both strands labeled with BrdU. Using a differential stain, chromatids with the analog in both strands stain less brightly than chromatids with BrdU in only one strand. As a result, SCEs are readily detectable if they occur. In **Figure 14,** numerous instances of SCE events

are clearly evident. These sister chromatids are sometimes referred to as **harlequin chromosomes** because of their patchlike appearance.

The significance of SCEs is still uncertain, but several observations have generated great interest in this phenomenon. We know, for example, that agents that induce chromosome damage (viruses, X rays, ultraviolet light, and certain chemical mutagens) increase the frequency of SCEs. The frequency of SCEs is also elevated in **Bloom syndrome,** a human disorder caused by a mutation in the *BLM* gene on chromosome 15. This rare, recessively inherited disease is characterized

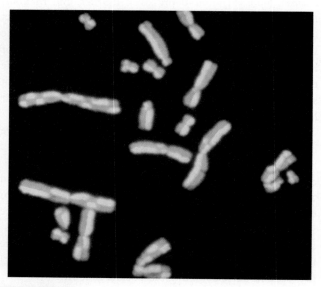

FIGURE 14 Demonstration of sister chromatid exchanges (SCEs) in mitotic chromosomes. Sometimes called harlequin chromosomes because of the alternating patterns they exhibit, sister chromatids containing the thymidine analog BrdU are seen to fluoresce *less* brightly where they contain the analog in both DNA strands than when they contain the analog in only one strand. These chromosomes were stained with 33258-Hoechst reagent and acridine orange and then viewed under fluorescence microscopy.

by prenatal and postnatal retardation of growth, a great sensitivity of the facial skin to the sun, immune deficiency, a predisposition to malignant and benign tumors, and abnormal behavior patterns. The chromosomes from cultured leukocytes, bone marrow cells, and fibroblasts derived from homozygotes are very fragile and unstable compared to those of homozygous and heterozygous normal individuals. Increased breaks and rearrangements between nonhomologous chromosomes are observed in addition to excessive amounts of sister chromatid exchanges. Work by James German and colleagues suggests that the *BLM* gene encodes an enzyme called **DNA helicase**, which is best known for its role in DNA replication .

ESSENTIAL POINT ■ ■ ■

Recombination events between sister chromatids in mitosis, referred to as sister chromatid exchanges (SCEs), occur at an elevated frequency in the human disorder, Bloom syndrome.

7 | Did Mendel Encounter Linkage?

We conclude by examining a modern-day interpretation of the experiments that form the cornerstone of transmission genetics—Mendel's crosses with garden peas.

Some observers believe that Mendel had extremely good fortune in his classic experiments. He did not encounter apparent linkage relationships between the seven mutant characters in any of his crosses. Had Mendel obtained highly variable data characteristic of linkage and crossing over, these unorthodox ratios might have hindered his successful analysis and interpretation.

The article by Stig Blixt, reprinted in its entirety in the following box, demonstrates the inadequacy of this hypothesis. As we shall see, some of Mendel's genes were indeed linked. We leave it to Stig Blixt to enlighten you as to why Mendel did not detect linkage.

Why Didn't Gregor Mendel Find Linkage?

It is quite often said that Mendel was very fortunate not to run into the complication of linkage during his experiments. He used seven genes, and the pea has only seven chromosomes. Some have said that had he taken just one more, he would have had problems. This, however, is a gross oversimplification. The actual situation, most probably, is shown in **Table 1**. This shows that Mendel worked with three genes in chromosome 4, two genes in chromosome 1, and one gene in each of chromosomes 5 and 7. It seems at first glance that, out of the 21 dihybrid combinations Mendel theoretically could have studied, no fewer than four (that is, *a–i, v–fa, v–le, fa–le,*) ought to have resulted in linkages. However, as found in hundreds of crosses and shown by the genetic map of the pea, *a* and *i* in chromosome 1 are so distantly located on the chromosome that no linkage is normally detected. The same is true for *v* and *le* on the one hand, and *fa* on the other, in chromosome 4. This leaves *v–le*, which ought to have shown linkage.

Mendel, however, seems not to have published this particular combination and thus, presumably, never made the appropriate cross to obtain both genes segregating simultaneously. It is therefore not so astonishing that Mendel did not run into the complication of linkage, although he did not avoid it by choosing one gene from each chromosome.

STIG BLIXT
Weibullsholm Plant Breeding Institute, Landskrona, Sweden, and Centro Energia Nucleate na Agricultura, Piracicaba, SP, Brazil

Source: Reprinted by permission from *Nature,* Vol. 256, p. 206. © 1975 Macmillan Magazines Limited.

TABLE 1 Relationship between Modern Genetic Terminology and Character Pairs Used by Mendel

Character Pair Used by Mendel	Alleles in Modern Terminology	Located in Chromosome
Seed color, yellow–green	*I–i*	1
Seed coat and flowers, colored–white	*A–a*	1
Mature pods, smooth expanded–wrinkled indented	*V–v*	4
Inflorescences, from leaf axis–umbellate in top of plant	*Fa–fa*	4
Plant height, 0.5–1m	*Le–le*	4
Unripe pods, green–yellow	*Gp–gp*	5
Mature seeds, smooth–wrinkled	*R–r*	7

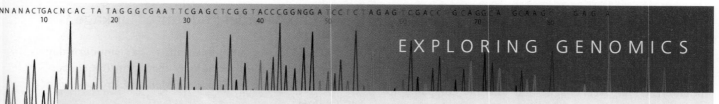

Human Chromosome Maps on the Internet

 Study Area: *Exploring Genomics*

I n this chapter textwe discussed how recombination data can be analyzed to develop chromosome maps based on linkage. Although linkage analysis and chromosome mapping continue to be important approaches in genetics, chromosome maps are increasingly being developed for many species using genomics techniques to sequence entire chromosomes. As a result of the Human Genome Project, maps of human chromosomes are now freely available on the Internet. With the click of a mouse you can have immediate access to an incredible wealth of information. In this exercise we will explore the **National Center for Biotechnology Information (NCBI) Genes and Disease** Web site to learn more about human chromosome maps.

■ NCBI Genes and Disease

The NCBI Web site is an outstanding resource for genome data. Here we explore the Genes and Disease site, which presents human chromosome maps that show the locations of specific disease genes.

1. Access the Genes and Disease site at http://www.ncbi.nlm.nih.gov/books/NBK22183/

2. Under contents, click on "chromosome maps" to see a page with an image of a karyotype of human chromosomes. Click on a chromosome in the chromosome map image, scroll down the page to view a chromosome or click on a chromosome listed on the right side of the page. For example, click on chromosome 7. Notice that the number of genes on the chromosomes and the number of base pairs the chromosome contains are displayed above the image.

3. Look again at chromosome 7. At first you might think there are only five disease genes on this chromosome because the initial view shows only selected disease genes. However, if you click the "MapViewer" link for the chromosome (just above the drawing), you will see detailed information about the chromosome, including a complete "Master Map" of the genes it contains, including the symbols used in naming genes:

Gene Symbols: Clicking on the gene symbols takes you to the **NCBI Entrez Gene database**, a searchable tool for information on genes in the NCBI database. Links: The items in the "Links" column provide access to OMIM (Online Mendelian Inheritance in Man) data for a particular gene, as well as to protein information (*pn*) and lists of homologous genes (*hm*; these are other genes that have similar sequences).

4. Click on the links in MapViewer to learn more about a gene of interest.

5. Scan the chromosome maps in Map Viewer until you see one of the genes listed as a "hypothetical gene or protein."
   a. What does it mean if a gene or protein is referred to as hypothetical?
   b. What information do you think genome scientists use to assign a gene locus for a gene encoding a hypothetical protein?

Visit the **NCBI Map Viewer** homepage (www.ncbi.nlm.nih.gov/mapview) for an excellent database containing chromosome maps for a wide variety of different organisms. Search this database to learn more about chromosome maps for an organism you are interested in.

CASE STUDY » Links to autism

As parents of an autistic child, entering a research study seemed to be a way of not only educating themselves about their son's condition, but also of furthering research into this complex, behaviorally defined disorder. Researchers told the couple that there is a strong genetic influence for autism since the concordance rate in identical twins is about 75 percent and only about 5 percent in fraternal twins. In addition, researchers have identified interactions among at least ten candidate genes distributed among chromosomes 3, 7, 15, and 17. Generally unaware of the principles of basic genetics, the couple asked a number of interesting questions. How would you respond to them?

1. What is a "candidate" gene?
2. Since several candidate genes must be on the same chromosome, will they always be transmitted as a block of harmful genetic information to future offspring?
3. With such an apparently complex genetic condition, what is the likelihood that our next child will also be autistic?
4. Is prenatal diagnosis possible during future pregnancies?

INSIGHTS AND SOLUTIONS

1. In rabbits, black color (*B*) is dominant to brown (*b*), while full color (*C*) is dominant to *chinchilla* (*c^ch*). The genes controlling these traits are linked. Rabbits that are heterozygous for both traits and express black, full color are crossed to rabbits that express brown, chinchilla with the following results:

31	brown, chinchilla	34	black, full
16	brown, full	19	black, chinchilla

Determine the arrangement of alleles in the heterozygous parents and the map distance between the two genes.

Solution: This is a two-point map problem, where the two most prevalent reciprocal phenotypes are the noncrossovers. The less frequent reciprocal phenotypes arise from a single crossover. The arrangement of alleles is derived from the noncrossover phenotypes because they enter gametes intact.

The single crossovers give rise to 35/100 offspring (35 percent). Therefore, the distance between the two genes is 35 mu.

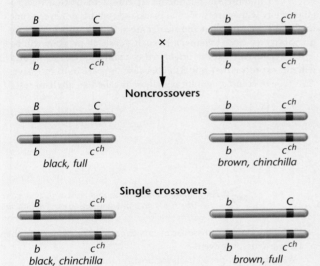

2. In *Drosophila*, *Lyra* (*Ly*) and *Stubble* (*Sb*) are dominant mutations located at locus 40 and 58, respectively, on chromosome III. A recessive mutation with bright red eyes is discovered and shown also to be located on chromosome III. A map is obtained by crossing a female who is heterozygous for all three mutations to a male that is homozygous for the *bright-red* mutation (which we will call *br*), and the data in the table are generated. Determine the location of the *br* mutation on chromosome III.

Phenotype			Number
(1) *Ly*	*Sb*	*br*	404
(2) +	+	+	422
(3) *Ly*	+	+	18
(4) +	*Sb*	*br*	16
(5) *Ly*	+	*br*	75
(6) +	*Sb*	+	59
(7) *Ly*	*Sb*	+	4
(8) +	+	*br*	2
Total			= 1000

Solution: First, determine the arrangement of the alleles on the homologs of the heterozygous crossover parent (the female in this case). To do this, locate the most frequent reciprocal phenotypes, which arise from the noncrossover gametes—these are phenotypes (1) and (2). Each phenotype represents the arrangement of alleles on one of the homologs. Therefore, the arrangement is

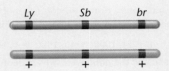

Second, find the correct sequence of the three loci along the chromosome. This is done by determining which sequence yields the observed double-crossover phenotypes, which are the least frequent reciprocal phenotypes (7 and 8). If the sequence is correct as written, then the double crossover depicted here,

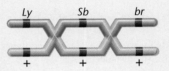

will yield *Ly* + *br* and + *Sb* + as phenotypes. Inspection shows that these categories (5 and 6) are actually single crossovers, not double crossovers. Therefore, the sequence is incorrect as written. Only two other sequences are possible: The *br* gene is either to the left of *Ly* (Case A), or it is between *Ly* and *Sb* (Case B).

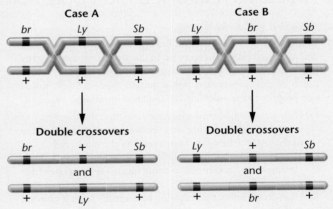

Comparison with the actual data shows that Case B is correct. The double-crossover gametes yield flies that express *Ly* and *Sb* but not *br*, or express *br* but not *Ly* and *Sb*. Therefore, the correct arrangement and sequence are as follows.

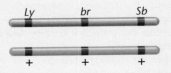

Once this sequence is found, determine the location of *br* relative to *Ly* and *Sb*. A single crossover between *Ly* and *br*, as shown here,

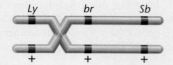

yields flies that are *Ly + +* and *+br Sb* (phenotypes 3 and 4). Therefore, the distance between the *Ly* and *br* loci is equal to

$$(18 + 16 + 4 + 2)/1000 = 40/1000 = 0.04 = 4 \text{ mu}$$

Remember to add the double crossovers because they represent two single crossovers occurring simultaneously. You need to know the frequency of all crossovers between *Ly* and *br*, so they must be included.

Similarly, the distance between the *br* and *Sb* loci is derived mainly from single crossovers between them.

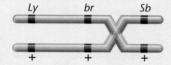

This event yields *Ly br+* and *+ + Sb* phenotypes (phenotypes 5 and 6). Therefore, the distance equals

$$(75 + 59 + 4 + 2)/1000 = 140/1000 = 0.14 = 14 \text{ mu}$$

The final map shows that *br* is located at locus 44, since *Lyra* and *Stubble* are known.

3. In reference to Problem 2, a student predicted that the mutation was actually the known mutation *scarlet* located at locus 44.0. Suggest an experimental cross that would confirm this prediction.

Solution: Since the *scarlet* locus is identical to the experimental assignment, it is reasonable to hypothesize that the *bright-red* eye mutation is an allele at the *scarlet* locus.

To test this hypothesis, you could perform complementation analysis by crossing females expressing the *bright-red* mutation with known *scarlet* males. If the two mutations are alleles, no complementation will occur and all progeny will reveal a *bright-red* mutant eye phenotype. If complementation occurs, all progeny will express normal brick-red (wild-type) eyes, since the bright-red mutation and *scarlet* are at different loci (they are probably very close together). In such a case, all progeny will be heterozygous at both the *bright-red* eye and the *scarlet* loci and will not express either mutation because they are both recessive. This type of complementation analysis is called an **allelism test.**

PROBLEMS AND DISCUSSION QUESTIONS

 For activities, animations, and review quizzes, go to the study area at www.masteringgenetics.com

HOW DO WE KNOW?

1. In this chapter, we focused on linkage, chromosomal mapping, and many associated phenomena. In the process, we found many opportunities to consider the methods and reasoning by which much of this information was acquired. What answers would you propose to the following fundamental questions?

   (a) How was it established experimentally that the frequency of recombination (crossing over) between two genes is related to the distance between them along the chromosome?

   (b) How do we know that specific genes are linked on a single chromosome, in contrast to being located on separate chromosomes?

   (c) How do we know that crossing over results from a physical exchange between chromatids?

   (d) How do we know that sister chromatids undergo recombination during mitosis?

2. What is the significance of crossing over (which leads to genetic recombination) to the process of evolution?

3. Describe the cytological observation that suggests that crossing over occurs during the first meiotic prophase.

4. Why does more crossing over occur between two distantly linked genes than between two genes that are very close together on the same chromosome?

5. Why is a 50 percent recovery of single-crossover products the upper limit, even when crossing over *always* occurs between two linked genes?

6. Why are double-crossover events expected less frequently than single-crossover events?

7. What is the proposed basis for positive interference?

8. What three essential criteria must be met in order to execute a successful mapping cross?

9. The genes *dumpy* wings (*dp*), *clot* eyes (*cl*), and *apterous* wings (*ap*) are linked on chromosome II of *Drosophila*. In a series of two-point mapping crosses, the genetic distances shown below were determined. What is the sequence of the three genes?

dp–ap	42
dp–cl	3
ap–cl	39

10. Colored aleurone in the kernels of corn is due to the dominant allele *R*. The recessive allele *r*, when homozygous, produces colorless aleurone. The plant color (not kernel color) is controlled by another gene with two alleles, *Y* and *y*. The dominant *Y* allele results in green color, whereas the homozygous presence of the recessive *y* allele causes the plant to appear yellow. In a testcross between a plant of unknown genotype and phenotype and a plant that is homozygous recessive for both traits, the following progeny were obtained:

colored, green	88
colored, yellow	12
colorless, green	8
colorless, yellow	92

Explain how these results were obtained by determining the exact genotype and phenotype of the unknown plant, including the precise association of the two genes on the homologs (i.e., the arrangement).

11. In the cross shown here, involving two linked genes, *ebony* (*e*) and *claret* (*ca*), in *Drosophila*, where crossing over does not occur in males, offspring were produced in a (2 + :1 *ca* :1 *e*) phenotypic ratio:

$$\frac{e \quad ca^+}{e^+ \quad ca} \quad \times \quad \frac{e \quad ca^+}{e^+ \quad ca}$$

These genes are 30 mu apart on chromosome III. What did crossing over in the female contribute to these phenotypes?

12. In a series of two-point map crosses involving five genes located on chromosome II in *Drosophila,* the following recombinant (single-crossover) frequencies were observed:

pr–adp	29
pr–vg	13
pr–c	21
pr–b	6
adp–b	35
adp–c	8
adp–vg	16
vg–b	19
vg–c	8
c–b	27

(a) If the *adp* gene is present near the end of chromosome II (locus 83), construct a map of these genes.

(b) In another set of experiments, a sixth gene (*d*) was tested against *b* and *pr,* and the results were *d − b* = 17% and *d − pr* = 23%. Predict the results of two-point maps between *d* and *c, d* and *vg,* and *d* and *adp.*

13. Two different female *Drosophila* were isolated, each heterozygous for the autosomally linked genes *black* body (*b*), *dachs* tarsus (*d*), and *curved* wings (*c*). These genes are in the order *d–b–c*, with *b* closer to *d* than to *c*. Shown in the following table is the genotypic arrangement for each female, along with the various gametes formed by both. Identify which categories are noncrossovers (NCO), single crossovers (SCO), and double crossovers (DCO) in each case. Then, indicate the relative frequency with which each will be produced.

Female A		Female B	
$\dfrac{d\ \ b\ \ +}{+\ \ +\ \ c}$		$\dfrac{d\ \ +\ \ +}{+\ \ b\ \ c}$	
	↓ *Gamete formation*	↓	
(1) d b c	(5) d + +	(1) d b +	(5) d b c
(2) + + +	(6) + b c	(2) + + c	(6) + + +
(3) + + c	(7) d + c	(3) d + c	(7) d + +
(4) d b +	(8) + b +	(4) + b +	(8) + b c

14. In *Drosophila,* a cross was made between females expressing the three X-linked recessive traits, *scute* bristles (*sc*), *sable* body (*s*), and *vermilion* eyes (*v*), and wild-type males. All females were wild type in the F₁, while all males expressed all three mutant traits. The cross was carried to the F₂ generation and 1000 offspring were counted, with the results shown in the following table. No

determination of sex was made in the F₂ data. (a) Using proper nomenclature, determine the genotypes of the P₁ and F₁ parents. (b) Determine the sequence of the three genes and the map distance between them. (c) Are there more or fewer double crossovers than expected? (d) Calculate the coefficient of coincidence; does this represent positive or negative interference?

Phenotype			Offspring
sc	s	v	314
+	+	+	280
+	s	v	150
sc	+	+	156
sc	+	v	46
+	s	+	30
sc	s	+	10
+	+	v	14

15. A cross in *Drosophila* involved the recessive, X-linked genes *yellow* body (*y*), *white* eyes (*w*), and *cut* wings (*ct*). A yellow-bodied, white-eyed female with normal wings was crossed to a male whose eyes and body were normal, but whose wings were cut. The F₁ females were wild type for all three traits, while the F₁ males expressed the yellow-body, white-eye traits. The cross was carried to F₂ progeny, and only male offspring were tallied. On the basis of the data shown here, a genetic map was constructed. (a) Diagram the genotypes of the F₁ parents. (b) Construct a map, assuming that *w* is at locus 1.5 on the X chromosome. (c) Were any double-crossover offspring expected? (d) Could the F₂ female offspring be used to construct the map? Why or why not?

Phenotype			Male Offspring
y	+	ct	9
+	w	+	6
y	w	ct	90
+	+	+	95
+	+	ct	424
y	w	+	376
y	+	+	0
+	w	ct	0

16. In *Drosophila*, *Dichaete* (*D*) is a mutation on chromosome III with a dominant effect on wing shape. It is lethal when homozygous. The genes *ebony* body (*e*) and *pink* eye (*p*) are recessive mutations on chromosome III. Flies from a Dichaete stock were crossed to homozygous ebony, pink flies, and the F₁ progeny with a Dichaete phenotype were backcrossed to the ebony, pink homozygotes. Using the results of this backcross shown in the following table, (a) diagram the cross, showing the genotypes of the parents and offspring of both crosses. (b) What is the sequence and interlocus distance between these three genes?

Phenotype	Number
Dichaete	401
ebony, pink	389
Dichaete, ebony	84
pink	96
Dichaete, pink	2
ebony	3
Dichaete, ebony, pink	12
wild type	13

17. *Drosophila* females homozygous for the third chromosomal genes *pink* eye (*p*) and *ebony* body (*e*) were crossed with males homozygous for the second chromosomal gene *dumpy* wings (*dp*). Because these genes are recessive, all offspring were wild type (normal). F_1 females were testcrossed to triply recessive males. If we assume that the two linked genes (*p* and *e*) are 20 mu apart, predict the results of this cross. If the reciprocal cross were made (F_1 males—where no crossing over occurs—with triply recessive females), how would the results vary, if at all?

18. In *Drosophila*, the two mutations *Stubble* bristles (*Sb*) and *curled* wings (*cu*) are linked on chromosome III. *Sb* is a dominant gene that is lethal in a homozygous state, and *cu* is a recessive gene. If a female of the genotype

$$\frac{Sb \quad cu}{+ \quad +}$$

is to be mated to detect recombinants among her offspring, what male genotype would you choose as her mate?

19. A female of genotype

$$\frac{a \quad b \quad c}{+ \quad + \quad +}$$

produces 100 meiotic tetrads. Of these, 68 show no crossover events. Of the remaining 32, 20 show a crossover between *a* and *b*, 10 show a crossover between *b* and *c*, and 2 show a double crossover between *a* and *b* and between *b* and *c*. Of the 400 gametes produced, how many of each of the eight different genotypes will be produced? Assuming the order a–b–c and the allele arrangement shown above, what is the map distance between these loci?

20. In a plant, fruit color is either red or yellow, and fruit shape is either oval or long. Red and oval are the dominant traits. Two plants, both heterozygous for these traits, were testcrossed, with the results shown in the following table. Determine the location of the genes relative to one another and the genotypes of the two parental plants.

Phenotype	Progeny Plant A	Plant B
red, long	46	4
yellow, oval	44	6
red, oval	5	43
yellow, long	5	47
Total	100	100

21. *Drosophila melanogaster* has one pair of sex chromosomes (XX or XY) and three autosomes (chromosomes II, III, and IV). A genetics student discovered a male fly with very short (*sh*) legs. Using this male, the student was able to establish a pure-breeding stock of this mutant and found that it was recessive. She then incorporated the mutant into a stock containing the recessive gene *black* (*b*, body color, located on chromosome II) and the recessive gene *pink* (*p*, eye color, located on chromosome III). A female from the homozygous black, pink, short stock was then mated to a wild-type male. The F_1 males of this cross were all wild type and were then backcrossed to the homozygous *b, p, sh* females. The F_2 results appeared as shown in the following table, and no other phenotypes were observed. (a) Based on these results, the student was able to assign *sh* to a linkage group (a chromosome). Determine which chromosome, and include step-by-step reasoning. (b) The student repeated the experiment, making the reciprocal cross: F_1 females backcrossed to homozygous *b, p, sh* males. She observed that 85 percent of the offspring fell into the given classes, but that 15 percent of the offspring were equally divided among *b+p, b++, +shp,* and *+sh+* phenotypic

males and females. How can these results be explained, and what information can be derived from these data?

Phenotype	Female	Male
wild	63	59
pink*	58	65
black, short	55	51
black, pink, short	69	60

*Pink indicates that the other two traits are wild type (normal). Similarly, black, short offspring are wild type for eye color.

22. In *Drosophila*, a female fly is heterozygous for three mutations: *Bar* eyes (*B*), *miniature* wings (*m*), and *ebony* body (*e*). (Note that *Bar* is a dominant mutation.) The fly is crossed to a male with normal eyes, miniature wings, and ebony body. The results of the cross are shown in the following list. Interpret the results of this cross. If you conclude that linkage is involved between any of the genes, determine the map distance(s) between them.

miniature	111	Bar	117
wild type	29	Bar, miniature	26
Bar, ebony	101	ebony	35
Bar, miniature, ebony	31	miniature, ebony	115

23. An organism of the genotype *AaBbCc* was testcrossed to a triply recessive organism (*aabbcc*). The genotypes of the progeny are in the following table.

AaBbCc	20	*AaBbcc*	20
aabbCc	20	*aabbcc*	20
AabbCc	5	*Aabbcc*	5
aaBbCc	5	*aaBbcc*	5

(a) Assuming simple dominance and recessiveness in each gene pair, if these three genes were all assorting independently, how many genotypic and phenotypic classes would result in the offspring, and in what proportion?

(b) Answer part (a) again, assuming the three genes are so tightly linked on a single chromosome that no crossover gametes were recovered in the sample of offspring.

(c) What can you conclude from the *actual* data about the location of the three genes in relation to one another?

24. Based on our discussion of the potential inaccuracy of mapping (see Figure 12), would you revise your answer to Problem 23(c)? If so, how?

25. In Creighton and McClintock's experiment demonstrating that crossing over involves physical exchange between chromosomes (see Section 6), explain the importance of the cytological markers (the translocated segment and the chromosome knob) in the experimental rationale.

26. Traditional gene mapping has been applied successfully to a variety of organisms including yeast, fungi, maize, and *Drosophila*. However, human gene mapping has only recently shared a similar spotlight. What factors have delayed the application of traditional gene-mapping techniques in humans?

27. DNA markers have greatly enhanced the mapping of genes in humans. What are DNA markers, and what advantage do they confer?

28. Are sister chromatid exchanges effective in producing genetic variability in an individual? in the offspring of individuals?

SOLUTIONS TO SELECTED PROBLEMS AND DISCUSSION QUESTIONS

Answers to Now Solve This

1. **(a)** 1/4 *AaBb*   1/4 *Aabb*   1/4 *aaBb*   1/4 *aabb*

   (b) 1/4 *AaBb*   1/4 *Aabb*   1/4 *aaBb*   1/4 *aabb*

   (c) If the arrangement is *AB/ab* × *ab/ab*
   then the two types of offspring will be as follows:

               1/2 *Ab/ab*   1/2 *aB/ab*

   If, however, *A* and *B* are not coupled, then the symbolism would be *Ab/aB* × *aabb*.
   The offspring would occur as follows:

               1/2 *Ab/ab*   1/2 *aB/ab*

2. The most frequent classes are *PZ* and *pz*. These classes represent the parental (noncrossover) groups, which indicates that the original parental arrangement in the testcross was *PZ/pz* × *pz/pz*. Adding the crossover percentages together (6.9 + 7.1) gives 14 percent, which would be the map distance between the two genes.

3. Examine the progeny list to see which types are not present. In this case, the double crossover classes are the following: + + *c* and *a b* +

   (a, b) Gene *b* is in the middle and the arrangement is as follows.

           + *b c/a* + +

       a − *b* = 7 map units   *b* − *c* = 2 map units

   (c) The progeny phenotypes that are missing are + + *c* and *a b* +, of which, from 1000 offspring, 1.4 (0.07 × 0.02 × 1000) would be expected. Perhaps by chance or some other unknown selective factor, they were not observed.

Solutions to Problems and Discussion Questions

4. With some qualification, especially around the centromeres and telomeres, one can say that crossing over is somewhat randomly distributed over the length of the chromosome. Two loci that are far apart are more likely to have a crossover between them than two loci that are close together.

6. If the probability of one event is 1/X, the probability of two events occurring at the same time will be $1/X^2$.

10. The heterozygous parent in the test cross is *RY/ry* × *ry/ry* with the two dominant alleles on one chromosome and the two recessives on the homolog. The map distance would be 10 map units between the *R* and *Y* loci.

12. The map for parts **(a)** and **(b)** is the following:

    d...........*b*.........*pr*.........*vg*........*c*.........*adp*
       31   48   54   67   75   83

    Map units

    The expected map units between *d* and *c* would be 44, between *d* and *vg* 36, and between *d* and *adp* 52. However, because there is a theoretical maximum of 50 map units possible between two loci in any one cross, that distance would be below the 52 determined by simple subtraction.

14. **(a)** P$_1$: *sc s v/sc s v* × + + +/Y
         F$_1$: + + +/*sc s v* × *sc s v*/Y

    (b) The map distances are determined by first writing the proper arrangement and sequence of genes, and then computing the distances between each set of genes.

               sc v s
                + + +

sc − *v* = 33 percent (map units)
  v − *s* = 10 percent (map units)

(c, d) The coefficient of coincidence = 0.727, which indicates that there were fewer double crossovers than expected; therefore, positive chromosomal interference is present.

16. **(a)** Represent the *Dichaete* gene as an uppercase letter because it is dominant. At this point, the gene sequence is not given.

    P$_1$: *D* + +/ + + + × + *e p*/ + *e p*
    F$_1$: *D* + +/ + *e p* × + *e p*/ + *e p*

F$_2$:	*D* + +/ + *e p*	Dichaete
	+ *e p*/ + *e p*	ebony, pink
	D e +/ + *e p*	Dichaete, ebony
	+ + *p*/ + *e p*	pink
	D + *p*/ + *e p*	Dichaete, pink
	+ *e* +/ + *e p*	ebony
	D e p/ + *e p*	Dichaete, ebony, pink
	+ + +/ + *e p*	wild type

    (b) F$_1$: *D* + +/ + *p e* × + *p e*/ + *p e*

      D − *p* = 3.0 map units   *p* − *e* = 18.5 map units

18. One would use the typical testcross arrangement with the *curled* gene so the male parent arrangement would be + *cu*/ + *cu*.

20. Assign the following symbols, for example:

        R = Red   *r* = yellow
        O = Oval   *o* = long

    Progeny A: *Ro/rO* × *rroo* = 10 map units
    Progeny B: *RO/ro* × *rroo* = 10 map units

22. Begin with a set of symbols as indicated below:

      B^+ = wild eye shape   *B* = Bar eye shape
      m^+ = wild wings   *m* = miniature wings
      e^+ = wild body color   *e* = ebony body color

    Superficially, the cross would be as follows:

      B^+B m^+m e^+e × B^+? *m*? *e*

    (The *?* is used at this point to indicate that we have no information allowing us to decide whether any of the alleles in the male are X-linked.) The arrangement is as indicated in the following: B m^+/B^+m; *e*+/*e*

    Notice that a semicolon is used to indicate that the *ebony* locus is on a different chromosome. At this point and without prior knowledge, we still don't know whether any of the genes are X-linked; however, it is of no consequence to the solution of the problem. (In actuality, both *B* and *m* loci are X-linked.) To determine the map distances (again, *ebony* is out of the mapping picture at this point because it is not linked to either *B* or *m*):

        111 + 115 = 226 = parental
        117 + 101 = 218 = parental
        26 +   31 = 57 = crossover
        29 +   35 = 64 = crossover

    Mapping the distance between *B* and *m* would be as follows:

      (57 + 64)/(226 + 218 + 57 + 64) × 100 =
      121/565 × 100 = 21.4 map units

We would conclude that the *ebony* locus is either far away from *B* and *m* (50 map units or more) or on a different chromosome. In fact, *ebony* is on a different chromosome.

24. Since the genetic map is more accurate when relatively small distances are covered and when large numbers of offspring are scored, this map would probably not be too accurate with such a small sample size.

26. In contrast to the other organisms mentioned, a single human mating pair produces relatively few offspring and the haploid number of chromosomes is relatively high (23), so there are rather small numbers of identifiable genes per chromosome. In addition, accurate medical records are often difficult to obtain and the life cycle is relatively long.

28. Because sister chromatids are genetically identical (with the exception of rare new mutations), crossing over between sisters provides no increase in genetic variability.

CREDITS

Credits are listed in order of appearance.

Photo

CO, James Kezer C/O Stanley Sessions; F-14, Dr. Sheldon Wolff & Judy Bodycote/Laboratory of Radiobiology and Environmental Health, University of California, San Francisco

Text

Source: *A History of Genetics*, by Alfred H. Sturtevant. New York: Harper & Row, 1965, Table "Molecular structure of nucleic acids:

A structure for deoxyribose nucleic acid," by Crick and Watson, from *Nature Journal*, 171.4356, p. 737-38. Copyright © 1953 by Macmillan Magazines Limited. Reprinted with permission.

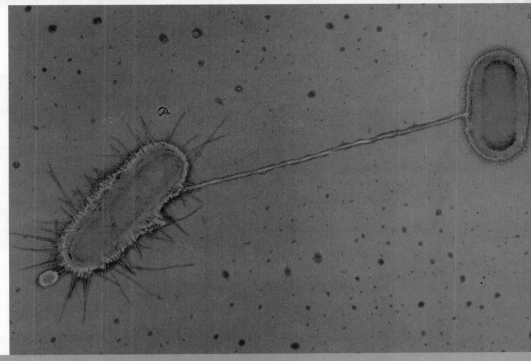

An electron micrograph showing the sex pilus between two conjugating *E. coli* cells.

Genetic Analysis and Mapping in Bacteria and Bacteriophages

CHAPTER CONCEPTS

- Bacterial genomes are most often contained in a single circular chromosome.

- Bacteria have developed numerous ways in which they can exchange and recombine genetic information between individual cells, including conjugation, transformation, and transduction.

- The ability to undergo conjugation and to transfer the bacterial chromosome from one cell to another is governed by the presence of genetic information contained in the DNA of a "fertility," or F factor.

- The F factor can exist autonomously in the bacterial cytoplasm as a plasmid, or it can integrate into the bacterial chromosome, where it facilitates the transfer of the host chromosome to the recipient cell, leading to genetic recombination.

- Genetic recombination during conjugation provides a means of mapping bacterial genes.

- Bacteriophages are viruses that have bacteria as their hosts.

- During infection of the bacterial host, bacteriophage DNA is injected into the host cell, where it is replicated and directs the reproduction of the bacteriophage.

I n this chapter, we discuss the analysis and mapping of genes in **bacteria** (prokaryotes) and **bacteriophages**, viruses that use bacteria as their hosts. The study of bacteria and bacteriophages has been essential to the accumulation of knowledge in many areas of genetic study. For example, much of what we know about molecular genetics, recombinational phenomena, and gene structure was initially derived from experimental work with them. Furthermore, our extensive knowledge of bacteria and their resident plasmids has led to their widespread use in DNA cloning and other recombinant DNA studies.

Bacteria and their viruses are especially useful research organisms in genetics for several reasons. They have extremely short reproductive cycles—literally hundreds of generations, giving rise to billions of genetically identical bacteria or phages, can be produced in short periods of time. Furthermore, they can be studied in pure cultures. That is, a single species or mutant strain of bacteria or one type of virus can be isolated and investigated independently of other similar organisms.

In this chapter, we focus on genetic recombination and chromosome mapping. Complex processes have evolved in bacteria and bacteriophages that facilitate the transfer of

From Chapter 8 of *Essentials of Genetics, Eighth edition*, William S. Klug, Michael R. Cummings, Charlotte A. Spencer, Michael A. Palladino. ©2013 by Pearson Education, Inc. All rights reserved.

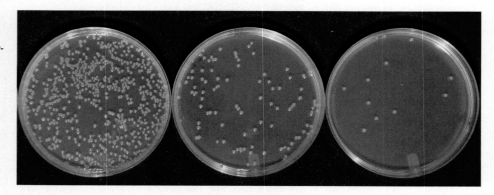

FIGURE 1 Results of the serial dilution technique and subsequent culture of bacteria. Each dilution varies by a factor of 10. Each colony is derived from a single bacterial cell.

genetic information between individual cells within populations. As we shall see, these processes are the basis for the chromosome mapping analysis that forms the cornerstone of molecular genetic investigations of bacteria and the viruses that invade them.

1 Bacteria Mutate Spontaneously and Are Easily Cultured

It has long been known that pure cultures of bacteria give rise to cells that exhibit heritable variation, particularly with respect to growth under unique environmental conditions. Mutant cells that arise spontaneously in otherwise pure cultures can be isolated and established independently from the parent strain by using established selection techniques. As a result, mutations for almost any desired characteristic can now be isolated. Because bacteria and viruses usually contain only a single chromosome and are therefore haploid, all mutations are expressed directly in the descendants of mutant cells, adding to the ease with which these microorganisms can be studied.

Bacteria are grown in a liquid culture medium or in a petri dish on a semisolid agar surface. If the nutrient components of the growth medium are simple and consist only of an organic carbon source (such as glucose or lactose) and a variety of ions, including Na^+, K^+, Mg^{2+}, Ca^{2+}, and NH_4^+, present as inorganic salts, it is called **minimal medium.** To grow on such a medium, a bacterium must be able to synthesize all essential organic compounds (e.g., amino acids, purines, pyrimidines, sugars, vitamins, and fatty acids). A bacterium that can accomplish this remarkable biosynthetic feat—one that we ourselves cannot duplicate—is a **prototroph.** It is said to be wild-type for all growth requirements. On the other hand, if a bacterium loses the ability to synthesize one or more organic components through mutation, it is an **auxotroph.** For example, if a bacterium loses the ability to make histidine, then this amino acid must be added as a supplement to the minimal medium for growth to occur. The resulting bacterium is designated as an his^- auxotroph, in contrast to its prototrophic his^+ counterpart.

To study bacterial growth quantitatively, an inoculum of bacteria is placed in liquid culture medium. Cells grown in liquid medium can be quantified by transferring them to a semisolid medium in a petri dish. Following incubation and many divisions, each cell gives rise to a visible colony on the surface of the medium. If the number of colonies is too great to count, then a series of successive dilutions (a technique called **serial dilution**) of the original liquid culture is made and plated, until the colony number is reduced to the point where it can be counted (**Figure 1**). This technique allows the number of bacteria present in the original culture to be calculated.

As an example, let's assume that the three dishes in Figure 1 represent serial dilutions of 10^{-3}, 10^{-4}, and 10^{-5} (from left to right). We need only select the dish in which the number of colonies can be counted accurately. Because each colony arose from a single bacterium, the number of colonies multiplied by the dilution factor represents the number of bacteria in each milliliter of the initial inoculum used to start the serial dilutions. In Figure 1, the rightmost dish has 12 colonies. The dilution factor for a 10^{-5} dilution is 10^5. Therefore, the initial number of bacteria was 12×10^5 per mL.

2 Genetic Recombination Occurs in Bacteria

Development of techniques that allowed the identification and study of bacterial mutations led to detailed investigations of the transfer of genetic information between individual organisms. As we shall see, as with meiotic crossing over in eukaryotes, the process of genetic recombination in bacteria provided the basis for the development of chromosome mapping methodology. It is important to note at the outset of our discussion that the term *genetic recombination*, as applied to bacteria, refers to the *replacement* of one or more genes present in the chromosome of one cell with those from the chromosome of a genetically distinct cell. While this is somewhat different from our use of the term in eukaryotes—where it describes *crossing over resulting in a reciprocal exchange*—the overall effect is the same: Genetic information is transferred and it results in an altered genotype.

We will discuss three processes that result in the transfer of genetic information from one bacterium to another: *conjugation, transformation,* and *transduction.* Collectively, knowledge of these processes has helped us understand the origin of genetic variation between members of the same bacterial species, and in some cases, between members of different species. When transfer of genetic information occurs between members of the same species, the term **vertical gene transfer** applies. When transfer occurs between members of related, but distinct bacterial species, the term **horizontal gene transfer** is used. The horizontal gene transfer process has played a significant role in the evolution of bacteria. Often, the genes discovered to be involved in horizontal transfer are those that also confer survival advantages to the recipient species. For example, one species may transfer antibiotic resistance genes to

another species. Or genes conferring enhanced pathogenicity may be transferred. Thus, the potential for such transfer is a major concern in the medical community. In addition, horizontal gene transfer has been a major factor in the process of speciation in bacteria. Many, if not most, bacterial species have been the recipient of genes from other species.

Conjugation in Bacteria: The Discovery of F⁺ and F⁻ Strains

Studies of bacterial recombination began in 1946, when Joshua Lederberg and Edward Tatum showed that bacteria undergo **conjugation,** a process by which genetic information from one bacterium is transferred to and recombined with that of another bacterium. Their initial experiments were performed with two multiple auxotrophs (nutritional mutants) of *E. coli* strain K12. As shown in **Figure 2,** strain A required

FIGURE 2 Genetic recombination of two auxotrophic strains producing prototrophs. Neither auxotroph grows on minimal medium, but prototrophs do, suggesting that genetic recombination has occurred.

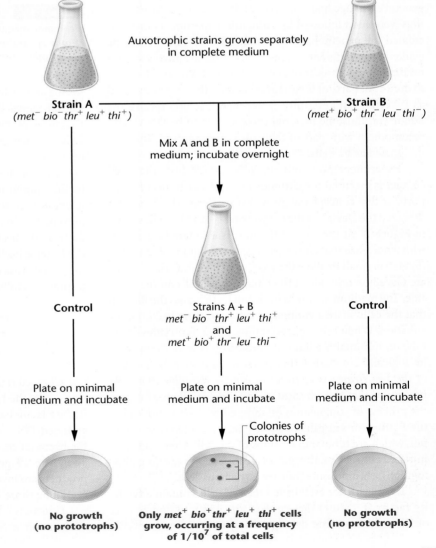

methionine (met) and biotin (bio) in order to grow, whereas strain B required threonine (thr), leucine (leu), and thiamine (thi). Neither strain would grow on minimal medium. The two strains were first grown separately in supplemented media, and then cells from both were mixed and grown together for several more generations. They were then plated on minimal medium. Any cells that grew on minimal medium were prototrophs. It is highly improbable that any of the cells containing two or three mutant genes would undergo spontaneous mutation simultaneously at two or three independent locations to become wild-type cells. Therefore, the researchers assumed that any prototrophs recovered must have arisen as a result of some form of genetic exchange and recombination between the two mutant strains.

In this experiment, prototrophs were recovered at a rate of $1/10^7$ (or 10^{-7}) cells plated. The controls for this experiment involved separate plating of cells from strains A and B on minimal medium. No prototrophs were recovered. Based on these observations, Lederberg and Tatum proposed that genetic exchange had occurred. Lederberg and Tatum's findings were soon followed by numerous experiments that elucidated the genetic basis of conjugation. It quickly became evident that different strains of bacteria are involved in a unidirectional transfer of genetic material. When cells serve as donors of parts of their chromosomes, they are designated as **F$^+$ cells** (F for "fertility"). Recipient bacteria receive the donor chromosome material (now known to be DNA), and recombine it with part of their own chromosome. They are designated as **F$^-$ cells.**

Experimentation subsequently established that cell contact is essential for chromosome transfer to occur. Support for this concept was provided by Bernard Davis, who designed the Davis U-tube for growing F$^+$ and F$^-$ cells shown in **Figure 3**. At the base of the tube is a sintered glass filter with a pore size that allows passage of the liquid medium but that is too small to allow the passage of bacteria. The F$^+$ cells are placed on one side of the filter, and F$^-$ cells on the other side. The medium is moved back and forth across the filter so that the cells share a common medium during bacterial incubation. When Davis plated samples from both sides of the tube on minimal medium, no prototrophs were found, and he logically concluded that *physical contact between cells of the two strains is essential to genetic recombination.* We now know that this physical interaction is the initial step in the process of conjugation established by a structure called the **F pilus** (or **sex pilus**; pl. pili). Bacteria often have many pili, which are tubular extensions of the cell. After contact is initiated between mating pairs (see the chapter opening photograph), chromosome transfer is then possible.

Later evidence established that F$^+$ cells contain a **fertility factor (F factor)** that confers the ability to donate part of their chromosome during conjugation. Experiments by

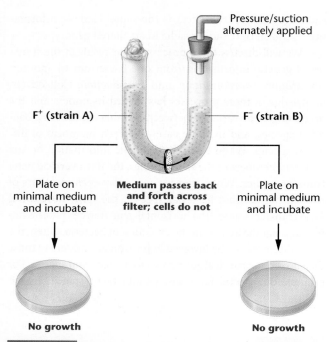

FIGURE 3   When strain A and B auxotrophs are grown in a common medium but separated by a filter, as in this Davis U-tube apparatus, no genetic recombination occurs and no prototrophs are produced.

Joshua and Esther Lederberg and by William Hayes and Luca Cavalli-Sforza showed that certain conditions eliminate the F factor in otherwise fertile cells. However, if these "infertile" cells are then grown with fertile donor cells, the F factor is regained.

The conclusion that the F factor is a mobile element is further supported by the observation that, following conjugation and genetic recombination, recipient cells always become F$^+$. Thus, in addition to the *rare* cases of gene transfer from the bacterial chromosome (genetic recombination), the F factor itself is passed to *all* recipient cells. On this basis, the initial cross of Lederberg and Tatum (see Figure 2) is interpreted as follow:

Strain A		Strain B
F$^+$	$\times$	F$^-$
Donor		**Recipient**

Characterization of the F factor confirmed these conclusions. Like the bacterial chromosome, though distinct from it, the F factor has been shown to consist of a circular, double-stranded DNA molecule, equivalent to about 2 percent of the bacterial chromosome (about 100,000 nucleotide pairs). There are 19 genes contained within the F factor whose products are involved in the transfer of genetic information, excluding those involved in the formation of the sex pilus.

Geneticists believe that transfer of the F factor during conjugation involves separation of the two strands of its double helix and movement of one of the two strands into the recipient cell. Both strands, one moving across the conjugation tube and one remaining in the donor cell, are replicated. The result is that both the donor *and* the recipient cells become F^+. This process is diagrammed in **Figure 4**.

To summarize, an *E. coli* cell may or may not contain the F factor. When this factor is present, the cell is able to form a sex pilus and potentially serve as a donor of genetic information. During conjugation, a copy of the F factor is almost always transferred from the F^+ cell to the F^- recipient, converting the recipient to the F^+ state. The question remained as to exactly why such a low proportion of cells involved in these matings (10^{-7}) also results in genetic recombination. The answer awaited further experimentation.

As you soon shall see, the F factor is in reality an autonomous genetic unit called a *plasmid*. However, in our historical coverage of its discovery, we will continue to refer to it as a factor.

ESSENTIAL POINT ■ ■ ■

Conjugation may be initiated by a bacterium housing a plasmid called the F factor in its cytoplasm, making it a donor (F^+) cell. Following conjugation, the recipient (F^-) cell receives a copy of the F factor and is converted to the F^+ status.

Hfr Bacteria and Chromosome Mapping

Subsequent discoveries not only clarified how genetic recombination occurs but also defined a mechanism by which the *E. coli* chromosome could be mapped. Let's address chromosome mapping first.

In 1950, Cavalli-Sforza treated an F^+ strain of *E. coli* K12 with nitrogen mustard, a chemical known to induce mutations. From these treated cells, he recovered a genetically altered strain of donor bacteria that underwent recombination at a rate of $1/10^4$ (or 10^{-4}), 1000 times more frequently than the original F^+ strains. In 1953, Hayes isolated another strain that

FIGURE 4 An $F^+ \times F^-$ mating demonstrating how the recipient F^- cell converts to F^+. During conjugation, the DNA of the F factor replicates with one new copy entering the recipient cell, converting it to F^+. Newly replicated DNA is depicted by a lighter shade of blue as the F factor is transferred.

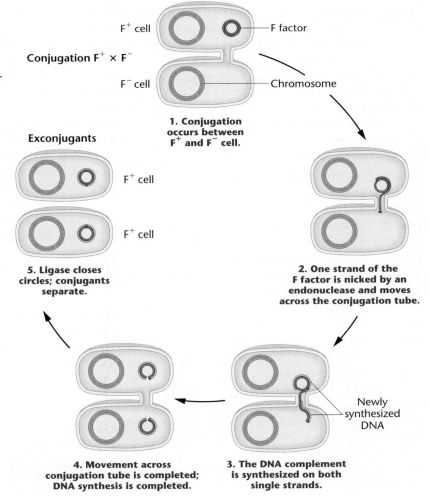

Conjugation $F^+ \times F^-$

F^+ cell — F factor

F^- cell — Chromosome

1. Conjugation occurs between F^+ and F^- cell.

2. One strand of the F factor is nicked by an endonuclease and moves across the conjugation tube.

Newly synthesized DNA

3. The DNA complement is synthesized on both single strands.

4. Movement across conjugation tube is completed; DNA synthesis is completed.

Exconjugants

F^+ cell

F^+ cell

5. Ligase closes circles; conjugants separate.

demonstrated an elevated frequency. Both strains were designated **Hfr,** for **high-frequency recombination.** Because Hfr cells behave as donors, they are a special class of F$^+$ cells.

Another important difference was noted between Hfr strains and the original F$^+$ strains. If the donor is from an Hfr strain, recipient cells, though sometimes displaying genetic recombination, never become Hfr; that is, they remain F$^-$. In comparison, then,

$$F^+ \times F^- \longrightarrow F^+ \quad \text{(low rate of recombination)}$$

$$Hfr \times F^- \longrightarrow F^- \quad \text{(higher rate of recombination)}$$

Perhaps the most significant characteristic of Hfr strains is the *nature of recombination.* In any given strain, certain genes are more frequently recombined than others, and some not at all. This *nonrandom* pattern was shown to vary between Hfr strains. Although these results were puzzling, Hayes interpreted them to mean that some physiological alteration of the F factor had occurred, resulting in the production of Hfr strains of *E. coli.*

In the mid-1950s, experimentation by Ellie Wollman and François Jacob elucidated the difference between Hfr and F$^+$ strains and showed how Hfr strains allow genetic mapping of the *E. coli* chromosome. In their experiments, Hfr and F$^-$ strains with suitable marker genes were mixed, and recombination of specific genes was assayed at different times. To accomplish this, a culture containing a mixture of an Hfr and an F$^-$ strain was first incubated, and samples were removed at various intervals and placed in a blender. The shear forces in the blender separated conjugating bacteria so that the transfer of the chromosome was terminated. The cells were then assayed for genetic recombination.

This process, called the **interrupted mating technique,** demonstrated that specific genes of a given Hfr strain were transferred and recombined sooner than others. The graph in **Figure 5** illustrates this point. During the first 8 minutes after the two strains were mixed, no genetic recombination was detected. At about 10 minutes, recombination of the *azi*^{R} gene was detected, but no transfer of the *ton*^{s}, *lac*$^+$, or *gal*$^+$ genes was noted. By 15 minutes, 50 percent of the recombinants were *azi*^{R}, and 15 percent were *ton*^{s}; but none were *lac*$^+$ or *gal*$^+$. Within 20 minutes, the *lac*$^+$ was found among the recombinants; and within 25 minutes, *gal*$^+$ was also being transferred. Wollman and Jacob had demonstrated *an ordered transfer of genes* that correlated with the length of time conjugation proceeded.

It appeared that the chromosome of the Hfr bacterium was transferred linearly and that the gene order and distance between genes, as measured in minutes, could be predicted from such experiments **(Figure 6)**. This information served as the basis for the first genetic map of the *E. coli* chromosome. Minutes in bacterial mapping are similar to map units in eukaryotes.

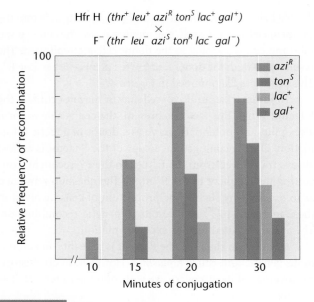

FIGURE 5 The progressive transfer during conjugation of various genes from a specific Hfr strain of *E. coli* to an F$^-$ strain. Certain genes (*azi* and *ton*) transfer more quickly than others and recombine more frequently. Others (*lac* and *gal*) take longer to transfer, and recombinants are found at a lower frequency.

Wollman and Jacob then repeated the same type of experiment with other Hfr strains, obtaining similar results with one important difference. Although genes were always transferred linearly with time, as in their original experiment, the order in which genes entered seemed to vary from Hfr strain to Hfr strain **[Figure 7(a)]**. When they reexamined the entry rate of genes, and thus the genetic maps for each strain, a definite pattern emerged. The major difference between each strain was simply the point of origin (*O*) and the direction in which entry proceeded from that point **[Figure 7(b)]**.

To explain these results, Wollman and Jacob postulated that the *E. coli* chromosome is circular (a closed circle,

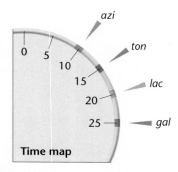

FIGURE 6 A time map of the genes studied in the experiment depicted in Figure 5.

(a)

Hfr strain	(earliest)	Order of transfer							(latest)
H	thr –	leu –	azi –	ton –	pro –	lac –	gal –	thi	
1	leu –	thr –	thi –	gal –	lac –	pro –	ton –	azi	
2	pro –	ton –	azi –	leu –	thr –	thi –	gal –	lac	
7	ton –	azi –	leu –	thr –	thi –	gal –	lac –	pro	

(b)

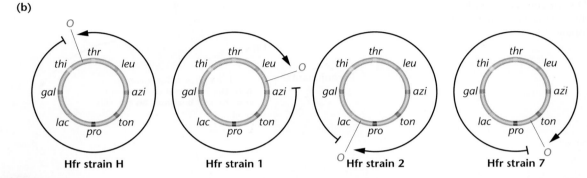

FIGURE 7 (a) The order of gene transfer in four Hfr strains, suggesting that the *E. coli* chromosome is circular. (b) The point where transfer originates (*O*) is identified in each strain. The origin is determined by the point of integration into the chromosome of the F factor, and the direction of transfer is determined by the orientation of the F factor as it integrates. The arrowheads indicate the points of initial transfer.

with no free ends). If the point of origin (*O*) varies from strain to strain, a different sequence of genes will be transferred in each case. But what determines *O*? They proposed that *in various Hfr strains, the F factor integrates into the chromosome at different points and that its position determines the site of O.* A case of integration is shown in step 1 of **Figure 8**. During conjugation between an Hfr and an F⁻ cell, the position of the F factor determines the initial point of transfer (steps 2 and 3). Those genes adjacent to *O* are transferred first, and the F factor becomes the last part that can be transferred (step 4). However, conjugation rarely, if ever, lasts long enough to allow the entire chromosome to pass across the conjugation tube (step 5). *This proposal explains why recipient cells, when mated with Hfr cells, remain F⁻.*

Figure 8 also depicts the way in which the two strands making up a DNA molecule behave during transfer, allowing for the entry of one strand of DNA into the recipient (see step 3). Following replication, the entering DNA now has the potential to recombine with its homologous region of the host chromosome. The DNA strand that remains in the donor also undergoes replication.

Use of the interrupted mating technique with different Hfr strains allowed researchers to map the entire *E. coli* chromosome. Mapped in time units, strain K12 (or *E. coli* K12) was shown to be 100 minutes long. While modern genome analysis of the *E. coli* chromosome has now established the presence of just over 4000 protein-coding sequences, this

original mapping procedure established the location of approximately 1000 genes.

ESSENTIAL POINT ▪ ▪ ▪

When the F factor is integrated into the donor cell chromosome (making it Hfr), the donor chromosome moves unidirectionally into the recipient, initiating recombination and providing the basis for time mapping of the bacterial chromosome.

NOW SOLVE THIS

1 When the interrupted mating technique was used with five different strains of Hfr bacteria, the following orders of gene entry and recombination were observed. On the basis of these data, draw a map of the bacterial chromosome. Do the data support the concept of circularity?

Hfr Strain	Order				
1	T	C	H	R	O
2	H	R	O	M	B
3	M	O	R	H	C
4	M	B	A	K	T
5	C	T	K	A	B

▪ **Hint** *This problem involves an understanding of how the bacterial chromosome is transferred during conjugation, leading to recombination and providing data for mapping. The key to its solution is to understand that chromosome transfer is strain-specific and depends on where in the chromosome, and in which orientation, the F factor has integrated.*

FIGURE 8 Conversion of F⁺ to an Hfr state occurs by integrating the F factor into the bacterial chromosome. The point of integration determines the origin (O) of transfer. During conjugation, an enzyme nicks the F factor, now integrated into the host chromosome, initiating transfer of the chromosome at that point. Conjugation is usually interrupted prior to complete transfer. Only the A and B genes are transferred to the F⁻cell, which may recombine with the host chromosome. Newly replicated DNA of the chromosome is depicted by a lighter shade of orange.

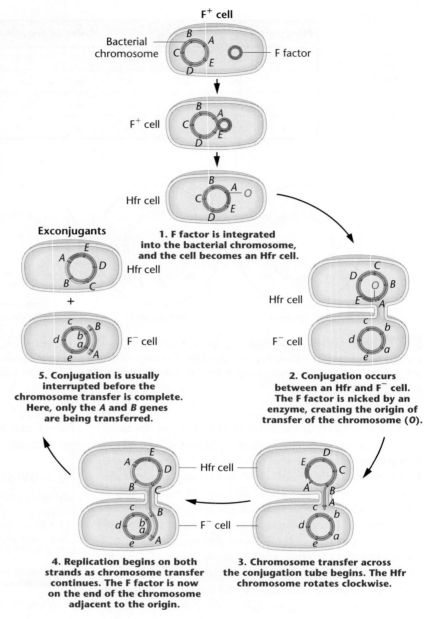

1. F factor is integrated into the bacterial chromosome, and the cell becomes an Hfr cell.

2. Conjugation occurs between an Hfr and F⁻ cell. The F factor is nicked by an enzyme, creating the origin of transfer of the chromosome (O).

3. Chromosome transfer across the conjugation tube begins. The Hfr chromosome rotates clockwise.

4. Replication begins on both strands as chromosome transfer continues. The F factor is now on the end of the chromosome adjacent to the origin.

5. Conjugation is usually interrupted before the chromosome transfer is complete. Here, only the A and B genes are being transferred.

Recombination in F⁺ × F⁻ Matings: A Reexamination

The preceding experiment helped geneticists better understand how genetic recombination occurs during F⁺ × F⁻ matings. Recall that recombination occurs much less frequently than in Hfr × F⁻ matings and that random gene transfer is involved. The current belief is that when F⁺ and F⁻ cells are mixed, conjugation occurs readily and each F⁻ cell involved in conjugation with an F⁺ cell receives a copy of the F factor, *but no genetic recombination occurs.* However, at an extremely low frequency in a population of F⁺ cells, the F factor integrates spontaneously from the cytoplasm to a random point in the bacterial chromosome, converting the F⁺ cell to the Hfr state, as we saw in Figure 8. Therefore, in F⁺ × F⁻ matings, the extremely low frequency of genetic recombination (10^{-7}) is attributed to the rare, newly formed Hfr cells, which then undergo conjugation with F⁻ cells. Because the point of integration of the F factor is random, the gene or genes that are transferred by any newly formed Hfr donor *will also appear to be random within the larger F⁺/F⁻ population.* The recipient bacterium will appear as a recombinant but will remain F⁻. If it subsequently undergoes conjugation with an F⁺ cell, it will then be converted to F⁺.

FIGURE 9 Conversion of an Hfr bacterium to F′ and its subsequent mating with an F⁻ cell. The conversion occurs when the F factor loses its integrated status. During excision from the chromosome, it carries with it one or more chromosomal genes (*A* and *E*). Following conjugation with an F⁻ cell, the recipient cell becomes partially diploid and is called a merozygote; it also behaves as an F⁺ donor cell.

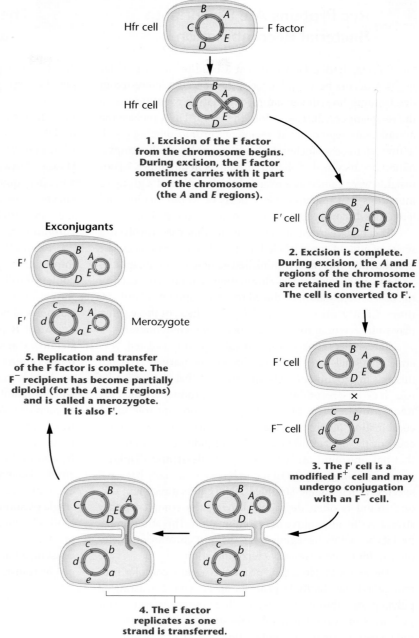

The F′ State and Merozygotes

In 1959, during experiments with Hfr strains of *E. coli*, Edward Adelberg discovered that the F factor could lose its integrated status, causing the cell to revert to the F⁺ state (**Figure 9**, step 1). When this occurs, the F factor frequently carries several adjacent bacterial genes along with it (step 2). Adelberg labeled this condition **F′** to distinguish it from F⁺ and Hfr. F′, like Hfr, is thus another special case of F⁺, but this conversion is from Hfr to F′.

The presence of bacterial genes within a cytoplasmic F factor creates an interesting situation. An F′ bacterium behaves like an F⁺ cell by initiating conjugation with F⁻ cells (Figure 9, step 3). When this occurs, the F factor, containing chromosomal genes, is transferred to the F⁻ cell (step 4). As a result, whatever chromosomal genes are part of the F factor are now present as duplicates in the recipient cell (step 5) because the recipient still has a complete chromosome. This creates a partially diploid cell called a **merozygote.** Pure cultures of F′ merozygotes can be established. They have been extremely useful in the study of bacterial genetics, particularly in genetic regulation.

3   Rec Proteins Are Essential to Bacterial Recombination

Once researchers established that a unidirectional transfer of DNA occurs between bacteria, they became interested in determining how the actual recombination event occurs in the recipient cell. Just how does the donor DNA replace the homologous region in the recipient chromosome? As with many systems, the biochemical mechanism by which recombination occurs was deciphered through genetic studies. Major insights were gained as a result of the isolation of a group of mutations that impaired the process of recombination and led to the discovery of *rec* (for recombination) genes.

The first relevant observation in this case involved a series of mutant genes labeled *recA, recB, recC,* and *recD.* The first mutant gene, *recA,* diminished genetic recombination in bacteria 1000-fold, nearly eliminating it altogether; each of the other *rec* mutations reduced recombination by about 100 times. Clearly, the normal wild-type products of these genes play some essential role in the process of recombination.

Researchers looked for, and subsequently isolated, several functional gene products present in normal cells but missing in *rec* mutant cells and showed that they played a role in genetic recombination. The first product is called the **RecA protein.**[*] This protein plays an important role in recombination involving either a single-stranded DNA molecule or the linear end of a double-stranded DNA molecule that has unwound. As it turns out, **single-strand displacement** is a common form of recombination in many bacterial species. When double-stranded DNA enters a recipient cell, one strand is often degraded, leaving the complementary strand as the only source of recombination. This strand must find its homologous region along the host chromosome, and once it does, RecA facilitates recombination.

The second related gene product is a more complex protein called the **RecBCD protein,** an enzyme consisting of polypeptide subunits encoded by three other *rec* genes. This protein is important when double-stranded DNA serves as the source of genetic recombination. RecBCD unwinds the helix, facilitating recombination that involves RecA. These discoveries have extended our knowledge of the process of recombination considerably and underscore the value of isolating mutations, establishing their phenotypes, and determining the biological role of the normal, wild-type genes. The model of recombination based on the *rec* discoveries also applies to eukaryotes: Eukaryotic proteins similar to RecA have been isolated and studied.

---

[*]Note that the names of bacterial genes use lowercase letters and are italicized, while the names of the corresponding gene products begin with capital letters and are not italicized. For example, the *recA* gene encodes the RecA protein.

4   The F Factor Is an Example of a Plasmid

The preceding sections introduced the extrachromosomal heredity unit required for conjugation called the F factor. When it exists autonomously in the bacterial cytoplasm, the F factor is composed of a double-stranded closed circle of DNA. These characteristics place the F factor in the more general category of genetic structures called **plasmids** [Figure 10(a)]. These structures contain one or more genes and often, quite a few. Their replication depends on the same enzymes that replicate the chromosome of the host cell, and they are distributed to daughter cells along with the host chromosome during cell division.

Plasmids are generally classified according to the genetic information specified by their DNA. The F factor plasmid confers fertility and contains the genes essential for sex pilus formation, on which genetic recombination depends. Other examples of plasmids include the R and Col plasmids.

Most **R plasmids** consist of two components: the **resistance transfer factor (RTF)** and one or more **r-determinants** [Figure 10(b)]. The RTF encodes genetic information essential to transferring the plasmid between bacteria, and the r-determinants are genes that confer resistance to antibiotics or mercury. While RTFs are similar in a variety of plasmids from different bacterial species, r-determinants are specific for resistance to one class of antibiotic and vary widely. Resistance to tetracycline, streptomycin, ampicillin, sulfonamide, kanamycin, or chloramphenicol is most frequently encountered. Sometimes several r-determinants occur in a single plasmid, conferring multiple resistance to several antibiotics [Figure 10(b)]. Bacteria bearing these plasmids are of great medical significance not only because of their multiple resistance but because of the ease with which the plasmids can be transferred to other bacteria.

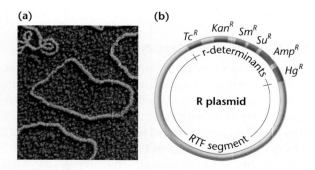

FIGURE 10   (a) Electron micrograph of a plasmid isolated from *E. coli.* (b) An R plasmid containing resistance transfer factors (RTFs) and multiple r-determinants (Tc, tetracycline; Kan, kanamycin; Sm, streptomycin; Su, sulfonamide; Amp, ampicillin; and Hg, mercury).

The first known case of such a plasmid occurred in Japan in the 1950s in the bacterium *Shigella*, which causes dysentery. In hospitals, bacteria were isolated that were resistant to as many as five of the above antibiotics. Obviously, this phenomenon represents a major health threat. Fortunately, a bacterial cell sometimes contains r-determinant plasmids but no RTF. Although such a cell is resistant, it cannot transfer the genetic information for resistance to recipient cells. The most commonly studied plasmids, however, contain the RTF as well as one or more r-determinants.

The **Col plasmid,** ColE1 (derived from *E. coli*), is clearly distinct from the R plasmid. It encodes one or more proteins that are highly toxic to bacterial strains that do not harbor the same plasmid. These proteins, called **colicins,** can kill neighboring bacteria, and bacteria that carry the plasmid are said to be *colicinogenic*. Present in 10 to 20 copies per cell, a gene in the Col plasmid encodes an immunity protein that protects the host cell from the toxin. Unlike an R plasmid, the Col plasmid is not usually transmissible to other cells.

Interest in plasmids has increased dramatically because of their role in recombinant DNA research. Specificgenes from any source can be inserted into a plasmid, which may then be inserted into a bacterial cell. As the altered cell replicates its DNA and undergoes division, the foreign gene is also replicated, thus cloning the foreign genes.

ESSENTIAL POINT ■　■　■

Plasmids, such as the F factor, are autonomously replicating DNA molecules found in the bacterial cytoplasm, sometimes containing unique genes conferring antibiotic resistance as well as the genes necessary for plasmid transfer during conjugation.

5　Transformation Is Another Process Leading to Genetic Recombination in Bacteria

Transformation provides another mechanism for recombining genetic information in some bacteria. Small pieces of extracellular DNA are taken up by a living bacterium, potentially leading to a stable genetic change in the recipient cell. We discuss transformation herebecause in those bacterial species where it occurs, the process can be used to map bacterial genes, though in a more limited way than conjugation. Transformation has also played a central role in experiments proving that DNA is the genetic material.

The process of transformation consists of numerous steps divided into two categories: (1) entry of DNA into a recipient cell and (2) recombination of the donor DNA with its homologous region in the recipient chromosome. In a population of bacterial cells, only those in the particular physiological state of **competence** take up DNA. Entry is thought to occur at a limited number of receptor sites on the surface of the bacterial cell. Passage into the cell is an active process that requires energy and specific transport molecules. This model is supported by the fact that substances that inhibit energy production or protein synthesis in the recipient cell also inhibit the transformation process.

Soon after entry, one of the two strands of the double helix is digested by nucleases, leaving only a single strand to participate in transformation. The surviving strand of DNA then aligns with its complementary region of the bacterial chromosome. In a process involving several enzymes, the segment replaces its counterpart in the chromosome, which is excised and degraded. For recombination to be detected, the transforming DNA must be derived from a different strain of bacteria that bears some genetic variation, such as a mutation. Once it is integrated into the chromosome, the recombinant region contains one host strand (present originally) and one mutant strand. Because these strands are from different sources, this helical region is referred to as a **heteroduplex.** Following one round of DNA replication, one chromosome is restored to its original configuration, and the other contains the mutant gene. Following cell division, one untransformed cell (nonmutant) and one transformed cell (mutant) are produced.

Transformation and Linked Genes

In early transformation studies, the most effective exogenous DNA contained 10,000–20,000 nucleotide pairs, a length sufficient to encode several genes.[*] Genes adjacent to or very close to one another on the bacterial chromosome can be carried on a single segment of this size. Consequently, a single transfer event can result in the cotransformation of several genes simultaneously. Genes that are close enough to each other to be cotransformed are *linked*. In contrast to *linkage groups* in eukaryotes, which consist of all genes on a single chromosome, note that here *linkage* refers to the proximity of genes that permits cotransformation (i.e., the genes are next to, or close to, one another).

If two genes are not linked, simultaneous transformation occurs only as a result of two independent events involving two distinct segments of DNA. As in double crossovers in eukaryotes, the probability of two independent events occurring simultaneously is equal to the product of the individual probabilities. Thus, the frequency of two unlinked genes being transformed simultaneously is much lower than if they are linked.

---

[*]Today, we know that a 2000 nucleotide pair length of DNA is highly effective in gene cloning experiments.

NOW SOLVE THIS

2 In a transformation experiment involving a recipient bacterial strain of genotype a^-b^-, the following results were obtained. What can you conclude about the location of the a and b genes relative to each other?

	Transformants (%)		
Transforming DNA	$a^+ b^-$	$a^- b^+$	$a^+ b^+$
a^+b^+	3.1	1.2	0.04
a^+b^- and a^-b^+	2.4	1.4	0.03

■ **Hint** *This problem involves an understanding of how transformation can be used to determine if bacterial genes are closely "linked." You are asked to predict the location of two genes relative to one another. The key to its solution is to understand that cotransformation (of two genes) occurs according to the laws of probability. Two "unlinked" genes are transformed only as a result of two independent events. In such a case, the probability of that occurrence is equal to the product of the individual probabilities.*

6 Bacteriophages Are Bacterial Viruses

Bacteriophages, or **phages** as they are commonly known, are viruses that have bacteria as their hosts. During their reproduction, phages can be involved in still another mode of bacterial genetic recombination called *transduction*. To understand this process, we must consider the genetics of bacteriophages, which themselves undergo recombination.

A great deal of genetic research has been done using bacteriophages as a model system, making them a worthy subject of discussion. In this section, we will first examine the structure and life cycle of one type of bacteriophage. We then discuss how these phages are studied during their infection of bacteria. Finally, we contrast two possible modes of behavior once the initial phage infection occurs. This information is background for our discussion of *transduction* and *bacteriophage recombination*.

Phage T4: Structure and Life Cycle

Bacteriophage T4 is one of a group of related bacterial viruses referred to as T-even phages. It exhibits the intricate structure shown in **Figure 11**. The phage T4's genetic material (DNA)

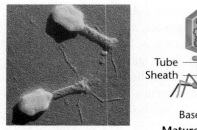 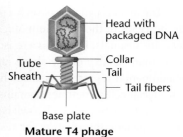

Head with packaged DNA — Collar — Tail — Tail fibers — Tube — Sheath — Base plate

Mature T4 phage

FIGURE 11 The structure of bacteriophage T4 includes an icosahedral head filled with DNA, a tail consisting of a collar, tube, sheath, base plate, and tail fibers. During assembly, the tail components are added to the head, and then tail fibers are added.

is contained within an icosahedral (a polyhedron with 20 faces) protein coat, making up the head of the virus. The DNA is sufficient in quantity to encode more than 150 average-sized genes. The head is connected to a tail that contains a collar and a contractile sheath surrounding a central core. Tail fibers, which protrude from the tail, contain binding sites in their tips that specifically recognize unique areas of the outer surface of the cell wall of the bacterial host, *E. coli*.

The life cycle of phage T4 (**Figure 12**) is initiated when the virus binds by adsorption to the bacterial host cell. Then, an ATP-driven contraction of the tail sheath causes the central core to penetrate the cell wall. The DNA in the head is extruded, and it moves across the cell membrane into the bacterial cytoplasm. Within minutes, all bacterial DNA, RNA, and protein synthesis in the host cell is inhibited, and synthesis of viral molecules begins. At the same time, degradation of the host DNA is initiated.

A period of intensive viral gene activity characterizes infection. Initially, phage DNA replication occurs, leading to a pool of viral DNA molecules. Then, the components of the head, tail, and tail fibers are synthesized. The assembly of mature viruses is a complex process that has been well studied by William Wood, Robert Edgar, and others. Three sequential pathways occur: (1) DNA packaging as the viral heads are assembled, (2) tail assembly, and (3) tail fiber assembly. Once DNA is packaged into the head, it combines with the tail components, to which tail fibers are added. Total construction is a combination of self-assembly and enzyme-directed processes.

When approximately 200 new viruses have been constructed, the bacterial cell is ruptured by the action of the enzyme lysozyme (a phage gene product), and the mature phages are released from the host cell. The new phages infect other available bacterial cells, and the process repeats itself over and over again.

The Plaque Assay

Bacteriophages and other viruses have played a critical role in our understanding of molecular genetics. During infection of bacteria, enormous quantities of bacteriophages

FIGURE 12 Life cycle of bacteriophage T4.

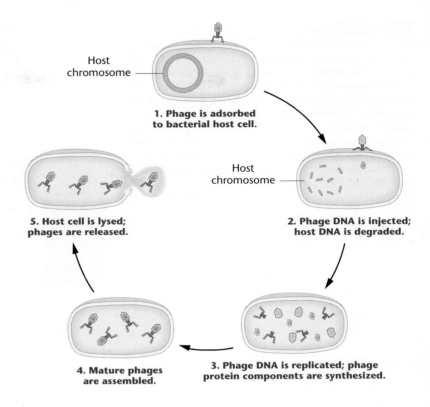

1. Phage is adsorbed to bacterial host cell.

Host chromosome

Host chromosome

2. Phage DNA is injected; host DNA is degraded.

5. Host cell is lysed; phages are released.

4. Mature phages are assembled.

3. Phage DNA is replicated; phage protein components are synthesized.

can be obtained for investigation. Often, over 10^{10} viruses are produced per milliliter of culture medium. Many genetic studies rely on the ability to quantify the number of phages produced following infection under specific culture conditions. The **plaque assay** is a routinely used technique, which is invaluable in mutational and recombinational studies of bacteriophages.

This assay is shown in **Figure 13**, where actual plaque morphology is also illustrated. A serial dilution of the original virally infected bacterial culture is performed first. Then, a 0.1-mL sample (an *aliquot*) from a dilution is added to melted nutrient agar (about 3 mL) into which a few drops of a healthy bacterial culture have been added. The solution is then poured evenly over a base of solid nutrient agar in a petri dish and allowed to solidify before incubation. A clear area called a **plaque** occurs wherever a single virus initially infected one bacterium in the culture (the lawn) that has grown up during incubation. The plaque represents clones of the single infecting bacteriophage, created as reproduction cycles are repeated. If the dilution factor is too low, the plaques are plentiful, and they will fuse, lysing the entire lawn—which has occurred in the 10^{-3} dilution of Figure 13. On the other hand, if the dilution factor is increased, plaques can be counted and the density of viruses in the initial culture can be estimated as

$$\text{initial phage density} = (\text{plaque number/mL}) \times (\text{dilution factor})$$

Using the results shown in Figure 13, 23 phage plaques are derived from the 0.1-mL aliquot of the 10^{-5} dilution.

Therefore, we estimate that there are 230 phages/mL *at this dilution* (since the initial aliquot was 0.1 mL). The initial phage density in the undiluted sample, factoring in the 10^{-5} dilution, is then calculated as

$$\text{initial phage density} = (230/\text{mL}) \times (10^5) = 230 \times 10^5/\text{mL}$$

Because this figure is derived from the 10^{-5} dilution, we can also estimate that there will be only 0.23 phage/0.1 mL in the 10^{-7} dilution. Thus, when 0.1 mL from this tube is assayed, it is predicted that no phage particles will be present. This possibility is borne out in Figure 13, which depicts an intact lawn of bacteria lacking any plaques. The dilution factor is simply too great.

ESSENTIAL POINT

Bacteriophages (viruses that infect bacteria) demonstrate a well-defined life cycle where they reproduce within the host cell and can be studied using the plaque assay.

Lysogeny

Infection of a bacterium by a virus does not always result in viral reproduction and lysis. As early as the 1920s, it was known that a virus can enter a bacterial cell and coexist with it. The precise molecular basis of this relationship is now well understood. Upon entry, the viral DNA is integrated into the bacterial chromosome instead of replicating in the bacterial cytoplasm, a step that characterizes the developmental stage referred to as **lysogeny.** Subsequently, each time the bacterial

FIGURE 13 A plaque assay for bacteriophage analysis. Serial dilutions of a bacterial culture infected with bacteriophages are first made. Then three of the dilutions (10^{-3}, 10^{-5}, and 10^{-7}) are analyzed using the plaque assay technique. Each plaque represents the initial infection of one bacterial cell by one bacteriophage. In the 10^{-3} dilution, so many phages are present that all bacteria are lysed. In the 10^{-5} dilution, 23 plaques are produced. In the 10^{-7} dilution, the dilution factor is so great that no phages are present in the 0.1-mL sample, and thus no plaques form. From the 0.1-mL sample of the 10^{-5} dilution, the original bacteriophage density is calculated to be $23 \times 10 \times 10^5$ phages/mL (23×10^6, or 2.3×10^7). The photograph shows phage plaques on a lawn of *E. coli*.

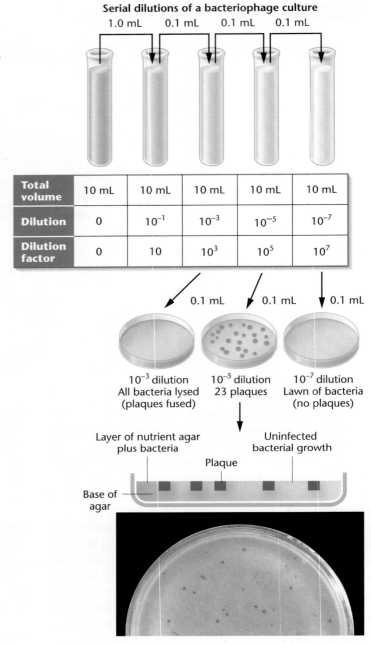

Serial dilutions of a bacteriophage culture

Total volume	10 mL	10 mL	10 mL	10 mL	10 mL
Dilution	0	10^{-1}	10^{-3}	10^{-5}	10^{-7}
Dilution factor	0	10	10^3	10^5	10^7

10^{-3} dilution
All bacteria lysed
(plaques fused)

10^{-5} dilution
23 plaques

10^{-7} dilution
Lawn of bacteria
(no plaques)

Layer of nutrient agar plus bacteria

Uninfected bacterial growth

Plaque

Base of agar

chromosome is replicated, the viral DNA is also replicated and passed to daughter bacterial cells following division. No new viruses are produced, and no lysis of the bacterial cell occurs. However, under certain stimuli, such as chemical or ultraviolet light treatment, the viral DNA loses its integrated status and initiates replication, phage reproduction, and lysis of the bacterium.

Several terms are used to describe this relationship. The viral DNA that integrates into the bacterial chromosome is called a **prophage.** Viruses that either lyse the cell or behave as a prophage are **temperate phages.** Those that only lyse the cell are referred to as **virulent phages.** A bacterium harboring a prophage is said to be **lysogenic;** that is, it is capable of being lysed as a result of induced viral reproduction. The viral DNA, which can either replicate in the bacterial cytoplasm or become integrated into the bacterial chromosome, is thus classified as an **episome**, meaning a genetic molecule that can replicate either in the cytoplasm of a cell or as part of its chromosome.

ESSENTIAL POINT ■ ■ ■

Bacteriophages can be lytic, meaning they infect the host cell, reproduce, and then lyse it, or in contrast, they can lysogenize the host cell, where they infect it and integrate their DNA into the host chromosome, but do not reproduce.

7 | Transduction Is Virus-Mediated Bacterial DNA Transfer

In 1952, Norton Zinder and Joshua Lederberg were investigating possible recombination in the bacterium *Salmonella typhimurium*. Although they recovered prototrophs from mixed cultures of two different auxotrophic strains, investigation revealed that recombination was occurring in a manner different from that attributable to the presence of an F factor, as in *E. coli*. What they had discovered was a process of bacterial recombination mediated by bacteriophages and now called **transduction.**

The Lederberg–Zinder Experiment

Lederberg and Zinder mixed the *Salmonella* auxotrophic strains LA-22 and LA-2 together, and when the mixture was plated on minimal medium, they recovered prototrophic cells. The LA-22 strain was unable to synthesize the amino acids phenylalanine and tryptophan (phe^-, trp^-), and LA-2 could not synthesize the amino acids methionine and histidine (met^-, his^-). Prototrophs (phe^+, trp^+, met^+, his^+) were recovered at a rate of about $1/10^5$ (10^{-5}) cells.

Although these observations at first suggested that the recombination involved was the type observed earlier in conjugative strains of *E. coli*, experiments using the Davis U-tube soon showed otherwise **(Figure 14)**. The two auxotrophic strains were separated by a sintered glass filter, thus preventing cell contact but allowing growth to occur in a common medium. Surprisingly, when samples were removed from both sides of the filter and plated independently on minimal medium, prototrophs *were* recovered, but only from the side of the tube containing LA-22 bacteria. Recall that if conjugation were responsible, the conditions in the Davis U-tube would be expected to *prevent* recombination altogether (see Figure 3).

Since LA-2 cells appeared to be the source of the new genetic information (phe^+ and trp^+) how that information crossed the filter from the LA-2 cells to the LA-22 cells, allowing recombination to occur, was a mystery. The unknown source was designated simply as a *filterable agent* (FA).

Three observations were used to identify the FA:

1. The FA was produced by the LA-2 cells only when they were grown in association with LA-22 cells. If LA-2 cells were grown independently and that culture medium was then added to LA-22 cells, recombination did not occur. Therefore, LA-22 cells play some role in the production of FA by LA-2 cells and do so only when they share a common growth medium.

2. The addition of DNase, which enzymatically digests DNA, did not render the FA ineffective. Therefore, the FA is not naked DNA, ruling out transformation.

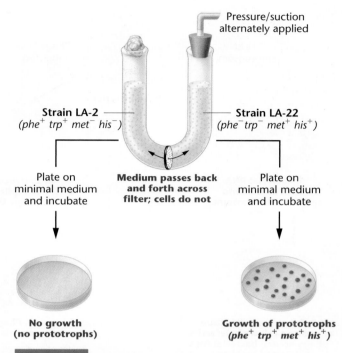

Strain LA-2 (phe^+ trp^+ met^- his^-)   **Strain LA-22** (phe^- trp^- met^+ his^+)

Pressure/suction alternately applied

Plate on minimal medium and incubate   Medium passes back and forth across filter; cells do not   Plate on minimal medium and incubate

No growth (no prototrophs)   **Growth of prototrophs** (phe^+ trp^+ met^+ his^+)

FIGURE 14   The Lederberg–Zinder experiment using *Salmonella*. After placing two auxotrophic strains on opposite sides of a Davis U-tube, Lederberg and Zinder recovered prototrophs from the side with the LA-22 strain, but not from the side containing the LA-2 strain.

3. The FA could not pass across the filter of the Davis U-tube when the pore size was reduced below the size of bacteriophages.

Aided by these observations and aware that temperate phages can lysogenize *Salmonella*, researchers proposed that the genetic recombination event is mediated by bacteriophage P22, present initially as a prophage in the chromosome of the LA-22 *Salmonella* cells. They hypothesized that P22 prophages rarely enter the vegetative or lytic phase, reproduce, and are released by the LA-22 cells. Such P22 phages, being much smaller than a bacterium, then cross the filter of the U-tube and subsequently infect and lyse some of the LA-2 cells. In the process of lysis of LA-2, these P22 phages occasionally package a region of the LA-2 chromosome in their heads. If this region contains the phe^+ and trp^+ genes and the phages subsequently pass back across the filter and infect LA-22 cells, these newly lysogenized cells will behave as prototrophs. This process of transduction, whereby bacterial recombination is mediated by bacteriophage P22, is diagrammed in **Figure 15**.

ESSENTIAL POINT ■ ■ ■

Transduction is virus-mediated bacterial DNA transfer.

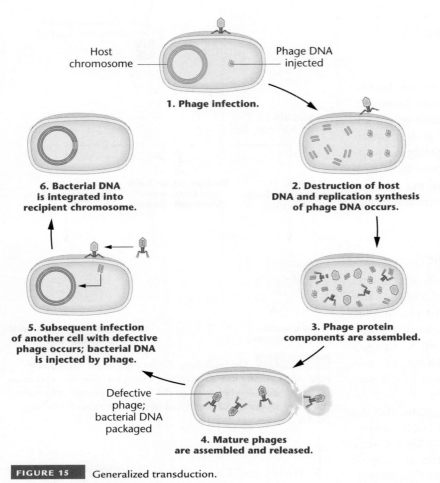

Host chromosome

Phage DNA injected

1. Phage infection.

6. Bacterial DNA is integrated into recipient chromosome.

2. Destruction of host DNA and replication synthesis of phage DNA occurs.

5. Subsequent infection of another cell with defective phage occurs; bacterial DNA is injected by phage.

3. Phage protein components are assembled.

Defective phage; bacterial DNA packaged

4. Mature phages are assembled and released.

FIGURE 15 Generalized transduction.

The Nature of Transduction

Further studies have revealed the existence of transducing phages in other species of bacteria. For example, *E. coli* can be transduced by phages P1 and λ and *B. subtilis* and *Pseudomonas aeruginosa* can be transduced by phages SPO1 and F116, respectively. The details of several different modes of transduction have also been established. Even though the initial discovery of transduction involved a temperate phage and a lysogenized bacterium, the same process can occur during the normal lytic cycle. Sometimes a small piece of bacterial DNA is packaged *along with* the viral chromosome so that the transducing phage contains both viral and bacterial DNA. In such cases, only a few bacterial genes are present in the transducing phage. However, when *only* bacterial DNA is packaged, regions as large as 1 percent of the bacterial chromosome can become enclosed in the viral head. In either case, the ability to infect a host cell is unrelated to the type of DNA in the phage head, making transduction possible.

When bacterial rather than viral DNA is injected into the bacterium, it either remains in the cytoplasm or recombines with the homologous region of the bacterial chromosome.

If the bacterial DNA remains in the cytoplasm, it does not replicate but is transmitted to one progeny cell following each division. When this happens, only a single cell, partially diploid for the transduced genes, is produced—a phenomenon called *abortive transduction*. If the bacterial DNA recombines with its homologous region of the bacterial chromosome, *complete transduction* occurs, where the transduced genes are replicated as part of the chromosome and passed to all daughter cells.

Both abortive and complete transduction are subclasses of the broader category of *generalized transduction,* which is characterized by the random nature of DNA fragments and genes that are transduced. Each fragment of the bacterial chromosome has a finite but small chance of being packaged in the phage head. Most cases of generalized transduction are of the abortive type; some data suggest that complete transduction occurs 10 to 20 times less frequently. In contrast to generalized transduction, *specialized transduction* occurs when transfer of bacterial DNA is not random, but instead, only strain-specific genes are transduced. This occurs when the DNA representing a temperate bacteriophage breaks out of the host chromosome, bringing with it bacterial DNA on either of its ends.

Transduction and Mapping

Like transformation, generalized transduction was used in linkage and mapping studies of the bacterial chromosome. The fragment of bacterial DNA involved in a transduction event is large enough to include numerous genes. As a result, two genes that closely align (are linked) on the bacterial chromosome can be simultaneously transduced, a process called **cotransduction.** Two genes that are not close enough to one another along the chromosome to be included on a single DNA fragment require two independent events to be transduced into a single cell. Since this occurs with a much lower probability than cotransduction, linkage can be determined.

By concentrating on two or three linked genes, transduction studies can also determine the precise order of these genes. The closer linked genes are to each other, the greater the frequency of cotransduction. Mapping studies involving three closely aligned genes can thus be executed, and the analysis of such an experiment is predicated on the same rationale underlying other mapping techniques.

8 Bacteriophages Undergo Intergenic Recombination

Around 1947, several research teams demonstrated that genetic recombination can be detected in bacteriophages. This led to the discovery that gene mapping can be performed in these viruses. Such studies relied on finding numerous phage mutations that could be visualized or assayed. As in bacteria and eukaryotes, these mutations allow genes to be identified and followed in mapping experiments. Thus, before considering recombination and mapping in these bacterial viruses, we briefly introduce several of the mutations that were studied.

Bacteriophage Mutations

Phage mutations often affect the morphology of the plaques formed following lysis of bacterial cells. For example, in 1946, Alfred Hershey observed unusual T2 plaques on plates of *E. coli* strain B. Normal T2 plaques are small and have a clear center surrounded by a diffuse (nearly invisible) halo, but the unusual plaques were larger and possessed a distinctive outer perimeter (**Figure 16**). When the viruses were isolated from these plaques and replated on *E. coli* B cells, the resulting plaque appearance was identical. Thus, the plaque phenotype was an inherited trait resulting from the reproduction of mutant phages. Hershey named the mutant *rapid lysis* (*r*) because the plaques were larger, apparently resulting from a more rapid or more efficient life cycle of the phage. We now know that in wild-type phages, reproduction is inhibited once a particular-sized plaque has been formed. The *r* mutant T2 phages overcome this inhibition, producing larger plaques.

Salvador Luria discovered another bacteriophage mutation, *host range* (*h*). This mutation extends the range of bacterial hosts that the phage can infect. Although wild-type T2 phages can infect *E. coli* B (a unique strain), they normally cannot attach or be adsorbed to the surface of *E. coli* B-2 (a different strain). The *h* mutation, however, provides the basis for adsorption and subsequent infection of *E. coli* B-2. When grown on a mixture of *E. coli* B and B-2, the center of the *h* plaque appears much darker than the *h*⁺ plaque (Figure 16).

Mapping in Bacteriophages

Genetic recombination in bacteriophages was discovered during **mixed infection experiments** in which two distinct mutant strains were allowed to *simultaneously* infect the same bacterial culture. These studies were designed so that the number of viral particles sufficiently exceeded the number of bacterial cells to ensure simultaneous infection of most cells by both viral strains. If two loci are involved, recombination is referred to as intergenic.

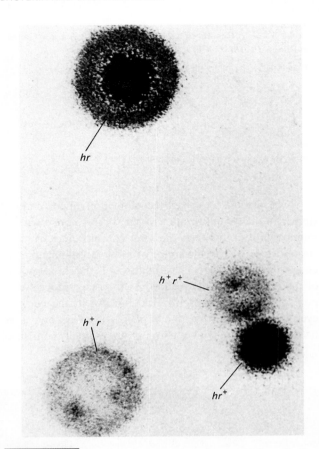

FIGURE 16 Plaque morphology phenotypes observed following simultaneous infection of *E. coli* by two strains of phage T2, *h*⁺*r* and *hr*⁺. In addition to the parental genotypes, recombinant plaques *hr* and *h*⁺*r*⁺ are shown.

For example, in one study using the T2/*E. coli* system, the parental viruses were of either the *h*⁺*r* (wild-type host range, rapid lysis) or *hr*⁺ (extended host range, normal lysis) genotype. If no recombination occurred, these two parental genotypes would be the only expected phage progeny. However, the recombinants *h*⁺*r*⁺ and *hr* were detected in addition to the parental genotypes (see Figure 16). As with eukaryotes, the percentage of recombinant plaques divided by the total number of plaques reflects the relative distance between the genes.

recombinational frequency = (h⁺r⁺ + hr)/total plaques × 100

Sample data for the *h* and *r* loci are shown in **Table 1**.

Similar recombinational studies have been conducted with numerous mutant genes in a variety of bacteriophages. Data are analyzed in much the same way as in eukaryotic mapping experiments. Two- and three-point mapping crosses are possible, and the percentage of recombinants in the total number of phage progeny is calculated. This value is proportional to the relative distance between two genes along the DNA molecule constituting the chromosome.

TABLE 1   Results of a Cross Involving the *h* and *r* Genes in Phage T2   ($hr^+ \times h^+r$)

Genotype	Plaques	Designation
h r^+	42 ⎫	
h^+ *r*	34 ⎭	Parental progency 76%
h^+ r^+	12 ⎫	
h *r*	12 ⎭	Recombinants 24%

Source: Data derived from Hershey and Rotman (1949).

Investigations into phage recombination support a model similar to that of eukaryotic crossing over—a breakage and reunion process between the viral chromosomes. A fairly clear picture of the dynamics of viral recombination has emerged. After the early phase of infection, the chromosomes of the phages begin replication. As this stage progresses, a pool of phage chromosomes accumulates in the bacterial cytoplasm. If double infection by phages of two genotypes has occurred, then the pool of chromosomes initially consists of the two parental types. Genetic exchange between these two types will occur before, during, and after replication, producing recombinant chromosomes.

In the case of the h^+r and hr^+ example discussed here, recombinant h^+r^+ and hr chromosomes are produced. Each of these chromosomes can undergo replication, with new replicates exchanging with each other and with parental chromosomes. Furthermore, recombination is not restricted to exchanges between two chromosomes—three or more can be involved simultaneously. As phage development progresses, chromosomes are randomly removed from the pool and packed into the phage head, forming mature phage particles. Thus, a variety of parental and recombinant genotypes are represented in progeny phages.

ESSENTIAL POINT ■   ■   ■

Various mutant phenotypes, including ones involving plaque morphology and host range, have been studied in bacteriophages and have served as the basis for mapping in these viruses.

GENETICS, TECHNOLOGY, AND SOCIETY

From Cholera Genes to Edible Vaccines

Using an expanding toolbox of molecular genetic tools, scientists are tackling some of the most serious bacterial diseases affecting our species. Our ability to clone bacterial genes and transfer them into other organisms is leading directly to exciting new treatments. The story of edible vaccines for the treatment of cholera illustrates how genetic engineering is being applied to control a serious human disease.

Cholera is caused by *Vibrio cholerae*, a curved, rod-shaped bacterium found in rivers and oceans. Most genetic strains of *V. cholerae* are harmless; only a few are pathogenic. Infection occurs when a person drinks water or eats food contaminated with pathogenic *V. cholerae*. Once in the digestive system, these bacteria colonize the small intestine and produce proteins called enterotoxins that invade the mucosal cells lining the intestine. This triggers a massive secretion of water and dissolved salts, resulting in violent diarrhea, severe dehydration, muscle cramps, lethargy, and often death. The enterotoxin consists of two polypeptides, called the A and B subunits, encoded by two separate genes.

Cholera remains a leading cause of human deaths throughout the Third World, where basic sanitation is lacking and water supplies are often contaminated. For example, in July 1994, 70,000 cases of cholera leading to 12,000 fatalities were reported among the Rwandans crowded into refugee camps in Goma, Zaire. And after an absence of over 100 years, cholera reappeared in Latin America in 1991, spreading from Peru to Mexico and claiming more than 10,000 lives. Following the 2010 earthquakes in Haiti, a severe cholera outbreak spread through the country, claiming more than 4000 lives.

A new gene-based technology is emerging to attack cholera. This technology centers on genetically engineered plants that act as vaccines. Scientists introduce a cloned gene—such as a gene encoding a bacterial protein—into the plant genome. The transgenic plant produces the new gene product, and immunity is acquired when an animal eats the plant. The gene product in the plant acts as an antigen, stimulating the production of antibodies to protect against bacterial infection or the effects of their toxins. Since the B subunit of the cholera enterotoxin binds to intestinal cells, research has focused on using this polypeptide as the antigen, with the hope that antibodies against it will prevent toxin binding and render the bacteria harmless.

Leading the efforts to develop an edible vaccine are Charles Arntzen and associates at Cornell University. To test the system, they are using the B subunit of an *E. coli* enterotoxin, which is similar in structure and immunological properties to the cholera protein. Their first step was to obtain the DNA clone of the gene encoding the B subunit and to attach it to a promoter that would induce transcription in all tissues of the plant. Second, the researchers introduced the hybrid gene into potato plants by means of *Agrobacterium*-mediated transformation. The engineered plants expressed their new gene and produced the enterotoxin B subunit. Third, they fed mice a few grams of the genetically engineered tubers. Arntzen's group found that the mice produced specific antibodies against the B subunit and secreted them into the small intestine. When they fed purified enterotoxin to the mice, the mice were protected from its effects and did not develop the symptoms of cholera. In clinical trials conducted using humans in 1998, almost all of the volunteers developed an immune response, and none experienced adverse side effects.

Arntzen's experiments have served as models for other research efforts involving edible vaccines. Currently, scientists are developing edible vaccines against bacterial diseases such as anthrax and tetanus, as well as viral diseases such as rabies, AIDS, and measles.

Your Turn

Take time, individually or in groups, to answer the following questions. Investigate the references and links to help you understand some of the issues that surround the development and uses of edible vaccines.

1. What are the latest research developments on edible vaccines for cholera?

A source of information is the PubMed Web site (http://www.ncbi.nlm.nih.gov/sites/entrez?db=PubMed).

2. Several oral vaccines against cholera are currently available. Given the availability of these vaccines, why do you think that scientists are also developing edible vaccines for cholera? Which vaccine type would you choose for vaccinating populations at risk for cholera?
Read about these cholera vaccines on the World Health Organization Web site at http://www.who.int/immunization/topics/Cholera/en/index.html

3. Cholera is spread through ingestion of contaminated water and food. Despite its severity, cholera patients can be effectively treated by oral rehydration. Cholera becomes a major health problem only when proper sanitation and medical treatment are lacking. Given these facts, how much research funding do you think we should spend to develop cholera vaccines, relative to funds spent on im-

proved sanitation, water treatment, and education about treatments?
A discussion of sanitation vs. cholera vaccination can be found on the Web site of the Integrated Regional Information Network at http://www.irinnews.org/Report/84386/GLOBAL-Sanitation-vs-vaccination-in-cholera-control

4. One of the problems associated with edible vaccines is the public's concern about genetically modified organisms (GMOs). Attitudes vary from outright moral opposition to GMOs to concern about the potential environmental hazards associated with growing transgenic plants. How do you feel about the use of GMOs? What do you think are the most valid arguments for and against them, and why?
Scientists also debate these issues. One such debate is presented in the article Arntzen, C. J., et al., 2003, GM crops: Science, politics and communication, *Nature Rev Genet* 4: 839–843.

CASE STUDY » To treat or not to treat

A 4-month-old infant had been running a moderate fever for 36 hours, and a nervous mother made a call to her pediatrician. Examination and testing revealed no outward signs of infection or cause of the fever. The anxious mother asked the pediatrician about antibiotics, but the pediatrician recommended watching the infant carefully for two days before making a decision. He explained that decades of rampant use of antibiotics in medicine and agriculture had caused a worldwide surge in bacteria that are now resistant to such drugs. He also said that the reproductive behavior of bacteria allows them to exchange antibiotic resistance traits with a wide range of other disease-causing bacteria, and that many strains are now resistant to multiple antibiotics. The physician's information raises several interesting questions.

1. Was the physician correct in saying that bacteria can share resistance?
2. Where do bacteria carry antibiotic resistance genes, and how are they exchanged?
3. If the infant was given an antibiotic as a precaution, how might it contribute to the production of resistant bacteria?
4. Aside from hospitals, where else would infants and children come in contact with antibiotic-resistant strains of bacteria? Does the presence of such bacteria in the body always mean an infection?

INSIGHTS AND SOLUTIONS

1. Time mapping is performed in a cross involving the genes *his*, *leu*, *mal*, and *xyl*. The recipient cells are auxotrophic for all four genes. After 25 minutes, mating is interrupted, with the results in recipient cells shown below. Diagram the positions of these genes relative to the origin (*O*) of the F factor and to one another.

(a) 90% are *xyl*⁺

(b) 80% are *mal*⁺

(c) 20% are *his*⁺

(d) None are *leu*⁺

Solution: The *xyl* gene is transferred most frequently, so it is closest to *O* (very close). The *mal* gene is next and reasonably close

to *xyl*, followed by the more distant *his* gene. The *leu* gene is far beyond these three, since no recovered recombinants include it. The diagram shows these relative locations along a piece of the circular chromosome.

2. In four Hfr strains of bacteria, all derived from an original F⁺ culture grown over several months, a group of hypothetical genes is studied and shown to transfer in the orders shown in the following table. (a) Assuming *b* is the first gene along the chromosome, determine the sequence of all genes shown. (b) One strain

creates an apparent dilemma. Which one is it? Explain why the dilemma is only apparent, not real.

Hfr Strain	Order of Transfer
1	e r i u m b
2	u m b a c t
3	c t e r i u
4	r e t c a b

Solution:

(a) The sequence is found by overlapping the genes in each strain.

Strain 2	u m b a c t
Strain 3	c t e r i u
Strain 1	e r i u m b

Starting with *b* in strain 2, the gene sequence is *bacterium*.

(b) Strain 4 creates a dilemma, which is resolved when we realize that the F factor is integrated in the opposite orientation. Thus, the genes enter in the opposite sequence, starting with gene *r*.

$$\overrightarrow{retcab}$$

PROBLEMS AND DISCUSSION QUESTIONS

 For activities, animations, and review quizzes, go to the study area at www.masteringgenetics.com

HOW DO WE KNOW?

1. In this chapter, weWe have focused on genetic systems present in bacteria and the viruses that use bacteria as hosts (bacteriophages). In particular, we discussed mechanisms by which bacteria and their phages undergo genetic recombination, the basis of chromosome mapping. Based on your knowledge of these topics, answer several fundamental questions:

   (a) How do we know that bacteria undergo genetic recombination, allowing the transfer of genes from one organism to another?

   (b) How do we know that conjugation leading to genetic recombination between bacteria involves cell contact, which precedes the transfer of genes from one bacterium to another?

   (c) How do we know that during transduction bacterial cell-to-cell contact is not essential?

   (d) How do we know that intergenic exchange occurs in bacteriophages?

2. Distinguish among the three modes of recombination in bacteria.
3. With respect to F$^+$ and F$^-$ bacterial matings,

   (a) How was it established that physical contact was necessary?

   (b) How was it established that chromosome transfer was unidirectional?

   (c) What is the genetic basis of a bacterium being F$^+$?

4. List all of the differences between F$^+ \times$ F$^-$ and Hfr $\times$ F$^-$ bacterial crosses.
5. List all of the differences between F$^+$, F$^-$, Hfr, and F$'$ bacteria.
6. Describe the basis for chromosome mapping in the Hfr $\times$ F$^-$ crosses.
7. Why are the recombinants produced from an Hfr $\times$ F$^-$ cross rarely, if ever, F$^+$?
8. Describe the origin of F$'$ bacteria and merozygotes.
9. Describe the mechanism of transformation.
10. The bacteriophage genome consists primarily of genes encoding proteins that make up the head, collar and tail, and tail fibers. When these genes are transcribed following phage infection, how are these proteins synthesized, since the phage genome lacks genes essential to ribosome structure?
11. Describe the temporal sequence of the bacteriophage life cycle.
12. In the plaque assay, what is the precise origin of a single plaque?
13. In the plaque assay, exactly what makes up a single plaque?

14. A plaque assay is performed beginning with 1.0 mL of a solution containing bacteriophages. This solution is serially diluted three times by taking 0.1 mL and adding it to 9.9 mL of liquid medium. The final dilution is plated and yields 17 plaques. What is the initial density of bacteriophages in the original 1.0 mL?
15. Describe the difference between the lytic cycle and lysogeny when bacteriophage infection occurs.
16. Define the term prophage.
17. Explain the observations that led Zinder and Lederberg to conclude that the prototrophs recovered in their transduction experiments were not the result of Hfr-mediated conjugation.
18. Differentiate between generalized and specialized transduction.
19. Two theoretical genetic strains of a virus ($a^- b^- c^-$ and $a^+ b^+ c^+$) are used to simultaneously infect a culture of host bacteria. Of 10,000 plaques scored, the genotypes in the following table were observed. Determine the genetic map of these three genes on the viral chromosome.

$a^+ b^+ c^+$	4100	$a^- b^+ c^-$	160
$a^- b^- c^-$	3990	$a^+ b^- c^+$	140
$a^+ b^- c^-$	740	$a^- b^- c^+$	90
$a^- b^+ c^+$	670	$a^+ b^+ c^-$	110

20. Describe the conditions under which genetic recombination may occur in bacteriophages.
21. If a single bacteriophage infects one *E. coli* cell present in a culture of bacteria and, upon lysis, yields 200 viable viruses, how many phages will exist in a single plaque if three more lytic cycles occur?
22. A phage-infected bacterial culture was subjected to a series of dilutions, and a plaque assay was performed in each case, with the following results. What conclusion can be drawn in the case of each dilution?

	Dilution Factor	Assay Results
(a)	10^4	All bacteria lysed
(b)	10^5	14 plaques
(c)	10^6	0 plaques

SOLUTIONS TO SELECTED PROBLEMS AND DISCUSSION QUESTIONS

Answers to Now Solve This

1. | *Hfr Strain* | *Order* |
   |---|---|
   | 1 | *t c h r o* |
   | 2 | *h r o m b* |
   | 3 | *<<c h r o m* |
   | 4 | *m b a k t>>* |
   | 5 | *<<b a k t c* |

   Overall: $\boxed{t\ c\ h\ r\ o\ m\ b\ a\ k\ t}\ c$

2. In the first dataset, the transformation of each locus, a^+ and b^+, occurs at a frequency of 0.031 and 0.012, respectively. To determine whether there is linkage, one would determine whether the frequency of double transformants a^+b^+ is greater than that expected by a multiplication of the two independent events. Multiplying 0.031×0.012 gives 0.00037, or approximately 0.04 percent. From this information, one would consider no linkage between these two loci. Notice that this frequency is approximately the same as the frequency in the second experiment, where the loci are transformed independently.

Solutions to Problems and Discussion Questions

4. In an $F^+ \times F^-$ cross, the transfer of the F factor produces a recipient bacterium that is F^+. Any gene may be transferred, and the frequency of transfer is relatively low. Crosses that are Hfr $\times$ F^- produce recombinants at a higher frequency than the $F^+ \times F^-$ cross. The transfer is oriented (nonrandom) and the recipient cell remains F^-.

8. The F^+ element can enter the host bacterial chromosome, and upon returning to its independent state, it may pick up a piece of a bacterial chromosome. When transferred to a bacterium with a complete chromosome, a partial diploid, or merozygote, is formed.

10. The phage not only lacks genes for ribosomal construction, but also contains no ribosomes. Upon infection, phage genes are transcribed, and the transcripts are translated using bacterial ribosomes.

12. A single plaque originates from the replicative activity of a single bacteriophage.

14. Assuming the typical introduction of 0.1 ml of the phage suspension to the bacterial solution, since 17 plaques were formed, the initial density of bacteriophage suspension would be 1.7×10^8 phage/ml.

20. Viral recombination occurs when there is a sufficiently high number of infecting viruses so that there is a high likelihood that more than one variant of phage will infect a given bacterium. Under this condition, phage chromosomes can recombine by crossing over.

22. (a) Remembering that 0.1 ml is typically used in the plaque assay, the initial concentration of phage per milliliter is greater than 10^5.

   (b) Remembering that 0.1 ml is typically used in the plaque assay, the initial concentration of phage per milliliter is around 140×10^5 or 1.4×10^7.

   (c) Remembering that 0.1 ml is typically used in the plaque assay, the initial concentration of phage is less than 10^7. Coupling this information with the calculations in part (b) above, it would appear that the initial concentration of phage is around 1×10^7, and that the failure to obtain plaques in this portion of the experiment is expected and due to sampling error.

CREDITS

Credits are listed in order of appearance.

Photo

CO, Charles C. Brinton, Jr., University of Pittsburgh; F-1, Experiment performed and photographed by L. Brent Selinger, Department of Biological Sciences, University of Lethbridge, Alberta,

Canada: Pearson Education Benjamin Cummings; F-10 (a) Dr. Gopal Murti/Science Photo Library/Photo Researchers, Inc.; F-11 (a) M. Wurtz/Biozentrum, University of Basel/Science Photo Library/Photo Researchers, Inc.; F-13, Christine Case, F-16, Dr. William S. Klug

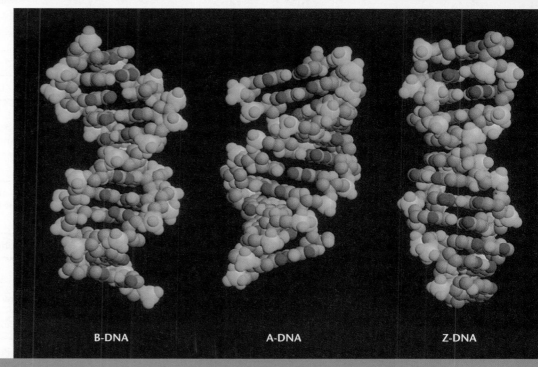

Computer-generated space-filling models of alternative forms of DNA.

B-DNA   A-DNA   Z-DNA

DNA Structure and Analysis

CHAPTER CONCEPTS

- With the exception of some viruses, DNA serves as the genetic material in all living organisms on Earth.

- According to the Watson–Crick model, DNA exists in the form of the right-handed double helix.

- The strands of the double helix are antiparallel and held together by hydrogen bonding between complementary nitrogenous bases.

- The structure of DNA provides the basis for storing and expressing genetic information.

- RNA has many similarities to DNA but exists mostly as a single-stranded molecule.

- In some viruses, RNA serves as the genetic material.

- Numerous techniques have been developed to facilitate the analysis of nucleic acids; many of these techniques are based on detection of the complementarity of nitrogenous bases.

U p to this point in the text, we have described chromosomes as structures containing genes that control phenotypic traits that are transmitted through gametes to future offspring. Logically, genes must contain some sort of information that, when passed to a new generation, influences the form and characteristics of each individual. We refer to that information as the **genetic material.**

Logic also suggests that this same information in some way directs the many complex processes that lead to an organism's adult form.

Until 1944, it was not clear what chemical component of the chromosome makes up genes and constitutes the genetic material. Because chromosomes were known to have both a nucleic acid and a protein component, both were candidates. In 1944, however, direct experimental evidence emerged showing that the nucleic acid DNA serves as the informational basis for heredity.

Once the importance of DNA in genetic processes was realized, work intensified with the hope of discerning not only the structural basis of this molecule but also the relationship of its structure to its function. Between 1944 and 1953, many scientists sought information that might answer the most significant and intriguing question in the history of biology: How does DNA serve as the genetic basis for the living process? Researchers believed the answer depended strongly on the chemical structure of the DNA molecule, given the complex but orderly functions ascribed to it.

These efforts were rewarded in 1953 when James Watson and Francis Crick set forth their hypothesis for the double-helical nature of DNA. The assumption that the molecule's functions would be clarified more easily once its general structure was determined proved to be correct. In this chapter, we initially review the evidence that DNA is the genetic material and then discuss the elucidation of its structure.

From Chapter 9 of *Essentials of Genetics,, Eighth edition,* William S. Klug, Michael R. Cummings, Charlotte A. Spencer, Michael A. Palladino. ©2013 by Pearson Education, Inc. All rights reserved.

1 The Genetic Material Must Exhibit Four Characteristics

For a molecule to serve as the genetic material, it must possess four major characteristics: **replication, storage of information, expression of information,** and **variation by mutation.** Replication of the genetic material is one facet of the cell cycle, a fundamental property of all living organisms. Once the genetic material of cells replicates and is doubled in amount, it must then be partitioned equally into daughter cells. During the formation of gametes, the genetic material is also replicated but is partitioned so that each cell gets only one-half of the original amount of genetic material—the process of meiosis. Although the products of mitosis and meiosis differ, both of these processes are part of the more general phenomenon of cellular reproduction.

Storage of information requires the molecule to act as a repository of genetic information that may or may not be expressed by the cell in which it resides. It is clear that while most cells contain a complete copy of the organism's genome, at any point in time they express only a part of this genetic potential. For example, in bacteria many genes "turn on" in response to specific environmental conditions and "turn off" when conditions change. In vertebrates, skin cells may display active melanin genes but never activate their hemoglobin genes; in contrast, digestive cells activate many genes specific to their function but do not activate their melanin genes.

Expression of the stored genetic information is the basis of the process of **information flow** within the cell (**Figure 1**). The initial event is the **transcription** of DNA, in which three main types of RNA molecules are synthesized: messenger RNA (mRNA), ribosomal RNA (rRNA), and transfer RNA (tRNA). Of these, mRNAs are translated into proteins. Each mRNA is the product of a specific gene and directs the synthesis of a different protein. In **translation,** the chemical information in mRNA directs the construction of a chain of amino acids, called a polypeptide, which then folds into a protein. Collectively, these processes form the **central dogma of molecular genetics:** "DNA makes RNA, which makes proteins."

The genetic material is also the source of variation among organisms through the process of mutation. If a mutation—a change in the chemical composition of DNA—occurs, the alteration is reflected during transcription and translation, affecting the specific protein. If a mutation is present in gametes, it may be passed to future generations and, with time, become distributed throughout the population. Genetic variation, which also includes alterations of chromosome number and rearrangements within and between chromosomes, provides the raw material for the process of evolution.

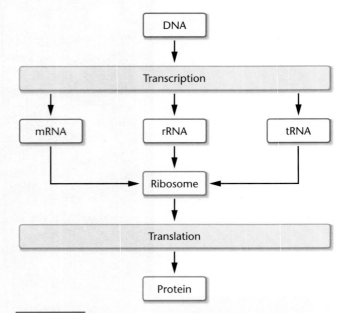

FIGURE 1 Simplified view of information flow (the central dogma) involving DNA, RNA, and proteins within cells.

2 Until 1944, Observations Favored Protein as the Genetic Material

The idea that genetic material is physically transmitted from parent to offspring has been accepted for as long as the concept of inheritance has existed. Beginning in the late nineteenth century, research into the structure of biomolecules progressed considerably, setting the stage for describing the genetic material in chemical terms. Although both proteins and nucleic acid were major candidates for the role of the genetic material, until the 1940s many geneticists favored proteins. This is not surprising because a diversity of proteins was known to be abundant in cells, and much more was known about protein chemistry.

DNA was first studied in 1868 by a Swiss chemist, Friedrich Miescher. He isolated cell nuclei and derived an acid substance containing DNA that he called **nuclein.** As investigations progressed, however, DNA, which was shown to be present in chromosomes, seemed to lack the chemical diversity necessary to store extensive genetic information. This conclusion was based largely on Phoebus A. Levene's observations in 1910 that DNA contained approximately equal amounts of four similar molecules called *nucleotides.* Levene postulated incorrectly that identical groups of these four components were repeated over and over, which was the basis of his **tetranucleotide hypothesis** for DNA structure. Attention was thus directed away from DNA, favoring proteins. However, in the 1940s, Erwin Chargaff showed that Levene's proposal was incorrect when he demonstrated that most organisms do not contain precisely equal proportions of the four nucleotides. We shall see later that the structure of DNA accounts for Chargaff's observations.

ESSENTIAL POINT ■ ■ ■

Although both proteins and nucleic acids were initially considered as possible candidates, proteins were initially favored to serve as the genetic material.

TABLE 1 Strains of *Diplococcus pneumoniae* Used by Frederick Griffith in His Original Transformation Experiments

Serotype	Colony Morphology	Capsule	Virulence
II*R*	Rough	Absent	Avirulent
III*S*	Smooth	Present	Virulent

3 | Evidence Favoring DNA as the Genetic Material Was First Obtained during the Study of Bacteria and Bacteriophages

Oswald Avery, Colin MacLeod, and Maclyn McCarty's 1944 publication on the chemical nature of a "transforming principle" in bacteria was the initial event that led to the acceptance of DNA as the genetic material. Their work, along with subsequent findings of other research teams, constituted the first direct experimental proof that DNA, and not protein, is the biomolecule responsible for heredity. It marked the beginning of the era of molecular genetics, a period of discovery in biology that made biotechnology feasible and has moved us closer to understanding the basis of life. The impact of the initial findings on future research and thinking paralleled that of the publication of Darwin's theory of evolution and the subsequent rediscovery of Mendel's postulates of transmission genetics. Together, these events constitute the three great revolutions in biology.

Transformation Studies

The research that provided the foundation for Avery, MacLeod, and McCarty's work was initiated in 1927 by Frederick Griffith, a medical officer in the British Ministry of Health. He experimented with several different strains of the bacterium *Diplococcus pneumoniae*.* Some were **virulent strains,** which cause pneumonia in certain vertebrates (notably humans and mice), while others were **avirulent strains,** which do not cause illness.

The difference in virulence depends on the existence of a polysaccharide capsule; virulent strains have this capsule, whereas avirulent strains do not. The nonencapsulated bacteria are readily engulfed and destroyed by phagocytic cells in the animal's circulatory system. Virulent bacteria, which possess the polysaccharide coat, are not easily engulfed; they multiply and cause pneumonia.

The presence or absence of the capsule causes a visible difference between colonies of virulent and avirulent strains. Encapsulated bacteria form **smooth colonies** (*S*) with a shiny

surface when grown on an agar culture plate; nonencapsulated strains produce **rough colonies** (*R*) **(Figure 2).** Thus, virulent and avirulent strains are easily distinguished by standard microbiological culture techniques.

Each strain of *Diplococcus* may be one of dozens of different types called **serotypes.** The specificity of the serotype is due to the detailed chemical structure of the polysaccharide constituent of the thick, slimy capsule. Serotypes are identified by immunological techniques and are usually designated by Roman numerals. Griffith used the avirulent type II*R* and the virulent type III*S* in his critical experiments. **Table 1** summarizes the characteristics of these strains.

Griffith knew from the work of others that only living virulent cells produced pneumonia in mice. If heat-killed virulent bacteria were injected into mice, no pneumonia resulted, just as living avirulent bacteria failed to produce the disease. Griffith's critical experiment (Figure 2) involved injecting mice with living II*R* (avirulent) cells combined with heat-killed III*S* (virulent) cells. Since neither cell type caused death in mice when injected alone, Griffith expected that the double injection would not kill the mice. But, after five days, all of the mice that had received both types of cells were dead. Paradoxically, analysis of their blood revealed large numbers of living type III*S* bacteria.

As far as could be determined, these III*S* bacteria were identical to the III*S* strain from which the heat-killed cell preparation had been made. Control mice, injected only with living avirulent II*R* bacteria, did not develop pneumonia and remained healthy. This ruled out the possibility that the avirulent II*R* cells simply changed (or mutated) to virulent III*S* cells in the absence of the heat-killed III*S* bacteria. Instead, some type of interaction had taken place between living II*R* and heat-killed III*S* cells.

Griffith concluded that the heat-killed III*S* bacteria somehow converted live avirulent II*R* cells into virulent III*S* cells. Calling the phenomenon **transformation,** he suggested that the **transforming principle** might be some part of the polysaccharide capsule or a compound required for capsule synthesis, although the capsule alone did not cause pneumonia. To use Griffith's term, the transforming principle from the dead III*S* cells served as a "pabulum" for the II*R* cells.

Griffith's work led others to explore the phenomenon of transformation. By 1931, Henry Dawson and his coworkers showed that transformation could occur *in vitro* (in a test

---
*This organism is now named *Streptococcus pneumoniae*.

FIGURE 2 Griffith's transformation experiment. The photographs show bacterial cells that exhibit capsules (type IIIS) or not (type IIR).

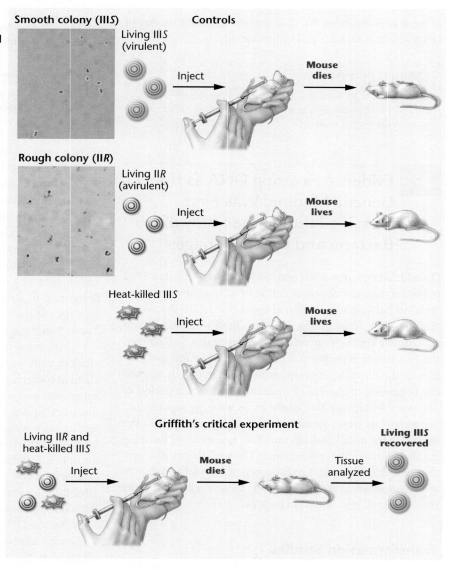

Smooth colony (IIIS)

Living IIIS (virulent) — Inject — Mouse dies

Controls

Rough colony (IIR)

Living IIR (avirulent) — Inject — Mouse lives

Heat-killed IIIS — Inject — Mouse lives

Griffith's critical experiment

Living IIR and heat-killed IIIS — Inject — Mouse dies — Tissue analyzed — Living IIIS recovered

tube containing only bacterial cells). That is, injection into mice was not necessary for transformation to occur. By 1933, Lionel J. Alloway had refined the *in vitro* experiments using extracts from *S* cells added to living *R* cells. The soluble filtrate from the heat-killed IIIS cells was as effective in inducing transformation as were the intact cells. Alloway and others did not view transformation as a genetic event, but rather as a physiological modification of some sort. Nevertheless, the experimental evidence that a chemical substance was responsible for transformation was quite convincing.

Then, in 1944, after ten years of work, Avery, MacLeod, and McCarty published their results in what is now regarded as a classic paper in the field of molecular genetics. They reported that they had obtained the transforming principle in a highly purified state and that beyond reasonable doubt it was DNA.

The details of their work are illustrated in **Figure 3**. The researchers began their isolation procedure with large quantities (50–75 L) of liquid cultures of type IIIS virulent

cells. The cells were centrifuged, collected, and heat-killed. Following various chemical treatments, a soluble filtrate was derived from these cells, which retained the ability to induce transformation of type IIR avirulent cells. The soluble filtrate was treated with a protein-digesting enzyme, called a protease, and an RNA-digesting enzyme, called **ribonuclease.** Such treatment destroyed the activity of any remaining protein and RNA. Nevertheless, transforming activity still remained. They concluded that neither protein nor RNA was responsible for transformation. The final confirmation came with experiments using crude samples of the DNA-digesting enzyme **deoxyribonuclease,** isolated from dog and rabbit sera. Digestion with this enzyme destroyed transforming activity present in the filtrate; thus, Avery and his coworkers were certain that the active transforming principle in these experiments was DNA.

The great amount of work, the confirmation and reconfirmation of the conclusions, and the logic of the experimental design involved in the research of these three scientists are

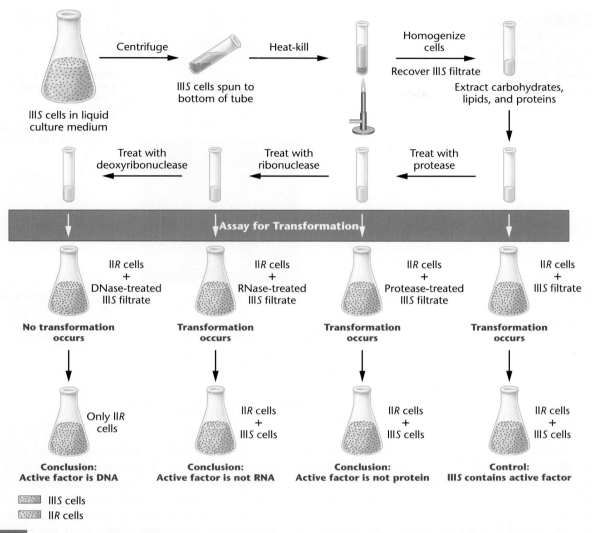

FIGURE 3   Summary of Avery, MacLeod, and McCarty's experiment demonstrating that DNA is the transforming principle.

truly impressive. Their conclusion in the 1944 publication, however, was stated very simply: "The evidence presented supports the belief that a nucleic acid of the desoxyribose* type is the fundamental unit of the transforming principle of *Pneumococcus* type III."

The researchers also immediately recognized the genetic and biochemical implications of their work. They suggested that the transforming principle interacts with the IIR cell and gives rise to a coordinated series of enzymatic reactions that culminates in the synthesis of the type IIIS capsular polysaccharide. They emphasized that, once transformation occurs, the capsular polysaccharide is produced in successive generations. Transformation is therefore heritable, and the process affects the genetic material.

Transformation has now been shown to occur in *Hemophilus influenzae, Bacillus subtilis, Shigella paradysenteriae,* and *Escherichia coli,* among many other microorganisms.

Transformation of numerous genetic traits other than colony morphology has been demonstrated, including those that resist antibiotics. These observations further strengthened the belief that transformation by DNA is primarily a genetic event rather than simply a physiological change.

The Hershey–Chase Experiment

The second major piece of evidence supporting DNA as the genetic material was provided during the study of the bacterium *E. coli* and one of its infecting viruses, **bacteriophage T2.** Often referred to simply as a **phage,** the virus consists of a protein coat surrounding a core of DNA. Electron micrographs reveal that the phage's external structure is composed of a hexagonal head plus a tail. **Figure 4** shows the life cycle of a T-even bacteriophage such as T2, as it was known in 1952. Recall that the phage adsorbs to the bacterial cell and that some component of the phage enters the bacterial cell.

---

*Desoxyribose is now spelled deoxyribose.

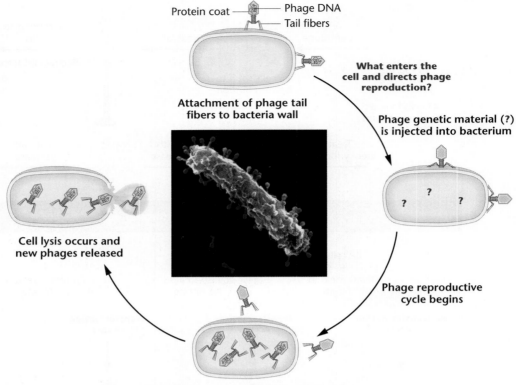

FIGURE 4 Life cycle of a T-even bacteriophage. The electron micrograph shows an *E. coli* cell during infection by numerous T2 phages (shown in blue).

Protein coat — Phage DNA
Tail fibers

Attachment of phage tail fibers to bacteria wall

What enters the cell and directs phage reproduction?

Phage genetic material (?) is injected into bacterium

Phage reproductive cycle begins

Cell lysis occurs and new phages released

Components accumulate; assembly of mature phages occurs

Following infection, the viral information "commandeers" the cellular machinery of the host and undergoes viral reproduction. In a reasonably short time, many new phages are constructed and the bacterial cell is lysed, releasing the progeny viruses.

In 1952, Alfred Hershey and Martha Chase published the results of experiments designed to clarify the events leading to phage reproduction. Several of the experiments clearly established the independent functions of phage protein and nucleic acid in the reproduction process of the bacterial cell. Hershey and Chase knew from this existing data that:

1. T2 phages consist of approximately 50 percent protein and 50 percent DNA.

2. Infection is initiated by adsorption of the phage by its tail fibers to the bacterial cell.

3. The production of new viruses occurs within the bacterial cell.

It appeared that some molecular component of the phage, DNA and/or protein, entered the bacterial cell and directed viral reproduction. Which was it?

Hershey and Chase used radioisotopes to follow the molecular components of phages during infection. Both ^{32}P and ^{35}S, radioactive forms of phosphorus and sulfur, respectively, were used. DNA contains phosphorus but not sulfur, so ^{32}P effectively labels DNA. Because proteins contain sulfur, but not phosphorus, ^{35}S labels protein. *This is a key point in the experiment.* If *E. coli* cells are first grown in the presence of *either* ^{32}P *or* ^{35}S and then infected with T2 viruses, the progeny phage will have *either* a labeled DNA core *or* a labeled protein coat, respectively. These radioactive phages can be isolated and used to infect unlabeled bacteria (**Figure 5**).

When labeled phage and unlabeled bacteria were mixed, an adsorption complex was formed as the phages attached their tail fibers to the bacterial wall. These complexes were isolated and subjected to a high shear force by placing them in a blender. This force stripped off the attached phages, which were then analyzed separately (Figure 5). By tracing the radioisotopes, Hershey and Chase were able to demonstrate that most of the ^{32}P-labeled DNA had transferred into the bacterial cell following adsorption; on the other hand, almost all of the ^{35}S-labeled protein remained outside the bacterial cell and was recovered in the phage "ghosts" (empty phage coats) after the blender treatment. Following separation, the bacterial cells, which now contained viral DNA, were eventually lysed as new phages were produced. These progeny contained ^{32}P but not ^{35}S.

Hershey and Chase interpreted these results as indicating that the protein of the phage coat remains outside the host cell and is not involved in the production of new phages. On the other hand, and most important, phage DNA enters the host cell and directs phage reproduction. Hershey and

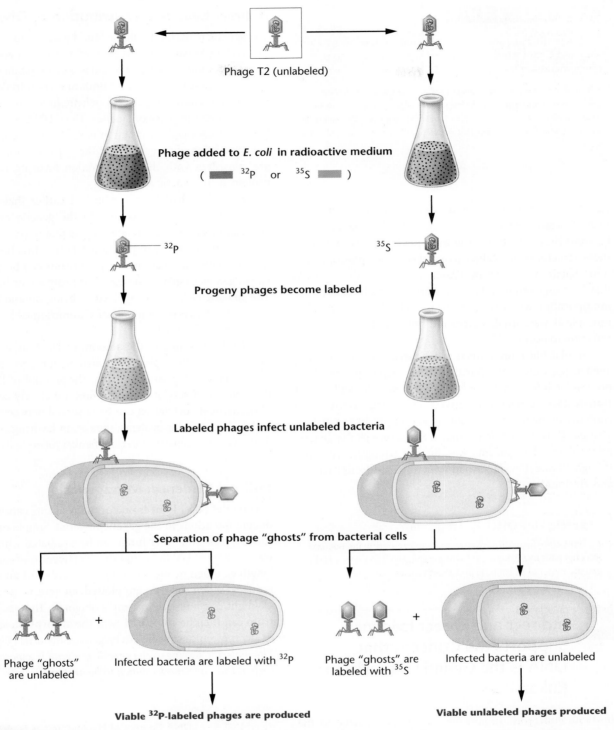

FIGURE 5 Summary of the Hershey–Chase experiment demonstrating that DNA, not protein, is responsible for directing the reproduction of phage T2 during the infection of *E. coli*.

Chase had demonstrated that the genetic material in phage T2 is DNA, not protein.

These experiments, along with those of Avery and his colleagues, provided convincing evidence that DNA is the molecule responsible for heredity. This conclusion has since served as the cornerstone of the field of molecular genetics.

Transfection Experiments

During the eight years following publication of the Hershey–Chase experiment, additional research with bacterial viruses provided even more solid proof that DNA is the genetic material. In 1957, several reports demonstrated that if *E. coli* is treated with the enzyme lysozyme, the outer wall of the

1 Would an experiment similar to that performed by Hershey and Chase work if the basic design were applied to the phenomenon of transformation? Explain why or why not.

■ **Hint** *This problem involves an understanding of the protocol of the Hershey–Chase experiment as applied to the investigation of transformation. The key to its solution is to remember that in transformation, exogenous DNA enters the soon-to-be transformed cell and that no cell-to-cell contact is involved in the process.*

cell can be removed without destroying the bacterium. Enzymatically treated cells are naked, so to speak, and contain only the cell membrane as the outer boundary of the cell; these structures are called **protoplasts** (or **spheroplasts**). John Spizizen and Dean Fraser independently reported that by using protoplasts, they were able to initiate phage multiplication with disrupted T2 particles. That is, provided protoplasts were used, a virus did not have to be intact for infection to occur.

Similar but more refined experiments were reported in 1960 using only DNA purified from bacteriophages. This process of infection by only the viral nucleic acid, called **transfection,** proves conclusively that phage DNA alone contains all the necessary information for producing mature viruses. Thus, the evidence that DNA serves as the genetic material in all organisms was further strengthened, even though all direct evidence had been obtained from bacterial and viral studies.

ESSENTIAL POINT ■ ■ ■

By 1952, transformation studies and experiments using bacteria infected with bacteriophages strongly suggested that DNA is the genetic material in bacteria and most viruses.

4 Indirect and Direct Evidence Supports the Concept that DNA Is the Genetic Material in Eukaryotes

In 1950, eukaryotic organisms were not amenable to the types of experiments that used bacteria and viruses to demonstrate that DNA is the genetic material. Nevertheless, it was generally assumed that the genetic material would be a universal substance and also serve this role in eukaryotes. Initially, support for this assumption relied on several circumstantial observations that, taken together, indicated that DNA is also the genetic material in eukaryotes. Subsequently, direct evidence established unequivocally the central role of DNA in genetic processes.

Indirect Evidence: Distribution of DNA

The genetic material should be found where it functions—in the nucleus as part of chromosomes. Both DNA and protein fit this criterion. However, protein is also abundant in the cytoplasm, whereas DNA is not. Both mitochondria and chloroplasts are known to perform genetic functions, and DNA is also present in these organelles. Thus, DNA is found only where primary genetic function is known to occur. Protein, however, is found everywhere in the cell. These observations are consistent with the interpretation favoring DNA over protein as the genetic material.

Because it had been established earlier that chromosomes within the nucleus contain the genetic material, a correlation was expected between the ploidy (n, $2n$, etc.) of cells and the quantity of the molecule that functions as the genetic material. Meaningful comparisons can be made between gametes (sperm and eggs) and somatic or body cells. The somatic cells are recognized as being diploid ($2n$) and containing twice the number of chromosomes as gametes, which are haploid (n).

Table 2 compares the amount of DNA found in haploid sperm and the diploid nucleated precursors of red blood cells from a variety of organisms. The amount of DNA and the number of sets of chromosomes are closely correlated. No consistent correlation can be observed between gametes and diploid cells for proteins, thus again favoring DNA over proteins as the genetic material of eukaryotes.

Indirect Evidence: Mutagenesis

Ultraviolet (UV) light is one of many agents capable of inducing mutations in the genetic material. Simple organisms such as yeast and other fungi can be irradiated with various wavelengths of UV light, and the effectiveness of each wavelength can then be measured by the number of mutations it induces. When the data are plotted, an **action spectrum** of UV light as a mutagenic agent is obtained. This action spectrum can then be compared with the **absorption spectrum** of any molecule suspected to be genetic material (**Figure 6**). *The molecule serving as the genetic material is expected to absorb at the wavelengths shown to be mutagenic.*

TABLE 2 **DNA Content of Haploid Versus Diploid Cells of Various Species (in picograms)***

Organism	*n*	*2n*
Human	3.25	7.30
Chicken	1.26	2.49
Trout	2.67	5.79
Carp	1.65	3.49
Shad	0.91	1.97

*Sperm (*n*) and nucleated precursors to red blood cells (*2n*) were used to contrast ploidy levels.

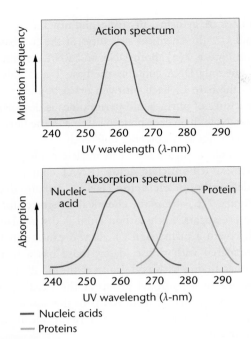

FIGURE 6 Comparison of the action spectrum, which determines the most effective mutagenic UV wavelength, and the absorption spectrum, which shows the range of wavelengths where nucleic acids and proteins absorb UV light.

UV light is most mutagenic at the wavelength (λ) of about 260 nanometers (nm), and both DNA and RNA strongly absorb UV light at 260 nm. On the other hand, protein absorbs most strongly around 280 nm, yet no significant mutagenic effects are observed at this wavelength. This indirect evidence also supports the idea that a nucleic acid is the genetic material and tends to exclude protein.

Direct Evidence: Recombinant DNA Studies

Although the circumstantial evidence described earlier does not constitute direct proof that DNA is the genetic material in eukaryotes, these observations spurred researchers to forge ahead, basing their work on this hypothesis. Today, there is no doubt of the validity of this conclusion. DNA *is* the genetic material in eukaryotes. The strongest evidence is provided by molecular analysis utilizing **recombinant DNA technology.** In this procedure, segments of eukaryotic DNA corresponding to specific genes are isolated and spliced into bacterial DNA. This complex can then be inserted into a bacterial cell, and its genetic expression is monitored. If a eukaryotic gene is introduced, the presence of the corresponding eukaryotic protein product demonstrates directly that this DNA is functional in the bacterial cell. This has been shown to be the case in countless instances. For example, the products of the human genes specifying insulin and interferon are produced by bacteria after they have incorporated the human genes that encode these proteins. As the bacterium divides, the eukaryotic DNA replicates along with the host DNA and is distributed to the daughter cells, which also express the human genes and synthesize the corresponding proteins.

The availability of vast amounts of DNA coding for specific genes, derived from recombinant DNA research, has led to other direct evidence that DNA serves as the genetic material. Work done in the laboratory of Beatrice Mintz demonstrated that DNA encoding the human β-globin protein, when microinjected into a fertilized mouse egg, is later found to be present and expressed in adult mouse tissue, and it is transmitted to and expressed in that mouse's progeny. These mice are examples of **transgenic animals**, which are now commonplace in genetic research. They clearly demonstrate that DNA meets the requirement of expression of genetic information in eukaryotes.

> **ESSENTIAL POINT** ■ ■ ■
>
> Although initially only indirect observations supported the hypothesis that DNA controls inheritance in eukaryotes, subsequent studies involving recombinant DNA techniques and transgenic mice provided direct experimental evidence that the eukaryotic genetic material is DNA.

5 RNA Serves as the Genetic Material in Some Viruses

Some viruses contain an RNA core rather than a DNA core. In these viruses, it appears that RNA serves as the genetic material—an exception to the general rule that DNA performs this function. In 1956, it was demonstrated that when purified RNA from **tobacco mosaic virus (TMV)** was spread on tobacco leaves, the characteristic lesions caused by viral infection subsequently appeared on the leaves. Thus, it was concluded that RNA is the genetic material of this virus.

In 1965 and 1966, Norman Pace and Sol Spiegelman demonstrated further that RNA from the phage Qβ can be isolated and replicated *in vitro*. Replication depends on an enzyme, **RNA replicase,** which is isolated from host *E. coli* cells following normal infection. When the RNA replicated *in vitro* is added to *E. coli* protoplasts, infection and viral multiplication (transfection) occur. Thus, RNA synthesized in a test tube serves as the genetic material in these phages by directing the production of all the components necessary for viral reproduction.

One other group of RNA-containing viruses bears mention. These are the **retroviruses,** which replicate in an unusual way. Their RNA serves as a template for the synthesis of the complementary DNA molecule. The process, **reverse transcription**, occurs under the direction of an RNA-dependent DNA polymerase enzyme called **reverse transcriptase.** This DNA intermediate can be incorporated into the genome of the host cell, and when the host DNA is transcribed, copies of the original retroviral RNA chromosomes are produced. Retroviruses include the human immunodeficiency virus (HIV), which causes AIDS, as well as RNA tumor viruses.

6 The Structure of DNA Holds the Key to Understanding Its Function

Having established that DNA is the genetic material in all living organisms (except certain viruses), we turn now to the structure of this nucleic acid. In 1953, James Watson and Francis Crick proposed that the structure of DNA is in the form of a double helix. Their proposal was published in a short paper in the journal *Nature*. In a sense, this publication was the finish of a highly competitive scientific race to obtain what some consider to be the most significant finding in the history of biology. This race, as recounted in Watson's book *The Double Helix*, demonstrates the human interaction, genius, frailty, and intensity involved in the scientific effort that eventually led to elucidation of the DNA structure.

The data available to Watson and Crick, crucial to the development of their proposal, came primarily from two sources: (1) base composition analysis of hydrolyzed samples of DNA and (2) X-ray diffraction studies of DNA. Watson and Crick's analytical success can be attributed to model building that conformed to the existing data. If the correct solution to the structure of DNA is viewed as a puzzle, Watson and Crick, working in the Cavendish Laboratory in Cambridge, England, were the first to fit the pieces together successfully.

Nucleic Acid Chemistry

Before turning to this work, a brief introduction to nucleic acid chemistry is in order. This chemical information was well known to Watson and Crick during their investigation and served as the basis of their model building.

DNA is a nucleic acid, and nucleotides are the building blocks of all nucleic acid molecules. Sometimes called mononucleotides, these structural units have three essential components: a **nitrogenous base**, a **pentose sugar** (a five-carbon sugar), and a **phosphate group**. There are two kinds of nitrogenous bases: the nine-member double-ring **purines** and the six-member single-ring **pyrimidines.**

Two types of purines and three types of pyrimidines are found in nucleic acids. The two purines are **adenine** and **guanine,** abbreviated A and G, respectively. The three

pyrimidines are **cytosine, thymine,** and **uracil** (respectively, C, T, and U). The chemical structures of the five bases are shown in **Figure 7(a)**. Both DNA and RNA contain A, G, and C, but only DNA contains the base T and only RNA contains the base U. Each nitrogen or carbon atom of the ring structures of purines and pyrimidines is designated by a number. Note that corresponding atoms in the purine and pyrimidine rings are numbered differently.

The pentose sugars found in nucleic acids give them their names. **Ribonucleic acids (RNA)** contain **ribose,** while **deoxyribonucleic acids (DNA)** contain **deoxyribose. Figure 7(b)** shows the chemical structures for these two pentose sugars. Each carbon atom is distinguished by a number with a prime sign ($'$). As you can see in Figure 7(b), compared with ribose, deoxyribose has a hydrogen atom rather than a hydroxyl group at the (C-2$'$) position. The absence of a hydroxyl group at the (C-2$'$) position thus distinguishes DNA from RNA. In the absence of the (C-2$'$) hydroxyl group, the sugar is more specifically named **2-deoxyribose.**

If a molecule is composed of a purine or pyrimidine base and a ribose or deoxyribose sugar, the chemical unit is called a **nucleoside.** If a phosphate group is added to the nucleoside, the molecule is now called a **nucleotide.** Nucleosides and nucleotides are named according to the specific nitrogenous base (A, G, C, T, or U) that is part of the molecule. The structure of a nucleotide and the nomenclature used in naming DNA nucleotides and nucleosides are shown in **Figure 8**.

The bonding between components of a nucleotide is highly specific. The (C-1$'$) atom of the sugar is involved in the chemical linkage to the nitrogenous base. If the base is a purine, the N-9 atom is covalently bonded to the sugar; if the base is a pyrimidine, the N-1 atom bonds to the sugar. In deoxyribonucleotides, the phosphate group may be bonded to the (C-2$'$), (C-3$'$), or (C-5$'$) atom of the sugar. The (C-5$'$)-phosphate configuration is shown in Figure 8. It is by far the most prevalent one in biological systems and the one found in DNA and RNA.

Nucleotides are also described by the term **nucleoside monophosphate (NMP).** The addition of one or two phosphate groups results in **nucleoside diphosphates (NDP)** and **triphosphates (NTP),** respectively, as shown in **Figure 9**. The triphosphate form is significant because it is the precursor molecule during nucleic acid synthesis within the cell. In addition, **adenosine triphosphate (ATP)** and **guanosine triphosphate (GTP)** are important in cell bioenergetics because of the large amount of energy involved in adding or removing the terminal phosphate group. The hydrolysis of ATP or GTP to ADP or GDP and inorganic phosphate (P_i) is accompanied by the release of a large amount of energy in the cell. When these chemical conversions are coupled to other reactions, the energy produced is

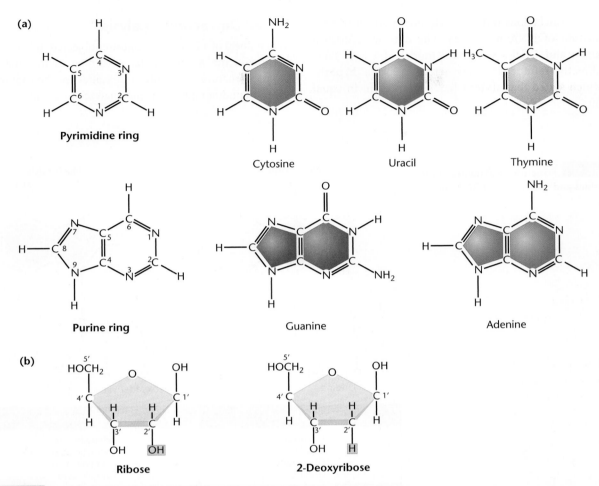

FIGURE 7 (a) Chemical structures of the pyrimidines and purines that serve as the nitrogenous bases in RNA and DNA. (b) Chemical ring structures of ribose and 2-deoxyribose, which serve as the pentose sugars in RNA and DNA, respectively.

used to drive them. As a result, ATP and GTP are involved in many cellular activities.

The linkage between two mononucleotides involves a phosphate group linked to two sugars. A **phosphodiester bond** is formed as phosphoric acid is joined to two alcohols (the hydroxyl groups on the two sugars) by an ester linkage on both sides. **Figure 10**, shows the resultant phosphodiester bond in DNA. Each structure has a (C-3′) end and a (C-5′) end. Two joined nucleotides form a dinucleotide; three nucleotides, a trinucleotide; and so forth. Short chains consisting of up to 20 nucleotides or so are called **oligonucleotides;** longer chains are **polynucleotides.**

Long polynucleotide chains account for the large molecular weight of DNA and explain its most important property—storage of vast quantities of genetic information. If each nucleotide position in this long chain can be occupied by any one of four nucleotides, extraordinary variation is possible. For example, a polynucleotide only 1000 nucleotides in length can be arranged 4^{1000} different ways, each one different from all other possible sequences. This potential variation in molecular structure is essential if DNA is to store the vast amounts of chemical information necessary to direct cellular activities.

Base-Composition Studies

Between 1949 and 1953, Erwin Chargaff and his colleagues used chromatographic methods to separate the four bases in DNA samples from various organisms. Quantitative methods were then used to determine the amounts of the four nitrogenous bases from each source. On the basis of these data, the following conclusions may be drawn:

1. The amount of adenine residues is proportional to the amount of thymine residues in DNA. Also, the amount of guanine residues is proportional to the amount of cytosine residues.

2. Based on this proportionality, the sum of the purines (A + G) equals the sum of the pyrimidines (C + T).

3. The percentage of (G + C) does not necessarily equal the percentage of (A + T). Instead, this ratio varies greatly between different organisms.

These conclusions indicate a definite pattern of base composition of DNA molecules. The data were critical to Watson and Crick's successful model of DNA. They also directly refuted Levene's tetranucleotide hypothesis, which stated that all four bases are present in equal amounts.

X-Ray Diffraction Analysis

When fibers of a DNA molecule are subjected to X-ray bombardment, the X rays scatter according to the molecule's atomic structure. The pattern of scatter can be captured as spots on photographic film and analyzed, particularly for the

FIGURE 8 Structures and names of the nucleosides and nucleotides of RNA and DNA.

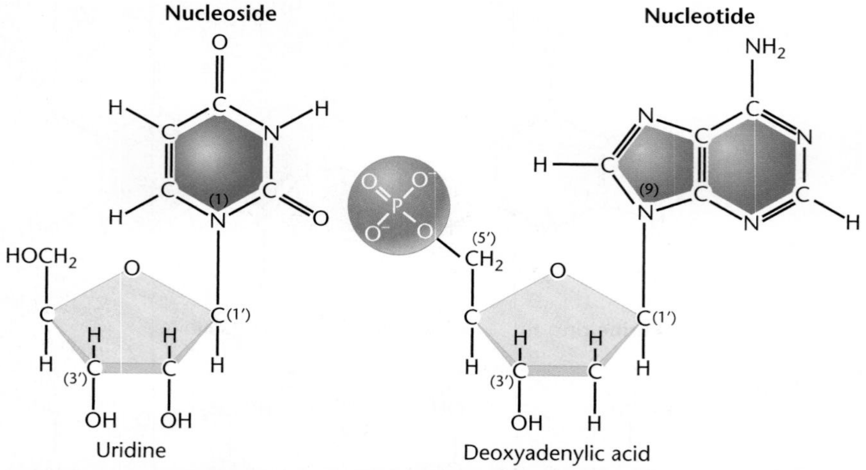

Nucleoside

Uridine

Nucleotide

Deoxyadenylic acid

Ribonucleosides	Ribonucleotides
Adenosine	Adenylic acid
Cytidine	Cytidylic acid
Guanosine	Guanylic acid
Uridine	Uridylic acid
Deoxyribonucleosides	**Deoxyribonucleotides**
Deoxyadenosine	Deoxyadenylic acid
Deoxycytidine	Deoxycytidylic acid
Deoxyguanosine	Deoxyguanylic acid
Deoxythymidine	Deoxythymidylic acid

FIGURE 9 Basic structures of nucleoside diphosphates and triphosphates. Deoxythymidine diphosphate and adenosine triphosphate are diagrammed here.

Deoxynucleoside diphosphate (dNDP)

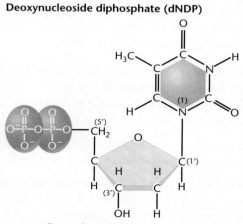

Deoxythymidine diphosphate (dTDP)

Ribonucleoside triphosphate (NTP)

Adenosine triphosphate (ATP)

(a)

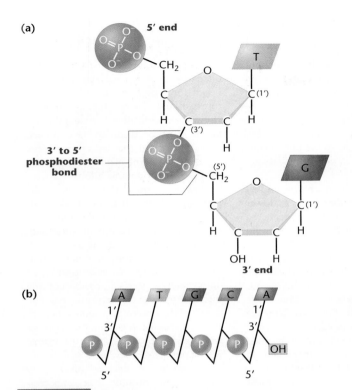

(b)

FIGURE 10 (a) Linkage of two nucleotides by the formation of a C–3′ to C–5′ (3′–5′) phosphodiester bond, producing a dinucleotide. (b) Shorthand notation for a polynucleotide chain.

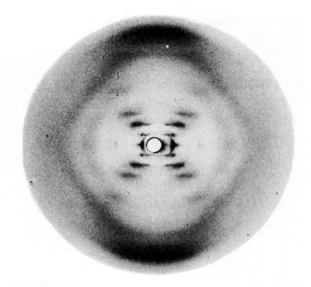

FIGURE 11 X-ray diffraction photograph of purified DNA fibers. The strong arcs on the periphery show closely spaced aspects of the molecule, providing an estimate of the periodicity of nitrogenous bases, which are 3.4 Å apart. The inner cross pattern of spots shows the grosser aspect of the molecule, indicating its helical nature.

overall shape of and regularities within the molecule. This process, **X-ray diffraction analysis,** was applied successfully to the study of protein structure by Linus Pauling and other chemists. The technique had been attempted on DNA as early as 1938 by William Astbury. By 1947, he had detected a periodicity of 3.4 angstroms (Å)[*] within the structure of the molecule, which suggested to him that the bases were stacked like coins on top of one another.

Between 1950 and 1953, Rosalind Franklin, working in the laboratory of Maurice Wilkins, obtained improved X-ray data from more purified samples of DNA **(Figure 11)**. Her work confirmed the 3.4 Å periodicity seen by Astbury and suggested that the structure of DNA was some sort of helix. However, she did not propose a definitive model. Pauling had analyzed the work of Astbury and others and proposed incorrectly that DNA is a triple helix.

The Watson–Crick Model

Watson and Crick published their analysis of DNA structure in 1953. By building models under the constraints of the information just discussed, they proposed the double-helical form of DNA shown in **Figure 12(a)**. This model has the following major features:

1. Two long polynucleotide chains are coiled around a central axis, forming a right-handed double helix.

2. The two chains are antiparallel; that is, their (C-5′) to (C-3′) orientations run in opposite directions.

3. The bases of both chains are flat structures, lying perpendicular to the axis; they are "stacked" on one another, 3.4 Å (0.34 nm) apart, and located on the inside of the structure.

4. The nitrogenous bases of opposite chains are *paired* as the result of hydrogen bonds; in DNA, only A═T and G≡C pairs occur.

5. Each complete turn of the helix is 34 Å (3.4 nm) long; thus, 10 bases exist per turn in each chain.

6. In any segment of the molecule, alternating larger **major grooves** and smaller **minor grooves** are apparent along the axis.

7. The double helix measures 20 Å (2.0 nm) in diameter.

The nature of *base pairing* (point 4 above) is the most genetically significant feature of the model. Before we discuss it in detail, several other important features warrant emphasis. First, the antiparallel nature of the two chains is a key part of the double helix model. While one chain runs in the 5′-to-3′ orientation (what seems right side up to us), the other chain goes in the 3′-to-5′ orientation (and thus appears upside down). This is illustrated in **Figure 12(b)** and **(c)**. Given the constraints of the bond angles of the various nucleotide com-

---

[*]Today, measurement in nanometers (nm) is favored (1 nm = 10 Å).

219

Molecular Structure of Nucleic Acids: A Structure for Deoxyribose Nucleic Acid

We wish to suggest a structure for the salt of deoxyribose nucleic acid (DNA). This structure has novel features which are of considerable biological interest. A structure for nucleic acid has already been proposed by Pauling and Corey.[1] They kindly made their manuscript available to us in advance of publication.

Their model consists of three intertwined chains, with the phosphates near the fibre axis, and the bases on the outside. In our opinion, this structure is unsatisfactory for two reasons: (1) We believe that the material which gives the X-ray diagrams is the salt, not the free acid. Without the acidic hydrogen atoms it is not clear what forces would hold the structure together, especially as the negatively charged phosphates near the axis will repel each other. (2) Some of the van der Waals distances appear to be too small.

Another three-chain structure has also been suggested by Fraser (in the press). In his model the phosphates are on the outside and the bases on the inside, linked together by hydrogen bonds. This structure as described is rather ill-defined, and for this reason we shall not comment on it.

We wish to put forward a radically different structure for the salt of deoxyribose nucleic acid. This structure has two helical chains each coiled round the same axis. We have made the usual chemical assumptions, namely, that each chain consists of phosphate diester groups joining β-D-deoxyribofuranose residues with 3′, 5′ linkages. The two chains (but not their bases) are related by a dyad perpendicular to the fibre axis. Both chains follow right-handed helices, but owing to the dyad the sequences of the atoms in the two chains run in opposite directions. Each chain loosely resembles Furberg's[2] model No. 1; that is, the bases are on the inside of the helix and the phosphates on the outside. The configuration of the sugar and the atoms near it is close to Furberg's "standard configuration," the sugar being roughly perpendicular to the attached base. There is a residue on each chain every 3.4 Å in the z-direction. We have assumed an angle of between adjacent residues in the same chain, so that the structure repeats after 10 residues on each chain, that is, after 34 Å. The distance of a phosphate atom from the fibre axis is 10 Å. As the phosphates are on the outside, cations have easy access to them.

The structure is an open one, and its water content is rather high. At lower water content we would expect the bases to tilt so that the structure could become more compact.

The novel feature of the structure is the manner in which the two chains are held together by the purine and pyrimidine bases. The planes of the bases are perpendicular to the fibre axis. They are joined together in pairs, a single base from one chain being hydrogen-bonded to a single base from the other chain, so that the two lie side by side with identical z-coordinates. One of the pair must be a purine and the other a pyrimidine for bonding to occur. The hydrogen bonds are made as follows: purine position 1 to pyrimidine position 1; purine position 6 to pyrimidine position 6.

If it is assumed that the bases only occur in the structure in the most plausible tautomeric forms (that is, with the keto rather than the enol configuration), it is found that only specific pairs of bases can bond together. These pairs are: adenine (purine) with thymine (pyrimidine), and guanine (purine) with cytosine (pyrimidine).

In other words, if an adenine forms one member of a pair, on either chain, then on these assumptions the other member must be thymine; similarly for guanine and cytosine. The sequence of bases on a single chain does not appear to be restricted in any way. However, if only specific pairs of bases can be formed, it follows that if the sequence of bases on one chain is given, then the sequence on the other chain is automatically determined.

It has been found experimentally[3,4] that the ratio of the amounts of adenine to thymine, and the ratio of guanine to cytosine, are always very close to unity for deoxyribose nucleic acid.

It is probably impossible to build this structure with a ribose sugar in place of deoxyribose, as the extra oxygen atom would make too close a van der Waals contact.

The previously published X-ray data[5,6] on deoxyribose nucleic acid are insufficient for a rigorous test of our structure. So far as we can tell, it is roughly compatible with the experimental data, but it must be regarded as unproved until it has been checked against more exact results. Some of these are given in the following communications. We were not aware of the details of the results presented there when we devised our structure, which rests mainly though not entirely on published experimental data and stereochemical arguments.

It has not escaped our notice that the specific pairing we have postulated immediately suggests a possible copying mechanism for the genetic material. Full details of the structure, including the conditions assumed in building it, together with a set of co-ordinates for the atoms, will be published elsewhere.

We are much indebted to Dr. Jerry Donohue for constant advice and criticism, especially on interatomic distances. We have also been stimulated by a knowledge of the general nature of the unpublished experimental results and ideas of Dr. M. H. F. Wilkins, Dr. R. E. Franklin and their co-workers at King's College, London. One of us (J. D. W.) has been aided by a fellowship from the National Foundation for Infantile Paralysis.

J. D. WATSON
F. H. C. CRICK
Medical Research Council Unit for the Study of the Molecular Structure of Biological Systems, Cavendish Laboratory, Cambridge, England

[1] Pauling L., and Corey, R. B., Nature 171, 346 (1953); Proc. U.S. Nat. Acad. Sci., 39, 84 (1953).

[2] Furberg, S., Acta Chem. Scand., 6, 634 (1952).

[3] Chargaff, E., for references see Zamenhof, S., Brawerman, G., and Chargaff, E., Biochim. et Biophys. Acta, 9, 402 (1952).

[4] Wyatt, G. R., J. Gen. Physiol., 36, 201 (1952).

[5] Astbury, W. T., Symp. Soc. Exp. Biol. 1, Nucleic Acid, 66 (Camb. Univ. Press, 1947).

[6] Wilkins, M. H. F., and Randall, J. T., Biochim. et Biophys. Acta, 10, 192 (1953).

Source: J. D. Watson and F. H. C. Crick, Molecular structure of nucleic acids: A structure of deoxyribose nucleic acid. Reprinted by permission from Nature, 171, No. 4356, pp. 737–738. Copyright © 1953 Macmillan Journals Limited.

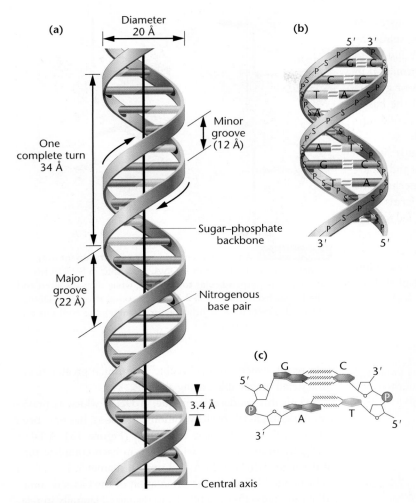

(a)

Diameter
20 Å

One
complete turn
34 Å

Minor
groove
(12 Å)

Major
groove
(22 Å)

Sugar–phosphate
backbone

Nitrogenous
base pair

3.4 Å

Central axis

(b)

5' 3'

3' 5'

(c)

G C
5' 3'
P
A T
3' 5'
P

FIGURE 12 (a) The DNA double helix as proposed by Watson and Crick. The ribbonlike strands constitute the sugar–phosphate backbones, and the horizontal rungs constitute the nitrogenous base pairs, of which there are 10 per complete turn. The major and minor grooves are shown. The solid vertical bar represents the central axis. (b) A detailed view labeled with the bases, sugars, phosphates, and hydrogen bonds of the helix. (c) A demonstration of the antiparallel nature of the helix and the horizontal stacking of the bases.

The specific A=T and G≡C base pairing is the basis for **complementarity**. This term describes the chemical affinity provided by hydrogen bonding between the bases. As we shall see, complementarity is very important in DNA replication and gene expression.

It is appropriate to inquire into the nature of a hydrogen bond and to ask whether it is strong enough to stabilize the helix. A **hydrogen bond** is a very weak electrostatic attraction between a covalently bonded hydrogen atom and an atom with an unshared electron pair. The hydrogen atom assumes a partial positive charge, while the unshared electron pair—characteristic of covalently bonded oxygen and nitrogen atoms—assumes a partial negative charge. These opposite charges are responsible for the weak chemical attractions. As oriented in the double helix, adenine forms two hydrogen bonds with thymine, and guanine forms three hydrogen bonds with cytosine. Although two or three individual hydrogen bonds are energetically very weak, 2000 to 3000 bonds in tandem (typical of two long polynucleotide chains) provide great stability to the helix.

Another stabilizing factor is the arrangement of sugars and bases along the axis. In the Watson–Crick model, the *hydrophobic* ("water-fearing") nitrogenous bases are stacked almost horizontally on the interior of the axis and are thus shielded from water. The *hydrophilic* ("water-loving") sugar–phosphate backbone is on the outside of the axis, where both components can interact with water. These molecular arrangements provide significant chemical stabilization to the helix.

ponents, the double helix could not be constructed easily if both chains ran parallel to one another.

The key to the model proposed by Watson and Crick is the specificity of base pairing. Chargaff's data suggested that the amounts of A equaled T and that the amounts of G equaled C. Watson and Crick realized that if A pairs with T and C pairs with G, this would account for these proportions and that such pairing could occur as a result of hydrogen bonding between base pairs [Figure 12(c)], providing the chemical stability necessary to hold the two chains together. Arranged in this way, both major and minor grooves become apparent along the axis. Further, a purine (A or G) opposite a pyrimidine (T or C) on each "rung of the spiral staircase" of the proposed double helix accounts for the 20 Å (2 nm) diameter suggested by X-ray diffraction studies.

A more recent and accurate analysis of the form of DNA that served as the basis for the Watson–Crick model has revealed a minor structural difference between the substance and the model. A precise measurement of the number of base pairs per turn has demonstrated a value of 10.4, rather than the 10.0 predicted by Watson and Crick. In the classic model, each base pair is rotated 36° around the helical axis relative to the adjacent base pair, but the new finding requires a rotation of 34.6°. This results in slightly more than 10 base pairs per 360° turn.

The Watson–Crick model had an immediate effect on the emerging discipline of molecular biology. Even in their initial 1953 article, the authors noted, "It has not escaped our notice that the specific pairing we have postulated immediately suggests a possible copying mechanism for the genetic material." Two months later, Watson and Crick pursued this idea in a second article in *Nature,* suggesting a specific mechanism of replication of DNA—the **semiconservative mode of replication.** The second article alluded to two new concepts: (1) the storage of genetic information in the sequence of the bases, and (2) the mutations or genetic changes that would result from an alteration of the bases. These ideas have received vast amounts of experimental support since 1953 and are now universally accepted.

Watson and Crick's synthesis of ideas was highly significant with regard to subsequent studies of genetics and biology. The nature of the gene and its role in genetic mechanisms could now be viewed and studied in biochemical terms. Recognition of their work, along with that of Wilkins, led to their receipt of the Nobel Prize in Physiology or Medicine in 1962. Unfortunately, Rosalind Franklin had died in 1958 at the age of 37, making her contributions ineligible for consideration since the award is not given posthumously. The Nobel Prize was to be one of many such awards bestowed for work in the field of molecular genetics.

ESSENTIAL POINT ■ ■ ■

As proposed by Watson and Crick, DNA exists in the form of a right-handed double helix composed of two long antiparallel polynucleotide chains held together by hydrogen bonds formed between complementary, nitrogenous base pairs.

NOW SOLVE THIS

2 In sea urchin DNA, which is double-stranded, 17.5 percent of the bases were shown to be cytosine (C). What percentages of the other three bases are expected to be present in this DNA?

■ **Hint** *This problem asks you to extrapolate from one measurement involving a unique DNA molecule to three other values characterizing the molecule. The key to its solution is to understand the base-pairing rules in the Watson–Crick model of DNA.*

7 Alternative Forms of DNA Exist

Under different conditions of isolation, several conformational forms of DNA have been recognized. At the time Watson and Crick performed their analysis, two forms—**A-DNA** and **B-DNA**—were known. Watson and Crick's analysis was based on X-ray studies of B-DNA performed by Franklin, which is present under aqueous, low-salt conditions and is believed to be the biologically significant conformation.

While DNA studies around 1950 relied on the use of X-ray diffraction, more recent investigations have been performed using **single-crystal X-ray analysis**. The earlier studies achieved limited resolution of about 5 Å, but single crystals diffract X rays at about 1 Å, near atomic resolution.

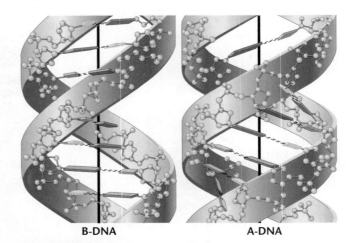

B-DNA   **A-DNA**

FIGURE 13   An artist's rendering showing the orientation of the base pairs of B-DNA and A-DNA. Note that in B-DNA the base pairs are perpendicular to the helix, while they are tilted and pulled away from the helix in A-DNA. In comparison to B-DNA, A-DNA also displays a slightly greater diameter and contains one more base pair per turn.

As a result, every atom is "visible" and much greater structural detail is available during analysis.

With this modern technique, A-DNA, which is prevalent under high-salt or dehydration conditions, has now been scrutinized. In comparison to B-DNA **(Figure 13)**, A-DNA is slightly more compact, with 11 bp in each complete turn of the helix, which is 23 Å (2.3 nm) in diameter. It is also a right-handed helix, but the orientation of the bases is somewhat different—they are tilted and displaced laterally in relation to the axis of the helix. As a result, the appearance of the major and minor grooves is modified. It seems doubtful that A-DNA occurs *in vivo* (under physiological conditions).

Other forms of DNA (e.g., C-, D-, E-, and most recently, P-DNA) are now known, but it is **Z-DNA** that has drawn the most attention. Discovered by Andrew Wang, Alexander Rich, and their colleagues in 1979 when they examined a small synthetic DNA fragment containing only G≡C base pairs, Z-DNA takes on the rather remarkable configuration of a *left-handed double helix*. Like A- and B-DNA, Z-DNA consists of two antiparallel chains held together by Watson–Crick base pairs. The three forms of DNA are shown in the chapter opening photograph. Beyond these characteristics, Z-DNA is quite different. The left-handed helix is 18 Å (1.8 nm) in diameter, contains 12 bp per turn, and assumes a zigzag conformation (hence its name). The major groove present in B-DNA is nearly eliminated in Z-DNA.

Speculation abounds over the possibility that regions of Z-DNA exist in the chromosomes of living organisms. The unique helical arrangement could provide an important recognition point for the interaction with other molecules. However, it is still not clear whether Z-DNA occurs *in vivo*.

Still other forms of DNA have been studied, including P-DNA, named after Linus Pauling. It is produced by artificial "stretching" of DNA, creating a longer, more narrow version with the phosphate groups on the interior.

8 The Structure of RNA Is Chemically Similar to DNA, but Single-Stranded

The structure of RNA molecules resembles DNA, with several important exceptions. Although RNA also has nucleotides linked with polynucleotide chains, the sugar ribose replaces deoxyribose, and the nitrogenous base uracil replaces thymine. Another important difference is that most RNA is single-stranded, although there are two important exceptions. First, RNA molecules sometimes fold back on themselves to form double-stranded regions of complementary base pairs. Second, some animal viruses that have RNA as their genetic material contain double-stranded helices.

As established earlier (see Figure 1), three major classes of cellular RNA molecules function during the expression of genetic information: **ribosomal RNA (rRNA), messenger RNA (mRNA),** and **transfer RNA (tRNA).** These molecules all originate as complementary copies of deoxyribonucleotide sequences of DNA. Because uracil replaces thymine in RNA, uracil is complementary to adenine during transcription and RNA base pairing.

Different RNAs are distinguished by their sedimentation behavior in a centrifugal field. Sedimentation behavior depends on a molecule's density, mass, and shape, and its measure is called the **Svedberg coefficient** (S). Although higher S values almost always designate molecules of greater molecular weight, the correlation is not direct; that is, a two-fold increase in molecular weight does not lead to a two-fold increase in S.

Ribosomal RNAs are generally the largest of these molecules and usually constitute about 80 percent of all RNA in the cell. Ribosomal RNAs are important structural components of **ribosomes,** which function as nonspecific workbenches where proteins are synthesized during translation. The various forms of rRNA found in prokaryotes and eukaryotes differ distinctly in size.

Messenger RNA molecules carry genetic information from the DNA of the gene to the ribosome. The mRNA molecules vary considerably in size, which reflects the variation in the size of the protein encoded by the mRNA as well as the gene serving as the template for transcription of mRNA.

Transfer RNA, the smallest class of these RNA molecules, carries amino acids to the ribosome during translation. Since more than one tRNA molecule interacts simultaneously with the ribosome, the molecule's smaller size facilitates these interactions.

Other unique RNAs exist that perform various genetic roles, especially in eukaryotes. For example, **telomerase RNA** is involved in DNA replication at the ends of chromosomes (the telomeres). **Small nuclear RNA (snRNA)** participates in processing mRNAs and **antisense RNA, microRNA (miRNA),** and **short interfering RNA (siRNA)** are involved in gene regulation.

ESSENTIAL POINT ■ ■ ■

The second category of nucleic acids important in genetic function is RNA, which is similar to DNA with the exceptions that it is usually single-stranded, the sugar ribose replaces the deoxyribose, and the pyrimidine uracil replaces thymine.

NOW SOLVE THIS

3 German measles results from an infection of the rubella virus, which can cause a multitude of health problems in newborns. What conclusions can you reach from a nucleic acid analysis of the virus that reveals an A + G/U + C ratio of 1.13?

■ **Hint** *This problem asks you to analyze information about the chemical composition of a nucleic acid serving as the genetic material of a virus. The key to its solution is to apply your knowledge of nucleic acid chemistry, in particular your understanding of base pairing.*

9 Many Analytical Techniques Have Been Useful during the Investigation of DNA and RNA

Since 1953, the role of DNA as the genetic material and the role of RNA in transcription and translation have been clarified through detailed analysis of nucleic acids. Several important methods of analysis are based on the unique nature of the hydrogen bond that is so integral to the structure of nucleic acids. For example, if DNA is subjected to heat, the double helix is denatured and unwinds. During unwinding, the viscosity of DNA decreases and UV absorption increases (called the **hyperchromic shift**). A melting profile, in which A260 is plotted against temperature, is shown for two DNA molecules in **Figure 14.** The midpoint of each curve is called the **melting temperature** T_m where 50 percent of the strands have unwound. The molecule with a higher T_m has a higher percentage of $G\equiv C$ base pairs than $A\!=\!T$ base pairs since $G\equiv C$ pairs share three hydrogen bonds compared to the two bonds between $A\!=\!T$ pairs.

The denaturation/renaturation of nucleic acids is the basis for one of the most useful techniques in molecular genetics—**molecular hybridization**. Provided that a reasonable degree of base complementarity exists between

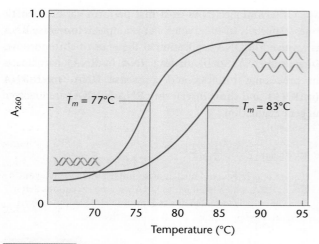

FIGURE 14 A melting profile shows the increase in UV absorption versus temperature (the hyperchromic effect) for two DNA molecules with different G≡C contents. The molecule with a melting point (T_m) of 83°C has a greater G≡C content than the molecule with a T_m of 77°C.

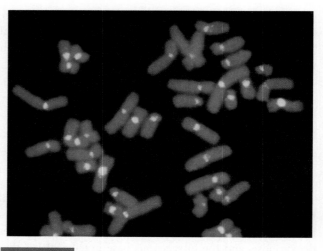

FIGURE 15 Fluorescent *in situ* hybridization (FISH) of human metaphase chromosomes. The probe, specific to centromeric DNA, produces a yellow fluorescence signal indicating hybridization. The red fluorescence is produced by propidium iodide counterstaining of chromosomal DNA.

any two nucleic acid strands, denaturation can be reversed whereby molecular hybridization is possible. Duplexes can be re-formed between DNA strands, even from different organisms, and between DNA and RNA strands. For example, an RNA molecule will hybridize with the segment of DNA from which it was transcribed. As a result, nucleic acid **probes** are often used to identify complementary sequences.

The technique can even be performed using the DNA present in chromosomal preparations as the "target" for hybrid formation. This process is called *in situ* **molecular hybridization.** Mitotic cells are first fixed to slides and then subjected to hybridization conditions. Single-stranded DNA or RNA is added (a probe), and hybridization is monitored. The nucleic acid that is added may be either radioactive or contain a fluorescent label to allow its detection. In the former case, autoradiography is used.

Figure 15 illustrates the use of a fluorescent label. A short fragment of DNA that is complementary to DNA in the chromosomes' centromere regions has been hybridized. Fluorescence occurs only in the centromere regions and thus identifies each one along its chromosome. Because fluorescence is used, the technique is known by the acronym **FISH** (**fluorescent *in situ* hybridization**). The use of this technique to identify chromosomal locations housing specific genetic information has been a valuable addition to geneticists' repertoire of experimental techniques.

Reassociation Kinetics and Repetitive DNA

In one extension of molecular hybridization procedures, the *rate* of reassociation of complementary single DNA strands is analyzed. This technique, **reassociation kinetics,** was first refined and studied by Roy Britten and David Kohne.

The DNA used in such studies is first fragmented into small pieces by shearing forces introduced during its isolation. The resultant DNA fragments have an average size of several hundred base pairs. The fragments are then dissociated into single strands (denatured) by heating, and when the temperature is lowered, reassociation is monitored. During reassociation, pieces of single-stranded DNA randomly collide. If they are complementary, a stable double strand is formed; if not, they separate and are free to encounter other DNA fragments. The process continues until all possible matches are made.

A great deal of information can be obtained from studies that compare the reassociation of DNA of different organisms. As genome size increases and there is more DNA, reassociation time is extended. Reassociation occurs more slowly in larger genomes because with random collisions it takes more time for all correct matches to be made.

When reassociation kinetics in eukaryotic organisms with much larger genome sizes was first studied, a surprising observation was made. Rather than requiring an extended reassociation time, *some* eukaryotic DNA reassociated even more rapidly than those derived from bacteria. The remaining DNA, as expected because of its complexity, took longer to reassociate.

Based on this observation, Britten and Kohne hypothesized that the rapidly reassociating fraction might represent **repetitive DNA sequences.** This interpretation would explain why these segments reassociate so rapidly—multiple copies of the same sequence are much more likely to make matches, thus reassociating more quickly than single copies. On the other hand, the remaining DNA segments consist of **unique DNA sequences,** present only once in the genome.

It is now known that repetitive DNA is prevalent in eukaryotic genomes and is key to our understanding of how genetic information is organized in chromosomes. Careful study has shown that various levels of repetition exist. In some cases, short DNA sequences are repeated over a million times. In other cases, longer sequences are repeated only a few times, or intermediate levels of sequence redundancy are present. In this chapter, we will simply point out that the discovery of repetitive DNA was one of the first clues that much of the DNA in eukaryotes is not contained in genes that encode proteins. We will develop and elaborate on this concept as we proceed with our coverage of the molecular basis of heredity.

Electrophoresis

Another technique essential to the analysis of nucleic acids is **electrophoresis.** This technique may be adapted to separate different-sized fragments of DNA and RNA chains and is invaluable in current research investigations in molecular genetics.

Electrophoresis separates the molecules in a mixture by causing them to migrate under the influence of an electric field. A sample is placed on a porous substance, such as a semisolid gel, which is then placed in a solution that conducts electricity. Mixtures of molecules with a similar charge–mass ratio but of different sizes will migrate at different rates through the gel based on their size. For example,

two polynucleotide chains of different lengths, such as 10 versus 20 nucleotides, are both negatively charged (based on the phosphate groups of the nucleotides) and will both move to the positively charged pole (the anode), but at different rates. Using a medium such as a **polyacrylamide gel** or an **agarose gel,** which can be prepared with various pore sizes, the *shorter chains migrate at a faster rate through the gel than larger chains* **(Figure 16)**. Once electrophoresis is complete, bands representing the variously sized molecules are identified either by autoradiography (if a component of the molecule is radioactive) or by use of a fluorescent dye that binds to nucleic acids. The resolving power is so great that polynucleotides that vary by just one nucleotide in length may be separated.

Electrophoretic separation of nucleic acids is at the heart of a variety of other commonly used research techniques. Of particular note are the various "blotting" techniques (e.g., Southern blots and Northern blots), as well as DNA sequencing methods.

ESSENTIAL POINT ▪ ▪ ▪

Various methods of analysis of nucleic acids, particularly molecular hybridization and electrophoresis, have led to studies essential to our understanding of genetic mechanisms.

FIGURE 16 Electrophoretic separation of a mixture of DNA fragments that vary in length. The photograph at the bottom right shows an autoradiogram derived from an agarose gel that reveals DNA bands.

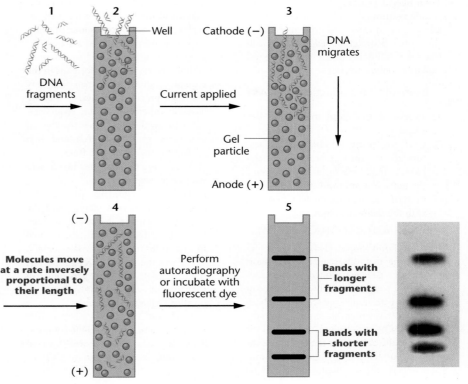

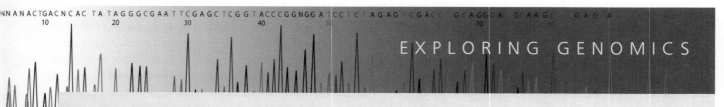

Introduction to Bioinformatics: BLAST

 Study Area: *Exploring Genomics*

I n this chapter, we focused on the structural details of DNA, the genetic material for living organisms. The explosion of DNA and protein sequence data that has occurred in the last 15 years has launched the field of *bioinformatics,* an interdisciplinary science that applies mathematics and computing technology to develop hardware and software for storing, sharing, comparing, and analyzing nucleic acid and protein sequence data.

A large number of sequence databases that make use of bioinformatics have been developed. An example is **GenBank** (www.ncbi.nlm.nih.gov/Genbank/index.html), which is the National Institutes of Health sequence database. This global resource, with access to databases in Europe and Japan, currently contains more than 148 billion base pairs of sequence data!

Here, we use an NCBI application called **BLAST, Basic Local Alignment Search Tool.** BLAST is an invaluable program for searching through GenBank and other databases to find DNA and protein-sequence similarities between cloned substances. It has many additional functions that we will explore in other exercises.

■ Exercise I – Introduction to BLAST

1. Access BLAST from the NCBI Web site at www.ncbi.nlm.nih.gov/BLAST.
2. Click on "nucleotide blast." This feature allows you to search DNA databases to look for a similarity between a sequence you enter and other sequences in the database. Do a nucleotide search with the following sequence:

   CCAGAGTCCAGCTGCTGCT CATA CTACTGATACTGCTGGG

3. Imagine that this sequence is a short part of a gene you cloned in your laboratory. You want to know if this gene or others with similar sequences have been discovered. Enter this sequence into the "Enter Query Sequence" text box at the top of the page. Near the bottom of the page, under the "Program Selection" category, choose "blastn"; then click on

the "BLAST" button at the bottom of the page to run the search. It may take several minutes for results to be available because BLAST is using powerful algorithms to scroll through billions of bases of sequence data! A new page will appear with the results of your search.

4. Near the top of this page you will see a table showing significant matches to the sequence you searched with (called the query sequence). BLAST determines significant matches based on statistical measures that consider the length of the query sequence, the number of matches with sequences in the database, and other factors. Significant alignments, regions of significant similarity in the query and subject sequences, typically have E values less than 1.0.
5. The top part of the table lists matches to transcripts (mRNA sequences), and the lower part lists matches to genomic DNA sequences, in order of highest to lowest number of matches. Use the "Links" column to the far right of this table to explore gene and chromosome databases relevant to the matched sequences.
6. Alignments are indicated by horizontal lines. BLAST adjusts for gaps in the sequences, that is, for areas that may not align precisely because of missing bases in otherwise similar sequences. Scroll below the table to see the aligned sequences from this search, and then answer the following questions:

   a. What were the top three matches to your query sequence?

   b. For each alignment, BLAST also indicates the percent *identity* and the number of gaps in the match between the query and subject sequences (shown in the column under "Max ident"). What was the percent identity for the top three matches? What percentage of each aligned sequence showed gaps indicating sequence differences?

   c. Click on the links for the first matched sequence (far-right column). These will take you to a wealth of information, including the size of the

sequence; the species it was derived from; a PubMed-linked chronology of research publications pertaining to this sequence; the complete sequence; and if the sequence encodes a polypeptide, the predicted amino acid sequence coded by the gene. Skim through the information presented for this gene. What is the gene's function?

7. A BLAST search can also be done by entering the *accession number* for a sequence, which is a unique identifying number assigned to a sequence before it can be put into a database. For example, search with the accession number NM_007305. What did you find?
8. Run a BLAST search using the sequences or accession numbers listed below. In each case, after entering the accession number or sequence in the "Enter Query Sequence" box, go to the "Choose Search Set" box and click on the "Others" button for database.

   Then go to the "Program Selection" box and click "megablast" before running your search. These features will allow you to align the query sequence with similar genes from a number of other species. When each search is completed, explore the information BLAST provides so that you can identify and learn about the gene encoded by the sequence.

   a. NM_001006650. What is the top sequence that aligns with the query sequence of this accession number and shows 100 percent sequence identity?

   b. DQ991619. What gene is encoded by this sequence?

   c. NC_007596. What living animal has a sequence similar to this one?

CASE STUDY » Zigs and zags of the smallpox virus

Smallpox, a once highly lethal contagious disease, has been eradicated worldwide. However, research continues with stored samples of variola, the smallpox virus, because it is a potential weapon in bioterrorism. Human cells protect themselves from the variola virus (and other viruses) by activating genes that encode protective proteins. It has recently been discovered that in response to variola, human cells create small transitory stretches of Z-DNA at sites that regulate these genes. The smallpox virus can bypass this cellular defense mechanism by specifically targeting the segments of Z-DNA and inhibiting the synthesis of the protective proteins. This discovery raises some interesting questions:

1. What is unique about Z-DNA that might make it a specific target during viral infection?
2. How might the virus target host-cell Z-DNA formation to block the synthesis of antiviral proteins?
3. To study the interaction between viral proteins and Z-DNA, how could Z-DNA-forming DNA be synthesized in the lab?
4. How could this research lead to the development of drugs to combat infection by variola and related viruses?

INSIGHTS AND SOLUTIONS

This chapter recounts some of the initial experimental analyses that launched the era of molecular genetics. Quite fittingly, then, our "Insights and Solutions" section shifts its emphasis from problem solving to experimental rationale and analytical thinking.

1. Based strictly on the transformation analysis of Avery, MacLeod, and McCarty, what objection might be made to the conclusion that DNA is the genetic material? What other conclusion might be considered?

Solution: Based solely on their results, we could conclude that DNA is essential for transformation. However, DNA might have been a substance that caused capsular formation by converting nonencapsulated cells *directly* to cells with a capsule. That is, DNA may simply have played a catalytic role in capsular synthesis, leading to cells that display smooth, type III colonies.

2. What observations argue against this objection?

Solution: First, transformed cells pass the trait on to their progeny cells, thus supporting the conclusion that DNA is responsible for heredity, not for the direct production of polysaccharide coats. Second, subsequent transformation studies over the next five years showed that other traits, such as antibiotic resistance, could be transformed. Therefore, the transforming factor has a broad general effect, not one specific to polysaccharide synthesis.

3. If RNA were the universal genetic material, how would this have affected the Avery experiment and the Hershey–Chase experiment?

Solution: In the Avery experiment, ribonuclease (RNase), rather than deoxyribonuclease (DNase), would have eliminated transformation. Had this occurred, Avery and his colleagues would have concluded that RNA was the transforming factor. Hershey and Chase would have obtained identical results, since ^{32}P would also label RNA but not protein.

PROBLEMS AND DISCUSSION QUESTIONS

 For activities, animations, and review quizzes, go to the study area at www.masteringgenetics.com

HOW DO WE KNOW?

1. In this chapter, we have focused on DNA, the molecule that stores genetic information in all living things. In particular, we discussed its structure and delved into how we analyze this molecule. Based on your knowledge of these topics, answer several fundamental questions:

    (a) How were we able to determine that DNA, and not some other molecule, serves as the genetic material in bacteria, bacteriophages, and eukaryotes?

    (b) How do we know that the structure of DNA is in the form of a right-handed double-helical molecule?

    (c) How do we know that in DNA G pairs with C and that A pairs with T as complementary strands are formed?

    (d) How do we know that repetitive DNA sequences exist in eukaryotes?

2. The functions ascribed to the genetic material are replication, expression, storage, and mutation. What does each of these terms mean?

3. Discuss the reasons why proteins were generally favored over DNA as the genetic material before 1940. What was the role of the tetranucleotide hypothesis in this controversy?

4. Contrast the various contributions made to our understanding of transformation by Griffith, Alloway, and Avery.

5. When Avery and his colleagues had obtained what was concluded to be the transforming factor from the IIIS virulent cells, they treated the fraction with proteases, ribonuclease, and deoxyribonuclease, followed by the assay for retention or loss of transforming ability. What were the purpose and results of these experiments? What conclusions were drawn?

6. Why were ^{32}P and ^{35}S chosen in the Hershey–Chase experiment? Discuss the rationale and conclusions of this experiment.

7. Does the design of the Hershey–Chase experiment distinguish between DNA and RNA as the molecule serving as the genetic material? Why or why not?

8. What observations are consistent with the conclusion that DNA serves as the genetic material in eukaryotes? List and discuss them.

9. What are the exceptions to the general rule that DNA is the genetic material in all organisms? What evidence supports these exceptions?

10. Draw the chemical structure of the three components of a nucleotide, and then link them together. What atoms are removed from the structures when the linkages are formed?

11. How are the carbon and nitrogen atoms of the sugars, purines, and pyrimidines numbered?

12. Adenine may also be named 6-amino purine. How would you name the other four nitrogenous bases, using this alternative system? (O is oxy, and CH_3 is methyl.)

13. Draw the chemical structure of a dinucleotide composed of A and G. Opposite this structure, draw the dinucleotide composed of T and C in an antiparallel (or upside-down) fashion. Form the possible hydrogen bonds.

14. Describe the various characteristics of the Watson–Crick double helix model for DNA.

15. What evidence did Watson and Crick have at their disposal in 1953? What was their approach in arriving at the structure of DNA?

16. What might Watson and Crick have concluded, had Chargaff's data from a single source indicated the following base composition?

	A	T	G	C
%	29	19	21	31

Why would this conclusion be contradictory to Wilkins and Franklin's data?

17. How do covalent bonds differ from hydrogen bonds? Define base complementarity.

18. List three main differences between DNA and RNA.

19. What are the three major types of RNA molecules? How is each related to the concept of information flow?

20. What component of the nucleotide is responsible for the absorption of ultraviolet light? How is this technique important in the analysis of nucleic acids?

21. What is the physical state of DNA after being denatured by heat?

22. What is the hyperchromic effect? How is it measured? What does T_m imply?

23. Why is T_m related to base composition?

24. What is the chemical basis of molecular hybridization?

25. What did the Watson–Crick model suggest about the replication of DNA?

26. A genetics student was asked to draw the chemical structure of an adenine- and thymine-containing dinucleotide derived from DNA. His answer is shown below. The student made more than six major errors. One of them is circled, numbered 1, and explained. Find five others. Circle them, number them 2–6, and briefly explain each by following the example given.

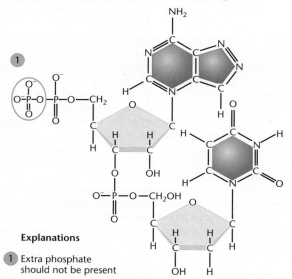

Explanations

1 Extra phosphate should not be present

27. A primitive eukaryote was discovered that displayed a unique nucleic acid as its genetic material. Analysis revealed the following observations:

(a) X-ray diffraction studies display a general pattern similar to DNA, but with somewhat different dimensions and more irregularity.

(b) A major hyperchromic shift is evident upon heating and monitoring UV absorption at 260 nm.

(c) Base-composition analysis reveals four bases in the following proportions:

Adenine = 8%   Hypoxanthine = 18%

Guanine = 37%   Xanthine = 37%

(d) About 75 percent of the sugars are deoxyribose, whereas 25 percent are ribose. Attempt to solve the structure of this molecule by postulating a model that is consistent with the foregoing observations.

28. One of the most common spontaneous lesions that occurs in DNA under physiological conditions is the hydrolysis of the amino group of cytosine, converting it to uracil. What would be the effect on DNA structure if a uracil group replaced cytosine?

29. In some organisms, cytosine is methylated at carbon 5 of the pyrimidine ring after it is incorporated into DNA. If a 5-methyl cytosine is then hydrolyzed, as described in Problem 28, what base will be generated?

30. *Newsdate: March 1, 2015.* A unique creature has been discovered during exploration of outer space. Recently, its genetic material has been isolated and analyzed, and has been found to be similar in some ways to DNA in chemical makeup. It contains in abundance the 4-carbon sugar erythrose and a molar equivalent of phosphate groups. In addition, it contains six nitrogenous bases: adenine (A), guanine (G), thymine (T), cytosine (C), hypoxanthine (H), and xanthine (X). These bases exist in the following relative proportion:

$$A = T = H \quad \text{and} \quad C = G = X$$

X-ray diffraction studies have established a regularity in the molecule and a constant diameter of about 30 Å.

Together, these data have suggested a model for the structure of this molecule. (a) Propose a general model of this molecule, and briefly describe it. (b) What base-pairing properties must exist for H and for X in the model? (c) Given the constant diameter of 30 Å, do you think *either* (i) both H and X are purines or both pyrimidines, *or* (ii) one is a purine and one is a pyrimidine?

31. You are provided with DNA samples from two newly discovered bacterial viruses. Based on the various analytical techniques discussed in this chapter, construct a research protocol that would be useful in characterizing and contrasting the DNA of both viruses. Indicate the type of information you hope to obtain for each technique included in the protocol.

32. During electrophoresis, DNA molecules can easily be separated according to size because all DNA molecules have the same charge–mass ratio and the same shape (long rod). Would you expect RNA molecules to behave in the same manner as DNA during electrophoresis? Why or why not?

33. Assume that you are interested in separating short (25–40 nucleotides) DNA molecules from a pool of longer molecules in the 900–1000 nucleotide range. You have two recipes for making your polyacrylamide gels; one recipe uses 12 percent acrylamide and would be considered a "hard gel," while the other uses 4 percent acrylamide and would be considered a loose gel. Which recipe would you consider using and why?

SOLUTIONS TO SELECTED PROBLEMS AND DISCUSSION QUESTIONS

Answers to Now Solve This

1. In theory, the general design would be appropriate in that some substance, if labeled, would show up in the progeny of transformed bacteria. However, since the amount of transforming DNA is extremely small compared to the genomic DNA of the recipient bacterium and its progeny, it would be technically difficult to assay for the labeled nucleic acid. In addition, it would be necessary to know that the small stretch of DNA that caused the genetic transformation was actually labeled.

2. Guanine = 17.5%, Adenine and Thymine both = 32.5%.

3. Assuming the value of 1.13 is statistically different from 1.00, one can conclude that rubella is a single-stranded RNA virus.

Solutions to Problems and Discussion Questions

8. The early evidence would be considered indirect in that at no time was there an experiment, like transformation in bacteria, in which genetic information in one organism was transferred to another using DNA. Rather, by comparing DNA content in various cell types (sperm and somatic cells) and observing that the *action* and *absorption* spectra of ultraviolet light were correlated, DNA was considered to be the genetic material. This suggestion was supported by the fact that DNA was shown to be the genetic material in bacteria and some phage. Direct evidence that DNA is the genetic material comes from a variety of observations, including gene transfer that has been facilitated by recombinant DNA techniques.

12. Examine the structures of the bases in the text. The other bases would be named as follows:

Guanine:	2-amino-6-oxypurine
Cytosine:	2-oxy-4-aminopyrimidine
Thymine:	2,4-dioxy-5-methylpyrimidine
Uracil:	2,4-dioxypyrimidine

16. Because in double-stranded DNA, A = T and G = C (within limits of experimental error), the data presented would have indicated a lack of pairing of these bases in favor of a single-stranded structure or some other nonhydrogen-bonded structure.

    Alternatively, from the data it would appear that A = C and T = G, which would negate the chance for typical hydrogen bonding since opposite charge relationships do not exist. Therefore, it is quite unlikely that a tight helical structure would form at all.

20. The nitrogenous bases of nucleic acids (nucleosides, nucleotides, and single- and double-stranded polynucleotides) absorb UV light maximally at wavelengths of 254 to 260 nm. Using this phenomenon, one can often determine the presence and concentration of nucleic acids in a mixture. Since proteins absorb UV light maximally at 280 nm, this is a relatively simple way of dealing with mixtures of biologically important molecules.

    UV absorption is greater in single-stranded molecules (hyperchromic shift) than in double-stranded structures. Therefore, by applying denaturing conditions, one can easily determine whether a nucleic acid is in the single- or double-stranded form. In addition, A-T rich DNA denatures more readily than G-C rich DNA. Therefore, one can estimate base content by denaturation kinetics.

22. *A hyperchromic effect* is the increased absorption of UV light as double-stranded DNA (or RNA, for that matter) is converted to single-stranded DNA. As illustrated in the text, the change in absorption is quite significant, with a structure of higher G-C content *melting* at a higher temperature than an A-T rich nucleic acid. If one monitors the UV absorption with a spectrophotometer during the melting process, the hyperchromic shift can be observed. The T_m is the point on the profile (temperature) at which half (50 percent) of the sample is denatured.

24. The reassociation of separate complementary strands of a nucleic acid, either DNA or RNA, is based on hydrogen bonds forming between A-T (or U) and G-C.

26. (1) As shown, the extra phosphate is not normally expected.
    (2) In the adenine ring, a nitrogen is at position 8 rather than position 9.
    (3) The bond from the C-1′ to the sugar should form with the N at position 9 (N-9) of the adenine.
    (4) The dinucleotide is a "deoxy" form; therefore, each C-2′ should not have a hydroxyl group. Notice the hydroxyl group at C-2′ on the sugar of the adenylic acid.
    (5) At the C-5 position on the thymine residue, there should be a methyl group.
    (6) There are too many bonds at the N-3 position on the thymine.
    (7) There are too few bonds at the C-5 of thymine.

28. Since cytosine pairs with guanine and uracil pairs with adenine, the result would be a base substitution of G:C to A:T after rounds of replication.

30. Without knowing the exact bonding characteristics of hypoxanthine or xanthine, it may be difficult to predict the likelihood of each pairing type. It is likely that both are of the same class (purine or pyrimidine) because the names of the molecules indicate a similarity. In addition, the diameter of the structure is constant, which, under the model to follow, would be expected. In fact, hypoxanthine and xanthine are both purines.

    Because there are equal amounts of A, T, and H, one could suggest that they are hydrogen bonded to each other; the same may be said for C, G, and X. Given the molar equivalence of erythrose and phosphate, an alternating sugar-phosphate-sugar backbone as in "earth-type" DNA would be acceptable. A model of a triple helix would be acceptable, since the diameter is constant. Given the chemical similarities to "earth-type" DNA, it is probable that the unique creature's DNA follows the same structural plan.

32. In comparing DNA migration to RNA, even though RNA molecules have the same charge-to-mass ratios, they can exist in a variety of shapes. Complementary intrastrand base pairing can make more compact structures compared to the more relaxed, open conformation. During electrophoresis, compact molecules migrate faster than relaxed, open structures. For electrophoretic size comparisons, RNA molecules must be denatured to eliminate secondary structural variables.

CREDITS

Credits are listed in order of appearance.

Photo

CO, Ken Eward/Science Source/Photo Researchers, Inc.; F-2 (a, b) Bruce Iverson, Photomicrography; F-4, Oliver Meckes/Max-Planck-Institut-Tubingen/Photo Researchers, Inc.; F-11, M.H.F. Wilkins; F-15, Ventana Medical Systems Inc.; F-16, Dr. William S. Klug

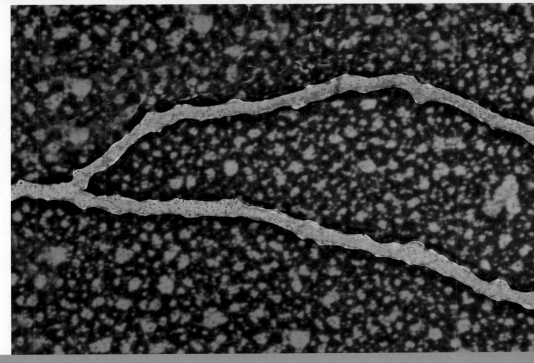

Transmission electron micrograph of human DNA from a HeLa cell, illustrating a replication fork characteristic of active DNA replication.

DNA Replication and Recombination

CHAPTER CONCEPTS

- Genetic continuity between parental and progeny cells is maintained by semiconservative replication of DNA, as predicted by the Watson–Crick model.

- Semiconservative replication uses each strand of the parent double helix as a template, and each newly replicated double helix includes one "old" and one "new" strand of DNA.

- DNA synthesis is a complex but orderly process, occurring under the direction of a myriad of enzymes and other proteins.

- DNA synthesis involves the polymerization of nucleotides into polynucleotide chains.

- DNA synthesis is similar in prokaryotes and eukaryotes, but more complex in eukaryotes.

- In eukaryotes, DNA synthesis at the ends of chromosomes (telomeres) poses a special problem, overcome by a unique RNA-containing enzyme, telomerase.

- Genetic recombination, an important process leading to the exchange of segments between DNA molecules, occurs under the direction of a group of enzymes.

Following Watson and Crick's proposal for the structure of DNA, scientists focused their attention on how this molecule is replicated. Replication is an essential function of the genetic material and must be executed precisely if genetic continuity between cells is to be maintained following cell division. It is an enormous, complex task. Consider for a moment that more than 3×10^9 (3 billion) base pairs exist within the human genome. To duplicate faithfully the DNA of just one of these chromosomes requires a mechanism of extreme precision. Even an error rate of only 10^{-6} (one in a million) will still create 3000 errors (obviously an excessive number) during each replication cycle of the genome. Although it is not error free, and much of evolution would not have occurred if it were, an extremely accurate system of DNA replication has evolved in all organisms.

As Watson and Crick noted in the concluding paragraph of their 1953 paper), their proposed model of the double helix provided the initial insight into how replication occurs. Called semiconservative replication, this mode of DNA duplication was soon to receive strong support from numerous studies of viruses, prokaryotes, and eukaryotes. Once the general *mode* of replication was clarified, research to determine the precise details of the *synthesis* of DNA intensified. What has since been discovered is that numerous enzymes and other proteins are needed to copy a DNA helix. Because of the complexity of the chemical events during synthesis, this subject remains an extremely active area of research.

In this chaptertext, we will discuss the general mode of replication, as well as the specific details of DNA synthesis. The research leading to such knowledge is another link in our understanding of life processes at the molecular level.

From Chapter 10 of *Essentials of Genetics, Eighth edition,* William S. Klug, Michael R. Cummings, Charlotte A. Spencer, Michael A. Palladino. ©2013 by Pearson Education, Inc. All rights reserved.

1 DNA Is Reproduced by Semiconservative Replication

Watson and Crick recognized that, because of the arrangement and nature of the nitrogenous bases, each strand of a DNA double helix could serve as a template for the synthesis of its complement **(Figure 1)**. They proposed that, if the helix were unwound, each nucleotide along the two parent strands would have an affinity for its complementary nucleotide. Thecomplementarity is due to the potential hydrogen bonds that can be formed. If thymidylic acid (T) were present, it would "attract" adenylic acid (A); if guanidylic acid (G) were present, it would attract cytidylic acid (C); likewise, A would attract T, and C would attract G. If these nucleotides were then covalently linked into polynucleotide chains along both templates, the result would be the production of two identical double strands of DNA. Each replicated DNA molecule would consist of one "old" and one "new" strand, hence the reason for the name **semiconservative replication.**

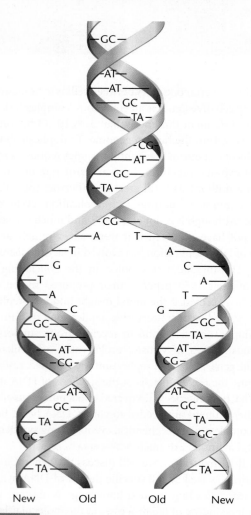

FIGURE 1 Generalized model of semiconservative replication of DNA. New synthesis is shown in blue.

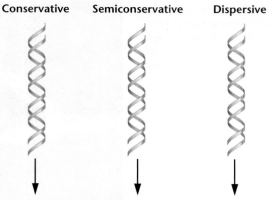

One round of replication—new synthesis is shown in blue

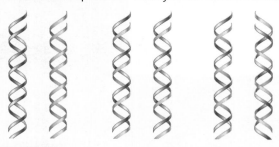

FIGURE 2 Results of one round of replication of DNA for each of the three possible modes by which replication could be accomplished.

Two other theoretical modes of replication are possible that also rely on the parental strands as a template **(Figure 2)**. In **conservative replication,** complementary polynucleotide chains are synthesized as described earlier. Following synthesis, however, the two newly created strands then come together and the parental strands reassociate. The original helix is thus "conserved."

In the second alternative mode, called **dispersive replication,** the parental strands are dispersed into two new double helices following replication. Hence, each strand consists of both old and new DNA. This mode would involve cleavage of the parental strands during replication. It is the most complex of the three possibilities and is therefore considered to be least likely to occur. It could not, however, be ruled out as an experimental model. Figure 2 shows the theoretical results of a single round of replication by each of the three different modes.

The Meselson–Stahl Experiment

In 1958, Matthew Meselson and Franklin Stahl published the results of an experiment providing strong evidence that semiconservative replication is the mode used by bacterial cells to produce new DNA molecules. They grew *E. coli* cells for many generations in a medium that had $^{15}NH_4Cl$ (ammonium chloride) as the only nitrogen source. A "heavy" isotope of nitrogen, ^{15}N contains one more neutron than the naturally occurring ^{14}N isotope; thus, molecules containing

^{15}N are more dense than those containing ^{14}N. Unlike radioactive isotopes, ^{15}N is stable. After many generations in this medium, almost all nitrogen-containing molecules in the *E. coli* cells, including the nitrogenous bases of DNA, contained the heavier isotope.

Critical to the success of this experiment, DNA containing ^{15}N can be distinguished from DNA containing ^{14}N. The experimental procedure involves the use of a technique referred to as **sedimentation equilibrium centrifugation** (also called buoyant density gradient centrifugation). Samples are forced by centrifugation through a density gradient of a heavy metal salt, such as cesium chloride. Molecules of DNA will reach equilibrium when their density equals the density of the gradient medium. In this case, ^{15}N-DNA will reach this point at a position closer to the bottom of the tube than will ^{14}N-DNA.

In this experiment **(Figure 3)**, uniformly labeled ^{15}N cells were transferred to a medium containing only ^{14}NH$_4$Cl. Thus, all "new" synthesis of DNA during replication contained only the "lighter" isotope of nitrogen. The time of transfer to the new medium was taken as time zero $t = 0$. The *E. coli* cells were allowed to replicate over several generations, with cell samples removed after each replication cycle. DNA

was isolated from each sample and subjected to sedimentation equilibrium centrifugation.

After one generation, the isolated DNA was present in only a single band of intermediate density—the expected result for semiconservative replication in which each replicated molecule was composed of one new ^{14}N-strand and one old ^{15}N-strand **(Figure 4)**. This result was not consistent with the prediction of conservative replication, in which two distinct bands would occur; thus this mode may be ruled out.

After two cell divisions, DNA samples showed two density bands—one intermediate band and one lighter band corresponding to the ^{14}N position in the gradient. Similar results occurred after a third generation, except that the proportion of the lighter band increased. This was again consistent with the interpretation that replication is semiconservative.

You may have realized that a molecule exhibiting intermediate density is also consistent with dispersive replication. However, Meselson and Stahl also ruled out this mode of replication on the basis of two observations. First, after the first generation of replication in an ^{14}N-containing medium, they isolated the hybrid molecule and heat denatured it. When the densities of the single strands of the hybrid were determined, they exhibited *either* an ^{15}N profile *or* an ^{14}N

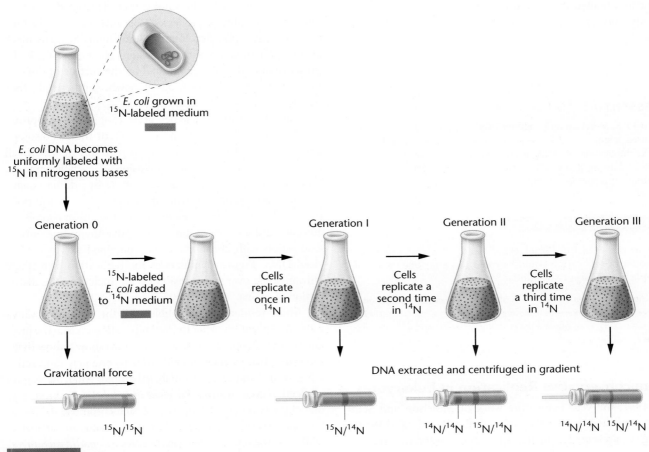

FIGURE 3 The Meselson–Stahl experiment.

FIGURE 4 The expected results of two generations of semiconservative replication in the Meselson–Stahl experiment.

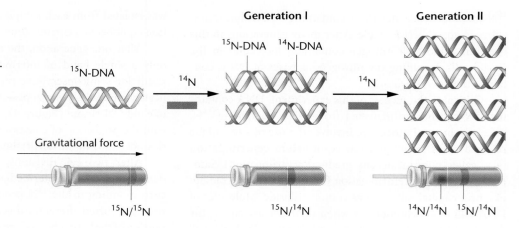

profile, but *not* an intermediate density. This observation is consistent with the semiconservative mode but inconsistent with the dispersive mode.

Furthermore, if replication were dispersive, *all* generations after $t = 0$ would demonstrate DNA of an intermediate density. In each generation after the first, the ratio of $^{15}N/^{14}N$ would decrease, and the hybrid band would become lighter and lighter, eventually approaching the ^{14}N band. This result was not observed. The Meselson–Stahl experiment provided conclusive support for semiconservative replication in bacteria and tended to rule out both the conservative and dispersive modes.

ESSENTIAL POINT ■ ■ ■

In 1958, Meselson and Stahl resolved the question of which of three potential modes of replication is utilized by *E. coli* during the duplication of DNA in favor of semiconservative replication, showing that newly synthesized DNA consists of one old strand and one new strand.

NOW SOLVE THIS

1 In the Meselson–Stahl experiment, which of the three modes of replication could be ruled out after one round of replication? after two rounds?

■ **Hint** *This problem involves an understanding of the nature of the experiment as well as the difference between the three possible modes of replication. The key to its solution is to determine which mode will not create "hybrid" helices after one round of replication.*

Semiconservative Replication in Eukaryotes

In 1957, the year before the work of Meselson and Stahl was published, J. Herbert Taylor, Philip Woods, and Walter Hughes presented evidence that semiconservative replication also occurs in eukaryotic organisms. They experimented

with root tips of the broad bean *Vicia faba*, which are an excellent source of dividing cells. These researchers were able to monitor the process of replication by labeling DNA with ^{3}H-thymidine, a radioactive precursor of DNA, and performing autoradiography.

Autoradiography is a common technique that, when applied cytologically, pinpoints the location of a radioisotope in a cell. In this procedure, a photographic emulsion is placed over a histological preparation containing cellular material (root tips, in this experiment), and the preparation is stored in the dark. The slide is then developed, much as photographic film is processed. Because the radioisotope emits energy, upon development the emulsion turns black at the approximate point of emission. The end result is the presence of dark spots or "grains" on the surface of the section, identifying the location of newly synthesized DNA within the cell.

Taylor and his colleagues grew root tips for approximately one generation in the presence of the radioisotope and then placed them in unlabeled medium in which cell division continued. At the conclusion of each generation, they arrested the cultures at metaphase by adding colchicine (a chemical derived from the crocus plant that poisons the spindle fibers) and then examined the chromosomes by autoradiography. They found radioactive thymidine only in association with chromatids that contained newly synthesized DNA. **Figure 5** illustrates the replication of a single chromosome over two division cycles, including the distribution of grains.

These results are compatible with the semiconservative mode of replication. After the first replication cycle in the presence of the isotope, both sister chromatids show radioactivity, indicating that each chromatid contains one new radioactive DNA strand and one old unlabeled strand. After the second replication cycle, *which takes place in unlabeled medium,* only one of the two sister chromatids of each chromosome should be radioactive because half of the parent strands are unlabeled. With only the minor exceptions of *sister chromatid exchanges*, this result was observed.

FIGURE 5 The Taylor–Woods–Hughes experiment, demonstrating the semiconservative mode of replication of DNA in the root tips of *Vicia faba*. A portion of the plant is shown in the top photograph. (a) An unlabeled chromosome proceeds through the cell cycle in the presence of ^{3}H-thymidine. As it enters mitosis, both sister chromatids of the chromosome are labeled, as shown, by autoradiography. After a second round of replication (b), this time in the absence of ^{3}H-thymidine, only one chromatid of each chromosome is expected to be surrounded by grains. Except where a reciprocal exchange has occurred between sister chromatids (c), the expectation was upheld. The micrographs are of the actual autoradiograms obtained in the experiment.

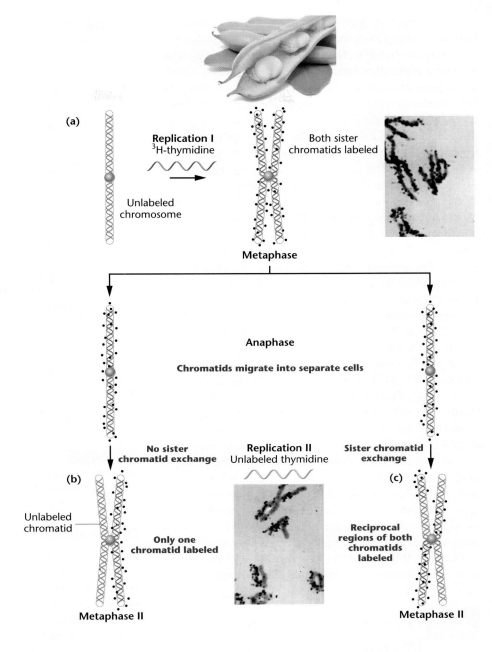

Together, the Meselson–Stahl experiment and the experiment by Taylor, Woods, and Hughes soon led to the general acceptance of the semiconservative mode of replication. Later studies with other organisms reached the same conclusion and also strongly supported Watson and Crick's proposal for the double helix model of DNA.

ESSENTIAL POINT

Taylor, Woods, and Hughes demonstrated semiconservative replication in eukaryotes using the root tips of the broad bean as the source of dividing cells.

Origins, Forks, and Units of Replication

To enhance our understanding of semiconservative replication, let's briefly consider a number of relevant issues. The first concerns the **origin of replication.** Where along the chromosome is DNA replication initiated? Is there only a single origin, or does DNA synthesis begin at more than one point? Is any given point of origin random, or is it located at a specific region along the chromosome? Second, once replication begins, does it proceed in a single direction or in both directions away from the origin? In other words, is replication **unidirectional** or **bidirectional**?

To address these issues, we need to introduce two terms. First, at each point along the chromosome where replication is occurring, the strands of the helix are unwound, creating what is called a **replication fork.** Such a fork will initially appear at the point of origin of synthesis and then move along the DNA duplex as replication proceeds. If replication is bidirectional, two such forks will be present, migrating in opposite directions away from the origin. The second term refers to the length of DNA that is replicated following one initiation event at a single origin. This is a unit referred to as the **replicon.**

The evidence is clear regarding the origin and direction of replication. John Cairns tracked replication in *E. coli,* using radioactive precursors of DNA synthesis and autoradiography. He was able to demonstrate that in *E. coli* there is only a single region, called *oriC,* where replication is initiated. The presence of only a single origin is characteristic of bacteria, which have only one circular chromosome. Since DNA synthesis in bacteriophages and bacteria originates at a single point, the entire chromosome constitutes one replicon. In *E. coli,* the replicon consists of the entire genome of 4.6 Mb (4.6 million base pairs).

Figure 6 illustrates Cairns's interpretation of DNA replication in *E. coli.* This interpretation and the accompanying micrograph do not answer the question of unidirectional versus bidirectional synthesis. However, other results, derived from studies of bacteriophage lambda, have demonstrated that replication is bidirectional, moving away from *oriC* in both directions. Figure 6 therefore interprets Cairns's work with that understanding. Bidirectional replication creates two replication forks that migrate farther and farther apart as replication proceeds. These forks eventually merge, as semiconservative replication of the entire chromosome is completed, at a termination region, called *ter.*

Later in this chapter,text we will see that in eukaryotes, each chromosome contains multiple points of origin.

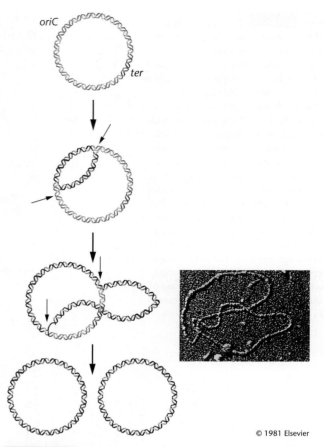

© 1981 Elsevier

FIGURE 6 Bidirectional replication of the *E. coli* chromosome. The thin black arrows identify the advancing replication forks. The micrograph is of a bacterial chromosome in the process of replication, comparable to the figure next to it.

2 DNA Synthesis in Bacteria Involves Five Polymerases, as Well as Other Enzymes

To say that replication is semiconservative and bidirectional describes the overall *pattern* of DNA duplication and the association of finished strands with one another once synthesis is completed. However, it says little about the more complex issue of how the actual *synthesis* of long complementary polynucleotide chains occurs on a DNA template. Like most questions in molecular biology, this one was first studied using microorganisms. Research on DNA synthesis began about the same time as the Meselson–Stahl work, and the topic is still an active area of investigation. What is most apparent in this research is the tremendous complexity of the biological synthesis of DNA.

DNA Polymerase I

Studies of the enzymology of DNA replication were first reported by Arthur Kornberg and colleagues in 1957. They isolated an enzyme from *E. coli* that was able to direct DNA synthesis in a cell-free (*in vitro*) system. The enzyme is called **DNA polymerase I,** because it was the first of several similar enzymes to be isolated.

Kornberg determined that there were two major requirements for *in vitro* DNA synthesis under the direction of DNA polymerase I: (1) all four deoxyribonucleoside triphosphates (dNTPs) and (2) template DNA. If any one of the four deoxyribonucleoside triphosphates was omitted from the reaction, no measurable synthesis occurred. If derivatives of these precursor molecules other than the nucleoside triphosphate were used (nucleotides or nucleoside diphosphates), synthesis also did not occur. If no template DNA was added, synthesis of DNA occurred but was reduced greatly.

Most of the synthesis directed by Kornberg's enzyme appeared to be exactly the type required for semiconservative replication. The reaction is summarized in **Figure 7,** which depicts the addition of a single nucleotide. The enzyme has

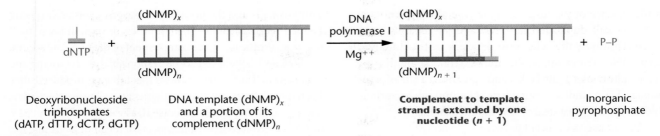

FIGURE 7 The chemical reaction catalyzed by DNA polymerase I. During each step, a single nucleotide is added to the growing complement of the DNA template using a nucleoside triphosphate as the substrate. The release of inorganic pyrophosphate drives the reaction energetically.

since been shown to consist of a single polypeptide containing 928 amino acids.

The way in which each nucleotide is added to the growing chain is a function of the specificity of DNA polymerase I. As shown in **Figure 8**, the precursor dNTP contains the three phosphate groups attached to the 5′-carbon of deoxyribose. As the two terminal phosphates are cleaved during synthesis, the remaining phosphate attached to the 5′-carbon is covalently linked to the 3′-OH group of the deoxyribose to which it is added. Thus, **chain elongation** occurs in the **5′ to 3′ direction** by the addition of one nucleotide at a time to the growing 3′ end. Each step provides a newly exposed 3′-OH group that can participate in the next addition of a nucleotide as DNA synthesis proceeds.

Having isolated DNA polymerase I and demonstrated its catalytic activity, Kornberg next sought to demonstrate the accuracy, or fidelity, with which the enzyme replicated the DNA template. Because technology for ascertaining the nucleotide sequences of the template and newly synthesized strand was not yet available in 1957, he initially had to rely on several indirect methods.

One of Kornberg's approaches was to compare the nitrogenous base compositions of the DNA template with those of the recovered DNA product. Using several sources of DNA (phage T2, *E. coli*, and calf thymus), he discovered that, within experimental error, the base composition of each product agreed with the template DNA used. This suggested that the templates were replicated faithfully.

ESSENTIAL POINT

Arthur Kornberg isolated the enzyme DNA polymerase I from *E. coli* and showed that it is capable of directing *in vitro* DNA synthesis, provided that a template and precursor nucleoside triphosphates were supplied.

DNA Polymerase II, III, IV, and V

While DNA polymerase I clearly directs the synthesis of DNA, a serious reservation about the enzyme's true biological role was raised in 1969. Paula DeLucia and John Cairns discovered a mutant strain of *E. coli* that was deficient in polymerase I activity. The mutation was designated *polA1*.

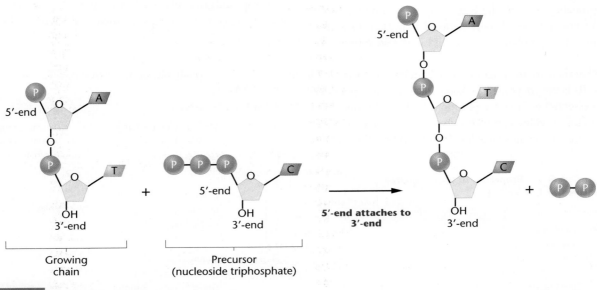

FIGURE 8 Demonstration of 5′ to 3′ synthesis of DNA.

In the absence of the functional enzyme, this mutant strain of *E. coli* still duplicated its DNA and successfully reproduced. However, the cells were deficient in their ability to repair DNA. For example, the mutant strain is highly sensitive to ultraviolet light (UV) and radiation, both of which damage DNA and are mutagenic. Nonmutant bacteria are able to repair a great deal of UV-induced damage.

These observations led to two conclusions:

1. At least one other enzyme that is responsible for replicating DNA *in vivo* is present in *E. coli* cells.

2. DNA polymerase I serves a secondary function *in vivo,* now believed to be critical to the maintenance of fidelity of DNA synthesis.

To date, four other unique DNA polymerases have been isolated from cells lacking polymerase I activity and from normal cells that contain polymerase I. **Table 1** contrasts several characteristics of DNA polymerase I with **DNA polymerase II and III**. Although none of the three can *initiate* DNA synthesis on a template, all three can *elongate* an existing DNA strand, called a **primer**.

All the DNA polymerase enzymes are large proteins exhibiting a molecular weight in excess of 100,000 Daltons (Da). All three possess 3′ to 5′ exonuclease activity, which means that they have the potential to polymerize in one direction and then pause, reverse their direction, and excise nucleotides just added. As we will discuss later , this activity provides a capacity to proofread newly synthesized DNA and to remove and replace incorrect nucleotides.

DNA polymerase I also demonstrates 5′ to 3′ exonuclease activity. This activity allows the enzyme to excise nucleotides, starting at the end at which synthesis begins and proceeding in the same direction of synthesis. Two final observations probably explain why Kornberg isolated polymerase I and not polymerase III: polymerase I is present in greater amounts than is polymerase III, and it is also much more stable.

What then are the roles of the polymerases *in vivo*? Polymerase III is the enzyme responsible for the 5′ to 3′ polymerization essential to *in vivo* replication. Its 3′ to 5′ exonuclease activity also provides a proofreading function that is activated when it inserts an incorrect nucleotide. When this occurs,

synthesis stalls and the polymerase "reverses course," excising the incorrect nucleotide. Then, it proceeds back in the 5′ to 3′ direction, synthesizing the complement of the template strand. Polymerase I is believed to be responsible for removing the primer, as well as for the synthesis that fills gaps produced after this removal. Its exonuclease activities also allow for its participation in DNA repair. Polymerase II, as well as **polymerase IV and V,** are involved in various aspects of repair of DNA that has been damaged by external forces, such as ultraviolet light. Polymerase II is encoded by a gene whose transcription is activated by disruption of DNA synthesis at the replication fork.

ESSENTIAL POINT ■ ■ ■

The discovery of the *polA1* mutant strain of *E. coli,* capable of DNA replication despite its lack of polymerase I activity, cast doubt on the enzyme's hypothesized *in vivo* replicative function. Polymerase III has been identified as the enzyme responsible for DNA replication *in vivo.*

We end this section by emphasizing the complexity of the DNA polymerase III molecule. In contrast to DNA polymerase I, which is but a single polypeptide, the active form of DNA polymerase III is a **holoenzyme**—a complex enzyme made up of multiple subunits. Polymerase III consists of ten kinds of polypeptide subunits **(Table 2)** and has a molecular weight of 900,000 Da. The largest subunit, α, has a molecular weight of 140,000 Da and, along with subunits ε and θ, constitutes a **core enzyme** responsible for the polymerization activity. The α subunit is responsible for nucleotide polymerization on the template strands, whereas the ε subunit of the core enzyme possesses the 3′ to 5′ exonuclease activity. A single DNA polymerase III holoenzyme contains, along with other components, two core enzymes combined into a dimer.

A second group of five subunits (γ, δ, δ', χ, and ψ) forms what is called the γ complex, which is involved in "loading"

TABLE 1   Properties of Bacterial DNA Polymerases I, II, and III

Properties	I	II	III
Initiation of chain synthesis	−	−	−
5′−3′ polymerization	+	+	+
3′−5′ exonuclease activity	+	+	+
5′−3′ exonuclease activity	+	−	−
Molecules of polymerase/cell	400	?	15

TABLE 2   Subunits of the DNA Polymerase III Holoenzyme

Subunit	Function	Groupings
α	5′−3′ polymerization	Core enzyme: Elongates polynucleotide chain and proofreads
ε	3′−5′ exonuclease	
θ	Core assembly	
γ		
δ	Loads enzyme on template (serves as clamp loader)	γ complex
δ'		
χ		
ψ		
β	Sliding clamp structure (processivity factor)	
τ	Dimerizes core complex	

the enzyme onto the template at the replication fork. This enzymatic function requires energy and is dependent on the hydrolysis of ATP. The β subunit serves as a "clamp" and prevents the core enzyme from falling off the template during polymerization. Finally, the τ subunit functions to dimerize two core polymerases facilitating simultaneous synthesis of both strands of the helix at the replication fork. The holoenzyme and several other proteins at the replication fork together form a huge complex (nearly as large as a ribosome) known as the **replisome.** We consider the function of DNA polymerase III in more detail later in this chapter.

3 | Many Complex Issues Must Be Resolved during DNA Replication

We have thus far established that in bacteria and viruses replication is semiconservative and bidirectional along a single replicon. We also know that synthesis is catalyzed by DNA polymerase III and occurs in the 5′ to 3′ direction. Bidirectional synthesis creates two replication forks that move in opposite directions away from the origin of synthesis. As we can see from the following list, many issues remain to be resolved in order to provide a comprehensive understanding of DNA replication:

1. The helix must undergo localized unwinding, and the resulting "open" configuration must be stabilized so that synthesis may proceed along both strands.

2. As unwinding and subsequent DNA synthesis proceed, increased coiling creates tension further down the helix, which must be reduced.

3. A primer of some sort must be synthesized so that polymerization can commence under the direction of DNA polymerase III. Surprisingly, RNA, not DNA, serves as the primer.

4. Once the RNA primers have been synthesized, DNA polymerase III begins to synthesize the DNA complement of both strands of the parent molecule. Because the two strands are antiparallel to one another, continuous synthesis in the direction that the replication fork moves is possible along only one of the two strands. On the other strand, synthesis must be discontinuous and thus involves a somewhat different process.

5. The RNA primers must be removed prior to completion of replication. The gaps that are temporarily created must be filled with DNA complementary to the template at each location.

6. The newly synthesized DNA strand that fills each temporary gap must be joined to the adjacent strand of DNA.

7. While DNA polymerases accurately insert complementary bases during replication, they are not perfect, and, occasionally, incorrect nucleotides are added to the growing strand. A proofreading mechanism that also corrects errors is an integral process during DNA synthesis.

As we consider these points, examine Figures 9, 10, and 11 to see how each issue is resolved. Figure 12 summarizes the model of DNA synthesis.

Unwinding the DNA Helix

As discussed earlier, there is a single point of origin along the circular chromosome of most bacteria and viruses at which DNA synthesis is initiated. This region of the *E. coli* chromosome has been particularly well studied. Called **oriC,** it consists of 245 nucleotide pairs characterized by repeating sequences of 9 and 13 bases (called **9mers** and **13mers**). One particular protein, called **DnaA** (because it is encoded by the gene called *dnaA*), is responsible for the initial step in unwinding the helix. A number of subunits of the DnaA protein bind to each of several 9mers. This step facilitates the subsequent binding of **DnaB** and **DnaC** proteins that further open and destabilize the helix. Proteins such as these, which require the energy supplied by the hydrolysis of ATP in order to break hydrogen bonds and denature the double helix, are called **helicases.** Other proteins, called **single-stranded binding proteins (SSBPs),** stabilize this open conformation.

As unwinding proceeds, a coiling tension is created ahead of the replication fork, often producing **supercoiling.** In circular molecules, supercoiling may take the form of added twists and turns of the DNA, much like the coiling you can create in a rubber band by stretching it out and then twisting one end. Such supercoiling can be relaxed by **DNA gyrase,** a member of a larger group of enzymes referred to as **DNA topoisomerases.** The gyrase makes either single- or double-stranded "cuts" and also catalyzes localized movements that have the effect of "undoing" the twists and knots created during supercoiling. The strands are then resealed. These various reactions are driven by the energy released during ATP hydrolysis.

Together, the DNA, the polymerase complex, and associated enzymes make up an array of molecules that participate in DNA synthesis and are part of what we have previously called the *replisome.*

ESSENTIAL POINT ▪ ▪ ▪

During the initiation of DNA synthesis, the double helix unwinds, forming a replication fork at which synthesis begins. Proteins stabilize the unwound helix and assist in relaxing the coiling tension created ahead of the replication fork.

Initiation of DNA Synthesis Using an RNA Primer

Once a small portion of the helix is unwound, what else is needed to initiate synthesis? As we have seen, DNA polymerase III requires a primer with a free 3′-hydroxyl group in order to elongate a polynucleotide chain. Since none is available in a circular chromosome, this absence prompted researchers to investigate how the first nucleotide could be added. It is now clear that RNA serves as the primer that initiates DNA synthesis.

A short segment of RNA (about 10 to 12 ribonucleotides long), complementary to DNA, is first synthesized on the DNA template. Synthesis of the RNA is directed by a form of RNA polymerase called **primase,** which does not require a free 3′ end to initiate synthesis. It is to this short segment of RNA that DNA polymerase III begins to add deoxyribonucleotides, initiating DNA synthesis. A conceptual diagram of initiation on a DNA template is shown in **Figure 9**. Later, the RNA primer is clipped out and replaced with DNA. This is thought to occur under the direction of DNA polymerase I. Recognized in viruses, bacteria, and several eukaryotic organisms, RNA priming is a universal phenomenon during the initiation of DNA synthesis.

ESSENTIAL POINT ▨ ▨ ▨

DNA synthesis is initiated at specific sites along each template strand by the enzyme primase, resulting in short segments of RNA that provide suitable 3′ ends upon which DNA polymerase III can begin polymerization.

Continuous and Discontinuous DNA Synthesis

We must now revisit the fact that the two strands of a double helix are **antiparallel** to each other—that is, one runs in the 5′–3′ direction, while the other has the opposite 3′–5′ polarity. Because DNA polymerase III synthesizes DNA in only the 5′–3′ direction, synthesis along an advancing replication fork occurs in one direction on one strand and in the opposite direction on the other.

As a result, as the strands unwind and the replication fork progresses down the helix (**Figure 10**), only one strand

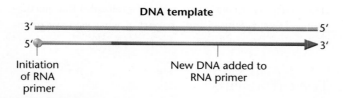

DNA template

3′ ════════════════════════ 5′
5′ ●━━━━━━━━━━━━━━━━━━▶ 3′

Initiation
of RNA
primer

New DNA added to
RNA primer

FIGURE 9 The initiation of DNA synthesis. A complementary RNA primer is first synthesized, to which DNA is added. All synthesis is in the 5′ to 3′ direction. Eventually, the RNA primer is replaced with DNA under the direction of DNA polymerase I.

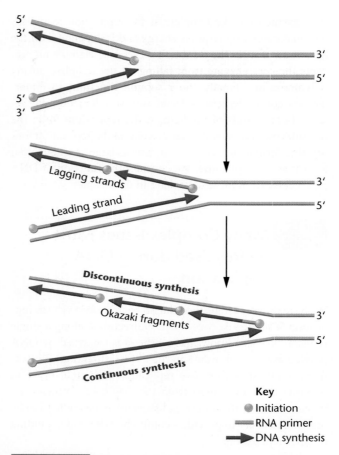

Key
● Initiation
━━ RNA primer
▶ DNA synthesis

FIGURE 10 Opposite polarity of synthesis along the two strands of DNA is necessary because they run antiparallel to one another, and because DNA polymerase III synthesizes in only one direction (5′ to 3′). On the lagging strand, synthesis must be discontinuous, resulting in the production of Okazaki fragments. On the leading strand, synthesis is continuous. RNA primers are used to initiate synthesis on both strands.

can serve as a template for **continuous DNA synthesis.** This newly synthesized DNA is called the **leading strand.** As the fork progresses, many points of initiation are necessary on the opposite DNA template, resulting in **discontinuous DNA synthesis** of the **lagging strand.**[*]

Evidence supporting the occurrence of discontinuous DNA synthesis was first provided by Reiji and Tuneko Okazaki. They discovered that when bacteriophage DNA is replicated in *E. coli,* some of the newly formed DNA that is hydrogen bonded to the template strand is present as small fragments containing 1000 to 2000 nucleotides. RNA primers are part of each such fragment. These pieces, now called **Okazaki fragments,** are converted into longer and longer DNA strands of higher molecular weight as synthesis proceeds.

---

[*]Because DNA synthesis is continuous on one strand and discontinuous on the other, the term **semidiscontinuous synthesis** is sometimes used to describe the overall process.

Discontinuous synthesis of DNA requires enzymes that both remove the RNA primers and unite the Okazaki fragments into the lagging strand. As we have noted, DNA polymerase I removes the primers and replaces the missing nucleotides. Joining the fragments is the work of **DNA ligase,** which is capable of catalyzing the formation of the phosphodiester bond that seals the nick between the discontinuously synthesized strands. The evidence that DNA ligase performs this function during DNA synthesis is strengthened by the observation of a ligase-deficient mutant strain (*lig*) of *E. coli,* in which a large number of unjoined Okazaki fragments accumulate.

Concurrent Synthesis Occurs on the Leading and Lagging Strands

Given the model just discussed, we might ask how DNA polymerase III synthesizes DNA on both the leading and lagging strands, which are antiparallel to one another. As the model depicted in **Figure 11** illustrates, if the lagging strand is spooled out forming a loop, nucleotide polymerization can occur on both template strands. This is accomplished under the direction of the dimer of the enzyme, which is held together by the γ complex (the sliding clamp loader). After the synthesis of 1000 to 2000 nucleotides on the lagging strand, the monomer of the enzyme will encounter a completed Okazaki fragment, at which point it must be released from the lagging strand. A new loop of the lagging template strand is spooled out, and the process is repeated. The end result is that looping inverts the orientation of the template but not the direction of actual synthesis at the replication fork.

One final consideration involves the role of the sliding clamp structure, a donut-like monomer that is found associated with each polymerase unit at the replication fork (Figure 12 and Table 2). This clamp is critical in maintaining what is called the **processivity** of the enzyme—that is, the number of nucleotides that may be added during synthesis before the

enzyme dissociates from the template. The clamp facilitates high processivity (of particular importance to synthesis on the leading strand), and is considered critical to the rapid *in vivo* rate of synthesis by DNA polymerase.

ESSENTIAL POINT ■ ■ ■

Concurrent DNA synthesis occurs continuously on the leading strand and discontinuously on the opposite lagging strand, resulting in short Okazaki fragments that are later joined by DNA ligase.

NOW SOLVE THIS

2 An alien organism was investigated. When DNA replication was studied, a unique feature was apparent: No Okazaki fragments were observed. Create a model of DNA that is consistent with this observation.

■ **Hint** *This problem involves an understanding of the process of DNA synthesis in prokaryotes, as depicted in Figure 12. The key to its solution is to consider why Okazaki fragments are observed during DNA synthesis and how their formation relates to DNA structure, as described in the Watson–Crick model.*

Proofreading and Error Correction Occur during DNA Replication

The immediate purpose of DNA replication is the synthesis of a new strand that is precisely complementary to the template strand at each nucleotide position. Although the action of DNA polymerases is very accurate, synthesis is not perfect and a noncomplementary nucleotide is occasionally inserted erroneously. To compensate for such inaccuracies, the DNA polymerases all possess 3′ to 5′ exonuclease activity. This property imparts the potential for them to detect and excise a mismatched nucleotide (in the 3′ to 5′ direction). Once the mismatched nucleotide is removed, 5′ to 3′ synthesis can again proceed. This process, called **proofreading,** increases

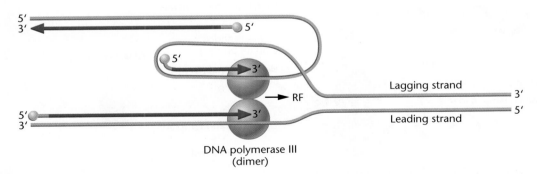

FIGURE 11   Illustration of how concurrent DNA synthesis may be achieved on both the leading and lagging strands at a single replication fork (RF). The lagging template strand is "looped" in order to invert the physical direction of synthesis, but not the biochemical direction. The enzyme functions as a dimer, with each core enzyme achieving synthesis on one or the other strand.

the fidelity of synthesis by a factor of about 100. In the case of the holoenzyme form of DNA polymerase III, the epsilon (ε) subunit is directly involved in the proofreading step. In strains of *E. coli* with a mutation that has rendered the ε subunit nonfunctional, the error rate (the mutation rate) during DNA synthesis is increased substantially.

4 A Coherent Model Summarizes DNA Replication

We can now combine the various aspects of DNA replication occurring at a single replication fork into a coherent model, as shown in **Figure 12**. At the advancing fork, a helicase is unwinding the double helix. Once unwound, single-stranded binding proteins associate with the strands, preventing the re-formation of the helix. In advance of the replication fork, DNA gyrase functions to diminish the tension created as the helix supercoils. Each half of the dimeric polymerase is a core enzyme bound to one of the template strands by a β-subunit sliding clamp. Continuous synthesis occurs on the leading strand, while the lagging strand must loop out and around the polymerase in order for simultaneous (concurrent) synthesis to occur on both strands. Not shown in the figure, but essential to replication on the lagging strand, is the action of DNA polymerase I and DNA ligase, which together replace the RNA primers with DNA and join the Okazaki fragments, respectively.

Because the investigation of DNA synthesis is still an extremely active area of research, this model will no doubt be extended in the future. In the meantime, it provides a summary of DNA synthesis against which genetic phenomena can be interpreted.

5 Replication Is Controlled by a Variety of Genes

Much of what we know about DNA replication in viruses and bacteria is based on genetic analysis of the process. For example, we have already discussed studies involving the *polA1* mutation, which revealed that DNA polymerase I is not the major enzyme responsible for replication. Many other mutations interrupt or seriously impair some aspect of replication, such as the ligase-deficient and the proofreading-deficient mutations mentioned previously. Because such mutations are lethal ones, genetic analysis frequently uses **conditional mutations,** which are expressed under one condition but not under a different condition. For example, a **temperature-sensitive mutation** may not be expressed at a particular *permissive* temperature. When mutant cells are grown at a *restrictive* temperature, the mutant phenotype is expressed and can be studied. By examining the effect of the loss of function associated with the mutation, the investigation of such temperature-sensitive mutants can provide insight into the product and the associated function of the normal, nonmutant gene.

As shown in **Table 3**, a variety of genes in *E. coli* specify the subunits of the DNA polymerases and encode products involved in specification of the origin of synthesis, helix unwinding and stabilization, initiation and priming, relaxation of supercoiling, repair, and ligation. The discovery of such a large group of genes attests to the complexity of the process of replication, even in the relatively simple prokaryote. Given the enormous quantity of DNA that must be unerringly replicated in a very brief time, this level of complexity is not unexpected. As we will see in the next section, the process is even more involved and therefore more difficult to investigate in eukaryotes.

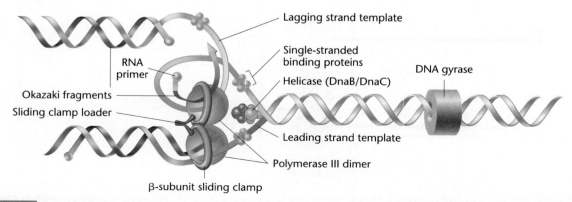

FIGURE 12 Summary of DNA synthesis at a single replication fork. Various enzymes and proteins essential to the process are shown.

TABLE 3	Some of the Various *E. coli* Genes and Their Products or Role in Replication
Gene	**Product or Role**
polA	DNA polymerase I
polB	DNA polymerase II
dnaE,N, Q, X, Z	DNA polymerase III subunits
dnaG	Primase
dnaA, I, P	Initiation
dnaB, C	Helicase at *oriC*
gyrA, B	Gyrase subunits
lig	DNA ligase
rep	DNA helicase
ssb	Single-stranded binding proteins
rpoB	RNA polymerase subunit

6 — Eukaryotic DNA Replication Is Similar to Replication in Prokaryotes, but Is More Complex

Eukaryotic DNA replication shares many features with replication in bacteria. In both systems, double-stranded DNA is unwound at replication origins, replication forks are formed, and bidirectional DNA synthesis creates leading and lagging strands from single-stranded DNA templates under the direction of DNA polymerase. Eukaryotic polymerases have the same fundamental requirements for DNA synthesis as do bacterial polymerases: four deoxyribonucleoside triphosphates, a template, and a primer. However, because eukaryotic cells contain much more DNA, this DNA is complexed with nucleosomes, and because eukaryotic chromosomes are linear rather than circular, eukaryotic DNA synthesis is more complicated. In this section, we will describe some of the ways in which eukaryotes deal with this added complexity.

Initiation of Replication at Multiple Replication Origins

The most obvious difference between eukaryotic and prokaryotic DNA replication is that eukaryotic replication must deal with greater amounts of DNA. For example, yeast cells contain three times as much DNA, and *Drosophila* cells contain 40 times as much as *E. coli* cells. In addition, eukaryotic DNA polymerases synthesize DNA at a rate 25 times slower (about 2000 nucleotides per minute) than that in prokaryotes. Under these conditions, replication from a single origin on a typical eukaryotic chromosome would take days to complete! However, replication of entire eukaryotic genomes is usually accomplished in a matter of minutes to hours.

To facilitate the rapid synthesis of large quantities of DNA, eukaryotic chromosomes contain multiple replication

origins. Yeast genomes contain between 250 and 400 origins, and mammalian genomes have as many as 25,000. Multiple origins are visible under the electron microscope as "replication bubbles" that form as the DNA helix opens up, each bubble providing two potential replication forks (**Figure 13**). Origins in yeast, called **autonomously replicating sequences (ARSs)**, consist of an approximately 120 base-pair unit containing a **consensus sequence** (meaning a sequence that is the same, or nearly the same, in all yeast ARSs) of 11 base pairs. Origins in mammalian cells appear to be unrelated to specific sequence motifs and may be defined more by chromatin structure over a 6–55 kb region.

Eukaryotic replication origins not only are the sites of replication initiation, but also control the timing of DNA replication. These regulatory functions are carried out by a complex of more than 20 proteins, called the **prereplication complex (pre-RC)**, which assembles at replication origins. In the early G1 phase of the cell cycle, replication origins are recognized by a six-protein complex known as an **origin recognition complex (ORC)**, which tags the origin as a site of initiation. Throughout the G1 phase of the cell cycle, other proteins associate with the ORC to form the pre-RC. The presence of a pre-RC at an origin "licenses" that origin for replication. Once DNA polymerases initiate synthesis at the origin, the pre-RC is disrupted and does not reassemble again until the G1 phase of the next cell cycle. This is an important mechanism because it distinguishes segments of DNA that have completed replication from segments of unreplicated DNA, thus maintaining orderly and efficient replication. It ensures that replication occurs only once along each stretch of DNA during each cell cycle.

The initiation of DNA replication is also regulated at the pre-RC. A number of cell-cycle kinases that phosphorylate replication proteins, along with helicases that unwind DNA,

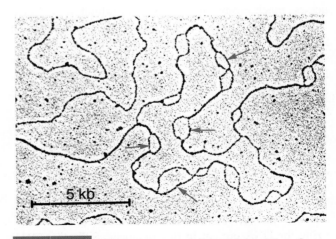

FIGURE 13 A demonstration of the multiple origins of replication along a eukaryotic chromosome. Each origin is apparent as a replication bubble along the axis of the chromosome. Arrows identify some of these replication bubbles.

5 kb

associate with the pre-RC and are essential for initiation. The kinases are activated in S phase, at which time they phosphorylate other proteins that trigger the initiation of DNA replication. The end result is the unwinding of DNA at the replication forks, the stabilization of single-stranded DNA, the association of DNA polymerases with the origins, and the initiation of DNA synthesis.

Multiple Eukaryotic DNA Polymerases

To accommodate the increased number of replicons, eukaryotic cells contain many more DNA polymerase molecules than do bacterial cells. For example, a single *E. coli* cell contains about 15 molecules of DNA polymerase III, but mammalian cells contain tens of thousands of DNA polymerase molecules.

Eukaryotes also utilize a larger array of different DNA polymerase types than do prokaryotes. The human genome contains genes that encode at least 14 different DNA polymerases (abbreviated as Pol), only three of which are involved in the majority of nuclear genome DNA replication.

Pol α and δ, as well as ε, are the major forms of the enzyme involved in initiation and elongation during nuclear DNA synthesis, so we will concentrate our discussion on these. Two of the four subunits of the Pol α enzyme synthesize RNA primers on both the leading and lagging strands. After the RNA primer reaches a length of about 10 ribonucleotides, another subunit adds 20 to 30 complementary deoxyribonucleotides. Pol α is said to possess low processivity, a term that refers to the strength of the association between the enzyme and its substrate. Thus, the length of DNA synthesized before the enzyme dissociates from the template is shorter. Once the primer is in place, an event known as **polymerase switching** occurs, whereby Pol α dissociates from the template and is replaced by Pol δ and ε. These enzymes extend the primers on opposite strands of DNA, possess much greater processivity, and exhibit 3' to 5' exonuclease activity, thus having the potential to proofread. Pol ε synthesizes DNA on the leading strand, and Pol δ synthesizes the lagging strand. Both Pol δ and ε participate in other DNA synthesizing events in the cell, including several types of DNA repair and recombination. All three enzymes are essential for viability.

As in prokaryotic DNA replication, the final stages in eukaryotic DNA replication involve replacing the RNA primers with DNA and ligating the Okazaki fragments on the lagging strand. In eukaryotes, the Okazaki fragments are about ten times smaller (100 to 150 nucleotides) than in prokaryotes.

Included in the remainder of DNA-replicating enzymes is Pol γ, which is found exclusively in mitochondria, synthesizing the DNA present in that organelle. Still other DNA polymerases are involved in DNA repair and replication through regions of the DNA template that contain damage or distortions.

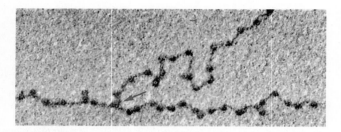

FIGURE 14 An electron micrograph of a eukaryotic replicating fork demonstrating the presence of histone-protein-containing nucleosomes on both branches.

Replication through Chromatin

One of the major differences between prokaryotic and eukaryotic DNA is that eukaryotic DNA is complexed with DNA-binding proteins, existing in the cell as chromatin. Chromatin consists of regularly repeating units called nucleosomes, each of which consists of about 200 base pairs of DNA complexed with eight histone protein molecules **(Figure 14)**. Before polymerases can begin synthesis, nucleosomes and other DNA-binding proteins must be stripped away or otherwise modified to allow the passage of replication proteins. As DNA synthesis proceeds, the histones and nonhistone proteins must rapidly reassociate with the newly formed duplexes, reestablishing the characteristic nucleosome pattern.

In order to re-create nucleosomal chromatin on replicated DNA, the synthesis of new histone proteins is tightly coupled to DNA synthesis during the S phase of the cell cycle. Research data suggest that nucleosomes are disrupted just ahead of the replication fork and that the preexisting histone proteins assemble with newly synthesized histone proteins into new nucleosomes. The new nucleosomes are assembled behind the replication fork, onto the two daughter strands of DNA. Assembly is carried out by **chromatin assembly factors (CAFs)** that move along with the replication fork.

> **ESSENTIAL POINT** ■ ■ ■
>
> DNA replication in eukaryotes is more complex than replication in prokaryotes, using multiple replication origins, multiple forms of DNA polymerases, and factors that disrupt and assemble nucleosomal chromatin.

7 The Ends of Linear Chromosomes Are Problematic during Replication

A final difference between prokaryotic and eukaryotic DNA synthesis stems from the structural differences in their chromosomes. Unlike the closed, circular DNA of bacteria and most bacteriophages, eukaryotic chromosomes are linear. During replication, a special problem arises at the "ends" of these linear molecules.

Eukaryotic chromosomes end in distinctive sequences called **telomeres** that help preserve the integrity and stability of the chromosome. Telomeres are necessary because the double-stranded "ends" of DNA molecules at the termini of linear chromosomes potentially resemble the **double-stranded breaks (DSBs)** that can occur when a chromosome becomes fragmented internally. In such cases, the double-stranded loose ends can fuse to other such ends; if they don't fuse, they are vulnerable to degradation by nucleases. Either outcome can lead to problems. Telomeres are believed to create inert chromosome ends, protecting intact eukaryotic chromosomes from improper fusion or degradation.

Telomere Structure

We could speculate that there must be something unique about the DNA sequence or the proteins that bind to it that confers this protective property to telomeres. Indeed, this has been shown to be the case. First discovered by Elizabeth Blackburn and Joe Gall in their study of micronuclei—the smaller of two nuclei in the ciliated protozoan *Tetrahymena*—the DNA at the protozoan's chromosome ends consists of the short tandem repeating sequence TTGGGG. This sequence is present many times on one of the two DNA strands making up each telomere. This strand is referred to as the G-rich strand, in contrast to its complementary strand, the so-called C-rich strand, which displays the repeated sequence AACCCC. In a similar way, all vertebrates contain the sequence TTAGGG at the ends of G-rich strands, repeated several thousand times in somatic cells. Since each linear chromosome ends with two DNA strands running antiparallel to one another, one strand has a 3'-ending and the other has a 5'-ending. It is the 3'-strand that is the G-rich one. This has special significance during telomere replication.

But first, let's describe how this tandemly repeated DNA confers inertness to the chromosome ends. One model is based on the discovery that the 3'-ending G-rich strand extends as an overhang, lacking a complement, and thus forms a single-stranded tail at the terminus of each telomere. In *Tetrahymena*, this tail is only 12 to 16 nucleotides long. However, in vertebrates, it may be several hundred nucleotides long. The final conformation of these tails has been correlated with chromosome inertness. Though not considered complementary in the same way as A-T and G-C base pairs are, G-containing nucleotides are nevertheless capable of base pairing with one another when several are aligned opposite another G-rich sequence. Thus, the G-rich single-stranded tails are capable of looping back on themselves, forming multiple G-G hydrogen bonds to create what are referred to as **G-quartets.** The resulting loops at the chromo-

some ends (called **t-loops**) are much like those created when you tie your shoelaces into a bow. It is believed that these structures, in combination with specific proteins that bind to them, essentially close off the ends of chromosomes and make them inert.

Replication at the Telomere

Now let's consider the problem that semiconservative replication poses at the end of a double-stranded DNA molecule. Although 5' to 3' synthesis on the leading-strand template may proceed to the end, a difficulty arises on the lagging strand once the final RNA primer is removed (**Figure 15**). Normally, the newly created gap would be filled in starting with the addition of a nucleotide to the adjacent 3'-OH group [the group to the right of gap (a) in Figure 15]. However, since the final gap [gap (b) in Figure 15] is at the end of the strand being synthesized, there is no Okazaki fragment present to provide the needed 3'-OH group. Thus, in the situation depicted in Figure 15, a gap remains on the

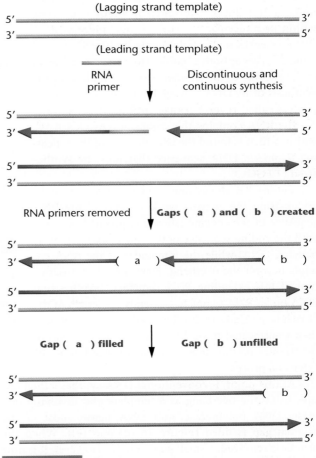

FIGURE 15 Diagram illustrating the difficulty encountered during the replication of the ends of linear chromosomes. A gap (b) is left following synthesis on the lagging strand.

lagging strand produced in each successive round of synthesis, shortening the double-stranded end of the chromosome by the length of the RNA primer. With each round of replication, the shortening becomes more severe in each daughter cell, eventually *extending beyond the telomere to potentially delete gene-coding regions.*

A unique eukaryotic enzyme called **telomerase,** first discovered by Elizabeth Blackburn and Carol Greider in studies of *Tetrahymena,* has helped us understand the solution to the problem of telomere shortening. As noted earlier, telomeric DNA in eukaryotes is always found to consist of many short, repeated nucleotide sequences, with the G-rich strand overhanging in the form of a single-stranded tail. In *Tetrahymena* the tail contains several repeats of the sequence 5′-TTGGGG-3′. As we will see, telomerase is capable of adding several more repeats of this six-nucleotide sequence to the 3′-end of the G-rich strand (using 5′−3′ synthesis). Detailed investigation by Blackburn and Greider of how the *Tetrahymena* telomerase enzyme accomplishes this synthesis yielded an extraordinary finding. The enzyme is highly unusual in that it is a **ribonucleoprotein,** containing within its molecular structure a short piece of RNA that is essential to its catalytic activity. The RNA component serves as both a "guide" (to proper attachment of the enzyme to the telomere) and a template for the synthesis of its DNA complement, the latter being a process called **reverse transcription.** In *Tetrahymena,* the RNA contains the sequence AACCCCAAC, within which is found the complement of the repeating telomeric DNA sequence that must be synthesized (TTGGGG).

Figure 16 shows one model of how researchers envision the enzyme working. Part of the RNA sequence of the enzyme (shown in green) base-pairs with the ending sequence of the single-stranded overhanging DNA, while the remainder of the RNA extends beyond the overhang. Next, reverse transcription of this extending RNA sequence—synthesizing DNA on an RNA template—extends the length of the G-rich lagging strand. It is believed that the enzyme is then translocated toward the (newly formed) end of the strand, and the same events are repeated, continuing the extension process.

At this point, if conventional DNA synthesis then ensues using the overhang as a template and involving primase, DNA polymerase, and DNA ligase, most of the original gap is filled [Figure 16 (d) and (e)]. When the primer is removed, a small gap remains. However, it is now found well beyond the original end of the chromosome,

(a) Telomerase binds to 3′ G-rich tail

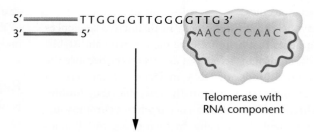

Telomerase with RNA component

(b) Telomeric DNA is synthesized on G-rich tail

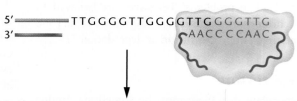

(c) Telomerase is translocated and steps (a) and (b) are repeated

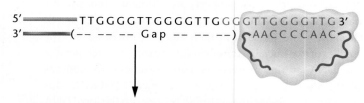

(d) Telomerase released; primase and DNA polymerase fill gap

(e) Primer removed; gap sealed by DNA ligase

FIGURE 16 The predicted solution to the problem posed in Figure 15. The enzyme telomerase (with its RNA component shown in green) directs synthesis of repeated TTGGGG sequences, resulting in the formation of an extended 3′-overhang. This facilitates DNA synthesis on the opposite strand, filling in the gap that would otherwise be created on the ends of linear chromosomes during each replication cycle.

thus preventing any shortening. Another model suggests that the DNA extension, created by telomerase, facilitates DNA synthesis on the opposite C-rich strand. In this model, the single-stranded extension loops back on itself, providing the 3′-OH group necessary for initiation of synthesis to fill the gap.

Telomerase function has now been found in all eukaryotes studied. Telomeric DNA sequences have been highly conserved throughout evolution, reflecting the critical function of telomeres. As mentioned earlier, in humans, the telomeric DNA sequence on the lagging strand that is repeated is 5′-TTAGGG-3′, differing from *Tetrahymena* by only one nucleotide.

In most eukaryotic somatic cells, telomerase is not active, and thus, with each cell division, the telomeres of each chromosome do shorten. After many divisions, the telomere may be seriously eroded, causing the cell to lose the capacity for further division. Malignant cells, on the other hand, maintain telomerase activity and in this way are immortalized. In the "Genetics, Technology, and Society" feature at the end of this chapter, we will see that telomere shortening, in the absence of telomerase in somatic cells, has been linked to a molecular mechanism involved in cellular aging.

ESSENTIAL POINT　■　■　■

Replication at the ends of linear chromosomes in eukaryotes poses a special problem that can be solved by the presence of telomeres and by a unique RNA-containing enzyme called telomerase.

---

8 DNA Recombination, Like DNA Replication, Is Directed by Specific Enzymes

The process of crossing over between homologs depends on the breakage and rejoining of the DNA strands, and results in the exchange of genetic information between DNA molecules. Now that we have discussed the chemistry and replication of DNA, we can consider how recombination occurs at the molecular level. In general, our discussion pertains to genetic exchange between any two homologous double-stranded DNA molecules, whether they are viral or bacterial chromosomes or eukaryotic homologs during meiosis. Genetic exchange at equivalent positions along two chromosomes with substantial DNA sequence homology is referred to as **general,** or **homologous recombination.**

Several models attempt to explain homologous recombination, but they all have certain features in common. First, all are based on proposals initally put forth independently by Robin Holliday and Harold L. K. Whitehouse in 1964. Second, they all depend on the complementarity between DNA strands to explain the precision of the exchange. Finally, each model relies on a series of enzymatic processes in order to accomplish genetic recombination.

One such model is shown in **Figure 17**. It begins with two paired DNA duplexes, or homologs [Step (a)], in each of which an endonuclease introduces a single-stranded nick at an identical position [Step (b)]. The internal strand endings produced by these cuts are then displaced and subsequently pair with their complements on the opposite duplex [Step (c)]. Next, a ligase seals the loose ends [Step (d)], creating hybrid duplexes called **heteroduplex DNA molecules,** held together by a cross-bridge structure. The position of this cross bridge can then move down the chromosome by a process referred to as **branch migration** [Step (e)], which occurs as a result of a zipperlike action as hydrogen bonds are broken and then re-formed between complementary bases of the displaced strands of each duplex. This migration yields an increased length of heteroduplex DNA on both homologs.

If the duplexes bend [Step (f)] and the bottom portion shown in the figure rotates 180° [Step (g)], an intermediate planar structure called a χ (chi) form—or **Holliday structure**—is created. If the two strands on opposite homologs previously uninvolved in the exchange are now nicked by an endonuclease [Step (h)] and ligation occurs as in Step (i), two recombinant duplexes are created. Note that the arrangement of alleles is altered as a result of this recombination.

Whereas the model above involves *single-stranded breaks,* other recombination models have been proposed that involve *double-stranded breaks* in one of the DNA double helices. In these models, endonucleases remove nucleotides at the breakpoint, creating 3′ overhangs on each strand. One of the broken strands invades the intact double helix of the other homolog, and both strands line up with the intact homolog. DNA repair synthesis then fills all gaps, and two Holliday junctions are formed. Endonuclease cleavages and ligations finalize the exchange. The end result is the same as our original model: genetic exchange occurs during crossing over in meiotic tetrads. A similar mechanism is thought to occur when cells repair double-stranded breaks in chromosomes. Such damage can occur from numerous causes, including the energy of ionizing radiation.

As with DNA replication, the processes involved in DNA recombination require the activities of numerous enzymes and other proteins. Mutations in genes encoding these proteins may cause defects in recombination, as well as in DNA repair and replication. One of the key proteins involved in *E. coli* recombination is the **RecA protein.** This molecule promotes the exchange of reciprocal single-stranded DNA molecules as occurs in Step (c) of the model. RecA also enhances the hydrogen-bond formation during strand displacement, thus initiating heteroduplex formation. The **RecB, RecC,** and **RecD proteins** can cleave DNA strands and help unwind the duplex. Other proteins are involved in branch migration and resolution of Holliday structures. DNA replication proteins, such as DNA polymerases, DNA

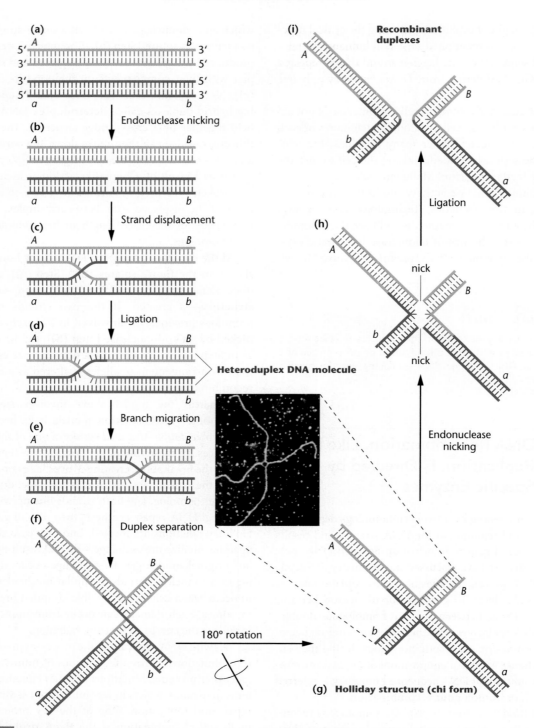

(a)

(b) Endonuclease nicking

(c) Strand displacement

(d) Ligation

Heteroduplex DNA molecule

(e) Branch migration

(f) Duplex separation

180° rotation

(g) Holliday structure (chi form)

(h) nick   nick

Ligation

Endonuclease nicking

(i) Recombinant duplexes

FIGURE 17 Model depicting how genetic recombination can occur as a result of the breakage and rejoining of heterologous DNA strands. Each stage is described in the text. The electron micrograph shows DNA in a X-form structure similar to the diagram in (g); the DNA is an extended Holliday structure, derived from the *Col*E1 plasmid of *E. coli. David Dressler, Oxford University, England.*

ligase, gyrases, and single-stranded binding proteins, are also involved in DNA recombination and repair.

ESSENTIAL POINT ■   ■   ■

Homologous recombination between DNA molecules relies on precise alignment of homologs and the actions of a series of enzymes that can cut, realign, and reseal DNA strands.

GENETICS, TECHNOLOGY, AND SOCIETY

Telomeres: The Key to Immortality?

Humans, like all multicellular organisms, grow old and die. As we age, our immune systems become less efficient, wound healing is impaired, and tissues lose resilience. It has always been a mystery why we go through these age-related declines and why each species has a characteristic finite life span. Why do we grow old? Can we reverse this march to mortality? Some recent discoveries suggest that the answers to these questions may lie at the ends of our chromosomes.

The study of human aging begins with a study of human cells growing in culture dishes. Like the organisms from which the cells are taken, cells in culture have a finite life span. This *replicative senescence* was first noted by Leonard Hayflick in the 1960s. He reported that normal human fibroblasts lose their ability to grow and divide after about 50 cell divisions. These senescent cells remain metabolically active but can no longer proliferate. Eventually, they die. Although we don't know whether cellular senescence directly causes organismal aging, the evidence is suggestive. For example, cells derived from young people undergo more divisions than those from older people; cells from short-lived species stop growing after fewer divisions than those from longer-lived species; and cells from patients with premature aging syndromes undergo fewer divisions than those from normal patients.

Another characteristic of aging cells involves their telomeres. In most mammalian somatic cells, telomeres become shorter with each DNA replication because DNA polymerase cannot synthesize new DNA at the ends of each parent strand. However, as discussed in detail in this chapter**text**, cells that undergo extensive proliferation, like embryonic cells, germ cells, and adult stem cells, maintain their telomere length by using *telomerase*—a remarkable RNA-containing enzyme that adds telomeric DNA sequences onto the ends of linear chromosomes. However, most somatic cells in adult organisms

do not proliferate and do not contain active telomerase.

Could we gain perpetual youth and vitality by increasing our telomere lengths? Studies suggest that it may be possible to reverse senescence by artificially increasing the amount of telomerase in our cells. When investigators introduced cloned telomerase genes into normal human cells in culture, telomeres lengthened, and the cells continued to grow past their typical senescence point. These studies suggest that some of the atrophy of tissues that accompanies old age could be reversed by activating telomerase genes. However, before we use telomerase to achieve immortality, we need to consider a potential serious side effect—cancer.

Although normal cells shorten their telomeres and undergo senescence after a specific number of cell divisions, cancer cells do not. More than 80 percent of human tumor cells contain telomerase activity, maintain telomeres, and achieve immortality. Those that do not contain active telomerase use a less well understood mechanism known as ALT (for "alternative lengthening of telomeres").

These observations have motivated scientists to devise new cancer therapies based on the idea that agents that inhibit telomerase might destroy cancer cells by allowing telomeres to shorten, thereby forcing the cells into senescence. Because most normal human cells do not express telomerase, such a therapy might target tumor cells and be less toxic than most current anticancer drugs. Many such anti-telomerase drugs are currently under development, and some are in clinical trials.

Will a deeper understanding of telomeres allow us to both arrest cancers *and* reverse the descent into old age? Time will tell.

Your Turn

Take time, individually or in groups, to answer the following questions. Investigate the references and links to help you

understand some of the research on telomeres, aging, and cancer.

1. How might our knowledge about telomeres and telomerase be applied to anti-aging strategies? Are such strategies or therapies being developed?

*Sources of information can be obtained by using the PubMed Web site (*http://www.ncbi.nlm.nih.gov/pubmed*).*

2. One anti-telomerase drug, called GRN163L, is being developed by Geron Corporation as a treatment for cancer. How does GRN163L work? What is the current status of GRN163L clinical trials? What are some possible side effects for anti-telomerase drugs?

Read about this drug and its clinical trials on the Geron Web site at http://www.geron.com. *Search on PubMed for scientific papers dealing with GRN163L's anticancer effects.*

3. People suffering from chronic stress appear to have more health problems and to age prematurely. Is there any evidence that chronic stress, poor health, and telomere length are linked? How might stress affect telomere length or vice versa?

Some recent papers suggest how these phenomena may be linked. One such paper is Epel, E. S., et al. 2004. Accelerated telomere shortening in response to life stress. Proc. Natl. Acad. Sci. USA 101(49): 17312–17315.

4. In 2006, the Lasker Award for Basic Medical Research was awarded to Drs. Elizabeth Blackburn, Carol Greider, and Jack Szostak, who subsequently were awarded the 2009 Nobel Prize in Physiology or Medicine. How did the intersections of people, ideas, and good fortune lead to their discovery of telomerase and its role in aging and cancer? What is the future for this research?

Listen to interviews with these scientists, in which they tell their stories about their research and where they see the field going, at http://www.laskerfoundation.org/2006videoawards.

CASE STUDY » At loose ends

A researcher was asked if his work on human telomere replication was related to any genetic disorders. He replied that one might think that any such mutation would be lethal during early development, but in fact a rare human genetic disorder affecting telomeres is known. This disorder, dyskeratosis congenita (DKC), is associated with mutations in the protein subunits of telomerase, the enzyme responsible for replicating the ends of eukaryotic chromosomes. Initial symptoms appear in tissues derived from rapidly dividing cells, including the skin, nails, and bone marrow, and first affect children between the ages of 5 and 15 years.

This disorder raises several interesting questions.

1. How could such individuals survive?
2. Why are the tissues derived from rapidly dividing cells initially affected?
3. Is this disorder likely to impact the life span?
4. Would you predict that mutations in the RNA component of telomerase might also cause DKC?

INSIGHTS AND SOLUTIONS

1. Predict the theoretical results of conservative and dispersive replication of DNA under the conditions of the Meselson–Stahl experiment. Follow the results through two generations of replication after cells have been shifted to an ^{14}N-containing medium, using the following sedimentation pattern.

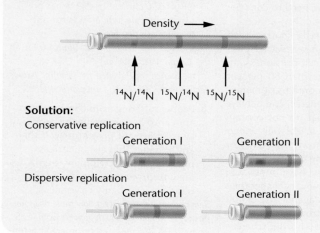

Density ⟶

$^{14}N/^{14}N$ $^{15}N/^{14}N$ $^{15}N/^{15}N$

Solution:
Conservative replication

Generation I   Generation II

Dispersive replication

Generation I   Generation II

2. Mutations in the *dnaA* gene of *E. coli* are lethal and can only be studied following the isolation of conditional, temperature-sensitive mutations. Such mutant strains grow nicely and replicate their DNA at the permissive temperature of 18°C, but they do not grow or replicate their DNA at the restrictive temperature of 37°C. Two observations were useful in determining the function of the DnaA protein product. First, *in vitro* studies using DNA templates that have unwound do not require the DnaA protein. Second, if intact cells are grown at 18°C and are then shifted to 37°C, DNA synthesis continues at this temperature until one round of replication is completed and then stops. What do these observations suggest about the role of the *dnaA* gene product?

Solution: At 18°C (the permissive temperature), the mutation is not expressed and DNA synthesis begins. Following the shift to the restrictive temperature, the already initiated DNA synthesis continues, but no new synthesis can begin. Because the DnaA protein is not required for synthesis of unwound DNA, these observations suggest that, *in vivo*, the DnaA protein plays an essential role in DNA synthesis by interacting with the intact helix and somehow facilitating the localized denaturation necessary for synthesis to proceed.

PROBLEMS AND DISCUSSION QUESTIONS

 For activities, animations, and review quizzes, go to the study area at www.masteringgenetics.com

HOW DO WE KNOW?

1. In this chapter, we focused on how DNA is replicated and synthesized. In particular, we elucidated the general mechanism of replication and described how DNA is synthesized when it is copied. Based on your study of these topics, answer the following fundamental questions:

   (a) What is the experimental basis for concluding that DNA replicates semiconservatively in both prokaryotes and eukaryotes?

   (b) How was it demonstrated that DNA synthesis occurs under the direction of DNA polymerase III and not polymerase I?

   (c) How do we know that *in vivo* DNA synthesis occurs in the 5′ to 3′ direction?

   (d) How do we know that DNA synthesis is discontinuous on one of the two template strands?

   (e) What observations reveal that a "telomere problem" exists during eukaryotic DNA replication, and how did we learn of the solution to this problem?

2. Compare conservative, semiconservative, and dispersive modes of DNA replication.

3. Describe the role of ^{15}N in the Meselson–Stahl experiment.

4. Predict the results of the experiment by Taylor, Woods, and Hughes if replication were (a) conservative and (b) dispersive.

5. This question was intentionally excluded from this edition.

6. What are the requirements for *in vitro* synthesis of DNA under the direction of DNA polymerase I?

7. In Kornberg's initial experiments, it was rumored that he grew *E. coli* in Anheuser-Busch beer vats. (He was working at

Washington University in St. Louis.) Why do you think this might have been helpful to the experiment?

8. How did Kornberg assess the fidelity of DNA polymerase I in copying a DNA template?

9. Which characteristics of DNA polymerase I raised doubts that its *in vivo* function is the synthesis of DNA leading to complete replication?

10. Kornberg showed that nucleotides are added to the 3'-end of each growing DNA strand. In what way does an exposed 3'-OH group participate in strand elongation?

11. What was the significance of the *polA1* mutation?

12. Summarize and compare the properties of DNA polymerase I, II, and III.

13. List and describe the function of the ten subunits constituting DNA polymerase III. Distinguish between the holoenzyme and the core enzyme.

14. Distinguish between (a) unidirectional and bidirectional synthesis, and (b) continuous and discontinuous synthesis of DNA.

15. List the proteins that unwind DNA during *in vivo* DNA synthesis. How do they function?

16. Define and indicate the significance of (a) Okazaki fragments, (b) DNA ligase, and (c) primer RNA during DNA replication.

17. Outline the current model for DNA synthesis.

18. Why is DNA synthesis expected to be more complex in eukaryotes than in bacteria? How is DNA synthesis similar in the two types of organisms?

19. If the analysis of DNA from two different microorganisms demonstrated very similar base compositions, are the DNA sequences of the two organisms also nearly identical?

20. Suppose that *E. coli* synthesizes DNA at a rate of 100,000 nucleotides per minute and takes 40 minutes to replicate its chromosome. (a) How many base pairs are present in the entire *E. coli* chromosome? (b) What is the physical length of the chromosome in its helical configuration—that is, what is the circumference of the chromosome if it were opened into a circle?

21. Several temperature-sensitive mutant strains of *E. coli* display the following characteristics. Predict what enzyme or function is being affected by each mutation.

    (a) Newly synthesized DNA contains many mismatched base pairs.

    (b) Okazaki fragments accumulate, and DNA synthesis is never completed.

    (c) No initiation occurs.

    (d) Synthesis is very slow.

    (e) Supercoiled strands remain after replication, which is never completed.

22. Reiji and Tuneko Okazaki conducted a now classic experiment in 1968 in which they discovered a population of short fragments synthesized during DNA replication. They introduced a short pulse of ^{3}H-thymidine into a culture of *E. coli* and extracted DNA from the cells at various intervals. In analyzing the DNA after centrifugation in denaturing gradients, they noticed that as the interval between the time of ^{3}H-thymidine introduction and the time of centrifugation increased, the proportion of short strands decreased and more labeled DNA was found in larger strands. What would account for this observation?

23. Many of the gene products involved in DNA synthesis were initially defined by studying mutant *E. coli* strains that could not synthesize DNA. (a) The *dnaE* gene encodes the α subunit of DNA polymerase III. What effect is expected from a mutation in this gene? How could the mutant strain be maintained? (b) The *dnaQ* gene encodes the ε subunit of DNA polymerase. What effect is expected from a mutation in this gene?

24. In 1994, telomerase activity was discovered in human cancer cell lines. Although telomerase is not active in human somatic tissue, this discovery indicated that humans do contain the genes for telomerase proteins and telomerase RNA. Since inappropriate activation of telomerase can cause cancer, why do you think the genes coding for this enzyme have been maintained in the human genome throughout evolution? Are there any types of human body cells where telomerase activation would be advantageous or even necessary? Explain.

25. The genome of *D. melanogaster* consists of approximately 1.7×10^8 base pairs. DNA synthesis occurs at a rate of 30 base pairs per second. In the early embryo, the entire genome is replicated in five minutes. How many bidirectional origins of synthesis are required to accomplish this feat?

26. Assume a hypothetical organism in which DNA replication is conservative. Design an experiment similar to that of Taylor, Woods, and Hughes that will unequivocally establish this fact. Using the format established in Figure 5, draw sister chromatids and illustrate the expected results depicting this mode of replication.

27. Given the following diagram, assume that the phase G1 chromosome on the left underwent one round of replication in ^{3}H-thymidine and that the metaphase chromosome on the right had both chromatids labeled. Which of the replicative models (conservative, dispersive, semiconservative) could be eliminated by this observation?

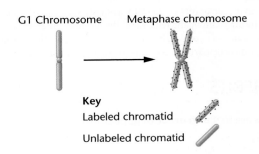

G1 Chromosome   Metaphase chromosome

Key
Labeled chromatid
Unlabeled chromatid

SOLUTIONS TO SELECTED PROBLEMS AND DISCUSSION QUESTIONS

Answers to Now Solve This

1. After one round of replication in the ^{14}N medium, the conservative scheme can be ruled out. After one round of replication in ^{14}N under a dispersive model, the DNA is of intermediate density, just as it is in the semiconservative model. However, in the next round of replication in ^{14}N medium, the density of the DNA is between the intermediate and "light" densities and therefore could be ruled out.

2. If the DNA contained parallel strands in the double helix and the polymerase were able to accommodate such parallel strands, there would be continuous synthesis and no Okazaki fragments. Several other possibilities exist. For example, if the DNA existed only as a single strand, the same results would occur.

Solutions to Problems and Discussion Questions

4. **(a)** Under a conservative scheme, all of the newly labeled DNA will go to one sister chromatid, while the other sister chromatid will remain unlabeled. In contrast to a semiconservative scheme, the first replicative round would produce one sister chromatid that has label on both strands of the double helix.

   (b) Under a dispersive scheme, all of the newly labeled DNA will be interspersed with unlabeled DNA. Because these preparations (metaphase chromosomes) are highly coiled and condensed structures derived from the "spread-out" form at interphase (which includes the S phase), it is impossible to detect the areas where label is not found. Rather, both sister chromatids would appear as evenly labeled structures.

6. The *in vitro* replication requires a DNA template, a primer to give a double-stranded portion, a divalent cation (Mg^{2+}), and all four of the deoxyribonucleoside triphosphates: dATP, dCTP, dTTP, and dGTP. The lowercase "d" refers to the deoxyribose sugar.

8. Several analytical approaches showed that the products of DNA polymerase I were probably copies of the template DNA. *Base composition* was used initially to compare both templates and products. Within experimental error, those data strongly suggested that the DNA replicated faithfully.

10. Each precursor (dNTP) to DNA synthesis is added to the 3′ end of a growing chain by the removal of the terminal phosphates and the formation of a covalent bond. The 3′-OH provided by the 3′ nucleotide directly participates in the formation of that covalent bond.

14. Given a stretch of double-stranded DNA, one could initiate synthesis at a given point and replicate strands either in one direction only (unidirectional) or in both directions (bidirectional). Notice that in the text the synthesis of complementary strands occurs in a *continuous* 5′>3′ mode on the leading strand in the direction of the replication fork, and in a *discontinuous* 5′>3′ mode on the lagging strand opposite the direction of the replication fork. Such discontinuous replication forms Okazaki fragments.

18. Eukaryotic DNA is replicated in a manner that is very similar to that of *E. coli*. Synthesis is bidirectional, continuous on one strand, and discontinuous on the other, and the requirements of synthesis (four deoxyribonucleoside triphosphates, divalent cation, template, and primer) are the same. Okazaki fragments of eukaryotes are about one-tenth the size of those in bacteria.

   Because there is a much greater amount of DNA to be replicated and DNA replication is slower, there are multiple initiation sites for replication in eukaryotes (and increased DNA polymerase per cell) in contrast to the single replication origin in prokaryotes. Replication occurs at different sites during different intervals of the S phase. The proposed functions of four DNA polymerases are described in the text. Because most eukaryotic chromosomes are linear, enzymes such as telomerase are needed to replicate the telomeres, or ends of chromosomes.

20. **(a)** In *E. coli*, 100 kb are added to each growing chain per minute. Therefore, the chain should be about 4,000,000 bp.

   (b) Given $(4 \times 10^6 \text{ bp}) \times 0.34 \text{ nm/bp} =$

   $$1.36 \times 10^6 \text{ nm or } 1.3 \text{ mm}$$

22. As time passes, the Okazaki fragments that are synthesized on the lagging strand are joined by DNA ligase so that larger strands are formed which form their own higher molecular weight peak.

24. Telomerase activity is present in germ-line tissue to maintain telomere length from one generation to the next. In other words, telomeres cannot shorten indefinitely without eventually eroding genetic information.

26. If replication is conservative, the first autoradiograms (see metaphase I in the text) would have label distributed on only one side (chromatid) of the metaphase chromosome.

CREDITS

Credits are listed in order of appearance.

Photo

CO, Dr. Gopal Murti/Science Photo Library/Photo Researchers, Inc.; F-5 (a1) Imageman/Shutterstock.com, F-5 (a2) Figure from "Molecular Genetics," Pt. 1 pp. 74–75, J.H. Taylor (ed). Copyright © 1963 and renewed 1991, reproduced with permission from Elsevier Science Ltd.; F10-5 (b) Figure from "Molecular Genetics," Pt. 1 pp. 74–75, J.H. Taylor (ed). Copyright © 1963 and renewed

1991, reproduced with permission from Elsevier Science Ltd.; F-6, Reprinted from CELL, Vol. 25, 1981, pp. 659, Sundin and Varshavsky, (1 figure), with permission from Elsevier Science. Courtesy of A. Varshavsky; F-13, David S. Hogness; F-14, Dr. Harold Weintraub, Howard Hughes Medical Institute, Fred Hutchinson Cancer Center/"Essential Molecular Biology" 2e, Freifelder & Malachinski, Jones & Bartlett, Fig. 7-24, pp. 141; F-17, David Dressler, Oxford University, England

Text

Source: Lasker Foundation.

Chromosome Structure and DNA Sequence Organization

From Chapter 11 of *Essentials of Genetics, Eighth edition,* William S. Klug, Michael R. Cummings, Charlotte A. Spencer, Michael A. Palladino. © 2013 by Pearson Education, Inc. All rights reserved.

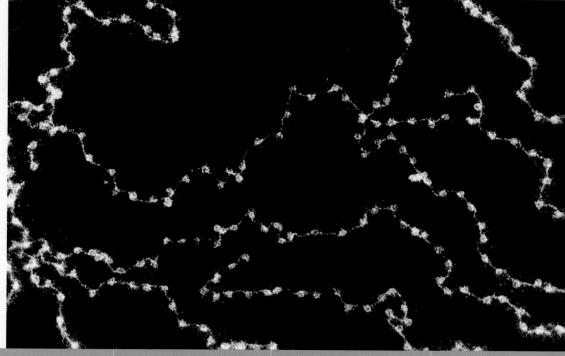

A chromatin fiber viewed using a scanning transmission electron microscope (STEM)

Chromosome Structure and DNA Sequence Organization

CHAPTER CONCEPTS

- Genetic information in viruses, bacteria, mitochondria, and chloroplasts is most often contained in a short, circular DNA molecule, relatively free of associated proteins.

- Eukaryotic cells, in contrast to viruses and bacteria, contain relatively large amounts of DNA organized into nucleosomes and present during most of the cell cycle as chromatin fibers.

- Uncoiled chromatin fibers characteristic of interphase coil up and condense into chromosomes during eukaryotic cell division.

- Eukaryotic genomes are characterized by both unique and repetitive DNA sequences.

- Eukaryotic genomes consist mostly of noncoding DNA sequences.

Once geneticists understood that DNA houses genetic information, it became very important to determine how DNA is organized into genes and how these basic units of genetic function are organized into chromosomes. In short, the major question had to do with how the genetic material was organized as it makes up the genome of organisms. There has been much interest in this question because knowledge of the organization of the genetic material and associated molecules is important to understanding many other areas of genetics. For example, the way in which the genetic information is stored, expressed, and regulated must be related to the molecular organization of the genetic molecule, DNA.

In this chapter, we focus on the various ways DNA is organized into chromosomes. These structures have been studied using numerous techniques, instruments, and approaches, including analysis by light microscopy and electron microscopy. More recently, molecular analysis has provided significant insights into chromosome organization. In the first half of the chapter, after surveying what we know about chromosomes in viruses and bacteria, we examine the large specialized eukaryotic structures called polytene and lampbrush chromosomes. Then, in the second half, we discuss how eukaryotic chromosomes are organized at the molecular level—for example, how DNA is complexed with proteins to form chromatin and how the chromatin fibers characteristic of interphase are condensed into chromosome structures visible during mitosis and meiosis. We conclude the chapter by examining certain aspects of DNA sequence organization in eukaryotic genomes.

1   Viral and Bacterial Chromosomes Are Relatively Simple DNA Molecules

The chromosomes of viruses and bacteria are much less complicated than those of eukaryotes. They usually consist of a single nucleic acid molecule, unlike the multiple chromosomes comprising the genome of higher forms. Compared to eukaryotes, the chromosomes contain much less genetic information and the DNA is not as extensively bound to proteins. These characteristics have greatly simplified analysis, and we now have a fairly comprehensive view of the structure of viral and bacterial chromosomes.

The chromosomes of viruses consist of a nucleic acid molecule—either DNA or RNA—that can be either single- or double-stranded. They can exist as circular structures (closed loops), or they can take the form of linear molecules. For example, the single-stranded DNA of the **ϕX174 bacteriophage** and the double-stranded DNA of the **polyoma virus** are closed loops housed within the protein coat of the mature viruses. The **bacteriophage lambda (λ)**, on the other hand, possesses a linear double-stranded DNA molecule prior to infection, which closes to form a ring upon its infection of the host cell. Still other viruses, such as the T-even series of bacteriophages, have linear double-stranded chromosomes of DNA that do not form circles inside the bacterial host. Thus, circularity is not an absolute requirement for replication in viruses.

Viral nucleic acid molecules have been seen with the electron microscope. **Figure 1** shows a mature bacteriophage λ and its double-stranded DNA molecule in the circular configuration. One constant feature shared by viruses, bacteria, and eukaryotic cells is the ability to package an exceedingly long DNA molecule into a relatively small volume. In λ, the DNA is 17 μm long and must fit into the phage head, which is less than 0.1 μm on any side. **Table 1** compares the length of the chromosomes of several viruses to the size of their head structure. In each case, a similar packaging feat must be accomplished. Compare the dimensions given for phage T2 with the micrograph of both the DNA and the viral particle shown in **Figure 2**. Seldom does the space available in the head of a virus exceed the chromosome volume by more than a factor of two. In many cases, almost all of the space is filled, indicating nearly perfect packing. Once packed within the head, the genetic material is functionally inert until it is released into a host cell.

Bacterial chromosomes are also relatively simple in form. They generally consist of a double-stranded DNA molecule, compacted into a structure sometimes referred to as the **nucleoid**. *Escherichia coli,* the most extensively studied bacterium, has a large circular chromosome measuring approximately 1200 μm (1.2 mm) in length that may occupy up to one-third of the volume of the cell. When the cell is gently lysed and the chromosome is released, it can be visualized under the electron microscope (**Figure 3**).

DNA in bacterial chromosomes is associated with several types of **DNA-binding proteins.** Two, called **HU** and **H1 proteins,** are small but abundant in the cell and contain a high percentage of positively charged amino acids that can bond ionically to the negative charges of the phosphate groups in DNA. Although these proteins are structurally similar to molecules called histones that are associated with eukaryotic DNA, they clearly do not compact DNA to the extent of histones, since the bacterial chromosome is *not* functionally inert and can be readily replicated and transcribed. Nor are these

(a)   (b)

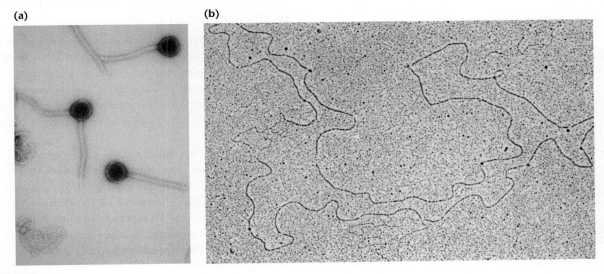

FIGURE 1   Electron micrographs of phage λ (a) and the DNA that was isolated from it (b). The chromosome is 17 μm long. Note that the phages are magnified about five times more than the DNA.

TABLE 1	The Genetic Material of Representative Viruses and Bacteria				
		Nucleic Acid			Overall Size of Viral
Organism		Type	SS or DS*	Length (μm)	Head or Bacteria (μm)
Viruses	φX174	DNA	SS	2.0	0.025 × 0.025
	Tobacco mosaic virus	RNA	SS	3.3	0.30 × 0.02
	Phage λ	DNA	DS	17.0	0.07 × 0.07
	T2 phage	DNA	DS	52.0	0.07 × 0.10
Bacteria	*Haemophilus influenzae*	DNA	DS	832.0	1.00 × 0.30
	Escherichia coli	DNA	DS	1200.0	2.00 × 0.50

*SS = single-stranded, DS = double-stranded.

proteins essential since null mutations in the genes encoding them are not lethal.

ESSENTIAL POINT ■ ■ ■

In contrast to eukaryotes, bacteriophage and bacterial chromosomes are largely devoid of associated proteins, are of much smaller size, and most often consist of circular DNA.

NOW SOLVE THIS

1 In bacteriophages and bacteria, the DNA is almost always organized into circular (closed loops) chromosomes. Phage λ is an exception, maintaining its DNA in a linear chromosome within the viral particle. However, as soon as this DNA is injected into a host cell, it circularizes before replication begins. What advantage exists in replicating circular DNA molecules compared to linear molecules, characteristic of eukaryotic chromosomes?

■ **Hint** *This problem involves an understanding of eukaryotic DNA replication. The key to its solution is to consider why the enzyme telomerase is essential in eukaryotic DNA replication, and why bacterial and viral chromosomes can be replicated without encountering the "telomere problem."*

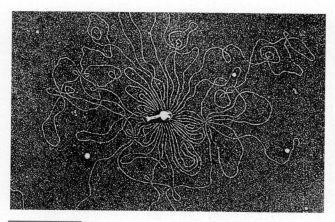

FIGURE 2 Electron micrograph of bacteriophage T2, which has had its DNA released by osmotic shock. The chromosome is 52 μm long.

2 | Mitochondria and Chloroplasts Contain DNA Similar to Bacteria and Viruses

That both **mitochondria** and **chloroplasts** contain their own DNA and a system for expressing genetic information was first suggested by the discovery of mutations and the resultant inheritance patterns in plants, yeast, and other fungi. Because both mitochondria and chloroplasts are inherited through the maternal cytoplasm in most organisms, and because each of the above-mentioned examples of mutations could be linked hypothetically to the altered function of either chloroplasts or mitochondria, geneticists set out to look for more direct evidence of DNA in these organelles. Not only was unique DNA found to be a normal component of both mitochondria and chloroplasts, but careful examination of the nature of this genetic information revealed a remarkable similarity to that found in viruses and bacteria.

Molecular Organization and Gene Products of Mitochondrial DNA

Extensive information is also available concerning the structure and gene products of **mitochondrial DNA (mtDNA).** In most eukaryotes, mtDNA exists as a double-stranded, closed circle **(Figure 4)** that is free of the chromosomal proteins characteristic of eukaryotic chromosomal DNA. An exception is found in some ciliated protozoans, in which the DNA is linear.

In size, mtDNA varies greatly among organisms. In a variety of animals, including humans, mtDNA consists of about 16,000 to 18,000 bp (16 to 18 kb). However, yeast (*Saccharomyces*) mtDNA consists of 75 kb. Plants typically exceed this amount—367 kb is present in mitochondria in the mustard plant, *Arabidopsis*. Vertebrates have 5 to 10 such DNA molecules per organelle, whereas plants have 20 to 40 copies per organelle.

There are several other noteworthy aspects of mtDNA. With only rare exceptions, *introns* (noncoding regions

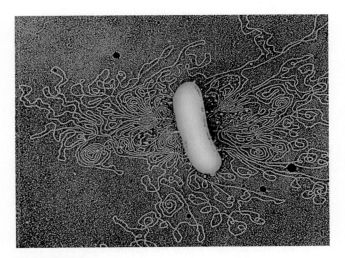

FIGURE 3 Electron micrograph of the bacterium *E. coli,* which has had its DNA released by osmotic shock. The chromosome is 1200 μm long.

within genes) are absent from mitochondrial genes, and gene repetitions are seldom present. Nor is there usually much in the way of intergenic spacer DNA. This is particularly true in species whose mtDNA is fairly small in size, such as humans. In *Saccharomyces*, with a much larger mtDNA molecule, introns and intergenic spacer DNA account for much of the excess DNA. The expression of mitochondrial genes uses several modifications of the otherwise standard genetic code. Also of interest is the fact that replication in mitochondria is dependent on enzymes encoded by nuclear DNA.

Another interesting observation is that in vertebrate mtDNA, the two strands vary in density, as revealed by centrifugation. This provides researchers with a way to isolate the strands for study, designating one heavy (H) and the other light (L). While most of the mitochondrial genes are encoded by the H strand, several are encoded by the complementary L strand.

Molecular Organization and Gene Products of Chloroplast DNA

Chloroplasts provide the photosynthetic function specific to plants. Like mitochondria, they contain an autonomous genetic system distinct from that found in the nucleus and cytoplasm, which has as its foundation a unique DNA molecule (**cpDNA**). **Chloroplast DNA,** shown in **Figure 5**, is fairly uniform in size among different organisms, ranging between 100 and 225 kb in length. It shares many similarities to DNA found in prokaryotic cells. It is circular and double-stranded, and it is free of the associated proteins characteristic of eukaryotic DNA.

The size of cpDNA is much larger than that of mtDNA. To some extent, this can be accounted for by a larger number of genes. However, most of the difference appears to be due to the presence in cpDNA of many long noncoding nucleotide sequences both between and within genes, the latter being called **introns.** Duplications of many DNA sequences are also present.

In the green alga *Chlamydomonas*, there are about 75 copies of the chloroplast DNA molecule per organelle. Each copy consists of a length of DNA that contains 195,000 base pairs (195 kb). In higher plants, such as the sweet pea, multiple copies of the DNA molecule are also present in each organelle, but the molecule is considerably smaller (134 kb) than that in *Chlamydomonas*.

ESSENTIAL POINT ■ ■ ■

Mitochondria and chloroplasts contain DNA that is remarkably similar in form and appearance to some bacterial and bacteriophage DNA.

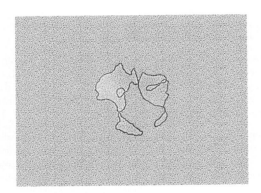

FIGURE 4 Electron micrograph of mitochondrial DNA (mtDNA) derived from *Xenopus laevis*.

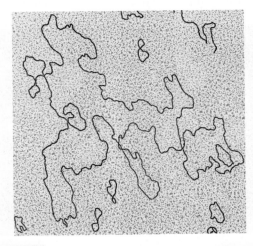

FIGURE 5 Electron micrograph of chloroplast DNA obtained from lettuce.

257

3   Specialized Chromosomes Reveal Variations in the Organization of DNA

We now consider two cases of genetic organization that demonstrate the specialized forms that eukaryotic chromosomes can take. Both types—*polytene chromosomes* and *lampbrush chromosomes*—are so large that their organization was discerned using light microscopy long before we understood how mitotic chromosomes form from interphase chromatin. The study of these chromosomes provided many of our initial insights into the arrangement and function of the genetic information. It is important to note that polytene and lampbrush chromosomes are unusual and not typically found in most eukaryotic cells, but the study of their structure has revealed many common themes of chromosome organization.

Polytene Chromosomes

Giant **polytene chromosomes** are found in various tissues (salivary, midgut, rectal, and malpighian excretory tubules) in the larvae of some flies and in several species of protozoans and plants. Such structures, first observed by E. G. Balbiani in 1881, provided a model system for subsequent investigations of chromosomes. What is particularly intriguing about polytene chromosomes is that they can be seen in the nuclei of interphase cells.

Each polytene chromosome is 200 to 600 μm long, and when they are observed under the light microscope, they reveal a linear series of alternating bands and interbands (**Figure 6**). The banding pattern is distinctive for each chromosome in any given species. Individual bands are sometimes called **chromomeres,** a generalized term describing lateral condensations of material along the axis of a chromosome.

Extensive study using electron microscopy and radioactive tracers led to an explanation for the unusual appearance of these chromosomes. First, polytene chromosomes represent paired homologs. This is highly unusual because they are present in somatic cells, where in most organisms, chromosomal material is normally dispersed as chromatin and homologs are not paired. Second, their large size and distinctiveness result from the many DNA strands that compose them. The DNA of these paired homologs undergoes many rounds of replication, *but without strand separation or cytoplasmic division.* As replication proceeds, chromosomes contain 1000 to 5000 DNA strands that remain in precise parallel alignment with one another, giving rise to the distinctive band pattern along the axis of the chromosome.

The presence of bands on polytene chromosomes was initially interpreted as the visible manifestation of individual genes. The discovery that the strands present in bands undergo localized uncoiling during genetic activity further strengthened this view. Each such uncoiling event results in what is called a **puff** because of its appearance (**Figure 7**). That puffs are visible manifestations of gene activity (transcription that produces RNA) is evidenced by their high rate of incorporation of radioactively labeled RNA precursors, as assayed by autoradiography. Bands that are not extended into puffs incorporate fewer radioactive precursors or none at all.

The study of bands during development in insects such as *Drosophila* and the midge fly *Chironomus* reveals *differential gene activity.* A characteristic pattern of band formation that is equated with gene activation is observed as development proceeds. Despite attempts to resolve the issue, it is not yet clear how many genes are contained in each band. However, we do know that in *Drosophila,* which contains about 15,000 genes, there are approximately 5000 bands. Interestingly, a band may contain up to 10^7 base pairs of DNA, enough to encode 50 to 100 average-size genes.

NOW SOLVE THIS

2 After salivary gland cells from *Drosophila* are isolated and cultured in the presence of radioactive thymidylic acid, autoradiography is performed, revealing the presence of thymidine within polytene chromosomes. Predict the distribution of the grains along the chromosomes.

■ **Hint** *This problem involves an understanding of the organization of DNA in polytene chromosomes. The key to its solution is to be aware that ^{3}H-thymidine, as a molecular tracer, will only be incorporated into DNA during its replication.*

Lampbrush Chromosomes

Another specialized chromosome that has given us insight into chromosomal structure is the **lampbrush chromosome,** so named because it resembles the brushes used to clean

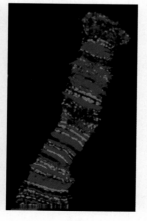

FIGURE 6   Polytene chromosomes derived from larval salivary gland cells of *Drosophila.*

FIGURE 7   Photograph of a puff within a polytene chromosome. The diagram depicts the uncoiling of strands within a band (B) region to produce a puff (P) in polytene chromosomes. Interband regions (IB) are also labeled.

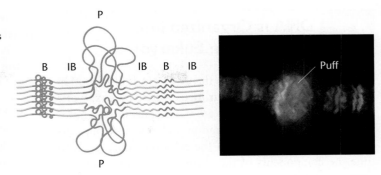

kerosene-lamp chimneys in the nineteenth century. Lampbrush chromosomes were first discovered in 1892 in the oocytes of sharks and are now known to be characteristic of most vertebrate oocytes as well as the spermatocytes of some insects. Therefore, they are meiotic chromosomes. Most experimental work has been done with material taken from amphibian oocytes.

These unique chromosomes are easily isolated from oocytes in the first prophase stage of meiosis, where they are active in directing the metabolic activities of the developing cell. The homologs are seen as synapsed pairs held together by chiasmata. However, instead of condensing, as most meiotic chromosomes do, lampbrush chromosomes often extend to lengths of 500 to 800 μm. Later in meiosis, they revert to their normal length of 15 to 20 μm. Based on these observations, lampbrush chromosomes are interpreted as extended, uncoiled versions of the normal meiotic chromosomes.

The two views of lampbrush chromosomes in **Figure 8** provide significant insights into their morphology. Part (a) shows the meiotic configuration under the light microscope. The linear axis of each structure contains a large number of condensed areas, and as with polytene chromosomes, these are referred to as *chromomeres*. Emanating from each chromomere is a pair of lateral loops, which give the chromosome its distinctive appearance. In part (b), the scanning electron micrograph (SEM) reveals adjacent loops present along one of the two axes of the chromosome. As with bands in polytene chromosomes, much more DNA is present in each loop than is needed to encode a single gene. Such an SEM provides a clear view of the chromomeres and the chromosomal fibers emanating from them. Each chromosomal loop is thought to be composed of one DNA double helix, while the central axis is made up of two DNA helices. This hypothesis is consistent with the belief that each meiotic chromosome is composed of a pair of sister chromatids. Studies

using radioactive RNA precursors reveal that the loops are active in the synthesis of RNA. The lampbrush loops, in a manner similar to puffs in polytene chromosomes, represent DNA that has been reeled out from the central chromomere axis during transcription.

ESSENTIAL POINT ▪ ▪ ▪

Polytene and lampbrush chromosomes are examples of specialized structures that extended our knowledge of genetic organization and function well in advance of the technology available to the modern-day molecular biologist.

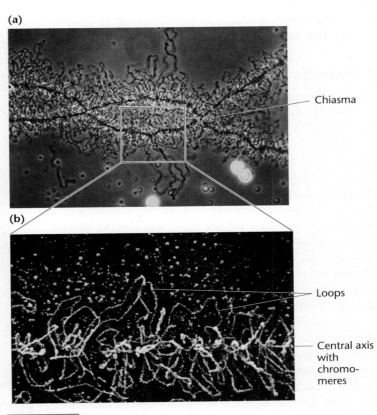

FIGURE 8   Lampbrush chromosomes derived from amphibian oocytes. Part (a) is a photomicrograph; part (b) is a scanning electron micrograph.

4 DNA Is Organized into Chromatin in Eukaryotes

We now turn our attention to the way DNA is organized in eukaryotic chromosomes. Our focus will be on eukaryotic cells, in which chromosomes are visible only during mitosis. After chromosome separation and cell division, cells enter the interphase stage of the cell cycle, during which time the components of the chromosome uncoil and are present in the form referred to as **chromatin.** While in interphase, the chromatin is dispersed in the nucleus, and the DNA of each chromosome is replicated. As the cell cycle progresses, most cells reenter mitosis, whereupon chromatin coils into visible chromosomes once again. This condensation represents a length contraction of some 10,000 times for each chromatin fiber.

The organization of DNA during the transitions just described is much more intricate and complex than in viruses or bacteria, which never exhibit a process similar to mitosis. This is due to the greater amount of DNA per chromosome, as well as the presence of a large number of proteins associated with eukaryotic DNA. For example, while DNA in the *E. coli* chromosome is 1200 μm long, the DNA in each human chromosome ranges from 19,000 to 73,000 μm in length. In a single human nucleus, all 46 chromosomes contain sufficient DNA to extend to more than 2 meters. This genetic material, along with its associated proteins, is contained within a nucleus that usually measures about 5 to 10 μm in diameter.

Chromatin Structure and Nucleosomes

As we have seen, the genetic material of viruses and bacteria consists of strands of DNA or RNA that are nearly devoid of proteins. In eukaryotic chromatin, a substantial amount of protein is associated with the chromosomal DNA in all phases of the eukaryotic cell cycle. The associated proteins are divided into basic, positively charged **histones** and less positively charged nonhistones. The histones clearly play the most essential structural role of all the proteins associated with DNA. There are five types, and they all contain large amounts of the positively charged amino acids lysine and arginine. This makes it possible for them to bond electrostatically to the negatively charged phosphate groups of nucleotides in DNA. Recall that a similar interaction has been proposed for several bacterial proteins.

The general model for chromatin structure is based on the assumption that chromatin fibers, composed of DNA and protein, undergo extensive coiling and folding as they are condensed within the cell nucleus. X-ray diffraction studies confirm that histones play an important role in chromatin structure. Chromatin produces regularly spaced diffraction rings, suggesting that repeating structural units occur along the chromatin axis. If the histone molecules are chemically removed from chromatin, the regularity of this diffraction pattern is disrupted.

A basic model for chromatin structure was worked out in the mid-1970s. Several observations were particularly relevant to the development of this model:

1. Digestion of chromatin by certain endonucleases, such as micrococcal nuclease, yields DNA fragments that are approximately 200 bp in length or multiples thereof. This demonstrates that enzymatic digestion is not random, for if it were, we would expect a wide range of fragment sizes. Thus, chromatin consists of some type of repeating unit, each of which is protected from enzymatic cleavage, except where any two units are joined. It is the area between units that is attacked and cleaved by the endonuclease.

2. Electron microscopic observations of chromatin reveal that chromatin fibers are composed of linear arrays of spherical particles (**Figure 9**). Discovered by Ada and Donald Olins, the particles occur regularly along the axis of a chromatin strand and resemble beads on a string. This conforms nicely to the earlier observation, which suggests the existence of repeating units. These particles, initially referred to as *v*-bodies (*v* is the Greek letter nu), are now called **nucleosomes.**

3. Studies of precise interactions of histone molecules and DNA in the nucleosomes constituting chromatin show

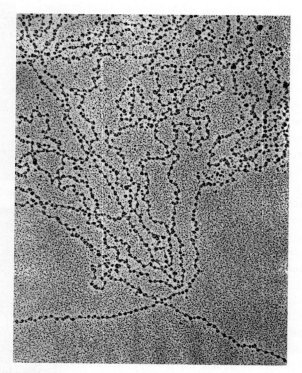

FIGURE 9   An electron micrograph revealing nucleosomes appearing as "beads on a string" along chromatin strands derived from *Drosophila melanogaster.*

that histones H2A, H2B, H3, and H4 occur as two types of tetramers, $(H2A)_2 \bullet (H2B)_2$ and $(H3)_2 \bullet (H4)_2$. Roger Kornberg predicted that each repeating nucleosome unit consists of one of each tetramer (creating an octamer) in association with about 200 bp of DNA. Such a structure is consistent with previous observations and provides the basis for a model that explains the interaction of histones and DNA in chromatin.

4. When nuclease digestion time is extended, some of the 200 bp of DNA are removed from the nucleosome, creating a **nucleosome core particle** consisting of 147 bp. The DNA lost in this prolonged digestion is responsible for linking nucleosomes together. This linker DNA is associated with the fifth histone, H1.

On the basis of this information, as well as on X-ray and neutron-scattering analyses of crystallized core particles by John T. Finch, Aaron Klug, and others, a detailed model of

the nucleosome was put forward in 1984, providing a basis for predicting chromatin structure and its condensation into chromosomes. In this model, illustrated in **Figure 10**, a 147-bp length of the 2-nm-diameter DNA molecule coils around an octamer of histones in a left-handed superhelix that completes about 1.7 turns per nucleosome. Each nucleosome, ellipsoidal in shape, measures about 11 nm at its longest point [**Figure 10(a)**]. Significantly, the formation of the nucleosome represents the first level of packing, whereby the DNA helix is reduced to about one-third of its original length by winding around the histones.

In the nucleus, the chromatin fiber seldom, if ever, exists in the extended form described in the previous paragraph (that is, as an extended chain of nucleosomes). Instead, the 11-nm-diameter fiber is further packed into a thicker, 30-nm-diameter structure that was initially called a *solenoid* [**Figure 10(b)**]. This thicker structure, which is dependent on the presence of histone H1, consists of numerous nucleo-

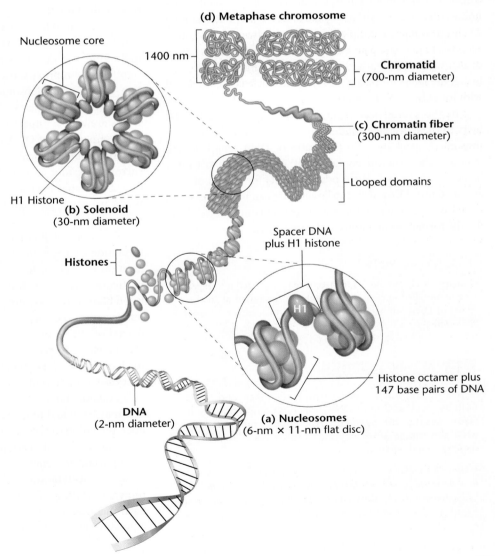

FIGURE 10 General model of the association of histones and DNA to form nucleosomes, illustrating the way in which each thickness of fiber may be coiled into a more condensed structure, ultimately producing a metaphase chromosome.

Nucleosome core

H1 Histone

(b) Solenoid (30-nm diameter)

Histones

DNA (2-nm diameter)

(d) Metaphase chromosome

1400 nm

Chromatid (700-nm diameter)

(c) Chromatin fiber (300-nm diameter)

Looped domains

Spacer DNA plus H1 histone

H1

Histone octamer plus 147 base pairs of DNA

(a) Nucleosomes (6-nm × 11-nm flat disc)

somes coiled around and stacked upon one another, creating a second level of packing. This provides a six-fold increase in compaction of the DNA. It is this structure that is characteristic of an uncoiled chromatin fiber in interphase of the cell cycle. In the transition to the mitotic chromosome, still further compaction must occur. The 30-nm structures are folded into a series of *looped domains,* which further condense the chromatin fiber into a structure that is 300 nm in diameter [**Figure 10(c)**]. These *coiled chromatin fibers* are then compacted into the chromosome arms that constitute a chromatid, one of the longitudinal subunits of the metaphase chromosome [**Figure 10(d)**]. While Figure 10 shows the chromatid arms to be 700 nm in diameter, this value undoubtedly varies among different organisms. At a value of 700 nm, a pair of sister chromatids comprising a chromosome measures about 1400 nm.

The importance of the organization of DNA into chromatin and chromatin into mitotic chromosomes can be illustrated by considering a human cell that stores its genetic material in a nucleus that is about 5 to 10 μm in diameter. The haploid genome contains 3.2×10^9 base pairs of DNA distributed among 23 chromosomes. The diploid cell contains twice that amount. At 0.34 nm per base pair, this amounts to an enormous length of DNA (as stated earlier, to more than 2 m). One estimate is that the DNA inside a typical human nucleus is complexed with roughly 2.5×10^7 nucleosomes.

In the overall transition from a fully extended DNA helix to the extremely condensed status of the mitotic chromosome, a packing ratio (the ratio of DNA length to the length of the structure containing it) of about 500 to 1 must be achieved. In fact, our model accounts for a ratio of only about 50 to 1. Obviously, the larger fiber can be further bent, coiled, and packed to achieve even greater condensation during the formation of a mitotic chromosome.

ESSENTIAL POINT ■ ■ ■

Eukaryotic chromatin is a nucleoprotein organized into repeating units called nucleosomes, which are composed of 200 base pairs of DNA, an octamer of four types of histones, plus one linker histone.

NOW SOLVE THIS

3 If a human nucleus is 10 μm in diameter, and it must hold as much as 2 m of DNA, which is complexed into nucleosomes that during full extension are 11 nm in diameter, what percentage of the volume of the nucleus does the genetic material occupy?

■ **Hint** *This problem asks you to make some numerical calculations in order to see just how "filled" the eukaryotic nucleus is with a diploid amount of DNA. The key to its solution is the use of the formula* $V = (4/3)\pi r^3$, *which calculates the volume of a sphere.*

Chromatin Remodeling

As with many significant findings in genetics, the study of nucleosomes has answered some important questions, but at the same time it has also led us to new ones. For example, in the preceding discussion, we established that histone proteins play an important structural role in packaging DNA into the nucleosomes that make up chromatin. While solving the structural problem of how to organize a huge amount of DNA within the eukaryotic nucleus, a new problem was apparent: *the chromatin fiber, when complexed with histones and folded into various levels of compaction, makes the DNA inaccessible to interaction with important nonhistone proteins.* For example, the proteins that function in enzymatic and regulatory roles during the processes of replication and gene expression must interact directly with DNA. To accommodate these protein–DNA interactions, chromatin must be induced to change its structure, a process called **chromatin remodeling.** In the case of replication and gene expression, chromatin must relax its compact structure but be able to reverse the process during periods of inactivity.

Insights into how different states of chromatin structure may be achieved were forthcoming in 1997, when Timothy Richmond and members of his research team were able to significantly improve the level of resolution in X-ray diffraction studies of nucleosome crystals (from 7 Å in the 1984 studies to 2.8 Å in the 1997 studies). At this resolution, most atoms are visible, thus revealing the subtle twists and turns of the superhelix of DNA that encircles the histones. Recall that the double-helical ribbon represents 147 bp of DNA surrounding four pairs of histone proteins. This configuration is repeated over and over in the chromatin fiber and is the principal packaging unit of DNA in the eukaryotic nucleus.

The work of Richmond and colleagues, extended to a resolution of 1.9 Å in 2003, has revealed the details of the location of each histone entity within the nucleosome. Of particular interest to chromatin remodeling is that unstructured **histone tails** are not packed into the folded histone domains within the core of the nucleosome. For example, tails devoid of any secondary structure extending from histones H3 and H2B protrude through the minor groove channels of the DNA helix. The tails of histone H4 appear to make a connection with adjacent nucleosomes. Histone tails also provide potential targets for a variety of chemical modifications that may be linked to genetic functions along the chromatin fiber, including the regulation of gene expression.

Potential modifications of histones, now recognized as important to genetic function, include the chemical processes of **acetylation, methylation, ubiquitination,** and **phosphorylation** of amino acids that are part of histones. Such chemical modifications are believed to be related to gene regulation as well as the cycle of chromatin unfolding and condensation that occurs during and after DNA replication. Although a great deal more work must be done to elucidate the specific involvement of chromatin remodeling during genetic processes, it is now clear that the dynamic forms in which chromatin exists are vitally important to the way that all genetic processes directly involving DNA are executed.

We will also return to a more detailed discussion of the role of chromatin remodeling when we consider the phenomenon of epigenetics, the study of modifications of an organism's genetic and phenotypic expression that are *not* attributable to alteration of the DNA sequence making up a gene.

Heterochromatin

Although we know that the DNA of the eukaryotic chromosome consists of one continuous double-helical fiber along its entire length, we also know that the whole chromosome is not structurally uniform from end to end. In the early part of the twentieth century, it was observed that some parts of the chromosome remain condensed and stain deeply during interphase, while most parts are uncoiled and do not stain. In 1928, the terms **euchromatin** and **heterochromatin** were coined to describe the parts of chromosomes that are uncoiled and those that remain condensed, respectively.

Subsequent investigation revealed a number of characteristics that distinguish heterochromatin from euchromatin. Heterochromatic areas are genetically inactive because they either lack genes or contain genes that are repressed. Also, heterochromatin replicates later during the S phase of the cell cycle than euchromatin does. The discovery of heterochromatin provided the first clues that parts of eukaryotic chromosomes do not always encode proteins. Instead, some chromosome regions are thought to be involved in maintenance of the chromosome's structural integrity and in other functions, such as chromosome movement during cell division.

The presence of heterochromatin is unique to and characteristic of the genetic material of eukaryotes. In some cases, whole chromosomes are heterochromatic. A case in point is the mammalian Y chromosome, much of which is genetically inert. The inactivated X chromosome in mammalian females is condensed into an inert heterochromatic Barr body. In some species, such as mealy bugs, all chromosomes of one entire haploid set are heterochromatic.

When certain heterochromatic areas from one chromosome are translocated to a new site on the same or another nonhomologous chromosome, genetically active areas sometimes become genetically inert if they lie adjacent to the translocated heterochromatin. This influence on existing euchromatin is one example of what is more generally referred to as a **position effect.** That is, the position of a gene or group of genes relative to all other genetic material may affect their expression.

Chromosome Banding

Until about 1970, mitotic chromosomes viewed under the light microscope could be distinguished only by their relative sizes and the positions of their centromeres. In karyotypes, two or more chromosomes are often visually indistinguishable from one another. Numerous cytological procedures, referred to as **chromosome banding,** have now made it possible to distinguish such chromosomes from one another as a result of differential staining along the longitudinal axis of mitotic chromosomes.

The most useful of these techniques, called **G-banding** (see Figure 2), involves the digestion of the mitotic chromosomes with the proteolytic enzyme trypsin, followed by Giemsa staining. This procedure stains regions of DNA that are rich in A=T base pairs. Another technique, called **C-banding,** uses chromosome preparations that are heat denatured. Subsequent Giemsa staining reveals only the heterochromatic regions of the centromeres.

These, and other chromosome-banding techniques, reflect the heterogeneity and complexity of the chromosome along its length. So precise is the banding pattern that when a segment of one chromosome has been translocated to another chromosome, its origin can be determined with great precision.

> **ESSENTIAL POINT**
>
> Heterochromatin, prematurely condensed in interphase and for the most part genetically inert, is illustrated by the centromeric and telomeric regions of eukaryotic chromosomes, the Y chromosome, and the Barr body.

5 Eukaryotic Genomes Demonstrate Complex Sequence Organization Characterized by Repetitive DNA

Thus far, we have looked at how DNA is organized into chromosomes in bacteriophages, bacteria, and eukaryotes. We now begin an examination of what we know about the organization of DNA sequences within the chromosomes making up an organism's genome, placing our emphasis on eukaryotes.

263

In addition to single copies of unique DNA sequences that comprise genes, a great deal of the DNA sequences within chromosomes are repetitive in nature and that various levels of repetition occur within the genome of organisms. Many studies have now provided insights into **repetitive DNA,** demonstrating various classes of these sequences and their organization within the genome. **Figure 11** outlines the various categories of repetitive DNA. Some functional genes are present in more than one copy and are therefore repetitive in nature. However, the majority of repetitive sequences are nongenic, and in fact, most serve no known function. We explore three main categories: (1) heterochromatin found associated with centromeres and making up telomeres, (2) tandem repeats of both short and long DNA sequences, and (3)

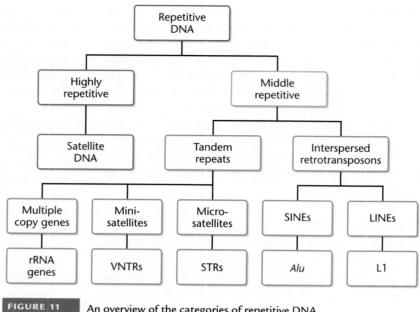

FIGURE 11 An overview of the categories of repetitive DNA.

transposable sequences that are interspersed throughout the genome of eukaryotes.

Repetitive DNA and Satellite DNA

The nucleotide composition of the DNA (e.g., the percentage of $G \equiv C$ versus $A = T$ pairs) of a particular species is reflected in its density, which can be measured with sedimentation equilibrium centrifugation. When eukaryotic DNA is analyzed in this way, the majority is present as a single main band, or peak, of fairly uniform density. However, one or more additional peaks represent DNA that differs slightly in density. This component, called **satellite DNA,** represents a variable proportion of the total DNA, depending on the species. A profile of main-band and satellite DNA from the mouse is shown in **Figure 12.** By contrast, prokaryotes contain only main-band DNA.

The significance of satellite DNA remained an enigma until the mid-1960s, when Roy Britten and David Kohne developed the technique for measuring the reassociation kinetics of DNA that had previously been dissociated into single strands. They demonstrated that certain portions of DNA reassociated more rapidly than others. They concluded that rapid reassociation was characteristic of multiple DNA fragments composed of identical or nearly identical nucleotide sequences—the basis for the descriptive term **repetitive DNA.**

When satellite DNA is subjected to analysis by reassociation kinetics, it falls into the category of **highly repetitive DNA,** which is known to consist of relatively short sequences repeated a large number of times. Further evidence suggests that these sequences are present as tandem repeats clustered in very specific chromosomal areas known to be

heterochromatic—the regions flanking centromeres. This was discovered in 1969 when several researchers, including Mary Lou Pardue and Joe Gall, applied *in situ* **molecular hybridization** to the study of satellite DNA. This technique involves the molecular hybridization between an isolated fraction of radioactively labeled DNA or RNA probes and the DNA contained in the chromosomes of a cytological preparation. Following the hybridization procedure, autoradiography is performed to locate the chromosome areas complementary to the fraction of DNA or RNA.

Pardue and Gall demonstrated that radioactive probes made from mouse satellite DNA hybridize with the DNA of centromeric regions of mouse mitotic chromosomes, which are all telocentric **(Figure 13).** Several conclusions were

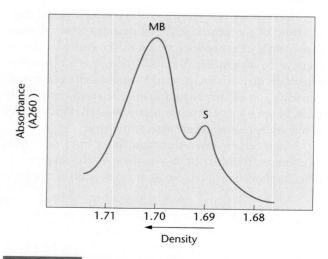

FIGURE 12 Separation of main-band (MB) and satellite (S) DNA from the mouse, using ultracentrifugation in a CsCl gradient.

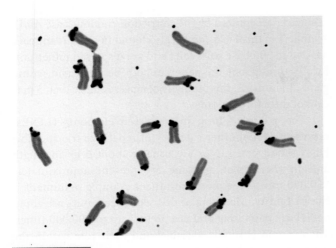

FIGURE 13 *In situ* molecular hybridization between RNA transcribed from mouse satellite DNA and mitotic chromosomes. The grains in the autoradiograph localize the chromosome regions (the centromeres) containing satellite DNA sequences.

drawn: Satellite DNA differs from main-band DNA in its molecular composition, as established by buoyant density studies. It is composed of repetitive sequences. Finally, satellite DNA is found in the heterochromatic centromeric regions of chromosomes.

Centromeric DNA Sequences

The separation of homologs during mitosis and meiosis depends on **centromeres,** described cytologically as the *primary constrictions* along eukaryotic chromosomes. In this role, it is believed that the DNA sequence contained within the centromere is critical. Careful analysis has confirmed this prediction. The minimal region of the centromere that supports the function of chromosomal segregation is designated the **CEN region.** Within this heterochromatic region of the chromosome, the DNA binds a platform of proteins, which in multicellular organisms includes the **kinetochore** that binds to the spindle fiber during division (see Figure 8).

The CEN regions of the yeast *Saccharomyces cerevisiae* were the first to be studied. Each centromere serves an identical function, so it is not surprising that CENs from different chromosomes were found to be remarkably similar in their organization. The CEN region of yeast chromosomes consists of about 120 bp. Mutational analysis suggests that portions near the 3′-end of this DNA region are most critical to centromere function since mutations in them, but not those nearer the 5′-end, disrupt centromere function. Thus, the DNA of this region appears to be essential to the eventual binding to the spindle fiber (yeast do not have kinetechores).

Centromere sequences of multicellular eukaryotes are much more extensive than in yeast and vary considerably in

size. Such sequences, absent from yeast but characteristic of most multicellular organisms, vary considerably in size. For example, in *Drosophila* the CEN region is found within some 200 to 600 kb of DNA, much of which is highly repetitive. Recall from our prior discussion that highly repetitive satellite DNA is localized in the centromere regions of mice. In humans, one of the most recognized satellite DNA sequences is the **alphoid family,** found mainly in the centromere regions. Alphoid sequences, each about 170 bp in length, are present in tandem arrays of up to 1 million base pairs. Embedded within this repetitive DNA are more specific sequences that are critical to centromere function.

Telomeric DNA Sequences

We now return to a consideration of a region of the chromosome, the **telomere**—the structure that "caps" the ends of linear eukaryotic chromosomes. DNA sequences in these structures function to maintain the stability of chromosomes by rendering chromosome ends inert in interactions with other chromosome ends and with enzymes that use double-stranded DNA ends as substrates (such as repair enzymes). As with centromeres, the analysis of telomeres was first approached by investigating smaller chromosomes of simple eukaryotes, such as protozoans and yeast.

Telomeric DNA sequences consist of short tandem repeats. It is this group of repetitive sequences that contributes to the stability and integrity of the chromosome. In the ciliate *Tetrahymena,* more than 50 tandem repeats of the hexanucleotide sequence 5′-TTGGGG-3′ occur. In all vertebrates, including humans, the sequence 5′-TTAGGGG-3′ is repeated many times. The number of copies making up a telomere varies in different organisms, and there may be as many as 1000 repeats in some species.

A central question concerns the role of these repeat DNA sequences in telomere function. Recent findings have established that the sequences are transcribed and that the RNA product, called **TERRA (telomere repeat-containing RNA),** is an integral component of the telomere, contributing to its heterochromatic nature by facilitating methylation of the histone H3K9. In addition, these sequences have been shown to regulate telomerase, the RNA-containing enzyme that replicates the telomere. This is possible because TERRA sequences are complementary to those of the RNA component of *telomerase,* which provides the template for the synthesis of telomeric DNA. Serving as a telomerase ligand, TERRA acts as an inhibitor of telomerase.

Learning about telomerase regulation is of great interest, since in multicellular organisms, including humans, telomerase is active in germ-line cells but is inactive in somatic cells. And, in human cancer cells, which have become immortalized, the transition to malignancy appears to require the activation of telomerase in order to overcome the normal senescence associated with chromosome shortening.

Middle Repetitive Sequences: VNTRs and STRs

A brief look at still another prominent category of repetitive DNA sheds additional light on the organization of the eukaryotic genome. In addition to highly repetitive DNA, which constitutes about 5 percent of the human genome (and 10 percent of the mouse genome), a second category, **middle (or moderately) repetitive DNA,** is fairly well characterized. Because we now know a great deal about the human genome, we will use our own species to illustrate this category of DNA in genome organization.

Although middle repetitive DNA does include some duplicated genes (such as those encoding ribosomal RNA), most prominent in this category are either noncoding tandemly repeated sequences or noncoding interspersed sequences. No function has been ascribed to these components of the genome. An example is DNA described as **variable number tandem repeats (VNTRs).** These repeating DNA sequences may be 15 to 100 bp long and are found within and between genes. Many such clusters are dispersed throughout the genome and are often referred to as **minisatellites.**

The number of tandem copies of each specific sequence at each location varies from one individual to the next, creating localized regions of 1,000 to 20,000 bp (1–20 kb) in length. The variation in size (length) of these regions between individual humans was originally the basis for the forensic technique referred to as **DNA fingerprinting.**

Another group of tandemly repeated sequences consists of di-, tri-, tetra-, and pentanucleotides, also referred to as **microsatellites** or **short tandem repeats (STRs).** Like VNTRs, they are dispersed throughout the genome and vary among individuals in the number of repeats present at any site. For example, in humans, the most common microsatellite is the dinucleotide $(CA)_n$, where n equals the number of repeats. Most commonly, n is between 5 and 50. These clusters have served as useful molecular markers for genome analysis.

Repetitive Transposed Sequences: SINEs and LINEs

Still another category of repetitive DNA consists of sequences that are interspersed individually throughout the genome, rather than being tandemly repeated. They can be either short or long, and many have the added distinction of being similar to **transposable sequences,** which are mobile and can potentially relocate within the genome. A large portion of the human genome is composed of such sequences.

For example, **short interspersed elements,** called **SINEs,** are less than 500 base pairs long and may be present 500,000 times or more in the human genome. The best characterized human SINE is a set of closely related sequences called the *Alu* family (the name is based on the presence of DNA

sequences recognized by the restriction endonuclease *Alu*I). Members of this DNA family, also found in other mammals, are 200 to 300 base pairs long and are dispersed rather uniformly throughout the genome, both between and within genes. In humans, this family encompasses more than 5 percent of the entire genome.

The group of **long interspersed elements (LINEs)** represents yet another category of repetitive transposable DNA sequences. LINEs are usually about 6 kb in length and in the human genome are present approximately 850,000 times. The most prominent example in humans is the **L1 family.** Members of this sequence family are about 6400 base pairs long and are present up to 100,000 times. Their 5′-end is highly variable, and their role within the genome has yet to be defined.

The general mechanism for transposition of L1 elements is now clear. The L1 DNA sequence is first transcribed into an RNA molecule. The RNA then serves as the template for the synthesis of the DNA complement using the enzyme *reverse transcriptase.* This enzyme is encoded by a portion of the L1 sequence. The new L1 copy then integrates into the DNA of the chromosome at a new site. Because of the similarity of this transposition mechanism to that used by retroviruses, LINEs are referred to as **retrotransposons.**

SINEs and LINEs represent a significant portion of human DNA. SINEs constitute about 13 percent of the human genome, whereas LINEs constitute up to 21 percent. Within both types of elements, repeating sequences of DNA are present in combination with unique sequences.

Middle Repetitive Multiple-Copy Genes

In some cases, middle repetitive DNA includes functional genes present tandemly in multiple copies. For example, many copies exist of the genes encoding ribosomal RNA. *Drosophila* has 120 copies per haploid genome. Single genetic units encode a large precursor molecule that is processed into the 5.8*S*, 18*S*, and 28*S* rRNA components. In humans, multiple copies of this gene are clustered on the p arm of the acrocentric chromosomes 13, 14, 15, 21, and 22. Multiple copies of the genes encoding 5*S* rRNA are transcribed separately from multiple clusters found together on the terminal portion of the p arm of chromosome 1.

ESSENTIAL POINT ■ ■ ■

Eukaryotic genomes demonstrate complex sequence organization characterized by numerous categories of repetitive DNA, consisting of either tandem repeats clustered in various regions of the genome or single sequences repeatedly interspersed at random throughout the genome.

6 The Vast Majority of a Eukaryotic Genome Does Not Encode Functional Genes

Given the preceding information concerning various forms of repetitive DNA in eukaryotes, it is of interest to pose an important question: *What proportion of the eukaryotic genome actually encodes functional genes?*

We have seen that, taken together, the various forms of highly repetitive and moderately repetitive DNA comprise a substantial portion of the human genome. In addition to repetitive DNA, a large amount of the DNA consists of single-copy sequences as defined by reassociation kinetic analysis that appear to be noncoding. Included are many instances of what we call **pseudogenes.** These are DNA sequences representing evolutionary vestiges of duplicated copies of genes that have undergone significant mutational alteration. As a result, although they show some homology to their parent gene, they are usually not transcribed because of insertions and deletions throughout their structure.

Although the proportion of the genome consisting of repetitive DNA varies among organisms, one feature seems to be shared: *Only a very small part of the genome actually codes for proteins.* For example, the 20,000 to 30,000 genes encoding proteins in the sea urchin occupy less than 10 percent of the genome. In *Drosophila*, only 5 to 10 percent of the genome is occupied by genes coding for proteins. In humans, it appears that the estimated 20,000 functional genes occupy less than 2 percent of the total DNA sequence making up the genome.

Study of the various forms of repetitive DNA has significantly enhanced our understanding of genome organization.

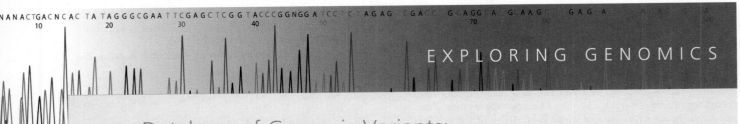

EXPLORING GENOMICS

Database of Genomic Variants: Structural Variations in the Human Genome

MG™ Study Area: *Exploring Genomics*

In this chapter, we focused on structural details of chromosomes and DNA sequence organization in chromosomes. Large segments of DNA and many genes can vary greatly in copy number due to duplications, creating **copy number variations (CNVs).** Many studies are underway to identify and map CNVs and to find possible disease conditions associated with them.

To date, over 2000 CNVs have been identified in the human genome, and estimates suggest there may be thousands more within human populations. In this Exploring Genomics exercise we will visit the **Database of Genomic Variants (DGV),** which provides a quickly expanding summary of structural variations in the human genome including CNVs.

■ Exercise I: Database of Genomic Variants

1. Access the DGV at http://projects. tcag.ca/variation. Click the "About the Project" tab to learn more about the purpose of the DGV.

2. Information in the DGV is easily viewed by clicking on a chromosome of interest using the "View Data by Chromosome" feature or by clicking on a chromosome using the "View Data by Genome" feature.

3. Click on a chromosome of interest to you using the "View Data by Chromosome" feature. A table will appear showing several columns of data including:

   ■ Locus: Shows the locus for the CNV, including the base pairs that span the variation.

   ■ Landmark: Shows different variations of CNVs for a particular locus.

   ■ Variation ID: Provides a unique identifying number for each variation.

   ■ Variation Type: Listed as "copy number." Most variations in this database are CNVs. Variations known to be insertions or deletions based on relatively small changes (a few bases) are labeled "InDel." Inversions labeled "Inv."

   ■ Cytoband: Indicates the chromosomal banding location for the variation.

   ■ Position (Mb): Shows the relative location in megabases (Mb) on the chromosome.

   ■ Known Genes in the Locus: Lists genes that are located in a particular CNV.

4. Let's analyze a particular group of CNVs. Many CNVs are unlikely to affect phenotype because they involve large areas of non–protein-coding or nonregulatory sequences. But gene-containing CNVs have been identified, including variants containing genes associated with Alzheimer's disease, Parkinson's disease, and other conditions.

Defensin (*DEF*) genes are part of a large family of highly duplicated genes.

(Cont. on the next page)

To learn more about *DEF* genes and CNVs, use the Keyword Search box to search for *DEF* genes (click "No" for the exact match button). A results page for the search will appear with a listing of CNVs. Click on one of the links shown. The top part of each report has graphs indicating the position of each CNV on a chromosome. Scroll down to the bottom of this page until you see "Known Genes." Click on the different *DEF* genes listed in the known genes category, which will take you to the National Center for Biotechnology Information (NCBI) Entrez site with a wealth of information about these genes so that you can answer the questions below. Do this for several *DEF*-containing CNVs on different chromosomes to find the information you will need for your answers.

a. On what chromosome(s) did you find CNVs containing *DEF* genes?
b. What did you learn about the function of *DEF* gene products? What do DEF proteins do?
c. Variations in *DEF* genotypes and *DEF* gene expression in humans have been implicated in a number of different human disease conditions. Give examples of the kinds of disorders affected by variations in *DEF* genotypes.
d. Explore the DGV to search a chromosome of interest to you and learn more about CNVs that have been mapped to that chromosome. Try the "View Data by Genome" feature that will show you maps of each chromosome indicating different variations. For CNVs (shown in blue), clicking on the CNV will take you to its locus on the chromosome.

CASE STUDY » Art inspires learning

A genetics student visiting a museum saw a painting by Goya showing a woman with a newborn baby in her lap that had very short arms and legs along with some facial abnormalities. Wondering whether this condition might be a genetic disorder, the student went online, learning that the baby might have Roberts syndrome (RBS), a rare autosomal recessive trait. She read that cells in RBS have mitotic errors, including the premature separation of centromeres and other heterochromatic regions of homologs in metaphase instead of anaphase. As a result, metaphase chromosomes have a rigid, or "railroad track" appearance. RBS has been shown to be caused by mutant alleles of the *ESCO2* gene, which functions during cell division.

The student wrote a list of questions to investigate in an attempt to better understand this condition. How would you answer these questions?

1. What do centromeres and other heterochromatic regions have in common that might cause this appearance?
2. What might be the role of the protein encoded by *ESCO2*, which in mutant form could cause these changes in mitotic chromosomes?
3. How could premature separation of centromeres cause the problems seen in RBS?

INSIGHTS AND SOLUTIONS

A previously undiscovered single-cell organism was found living at a great depth on the ocean floor. Its nucleus contained only a single linear chromosome with 7×10^6 nucleotide pairs of DNA coalesced with three types of histonelike proteins. Consider the following questions:

1. A short micrococcal nuclease digestion yielded DNA fractions of 700, 1400, and 2100 bp. Predict what these fractions represent. What conclusions can be drawn?

Solution: The chromatin fiber may consist of a variation of nucleosomes containing 700 bp of DNA. The 1400- and 2100-bp fractions, respectively, represent two and three linked nucleosomes. Enzymatic digestion may have been incomplete, leading to the latter two fractions.

2. The analysis of individual nucleosomes reveals that each unit contained one copy of each protein and that the short linker DNA contained no protein bound to it. If the entire chromosome consists of nucleosomes (discounting any linker DNA), how many are there, and how many total proteins are needed to form them?

Solution: Since the chromosome contains 7×10^6 bp of DNA, the number of nucleosomes, each containing 7×10^2 bp, is equal to

$$7 \times 10^6 / 7 \times 10^2 = 10^4 \text{ nucleosomes}$$

The chromosome contains 10^4 copies of each of the three proteins, for a total of 3×10^4 proteins.

3. Further analysis revealed the organism's DNA to be a double helix similar to the Watson–Crick model but containing 20 bp per complete turn of the right-handed helix. The physical size of the nucleosome was exactly double the volume occupied by that found in all other known eukaryotes, by virtue of increasing the distance along the fiber axis by a factor of two. Compare the degree of compaction of this organism's nucleosome to that found in other eukaryotes.

Solution: The unique organism compacts a length of DNA consisting of 35 complete turns of the helix (700 bp per nucleosome/20 bp per turn) into each nucleosome. The normal eukaryote compacts a length of DNA consisting of 20 complete turns of the helix (200 bp per nucleosome/10 bp per turn) into a nucleosome one-half the volume of that in the unique organism. The degree of compaction is therefore less in the unique organism.

PROBLEMS AND DISCUSSION QUESTIONS

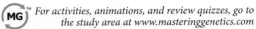

For activities, animations, and review quizzes, go to the study area at www.masteringgenetics.com

HOW DO WE KNOW?

1. In this chapter, we focused on how DNA is organized at the chromosomal level. Along the way, we found many opportunities to consider the methods and reasoning by which much of this information was acquired. From the explanations given in the chapter, what answers would you propose to the following fundamental questions:

   (a) How do we know that viral and bacterial chromosomes most often consist of circular DNA molecules devoid of protein?

   (b) What is the experimental basis for concluding that puffs in polytene chromosomes and loops in lampbrush chromosomes are areas of intense transcription of RNA?

   (c) How did we learn that eukaryotic chromatin exists in the form of repeating nucleosomes, each consisting of about 200 base pairs and an octamer of histones?

   (d) How do we know that satellite DNA consists of repetitive sequences and has been derived from regions of the centromere?

2. Contrast the sizes of the chromosomes of bacteriophage λ and T2 with those of *E. coli*. How does this relate to the relative size and complexity of the phages and bacteria?

3. Contrast the appearance and size of the DNA associated with mitochondria and chloroplasts.

4. Describe how giant polytene chromosomes are formed.

5. What genetic process is occurring in a puff of a polytene chromosome?

6. During what genetic process are lampbrush chromosomes present in vertebrates?

7. Why might we predict that the organization of eukaryotic genetic material will be more complex than that of viruses or bacteria?

8. Describe the sequence of research findings that led to the development of the model of chromatin structure.

9. What are the molecular composition and arrangement of the components in the nucleosome?

10. Describe the transitions that occur as nucleosomes are coiled and folded, ultimately forming a chromatid.

11. Provide a comprehensive definition of heterochromatin, and list as many examples as you can.

12. Contrast the various categories of repetitive DNA.

13. Define satellite DNA. Describe where it is found in the genome of eukaryotes and its role as part of chromosomes.

14. Describe the structure of LINE sequences. Why are LINEs referred to as retrotransposons?

15. Mammals contain a diploid genome consisting of at least 10^9 bp. If this amount of DNA is present as chromatin fibers, where each group of 200 bp of DNA is combined with 9 histones into a nucleosome and each group of 6 nucleosomes is combined into a solenoid, achieving a final packing ratio of 50, determine (a) the total number of nucleosomes in all fibers, (b) the total number of histone molecules combined with DNA in the diploid genome, and (c) the combined length of all fibers.

16. Assume that a viral DNA molecule is a 50-μm-long circular strand of a uniform 20 Å diameter. If this molecule is contained in a viral head that is a 0.08-μm-diameter sphere, will the DNA molecule fit into the viral head, assuming complete flexibility of the molecule? Justify your answer mathematically.

17. How many base pairs are in a molecule of phage T2 DNA that is 52 μm long?

18. While much remains to be learned about the role of nucleosomes and chromatin structure and function, recent research indicates that *in vivo* chemical modification of histones is associated with changes in gene activity. For example, Bernstein and others (2000. *Proc. Natl. Acad. Sci. (USA)* 97: 5340–5345) determined that acetylation of H3 and H4 is associated with 21.1 percent and 13.8 percent increase in yeast gene activity, respectively, and that yeast heterochromatin is hypomethylated relative to the genome average. Speculate on the significance of these findings in terms of nucleosome–DNA interactions and gene activity.

19. In an article entitled "*Nucleosome Positioning at the Replication Fork,*" Lucchini and others (2002. *EMBO* 20: 7294–7302) state, "both the 'old' randomly segregated nucleosomes as well as the 'new' assembled histone octamers rapidly position themselves (within seconds) on the newly replicated DNA strands." Given this statement, how would one compare the distribution of nucleosomes and DNA in newly replicated chromatin? How could one experimentally test the distribution of nucleosomes on newly replicated chromosomes?

20. The human genome contains approximately 10^6 copies of an *Alu* sequence, one of the best-studied classes of short interspersed elements (SINEs), per haploid genome. Individual *Alus* share a 282-nucleotide consensus sequence followed by a 3'-adenine-rich tail region (Schmid, 1998. *Nuc. Acids Res.* 26: 4541–4550). Given that there are approximately 3×10^9 bp per human haploid genome, about how many base pairs are spaced between each *Alu* sequence?

21. Below is a diagram of the general structure of the bacteriophage λ chromosome. Speculate on the mechanism by which it forms a closed ring upon infection of the host cell.

5'GGGCGGCGACCT—double-stranded region—3'
3'—double-stranded region—CCCGCCGCTGGA5'

SOLUTIONS TO SELECTED PROBLEMS AND DISCUSSION QUESTIONS

Answers to Now Solve This

1. By having a circular chromosome, no free ends present the problem of linear chromosomes, namely, complete replication of terminal sequences.

2. Since eukaryotic chromosomes are "multirepliconic" in that there are multiple replication forks along their lengths, one would expect to see multiple clusters of radioactivity.

3. Volume of the nucleus $= 4/3\pi r^3$

$$= 4/3 \times 3.14 \times (5 \times 10^3 \text{ nm})^3 = 5.23 \times 1011 \text{ nm}^3$$

Volume of the chromosome $= \pi r^2 \times$ length

$$= 3.14 \times 5.5 \text{ nm} \times 5.5 \text{ nm} \times (2 \times 10^9 \text{ nm})$$
$$= 1.9 \times 10^{11} \text{ nm}^3$$

Therefore, the percentage of the volume of the nucleus occupied by the chromatin is

$$1.9 \times 10^{11} \text{ nm}^3/5.23 \times 10^{11} \text{ nm}^3 \times 100$$
$$= \text{about 36.3 percent}$$

Solutions to Problems and Discussion Questions

2. Bacteriophage λ has a linear, double-stranded DNA while in the phage coat and upon infection closes to form a circular chromosome with a size of about 50 kb. T2 phage also has a linear, double-stranded DNA chromosome of less than 200 kb. *E. coli* has a circular, double-stranded DNA chromosome of about 4.2×10^3 kb. Both intact phages are about 1/150 the size of *E. coli*. Since phages are obligate parasites of bacteria, they are dependent on their hosts for the manufacture of materials for their replication. Bacteria contain all genetic information for metabolism, replication, and *de novo* synthesis of numerous life-supporting materials. Phages, on the other hand, contain relatively few genes, namely, those needed to adsorb, inject, and produce progeny using primarily bacterial materials.

6. Lampbrush chromosomes are typically present in vertebrate oocytes. They are also found in spermatocytes of some insects. They are found as diplotene stage structures and are active uncoiled versions of condensed meiotic chromosomes.

8. Digestion of chromatin with endonucleases, such as micrococcal nuclease, gives DNA fragments of approximately 200 base

pairs or multiples of such segments.

X-ray diffraction data indicated a regular spacing of DNA in chromatin. Regularly spaced bead-like structures (nucleosomes) were identified by electron microscopy.

10. As chromosome condensation occurs, a 300-Å fiber is formed. It appears to be composed of five or six nucleosomes coiled together. Such a structure is called a solenoid. These fibers form a series of loops that further condense into the chromatin fiber and are then coiled into chromosome arms making up each chromatid.

14. Long interspersed elements (LINEs) are repetitive transposable DNA sequences in humans. The most prominent family, designated **L1**, is about 6.4 kb each and is represented about 100,000 times. LINEs are often referred to as retrotransposons because their mechanism of transposition resembles that used by retroviruses.

16. Using the formula πr^2 for the area of a circle and $4/3\pi r^3$ for the volume of a sphere, the following calculations apply:

Volume of DNA:
$$3.14 \times 10 \text{ Å} \times 10 \text{ Å} \times (50 \times 10^4 \text{ Å}) = 1.57 \times 10^8 \text{ Å}^3$$

Volume of capsid:
$$4/3 (3.14 \times 400 \text{ Å} \times 400 \text{ Å} \times 400 \text{ Å}) = 2.67 \times 10^8 \text{ Å}^3$$

Because the capsid head has a greater volume than the volume of DNA, the DNA will fit into the capsid.

18. That nucleosomes are associated with chromatin during periods of gene activity raises the question of the possible roles they play in influencing not only chromosome structure but also gene function. The finding that natural chemical modification of nucleosomal components, as indicated in the question, increases gene activity suggests that changes in the binding of nucleosomes to DNA enable genes to be more accessible to factors that promote gene function. In addition, the finding that heterochromatin, containing fewer genes and more repressed genes, is undermethylated further supports the suggestion that histone modification is functionally related to changes in gene activity.

20. Dividing 3×10^9 base pairs by 10^6 gives an average of 3000 base pairs or 3 kb between *Alu* sequences.

CREDITS

Credits are listed in order of appearance.

Photo

CO, David Dressler, Oxford University, England; F-1 (a) Dr. M. Wurtz/Biozentrum, University of Basel/Science Photo Library/ Photo Researchers, Inc., F-1 (b) Dr. William S. Klug; F-2, Biology Pics/Photo Researchers, Inc.; F-3, Dr. Gopal Murti/Science Photo Library/Photo Researchers, Inc.; F-4, Don W. Fawcett/Kahri/Dawid/

Science Source/Photo Researchers, Inc.; F-5, Dr. Richard D. Kolodnar/ Dana-Farber Cancer Institute, F-6 (a) Harald Eggert, F-6 (b) The Company of Biologists, Ltd.; F-7 (b) John Ellison, Richardson Lab, Integrative Biology, The University of Texas at Austin; F-08 (a) Omikron/Photo Researchers, Inc.; F-08 (b) Dr. William S. Klug; F-9, Dr. William S. Klug; F-13, Dr. William S. Klug

The Genetic Code and Transcription

From Chapter 12 of *Essentials of Genetics, Eighth edition,* William S. Klug, Michael R. Cummings, Charlotte A. Spencer, Michael A. Palladino. © 2013 by Pearson Education, Inc. All rights reserved.

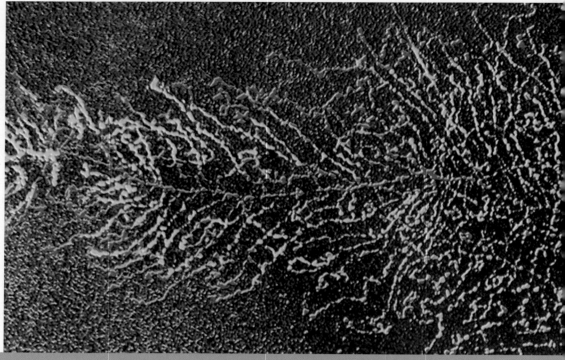

Electron micrograph visualizing the process of transcription.

The Genetic Code and Transcription

CHAPTER CONCEPTS

- Genetic information is stored in DNA using a triplet code that is nearly universal to all living things on Earth.

- The genetic code is initially transferred from DNA to RNA during the process of transcription.

- Once transferred from DNA to RNA, the genetic code exists as triplet codons, using the four ribonucleotides in RNA as the letters composing it.

- By using four different letters taken three at a time, 64 triplet sequences are possible. Most encode one of the 20 amino acids present in proteins, which serve as the end products of most genes.

- Several codons provide signals that initiate or terminate protein synthesis.

- The process of transcription is similar but more complex in eukaryotes compared to prokaryotes and bacteriophages that infect them.

The linear sequence of deoxyribonucleotides making up DNA ultimately dictates the components constituting proteins, the end product of most genes. The central question is how such information stored as a nucleic acid is decoded into a protein. **Figure 1** gives a simplified overview of how this transfer of information occurs. In the first step in gene expression, information on one of the two strands of DNA (the template strand) is transferred into an RNA complement through transcription. Once synthesized, this RNA acts as a "messenger" molecule bearing the coded information—hence its name, **messenger RNA (mRNA).** The mRNAs then associate with ribosomes, where decoding into proteins takes place.

In this chapter, we focus on the initial phases of gene expression by addressing two major questions. First, how is genetic information encoded? Second, how does the transfer from DNA to RNA occur, thus defining the process of transcription? As you shall see, ingenious analytical research established that the genetic code is written in units of three letters—ribonucleotides present in mRNA that reflect the stored information in genes. Most all triplet code words direct the incorporation of a specific amino acid into a protein as it is synthesized. As we can predict based on our prior discussion of the replication of DNA, transcription is also a complex process dependent on a major polymerase enzyme and a cast of supporting proteins. We will explore what is known about transcription in bacteria and then contrast this prokaryotic model with the differences found in eukaryotes. The information in this chapter provides a comprehensive picture of molecular genetics, which serves as the most basic foundation for understanding living organisms.

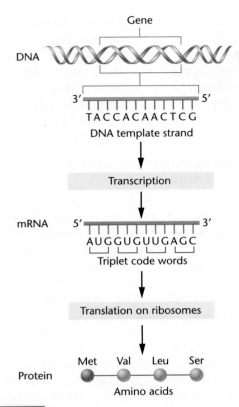

Gene

DNA

3′ _____ 5′

TACCACAACTCG

DNA template strand

↓

Transcription

↓

mRNA 5′ _____ 3′

AUGGUGUUGAGC

Triplet code words

↓

Translation on ribosomes

↓

Met   Val   Leu   Ser

Protein

Amino acids

FIGURE 1 Flowchart illustrating how genetic information encoded in DNA produces protein.

1 The Genetic Code Exhibits a Number of Characteristics

Before we consider the various analytical approaches that led to our current understanding of the genetic code, let's summarize the general features that characterize it.

1. The genetic code is written in linear form, using the ribonucleotide bases that compose mRNA molecules as "letters." The ribonucleotide sequence is derived from the complementary nucleotide bases in DNA.

2. Each "word" within the mRNA consists of three ribonucleotide letters, thus referred to as a triplet code. With several exceptions, each group of *three* ribonucleotides, called a codon, specifies *one* amino acid.

3. The code is **unambiguous**—each triplet specifies only a single amino acid.

4. The code is **degenerate;** that is, a given amino acid can be specified by more than one triplet codon. This is the case for 18 of the 20 amino acids.

5. The code contains one "start" and three "stop" signals, triplets that **initiate** and **terminate** translation.

6. No internal punctuation (such as a comma) is used in the code. Thus, the code is said to be **commaless.** Once translation of mRNA begins, the codons are read one after the other, with no breaks between them.

7. The code is **nonoverlapping.** Once translation commences, any single ribonucleotide at a specific location within the mRNA is part of only one triplet.

8. The code is nearly **universal.** With only minor exceptions, almost all viruses, prokaryotes, archaea, and eukaryotes use a single coding dictionary.

2 Early Studies Established the Basic Operational Patterns of the Code

In the late 1950s, before it became clear that mRNA is the intermediate that transfers genetic information from DNA to proteins, researchers thought that DNA itself might directly encode proteins during their synthesis. Because ribosomes had already been identified, the initial thinking was that information in DNA was transferred in the nucleus to the RNA of the ribosome, which served as the template for protein synthesis in the cytoplasm. This concept soon became untenable as accumulating evidence indicated the existence of an unstable intermediate template. The RNA of ribosomes, on the other hand, was extremely stable. As a result, in 1961 François Jacob and Jacques Monod postulated the existence of **messenger RNA (mRNA).** Once mRNA was discovered, it was clear that even though genetic information is stored in DNA, the code that is translated into proteins resides in RNA. The central question then was how only four letters—the four ribonucleotides—could specify 20 words (the amino acids).

The Triplet Nature of the Code

In the early 1960s, Sidney Brenner argued on theoretical grounds that the code had to be a triplet since three-letter words represent the minimal use of four letters to specify 20 amino acids. A code of four nucleotides, taken two at a time, for example, provides only 16 unique code words (4^2). A triplet code yields 64 words (4^3)—clearly more than the 20 needed—and is much simpler than a four-letter code (4^4), which specifies 256 words.

Experimental evidence supporting the triplet nature of the code was subsequently derived from research by Francis Crick and his colleagues. Using phage T4, they studied **frameshift mutations,** which result from the addition or deletion of one or more nucleotides within a gene and subsequently the mRNA transcribed by it. The gain or loss of letters shifts the *frame of reading* during translation. Crick

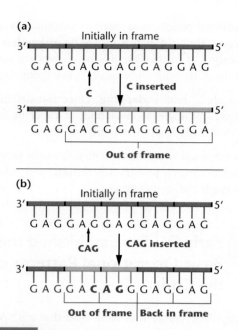

(a)

Initially in frame

G A G G A G G A G G A G G A G

C inserted / C

G A G G A C G G A G G A G G A

Out of frame

(b)

Initially in frame

G A G G A G G A G G A G G A G

CAG inserted / CAG

G A G G A C A G G G A G G A G

Out of frame | Back in frame

FIGURE 2 The effect of frameshift mutations on a DNA sequence with the repeating triplet sequence GAG. (a) The insertion of a single nucleotide shifts all subsequent triplet reading frames. (b) The insertion of three nucleotides changes only two triplets, but the frame of reading is then reestablished to the original sequence.

and his colleagues found that the gain or loss of one or two nucleotides caused a frameshift mutation, but when three nucleotides were involved, the frame of reading was reestablished **(Figure 2)**. This would not occur if the code was anything other than a triplet. This work also suggested that most triplet codes are not blank, but rather encode amino acids, supporting the concept of a degenerate code.

3 Studies by Nirenberg, Matthaei, and Others Deciphered the Code

In 1961, Marshall Nirenberg and J. Heinrich Matthaei deciphered the first specific coding sequences, which served as a cornerstone for the complete analysis of the genetic code.

Their success, as well as that of others who made important contributions to breaking the code, was dependent on the use of two experimental tools—an *in vitro* (**cell-free**) **protein-synthesizing system** and an enzyme, **polynucleotide phosphorylase,** which enabled the production of synthetic mRNAs. These mRNAs are templates for polypeptide synthesis in the cell-free system.

Synthesizing Polypeptides in a Cell-Free System

In the cell-free system, amino acids are incorporated into polypeptide chains. This *in vitro* mixture must contain the essential factors for protein synthesis in the cell: ribosomes, tRNAs, amino acids, and other molecules essential to translation. In order to follow (or trace) protein synthesis, one or more of the amino acids must be radioactive. Finally, an mRNA must be added, which serves as the template that will be translated.

In 1961, mRNA had yet to be isolated. However, use of the enzyme polynucleotide phosphorylase allowed the artificial synthesis of RNA templates, which could be added to the cell-free system. This enzyme, isolated from bacteria, catalyzes the reaction shown in **Figure 3**. Discovered in 1955 by Marianne Grunberg-Manago and Severo Ochoa, the enzyme functions metabolically in bacterial cells to degrade RNA. However, *in vitro*, with high concentrations of ribonucleoside diphosphates, the reaction can be "forced" in the opposite direction to synthesize RNA, as shown.

In contrast to RNA polymerase, polynucleotide phosphorylase does not require a DNA template. As a result, each addition of a ribonucleotide is random, based on the relative concentration of the four ribonucleoside diphosphates added to the reaction mixtures. The probability of the insertion of a specific ribonucleotide is proportional to the availability of that molecule, relative to other available ribonucleotides. *This point is absolutely critical to understanding the work of Nirenberg and others in the ensuing discussion.*

Together, the cell-free system and the availability of synthetic mRNAs provided a means of deciphering the ribonucleotide composition of various triplets encoding specific amino acids.

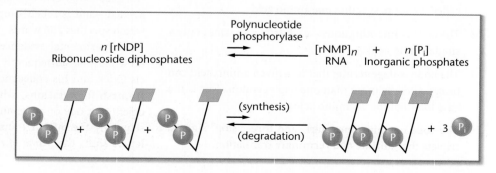

FIGURE 3 The reaction catalyzed by the enzyme polynucleotide phosphorylase. Note that the equilibrium of the reaction favors the degradation of RNA but can be "forced" in the direction favoring synthesis.

n [rNDP]
Ribonucleoside diphosphates

Polynucleotide phosphorylase

[rNMP]$_n$ + n [P$_i$]
RNA   Inorganic phosphates

(synthesis)

(degradation)

+ 3 P$_i$

The Use of Homopolymers

In their initial experiments, Nirenberg and Matthaei synthesized **RNA homopolymers,** each with only one type of ribonucleotide. Therefore, the mRNA added to the *in vitro* system was either UUUUUU . . . , AAAAAA . . . , CCCCCC . . . , or GGGGGG . . . . They tested each mRNA and were able to determine which, if any, amino acids were incorporated into newly synthesized proteins. To do this, the researchers labeled 1 of the 20 amino acids added to the *in vitro* system and conducted a series of experiments, each with a different radioactively labeled amino acid.

For example, in experiments using ^{14}C-phenylalanine (**Table 1**), Nirenberg and Matthaei concluded that the message poly U (polyuridylic acid) directed the incorporation of only phenylalanine into the homopolymer polyphenylalanine. Assuming the validity of a triplet code, they determined the first specific codon assignment—UUU codes for phenylalanine. Using similar experiments, they quickly found that AAA codes for lysine and CCC codes for proline. Poly G was not an adequate template, probably because the molecule folds back upon itself. Thus, the assignment for GGG had to await other approaches.

Note that the *specific triplet codon assignments* were possible only because homopolymers were used. This method yields only the *composition of triplets,* but since three identical letters can have only one possible sequence (e.g., UUU), the actual codons were identified.

Mixed Copolymers

With these techniques in hand, Nirenberg and Matthaei, and Ochoa and coworkers turned to the use of **RNA heteropolymers.** In this type of experiment, two or more different ribonucleoside diphosphates are added in combination to form the artificial message. The researchers reasoned that if they knew the relative proportion of each type of ribonucleoside diphosphate, they could predict the frequency of any particular triplet codon occurring in the synthetic mRNA. If they then added the mRNA to the cell-free system and ascertained the percentage of any particular amino acid present in the new protein, they could analyze the results and predict the composition (not the specific sequence) of triplets specifying particular amino acids.

This approach is shown in **Figure 4.** Suppose that A and C are added in a ratio of 1A:5C. The insertion of a ribo-

nucleotide at any position along the RNA molecule during its synthesis is determined by the ratio of A:C. Therefore, there is a 1/6 chance for an A and a 5/6 chance for a C to occupy each position. On this basis, we can calculate the frequency of any given triplet appearing in the message.

For AAA, the frequency is $(1/6)^3$, or about 0.4 percent. For AAC, ACA, and CAA, the frequencies are identical—that is, $(1/6)^2 (5/6)$, or about 2.3 percent for each triplet. Together, all three 2A:1C triplets account for 6.9 percent of the total three-letter sequences. In the same way, each of three 1A:2C triplets accounts for $(1/6) (5/6)^2$, or 11.6 percent (or a total of 34.8 percent); CCC is represented by $(5/6)^3$, or 57.9 percent of the triplets.

By examining the percentages of any given amino acid incorporated into the protein synthesized under the direction of this message, we can propose probable base compositions for each amino acid (Figure 4). Since proline appears 69 percent of the time, we could propose that proline is encoded by CCC (57.9 percent) and one triplet of 2C:1A (11.6 percent). Histidine, at 14 percent, is probably coded by one 2C:1A (11.6 percent) and one 1C:2A (2.3 percent). Threonine, at 12 percent, is likely coded by only one 2C:1A. Asparagine and glutamine each appear to be coded by one of the 1C:2A triplets, and lysine appears to be coded by AAA.

Using as many as all four ribonucleotides to construct the mRNA, the researchers conducted many similar experiments. Although determining the *composition* of the triplet code words for all 20 amino acids represented a significant breakthrough, the *specific sequences* of triplets were still unknown—other approaches were needed.

NOW SOLVE THIS

1 In a mixed copolymer experiment using polynucleotide phosphorylase, 3/4G:1/4C was used to form the synthetic message. The amino acid composition of the resulting protein was determined to be:

Glycine	36/64	(56 percent)
Alanine	12/64	(19 percent)
Arginine	12/64	(19 percent)
Proline	4/64	(6 percent)

From this information,

(a) indicate the percentage (or fraction) of the time each possible codon will occur in the message.

(b) determine one consistent base-composition assignment for the amino acids present.

(c) Once the wobble hypothesis has been discussed, return to this problem and predict as many specific codon assignments as possible.

■ **Hint** *This problem asks you to analyze a mixed copolymer experiment and to predict codon composition assignments for the amino acids encoded by the synthetic message. The key to its solution is to first calculate the proportion of each triplet codon in the synthetic RNA and then match these to the proportions of amino acids that are synthesized.*

TABLE 1   Incorporation of ^{14}C-phenylalanine into Protein

Artificial mRNA	Radioactivity (counts/min)
None	44
Poly U	39,800
Poly A	50
Poly C	38

Source: After Nirenberg and Matthaei (1961).

FIGURE 4 Results and interpretation of a mixed copolymer experiment where a ratio of 1A:5C is used (1/6A:5/6C).

Possible compositions	Possible triplets	Probability of occurrence of any triplet	Final %
3A	AAA	$(1/6)^3 = 1/216 = 0.4\%$	0.4
1C:2A	AAC ACA CAA	$(5/6)(1/6)^2 = 5/216 = 2.3\%$	$3 \times 2.3 = 6.9$
2C:1A	ACC CAC CCA	$(5/6)^2(1/6) = 25/216 = 11.6\%$	$3 \times 11.6 = 34.8$
3C	CCC	$(5/6)^3 = 125/216 = 57.9\%$	57.9
			100.0

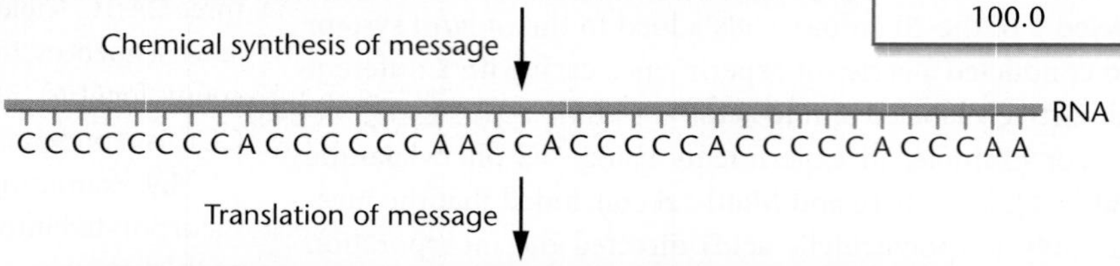

Chemical synthesis of message ↓

———————————————————————————————————— RNA
C C C C C C C C A C C C C C C A A C C A C C C C C C A C C C C C A C C C A A

Translation of message ↓

Percentage of amino acids in protein		Probable base-composition assignments
Lysine	<1	AAA
Glutamine	2	1C:2A
Asparagine	2	1C:2A
Threonine	12	2C:1A
Histidine	14	2C:1A, 1C:2A
Proline	69	CCC, 2C:1A

ESSENTIAL POINT ▪ ▪ ▪

The use of RNA homopolymers and mixed copolymers in a cell-free system allowed the determination of the composition, but not the sequence, of triplet codons designating specific amino acids.

The Triplet Binding Assay

It was not long before more advanced techniques were developed. In 1964, Nirenberg and Philip Leder developed the **triplet binding assay,** which led to specific assignments of triplets. The technique took advantage of the observation that ribosomes, when presented *in vitro* with an RNA sequence as short as three ribonucleotides, will bind to it and form a complex similar to that found *in vivo*. The triplet acts like a codon in mRNA, attracting the complementary sequence within tRNA **(Figure 5)**. The triplet sequence in tRNA that is complementary to a codon of mRNA is an **anticodon.**

Although it was not yet feasible to chemically synthesize long stretches of RNA, triplets of known sequence could be synthesized in the laboratory to serve as templates. All that was needed was a method to determine which tRNA–amino acid was bound to the triplet RNA–ribosome complex. The test system Nirenberg and Leder devised was quite simple. The amino acid to be tested was made radioactive, and a charged tRNA was produced. Because codon compositions were known, researchers could narrow the range of amino acids that should be tested for each specific triplet.

The radioactively charged tRNA, the RNA triplet, and ribosomes were incubated together and then passed through a nitrocellulose filter, which retains the larger ribosomes but not the other smaller components, such as unbound charged tRNA. If radioactivity is not retained on the filter, an incorrect amino acid has been tested. But if radioactivity remains on the filter, it is retained because the charged tRNA has bound to the triplet associated with the ribosome. When this occurs, a specific codon assignment can be made.

Work proceeded in several laboratories, and in many cases clear-cut, unambiguous results were obtained. **Table 2,** for example, shows 26 triplets assigned to 9 amino acids. However, in some cases, the degree of triplet binding was inefficient and assignments were not possible. Eventually, about 50 of the 64 triplets were assigned. These specific assignments of triplets to amino acids led to two major conclusions. First, the genetic code is **degenerate;** that is, one amino acid can be specified by more than one triplet. Second, the code is **unambiguous.** That is, a single triplet specifies only one amino acid. As you shall see later in this chapter, these conclusions have been upheld with only minor exceptions. The triplet binding technique was a major innovation in deciphering the genetic code.

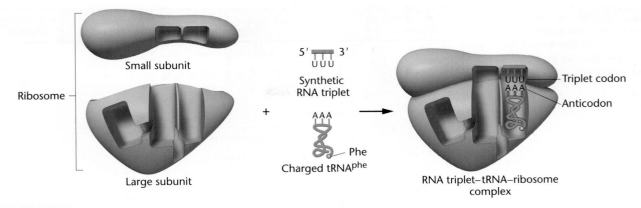

FIGURE 5 Illustration of the behavior of the components during the triplet binding assay. The synthetic UUU triplet RNA sequence acts as a codon, attracting the complementary AAA anticodon of the charged tRNA^phe, which together are bound by the subunits of the ribosome.

Repeating Copolymers

Yet another innovative technique used to decipher the genetic code was developed in the early 1960s by Gobind Khorana, who chemically synthesized long RNA molecules consisting of short sequences repeated many times. First, he created shorter sequences (e.g., di-, tri-, and tetranucleotides), which were then replicated many times and finally joined enzymatically to form the long polynucleotides. As shown in **Figure 6**, a dinucleotide made in this way is converted to an mRNA with two repeating triplets. A trinucleotide is converted to an mRNA with three potential triplets, depending on the point at which initiation occurs, and a tetranucleotide creates four repeating triplets.

When synthetic messages were added to a cell-free system, the predicted number of amino acids incorporated into polypeptides was upheld. Several examples are shown in **Table 3**. When the data were combined with those on composition assignment and triplet binding, specific assignments were possible.

One example of specific assignments made in this way will illustrate the value of Khorana's approach. Consider the following experiments in concert with one another: (1) The repeating *trinucleotide sequence* UUCUUCUUC . . . can be read as three possible repeating triplets—UUC, UCU, and CUU—depending on the initiation point. When placed in a cell-free translation system, three different polypeptide homopolymers—containing either phenylalanine (phe), serine (ser), or leucine (leu)—are produced. Thus, we know that each of the three triplets encodes one of the three amino acids, but we do not know which codes which; (2) On the other hand, the *repeating dinucleotide sequence* UCUCUCUC . . . produces the triplets UCU and CUC and, when used in an experiment, leads to the incorporation of leucine and serine into a polypeptide. Thus, the triplets UCU and CUC specify leucine and serine, but we still do not know which triplet specifies which amino acid. However, when considering both sets of results in concert, we can conclude that UCU, which is common to both experiments, must encode either leucine or serine but not phenylalanine. Thus, either CUU *or* UUC encodes leucine *or* serine, while the other encodes phenylalanine; (3) To derive more specific information, we can examine the results of using the repeating tetranucleotide sequence UUAC, which produces the triplets UUA, UAC, ACU, and CUU. The CUU triplet is one of the two in which we are interested. Three amino acids are incorporated by this experiment: leucine, threonine, and tyrosine. Because CUU must specify only serine or leucine, and because, of these two, only leucine appears in the resulting polypeptide, we may conclude that CUU specifies leucine. Once this assignment is established, we can logically determine all others. Of the two triplet pairs remaining (UUC and UCU from the first experiment *and* UCU and CUC from the second experiment), whichever triplet is common to both must encode serine. This is UCU. By elimination, UUC is determined to encode phenylalanine and CUC is determined to encode leucine. Thus, through painstaking logical analysis, four specific triplets encoding three different amino acids have been assigned from these experiments.

TABLE 2 Amino Acid Assignments to Specific Trinucleotides Derived from the Triplet Binding Assay

Trinucleotides	Amino Acid
AAA AAG	Lysine
AUG	Methionine
AUU AUC AUA	Isoleucine
CCG CCA CCU CCC	Proline
CUC CUA CUG CUU	Leucine
GAA GAG	Glutamic acid
UCA UCG UCU UCC	Serine
UGU UGC	Cysteine
UUA UUG	Leucine
UUU UUC	Phenylalanine

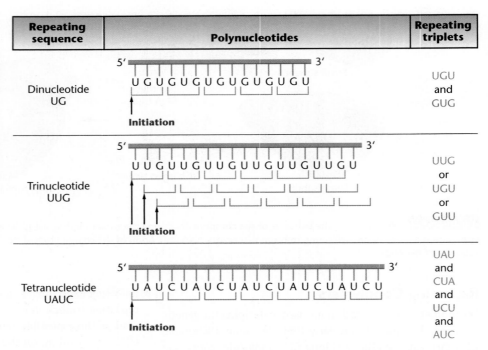

FIGURE 6 The conversion of di-, tri-, and tetranucleotides into repeating RNA copolymers. The triplet codons that are produced in each case are shown.

Repeating sequence	Polynucleotides	Repeating triplets
Dinucleotide UG	5′ U G U G U G U G U G U G U Initiation	UGU and GUG
Trinucleotide UUG	5′ U G U U G U U G U U G U U G U U G U Initiation	UUG or UGU or GUU
Tetranucleotide UAUC	5′ U A U C U A U C U A U C U A U C U A U C U Initiation	UAU and CUA and UCU and AUC

From these interpretations, Khorana reaffirmed the identity of triplets that had already been deciphered and filled in gaps left from other approaches. A number of examples are shown in Table 3. For example, the use of two tetranucleotide sequences, GAUA and GUAA, suggested that at least two triplets were *termination codons*. Khorana reached this conclusion because neither of these repeating sequences directed the incorporation of more than a few amino acids into a polypeptide, too few for him to detect. There are no triplets common to both messages, and both seemed to contain at least one triplet that terminates protein synthesis. Of the possible triplets in the poly–(GAUA) sequence shown in Table 3, UAG was later shown to be a termination codon.

ESSENTIAL POINT ■ ■ ■

Use of the triplet-binding assay and of repeating copolymers allowed the determination of the specific sequences of triplet codons designating specific amino acids.

NOW SOLVE THIS

2 When repeating copolymers are used to form synthetic mRNAs, dinucleotides produce a single type of polypeptide that contains only two different amino acids. On the other hand, using a trinucleotide sequence produces three different polypeptides, each consisting of only a single amino acid. Why? What will be produced when a repeating tetranucleotide is used?

■ Hint *This problem asks you to consider different outcomes of repeating copolymer experiments. The key to its solution is to be aware that when using a repeating copolymer of RNA, translation can be initiated at different ribonucleotides. You must simply determine the number of triplet codons produced by initiation at each of the different ribonucleotides.*

TABLE 3 Amino Acids Incorporated Using Repeated Synthetic Copolymers of RNA

Repeating Copolymer	Codons Produced	Amino Acids in Polypeptides
UG	UGU	Cysteine
	GUG	Valine
AC	ACA	Threonine
	CAC	Histidine
UUC	UUC	Phenylalanine
	UCU	Serine
	CUU	Leucine
AUC	AUC	Isoleucine
	UCA	Serine
	CAU	Histidine
UAUC	UAU	Tyrosine
	CUA	Leucine
	UCU	Serine
	AUC	Isoleucine
GAUA	GAU	None
	AGA	None
	UAG	None
	AUA	None

4 The Coding Dictionary Reveals the Function of the 64 Triplets

The various techniques used to decipher the genetic code have yielded a dictionary of 61 triplet codons assigned to amino acids. The remaining three triplets are termination signals and do not specify any amino acid.

Degeneracy and the Wobble Hypothesis

A general pattern of triplet codon assignments becomes apparent when we look at the genetic coding dictionary. **Figure 7** designates the assignments in a particularly illustrative form first suggested by Francis Crick.

Most evident is that the code is degenerate, as the early researchers predicted. That is, almost all amino acids are specified by two, three, or four different codons. Three amino acids (serine, arginine, and leucine) are each encoded by six different codons. Only tryptophan and methionine are encoded by single codons.

Also evident is the *pattern of degeneracy.* Most often, in a set of codons specifying the same amino acid, the first two letters are the same, with only the third differing. Crick discerned a pattern in the degeneracy at the third position, and in 1966, he postulated the **wobble hypothesis.**

Crick's hypothesis first predicted that the initial two ribonucleotides of triplet codes are more critical than the third in attracting the correct tRNA during translation. He postulated that hydrogen bonding at the third position of the codon–anticodon interaction is less constrained and need not adhere as specifically to the established base-pairing rules. The wobble hypothesis thus proposes a more flexible set of base-pairing rules at the third position of the codon **(Table 4)**.

This relaxed base-pairing requirement, or "wobble," allows the anticodon of a single form of tRNA to pair with more than one triplet in mRNA. Consistent with the wobble

TABLE 4	Codon–Anticodon Base-Pairing Rules
Base at First Position (5′ end) of tRNA	**Base at Third Position (3′ end) of mRNA**
A	U
C	G
G	C or U
U	A or G
I	A, U, or C

hypothesis and degeneracy, U at the first position (the 5′ end) of the tRNA anticodon may pair with A or G at the third position (the 3′ end) of the mRNA codon, and G may likewise pair with U or C. Inosine (I), one of the modified bases found in tRNA, may pair with C, U, or A. Applying these wobble rules, a minimum of about 30 different tRNA species is necessary to accommodate the 61 triplets specifying an amino acid. If nothing more, wobble can be considered a potential economy measure, provided that the fidelity of translation is not compromised. Current estimates are that 30 to 40 tRNA species are present in bacteria and up to 50 tRNA species exist in animal and plant cells.

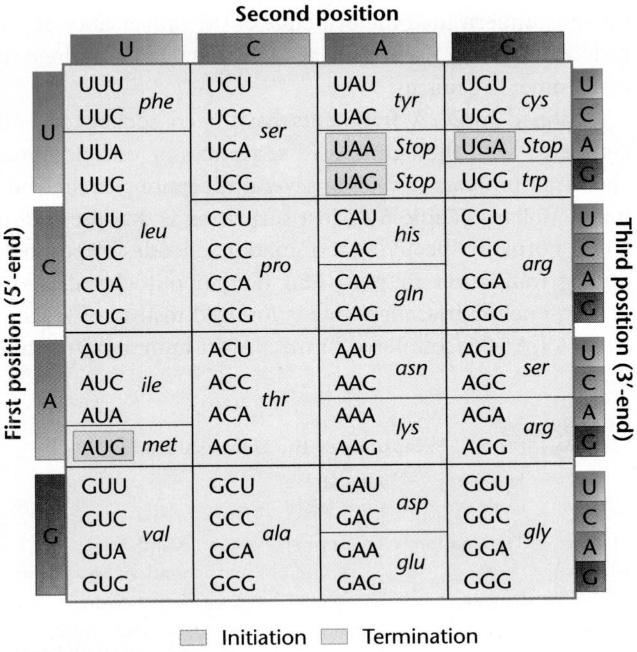

The Ordered Nature of the Code

Still another observation has become apparent in the pattern of codon sequences and their corresponding amino acids, leading to the description referred to as an **ordered genetic code.** Chemically similar amino acids often share one or two "middle" bases in the different triplets encoding them. For example, either U or C is often present in the second position of triplets that specify hydrophobic amino acids, including valine and alanine, among others. Two codons (AAA and AAG) specify the positively charged amino acid lysine. If only the middle letter of these codons is changed from A to G (AGA and AGG), the positively charged amino acid arginine is specified. Hydrophilic amino acids, such as serine and threonine, are specified by triplet codons, with G or C in the second position.

The end result of an "ordered" code is that it buffers the potential effect of mutation on protein function. While many mutations of the second base of triplet codons result in a change of one amino acid to another, the change is often to an amino acid with similar chemical properties. In such cases, protein function may not be noticeably altered.

Initiation and Termination

In contrast to the *in vitro* experiments discussed earlier, initiation of protein synthesis *in vivo* is a highly specific process. In bacteria, the initial amino acid inserted into all polypeptide chains is a modified form of methionine—

Second position

FIGURE 7 The coding dictionary. AUG encodes methionine, which initiates most polypeptide chains. All other amino acids except tryptophan, which is encoded only by UGG, are encoded by two to six triplets. The triplets UAA, UAG, and UGA are termination signals and do not encode any amino acids.

N-formylmethionine (fmet). Only one codon, AUG, codes for methionine, and it is sometimes called the **initiator codon.** However, when AUG appears internally in mRNA, rather than at an initiating position, unformylated methionine is inserted into the polypeptide chain. Rarely, another codon, GUG, specifies methionine during initiation, though it is not clear why this happens, since GUG normally encodes valine.

In bacteria, either the formyl group is removed from the initial methionine upon the completion of protein synthesis or the entire formylmethionine residue is removed. In eukaryotes, methionine is also the initial amino acid during polypeptide synthesis. However, it is not formylated.

As mentioned in the preceding section, three other triplets (UAG, UAA, and UGA) serve as **termination codons,** punctuation signals that do not code for any amino acid. They are not recognized by a tRNA molecule, and translation terminates when they are encountered. Mutations that produce any of the three triplets internally in a gene will also result in termination. Consequently, only a partial polypeptide has been synthesized when it is prematurely released from the ribosome. When such a change occurs in the DNA, it is called a **nonsense mutation.**

ESSENTIAL POINT ■ ■ ■

The complete coding dictionary reveals that of the 64 possible triplet codons, 61 encode the 20 amino acids found in proteins, while three triplets terminate translation.

---

5 The Genetic Code Has Been Confirmed in Studies of Bacteriophage MS2

The various aspects of the genetic code discussed thus far yield a fairly complete picture, suggesting that it is triplet in nature, degenerate, unambiguous, and commaless, and that it contains punctuation start and stop signals. That these features are correct was confirmed by analysis of the RNA-containing **bacteriophage MS2** by Walter Fiers and his coworkers.

MS2 is a bacteriophage that infects *E. coli.* Its nucleic acid (RNA) contains only about 3500 ribonucleotides, making up only three genes, specifying a coat protein, an RNA replicase, and a maturation protein. The small genome and a few gene products enabled Fiers and his colleagues to sequence the genes and their products. When the chemical constitution of these genes and their encoded proteins were compared, they were found to exhibit **colinearity.** That is, based on the coding dictionary, *the linear sequence of triplet codons corresponds precisely with the linear sequence of amino acids in each protein.* Punctuation was also confirmed. For example, in the coat protein gene, the codon for the first amino acids is AUG, the common initiator codon. The codon for the

last amino acid is followed by two consecutive termination codons, UAA and UAG. The analysis clearly showed that the genetic code in this virus was identical to that established experimentally in bacterial systems.

6 The Genetic Code Is Nearly Universal

Between 1960 and 1978, it was generally assumed that the genetic code would be found to be universal, applying equally to viruses, bacteria, archaea, and eukaryotes. Certainly, the nature of mRNA and the translation machinery seemed to be very similar in these organisms. For example, cell-free systems derived from bacteria can translate eukaryotic mRNAs. Poly U stimulates synthesis of polyphenylalanine in cell-free systems when the components are derived from eukaryotes. Many recent studies involving recombinant DNA technology reveal that eukaryotic genes can be inserted into bacterial cells, which are then transcribed and translated. Within eukaryotes, mRNAs from mice and rabbits have been injected into amphibian eggs and efficiently translated. For the many eukaryotic genes that have been sequenced, notably those for hemoglobin molecules, the amino acid sequence of the encoded proteins adheres to the coding dictionary established from bacterial studies.

However, several 1979 reports on the coding properties of DNA derived from mitochondria (**mtDNA**) of yeast and humans undermined the principle of the universality of the genetic language. Since then, mtDNA has been examined in many other organisms.

Cloned mtDNA fragments have been sequenced and compared with the amino acid sequences of various mitochondrial proteins, revealing several exceptions to the coding dictionary **(Table 5)**. Most surprising is that the codon UGA, normally specifying termination, encodes tryptophan during translation in yeast and human mitochondria. In yeast mitochondria, threonine is inserted instead of leucine when CUA is encountered in mRNA. In human mitochon-

TABLE 5 Exceptions to the Universal Code

Triplet	Normal Code Word	Altered Code Word	Source
UGA	Termination	Tryptophan	Human and yeast mitochondria; *Mycoplasma*
CUA	Leucine	Threonine	Yeast mitochondria
AUA	Isoleucine	Methionine	Human mitochondria
AGA	Arginine	Termination	Human mitochondria
AGG	Arginine	Termination	Human mitochondria
UAA	Termination	Glutamine	*Paramecium; Tetrahymena; Stylonychia*
UAG	Termination	Glutamine	*Paramecium*

dria, AUA, which normally specifies isoleucine, directs the internal insertion of methionine.

In 1985, several other exceptions to the standard coding dictionary were discovered in the bacterium *Mycoplasma capricolum* and in the nuclear genes of the protozoan ciliates *Paramecium, Tetrahymena,* and *Stylonychia.* For example, as shown in Table 5, one alteration converts the termination codon UGA to tryptophan, yet several others convert the normal termination codons UAA and UAG to glutamine. These changes are significant because both a prokaryote and several eukaryotes are involved, representing distinct species that have evolved separately over a long period of time.

Note the apparent pattern in several of the altered codon assignments. The change in coding capacity involves only a shift in recognition of the third, or wobble, position. For example, AUA specifies isoleucine in the cytoplasm and methionine in the mitochondrion, but in the cytoplasm, methionine is specified by AUG. Similarly, UGA calls for termination in the cytoplasm, but it specifies tryptophan in the mitochondrion; in the cytoplasm, tryptophan is specified by UGG. It has been suggested that such changes in codon recognition may represent an evolutionary trend toward reducing the number of tRNAs needed in mitochondria; only 22 tRNA species are encoded in human mitochondria, for example. However, until more examples are found, the differences must be considered to be exceptions to the previously established general coding rules.

7 | Different Initiation Points Create Overlapping Genes

Earlier we stated that the genetic code is nonoverlapping—each ribonucleotide in an mRNA is part of only one codon. However, this characteristic of the code does not rule out the possibility that a single mRNA may have multiple initiation points for translation. If so, these points could theoretically create several different reading frames within the same mRNA, thus specifying more than one polypeptide and leading to the concept of **overlapping genes.**

That this might actually occur in some viruses was suspected when phage φX174 was carefully investigated. The circular DNA chromosome consists of 5386 nucleotides, which should encode a maximum of 1795 amino acids, sufficient for five or six proteins. However, this small virus in fact synthesizes 11 proteins consisting of more than 2300 amino acids. A comparison of the nucleotide sequence of the DNA and the amino acid sequences of the polypeptides synthesized has clarified the apparent paradox. At least four cases of multiple initiation have been discovered, creating overlapping genes.

For example, in one case, the coding sequences for the initiation of two polypeptides are found at separate positions within the reading frame that specifies the sequence of a third polypeptide. In one case, seven different polypeptides may be created from a DNA sequence that might otherwise have specified only three polypeptides.

A similar situation has been observed in other viruses, including phage G4 and the animal virus SV40. Like φX174, phage G4 contains a circular single-stranded DNA molecule. The use of overlapping reading frames optimizes the use of a limited amount of DNA present in these small viruses. However, such an approach to storing information has a distinct disadvantage in that a single mutation may affect more than one protein and thus increase the chances that the change will be deleterious or lethal.

8 | Transcription Synthesizes RNA on a DNA Template

Even while the genetic code was being studied, it was quite clear that proteins were the end products of many genes. Thus, while some geneticists attempted to elucidate the code, other research efforts focused on the nature of genetic expression. The central question was how DNA, a nucleic acid, could specify a protein composed of amino acids.

The complex multistep process begins with the transfer of genetic information stored in DNA to RNA. The process by which RNA molecules are synthesized on a DNA template is called **transcription.** It results in an mRNA molecule complementary to the gene sequence of one of the double helix's two strands. Each triplet codon in the mRNA is, in turn, complementary to the anticodon region of its corresponding tRNA as the amino acid is correctly inserted into the polypeptide chain during translation. The significance of transcription is enormous, for it is the initial step in the process of **information flow** within the cell. The idea that RNA is involved as an intermediate molecule in the process of information flow between DNA and protein was suggested by the following observations:

1. DNA is, for the most part, associated with chromosomes in the nucleus of the eukaryotic cell. However, protein synthesis occurs in association with ribosomes located outside the nucleus in the cytoplasm. Therefore, DNA does not appear to participate directly in protein synthesis.

2. RNA is synthesized in the nucleus of eukaryotic cells, where DNA is found, and is chemically similar to DNA.

3. Following its synthesis, most RNA migrates to the cytoplasm, where protein synthesis (translation) occurs.

4. The amount of RNA is generally proportional to the amount of protein in a cell.

Collectively, these observations suggested that genetic information, stored in DNA, is transferred to an RNA intermediate, which directs the synthesis of proteins. As with most new ideas in molecular genetics, the initial supporting evidence was based on experimental studies of bacteria and their phages. It was clearly established that during initial infection, RNA synthesis preceded phage protein synthesis and that the RNA is complementary to phage DNA.

The results of these experiments agree with the concept of a messenger RNA (mRNA) being made on a DNA template and then directing the synthesis of specific proteins in association with ribosomes. This concept was formally proposed by François Jacob and Jacques Monod in 1961 as part of a model for gene regulation in bacteria. Since then, mRNA has been isolated and studied thoroughly. There is no longer any question about its role in genetic processes.

9 RNA Polymerase Directs RNA Synthesis

To establish that RNA can be synthesized on a DNA template, it was necessary to demonstrate that there is an enzyme capable of directing this synthesis. By 1959, several investigators, including Samuel Weiss, had independently isolated such a molecule from rat liver. Called **RNA polymerase,** it has the same general substrate requirements as does DNA polymerase, the major exception being that the substrate nucleotides contain the ribose rather than the deoxyribose form of the sugar. Unlike DNA polymerase, no primer is required to initiate synthesis; the initial base remains as a nucleoside triphosphate (NTP). The overall reaction summarizing the synthesis of RNA on a DNA template can be expressed as

$$n(\text{NTP}) \xrightarrow[\text{enzyme}]{\text{DNA}} (\text{NMP})_n + n\,(\text{PP}_i)$$

As this equation reveals, nucleoside triphosphates (NTPs) are substrates for the enzyme, which catalyzes the polymerization of nucleoside monophosphates (NMPs), or nucleotides, into a polynucleotide chain (NMP)n. Nucleotides are linked during synthesis by 3′-to-5′ phosphodiester bonds (see Figure 10). The energy created by cleaving the triphosphate precursor into the monophosphate form drives the reaction, and inorganic pyrophosphates (PP$_i$) are produced.

A second equation summarizes the sequential addition of each ribonucleotide as the process of transcription progresses

$$(\text{NMP})n + \text{NTP} \xrightarrow[\text{enzyme}]{\text{DNA}} (\text{NMP})_{n+1} + \text{PP}_i$$

As this equation shows, each step of transcription involves the addition of one ribonucleotide (NMP) to the growing polyribonucleotide chain (NMP)$_{n+1}$, using a nucleoside triphosphate (NTP) as the precursor.

RNA polymerase from *E. coli* has been extensively characterized and shown to consist of subunits designated α, β, β′, and σ. The active form of the enzyme, the **holoenzyme,** contains the subunits α$_2$, β, β′, and σ and has a molecular weight of almost 500,000 Da. Of these subunits, it is the β and β′ polypeptides that provide the catalytic basis and active site of transcription. Still another subunit, the σ **(sigma) factor,** plays a regulatory function in the initiation of RNA transcription.

Although there is but a single form of the enzyme in *E. coli,* there are several different σ factors, creating variations of the polymerase holoenzyme. On the other hand, eukaryotes display three distinct forms of RNA polymerase, each consisting of a greater number of polypeptide subunits than in bacteria.

ESSENTIAL POINT ■ ■ ■

Transcription—the initial step in gene expression—is the synthesis, under the direction of RNA polymerase, of a strand of RNA complementary to a DNA template.

NOW SOLVE THIS

3 The following represent deoxyribonucleotide sequences in the template strand of DNA:

Sequence 1: 5′-CTTTTTTGCCAT-3′

Sequence 2: 5′-ACATCAATAACT-3′

Sequence 3: 5′-TACAAGGGTTCT-3′

(a) For each strand, determine the mRNA sequence that would be derived from transcription.

(b) Using Figure 7, determine the amino acid sequence that is encoded by these mRNAs.

(c) For Sequence 1, what is the sequence of the partner DNA strand?

■ **Hint** *This problem asks you to consider the outcome of the transfer of complementary information from DNA to RNA and to determine the amino acids encoded by this information. The key to its solution is to remember that in RNA, uracil is complementary to adenine, and that while DNA stores genetic information in the cell, the code that is translated is contained in the RNA complementary to the template strand of DNA making up a gene.*

Promoters, Template Binding, and the σ Subunit

Transcription results in the synthesis of a single-stranded RNA molecule complementary to a region along only one strand of the DNA double helix. For simplicity, let's call the transcribed DNA strand the **template strand** and its complement the **partner strand.**

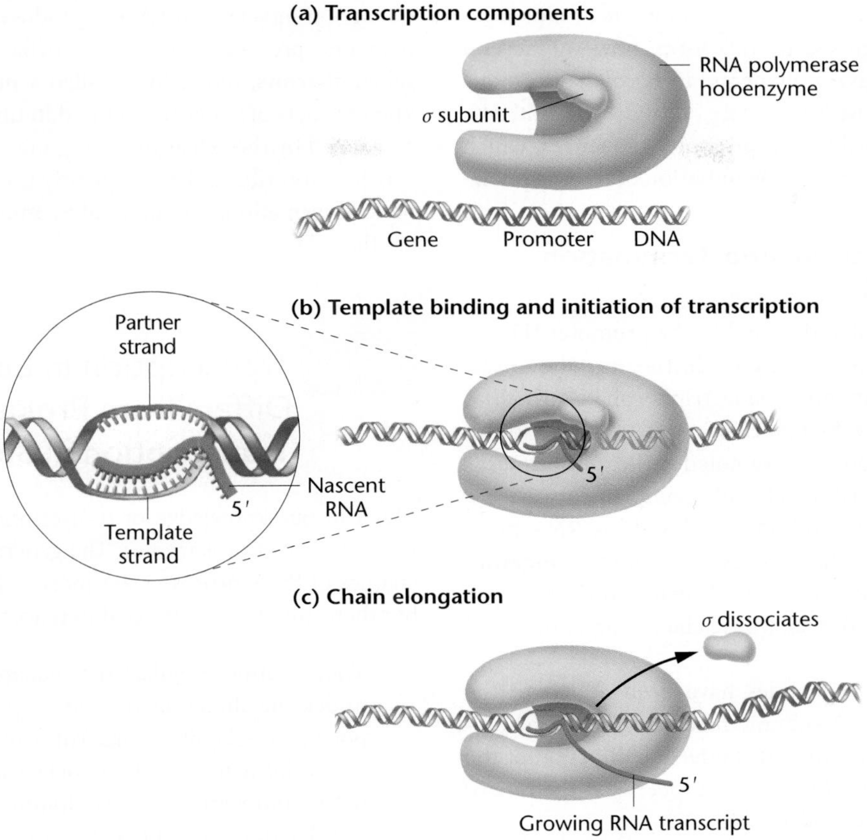

(a) Transcription components

(b) Template binding and initiation of transcription

(c) Chain elongation

FIGURE 8 The early stages of transcription in prokaryotes, showing (a) the components of the process; (b) template binding at the −10 site involving the σ subunit of RNA polymerase and subsequent initiation of RNA synthesis; and (c) chain elongation, after the σ subunit has dissociated from the transcription complex and the enzyme moves along the DNA template.

The initial step is **template binding (Figure 8)**. In bacteria, the site of this initial binding is established when the RNA polymerase σ subunit recognizes specific DNA sequences called **promoters.** These regions are located in the region upstream (5′) from the point of initial transcription of a gene. It is believed that the enzyme "explores" a length of DNA until it recognizes the promoter region and binds to about 60 nucleotide pairs of the helix, 40 of which are upstream from the point of initial transcription. Once this occurs, the helix is denatured or unwound locally, making the DNA template accessible to the action of the enzyme. The point at which transcription actually begins is called the **transcription start site.**

The importance of promoter sequences cannot be overemphasized. They govern the efficiency of the initiation of transcription. In bacteria, both strong promoters and weak promoters have been discovered. Because the interaction of promoters with RNA polymerase governs transcription, the nature of the binding between them is at the heart of discussions concerning genetic regulation. While we will later pursue more detailed information involving promoter–enzyme interactions, we must address two points here.

The first point is the concept of **consensus sequences** of DNA. These sequences are similar (homologous) in different genes of the same organism or in one or more genes of related organisms. Their conservation throughout evolution attests to the critical nature of their role in biological processes. Two such sequences have been found in bacterial promoters. One, TATAAT, is located 10 nucleotides upstream from the site of initial transcription (the −10 region, or **Pribnow box**). The other, TTGACA, is located 35 nucleotides upstream (the −35 region). Mutations in either region diminish transcription, often severely.

Sequences such as these are said to be *cis*-**acting elements.** Use of the term *cis* is drawn from organic chemistry nomenclature, meaning "next to" or on the same side as, in contrast to being "across from," or *trans,* to other functional groups. In molecular genetics, then, *cis*-elements are adjacent parts of the same DNA molecule. This is in contrast to *trans*-**acting factors**, molecules that bind to these DNA elements. As we will soon see, in most eukaryotic genes studied, a consensus sequence comparable to that in the −10 region has been recognized. Because it is rich in adenine and thymine residues, it is called the **TATA box.**

The second point involves the σ subunit in bacteria. The major form is designated as σ^{70}, based on its molecular weight of 70 kilodaltons (kDa). The promoters of most bacterial genes are recognized by this form; however, several alternative forms of RNA polymerase in *E. coli* have unique σ subunits associated with them (e.g., σ^{32}, σ^{54}, σ^{S}, and σ^{E}). Each form recognizes different promoter sequences, which in turn provides specificity to the initiation of transcription.

Initiation, Elongation, and Termination of RNA Synthesis

Once it has recognized and bound to the promoter [Figure 8(b)], RNA polymerase catalyzes **initiation,** the insertion of the first 5′-ribonucleoside triphosphate, which is complementary to the first nucleotide at the start site of the DNA template strand. As we noted earlier, no primer is required. Subsequent ribonucleotide complements are inserted and linked by phosphodiester bonds as RNA polymerization proceeds. This process of **chain elongation** [Figure 8(c)] continues in a 5′ to 3′ extension, creating a temporary DNA/RNA duplex whose chains run antiparallel to one another.

After a few ribonucleotides have been added to the growing RNA chain, the σ subunit dissociates from the holoenzyme and elongation proceeds under the direction of the core enzyme. In *E. coli,* this process proceeds at the rate of about 50 nucleotides/second at 37°C.

Eventually, the enzyme traverses the entire gene until it encounters a specific nucleotide sequence that acts as a termination signal. The termination sequences, about 40 base pairs in length, are extremely important in prokaryotes because of the close proximity of the end of one gene and the upstream sequences of the adjacent gene. An interesting aspect of termination in bacteria is that the termination sequence alluded to above is actually transcribed into RNA. The unique sequence of nucleotides in this termination region causes the newly formed transcript to fold back on itself, forming what is called a **hairpin secondary structure,** held together by hydrogen bonds. The hairpin is important to termination. In some cases, the termination of synthesis is dependent on the **termination factor, ρ (rho)**—a large hexameric protein that physically interacts with the growing RNA transcript.

At the point of termination, the transcribed RNA molecule is released from the DNA template and the core polymerase enzyme dissociates. The synthesized RNA molecule is precisely complementary to a DNA sequence representing the template strand of a gene. Wherever an A, T, C, or G residue existed, a corresponding U, A, G, or C residue, respectively, is incorporated into the RNA molecule. These RNA molecules ultimately provide the information leading to the synthesis of all proteins present in the cell.

In bacteria, groups of genes whose products are related are often clustered along the chromosome. In many such cases, they are contiguous, and all but the last gene lack the encoded signals for termination. The result is that during transcription, a large mRNA is produced that encodes more than one protein. Since genes in bacteria are sometimes called **cistrons,** the RNA is called a **polycistronic mRNA.** The products of genes transcribed in this fashion are usually all needed by the cell at the same time, so this is an efficient way to transcribe and subsequently translate the needed genetic information. In eukaryotes, **monocistronic mRNAs** are the rule.

10 Transcription in Eukaryotes Differs from Prokaryotic Transcription in Several Ways

Much of our knowledge of transcription has been derived from studies of prokaryotes. The general aspects of the mechanics of these processes are mostly similar in eukaryotes, but there are several notable differences:

1. Transcription in eukaryotes occurs within the nucleus under the direction of three separate forms of RNA polymerase. Unlike prokaryotes, in eukaryotes the RNA transcript is not free to associate with ribosomes prior to the completion of transcription. For the mRNA to be translated, it must move out of the nucleus into the cytoplasm.

2. Initiation of transcription of eukaryotic genes requires that the compact chromatin fiber, characterized by nucleosome coiling, must be uncoiled and the DNA made accessible to RNA polymerase and other regulatory proteins. This transition is referred to as **chromatin remodeling,** reflecting the dynamics involved in the conformational change that occurs as the DNA helix is opened.

3. Initiation and regulation of transcription entail a more extensive interaction between *cis*-acting DNA sequences and *trans*-acting protein factors involved in stimulating and initiating transcription. Eukaryotic RNA polymerases, for example, rely on *transcription factors* (*TFs*) to scan and bind to DNA. In addition to promoters, other control units, called *enhancers* and *silencers*, may be located in the 5′ regulatory region upstream from the initiation point, but they have also been found within the gene or even in the 3′ downstream region, beyond the coding sequence.

4. Alteration of the primary RNA transcript to produce mature eukaryotic mRNA involves many complex stages referred to generally as "processing." An initial processing step involves the addition of a 5′ cap and a 3′ tail to most transcripts destined to become mRNAs. Other extensive modifications occur to the internal nucleotide

sequence of eukaryotic RNA transcripts that eventually serve as mRNAs. The initial (or primary) transcripts are most often much larger than those that are eventually translated. Sometimes called **pre-mRNAs,** they are part of a group of molecules found only in the nucleus—a group referred to collectively as **heterogeneous nuclear RNA (hnRNA).** Such RNA molecules are of variable but large size and are complexed with proteins, forming **heterogeneous nuclear ribonucleoprotein particles (hnRNPs).** Only about 25 percent of hnRNA molecules are converted to mRNA. In those that are converted, substantial amounts of the ribonucleotide sequence are excised, and the remaining segments are spliced back together prior to nuclear export and translation. This phenomenon has given rise to the concepts of **split genes** and **splicing** in eukaryotes.

In the remainder of this chapter, we will look at the basic details of transcription in eukaryotic cells. The process of transcription is highly regulated, determining which DNA sequences are copied into RNA and when and how frequently they are transcribed.

Initiation of Transcription in Eukaryotes

Eukaryotic RNA polymerase exists in three unique forms, each of which transcribes different types of genes, as indicated in **Table 6**. Each enzyme is larger and more complex than the single prokaryotic polymerase. For example, in yeast, the holoenzyme consists of two large subunits and ten smaller subunits.

In regard to the initial template-binding step and promoter regions, most is known about **RNA polymerase II (RNP II)**, which is responsible for the transcription of a wide range of genes in eukaryotes. The activity of RNP II is dependent on both *cis*-acting elements surrounding the gene itself and a number of *trans*-acting transcription factors that bind to these DNA elements (we will return to the topic of transcription factors below). At least four *cis*-acting DNA elements regulate the initiation of transcription by RNP II. The first of these elements, the **core-promoter,** determines where RNP II binds to the DNA and where it begins copying the DNA into RNA. The other three types of regulatory DNA sequences, called **proximal-promoter elements, enhancers,** and **silencers,** influence the efficiency or the rate

of transcription initiation by RNP II. Recall that in prokaryotes, the DNA sequence recognized by RNA polymerase is also called the promoter. In eukaryotes, however, transcriptional initiation is controlled by a larger number of *cis*-acting DNA elements.

In many eukaryotic genes, a *cis*-acting core-promoter element is the **TATA box** (or the **Goldberg-Hogness box).** Located about 30 nucleotide pairs upstream (−30) from the start point of transcription, TATA boxes share a consensus sequence TATA^A/$_T$AAR, where R indicates any purine nucleotide. The sequence and function of TATA boxes are analogous to those found in the −10 promoter region of prokaryotic genes. However, recall that in prokaryotes, RNA polymerase binds directly to the −10 promoter region. As we will see below, this is not the case in eukaryotes. A wide range of core-promoter and proximal-promoter elements are also found within eukaryotic gene-regulatory regions, and each can have an effect on the efficiency of transcription initiation from the start site.

Heterogeneous Nuclear RNA and Its Processing: Caps and Tails

In bacteria the base sequence of DNA is transcribed into an mRNA that is immediately and directly translated into the amino acid sequence as dictated by the genetic code. In contrast, eukaryotic RNA transcripts require significant alteration before they are transported to the cytoplasm and translated. By 1970, accumulating evidence showed that eukaryotic mRNA is transcribed initially as a precursor molecule much larger than that which is translated into protein. This notion was based on the observation by James Darnell and his coworkers of the large **heterogeneous nuclear RNA (hnRNA)** in mammalian nuclei that contained nucleotide sequences common to the smaller mRNA molecules present in the cytoplasm. They proposed that the initial transcript of a gene results in a large RNA molecule that must be processed in the nucleus before it appears in the cytoplasm as a mature mRNA molecule. The various processing steps, discussed in the sections that follow, are summarized in **Figure 9**.

An important **posttranscriptional modification** of eukaryotic RNA transcripts destined to become mRNAs occurs at the 5′ end of these molecules, where a **7-methylguanosine (7-mG)** cap is added. The cap is added even before synthesis of the initial transcript is complete and appears to be important to subsequent processing within the nucleus. The cap also protects the 5′ end of the molecule from nuclease attack. Subsequently, it may be involved in the transport of mature mRNAs across the nuclear membrane into the cytoplasm and in the initiation of translation of the mRNA into protein. The cap is fairly complex and is distinguished by the unique 5′-5′ bonding that connects it to the initial ribonucleotide

TABLE 6	RNA Polymerases in Eukaryotes	
Form	Product	Location
I	rRNA	Nucleolus
II	mRNA, snRNA	Nucleoplasm
III	5S rRNA, tRNA	Nucleoplasm

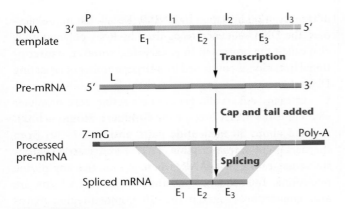

FIGURE 9 Posttranscriptional RNA processing in eukaryotes. Transcription produces a pre-mRNA containing a leader sequence (L), several introns (I), and several exons (E), as identified in the DNA template strand. This is processed by the addition of a 5'7-mG cap and a 3'-poly-A tail. The introns are then spliced out and the exons joined to create the mature mRNA. While the above figure depicts these steps sequentially, in some eukaryotic transcripts, splicing actually occurs in introns before transcription is complete and the poly-A tail has been added, leading to the concept of *cotranscriptional splicing*.

of the RNA. Some eukaryotes also acquire a methyl group (CH_3) at the 2'-carbon of the ribose sugars of the first two ribonucleotides of the RNA.

Further insights into the processing of RNA transcripts during the maturation of mRNA came from the discovery that both pre-RNAs and mRNAs contain at their 3' end a stretch of as many as 250 adenylic acid residues. This **poly-A tail** is added after the 3' end of the initial transcript is cleaved enzymatically at a position some 10 to 35 ribonucleotides from a highly conserved AAUAAA sequence. Poly A has now been found at the 3' end of almost all mRNAs studied in a variety of eukaryotic organisms. In fact, poly-A tails have also been detected in some prokaryotic mRNAs. The exceptions in eukaryotes seem to be the RNAs that encode the histone proteins.

While the AAUAAA sequence is not found on all eukaryotic transcripts, it appears to be essential to those that have it. If the sequence is changed as a result of a mutation, those transcripts that would normally have it cannot add the poly-A tail. In the absence of this tail, these RNA transcripts are rapidly degraded. Both the 5' cap and the 3' poly-A tail are critical if an mRNA transcript is to be transported to the cytoplasm and translated.

ESSENTIAL POINT

The process of creating the initial transcript during transcription is more complex in eukaryotes than in prokaryotes, including the addition of a 5' 7-mG cap and a 3' poly-A tail, to the pre-mRNA.

11 The Coding Regions of Eukaryotic Genes Are Interrupted by Intervening Sequences Called Introns

One of the most exciting findings in molecular genetics occurred in 1977, when Susan Berget, Philip Sharp, and Richard Roberts presented direct evidence that the genes of animal viruses contain *internal* nucleotide sequences that are not expressed in the amino acid sequence of the proteins they encode. These internal DNA sequences are represented in initial RNA transcripts, but they are removed before the mature mRNA is translated (Figure 9). Such nucleotide segments are called intervening sequences, and the genes that contain them are split genes. DNA sequences that are not represented in the final mRNA product are also called introns ("int" for intervening), and those retained and expressed are called exons ("ex" for expressed). Splicing involves the removal of the corresponding ribonucleotide sequences representing introns as a result of an excision process and the rejoining of the regions representing exons.

Similar discoveries were soon made in many other eukaryotic genes. Two approaches have been most fruitful for this purpose. The first involves the molecular hybridization of purified, functionally mature mRNAs with DNA containing the genes from which the RNA was originally transcribed. Hybridization between nucleic acids that are not perfectly complementary results in heteroduplexes, in which introns present in the DNA but absent in the mRNA loop out and remain unpaired. Such structures can be visualized with the electron microscope, as shown in **Figure 10**. The chicken ovalbumin complex shown in the figure is a heteroduplex with seven loops (A through G), representing seven introns whose sequences are present in the DNA but not in the final mRNA.

The second approach provides more specific information. It involves a direct comparison of nucleotide sequences of DNA with those of mRNA and their correlation with amino acid sequences. Such an approach allows the precise identification of all intervening sequences.

Thus far, most eukaryotic genes have been shown to contain introns (**Figure 11**). One of the first so identified was the β-globin gene in mice and rabbits, studied independently by Philip Leder and Richard Flavell. The mouse gene contains an intron 550 nucleotides long, beginning immediately after the codon specifying the 104th amino acid. In the rabbit, there is an intron of 580 base pairs near the codon for the 110th amino acid. In addition, another intron of about 120 nucleotides exists earlier in both genes. Similar introns have been found in the β-globin gene in all mammals examined.

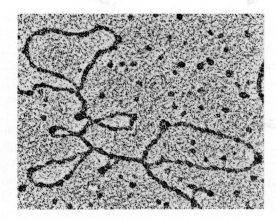

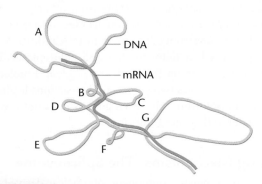

FIGURE 10 An electron micrograph and interpretive drawing of the hybrid molecule (heteroduplex) formed between the template DNA strand of the chicken ovalbumin gene and the mature ovalbumin mRNA. Seven DNA introns, labeled A–G, produce unpaired loops.

The ovalbumin gene of chickens has been extensively characterized by Bert O'Malley in the United States and Pierre Chambon in France. As shown in Figure 11, the gene contains seven introns. In fact, the majority of the gene's DNA sequence is composed of introns and is thus "silent." The initial RNA transcript is nearly three times the length of the mature mRNA. Compare the ovalbumin gene in Figures 10 and 11. Can you match the unpaired loops in Figure 10 with the order of introns specified in Figure 11?

The list of genes containing intervening sequences is long. In fact, few eukaryotic genes seem to lack introns. An extreme example of the number of introns in a single gene is provided by the gene coding for one of the subunits of collagen, the major connective tissue protein in vertebrates. The *pro-α-2(1)* collagen gene contains 50 introns. The precision of cutting and splicing that occurs must be extraordinary if errors are not to be introduced into the mature mRNA. Equally noteworthy is the difference between the size of a typical gene and the size of the final mRNA transcribed from it once introns are removed. As shown in **Table 7**, only about 15 percent of the collagen gene consists of exons that finally appear in mRNA. For other proteins, an even more extreme picture emerges. Only about 8 percent of the albumin gene remains to be translated, and in the largest human gene known, dystrophin (which is the protein product absent in Duchenne muscular dystrophy), less than 1 percent of the gene sequence is retained in the mRNA. Two other human genes are also contrasted in Table 7.

Although the vast majority of eukaryotic genes examined thus far contain introns, there are several exceptions. Notably, the genes coding for histones and for interferon appear to contain no introns. It is not clear why or how the genes encoding these molecules have been maintained throughout evolution without acquiring the extraneous information characteristic of almost all other genes.

Splicing Mechanisms: Self-Splicing RNAs

The discovery of split genes led to intensive attempts to elucidate the mechanism by which introns of RNA are excised and exons are spliced back together. A great deal of progress has already been made, relying heavily on *in vitro* studies. Interestingly, it appears that somewhat different mechanisms exist for different types of RNA, as well as for RNAs produced in mitochondria and chloroplasts.

We might envision the simplest possible mechanism for removing an intron to be as illustrated in Figure 9. After an endonucleolytic "cut" is made at each end of an intron, the intron is removed, and the terminal ends of the adjacent

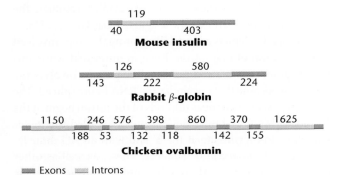

FIGURE 11 Intervening sequences in various eukaryotic genes. The numbers indicate the number of nucleotides present in various intron and exon regions.

TABLE 7 Comparing Human Gene Size, mRNA Size, and the Number of Introns

Gene	Gene Size (kb)	mRNA Size (kb)	Number of Introns
Insulin	1.7	0.4	2
Collagen [pro-α-2(1)]	38.0	5.0	51
Albumin	25.0	2.1	14
Phenylalanine hydroxylase	90.0	2.4	12
Dystrophin	2000.0	17.0	79

exons are ligated by an enzyme (in short, the intron is snipped out, and the exon ends are rejoined). This is apparently what happens to the introns present in transfer RNAs (tRNAs) in bacteria. A specific endonuclease recognizes the intron termini and excises the intervening sequences. Then RNA ligase seals the exon ends to complete each splicing event. However, in the studies of all other RNAs—tRNA in higher eukaryotes and rRNAs and pre-mRNAs in all eukaryotes—precise excision of introns is much more complex and a much more interesting story.

Introns in eukaryotes can be categorized into several groups based on their splicing mechanisms. Group I, represented by introns that are part of the primary transcript of rRNAs, require no additional components for intron excision; the intron itself is the source of the enzymatic activity necessary for removal. This amazing discovery was made in 1982 by Thomas Cech and his colleagues during a study of the ciliate protozoan *Tetrahymena*. RNAs that are capable of catalytic activity are referred to as **ribozymes.** The self-excision process for Group I introns serves to illustrate this concept and is shown in **Figure 12.** Chemically, two

nucleophilic reactions take place—that is, reactions caused by the presence of electron-rich chemical species (in this case, they are *transesterification reactions*). The first is an interaction between guanosine, which acts as a cofactor in the reaction, and the primary transcript [Figure 12(a)]. The 3′-OH group of guanosine is transferred to the nucleotide adjacent to the 5′ end of the intron [Figure 12(b) and Figure 12(c)]. The second reaction involves the interaction of the newly acquired 3′-OH group on the left-hand exon and the phosphate on the 3′ end of the intron [Figure 12(c)]. The intron is spliced out and the two exon regions are ligated, leading to the mature RNA [Figure 12(d)].

Self-excision of group I introns, as described above, is now known to apply to pre-rRNAs from other protozoans besides *Tetrahymena*. Self-excision also seems to govern the removal of introns from the primary mRNA and tRNA transcripts produced in mitochondria and chloroplasts. These are referred to as group II introns. As in group I molecules, splicing here involves two autocatalytic reactions leading to the excision of introns. However, guanosine is not involved as a cofactor with group II introns.

Splicing Mechanisms: The Spliceosome

Introns are a major component of nuclear-derived pre-mRNA transcripts of eukaryotes. Compared to the group I and group II introns discussed above, those in nuclear-derived mRNA can be much larger—up to 20,000 nucleotides—and they are more plentiful. Their removal appears to require a much more complex mechanism. Nevertheless, we now have a good handle on the process.

Interestingly, the splicing reactions are mediated by a huge molecular complex called a **spliceosome,** which has now been identified in extracts of yeast as well as in mammalian cells. This structure is very large, 40*S* in yeast and 60*S* in mammals, being the same size as ribosomal subunits! One set of essential components of spliceosomes is a unique set of small nuclear RNAs (snRNAs). These RNAs are usually 100 to 200 nucleotides long or less and are complexed with proteins to form small nuclear ribonucleoproteins (snRNPs or snurps). Because they are rich in uridine residues, the snRNAs have been arbitrarily designated U1, U2, . . . , U6.

Figure 13 depicts a model illustrating the steps involved in the removal of one intron. Keep in mind that while this figure shows separate components, the process involves the huge spliceosome that envelopes the RNA being spliced. The nucleotide sequences near the ends of the intron begin at the 5′ end with a GU dinucleotide sequence, called the *donor sequence,* and terminate at the 3′ end with an AG dinucleotide, called the *acceptor sequence.* These, as well as other consensus sequences shared by introns, attract specific snRNAs of the spliceosome. For example, the snRNA U1 bears a nucleotide sequence that is complementary to the 5′-donor sequence end of the intron. Base pairing resulting

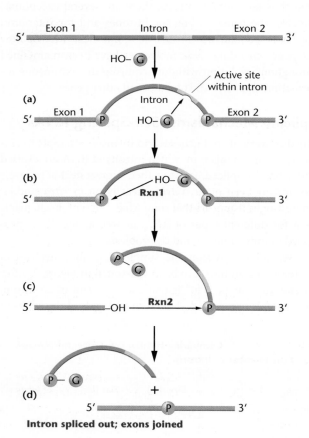

FIGURE 12 Splicing mechanism of pre-rRNA involving group I introns that are removed from the initial transcript. The process is one of self-excision involving two transesterification reactions.

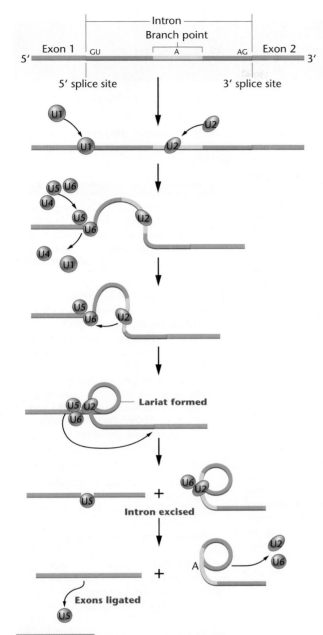

FIGURE 13 A model of the splicing mechanism involved during the removal of an intron from a pre-mRNA. Excision is dependent on various snRNAs (U1, U2, . . . , U6) that combine with proteins to form snurps, which are part of the spliceosome. The lariat structure in the intermediate stage is characteristic of this mechanism.

from this homology promotes the binding that represents the initial step in the formation of the spliceosome. After the other snRNPs (U2, U4, U5, and U6) are added, splicing commences. As with group I splicing, two transesterification reactions occur. The first involves the interaction of the 3'-OH group from an adenine (A) residue present within the branch point region of the intron. The A residue attacks the 5'-splice site, cutting the RNA chain. In a subsequent

step involving several other snRNPs, an intermediate structure is formed and the second reaction ensues, linking the cut 5' end of the intron to the A. This results in the formation of a characteristic loop structure called a *lariat,* which contains the excised intron. The exons are then ligated and the snRNPs are released.

The processing involved in splicing, which occurs within the nucleus, represents a potential regulatory step in gene expression in eukaryotes. For instance, several cases are known wherein introns present in pre-mRNAs *derived from the same gene* are spliced *in more than one way,* thereby yielding different collections of exons in the mature mRNA. Such alternative splicing yields a group of similar but nonidentical mRNAs that, upon translation, result in a series of related proteins called isoforms. Many examples have been encountered in organisms ranging from viruses to *Drosophila* to humans. Alternative splicing of pre-mRNAs represents a way of producing related proteins from a single gene, increasing the number of gene products that can be derived from an organism's genome.

ESSENTIAL POINT ■ ■ ■

The primary transcript in eukaryotes reflects the presence of intervening sequences, or introns, present in DNA, which must be spliced out to create the mature mRNA.

12 Transcription Has Been Visualized by Electron Microscopy

We conclude our coverage of transcription by referring you back to the chapter opening photograph, which is a striking visualization of transcription occurring in the oocyte nucleus of *Xenopus laevis,* the clawed frog. Note the central axis that runs horizontally from left to right and from which threads appear to be emanating vertically. This axis, appearing as a thin thread, is the DNA of most of one gene encoding ribosomal RNA (rDNA). Each of the emanating threads, which grows longer the farther to the right it is found, is an rRNA molecule being transcribed. What is apparent is that multiple copies of RNA polymerase have initiated transcription at a point near the left end and that transcription by each of them has proceeded to the right. Simultaneous transcription by many of these polymerases results in the electron micrograph that has captured an image of the entire process.

It is fascinating to visualize the process and to confirm our expectations based on the biochemical analysis of this process.

GENETICS, TECHNOLOGY, AND SOCIETY

Nucleic Acid-Based Gene Silencing: Attacking the Messenger

Standard chemotherapies for diseases such as cancer and AIDS are often accompanied by toxic side effects. Conventional therapeutic drugs affect both normal and diseased cells, with diseased or infected cells being only slightly more susceptible than the patient's normal cells. Scientists have long wished for a magic bullet that could seek out and destroy viruses or diseased cells, leaving normal cells alive and healthy. Over the last decade, a group of promising candidates has emerged, collectively described as *nucleic acid-based gene-silencing* drugs.

The two chief nucleic acid-based therapies currently being investigated are *antisense oligonucleotides* (*ASOs*) and *RNA interference* (*RNAi*). Both have been developed through an understanding of the molecular biology of gene expression: First, a single-stranded messenger RNA (mRNA) is copied from the template strand of the duplex DNA molecule; and second, the mRNA is complexed with ribosomes, and its coded information is translated into the amino acid sequence of a polypeptide.

Normally, a gene is transcribed into RNA from only one strand of the DNA duplex. The resulting RNA is known as *sense RNA*. However, it is possible for the other DNA strand to be copied into RNA, and this RNA, produced by transcription of the "wrong" strand of DNA, is called *antisense RNA*. When present together, sense and antisense RNA strands can form double-stranded duplex structures, the formation of which may affect the sense RNA in several ways.

In ASO technologies, scientists design single-stranded antisense DNA oligonucleotides (about 20 nucleotides long) of known sequence and then synthesize large amounts of these antisense nucleic acids *in vitro*. It is theoretically possible to treat cells with these synthetic antisense oligonucleotides, so that they enter the cells and bind to their target mRNAs.

The binding of antisense DNA to sense mRNA may physically block its translation. Alternatively, the degradation of the RNA may result. In either case, gene expression is blocked.

The antisense approach is exciting because of its potential specificity. Because an ASO has a sequence that specifically binds to a particular sense RNA, it should be possible to inhibit synthesis of the specific protein encoded by the sense RNA. If the protein is necessary for virus reproduction or cancer cell growth (but is not necessary in normal cells), the antisense oligonucleotide should have only therapeutic effects.

One ASO drug, which targets cytomegalovirus infection, is currently on the market. Others, which are designed to counteract cancers, Crohn's disease, HIV-1, and Hepatitis C, are in Phase II clinical trials.

The second gene-silencing approach, called RNAi technology, uses short double-stranded RNA molecules (~20–25 nucleotides long) with sequences complementary to specific mRNAs within cells. These are known as *short interfering RNAs* (*siRNAs*). These siRNA molecules may be synthesized *in vitro* or may be transcribed within a cell from cloned vectors that are introduced into cells.

Once within a cell's cytoplasm, the siRNA associates with an enzyme complex called an *RNA-induced silencing complex* (*RISC*), which is found within the cell's cytoplasm. One strand of the siRNA is cleaved within the RISC, and the other strand binds to a target mRNA that contains the complementary RNA sequence. When RISC and the siRNA are bound to the target mRNA, the RISC may cleave the target mRNA or may interfere with its translation.

RNAi clinical trials are being conducted to study its use in combating the eye disease, macular degeneration, with encouraging results. Other areas of high interest for RNAi-based treatments are cancers, diseases of the nervous system, and viral infections such as hepatitis B and HIV-1.

Your Turn

Take time, individually or in groups, to answer the following questions. Investigate the references and links to help you discuss some of the issues that surround the development and uses of antisense therapies.

1. What are some of the challenges in the use of ASOs as therapeutics? Do siRNAs share these challenges?

   A balanced discussion of antisense oligonucleotide drugs is presented in: Lebedeva, I. and Stein, C.A. 2001. Antisense oligonucleotides: Promise and reality. *Annu. Rev. Pharmacol.* 41:403–419. *RNAi drugs are critiqued in* Dykxhoorn, D. M. and Lieberman, J. 2006. Running interference: Prospects and obstacles to using small interfering RNAs as small molecule drugs. *Annu. Rev. Biomed. Eng.* 8:377–402.

2. Have any RNAi-based therapeutic drugs reached the market? What clinical trials for RNAi drugs are currently in progress?

   Information about clinical trials can be found at http://www.ClinicalTrials.gov. *A number of biotechnology companies, including Sirna Therapeutics, Alnylam Pharmaceuticals, and Opko Health, are developing RNAi-based therapeutics. Their Web sites also contain information about the RNAi drug pipelines and clinical trials.*

3. Studies in model organisms show that RNAi is effective in silencing genes involved in a wide range of infections and diseases. What do you think is the most promising use of RNAi as a therapeutic?

   To read about animal studies using siRNA gene silencing, see Dykxhoorn, D.M., et al. 2006. The silent treatment: siRNAs as small molecule drugs. *Gene Therapy* 13:541–552.

CASE STUDY » A drug that sometimes works

A 30-year-old woman with β-thalassemia, a recessively inherited genetic disorder caused by absence of the hemoglobin β chain, had been treated with blood transfusions since the age of 7. However, in spite of the transfusions, her health was declining. As an alternative treatment, her physician administered 5-azacytidine to induce transcription of the fetal β hemoglobin chain to replace her missing β chain. This drug activates gene transcription by removing methyl groups from DNA. Addition of methyl groups silences genes. However, the physician expressed concern that approximately 40 percent of all human genes are normally silenced by methylation. Nevertheless, after several weeks of 5-azacytidine treatment, the patient's condition improved dramatically. Although the treatment was successful, use of this drug raises several important questions.

1. Why was her physician concerned that a high percentage of human genes are transcriptionally silenced by methylation?
2. What genes might raise the greatest concern?
3. What criteria would you use when deciding to administer a drug such as 5-azacytidine?

INSIGHTS AND SOLUTIONS

1. Calculate how many triplet codons would be possible had evolution seized on six bases (three complementary base pairs) rather than four bases within the structure of DNA. Would six bases accommodate a two-letter code, assuming 20 amino acids and start and stop codons?

Solution: Six things taken three at a time will produce $(6)^3$ or 216 triplet codes. If the code was a doublet, there would be $(6)^2$ or 36 two-letter codes, more than enough to accommodate 20 amino acids and start and stop signals.

2. In a heteropolymer experiment using 1/2C:1/4A:1/4G, how many different triplets will occur in the synthetic RNA molecule? How frequently will the most frequent triplet occur?

Solution: There will be $(3)^3$ or 27 triplets produced. The most frequent will be CCC, present $(1/2)^3$ or 1/8 of the time.

3. In a regular copolymer experiment, where UUAC is repeated over and over, how many different triplets will occur in the synthetic RNA, and how many amino acids will occur in the polypeptide when this RNA is translated? (Consult Figure 7.)

Solution: The synthetic RNA will repeat four triplets—UUA, CUU, ACU, UAC—over and over. Because both UUA and CUU encode leucine, while ACU and UAC encode threonine and tyrosine, respectively, the polypeptides synthesized under the directions of this RNA would contain three amino acids in the repeating sequence leu-leu-thr-tyr.

4. Actinomycin D inhibits DNA-dependent RNA synthesis. This antibiotic is added to a bacterial culture where a specific protein is being monitored. Compared to a control culture, where no antibiotic is added, translation of the protein declines over a period of 20 minutes, until no further protein is made. Explain these results.

Solution: The mRNA, which is the basis for translation of the protein, has a lifetime of about 20 minutes. When actinomycin D is added, transcription is inhibited and no new mRNAs are made. Those already present support the translation of the protein for up to 20 minutes.

PROBLEMS AND DISCUSSION QUESTIONS

 For activities, animations, and review quizzes, go to the study area at www.masteringgenetics.com

HOW DO WE KNOW?

1. In this chapter, we focused on the genetic code and the transcription of genetic information stored in DNA into complementary RNA molecules. Along the way, we found many opportunities to consider the methods and reasoning by which much of this information was acquired. From the explanations given in the chapter, what answers would you propose to the following fundamental questions:

    (a) How did we determine the *compositions* of codons encoding specific amino acids?

    (b) How were the specific sequences of triplet codes determined experimentally?

    (c) How were the experimentally derived triplet codon assignments verified in studies using bacteriophage MS2?

    (d) How do we know that mRNA exists and serves as an intermediate between information encoded in DNA and its concomitant gene product?

    (e) How do we know that the initial transcript of a eukaryotic gene contains noncoding sequences that must be removed before accurate translation into proteins can occur?

2. Early proposals regarding the genetic code considered the possibility that DNA served directly as the template for polypeptide synthesis. In eukaryotes, what difficulties would such a system pose? What observations and theoretical considerations argue against such a proposal?

3. In studies of frameshift mutations, Crick, Barnett, Brenner, and Watts–Tobin found that either three nucleotide insertions or deletions restored the correct reading frame. (a) Assuming the

code is a triplet, what effect would the addition or loss of six nucleotides have on the reading frame? (b) If the code were a sextuplet (consisting of six nucleotides), would the reading frame be restored by the addition or loss of three, six, or nine nucleotides?

4. The mRNA formed from the repeating tetranucleotide UUAC incorporates only three amino acids, but the use of UAUC incorporates four amino acids. Why?

5. In studies using repeating copolymers, AC . . . incorporates threonine and histidine, and CAACAA . . . incorporates glutamine, asparagine, and threonine. What triplet code can definitely be assigned to threonine?

6. In a coding experiment using repeating copolymers (as shown in Table 3), the following data were obtained.

Copolymer	Codons Produced	Amino Acids in Polypeptide
AG	AGA, GAG	arg, glu
AAG	AGA, AAG, GAA	lys, arg, glu

AGG is known to code for arginine. Taking into account the wobble hypothesis, assign each of the four remaining different triplet codes to its correct amino acid.

7. In the triplet-binding assay technique, radioactivity remains on the filter when the amino acid corresponding to the experimental triplet is labeled. Explain the basis of this technique.

8. When the amino acid sequences of insulin isolated from different organisms were determined, some differences were noted. For example, alanine was substituted for threonine, serine was substituted for glycine, and valine was substituted for isoleucine at corresponding positions in the protein. List the single-base changes that could occur in triplets to produce these amino acid changes.

9. In studies of the amino acid sequence of wild-type and mutant forms of tryptophan synthetase in *E. coli,* the following changes have been observed:

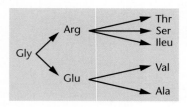

Determine a set of triplet codes in which only a single-nucleotide change produces each amino acid change.

10. Why doesn't polynucleotide phosphorylase (Ochoa's enzyme) synthesize RNA *in vivo*?

11. Refer to Table 1. Can you hypothesize why a mixture of (Poly U) + (Poly A) would not stimulate incorporation of ^{14}C-phenylalanine into protein?

12. Predict the amino acid sequence produced during translation of the short theoretical mRNA sequences below. (Note that the second sequence was formed from the first by a deletion of only one nucleotide.) What type of mutation gave rise to sequence 2?

Sequence 1: 5'-AUGCCGGAUUAUAGUUGA-3'

Sequence 2: 5'-AUGCCGGAUUAAGUUGA-3'

13. A short RNA molecule was isolated that demonstrated a hyperchromic shift indicating secondary structure. Its sequence was determined to be

5'-AGGCGCCGACUCUACU-3'

(a) Propose a two-dimensional model for this molecule.
(b) What DNA sequence would give rise to this RNA molecule through transcription?
(c) If the molecule were a tRNA fragment containing a CGA anticodon, what would the corresponding codon be?
(d) If the molecule were an internal part of a message, what amino acid sequence would result from it following translation? (Refer to the code chart in Figure 7.)

14. A glycine residue exists at position 210 of the tryptophan synthetase enzyme of wild-type *E. coli.* If the codon specifying glycine is GGA, how many single-base substitutions will result in an amino acid substitution at position 210, and what are they? How many will result if the wild-type codon is GGU?

15. Shown here is a theoretical viral mRNA sequence

5'-AUGCAUACCUAUGAGACCCUUGGA-3'

(a) Assuming that it could arise from overlapping genes, how many different polypeptide sequences can be produced? Using the chart in Figure 7, what are the sequences?
(b) A base substitution mutation that altered the sequence in part (a) eliminated the synthesis of all but one polypeptide. The altered sequence is shown below. Use Figure 7 to determine why it was altered.

5'-AUGCAUACCUAUGUGACCCUUGGA-3'

16. Most proteins have more leucine than histidine residues but more histidine than tryptophan residues. Correlate the number of codons for these three amino acids with this information.

17. Define the process of transcription. Where does this process fit into the central dogma of molecular genetics?

18. Describe the structure of RNA polymerase in bacteria. What is the core enzyme? What is the role of the σ subunit?

19. In a written paragraph, describe the abbreviated chemical reactions that summarize RNA polymerase-directed transcription.

20. Messenger RNA molecules are very difficult to isolate from prokaryotes because they are quickly degraded. Can you suggest a reason why this occurs? Eukaryotic mRNAs are more stable and exist longer in the cell than do prokaryotic mRNAs. Is this an advantage or a disadvantage for a pancreatic cell making large quantities of insulin?

21. One form of posttranscriptional modification of most eukaryotic RNA transcripts is the addition of a poly-A sequence at the 3' end. The absence of a poly-A sequence leads to rapid degradation of the transcript. Poly-A sequences of various lengths are also added to many prokaryotic RNA transcripts where, instead of promoting stability, they enhance degradation. In both cases, RNA secondary structures, stabilizing proteins, or degrading enzymes interact with poly-A sequences. Considering the activities of RNAs, what might be the general functions of 3'-polyadenylation?

22. In a mixed copolymer experiment, messages were created with either 4/5C:1/5A or 4/5A:1/5C. These messages yielded proteins with the amino acid compositions shown in the following table. Using these data, predict the most specific *coding composition* for each amino acid.

4/5C:1/5A		4/5A:1/5C	
Proline	63.0%	Proline	3.5%
Histidine	13.0%	Histidine	3.0%
Threonine	16.0%	Threonine	16.6%
Glutamine	3.0%	Glutamine	13.0%
Asparagine	3.0%	Asparagine	13.0%
Lysine	0.5%	Lysine	50.0%
	98.5%		99.1%

23. Shown in this problem are the amino acid sequences of the wild type and three mutant forms of a short protein.
    (a) Using Figure 7, predict the type of mutation that created each altered protein.
    (b) Determine the specific ribonucleotide change that led to the synthesis of each mutant protein.
    (c) The wild-type RNA consists of nine triplets. What is the role of the ninth triplet?
    (d) For the first eight wild-type triplets, which, if any, can you determine specifically from an analysis of the mutant proteins? In each case, explain why or why not.
    (e) Another mutation (mutant 4) is isolated. Its amino acid sequence is unchanged, but mutant cells produce abnormally low amounts of the wild-type proteins. As specifically as you can, predict where this mutation exists in the gene.

Wild type:	met-trp-tyr-arg-gly-ser-pro-thr
Mutant 1:	met-trp
Mutant 2:	met-trp-his-arg-gly-ser-pro-thr
Mutant 3:	met-cys-ile-val-val-val-gln-his

24. Alternative splicing is a common mechanism for eukaryotes to expand their repertoire of gene functions. Studies by Xu and colleagues (2002. *Nuc. Acids Res.* 30: 3754–3766) indicate that approximately 50 percent of human genes use alternative splicing, and approximately 15 percent of disease-causing mutations involve aberrant alternative splicing. Different tissues show remarkably different frequencies of alternative splicing, with the brain accounting for approximately 18 percent of such events.
    (a) Define alternative splicing and speculate on the evolutionary strategy alternative splicing offers to organisms.
    (b) Why might some tissues engage in more alternative splicing than others?

25. The genetic code is degenerate. Amino acids are encoded by either 1, 2, 3, 4, or 6 triplet codons. (See Figure 7.) An interesting question is whether the number of triplet codes for a given amino acid is in any way correlated with the frequency with which that amino acid appears in proteins. That is, is the genetic code optimized for its intended use? Some approximations of the frequency of appearance of nine amino acids in proteins in *E. coli* are

Amino Acid	Percentage
Met	2
Cys	2
Gln	5
Pro	5
Arg	5
Ile	6
Glu	7
Ala	8
Leu	10

(a) Determine how many triplets encode each amino acid.
(b) Devise a way to graphically compare the two sets of information (data).
(c) Analyze your data to determine what, if any, correlations can be drawn between the relative frequency of amino acids making up proteins and the number of codons for each. Write a paragraph that states your specific and general conclusions.
(d) How would you proceed with your analysis if you wanted to pursue this problem further?

SOLUTIONS TO SELECTED PROBLEMS AND DISCUSSION QUESTIONS

Answers to Now Solve This

1. (a) The way to determine the fraction of each triplet that will occur with a random incorporation system is to determine the likelihood that each base will occur in each position of the codon (first, second, third), and then multiply the individual probabilities (fractions) for a final probability (fraction).

$$GGG = 3/4 \times 3/4 \times 3/4 = 27/64$$
$$GGC = 3/4 \times 3/4 \times 1/4 = 9/64$$
$$GCG = 3/4 \times 1/4 \times 3/4 = 9/64$$
$$CGG = 1/4 \times 3/4 \times 3/4 = 9/64$$

$$CCG = 1/4 \times 1/4 \times 3/4 = 3/64$$
$$CGC = 1/4 \times 3/4 \times 1/4 = 3/64$$
$$GCC = 3/4 \times 1/4 \times 1/4 = 3/64$$
$$CCC = 1/4 \times 1/4 \times 1/4 = 1/64$$

(b) Glycine:
GGG and one G_2C (adds up to 36/64)
Alanine:
one G_2C and one C_2G (adds up to 12/64)
Arginine:
one G_2C and one C_2G (adds up to 12/64)
Proline:
one C_2G and CCC (adds up to 4/64)

(c) With the wobble hypothesis, variation can occur in the third position of each codon. Below are possible unordered codon assignments.

Glycine: GGG, GGC
Alanine: CGG, GCC, CGC, GCG
Arginine: GCG, GCC, CGC, CGG
Proline: CCC, CCG

2. Because of a triplet code, a trinucleotide sequence will, once initiated, remain in the same reading frame and produce the same code all along the sequence regardless of the initiation site. If a tetranucleotide is used, such as ACGUACGUACGU...:

Codons:	ACG	UAC	GUA	CGU	ACG
Amino acids:	thr	tyr	val	arg	thr
	CGU	ACG	UAC	GUA	CGU
	arg	thr	tyr	val	arg
	GUA	CGU	ACG	UAC	GUA
	val	arg	thr	tyr	val
	UAC	GUA	CGU	ACG	UAC
	tyr	val	arg	thr	tyr

Notice that the sequences are the same except that the starting amino acid changes.

3. Apply complementary bases, substituting U for T:
(a) Sequence 1: 3'-GAAAAAACGGUA-5'
Sequence 2: 3'-UGUAGUUAUUGA-5'
Sequence 3: 3'-AUGUUCCCAAGA-5'

(b) Sequence 1: *met-ala-lys-lys*
Sequence 2: *ser-tyr-[ter]*
Sequence 3: *arg-thr-leu-val*

(c) Apply complementary bases: 3'-GAAAAAACGGTA-5'

Solutions to Problems and Discussion Questions

2. In eukaryotes, protein synthesis occurs primarily in the cytoplasm, far from the location of DNA and the encoded information. In addition, while some of the basic amino acids would

be able to associate directly with DNA, the acidic amino acids would be unable to do so. Thus some sort of "adaptor" system was needed for DNA to direct amino acid assembly.

4. The UUACUUACUUAC tetranucleotide sequence will produce the following triplets depending on the initiation point: UUA = leu; UAC = tyr; ACU = thr; CUU = leu. Notice that because of the degenerate code, two codons correspond to the amino acid leucine.

The UAUCUAUCUAUC tetranucleotide sequence will produce the following triplets depending on the initiation point: UAU = tyr; AUC = ile; UCU = ser; CUA = leu. Notice that in this case, degeneracy is not revealed and all the codons produce unique amino acids.

6. Given that AGG = arg, information from the AG copolymer indicates that AGA also codes for arg and that GAG must therefore code for glu. Coupling this information with that of the AAG copolymer, one can see that GAA must also code for glu and AAG must code for lys.

8. List the substitutions, then from the code table apply the codons to the original amino acids. Select codons that provide single base changes.

Original		Substitutions
threonine	----->	*alanine*
_AC (U, C, A, or G)		_GC (U, C, A, or G)
glycine	----->	*serine*
_GG (U or C)		_AG (U or C)
isoleucine	----->	*valine*
_AU (U, C, or A)		_GU (U, C, or A)

10. The enzyme generally functions in the degradation of RNA; however, in an *in vitro* environment, with high concentrations of the ribonucleoside diphosphates, the direction of the reaction can be forced toward polymerization. *In vivo*, the concentration of ribonucleoside diphosphates is low and the degradative process is favored.

12. Applying the coding dictionary, the following sequences are "decoded":

Sequence 1: met-pro-asp-tyr-ser-(term)
Sequence 2: met-pro-asp-(term)

The 12th base (a uracil) is deleted from sequence #1, thereby causing a frameshift mutation that introduced a terminating triplet UAA.

16. By examining the coding dictionary, one will notice that the number of codons for each particular amino acid (synonyms) is directly related to the frequency of amino acid incorporation stated in the problem.

20. While some folding (from complementary base pairing) may occur with mRNA molecules, they generally exist as single-stranded structures that are quite labile. Eukaryotic mRNAs are generally processed such that the 5' end is "capped" and the 3' end has a considerable string of adenine bases. It is thought that these features protect the mRNAs from degradation. Such stability of eukaryotic mRNAs probably evolved with the differentiation of nuclear and cytoplasmic functions. Because prokaryotic cells exist in a more unstable environment (nutritionally and physically, for example) than many cells of multicellular organisms, rapid genetic response to environmental change

is likely to be adaptive. To accomplish such rapid responses, a labile gene product (mRNA) is advantageous. A pancreatic cell, which is developmentally stable and exists in a relatively stable environment, could produce more insulin on stable mRNAs for a given transcriptional rate.

22. Proline: C_3 and one of the C_2A triplets
    Histidine: one of the C_2A triplets
    Threonine: one C_2A triplet and one A_2C triplet
    Glutamine: one of the A_2C triplets
    Asparagine: one of the A_2C triplets
    Lysine: A_3

24. **(a, b)** Alternative splicing occurs when pre-mRNAs are spliced in more than one way to yield various combinations of exons in the final mRNA product. Upon translation of a group of alternatively spliced mRNAs, a series of related proteins, called isoforms, are produced. It is likely that alternative splicing evolved to provide a variety of functionally related proteins in a particular tissue from one original source. In other words, varieties of similar proteins can be produced by alternative splicing rather than by independent evolution.

Some tissues might be more prone to develop alternative splicing if they depend on a number of related protein functions. In addition, if genes found in certain tissues have more exons in their active genes, alternative splicing would be expected. The use of alternative splicing to generate varieties of products is also known to exist at different developmental stages. In such cases, as development occurs, different splicing mechanisms occur.

CREDITS

Credits are listed in order of appearance.

Photo

CO, Prof. Oscar L. Miller/Science Photo Library/Photo Researchers, Inc.; F-10 Bert W. O'Malley, M.D., Baylor College of Medicine

Translation and Proteins

From Chapter 13 of *Essentials of Genetics, Eighth edition*, William S. Klug, Michael R. Cummings, Charlotte A. Spencer, Michael A. Palladino. © 2013 by Pearson Education, Inc. All rights reserved.

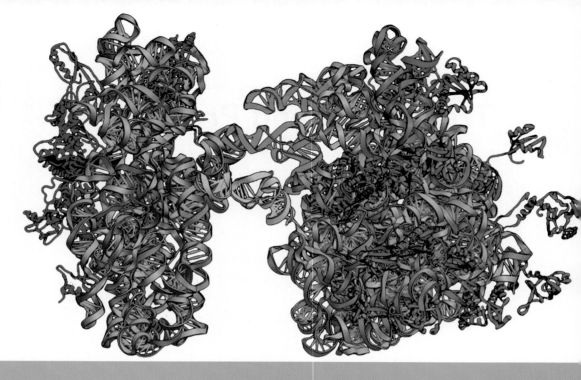

Crystal structure of a *Thermus thermophilus* 70S ribosome containing three bound transfer RNAs.

Translation and Proteins

CHAPTER CONCEPTS

- The ribonucleotide sequence of messenger RNA (mRNA) reflects genetic information stored in DNA that makes up genes and corresponds to the amino acid sequences in proteins encoded by those genes.

- The process of translation decodes the information in mRNA, leading to the synthesis of polypeptide chains.

- Translation involves the interactions of mRNA, tRNA, ribosomes, and a variety of translation factors essential to the initiation, elongation, and termination of the polypeptide chain.

- Proteins, the final product of most genes, achieve a three-dimensional conformation that is based on the primary amino acid sequences of the polypeptide chains making up each protein.

- The function of any protein is closely tied to its three-dimensional structure, which can be disrupted by mutation.

In A genetic code stores information in the form of triplet nucleotides in DNA and this information is initially expressed through the process of transcription into a messenger RNA that is complementary to one strand of the DNA helix. However, in most instances, the final product of gene expression is a polypeptide chain consisting of a linear series of amino acids whose sequence has been prescribed by the genetic code. In this

, we will examine how the information present in mRNA is utilized to create polypeptides, which then fold into protein molecules. We will also review the evidence confirming that proteins are the end products of gene expression, and we will briefly discuss the various levels of protein structure, diversity, and function. This information extends our understanding of gene expression and provides an important foundation for interpreting how the mutations that arise in DNA can result in the diverse phenotypic effects observed in organisms.

1   Translation of mRNA Depends on Ribosomes and Transfer RNAs

Translation of mRNA is the biological polymerization of amino acids into polypeptide chains. This process occurs only in association with ribosomes, which serve as nonspecific workbenches. The central question in translation is how triplet ribonucleotides of mRNA direct specific amino acids into their correct position in the polypeptide. This question was answered once **transfer RNA (tRNA)** was discovered. This class of molecules adapts specific triplet codons in mRNA to their correct amino acids. The *adaptor hypothesis* for the role of tRNA was postulated by Francis Crick in 1957.

In association with a ribosome, mRNA presents a triplet codon that calls for a specific amino acid. A specific tRNA molecule contains within its nucleotide sequence three consecutive ribonucleotides complementary to the codon, called the **anticodon,** which can base-pair with the codon. Another region of this tRNA is covalently bonded to its corresponding amino acid.

Hydrogen bonding of tRNAs to mRNA holds the amino acids in proximity so that a peptide bond can be formed. This process occurs over and over as mRNA runs through the ribosome and amino acids are polymerized into a polypeptide. Before we discuss the actual process of translation, let's first consider the structures of the ribosome and tRNA.

Ribosomal Structure

Because of its essential role in the expression of genetic information, the **ribosome** has been extensively analyzed. One bacterial cell contains about 10,000 ribosomes, and a eukaryotic cell contains many times more. Electron microscopy reveals that the bacterial ribosome is about 40 nm at its largest dimension and consists of two subunits, one large and one small. Both subunits consist of one or more molecules of rRNA and an array of **ribosomal proteins.** When the two subunits are associated with each other in a single ribosome, the structure is sometimes called a **monosome.**

The main differences between prokaryotic and eukaryotic ribosomes are summarized in **Figure 1**. The subunit and rRNA components are most easily isolated and characterized on the basis of their sedimentation behavior in sucrose gradients (their rate of migration, or *Svedberg coefficient S*, which is a reflection of their density, mass, and shape). In prokary-

otes, the monosome is a 70S particle; in eukaryotes, it is approximately 80S. Sedimentation coefficients, which reflect the variable rate of migration of different-sized particles and molecules, are not additive. For example, the prokaryotic 70S monosome consists of a 50S and a 30S subunit, and the eukaryotic 80S monosome consists of a 60S and a 40S subunit.

The larger subunit in prokaryotes consists of a 23S rRNA molecule, a 5S rRNA molecule, and 31 ribosomal proteins. In the eukaryotic equivalent, a 28S rRNA molecule is accompanied by a 5.8S and 5S rRNA molecule and 46 proteins. The smaller prokaryotic subunits consist of a 16S rRNA component and 21 proteins. In the eukaryotic equivalent, an 18S rRNA component and 33 proteins are found. The approximate molecular weights (MWs) and the number of nucleotides of these components are also shown in Figure 1.

It is now clear that the RNA molecules perform the all-important catalytic functions associated with translation. The many proteins, whose functions were long a mystery, are thought to promote the binding of the various molecules involved in translation and, in general, to fine-tune the process. This conclusion is based on the observation that some of the catalytic functions in ribosomes still occur in experiments involving "ribosomal protein-depleted" ribosomes.

Molecular hybridization studies have established the degree of redundancy of the genes coding for the rRNA components. The *E. coli* genome contains seven copies of a single sequence that encodes all three components—23S, 16S, and 5S. The initial transcript of each set of these genes produces a 30S RNA molecule that is enzymatically cleaved into these small-

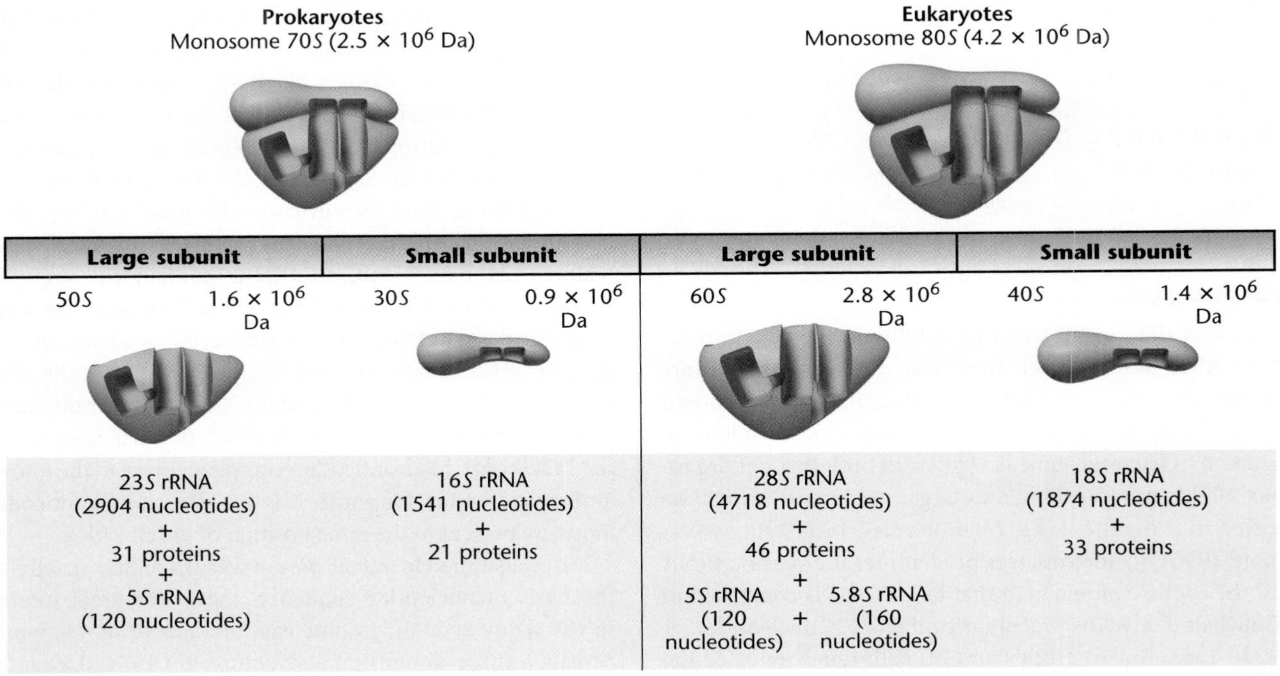

Prokaryotes Monosome 70S (2.5×10^6 Da)				**Eukaryotes** Monosome 80S (4.2×10^6 Da)			
Large subunit		**Small subunit**		**Large subunit**		**Small subunit**	
50S	1.6×10^6 Da	30S	0.9×10^6 Da	60S	2.8×10^6 Da	40S	1.4×10^6 Da
23S rRNA (2904 nucleotides) + 31 proteins + 5S rRNA (120 nucleotides)		16S rRNA (1541 nucleotides) + 21 proteins		28S rRNA (4718 nucleotides) + 46 proteins + 5S rRNA 5.8S rRNA (120 + (160 nucleotides) nucleotides)		18S rRNA (1874 nucleotides) + 33 proteins	

FIGURE 1   A comparison of the components of prokaryotic and eukaryotic ribosomes.

er components. Coupling of the genetic information encoding these three rRNA components ensures that after multiple transcription events, equal quantities of all three will be present as ribosomes are assembled.

In eukaryotes, many more copies of a sequence encoding the 28S, 18S, and 5.8S components are present. In *Drosophila*, approximately 120 copies per haploid genome are each transcribed into a molecule of about 34S. This molecule is then processed into the 28S, 18S, and 5.8S rRNA species. These species are homologous to the three rRNA components of *E. coli*. In *Xenopus laevis*, over 500 copies of the 34S component are present per haploid genome. In mammalian cells, the initial transcript is even larger at 45S.

The rRNA genes, called **rDNA,** are part of the moderately repetitive DNA fraction and are present in clusters at various chromosomal sites. Each cluster in eukaryotes consists of tandem repeats, with each unit separated by a noncoding spacer DNA sequence. In humans, these gene clusters have been localized near the ends of chromosomes 13, 14, 15, 21, and 22. The unique 5S rRNA component of eukaryotes is not part of this larger transcript. Instead, genes coding for the 5S ribosomal component are distinct and located separately. In humans, a gene cluster encoding 5S rRNA has been located on chromosome 1.

Despite their detailed knowledge of the structure and genetic origin of the ribosomal components, a complete understanding of the function of these components has eluded geneticists. This is not surprising; the ribosome is the largest and perhaps the most intricate of all cellular structures. For example, the bacterial monosome has a combined molecular weight of 2.5 million Da!

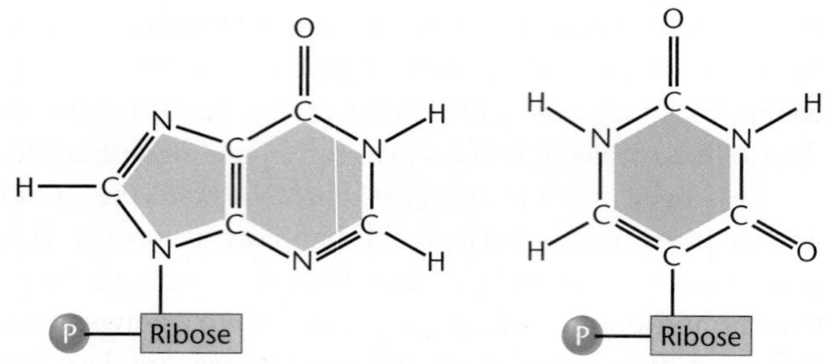

ESSENTIAL POINT

Translation is the synthesis of polypeptide chains under the direction of mRNA in association with ribosomes.

tRNA Structure

Because of their small size and stability in the cell, transfer RNAs (tRNAs) have been investigated extensively and are the best-characterized RNA molecules. They are composed of only 75 to 90 nucleotides, displaying a nearly identical structure in bacteria and eukaryotes. In both types of organisms, tRNAs are transcribed as larger precursors, which are cleaved into mature 4S tRNA molecules. In *E. coli,* for example, tRNA[Tyr] (the superscript identifies the specific tRNA and the cognate amino acid that binds to it) is composed of 77 nucleotides, yet its precursor contains 126 nucleotides.

In 1965, Robert Holley and his colleagues reported the complete sequence of tRNA[Ala] isolated from yeast. Of great interest was their finding that a number of nucleotides are

FIGURE 2 Ribonucleotides containing two unusual nitrogenous bases found in transfer RNA.

unique to tRNA, containing a so-called modified base. Two of these nucleotides, inosinic acid and pseudouridylic acid, are illustrated in **Figure 2**. These modified structures are created *after* transcription of tRNA, illustrating the more general concept of **posttranscriptional modification.** While it is still not clear why such modified bases are created, it is believed that their presence enhances hydrogen bonding efficiency during translation.

Holley's sequence analysis led him to propose the two-dimensional **cloverleaf model of tRNA.** It was known that tRNA demonstrates a secondary structure due to base pairing. Holley discovered that he could arrange the linear model in such a way that several stretches of base pairing would result. This arrangement created a series of paired stems and unpaired loops resembling the shape of a cloverleaf. Loops consistently contained modified bases that did not generally form base pairs. Holley's model is shown in **Figure 3**.

The triplets GCU, GCC, and GCA specify alanine; therefore, Holley looked for an anticodon sequence complementary to one of these codons in his tRNA[Ala] molecule. He found it in the form of CGI (the 3′ to 5′ direction) in one loop of the cloverleaf. The nitrogenous base I (inosinic acid) can form hydrogen bonds with U, C, or A, the third members of the alanine triplets. Thus, the **anticodon loop** was established.

Studies of other tRNA species reveal many constant features. At the 3′ end, all tRNAs contain the sequence (…pCpCpA-3′). This is the end of the molecule where the amino acid is covalently joined to the terminal adenosine residue. All tRNAs contain the nucleotide (5′-Gp…) at the other end of the molecule. In addition, the lengths of various stems and loops are very similar. Each tRNA that has been examined also contains an anticodon complementary to the known amino acid codon for which it is specific, and all anticodon loops are present in the same position of the cloverleaf.

Because the cloverleaf model was predicted strictly on the basis of nucleotide sequence, there was great interest in the X-ray crystallographic examination of tRNA, which reveals a three-dimensional structure. By 1974, Alexander Rich and his colleagues in the United States, and J. Roberts, B. Clark, Aaron Klug, and their colleagues in England had

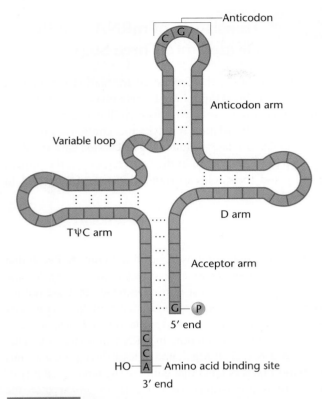

FIGURE 3 Holley's two-dimensional cloverleaf model of transfer RNA. Blocks represent nitrogenous bases.

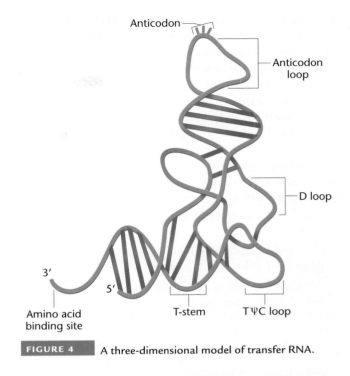

FIGURE 4 A three-dimensional model of transfer RNA.

succeeded in crystallizing tRNA and performing X-ray crystallography at a resolution of 3 Å. At this resolution, the pattern formed by individual nucleotides is discernible.

As a result of these studies, a complete three-dimensional model of tRNA was proposed, as shown in **Figure 4**. At one end of the molecule is the anticodon loop and stem, and at the other end is the 3′-acceptor region where the amino acid is bound. Geneticists speculate that the shapes of the intervening loops may be recognized by the specific enzymes responsible for adding amino acids to tRNAs—a subject to which we now turn our attention.

Charging tRNA

Before translation can proceed, the tRNA molecules must be chemically linked to their respective amino acids. This activation process, called **charging,** occurs under the direction of enzymes called **aminoacyl tRNA synthetases.** There are 20 different amino acids, so there must be at least 20 different tRNA molecules and as many different enzymes. In theory, because there are 61 triplets that encode amino acids, there could be 61 specific tRNAs and enzymes. However, because of the ability of the third member of a triplet code to "wobble," it is now thought that there are only 31 different tRNAs. It is also believed that there are only 20 synthetases, one for each amino acid, regardless of the greater number of corresponding tRNAs.

The charging process is outlined in **Figure 5**. In the initial step, the amino acid is converted to an activated form,

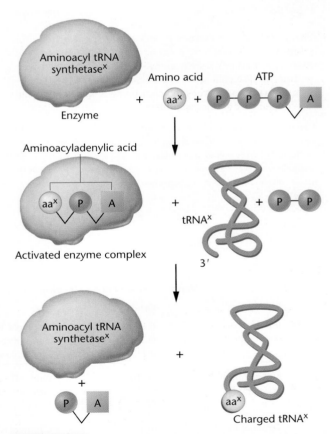

FIGURE 5 Steps involved in charging tRNA. The superscript x denotes that only the corresponding specific tRNA and specific aminoacyl tRNA synthetase enzyme are involved in the charging process for each amino acid.

301

reacting with ATP to create an **aminoacyladenylic acid.** A covalent linkage is formed between the 5′-phosphate group of ATP and the carboxyl end of the amino acid. This molecule remains associated with the synthetase enzyme, forming a complex that then reacts with a specific tRNA molecule. During this next step, the amino acid is transferred to the appropriate tRNA and bonded covalently to the adenine residue at the 3′ end. The charged tRNA may now participate directly in protein synthesis. Aminoacyl tRNA synthetases are highly specific enzymes because they recognize only one amino acid and the subset of corresponding tRNAs called **isoaccepting tRNAs.** Accurate charging is crucial if fidelity of translation is to be maintained.

ESSENTIAL POINT ■ ■ ■

Translation depends on tRNA molecules that serve as adaptors between triplet codons in mRNA and the corresponding amino acids.

NOW SOLVE THIS

1 In 1962, F. Chapeville and others reported an experiment in which they isolated radioactive ^{14}C-cysteinyl-tRNA^{Cys} (charged tRNA^{Cys} + cysteine). They then removed the sulfur group from the cysteine, creating alanyl-tRNA^{Cys} (charged tRNA^{Cys} + alanine). When alanyl-tRNA^{Cys} was added to a synthetic mRNA calling for cysteine, but not alanine, a polypeptide chain was synthesized containing alanine. What can you conclude from this experiment?

■ **Hint** *This problem is concerned with establishing whether tRNA or the amino acid added to the tRNA during charging is responsible for attracting the charged tRNA to mRNA during translation. The key to its solution is the observation that in this experiment, when the triplet codon in mRNA calls for cysteine, alanine is inserted during translation, even though it is the "incorrect" amino acid.*

2 Translation of mRNA Can Be Divided into Three Steps

Like transcription, the process of translation can be best described by breaking it into discrete phases. We will consider three phases, each with its own illustration, but keep in mind that translation is a dynamic, continuous process. You should correlate the following discussion with the step-by-step characterization in the figures. Many of the protein factors and their roles in translation are summarized in **Table 1**.

Initiation

Initiation of translation is depicted in **Figure 6**. Recall that the ribosome serves as a nonspecific workbench for the translation process. Most ribosomes, when they are not involved in translation, are dissociated into their large and small subunits. Initiation of translation in *E. coli* involves the small ribosome subunit, an mRNA molecule, a specific charged tRNA, GTP, Mg^{2+}, and at least three proteinaceous **initiation factors (IFs)** that enhance the binding affinity of the various translational components. In prokaryotes, the initiation codon of mRNA (AUG) calls for the modified amino acid **formylmethionine (fmet).**

The small ribosomal subunit binds to IF1, and this complex then binds to mRNA (Step 1). In bacteria, this binding involves a sequence of up to six ribonucleotides (AGGAGG, not shown), which *precedes* the initial AUG start codon of mRNA. This sequence (containing only purines and called the **Shine-Dalgarno sequence**) base-pairs with a region of the 16S rRNA of the small ribosomal subunit, facilitating initiation.

Another initiation protein then enhances the binding of charged fmet-tRNA to the small subunit in response to the AUG triplet (Step 2). This step sets the reading frame so that

TABLE 1	Various Protein Factors Involved during Translation in *E. coli*		
Process	**Factor**	**Role**	
Initiation of translation	IF1	Stabilizes 30S subunit	
	IF2	Binds fmet-tRNA to 30S-mRNA complex; binds to GTP	
	IF3	Binds 30S subunit to mRNA	
Elongation of polypeptide	EF-Tu	Binds GTP; mediates aminoacyl-tRNA entry to the A site of ribosome	
	EF-Ts	Generates active EF-Tu	
	EF-G	Stimulates translocation; GTP-dependent	
Termination of translation and release of polypeptide	RF1	Catalyzes release of the polypeptide chain from tRNA and dissociation of the translocation complex; specific for UAA and UAG termination codons	
	RF2	Behaves like RF1; specific for UGA and UAA codons	
	RF3	Stimulates RF1 and RF2	

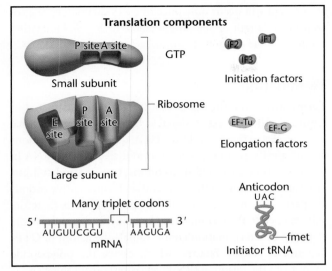

Translation components

Initiation of Translation

1. mRNA and IF1 bind to small subunit

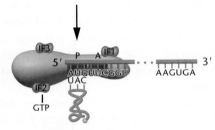

Initiation complex

2. IF2 and IF3 bind to complex; Initiator tRNA^{fmet} binds to mRNA codon in P site

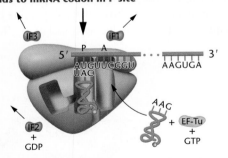

3. Large subunit binds to complex; IF factors are released; EF-Tu binds to tRNA, facilitating entry into A site

FIGURE 6   Initiation of translation. The components are depicted at the top of the figure.

all subsequent groups of three ribonucleotides are translated accurately. This aggregate represents the **initiation complex,** which then combines with the large ribosomal subunit. At

this point, a molecule of GTP is hydrolyzed, providing the required energy for the release of the initiation factors (Step 3).

Elongation

The second phase of translation, elongation, is depicted in **Figure 7.** Once both subunits of the ribosome are assembled with the mRNA, binding sites for two charged tRNA molecules are formed. These are the **P (peptidyl) site** and the **A (aminoacyl) site.** The charged initiator tRNA binds to the P site, provided that the AUG triplet of mRNA is in the corresponding position of the small subunit.

The lengthening of the growing polypeptide chain by one amino acid is called **elongation.** The sequence of the second triplet in mRNA dictates which charged tRNA molecule will become positioned at the A site (Step 1). Once tRNA is present, a reaction occurs within the large subunit of the ribosome, catalyzing the formation of the peptide bond that links the two amino acids together. At the same time, the covalent bond between the tRNA occupying the P site and its amino acid is hydrolyzed (broken). The dipeptide remains attached to the end of the tRNA still residing in the A site (Step 2). These reactions were initially believed to be catalyzed by an enzyme called **peptidyl transferase,** embedded in but never isolated from the large subunit of the ribosome. However, it is now clear that this catalytic activity is a function of the 23S rRNA of the large subunit, perhaps in conjunction with one or more of the ribosomal proteins. In such a case, we refer to the complex as a **ribozyme,** recognizing the catalytic role that RNA plays in the process.

Before elongation can be repeated, the tRNA attached to the P site, which is now uncharged, must be released from the large subunit. The uncharged tRNA moves transiently through a third site on the ribosome, called the **E (exit) site.** The entire *mRNA-tRNA-aa2-aa1 complex* then shifts in the direction of the P site by a distance of three nucleotides (Step 3). This event requires several protein **elongation factors (EFs)** as well as the energy derived from hydrolysis of GTP. The result is that the third triplet of mRNA is now in a position to accept another specific charged tRNA into the A site (Step 4). One simple way to distinguish the two sites is to remember that, *following the shift,* the P site (P for peptidyl) contains a tRNA attached to a peptide chain, whereas the A site (A for aminoacyl) contains a tRNA with an amino acid attached.

These elongation events are repeated over and over (Steps 4 and 5). An additional amino acid is added to the growing polypeptide chain each time the mRNA advances through the ribosome. Once a polypeptide chain of reasonable size is assembled (about 30 amino acids), it begins to emerge from the base of the large subunit, as illustrated in Step 6. A tunnel exists within the large subunit, from which the elongating polypeptide emerges.

As we have seen, the role of the small subunit during elongation is one of "decoding" the triplets present in mRNA, whereas the role of the large subunit is peptide bond synthesis. The efficiency of the process is remarkably high. The observed error rate is only about 10^{-4}. At this rate, an incorrect amino acid will occur only once in every 20 polypeptides of an average length of 500 amino acids. In *E. coli,* elongation occurs at a rate of about 15 amino acids per second at 37°C.

Termination

Termination, the third phase of translation, is depicted in **Figure 8.** The process is signaled by one or more of three triplet codes in the A site: UAG, UAA, or UGA. These codons do not specify an amino acid, nor do they call for a tRNA in the A site. They are called **stop codons, termination codons,** or **nonsense codons.** Often, several such consecutive codons are part of an mRNA. The finished polypeptide is therefore still attached to the terminal tRNA at the P site, and the A site is empty. The termination codon signals the action of **GTP-dependent release factors,** which cleave the polypeptide chain from the terminal tRNA, releasing it from the translation complex (Step 1). Then, the tRNA is released from the ribosome, which then dissociates into its subunits (Step 2). If a termination codon should appear in the middle of an mRNA molecule as a result of mutation, the same process occurs, and the polypeptide chain is prematurely terminated.

Polyribosomes

As elongation proceeds and the initial portion of mRNA has passed through the ribosome, this mRNA is free to associate with another small subunit to form a second initiation complex. This process can be repeated several times with a single mRNA and results in what are called **polyribosomes,** or just **polysomes.**

Polyribosomes can be isolated and analyzed following a gentle lysis of cells. The photos in **Figure 9** show these complexes as seen under an electron microscope. In **Figure 9(a),** you can see the thin lines of mRNA between the individual ribosomes. The micrograph in **Figure 9(b)** is even more remarkable, for it shows the polypeptide chains emerging from the ribosomes during translation. The for-

Elongation during Translation

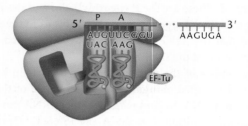

1. Second charged tRNA has entered A site, facilitated by EF-Tu; first elongation step commences

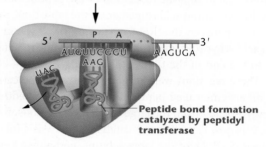

Peptide bond formation catalyzed by peptidyl transferase

2. Peptide bond forms; uncharged tRNA moves to the E site and subsequently out of the ribosome; the mRNA has been translocated three bases to the left, causing the tRNA bearing the dipeptide to shift into the P site

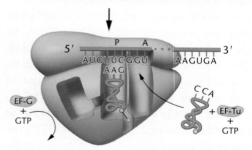

3. The first elongation step is complete, facilitated by EF-G. The third charged tRNA is ready to enter the A site

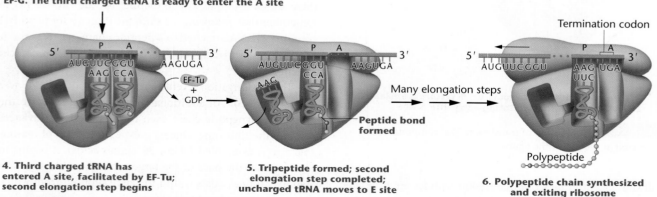

4. Third charged tRNA has entered A site, facilitated by EF-Tu; second elongation step begins

5. Tripeptide formed; second elongation step completed; uncharged tRNA moves to E site

6. Polypeptide chain synthesized and exiting ribosome

FIGURE 7 Elongation of the growing polypeptide chain during translation.

Termination of Translation

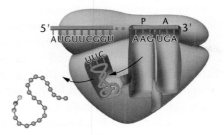

1. tRNA and polypeptide chain released

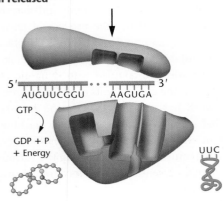

GTP

GDP + P + Energy

2. GTP-dependent termination factors stimulate the release of tRNA and the dissociation of the ribosomal subunits. Polypeptide folding occurs

FIGURE 8 Termination of the process of translation.

mation of polysome complexes represents an efficient use of the components available for protein synthesis during a particular unit of time. Using the analogy of a song recorded on a tape and a tape recorder, in polysome complexes one tape (mRNA) would be played simultaneously by several recorders (the ribosomes), but at any given moment, each recorder would be playing a different part of the song (the polypeptide being synthesized in each ribosome).

ESSENTIAL POINT ■ ■ ■

Translation, like transcription, is subdivided into the stages of initiation, elongation, and termination and relies on base-pairing affinities between complementary nucleotides.

3 High-Resolution Studies Have Revealed Many Details about the Functional Prokaryotic Ribosome

Our knowledge of the process of translation and the structure of the ribosome is based primarily on biochemical and genetic observations, in addition to the visualization of ribosomes under the electron microscope. To confirm and refine this information, the next step is to examine the ribosome at even higher levels of resolution. For example, X-ray diffraction analysis of ribosome crystals is one way to achieve this. However, because of its tremendous size and the complexity of molecular interactions occurring in the functional ribosome, it was extremely difficult to obtain the crystals necessary to perform X-ray diffraction studies. Nevertheless, great strides have been made over the past decade. First, the individual ribosomal subunits were crystallized and examined in several laboratories, most prominently that of V. Ramakrishnan. Then, the crystal structure of the intact 70S ribosome, complete with

(a)

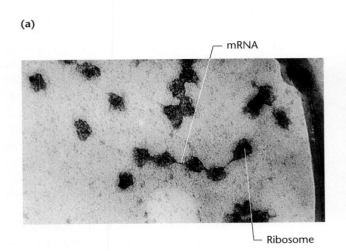

mRNA

Ribosome

(b)

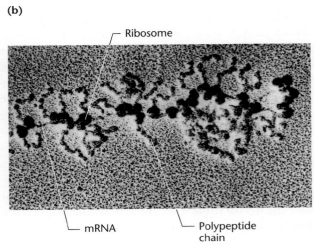

Ribosome

mRNA

Polypeptide chain

FIGURE 9 Polyribosomes as seen under the electron microscope. Those in (a) were derived from rabbit reticulocytes engaged in the translation of hemoglobin mRNA. The polyribosomes in (b) were taken from the giant salivary gland cells of the midgefly, *Chironomus thummi*. Note that the nascent polypeptide chains are apparent as they emerge from each ribosome. Their length increases as translation proceeds from left (5′) to right (3′) along the mRNA.

associated mRNA and tRNAs, was examined by Harry Noller and colleagues. In essence, the entire translational complex was seen at the atomic level. Both Ramakrishnan and Noller derived the ribosomes from the bacterium *Thermus thermophilus.*

Many noteworthy observations have come from these investigations. For example, the shape of the ribosome changes during different functional states, attesting to the dynamic nature of the process of translation. A great deal has also been learned about the location of the RNA components of the subunits. About one-third of the 16S RNA is responsible for producing a flat projection, referred to as the *platform,* within the smaller 30S subunit, and it modulates movement of the mRNA–tRNA complex during translocation. One of the models based on Noller's findings is shown in the opening photograph of this .

Crystallographic analysis also supports the concept that RNA is the real "player" in the ribosome during translation. The interface between the two subunits, considered to be the location in the ribosome where polymerization of amino acids occurs, is composed almost exclusively of RNA. In contrast, the numerous ribosomal proteins are found mostly on the periphery of the ribosome. These observations confirm what has been predicted on genetic grounds—the catalytic steps that join amino acids during translation occur under the direction of RNA, not proteins.

Another interesting finding involves the actual location of the various sites predicted to house tRNAs during translation. All three sites (A, P, and E), have been identified in X-ray diffraction studies, and in each case, the RNA of the ribosome makes direct contact with the various loops and domains of the tRNA molecule. This observation helps us understand why the distinctive three-dimensional conformation that is characteristic of all tRNA molecules has been preserved throughout evolution.

Still another noteworthy observation is that the intervals between the A, P, and E sites are at least 20 Å, and perhaps as much as 50 Å, wide, thus defining the atomic distance that the tRNA molecules must shift during each translocation event. This is considered a fairly large distance relative to the size of the tRNAs themselves. Further analysis has led to the identification of molecular (RNA–protein) bridges existing between the three sites and apparently involved in the translocation events. These observations provide us with a much more complete picture of the dynamic changes that must occur within the ribosome during translation. A final observation takes us back almost 50 years, to when Francis Crick proposed the wobble hypothesis. The Ramakrishnan group has identified the precise location along the 16S rRNA of the 30S subunit involved in the decoding step that connects mRNA to the proper tRNA. At this location, two particular nucleotides of the 16S rRNA actually flip out and probe the codon: anticodon region, and are believed to check for accuracy of

base pairing during this interaction. According to the wobble hypothesis, the stringency of this step is high for the first two base pairs but less so for the third (or wobble) base pair.

As our knowledge of the translation process in prokaryotes has continued to grow, a remarkable study was reported in 2010 by Niels Fischer and colleagues. Using a unique high-resolution approach—the technique of time-resolved single particle cryo-electron microscopy (cryo-EM)—the 70S *E. coli* ribosome was captured and examined while in the process of translation at a resolution of 5.5 Å. In this work, over two million images were obtained and computationally analyzed, establishing a temporal snapshot of the trajectories of tRNA during the process of translocation. This research team examined how tRNA is translocated during elongation of the polypeptide chain. They demonstrated that the trajectories are coupled with dynamic conformational changes in the components of the ribosome. Surprisingly, the work has revealed that during translation, the ribosome behaves as a complex molecular machine *powered by Brownian movement driven by thermal energy.* That is, the energetic requirement for achieving the various conformational changes essential to translocation are inherent to the ribosome itself.

Numerous questions about ribosome structure and function still remain. In particular, the precise role of the many ribosomal proteins is yet to be clarified. Nevertheless, the models that are emerging from the above research provide us with a much better understanding of the mechanism of translation.

4   Translation Is More Complex in Eukaryotes

The general features of the model we just discussed were initially derived from investigations of the translation process in bacteria. As we have seen (Figure 13–1), one main difference between prokaryotes and eukaryotes is that in eukaryotes, translation occurs on larger ribosomes whose rRNA and protein components are more complex. Interestingly, prokaryotic and eukaryotic rRNAs do share what is called a *core sequence,* but in eukaryotes, they are lengthened by the addition of *expansion sequences (ES),* which presumably impart added functionality. Another significant distinction is that whereas transcription and translation are coupled in prokaryotes, in eukaryotes these two processes are separated both spatially and temporally. In eukaryotic cells, transcription occurs in the nucleus and translation in the cytoplasm. This separation provides multiple opportunities for regulation of genetic expression in eukaryotic cells.

A number of aspects of the initiation of translation vary in eukaryotes. Three differences center on the mRNA that is being translated. First, the 5' end of mRNA is capped with a **7-methylguanosine (7-mG)** residue at maturation . The presence of the cap, absent in prokaryotes, is essential for efficient initiation of translation. A second difference is that many mRNAs contain a purine (A or G) three bases upstream from the AUG initiator codon, which is followed by a G (A/GNN<u>A</u>UGG). Named after its discoverer, Marilyn Kozak, its presence in eukaryotes is considered to increase the efficiency of translation by interacting with the initator tRNA. This **Kozak sequence** is considered analogous to the *Shine-Dalgarno* sequence found in the upstream region of prokaryotic mRNAs. Above, N depicts any base.

Third, eukaryotic mRNAs require the posttranscriptional addition of a **poly-A tail** on their 3'end; that is, they are *polyadenylated*. In the absence of poly A, these potential messages are rapidly degraded in the cytoplasm. Interestingly, histone mRNAs serve as an exception and are not polyadenylated. Still another difference related to initiation of translation is that in eukaryotes the amino acid formylmethionine is not required as it is in prokaryotes. However, the AUG triplet, which encodes methionine, is essential to the formation of the translational complex, and a unique transfer RNA ($tRNA_i^{Met}$) is used during initiation.

Still other differences are noteworthy. Eukaryotic mRNAs are much longer lived than are their prokaryotic counterparts. Most exist for hours rather than minutes prior to degradation by nucleases in the cell; thus they remain available much longer to orchestrate protein synthesis. And, during translation, protein factors similar to those in prokaryotes guide the initiation, elongation, and termination of translation in eukaryotes. Many of these eukaryotic factors are clearly homologous to their counterparts in prokaryotes. However, a greater number of factors are usually required during each step, and some are more complex than in prokaryotes. Finally, recall that in eukaryotes, many, but not all, of the cell's ribosomes are found in association with the membranes that make up the endoplasmic reticulum (forming the rough ER). Such membranes are absent from the cytoplasm of prokaryotic cells. This association in eukaryotes facilitates the secretion of newly synthesized proteins from the ribosomes directly into the channels of the endoplasmic reticulum. Recent studies using electron microscopy have established how this occurs. A *tunnel* in the large subunit of the ribosome begins near the point where the two subunits interface and exits near the back of the large subunit. The location of the tunnel within the large subunit is the basis for the belief that it provides the conduit for the movement of the newly synthesized polypeptide chain out of the ribosome. In studies in yeast, newly synthesized polypeptides enter the ER through a membrane channel formed by a specific protein, Sec61. This channel is perfectly aligned with the exit point of the ribosomal tunnel. In prokaryotes, the polypeptides are released by the ribosome directly into the cytoplasm.

---

5 The Initial Insight that Proteins Are Important in Heredity Was Provided by the Study of Inborn Errors of Metabolism

Let's consider how we know that proteins are the end products of genetic expression. The first insight into the role of proteins in genetic processes was provided by observations made by Sir Archibald Garrod and William Bateson early in the twentieth century. Garrod was born into an English family of medical scientists. His father was a physician with a strong interest in the chemical basis of rheumatoid arthritis, and his eldest brother was a leading zoologist in London. It is not surprising, then, that as a practicing physician, Garrod became interested in several human disorders that seemed to be inherited. Although he also studied albinism and cystinuria, we shall describe his investigation of the disorder **alkaptonuria.** Individuals afflicted with this disorder have an important metabolic pathway blocked. As a result, they cannot metabolize the alkapton 2,5-dihydroxyphenylacetic acid, also known as homogentisic acid. Homogentisic acid accumulates in cells and tissues and is excreted in the urine. The molecule's oxidation products are black and easily detectable in the diapers of newborns. The products tend to accumulate in cartilaginous areas, causing the ears and nose to darken. The deposition of homogentisic acid in joints leads to a benign arthritic condition. This rare disease is not serious, but it persists throughout an individual's life.

Garrod studied alkaptonuria by looking for patterns of inheritance of this benign trait. Eventually he concluded that it was genetic in nature. Of 32 known cases, he ascertained that 19 were confined to seven families, with one family having four affected siblings. In several instances, the parents were unaffected but known to be related as first cousins, and therefore **consanguine,** a term describing individuals descended from a common recent ancestor. Parents who are so related have a higher probability than unrelated parents of producing offspring that express recessive traits because such parents are both more likely to be heterozygous for some of the same recessive traits. Garrod concluded that this inherited condition was the result of an alternative mode of metabolism, thus implying that hereditary information controls chemical reactions in the body. While *genes* and *enzymes* were not familiar terms during Garrod's time, he used the corresponding concepts of *unit factors* and *ferments*. Garrod published his initial observations in 1902.

Only a few geneticists, including Bateson, were familiar with or referred to Garrod's work. Garrod's ideas fit nicely with Bateson's belief that inherited conditions are caused by the lack of some critical substance. In 1909, Bateson published *Mendel's Principles of Heredity,* in which he linked Garrod's ferments with heredity. However, for almost 30 years, most geneticists failed to see the relationship between genes and enzymes. Garrod and Bateson, like Mendel, were ahead of their time.

6 Studies of *Neurospora* Led to the One-Gene:One-Enzyme Hypothesis

In two separate investigations beginning in 1933, George Beadle provided the first convincing experimental evidence that genes are directly responsible for the synthesis of enzymes. The first investigation, conducted in collaboration with Boris Ephrussi, involved *Drosophila* eye pigments. Together, they confirmed that mutant genes that alter the eye color of fruit flies could be linked to biochemical errors that, in all likelihood, involved the loss of enzyme function. Encouraged by these findings, Beadle then joined with Edward Tatum to investigate nutritional mutations in the pink bread mold *Neurospora crassa.* This investigation led to the **one-gene:one-enzyme hypothesis.**

Analysis of *Neurospora* Mutants by Beadle and Tatum

In the early 1940s, Beadle and Tatum chose to work with *Neurospora* because much was known about its biochemistry and because mutations could be induced and isolated with relative ease. By inducing mutations, they produced strains that had genetic blocks of reactions essential to the growth of the organism.

Beadle and Tatum knew that this mold could manufacture nearly everything necessary for normal development. For example, using rudimentary carbon and nitrogen sources, this organism can synthesize nine water-soluble vitamins, 20 amino acids, numerous carotenoid pigments, and all essential purines and pyrimidines. Beadle and Tatum irradiated asexual conidia (spores) with X rays to increase the frequency of mutations and allowed them to be grown on "complete" medium containing all the necessary growth factors (e.g., vitamins and amino acids). Under such growth conditions, a mutant strain unable to grow on minimal medium was able to grow by virtue of supplements present in the enriched complete medium. All the cultures were then transferred to minimal medium. If growth occurred on the minimal medium, the organisms were able to synthesize all the necessary growth factors themselves, and

the researchers concluded that the culture did not contain a nutritional mutation. If no growth occurred on minimal medium, they concluded that the culture contained a nutritional mutation, and the only task remaining was to determine its type. These results are shown in **Figure 10(a)**.

Many thousands of individual spores from this procedure were isolated and grown on complete medium. In subsequent tests on minimal medium, many cultures failed to grow, indicating that a nutritional mutation had been induced. To identify the mutant type, the mutant strains were then tested on a series of different minimal media [**Figure 10(b)**], each containing groups of supplements, and subsequently on media containing single vitamins, purines, pyrimidines, or amino acids [**Figure 10(c)**] until one specific supplement that permitted growth was found. Beadle and Tatum reasoned that *the supplement that restored growth would be the molecule that the mutant strain could not synthesize.*

The first mutant strain they isolated required vitamin B_6 (pyridoxine) in the medium, and the second required vitamin B_1 (thiamine). Using the same procedure, Beadle and Tatum eventually isolated and studied hundreds of mutants deficient in the ability to synthesize other vitamins, amino acids, or other substances.

The findings derived from testing over 80,000 spores convinced Beadle and Tatum that genetics and biochemistry have much in common. It seemed likely that each nutritional mutation caused the loss of the enzymatic activity that facilitated an essential reaction in wild-type organisms. It also appeared that a mutation could be found for nearly any enzymatically controlled reaction. Beadle and Tatum had thus provided sound experimental evidence for the hypothesis that *one gene specifies one enzyme,* an idea alluded to over 30 years earlier by Garrod and Bateson. With modifications, this concept was to become another major principle of genetics.

ESSENTIAL POINT ■ ■ ■
Beadle and Tatum's work with nutritional mutations in *Neurospora* led them to propose that one gene encodes one enzyme.

7 Studies of Human Hemoglobin Established that One Gene Encodes One Polypeptide

The one-gene:one-enzyme hypothesis that was developed in the early 1940s was not immediately accepted by all geneticists. This is not surprising because it was not yet clear how mutant enzymes could cause variation in many phenotypic traits. For example, *Drosophila* mutants demonstrate altered eye size, wing shape, wing-vein pattern, and so on. Plants

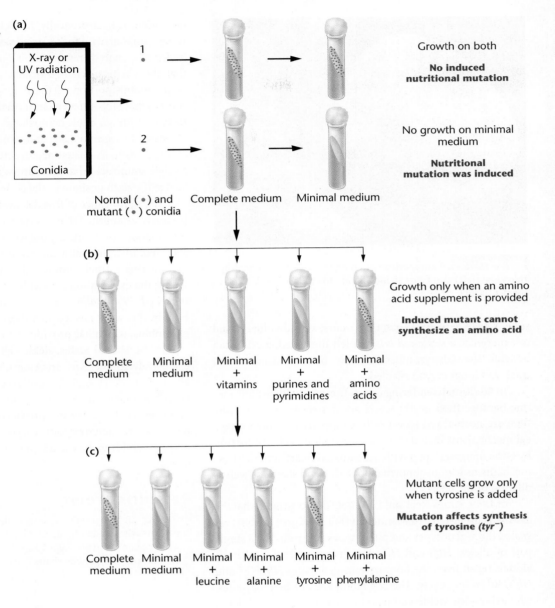

FIGURE 10 Induction, isolation, and characterization of a nutritional auxotrophic mutation in *Neurospora*. (a) Most conidia are not affected, but one conidium (shown in red) contains a mutation. In (b) and (c), the precise nature of the mutation is established and found to involve the biosynthesis of tyrosine.

(a)

X-ray or UV radiation

Conidia

Normal (•) and mutant (•) conidia

Complete medium

Minimal medium

1 — Growth on both

No induced nutritional mutation

2 — No growth on minimal medium

Nutritional mutation was induced

(b)

Complete medium   Minimal medium   Minimal + vitamins   Minimal + purines and pyrimidines   Minimal + amino acids

Growth only when an amino acid supplement is provided

Induced mutant cannot synthesize an amino acid

(c)

Complete medium   Minimal medium   Minimal + leucine   Minimal + alanine   Minimal + tyrosine   Minimal + phenylalanine

Mutant cells grow only when tyrosine is added

Mutation affects synthesis of tyrosine (tyr⁻)

exhibit mutant varieties of seed texture, height, and fruit size. How an inactive mutant enzyme could result in such phenotypes puzzled many geneticists.

Two factors soon modified the one-gene:one-enzyme hypothesis. First, although *nearly all enzymes are proteins, not all proteins are enzymes*. As the study of biochemical genetics progressed, it became clear that all proteins are specified by the information stored in genes, leading to the more accurate phraseology, **one-gene:one-protein hypothesis.** Second, proteins often show a substructure consisting of two or more polypeptide chains. This is the basis of the quaternary protein structure, which we will discuss later in this text.

Because each distinct polypeptide chain is encoded by a separate gene, a more accurate statement of Beadle and Tatum's basic tenet is **one-gene:one-polypeptide chain hypothesis.** These modifications of the original hypothesis became apparent during the analysis of hemoglobin structure in individuals afflicted with sickle-cell anemia.

Sickle-cell Anemia

The first direct evidence that genes specify proteins other than enzymes came from work on mutant hemoglobin molecules found in humans afflicted with the disorder **sickle-cell anemia.** Affected individuals have erythrocytes that, under low oxygen tension, become elongated and curved because of the polymerization of hemoglobin. The sickle shape of these erythrocytes is in contrast to the biconcave disc shape characteristic in unaffected individuals (**Figure 11**). Those with the disease suffer attacks when red blood cells aggregate in the venous side of capillary systems, where oxygen tension is very low. As a result, a variety of tissues are deprived of oxygen and

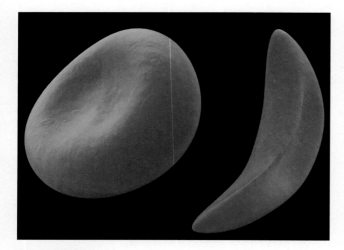

FIGURE 11 A comparison of an erythrocyte from a healthy individual (left) and from an individual afflicted with sickle-cell anemia (right).

suffer severe damage. When this occurs, an individual is said to experience a *sickle-cell crisis.* If left untreated, a crisis can be fatal. The kidneys, muscles, joints, brain, gastrointestinal tract, and lungs can be affected.

In addition to suffering crises, these individuals are anemic because their erythrocytes are destroyed more rapidly than are normal red blood cells. Compensatory physiological mechanisms include increased red blood cell production by bone marrow, along with accentuated heart action. These mechanisms lead to abnormal bone size and shape, as well as dilation of the heart.

In 1949, James Neel and E. A. Beet demonstrated that the disease is inherited as a Mendelian trait. Pedigree analysis revealed three genotypes and phenotypes controlled by a single pair of alleles, Hb^A and Hb^S. Unaffected and affected individuals result from the homozygous genotypes $Hb^A Hb^A$ and $Hb^S Hb^S$, respectively. The red blood cells of heterozygotes, who exhibit the **sickle-cell trait** but not the disease, undergo much less sickling because over half of their hemoglobin is normal. Although they are largely unaffected, heterozygotes are "carriers" of the defective gene, which is transmitted on average to 50 percent of their offspring.

In the same year, Linus Pauling and his coworkers provided the first insight into the molecular basis of the disease. They showed that hemoglobins isolated from diseased and normal individuals differ in their rates of electrophoretic migration. In this technique, charged molecules migrate in an electric field. If the net charge of two molecules is different, their rates of migration will be different. On this basis, Pauling and his colleagues concluded that a chemical difference exists between normal (**HbA**) and sickle-cell (**HbS**) hemoglobin.

Pauling's findings suggested two possibilities. It was known that hemoglobin consists of four nonproteinaceous, iron-containing *heme groups* and a *globin portion* that contains four polypeptide chains. The alteration in net charge in HbS

had to be due, theoretically, to a chemical change in one of these components. Work carried out between 1954 and 1957 by Vernon Ingram resolved this question. He demonstrated that the chemical change occurs in the primary structure of the globin portion of the hemoglobin molecule. Ingram showed that HbS differs in amino acid composition compared to HbA. Human adult hemoglobin contains two identical α chains of 141 amino acids and two identical β chains of 146 amino acids in its quaternary structure. Analysis revealed just a single amino acid change: valine was substituted for glutamic acid at the sixth position of the β chain (**Figure 12**).

The significance of this discovery has been multifaceted. It clearly establishes that a single gene provides the genetic information for a single polypeptide chain. Studies of HbS also demonstrate that a mutation can affect the phenotype by directing a single amino acid substitution. Also, by providing the explanation for sickle-cell anemia, the concept of *inherited molecular disease* was firmly established. Finally, this work has led to a thorough study of human hemoglobins, which has provided valuable genetic insights.

In the United States, sickle-cell anemia is found almost exclusively in the African-American population. It affects about 1 in every 625 African-American infants. Currently, about 50,000 to 75,000 individuals are afflicted. In 1 of about every 145 African-American married couples, both partners are heterozygous carriers. In these cases, each of their children has a 25 percent chance of having the disease.

ESSENTIAL POINT

Pauling and Ingram's investigations of hemoglobin from patients with sickle-cell anemia led to the modification of the one-gene:one-enzyme hypothesis to indicate that one gene encodes one polypeptide chain.

NOW SOLVE THIS

2 HbS results from the substitution of valine for glutamic acid at the number 6 position in the β chain of human hemoglobin. HbC is the result of a change at the same position in the β chain, but in this case lysine replaces glutamic acid. Return to the genetic code table (Figure 7) and determine whether single-nucleotide changes can account for these mutations. Then view Figure 13 and examine the R groups in the amino acids glutamic acid, valine, and lysine. Describe the chemical differences between the three amino acids. Predict how the changes might alter the structure of the molecule and lead to altered hemoglobin function.

■ **Hint** *This problem asks you to consider the potential impact of several amino acid substitutions that result from mutations in one of the genes encoding one of the chains making up human hemoglobin. The key to its solution is to consider and compare the structure of the three amino acids (glutamic acid, lysine, and valine) and their net charge (see Figure 13).*

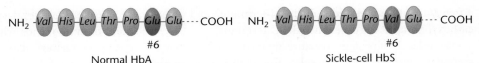

FIGURE 12 A comparison of the amino acid sequence of the β chain found in HbA and HbS.

NH₂ –Val–His–Leu–Thr–Pro–Glu–Glu- ---COOH   NH₂ –Val–His–Leu–Thr–Pro–Val–Glu- --- COOH

#6   #6

Normal HbA   Sickle-cell HbS

Partial amino acid sequences of β chains

8 Variation in Protein Structure Is the Basis of Biological Diversity

Having established that the genetic information is stored in DNA and influences cellular activities through the proteins it encodes, we turn now to a brief discussion of protein structure. How can these molecules play such a critical role in determining the complexity of cellular activities? As we shall see, the fundamental aspects of the structure of proteins provide the basis for incredible complexity and diversity. At the outset, we should differentiate between **polypeptides** and **proteins.** Both are molecules composed of amino acids. They differ, however, in their state of assembly and functional capacity. Polypeptides are the precursors of proteins. As it is assembled on the ribosome during translation, the molecule is called a *polypeptide.* When released from the ribosome following translation, a polypeptide folds up and assumes a higher order of structure. When this occurs, a three-dimensional conformation emerges. In many cases, several polypeptides interact to produce this conformation. When the final conformation is achieved, the molecule is now fully functional and is appropriately called a *protein.* Its three-dimensional conformation is essential to the function of the molecule.

The polypeptide chains of proteins, like nucleic acids, are linear nonbranched polymers. There are 20 commonly occurring amino acids that serve as the subunits (the building blocks) of proteins. Each amino acid has a **carboxyl group,** an **amino group,** and an **R (radical) group** (a side chain) bound covalently to a **central carbon (C) atom.** The R group gives each amino acid its chemical identity exhibiting a variety of configurations that can be divided into four main classes: *nonpolar* (hydrophobic), *polar* (hydrophilic), *positively charged,* and *negatively charged.* **Figure 13** shows the chemical structure of an amino acid and one example from each of these categories. Because polypeptides are often long polymers and because each position may be occupied by any 1 of the 20 amino acids with their unique chemical properties, enormous variation in chemical conformation and activity is possible. For example, if an average polypeptide is composed of 200 amino acids (molecular weight of about 20,000 Da), 20^{200} different molecules, each with a unique sequence, can be created using the 20 different building blocks.

Around 1900, German chemist Emil Fischer determined the manner in which the amino acids are bonded together. He showed that the amino group of one amino acid

reacts with the carboxyl group of another amino acid during a dehydration reaction, releasing a molecule of H_2O. The resulting covalent bond is a **peptide bond.** Two amino acids linked together constitute a **dipeptide,** three a **tripeptide,** and so on. Once 10 or more amino acids are linked by peptide bonds, the chain is referred to as a polypeptide. Generally, no matter how long a polypeptide is, it will contain

Nonpolar: Hydrophobic

Valine (Val, V)

Polar: Hydrophilic

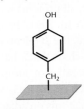

Tyrosine (Tyr, Y)

Polar: positively charged (basic)

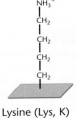

Lysine (Lys, K)

Polar: negatively charged (acidic)

Glutamic acid (Glu, E)

Amino group

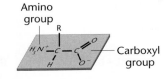

Carboxyl group

Amino acid structure

FIGURE 13 Chemical structure of an amino acid as well as an example of the R group characterizing each of the four categories of amino acids. Each amino acid has two abbreviations, often based on the first three letters of its name; for example, valine is designated either Val or V.

a free amino group at one end (the N-terminus) and a free carboxyl group at the other end (the C-terminus).

Four levels of protein structure are recognized: primary, secondary, tertiary, and quaternary. The sequence of amino acids in the linear backbone of the polypeptide constitutes its **primary structure.** It is specified by the sequence of deoxyribonucleotides in DNA via an mRNA intermediate. The primary structure of a polypeptide helps determine the specific characteristics of the higher orders of organization as a protein is formed.

Secondary structures are certain regular or repeating configurations in space assumed by amino acids lying close to one another in the polypeptide chain. In 1951, Linus Pauling and Robert Corey predicted, on theoretical grounds, an **α helix** as one type of secondary structure. The α-helix model **[Figure 14(a)]** has since been confirmed by X-ray crystallographic studies. The helix is composed of a spiral chain of amino acids stabilized by hydrogen bonds.

The side chains (the R groups) of amino acids extend outward from the helix, and each amino acid residue occupies a vertical distance of 1.5 Å in the helix. There are 3.6 residues per turn. While left-handed helices are theoretically possible, all proteins seen with an α helix are right-handed.

Also in 1951, Pauling and Corey proposed a second structure, the **β-pleated sheet.** In this model, a single-polypeptide chain folds back on itself, or several chains run in either parallel or antiparallel fashion next to one another. Each such structure is stabilized by hydrogen bonds formed between atoms on adjacent chains **[Figure 14(b)].** A zigzagging plane is formed in space with adjacent amino acids 3.5 Å apart. As a general rule, most proteins demonstrate a mixture of α-helix and β-pleated-sheet structures.

While the secondary structure describes the arrangement of amino acids within certain areas of a polypeptide chain, the **tertiary structure** defines the three-dimensional conformation of the entire chain in space. Each protein twists and turns and loops around itself in a very particular fashion, characteristic of the specific protein. A model of the three-dimensional tertiary structure of the respiratory pigment myoglobin is shown in **Figure 15.**

The three-dimensional conformation achieved by any protein is a product of the *primary structure* of the polypeptide. As the polypeptide is folded, the most thermodynamically stable conformation is created. This level of organization is essential because the specific function of any protein is directly related to its tertiary structure.

The concept of **quaternary structure** applies to those proteins composed of more than one polypeptide chain and indicates the position of the various chains in relation to one another. Hemoglobin, a protein consisting of four polypeptide chains, has been studied in great detail. Most enzymes, including DNA and RNA polymerase, demonstrate quaternary structure.

ESSENTIAL POINT ■   ■   ■

Proteins, the end products of genes, demonstrate four levels of structural organization that together describe their three-dimensional conformation, which is the basis of each molecule's function.

Protein Folding and Misfolding

It was long thought that **protein folding** was a spontaneous process whereby a linear molecule exiting the ribosome achieved a three-dimensional, thermodynamically stable

FIGURE 14 (a) The right-handed α helix, which represents one form of secondary structure of a polypeptide chain. (b) The β-pleated sheet, an alternative form of secondary structure of polypeptide chains. To maintain clarity, not all atoms are shown.

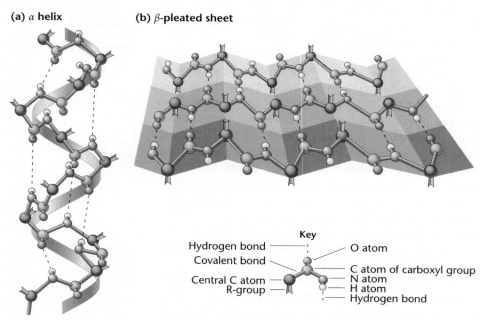

(a) α helix

(b) β-pleated sheet

Key

Hydrogen bond
Covalent bond
Central C atom
R-group

O atom
C atom of carboxyl group
N atom
H atom
Hydrogen bond

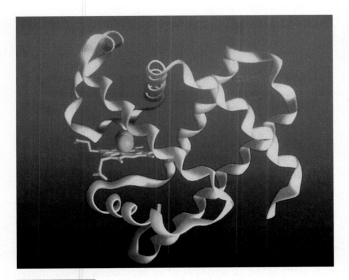

FIGURE 15 The tertiary level of protein structure in a respiratory pigment, myoglobin. The bound oxygen atom is shown in red.

conformation based solely on the combined chemical properties inherent in the amino acid sequence. This indeed is the case for many proteins. However, numerous studies have shown that for other proteins, correct folding is dependent on members of a family of molecules called **chaperones.** Chaperones are themselves proteins (sometimes called *molecular chaperones* or *chaperonins*) that function by mediating the folding process by excluding the formation of alternative, incorrect patterns. While they may bind to the protein in question, like enzymes, they do not become part of the final product. Initially discovered in *Drosophila*, in which they are called **heat-shock proteins,** chaperones are ubiquitous, having now been discovered in all organisms. They are even present in mitochondria and chloroplasts.

In eukaryotic cells, chaperones are particularly important when translation occurs on membrane-bound ribosomes, where the newly translated polypeptide is extruded into the lumen of the endoplasmic reticulum. Even in their presence, misfolding may still occur, and one more system of "quality control" exists. As misfolded proteins are transported out of the endoplasmic reticulum to the cytoplasm, they are "tagged" by another class of small proteins called **ubiquitins.** The protein–ubiquitin complex moves to a cellular structure called the **proteasome,** within which the ubiquitin is released and the misfolded proteins are degraded by proteases.

Protein folding is a critically important process, not only because misfolded proteins may be nonfunctional, but also because improperly folded proteins can accumulate and be detrimental to cells and the organisms that contain them. For example, a group of transmissible brain disorders in mammals—**scrapie** in sheep, **bovine spongiform encephalopathy (mad cow disease)** in cattle, and **Creutzfeldt–Jakob disease** in humans—are caused by the presence in the brain of **prions,** which are aggregates of a misfolded protein.

The misfolded protein (called PrP^{Sc}) is an altered version of a normal cellular protein (called PrP^{C}) synthesized in neurons and found in the brains of all adult animals. The difference between PrP^{C} and PrP^{Sc} lies in their secondary protein structures. Normal, noninfectious PrP^{c} folds into an α helix, whereas infectious PrP^{Sc} folds into a β-pleated sheet. When an abnormal PrP^{Sc} molecule contacts a PrP^{C} molecule, the normal protein refolds into the abnormal conformation. The process continues as a chain reaction, with potentially devastating results—the formation of prion particles that eventually destroy the brain. Hence, this group of disorders can be considered diseases of secondary protein structure.

Currently, many laboratories are studying protein folding and misfolding, particularly as related to genetics. Numerous inherited human disorders are caused by misfolded proteins that form abnormal aggregates. **Sickle-cell anemia,** discussed earlier in this text, is a case in point, where the β chains of hemoglobin are altered as the result of a single amino acid change, causing the molecules to aggregate within erythrocytes, with devastating results. An autosomal dominant inherited form of **Creutzfeldt–Jakob disease** is known in which the mutation alters the PrP amino acid sequence, leading to prion formation. And various progressive neurodegenerative diseases such as **Huntington disease, Alzheimer disease,** and **Parkinson disease** are linked to the formation of abnormal protein aggregates in the brain. Huntington disease is inherited as an autosomal dominant trait, whereas less clearly defined genetic components are associated with Alzheimer and Parkinson diseases.

9 Proteins Function in Many Diverse Roles

The essence of life on Earth rests at the level of diverse cellular function. One can argue that DNA and RNA simply serve as vehicles to store and express genetic information. However, proteins are at the heart of cellular function. And it is the capability of cells to assume diverse structures and functions that distinguishes most eukaryotes from less evolutionarily advanced organisms such as bacteria. Therefore, an introductory understanding of protein function is critical to a complete view of genetic processes.

Proteins are the most abundant macromolecules found in cells. As the end products of genes, they play many diverse roles. For example, the respiratory pigments **hemoglobin** and **myoglobin,** discussed earlier in the text, transport oxygen, which is essential for cellular metabolism. **Collagen** and **keratin** are structural proteins associated with the skin, connective tissue, and hair of organisms. **Actin** and **myosin** are contractile proteins, found in abundance in muscle tissue, while **tubulin** is the basis of the function of microtubules in mitotic and meiotic spindles. Still other examples are the

immunoglobulins, which function in the immune system of vertebrates; **transport proteins,** involved in the movement of molecules across membranes; some of the hormones and their receptors, which regulate various types of chemical activity; **histones**, which bind to DNA in eukaryotic organisms; and **transcription factors** that regulate gene expression.

Nevertheless, the most diverse and extensive group of proteins (in terms of function) are the enzymes, to which we have referred throughout this text. Enzymes specialize in catalyzing chemical reactions within living cells. Like all catalysts, they increase the rate at which a chemical reaction reaches equilibrium, but they do not alter the end-point of the chemical equilibrium. Their remarkable, highly specific catalytic properties largely determine the metabolic capacity of any cell type and provide the underlying basis of what we refer to as biochemistry. The specific functions of many enzymes involved in the genetic and cellular processes of cells are described throughout the text.

> ### ESSENTIAL POINT ■ ■ ■
>
> Of the myriad functions performed by proteins, the most influential role belongs to enzymes, which serve as highly specific biological catalysts that play a central role in the production of all classes of molecules in living systems.

Protein Domains Impart Function

We conclude this by briefly discussing the important finding that regions made up of specific amino acid sequences are associated with specific functions in protein molecules. Such sequences, usually between 50 and 300 amino acids, constitute **protein domains** and represent modular portions of the protein that fold into stable, unique conformations independently of the rest of the molecule. Different domains impart different functional capabilities. Some proteins contain only a single domain, while others contain two or more.

The significance of domains resides in the tertiary structures of proteins. Each domain can contain a mixture of secondary structures, including α helices and β-pleated sheets. The unique conformation of a given domain imparts a specific function to the protein. For example, a domain may serve as the catalytic site of an enzyme, or it may impart an ability to bind to a specific ligand. Thus, discussions of proteins may mention *catalytic domains, DNA-binding domains,* and so on. In short, a protein must be seen as being composed of a series of structural and functional modules. Obviously, the presence of multiple domains in a single protein increases the versatility of each molecule and adds to its functional complexity.

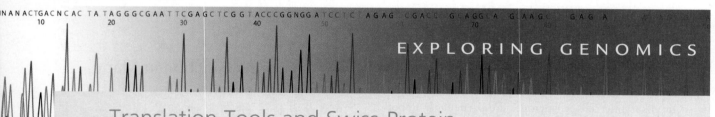

EXPLORING GENOMICS

Translation Tools and Swiss-Protein for Studying Protein Sequences

 Study Area: *Exploring Genomics*

Many of the databases and bioinformatics programs we have used for "Exploring Genomics" exercises have focused on manipulating and analyzing DNA and RNA sequences. However, scientists working on various aspects of translation and protein structure and function also have a wide range of bioinformatics tools and databases at their disposal via the Internet.

We will use a program from **ExPASy (Expert Protein Analysis System)** to translate a segment of a gene into a possible polypeptide. We will then explore databases for learning more about this polypeptide.

■ Exercise I – Translating a Nucleotide Sequence and Analyzing a Polypeptide

ExPASy (Expert Protein Analysis System), hosted by the Swiss Institute of Bioinformatics, provides a wealth of resources for *proteomics,* the study of all the proteins expressed in a cell or tissue.

We will use a program from ExPASy called **Translate Tool** to translate a nucleotide sequence to a polypeptide sequence. Although many other programs are available on the Web for this purpose, ExPASy is one of the more student-friendly tools. Translate Tool allows you to make a predicted polypeptide sequence from a cloned gene

and then look for open reading frames and variations in possible polypeptides.

1. Go to the Companion Website for *Essentials of Genetics* and open the Exploring Genomics exercise for this chapter. At the Website we provide partial sequence for a human gene based on a complementary DNA (cDNA) sequence. . Before you translate this sequence in ExPASy, run a nucleotide–nucleotide BLAST search from the NCBI Web site (http://www.ncbi.nlm.nih.gov/BLAST) to identify the gene corresponding to this sequence.

2. Access the ExPASy Translate Tool program at http://us.expasy.org/tools/dna.html. Copy and paste the cDNA sequence into Translate Tool and click "Translate Sequence" to generate possible polypeptide sequences encoded by this cDNA.

3. Review the translation results and then answer the following questions:
a. Did Translate Tool provide one or multiple possible polypeptide sequences?
b. If the translation results showed multiple polypeptide sequences, what does this mean? Explain.
c. Refer to Figure 12. Based on this figure, which reading frame generated by Translate Tool appears to be correct?

CASE STUDY » Lost in translation

A recessively inherited brain disorder called vanishing white matter (VWM) was first described in the 1990s using magnetic resonance imaging (MRI). Affected individuals show neurological deterioration early in childhood, dying soon after diagnosis, or they may express a slow progressive form of the disease. VWM is caused by mutations in any of the five genes that encode the protein subunits of the translation initiation factor 2B. This factor helps position the ribosome on the mRNA during the initiation of translation. It is known that other cells that rapidly synthesize large amounts of protein, such as insulin-secreting cells, are also affected. Many questions about this disorder remain to be answered.

1. Given the two forms of the disease, discuss the possible nature of VWM mutations. Why do you think that this is a recessive rather than a dominant mutation?
2. Why might some cells in the body be more susceptible to mutations in genes encoding factor 2B than other cells?
3. How could we study the cellular and molecular aspects of VWM?

INSIGHTS AND SOLUTIONS

1. As an extension of Beadle and Tatum's work with *Neurospora*, it is possible to study multiple mutations whose impact is on the same biochemical pathway. The growth responses in the following chart were obtained using four mutant strains of *Neurospora* and the chemically related compounds A, B, C, and D. None of the mutants grow on minimal medium. Draw all possible conclusions from these data.

| | **Growth Supplement** | | | |
Mutation	A	B	C	D
1	−	−	−	−
2	+	+	−	+
3	+	+	−	−
4	−	+	−	−

Solution: Nothing can be concluded about mutation *1* except that it lacks some essential growth factor, perhaps even unrelated to the biochemical pathway represented by mutations *2, 3,* and *4*. Nor can anything be concluded about compound C. If it is involved

in the pathway, it is a product that was synthesized prior to compounds A, B, and D.

We now analyze these three compounds and the control of their synthesis by the enzymes encoded by mutations *2, 3,* and *4*. Because product B allows growth in all three cases, it may be considered the "end product"—it bypasses the block in all three instances. Using similar reasoning, product A precedes B in the pathway because it bypasses the block in two of the three steps, and product D precedes B yielding a partial solution

$$C(?) \longrightarrow D \longrightarrow A \longrightarrow B$$

Now let's determine which mutations control which steps. Since mutation *2* can be alleviated by products D, B, and A, it must control a step prior to all three products, perhaps the direct conversion to D (although we cannot be certain). Mutation *3* is alleviated by B and A, so its effect must precede them in the pathway. Thus, we assign it as controlling the conversion of D to A. Likewise, we can assign mutation *4* to the conversion of A to B, leading to a more complete solution

$$C(?) \xrightarrow{2(?)} D \xrightarrow{3} A \xrightarrow{4} B$$

PROBLEMS AND DISCUSSION QUESTIONS

 For activities, animations, and review quizzes, go to the study area at www.masteringgenetics.com

HOW DO WE KNOW?

1. In this text, we focused on the translation of mRNA into proteins as well as on protein structure and function. Along the way, we found many opportunities to consider the methods and reasoning by which much of this information was acquired. From the explanations given in the chapter, what answers would you propose to the following fundamental questions:

(a) What experimentally derived information led to Holley's proposal of the two-dimensional cloverleaf model of tRNA?
(b) What experimental information verifies that certain codons in mRNA specify chain termination during translation?
(c) How do we know, based on studies of *Neurospora* nutritional mutations, that one gene specifies one enzyme?
(d) On what basis have we concluded that proteins are the end products of genetic expression?

2. List and describe the role of all molecular constituents present in a functional polyribosome.

3. Contrast the roles of tRNA and mRNA during translation, and list all enzymes that participate in the transcription and translation processes.

4. Francis Crick proposed the adaptor hypothesis for the function of tRNA. Why did he choose that description?

5. During translation, what molecule bears the anticodon? The codon?

6. The α chain of eukaryotic hemoglobin is composed of 141 amino acids. What is the minimum number of nucleotides in an mRNA coding for this polypeptide chain?

7. Summarize the steps involved in charging tRNAs with their appropriate amino acids.

8. Each transfer RNA requires at least four specific recognition sites that must be inherent in its tertiary protein structure in order for it to carry out its role. What are these sites?

9. Explain why the one-gene:one-enzyme hypothesis is no longer considered to be totally accurate.

10. Hemoglobin is a tetramer consisting of two α and two β chains. What level of protein structure is described in the above statement?

11. Using sickle-cell anemia as a basis, describe what is meant by a genetic or inherited molecular disease. What are the similarities and dissimilarities between this type of a disorder and a disease caused by an invading microorganism?

12. Describe the genetic and molecular basis of sickle-cell anemia.

13. Assuming that each nucleotide is 0.34 nm long in mRNA, how many triplet codes can simultaneously occupy space in a ribosome that is 20 nm in diameter?

14. Consider the following question: Certain mutations called *amber* in bacteria and viruses result in premature termination of polypeptide chains during translation. Many *amber* mutations have been detected at different points along the gene that codes for a head protein in phage T4. How might this system be further investigated to demonstrate and support the concept of colinearity?

15. In your opinion, which of the four levels of protein organization is the most critical to a protein's function? Defend your choice.

16. List and describe the function of as many nonenzymatic proteins as you can that are unique to eukaryotes.

17. How does an enzyme function? Why are enzymes essential for living organisms?

18. Shown in the following table are several amino acid substitutions in the α and β chains of human hemoglobin.

Hb Type	Normal Amino Acid	Substituted Amino Acid
HbJ Toronto	Ala	Asp (α-5)
HbJ Oxford	Gly	Asp (α-15)
Hb Mexico	Gln	Glu (α-54)
Hb Bethesda	Tyr	His (β-145)
Hb Sydney	Val	Ala (β-67)
HbM Saskatoon	His	Tyr (β-63)

19. Three independently assorting genes are known to control the biochemical pathway below that provides the basis for flower color in a hypothetical plant

$$\text{colorless} \xrightarrow{A-} \text{yellow} \xrightarrow{B-} \text{green} \xrightarrow{C-} \text{speckled}$$

Homozygous recessive mutations, which disrupt enzyme function controlling each step, are known. Determine the phenotypic results in the F1 and F2 generations resulting from the P1 crosses involving true-breeding plants given here.

(a) speckled	(*AABBCC*)	×	yellow	(*AAbbCC*)
(b) yellow	(*AAbbCC*)	×	green	(*AABBcc*)
(c) colorless	(*aaBBCC*)	×	green	(*AABBcc*)

20. How would the results in cross (a) of Problem 19 vary if genes A and B were linked with no crossing over between them? How would the results of cross (a) vary if genes A and B were linked and 20 map units apart?

21. A series of mutations in the bacterium *Salmonella typhimurium* results in the requirement of either tryptophan or some related molecule in order for growth to occur. From the data shown here, suggest a biosynthetic pathway for tryptophan.

Mutation	Minimal Medium	Anthranilic Acid	Indole Glycerol Phosphate	Indole	Tryptophan
trp-8	−	+	+	+	+
trp-2	−	−	+	+	+
trp-3	−	−	−	+	+
trp-1	−	−	−	−	+

Growth Supplement

22. The emergence of antibiotic-resistant strains of *Enterococci* and transfer of resistant genes to other bacterial pathogens have highlighted the need for new generations of antibiotics to combat serious infections. To grasp the range of potential sites for the action of existing antibiotics, sketch the components of the translation machinery (e.g., see Step 3 of Figure 6), and using a series of numbered pointers, indicate the specific location for the action of the antibiotics shown in the following table.

Antibiotic	Action
1. Streptomycin	Binds to 30S ribosomal subunit
2. Chloramphenicol	Inhibits the peptidyl transferase function of 70S ribosome
3. Tetracycline	Inhibits binding of charged tRNA to ribosome
4. Erythromycin	Binds to free 50S particle and prevents formation of 70S ribosome
5. Kasugamycin	Inhibits binding of tRNA^{fmet}
6. Thiostrepton	Prevents translocation by inhibiting EF-G

SOLUTIONS TO SELECTED PROBLEMS AND DISCUSSION QUESTIONS

Answers to Now Solve This

1. One can conclude that the tRNA and not the amino acid is involved in recognition of the codon.

2. With the codes for valine being GUU, GUC, GUA, and GUG, single base changes from glutamic acid's GAA and GAG can cause the glu>>>val switch. The same can be said for lysine with its AAG codon. The normal glutamic acid is a negatively charged amino acid, whereas valine carries no net charge and lysine is positively charged. Given these significant charge changes, one would predict some, if not considerable, influence on protein structure and function. Such changes could stem from internal changes in folding or interactions with other molecules in the RBC, especially other hemoglobin molecules.

Solutions to Problems and Discussion Questions

2. A functional polyribosome will contain the following components: mRNA, charged tRNA, large and small ribosomal subunits, elongation and perhaps initiation factors, peptidyl transferase, GTP, Mg^{2+}, nascent proteins, and possibly GTP-dependent release factors.

4. It was reasoned that there would not be sufficient affinity between amino acids and nucleic acids to account for protein synthesis. For example, acidic amino acids would not be attracted to nucleic acids. With an adaptor molecule, specific hydrogen bonding could occur between nucleic acids, and specific covalent bonding could occur between an amino acid and a nucleic acid tRNA.

6. Since there are three nucleotides that code for each amino acid, there would be 423 code letters (nucleotides), 426 including a termination codon. This assumes that other features such as the poly-A tail, the 5′ cap, promoter, and 5′ and 3′ untranslated sequences are omitted.

8. The four sites in tRNA that provide for specific recognition are the following: attachment of the specific amino acid, interaction with the aminoacyl tRNA synthetase, interaction with the ribosome, and interaction with the codon (anticodon).

10. The quaternary level results from the associations of individual polypeptide chains.

14. Having the precise intragenic location of mutations as well as the ability to isolate the products, especially mutant products, allows scientists to compare the locations of lesions within genes. Mutations occurring nearer the initiation site in a gene will produce proteins with defects near the N-terminus. In this problem, the lesions cause chain termination; therefore, the nearer the mutations to the 5′ end of the mRNA, the shorter the polypeptide product. Relating the position of the mutation with the length of the product establishes the colinear relationship.

18. All of the substitutions involve one base change.

20. Because cross (a) is essentially a monohybrid cross, there would be no difference in the results if crossing over occurred (or did not occur) between the *a* and *b* loci.

22.

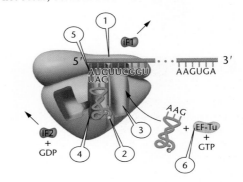

CREDITS

Credits are listed in order of appearance.

Photo

CO, American Association for the Advancement of Science; F-9 (a) Cold Spring Harbor Laboratory; F-9 (b) Elena Kiseleva; F-11, Sebastian Kaulitzki/Shutterstock.com; F-15, Kenneth Eward/BioGrafx/Science Source/Photo Researchers, Inc.

Text

Source: ExPASy

Pigment mutations within an ear of corn, caused by transposition of the *Ds* element.

Gene Mutation, DNA Repair, and Transposition

CHAPTER CONCEPTS

- Mutations comprise any change in the base–pair sequence of DNA.

- Mutation is a source of genetic variation and provides the raw material for natural selection. It is also the source of genetic damage that contributes to cell death, genetic diseases, and cancer.

- Mutations have a wide range of effects on organisms depending on the type of base-pair alteration, the location of the mutation within the chromosome, and the function of the affected gene product.

- Mutations can occur spontaneously as a result of natural biological and chemical processes, or they can be induced by external factors, such as chemicals or radiation.

- Single-gene mutations cause a wide variety of human diseases.

- Organisms rely on a number of DNA repair mechanisms to counteract mutations. These mechanisms range from proofreading and correction of replication errors to base excision and homologous recombination repair.

- Mutations in genes whose products control DNA repair lead to genome hypermutability, human DNA repair diseases, and cancers.

- Transposable elements move into and out of chromosomes, causing chromosome breaks and inducing mutations both within coding regions and in gene-regulatory regions.

The ability of DNA molecules to store, replicate, transmit, and decode information is the basis of genetic function. But equally important is the capacity of DNA to make mistakes. Without the variation that arises from changes in DNA sequences, there would be no phenotypic variability, no adaptation to environmental changes, and no evolution. Gene mutations are the source of most new alleles and are the origin of genetic variation within populations. On the downside, they are also the source of genetic changes that can lead to cell death, genetic diseases, and cancer.

Mutations also provide the basis for genetic analysis. The phenotypic variability resulting from mutations allows geneticists to identify and study the genes responsible for the modified trait. In genetic investigations, mutations act as identifying "markers" for genes so that they can be followed during their transmission from parents to offspring. Without phenotypic variability, classical genetic analysis would be impossible. For example, if all pea plants displayed a uniform phenotype, Mendel would have had no foundation for his research.

We will now explore mutations that occur primarily in the base–pair sequence of DNA within individual genes—**gene mutations.** We will also describe how the cell defends itself from such mutations using various mechanisms of DNA repair.

From Chapter 14 of *Essentials of Genetics, Eighth edition,* William S. Klug, Michael R. Cummings, Charlotte A. Spencer, Michael A. Palladino. © 2013 by Pearson Education, Inc. All rights reserved.

1 Gene Mutations Are Classified in Various Ways

A gene mutation can be defined as an alteration in DNA sequence. Any base-pair change in any part of a DNA molecule can be considered a mutation. A mutation may comprise a single base-pair substitution, a deletion or insertion of one or more base pairs, or a major alteration in the structure of a chromosome.

Mutations may occur within regions of a gene that code for protein or within noncoding regions of a gene such as introns and regulatory sequences. Mutations may or may not bring about a detectable change in phenotype. The extent to which a mutation changes the characteristics of an organism depends on where the mutation occurs and the degree to which the mutation alters the function of a gene product or a gene-regulatory region.

Mutations can occur in somatic cells or within germ cells. Those that occur in germ cells are heritable and are the basis for the transmission of genetic diversity and evolution, as well as genetic diseases. Those that occur in somatic cells are not transmitted to the next generation but may lead to altered cellular function or tumors.

Because of the wide range of types and effects of mutations, geneticists classify mutations according to several different schemes. These organizational schemes are not mutually exclusive. In this section, we outline some of the ways in which gene mutations are classified.

Spontaneous and Induced Mutations

Mutations can be classified as either spontaneous or induced, although these two categories overlap to some degree. **Spontaneous mutations** are changes in the nucleotide sequence of genes that appear to have no known cause. No specific agents are associated with their occurrence, and they are generally assumed to be accidental. Many of these mutations arise as a result of normal biological or chemical processes in the organism that alter the structure of nitrogenous bases. Often, spontaneous mutations occur during the enzymatic process of DNA replication, as we discuss later in this .

In contrast to spontaneous mutations, mutations that result from the influence of extraneous factors are considered to be **induced mutations.** Induced mutations may be the result of either natural or artificial agents. For example, radiation from cosmic and mineral sources and ultraviolet radiation from the sun are energy sources to which most organisms are exposed and, as such, may be factors that cause induced mutations. The earliest demonstration of the artificial induction of mutations occurred in 1927, when Hermann J. Muller reported that X rays could cause mutations in *Drosophila.* In 1928, Lewis J. Stadler reported that X rays had the same effect on barley. In addition to various forms of radiation, numerous natural and synthetic chemical agents are also mutagenic.

The **mutation rate** is defined as the likelihood that a gene will undergo a mutation in a single generation or in forming a single gamete. Several generalizations can be made regarding spontaneous mutation rates in organisms. First, the rate of spontaneous mutation is exceedingly low for all organisms. Second, the rate varies considerably between different organisms. Third, even within the same species, the spontaneous mutation rate varies from gene to gene.

Viral and bacterial genes undergo spontaneous mutation at an average of about 1 in 100 million (10^{-8}) replications or cell divisions. *Neurospora* exhibits a similar rate, but maize, *Drosophila,* and humans demonstrate rates several orders of magnitude higher. The genes studied in these groups average between 1/1,000,000 and 1/100,000 (10^{-6} and 10^{-5}) mutations per gamete formed. Some mouse genes are another order of magnitude higher in their spontaneous mutation rate, 1/100,000 to 1/10,000 (10^{-5} to 10^{-4}). It is not clear why such a large variation occurs in mutation rates. The variation between organisms may reflect the relative efficiencies of their DNA proofreading and repair systems. We will discuss these systems later in the text.

Classification Based on Location of Mutation

Mutations may be classified according to the cell type or chromosomal locations in which they occur. **Somatic mutations** are those occurring in any cell in the body except germ cells. **Germ-line mutations** are those occurring in gametes. **Autosomal mutations** are mutations within genes located on the autosomes, whereas **X-linked mutations** are those within genes located on the X chromosome.

Mutations arising in somatic cells are not transmitted to future generations. When a recessive autosomal mutation occurs in a somatic cell of a diploid organism, it is unlikely to result in a detectable phenotype. The expression of most such mutations is likely to be masked by expression of the wild-type allele within that cell. Somatic mutations will have a greater impact if they are dominant or, in males, if they are X-linked, since such mutations are most likely to be immediately expressed. Similarly, the impact of dominant or X-linked somatic mutations will be more noticeable if they occur early in development, when a small number of undifferentiated cells replicate to give rise to several differentiated tissues or organs. Dominant mutations that occur in cells of adult tissues are often masked by the activity of thousands upon thousands of nonmutant cells in the same tissue that perform the nonmutant function.

Mutations in gametes are of greater significance because they are transmitted to offspring as part of the germ line. They have the potential of being expressed in all cells of an offspring. Inherited dominant autosomal mutations will be expressed phenotypically in the first generation. X-linked recessive mutations arising in the gametes of a **homogametic** female may be expressed in hemizygous male offspring. This will occur provided that the male offspring receives the affected X chromosome. Because of heterozygosity, the occurrence of an autosomal recessive mutation in the gametes of either males or females (even one resulting in a lethal allele) may go unnoticed for many generations, until the resultant allele has become widespread in the population. Usually, the new allele will become evident only when a chance mating brings two copies of it together into the homozygous condition.

Classification Based on Type of Molecular Change

Geneticists often classify gene mutations in terms of the nucleotide changes that constitute the mutation. A change of one base pair to another in a DNA molecule is known as a **point mutation,** or **base substitution (Figure 1)**. A change of one nucleotide of a triplet within a protein–coding portion of a gene may result in the creation of a new triplet that codes for a different amino acid in the protein product. If this occurs, the mutation is known as a **missense mutation.** A second possible outcome is that the triplet will be changed into a stop codon, resulting in the termination of translation of the protein. This is known as a **nonsense mutation.** If the point mutation alters a codon but does not result in a change in the amino acid at that position in the protein (due to degeneracy of the genetic code), it can be considered a **silent mutation.**

You will often see two other terms used to describe base substitutions. If a pyrimidine replaces a pyrimidine or a purine replaces a purine, a **transition** has occurred. If a purine replaces a pyrimidine, or vice versa, a **transversion** has occurred.

Another type of change is the insertion or deletion of one or more nucleotides at any point within the gene. As illustrated in Figure 1, the loss or addition of a single nucleotide causes all of the subsequent three–letter codons to be changed. These are called **frameshift mutations** because the frame of triplet reading during translation is altered.

A frameshift mutation will occur when any number of bases are added or deleted, except multiples of three, which would reestablish the initial frame of reading. It is possible that one of the many altered triplets will be UAA, UAG, or UGA, the translation termination codons. When one of these triplets is encountered during translation, polypeptide synthesis is terminated at that point. Obviously, the results of frameshift mutations can be very severe, especially if they occur early in the coding sequence.

Classification Based on Phenotypic Effects

Depending on their type and location, mutations can have a wide range of phenotypic effects, from none to severe.

A **loss-of-function mutation** is one that reduces or eliminates the function of the gene product. Any type of mutation, ranging from a point mutation to deletion of the entire gene, may lead to a loss of function. Mutations that result in complete loss of function are known as **null mutations.** It is possible for a loss-of-function mutation to be either dominant or recessive. A dominant loss–of–function mutation may result from the presence of a defective protein product that binds to, or inhibits the action of, the normal gene product, which is also present in the same organism. A **gain-of-function** mutation results in a gene product with enhanced or new functions. This may be due to a change in the amino acid sequence of the protein that confers a new activity, or it may result from a mutation in a regulatory region of the gene, leading to expression of the gene at higher levels, or the synthesis of the gene product at abnormal times, or places. Most gain-of-function mutations are dominant.

The most easily observed mutations are those affecting a morphological trait. These mutations are known as **visible mutations** and are recognized by their ability to alter a normal or wild-type visible phenotype. For example, all of Mendel's pea characteristics and many genetic variations encountered in *Drosophila* fit this designation, since they cause obvious changes to the morphology of the organism.

Some mutations exhibit nutritional or biochemical effects. In bacteria and fungi, a typical **nutritional mutation**

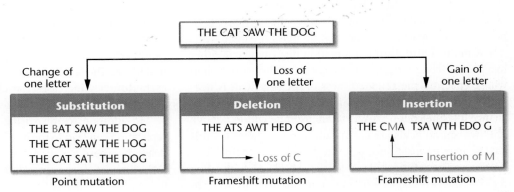

FIGURE 1 Analogy showing the effects of substitution, deletion, and insertion of one letter in a sentence composed of three-letter words to demonstrate point and frameshift mutations.

THE CAT SAW THE DOG

Change of one letter

Loss of one letter

Gain of one letter

Substitution

THE BAT SAW THE DOG
THE CAT SAW THE HOG
THE CAT SAT THE DOG

Point mutation

Deletion

THE ATS AWT HED OG

→ Loss of C

Frameshift mutation

Insertion

THE CMA TSA WTH EDO G

Insertion of M

Frameshift mutation

results in a loss of ability to synthesize an amino acid or vitamin. In humans, sickle-cell anemia and hemophilia are examples of diseases resulting from **biochemical mutations.** Although such mutations do not always affect morphological characters, they can have an effect on the well-being and survival of the affected individual.

Still another category consists of mutations that affect the behavior patterns of an organism. The primary effect of **behavioral mutations** is often difficult to analyze. For example, the mating behavior of a fruit fly may be impaired if it cannot beat its wings. However, the defect may be in the flight muscles, the nerves leading to them, or the brain, where the nerve impulses that initiate wing movements originate.

Another group of mutations—**regulatory mutations**—affect the regulation of gene expression. A mutation in a regulatory gene or a gene control region can disrupt normal regulatory processes and may inappropriately activate or inactivate expression of a gene. For example, a regulatory gene encodes a product that controls the transcription of the entire *lac* operon. Mutations within this regulatory gene can lead to the production of a regulatory protein with abnormal effects on the *lac* operon. Our knowledge of genetic regulation has been dependent on the study of such regulatory mutations.

It is also possible that a mutation may interrupt a process that is essential to the survival of the organism. In this case, it is referred to as a **lethal mutation.** For example, a mutant bacterium that has lost the ability to synthesize an essential amino acid will cease to grow and eventually will die when placed in a medium lacking that amino acid. Various inherited human biochemical disorders are also examples of lethal mutations. For example, Tay-Sachs disease and Huntington disease are caused by mutations that result in lethality, but at different points in the life cycle of humans.

Another interesting class of mutations are those whose expression depends on the environment in which the organism finds itself. Such mutations are called **conditional mutations** because the mutation is present in the genome of an organism but can be detected only under certain conditions. Among the best examples of conditional mutations are **temperature-sensitive mutations.** At a "permissive" temperature, the mutant gene product functions normally, but it loses its function at a different, "restrictive" temperature. Therefore, when the organism is shifted from the permissive to the restrictive temperature, the impact of the mutation becomes apparent.

A **neutral mutation** is a mutation that can occur either in a protein-coding region or in any part of the genome, and whose effect on the genetic fitness of the organism is negligible. For example, a neutral mutation within a gene may change a lysine codon (AAA) to an arginine codon (AGA). The two amino acids are chemically similar; therefore, this change may be insignificant to the function of the protein. Because eukaryotic genomes consist mainly of noncoding regions, the vast majority of mutations are likely to occur in

the large portions of the genome that do not contain genes. These may be considered neutral mutations, if they do not affect gene products or gene expression.

ESSENTIAL POINT ■ ■ ■

Mutations can be spontaneous or induced, somatic or germ-line, autosomal or X-linked. They can have many different effects on gene function, depending on the type of nucleotide changes that comprise the mutation. Phenotypic effects can range from neutral or silent, to loss of function or gain of function, to lethality.

NOW SOLVE THIS

1 If one spontaneous mutation occurs within a human egg cell genome, and this mutation changes an A to a T, what is the most likely effect of this mutation on the phenotype of an offspring that develops from this mutated egg?

■ **Hint** *This problem asks you to predict the effects of a single base-pair mutation on phenotype. The key to its solution involves an understanding of the organization of the human genome as well as the effects of mutations on development and on coding and noncoding regions of genes.*

2 | Spontaneous Mutations Arise from Replication Errors and Base Modifications

In this section, we will outline some of the processes that lead to spontaneous mutations. It is useful to keep in mind, however, that many of the DNA changes that occur during spontaneous mutagenesis also occur, at a higher rate, during induced mutagenesis.

DNA Replication Errors and Slippage

process of DNA replication is imperfect. Occasionally, DNA polymerases insert incorrect nucleotides during replication of a strand of DNA. Although DNA polymerases can correct most of these replication errors using their inherent 3′ to 5′ exonuclease proofreading capacity, misincorporated nucleotides may persist after replication. If these errors are not detected and corrected by DNA repair mechanisms, they may lead to mutations. Replication errors due to mispairing predominantly lead to point mutations. The fact that bases can take several forms, known as **tautomers,** increases the chance of mispairing during DNA replication, as we explain next.

In addition to mispairing and point mutations, DNA replication can lead to the introduction of small insertions or deletions. These mutations can occur when one strand of the DNA template loops out and becomes displaced during replication, or when DNA polymerase slips or stutters during replication. If a loop occurs in the template strand during replication, DNA polymerase may miss the looped-out nucleotides, and a small deletion in the new strand will be introduced. If DNA polymerase repeatedly introduces nucleotides that are not present in the template strand, an insertion of one or more nucleotides will occur, creating an unpaired loop on the newly synthesized strand. Insertions and deletions may lead to frameshift mutations, or amino acid insertions or deletions in the gene product.

Replication slippage can occur anywhere in the DNA but seems distinctly more common in regions containing repeated sequences. Repeat sequences are hot spots for DNA mutation and in some cases contribute to hereditary diseases, such as fragile-X syndrome and Huntington disease. The hypermutability of repeat sequences in noncoding regions of the genome is the basis for current methods of forensic DNA analysis.

Tautomeric Shifts

Purines and pyrimidines can exist in tautomeric forms—that is, in alternate chemical forms that differ by only a single proton shift in the molecule. The biologically important tautomers are the keto–enol forms of thymine and guanine and the amino–imino forms of cytosine and adenine. These shifts change the bonding structure of the molecule, allowing hydrogen bonding with noncomplementary bases. Hence, **tautomeric shifts** may lead to permanent base-pair changes and mutations. **Figure 2** compares normal base-pairing relationships with rare unorthodox pairings. Anomalous T≡G and C═A pairs, among others, may be formed.

A mutation occurs during DNA replication when a transiently formed tautomer in the template strand pairs with a noncomplementary base. In the next round of replication, the "mismatched" members of the base pair are separated, and each becomes the template for its normal complementary base. The end result is a point mutation **(Figure 3)**.

Depurination and Deamination

Some of the most common causes of spontaneous mutations are two forms of DNA base damage: depurination and deamination. **Depurination** is the loss of one of the nitrogenous bases in an intact double-helical DNA molecule. Most frequently, the base is a purine—either guanine or adenine. These bases may be lost if the glycosidic bond linking the 1′-C of the deoxyribose and the number 9 position of the purine ring is broken, leaving an **apurinic site** on one strand of the DNA. Geneticists estimate that thousands of such spontaneous lesions are formed daily in the DNA of mammalian cells in culture. If apurinic sites are not repaired, there will be no base at that position to act as a template during DNA replication. As a result, DNA polymerase may introduce a nucleotide at random at that site.

In **deamination,** an amino group in cytosine or adenine is converted to a keto group **(Figure 4)**. In these cases, cytosine is converted to uracil, and adenine is changed to hypoxanthine. The major effect of these changes is an alteration in the base-pairing specificities of these two bases during DNA replication. For example, cytosine normally pairs with guanine.

(a) Standard base-pairing arrangements

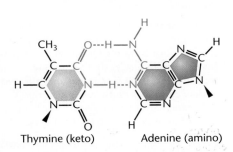

Thymine (keto)   Adenine (amino)

Cytosine (amino)   Guanine (keto)

(b) Anomalous base-pairing arrangements

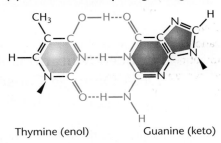

Thymine (enol)   Guanine (keto)

Cytosine (imino)   Adenine (amino)

FIGURE 2   Standard base-pairing relationships (a) compared with examples of the anomalous base-pairing that occurs as a result of tautomeric shifts (b). The long triangle indicates the point at which the base bonds to the pentose sugar.

323

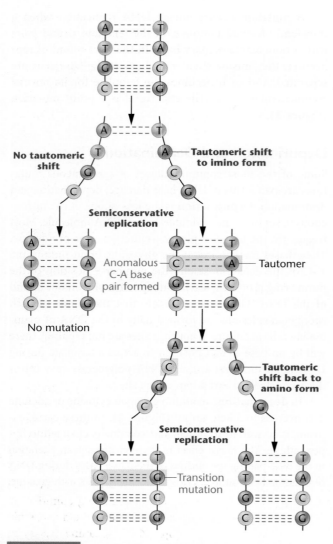

FIGURE 3 Formation of an A=T to G≡C transition mutation as a result of a tautomeric shift in adenine.

Following its conversion to uracil, which pairs with adenine, the original G≡C pair is converted to an A=U pair and then, in the next replication, is converted to an A=T pair. When adenine is deaminated, the original A=T pair is converted to a G≡C pair because hypoxanthine pairs naturally with cytosine. Deamination may occur spontaneously or as a result of treatment with chemical mutagens such as nitrous acid (HNO₂).

Oxidative Damage

DNA may also suffer damage from the by-products of normal cellular processes. These by-products include reactive oxygen species (electrophilic oxidants) that are generated during normal aerobic respiration. For example, superoxides (O_2^-), hydroxyl radicals ($\cdot OH$), and hydrogen peroxide (H_2O_2) are created during cellular metabolism and are constant threats to the integrity of DNA. Such **reactive oxidants,** also generated by exposure to high-energy radiation, can produce more than 100 different types of chemical modifications

in DNA, including modifications to bases, loss of bases, and single-stranded breaks.

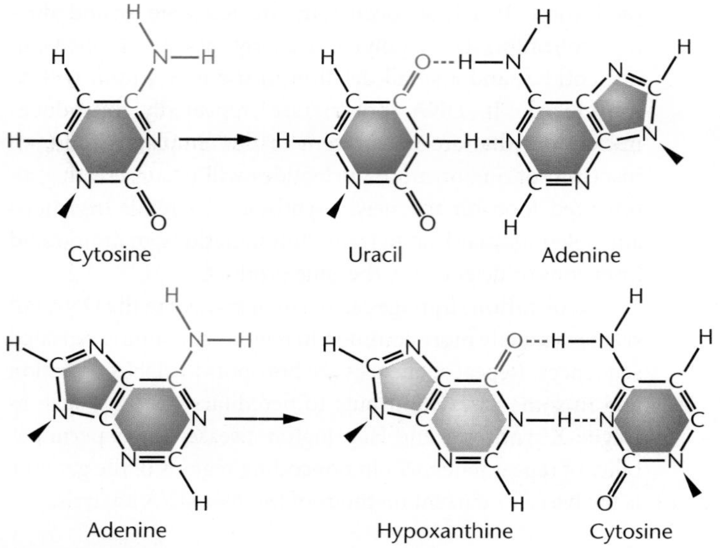

FIGURE 4 Deamination of cytosine and adenine, leading to new base pairing and mutation. Cytosine is converted to uracil, which base-pairs with adenine. Adenine is converted to hypoxanthine, which base-pairs with cytosine.

ESSENTIAL POINT ■ ■ ■

Spontaneous mutations occur in many ways, ranging from errors during DNA replication to changes in DNA base pairing caused by tautomeric shifts, depurination, deamination, and reactive oxidant damage.

NOW SOLVE THIS

2 One of the most famous cases of an X–linked recessive mutation in humans is that of hemophilia found in the descendants of Britain's Queen Victoria. The pedigree of the royal family indicates that Victoria was heterozygous for the trait; however, her father was not affected, and there is no evidence that her mother was a carrier. What are some possible explanations for how the mutation arose? What types of mutations could lead to the disease?

■ **Hint** *This problem asks you to determine the sources of new mutations. The key to its solution is to consider the ways in which mutations occur, the types of cells in which they can occur, and how they are inherited.*

3 Induced Mutations Arise from DNA Damage Caused by Chemicals and Radiation

Induced mutations are those that increase the rate of mutation above the spontaneous background. All cells on Earth are exposed to agents called **mutagens,** which have the potential to damage DNA and cause mutations. Some of these agents, such as some fungal toxins, cosmic rays, and ultraviolet light, are natural components of our environment. Others, including some industrial pollutants, medical X rays, and chemicals within tobacco smoke, can be considered as unnatural or human-made additions to our modern world. On the positive side, geneticists harness some mutagens to create mutations within model organisms such as yeast or *Drosophila*, as a first step in analyzing genes and gene functions. The mechanisms by which some of these natural and unnatural agents lead to mutations are outlined in this section.

Base Analogs

One category of mutagenic chemicals includes **base analogs,** compounds that can substitute for purines or pyrimidines during nucleic acid biosynthesis. For example, **5-bromo-uracil (5-BU),** a derivative of uracil, behaves as a thymine analog but is halogenated at the number 5 position of the pyrimidine ring. If 5-BU is chemically linked to deoxyribose, the nucleoside analog **bromodeoxyuridine (BrdU)**

is formed. **Figure 5** compares the structure of this analog with that of thymine. The presence of the bromine atom in place of the methyl group increases the probability that a tautomeric shift will occur. If 5-BU is incorporated into DNA in place of thymine and a tautomeric shift to the enol form occurs, 5-BU base-pairs with guanine. After one round of replication, an A=T to G≡C transition results. Furthermore, the presence of 5-BU within DNA increases the sensitivity of the molecule to ultraviolet (UV) light, which itself is mutagenic.

There are other base analogs that are mutagenic. For example, **2-amino purine (2-AP)** can act as an analog of adenine. In addition to its base-pairing affinity with thymine, 2-AP can also base-pair with cytosine, leading to possible transitions from A=T to G≡C following replication.

Alkylating, Intercalating, and Adduct-Forming Agents

A number of naturally occurring and human-made chemicals alter the structure of DNA and cause mutations. One type of chemical—the alkylating agent—donates alkyl groups such as CH_3 or CH_3CH_2, to amino or keto groups in nucleotides. Such alkylation alters the base pairing of nucleotides, leading to mutations. Ethylmethane sulfonate (EMS), for example, alkylates the keto groups in the number 6 position of guanine and in the number 4 position of thymine. As with base analogs, base-pairing affinities are altered, and transition mutations result. For example, 6-ethylguanine acts as an analog of adenine and pairs with thymine **(Figure 6)**.

Intercalating agents are chemicals that have dimensions and shapes that allow them to wedge between the base pairs of DNA. When bound between base pairs, intercalating agents cause base pairs to distort and DNA strands to unwind. These changes in DNA structure affect many functions including transcription, replication, and repair. Deletions and insertions occur during DNA replication and repair, leading to frameshift mutations.

Another group of chemicals that cause mutations are known as adduct-forming agents. A DNA adduct is a substance that covalently binds to DNA, altering its conformation and interfering with replication and repair. Two examples of adduct-forming substances are acetaldehyde (a component of cigarette smoke) and

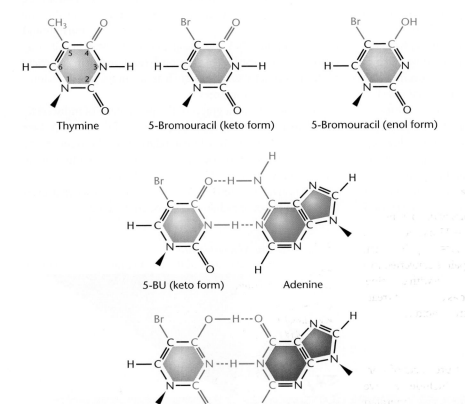

FIGURE 5 Similarity of 5-bromouracil (5-BU) structure to thymine structure. In the common keto form, 5-BU base-pairs normally with adenine, behaving as a thymine analog. In the rare enol form, it pairs anomalously with guanine.

FIGURE 6 Conversion of guanine to 6-ethylguanine by the alkylating agent ethylmethane sulfonate (EMS). The 6-ethylguanine base-pairs with thymine.

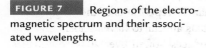

Guanine   6-Ethylguanine   Thymine

heterocyclic amines (HCAs). HCAs are cancer-causing chemicals that are created during the cooking of meats such as beef, chicken, and fish. Many HCAs covalently bind to guanine bases. At least 17 different HCAs have been linked to the development of cancers, such as those of the stomach, colon, and breast.

Ultraviolet Light

All electromagnetic radiation consists of energetic waves that we define by their different wavelengths (Figure 7). The full range of wavelengths is referred to as the **electromagnetic spectrum,** and the energy of any radiation in the spectrum varies inversely with its wavelength. Waves that are in the range of visible light and longer are benign when they interact with most organic molecules. However, waves of shorter length than visible light, being inherently more energetic, have the potential to disrupt organic molecules. As we know, purines and pyrimidines absorb **ultraviolet (UV) radiation** most intensely at a wavelength of about 260 nm. One major effect of UV radiation on DNA is the creation of **pyrimidine dimers**—chemical species consisting of two identical pyrimidines—particularly those consisting of two thymine residues on the same strand of DNA (Figure 8). The dimers distort the DNA conformation and inhibit normal replication. As a result, errors can be introduced in the base sequence of DNA during replication. When UV-induced dimerization is extensive, it is responsible (at least in part) for the killing effects of UV radiation on cells.

Ionizing Radiation

As noted above, the energy of radiation varies inversely with wavelength. Therefore, **X rays, gamma rays,** and **cosmic rays** are more energetic than UV radiation. As a result, they penetrate deeply into tissues, causing ionization of the molecules encountered along the way. Hence, this type of radiation is called **ionizing radiation.**

As X rays penetrate cells, stable molecules and atoms are transformed into **free radicals**—chemical species containing one or more unpaired electrons. Free radicals can directly or indirectly affect the genetic material, altering purines and pyrimidines in DNA, breaking phosphodiester bonds, disrupting the integrity of chromosomes, and producing a variety of chromosomal aberrations, such as deletions, translocations, and chromosomal fragmentation.

Figure 9 shows a graph of the percentage of induced X-linked recessive lethal mutations versus the dose of X rays administered. There is a linear relationship between X-ray dose and the induction of mutation; for each doubling of the dose, twice as many mutations are induced. Because the line intersects near the zero axis, this graph suggests that even very small doses of radiation may be mutagenic.

FIGURE 7 Regions of the electromagnetic spectrum and their associated wavelengths.

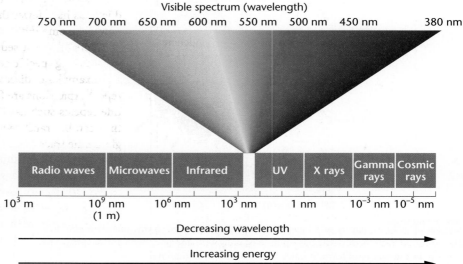

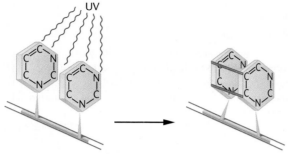

Dimer formed between adjacent thymidine
residues along a DNA strand

FIGURE 8 Induction of a thymine dimer by UV radiation,
leading to distortion of the DNA. The covalent crosslinks occur
between the atoms of the pyrimidine ring.

ESSENTIAL POINT ■ ■ ■

Mutations can be induced by many types of chemicals and radiation. These agents can damage both DNA bases and the sugar-phosphate backbone of DNA molecules.

NOW SOLVE THIS

3 The cancer drug melphalan is an alkylating agent of the mustard gas family. It acts in two ways: by causing alkylation of guanine bases and by cross linking DNA strands together. Describe two ways in which melphalan might kill cancer cells. What are two ways in which cancer cells could repair the DNA-damaging effects of melphalan?

■ **Hint** *This problem asks you to consider the effect of the alkylation of guanine on base pairing during DNA replication. The key to its solution is to consider the effects of mutations on cellular processes that allow cells to grow and divide. In Section 5, you will learn about the ways in which cells repair the types of mutations introduced by alkylating agents.*

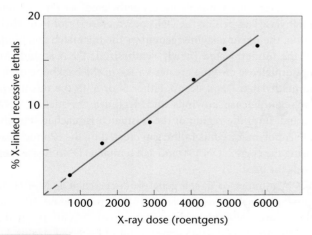

FIGURE 9 Plot of the percentage of X-linked recessive mutations induced in *Drosophila* by increasing doses of X rays. If extrapolated, the graph intersects the zero axis as shown by the dashed line.

4 Single-Gene Mutations Cause a Wide Range of Human Diseases

Although most human genetic diseases are polygenic—that is, caused by variations in several genes—even a single base-pair change in one of the approximately 20,000 human genes can lead to a serious genetic disorder. These monogenic diseases can be caused by many different types of single-gene mutations. **Table 1** lists some examples of the types of single-gene mutations that can lead to serious genetic diseases. A comprehensive database of human genes, mutations, and disorders is available in the Online Mendelian Inheritance in Man (OMIM) database. As of 2011, the OMIM database has catalogued over 3000 human phenotypes for which the molecular basis is known.

Geneticists estimate that approximately 30 percent of mutations that cause human diseases are single base-pair changes that create nonsense mutations. These mutations not only code for a prematurely terminated protein product, but also trigger rapid decay of the mRNA. Many more mutations are missense mutations that alter the amino acid sequence of a protein and frameshift mutations that alter the protein sequence and create internal nonsense codons. Other common disease-associated mutations affect the sequences of gene promoters, mRNA splicing signals, and other noncoding sequences that affect transcription, processing, and stability of mRNA or protein. One recent study showed that about 15 percent of all point mutations that cause human genetic diseases result in abnormal mRNA splicing. Approximately 85 percent of these splicing mutations alter the sequence of 5′ and 3′ splice signals. The remainder create new splice sites within the gene. Splicing defects often result in degradation of the abnormal mRNA or creation of abnormal protein products.

Another type of single-gene mutation is caused by expansions of **trinucleotide repeat sequences**—specific short DNA sequences that are repeated many times. Normal individuals have fewer than 30 repetitions of these sequences; however, some individuals appear to have abnormally large numbers of repeat sequences—often over 200—within and surrounding specific genes.

Examples of diseases associated with these trinucleotide repeat expansions are fragile-X syndrome . When trinucleotide repeats such as $(CAG)_n$ occur within a coding region, they can be translated into long tracks of glutamine. These glutamine tracks may cause the proteins to aggregate abnormally. When the repeats occur outside coding regions, but within the mRNA, it is thought that the mRNAs may act as "toxic" RNAs that bind to important regulatory proteins, sequestering them away from their normal functions in the cell. Another possible consequence of long trinucleotide repeats is that the regions of DNA that code for these repeats

TABLE 1   Examples of Human Disorders Caused by Single-Gene Mutations

Type of DNA Mutation	Disorder	Molecular Change
Missense	Achondroplasia	Glycine to arginine at position 380 of *FGFR2* gene
Nonsense	Marfan syndrome	Tyrosine to STOP codon at position 2113 of *fibrillin-1* gene
Insertion	Familial hypercholesterolemia	Various short insertions throughout the *LDLR* gene
Deletion	Cystic fibrosis	Three base-pair deletion of phenylalanine codon at position 508 of *CFTR* gene
Trinucleotide repeat expansions	Huntington disease	>40 repeats of (CAG) sequence in coding region of *Huntingtin* gene

may become abnormally methylated, leading to silencing of gene transcription.

The mechanisms by which the repeated sequences expand from generation to generation are still unclear. It is thought that expansions may result from errors that take place either during DNA replication or during DNA damage repair.

5   Organisms Use DNA Repair Systems to Counteract Mutations

Living systems have evolved a variety of elaborate repair systems that counteract both spontaneous and induced DNA damage. These **DNA repair** systems are absolutely essential to the maintenance of the genetic integrity of organisms and, as such, to the survival of organisms on Earth. The balance between mutation and repair results in the observed mutation rates of individual genes and organisms. Of foremost interest in humans is the ability of these systems to counteract genetic damage that would otherwise result in genetic diseases and cancer.

We now embark on a review of some systems of DNA repair. Since the field is expanding rapidly, our goal here is merely to survey the major approaches that organisms use to counteract genetic damage.

Proofreading and Mismatch Repair

Some of the most common types of mutations arise during DNA replication when an incorrect nucleotide is inserted by DNA polymerase. The enzyme in bacteria (**DNA polymerase III**) makes an error approximately once every 100,000 insertions, leading to an error rate of 10^{-5}. Fortunately, DNA polymerase proofreads each step, catching 99 percent of those errors. If an incorrect nucleotide is inserted during polymerization, the enzyme can recognize the error and "reverse" its direction. It then behaves as a 3′ to 5′ exonuclease, cutting out the incorrect nucleotide and replacing it with the correct one. This improves the efficiency of replication one hundredfold, creating only 1 mismatch in every 10^7 insertions, for a final error rate of 10^{-7}.

To cope with those errors that remain after proofreading, another mechanism, called **mismatch repair,** may be activated. During mismatch repair, the mismatch is detected, the incorrect nucleotide is removed, and the correct nucleotide is inserted in its place. But how does the repair system recognize which nucleotide is correct (on the template strand) and which nucleotide is incorrect (on the newly synthesized strand)? If the mismatch is recognized but no such discrimination occurs, the excision will be random, and the strand bearing the correct base will be clipped out 50 percent of the time. Hence, strand discrimination is a critical step.

The process of strand discrimination has been elucidated in some bacteria, including *E. coli,* and is based on **DNA methylation.** These bacteria contain an enzyme, **adenine methylase,** which recognizes the DNA sequence

$$5'\text{——GATC——}3'$$
$$3'\text{——CTAG——}5'$$

as a substrate, adding a methyl group to each of the adenine residues during DNA replication.

Following replication, the newly synthesized DNA strand remains temporarily unmethylated, as the adenine methylase lags behind the DNA polymerase. Prior to methylation, the repair enzyme recognizes the mismatch and binds to the unmethylated (newly synthesized) DNA strand. An **endonuclease** enzyme creates a nick in the backbone of the unmethylated DNA strand, either 5′ or 3′ to the mismatch. An **exonuclease** unwinds and degrades the nicked DNA strand, until the region of the mismatch is reached. Finally, DNA polymerase fills in the gap created by the exonuclease, using the correct DNA strand as a template. DNA ligase then seals the gap.

A number of *E. coli* gene products, such as Mut H, L, and S, as well as exonucleases, DNA polymerase III, and ligase, are involved in mismatch repair. Mutations in the *MutH, MutL,* and *MutS* genes result in bacterial strains deficient in mismatch repair. While the preceding mechanism is based on studies of *E. coli,* similar mechanisms involving homologous proteins exist in yeast and in mammals. Mutations in the human equivalents of the *MutS* and *MutL* genes of *E. coli* are associated with human hereditary nonpolyposis colon cancer. Mismatch repair defects are commonly found in other cancers, such as leukemias, lymphomas, and tumors of the ovary, prostate, and endometrium. Cells from these cancers show genome-wide increases in the rate of spontaneous mutation. The link between defective mismatch repair and cancer is supported by experiments with mice. Mice that are engineered to have deficiencies in mismatch repair genes accumulate large numbers of mutations and are cancer-prone.

Postreplication Repair and the SOS Repair System

Another DNA repair system, called **postreplication repair,** responds *after* damaged DNA has escaped repair and has failed to be completely replicated. As illustrated in **Figure 10,** when DNA bearing a lesion of some sort (such as a pyrimidine dimer) is being replicated, DNA polymerase may stall at the lesion and then skip over it, leaving an unreplicated gap on the newly synthesized strand. To correct the gap, the RecA protein directs a recombinational exchange with the corresponding region on the undamaged parental strand of the same polarity (the "donor" strand). When the undamaged segment of the donor strand DNA replaces the gapped segment, a gap is created on the donor strand. The gap can be filled by repair synthesis as replication proceeds. Because a recombinational event is involved in this type of DNA repair, it is considered to be a form of homologous recombination repair.

Still another repair pathway, the *E. coli* **SOS repair system,** also responds to damaged DNA, but in a different way. In the presence of a large number of unrepaired DNA mismatches and gaps, bacteria can induce the expression of about 20 genes (including *lexA, recA,* and *uvr*) whose products allow DNA replication to occur even in the presence of these lesions. This type of repair is a last resort to minimize DNA damage, hence its name. During SOS repair, DNA synthesis becomes error-prone, inserting random and possibly incorrect nucleotides in places that would normally stall DNA replication. As a result, SOS repair itself becomes mutagenic—although it may allow the cell to survive DNA damage that would otherwise kill it.

Photoreactivation Repair: Reversal of UV Damage

As illustrated in Figure 8, UV light is mutagenic as a result of the creation of pyrimidine dimers. UV-induced damage to *E. coli* DNA can be partially reversed if, following irradiation, the cells are exposed briefly to light in the blue range of the visible spectrum. The process is dependent on the activity of a protein called **photoreactivation enzyme**

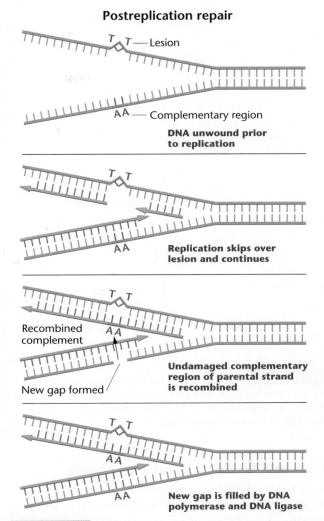

Postreplication repair

T T—Lesion

A A—Complementary region

DNA unwound prior to replication

Replication skips over lesion and continues

Recombined complement

New gap formed

Undamaged complementary region of parental strand is recombined

New gap is filled by DNA polymerase and DNA ligase

FIGURE 10 Postreplication repair occurs if DNA replication has skipped over a lesion such as a thymine dimer. Through the process of recombination, the correct complementary sequence is recruited from the parental strand and inserted into the gap opposite the lesion. The new gap is filled by DNA polymerase and DNA ligase.

(PRE). The enzyme's mode of action is to cleave the bonds between thymine dimers, thus directly reversing the effect of UV radiation on DNA **(Figure 11).** Although the enzyme will associate with a dimer in the dark, it must absorb a photon of light to cleave the dimer. In spite of its ability to reduce the number of UV-induced mutations, **photoreactivation repair** is not absolutely essential in *E. coli;* we know this because a mutation creating a null allele in the gene coding for PRE is not lethal. Nonetheless, the enzyme is detectable in many organisms, including bacteria, fungi, plants, and some vertebrates—although not in humans. Humans and other organisms that lack photoreactivation repair must rely on other repair mechanisms to reverse the effects of UV radiation.

Photoreactivation repair

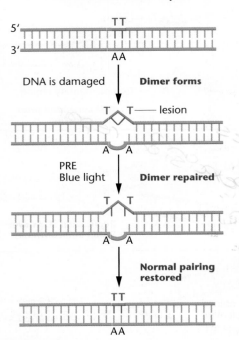

FIGURE 11   Damaged DNA repaired by photoreactivation repair. The bond creating the thymine dimer is cleaved by the photoreactivation enzyme (PRE), which must be activated by blue light in the visible spectrum.

Base excision repair

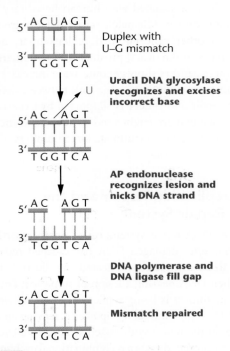

FIGURE 12   Base excision repair (BER) accomplished by uracil DNA glycosylase, AP endonuclease, DNA polymerase, and DNA ligase. Uracil is recognized as a noncomplementary base, excised, and replaced with the complementary base (C).

Base and Nucleotide Excision Repair

A number of light-independent DNA repair systems exist in all prokaryotes and eukaryotes. The basic mechanisms involved in these types of repair—collectively referred to as **excision repair** or cut-and-paste mechanisms—consist of the following three steps.

1. The distortion or error present on one of the two strands of the DNA helix is recognized and enzymatically clipped out by an endonuclease. Excisions in the phosphodiester backbone usually include a number of nucleotides adjacent to the error as well, leaving a gap on one strand of the helix.

2. A DNA polymerase fills in the gap by inserting nucleotides complementary to those on the intact strand, which it uses as a replicative template. The enzyme adds these nucleotides to the free 3′-OH end of the clipped DNA. In *E. coli*, this step is usually performed by DNA polymerase I.

3. DNA ligase seals the final "nick" that remains at the 3′-OH end of the last nucleotide inserted, closing the gap.

There are two types of excision repair: base excision repair and nucleotide excision repair. **Base excision repair**

(BER) corrects DNA that contains a damaged DNA base. The first step in the BER pathway in *E. coli* involves the recognition of the altered base by an enzyme called **DNA glycosylase.** There are a number of DNA glycosylases, each of which recognizes a specific base **(Figure 12)**. For example, the enzyme uracil DNA glycosylase recognizes the presence of uracil in DNA. DNA glycosylases first cut the glycosidic bond between the base and the sugar, creating an **apyrimidinic** or **apurinic site.** The sugar with the missing base is then recognized by an enzyme called **AP endonuclease.** The AP endonuclease makes a cut in the phosphodiester backbone at the apyrimidinic or apurinic site. Endonucleases then remove the deoxyribose sugar and the gap is filled by DNA polymerase and DNA ligase.

Although much has been learned about the mechanisms of BER in *E. coli*, BER systems have also been detected in eukaryotes from yeast to humans. Experimental evidence shows that both mouse and human cells that are defective in BER activity are hypersensitive to the killing effects of gamma rays and oxidizing agents.

Nucleotide excision repair (NER) pathways repair "bulky" lesions in DNA that alter or distort the double helix, such as the UV-induced pyrimidine dimers discussed previously. The NER pathway **(Figure 13)** was first discovered in *E. coli* by Paul Howard-Flanders and coworkers, who

Nucleotide excision repair

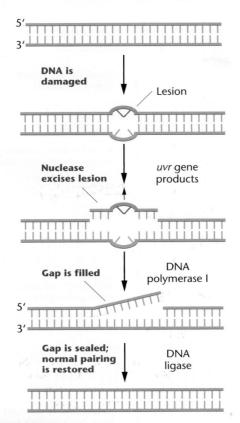

FIGURE 13 Nucleotide excision repair (NER) of a UV-induced thymine dimer. During repair, 13 nucleotides are excised in prokaryotes, and 28 nucleotides are excised in eukaryotes.

FIGURE 14 Two individuals with xeroderma pigmentosum. These XP patients show characteristic XP skin lesions induced by sunlight, as well as mottled redness (erythema) and irregular pigment changes to the skin, in response to cellular injury.

isolated several independent mutants that are sensitive to UV radiation. One group of genes was designated *uvr* (ultraviolet repair) and included the *uvrA, uvrB,* and *uvrC* mutations. In the NER pathway, the *uvr* gene products are involved in recognizing and clipping out lesions in the DNA. Usually, a specific number of nucleotides is clipped out around both sides of the lesion. In *E. coli,* usually a total of 13 nucleotides are removed, including the lesion. The repair is then completed by DNA polymerase I and DNA ligase, in a manner similar to that occurring in BER. The undamaged strand opposite the lesion is used as a template for the replication, resulting in repair.

Nucleotide Excision Repair and Xeroderma Pigmentosum in Humans

The mechanism of NER in eukaryotes is much more complicated than that in prokaryotes and involves many more proteins, encoded by about 30 genes. Much of what is known about the system in humans has come from detailed studies of individuals with **xeroderma pigmentosum (XP),** a rare

recessive genetic disorder that predisposes individuals to severe skin abnormalities, skin cancers, and a wide range of other symptoms including developmental and neurological defects. The condition is severe and may be lethal, although early detection and protection from sunlight can arrest it (**Figure 14**).

The repair of UV-induced lesions in XP has been investigated *in vitro,* using human fibroblast cell cultures derived from normal individuals and those with XP. (Fibroblasts are undifferentiated connective tissue cells.) The results of these studies suggest that the XP phenotype is caused by defects in NER pathways and by mutations in more than one gene.

In 1968, James Cleaver showed that cells from XP patients were deficient in DNA synthesis other than that occurring during chromosome replication—a phenomenon known as **unscheduled DNA synthesis.** Unscheduled DNA synthesis is elicited in normal cells by UV radiation. Because this type of synthesis is thought to represent the activity of DNA polymerization during NER, the lack of unscheduled DNA synthesis in XP patients suggested that XP may be a deficiency in NER.

The involvement of multiple genes in NER and XP was further investigated by studies using **somatic cell hybridization.** Fibroblast cells from any two unrelated XP patients, when grown together in tissue culture, can fuse, forming heterokaryons. A **heterokaryon** is a single cell with two nuclei from different organisms but a common cytoplasm. NER in the heterokaryon can be measured by the level of unscheduled DNA synthesis. If the mutation in each of the two XP cells occurs in the same gene, the heterokaryon, like the cells that fused to form it, will still be unable to undergo NER. This is because there is no normal copy of the relevant gene present in the heterokaryon. However, if NER does occur in the heterokaryon, the mutations in the two XP cells must have been present in two different genes. Hence, the two mutants are said to demonstrate **complementation**. Complementation occurs because the heterokaryon has at least one normal copy of each gene in the fused cell. By fusing XP cells from a large number of XP patients, researchers were able to determine how many genes contribute to the XP phenotype.

Based on these and other studies, XP patients have been divided into seven complementation groups, indicating that at least seven different genes are involved in excision repair in humans. Genes representing each of these complementation groups, *XPA* to *XPG* (*X*eroderma *P*igmentosum gene *A* to *G*), have now been identified, and a homologous gene for each has been identified in yeast. Approximately 20 percent of XP patients do not fall into any of the seven complementation groups. Cells from these patients often have mutations in the gene coding for DNA polymerase η and are defective in repair during DNA synthesis.

As a result of the study of defective genes in XP, a great deal is now known about how NER counteracts DNA damage in normal cells. The first step in humans is recognition of the damaged DNA by proteins encoded by the *XPC*, *XPE*, and *XPA* genes. These proteins then recruit the remainder of the repair proteins to the site of DNA damage. The *XPB* and *XPD* genes encode helicases, and the *XPF* and *XPG* genes encode nucleases. The excision repair complex containing these and other factors is responsible for the excision of an approximately 28-nucleotide-long fragment from the DNA strand that contains the lesion.

Double-Strand Break Repair in Eukaryotes

Thus far, we have discussed repair pathways that deal with damage or errors within one strand of DNA. We conclude our discussion of DNA repair by considering what happens when both strands of the DNA helix are cleaved—as a result of exposure to ionizing radiation, for example. These types of damage are extremely dangerous to cells, leading to chromosome rearrangements, cancer, or cell death. In this section, we will discuss double-strand breaks in eukaryotic cells.

Specialized forms of DNA repair, the **DNA double-strand break repair (DSB repair)** pathways, are activated and are responsible for reattaching two broken DNA strands. Recently, interest in DSB repair has grown because defects in these pathways are associated with X-ray hypersensitivity and immune deficiency. Such defects may also underlie familial disposition to breast and ovarian cancer. Several

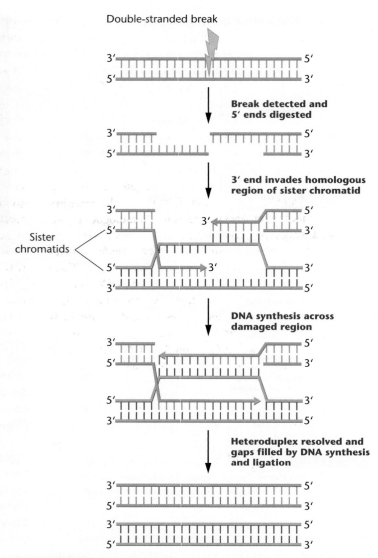

FIGURE 15 Steps in homologous recombination repair of double-stranded breaks.

human disease syndromes, such as Fanconi's anemia and ataxia telangiectasia, result from defects in DSB repair.

One pathway involved in double-strand break repair is **homologous recombination repair.** The first step in this process involves the activity of an enzyme that recognizes the double-strand break, then digests back the 5′ ends of the broken DNA helix, leaving overhanging 3′ ends (**Figure 15**). One 3′ overhanging end searches for a region of sequence complementarity on the sister chromatid and then invades the homologous DNA duplex, aligning the complementary sequences. Once aligned, DNA synthesis proceeds from the 3′ overhanging ends, using the undamaged DNA strands as templates. The interaction of two sister chromatids is necessary because, when both strands of one helix are broken, there is no undamaged parental DNA strand available to use as a source of the complementary template DNA sequence during repair. After DNA repair synthesis, the resulting heteroduplex molecule is resolved and the two chromatids separate.

DSB repair usually occurs during the late S or early G2 phase of the cell cycle, after DNA replication, a time when sister chromatids are available to be used as repair templates. Because an undamaged template is used during repair synthesis, homologous recombination repair is an accurate process.

A second pathway, called **nonhomologous end joining,** also repairs double-strand breaks. However, as the name implies, the mechanism does not recruit a homologous region of DNA during repair. This system is activated in G1, prior to DNA replication. End joining involves a complex of three proteins, including DNA-dependent protein kinase. These proteins bind to the free ends of the broken DNA, trim the ends, and ligate them back together. Because some nucleotide sequences are lost in the process of end joining, it is an error-prone repair system. In addition, if more than one chromosome suffers a double-strand break, the wrong ends could be joined together, leading to abnormal chromosome structures.

ESSENTIAL POINT ▪ ▪ ▪

Organisms counteract mutations by using a range of DNA repair systems. Errors in DNA synthesis can be repaired by proofreading, mismatch repair, and postreplication repair. DNA damage can be repaired by photoreactivation repair, SOS repair, base excision repair, nucleotide excision repair, and double-strand break repair.

NOW SOLVE THIS

4 Geneticists often use ethylmethane sulfonate (EMS) to induce mutations in *Drosophila*. Why is EMS a mutagen of choice for genetic research? What would be the effects of EMS in a strain of *Drosophila* lacking functional mismatch repair systems?

▪ **Hint** *This problem asks you to evaluate EMS as a useful mutagen and to determine its effects in the absence of DNA repair. The key to its solution is to consider the chemical effects of EMS on DNA. Also, consider the types of DNA repair that may operate on EMS-mutated DNA and the efficiency of these processes.*

---

6   The Ames Test Is Used to Assess the Mutagenicity of Compounds

There is great concern about the possible mutagenic properties of any chemical that enters the human body, whether through the skin, the digestive system, or the respiratory tract. Examples of synthetic chemicals that concern us are those found in air and water pollution, food preservatives, artificial sweeteners, herbicides, pesticides, and pharmaceutical products. Mutagenicity can be tested in various organisms, including fungi, plants, and cultured mammalian cells; however, one of the most common tests, which we describe here, uses bacteria.

The **Ames test** uses a number of different strains of the bacterium *Salmonella typhimurium* that have been selected for their ability to reveal the presence of specific types of mutations. For example, some strains are used to detect base-pair substitutions, and other strains detect various frameshift mutations. Each strain contains a mutation in one of the genes of the histidine operon. The mutant strains are unable to synthesize histidine (his^- strains) and therefore require histidine for growth **(Figure 16)**. The assay measures the frequency of reverse mutations that occur within the mutant gene, yielding wild-type bacteria (his^+ revertants). These *Salmonella* strains also have an increased sensitivity to mutagens due

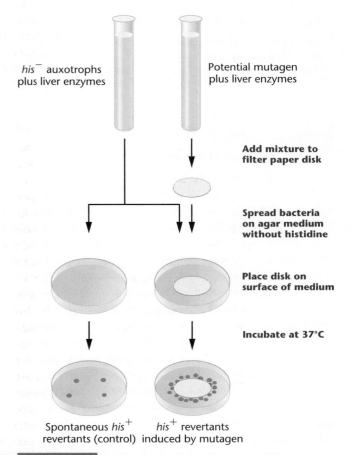

FIGURE 16   The Ames test, which screens compounds for potential mutagenicity.

to the presence of mutations in genes involved in DNA damage repair and synthesis of the lipopolysaccharide barrier that coats bacteria and protects them from external substances.

Many substances entering the human body are relatively innocuous until activated metabolically, usually in the liver, to a more chemically reactive product. Thus, the Ames test includes a step in which the test compound is incubated *in vitro* in the presence of a mammalian liver extract. Alternatively, test compounds may be injected into a mouse where they are modified by liver enzymes and then recovered for use in the Ames test.

In the initial use of Ames testing in the 1970s, a large number of known **carcinogens,** or cancer-causing agents, were examined, and more than 80 percent of these were shown to be strong mutagens. This is not surprising, as the transformation of cells to the malignant state occurs as a result of mutations. Although a positive response in the Ames test does not prove that a compound is carcinogenic, the test is useful as a preliminary screening device. The test is used extensively during the development of industrial and pharmaceutical chemical compounds.

ESSENTIAL POINT ■   ■   ■

The Ames test allows scientists to estimate the mutagenicity and cancer-causing potential of chemical agents by following the rate of mutation in specific strains of bacteria.

---

7 Transposable Elements Move within the Genome and May Create Mutations

Transposable elements, also known as **transposons** or "jumping genes," are DNA sequences that can move or transpose within and between chromosomes, inserting themselves into various locations within the genome.

Transposons are present in the genomes of all organisms from bacteria to humans. Not only are they ubiquitous, but they also comprise large portions of some eukaryotic genomes. For example, almost 50 percent of the human genome is derived from transposable elements. Some organisms with unusually large genomes, such as salamanders and barley, contain hundreds of thousands of copies of various types of transposable elements. Although the function of these elements is unknown, data from human genome sequencing suggest that some genes may have evolved from transposons and that transposons may help to modify and reshape the genome. Transposable elements are also valuable tools in genetic research. Geneticists harness transposons as

mutagens, as cloning tags, and as vehicles for introducing foreign DNA into model organisms.

In this section, we discuss transposable elements as naturally occurring mutagens. The movement of transposons from one place in the genome to another has the capacity to disrupt genes and cause mutations, as well as to create chromosomal damage such as double-strand breaks.

Insertion Sequences and Bacterial Transposons

There are two types of transposable elements in bacteria: insertion sequences and bacterial transposons. **Insertion sequences (IS elements)** can move from one location to another and, if they insert into a gene or gene-regulatory region, may cause mutations.

IS elements were first identified during analyses of mutations in the *gal* operon of *E. coli*. Researchers discovered that certain mutations in this operon were due to the presence of several hundred base pairs of extra DNA inserted into the beginning of the operon. Surprisingly, the segment of mutagenic DNA could spontaneously excise from this location, restoring wild-type function to the *gal* operon. Subsequent research revealed that several other DNA elements could behave in a similar fashion, inserting into bacterial chromosomes and affecting gene function.

IS elements are relatively short, not exceeding 2000 bp (2 kb). The first insertion sequence to be characterized in *E. coli*, IS1, is about 800 bp long. Other IS elements such as IS2, 3, 4, and 5 are about 1250 to 1400 bp in length. IS elements are present in multiple copies in bacterial genomes. For example, the *E. coli* chromosome contains five to eight copies of IS1, five copies each of IS2 and IS3, as well as copies of IS elements on plasmids such as F factors.

All IS elements contain two features that are essential for their movement. First, they contain a gene that encodes an enzyme called **transposase.** This enzyme is responsible for making staggered cuts in chromosomal DNA, into which the IS element can insert. Second, the ends of IS elements contain **inverted terminal repeats (ITRs).** ITRs are short segments of DNA that have the same nucleotide sequence as each other but are oriented in the opposite direction (**Figure 17**). Although Figure 17 shows that ITRs consist of only a few nucleotides, IS ITRs usually contain about 20 to 40 nucleotide pairs. ITRs are essential for transposition and act as recognition sites for the binding of the transposase enzyme.

Bacterial transposons (**Tn elements**) are larger than IS elements and contain protein-coding genes that are unrelated to their transposition. Some Tn elements, such as Tn10, are comprised of a drug-resistance gene flanked by two IS elements present in opposite orientations. The IS elements encode the transposase enzyme that is necessary for transpo-

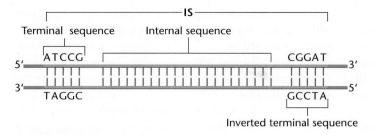

FIGURE 17 An insertion sequence (IS), shown in purple. The terminal sequences are perfect inverted repeats of one another.

sition of the Tn element. Other types of Tn elements, such as Tn3, have shorter inverted repeat sequences at their ends and encode their transposase enzyme from a transposase gene located in the middle of the Tn element. Like IS elements, Tn elements are mobile in both bacterial chromosomes and in plasmids, and can cause mutations if they insert into genes or gene-regulatory regions.

Tn elements are currently of interest because they can introduce multiple drug resistance onto bacterial plasmids. These plasmids, called **R factors,** may contain many Tn elements conferring simultaneous resistance to heavy metals, antibiotics, and other drugs. These elements can move from plasmids onto bacterial chromosomes and can spread multiple drug resistance between different strains of bacteria.

The *Ac–Ds* System in Maize

About 20 years before the discovery of transposons in bacteria, Barbara McClintock discovered mobile genetic elements in corn plants (maize). She did this by analyzing the genetic behavior of two mutations, *dissociation (Ds)* and *activator (Ac),* expressed in either the endosperm or aleurone layers. She then correlated her genetic observations with cytological examinations of the maize chromosomes. Initially, McClintock determined that *Ds* was located on chromosome 9. If *Ac* was also present in the genome, *Ds* induced breakage at a point on the chromosome adjacent to its own location. If chromosome breakage occurred in somatic cells during their development, progeny cells often lost part of the broken chromosome, causing a variety of phenotypic effects.

Subsequent analysis suggested to McClintock that both *Ds* and *Ac* elements sometimes moved to new chromosomal locations. While *Ds* moved only if *Ac* was also present, *Ac* was capable of autonomous movement. Where *Ds* came to reside determined its genetic effects—that is, it might cause chromosome breakage, or it might inhibit expression of a certain gene. In cells in which *Ds* caused a

gene mutation, *Ds* might move again, restoring the gene mutation to wild type.

Figure 18 illustrates the types of movements and effects brought about by *Ds* and *Ac* elements. In McClintock's original observation, pigment synthesis was restored in cells in which the *Ds* element jumped out of chromosome 9. McClintock concluded that the *Ds* and *Ac* genes were **mobile controlling elements.** We now commonly refer to them as transposable elements, a term coined by another great maize geneticist, Alexander Brink.

(a) In absence of *Ac*, *Ds* is not transposable.

(b) When *Ac* is present, *Ds* may be transposed.

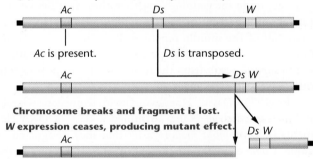

(c) *Ds* can move into and out of another gene.

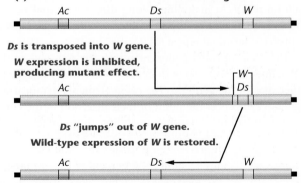

FIGURE 18 Effects of *Ac* and *Ds* elements on gene expression. (a) If *Ds* is present in the absence of *Ac*, there is normal expression of a distantly located hypothetical gene *W*. (b) In the presence of *Ac*, *Ds* may transpose to a region adjacent to *W*. *Ds* can induce chromosome breakage, which may lead to loss of a chromosome fragment bearing the *W* gene. (c) In the presence of *Ac*, *Ds* may transpose into the *W* gene, disrupting *W*-gene expression. If *Ds* subsequently transposes out of the *W* gene, *W*-gene expression may return to normal.

Several *Ac* and *Ds* elements have now been analyzed, and the relationship between the two elements has been clarified. The first *Ds* element studied (*Ds-a*) is nearly identical to *Ac* except for a 194-bp deletion within the transposase gene. The deletion of part of the transposase gene in the *Ds-a* element explains its dependence on the *Ac* element for transposition. Several other *Ds* elements have also been sequenced, and each contains an even larger deletion within the transposase gene. In each case, however, the ITRs are retained.

Although the significance of Barbara McClintock's mobile controlling elements was not fully appreciated following her initial observations, molecular analysis has since verified her conclusions. She was awarded the Nobel Prize in Physiology or Medicine in 1983.

Copia and *P* Elements in *Drosophila*

There are more than 30 families of transposable elements in *Drosophila melanogaster,* each of which is present in 20 to 50 copies in the genome. Together, these families constitute about 5 percent of the *Drosophila* genome and more than half of the middle repetitive DNA of this organism. One study suggests that 50 percent of all visible mutations in *Drosophila* result from the insertion of transposons into otherwise wild-type genes.

In 1975, David Hogness, David Finnegan, Gerald Rubin, and Michael Young identified a class of DNA elements in *Drosophila melanogaster* that they designated as **copia.** *Copia* elements are present in 10 to 100 copies in the genomes of *Drosophila* cells and may move to different chromosomal locations throughout the genome.

Each *copia* element consists of approximately 5000 to 8000 bp of DNA, including a long **direct terminal repeat (DTR)** sequence of 267 bp at each end. Within each DTR is an inverted terminal repeat (ITR) of 17 bp **(Figure 19)**.

Insertion of *copia* is dependent on the presence of the ITR sequences and seems to occur preferentially at specific target sites in the genome. The *copia* elements demonstrate regulatory effects at the point of their insertion in the chromosome. Certain mutations, including those affecting eye color and segment formation, are due to *copia* insertions within genes. For example, the eye-color mutation *white-apricot* is caused by an allele of the *white* gene, which contains a *copia* element within the gene. Transposition of the *copia* element out of the *white-apricot* allele can restore the allele to wild type.

Perhaps the most significant *Drosophila* transposable elements are the **P elements.** These were discovered while studying the phenomenon of **hybrid dysgenesis,** a condition characterized by sterility, elevated mutation rates, and chromosome rearrangements in the offspring of crosses between certain strains of fruit flies. Hybrid dysgenesis is caused by high rates of *P* element transposition in the germ line, in which transposons insert themselves into or near genes, thereby causing mutations. *P* elements range from 0.5 to 2.9 kb long, with 31-bp ITRs. Full-length *P* elements encode at least two proteins, one of which is the transposase enzyme, and another is a repressor protein that inhibits transposition. The transposase gene is expressed only in the germ line, accounting for the tissue specificity of *P* element transposition.

Mutations can arise from several kinds of insertional events. If a *P* element inserts into the coding region of a gene, it can terminate transcription of the gene and destroy normal gene expression. If it inserts into the promoter region of a gene, it can affect the level of expression of the gene. Insertions into introns can affect splicing or cause the premature termination of transcription.

Geneticists have harnessed *P* elements as tools for genetic analysis. One of the most useful applications of *P* elements is as vectors to introduce transgenes into *Drosophila*—a technique known as **germ-line transformation.** *P* elements are also used to generate mutations and to clone mutant genes. In addition, researchers are perfecting methods to target *P* element insertions to precise single-chromosomal sites, which should increase the precision of germ-line transformation in the analysis of gene activity.

Transposable Elements in Humans

The human genome, like that of other eukaryotes, is riddled with transposons. Recent genomic sequencing data reveal that approximately half of the human genome is comprised of transposable element DNA. In contrast, only about 2 percent of the human genome comprises gene coding-sequence DNA. major families of human transposons are the long interspersed elements and short interspersed elements (LINES and SINES, respectively). Together, they comprise over 30 percent of the human genome.

Although most human transposons appear to be inactive, the potential mobility and mutagenic effects of transposable elements have far-reaching implications for human genetics, as can be seen in a recent example of a transpo-

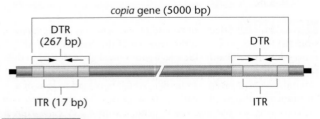

FIGURE 19 Structural organization of a *copia* transposable element in *Drosophila melanogaster,* showing the terminal repeats.

son "caught in the act." The case involves a male child with hemophilia. One cause of hemophilia is a defect in blood-clotting factor VIII, the product of an X-linked gene. Haig Kazazian and his colleagues found LINES inserted at two points within the patient's *factor VIII* gene. Researchers were interested in determining if one of the mother's X chromosomes also contained these specific LINES. If so, the unaffected mother would be heterozygous and pass the LINE-containing chromosome to her son. The surprising finding was that the LINE sequences were *not* present on either of her X chromosomes but *were* detected on chromosome 22 of both parents. This suggests that this mobile element may have transposed from one chromosome to another in the gamete-forming cells of the mother, prior to being transmitted to the son.

LINE insertions into the human *dystrophin* gene have resulted in at least two separate cases of Duchenne muscular dystrophy. In one case, a transposon inserted into exon 48, and in another case, a transposon inserted into exon 44; both led to frameshift mutations and premature termination of translation of the dystrophin protein. There are also reports that LINES have inserted into the *APC* and *c-myc* genes, leading to mutations that may have contributed to the development of some colon and breast cancers. In the latter cases, the transposition had occurred within one or a few somatic cells.

SINE insertions are also responsible for a number of cases of human disease. In one case, an *Alu* element integrated into the *BRCA2* gene, inactivating this tumor-suppressor gene and leading to a familial case of breast cancer. Other genes that have been mutated by *Alu* integrations are the *factor IX* gene (leading to hemophilia B), the *ChE* gene (leading to acholinesterasemia), and the *NF1* gene (leading to neurofibromatosis).

Transposons, Mutations, and Evolution

Transposons can have a wide range of effects on genes. The insertion of a transposon into the coding region of a gene may disrupt the gene's normal translation reading frame or may induce premature termination of translation of the mRNA transcribed from the gene. Many transposons contain their own transcription promoters and enhancers, as well as splice sites and polyadenylation signals. The presence of these transposon regulatory sequences can have effects on nearby genes. The insertion of a transposon containing polyadenylation or transcription termination signals into a gene's intron may bring about termination of the gene's transcription within the transposon. In addition, it can cause aberrant splicing of RNA transcribed from the gene. Insertions of a transposon into a gene's transcription

regulatory region may disrupt the gene's normal regulation or may cause the gene to be expressed differently as a result of the presence of the transposon's own transcription promoter or enhancer sequences. The presence of two or more identical transposons in a genome creates the potential for recombination between the transposons, leading to duplications, deletions, inversions, or chromosome translocations. Any of these rearrangements may bring about phenotypic changes or disease.

New germ-line transpositions are estimated to occur once in every 50 to 100 human births. Most of these do not cause disease or a change in phenotype; however, it is thought that about 0.2 percent of detectable human mutations may be due to transposon insertions. Other organisms appear to suffer more damage due to transposition. About 10 percent of new mouse mutations and 50 percent of *Drosophila* mutations are caused by insertions of transposons in or near genes.

Because of their ability to alter genes and chromosomes, transposons may contribute to the variability that underlies evolution. For example, the Tn elements of bacteria carry antibiotic resistance genes between organisms, conferring a survival advantage to the bacteria under certain conditions. Another example of a transposon's contribution to evolution is provided by *Drosophila* telomeres. LINE-like elements are present at the ends of *Drosophila* chromosomes, and these elements act as telomeres, maintaining the length of *Drosophila* chromosomes over successive cell divisions. Other examples of evolved transposons are the *RAG1* and *RAG2* genes in humans. These genes encode **recombinase** enzymes that are essential to the development of the immune system. These two genes appear to have evolved from transposons.

Transposons may also affect the evolution of genomes by altering gene-expression patterns in ways that are subsequently retained by the host. For example, the human *amylase* gene contains an enhancer that causes the gene to be expressed in the parotid gland. This enhancer evolved from transposon sequences that were inserted into the gene-regulatory region early in primate evolution. Other examples of gene-expression patterns that were affected by the presence of transposon sequences are T-cell-specific expression of the *CD8* gene and placenta-specific expression of the *leptin* and *CYP19* genes.

ESSENTIAL POINT ■ ■ ■

Transposable elements can move within a genome, creating mutations and altering gene expression. Besides creating mutations, transposons may contribute to evolution. Geneticists use transposons as a research tool to create mutations, clone genes, and introduce genes into model organisms.

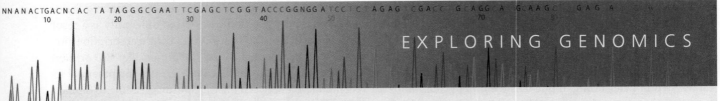

NN AN ACTGACNCAC TA TAGGGCGAA T TCGAGCTCGG TACCCGGNGG ATCCTC AGAG CGACC GCAGGCA GCAAGC GAGA

EXPLORING GENOMICS

Sequence Alignment to Identify a Mutation

MG Study Area: *Exploring Genomics*

In this , we examined the causes of different types of mutations and how mutations affect phenotype by altering the structure and function of proteins. The emergence of genomics, bioinformatics, and proteomics as key disciplines in modern genetics has provided geneticists with an unprecedented set of tools for identifying and analyzing mutations in gene and protein sequences.

In this exercise we will use the **ExPASy (Expert Protein Analysis System)** site, which is hosted by the Swiss Institute for Bioinformatics and provides a wealth of resources for studying proteins. Here we will use an ExPASy program called SIM (for "similarity" in sequence) to compare two polypeptide sequences so as to pinpoint a mutation. Once the mutation has been identified, you will learn more about the gene encoding these polypeptides and about a human disease condition associated with this gene.

■ Exercise I – Identifying a Missense Mutation Affecting a Protein in Humans

1. Begin this exercise by accessing the ExPASy site at http://web.expasy.org/ sim/. The SIM feature is an algorithm-based software program that allows us to compare multiple polypeptide

sequences by looking for amino acid similarity in the sequences.

2. Go to the Companion Web site for *Essentials of Genetics* and open the "Exploring Genomics" exercise for this chapter. At the Web site we provide amino acid sequences for polypeptides expressed in two different people (Person A and Person B). Note that the amino acid sequence is provided using the single letter code for each amino acid (see the accompanying table for amino acid names and single-letter codes).

3. Copy and paste each sequence into separate "SEQUENCE" text boxes in SIM. Use the Person A sequence for sequence 1 and the Person B sequence for sequence 2. Click the "User-entered sequence" button for each. Name the sequences Person A and Person B as appropriate. Submit the sequences for comparison and then answer the following questions:
   a. How many amino acids (called "residues" in the SIM results report) are in each polypeptide sequence that was analyzed?
   b. Look carefully at the alignment results. Can you find any differences in amino acid sequence when comparing these two polypeptides? What did you find?

■ Exercise II – Identifying the Genetic Basis for a Human Genetic Disease Condition

1. From the ExPASy site use the BLAST link (http://web.expasy.org/blast/) and run a protein BLAST (blastp) search to identify which polypeptide you have been studying. Explore the BLAST reports for the top three protein sequences that aligned with your query sequence by clicking on the link for each sequence. Pay particular attention to the "Comment" section of each report to help you answer the questions in the following part.

2. Now that you know what gene you are working with, go to PubMed (http:// www.ncbi.nlm.nih.gov/entrez/query .fcgi?db=PubMed) and search for a review article from the authors Vajo, Z., Francomano, C. A., and Wilkin, D. J.

3. Answer the following questions:
   a. What gene codes for the polypeptides have you been studying?
   b. What is the function of this protein?
   c. Based on what you learned from the alignment results you analyzed in Exercise I, the BLAST reports, and your PubMed search, what human disease is caused by the mutation you identified in Exercise I? Explain your answer and briefly describe phenotypes associated with the disease.

CASE STUDY » Genetic dwarfism

Seven months pregnant, an expectant mother was undergoing a routine ultrasound. While prior tests had been normal, this one showed that the limbs of the fetus were unusually short. The doctor suspected that the baby might have a genetic form of dwarfism called achondroplasia. He told her that the disorder was due to an autosomal dominant mutation and occurred with a frequency of about 1 in 25,000 births. The expectant mother had studied genetics in college and immediately raised several questions. How would you answer them?

1. How could her baby have a dominantly inherited disorder if there was no history of this condition on either side of the family?
2. Is the mutation more likely to have come from the mother or the father?
3. If this child has achondroplasia, would the chances increase that their next child would also have this disorder?
4. Could this disorder have been caused by X rays or ultrasounds she had earlier in pregnancy?

INSIGHTS AND SOLUTIONS

1. The base analog 2-amino purine (2-AP) substitutes for adenine during DNA replication, but it may base-pair with cytosine. The base analog 5-bromouracil (5-BU) substitutes for thymidine, but it may base-pair with guanine. Follow the double-stranded trinucleotide sequence shown here through three rounds of replication, assuming that, in the first round, both analogs are present and become incorporated wherever possible. Before the second and third round of replication, any unincorporated base analogs are removed. What final sequences occur?

Solution:

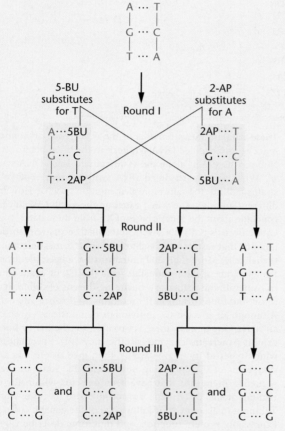

2. A rare dominant mutation expressed at birth was studied in humans. Records showed that six cases were discovered in 40,000 live births. Family histories revealed that in two cases, the mutation was already present in one of the parents. Calculate the spontaneous mutation rate for this mutation. What are some underlying assumptions that may affect our conclusions?

Solution: Only four cases represent a new mutation. Because each live birth represents two gametes, the sample size is from 80,000 meiotic events. The rate is equal to

$$\frac{4}{80,000} = \frac{1}{20,000} = 5 \times 10^{-5}$$

We have assumed that the mutant gene is fully penetrant and is expressed in each individual bearing it. If it is not fully penetrant, our calculation may be an underestimate because one or more mutations may have gone undetected. We have also assumed that the screening was 100 percent accurate. One or more mutant individuals may have been "missed," again leading to an underestimate. Finally, we assumed that the viability of the mutant and nonmutant individuals is equivalent and that they survive equally *in utero*. Therefore, our assumption is that the number of mutant individuals at birth is equal to the number at conception. If this were not true, our calculation would again be an underestimate.

3. Consider the following estimates:

   a. There are 7×10^9 humans living on this planet.

   b. Each individual has about 20,000 (0.2×10^5) genes.

   c. The average mutation rate at each locus is 10^{-5}.

How many spontaneous mutations are currently present in the human population? Assuming that these mutations are equally distributed among all genes, how many new mutations have arisen in each gene in the human population?

Solution: First, since each individual is diploid, there are two copies of each gene per person, each arising from a separate gamete. Therefore, the total number of spontaneous mutations is

$(2 \times 0.2 \times 10^5 \text{ genes/individual})$

     $\times (7 \times 10^9 \text{ individuals}) \times (10^{-5} \text{ mutations/gene})$

   $= (0.4 \times 10^5) \times (7 \times 10^9) \times (10^{-5}) \text{ mutations}$

   $= 2.8 \times 10^9 \text{ mutations in the population}$

$2.8 \times 10^9 \text{ mutations}/0.2 \times 10^5 \text{ genes}$

   $= 14 \times 10^4 \text{ mutations per gene in the population}$

PROBLEMS AND DISCUSSION QUESTIONS

 For activities, animations, and review quizzes, go to the study area at www.masteringgenetics.com

HOW DO WE KNOW?

1. In this chapter, we focused on how gene mutations arise and how cells repair DNA damage. In particular, we discussed spontaneous and induced mutations, DNA repair methods, and transposable elements. Based on your knowledge of these topics, answer several fundamental questions:

   (a) How do we know that mutations occur spontaneously?

   (b) How do we know that certain chemicals and wavelengths of radiation induce mutations in DNA?

   (c) How do we know that DNA repair mechanisms detect and correct the majority of spontaneous and induced mutations?

2. Discuss the importance of mutations in genetic studies.

3. Why would a mutation in a somatic cell of a multicellular organism escape detection?

4. Most mutations are thought to be deleterious. Why, then, is it reasonable to state that mutations are essential to the evolutionary process?

5. Why is a random mutation more likely to be deleterious than beneficial?

6. Most mutations in a diploid organism are recessive. Why?

7. What is meant by a conditional mutation?

8. Describe a tautomeric shift and how it may lead to a mutation.

9. Contrast and compare the mutagenic effects of deaminating agents, alkylating agents, and base analogs.

10. Why are frameshift mutations likely to be more detrimental than point mutations, in which a single pyrimidine or purine has been substituted?

11. Why are X rays more potent mutagens than UV radiation?

12. DNA damage brought on by a variety of natural and artificial agents elicits a wide variety of cellular responses. In addition to the activation of DNA repair mechanisms, there can be activation of pathways leading to apoptosis (programmed cell death) and cell-cycle arrest. Why would apoptosis and cell-cycle arrest often be part of a cellular response to DNA damage?

13. Contrast the various types of DNA repair mechanisms known to counteract the effects of UV radiation. What is the role of visible light in repairing UV-induced mutations?

14. Mammography is an accurate screening technique for the early detection of breast cancer in humans. Because this technique uses X rays diagnostically, it has been highly controversial. Can you explain why? What reasons justify the use of X rays for such a medical screening technique?

15. Describe how the Ames test screens for potential environmental mutagens. Why is it thought that a compound that tests positively in the Ames test may also be carcinogenic?

16. What genetic defects result in the disorder xeroderma pigmentosum (XP) in humans? How do these defects create the phenotypes associated with the disorder?

17. In a bacterial culture in which all cells are unable to synthesize leucine (leu^-), a potent mutagen is added, and the cells are allowed to undergo one round of replication. At that point, samples are taken, a series of dilutions is made, and the cells are plated on either minimal medium or minimal medium containing leucine. The first culture condition (minimal medium) allows the growth of only leu^+ cells, while the second culture condition (minimum medium with leucine added) allows the growth of all cells. The results of the experiment are as follows:

Culture Condition	Dilution	Colonies
Minimal medium	10^{-1}	18
Minimal + leucine	10^{-7}	6

What is the rate of mutation at the locus associated with leucine biosynthesis?

18. Human equivalents of bacterial DNA mismatch repair proteins are subject to mutational damage just as are other proteins. What evidence indicates that mutations in human DNA mismatch repair genes are related to certain forms of cancer?

19. A number of different types of mutations in the *HBB* gene can cause human β-thalassemia, a disease characterized by various levels of anemia. Many of these mutations occur within introns or in upstream noncoding sequences. Explain why mutations in these regions often lead to severe disease, although they may not directly alter the coding regions of the gene.

20. Some mutations that lead to diseases such as Huntington disease are caused by the insertion of trinucleotide repeats. Describe how the process of DNA replication could lead to expansions of trinucleotide repeat regions.

21. In maize, a *Ds* or *Ac* transposon can cause mutations in genes at or near the site of transposon insertion. It is possible for these elements to transpose away from their original site, causing a reversion of the mutant phenotype. In some cases, however, even more severe phenotypes appear, due to events at or near the mutant allele. What might be happening to the transposon or the nearby gene to create more severe mutations?

22. Presented here are hypothetical findings from studies of heterokaryons formed from seven human xeroderma pigmentosum cell strains:

	XP1	XP2	XP3	XP4	XP5	XP6	XP7
XP1	−						
XP2	−	−					
XP3	−	−	−				
XP4	+	+	+	−			
XP5	+	+	+	+	−		
XP6	+	+	+	+	−	−	
XP7	+	+	+	+	−	−	−

Note: " + " = complementation; " − " = no complementation

These data are measurements of the occurrence or nonoccurrence of unscheduled DNA synthesis in the fused heterokaryon. None of the strains alone shows any unscheduled DNA synthesis. What does unscheduled DNA synthesis represent? Which strains fall into the same complementation groups? How many different groups are revealed based on these data? What can we conclude about the genetic basis of XP from these data?

23. Cystic fibrosis (CF) is a severe autosomal recessive disorder in humans that results from a chloride ion–channel defect in epithelial cells. More than 500 mutations have been identified in the 24 exons of the responsible gene (*CFTR*, or cystic fibrosis transmembrane regulator), including dozens of different missense mutations, frameshift mutations, and splice-site defects. Although all affected CF individuals demonstrate chronic obstructive lung disease, there is variation in whether or not they exhibit pancreatic enzyme insufficiency (PI). Speculate as to which types of mutations are likely to give rise to less severe symptoms of CF, including only minor PI. Some of the 300 sequence alterations that have been detected within the exon regions of the *CFTR* gene do not give rise to cystic fibrosis. Taking into account your knowledge of the genetic code, gene expression, protein function, and mutation, describe why this might be so.

24. Electrophilic oxidants are known to create the modified base named 7,8-dihydro-8-oxoguanine (oxoG) in DNA. Whereas guanine base-pairs with cytosine, oxoG base-pairs with either cytosine or adenine.

    (a) What are the sources of reactive oxidants within cells that cause this type of base alteration?

    (b) Drawing on your knowledge of nucleotide chemistry, draw the structure of oxoG, and, below it, draw guanine. Opposite guanine, draw cytosine, including the hydrogen bonds that allow these two molecules to base-pair. Does the structure of oxoG, in contrast to guanine, provide any hint as to why it base-pairs with adenine?

    (c) Assume that an unrepaired oxoG lesion is present in the helix of DNA opposite cytosine. Predict the type of mutation that will occur following several rounds of replication.

    (d) Which DNA repair mechanisms might work to counteract an oxoG lesion? Which of these is likely to be most effective?

25. Skin cancer carries a lifetime risk nearly equal to that of all other cancers combined. Following is a graph (modified from Kraemer, 1997. *Proc. Natl. Acad. Sci. (USA)* 94: 11–14) depicting the age of onset of skin cancers in patients with or without XP, where cumulative percentage of skin cancer is plotted against age. The non-XP curve is based on 29,757 cancers surveyed by the National Cancer Institute, and the curve representing those with XP is based on 63 skin cancers from the Xeroderma Pigmentosum Registry.

   (a) Provide an overview of the information contained in the graph.

   (b) Explain why individuals with XP show such an early age of onset.

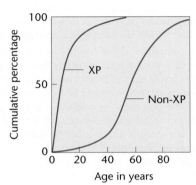

26. The initial discovery of IS elements in bacteria revealed the presence of an element upstream (5′) of three genes controlling galactose metabolism. All three genes were affected simultaneously, although there was only one IS insertion. Offer an explanation as to why this might occur.

27. It is estimated that about 0.2 percent of human mutations are due to transposon insertions and a much higher percentage of mutational damage is known to occur in some other organisms. In what way might transposon insertion contribute positively to evolution?

28. It has been noted that most transposons in humans and other organisms are located in noncoding regions of the genome—regions such as introns, pseudogenes, and stretches of particular types of repetitive DNA. There are several ways to interpret this observation. Describe two possible interpretations. Which interpretation do you favor? Why?

29. Two related forms of muscular dystrophy—Duchenne muscular dystrophy (DMD) and Becker muscular dystrophy (BMD)—are both recessive, X-linked, single-gene conditions caused by point mutations, deletions, and insertion in the *dystrophin* gene. Each mutated form of *dystrophin* is one allele. Of the two diseases, DMD is much more severe. Given your knowledge of mutations, the genetic code, and translation, propose an explanation for why the two disorders differ greatly in severity.

SOLUTIONS TO SELECTED PROBLEMS AND DISCUSSION QUESTIONS

Answers to Now Solve This

1. The phenotypic influence of any base change is dependent on a number of factors including, its location in coding or noncoding regions, its potential in dominance or recessiveness, and its interaction with other base sequences in the genome. If a base change is located in a non-coding region, there may be no influence on the phenotype, however, some non-coding regions in a traditional sense, may influence other genes and/or gene products. If a mutation occurs in a coding region acts as a full recessive, there should be no influence on the phenotype. If a mutant gene acts as a dominant, then there would be an influence on the phenotype. Some genes interact with other genes in a variety of ways which would be difficult to predict without additional information.

2. There are several ways that an unexpected mutant gene may enter a pedigree. If a gene is incompletely penetrant, it may be present in a population and only express itself under certain conditions. It is unlikely that the gene for hemophilia behaved in this manner. If a gene's expression is suppressed by another mutation in an individual, it is possible that offspring may inherit a given gene and not inherit its suppressor. Such offspring would have hemophila. Since all genetic variations must arise at some point, it is possible that the mutation in the queen Victoria family was new, arising in a cell of the father. Lastly, it is possible that the mother was heterozygous and by chance, no other individuals in her family were unlucky enough to receive the mutant gene.

3. Any agent that inhibits DNA replication, either directly or indirectly, through mutation and/or DNA crosslinking, will suppress the cell cycle and may be useful in cancer therapy. Since guanine alkylation often leads to mismatched bases, they can often be repaired by a variety of mismatched repair mechanisms. However, DNA crosslinking can be repaired by recombinational mechanisms; thus, for such agents to be successful in cancer

therapy, suppressors of DNA repair systems are often used in conjunction with certain cancer drugs. See: Wang, Z. et al. 2001. *J Nat'l Cancer Inst.* 93(19):1434-6.

4. Ethylmethane sulfonate (EMS) alkylates the keto groups at the 6th position of guanine and at the 4th position of thymine. In each case, base-pairing affinities are altered and transition mutations result. Altered bases are not readily repaired and once the transition to normal bases occurs through replication, such mutations avoid repair altogether.

Solutions to Problems and Discussion Questions

2. Mutations are the "windows" through which geneticists look at the normal function of genes, cells, and organisms. When a mutation occurs, it allows the investigator to formulate questions as to the function of the normal allele of that mutation. At a different level, mutations provide "markers" with which biologists can study the genetics and dynamics of populations.

4. It is true that *most* mutations are thought to be deleterious to an organism. A gene is a product of perhaps a billion or so years of evolution, and it is only natural to suspect that random changes will probably yield negative results. However, *all* mutations may not be deleterious. Those few, rare variations that are beneficial will provide a basis for possible differential propagation of the variation.

6. A diploid organism possesses at least two copies of each gene (except for "hemizygous" genes), and in most cases, the amount of product from one gene of each pair is sufficient for production of a normal phenotype.

8. Tautomeric shifts can result in mutations by allowing hydrogen bonding of normally noncomplementary bases so that incorrect nucleotide bases may be added during DNA replication.

10. Frameshift mutations are likely to change more than one amino acid in a protein product because as the reading frame is shifted, a different set of codons is generated. In addition, there is the possibility that a nonsense triplet could be introduced, thus causing

premature chain termination. If a single pyrimidine or purine has been substituted, then only one amino acid is influenced.

12. When DNA is damaged, mutations are likely. In many cases, such mutations are deleterious to the health of the organism. Several mechanisms have evolved to reduce the impact of such mutations; cell-cycle arrest to quarantine a cell line or allow DNA repair and programmed cell death (apoptosis). If damaged DNA cannot be repaired through cell-cycle arrest, programmed cell death is often activated to rid the cell population of mutant cell lines.

14. Because mammography involves the use of X rays and X rays are known to be mutagenic, it has been suggested that frequent mammograms may do harm. This subject is presently under considerable debate. At the 2002 World Health Organization conference in Barcelona, Spain, the conclusion was that "mammograms can prevent breast cancer deaths in one in 500 women ages 50 to 69."

16. *Xeroderma pigmentosum* is a form of human skin cancer caused by perhaps several rare autosomal genes, which interfere with the repair of damaged DNA. Since cancer is caused by mutations in several types of genes, interfering with DNA repair can enhance the occurrence of these types of mutations.

18. Mismatch repair defects are common in hereditary nonpolyposis colon cancer, leukemias, lymphomas, and tumors of the ovary, prostate, and endometrium. The link between mutant mismatch repair systems and cancer is seen in mice engineered to have defects in mismatch repair genes. Such mice are cancer-prone.

20. Replication slippage is a process that generates small deletions and insertions during DNA replication. While it can occur anywhere in the genome, it is most prevalent in regions already containing repeated sequences. Thus, repeated sequences are hypermutable.

22. Unscheduled DNA synthesis represents DNA repair. Complementation groups:

XP1	XP4	XP5
XP2		XP6
XP3		XP7

The groupings (complementation groups) indicate that there are at least three "genes" that form products necessary for unscheduled DNA synthesis. All of the cell lines that are in the same complementation group are defective in the same product.

26. It is probable that the IS occupied or interrupted normal function of a controlling region related to the *galactose* genes, which are in an operon with one controlling upstream element.

28. First, while less likely, one might suggest that transposons, for one reason or another, are more likely to insert in noncoding regions of the genome. One might also suggest that they are more stable in such regions. Second, and more likely, it is possible that transposons insert rather randomly and that selection eliminates those that have interrupted coding regions of the genome. Since such regions are more likely to influence the phenotype, selection is more likely to influence such regions.

CREDITS

Credits are listed in order of appearance.

Photo

CO, M.G. Neuffer, F-14, HADJ/SIPA/Newscom

Text

Source: ExPASy

Regulation of Gene Expression

From Chapter 15 of *Essentials of Genetics, Eighth edition,* William S. Klug, Michael R. Cummings, Charlotte A. Spencer, Michael A. Palladino. © 2013 by Pearson Education, Inc. All rights reserved.

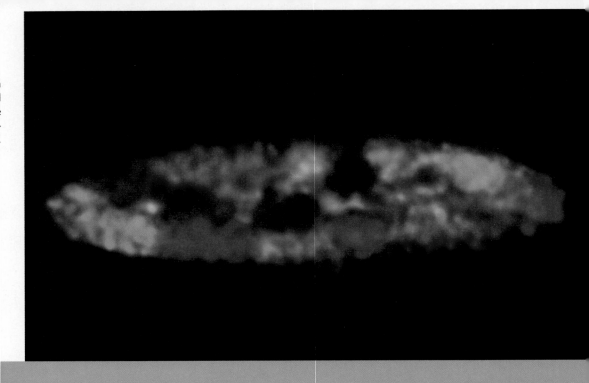

Regulation of Gene Expression

CHAPTER CONCEPTS

- Expression of genetic information is regulated by intricate regulatory mechanisms that control transcription, mRNA stability, translation, and posttranslational modifications.

- In prokaryotes, genes that encode proteins with related functions tend to be organized in clusters and are often under coordinated control. Such clusters, including their associated regulatory sequences, are called operons.

- Transcription within operons is either inducible or repressible and is often regulated by the metabolic substrate or end product of the pathway.

- Eukaryotic gene regulation is more complex than prokaryotic gene regulation.

- The organization of eukaryotic chromatin in the nucleus plays a role in regulating gene expression. Chromatin must be remodeled to provide access to regulatory DNA sequences within it.

- Eukaryotic transcription initiation requires the presence of transcription regulators at enhancer sites and general transcription complexes at promoter sites.

- Eukaryotic gene expression is also regulated at posttranscriptional steps, including splicing of pre-mRNA, mRNA stability, translation, and posttranslational processing.

We now consider one of the most fundamental questions in molecular genetics: *How is gene expression regulated?*

It is clear that not all genes are expressed at all times in all situations. For example, some proteins in the bacterium *E. coli* are present in as few as 5 to 10 molecules per cell, whereas others, such as ribosomal proteins and the many proteins involved in the glycolytic pathway, are present in as many as 100,000 copies per cell. Although many prokaryotic gene products are present continuously at low levels, the level of these products can increase dramatically when required. In multicellular eukaryotes, differential gene expression is also essential and is at the heart of embryonic development and maintenance of the adult state.

The activation and repression of gene expression is a delicate balancing act for an organism; expression of a gene at the wrong time, in the wrong cell type, or in abnormal amounts can lead to a deleterious phenotype, cancer, or cell death—even when the gene itself is normal.

In this chapter, we will explore the ways in which prokaryotic and eukaryotic organisms regulate gene expression. We will describe some of the fundamental components of gene regulation, including the *cis*-acting DNA elements and *trans*-acting factors that regulate transcription initiation. We will then explain how these components interact with each other and with other factors such as activators, repressors, and chromatin proteins. We will also consider the roles that posttranscriptional mechanisms play in the regulation of eukaryotic gene expression.

1 Prokaryotes Regulate Gene Expression in Response to Both External and Internal Conditions

Not only do bacteria respond metabolically to changes in their environment, but they also regulate gene expression in order to synthesize products required for a variety of normal cellular activities, including DNA replication, recombination, repair, and cell division. In the following sections, we will focus on prokaryotic gene regulation at the level of transcription, which is the predominant level of regulation in prokaryotes. Keep in mind, however, that posttranscriptional regulation also occurs in bacteria. We will defer discussion of posttranscriptional gene-regulatory mechanisms to subsequent sections dealing with eukaryotic gene expression.

The idea that microorganisms regulate the synthesis of gene products is not a new one. As early as 1900, it was shown that when lactose (a galactose and glucose-containing disaccharide) is present in the growth medium of yeast, the organisms synthesize enzymes required for lactose metabolism. When lactose is absent, the enzymes are not manufactured. Soon thereafter, investigators were able to generalize that bacteria also adapt to their environment, producing certain enzymes only when specific chemical substrates are present. Such enzymes are referred to as **inducible,** reflecting the role of the substrate, which serves as the **inducer** in enzyme production. In contrast, enzymes that are produced continuously, regardless of the chemical makeup of the environment, are called **constitutive.**

More recent investigation has revealed a contrasting system whereby the presence of a specific molecule inhibits gene expression. This is usually true for molecules that are end products of anabolic biosynthetic pathways. For example, the amino acid tryptophan can be synthesized by bacterial cells. If a sufficient supply of tryptophan is present in the environment or culture medium, it is energetically inefficient for the organism to synthesize the enzymes necessary for tryptophan production. A mechanism has evolved whereby tryptophan plays a role in repressing transcription of genes that encode the appropriate biosynthetic enzymes. In contrast to the inducible system controlling lactose metabolism, the system governing tryptophan expression is said to be **repressible.**

Regulation, whether it is inducible or repressible, may be under either **negative** or **positive control.** Under negative control, gene expression occurs *unless it is shut off by some form of a regulator molecule.* In contrast, under positive control, transcription occurs *only if a regulator molecule directly stimulates RNA production.* In theory, either type of control or a combination of the two can govern inducible or repressible systems.

2 Lactose Metabolism in *E. coli* Is Regulated by an Inducible System

Beginning in 1946, the studies of Jacques Monod (with later contributions by Joshua Lederberg, François Jacob, and André Lwoff) revealed genetic and biochemical insights into the mechanisms of lactose metabolism in bacteria. These studies explained how gene expression is repressed when lactose is absent, but induced when it is available. In the presence of lactose, concentrations of the enzymes responsible for lactose metabolism increase rapidly from a few molecules to thousands per cell. The enzymes responsible for lactose metabolism are thus *inducible,* and lactose serves as the *inducer.*

In prokaryotes, genes that code for enzymes with related functions (in this case, the genes involved with lactose metabolism) tend to be organized in clusters on the bacterial chromosome. In addition, transcription of these genes is often under the coordinated control of a single transcription regulatory region. The location of this regulatory region is almost always on the same DNA molecule and upstream of the gene cluster it controls. We refer to this type of regulatory region as a *cis*-acting site. *Cis*-acting regulatory regions bind molecules that control transcription of the gene cluster. Such molecules are called *trans*-acting molecules. Actions at the *cis*-acting regulatory site determine whether the genes are transcribed into RNA and thus whether the corresponding enzymes or other protein products are synthesized from the mRNA. Binding of a *trans*-acting molecule at a *cis*-acting site can regulate the gene cluster either negatively (by turning off transcription) or positively (by turning on transcription of genes in the cluster). In this section, we discuss how transcription of such bacterial gene clusters is coordinately regulated.

ESSENTIAL POINT ■ ■ ■

Research on the *lac* operon in *E. coli* pioneered our understanding of gene regulation in bacteria.

Structural Genes

As illustrated in **Figure 1**, three genes and an adjacent regulatory region constitute the **lactose,** or *lac,* **operon.** Together, the entire gene cluster functions in an integrated fashion to

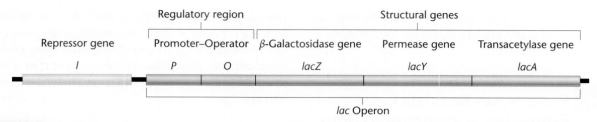

FIGURE 1 A simplified overview of the genes and regulatory units involved in the control of lactose metabolism. (This region of DNA is not drawn to scale.) A more detailed model is described later in this chapter.

provide a rapid response to the presence or absence of lactose.

Genes coding for the primary structure of the enzymes are called **structural genes.** There are three structural genes in the *lac* operon. The *lacZ* gene encodes **β-galactosidase,** an enzyme whose role is to convert the disaccharide lactose to the monosaccharides glucose and galactose **(Figure 2).** This conversion is essential if lactose is to serve as an energy source in glycolysis. The second gene, *lacY,* encodes the amino acid sequence of **permease,** an enzyme that facilitates the entry of lactose into the bacterial cell. The third gene, *lacA,* codes for the enzyme **transacetylase.** Although its physiological role is still not completely clear, it may be involved in the removal of toxic by-products of lactose digestion from the cell.

To study the genes encoding these three enzymes, researchers isolated numerous mutants, each of which eliminated the function of one of the enzymes. These mutants were first isolated and studied by Joshua Lederberg. Mutant cells that fail to produce active β-galactosidase (*lacZ⁻*) or

permease (*lacY⁻*) are unable to use lactose as an energy source and are collectively known as *lac⁻* mutants. Mapping studies by Lederberg established that all three genes are closely linked or contiguous to one another on the bacterial chromosome, in the order *Z–Y–A.* (See Figure 1.) All three genes are transcribed as a single unit, resulting in a polycistronic mRNA **(Figure 3).** This results in the coordinated regulation of all three genes, since a single messenger RNA is simultaneously translated into all three gene products.

The Discovery of Regulatory Mutations

How does lactose stimulate transcription of the *lac* operon and induce the synthesis of the related enzymes? A partial answer comes from studies using **gratuitous inducers,** chemical analogs of lactose such as the sulfur analog **isopropylthiogalactoside (IPTG),** shown in **Figure 4.** Gratuitous inducers behave like natural inducers, but they do not serve as substrates for the enzymes that are subsequently synthesized.

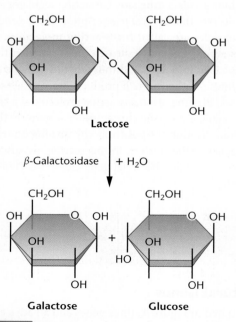

FIGURE 2 The catabolic conversion of the disaccharide lactose into its monosaccharide units, galactose and glucose.

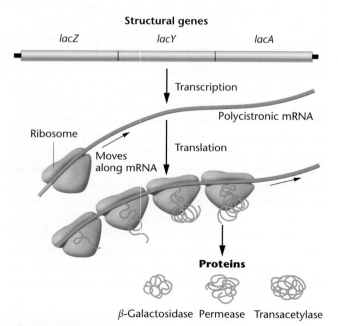

FIGURE 3 The structural genes of the *lac* operon are transcribed into a single polycistronic mRNA, which is translated simultaneously by several ribosomes into the three enzymes encoded by the operon.

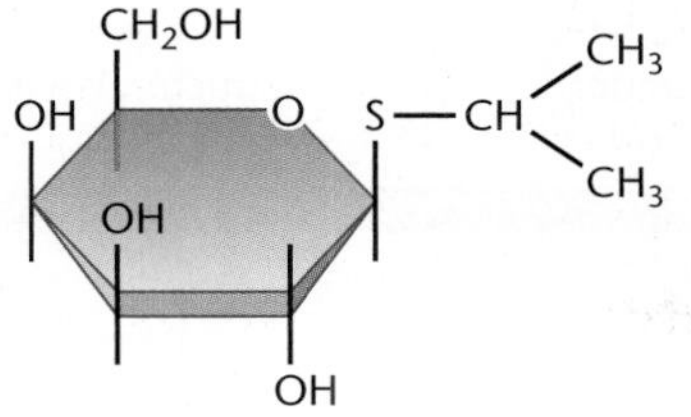

FIGURE 4 The gratuitous inducer isopropylthiogalactoside (IPTG).

What, then, is the role of lactose in gene regulation? The answer to this question required the study of a class of mutants called **constitutive mutants.** In cells bearing constitutive mutations, enzymes are produced regardless of the presence or absence of lactose. Studies of the constitutive mutation *lacI⁻* mapped the mutation to a site on the bacterial chromosome close to, but distinct from, the *lacZ, lacY,* and *lacA* genes. This mutation defined the *lacI* gene, which is appropriately called a **repressor gene.** Another set of constitutive mutations that produce identical effects to those of *lacI⁻* occur in a region immediately adjacent to the structural genes. This class of mutations, designated *lacO^C*, occur in the **operator region** of the operon. In both types of constitutive mutants, the enzymes are produced continuously, inducibility is eliminated, and gene regulation is lost.

The Operon Model: Negative Control

Around 1960, Jacob and Monod proposed a scheme involving negative control called the **operon model,** whereby a group of genes is regulated and expressed together as a unit. As we saw in Figure 1, the *lac* operon consists of the *Z, Y,* and *A* structural genes, as well as the adjacent sequences of DNA referred to as the *operator region.* They argued that the *lacI* gene regulates the transcription of the structural genes by producing a **repressor molecule,** and that the repressor is **allosteric,** meaning that it reversibly interacts with another molecule, causing both a conformational change in the repressor's three-dimensional shape and a change in its chemical activity. **Figure 5** illustrates the components of the *lac* operon as well as the action of the *lac* repressor in the presence and absence of lactose.

Jacob and Monod suggested that the repressor normally binds to the DNA sequence of the operator region. When it does so, it inhibits the action of RNA polymerase, effectively repressing the transcription of the structural genes [**Figure 5(b)**]. However, when lactose is present, the sugar binds to the repressor molecule and causes an allosteric conformational change. This change renders the repressor incapable of interacting with operator DNA [**Figure 5(c)**]. In the absence of the repressor–operator interaction, RNA polymerase transcribes the structural genes, and the enzymes necessary for lactose metabolism are produced. Because transcription oc-

curs only when the repressor *fails* to bind to the operator region, regulation is said to be under *negative control.*

To summarize, the operon model invokes a series of molecular interactions between proteins, inducers, and DNA to explain the efficient regulation of structural gene expression. In the absence of lactose, the enzymes encoded by the genes are not needed, and expression of genes encoding these enzymes is repressed. When lactose is present, it indirectly induces the transcription of the structural genes by interacting with the repressor.[*] If all lactose is metabolized, none is available to bind to the repressor, which is again free to bind to operator DNA and repress transcription.

Both the *I⁻* and *O^C* constitutive mutations interfere with these molecular interactions, allowing continuous transcription of the structural genes. In the case of the *I⁻* mutant, seen in **Figure 6(a)**, the repressor protein is altered or absent and cannot bind to the operator region, so the structural genes are always transcribed. In the case of the *O^C* mutant [**Figure 6(b)**], the nucleotide sequence of the operator DNA is altered and will not bind with a normal repressor molecule. The result is the same: the structural genes are always transcribed.

Genetic Proof of the Operon Model

The operon model leads to three major predictions that can be tested to determine its validity. The major predictions to be tested are that (1) the *I* gene produces a diffusible product; (2) the *O* region is involved in regulation but does not produce a product; and (3) the *O* region must be adjacent to the structural genes in order to regulate transcription.

The construction of partially diploid bacteria allows us to assess these assumptions, particularly those that predict *trans*-acting regulatory molecules. For example, the F plasmid may contain chromosomal genes, in which case it is designated F′. When an F⁻ cell acquires such a plasmid, it contains its own chromosome plus one or more additional genes present in the plasmid. This creates a host cell, called a **merozygote,** that is diploid for those genes. The use of such a plasmid makes it possible, for example, to introduce an *I⁺* gene into a host cell whose genotype is *I⁻* or to introduce an *O⁺* region into a host cell of genotype *O^C*. The Jacob–Monod operon model predicts how regulation should be affected in such cells. Adding an *I⁺* gene to an *I⁻* cell should restore inducibility because the normal wild-type repressor, which is a *trans*-acting factor, would be produced by the inserted *I⁺* gene. Adding an *O⁺* region to an *O^C* cell should have no effect on constitutive enzyme production, since regulation depends on the presence of an *O⁺* region immediately adjacent to the structural genes—that is, *O⁺* is a *cis*-acting regulator.

---

[*]Technically, allolactose, an isomer of lactose, is the inducer. When lactose enters the bacterial cell, some of it is converted to allolactose by the β-galactosidase enzyme.

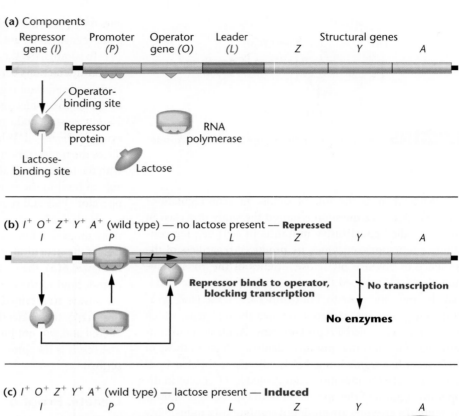

FIGURE 5 The components of the wild-type *lac* operon (a) and the response of the *lac* operon to the absence (b) and presence (c) of lactose. The Leader (L) sequence encodes a short region of mRNA that is 5′ of the AUG translation start codon and is not translated.

(a) Components

Repressor gene (*I*) · Promoter (*P*) · Operator gene (*O*) · Leader (*L*) · Structural genes *Z Y A*

Operator-binding site · Repressor protein · Lactose-binding site · RNA polymerase · Lactose

(b) $I^+ O^+ Z^+ Y^+ A^+$ (wild type) — no lactose present — **Repressed**

I P O L Z Y A

Repressor binds to operator, blocking transcription · **No transcription** · **No enzymes**

(c) $I^+ O^+ Z^+ Y^+ A^+$ (wild type) — lactose present — **Induced**

I P O L Z Y A

No binding occurs; transcription proceeds · Operator-binding region is altered when bound to lactose · **Transcription** · mRNA · **Translation** · **Enzymes**

The results of these experiments are shown in **Table 1**, where *Z* represents the structural genes. The inserted genes are listed after the designation F′. In both cases described here, the Jacob–Monod model is upheld (part B of Table 1). Part C shows the reverse experiments, where either an I^- gene or an O^C region is added to cells of normal inducible genotypes. As the model predicts, inducibility is maintained in these partial diploids.

Another prediction of the operon model is that certain mutations in the *I* gene should have the opposite effect of I^-. That is, instead of being constitutive because the repressor can't bind the operator, mutant repressor molecules should be produced that cannot interact with the inducer, lactose. As a result, the repressor would always bind to the operator sequence, and the structural genes would be permanently repressed **(Figure 7)**. If this were the case, the presence of an additional I^+ gene would have little or no effect on repression.

In fact, such a mutation, I^S, was discovered wherein the operon is "super-repressed," as shown in part D of Table 1. An additional I^+ gene does not effectively relieve repression of gene activity. These observations are consistent with the idea that the repressor contains separate DNA-binding domains and inducer-binding domains. The binding of lactose to the inducer-binding domain causes an allosteric change in the DNA-binding domain.

ESSENTIAL POINT ■ ■ ■

Genes involved in the metabolism of lactose are coordinately regulated by a negative control system that responds to the presence or absence of lactose.

FIGURE 6 The response of the *lac* operon in the absence of lactose when a cell bears either the *I*⁻ (a) or the *O*^C (b) mutation.

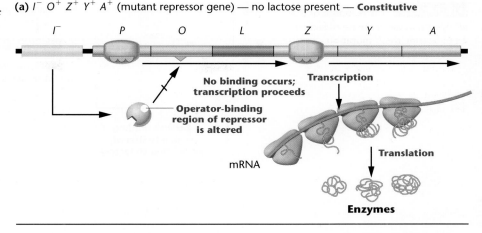

(a) *I*⁻ *O*⁺ *Z*⁺ *Y*⁺ *A*⁺ (mutant repressor gene) — no lactose present — **Constitutive**

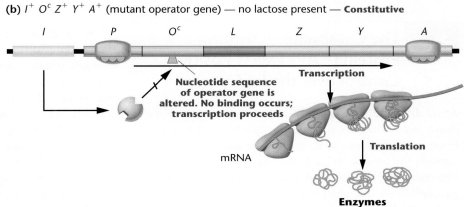

(b) *I*⁺ *O*^c *Z*⁺ *Y*⁺ *A*⁺ (mutant operator gene) — no lactose present — **Constitutive**

TABLE 1 A Comparison of Gene Activity (+ or −) in the Presence or Absence of Lactose for Various *E. coli* Genotypes

Genotype	Presence of β-Galactosidase Activity	
	Lactose Present	Lactose Absent
$I^+O^+Z^+$	+	−
A. $I^+O^+Z^-$	−	−
$I^-O^+Z^+$	+	+
$I^+O^cZ^+$	+	+
B. $I^-O^+Z^+/F'I^+$	+	−
$I^+O^cZ^+/F'O^+$	+	+
C. $I^+O^+Z^+/F'I^-$	+	−
$I^+O^+Z^+/F'O^c$	+	−
D. $I^SO^+Z^+$	−	−
$I^SO^+Z^+/F'I^+$	−	−

Note: In parts B to D, most genotypes are partially diploid, containing an F factor plus attached genes (F′).

Isolation of the Repressor

Although Jacob and Monod's operon theory succeeded in explaining many aspects of genetic regulation in prokaryotes, the nature of the repressor molecule was not known when their landmark paper was published in 1961. While they had assumed that the allosteric repressor was a protein, RNA was also a candidate because activity of the molecule required the ability to bind to DNA. A single *E. coli* cell contains no more than ten or so molecules of the *lac* repressor; therefore, direct chemical identification of ten molecules in a population of millions of proteins and RNAs in a single cell presented a tremendous challenge. Nevertheless, in 1966, Walter Gilbert and Benno Müller-Hill reported the isolation of the *lac* repressor. Once the repressor was purified, it was shown to have various characteristics of a protein. The isolation of the repressor thus confirmed the operon model, which had been put forward strictly on genetic grounds.

NOW SOLVE THIS

1 The *lac Z, Y,* and *A* structural genes are transcribed as a single polycistronic mRNA; however, each structural gene contains its own initiation and termination signals essential for translation. Predict what will happen when cells growing in the presence of lactose contain a deletion of one nucleotide (a) early in the *Z* gene and (b) early in the *A* gene.

■ **Hint** *This problem requires you to combine your understanding of the genetic expression of the lac operon, the genetic code, frameshift mutations, and termination of transcription. The key to its solution is to consider the effect of the loss of one nucleotide within a polycistronic mRNA.*

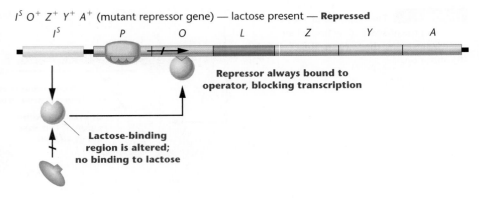

FIGURE 7 The response of the *lac* operon in the presence of lactose in a cell bearing the I^S mutation.

$I^S\ O^+\ Z^+\ Y^+\ A^+$ (mutant repressor gene) — lactose present — **Repressed**

Repressor always bound to operator, blocking transcription

Lactose-binding region is altered; no binding to lactose

3   The Catabolite-Activating Protein (CAP) Exerts Positive Control over the *lac* Operon

As we discussed previously, the role of β-galactosidase is to cleave lactose into its components, glucose and galactose. However, for galactose to be used by the cell, it also must be converted to glucose. What if the cell found itself in an environment that contained an ample amount of lactose *and* glucose? Given that glucose is the preferred carbon source for *E. coli,* it would not be energetically efficient for a cell to induce transcription of the *lac* operon, make β-galactosidase, and metabolize lactose, since what it really needs—glucose—is already present. As we shall see next, a molecule called the **catabolite-activating protein (CAP)** helps activate expression of the *lac* operon, but is able to inhibit expression when glucose is present. This inhibition is called **catabolite repression.**

To understand CAP and its role in regulation, let's backtrack for a moment. When the *lac* repressor is bound to the inducer, RNA polymerase transcribes the *lac* operon structural genes. Transcriptionis initiated as a result of the binding that occurs between RNA polymerase and the nucleotide sequence of the promoter region, found upstream (5′) from the initial coding sequences. Within the *lac* operon, the promoter is found between the *I* gene and the operator region (*O*). (See Figure 1.) Careful examination has revealed that RNA polymerase binding is never very efficient unless CAP is also present to facilitate the process.

The mechanism is summarized in **Figure 8.** In the absence of glucose and under inducible conditions, CAP exerts positive control by binding to the CAP site, facilitating RNA polymerase binding at the *lac* operon promoter and thus transcription. Therefore, for maximal transcription of the structural genes, the repressor must be bound by lactose (so as not to repress *lac* operon transcription), *and* CAP must be bound to the CAP-binding site.

What role does glucose play in inhibiting CAP binding? The answer involves still another molecule, **cyclic adenosine monophosphate (cAMP),** upon which CAP binding is dependent. In order to bind to the *lac* operon promoter, CAP must be bound to cAMP. The level of cAMP is itself dependent on an enzyme, **adenyl cyclase,** which catalyzes the conversion of ATP to cAMP.

The role of glucose in catabolite repression is to inhibit the activity of adenyl cyclase, causing a decline in the level of cAMP in the cell. Under this condition, CAP cannot form the cAMP–CAP complex that is essential to the positive control of transcription of the *lac* operon.

The structures of CAP and cAMP–CAP have been examined by using X-ray crystallography. CAP is a dimer that binds adjacent regions of a specific nucleotide sequence of the DNA making up the *lac* promoter. The cAMP–CAP complex, when bound to DNA, bends the DNA, causing it to assume a new conformation.

Binding studies in solution further clarify the mechanism of gene activation. Alone, neither cAMP–CAP nor RNA polymerase has a strong affinity to bind to *lac* promoter DNA, nor does either molecule have a strong affinity to bind to the other. However, when both are together in the presence of the *lac* promoter DNA, a tightly bound complex is formed, an example of what is called **cooperative binding.** The control conferred by the cAMP-CAP provides another illustration of how the regulation of one small group of genes can be fine-tuned by several simultaneous influences.

In contrast to the negative regulation conferred by the *lac* repressor, the action of cAMP–CAP constitutes positive regulation. Thus, a combination of positive and negative regulatory mechanisms determines transcription levels of the *lac* operon. Catabolite repression involving CAP has also been observed for other inducible operons, including those controlling the metabolism of galactose and arabinose.

(a) Glucose absent

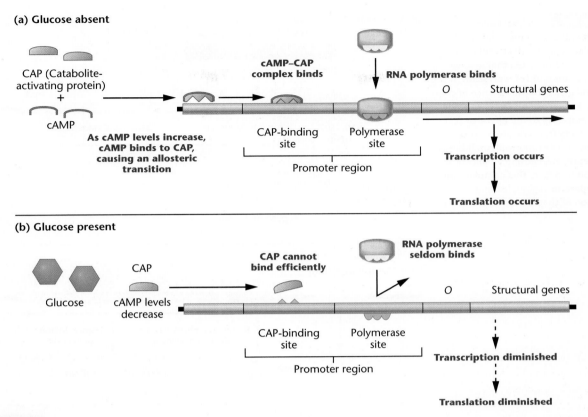

(b) Glucose present

FIGURE 8 Catabolite repression. (a) In the absence of glucose, cAMP levels increase, resulting in the formation of a cAMP–CAP complex, which binds to the CAP site of the promoter, stimulating transcription. (b) In the presence of glucose, cAMP levels decrease, cAMP–CAP complexes are not formed, and transcription is not stimulated.

NOW SOLVE THIS

2 Predict the level of gene expression of the *lac* operon, as well as the status of the *lac* repressor and the CAP protein, when bacterial growth media contain the following sugars: (a) no lactose or glucose, (b) lactose but no glucose, (c) glucose but no lactose, (d) both lactose and glucose.

■ **Hint** *This problem asks you to combine your knowledge of* lac *operon regulation with your understanding of how catabolite repression affects this regulation. The key to its solution is to keep in mind that regulation involving lactose is a negative control system, while regulation involving glucose and catabolite repression is a positive control system.*

ESSENTIAL POINT ■ ■ ■

The catabolite-activating protein (CAP) exerts positive control over *lac* gene expression by interacting with RNA polymerase at the *lac* promoter and by responding to the levels of cyclic AMP in the bacterial cell.

4 The Tryptophan (*trp*) Operon in *E. coli* Is a Repressible Gene System

Although inducible gene regulation had been known for some time, it was not until 1953 that Monod and colleagues discovered a repressible system. Studies on the biosynthesis of the essential amino acid tryptophan revealed that, if tryptophan is present in sufficient quantity in the growth medium, the enzymes necessary for its synthesis (such as **tryptophan synthase**) are not produced. It is energetically advantageous for bacteria to repress expression of genes involved in tryptophan synthesis when ample tryptophan is present in the growth medium.

Further investigation showed that enzymes encoded by five contiguous genes on the *E. coli* chromosome are involved in tryptophan synthesis. These genes are part of an operon and, in the presence of tryptophan, all are coordinately repressed, and none of the enzymes is produced. Because of the great similarity between this repression and the induction of enzymes for lactose metabolism, Jacob and

FIGURE 9 (a) The components involved in the regulation of the tryptophan operon. (b) Regulatory conditions are depicted that involve either activation or (c) repression of the structural genes. In the absence of tryptophan, an inactive repressor is made that cannot bind to the operator (*O*), thus allowing transcription to proceed. In the presence of tryptophan, it binds to the repressor, causing an allosteric transition to occur. This complex binds to the operator region, leading to repression of the operon.

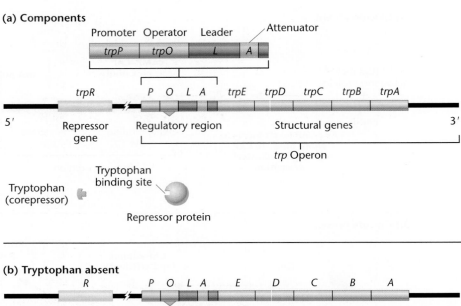

(a) Components

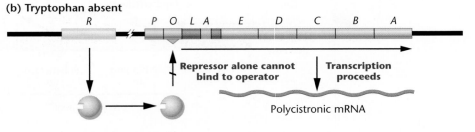

(b) Tryptophan absent

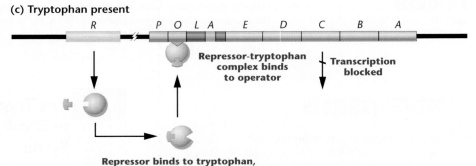

(c) Tryptophan present

Monod proposed a model of gene regulation resembling that of the *lac* system (**Figure 9**).

The model suggests the presence of a *normally inactive repressor* that alone cannot interact with the operator region of the operon. However, the repressor is an allosteric molecule that can bind to tryptophan. When tryptophan is present, the resultant complex of repressor and tryptophan attains a new conformation that binds to the operator, repressing transcription. Thus, when tryptophan, the end product of this anabolic pathway, is present, the operon is repressed and enzymes are not made. Since the regulatory complex inhibits transcription of the operon, this repressible system is under negative control. And as tryptophan participates in repression, it is referred to as a **corepressor** in this regulatory scheme.

Evidence for the *trp* Operon

Support for the concept of a repressible operon is based primarily on the isolation of two distinct categories of constitutive mutations. The first class, *trpR⁻*, maps at a considerable distance from the structural genes. This locus represents the gene coding for the repressor. Presumably, the mutation either inhibits the interaction of the repressor with tryptophan or inhibits repressor formation entirely. Whichever the case, no repression is present in cells with the *trpR⁻* mutation. As expected, if the *trpR⁺* gene encodes a functional repressor molecule, the presence of a copy of this gene will restore repressibility.

The second constitutive mutant is analogous to the O^C mutant of the lactose operon because it maps immediately adjacent to the structural genes. Furthermore, the addition of a wild-type operator gene into mutant cells (as an external element) does not restore repression. This is predictable if the mutant operator, which must be present in *cis*, no longer interacts with the repressor–tryptophan complex.

The entire *trp* operon has now been well defined, as shown in Figure 9. Five contiguous structural genes (*trp E, D, C, B,* and *A*) are transcribed as a polycistronic mRNA directing translation of the enzymes that catalyze the biosynthesis of tryptophan. As in the *lac* operon, a promoter region (*trpP*) represents the binding site for RNA polymerase, and an operator region (*trpO*) is the binding site for the repressor. In the absence of repressor binding, transcription initiates within the overlapping *trpP–trpO* region and proceeds along a **leader sequence** 162 nucleotides prior to the first structural gene (*trpE*). Within that leader sequence, still another regulatory site exists, called an *attenuator*, which we describe in the next section of this chapter. As we shall see, the attenuator is also an integral part of the control mechanism of the operon.

ESSENTIAL POINT ■ ■ ■

Unlike the inducible *lac* operon, the *trp* operon is repressible. In the presence of tryptophan, the repressor binds to the regulatory region of the *trp* operon and represses transcription initiation.

5 | Alterations to RNA Secondary Structure Also Contribute to Prokaryotic Gene Regulation

In the preceding sections of this chapter we focused on gene regulation brought about by DNA-binding regulatory proteins that interact with promoter and operator regions of the genes to be regulated. These regulatory proteins, such as the *lac* repressor and the CAP protein, act to decrease or increase transcription initiation from their target promoters by affecting the binding of RNA polymerase to the promoter.

Gene regulation in prokaryotes can also occur through the interactions of regulatory molecules with specific regions of a nascent mRNA, after transcription has been initiated. The binding of these regulatory molecules alters the secondary structure of the mRNA, leading to premature transcription termination or repression of translation. We will discuss two types of regulation by RNA secondary structure—attenuation and riboswitches. Both types of regulation help to fine-tune prokaryotic gene regulation and are used in addition to regulation of transcription initiation.

Transcription Attenuation

Charles Yanofsky, Kevin Bertrand, and their colleagues defined the mechanisms of bacterial attenuation. They observed that, when tryptophan is present and the *trp* operon is repressed, initiation of transcription still occurs at a low level, but is subsequently terminated at a point about 140 nucleotides along the transcript. They called this process **attenuation,** as it further diminishes expression of the operon. In contrast, when tryptophan is absent or present in very low concentrations, transcription is initiated but is *not* subsequently terminated, instead continuing beyond the leader sequence into the structural genes.

The site involved in attenuation is located 115 to 140 nucleotides into the leader sequence and is referred to as the **attenuator.** (See Figure 9.)

Yanofsky and colleagues presented a model to explain how attenuation occurs and is regulated. The initial DNA sequence that is transcribed gives rise to an mRNA sequence that has the potential to fold into two mutually exclusive stem-loop structures referred to as "hairpins." In the presence of excess tryptophan, the mRNA hairpin that is formed behaves as a **terminator** structure, and transcription is almost always terminated prematurely, just beyond the attenuator. On the other hand, if tryptophan is scarce, an alternative mRNA hairpin referred to as the **antiterminator hairpin** is formed. Transcription proceeds past the antiterminator hairpin region, and the entire mRNA is subsequently produced.

A key point in Yanofsky's model is that the leader transcript must be translated in order for the antiterminator hairpin to form. The leader transcript contains two triplets (UGG) that encode tryptophan, and these are present just downstream of the initial AUG sequence that signals the initiation of translation by ribosomes. When adequate tryptophan is present, charged tRNA^{trp} is present in the cell. As a result, ribosomes translate these UGG triplets, proceed through the attenuator, and allow the *terminator hairpin* to form. The terminator hairpin signals RNA polymerase to prematurely terminate transcription, and the operon is not transcribed. If cells are starved of tryptophan, charged tRNA^{trp} is unavailable. As a result, ribosomes "stall" during translation of the UGG triplets. The presence of ribosomes in this region of the mRNA interferes with the formation of the terminator hairpin, but allows the formation of the antiterminator hairpin within the leader transcript. As a result, transcription proceeds, leading to expression of the entire set of structural genes.

Many other bacterial operons use attenuation to control gene expression. These include operons that encode enzymes involved in the biosynthesis of amino acids such as threonine, histidine, leucine, and phenylalanine. As with the *trp* operon, attenuation occurs in a leader sequence that contains an attenuator region.

Riboswitches

Since the elucidation of attenuation in the *trp* operon, numerous cases of gene regulation that also depend on alternative forms of mRNA secondary structure have been documented. These involve what are called **riboswitches,** which are mRNA sequences (or elements), present in the 5'-untranslated region (5'-UTR) upstream from the coding sequences. These elements are capable of binding with small molecule ligands, such as metabolites, whose synthesis or activity is controlled by the genes encoded by the mRNA. Such binding causes a conformational change in one domain of the riboswitch element, which induces another change at a second RNA domain, most often creating a transcription *terminator structure.* This terminator structure interfaces directly with the transcriptional machinery and shuts it down.

Riboswitches can recognize a broad range of ligands, including amino acids, purines, vitamin cofactors, amino sugars, and metal ions, among others. They are widespread in bacteria. In *Bacillus subtilis,* for example, approximately 5 percent of this bacterium's genes are regulated by riboswitches. They are also found in archaea, fungi, and plants, and may prove to be present in animals as well.

The two important domains within a riboswitch are the ligand-binding site and the expression platform, which is capable of forming the terminator structure. **Figure 10** illustrates the principles involved in riboswitch control. The 5'-UTR of an mRNA is shown on the left side of the figure in the absence of the ligand (metabolite). RNA polymerase has transcribed the unbound ligand-binding site, and in the *default conformation,* the expression domain adopts an *antiterminator conformation.* Thus, transcription continues through the expression platform and into the coding region. On the right side of the figure, the presence of the ligand on the ligand-binding site induces an alternative conformation

in the expression platform, creating the *terminator conformation.* RNA polymerase is effectively blocked and transcription ceases.

ESSENTIAL POINT ■ ■ ■

Attenuation and riboswitches regulate gene expression by inducing alterations to mRNA secondary structure, leading to premature termination of transcription.

---

6 | Eukaryotic Gene Regulation Differs from That in Prokaryotes

In eukaryotes, gene expression is tightly controlled in order to express the required levels of gene products at specific times, in specific cell types, and in response to complex changes in the environment. Virtually all cells in a multicellular eukaryotic organism contain a complete genome; however, only a subset of genes is expressed in any particular cell type. Multicellular eukaryotes do not grow solely in response to the availability of nutrients. Instead, they regulate their growth and division to occur at appropriate places in the body and at appropriate times during development. The loss of gene regulation that controls normal cell growth and division may lead to developmental defects or cancer.

Eukaryotes employ a wide range of mechanisms for altering the expression of genes. In contrast to prokaryotic gene regulation, which occurs primarily at the level of transcription initiation, regulation of gene expression in eukaryotes can occur at many different levels. These include the initiation of transcription, mRNA modifications and stability, and the synthesis, modification, and stability of the protein product **(Figure 11)**.

FIGURE 10 Illustration of the mechanism of riboswitch regulation of gene expression, where the default position (left) is in the antiterminator conformation. Upon binding by the ligand, the mRNA adopts the terminator conformation (right).

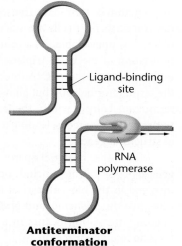

Antiterminator conformation

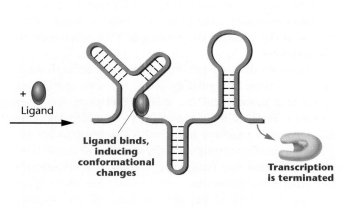

Terminator conformation

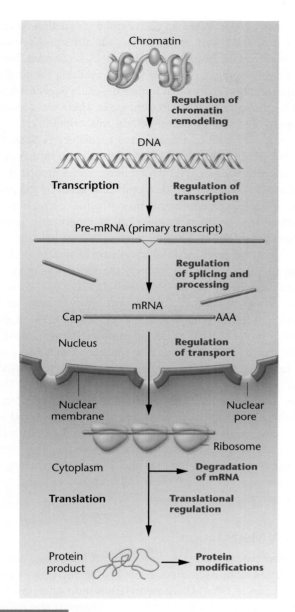

Chromatin

Regulation of chromatin remodeling

DNA

Transcription — Regulation of transcription

Pre-mRNA (primary transcript)

Regulation of splicing and processing

mRNA

Cap — AAA

Nucleus — Regulation of transport

Nuclear membrane — Nuclear pore

Ribosome

Cytoplasm → **Degradation of mRNA**

Translation — **Translational regulation**

Protein product → **Protein modifications**

FIGURE 11 Regulation can occur at any stage in the expression of genetic material in eukaryotes. All these forms of regulation affect the degree to which a gene is expressed.

Several features of eukaryotic cells make it possible for them to use more types of gene regulation than are possible in prokaryotic cells:

- Eukaryotic cells contain a much greater amount of DNA than do prokaryotic cells, and this DNA is associated with histones and other proteins to form highly compact chromatin structures within an enclosed nucleus. Eukaryotic cells modify this structural organization in order to influence gene expression.

- The mRNAs of most eukaryotic genes must be spliced, capped, and polyadenylated prior to transport from the nucleus. Each of these processes can be regulated in order

to influence the numbers and types of mRNAs available for translation.

- Genetic information in eukaryotes is carried on many chromosomes (rather than just one), and these chromosomes are enclosed within a double-membrane-bound nucleus. After transcription, transport of RNAs into the cytoplasm can be regulated in order to modulate the availability of mRNAs for translation.

- Eukaryotic mRNAs can have a wide range of half-lives ($t_{1/2}$). In contrast, the majority of prokaryotic mRNAs decay very rapidly. Rapid turnover of mRNAs allows prokaryotic cells to rapidly respond to environmental changes. In eukaryotes, the complement of mRNAs in each cell type can be more subtly manipulated by altering mRNA decay rates over a larger range.

- In eukaryotes, translation rates can be modulated, as well as the way proteins are processed, modified, and degraded.

In the following sections, we examine some of the major ways in which eukaryotic gene expression is regulated. As most eukaryotic genes are regulated, at least in part, at the transcriptional level, we will emphasize transcriptional control. In addition, we will limit our discussion to regulation of genes transcribed by RNA polymerase II. Three different RNA polymerases transcribe eukaryotic genes. RNA polymerase II transcribes all mRNAs and some small nuclear RNAs, whereas RNA polymerases I and III transcribe ribosomal RNAs, some small nuclear RNAs, and transfer RNAs. Transcription by each of these RNA polymerases is regulated differently, with RNA polymerase II having the most diverse and complex mechanisms.

7 Eukaryotic Gene Expression Requires Chromatin Modifications

Two structural features of eukaryotic genes distinguish them from the genes of prokaryotes. First, eukaryotic genes are situated on chromosomes that occupy a distinct location within the cell—the nucleus. This sequestering of genetic information in a discrete compartment allows the proteins that directly regulate transcription to be kept apart from those involved with translation and other aspects of cellular metabolism. Second, eukaryotic DNA is combined with histones and nonhistone proteins to form chromatin. Chromatin's basic structure is characterized by repeating units called nucleosomes that are wound into 30-nm fibers, which in turn form other, even more compact structures. The compactness of these chromatin structures is inhibitory to many processes, including transcription, replication, and DNA repair. In this section, we outline some of the ways in which eukaryotic cells use these structural features of eukaryotic genes to regulate their expression.

Chromosome Territories and Transcription Factories

Recent research has revealed that the interphase nucleus is not a bag of tangled chromosome arms, but has a highly organized structure. In the interphase nucleus, each chromosome occupies a discrete domain called a **chromosome territory** and stays separate from other chromosomes. Channels between chromosomes contain little or no DNA and are called **interchromosomal domains.**

Transcriptionally active genes appear to be cycled to the edge of chromosome territories at the border of the interchromosomal domains. Scientists hypothesize that this organization may bring actively expressed genes into closer association with transcription factors, or with other actively expressed genes, thereby facilitating their coordinated expression.

Another feature within the nucleus—the **transcription factory**—may also contribute to regulating gene expression. Transcription factories are nuclear sites at which most RNA polymerase II transcription occurs. These sites also contain the majority of active RNA polymerase and other transcription factors. By concentrating transcription proteins and actively transcribed genes in specific locations in the nucleus, the cell may enhance the expression of these genes.

Histone Modifications and Nucleosomal Chromatin Remodeling

Chromatin modification is an important step in gene regulation. Chromatin modification appears to be a prerequisite for transcription of some eukaryotic genes, although it can occur simultaneously with transcription of other genes.

Chromatin can be modified in two general ways. The first involves changes to nucleosomes, and the second involves modifications to DNA. In this subsection, we will discuss changes to the nucleosomal component of chromatin. In the next subsection, we present DNA modifications, specifically DNA methylation.

Nucleosomal chromatin can be modified in three ways. The first involves changes in nucleosome composition that can affect gene transcription. For example, most nucleosomes contain the normal histones H2A and H3. Some gene promoter regions may be flanked by nucleosomes containing variant histones, such as H2A.Z and H3.3. These variant nucleosomes help keep promoter regions free of normal repressive nucleosomes, thereby facilitating gene transcription.

A second mechanism of chromatin alteration involves histone modification. One such modification is acetylation, a chemical alteration of the histone component of nucleosomes that is catalyzed by histone acetyltransferase

enzymes (HATs). When an acetyl group is added to specific basic amino acids on a histone tail, the attraction between the basic histone protein and acidic DNA is lessened. These modifications make promoter regions available for binding to transcription factors that initiate the chain of events leading to gene transcription. In some cases, HATs are recruited to genes by the presence of certain transcription activator proteins that bind to transcription regulatory regions. Of course, what can be opened can also be closed. In that case, histone deacetylases (HDACs) remove acetyl groups from histone tails. HDACs can be recruited to genes by the presence of certain repressor proteins on regulatory regions. In addition to acetylation, histones can be modified in several other ways, including phosphorylation and methylation.

The third mechanism of chromatin alteration is chromatin remodeling which involves the repositioning or removal of nucleosomes on DNA, brought about by chromatin remodeling complexes. Repositioned nucleosomes make regions of the chromosome accessible to transcription regulatory proteins, such as transcription activators and RNA polymerase II. One of the best-studied remodeling complexes is the SWI/SNF complex. Remodelers such as SWI/SNF can act in several different ways (**Figure 12**). They may loosen the attachment between histones and DNA, resulting in the nucleosome sliding along the DNA and exposing regulatory regions. Alternatively, they may loosen the DNA strand from the nucleosome core, or they may cause reorganization of the internal nucleosome components. In all cases, the DNA is left transiently exposed to association with transcription factors and RNA polymerase.

DNA Methylation

Another type of change in chromatin that plays a role in gene regulation is the addition or removal of methyl groups to or from bases in DNA. **DNA methylation** most often involves cytosine. In the genome of any given eukaryotic species, approximately 5 percent of the cytosine residues are methylated. However, the extent of methylation can be tissue specific and can vary from less than 2 percent to more than 7 percent of cytosine residues.

Evidence of a role for methylation in eukaryotic gene expression is based on a number of observations. First, an inverse relationship exists between the degree of methylation and the degree of expression. Large transcriptionally inert regions of the genome, such as the inactivated X chromosome in mammalian female cells, are often heavily methylated. Second, methylation patterns are tissue specific and, once established, are heritable for all cells of that tissue. It appears that proper patterns of DNA methylation are essential for normal mammalian development. Undifferentiated embryonic cells that are not able to methylate DNA die when they are required to differentiate into specialized cell types.

(a) Alteration of DNA-histone contacts

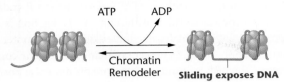

Sliding exposes DNA

(b) Alteration of the DNA path

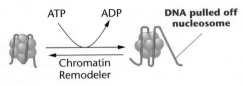

DNA pulled off nucleosome

(c) Remodeling of nucleosome core particle

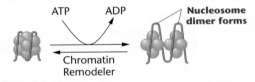

Nucleosome dimer forms

FIGURE 12 Three ways by which chromatin remodelers, such as the SWI/SNF complex, alter the association of nucleosomes with DNA. (a) The DNA–histone contacts may be loosened, allowing the nucleosomes to slide along the DNA, exposing DNA regulatory regions. (b) The path of the DNA around a nucleosome core particle may be altered. (c) Components of the core nucleosome particle may be rearranged, resulting in a modified nucleosome structure.

Perhaps the most direct evidence for the role of methylation in gene expression comes from studies using base analogs. The nucleotide **5-azacytidine** can be incorporated into DNA in place of cytidine during DNA replication. This analog cannot be methylated, causing the undermethylation of the sites where it is incorporated. The incorporation of 5-azacytidine into DNA changes the pattern of gene expression and stimulates expression of alleles on inactivated X chromosomes. In addition, the presence of 5-azacytidine in DNA can induce the expression of genes that would normally be silent in certain differentiated cells.

How might methylation affect gene regulation? Data from *in vitro* studies suggest that methylation can repress transcription by inhibiting the binding of transcription factors to DNA. Methylated DNA may also recruit repressive chromatin remodeling complexes to gene-regulatory regions.

ESSENTIAL POINT ■ ■ ■

Eukaryotic gene regulation at the level of chromatin may involve gene-specific chromatin remodeling, histone modifications, or DNA modifications. These modifications may either allow or inhibit access of promoters and enhancers to transcription factors, resulting in increased or decreased levels of transcription initiation.

NOW SOLVE THIS

3 Cancer cells often have abnormal patterns of chromatin modifications. In some cancers, the DNA repair genes *MLH1* and *BRCA1* are hypermethylated on their promoter regions. Explain how this abnormal methylation pattern could contribute to cancer.

■ **Hint** *This problem involves an understanding of the types of genes that are mutated in cancer cells. The key to its solution is to consider how methylation affects gene expression of cancer-related genes.*

8 Eukaryotic Transcription Requires Specific *Cis*-Acting Sites

As in prokaryotes, eukaryotic transcription regulation is controlled by *trans*-acting regulatory proteins that bind to specific *cis*-acting sites located in and around eukaryotic genes. Although these *cis*-acting sites do not, by themselves, regulate gene transcription, they are essential because they position regulatory proteins in regions where those proteins can act to stimulate or repress transcription of the associated gene. In this section, we will discuss some of these *cis*-acting DNA sequences including promoters, enhancers, and silencers.

Promoters

A **promoter** is a region of DNA that binds one or more proteins that regulate transcription initiation. Promoters are located immediately adjacent to the genes they regulate. They may be up to several hundred nucleotides in length and specify where transcription begins and the direction of transcription along the DNA. Within promoters are a number of **promoter elements**—short nucleotide sequences that bind specific regulatory factors.

There are two subcategories within eukaryotic promoters. First, the **core promoter** determines the accurate initiation of transcription by RNA polymerase II. Second, **proximal promoter elements** are those that modulate the efficiency of basal levels of transcription.

Recent bioinformatic research reveals that there is a great deal of diversity in eukaryotic core promoters in terms of both their structures and functions. Core promoters are now thought to be either *focused* or *dispersed*. **Focused promoters** specify transcription initiation at a single specific nucleotide (the transcription start site). In contrast, **dispersed promoters** direct initiation from a number of weak transcription start sites located over a 50- to 100-nucleotide region (**Figure 13**). Focused transcription initiation is the major type of initiation for most genes of lower eukaryotes, but for only about 30 percent of vertebrate genes. Focused promoters are usually associated with genes whose transcription levels are highly regulated, whereas dispersed promoters are associated with genes that are transcribed constitutively.

(a) Focused promoter

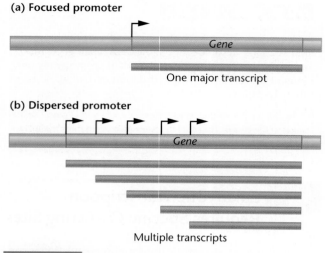

One major transcript

(b) Dispersed promoter

Multiple transcripts

FIGURE 13 Focused and dispersed promoters. Focused promoters (a) specify one specific transcription initiation site. Dispersed promoters (b) specify weak transcription initiation at multiple start site positions over an approximately 100-bp region. Transcription start sites and the directions of transcription are indicated with arrows.

Little is known about the DNA elements that make up dispersed promoters. These promoters are usually found within CG-rich regions, suggesting that chromatin modifications may influence initiation from these promoters. In addition, dispersed promoters are free of nucleosomes over a 150- to 200-nucleotide region.

Much more is known about the structure of focused promoters. These promoters are made up of one or more DNA sequence elements, as summarized in **Figure 14**. Each of these elements is found in only some core promoters, with no element being a universal component of all focused promoters.

The Inr element encompasses the transcription start site, from approximately nucleotides -2 to $+4$, relative to the start site. In humans, the Inr consensus sequence is $YYAN^A/_TYY$ (where Y indicates any pyrimidine nucleotide and N indicates any nucleotide). The transcription start site is the first A residue at $+1$. The TATA box element is located at

approximately -30 relative to the transcription start site and has the consensus sequence $TATA^A/_TAAR$ (where R indicates any purine nucleotide). Although the TATA box is a common element in prokaryotic and eukaryotic promoters, it is found in only about 15 percent of mammalian gene core promoters. The BRE is found in some core promoters at positions either immediately upstream or downstream from the TATA box. The MTE and DPE sequence motifs are located downstream of the transcription start site, at approximately $+18$ to $+27$ and $+28$ to $+32$ respectively.

Many promoters contain proximal promoter elements that act to increase the levels of basal transcription. For example, the **CAAT box** has the consensus sequence CAAT or CCAAT and is usually located about 70 to 80 base pairs upstream from the start site. Mutational analysis suggests that CAAT boxes (when present) are critical to the promoter's ability to initiate transcription. Mutations on either side of this element have no effect on transcription, whereas mutations within the CAAT sequence dramatically lower the rate of transcription. **Figure 15** summarizes the transcriptional effects of mutations in the CAAT box and other promoter elements. The **GC box** is another element often found in proximal promoter regions and has the consensus sequence GGGCGG. It is located, in one or more copies, at about position -110.

Enhancers and Silencers

Although eukaryotic promoter elements are essential for basal or low levels of transcription initiation, more dramatic changes in transcription initiation require the presence of other sequence elements known as enhancers and silencers.

Like promoters, **enhancers** are *cis* regulators because they function when adjacent to the structural genes they regulate. However, unlike promoters, enhancers can be located on either side of a gene, at some distance from the gene, or even within the gene. Enhancers are necessary for achieving the maximum level of transcription. In addition, enhancers are responsible for time- and tissue-specific gene expression. Thus, there is some degree of analogy between enhancers

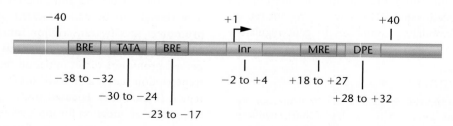

FIGURE 14 Core promoter elements found in focused promoters. Core promoter elements are usually located between -40 and $+40$ nucleotides, relative to the transcription start site, indicated as $+1$. None of these elements is universal, and a core promoter may contain only one, or several, of these elements. BRE is the TFIIB recognition element, TATA is the TATA box, Inr is the initiator element, MTE is the motif ten element, and DPE is the downstream promoter element.

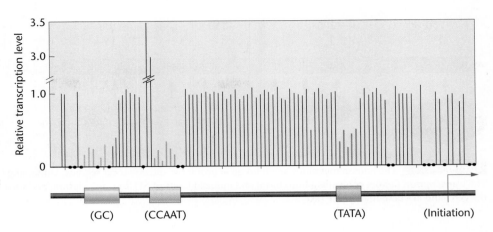

FIGURE 15 Summary of the effects on transcription levels of different point mutations in the promoter region of the β-globin gene. Each line represents the level of transcription produced in a separate experiment by a single-nucleotide mutation (relative to wild-type) at a particular location. Dots represent nucleotides for which no mutation was obtained. Note that mutations within specific elements of the promoter have the greatest effects on the level of transcription.

and operator regions in prokaryotes. However, enhancers are more complex in both structure and function.

Several features distinguish promoters from enhancers:

1. The position of an enhancer need not be limited in position; it will function whether it is upstream, downstream, or within the gene it regulates.

2. The orientation of an enhancer can be inverted without significant effect on its action.

3. If an enhancer is experimentally moved adjacent to a gene elsewhere in the genome, or if an unrelated gene is placed near an enhancer, the transcription of the newly adjacent gene is enhanced.

Another type of *cis*-acting transcription regulatory element, the **silencer,** acts upon eukaryotic genes to repress the level of transcription initiation. Silencers, like enhancers, are short DNA sequence elements that affect the rate of transcription initiated from an associated promoter. They often act in tissue- or temporal-specific ways to control gene expression.

ESSENTIAL POINT ■ ■ ■

Eukaryotic transcription regulation requires gene-specific promoter, enhancer, and silencer elements. The presence of these *cis*-acting regulatory sites can affect transcription in tissue- and temporal-specific ways.

9 Activators and Repressors Influence Transcription When Bound to *Cis*-Acting Sites

Eukaryotic promoters, enhancers, and silencers influence transcription initiation by acting as binding sites for transcription regulatory proteins. These transcription regulatory proteins, known as **transcription factors,** can have diverse and complicated effects on transcription. Some transcription factors increase the levels of transcription initiation and are

known as **activators,** whereas others reduce transcription levels and are known as **repressors.**

Some transcription factors are expressed in tissue-specific ways, regulating their target genes for tissue-specific levels of expression. In addition, some transcription factors are expressed in cells only at certain times during development or in response to external physiological signals. In some cases, a transcription factor that binds to a *cis*-acting site and regulates a certain gene may be present in a cell and may even bind to its appropriate *cis*-acting site but will only become active when modified structurally (for example, by phosphorylation or by binding to a coactivator such as a hormone). These modifications to transcription factors can also be regulated in tissue- or temporal-specific ways. In addition, different transcription factors may compete for binding to a given DNA sequence or to one of two overlapping sequences. In these cases, transcription factor concentrations and the strength with which each factor binds to DNA will dictate which factor binds. The same site may also bind different factors in different tissues. Finally, multiple transcription factors that bind to several different enhancers and promoter elements within a gene-regulatory region can interact with each other to fine-tune the levels and timing of transcription initiation.

The Human Metallothionein IIA Gene: Multiple *Cis*-Acting Elements and Transcription Factors

The **human metallothionein IIA gene** (*hMTIIA*) provides an example of how one gene can be transcriptionally regulated through the interplay of multiple promoter and enhancer elements and the transcription factors that bind to them. The product of the *hMTIIA* gene is a protein that binds to heavy metals such as zinc and cadmium, thereby protecting cells from the toxic effects of high levels of these metals. The gene is expressed at low levels in all cells but is transcriptionally induced to express at high levels when cells are exposed to heavy metals and steroid hormones such as glucocorticoids.

359

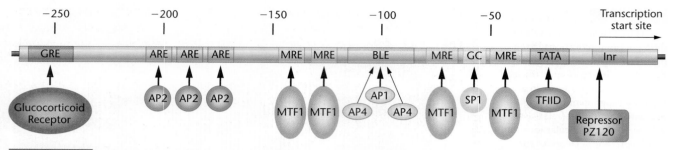

FIGURE 16 The human metallothionein IIA gene promoter and enhancer regions, containing multiple *cis*-acting regulatory sites. The transcription factors controlling both basal and induced levels of MTIIA transcription, and their binding sites, are indicated below the gene and are described in the text.

The *cis*-acting regulatory elements controlling transcription of the *hMTIIA* gene include promoter, enhancer, and silencer elements **(Figure 16)**. Each *cis*-acting element is a short DNA sequence that has specificity for binding to one or more transcription factors.

The *hMTIIA* gene contains the promoter elements TATA box and start site, which specify the start of transcription. The proximal promoter element, GC, binds the SP1 factor, which is present in most eukaryotic cells and stimulates transcription at low levels in most cells. Basal levels of expression are also regulated by the BLE (basal element) and ARE (AP factor response element) regions. These *cis*-elements bind the activator proteins 1, 2, and 4 (AP1, AP2, and AP4), which are present in various levels in different cell types and can be activated in response to extracellular growth signals. The BLE element contains overlapping binding sites for the AP1 and AP4 factors, providing some degree of selectivity in how these factors stimulate transcription of *hMTIIA* when bound to the BLE in different cell types. High levels of transcription induction are conferred by the MRE (metal response element) and GRE (glucocorticoid response element). The metal-inducible transcription factor (MTF1) binds to the MRE element in response to the presence of heavy metals. The glucocorticoid receptor protein binds to the GRE, but only when the receptor protein is also bound to the glucocorticoid steroid hormone. The glucocorticoid receptor is normally located in the cytoplasm of the cell. However, when glucocorticoid hormone enters the cytoplasm, it binds to the receptor and causes a conformational change that allows the receptor to enter the nucleus, bind to the GRE, and stimulate *hMTIIA* gene transcription. In addition to induction, transcription of the *hMTIIA* gene can be repressed by the actions of the repressor protein PZ120, which binds over the transcription start region.

The presence of multiple regulatory elements and transcription factors that bind to them allows the *hMTIIA* gene to be transcriptionally induced or repressed in response to subtle changes in both extracellular and intracellular conditions.

ESSENTIAL POINT ■ ■ ■

Transcription factors influence transcription rates by binding to *cis*-acting regulatory sites within or adjacent to a gene promoter.

---

10 ## Activators and Repressors Interact with General Transcription Factors and Affect Chromatin Structure

We have now discussed the first steps in eukaryotic transcription regulation: first, chromatin must be remodeled and modified to allow transcription proteins to bind to their specific *cis*-acting sites; second, transcription factors bind to *cis*-acting sites and bring about positive and negative effects on the transcription initiation rate—often in response to extracellular signals or in tissue- or time-specific ways. The next question is, how do these *cis*-acting regulatory elements and their DNA-binding factors act to influence transcription initiation? To answer this question, we must first discuss how eukaryotic RNA polymerase II and its basal transcription factors assemble at promoters.

Formation of the Transcription Pre-Initiation Complex

A number of proteins called **general transcription factors** are needed to initiate both basal-level and enhanced levels of transcription. These proteins assemble at the promoter in a specific order, forming a transcriptional **pre-initiation complex (PIC)** that in turn provides a platform for RNA polymerase to recognize and bind to the promoter. We will restrict our discussion of PIC formation to focused promoters with TATA boxes—the type of promoter for which the most information is available.

The general transcription factors and their interactions with the core promoter and RNA polymerase II are outlined

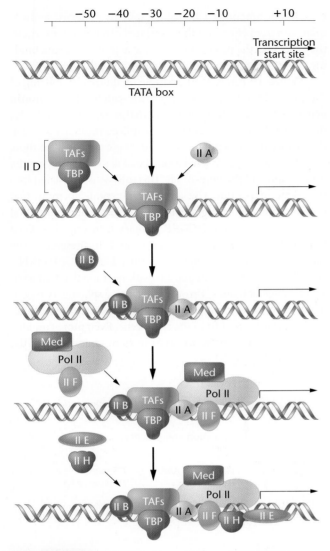

FIGURE 17 The assembly of general transcription factors required for the initiation of transcription by RNA polymerase II.

proceeds down the DNA template in an **elongation complex**. Several of the general transcription factors, specifically TFIID, TFIIE, TFIIH, and Mediator, remain on the core promoter to help set up the next PIC.

Interactions of the General Transcription Factors with Chromatin, Activators, and Repressors

One way in which activators may influence transcription is to bind to chromatin near a gene's promoter. Once bound, activators may recruit chromatin remodeling complexes. The remodeling complexes may open regions of the promoter for further interactions with the transcription machinery, such as the PIC.

A second way that activators may affect transcription is to make direct contacts with general transcription factors and either enhance or inhibit the ability of general transcription factors to associate with a promoter. Some activators, when bound to enhancers, interact with other proteins called coactivators in a complex known as an enhanceosome. The enhanceosome then makes contacts with one or more general transcription factors in the PIC, bending or looping out the intervening DNA **(Figure 18)**. By enhancing the rate of PIC assembly, or by influencing its stability, transcription activators stimulate the rate of transcription initiation.

In addition to chromatin remodeling and PIC formation, transcription activators may increase the rate of DNA unwinding within the gene and accelerate the release of RNA polymerase from the promoter into the transcribed region of the gene.

In contrast, transcription can be repressed by the actions of repressor proteins bound at silencer DNA elements. Repressors can inhibit the formation of a PIC, stimulate or recruit chromatin-remodeling proteins that create repressive

in **Figure 17**. The first step in the formation of a PIC is the binding of TFIID to one or more core promoter elements. TFIID is a multi-subunit complex that contains TBP (*TATA Binding Protein*) and approximately 13 proteins called TAFs (*TBP Associated Factors*). As its name implies, TBP binds to the TATA box. In addition, a subset of TAFs binds to Inr elements, as well as DPEs and MTEs. TFIIA interacts with TFIID and assists the binding of TFIID to the core promoter. Once TFIID has made contact with the core promoter, TFIIB binds to BRE elements on one or both sides of the TATA box. Once TFIID and TFIIB have bound the core promoter, the other general transcription factors interact with RNA polymerase II and help recruit it to the promoter. The fully formed PIC mediates the unwinding of promoter DNA at the start site and the transition of RNA polymerase II from transcription initiation to elongation. RNA polymerase then

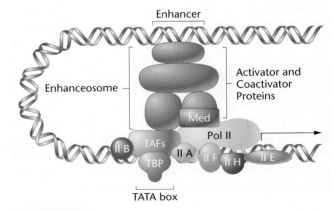

FIGURE 18 Formation of DNA loops allows factors that bind to an enhancer (or silencer) at a distance from a promoter to interact with general transcription factors in the pre-initiation complex and to regulate the level of transcription.

chromatin structures, or block the association of a gene's regulatory elements with activators, or with the PIC.

NOW SOLVE THIS

4 The hormone estrogen converts the estrogen receptor (ER) protein from an inactive molecule to an active transcription factor. The ER binds to *cis*-acting sites that act as enhancers, located near the promoters of a number of genes. In some tissues, the presence of estrogen appears to activate transcription of ER-target genes, whereas in other tissues, it appears to repress transcription of those same genes. Offer an explanation as to how this may occur.

■ Hint *This problem involves an understanding of how transcription enhancers and repressors work. The key to its solution is to consider the many ways that* trans*-acting factors can interact at enhancers to bring about changes in transcription initiation.*

11 Posttranscriptional Gene Regulation Occurs at All the Steps from RNA Processing to Protein Modification

Although transcriptional control is a major type of gene regulation in eukaryotes, **posttranscriptional regulation** also contributes to control of gene expression. Modification of eukaryotic nuclear RNA transcripts prior to translation includes the removal of noncoding introns, the precise splicing together of the remaining exons, and the addition of a cap at the mRNA's 5′ end and a poly-A tail at its 3′ end. The messenger RNA is then exported to the cytoplasm, where it is translated and degraded. Each of the mRNA processing steps can be regulated to control the quantity of functional mRNA available for synthesis of a protein product. In addition, the rate of translation, as well as the stability and activity of protein products, can be regulated. We will examine several mechanisms of posttranscriptional gene regulation that are especially important in eukaryotes—alternative splicing, mRNA stability, translation, and protein stability.

Alternative Splicing of mRNA

Alternative splicing can generate different forms of mRNA from identical pre-mRNA molecules, so that expression of one gene can give rise to a number of proteins with similar

or different functions. Changes in splicing patterns can have many different effects on the translated protein. For example, they can alter the protein's enzymatic activity, receptor-binding capacity, or protein localization in the cell.

Figure 19 presents an example of alternative splicing of the pre-mRNA transcribed from the **calcitonin/calcitonin gene-related peptide gene (*CT/CGRP* gene)**. In thyroid cells, the *CT/CGRP* primary transcript is spliced in such a way that the mature mRNA contains the first four exons only. In these cells, the exon 4 polyadenylation signal is used to process the mRNA and add the poly-A tail. This mRNA is translated into the calcitonin peptide, a 32-amino acid peptide hormone that is involved in regulating calcium. In the brain and peripheral nervous system, the *CT/CGRP* primary transcript is spliced to include exons 5 and 6, but not exon 4. In these cells, the exon 6 polyadenylation site is recognized. The *CGRP* mRNA encodes a 37-amino acid peptide with hormonal activities in a wide range of tissues. Through alternative splicing, two peptide hormones with different structures, locations, and functions are synthesized from the same gene. Even more complex alternative splicing patterns occur in some genes, such as the example in **Figure 20**.

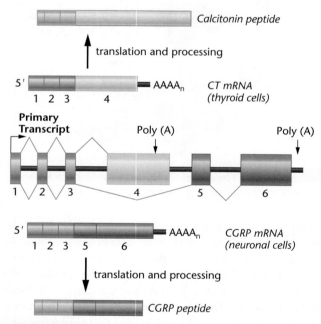

FIGURE 19 Alternative splicing of the *CT/CGRP* gene transcript. The primary transcript, which is shown in the middle of the diagram, contains six exons. The primary transcript can be spliced into two different mRNAs, both containing the first three exons but differing in their final exons. The *CT* mRNA contains exon 4, with polyadenylation occurring at the end of the fourth exon. The *CGRP* mRNA contains exons 5 and 6, and polyadenylation occurs at the end of exon 6. The *CT* mRNA is produced in thyroid cells. After translation, the resulting protein is processed into the calcitonin peptide. In contrast, the *CGRP* mRNA is produced in neuronal cells, and after translation, its protein product is processed into the CGRP peptide.

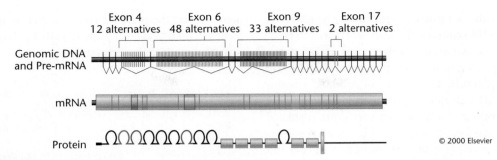

Exon 4
12 alternatives

Exon 6
48 alternatives

Exon 9
33 alternatives

Exon 17
2 alternatives

Genomic DNA
and Pre-mRNA

mRNA

Protein

© 2000 Elsevier

FIGURE 20 Alternative splicing of the *Dscam* gene mRNA. The *Dscam* gene encodes a protein that guides axon growth during development. Each mRNA will contain one of the 12 possible exons for exon 4 (red), one of the 48 possible exons for exon 6 (blue), one of the 33 possible exons for exon 9 (green), and one of the 2 possible exons for exon 17 (yellow). Counting all possible combinations of these exons, the *Dscam* gene could encode 38,016 different versions of the DSCAM protein.

Alternative splicing increases the number of proteins that can be made from each gene. As a result, the number of proteins that an organism can make—its **proteome**—is not the same as the number of genes in the genome, and protein diversity can exceed gene number by an order of magnitude. Alternative splicing is found in all metazoans but is especially common in vertebrates, including humans. It has been estimated that 30 to 60 percent of the genes in the human genome can undergo alternative splicing. Thus, humans can produce several hundred thousand different proteins (or perhaps more) from the approximately 20,000 genes in the haploid genome.

Mutations that affect regulation of splicing contribute to several genetic disorders. One of these disorders, **myotonic dystrophy (DM),** provides an example of how defects in alternative RNA splicing can lead to a wide range of symptoms. Myotonic dystrophy is the most common form of adult muscular dystrophy, affecting 1 in 8000 individuals. It is an autosomal dominant disorder that occurs in two forms—DM1 and DM2. Both of these diseases show a wide range of symptoms, including muscle wasting, **myotonia** (difficulty relaxing muscles), insulin resistance, behavior and cognitive defects, and cardiac muscle problems.

DM1 is caused by the expansion of the trinucleotide repeat CTG in the 3'-untranslated region of the ***DMPK* gene.** In unaffected individuals, the *DMPK* gene contains between 5 and 35 copies of the CTG repeat sequence, whereas in DM1 patients, the gene contains between 150 and 2000 copies. The severity of the symptoms is directly related to the number of copies of the repeat sequence. DM2 is caused by an expansion of the repeat sequence CCTG within the first intron of the ***ZNF9* gene.** Affected individuals may have up to 11,000 copies of the repeat sequence in the *ZNF9* intron. In DM2, the severity of symptoms is not related to the number of repeats.

Recently, scientists have discovered that DM1 and DM2 are caused not by changes in the protein products of the *DMPK* or *ZNF9* genes, but by the toxic effects of their repeat-containing RNAs. These RNAs accumulate and form inclusions within the nucleus. In the case of *ZNF9*, only the CCUG sequence repeat itself accumulates in the nucleus, as the remainder of the intron is degraded after splicing of the mRNA. It appears that the accumulated RNA repeats bind to, and sequester, proteins that would normally be involved in regulating the alternative splicing patterns of a large number of other RNAs. These RNAs include those whose products are required for the proper functioning of muscle and neural tissue. So far, scientists have discovered over 20 genes that are inappropriately spliced in the muscle, heart, and brain of DM1 patients. Often, the fetal splicing patterns occur in DM1 and DM2 patients, and the normal transitions to adult splicing patterns are lacking. Such defects in the regulation of RNA splicing are known as **spliceopathies.**

Control of mRNA Stability

The **steady-state level** of an mRNA is its amount in the cell as determined by a combination of the rate at which the gene is transcribed and the rate at which the mRNA is degraded. In turn, the steady-state level determines the amount of mRNA that is available for translation. All mRNA molecules are degraded at some point after their synthesis, but the lifetime of an mRNA, defined in terms of its **half-life,** or $t_{1/2}$, can vary widely between different mRNAs and can be regulated in response to the needs of the cell. Some mRNAs are degraded within minutes after their synthesis, whereas others can remain stable for hours, months, or even years (in the case of mRNAs stored in oocytes).

Regulation of mRNA stability is often linked with the process of translation. Several observations demonstrate this link between translation and mRNA stability. First, most mRNA molecules become stable in cells that are treated with translation inhibitors. Second, the presence of premature stop codons in the body of an mRNA, as well as premature translation termination, causes rapid degradation of mRNAs. Third, many of the ribonucleases and mRNA-binding proteins that affect mRNA stability associate with ribosomes.

363

Another way that an mRNA's half-life can be altered is through specific RNA sequence elements that recruit degrading or stabilizing complexes. One well-studied mRNA stability element is the adenosine-uracil rich element (ARE)—a stretch of ribonucleotides that consist of A and U ribonucleotides. These AU-rich elements are usually located in the 3′-untranslated regions of mRNAs that have short, regulated half-lives. These ARE-containing mRNAs encode proteins that are involved in cell growth or transcription control and need to be rapidly modulated in abundance. In cells that are not growing or require low levels of gene expression, specific complexes bind to the ARE elements of these mRNA molecules, bringing about shortening of the poly-A tail and rapid mRNA degradation. It is estimated that approximately 10 percent of mammalian mRNAs contain these instability elements.

Translational and Posttranslational Controls

In some cases, the translation of an mRNA can be regulated by the extent of the cell's requirement for the gene product. In some cases, specific regulatory proteins can bind to mRNA molecules, blocking their translation. The binding of these regulatory proteins can be controlled by their abundance in the cell or by their molecular modifications.

Protein levels and activities can also be controlled by regulation of their stabilities and by protein modifications. An example of regulated stability and modification is offered by the **p53 protein**. The p53 protein is essential to protect normal cells from the effects of DNA damage and other stresses. It is a transcription factor that increases the transcription of a number of genes whose products are involved in cell-cycle arrest, DNA repair, and programmed cell death. Under normal conditions, the levels of p53 protein are extremely low in cells, and the p53 that is present is inactive. When cells suffer DNA damage or metabolic stress, the levels of p53 protein increase dramatically, and p53 becomes an active transcription factor.

The changes in the abundance and activity of p53 are due to a combination of increased protein stability and modifications to the protein. In unstressed cells, p53 is bound by another protein called **Mdm2**. The Mdm2 protein binds to the p53 protein, blocking its ability to induce transcription. In addition, Mdm2 adds ubiquitin residues onto the p53 protein. **Ubiquitin** is a small protein that tags other proteins for degradation by proteolytic enzymes. The presence of ubiquitin on p53 results in p53 degradation. When cells are stressed, Mdm2 and p53 become modified by phosphorylation and acetylation, resulting in the release of Mdm2 from p53. As a consequence, p53 proteins become stable, the levels of p53 increase, and the protein is able to act as a transcription factor. An added level of control is that p53 is a transcription factor that induces the transcription of the *Mdm2*

gene. Hence, the presence of active p53 triggers a negative feedback loop that creates more Mdm2 protein, which rapidly returns p53 to its rare and inactive state.

ESSENTIAL POINT ■ ■ ■

Posttranscriptional gene regulation can involve alternative splicing of nascent RNA, changes in mRNA stability, translational control, and posttranslational modifications. These mechanisms may alter the type, quantity, or activity of a gene's protein product.

12 RNA-Induced Gene Silencing Controls Gene Expression in Several Ways

In the last several years, the discovery that small RNA molecules control gene expression has given rise to a new field of research. First discovered in plants, short RNA molecules, ~21 nucleotides long, are now known to regulate gene expression in the cytoplasm of both plants and animals by repressing translation and triggering the degradation of mRNAs. This form of sequence-specific posttranscriptional regulation is known as **RNA interference (RNAi)**. More recently, short RNAs have been shown to act in the nucleus to alter chromatin structure and bring about repression of transcription. Together, these phenomena are known as **RNA-induced gene silencing.** We will outline some basic features of RNA-induced gene silencing, while keeping in mind that this field is rapidly expanding and advancing. In addition, we will discuss some of the ways in which RNAi is being used in biotechnology and medicine.

RNAi was first discovered during laboratory research, in studies of plant and animal gene expression. In one research project, Andrew Fire and Craig Mello injected roundworm (*Caenorhabditis elegans*) cells with either single-stranded or double-stranded RNA molecules—both containing sequences complementary to the mRNA of the *unc-22* gene. Although they expected that the single-stranded antisense RNA molecules would suppress *unc-22* gene expression by binding to the endogenous sense mRNA, they were surprised to discover that the injection of double-stranded *unc-22* RNA was 10- to 100-fold more powerful in repressing expression of the *unc-22* mRNA. They studied the phenomenon further and published their results in the journal *Nature* in 1998. They reported that the presence of double-stranded RNA acts to degrade the mRNA if the mRNA is complementary in sequence to one strand of the double-stranded RNA. Only a few molecules of double-stranded RNA are needed to bring about the degradation of large amounts of mRNA. Fire and Mello's research opened up an entirely new and surprising branch of molecular biology, with far-reaching implications for practical applications.

For their insights into RNAi, they were awarded the Nobel Prize in Physiology or Medicine for 2006.

The Molecular Mechanisms of RNA-Induced Gene Silencing

Two types of short RNA molecules are involved in RNA-induced gene silencing: The **small interfering RNAs (siRNAs)** and the **microRNAs (miRNAs).** Although they arise from different sources, their mechanisms of action are similar. Both types of RNA are short, double-stranded molecules, between 21 and 24 ribonucleotides long.

The siRNAs are derived from longer RNA molecules that are linear, double-stranded, and located in the cell cytoplasm. In nature, these siRNA precursors arise within cells as a result of virus infection or the expression of transposons—both of which synthesize double-stranded RNA molecules as part of their life cycles. RNAi may be a method by which cells recognize these double-stranded RNAs and inactivate them, protecting the organism from external or internal assaults. Another source of siRNA molecules is in the research lab. Scientists are now able to introduce double-stranded RNAs into cells for research or therapeutic purposes. In the cytoplasm, double-stranded RNA molecules are recognized by an enzyme complex known as **Dicer** and are cleaved by Dicer into siRNAs.

The miRNAs are derived from single-stranded RNAs that are transcribed within the nucleus from the cell's own genome and that contain a double-stranded stem-loop structure. Nuclease enzymes within the nucleus recognize these stem-loop structures and cleave them from the longer single-stranded RNA. The stem-loop RNA fragments are exported from the nucleus into the cytoplasm where they are further processed by the Dicer complex into short, linear, double-stranded miRNAs. Over the last few years, scientists have discovered that significant amounts of eukaryotic genomes are transcribed by RNA polymerase II into RNA products that contain no open reading frames and are not translated into protein products. These RNAs are transcribed either from sequences within the introns of other protein-coding genes or from their own promoters. So far, more than 700 of these noncoding RNA genes have been discovered in the human genome. *Arabidopsis* has more than 130, and *C. elegans* has more than 100. This is probably an underestimate, and scientists speculate that eukaryotic genomes may contain thousands of genes that are transcribed into short noncoding RNAs, which may regulate the expression of more than half of all protein-coding genes.

How do these noncoding RNAs work to negatively regulate gene expression? The several different pathways involved in RNA-induced gene silencing are outlined in **Figure 21**.

The RNAi pathway takes several steps. First, siRNA or miRNA molecules associate with an enzyme complex called the **RNA-induced silencing complex (RISC).** Second, within the RISC, the short double-stranded RNA is denatured and the sense strand is degraded. Third, the RNA/RISC complex becomes a functional and highly specific agent of RNAi, seeking out mRNA molecules that are complementary to the antisense RNA contained in the RISC. At this point, RNAi can take one of two different pathways. If the antisense

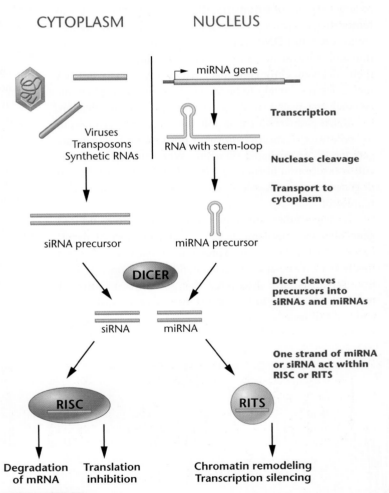

FIGURE 21 Mechanisms of gene regulation by RNA-induced gene silencing. The siRNA or miRNA precursors are processed into short double-stranded RNA molecules by the Dicer complex in the cytoplasm. They are then recognized by either the RISC complex or the RITS complex, and one strand is degraded. In the RNAi pathway, the RISC complex, guided by the antisense single-stranded RNA, recognizes target mRNA substrates, marking them for degradation or translation inhibition. In the transcription silencing pathway, the RITS complex acts in the nucleus by recognizing genomic DNA that is complementary to the single strands of the miRNAs or siRNAs. The RITS complex recruits chromatin remodeling proteins that modify chromatin and repress transcription.

RNA in the RISC is perfectly complementary to the mRNA, the RISC will cleave the mRNA. The cleaved mRNA is then degraded by ribonucleases. If the antisense RNA within the RISC is not exactly complementary to the mRNA, the RISC complex stays bound to the mRNA, interfering with the ability of ribosomes to translate the mRNA. Hence, RNAi can silence gene expression by affecting either mRNA stability or translation.

In addition to repressing mRNA translation and triggering mRNA degradation, siRNAs and miRNAs can also repress the transcription of specific genes and larger regions of the genome. They do this by associating with a different complex—the **RNA-induced initiation of transcription silencing complex (RITS)**. The antisense RNA strand within the RITS targets the RITS complex to specific gene promoters or larger regions of chromatin. RITS then recruits chromatin remodeling enzymes to these regions. These enzymes methylate histones and DNA, resulting in heterochromatin formation and subsequent transcriptional silencing. As a result of their effects on chromatin-mediated gene silencing, miRNA molecules are thought to be involved in epigenetic phenomena such as gene imprinting and X-chromosome inactivation.

RNAi pathways are also able to repress transcription in indirect ways. Transcription factor mRNAs are frequent targets for RNAi-mediated silencing. When the levels of specific transcription factors are reduced in a cell, transcription of genes whose expression depends on these factors is also repressed.

Recent studies are demonstrating that RNA-induced gene-silencing mechanisms operate during normal development and control the expression of batteries of genes involved in tissue-specific cellular differentiation. In addition, scientists have discovered that abnormal activities of miRNAs contribute to the occurrence of cancers, diabetes, and heart disease.

ESSENTIAL POINT ■　　■　　■

RNA-induced gene silencing affects the translatability or stability of mRNA, as well as transcription. It acts through the hybridization of small antisense RNAs to specific regions of an mRNA or a DNA region.

RNA-Induced Gene Silencing in Biotechnology and Medicine

Recently, geneticists have applied RNAi as a powerful research tool. RNAi technology allows investigators to create specific single-gene defects without having to induce inherited gene mutations. RNAi-mediated gene silencing is relatively specific and inexpensive, and it allows scientists to rapidly analyze gene function. Several dozen scientific supply companies now manufacture synthetic siRNA molecules of specific ribonucleotide sequences for use in research. These molecules can be introduced into cultured cells to knock out specific gene products.

In addition to its use in laboratory research, RNAi is being developed as a potential pharmaceutical agent. In theory, any disease caused by overexpression of a specific gene, or even normal expression of an abnormal gene product, could be attacked by therapeutic RNAi. Viral infections are obvious targets, and scientists have had promising results using RNAi in tissue cultures to reduce the severity of infection by several types of viruses such as HIV, influenza, and polio. In animal models, siRNA molecules have successfully treated virus infections, eye diseases, cancers, and inflammatory bowel disease.

New as it is, the science of RNAi holds powerful promise for molecular medicine.

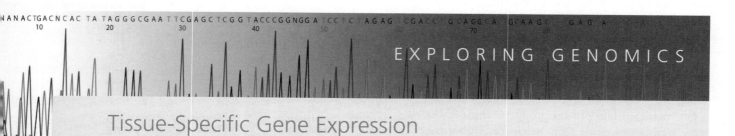

EXPLORING GENOMICS

Tissue-Specific Gene Expression

 Study Area: *Exploring Genomics*

In this chapter, we discussed how gene expression can be regulated in many complex ways. Recall that one aspect of gene-expression regulation we considered is the way promoter, enhancer, and silencer sequences can govern transcriptional initiation of genes to allow for tissue-specific gene expression. All cells and tissues of an organism possess the same genome, and many genes are expressed in all cell and tissue types. However, muscle cells, blood cells, and all other tissue types express genes that are largely tissue-specific (i.e., they have limited or no expression in other tissue types). In this exercise, we use BLAST to learn more about tissue-specific gene-expression patterns.

■ Exercise – Tissue-Specific Gene Expression

In this exercise, we return to the National Center for Biotechnology Information site

(NCBI) and use the search tool **BLAST, Basic Local Alignment Search Tool**.

1. Access BLAST from the NCBI Web site at http://www.ncbi.nlm.nih.gov/BLAST.
2. The following are GenBank accession numbers for four different genes that show tissue-specific expression patterns. You will perform your searches on these genes.

   NM_021588.1
   NM_00739.1
   AY260853.1
   NM_004917

3. For each gene, carry out a nucleotide BLAST search using the accession numbers for your sequence query. Because the accession numbers are for nucleotide sequences, be sure to use the "nucleotide blast" (blastn) program when running your searches. Once you enter "blastn," under the "Choose Search Set" category, you should set the database to "Others (nr etc.)," so that

you are not searching an organism-specific database.

4. Once your BLAST search results appear, look at the top alignments for each gene. The "Links" column (far right) contains colored boxes labeled U (for UniGene expression data), E (Gene Expression Profiles), and G (Gene Information).

   Some alignments will display an M box (which is a Genome Mapview). Each of these boxes will link you to information about the gene. The UniGene link will show you a UniGene report. For some genes, upon entering UniGene you may need to click a link above the gene name before retrieving a UniGene report. Be sure to explore the "Expression Profile" link under the "Gene Expression" category in each UniGene report. Expression profiles will show a table of gene-expression patterns in different tissues.

   Also explore the "GEO Profiles" link under the "Gene Expression" category

of the UniGene reports, when available. These links will take you to a number of gene-expression studies related to each gene of interest. Explore these resources for each gene, and then answer the following questions:

a. What is the identity of each sequence, based on sequence alignment? How do you know this?

b. What species was each gene cloned from?

c. Which tissue(s) are known to express each gene?

d. Does this gene show regulated expression during different times of development?

e. Which gene shows the most restricted pattern of expression by being expressed in the fewest tissues?

CASE STUDY » A mysterious muscular dystrophy

A man in his early 30s suddenly developed weakness in his hands and neck, followed a few weeks later by burning muscle pain—all symptoms of late-onset muscular dystrophy. His internist ordered genetic tests to determine whether he had one of the inherited muscular dystrophies, focusing on Becker muscular dystrophy, myotonic dystrophy Type I, and myotonic dystrophy Type II. These tests were designed to detect mutations in the related *dystrophin, DMPK,* and *ZNF9* genes. The testing ruled out Becker muscular dystrophy. While awaiting the results of the *DMPK,* and *ZNF9* gene tests, the internist explained that the possible mutations were due to expanded tri- and tetranucleotide repeats, but not in the protein-coding portion of the genes. She went on to say that the resulting disorders were due not to changes in the encoded proteins, which appear to be normal, but instead to

altered RNA splicing patterns, whereby the RNA splicing remnants containing the nucleotide repeats disrupt normal splicing of the transcripts of other genes. This discussion raises several interesting questions about the diagnosis and genetic basis of the disorders.

1. What is alternative splicing, where does it occur, and how could disrupting it affect the expression of the affected gene(s)?
2. What role might the expanded tri- and tetranucleotide repeats play in the altered splicing?
3. How does this contrast with other types of muscular dystrophy, such as Becker muscular dystrophy and Duchenne muscular dystrophy?

INSIGHTS AND SOLUTIONS

1. A theoretical operon (*theo*) in *E. coli* contains several structural genes encoding enzymes that are involved sequentially in the biosynthesis of an amino acid. Unlike the *lac* operon, in which the repressor gene is separate from the operon, the gene encoding the regulator molecule is contained within the *theo* operon. When the end product (the amino acid) is present, it combines with the regulator molecule, and this complex binds to the operator, repressing the operon. In the absence of the amino acid, the regulatory molecule fails to bind to the operator, and transcription proceeds.

Characterize this operon, then consider the following mutations, as well as the situation in which the wild-type gene is present along with the mutant gene in partially diploid cells (F′):

(a) Mutation in the operator region.

(b) Mutation in the promoter region.

(c) Mutation in the regulator gene.

In each case, will the operon be active or inactive in transcription, assuming that the mutation affects the regulation of the *theo* operon? Compare each response with the equivalent situation of the *lac* operon.

Solution: The *theo* operon is repressible and under negative control. When there is no amino acid present in the medium (or the environment), the product of the regulatory gene cannot bind to the operator region, and transcription proceeds under the direction of RNA polymerase. The enzymes necessary for synthesis of the amino acid are produced, as is the regulator molecule. If the amino acid *is* present, or is present after sufficient synthesis occurs, the amino acid binds to the regulator, forming a complex that interacts with the operator region, causing repression of transcription of the genes within the operon.

The *theo* operon is similar to the tryptophan system, except that the regulator gene is within the operon rather than separate from it. Therefore, in the *theo* operon, the regulator gene is itself regulated by the presence or absence of the amino acid.

(a) As in the *lac* operon, a mutation in the *theo* operator region inhibits binding with the repressor complex, and transcription occurs constitutively. The presence of an F′ plasmid bearing the wild-type allele would have no effect, since it is not adjacent to the structural genes.

(b) A mutation in the *theo* promoter region would no doubt inhibit binding to RNA polymerase and therefore inhibit transcription. This would also happen in the *lac* operon. A wild-type allele present in an F′ plasmid would have no effect.

(c) A mutation in the *theo* regulator gene, as in the *lac* system, may inhibit either its binding to the repressor or its binding to the operator gene. In both cases, transcription will be constitutive because the *theo* system is repressible. Both cases result in the failure of the regulator to bind to the operator, allowing transcription to proceed. In the *lac* system, failure to bind the corepressor lactose would permanently repress the system. The addition of a wild-type allele would restore repressibility, provided that this gene was transcribed constitutively.

2. Regulatory sites for eukaryotic genes are usually located within a few hundred nucleotides of the transcription start site, but they can be located up to several kilobases away. DNA sequence-specific binding assays have been used to detect and isolate protein factors present at low concentrations in nuclear extracts. In these experiments, short DNA molecules containing DNA-binding sequences are attached to material that is packed into a glass column, and nuclear extracts are passed over the column. The idea is that if proteins that specifically bind to the DNA sequence are present in the nuclear extract, they will bind to the DNA, and they can be recovered from the column after all other nonbinding material has been washed away. Once a DNA-binding protein has been isolated and identified, the problem is to devise a general method for screening cloned libraries for the genes encoding the DNA-binding factors. Determining the amino acid sequence of the protein and constructing synthetic oligonucleotide probes are time consuming and useful for only one factor at a time. Knowing the strong affinity for binding between the protein and its DNA-recognition sequence, how would you screen for genes encoding binding factors?

Solution: Several general strategies have been developed, and one of the most promising was devised by Steve McKnight's laboratory at the Fred Hutchinson Cancer Center. The cDNA isolated from cells expressing the binding factor is cloned into the lambda vector, gt11. Plaques of this library, containing proteins derived from expression of cDNA inserts, are adsorbed onto nitrocellulose filters and probed with double-stranded radioactive DNA corresponding to the binding site. If a fusion protein corresponding to the binding factor is present, it will bind to the DNA probe. After the unbound probe is washed off, the filter is subjected to autoradiography and the plaques corresponding to the DNA-binding proteins can be identified. An added advantage of this strategy is filter recycling by washing the bound DNA from the filters prior to their reuse. Such an ingenious procedure is similar to the colony-hybridization and plaque-hybridization procedures , and it provides a general method for isolating genes encoding DNA-binding factors.

PROBLEMS AND DISCUSSION QUESTIONS

 For activities, animations, and review quizzes, go to the study area at www.masteringgenetics.com

HOW DO WE KNOW ?

1. In this chapter, we have focused on how prokaryotic and eukaryotic organisms regulate the expression of genetic information. In particular, we discussed both transcriptional and posttranscriptional gene regulation. Based on your knowledge of these topics, answer several fundamental questions:

    (a) How do we know that bacteria regulate the expression of certain genes in response to the environment?

    (b) How do we know that bacterial gene clusters are often coordinately regulated by a regulatory region that must be located adjacent to the cluster?

    (c) What led researchers to conclude that a *trans*-acting repressor molecule regulates the *lac* operon?

    (d) How do we know that promoters and enhancers regulate transcription of eukaryotic genes?

    (e) How do we know that DNA methylation plays a role in the regulation of eukaryotic gene expression?

2. Describe which enzymes are required for lactose and tryptophan metabolism in bacteria when lactose and tryptophan, respectively, are (a) present and (b) absent.

3. Contrast positive versus negative regulation of gene expression.

4. Contrast the role of the repressor in an inducible system and in a repressible system.

5. Both attenuation and riboswitches rely on changes in the secondary structure of the leader regions of mRNA to regulate gene expression. Compare and contrast the specific mechanisms in these two types of regulation.

6. For the *lac* genotypes shown in the accompanying table, predict whether the structural gene (Z) is constitutive, permanently repressed, or inducible in the presence of lactose.

Genotype	Constitutive	Repressed	Inducible
$I^+O^+Z^+$			X
$I^-O^+Z^+$			
$I^+O^cZ^+$			
$I^-O^+Z^+/F'I^+$			
$I^+O^cZ^+/F'O^c$			
$I^sO^+Z^+$			
$I^sO^+Z^+/F'I^+$			

7. For the genotypes and conditions (lactose present or absent) shown in the accompanying table, predict whether functional enzymes, nonfunctional enzymes, or no enzymes are made.

Genotype	Condition	Functional Enzyme Made	Nonfunctional Enzyme Made	No Enzyme Made
$I^+O^+Z^+$	No lactose			X
$I^+O^cZ^+$	Lactose			
$I^-O^+Z^-$	No lactose			
$I^-O^+Z^-$	Lactose			
$I^-O^+Z^+/F'I^+$	No lactose			
$I^+O^cZ^+/F'O^+$	Lactose			
$I^+O^+Z^-/F'I^+O^+Z^+$	Lactose			
$I^-O^+Z^-/F'I^+O^+Z^+$	No lactose			
$I^sO^+Z^+/F'O^+$	No lactose			
$I^+O^cZ^+/F'O^+Z^+$	Lactose			

8. The locations of numerous $lacI^-$ and $lacI^S$ mutations have been determined within the DNA sequence of the $lacI$ gene. Among these, $lacI^-$ mutations were found to occur in the 5'-upstream region of the gene, while $lacI^S$ mutations were found to occur farther downstream in the gene. Are the locations of the two types of mutations within the gene consistent with what is known about the function of the repressor that is the product of the $lacI$ gene?

9. Explain why catabolite repression is used in regulating the lac operon and describe how it fine-tunes β-galactosidase synthesis.

10. Describe experiments that would confirm whether or not two transcription regulatory molecules act through the mechanism of cooperative binding.

11. Predict the level of genetic activity of the lac operon as well as the status of the lac repressor and the CAP protein under the cellular conditions listed in the accompanying table.

	Lactose	Glucose
(a)	−	−
(b)	+	−
(c)	−	+
(d)	+	+

12. Predict the effect on the inducibility of the lac operon of a mutation that disrupts the function of (a) the crp gene, which encodes the CAP protein, and (b) the CAP-binding site within the promoter.

13. Describe the role of attenuation in the regulation of tryptophan biosynthesis.

14. In a theoretical operon, genes A, B, C, and D represent the repressor gene, the promoter sequence, the operator gene, and the structural gene, *but not necessarily in that order*. This operon is concerned with the metabolism of a theoretic molecule (tm). From the data provided in the accompanying table, first decide whether the operon is inducible or repressible. Then assign A, B, C, and D to the four parts of the operon. Explain your rationale. (AE = active enzyme; IE = inactive enzyme; NE = no enzyme)

Genotype	tm Present	tm Absent
$A^+B^+C^+D^+$	AE	NE
$A^-B^+C^+D^+$	AE	AE
$A^+B^-C^+D^+$	NE	NE
$A^+B^+C^-D^+$	IE	NE
$A^+B^+C^+D^-$	AE	AE
$A^-B^+C^+D^+/F'A^+B^+C^+D^+$	AE	AE
$A^+B^-C^+D^+/F'A^+B^+C^+D^+$	AE	NE
$A^+B^+C^-D^+/F'A^+B^+C^+D^+$	AE + IE	NE
$A^-B^+C^+D^-/F'A^+B^+C^+D^+$	AE	NE

15. A bacterial operon is responsible for production of the biosynthetic enzymes needed to make the theoretical amino acid tisophane (tis). The operon is regulated by a separate gene, R, deletion of which causes the loss of enzyme synthesis. In the wild-type condition, when tis is present, no enzymes are made; in the absence of tis, the enzymes are made. Mutations in the operator gene (O^-) result in repression regardless of the presence of tis.

Is the operon under positive or negative control? Propose a model for (a) repression of the genes in the presence of tis in wild-type cells and (b) the mutations.

16. A marine bacterium is isolated and is shown to contain an inducible operon whose genetic products metabolize oil when it is encountered in the environment. Investigation demonstrates that the operon is under positive control and that there is a reg gene whose product interacts with an operator region (o) to regulate the structural genes designated sg.

In an attempt to understand how the operon functions, a constitutive mutant strain and several partial diploid strains were isolated and tested with the results shown here:

Host Chromosome	F′ Factor	Phenotype
wild type	none	inducible
wild type	reg gene from mutant strain	inducible
wild type	operon from mutant strain	constitutive
mutant strain	reg gene from wild type	constitutive

Draw all possible conclusions about the mutation as well as the nature of regulation of the operon. Is the constitutive mutation in the *trans*-acting reg element or in the *cis*-acting o operator element?

17. Why is gene regulation more complex in a multicellular eukaryote than in a prokaryote? Why is the study of this phenomenon in eukaryotes more difficult?

18. List and define the levels of gene regulation discussed in this chapter.

19. Distinguish between the *cis*-acting regulatory elements referred to as promoters and enhancers.

20. Is the binding of a transcription factor to its DNA recognition sequence necessary and sufficient for an initiation of transcription at a regulated gene? Explain your answer.

21. Compare the control of gene regulation in eukaryotes and prokaryotes at the level of initiation of transcription. How do the regulatory mechanisms work? What are the similarities and differences in these two types of organisms in terms of the specific components of the regulatory mechanisms? Address how the differences or similarities relate to the biological context of the control of gene expression.

22. Many promoter regions contain CAAT boxes with consensus sequences CAAT or CCAAT approximately 70 to 80 bases upstream from the transcription start site. How might one determine the influence of CAAT boxes on the transcription rate of a given gene?

23. Present an overview of RNA silencing achieved through RNA interference (RNAi) and microRNAs (miRNAs). How do the silencing processes begin and what major components participate?

24. Although it is customary to consider transcriptional regulation in eukaryotes as resulting from the positive or negative influence of different factors binding to DNA, a more complex picture is emerging. For instance, researchers have described the action of a transcriptional repressor (Net) that is regulated by nuclear export (Ducret et al., 1999. *Mol. and Cell. Biol.* 19: 7076–7087). Under neutral conditions, Net inhibits transcription of target genes; however, when phosphorylated, Net stimulates transcription of

target genes. When stress conditions exist in a cell (for example, from ultraviolet light or heat shock), Net is excluded from the nucleus, and target genes are transcribed. Devise a model (using diagrams) that provides a consistent explanation of these three conditions.

25. DNA methylation is commonly associated with a reduction of transcription. The following data come from a study of the impact of the location and extent of DNA methylation on gene activity in human cells. A bacterial gene, luciferase, was cloned next to eukaryotic promoter fragments that were methylated to various degrees, *in vitro*. The chimeric plasmids were then introduced into tissue culture cells, and the luciferase activity was assayed. These data compare the degree of expression of luciferase with differences in the location of DNA methylation (Irvine et al., 2002. *Mol. and Cell. Biol.* 22: 6689–6696). What general conclusions can be drawn from these data?

DNA Segment	Patch Size of Methylation (kb)	Number of Methylated CpGs	Relative Luciferase Expression
Outside transcription unit (0–7.6 kb away)	0.0	0	490X
	2.0	100	290X
	3.1	102	250X
	12.1	593	2X
Inside transcription unit	0.0	0	490X
	1.9	108	80X
	2.4	134	5X
	12.1	593	2X

26. The interphase nucleus appears to be a highly structured organelle with chromosome territories, interchromosomal compartments, and transcription factories. In cultured human cells, researchers have identified approximately 8000 transcription factories per cell, each containing an average of eight tightly associated RNA polymerase II molecules actively transcribing RNA. If each RNA polymerase II molecule is transcribing a different gene, how might such a transcription factory appear? Provide a simple diagram that shows eight different genes being transcribed in a transcription factory and include the promoters, structural genes, and nascent transcripts in your presentation.

27. It has been estimated that 30 to 60 percent of human genes produce alternatively spliced mRNA isoforms. In some cases, incorrectly spliced RNAs lead to human pathologies. Scientists have examined human cancer cells for splice-specific changes and found that many of the changes disrupt tumor-suppressor gene function (Xu and Lee, 2003. *Nucl. Acids Res.* 31: 5635–5643). In general, what would be the effects of splicing changes on these RNAs and the function of tumor-suppressor gene function? How might loss of splicing specificity be associated with cancer?

SOLUTIONS TO SELECTED PROBLEMS AND DISCUSSION QUESTIONS

Answers to Now Solve This

1. (a) It is likely that either premature chain termination of translation will occur (from the introduction of a nonsense triplet in a reading frame) or the normal chain termination will be ignored. Regardless, a mutant condition for the Z gene will be likely. If such a cell is placed on a lactose medium, it will be incapable of growth because β-galactosidase is not available.

   (b) If the deletion occurs early in the A gene, one might expect impaired function of the A gene product, but it will not influence the use of lactose as a carbon source.

2. (a) With no lactose and no glucose, the operon is off because the *lac* repressor is bound to the operator and although CAP is bound to its binding site, it will not override the action of the repressor.

   (b) With lactose added to the medium, the *lac* repressor is inactivated and the operon is transcribing the structural genes. With no glucose, the CAP is bound to its binding site, thus enhancing transcription.

   (c) With no lactose present in the medium, the *lac* repressor is bound to the operator region, and since glucose inhibits adenyl cyclase, the CAP protein will not interact with its binding site. The operon is therefore "off."

   (d) With lactose present, the *lac* repressor is inactivated; however, since glucose is also present, CAP will not interact with its binding site. Under this condition transcription is severely diminished and the operon can be considered to be "off."

3. Should hypermetylation occur in one of many DNA repair genes, the frequency of mutation would increase because the DNA repair system is compromised. The resulting increase in mutations might occur in tumor suppressor genes or proto-oncogenes.

4. General transcription factors associate with a promoter to stimulate transcription of a specific gene. Some *trans*-acting elements, when bound to enhancers, interact with coactivators to enhance transcription by forming an enhanceosome that stimulates transcription initiation. Transcription can be repressed when certain proteins bind to silencer DNA elements and generate repressive chromatin structures. The same molecule may bind to a different chromosomal regulatory site (enhancer or silencer), depending on the molecular environment of a given tissue type.

Solutions to Problems and Discussion Questions

2. The enzymes of the lactose operon are needed to break down and use lactose as an energy source. If lactose is the sole carbon source, the enzymes are synthesized to use that carbon source. With no lactose present, there is no "need" for the enzymes. The tryptophan operon contains structural genes for the *synthesis* of tryptophan. If there is little or no tryptophan in the medium, the tryptophan operon is "turned on" to manufacture tryptophan. If tryptophan is abundant in the medium, then there is no "need" for the operon to be manufacturing tryptophan synthetases.

4. In an *inducible system*, the repressor that normally interacts with the operator to inhibit transcription is inactivated by an *inducer*,

thus permitting transcription. In a *repressible system*, a normally inactive repressor is *activated* by a *corepressor*, thus enabling it (the activated repressor) to bind to the operator to inhibit transcription.

6. $I^+ \ O^+ \ Z^+ = $ **inducible** because a repressor protein can interact with the operator to turn off transcription.

$I^- \ O^+ \ Z^+ = $ **constitutive** because the repressor gene is mutant; therefore, no repressor protein is available.

$I^+ \ O^c \ Z^+ = $ **constitutive** because even though a repressor protein is made, it cannot bind with the mutant operator.

$I^- \ O^+ \ Z^+/F' \ I^+ = $ **inducible** because even though there is one mutant repressor gene, the other I^+ gene, on the F factor, produces a normal repressor protein that is diffusible and capable of interacting with the operon to repress transcription.

$I^+ \ O^c \ Z^+/F' \ O^c = $ **constitutive** because there is a constitutive operator (O^c) next to a normal Z gene. Constitutive synthesis of β-galactosidase will occur.

$I^s \ O^+ \ Z^+ = $ **repressed** because the product of the I^s gene is *insensitive* to the inducer lactose and thus cannot be inactivated. The repressor will continually interact with the operator and shut off transcription regardless of the presence or absence of lactose.

$I^s \ O^+ \ Z^+/F' \ I^+ = $ **repressed** because, as in the previous case, the product of the I^s gene is insensitive to the inducer lactose and thus cannot be inactivated. The repressor will continually interact with the operator and shut off transcription regardless of the presence or absence of lactose. The fact that there is a normal I^+ gene is of no consequence because once a repressor from I^s binds to an operator, the presence of normal repressor molecules will make no difference.

8. The mutations described are consistent with the structure of the lac repressor. The N-terminal portion of the repressor is involved in DNA binding, while the C-terminal portion is more involved in association with lactose and its analogs.

10. Generally, cooperative binding occurs when the final outcome is greater than the simple sum of its parts. In the case of transcription factors, each factor has little impact on transcription; however, when all components are present, a cooperative interaction (binding) occurs and a functional complex is made.

12. **(a)** Because activated CAP is a component of the cooperative binding of RNA polymerase to the *lac* promoter, absence of a functional *crp* would compromise the positive control exhibited by CAP.

(b) Without a CAP binding site there would be a reduction in the inducibility of the *lac* operon.

14. C codes for the **structural gene**. Because when B is mutant, no enzyme is produced, B must be the **promoter**. The A locus is the **operator** and the D locus is the **repressor** gene.

16. Oil stimulates the production of a protein, which turns on (positive control) genes to metabolize oil. The different results in strains #2 and #4 suggest a *cis*-acting system. Because the operon by itself (when mutant as in strain #3) gives constitutive synthesis of the structural genes, a *cis*-acting system is also supported. The *cis*-acting element is most likely part of the operon.

20. Transcription factors are proteins that are *necessary* for the initiation of transcription. However, they are not *sufficient* for the initiation of transcription. To be activated, RNA polymerase II requires a number of transcription factors.

22. Generally, one determines the influence of various regulatory elements by removing necessary elements or adding extra elements. In addition, examining the outcome of mutations within such elements often provides insight as to function.

24.

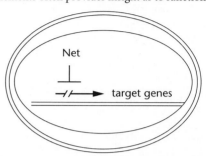

Neutral Conditions

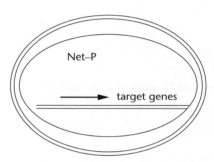

Phosphorylated Net

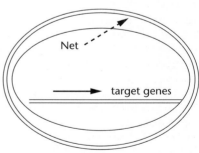

UV and Heat Shock

Sketches modified from Ducret et al. 1999 *Molecular and Cellular Biology* 19:7076–7087.

26. Below is a sketch of several RNA polymerase molecules (filled circles) in what might be a transcription factory. In this diagram there are eight RNA pol II molecules shown being transcribed. Nascent transcripts are shown extending from the RNA polymerase molecules. For simplicity, only one promoter is shown and one structural gene is shown.

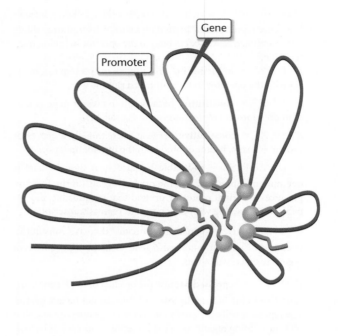

CREDITS

Credits are listed in order of appearance.

Photo

CO, T. Cremer/Dr. I. Solovei/Dr. F. Haberman/Biozentrum (LMU)

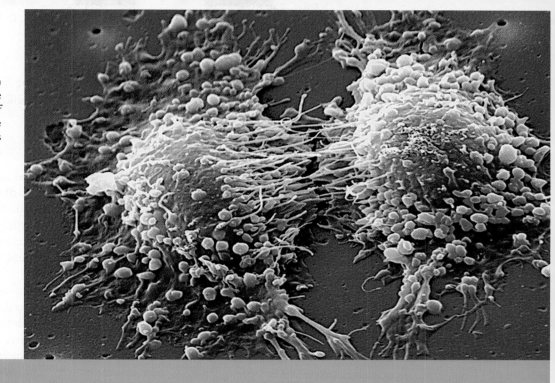

Colored scanning electron micrograph of two prostate cancer cells in the final stages of cell division (cytokinesis). The cells are still joined by strands of cytoplasm.

The Genetics of Cancer

CHAPTER CONCEPTS

- Cancer is a group of genetic diseases affecting fundamental aspects of cellular function, including DNA repair, the cell cycle, apoptosis, and signal transduction.

- Most cancer-causing mutations occur in somatic cells; only about 1 percent of cancers have a hereditary component.

- Mutations in cancer-related genes lead to abnormal proliferation and loss of control over how cells spread and invade surrounding tissues.

- The development of cancer is a multistep process requiring mutations in genes controlling many aspects of cell proliferation and metastasis.

- Cancer cells show high levels of genomic instability, leading to the accumulation of multiple mutations in cancer-related genes.

- Epigenetic effects such as DNA methylation and histone modifications may play significant roles in the development of cancers.

- Mutations in proto-oncogenes and tumor-suppressor genes contribute to the development of cancers.

- Oncogenic viruses stimulate cell proliferation and introduce oncogenes into infected cells.

- Environmental agents contribute to cancer by damaging DNA.

ancer is the leading cause of death in Western countries. It strikes people of all ages, and one out of three people will experience a cancer diagnosis sometime in his or her lifetime. Each year, more than 1 million cases of cancer are diagnosed in the United States, and more than 500,000 people die from the disease.

Over the last 30 years, scientists have discovered that cancer is a genetic disease at the somatic cell level, characterized by the presence of gene products derived from mutated or abnormally expressed genes. The combined effects of numerous abnormal gene products lead to the uncontrolled growth and spread of cancer cells. Although some mutated cancer genes may be inherited, most are created within somatic cells that then divide and form tumors. Completion of the Human Genome Project and numerous large-scale rapid DNA sequencing studies have opened the door to a wealth of new information about the mutations that trigger a cell to become cancerous. This new understanding of cancer genetics is also leading to new gene-specific treatments, some of which are now entering clinical trials. Some scientists predict that gene therapies will replace chemotherapies within the next 25 years.

The goal of this chapter is to highlight our current understanding of the nature and causes of cancer. As we will see, cancer is a genetic disease that arises from the accumulation of mutations in genes controlling many basic aspects of

From Chapter 16 of *Essentials of Genetics, Eighth edition,* William S. Klug, Michael R. Cummings, Charlotte A. Spencer, Michael A. Palladino. © 2013 by Pearson Education, Inc. All rights reserved.

cellular function. We will examine the relationship between genes and cancer, and consider how mutations, chromosomal changes, epigenetics, and environmental agents play roles in the development of cancer.

Cancer Is a Genetic Disease at the Level of Somatic Cells

Perhaps the most significant development in understanding the causes of cancer is the realization that cancer is a genetic disease. Genomic alterations that are associated with cancer range from single-nucleotide substitutions to large-scale chromosome rearrangements, amplifications, and deletions (Figure 1). However, unlike other genetic diseases, cancer is caused by mutations that occur predominantly in somatic cells. Only about 1 percent of cancers are associated with germ-line mutations that increase a person's susceptibility to certain types of cancer. Another important difference between cancer and other genetic diseases is that cancers rarely arise from a single mutation, but from the accumulation of mutations in many genes—as many as six to twelve. The mutations that lead to cancer affect multiple

(a)

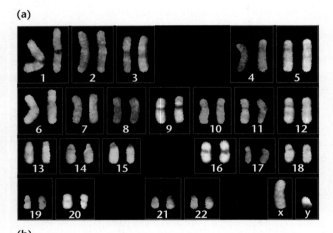

(b)

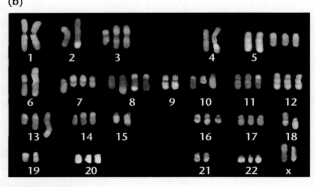

FIGURE 1 (a) Spectral karyotype of a normal cell. (b) Spectral karyotype of a cancer cell showing translocations, deletions, and aneuploidy—characteristic features of cancer cells.

cellular functions, including repair of DNA damage, cell division, apoptosis, cellular differentiation, migratory behavior, and cell–cell contact.

What Is Cancer?

Clinically, cancer is defined as a large number of complex diseases, up to a hundred, that behave differently depending on the cell types from which they originate. Cancers vary in their ages of onset, growth rates, invasiveness, prognoses, and responsiveness to treatments. However, at the molecular level, all cancers exhibit common characteristics that unite them as a family.

All cancer cells share two fundamental properties: (1) abnormal cell growth and division (**proliferation**), and (2) defects in the normal restraints that keep cells from spreading and invading other parts of the body (**metastasis**). In normal cells, these functions are tightly controlled by genes that are expressed appropriately in time and place. In cancer cells, these genes are either mutated or are expressed inappropriately.

It is this combination of uncontrolled cell proliferation and metastatic spread that makes cancer cells dangerous. When a cell simply loses genetic control over cell growth, it may grow into a multicellular mass, a **benign tumor.** Such a tumor can often be removed by surgery and may cause no serious harm. However, if cells in the tumor also acquire the ability to break loose, enter the bloodstream, invade other tissues, and form secondary tumors (**metastases**), they become malignant. **Malignant tumors** are difficult to treat and may become life threatening. As we will see later in the chapter, there are multiple steps and genetic mutations that convert a benign tumor into a dangerous malignant tumor.

ESSENTIAL POINT ■ ■ ■

Cancer cells show two fundamental properties: abnormal cell proliferation and a propensity to spread and invade other parts of the body (metastasis).

The Clonal Origin of Cancer Cells

Although malignant tumors may contain billions of cells, and may invade and grow in numerous parts of the body, all cancer cells in the primary and secondary tumors are clonal, meaning that they originated from a common ancestral cell that accumulated specific mutations. This is an important concept in understanding the molecular causes of cancer and has implications for its diagnosis.

Numerous data support the concept of cancer clonality. For example, reciprocal chromosomal translocations are characteristic of many cancers, including leukemias and lymphomas (two cancers involving white blood cells). Cancer cells from patients with **Burkitt's lymphoma** show reciprocal translocations between chromosome 8 (with translocation breakpoints at or near the *c-myc* gene) and chromosomes 2, 14, or 22 (with translocation breakpoints at or near one of the immunoglobulin genes). Each Burkitt's lymphoma patient exhibits unique breakpoints in his or her *c-myc* and immunoglobulin gene DNA sequences; however, all lymphoma cells within that patient contain identical translocation breakpoints. This demonstrates that all cancer cells in each case of Burkitt's lymphoma arise from a single cell, and this cell passes on its genetic aberrations to its progeny.

Another demonstration that cancer cells are clonal is their pattern of X-chromosome inactivation. Female humans are mosaic, with some cells containing an inactivated paternal X chromosome and other cells containing an inactivated maternal X chromosome. X-chromosome inactivation occurs early in development and takes place at random. All cancer cells within a tumor, both primary and metastatic, within one female individual, contain the same inactivated X chromosome. This supports the concept that all the cancer cells in that patient arose from a common ancestral cell.

ESSENTIAL POINT ■ ■ ■

Cancers are clonal, meaning that all cells within a tumor originate from a single cell that contained a number of mutations.

The Cancer Stem Cell Hypothesis

A concept that is related to the clonal origin of cancer cells is that of the cancer stem cell. Many scientists now believe that tumors are comprised of a mixture of cells, many of which do not proliferate. Those that do proliferate and give rise to all the cells within the tumor are known as cancer stem cells. Stem cells are cells that have the capacity for self-renewal—a process in which the stem cell divides unevenly, creating one daughter cell that goes on to differentiate into a mature cell type and one that remains a stem cell. The cancer stem cell hypothesis contrasts the random or stochastic model. This model predicts that every cell within a tumor has the potential to form a new tumor.

Although scientists still actively debate the existence of cancer stem cells, evidence is accumulating that cancer stem cells do exist, at least in some tumors. Cancer stem cells have been identified in leukemias as well as in solid tumors of the brain, breast, colon, ovary, pancreas, and prostate. It is still not clear what fraction of any tumor is comprised of cancer stem cells. For example, human acute myeloid leukemias contain less than 1 cancer stem cell in 10,000. In contrast, some solid tumors may contain as many as 40 percent cancer stem cells.

Scientists are also not sure about the origins of cancer stem cells. It is possible that they may arise from normal adult stem cells within a tissue, or they may be created from more differentiated cells that acquire properties similar to stem cells after accumulating numerous mutations.

Cancer As a Multistep Process, Requiring Multiple Mutations

Although we know that cancer is a genetic disease initiated by mutations that lead to uncontrolled cell proliferation and metastasis, a single mutation is not sufficient to transform a normal cell into a tumor-forming (tumorigenic), malignant cell. If it were sufficient, then cancer would be far more prevalent than it is. In humans, mutations occur spontaneously at a rate of about 10^{-6} mutations per gene, per cell division, mainly due to the intrinsic error rates of DNA replication. Because there are approximately 10^{16} cell divisions in a human body during a lifetime, a person might suffer up to 10^{10} mutations per gene somewhere in the body during his or her lifetime. However, only about one person in three will suffer from cancer.

The phenomenon of age-related cancer is another indication that cancer develops from the accumulation of several mutagenic events in a single cell. The incidence of most cancers rises exponentially with age. If a single mutation were sufficient to convert a normal cell to a malignant one, then cancer incidence would appear to be independent of age. The age-related incidence of cancer suggests that many independent mutations, occurring randomly, and with a low probability, are necessary before a cell is transformed into a malignant cancer cell. Another indication that cancer is a multistep process is the delay that occurs between exposure to **carcinogens** (cancer-causing agents) and the appearance of a cancer. For example, an incubation period of five to eight years separated exposure of people to the radiation of the atomic explosions at Hiroshima and Nagasaki and the onset of leukemias.

The multistep nature of cancer development is supported by the observation that cancers often develop in progressive steps, beginning with mildly aberrant cells and progressing to cells that are increasingly tumorigenic and malignant. This progressive nature of cancer is illustrated by the development of colon cancer, as discussed in Section 6.

Each step in tumorigenesis (the development of a malignant tumor) appears to be the result of two or more genetic alterations that release the cells progressively from the controls that normally operate on proliferation and malignancy. This observation suggests that the progressive genetic alterations that create a cancer cell confer selective advantages to the cell and are propagated through cell divisions during the creation of tumors.

Scientists are now applying some of the recent advances in DNA sequencing in order to identify all of the somatic mutations that occur during the development of a cancer cell. These studies compare the DNA sequences of genomes from cancer cells and normal cells derived from the same patient.

Data from these studies are revealing that tens of thousands of somatic mutations are present in cancer cells. Researchers believe that only a handful of these mutations—called driver mutations—give a growth advantage to a tumor cell. The remainder of the mutations may be acquired over time, perhaps as a result of the increased levels of DNA damage that accumulate in cancer cells, but these mutations have no direct contribution to the cancer phenotype. The total number of driver mutations that occur in any particular cancer is still unclear; however, scientists expect that the presence of fewer than a dozen mutated genes may be sufficient to create a cancer cell.

ESSENTIAL POINT ■ ■ ■

The development of cancer is a multistep process, requiring mutations in several cancer-related genes.

---

2 | Cancer Cells Contain Genetic Defects Affecting Genomic Stability, DNA Repair, and Chromatin Modifications

Cancer cells show higher than normal rates of mutation, chromosomal abnormalities, and genomic instability. Many researchers believe that the fundamental defect in cancer cells is a derangement of the cells' normal ability to repair DNA damage. This loss of genomic integrity leads to a general increase in the mutation rate for every gene in the genome, including specific genes that control aspects of cell proliferation and cell–cell contact. In turn, the accumulation of mutations in genes controlling these processes leads to cancer. The high level of genomic instability seen in cancer cells is known as the **mutator phenotype.** In addition, recent research has revealed that cancer cells contain aberrations in the types and locations of chromatin modifications, particularly DNA methylation patterns.

Genomic Instability and Defective DNA Repair

Genomic instability in cancer cells is characterized by the presence of gross defects such as translocations, aneuploidy, chromosome loss, DNA amplification, and chromosome deletions (Figure 1 and 2). Cancer cells that are grown in cultures in the lab also show a great deal of genomic instability—duplicating, losing, and translocating chromosomes or parts of chromosomes. Often cancer cells show specific chromosomal defects that are used to diagnose the type and stage of the cancer. For example, leukemic white blood cells from patients with **chronic myelogenous leukemia (CML)** bear a specific translocation, in

which the *c-ABL* gene on chromosome 9 is translocated into the *BCR* gene on chromosome 22. This translocation creates a structure known as the **Philadelphia chromosome (Figure 3).** The *BCR-ABL* fusion gene codes for a chimeric BCR-ABL protein. The normal ABL protein is a **protein kinase** that acts within signal transduction pathways, transferring growth factor signals from the external environment to the nucleus. The BCR-ABL protein is an abnormal signal transduction molecule in CML cells, which stimulates these cells to proliferate even in the absence of external growth signals.

In keeping with the concept of the cancer mutator phenotype, a number of inherited cancers are caused by defects in genes that control DNA repair. For example, xeroderma pigmentosum (XP) is a rare hereditary disorder that is characterized by extreme sensitivity to ultraviolet light and other carcinogens. Patients with XP often develop skin cancer. Cells from patients with XP are defective in nucleotide excision repair, with mutations appearing in any one of seven genes whose products are necessary to carry out DNA repair. XP cells are impaired in their ability to repair DNA lesions such as thymine dimers induced by UV light.

Another hereditary cancer, **hereditary nonpolyposis colorectal cancer (HNPCC),** is also caused by mutations in genes controlling DNA repair. HNPCC is an autosomal dominant syndrome, affecting about one in every 200 to 1000 people. Patients affected by HNPCC have an increased risk of developing colon, ovary, uterine, and kidney cancers. Cells from patients with HNPCC show higher than normal mutation rates and genomic instability. At least eight genes are associated with HNPCC, and four of these genes control aspects of DNA mismatch repair. Inactivation of any of these

(a) Double minutes   **(b)** Heterogeneous staining region

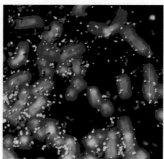 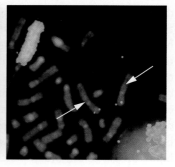

FIGURE 2   DNA amplifications in neuroblastoma cells.
(a) Two cancer genes (*MYCN* in red and *MDM2* in green) are amplified as small DNA fragments that remain separate from chromosomal DNA within the nucleus. These units of amplified DNA are known as double minute chromosomes. Normal chromosomes are stained blue. (b) Multiple copies of the *MYCN* gene are amplified within one large region called a heterogeneous staining region (green). Single copies of the *MYCN* gene are visible as green dots at the ends of the normal parental chromosomes (white arrows). Normal chromosomes are stained red.

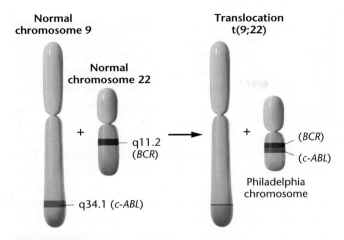

Normal chromosome 9

Normal chromosome 22

Translocation t(9;22)

q11.2 (*BCR*)

q34.1 (*c-ABL*)

(*BCR*)

(*c-ABL*)

Philadelphia chromosome

FIGURE 3 A reciprocal translocation involving the long arms of chromosomes 9 and 22 results in the formation of a characteristic chromosome, the Philadelphia chromosome, which is associated with chronic myelogenous leukemia (CML). The t(9;22) translocation results in the fusion of the *c-ABL* proto-oncogene on chromosome 9 with the *BCR* gene on chromosome 22. The fusion protein is a powerful hybrid molecule that allows cells to escape control of the cell cycle, contributing to the development of CML.

four genes—*MSH2, MSH6, MLH1,* and *MLH3*—causes a rapid accumulation of genomewide mutations and the subsequent development of colorectal and other cancers.

The observation that hereditary defects in genes controlling nucleotide excision repair and DNA mismatch repair lead to high rates of cancer lends support to the idea that the mutator phenotype is a significant contributor to the development of cancer.

Chromatin Modifications and Cancer Epigenetics

The field of cancer epigenetics is providing new perspectives on the genetics of cancer. Cancer cells contain major alterations in DNA methylation. Overall, there is much less DNA methylation in cancer cells than in normal cells. At the same time, the promoters of some genes are hypermethylated in cancer cells. These changes are thought to result in the release of transcription repression for some genes, while repressing transcription of other genes that would regulate normal cellular functions such as DNA repair and cell-cycle control.

Histone modifications are also disrupted in cancer cells. Genes that encode histone modifying enzymes are often mutated or aberrantly expressed in cancer cells. The large numbers of epigenetic abnormalities in tumors have prompted some scientists to speculate that there may be more epigenetic defects in cancer cells than there are gene mutations. In addition, because epigenetic modifications are reversible,

cancers may lend themselves to epigenetic-based therapies. Although the field of cancer epigenetics is still in its infancy, it has already provided major insights into tumorigenesis as well as new clinical applications.

ESSENTIAL POINT

Cancer cells show high rates of mutation, chromosomal abnormalities, genomic instability, and abnormal patterns of chromatin modifications.

NOW SOLVE THIS

1 In chronic myelogenous leukemia (CML), leukemic blood cells can be distinguished from other cells of the body by the presence of a functional BCR-ABL hybrid protein. Explain how this characteristic provides an opportunity to develop a treatment for CML.

■ **Hint** *This problem asks you to imagine a therapy that is based on the unique genetic characteristics of CML leukemic cells. The key to its solution is to remember that the BCR-ABL fusion protein is found only in CML white blood cells and that this unusual protein has a specific function thought to directly contribute to the development of CML. To help you answer this problem, you may wish to learn more about the cancer drug Gleevec (see http://www.cancer .gov/newscenter/qa/2001/gleevecqa).*

3 Cancer Cells Contain Genetic Defects Affecting Cell-Cycle Regulation and Apoptosis

One of the fundamental aberrations in all cancer cells is a loss of control over cell proliferation. Although some cells, such as epidermal cells of the skin or blood-forming cells in the bone marrow, continue to grow and divide throughout the organism's lifetime, most cells in adult multicellular organisms remain in a nondividing, quiescent, and differentiated state. The most extreme examples of nonproliferating cells are nerve cells, which divide little, if at all, even to replace damaged tissue. In contrast, cells in some differentiated tissues, such as those in the liver and kidney, are able to grow and divide when stimulated by extracellular signals and growth factors. In this way, multicellular organisms are able to replace dead and damaged tissue. The growth and differentiation of cells must be strictly regulated; otherwise, the integrity of organs and tissues would be compromised by the presence of inappropriate types and quantities of cells. Normal regulation over cell proliferation involves a large number of gene products that control steps in the cell cycle, programmed cell death, and the response of cells to external growth signals. In cancer cells, many of the genes that control these functions are mutated or abnormally expressed, leading to uncontrolled cell proliferation.

The Cell Cycle and Signal Transduction

Thecellular events that occur in sequence from one cell division to the next comprise the **cell cycle (Figure 4)**.

The regulation of cells entering and leaving the G0 (quiescent) phase of the cell cycle is particularly relevant to understanding cancer. Most differentiated cells in multicellular organisms can remain in this G0 phase indefinitely. In contrast, cancer cells are unable to enter G0, and instead, they continuously cycle. Their *rate* of proliferation is not necessarily any greater than that of normal proliferating cells; however, they are not able to become quiescent at the appropriate time or place.

Cells in G0 can often be stimulated to reenter the cell cycle by external growth signals. These signals are delivered to the cell by molecules such as growth factors and hormones that bind to cell-surface receptors, which then relay the signal from the plasma membrane to the cytoplasm. The process of transmitting growth signals from the external environment to the cell nucleus is known as **signal transduction**. Ultimately, signal transduction initiates a program of gene expression that stimulates the cell out of G0 back into the cell cycle. Cancer cells often have defects in signal transduction pathways. Sometimes, abnormal signal transduction molecules send continuous growth signals to the nucleus even in the absence of external growth signals. In addition, malignant cells may not respond to external signals from surrounding cells—signals that would normally inhibit cell proliferation within a mature tissue.

Cell-Cycle Control and Checkpoints

In normal cells, progress through the cell cycle is tightly regulated, and each step must be completed before the next step can begin. There are at least three distinct points in the cell cycle at which the cell monitors external signals and internal equilibrium before proceeding to the next stage. These are the **G1/S**, the **G2/M**, and **M checkpoints,** as indicated in Figure 4. At the G1/S checkpoint, the cell monitors its size and determines whether its DNA has been damaged. If the cell has not achieved an adequate size, or if the DNA has been damaged, further progress through the cell cycle is halted until these conditions are corrected. If cell size and DNA integrity are normal, the G1/S checkpoint is traversed, and the cell proceeds to S phase. The second important checkpoint is the G2/M checkpoint, where physiological conditions in the cell are monitored prior to mitosis. If DNA replication or repair of any DNA damage has not been completed, the cell cycle arrests until these processes are complete. The third major checkpoint occurs during mitosis and is called the M checkpoint. At this checkpoint, both the successful formation of the spindle-fiber system and the attachment of spindle fibers to the kinetochores associated with the centromeres are monitored. If spindle fibers are not properly formed or attachment is inadequate, mitosis is arrested.

In addition to regulating the cell cycle at checkpoints, the cell controls progress through the cell cycle by means of two classes of proteins: **cyclins** and **cyclin-dependent kinases (CDKs)**. The cell synthesizes and destroys cyclin proteins in a precise pattern during the cell cycle **(Figure 5)**. When a cyclin is present, it binds to a specific CDK, triggering activity of the CDK/cyclin complex. The CDK/cyclin complex then selectively phosphorylates and activates other proteins that in turn bring about the changes necessary to advance the cell through the cell cycle. For example, in G1 phase, CDK4/cyclin D complexes activate proteins that stimulate transcription of genes whose products (such as DNA polymerase δ and DNA ligase) are required for DNA replication during S phase. Another CDK/cyclin complex, CDK1/cyclin B, phosphorylates a number of proteins that bring about the events of early mitosis, such as nuclear membrane breakdown, chromosome condensation, and cytoskeletal reorganization. Mitosis can only be completed, however, when cyclin B is degraded and the protein phosphorylations characteristic of M phase are reversed. Although a large number of different protein kinases exist in cells, only a few are involved in cell-cycle regulation.

Mutation or misexpression of any of the genes controlling the cell cycle can contribute to the development of cancer. For example, if genes that control the G1/S or G2/M checkpoints are mutated, the cell may continue to grow and divide without repairing DNA damage. As these cells continue to divide, they accumulate mutations in genes whose products control cell proliferation or metastasis. Similarly, if genes that control progress through the cell cycle, such as those that encode the cyclins, are expressed at the wrong time or at incorrect levels, the cell may grow and divide continuously and may be unable to exit the cell cycle into G0.

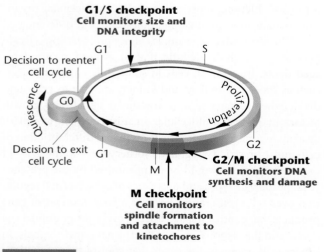

FIGURE 4 Checkpoints and proliferation decision points monitor the progress of the cell through the cell cycle.

FIGURE 5 Relative expression times and amounts of cyclins during the cell cycle. Cyclin D1 accumulates early in G1 and is expressed at a constant level through most of the cycle. Cyclin E accumulates in G1, reaches a peak, and declines by mid-S phase. Cyclin D2 begins accumulating in the last half of G1, reaches a peak just after the beginning of S, and then declines by early G2. Cyclin A appears in late G1, accumulates through S phase, peaks at the G2/M transition, and is rapidly degraded. Cyclin B peaks at the G2/M transition and declines rapidly in M phase.

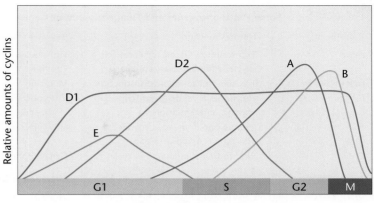

Phases of the cell cycle

The result in both cases is that the cell loses control over proliferation and is on its way to becoming cancerous.

Control of Apoptosis

As already described, if DNA replication, repair, or chromosome assembly is aberrant, normal cells halt their progress through the cell cycle until the condition is corrected. This reduces the number of mutations and chromosomal abnormalities that accumulate in normal proliferating cells. However, if DNA or chromosomal damage is so severe that repair is impossible, the cell may initiate a second line of defense—a process called **apoptosis,** or **programmed cell death.** Apoptosis is a genetically controlled process whereby the cell commits suicide. Besides its role in preventing cancer, apoptosis is also initiated during normal multicellular development in order to eliminate certain cells that do not contribute to the final adult organism. The steps in apoptosis are the same for damaged cells and for cells being eliminated during development: nuclear DNA becomes fragmented, internal cellular structures are disrupted, and the cell dissolves into small spherical structures known as apoptotic bodies. In the final step, the apoptotic bodies are engulfed by the immune system's phagocytic cells.

By removing damaged cells, programmed cell death reduces the number of mutations that are passed to the next generation, including those in cancer-causing genes. The same genes that control cell-cycle checkpoints can trigger apoptosis. These genes are mutated in many cancers. As a result of the mutation or inactivation of these checkpoint genes, the cell is unable to repair its DNA or undergo apoptosis. This inability leads to the accumulation of even more mutations in genes that control growth, division, and metastasis.

ESSENTIAL POINT

Cancer cells have defects in cell-cycle progression, checkpoint controls, and programmed cell death.

4 Proto-Oncogenes and Tumor-Suppressor Genes Are Altered in Cancer Cells

Two general categories of cancer-causing genes are mutated or misexpressed in cancer cells—the proto-oncogenes and the tumor-suppressor genes (Table 1). **Proto-oncogenes** encode transcription factors that stimulate expression of other genes, signal transduction molecules that stimulate cell division, and cell-cycle regulators that move the cell through the cell cycle. Their products are important for normal cell functions, especially cell growth and division. When normal cells become quiescent and cease division, they repress the expression of most proto-oncogenes or the activity of their products. In cancer cells, one or more proto-oncogenes are altered in such a way that their activities cannot be controlled in a normal fashion. This is sometimes due to a mutation in the proto-oncogene resulting in a protein product that acts abnormally. In other cases, proto-oncogenes are overexpressed or expressed at an incorrect time. If the proto-oncogene is continually in an "on" state, its product may constantly stimulate the cell to divide. When a proto-oncogene is mutated or aberrantly expressed and contributes to the development of cancer, it is known as an **oncogene**—a cancer-causing gene. Oncogenes are proto-oncogenes that have experienced a gain-of-function alteration. As a result, only one allele of a proto-oncogene needs to be mutated or misexpressed in order to trigger uncontrolled growth. Hence, oncogenes confer a dominant cancer phenotype.

Tumor-suppressor genes are genes whose products normally regulate cell-cycle checkpoints or initiate the process of apoptosis. In normal cells, proteins encoded by tumor-suppressor genes halt progress through the cell cycle in response to DNA damage or growth-suppression signals from the extracellular environment. When tumor-suppressor genes are mutated or inactivated, cells are unable to respond normally to cell-cycle checkpoints, or are unable to undergo

TABLE 1	Some Proto-oncogenes and Tumor-suppressor Genes		
Proto-oncogene	**Normal Function**	**Alteration in Cancer**	**Associated Cancers**
c-myc	Transcription factor, regulates cell cycle, differentiation, apoptosis	Translocation, amplification, point mutations	Lymphomas, leukemias, lung cancer, many types
c-kit	Tyrosine kinase, signal transduction	Mutation	Sarcomas
RARα	Hormone-dependent transcription factor, differentiation	Chromosomal translocations with PML gene, fusion product	Acute promyelocytic leukemia
E6	Human papillomavirus encoded oncogene, inactivates p53	HPV infection	Cervical cancer
Cyclins	Bind to CDKs, regulate cell cycle	Gene amplification, overexpression	Lung, esophagus, many types
Tumor Suppressor	**Normal Function**	**Alteration in Cancer**	**Associated Cancers**
RB1	Cell-cycle checkpoints, binds E2F	Mutation, deletion, inactivation by viral oncogene products	Retinoblastoma, osteosarcoma, many types
APC	Cell–cell interaction	Mutation	Colorectal cancers, brain, thyroid
P53	Transcription regulation	Mutation, deletion, viruses	Many types
BRCA1, BRCA2	DNA repair	Point mutations	Breast, ovarian, prostate cancers

programmed cell death if DNA damage is extensive. This leads to a further increase in mutations and to the inability of the cell to leave the cell cycle when it should become quiescent. When both alleles of a tumor-suppressor gene are inactivated, and other changes in the cell keep it growing and dividing, the cell may become tumorigenic.

The following are examples of proto-oncogenes and tumor-suppressor genes that contribute to cancer when mutated. More than 400 oncogenes and tumor-suppressor genes are now known, and more will likely be discovered as cancer research continues.

The *ras* Proto-Oncogenes

Some of the most frequently mutated genes in human tumors are those in the **ras gene family**. These genes are mutated in more than 30 percent of human tumors. The *ras* gene family encodes signal transduction molecules that are associated with the cell membrane and regulate cell growth and division. Ras proteins normally transmit signals from the cell membrane to the nucleus, stimulating the cell to divide in response to external growth factors. Ras proteins alternate between an inactive (switched off) and an active (switched on) state by binding either guanosine diphosphate (GDP) or guanosine triphosphate (GTP). When a cell encounters a growth factor (such as platelet-derived growth factor or epidermal growth factor), growth factor receptors on the cell membrane bind to the growth factor, resulting in activated Ras. The active, GTP-bound form of Ras then sends its signals through cascades of protein phosphorylations in the cytoplasm. The end-point of these cascades is activation of nuclear transcription factors that stimulate expression of genes whose products drive the cell from quiescence into the cell cycle. Once Ras has sent its signals to the nucleus, it hydrolyzes GTP to GDP and becomes inactive. Mutations

that convert the proto-oncogene *ras* to an oncogene prevent the Ras protein from hydrolyzing GTP to GDP and hence freeze the Ras protein into its "on" conformation, constantly stimulating the cell to divide.

> **ESSENTIAL POINT** ■ ■ ■
>
> Proto-oncogenes are normal genes that promote cell growth and division. When proto-oncogenes are mutated or misexpressed in cancer cells, they are known as oncogenes.

The *p53* Tumor-Suppressor Gene

The most frequently mutated gene in human cancers—mutated in more than 50 percent of all cancers—is the **p53 gene**. This gene encodes a nuclear protein that acts as a transcription factor, repressing or stimulating transcription of more than 50 different genes.

Normally, the p53 protein is synthesized continuously but is rapidly degraded and therefore is present in cells at low levels. Several types of cellular stress events bring about rapid increases in the nuclear levels of activated p53 protein. These include chemical damage to DNA, double-stranded breaks in DNA induced by ionizing radiation, and the presence of DNA-repair intermediates generated by exposure of cells to ultraviolet light. In response to these signals, p53 protein becomes more stable and more transcriptionally active.

The p53 protein initiates two different responses to DNA damage: either arrest of the cell cycle followed by DNA repair, or apoptosis and cell death if DNA cannot be repaired. Both of these responses are accomplished by p53 acting as a transcription factor that stimulates or represses the expression of genes involved in each response.

In normal cells, p53 can arrest the cell cycle at the G1/S and G2/M checkpoints, as well as retarding the progression of the cell through S phase. It accomplishes this by inhibiting cyclin/CDK complexes and regulating the transcription of other genes involved in these phases of the cell cycle.

Activated p53 can also instruct a damaged cell to commit suicide by apoptosis. It does so by activating the transcription of genes whose products control this process. In cancer cells that lack functional p53, these gene products are not synthesized and apoptosis may not occur.

Hence, cells lacking functional p53 are unable to arrest at cell-cycle checkpoints or to enter apoptosis in response to DNA damage. As a result, they move unchecked through the cell cycle, regardless of the condition of the cell's DNA. Cells lacking p53 have high mutation rates and accumulate the types of mutations that lead to cancer. Because of the importance of the *p53* gene to genomic integrity, it is often referred to as the "guardian of the genome."

The *RB1* Tumor-Suppressor Gene

The loss or mutation of the *RB1* (**retinoblastoma 1**) tumor-suppressor gene contributes to the development of many cancers, including those of the breast, bone, lung, and bladder. The *RB1* gene was identified originally as a result of studies on **retinoblastoma,** an inherited disorder in which tumors develop in the eyes of young children. Retinoblastoma occurs with a frequency of about 1 in 15,000 individuals. In the familial form of the disease, individuals inherit one mutated allele of the *RB1* gene and have an 85 percent chance of developing retinoblastomas as well as an increased chance of developing other cancers. All somatic cells of patients with hereditary retinoblastoma contain one mutated allele of the *RB1* gene. However, it is only when the second normal allele of the *RB1* gene is lost or mutated in certain retinal cells that retinoblastoma develops. In individuals who do not have this hereditary condition, retinoblastoma is extremely rare, as it requires at least two separate somatic mutations in a retinal cell in order to inactivate both copies of the *RB1* gene.

The **retinoblastoma protein (pRB)** is a tumor-suppressor protein that controls the G1/S cell-cycle checkpoint. The pRB protein is found in the nuclei of all cell types at all stages of the cell cycle. However, its activity varies throughout the cell cycle, depending on its phosphorylation state. When cells are in the G0 phase of the cell cycle, the pRB protein is nonphosphorylated and binds to transcription factors such as E2F, inactivating them (**Figure 6**). When the cell is stimulated by growth factors, it enters G1 and approaches S phase. Throughout the G1 phase, the pRB protein becomes phosphorylated by the CDK4/cyclin D1 complex. Phosphorylated pRB releases its bound regulatory proteins. When E2F and other regulators are released by pRB, they are free to induce the expression of over 30 genes whose products are required for the transition from G1 into S phase. After cells traverse S, G2, and M phases, pRB reverts to a nonphos-

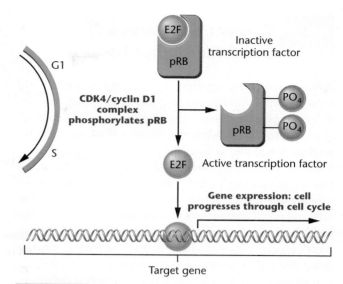

FIGURE 6 In the nucleus during G0 and early G1, pRB interacts with and inactivates transcription factor E2F. As the cell moves from G1 to S phase, a CDK4/cyclinD1 complex forms and adds phosphate groups to pRB. As pRB becomes phosphorylated, E2F is released and becomes transcriptionally active, allowing the cell to pass through S phase. Phosphorylation of pRB is transitory; as CDK/cyclin complexes are degraded and the cell moves through the cell cycle to early G1, pRB phosphorylation declines, allowing pRB to reassociate with E2F.

phorylated state, binds to regulatory proteins such as E2F, and keeps them sequestered until required for the next cell cycle. In normal quiescent cells, the presence of the pRB protein prevents passage into S phase. In many cancer cells, including retinoblastoma cells, both copies of the *RB1* gene are defective, inactive, or absent, and progression through the cell cycle is not regulated.

ESSENTIAL POINT

Tumor-suppressor genes normally regulate cell-cycle checkpoints and apoptosis. When tumor-suppressor genes are mutated or inactivated, cells cannot correct DNA damage. This leads to accumulations of mutations that may cause cancer.

NOW SOLVE THIS

2 People with a genetic condition known as Li-Fraumeni syndrome inherit one mutant copy of the *p53* gene. These people have a high risk of developing a number of different cancers, such as breast cancer, leukemia, and brain tumors. Explain how mutations in one cancer-related gene can give rise to such a diverse range of tumors.

■ **Hint** *This problem involves an understanding of how tumor-suppressor genes regulate cell growth and behavior. The key to its solution is to consider which cellular functions are regulated by the p53 protein and how the absence of p53 could affect each of these functions. Also, read about loss of heterozygosity in Section 6.*

5 Cancer Cells Metastasize and Invade Other Tissues

As discussed at the beginning of this chapter, uncontrolled growth alone is insufficient to create a malignant and life-threatening cancer. Cancer cells must also acquire the features of metastasis, which include the ability to disengage from the original tumor site, to enter the blood or lymphatic system, to invade surrounding tissues, and to develop into secondary tumors. In order to leave the site of the primary tumor and invade other tissues, tumor cells must dissociate from other cells and digest components of the **extracellular matrix** and **basal lamina,** which normally surround and separate the body's tissues. The extracellular matrix and basal lamina are composed of proteins and carbohydrates. They form the scaffold for tissue growth and inhibit the migration of cells.

The ability to invade the extracellular matrix is also a property of some normal cell types. For example, implantation of the embryo in the uterine wall during pregnancy requires cell migration across the extracellular matrix. In addition, white blood cells reach sites of infection by penetrating capillary walls. The mechanisms of invasion are probably similar in these normal cells and in cancer cells. The difference is that, in normal cells, the invasive ability is tightly regulated, whereas in tumor cells, this regulation has been lost.

Metastasis is controlled by a large number of genes, including those that encode cell-adhesion molecules, cytoskeleton regulators, and proteolytic enzymes. For example, epithelial tumors have a lower than normal level of the **E-cadherin glycoprotein,** which is responsible for cell–cell adhesion in normal tissues. Also, proteolytic enzymes such as **metalloproteinases** are present at higher than normal levels in highly malignant tumors and are not susceptible to the normal controls conferred by regulatory molecules such as **tissue inhibitors of metalloproteinases (TIMPs).** It has been shown that a tumor's level of aggressiveness correlates positively with the tumor's levels of proteolytic enzymes.

Like the tumor-suppressor genes of primary cancers, **metastasis-suppressor genes** are mutated or disrupted in metastatic tumors. Less than a dozen of these metastasis-suppressor genes have been identified so far, but all appear to affect the growth of metastatic tumors and not the primary tumor. The expression of these genes is often reduced by epigenetic mechanisms rather than by mutation. This observation provides hope that researchers can develop anti-metastasis therapies that target the epigenetic silencing of metastasis-suppressor genes.

ESSENTIAL POINT ■ ■ ■

The ability of cancer cells to metastasize requires defects in gene products that control a number of functions such as cell adhesion, proteolysis, and tissue invasion.

6 Predisposition to Some Cancers Can Be Inherited

Although the vast majority of human cancers are sporadic, a small fraction (1 to 2 percent) have a hereditary or familial component. At present, about 50 forms of hereditary cancer are known (Table 2).

Most inherited cancer-susceptibility alleles, though transmitted in a Mendelian dominant fashion, are not sufficient in themselves to trigger development of a cancer. At least one other somatic mutation in the other copy of the gene must occur in order to drive a cell toward tumorigenesis. In addition, mutations in still other genes are usually necessary to fully express the cancer phenotype. As mentioned earlier, inherited mutations in the *RB1* gene predispose individuals to developing various cancers. Although the normal somatic cells of these patients are heterozygous for the *RB1* mutation, cells within their tumors contain mutations in both copies of the gene. The phenomenon whereby the second, wild-type, allele is mutated in a tumor is known as **loss of heterozygosity.** Although loss of heterozygosity is an essential first step in expression of these inherited cancers, further mutations in other proto-oncogenes and tumor-suppressor genes are necessary for the tumor cells to become fully malignant.

The development of hereditary colon cancer illustrates how inherited mutations in one allele of a gene contribute only one step in the multistep pathway leading to malignancy.

About 1 percent of colon cancer cases result from a genetic predisposition to cancer known as **familial adenomatous polyposis (FAP).** In FAP, individuals inherit one mutant copy of the *APC* (**adenomatous polyposis**) **gene** located on the long arm of chromosome 5. Mutations include deletions, frameshift, and point mutations. The

TABLE 2   Some Inherited Predispositions to Cancer

Tumor Predisposition Syndromes	Gene Affected
Early-onset familial breast cancer	BRCA1
Familial adenomatous polyposis	APC
Familial melanoma	CDKN2
Gorlin syndrome	PTCH1
Hereditary nonpolyposis colon cancer	MSH2, 6
Li-Fraumeni syndrome	p53
Multiple endocrine neoplasia, type 1	MEN1
Multiple endocrine neoplasia, type 2	RET
Neurofibromatosis, type 1	NF1
Neurofibromatosis, type 2	NF2
Retinoblastoma	RB1
Von Hippel–Lindau syndrome	VHL
Wilms tumor	WT1

normal function of the *APC* gene product is to act as a tumor suppressor controlling cell–cell contact and growth inhibition by interacting with the β-catenin protein. The presence of a heterozygous *APC* mutation causes the epithelial cells of the colon to partially escape cell-cycle control, and the cells divide to form small clusters of cells called **polyps** or adenomas. People who are heterozygous for this condition develop hundreds to thousands of colon and rectal polyps early in life. Although it is not necessary for the second allele of the *APC* gene to be mutated in polyps at this stage, in the majority of cases, the second *APC* allele becomes mutant in a later stage of cancer development. The relative order of mutations in the development of FAP is shown in **Figure 7**.

The second mutation in polyp cells that contain an *APC* gene mutation occurs in the *ras* proto-oncogene. The combined *APC* and *ras* gene mutations bring about the development of intermediate adenomas. Cells within these adenomas have defects in normal cell differentiation. In addition, these cells will grow in culture and are not growth-inhibited by contact with other cells—a process known as **transformation.** The third step toward malignancy requires loss of function of both alleles of the *DCC* (*d*eleted in *c*olon *c*ancer) gene. The *DCC* gene product is thought to be involved with cell adhesion and differentiation. Mutations in both *DCC* alleles result in the formation of late-stage adenomas with a number of finger-like outgrowths (villi). When late adenomas progress to cancerous adenomas, they usually suffer loss of functional *p53* genes. The final steps toward malignancy involve mutations in an unknown number of genes associated with metastasis.

ESSENTIAL POINT ■ ■ ■

Inherited mutations in cancer-susceptibility genes are not sufficient to trigger cancer. Other somatic mutations in proto-oncogenes or tumor-suppressor genes are necessary for the development of hereditary cancers.

NOW SOLVE THIS

3 Although tobacco smoking is responsible for a large number of human cancers, not all smokers develop cancer. Similarly, some people who inherit mutations in the tumor-suppressor genes *p53* or *RB1* never develop cancer. Explain these observations.

■ **Hint** *This problem asks you to consider the reasons why only some people develop cancer as a result of environmental factors or mutations in tumor-suppressor genes. The key to its solution is to consider the steps involved in the development of cancer.*

7 Viruses Contribute to Cancer in Both Humans and Animals

It is thought that about 15 percent of human cancers are associated with viruses, making virus infection the second greatest risk factor for cancer, next to tobacco smoking. The most significant contributors to virus-induced cancers are described in **Table 3**. Like other risk factors for cancer, including hereditary predisposition to certain cancers, virus infection alone is not sufficient to trigger human cancers. Other factors, including DNA damage or the accumulation of mutations in one or more of a cell's oncogenes and tumor-suppressor genes, are required to move a cell down the multistep pathway to cancer.

DNA viruses are viruses whose genomes consist of double-stranded DNA. These types of viruses contribute to the development of cancers in a variety of ways. Because viruses are comprised solely of a nucleic acid genome surrounded by a protein coat, they must utilize the host cell's biosynthetic machinery in order to reproduce themselves. To access the host's DNA-synthesizing enzymes, viruses require the host cell to be in an actively growing state. Thus, many DNA viruses contain genes encoding products that stimulate the cell cycle. These products often interact with tumor-suppressor

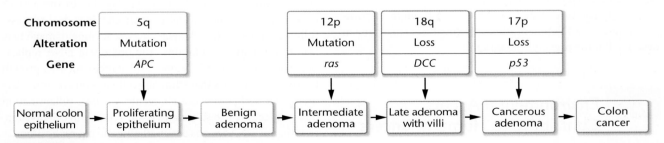

FIGURE 7   A model for the multistep development of colon cancer. The first step is the loss or inactivation of one allele of the *APC* gene on chromosome 5. In FAP cases, one mutant *APC* allele is inherited. Subsequent mutations involving genes on chromosomes 12, 17, and 18 in cells of benign adenomas can lead to a malignant transformation that results in colon cancer. Although the mutations on chromosomes 12, 17, and 18 usually occur at a later stage than those involving chromosome 5, the sum of changes is more important than the order in which they occur.

TABLE 3 Viruses Associated with Human Cancers

Virus Type	Virus Name	Associated Cancers
DNA Viruses	Epstein-Barr virus	Burkitt's lymphoma, B and T cell lymphomas
	Human papilloma virus 16, 18	Cervical cancer
	Hepatitis B virus	Hepatocellular carcinoma
	Human herpesvirus-8	Kaposi sarcoma, lymphoma
RNA Viruses	Human T lymphotrophic Virus type 1	Adult T-cell leukemia
	Hepatitis C virus	Hepatocellular carcinoma

proteins, inactivating them. If the host cell survives the infection, it may lose control of the cell cycle and begin its journey to carcinogenesis.

RNA viruses are those whose genomes consist of single or double-stranded RNA molecules. A type of RNA virus known as a **retrovirus** can contribute to the development of cancers in both animals and humans in three different ways. First, the viral genome, after being converted from RNA to DNA, may integrate by chance near one of the cell's normal proto-oncogenes. The strong promoters and enhancers in the viral genome then stimulate high levels or inappropriate timing of transcription of the proto-oncogene, leading to host-cell proliferation. Second, a retrovirus may pick up a copy of a host proto-oncogene during a previous infection and integrate it into its own genome. The cellular proto-oncogene may be mutated during the process of transfer into the virus, or it may be expressed at abnormal levels because it is now under the control of viral promoters. Retroviruses that carry these cell-derived oncogenes can infect and transform normal cells into tumor cells, and are known as acute transforming retroviruses. Third, a retrovirus may contain a normal viral gene whose product can either stimulate the cell cycle or act as a gene-expression regulator for both cellular and viral genes. As a result, expression of such a viral gene may lead to inappropriate cell growth or to abnormal expression of cancer-related cellular genes.

The retroviruses listed in Table 3 are thought to stimulate cancer development through the third mechanism described in the previous paragraph.

8   Environmental Agents Contribute to Human Cancers

Any substance or event that damages DNA has the potential to be carcinogenic. Unrepaired or inaccurately repaired DNA introduces mutations, which, if they occur in proto-oncogenes or tumor-suppressor genes, can lead to abnormal

regulation of the cell cycle or disruption of controls over apoptosis or metastasis.

Our environment, both natural and human-made, contains abundant carcinogens. These include chemicals, radiation, some viruses, and chronic infections. Perhaps the most significant carcinogen in our environment is tobacco smoke, which contains at least 60 chemicals that interact with DNA and cause mutations. Epidemiologists estimate that about 30 percent of human cancer deaths are associated with cigarette smoking.

Diet is often implicated in the development of cancer. Consumption of red meat and animal fat is associated with some cancers, such as colon, prostate, and breast cancer. The mechanisms by which these substances may contribute to carcinogenesis are not clear but may involve stimulation of cell division through hormones or creation of carcinogenic chemicals during cooking. Alcohol may cause inflammation of the liver and contribute to liver cancer.

Although most people perceive the human-made, industrial environment to be a highly significant contributor to cancer, it may account for only a small percentage of total cancers, and only in special situations. Some of the most mutagenic agents, and hence potentially the most carcinogenic, are natural substances and natural processes. For example, **aflatoxin,** a component of a mold that grows on peanuts and corn, is one of the most carcinogenic chemicals known. Most chemical carcinogens, such as **nitrosamines,** are components of synthetic substances and are found in some preserved meats; however, many are naturally occurring. For example, natural pesticides and antibiotics found in plants may be carcinogenic, and the human body itself creates alkylating agents in the acidic environment of the gut. Nevertheless, these observations do not diminish the serious cancer risks to specific populations who are exposed to human-made carcinogens such as synthetic pesticides or asbestos.

DNA lesions brought about by natural radiation (such as X rays, ultraviolet light), natural dietary substances, and substances in the external environment contribute mutations that lead to cancer. In addition, normal metabolism creates oxidative end products that can damage DNA, proteins, and lipids. It is estimated that the human body suffers about 10,000 damaging DNA lesions per day due to the actions of oxygen free radicals. DNA repair enzymes deal successfully with most of this damage; however, some damage may lead to permanent mutations. The process of DNA replication itself is mutagenic. Hence, substances such as growth factors or hormones that stimulate cell division are ultimately mutagenic and perhaps carcinogenic. Chronic inflammation due to infection also stimulates tissue repair and cell division, resulting in DNA lesions accumulating during replication.

ESSENTIAL POINT ■   ■   ■

The most significant environmental factors that affect human cancers are tobacco smoke, viruses, diet, and natural radiation.

GENETICS, TECHNOLOGY, AND SOCIETY

Breast Cancer: The Double-Edged Sword of Genetic Testing

These are exhilarating times for genetics and biotechnology. The prospect of using genetics to prevent and cure a wide range of diseases is exciting. However, in our enthusiasm, we often forget that these new technologies still have significant limitations and profound ethical complexities. The story of genetic testing for breast cancer illustrates how we must temper our high expectations with respect for uncertainty.

Breast cancer is the most common cancer among women. A woman's lifetime risk of developing breast cancer is about 12 percent. Each year, more than 200,000 new cases are diagnosed in the United States. Breast cancer is not limited to women; about 1400 men are also diagnosed with the disease each year.

Approximately 5 to 10 percent of breast cancers are familial, a category defined by the early onset of the disease and the appearance of several cases of breast or ovarian cancer among near blood relatives. In 1994, two genes were identified that show linkage to familial breast cancers: *BRCA1* and *BRCA2*. In normal cells, these two genes appear to encode tumor-suppressor proteins that are involved in repairing damaged DNA. Women with germ-line mutations in *BRCA1* or *BRCA2* have a 36 to 85 percent lifetime risk of developing breast cancer and a 16 to 60 percent risk of developing ovarian cancer. Men with germ-line mutations in *BRCA2* have a 6 percent lifetime breast cancer risk—a hundredfold increase over the general male population.

BRCA1 and *BRCA2* genetic tests are available, and these detect over 2000 different mutations that are known to occur within the coding regions of these genes. Many patients at risk for familial breast cancer opt to undergo genetic testing. These patients feel that test results could motivate them to take steps to prevent breast or ovarian cancers, guide them in childbearing decisions, and provide information concerning the risk to close relatives. But all these potential benefits are fraught with uncertainties.

A woman whose *BRCA* test results are negative may feel relieved and assume that she is not subject to familial breast cancer. However, her risk of developing breast cancer is still 12 percent (the population risk), and she should continue to monitor herself for the disease. Also, a negative *BRCA* genetic test does not eliminate the possibility that she carries an inherited mutation in another gene that increases breast cancer risk or that her *BRCA1* or *BRCA2* gene mutations exist in regions of the genes that are inaccessible to current genetic tests.

A woman whose test results are positive faces difficult choices. Her treatment options consist of close monitoring, prophylactic mastectomy or oophorectomy (removal of breasts and ovaries, respectively), or taking prophylactic drugs such as tamoxifen. Prophylactic surgery reduces her risk but does not eliminate it, as cancers can still occur in tissues that remain after surgery. Drugs such as tamoxifen are helpful but have serious side effects. Genetic tests also affect the patient's entire family. People often experience fear, anxiety, and guilt on learning that they are carriers of a genetic disease. Confidentiality is also a major concern. Patients fear that their genetic test results may be leaked to insurance companies or employers, jeopardizing their prospects for jobs or affordable health and life insurance.

The unanswered scientific and ethical questions about *BRCA1* and *BRCA2* genetic testing are many and important. As we develop genetic tests for more and more diseases over the next few decades, our struggle with these kinds of questions will continue.

Your Turn

Take time, individually or in groups, to answer the following questions. Investigate the references and links, to help you understand some of the issues that surround genetic testing for breast cancer.

1. New genomics research is rapidly identifying genes linked to human diseases, including breast cancer. How many genes are now thought to be involved in familial breast cancer? Are genetic tests available to detect mutations in these genes?
   Search for recent scientific data on breast cancer susceptibility genes by using the PubMed Web site (http://www.ncbi.nlm.nih.gov/pubmed), .

2. In 2008, the U.S. Congress passed a bill designed to eliminate discrimination by employers or health insurance companies, based on a person's genetic test results. Do you think that this act will be effective? What problems exist with the act, and how might they be solved?
   Read news articles about the Genetic Information Nondiscrimination Act by searching the New York Times Web site at: http://www.nytimes.com.

3. What are the rates of "false negatives" in *BRCA1/2* gene tests—that is, the number of high-risk patients who do not show BRCA1/2 gene mutations in genetic tests but still have genetic defects that lead to breast cancer?
   This topic is discussed in Stokstad, E. 2006. Genetic screen misses mutations in women at high risk of breast cancer. *Science* 311: 1847.

4. If your family was at risk for familial breast cancer, would you opt to take the BRCA1/2 genetic tests? What actions would you take if you received a positive test result? How would you feel about such a result?
   Two helpful sources are: (a) Surbone, A. 2001. Ethical implications of genetic testing for breast cancer susceptibility. *Crit. Rev. in Onc./Hem.* 40: 149–157. *(b) Genetic Testing for Breast and Ovarian Cancer Risk: It's your choice. National Cancer Institute Fact Sheet.* http://www.cancer.gov/cancertopics/genetics/genetic-testing-for-breast-and-ovarian-cancer-risk.pdf.

CASE STUDY » I thought it was safe

A middle-aged woman taking the breast cancer drug tamoxifen for ten years became concerned when she saw a news report with disturbing information. In some women, the drug made their cancer more aggressive and more likely to spread. Other women with breast cancer, the report stated, do not respond to tamoxifen at all, and 30 to 40 percent of women who take the drug eventually become resistant to chemotherapy. The woman contacted her oncologist to ask some questions:

1. How can some people react one way to a cancer treatment and others react a different way?
2. Why do most cancers eventually become resistant to a specific chemotherapeutic drug?
3. Why does it seem that some drugs are thought to be safe one day and declared unsafe the next day?

INSIGHTS AND SOLUTIONS

1. In disorders such as retinoblastoma, a mutation in one allele of the *RB1* gene can be inherited from the germ line, causing an autosomal dominant predisposition to the development of eye tumors. To develop tumors, a somatic mutation in the second copy of the *RB1* gene is necessary, indicating that the mutation itself acts as a recessive trait. Given that the first mutation can be inherited, in what ways can a second mutational event occur?

Solution: In considering how this second mutation arises, we must look at several types of mutational events, including changes in nucleotide sequence and events that involve whole chromosomes or chromosome parts. Retinoblastoma results when both copies of the *RB1* locus are lost or inactivated. With this in mind, you must first list the phenomena that can result in a mutational loss or the inactivation of a gene.

One way the second *RB1* mutation can occur is by a nucleotide alteration that converts the remaining normal *RB1* allele to a mutant form. This alteration can occur through a nucleotide substitution or through a frameshift mutation caused by the insertion or deletion of nucleotides during replication. A second mechanism involves the loss of the chromosome carrying the normal allele. This event would take place during mitosis, resulting in chromosome 13 monosomy and leaving the mutant copy of the gene as the only *RB1* allele. This mechanism does not necessarily involve loss of the entire chromosome; deletion of the long arm (*RB1* is on 13q) or an interstitial deletion involving the *RB1* locus and some surrounding material would have the same result. Alternatively, a chromosome aberration involving loss of the normal copy of the *RB1* gene might be followed by duplication of the chromosome carrying the mutant allele. Two copies of chromosome 13 would be restored to the cell, but the normal *RB1* allele would not be present. Finally, a recombination event followed by chromosome segregation could produce a homozygous combination of mutant *RB1* alleles.

2. Proto-oncogenes can be converted to oncogenes in a number of different ways. In some cases, the proto-oncogene itself becomes amplified up to hundreds of times in a cancer cell. An example is the *cyclin D1* gene, which is amplified in some cancers. In other cases, the proto-oncogene may be mutated in a limited number of specific ways leading to alterations in the gene product's structure. The *ras* gene is an example of a proto-oncogene that becomes oncogenic after suffering point mutations in specific regions of the gene. Explain why these two proto-oncogenes (*cyclin D1* and *ras*) undergo such different alterations in order to convert them into oncogenes.

Solution: The first step in solving this question is to understand the normal functions of these proto-oncogenes and to think about how either amplification or mutation would affect each of these functions.

The cyclin D1 protein regulates progression of the cell cycle from G1 into S phase, by binding to CDK4 and activating this kinase. The cyclin D1/CDK4 complex phosphorylates a number of proteins including pRB, which in turn activates other proteins in a cascade that results in transcription of genes whose products are necessary for DNA replication in S phase. The simplest way to increase the activity of cyclin D1 would be to increase the number of cyclin D1 molecules available for binding to the cell's endogenous CDK4 molecules. This can be accomplished by several mechanisms, including amplification of the *cyclin D1* gene. In contrast, a point mutation in the *cyclin D1* gene would most likely interfere with the ability of the cyclin D1 protein to bind to CDK4; hence, mutations within the gene would probably repress cell-cycle progression rather than stimulate it.

The *ras* gene product is a signal transduction protein that operates as an on/off switch in response to external stimulation by growth factors. It does so by binding either GTP (the "on" state) or GDP (the "off" state). Oncogenic mutations in the *ras* gene occur in specific regions that alter the ability of the Ras protein to exchange GDP for GTP. Oncogenic Ras proteins are locked in the "on" conformation, bound to GTP. In this way, they constantly stimulate the cell to divide. An amplification of the *ras* gene would simply provide more molecules of normal Ras protein, which would still be capable of on/off regulation. Hence, simple amplification of *ras* would less likely be oncogenic.

PROBLEMS AND DISCUSSION QUESTIONS

For activities, animations, and review quizzes, go to the study area at www.masteringgenetics.com

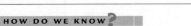

1. In this chapter, we focused on cancer as a genetic disease. In particular, we discussed the relationship between cancer, the cell cycle, and mutations in proto-oncogenes and tumor-suppressor genes. Based on your knowledge of these topics, answer several fundamental questions:

(a) How do we know that malignant tumors arise from a single cell that contains mutations?
(b) How do we know that cancer development requires more than one mutation?
(c) How do we know that cancer cells contain defects in DNA repair?

2. As a genetic counselor, you are asked to assess the risk for a couple with a family history of retinoblastoma who are thinking about having children. Both the husband and wife are phenotypically normal, but the husband has a sister with familial retinoblastoma in both eyes. What is the probability that this couple will have a child with retinoblastoma? Are there any tests that you could recommend to help in this assessment?

3. What events occur in each phase of the cell cycle? Which phase is most variable in length?

4. Where are the major regulatory points in the cell cycle?

5. List the functions of kinases and cyclins, and describe how they interact to cause cells to move through the cell cycle.

6. (a) How does pRB function to keep cells at the G1 checkpoint? (b) How do cells get past the G1 checkpoint to move into S phase?

7. What is the difference between saying that cancer is inherited and saying that the predisposition to cancer is inherited?

8. What is apoptosis, and under what circumstances do cells undergo this process?

9. Define tumor-suppressor genes. Why is a mutation in a single copy of a tumor-suppressor gene expected to behave as a recessive gene?

10. A genetic variant of the retinoblastoma protein, called PSM-RB (phosphorylation site mutated RB), is not able to be phosphorylated by the action of CDK4/cyclinD1 complex. Explain why PSM-RB is said to have a constitutive growth-suppressing action on the cell cycle.

11. Part of the Ras protein is associated with the plasma membrane, and part extends into the cytoplasm. How does the Ras protein transmit a signal from outside the cell into the cytoplasm? What happens in cases where the *ras* gene is mutated?

12. If a cell suffers damage to its DNA while in S phase, how can this damage be repaired before the cell enters mitosis?

13. Distinguish between oncogenes and proto-oncogenes. In what ways can proto-oncogenes be converted to oncogenes?

14. Of the two classes of genes associated with cancer, tumor-suppressor genes and oncogenes, mutations in which group can be considered gain-of-function mutations? In which group are the loss-of-function mutations? Explain.

15. How do translocations such as the Philadelphia chromosome contribute to cancer?

16. Given that cancers can be environmentally induced and that some environmental factors are the result of lifestyle choices such as smoking, sun exposure, and diet, what percentage of the money spent on cancer research do you think should be devoted to research and education on preventing cancer rather than on finding cancer cures?

17. Explain why some viruses that cause cancer contain genes whose products interact with tumor-suppressor proteins.

18. How do normal cells protect themselves from accumulating mutations in genes that could lead to cancer? How do cancer cells differ from normal cells in these processes?

19. Describe the difference between an acute transforming virus and a virus that does not cause tumors.

20. Explain how environmental agents such as chemicals and radiation cause cancer.

21. Radiotherapy (treatment with ionizing radiation) is one of the most effective current cancer treatments. It works by damaging DNA and other cellular components. In which ways could radiotherapy control or cure cancer, and why does radiotherapy often have significant side effects?

22. Genetic tests that detect mutations in the *BRCA1* and *BRCA2* oncogenes are widely available. These tests reveal a number of mutations in these genes—mutations that have been linked to familial breast cancer. Assume that a young woman in a suspected breast cancer family takes the *BRCA1* and *BRCA2* genetic tests and receives negative results. That is, she does not test positive for the mutant alleles of *BRCA1* or *BRCA2*. Can she consider herself free of risk for breast cancer?

23. Explain the apparent paradox that both hypermethylation and hypomethylation of DNA are found in the same cancer cell.

24. Explain the differences between a benign and malignant tumor.

25. As part of a cancer research project, you have discovered a gene that is mutated in many metastatic tumors. After determining the DNA sequence of this gene, you compare the sequence with those of other genes in the human genome sequence database. Your gene appears to code for an amino acid sequence that resembles sequences found in some serine proteases. Conjecture how your new gene might contribute to the development of highly invasive cancers.

26. A study by Bose and colleagues (1998. *Blood* 92: 3362–3367) and a previous study by Biernaux and others (1996. *Bone Marrow Transplant* 17: (Suppl. 3) S45–S47) showed that *BCR-ABL* fusion gene transcripts can be detected in 25 to 30 percent of healthy adults who do not develop chronic myelogenous leukemia (CML). Explain how these individuals can carry a fusion gene that is transcriptionally active and yet do not develop CML.

27. Those who inherit a mutant allele of the *RB1* gene are at risk for developing a bone cancer called osteosarcoma. You suspect that in these cases, osteosarcoma requires a mutation in the second *RB1* allele, and you have cultured some osteosarcoma cells and obtained a cDNA clone of a normal human *RB1* gene. A colleague sends you a research paper revealing that a strain of cancer-prone mice develops malignant tumors when injected with osteosarcoma cells, and you obtain these mice. Using these three resources, what experiments would you perform to determine (a) whether osteosarcoma cells carry two *RB1* mutations, (b) whether osteosarcoma cells produce any pRB protein, and (c) if the addition of a normal *RB1* gene will change the cancer-causing potential of osteosarcoma cells?

28. The following table shows neutral polymorphisms found in control families (those with no increased frequency of breast and ovarian cancer). Examine the data in the table and answer the following questions:

(a) What is meant by a neutral polymorphism?

(b) What is the significance of this table in the context of examining a family or population for *BRCA1* mutations that predispose an individual to cancer?

(c) Is the PM2 polymorphism likely to result in a neutral missense mutation or a silent mutation?

(d) Answer part (c) for the PM3 polymorphism.

Neutral Polymorphisms in *BRCA1*

Name	Codon Location	Base in Codon[†]	Frequency in Control Chromosomes[*]			
			A	C	G	T
PM1	317	2	152	0	10	0
PM6	878	2	0	55	0	100
PM7	1190	2	109	0	53	0
PM2	1443	3	0	115	0	58
PM3	1619	1	116	0	52	0

[*]The number of chromosomes with a particular base at the indicated polymorphic site (A, C, G, or T) is shown.
[†]Position 1, 2, or 3 of the codon.

SOLUTIONS TO SELECTED PROBLEMS AND DISCUSSION QUESTIONS

Answers to Now Solve This

1. Several approaches are used to combat CML. One includes the use of a tyrosine kinase inhibitor that binds competitively to the ATP binding site of ABL kinase, thereby inhibiting phosphorylation of BCR-ABL and preventing the activation of additional signaling pathways. In addition, real-time quantitative reverse transcription-polymerase chain reaction (Q-RT-PCR) allows one to monitor drug responses of cell populations in patients so that less toxic and more effective treatments are possible. Being able to distinguish leukemic cells from healthy cells allows one to not only target therapy to specific cell populations, but it also allows for the quantification of responses to therapy. Because such cells produce a hybrid protein, it may be possible to develop a therapy, perhaps an immunotherapy, based on the uniqueness of the BCR/ABL protein.

2. *p53* is a tumor suppressor gene that protects cells from multiplying with damaged DNA. It is present in its mutant state in more than 50 percent of all tumors. Since the immediate control of a critical and universal cell cycle checkpoint is mediated by *p53*, mutation will influence a wide range of cell types. *p53*'s action is not limited to specific cell types.

3. Cancer is a complex alteration in normal cell cycle controls. Even if a major "cancer-causing" gene is transmitted, other genes, often new mutations, are usually necessary in order to drive a cell towards tumor formation. Full expression of the cancer phenotype is likely to be the result of interplay among a variety of genes and therefore show variable penetrance and expressivity.

Solutions to Problems and Discussion Questions

2. The overall probability of the child having RB (using this logic) would be $0.025 \times 0.90 = 0.0225$, or just over 2 percent (or about 1 in 50). To test the presence of the RB gene in the husband, it is possible in some forms of RB to identify (by molecular probes) a defective or missing DNA segment. Otherwise, one might attempt to assay the RB product in cells to see if it is present and functional at normal levels.

4. The major regulatory points of the cell cycle include the following
   (a) late G1 (G1/S)
   (b) the border between G2 and mitosis (G2/M)
   (c) mitosis (M)

6. The retinoblastoma gene (*RB1*), located on chromosome 13, encodes a protein designated pRB. Cells progress through the G1/S transition when pRB is phosphorylated and CDK4 binds to cyclin D. In the absence of phosphorylation of pRB, it binds to members of the E2F family of transcription factors, which controls the expression of genes required to move the cell from G1 to S. When E2F and other regulators are released by pRB, they are free to induce the expression of over 30 genes whose products are required for the transition from G1 into S phase. After cells traverse S, G2, and M phases, pRB reverts to a nonphosphorylated state, binds to regulatory proteins such as E2F, and keeps them sequestered until required for the next cell cycle.

8. Apoptosis is a natural process involved in morphogenesis and a protective mechanism against cancer formation. During apoptosis, nuclear DNA becomes fragmented, cellular structures are disrupted, and the cells are dissolved. Caspases are involved in the initiation and progress of apoptosis.

10. The nonphosphorylated form of pRB binds to transcription factors such as E2F, causing inactivation and suppression of the cell cycle. Phosphorylation of pRB activates the cell cycle by releasing transcription factors (E2F) to advance the cell cycle.

With the phosphorylation site inactivated in the PSM-RB form, phosphorylation cannot occur, thereby leaving the cell cycle in a suppressed state.

12. Various kinases can be activated by breaks in DNA. One kinase, called ATM, and/or a kinase called Chk2 phosphorylates BRCA1 and p53. The activated p53 arrests replication during the S phase to facilitate DNA repair. The activated BRCA1 protein, in conjunction with BRCA2, mRAD51, and other nuclear proteins, is involved in repairing the DNA.

14. In the mutant state (oncogenes) induce or maintain uncontrolled cell division; that is, there is a gain of function. Generally, this gain of function takes the form of increased or abnormally continuous gene output. On the other hand, loss of function is generally attributed to mutations in tumor-suppressor genes, which function to halt passage through the cell cycle. When such genes are mutant, they have lost their capacity to halt the cell cycle. Such mutations are generally recessive.

16. It is less expensive, in terms of both human suffering and money, to seek preventive measures for as many diseases as possible. However, having gained some understanding of the mechanisms of disease, in this case cancer, it must also be stated that no matter what preventive measures are taken, it will be impossible to completely eliminate disease from the human population. It is extremely important, however, that we increase efforts to educate and protect the human population from as many hazardous environmental agents as possible.

18. Normal cells are often capable of withstanding mutational assault because they have checkpoints and DNA repair mechanisms in place. When such mechanisms fail, cancer may be a result. Through mutation, such protective mechanisms are compromised in cancer cells, and as a result they show higher than normal rates of mutation, chromosomal abnormalities, and genomic instability.

20. Certain environmental agents such as chemicals and X rays cause mutations. Since genes control the cell cycle, mutations in cell-cycle control genes, or those that impact cell cycle control, can lead to cancer.

22. No, she will still have the general population risk of about 10 percent. In addition, it is possible that genetic tests will not detect all breast cancer mutations.

24. A benign tumor is a multicellular cell mass that is usually localized to a given anatomical site. Malignant tumors are those generated by cells that have migrated to one or more secondary sites. Malignant tumors are more difficult to treat and can be life-threatening.

26. As with many forms of cancer, a single gene alteration is not the only requirement. The authors (Bose et al.) state "but only infrequently do the cells acquire the additional changes necessary to produce leukemia in humans." Some studies indicate that variations (often deletions) in the region of the breakpoints may influence expression of CML.

28. (a, b) Even though there are changes in the *BRCA1* gene, they do not always have physiological consequences. Such neutral polymorphisms make screening difficult in that one cannot always be certain that a mutation will cause problems for the patient.
   (c) The polymorphism in *PM2* is probably a silent mutation because the third base of the codon is involved.
   (d) The polymorphism in *PM3* is probably a neutral missense mutation because the first base is involved. However, because there is some first codon position degeneracy, it is possible for the mutation to be silent.

CREDITS

Credits are listed in order of appearance.

Photo

CO, SPL/Photo Researchers, Inc.; F-1 (a, b) Courtesy of Hesed M. Padilla-Nash, Antonio Fargiano, and Thomas Ried. Affiliation is Section of Cancer Genomics, Genetics Branch, Center for Research, National Cancer Institute, National Institutes of Health, Bethesda, MD 20892; F-2 (a, b) Courtesy of Professor Manfred Schwab, DKFZ, Heidelberg, Germany

Recombinant DNA Technology

From Chapter 17 of *Essentials of Genetics, Eighth edition,* William S. Klug, Michael R. Cummings, Charlotte A. Spencer, Michael A. Palladino. © 2013 by Pearson Education, Inc. All rights reserved.

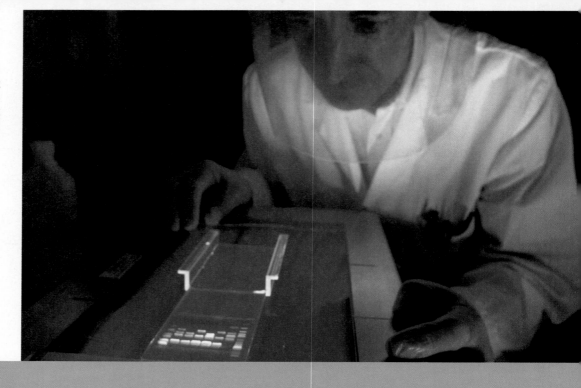

An agarose gel containing separated DNA fragments stained with the DNA-binding dye ethidium bromide and visualized under ultraviolet light.

Recombinant DNA Technology

CHAPTER CONCEPTS

- Recombinant DNA technology creates combinations of DNA sequences from different sources.

- A common application of recombinant DNA technology is to clone a DNA segment of interest.

- For some cloning applications, specific DNA segments are inserted into vectors to create recombinant DNA molecules that are transferred into eukaryotic or prokaryotic host cells such as bacteria, where the recombinant DNA replicates as the host cells divide.

- DNA libraries are collections of cloned DNA.

- DNA segments can be amplified quickly and cloned millions of times using the polymerase chain reaction (PCR).

- DNA, RNA, and proteins can be analyzed using a range of molecular techniques.

- DNA sequencing reveals the nucleotide composition of DNA, and major developments in sequencing technologies have advanced many areas of modern genetics research particularly genomics.

- Recombinant DNA technology has revolutionized our ability to investigate the genomes of diverse species and led to the modern revolution in genomics.

Researchers of the mid- to late 1970s developed various techniques to create, replicate, and analyze **recombinant DNA** molecules—DNA created by joining together pieces of DNA from different sources. The methods used to copy or **clone** DNA, called **recombinant DNA technology** and often known as "gene splicing" in the early days, were a major advance in research in molecular biology and genetics, allowing scientists to isolate and study specific DNA sequences. For their contributions to the development of this technology, Daniel Nathans, Hamilton Smith, and Werner Arber were awarded the 1978 Nobel Prize in Physiology or Medicine.

The power of recombinant DNA technology is astonishing, enabling geneticists to identify and isolate a single gene or DNA segment of interest from a genome. Through cloning, large quantities of identical copies, or clones, of this specific DNA molecule can be produced. Cloned DNA can then be manipulated for numerous purposes, including research into the structure and organization of the DNA, studying gene expression, and producing important commercial products from the protein encoded by a gene. The fundamental techniques involved in recombinant DNA technology subsequently led to the field of genomics, enabling scientists to sequence and analyze entire genomes. In this chapter, we review basic methods of recombinant DNA technology used to isolate, replicate, and analyze DNA.

1 Recombinant DNA Technology Began with Two Key Tools: Restriction Enzymes and DNA Cloning Vectors

Although natural genetic processes such as crossing over produce recombined DNA molecules, the term *recombinant DNA* is generally reserved for molecules produced by artificially joining DNA obtained from different sources. The methods used to create these molecules were derived largely from nucleic acid biochemistry, coupled with genetic techniques developed for the study of bacteria and viruses.

We will begin our discussion of recombinant DNA technology by considering two important tools used to construct and amplify recombinant DNA molecules: DNA-cutting enzymes called **restriction enzymes** and **DNA cloning vectors**. The use of restriction enzymes and cloning vectors was largely responsible for advancing the field of molecular biology because of a wide range of techniques that are based on recombinant DNA technology.

Restriction Enzymes Cut DNA at Specific Recognition Sequences

Restriction enzymes are produced by bacteria as a defense mechanism against infection by viruses. They restrict or prevent viral infection by degrading the DNA of invading viruses.

More than 3500 restriction enzymes have been identified, and over 250 are commercially produced and available for use by researchers. A restriction enzyme recognizes and binds to DNA at a specific nucleotide sequence called a **restriction site (Figure 1)**. The enzyme then cuts both strands of the DNA within that sequence by cleaving the phosphodiester backbone of DNA. Scientists commonly refer to this as "digestion" of DNA. The usefulness of restriction enzymes in cloning derives from their ability to accurately and reproducibly cut genomic DNA into fragments. Restriction enzymes represent sophisticated molecular scissors for cutting DNA into fragments of desired sizes. Restriction sites are present randomly in the genome. The actual fragment sizes produced by DNA digestion with a given restriction enzyme vary because the number and location of recognition sequences are not always distributed randomly in DNA.

Most restriction sites exhibit a form of symmetry described as a **palindrome**: the nucleotide sequence reads the same on both strands of the DNA when read in the 5′ to 3′ direction. Each restriction enzyme recognizes its particular restriction site and cuts the DNA in a characteristic cleavage pattern. The most common restriction sites are four or six nucleotides long, but some contain eight or more nucleotides. Enzymes such as *Eco*RI and *Hind*III make offset cuts in the DNA strands, thus producing fragments with single-stranded overhanging ends called **cohesive ends**, while others such as *Alu*I and *Bal*I cut both strands at the same nucleotide pair, producing DNA fragments with double-stranded ends called **blunt-end** fragments. Four common restriction enzymes, their restriction sites, and source microbes are indicated in Figure 1.

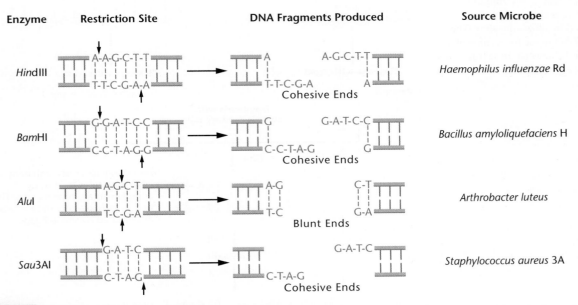

Enzyme	Restriction Site	DNA Fragments Produced	Source Microbe

FIGURE 1 Common restriction enzymes with their restriction sites, DNA cutting patterns, and sources. Arrows indicate the location in the DNA cut by each enzyme.

One of the first restriction enzymes to be identified was isolated from *Escherichia coli* strain R and was designated *Eco*RI. DNA fragments produced by *Eco*RI digestion (Figure 2) have cohesive ends or so-called sticky ends because they can base-pair with complementary single-stranded ends on other DNA fragments cut using *Eco*RI. When mixed together, single-stranded ends of DNA fragments from different sources cut with the same restriction enzyme can **anneal**, or stick together, by hydrogen bonding of complementary base pairs in single-stranded ends. Addition of the enzyme **DNA ligase** to DNA fragments will seal the phosphodiester backbone through covalent bonding, thus forming recombinant DNA molecules (Figure 2).

ESSENTIAL POINT ■ ■ ■

The development of recombinant DNA technology was made possible by the discovery of proteins called restriction enzymes, which cut DNA at specific recognition sites, producing fragments that can be joined with other DNA fragments to form recombinant DNA molecules.

DNA Vectors Accept and Replicate DNA Molecules to Be Cloned

Scientists recognized that DNA fragments produced by restriction enzyme digestion could be copied or cloned if they had a technique for replicating the fragments. The second key tool that allowed DNA cloning was the development of DNA **cloning vectors**—DNA molecules that accept DNA fragments and replicate inserted DNA fragments when placed into host cells.

Many different vectors are available for cloning. Although they differ in terms of the host cells they are able to enter and in the size of inserts they can carry, most DNA vectors have several key properties.

- A vector contains several restriction sites that allow insertion of the DNA fragments to be cloned.

- Vectors must be introduced into host cells to allow for independent replication of the vector DNA and any DNA fragment it carries.

- To distinguish host cells that have taken up vectors from host cells that have not, the vector should carry a **selectable marker gene** (usually an antibiotic resistance gene or the gene for an enzyme absent from the host cell).

- Many vectors incorporate specific sequences that allow for sequencing inserted DNA.

Bacterial Plasmid Vectors

Genetically modified bacterial **plasmids** were the first vectors developed and are still used widely for cloning. Plasmid cloning vectors were derived from naturally occurring plasmids. Plasmidsare extrachromosomal, double-stranded DNA molecules that replicate independently from the chromosomes within bacterial cells [Figure 3(a)]. Plasmids have been extensively modified by genetic engineering to serve as cloning vectors. Many commercially prepared plasmids are readily available with a range of useful features [Figure 3(b)]. Plasmids are introduced into bacteria by the process of **transformation** . Two main techniques are widely used for bacterial transformation. One approach involves

FIGURE 2 DNA from different sources is cleaved with *Eco*RI and mixed to allow annealing. The enzyme DNA ligase forms phosphodiester bonds between these fragments to create an intact recombinant DNA molecule.

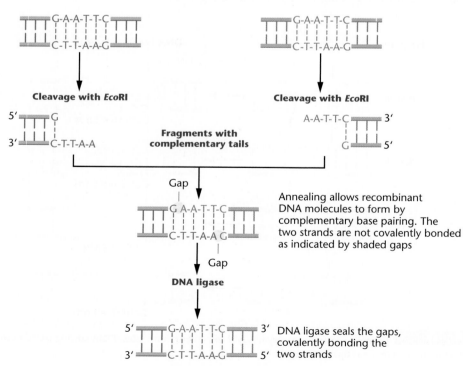

(a)

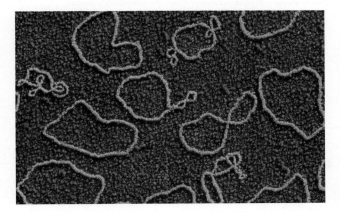

(b)

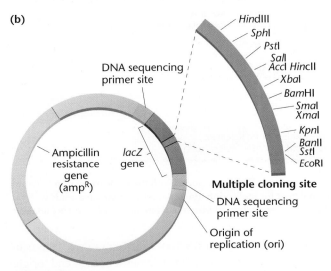

FIGURE 3 (a) A color-enhanced electron micrograph of circular plasmid molecules isolated from *E. coli*. Genetically engineered plasmids are used as vectors for cloning DNA. (b) A diagram of a typical DNA cloning plasmid.

cut once within the multiple cloning site to produce a linear vector. DNA restriction fragments from the DNA to be cloned are added to the linearized vector in the presence of DNA ligase. Sticky ends of DNA fragments anneal, joining the DNA to be cloned with the plasmid. DNA ligase is then used to create phosphodiester bonds to seal nicks in the DNA backbone, thus producing recombinant DNA, which is then introduced into bacterial host cells by transformation. Once inside the cell, plasmids replicate quickly to produce multiple copies.

However, when cloning DNA using plasmids, not all plasmids will incorporate DNA to be cloned. For example, a plasmid cut with a particular restriction enzyme can close back on itself (self-ligation) if cut ends of the plasmid rejoin. Obviously then, such nonrecombinant plasmids are not desired. Also, during transformation, not all host cells will take up plasmids. Therefore it is important that bacterial cells containing recombinant DNA can be readily identified in a cloning experiment. One way this is accomplished is through the use of selectable marker genes described earlier. Genes that provide resistance to antibiotics such as ampicillin and genes such as the *lacZ* gene are very effective selectable marker genes. **Figure 5** provides an example of how these genes can be used to identify bacteria containing recombinant plasmids. This process is referred to as **"blue-white" selection** for a reason that will soon become obvious. In blue-white selection a plasmid is used that contains the *lacZ* gene incorporated into the multiple cloning site. The *lacZ* gene encodes β-galactosidase, an enzyme that cleaves the disaccharide lactose into its component monosaccharides glucose and galactose.

Using this approach, one can easily identify transformed bacterial cells containing recombinant or nonrecombinant plasmids. If a DNA fragment is inserted anywhere in the multiple cloning site, the *lacZ* gene is disrupted and will not produce functional copies of β-galactosidase. Transformed bacteria in this experiment are plated on agar plates that contain an antibiotic—ampicillin in this case. Nontransformed bacteria cannot grow well on these plates because they do not have the *amp*R gene and will be prevented from growing by the antibiotic. But these agar plates also contain a substance called X-gal (technically 5-bromo-4-chloro-3-indolyl-β-D-galactopyranoside). X-gal is similar to lactose in structure. It can be cleaved by β-galactosidase, and when this happens it turns blue. As a result, bacterial cells carrying nonrecombinant plasmids (those that have closed up on themselves and do not contain inserted DNA) have a functional *lacZ* gene and produce β-galactosidase, which cleaves X-gal in the medium, and these cells turn blue. Recombinant bacteria with plasmids containing an inserted DNA fragment form white colonies on X-gal medium because the plasmids in these cells are not producing functional β-galactosidase (Figure 5). Bacteria in these white colonies are clones of each other—genetically identical cells with copies of recombinant plasmids. White colonies can be transferred to flasks of bacterial culture broth

treating cells with calcium ions and using a brief heat shock to pulse DNA into cells. The other technique, called **electroporation**, uses a brief, but high-intensity, pulse of electricity to move DNA into bacterial cells.

Only one or a few plasmids generally enter a bacterial host cell by transformation. Because plasmids have an origin of replication (*ori*) that allows for plasmid replication, many plasmids can increase their copy number to produce several hundred copies in a single host cell. These plasmids greatly enhance the number of DNA clones that can be produced. Plasmid vectors have also been genetically engineered to contain a number of restriction sites for commonly used restriction enzymes in a region called the **multiple cloning site**. Multiple cloning sites allow scientists to clone a range of different fragments generated by many commonly used restriction enzymes.

Cloning DNA with a plasmid generally begins by cutting both the plasmid DNA and the DNA to be cloned with the same restriction enzyme **(Figure 4)**. Typically the plasmid is

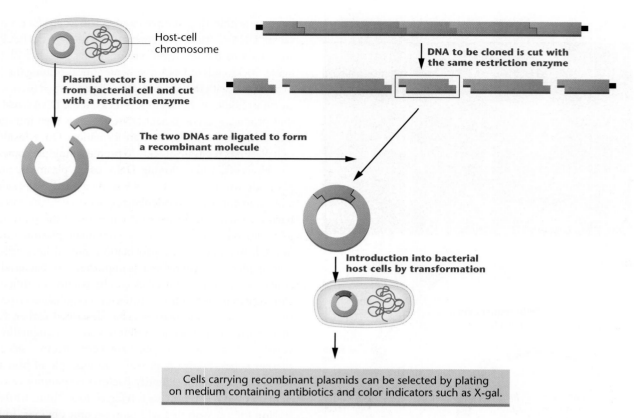

Host-cell chromosome

Plasmid vector is removed from bacterial cell and cut with a restriction enzyme

DNA to be cloned is cut with the same restriction enzyme

The two DNAs are ligated to form a recombinant molecule

Introduction into bacterial host cells by transformation

Cells carrying recombinant plasmids can be selected by plating on medium containing antibiotics and color indicators such as X-gal.

FIGURE 4 Cloning with a plasmid vector involves cutting both plasmid and the DNA to be cloned with the same restriction enzyme. The DNA to be cloned is spliced into the vector and transferred to a bacterial host for replication. Bacterial cells carrying plasmids with DNA inserts can be identified by selection and then isolated. The cloned DNA is then recovered from the bacterial host for further analysis.

and grown in large quantities, after which it is relatively easy to isolate recombinant plasmids from these cells.

Plasmids are still the workhorses for many applications of recombinant DNA technology, but they have a limitation: because they are small, they can only accept inserted pieces of DNA up to about 25 kilobases (kb) in size and most plasmids can often only accept substantially smaller pieces. Therefore as recombinant DNA technology has developed and it has become desirable to clone large pieces of DNA, other vectors have been developed primarily for their ability to accept larger pieces of DNA and because they can be used with other types of host cells beside bacteria.

Other Types of Cloning Vectors

Phage vector systems were among the earliest vectors used in addition to plasmids. These included genetically modified strains of **λ phage.** Phage vectors were popular for quite some time and are still in use today because they can carry inserts up to 45 kb, more than twice as long as DNA inserts in most plasmid vectors.

Bacterial artificial chromosomes (BACs) and **yeast artificial chromosomes (YACs)** are two other examples of vectors that can be used to clone large fragments of DNA. BACs are essentially very large but low copy number (typically one or two copies/bacterial cell) plasmids that can

accept DNA inserts in the 100- to 300-kb range. Like natural chromosomes, a YAC has telomeres at each end, origins of replication, and a centromere. Yeast chromosomes range in size from 230 kb to over 1900 kb, making it possible to clone DNA inserts from 100 to 1000 kb in YACs. The ability to clone large pieces of DNA in these vectors makes them an important tool in genome sequencing projects, including the Human Genome Project .

Expression vectors are designed to ensure mRNA expression of a cloned gene with the purpose of producing many copies of the gene's encoded protein in a host cell. For many research applications that involve studies of protein structure and function, producing a recombinant protein in bacteria (or other host cells) and purifying the protein is a routine approach, although it is not always easy to properly express a protein that maintains its biological function. The biotechnology industry also relies heavily on expression vectors to produce commercially valuable protein products from cloned genes.

Introducing genes into plants is a common application that can be done in many ways. One widely used approach to insert genes into plant cells involves the soil bacterium *Rhizobium radiobacter,* which infects plant cells and produces tumors (called crown galls) in many species of plants. Formerly *Agrobacterium tumefaciens,* this bacterium was renamed based on genomic analysis. *Rhizobium* contains a plasmid called the **Ti plasmid.**

Restriction sites in Ti plasmids can be used to insert foreign DNA, and recombinant vectors are introduced into *Rhizobium* by transformation. Tumor-inducing genes from Ti plasmids are removed from the vector so that the recombinant vector does not result in tumor production. *Rhizobium* containing recombinant DNA is mixed with plant cells (not all types of plant cells can be infected by *Rhizobium*). Once inside the cell, the foreign DNA is inserted into the host cell chromosome by the Ti plasmid. Plant cells carrying a recombinant Ti plasmid are grown to produce a mature plant.

ESSENTIAL POINT ■ ■ ■ ■

Vectors replicate autonomously in host cells and facilitate the cloning and manipulation of newly created recombinant DNA molecules.

NOW SOLVE THIS

1 An ampicillin-resistant, tetracycline-resistant plasmid, pBR322, is cleaved with *Pst*I, which cleaves within the ampicillin resistance gene. The cut plasmid is ligated with *Pst*I-digested *Drosophila* DNA to prepare a genomic library, and the mixture is used to transform *E. coli* K12.

(a) Which antibiotic should be added to the medium to select cells that have incorporated a plasmid?

(b) If recombinant cells were plated on medium containing ampicillin or tetracycline and medium with both antibiotics, on which plates would you expect to see growth of bacteria containing plasmids with *Drosophila* DNA inserts?

(c) How can you explain the presence of colonies that are resistant to both antibiotics?

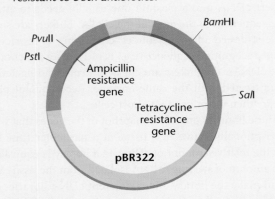

■ **Hint** *This problem involves an understanding of antibiotic selectable marker genes in plasmids and antibiotic DNA selection for identifying bacteria transformed with recombinant plasmid DNA. The key to its solution is to recognize that inserting foreign DNA into the plasmid vector disrupts one of the antibiotic resistance genes in the plasmid.*

FIGURE 5 In blue-white selection procedures, DNA inserted into the multiple cloning site of a plasmid disrupts the *lacZ* gene so that bacteria containing recombinant DNA are unable to metabolize X-gal. As a result, white colonies allow direct identification of bacterial colonies carrying cloned DNA inserts. Photo of a Petri dish showing the growth of bacterial cells after uptake of recombinant plasmids. Cells in blue colonies contain vectors without cloned DNA inserts, whereas cells in white colonies contain vectors carrying DNA inserts.

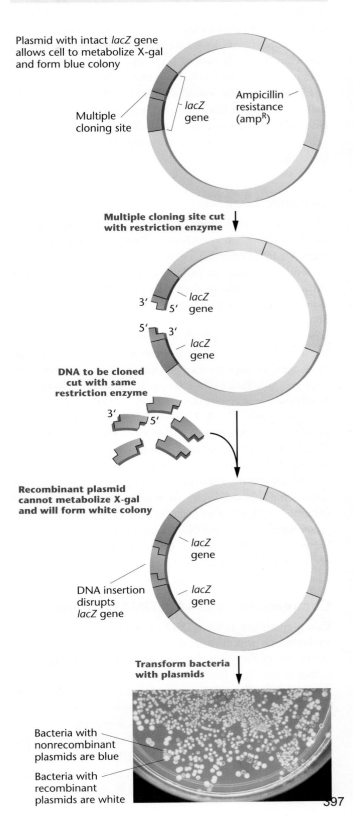

Plasmid with intact *lacZ* gene allows cell to metabolize X-gal and form blue colony

Multiple cloning site

lacZ gene

Ampicillin resistance (amp^R)

Multiple cloning site cut with restriction enzyme

3′ 5′ *lacZ* gene

5′ 3′ *lacZ* gene

DNA to be cloned cut with same restriction enzyme

3′ 5′

Recombinant plasmid cannot metabolize X-gal and will form white colony

lacZ gene

DNA insertion disrupts *lacZ* gene

lacZ gene

Transform bacteria with plasmids

Bacteria with nonrecombinant plasmids are blue

Bacteria with recombinant plasmids are white

2 | DNA Libraries Are Collections of Cloned Sequences

In the cloning discussions we have had so far, we have described *how* DNA can be inserted into vectors and cloned—a relatively straightforward process—but we have not discussed how one knows what particular DNA sequence they have cloned. Simply cutting DNA and inserting into vectors does not tell you *what* DNA sequences have been cloned.

During the first several decades of DNA cloning, scientists created **DNA libraries**, which represent a collection of cloned DNA samples. There are two main types of libraries, genomic DNA libraries and complementary DNA (cDNA) libraries.

Genomic Libraries

Ideally, a **genomic library** consists of many overlapping fragments of the genome, with at least one copy of every DNA sequence in an organism's genome which in summary span the entire genome. In making a genomic library, DNA is extracted from cells or tissues and cut with restriction enzymes, and the resulting fragments are inserted into vectors. Since vectors such as plasmids can carry only a few thousand base pairs of inserted DNA, selecting the vector so that the library contains the whole genome in the smallest number of clones is an important consideration. When working with large genomes such as the Human Genome, YACs were commonly used to accommodate large sizes of DNA necessary to span the approximately 3 billion bp of DNA in the human genome. Because genomic DNA is the foreign DNA introduced into vectors, genomic libraries contain coding and noncoding segments of DNA such as introns.

whole-genome shotgun cloning approaches and new sequencing methodologies are readily replacing traditional genomic DNA libraries because they effectively allow one to sequence all of the DNA fragments in a genomic DNA sample without the need for inserting DNA fragments into vectors and cloning them in host cells.

Complementary DNA (cDNA) Libraries

Complementary DNA (cDNA) libraries offer certain advantages over genomic libraries and continue to be a useful methodology for gene cloning. This is primarily because a cDNA library contains DNA copies—called cDNA—which are made from the mRNA molecules of a cell population and therefore represent the genes being expressed in the cells at the time the library was made. cDNA is complementary to the nucleotide sequence of the mRNA, and so, unlike a genomic library, which contains all of the DNA in a genome—gene coding and noncoding sequences—a cDNA library contains only expressed genes. As a result, cDNA libraries have been particularly useful for identifying and studying genes expressed in certain cells or tissues under certain conditions. For instance, one can use these libraries to compare expressed genes from normal tissues and diseased tissues. This approach has been valuable for identifying genes involved in cancer formation, such as genes that contribute to progression from a normal cell to a cancer cell and genes involved in cancer cell metastasis (spreading).

A cDNA library is prepared by isolating mRNA from a population of cells of interest, typically cells that express an abundance of mRNA for the genes to be cloned. The next step in making a cDNA library is to mix mRNAs with oligo-dT primers—short, single-stranded sequences of T nucleotides that anneal to the poly-A tail **(Figure 6)**. The enzyme **reverse transcriptase** extends the oligo-dT primer and synthesizes a complementary DNA copy of the mRNA sequence, creating an mRNA–DNA double-stranded hybrid molecule. The RNA in the hybrid molecule can be enzymatically digested or chemically degraded and another, opposing strand of DNA is synthesized by DNA polymerase. Alternatively, cDNA can also be synthesized using primers that bind randomly to mRNA and the first strand of cDNA to eventually create double-stranded cDNA from the original mRNA strand.

The cDNA molecules are subsequently inserted into vectors, usually plasmids. Because one typically wouldn't know what restriction enzymes could be used to cut this DNA, to insert cDNA into a plasmid one usually needs to attach linker sequences to the ends of the cDNA. Linkers are short double-stranded oligonucleotides containing a restriction enzyme recognition sequence (e.g., *Eco*RI). After attachment to the cDNAs, the linkers are cut with *Eco*RI and ligated to vectors treated with the same enzyme. Recombinant plasmids are then introduced into host cells.

These libraries provide a snapshot of the genes that were transcriptionally active in a tissue at a particular time because the relative amount of cDNA in a library is equivalent to the amount of starting mRNA isolated from the tissue and used to make the library. Many different cDNA libraries are available from cells and tissues in specific stages of development, different organs such as brain, muscle, and kidney, and tissues from different disease states such as cancer.

Specific Genes Can Be Recovered from a Library by Screening

Genomic and cDNA libraries often consist of several hundred thousand different DNA clones, much like a large book library may have many books but only a few of interest to your studies in genetics. So how can libraries be used to locate a specific gene of interest in a library? To find a specific gene, we need to identify and isolate only the clone or clones

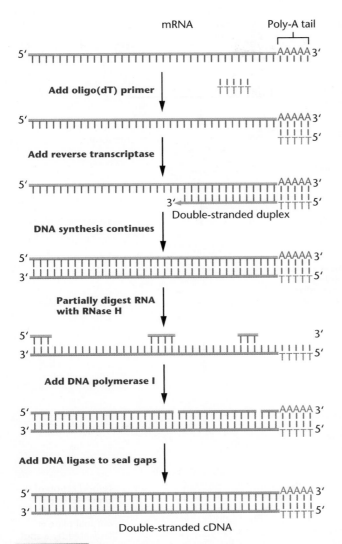

mRNA   Poly-A tail

5′ [_____] AAAAA 3′

Add oligo(dT) primer   │││││
                                            TTTTT

5′ [_____] AAAAA 3′
                                    │││││ 5′
                                    TTTTT

Add reverse transcriptase

5′ [_____] AAAAA 3′
   3′ [_____] │││││ 5′
                                    TTTTT
                Double-stranded duplex

DNA synthesis continues

5′ [_____] AAAAA 3′
3′ [_____] │││││ 5′
                                    TTTTT

Partially digest RNA with RNase H

5′ │││   │││   │││   3′
3′ [_____] │││││ 5′
                                    TTTTT

Add DNA polymerase I

5′ [___] [_____] [_____] AAAAA 3′
3′ [_____] │││││ 5′
                                    TTTTT

Add DNA ligase to seal gaps

5′ [_____] AAAAA 3′
3′ [_____] │││││ 5′
                                    TTTTT

Double-stranded cDNA

FIGURE 6   Producing cDNA from mRNA. Because many eukaryotic mRNAs have a poly-A tail of variable length at one end, a short oligo-dT molecule annealed to this tail serves as a primer for the enzyme reverse transcriptase. Reverse transcriptase uses the mRNA as a template to synthesize a complementary DNA strand (cDNA) and forms an mRNA/cDNA double-stranded duplex. The mRNA is digested with the enzyme RNAse H, producing gaps in the RNA strand. The 3′ ends of the remaining RNA serve as primers for DNA polymerase I, which synthesizes a second DNA strand. The result is a double-stranded cDNA molecule that can be cloned into a suitable vector or used directly as a probe for library screening.

containing that gene. We must also determine whether a given clone contains all or only part of the gene we are studying. We can sort through a library and isolate specific genes of interest by an approach that is called **library screening.**

Often, probes are used to screen a library. A **probe** is any DNA or RNA sequence that is complementary to some part of a cloned sequence present in the library—the target gene or sequence to be identified. A probe is also labeled or tagged in different ways so that it can be identified. When used in a hybridization reaction, the probe binds to any complementary DNA sequences present in one or more clones. Probes can be labeled with radioactive isotopes, or increasingly, they are labeled with nonradioactive compounds that undergo chemical or color reactions to indicate the location of a specific clone in a library. Probes are derived from a variety of sources—often related genes isolated from another species can be used if enough of the DNA sequence is conserved. For example, genes cloned in rats, mice, or even *Drosophila* that have conserved sequence similarity to human genes have been used as probes to identify human genes during library screening.

To screen a library with a probe, bacterial clones from the library are grown on nutrient agar plates, where they form hundreds or thousands of colonies (**Figure 7**). A replica of the colonies on each plate is made by gently pressing a nylon or nitrocellulose membrane onto the plate's surface; this transfers the pattern of bacterial colonies from the plate to the membrane. The membrane is processed to lyse the bacterial cells, denature the double-stranded DNA released from the cells into single strands, and bind these strands to the membrane. The DNA on the membrane is incubated with a labeled nucleic acid probe. First, the probe is heated and quickly cooled to form single-stranded molecules, and then it is added to a solution containing the membrane. If the nucleotide sequence of any DNA on the membrane is complementary to the probe, a double-stranded DNA–DNA hybrid molecule will form (one strand from the probe and the other from the cloned DNA on the membrane). This step is called **hybridization.**

After incubation of the probe and the membrane, unbound probe molecules are washed away, and the membrane is assayed to detect the hybrid molecules that remain. If a radioactive probe has been used, the membrane is overlaid with a piece of X-ray film. Radioactivity from the probe molecules bound to DNA on the membrane will expose the film, producing dark spots on the film. These spots represent colonies on the plate containing the cloned gene of interest (Figure 7). The positions of spots on the film are used as a guide to identify and recover the corresponding colonies on the plate. The cloned DNA they contain can be used in further experiments. With some nonradioactive probes, a chemical reaction emits photons of light (chemiluminescence) to expose the photographic film and reveal the location of colonies carrying the gene of interest.

As we have discussed here, libraries enable scientists to clone DNA and then identify individual genes in the library. Cloning DNA from libraries is still a technique with valuable applications. But the basic methods of recombinant DNA technology were the foundation for the development of powerful techniques for whole-genome cloning and sequencing in which entire

FIGURE 7 Screening a library constructed using a plasmid vector to recover a specific gene. The library, present in bacteria on Petri plates, is overlaid with a DNA-binding membrane, and colonies are transferred to the membrane. Colonies on the membrane are lysed, and the DNA is denatured to single strands. The membrane is placed in a hybridization bag along with buffer and a labeled single-stranded DNA probe. During incubation, the probe forms a double-stranded hybrid with any complementary sequences on the membrane. The membrane is removed from the bag and washed to remove excess probe. Hybrids are detected by placing a piece of X-ray film over the membrane and exposing it for a short time. The film is developed, and hybridization events are visualized as spots on the film. Colonies containing the insert that hybridized to the probe are identified from the orientation of the spots. Cells are picked from this colony for growth and further analysis.

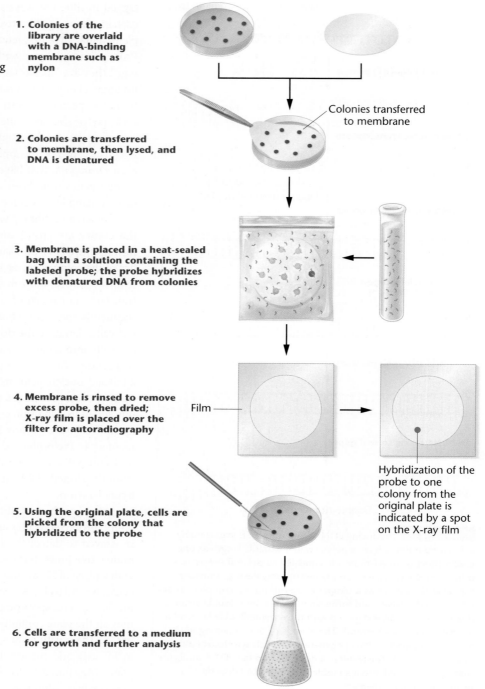

1. Colonies of the library are overlaid with a DNA-binding membrane such as nylon

Colonies transferred to membrane

2. Colonies are transferred to membrane, then lysed, and DNA is denatured

3. Membrane is placed in a heat-sealed bag with a solution containing the labeled probe; the probe hybridizes with denatured DNA from colonies

4. Membrane is rinsed to remove excess probe, then dried; X-ray film is placed over the filter for autoradiography

Film

Hybridization of the probe to one colony from the original plate is indicated by a spot on the X-ray film

5. Using the original plate, cells are picked from the colony that hybridized to the probe

6. Cells are transferred to a medium for growth and further analysis

genomes are sequenced without creating libraries. Increasingly whole-genome sequencing is rapidly replacing many traditional recombinant DNA techniques for gene cloning such as DNA libraries.

ESSENTIAL POINT

DNA libraries are collections of cloned DNA that can be screened to identify and isolate specific sequences of interest.

3 The Polymerase Chain Reaction Is a Powerful Technique for Copying DNA

Therecombinant DNA techniques developed in the early 1970s gave birth to the biotechnology industry because these methods enabled scientists to clone human genes, such as the insulin gene, whose protein product could be used for therapeutic purposes. However, cloning DNA using vectors and host cells is labor intensive and time consuming. In 1986, another technique, called the **polymerase chain reaction (PCR)**, was developed. This advance revolutionized recombinant DNA methodology and further accelerated the pace of biological research. In 1993, the Nobel Prize in Chemistry was awarded to Kary Mullis, who developed the technique.

PCR is a rapid method of DNA cloning that, in many cases, eliminates the need to use host cells for cloning. PCR is also a method of choice for many other research techniques in molecular biology and for applications such as genetic testing and forensics.

PCR can amplify target DNA sequences that are initially present in very small quantities in a population of other DNA molecules. When using PCR to clone DNA, double-stranded target DNA to be cloned is placed in a tube with DNA polymerase, Mg^{+2} (as an important cofactor for DNA polymerase), and the four deoxyribonucleoside triphosphates. In addition, as a prerequisite for PCR, some information about the nucleotide sequence of the target DNA is required. This sequence information is used to synthesize two oligonucleotide **primers**: short (typically about 20 nt long), single-stranded DNA sequences, one complementary to the 5′ end of one strand of target DNA to be amplified and another primer complementary to the opposing strand of target DNA at its 3′ end. When added to a sample of double-stranded DNA that has been denatured into single strands, the primers bind to complementary nucleotides flanking the sequence to be cloned. DNA polymerase can then extend the 3′ end of each primer to synthesize second strands of the target DNA. Therefore, one complete reaction process, called a **cycle**, doubles the number of DNA molecules in the reaction **(Figure 8)**. Repetition of the process produces large numbers of copies of the DNA very quickly. If desired, the products of PCR can be cloned into plasmid vectors for further use.

The amount of amplified DNA produced is theoretically limited only by the number of times these cycles are repeated, although several factors prevent PCR reactions from amplifying very long stretches of DNA or amplifying DNA indefinitely. Most routine PCR applications involve a series of three reaction steps in a cycle. These three steps are as follows:

1. **Denaturation:** The double-stranded target DNA to be cloned is *denatured* into single strands. The DNA can come from many sources, including genomic DNA, mummified remains, fossils, or forensic samples such as dried blood or semen, single hairs, or dried samples from medical records. Heating to 92–95°C for about 1 minute denatures the double-stranded DNA into single strands.

2. **Hybridization/Annealing:** The temperature of the reaction is lowered to a temperature between 45°C and 65°C, which causes primer binding (also called hybridization or annealing) to the denatured, single-stranded target DNA. Because primers are complementary to sequences flanking the target DNA they serve as starting points for DNA polymerase to synthesize new DNA strands complementary to the target DNA.

3. **Extension:** The reaction temperature is adjusted to between 65°C and 75°C, and DNA polymerase uses the primers as a starting point to synthesize new DNA strands by adding nucleotides to the ends of the primers in a 5′ to 3′ direction.

PCR is a chain reaction because the number of new DNA strands is doubled in each cycle, and the new strands, along with the old strands, serve as templates in the next cycle. Each cycle usually takes several minutes and can be repeated immediately, so that in less than 3 hours, 25 to 30 cycles result in over a million-fold increase in the amount of DNA (Figure 8). This process is automated by instruments called *thermocyclers,* or simply PCR machines, that can be programmed to carry out a predetermined number of cycles. A key requirement for PCR is the type of DNA polymerase used in PCR reactions. PCR reactions rely on thermostable forms of DNA polymerase capable of withstanding multiple heating and cooling cycles without significant loss of activity as a result of denaturation. PCR became a major tool when DNA polymerase was isolated from *Thermus aquaticus,* a bacterium living in the hot springs of Yellowstone National Park. Called *Taq Polymerase,* this enzyme is capable of tolerating extreme temperature changes and was the first thermostable polymerase used for PCR.

PCR-based DNA cloning has several advantages over library cloning approaches. PCR is rapid and can be carried out in a few hours, rather than the days required for making and screening DNA libraries. PCR is also very sensitive and amplifies specific DNA sequences from small amounts of DNA samples. A wide variety of PCR-based techniques involve different variations of the basic technique described here.

Limitations of PCR

Although PCR is a valuable technique, it does have limitations: some information about the nucleotide sequence of the target DNA must be known to synthesize primers. Minor contamination of a sample with DNA from other sources can cause problems. For example, cells shed from the skin of a researcher can contaminate samples gathered from a crime scene or taken from fossils, making it difficult to obtain accurate results. Normally, DNA polymerase in a PCR reaction only extends primers for relatively short distances and does

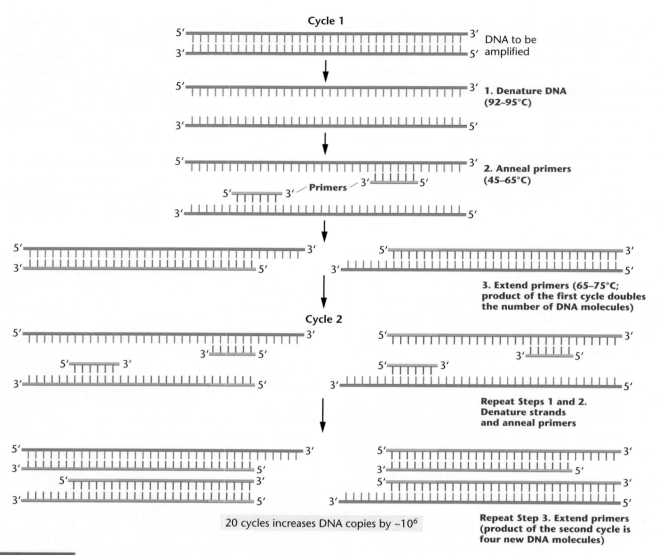

Cycle 1

5′ ————————————————— 3′ DNA to be
3′ ————————————————— 5′ amplified

5′ ————————————————— 3′ **1. Denature DNA**
(92–95°C)
3′ ————————————————— 5′

5′ ————————————————— 3′ **2. Anneal primers**
(45–65°C)
Primers
3′ ————————————————— 5′

3. Extend primers (65–75°C;
product of the first cycle doubles
the number of DNA molecules)

Cycle 2

Repeat Steps 1 and 2.
Denature strands
and anneal primers

20 cycles increases DNA copies by ~10⁶

Repeat Step 3. Extend primers
(product of the second cycle is
four new DNA molecules)

FIGURE 8 In the polymerase chain reaction (PCR), the target DNA is denatured into single strands; each strand is then annealed to short, complementary primers. DNA polymerase extends the primers in the 5′ to 3′ direction, using the single-stranded DNA as a template. The result after one round of replication is a doubling of DNA molecules to create two newly synthesized double-stranded DNA molecules. Repeated cycles of PCR can quickly amplify the original DNA sequence more than a millionfold. *Note:* Shown here is a relatively short sequence of DNA being amplified. Typically, much longer segments of DNA are used for PCR, and the primers bind somewhere within the DNA molecule and not so close to the end of the actual molecule.

not continue processively until it reaches the other end of long template strands of DNA. Because of this characteristic, scientists often use PCR to amplify pieces of DNA that are relatively short, several thousand nucleotides in length, which is fine for most routine applications.

Applications of PCR

Cloning DNA by PCR has been one of the most widely used techniques in genetics and molecular biology for over 20 years. PCR and its variations have many other applications, and PCR is one of the most versatile techniques in modern genetics research. Gene-specific primers provide a way of

using PCR for screening mutations involved in genetic disorders. PCR is a key diagnostic methodology for the detection of bacteria and viruses (such as hepatitis or HIV) in humans, and pathogenic bacteria such as *E. coli* and *Staphylococcus aureus* in contaminated food. PCR techniques are particularly advantageous when studying samples from single cells, fossils, or a crime scene, where a single hair or even a saliva-moistened postage stamp is the source of the DNA. Using PCR, researchers can also explore uncharacterized DNA regions adjacent to known regions and even sequence DNA. PCR has been used to enforce the worldwide ban on the sale of certain whale products and to settle arguments about the pedigree background of purebred dogs.

One commonly used PCR method is called **reverse transcription PCR (RT-PCR)**. RT-PCR is a powerful methodology for studying gene expression, that is, mRNA production by cells or tissues. In RT-PCR, RNA is isolated from cells or tissues to be studied, and reverse transcriptase is used to generate double-stranded cDNA molecules, as described earlier when we discussed preparation of cDNA libraries. This reaction is followed by PCR to amplify cDNA with a set of primers specific for the gene of interest. Amplified cDNA fragments are then separated on an agarose gel. Because the amount of amplified cDNA in an RT-PCR reaction is based on the relative number of mRNA molecules in the starting reaction, RT-PCR can be used to evaluate relative levels of gene expression in different samples. RT-PCR is more sensitive than conventional cDNA preparation and is a powerful tool for identifying mRNAs that may be present in only one or two copies per cell.

One of the most valuable modern applications of PCR involves a method called **quantitative real-time PCR** or simply **qPCR**. This approach makes it possible to determine the amount of PCR product made during an experiment, which enables researchers to quantify amplification reactions as they occur in "real time" without having to run a gel.

ESSENTIAL POINT ■ ■ ■

PCR allows DNA to be amplified, or copied, without cloning and is a rapid and sensitive method with wide-ranging applications.

NOW SOLVE THIS

2 Question 12 refers to creating a genomic DNA library from the African okapi. If you were the first person in the world attempting to use PCR to amplify particular genes from this library, what strategies might you use to design PCR primers for your experiments if the okapi genome has yet to be sequenced?

■ **Hint** *This problem asks you to design PCR primers to amplify the β-globin gene from a species whose genome you just sequenced. The key to its solution is to remember that you have at your disposal sequence data for the human β-globin gene and to consider that PCR experiments require the use of primers that bind to complementary bases in the DNA to be amplified.*

---

4 | Molecular Techniques for Analyzing DNA

The identification of genes and other DNA sequences plays a very powerful role in analyzing genomic structure and function. In addition, a wide range of molecular techniques is available to geneticists, molecular biologists, and almost anyone who does research involving DNA and RNA, particularly

those who study the structure, expression, and regulation of genes. Here we consider some of the most commonly used molecular methods that provide information about the organization and function of cloned sequences.

Restriction Mapping

Historically, one of the first steps in characterizing a DNA clone was the construction of a **restriction map**. A restriction map establishes the number of, order of, and distances between restriction sites along a cloned segment of DNA, thus providing information about the length of the cloned insert and the location of restriction sites within the clone. Information from these maps can be valuable when trying to reclone fragments of a gene for different applications such as making a probe.

Before DNA sequencing and bioinformatics became popular, restriction maps were created experimentally by cutting DNA with different restriction enzymes and separating DNA fragments by gel electrophoresis, a method that separates fragments by size, with the smallest pieces moving farthest through the gel **(Figure 9)**. The fragments form a series of bands that can be visualized by staining the DNA with ethidium bromide and illuminating it with ultraviolet light. The cutting pattern of fragments generated can then be interpreted to determine the location of restriction sites for different enzymes.

Because of recent advances in DNA sequencing and the use of bioinformatics, most restriction maps are now created by using software to identify restriction enzyme cutting sites in sequenced DNA. The Exploring Genomics exercise in this chapter involves a Web site, Webcutter, commonly used for generating restriction maps. Restriction digestion of cloned

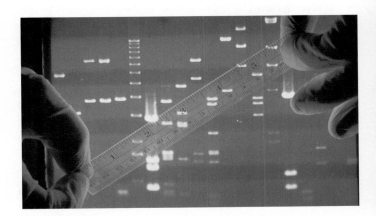

FIGURE 9 An agarose gel containing separated DNA fragments stained with the DNA-binding dye (ethidium bromide) and visualized under ultraviolet light. Smaller fragments migrate faster and farther than do larger fragments, resulting in the distribution shown. Molecular techniques involving agarose gel electrophoresis are routinely used in a wide range of applications.

DNA can still play an important role in chromosome analysis. In addition, if a restriction site maps close to a mutant allele, this site can be used as a marker in genetic testing to identify carriers of recessively inherited disorders or to prenatally diagnose a fetal genotype.

Nucleic Acid Blotting

Several of the techniques described in this chapter rely on hybridization between complementary nucleic acid (DNA or RNA) molecules. One of the most widely used methods for detecting such hybrids is called Southern blotting (after Edwin Southern, who devised it). The **Southern blot** method can be used to identify which clones in a library contain a given DNA sequence (such as ribosomal DNA, a β-globin gene, etc.) and to characterize the size of the fragments. Southern blots can also be used to identify fragments carrying specific genes in genomic DNA digested with a restriction enzyme. Fragments of genomic clones isolated by Southern blots can in turn be isolated and recloned, providing a way to isolate parts of a gene. Southern blotting has also been a valuable tool for identifying the number of copies of a particular sequence or gene that are present in a genome, and Southern blotting is still used in forensic applications.

Southern blotting has two components: separation of DNA fragments by gel electrophoresis and hybridization of the fragments using labeled probes. As shown in Figure 9, gel electrophoresis can be used to characterize the number of fragments produced by restriction digestion and to estimate their molecular weights. However, restriction digests of large genomes—such as the human genome, with more than 3 billion nucleotides—produce so many different fragments that they will run together on a gel to produce a continuous smear. The identification of specific fragments in these cases is accomplished by hybridization of these fragments to specific probe sequences.

The DNA to be characterized by Southern blot hybridization can come from several sources, including clones selected from a library or genomic DNA. To make a Southern blot, DNA is often cut into fragments with one or more restriction enzymes, and the fragments are separated by gel electrophoresis (**Figure 10**). In preparation for hybridization, the DNA in the gel is denatured with alkaline treatment to form single-stranded fragments. The gel is then overlaid with a DNA-binding membrane, usually nylon. Transfer of the DNA fragments to the membrane is accomplished by placing the membrane and gel on a wick sitting in a buffer solution. Layers of paper towels or blotting paper are placed on top of the membrane and held in place with a weight. Capillary action draws buffer up through the gel, transferring the DNA fragments from the gel to the membrane.

The membrane is placed in a heat-sealed bag with a labeled, single-stranded DNA probe for hybridization. For many years Southern blots were primarily carried out with radioactively labeled probes, but as was mentioned when describing library screening, probes that use fluorescent and chemiluminescent labels are now commonly used. DNA fragments on the membrane that are complementary to the probe's nucleotide sequence bind to the probe to form double-stranded hybrids. Excess probe is then washed away, and the hybridized fragments are visualized on a piece of film (Figure 10 and **11**).

To produce Figure 11, researchers cut samples of genomic DNA with several restriction enzymes. The pattern of fragments obtained for each restriction enzyme is shown in Figure 11(a). A Southern blot of this gel is illustrated in Figure 11(b). The probe hybridized to complementary sequences, identifying fragments of interest.

Southern blotting led to the development of other blotting approaches. RNA blotting was subsequently called **Northern blot analysis** or simply **Northern blotting,** and following a naming scheme that correlates with the directionality of a compass, a related technique involving proteins is known as **Western blotting.** Western blotting is a widely used technique for analyzing proteins. Thus, part of the historical significance of Southern blotting is that it led to the development of other very important blotting methods that are key tools for studying nucleic acids and proteins.

To determine whether a gene is actively being expressed in a given cell or tissue type, Northern blotting probes for the presence of mRNA complementary to a cloned gene (**Figure 12**). To do this, mRNA is extracted from a specific cell or tissue type, separated by gel electrophoresis, and RNA is transferred to a membrane. The membrane is then exposed to a labeled single-stranded DNA probe derived from a cloned copy of the gene. If mRNA complementary to the DNA probe is present, the complementary sequences will hybridize and be detected as a band on the film. Northern blots provide information about the expression of specific genes and are used to study patterns of gene expression.

Northern blots also detect alternatively spliced mRNAs (multiple types of transcripts derived from a single gene) and are used to derive other information about transcribed mRNAs. Northern blots can be used to measure the size of a gene's mRNA transcripts, and by measuring band density, an estimate of the relative transcriptional activity of the gene can be made. Thus, Northern blots characterize and quantify the transcriptional activity of genes in different samples. Northern blots are still used to study RNA expression, but because PCR-based techniques are faster and more sensitive than blotting methods, techniques such as RT-PCR are often the preferred approach, particularly for measuring changes in gene expression.

Finally, **fluorescent in situ hybridization (FISH)** is a powerful tool that involves hybridizing a probe directly to a chromosome or RNA without blotting . FISH can be carried out with isolated chromosomes on a slide or directly *in situ* in tissue sections or entire organisms, particularly when embryos are used for various studies in developmental genet-

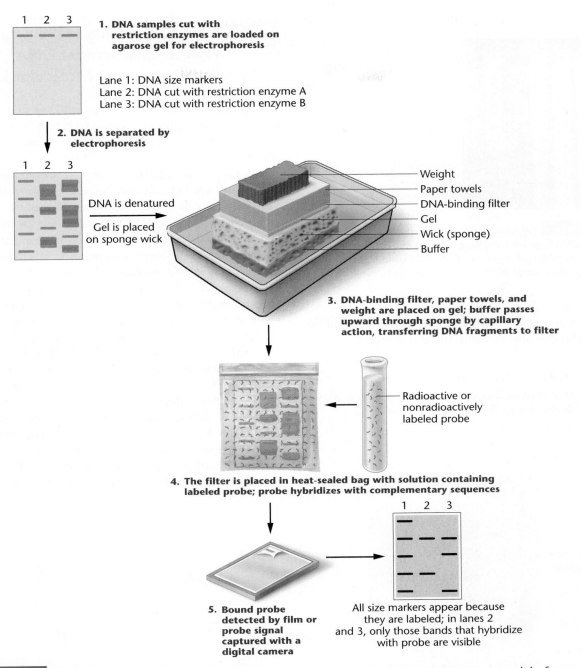

1. **DNA samples cut with restriction enzymes are loaded on agarose gel for electrophoresis**

Lane 1: DNA size markers
Lane 2: DNA cut with restriction enzyme A
Lane 3: DNA cut with restriction enzyme B

2. **DNA is separated by electrophoresis**

DNA is denatured

Gel is placed on sponge wick

Weight
Paper towels
DNA-binding filter
Gel
Wick (sponge)
Buffer

3. **DNA-binding filter, paper towels, and weight are placed on gel; buffer passes upward through sponge by capillary action, transferring DNA fragments to filter**

Radioactive or nonradioactively labeled probe

4. **The filter is placed in heat-sealed bag with solution containing labeled probe; probe hybridizes with complementary sequences**

5. **Bound probe detected by film or probe signal captured with a digital camera**

All size markers appear because they are labeled; in lanes 2 and 3, only those bands that hybridize with probe are visible

FIGURE 10 In Southern blotting, samples of the DNA to be probed are digested with restriction enzymes and the fragments are separated by gel electrophoresis. Then, the gel is placed on a wick that is in contact with a buffer solution and covered with a DNA-binding membrane. Layers of paper towels or blotting paper are placed on top of the membrane and held in place with a weight. Capillary action draws the buffer through the gel, transferring the pattern of DNA fragments from the gel to the membrane. The DNA fragments on the membrane are then denatured into single strands and hybridized with a labeled DNA probe. The membrane is washed to remove excess probe and overlaid with a piece of X-ray film or probe fluorescence is detected by a digital camera.

ics. For example, in developmental studies, one can identify which cell types in an embryo express different genes during different stages of development.

For *in situ* hybridization, a probe for a specific sequence is labeled with nucleotides tagged with a particular dye that will fluoresce under fluorescent light. When hybridized to a chromosome, for example, the probe can reveal the specific location of a gene on a particular chromosome. Variations of the FISH technique are also being used to produce karyotypes (sometimes called **spectral karyotypes**) in which individual chromosomes can be detected using probes labeled with dyes that will fluoresce at different wavelengths . Spectral karyotyping has proven to be extremely valuable for detecting deletions, translocations, duplications,

(a)

(b)

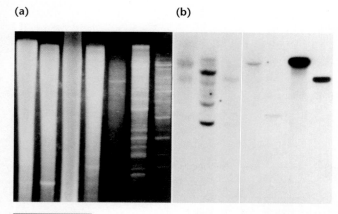

FIGURE 11 (a) Agarose gel stained with ethidium bromide to show DNA fragments. (b) Exposed X-ray film of a Southern blot prepared from the gel in part (a). Only those bands containing DNA sequences complementary to the probe show hybridization.

and other anomalies in chromosome structure, and for detecting chromosomal abnormalities in cancer cells.

ESSENTIAL POINT ■ ■ ■

Once cloned, DNA sequences can be analyzed through a variety of methods that involve hybridization techniques.

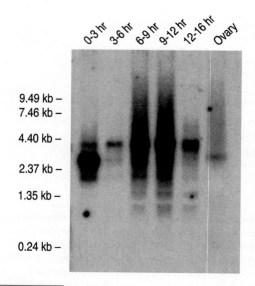

FIGURE 12 Northern blot analysis of *dfmr1* gene expression in *Drosophila* ovaries and embryos. A *dfmr1* transcript of approximately 2.8 kb is present in ovaries and 0- to 3-hr-old embryos. The *dfmr1* transcript peaks in abundance between 9 and 12 hr of embryonic development, when it measures 4.0 kb. These data suggest that *dfmr1* gene expression may be regulated at the levels of transcription or transcript processing during embryogenesis. The *dfmr1* gene is a homolog of the human *FMR1* gene. Loss-of-function mutations in *FMR1* result in human fragile-X mental retardation.

5 DNA Sequencing Is the Ultimate Way to Characterize DNA Structure at the Molecular Level

In a sense, a cloned DNA molecule or any DNA from a single gene to an entire genome, is completely characterized at the molecular level only when its nucleotide sequence is known. The ability to sequence DNA has greatly enhanced our understanding of genome organization and increased our knowledge of gene structure, function, and mechanisms of regulation.

Historically the most commonly used method of DNA sequencing was developed by Fred Sanger and his colleagues and is known as **dideoxynucleotide chain-termination sequencing** or simply **Sanger sequencing.** In this technique, a double-stranded DNA molecule to be sequenced is converted to single strands that are used as a template for synthesizing a series of complementary strands. The DNA to be sequenced is mixed with a primer that is complementary to the target DNA or vector along with DNA polymerase, and four deoxyribonucleotide triphosphates (dATP, dCTP, dGTP, and dTTP).

The key to the Sanger technique is the addition of a small amount of one modified deoxyribonucleotide (**Figure 13**), called a **dideoxynucleotide** (abbreviated ddNTP). Notice that dideoxynucleotides have a 3′ hydrogen instead of a 3′ hydroxyl group. Dideoxynucleotides are called chain-termination nucleotides because they lack the 3′ oxygen required to form a phosphodiester bond with another nucleotide. Thus when ddNTPs are included in a reaction

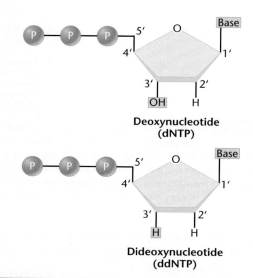

FIGURE 13 Deoxynucleotides (top) have an OH group at the 3′ position in the deoxyribose molecule. Dideoxynucleotides (bottom) lack an OH group and have only hydrogen (H) at this position.

as DNA synthesis takes place, the polymerase occasionally inserts a dideoxynucleotide instead of a deoxyribonucleotide into a growing DNA strand. Since the dideoxynucleotide does not have a 3'-OH group, it cannot form a 3' bond with another nucleotide, and DNA synthesis terminates because DNA polymerase cannot add new nucleotides to a ddNTP. The Sanger reaction takes advantage of this key modification.

For example, in **Figure 14**, notice that the shortest fragment generated is a sequence that has added ddCTP

to the 3' end of the primer and the chain has terminated. Over time as the reaction proceeds, eventually there will be a ddNTP inserted at every location in the newly synthesized DNA so that each strand synthesized differs in length by one nucleotide and is terminated by a ddNTP. This allows for separation of these DNA fragments by gel electrophoresis, which can then be used to determine the sequence.

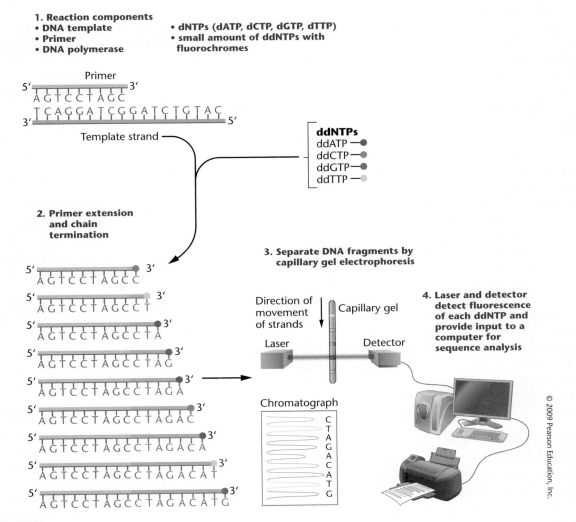

FIGURE 14 Computer-automated DNA sequencing using the chain-termination (Sanger) method. (1) A primer is annealed to a sequence adjacent to the DNA being sequenced (usually near the multiple cloning site of a cloning vector). (2) A reaction mixture is added to the primer–template combination. This includes DNA polymerase, the four dNTPs, and small molar amounts of dideoxynucleotides (ddNTPs) labeled with fluorescent dyes. All four ddNTPs are added to the same tube, and during primer extension, all possible lengths of chains are produced. During primer extension, the polymerase occasionally (randomly) inserts a ddNTP instead of a dNTP, terminating the synthesis of the chain because the ddNTP does not have the OH group needed to attach the next nucleotide. Over the course of the reaction, all possible termination sites will have a ddNTP inserted. The products of the reaction are added to a single lane on a capillary gel, and the bands are read by a detector and imaging system. The sequence is obtained by extension of the primer and is read from the newly synthesized strand, not the template strand. Thus, the sequence obtained begins with 5'-CTAGACATG-3'. Automated, robotic machines, such as those used in the Human Genome Project, sequence several hundred thousand nucleotides in a 24-hour period and then store and analyze the data automatically.

When the Sanger technique was first developed, it was done quite differently than is depicted in Figure 14. For example, four separate reaction tubes, each with a different single ddNTP, were used. These reactions also typically used either a radioactively labeled primer or a radioactively labeled ddNTP, polyacrylamide gel electrophoresis to separate fragments, and film detection (autoradiography) of radioactively labeled DNA. This original approach could typically read about 800 bases of 100 DNA molecules simultaneously. Dramatic modifications of the Sanger technique have led to technologies that allow sequencing reactions to occur in a single tube in which each of the four ddNTPs is labeled with a different-colored fluorescent dye (Figure 14). Reaction products are separated through a single, ultrathin-diameter polyacrylamide tube gel called a capillary gel (capillary gel electrophoresis). As DNA fragments move through the gel, they are scanned with a laser. The laser stimulates fluorescent dyes on each DNA fragment, which then emit different wavelengths of light for each ddNTP. Emitted light is captured by a detector that amplifies and feeds this information into a computer to convert the light patterns into a DNA sequence. The data are represented as a series of colored peaks, each corresponding to one nucleotide in the sequence.

Since the early 1990s, DNA sequencing has largely been performed through computer-automated Sanger-reaction-based technology and is referred to as **computer-automated high-throughput DNA sequencing.** Such systems generate relatively large amounts of sequence DNA at greatly decreased costs compared to the original Sanger method. Automated DNA sequencers often contain multiple capillary gels (as many as 96) that are several feet long and can process several thousand bases of sequences so that many of these instruments made it possible to generate over 2 million bp of sequences in a day! These systems became essential for enabling the rapidly accelerating progress of the Human Genome Project.

Sequencing Technologies Have Progressed Rapidly

DNA sequencing technologies have undergone an incredible evolution to dramatically improve sequencing capabilities. New innovations in sequencing technology have developed quickly in the past few years. When it comes to sequencing entire genomes, however, Sanger sequencing technologies are outdated. Even computer-automated DNA sequencing is simply not high enough to support the growing demand for genomic data. This demand is being driven in large part by personalized genomics and the desire to analyze the genomes of many other species.

Advances in genomics have spurred a demand for sequencers that are faster and capable of generating millions of bases of DNA sequences in a relatively short time, leading to the development of **next-generation sequencing (NGS) technologies.** Such technologies dispense with the Sanger technique and capillary electrophoresis methods in favor of sophisticated, parallel formats (simultaneous reaction formats) that use state-of-the art fluorescence imaging techniques. NGS technologies are providing an unprecedented capacity for generating massive amounts of DNA sequence data rapidly (up to 200 times faster than Sanger approaches in some cases!) and at dramatically reduced costs per base.

The company Applied Biosystems (ABI) has developed an approach called **SOLiD (supported oligonucleotide ligation and detection)** that can produce 6 *gigabases* of sequence data per run! The SOLiD method combines a variety of approaches to sequence DNA fragments that are linked to beads and amplified.

Third-generation sequencers are already available in some laboratories and there is every reason to believe these instruments will be widely used in the next few years. For example, one promising approach on the immediate horizon involves the use of nanotechnology by pushing single-stranded DNA fragments into nanopores and then cleaving off individual bases to produce a signal that can be captured. This method does not involve DNA amplification or fluorescent tags and thus provides direct sequencing of the DNA in a single strand.

The genomics research community has embraced NGS technologies. Which approaches will eventually emerge as the sequencing methods of choice for the short-term future is unclear, but what is clear is that the landscape of sequencing capabilities has dramatically changed for the better. Never before have scientists had the ability to generate so much sequence data so quickly.

The principal techniques of recombinant DNA technology, particularly DNA cloning, were essential for making genome projects possible, but no technology has had a greater influence on our ability to study genomes than DNA sequencing. Rapid advances in sequencing technology that pushed the capabilities of computer-automated sequencing were driven by the demands of genome scientists (particularly those working on the Human Genome Project) to rapidly generate more sequence with greater accuracy and at lower cost.

ESSENTIAL POINT ■ ■ ■

DNA sequencing technologies are changing rapidly. Next-generation methods generate large amounts of sequence data relatively inexpensively and in short time.

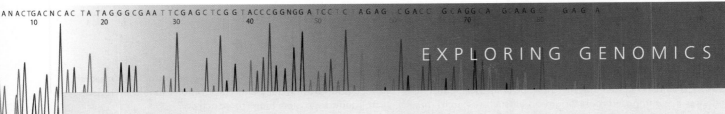

EXPLORING GENOMICS

Manipulating Recombinant DNA: Restriction Mapping and Designing PCR Primers

 Study Area: *Exploring Genomics*

As you learned in this chapter, restriction enzymes are sophisticated "scissors" that molecular biologists use to cut DNA, and they are routinely used in genetics and molecular biology laboratories for recombinant DNA experiments. A wide variety of online tools are available to assist scientists working with recombinant DNA for different applications, such as restriction mapping and designing primers for PCR experiments. Here we explore **Webcutter** and **Primer3**, two sites that make recombinant DNA experiments much easier.

■ Exercise I—Creating a Restriction Map in Webcutter

Suppose you had cloned and sequenced a gene and you wanted to design a probe approximately 600 bp long that could be used to analyze expression of this gene in different human tissues by Northern blot analysis. Not too long ago, you had primarily two ways to approach this task. You could digest the cloned DNA with whatever restriction enzymes were in your freezer, then run agarose gels and develop restriction maps in the hope of identifying cutting sites that would give you the size fragment you wanted. Or you could scan the sequence with your eyes, looking for restriction sites of interest—a very time-consuming and eye-straining effort!

Internet sites such as Webcutter take the guesswork out of developing restriction maps and make it relatively easy to design experiments for manipulating recombinant DNA. In this exercise, you will use Webcutter to create a restriction map of human DNA with the enzymes *Eco*RI, *Bam*HI, and *Pst*I.

1. Access **Webcutter** at http://rna.lundberg.gu.se/cutter2. Go to the Companion Website for *Essentials of Genetics* and open the Exploring Genomics exercise for this chapter. Copy the sequence of cloned human DNA from this exercise and paste it into the text box in Webcutter.

2. Scroll down to "Please indicate which enzymes to include in the analysis." Click the button indicating "Use only the following enzymes." Select the restriction enzymes *Eco*RI, *Bam*HI, and *Pst*I from the list provided, then click "Analyze sequence." (*Note:* Use the command, control, or shift key to select multiple restriction enzymes.)

3. After examining the results provided by Webcutter, create a table showing the number of cutting sites for each enzyme and the fragment sizes that would be generated by digesting with each enzyme. Draw a restriction map indicating cutting sites for each enzyme with distances between each site and the total size of this piece of human DNA.

■ Exercise II—Designing a Recombinant DNA Experiment

Now that you have created a restriction map of your piece of human DNA, you need to ligate the DNA into a plasmid DNA vector that you can use to make your probe (molecular biologists often refer to this as subcloning). To do this, you will need to determine which restriction enzymes would best be suited for cutting both the plasmid and the human DNA.

1. Referring back to the Companion Website and the Exploring Genomics exercise for this chapter, copy the plasmid DNA sequence from Exercise II into the text box in Webcutter and identify cutting sites for the same enzymes you used in Exercise I. Then answer the following questions:

a. What is the total size of the plasmid DNA analyzed in Webcutter?

b. Which enzyme(s) could be used in a recombinant DNA experiment to ligate the plasmid to the *largest* DNA fragment from the human gene? Briefly explain your answer.

c. What size recombinant DNA molecule will be created by ligating these fragments?

d. Draw a simple diagram showing the cloned DNA inserted into the plasmid and indicate the restriction-enzyme cutting site(s) used to create this recombinant plasmid.

2. As you prepare to carry out this subcloning experiment, you find that the expiration dates on most of your restriction enzymes have long since passed. Rather than run an experiment with old enzymes, you decide to purchase new enzymes. Fortunately, a site called **REBASE®: The Restriction Enzyme Database** can help you. Over 300 restriction enzymes are commercially available rather inexpensively, but scientists are always looking for ways to stretch their research budgets as far as possible. REBASE is excellent for locating enzyme suppliers and enzyme specifics, particularly if you need to work with an enzyme that you are unfamiliar with. Visit **REBASE®** at http://rebase.neb.com/rebase/rebase.html to identify companies that sell the restriction enzyme(s) you need for this experiment.

CASE STUDY » Should we worry about recombinant DNA technology?

Early in the 1970s, when recombinant DNA research was first developed, scientists realized that there may be unforeseen dangers, and after a self-imposed moratorium on all such research, they developed and implemented a detailed set of safety protocols for the construction, storage, and use of genetically modified organisms. These guidelines then formed the basis of regulations adopted by the federal government. Over time, safer methods were developed, and these stringent guidelines were gradually relaxed or in many cases, eliminated altogether. Now, however, the specter of bioterrorism has re-focused attention on the potential misuses of recombinant DNA technology. For example, individuals or small groups might use the information in genome databases coupled with recombinant DNA technology to construct or reconstruct agents of disease, such as the smallpox virus or the deadly influenza virus.

1. Do you think that the question of recombinant DNA research regulation by university and corporations should be revisited to monitor possible bioterrorist activity?

2. Should freely available access to genetic databases, including genomes, and gene or protein sequences be continued, or should it be restricted to individuals who have been screened and approved for such access?

3. Forty years after its development, the use of recombinant DNA technology is widespread and is found even in many middle school and high school biology courses. Are there some aspects of gene splicing that might be dangerous in the hands of an amateur?

INSIGHTS AND SOLUTIONS

1. The recognition sequence for the restriction enzyme *Sau*3AI is GATC (see Figure 1). In the recognition sequence for the enzyme *Bam*HI—GGATCC—the four internal bases are identical to the *Sau*3AI sequence. The single-stranded ends produced by the two enzymes are identical. Suppose you have a cloning vector that contains a *Bam*HI recognition sequence and you also have foreign DNA that was cut with *Sau*3AI.

(a) Can this DNA be ligated into the *Bam*HI site of the vector, and if so, why? (b) Can the DNA segment cloned into this sequence be cut from the vector with *Sau*3AI? With *Bam*HI? What potential problems do you see with the use of *Bam*HI?

Solution:

(a) DNA cut with *Sau*3AI can be ligated into the vector's *Bam*HI cutting site because the single-stranded ends generated by the two enzymes are identical. **(b)** The DNA can be cut from the vector with *Sau*3AI because the recognition sequence for this enzyme (GATC) is maintained on each side of the insert. Recovering the

cloned insert with *Bam*HI is more problematic. In the ligated vector, the conserved sequences are GGATC (left) and GATCC (right). The correct base for recognition by *Bam*HI will *follow* the conserved sequence (to produce GGATCC on the left) only about 25 percent of the time, and the correct base will *precede* the conserved sequence (and produce GGATCC on the right) about 25 percent of the time as well. Thus, *Bam*HI will be able to cut the insert from the vector ($0.25 \times 0.25 = 0.0625$), or only about 6 percent, of the time.

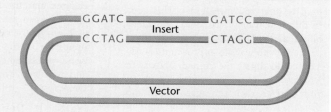

PROBLEMS AND DISCUSSION QUESTIONS

 For activities, animations, and review quizzes, go to the study area at www.masteringgenetics.com

HOW DO WE KNOW?

1. In this chapter we focused on how specific DNA sequences can be copied, identified, characterized, and sequenced. At the same time, we found many opportunities to consider the methods and reasoning underlying these techniques. From the explanations given in the chapter, what answers would you propose to the following fundamental questions?

   (a) In a recombinant DNA cloning experiment, how can we determine whether DNA fragments of interest have been incorporated into plasmids and, once host cells are transformed, which cells contain recombinant DNA?

   (b) When using DNA libraries to clone genes, what combination of techniques is used to identify a particular gene of interest?

   (c) What steps make PCR a chain reaction that can produce millions of copies of a specific DNA molecule in a matter of hours without using host cells?

   (d) How has DNA sequencing technology evolved in response to the emerging needs of genome scientists?

2. What roles do restriction enzymes, vectors, and host cells play in recombinant DNA studies?

3. What role does DNA ligase perform in a DNA cloning experiment? How does the action of DNA ligase differ from the function of restriction enzymes?

4. Although many cloning applications involve introducing recombinant DNA into bacterial host cells, many other cell types are also used as hosts for recombinant DNA. Why?

5. Using DNA sequencing on a cloned DNA segment, you recover the nucleotide sequence shown below. Does this segment contain a palindromic recognition sequence for a restriction enzyme? If so, what is the double-stranded sequence of the palindrome, and what enzyme would cut at this sequence? (Consult Figure 1 for a list of restriction sites.)

5′-GTCATAGGATCCGTA-3′

6. Restriction sites are palindromic; that is, they read the same in the 5′ to 3′ direction on each strand of DNA. What is the advantage of having restriction sites organized in this way?

7. List the advantages and disadvantages of using plasmids as cloning vectors.

8. The introduction of genes into plants is a common practice that has generated not only a host of genetically modified foodstuffs, but also significant worldwide controversy. Interestingly, a tumor-inducing plasmid is often used to produce genetically modified plants. Is the use of a tumor-inducing plasmid the source of such controversy?

9. In the context of recombinant DNA technology, of what use is a probe?

10. If you performed a PCR experiment starting with only one copy of double-stranded DNA, approximately how many DNA molecules would be present in the reaction tube after 15 cycles of amplification?

11. In a control experiment, a plasmid containing a HindIII recognition sequence within a kanamycin resistance gene is cut with HindIII, re-ligated, and used to transform E. coli K12 cells. Kanamycin-resistant colonies are selected, and plasmid DNA from these colonies is subjected to electrophoresis. Most of the colonies contain plasmids that produce single bands that migrate at the same rate as the original intact plasmid. A few colonies, however, produce two bands, one of original size and one that migrates much higher in the gel. Diagram the origin of this slow band as a product of ligation.

12. You have just created the world's first genomic library from the African okapi, a relative of the giraffe. No genes from this genome have been previously isolated or described. You wish to isolate the gene encoding the oxygen-transporting protein β-globin from the okapi library. This gene has been isolated from humans, and its nucleotide sequence and amino acid sequence are available in databases. Using the information available about the human β-globin gene, what two strategies can you use to isolate this gene from the okapi library?

13. What advantages do cDNA libraries provide over genomic DNA libraries? Describe cloning applications where the use of a genomic library is necessary to provide information that a cDNA library cannot.

14. Although the capture and trading of great apes has been banned in 112 countries since 1973, it is estimated that about 1000 chimpanzees are removed annually from Africa and smuggled into Europe, the United States, and Japan. This illegal trade is often disguised by simulating births in captivity. Until recently, genetic identity tests to uncover these illegal activities were not used because of the lack of highly polymorphic markers (markers that vary from one individual to the next) and the difficulties of obtaining chimpanzee blood samples. A study was reported in which DNA samples were extracted from freshly plucked chimpanzee hair roots and used as templates for PCR. The primers used in these studies flank highly polymorphic sites in human DNA that result from variable numbers of tandem nucleotide

repeats. Several offspring and their putative parents were tested to determine whether the offspring were "legitimate" or the product of illegal trading. The data are shown in the following Southern blot.

Examine the data carefully and choose the best conclusion.

(a) None of the offspring is legitimate.

(b) Offspring B and C are not the products of these parents and were probably purchased on the illegal market. The data are consistent with offspring A being legitimate.

(c) Offspring A and B are products of the parents shown, but C is not and was therefore probably purchased on the illegal market.

(d) There are not enough data to draw any conclusions. Additional polymorphic sites should be examined.

(e) No conclusion can be drawn because "human" primers were used.

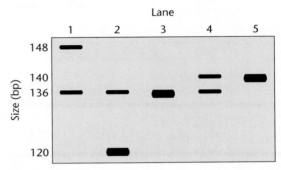

Lane 1: father chimpanzee
Lane 2: mother chimpanzee
Lanes 3–5: putative offspring A, B, C

15. List the steps involved in screening a genomic library. What must be known before starting such a procedure? What are the potential problems with such a procedure, and how can they be overcome or minimized?

16. In a typical PCR reaction, describe what is happening in stages occurring at temperature ranges (a) 90–95°C, (b) 50–70°C, and (c) 70–75°C?

17. We usually think of enzymes as being most active at around 37°C, yet in PCR the DNA polymerase is subjected to multiple exposures of relatively high temperatures and seems to function appropriately at 70–75°C. What is special about the DNA polymerizing enzymes typically used in PCR?

18. How are dideoxynucleotides (ddNTPs) structurally different from deoxynucleotides (dNTPs), and how does this structural difference make ddNTPs valuable in chain-termination methods of DNA sequencing?

19. How is fluorescent in situ hybridization (FISH) used to produce a spectral karyotype?

20. The gel presented here shows the pattern of bands of fragments produced with several restriction enzymes. The enzymes used are identified above and below the gel, and six possible restriction maps are shown in the column to the right.

One of the six restriction maps shown is consistent with the pattern of bands shown in the gel.

(a) From your analysis of the pattern of bands on the gel, select the correct map and explain your reasoning.

(b) In a Southern blot prepared from this gel, the highlighted bands (pink) hybridized with the gene pep. Where is the pep gene located?

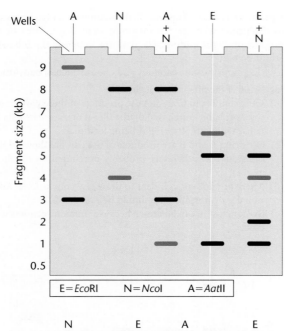

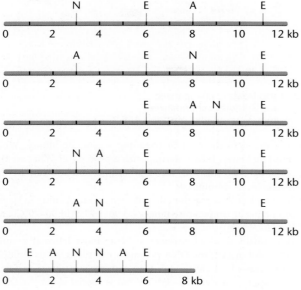

E = *Eco*RI   N = *Nco*I   A = *Aat*II

21. A widely used method for calculating the annealing temperature for a primer used in PCR is 5 degrees below the T_m (°C), which is computed by the equation $81.5 + 0.41 (\%GC) - (675/N)$, where %GC is the percentage of GC nucleotides in the oligonucleotide and N is the length of the oligonucleotide. Notice from the formula that both the GC content *and* the length of the oligonucleotide are variables. Assuming you have the following oligonucleotide as a primer, compute the annealing temperature for PCR. What is the relationship between T_m (°C) and %GC? Why? (*Note:* In reality, this computation provides only a starting point for empirical determination of the most useful annealing temperature.)

5'-TTGAAAATATTTCCCATTGCC-3'

22. The U.S. Department of Justice has established a database that catalogs PCR amplification products from short tandem repeats of the Y (Y-STRs) chromosome in humans. The database contains polymorphisms of five U.S. ethnic groups (African Americans, European Americans, Hispanics, Native Americans, and Asian Americans) as well as the worldwide population.

(a) Given that STRs are repeats of varying lengths, for example $(TCTG)_{9-17}$ or $(TAT)_{6-14}$, explain how PCR could reveal differences (polymorphisms) among individuals. How could the Department of Justice make use of those differences?

(b) Y-STRs from the nonrecombining region of the Y chromosome (NRY) have special relevance for forensic purposes. Why?

(c) What would be the value of knowing the ethnic population differences for Y-STR polymorphisms?

(d) For forensic applications, the probability of a "match" for a crime scene DNA sample and a suspect's DNA often culminates in a guilty or innocent verdict. How is a "match" determined, and what are the uses and limitations of such probabilities?

SOLUTIONS TO SELECTED PROBLEMS AND DISCUSSION QUESTIONS

Answers to Now Solve This

1. (a) Because the *Drosophila* DNA has been cloned into the *Pst*I site in the ampicillin resistance gene of the plasmid, the gene will be mutated and any bacterium with the recombinant plasmid will be ampicillin sensitive. The tetracycline resistance gene remains active, however. Bacteria that have been transformed with the recombinant plasmid will be resistant to tetracycline and therefore tetracycline should be added to the medium.

(b) Colonies that grow on a tetracycline medium only should contain the insert. Those bacteria which do not grow on the ampicillin medium probably contain the *Drosophila* DNA insert.

(c) Resistance to both antibiotics by a transformed bacterium could be explained in several ways. First, if cleavage with the *Pst*I was incomplete, then no change in biological properties of the uncut plasmids would be expected. Also, it is possible that the cut ends of the plasmid were ligated together in the original form with no insert.

2. Because the African okapi is a mammal (relative of the giraffe) it will have many sequences in common with those of humans and other mammals that have been sequenced. Using the human nucleotide sequence, for example, one can produce primers that are likely to be useful for isolating particular genes. If primers identical to humans are not successful, then a series of degenerate primers might be used.

Solutions to Problems and Discussion Questions

4. While bacteria are commonly used in cloning, other cell types are also very useful, such as yeast, mammalian, etc. Bacteria are prokaryotes and as such do not process transcripts as do eukaryotes, therefore, there is often an advantage to using a eukaryotic host. In addition, one might be interested in the influence of a specific DNA segment in a specific host environment, thus necessitating the use of a variety of hosts.

6. The simple answer to this question is to assume that one is asking about the advantage to the scientist of having restriction enzyme sites recognize palindromic sites. In this case the answer would be that single-stranded overhanging ends are often generated, which allow DNA from different sources cut with the same restriction enzyme to generate complementary overhangs, which can anneal to form recombinant molecules.

   If one considers the question from a bacterial standpoint, the answer is much more involved. In fact, bacterial chromosomes actually have fewer palindromic sites than expected based on chance. This adaptation stems from the fact that restriction sites cleave at palindromic sequences and one way to keep them from cleaving the host DNA is to evolve away from such sequences.

8. No. The tumor-inducing plasmid (Ti) that is used to produce genetically modified plants is specific for the bacterium *Agrobacterium tumifaciens* which causes tumors in many plant species. There is no danger that this tumor-inducing plasmid will cause tumors in humans.

10. The total number of molecules after 15 cycles would be 16,384 or $(2)^{14}$.

12. Using the human nucleotide sequence, one can produce a probe to screen the library of the African okapi. Second, one can use the amino acid sequence and the genetic code to generate a complementary DNA probe for screening of the library. The probe is used, through hybridization, to identify the DNA that is complementary to the probe and allow one to identify the library clone containing the DNA of interest. Cells with the desired clone are then picked from the original plate and the plasmid is isolated from the cells.

14. Option (b) fits the expectation because the thick band in the offspring probably represents the bands at approximately the same position in both parents. The likelihood of such a match is expected to be low in the general population.

16. (a) Heating to 90–95°C denatures the double-stranded DNA so that it dissociates into single strands. It usually takes about five minutes, depending on the length and GC content of the DNA.

(b) Lowering the temperature to 50–70°C allows the primers to bind to the denatured DNA.

(c) Bringing the temperature to 70–75°C allows the heat stable DNA polymerase an opportunity to extend the primers by adding nucleotides to the 3′ ends of each growing strand. Each PCR is designed with specific temperatures (not ranges) based on the characteristics of the DNAs (template and primers).

20. (a) The overall size of the fragment is 12 kb. From the A + N digest, sites A and N must be 1 kb apart. N must be 2 kb from an E site. Pattern #5 is the likely choice. Notice that digest A + N breaks up the 6 kb E fragment.

(b) By drawing lines though sections that hybridize to the probe, one can see that the only place of consistent overlap to the probe is the 1 kb fragment between A and N.

22. (a) Short tandem repeats of the Y chromosome (Y-STRs) vary considerably among individuals and populations. By amplifying Y-STRs by PCR and separating the amplified products by electrophoresis, one can genotypically type an individual as one does with a standard fingerprint. Because tissue samples are often left at the scene of a violent crime, DNA fingerprints are sometimes more available than standard fingerprints. Linking an individual with the time and place of a significant event has multiple forensic applications. Eliminating an individual as a suspect also has important forensic applications.

(b) The nonrecombining region of the Y is maintained strictly in the male population. Of special relevance in forensic applications would be the elimination of half the population (females) from a suspect group.

(c) Because different ethnic groups show different levels of Y-STR polymorphism, different final probabilities occur as products of individual probabilities. Since these probabilities are used to match individuals in forensics, ethnic variations must be taken under consideration.

(d) While there are many potential uses of DNA samples, generally a "match" is determined by multiplying the occurrence probabilites of each haplotype to arrive at the overall probability (product) of a genotype occurring in a population. If an individual's genotype matches that found in DNA at a crime scene, depending on the frequecies of the haplotypes, one might be able to say that the individual was at the crime scene. However, contamination, inappropriate genotyping, and laboratory expertise may give both false positive or negative results. Identical twins will have identical DNA fingerprints and may complicate forensic applications.

CREDITS

Credits are listed in order of appearance.

Photo

CO, Pascal Goetgheluck/Photo Researchers, Inc.; F-3 (a) Gopal Murti/Photo Researchers, Inc.; F-5, Michael Gabridge/Custom Medical Stock Photo; F-9, Custom Medical Stock Photo; F-11 (a, b) Proceedings of the National Academy of Sciences; F-12, American Society for Cell Biology

Text

Source: Webcutter., Source: REBASE

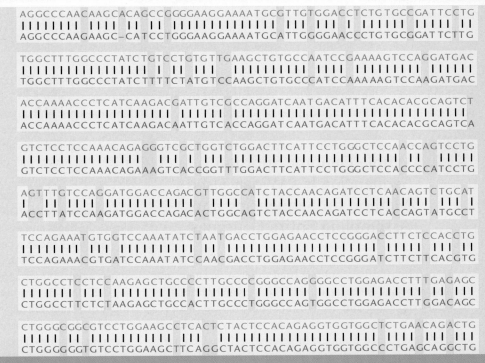

Alignment comparing DNA sequence for the leptin gene from dogs (top) and humans (bottom). Vertical lines and shaded boxes indicate identical bases. *LEP* encodes a hormone that functions to suppress appetite. This type of analysis is a common application of bioinformatics and a good demonstration of comparative genomics.

Genomics, Bioinformatics, and Proteomics

CHAPTER CONCEPTS

■ Genomics applies recombinant DNA, DNA sequencing methods, and bioinformatics to sequence, assemble, and analyze genomes.

■ Disciplines in genomics encompass several areas of study, including structural and functional genomics, comparative genomics, and metagenomics, and have led to an "omics" revolution in modern biology.

■ Bioinformatics merges information technology with biology and mathematics to store, share, compare, and analyze nucleic acid and protein sequence data.

■ The Human Genome Project has greatly advanced our understanding of the organization, size, and function of the human genome.

■ Ten years after completion of the Human Genome Project, a new era of genomics studies is providing deeper insights into the human genome.

■ Comparative genomics analysis of model prokaryotes and eukaryotes has revealed similarities and differences in genome size and organization.

■ Metagenomics is the study of genomes from environmental samples and is valuable for identifying microbial genomes.

■ Transcriptome analysis provides insight into patterns of gene expression and gene-regulatory activity of a genome.

■ Proteomics focuses on the protein content of cells and on the structures, functions, and interactions of proteins.

■ Systems biology approaches attempt to uncover complex interactions among genes, proteins, and other cellular components.

The term **genome,** meaning the complete set of DNA in a single cell of an organism, was coined in 1920, at a time when geneticists began to turn from the study of individual genes to a focus on the larger picture.

In 1977, as recombinant DNA-based techniques were developed, Fred Sanger and colleagues began the field of **genomics,** the study of genomes, by using a newly developed method of DNA sequencing to sequence the 5400-nucleotide genome of the virus ϕX174. Other viral genomes were sequenced in short order, but even this technology was slow and labor-intensive, limiting its use to small genomes. During the next three decades, the development of computer-automated DNA sequencing methods made it possible to consider sequencing the larger and more complex genomes of eukaryotes, including the 3.1 billion nucleotides that comprise the human genome. Recombinant DNA technologies coupled with the advent of computer-automated DNA sequencing methods, bioinformatics, and now next-generation sequencing technologies are responsible for rapidly accelerating the field of genomics. Genomic technologies have developed so quickly that modern biological research is currently

From Chapter 18 of *Essentials of Genetics, Eighth edition,* William S. Klug, Michael R. Cummings, Charlotte A. Spencer, Michael A. Palladino. ©2013 by Pearson Education, Inc. All rights reserved.

experiencing a genomics revolution. In this chapter, we will examine basic technologies used in genomics and then discuss examples of different disciplines of genomics.

1   Whole-Genome Shotgun Sequencing Is a Widely Used Method for Sequencing and Assembling Entire Genomes

RecombinantDNA technology made it possible to generate DNA libraries that could be used to identify, clone, and sequence specific genes of interest. But a primary limitation of library screening and even of most polymerase chain reaction (PCR) approaches is that they typically can identify only relatively small numbers of genes at a time. Genomics allows the sequencing of entire genomes. **Structural genomics** focuses on sequencing genomes and analyzing nucleotide sequences to identify genes and other important sequences such as gene-regulatory regions.

The most widely used strategy for sequencing and assembling an entire genome involves variations of a method called **whole-genome shotgun sequencing.** In simple terms, this technique is analogous to you and a friend taking your

respective copies of this textand randomly ripping the pages into strips about 5 to 7 inches long. Each page represents a chromosome, and all of the letters in the entire text are the "genome." Then, you and your friend would go through the painstaking task of comparing the pieces of paper to find places that match, thus creating overlapping sentences—areas where there are similar sentences on different pieces of paper. Eventually, in theory, many of the strips containing matching sentences would overlap in ways that you could use to reconstruct the pages and assemble the order of the entire text.

Figure 1 shows a basic overview of whole-genome shotgun sequencing. First, an entire chromosome is cut into short, overlapping fragments, either by mechanically shearing the DNA in various ways (such as excessive heat treatment or sonication in which sonic energy is used to break DNA) or by using restriction enzymes to cleave the DNA at different locations. For simplicity, here we present a basic example of DNA shearing using restriction enzymes. Different restriction enzymes can be used so that chromosomes are cut at different sites; or sometimes, *partial digests* of DNA using the same restriction enzyme are used. With partial digests, DNA is incubated with restriction enzymes for only a short period of time, so that not every site in a particular sequence is cut to completion by an individual enzyme. Either way, restriction digests of whole chromosomes gener-

FIGURE 1   An overview of whole-genome shotgun sequencing and assembly. This approach shows one strategy that involves using restriction enzymes to digest genomic DNA into contigs, which are then sequenced and aligned using bioinformatics to identify overlapping fragments based on sequence identity. Notice that *Eco*RI digestion of the portion of DNA depicted here produces two fragments (contigs 1, 2–4), whereas digestion with *Bam*HI produces three fragments (contigs 1–2, 3, 4).

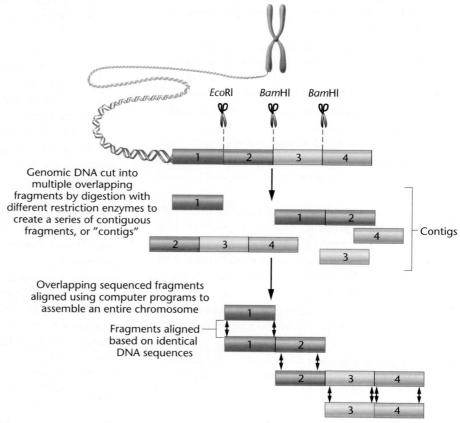

ate thousands to millions of overlapping DNA fragments. For example, a 6-bp cutter such as *Eco*RI creates about 700,000 fragments when used to digest the human genome! Because these overlapping fragments are adjoining segments that collectively form one continuous DNA molecule within a chromosome, they are called **contiguous fragments,** or **"contigs."**

In the next section, we will discuss the importance of bioinformatics to genomics. One of the earliest bioinformatics applications to be developed for genomic purposes was the use of algorithm-based software programs for creating a DNA sequence **alignment,** in which similar sequences of bases, such as contigs, are lined up for comparison. Alignment identifies overlapping sequences, allowing scientists to reconstruct their order in a chromosome. **Figure 2** shows an example of contig alignment and assembly for a portion of human chromosome 2. For simplicity, this figure shows relatively short sequences for each contig, which in actuality would be much longer. The figure is also simplified in that, in actual alignments, assembled sequences do not always overlap only at their ends.

The whole-genome shotgun sequencing method was developed by J. Craig Venter and colleagues at The Institute for Genome Research (TIGR). In 1995, TIGR scientists used this approach to sequence the 1.83-million-bp genome of the bacterium *Haemophilus influenzae*. This was the first com-

pleted genome sequence from a free-living organism, and it demonstrated "proof of concept" that shotgun sequencing could be used to sequence an entire genome. Even after the genome for *H. influenzae* was sequenced, many scientists were skeptical that a shotgun approach would work on the larger genomes of eukaryotes. But variations of shotgun approaches are now the predominant methods for sequencing genomes, including those of *Drosophila,* dog, several hundred species of bacteria, humans, and many other organisms, as you will read about later in this chapter.

Cutting a genome into contigs is not particularly difficult; however, a primary hurdle that had to be overcome to advance whole-genome sequencing was the question of how to sequence millions or billions of base pairs in a timely and cost-effective way. This was a challenge for scientists working on the Human Genome Project. The major technological breakthrough that made genomics possible was the development of computer-automated sequencers.

Computer-**automated DNA sequencing instruments** designed for so-called **high-throughput sequencing,** can process millions of base pairs in a day. Next-generation sequencers now enable genome scientists to produce even larger amounts of sequence DNA with improved accuracy and reduced cost.

ESSENTIAL POINT   ■   ■   ■

Whole-genome shotgun sequencing enables scientists to assemble sequence maps of entire genomes.

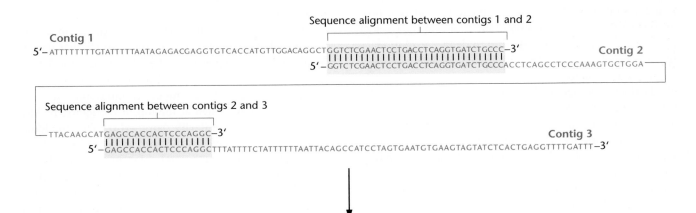

Sequence alignment between contigs 1 and 2

Contig 1
5′−ATTTTTTTTGTATTTTTAATAGAGACGAGGTGTCACCATGTTGGACAGGCTGGTCTCGAACTCCTGACCTCAGGTGATCTGCCC−3′   **Contig 2**
5′−GGTCTCGAACTCCTGACCTCAGGTGATCTGCCCACCTCAGCCTCCCAAAGTGCTGGA−

Sequence alignment between contigs 2 and 3

−TTACAAGCATGAGCCACCACTCCCAGGC−3′   **Contig 3**
5′−GAGCCACCACTCCCAGGCTTTATTTTCTATTTTTTAATTACAGCCATCCTAGTGAATGTGAAGTAGTATCTCACTGAGGTTTTGATTT−3′

Assembled sequence of a partial segment of chromosome 2 based on alignment of three contigs

5′−ATTTTTTTTGTATTTTTAATAGAGACGAGGTGTCACCATGTTGGACAGGCTGGTCTCGAACTCCTGACCTCAGGTGATCTGCCCACCTCAGCCTCCCAAAGTGCTGGA
TTACAAGCATGAGCCACCACTCCCAGGCTTTATTTTCTATTTTTTAATTACAGCCATCCTAGTGAATGTGAAGTAGTATCTCACTGAGGTTTTGATTT−3′

FIGURE 2   DNA sequence alignment of contigs on human chromosome 2. Single-stranded DNA for three different contigs from human chromosome 2 is shown in blue, red, or green. The actual sequence from chromosome 2 is shown, but in reality, contig alignment involves fragments that are several thousand bases in length. Alignment of the three contigs allows a portion of chromosome 2 to be assembled. Alignment of all contigs for a particular chromosome would result in assembly of a completely sequenced chromosome.

2 DNA Sequence Analysis Relies on Bioinformatics Applications and Genome Databases

Genomics necessitated the rapid development of **bioinformatics,** the use of computer hardware and software and mathematics applications to organize, share, and analyze data related to gene structure, gene sequence and expression, and protein structure and function. However, even before whole-genome sequencing projects had been initiated, a large amount of sequence information from a range of different organisms was accumulating as a result of gene cloning by recombinant DNA techniques. Scientists around the world needed databases that could be used to store, share, and obtain the maximum amount of information from protein and DNA sequences. Thus, bioinformatics software was already being widely used to compare and analyze DNA sequences and to create private and public databases. Once genomics emerged as a new approach for analyzing DNA, however, bioinformatics became even more important than before. Today, it is a dynamic area of biological research, providing new career opportunities for anyone interested in merging an understanding of biological data with information technology, mathematics, and statistical analysis.

Among the most important applications of bioinformatics are to compare DNA sequences, as in contig alignment, discussed in the previous section; to identify genes in a genomic DNA sequence; to find gene-regulatory regions, such as promoters and enhancers; to identify structural sequences, such as telomeric sequences, in chromosomes; to predict the amino acid sequence of a putative polypeptide encoded by a cloned gene sequence; to analyze protein structure and predict protein functions on the basis of identified domains and motifs; and to deduce evolutionary relationships between genes and organisms on the basis of sequence information.

High-throughput DNA sequencing techniques were developed nearly simultaneously with the expansion of the Internet. As genome data accumulated, many DNA sequence databases became freely available online. Databases are essential for archiving and sharing data with other researchers and with the public. One of the most important genomic databases, called **GenBank,** is maintained by the National Center for Biotechnology Information (NCBI) in Washington, D.C., and is the largest publicly available database of DNA sequences. GenBank shares and acquires data from databases in Japan and Europe; it contains more than 100 billion bases of sequence data from over 100,000 species; and it doubles in size roughly every 14–18 months! The Human Genome Nomenclature Committee, supported by the NIH, establishes rules for assigning names and symbols to newly cloned human genes. As sequences are identified and genes are named, each sequence deposited into GenBank is provided with an **accession number** that scientists can use to access and retrieve that sequence for analysis.

The NCBI is an invaluable source of public access databases and bioinformatics tools for analyzing genome data. In the Exploring Genomics feature for this chapter, you will use NCBI and GenBank to compare and align contigs to assemble a chromosome segment.

Annotation to Identify Gene Sequences

Although genome projects generate tremendous amounts of DNA sequence information, these data are of little use until they have been analyzed and interpreted. Thus, after a genome has been sequenced and compiled, scientists are faced with the task of identifying gene-regulatory sequences and other sequences of interest in the genome so that gene maps can be developed. This process, called **annotation,** relies heavily on bioinformatics, and a wealth of different software tools are available to carry it out.

One initial approach to annotating a sequence is to compare the newly sequenced genomic DNA to the known sequences already stored in various databases. The NCBI provides access to **BLAST (Basic Local Alignment Search Tool),** a very popular software application for searching through banks of DNA and protein sequence data. Using BLAST, we can compare a segment of genomic DNA to sequences throughout major databases such as GenBank to identify portions that align with or are the same as existing sequences.

Figure 3 shows a representative example of a sequence alignment based on a BLAST search. Here a 280-bp chromosome 12 contig from the rat was used to search a mouse database to determine whether a sequence in the rat contig matched a known gene in mice. Notice that the rat contig (the query sequence in the BLAST search) aligned with base pairs 174,612 to 174,891 of mouse chromosome 8. The accession number for the mouse chromosome sequence, NT_039455.6, is indicated at the top of the figure. BLAST searches calculate a **similarity score**—also called the **identity value**—determined by the sum of identical matches between aligned sequences divided by the total number of bases aligned. Gaps, indicating missing bases in the two sequences, are usually ignored in calculating similarity scores. The aligned rat and mouse sequences were 93 percent similar and showed no gaps in the alignment.

Because this mouse sequence on chromosome 8 is known to contain an insulin receptor gene (encoding a protein that binds the hormone insulin), it is highly likely that the rat contig sequence also contains an insulin receptor gene. We will return to the topic of similarity in Sections 3 and 6, where we consider how similarity between gene sequences can be used

ref | NT_039455.6 | Mm8_39495_36
Mus musculus chromosome 8 genomic contig, strain C57BL/6J
Features in this part of subject sequence: insulin receptor
Score = 418 bits (226), Expect = 2e-114
Identities = 262/280 (93%), Gaps = 0/280 (0%)

```
Query    1       CAGGCCATCCCGAAAGCGAAGATCCCTTGAAGAGGTGGGCAATGTGACAGCCACTACACC   60
                 |||||||||||||||||||||||||||||||||||||||||| |||||||||||| ||||
Sbjct    174891  CAGGCCATCCCGAAAGCGAAGATCCCTTGAAGAGGTGGGGAATGTGACAGCCACCACACT   174832

Query    61      CACACTTCCAGATTTTCCCAACATCTCCTCCACCATCGCGCCCACAAGCCACGAAGAGCA   120
                 |||||||||||||| |||||| ||||||| || ||||| || ||||| || || ||||||
Sbjct    174831  CACACTTCCAGATTTCCCCAACGTCTCCTCTACCATTGTGCCCACAAGTCAGGAGGAGCA   174772

Query    121     CAGACCATTTGAGAAAGTAGTAAACAAGGAGTCACTTGTCATCTCTGGCCTGAGACACTT   180
                 |||| |||||||||||||| || || |||||||||||||||||||||||||||||||||||
Sbjct    174771  CAGGCCATTTGAGAAAGTGGTGAACAAGGAGTCACTTGTCATCTCTGGCCTGAGACACTT   174712

Query    181     CACTGGGTACCGCATTGAGCTGCAGGCATGCAATCAGGACTCCCCAGAAGAGAGGTGCAG   240
                 |||||||||||||||||||||||||||||||||||||||| || ||||||||| |||||||
Sbjct    174711  CACTGGGTACCGCATTGAGCTGCAGGCATGCAATCAAGATTCCCCAGATGAGAGGTGCAG   174652

Query    241     CGTGGCTGCCTACGTCAGTGCCCGGACCATGCCTGAAGGT   280
                 ||||||||||| |||||||||||||||||||||||||||||
Sbjct    174651  TGTGGCTGCCTACGTCAGTGCCCGGACCATGCCTGAAGGT   174612
```

FIGURE 3 BLAST results showing a 280-base sequence of a chromosome 12 contig from rat (*Rattus norvegicus,* the "query") aligned with a portion of chromosome 8 from mice (*Mus musculus,* the "subject") that contains a partial sequence for the insulin receptor gene. Vertical lines indicate exact matches. The rat contig sequence was used as a query sequence to search a mouse database in GenBank. Notice that the two sequences show 93 percent identity, strong evidence that this rat contig sequence contains a gene for the insulin receptor.

to infer function and to identify evolutionarily related genes through comparative genomics.

Hallmark Characteristics of a Gene Sequence Can Be Recognized during Annotation

A major limitation of the annotation approach just described is that it works only if similar gene sequences are already in a database. Fortunately, it is not the only way to identify genes. Whether the genome under study is from a eukaryote or a prokaryote, several hallmark characteristics of genes can be searched for using bioinformatics software (**Figure 4**). For instance, gene-regulatory sequences found upstream of genes are marked by identifiable sequences such as promoters, enhancers, and silencers. Recall that TATA box, GC box, and CAAT box sequences are often present in the promoter region of eukaryotic genes. Recall also that splice sites between **exons** and **introns** contain a predictable sequence (most introns begin with CT and end with AG) and such splice site sequences are important for determining intron and exon boundaries. Interestingly, current estimates indicate that only 6 percent of human genes are transcribed from a single, linear stretch of DNA that does not contain any introns.

Annotation is intended to reveal identifiable features that provide clues to the presence of a protein-coding gene. Protein-coding genes often contain one or more **open reading frames (ORFs),** sequences of triplet nucleotides that, after transcription and mRNA splicing, are translated into the amino acid sequence of a protein. ORFs typically begin with an initiation sequence, usually ATG, which transcribes into the AUG start codon of an mRNA molecule, and end with a termination sequence, TAA, TAG, or TGA, which correspond to the stop codons of UAA, UAG, and UGA in mRNA.

Downstream elements, such as termination sequences and well-defined sequences at the end of a gene, where a polyadenylation sequence signals the addition of a poly-A tail to the 3′ end of a mRNA transcript are also important for annotation (Figure 4). Annotation can sometimes be a little bit easier for prokaryotic genes than for eukaryotic genes because there are no introns in prokaryotic genes. Gene-prediction programs are used to annotate sequences. These programs incorporate search elements for many of the criteria mentioned above and have become invaluable applications of bioinformatics.

ESSENTIAL POINT ■ ■ ■

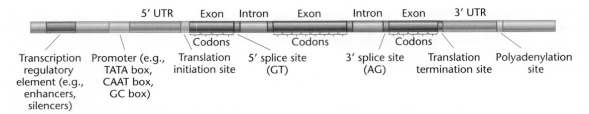

5' UTR   Exon   Intron   Exon   Intron   Exon   3' UTR

Codons   Codons   Codons

Transcription | Promoter (e.g., | Translation | 5' splice site | 3' splice site | Translation | Polyadenylation
regulatory | TATA box, | initiation site | (GT) | (AG) | termination site | site
element (e.g., | CAAT box, | | | | |
enhancers, | GC box) | | | | |
silencers) | | | | | |

FIGURE 4   Characteristics of a protein-coding gene that can be used during annotation to identify a gene in an unknown sequence of genomic DNA. Most eukaryotic genes are organized into coding segments (exons) and noncoding segments (introns). When annotating a genome sequence to determine whether it contains a gene, it is necessary to distinguish between introns and exons, gene-regulatory sequences, such as promoters and enhancers, untranslated regions (UTRs), and gene termination sequences.

NOW SOLVE THIS

1 In a sequence encompassing 99.4 percent of the euchromatic regions of human chromosome 1, Gregory et al. (Gregory, S.G. et al., *Nature*, 441: 315–321, 2006) identified 3141 genes.

(a) How does one identify a gene within a raw sequence of bases in DNA?

(b) What features of a genome are used to verify likely gene assignments?

(c) Given that chromosome 1 contains approximately 8 percent of the human genome, and assuming that there are approximately 20,000 genes, would you consider chromosome 1 to be "gene rich"?

■ **Hint** *This problem involves a basic understanding of bioinformatics and gene annotation approaches to determine how potential gene sequences can be identified in a stretch of sequenced DNA.*

Bioinformatics can be used for sequence annotation to identify protein-coding DNA sequences and noncoding sequences such as regulatory elements.

3   Functional Genomics Attempts to Identify Potential Functions of Genes and Other Elements in a Genome

Reading a genome sequence is a surefire cure for insomnia. What is exciting is not the sequence of the nucleotides but the information that the sequence contains. As the term suggests, **functional genomics** is the study of gene functions, based on the resulting RNAs or proteins they encode, and considers the functions of other components of the genome, such as gene-regulatory elements. Functional genomics can involve experimental approaches to confirm or refute computational predictions about genome functions (such as the number of protein-coding genes), and it also considers how genes are expressed and the regulation of gene expression.

Predicting Gene and Protein Functions by Sequence Analysis

Some newly identified genomic sequences may already have had functions assigned to their genes by classic methods such as mutagenesis and linkage mapping, but many other genes that have been sequenced have not yet been correlated with a function. One approach to assigning functions to genes is to use sequence similarity searches, as described in the previous section. Programs such as BLAST are used to search through databases to find alignments between the newly sequenced genome and genes that have already been identified, either in the same or in different species. You were introduced to this approach for predicting gene function in Figure 3, when we demonstrated how sequence similarity to the mouse gene was used to identify a gene in a rat contig as the insulin receptor gene. Inferring gene function from similarity searches is based on a relatively simple idea. If a genome sequence shows statistically significant similarity to the sequence of a gene whose function is known, then it is likely that the genome sequence encodes a protein with a similar or related function.

Another major benefit of similarity searches is that they are often able to identify **homologous genes,** genes that are evolutionarily related. After the human genome was sequenced, many ORFs in it were identified as protein-coding genes based on their alignment with related genes of known function in other species. As an example, **Figure 5** compares portions of the human leptin gene (*LEP*) with its homolog in mice (*ob/Lep*). These two genes are over 85 percent identical in sequence. The leptin gene was first discovered in mice. The match between the *LEP*-containing DNA sequence in humans and the mouse homolog sequence confirms the identity and leptin-coding function of this gene in human genomic DNA.

As an interesting aside, the leptin gene (also called *ob,* for obesity, in mice) is highly expressed in fat cells (adipocytes). This gene produces the protein hormone leptin, which targets cells in the brain to suppress appetite. Knockout mice lacking a functional *ob* gene grow dramatically overweight. A similar phenotype has been observed in small numbers of humans with particular mutations in

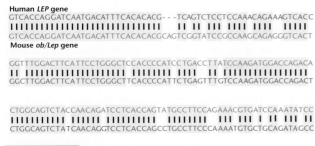

Human *LEP* gene

GTCACCAGGATCAATGACATTTCACACACG- - -TCAGTCTCCTCCAAACAGAAAGTCACC
||||||||||||||||||||||||||||| || || ||| |||| |||| |||| |||||
GTCACCAGGATCAATGACATTTCACACACGCAGTCGGTATCCGCCAAGCAGAGGGTCACT

Mouse *ob/Lep* gene

GGTTTGGACTTCATTCCTGGGCTCCACCCCATCCTGACCTTATCCAAGATGGACCAGACA
|| |||||||||||||||||||||| ||||||| |||| || |||||||||||||||||
GGCTTGGACTTCATTCCTGGGCTTCACCCCATTCTGAGTTTGTCCAAGATGGACCAGACT

CTGGCAGTCTACCAACAGATCCTCACCAGTATGCCTTCCAGAAACGTGATCCAAATATCC
||||||||||| ||||||| |||||||||| ||||||||| ||| ||| | || ||| ||
CTGGCAGTCTATCAACAGGTCCTCACCAGCCTGCCTTCCCAAAATGTGCTGCAGATAGCC

FIGURE 5 Comparison of the human *LEP* and mouse *ob/Lep* genes. Partial sequences for these homologs are shown with the human *LEP* gene on top and the mouse *ob/Lep* gene sequence below it. Notice from the number of identical nucleotides, indicated by vertical lines, that the nucleotide sequence for these two genes is very similar. Gaps are indicated by horizontal dashes.

LEP. Although it is important to note that weight control is not regulated by a single gene, the discovery of leptin has provided significant insight into lipid metabolism and weight disorders in humans. Further studies on leptin will be important for understanding more about the genetics of weight disorders.

If homologous genes in different species are thought to have descended from a gene in a common ancestor, the genes are known as **orthologs.** In Section 6 we will consider the globin gene family. Mouse and human β-globin genes are orthologs evolved from a common ancestor. Homologous genes in the same species are called **paralogs.** The α- and β-globin subunits in humans are paralogs resulting from a gene-duplication event. Paralogs often have similar or identical functions.

Predicting Function from Structural Analysis of Protein Domains and Motifs

When a gene sequence is used to predict a polypeptide sequence, the polypeptide can be analyzed for specific structural domains and motifs. Identification of **protein domains,** such as ion channels, membrane-spanning regions, DNA-binding regions, secretion and export signals, and other structural aspects of a polypeptide that are encoded by a DNA sequence, can in turn be used to predict protein function. Thestructures of many DNA-binding proteins have characteristic patterns, or **motifs,** such as the helix-turn-helix, leucine zipper, or zinc-finger motifs. These motifs can often easily be searched for using bioinformatics software, and their identification in a sequence is a common strategy for inferring the possible functions of a protein.

ESSENTIAL POINT ■ ■ ■
Functional genomics predicts gene function based on sequence analysis.

4 The Human Genome Project Revealed Many Important Aspects of Genome Organization in Humans

Now that you have a general idea of the strategies used for analyzing a genome, let's look at the largest genomics project completed to date. The **Human Genome Project (HGP)** was a coordinated international effort to determine the sequence of the human genome and to identify all the genes it contains. It has produced a plethora of information, much of which is still being analyzed and interpreted. What is clear so far, from all the different kinds of genomes sequenced, is that humans and all other species share a common set of genes essential for cellular function and reproduction, confirming that all living organisms arose from a common ancestor.

Origins of the Project

The publicly funded Human Genome Project began in 1990 under the direction of Dr. James Watson, the co-discoverer of the double-helix structure of DNA. Eventually, the public project was led by Dr. Francis Collins, who had previously led a research team involved in identifying the *CFTR* gene as the cause of cystic fibrosis. In the United States, the Collins-led HGP was coordinated by the Department of Energy and the National Center of Human Genome Research, a division of the National Institutes of Health. It established a 15-year plan with a proposed budget of $3 billion to identify all human genes, originally thought to number between 80,000 and 100,000, to sequence and map them all, and to sequence the approximately 3 billion base pairs thought to comprise the 24 chromosomes (22 autosomes, plus X and Y) in humans. Other primary goals of the HGP included the following:

- To establish functional categories for all human genes

- To analyze genetic variations between humans, including the identification of single-nucleotide polymorphisms (SNPs)

- To map and sequence the genomes of several model organisms used in experimental genetics, including *E. coli, S. cerevisiae, C. elegans, D. melanogaster,* and *M. musculus* (mouse)

- To develop new sequencing technologies, such as high-throughput computer-automated sequencers, in order to facilitate genome analysis

- To disseminate genome information among both scientists and the general public

Lastly, to deal with the impact that genetic information would have on society, the HGP set up the **ELSI program** (standing for Ethical, Legal, and Social Implications) to consider ethical, legal, and social issues arising from the HGP and to ensure that personal genetic information would be safeguarded and not used in discriminatory ways.

The HGP grew into an international effort as scientists from 18 countries became involved in the project. Much of the work was carried out by the International Human Genome Sequence Consortium, involving nearly 3000 scientists working at 20 centers in six countries (China, France, Germany, Great Britain, Japan, and the United States).

In 1999, a privately funded human genome project led by J. Craig Venter at **Celera Genomics** (aptly named from a word meaning "swiftness") was announced. Celera's goal was to use whole-genome shotgun sequencing and computer-automated high-throughput DNA sequencers to sequence the human genome more rapidly than HGP. The public project had proposed using a clone-by-clone approach to sequence the genome. Recall that Venter and colleagues had proven the potential of shotgun sequencing in 1995 when they completed the genome for *H. influenzae*. Celera's announcement set off an intense competition between the two teams, which both aspired to be first with the human genome sequence. This contest eventually led to the HGP finishing ahead of schedule and under budget after scientists from the public project began to use high-throughput sequencers and whole-genome sequencing strategies as well.

Major Features of the Human Genome

In June 2000, the leaders of the public and private genome projects met at the White House with President Clinton and jointly announced the completion of a draft sequence of the human genome. In February 2001, they each published an analysis covering about 96 percent of the euchromatic region of the genome. The remaining work of completing the sequence included filling in gaps clustered around centromeres, telomeres, and repetitive sequences, correcting misaligned segments, and re-sequencing portions of the genome to ensure accuracy. In 2003, genome sequencing and error-fixing were deemed sufficient to pass the international project's definition of completion—that it contained fewer than 1 error per 10,000 nucleotides and that it covered 95 percent of the gene-containing portions of the genome. Yet even at the time of "completion," there were still some 350 gaps in the sequence that continue to be worked on.

And of course the HGP did not sequence the genome of every person on Earth. The assembled genomes largely consist of haploid genomes pooled from different individuals so that they provide a *reference genome* representative of major, common elements of a human genome widely shared among populations of humans. Examples of major features of the human genome are summarized in **Table 1**. As you can see in this table, many unexpected observations have provided us with major new insights. The genome is not static! Genome variations, including the abundance of repetitive sequences scattered throughout the genome, verify that the genome is indeed dynamic, revealing many evolutionary examples of sequences that have changed in structure and

---

TABLE 1 Major Features of the Human Genome

- The human genome contains 3.1 billion nucleotides, but protein-coding sequences make up only about 2 percent of the genome.
- The genome sequence is ~99.9 percent similar in individuals of all nationalities. Single-nucleotide polymorphisms (SNPs) and copy number variations (CNVs) account for genome diversity from person to person.
- The genome is dynamic. At least 50 percent of the genome is derived from transposable elements, such as LINE and *Alu* sequences, and other repetitive DNA sequences.
- The human genome contains approximately 20,000 protein-coding genes, far fewer than the predicted number of 80,000–100,000 genes.
- The average size of a human gene is ~25 kb, including gene-regulatory regions, introns, and exons. On average, mature mRNAs produced by human genes are ~3000 nt long.
- Many human genes produce more than one protein through alternative splicing, thus enabling human cells to produce a much larger number of proteins (perhaps as many as 200,000) from only ~20,000 genes.
- More than 50 percent of human genes show a high degree of sequence similarity to genes in other organisms; however, more than 40 percent of the genes identified have no known molecular function.
- Genes are not uniformly distributed on the 24 human chromosomes. Gene-rich clusters are separated by gene-poor "deserts" that account for 20 percent of the genome. These deserts correlate with G bands seen in stained chromosomes. Chromosome 19 has the highest gene density, and chromosome 13 and the Y chromosome have the lowest gene densities.
- Chromosome 1 contains the largest number of genes, and the Y chromosome contains the smallest number.
- Human genes are larger and contain more and larger introns than genes in the genomes of invertebrates, such as *Drosophila*. The largest known human gene encodes dystrophin, a muscle protein. This gene, associated in mutant form with muscular dystrophy, is 2.5 Mb in length , larger than many bacterial chromosomes. Most of this gene is composed of introns.
- The number of introns in human genes ranges from 0 (in histone genes) to 234 (in the gene for *titin*, which encodes a muscle protein).

location. In many ways, the HGP has revealed just how little we know about our genome.

Two of the biggest surprises discovered by the HGP were that less than 2 percent of the genome codes for proteins and that there are only around 20,000 protein-coding genes. Recall that the number of genes had originally been estimated to be about 100,000, based in part on a prediction that human cells produce about 100,000 proteins. At least half of the genes show sequence similarity to genes shared by many other organisms, and as you will learn in Section 7, a majority of human genes are similar in sequence to genes from closely related species such as chimpanzees. The exact number of human genes is still not certain. One reason is that it is unclear whether or not many of the presumed genes produce functional proteins. Genome scientists continue to annotate the genome, and as mentioned earlier, functional genomics studies have important roles in determining whether or not computational predictions about the number of protein-coding and non–protein-coding genes are accurate.

The number of genes is much lower than the number of predicted proteins in part because many genes code for multiple proteins through **alternative splicing.** Recall that alternative splicing patterns can generate multiple mRNA molecules, and thus multiple proteins, from a single gene through different combinations of intron–exon splicing arrangements. Initial estimates suggested that over 50 percent of human genes undergo alternative splicing to produce mul-

tiple transcripts and multiple proteins. Recent studies suggest that ~94–95 percent of human pre-mRNAs contain multiple exons that are processed to produce multiple transcripts and potentially multiple different protein products. Clearly, alternative splicing produces an incredible diversity of proteins beyond simple predictions based on the number of genes in the human genome.

Functional categories have been assigned for human genes, primarily on the basis of (1) functions determined previously (for example, from recombinant DNA cloning of human genes and known mutations involved in human diseases), (2) comparison to known genes and predicted protein sequences from other species, and (3) predictions based on annotation and analysis of protein functional domains and motifs **(Figure 6)**. Although functional categories and assignments continue to be revised, the functions of over 40 percent of human genes remain unknown. Determining human gene functions, deciphering complexities of gene-expression regulation and gene interaction, and uncovering the relationships between human genes and phenotypes are among the many challenges for genome scientists.

The HGP has also shown us that in all humans, regardless of racial and ethnic origins, the genomic sequence is approximately 99.9 percent the same. Mostgenetic differences between humans result from **single-nucleotide polymorphisms (SNPs)** and **copy number variations (CNVs).** Recall that SNPs are single base changes in the genome and variations of many SNPs are associated with disease condi-

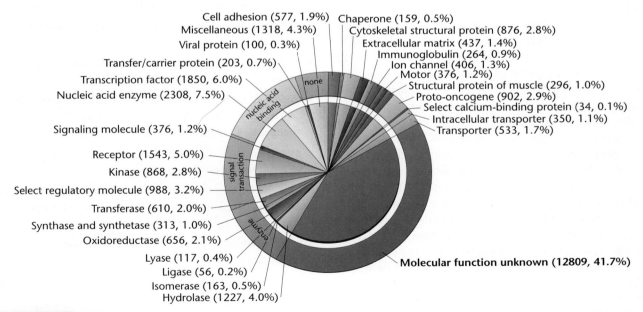

FIGURE 6 A preliminary list of the functional categories to which genes in the human genome have been assigned on the basis of similarity to proteins of known function. Among the most common genes are those involved in nucleic acid metabolism (7.5 percent of all genes identified), transcription factors (6.0 percent), receptors (5 percent), hydrolases (4 percent), protein kinases (2.8 percent), and cytoskeletal structural proteins (2.8 percent). A total of 12,809 predicted proteins (41 percent) have unknown functions, indicative of the work that is still needed to fully decipher our genome.

tions. For example, SNPs cause sickle-cell anemia and cystic fibrosis.

After the draft sequence of the human genome was completed, it initially appeared that most genetic variations between individuals (the 0.1 percent differences) were due to SNPs. While SNPs are important contributing factors to genome variation, structural differences such as deletions, duplications, inversions, and CNVs, which can span millions of bp of DNA, play much more important roles in genome variation than previously thought. Recall that CNVs are duplications or deletions of relatively large sections of DNA on the order of several hundred or several thousand base pairs. Many of the CNVs that vary the most among genomes appear to be at least 1 kilobase.

Although most human DNA is present in two copies per cell, one from each parent, CNVs are segments of DNA that are duplicated or deleted, resulting in variations in the number of copies of a DNA segment inherited by individuals. In some cases CNVs are major deletions removing entire genes; other deletions affect gene function by frameshifts in the reading code. CNV sequences that are duplicated can result in overexpression of a particular gene, yet many deleted and duplicated CNVs do not present clearly identifiable phenotypes.

Current estimates of the number of CNVs in an individual genome range from about 12 CNVs to perhaps 4–5 dozen per person. Some studies estimate that there may be as many as 1500 CNVs greater than 1 kb among the human genome. Other studies claim there are more than 1.5 million deletions of less than 100 bp that contribute to genome variation between individuals.

It is now possible to access databases and other sites on the Internet that display maps for all human chromosomes. **Figure 7(a)** displays a partial gene map for chromosome 12 that was taken from an NCBI database called Map Viewer. This image shows an ideogram, or cytogenetic map, of chromosome 12. To the right of the ideogram is a column showing the contigs (arranged lying vertically) that were aligned to sequence this chromosome. The Hs UniG column displays a histogram representation of gene density on chromosome 12. Notice that relatively few genes are located near the centromere. Gene symbols, loci, and gene names (by description) are provided for selected genes; in this figure only 20 genes are shown. When accessing these maps on the Internet, one can magnify, or zoom in on, each region of the chromosome, revealing all genes mapped to a particular area.

You can see that most of the genes listed here have been assigned descriptions based on the functions of their products, some of which are transmembrane proteins, some enzymes such as kinases, some receptors, including several involved in olfaction, and so on. Other genes are described in terms of hypothetical products; they are presumed to be genes based on the presence of ORFs, but their function remains unknown [Figure 7(a)].

The HGP's most valuable contribution will perhaps be the identification of disease genes and the development of new treatment strategies as a result. Thus, extensive maps have been developed for genes implicated in human disease conditions. The disease gene map of chromosome 21 shown in **Figure 7(b)** indicates genes involved in amyotrophic lateral sclerosis (ALS), Alzheimer disease, cataracts, deafness, and several different cancers.

ESSENTIAL POINT ■ ■ ■

The Human Genome Project revealed many surprises about human genetics, including gene number, the high degree of DNA sequence similarity between individuals and between humans and other species, and showed that many genes encode multiple proteins.

5 The "Omics" Revolution Has Created a New Era of Biological Research

The Human Genome Project and the development of genomics techniques have been largely responsible for launching a new era of biological research—the era of "omics." It seems that every year, more areas of biological research are being described as having an omics connection. Some examples of "omics" are

- proteomics—the analysis of all the proteins in a cell or tissue

- metabolomics—the analysis of proteins and enzymatic pathways involved in cell metabolism

- glycomics—the analysis of the carbohydrates of a cell or tissue

- toxicogenomics—the analysis of the effects of toxic chemicals on genes, including mutations created by toxins and changes in gene expression caused by toxins

- metagenomics—the analysis of genomes of organisms collected from the environment

- pharmacogenomics—the development of customized medicine based on a person's genetic profile for a particular condition

- transcriptomics—the analysis of all expressed genes in a cell or tissue

We will consider several of these genomics disciplines in other parts of this chapter.

As evidence of the impact of genomics, a new field of nutritional science called nutritional genomics, or **nutrigenomics,** has emerged. Nutrigenomics focuses on understanding the interactions between diet and genes. We have all had routine medical tests for blood pressure, blood sugar levels,

(a)

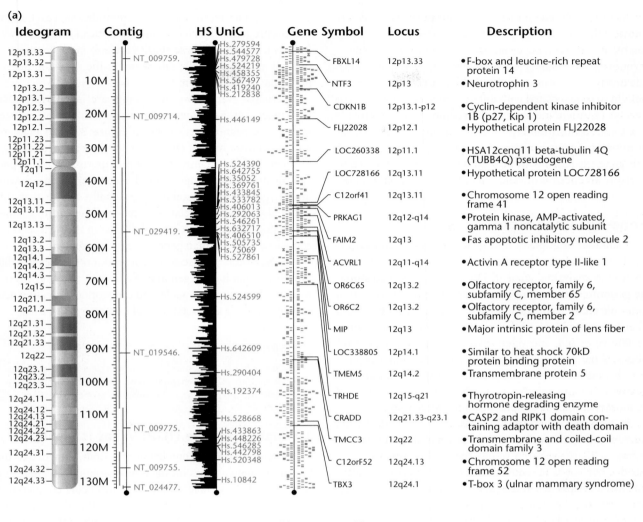

(b)

Chromosome 21
50 million bases

Coxsackie and adenovirus receptor
Amyloidosis cerebroarterial, Dutch type
Alzheimer disease, APP-related
Schizophrenia, chronic
Usher syndrome, autosomal recessive

Amyotrophic lateral sclerosis
Oligomycin sensitivity
Jervell and Lange-Nielsen syndrome
Long QT syndrome
Down syndrome cell adhesion molecule

Homocystinuria
Cataract, congenital, autosomal dominant
Deafness, autosomal recessive
Myxovirus (influenza) resistance
Leukemia, acute myeloid

Myeloproliferative syndrome, transient
Leukemia transient of Down syndrome

Enterokinase deficiency
Multiple carboxylase deficiency
T-cell lymphoma invasion and metastasis

Mycobacterial infection, atypical
Down syndrome (critical region)
Autoimmune polyglandular disease, type 1

Bethlem myopathy
Epilepsy, progressive myoclonic
Holoprosencephaly, alobar
Knobloch syndrome
Hemolytic anemia
Breast cancer
Platelet disorder, with myeloid malignancy

FIGURE 7 (a) A gene map for chromosome 12 from the NCBI database Map Viewer. (b) Partial map of disease genes on human chromosome 21. Maps such as this depict genes thought to be involved in human genetic disease conditions.

and heart rate. Based on these tests, your physician may recommend that you change your diet and exercise more to lose weight, or that you reduce your intake of sodium to help lower your blood pressure. Now several companies claim to provide nutrigenomics tests that analyze your genome for genes thought to be associated with different medical conditions or aspects of nutrient metabolism. The companies then provide a customized nutrition report, recommending diet changes for improving your health and preventing illness, based on your genes! It remains to be seen whether this approach as currently practiced is of valid scientific or nutritional value.

Stone-Age Genomics

In yet another example of how genomics has taken over areas of DNA analysis, a number of labs around the world are involved in analyzing "ancient" DNA. These so-called **stone-age genomics** studies are generating fascinating data from miniscule amounts of ancient DNA obtained from bone and other tissues such as hair that are tens of thousands to about 100,000 years old, and often involve samples from extinct species. Analysis of DNA from a 2400-year-old Egyptian mummy, bison, mosses, platypus, mammoths, Pleistocene-age cave bears and polar bears, and Neanderthals are some of the most prominent examples of stone-age genomics.

In 2005, researchers from McMaster University in Canada and Pennsylvania State University published about 13 million bp from a 27,000-year-old woolly mammoth. This study revealed a ~98.5 percent sequence identity between mammoths and African elephants. Subsequent studies by other scientists have used whole-genome shotgun sequencing of mitochondrial and nuclear DNA from Siberian mammoths to provide data on the mammoth genome. These studies suggest that the mammoth genome differs from the African elephant by as little as 0.6 percent. These studies are also great demonstrations of how stable DNA can be under the right conditions, particularly when frozen.

Perhaps even more intriguing are similarities that have been revealed between the mammoth and human genomes. For example, as shown in **Figure 8**, when the gene sequences from human chromosomes were aligned with sequences from the mammoth genome, approximately 50 percent of mammoth genes showed sequence alignment with human genes on autosomes. Incidentally, notice that this figure also shows the relative number of genes from James Watson's genome (which we will discuss in the next section) compared to the human genome reference sequence.

In Section 6 we discuss recent work on the Neanderthal genome. Obtaining the genome of a human ancestor this old was previously unimaginable, and this work is providing new insights into our understanding of human evolution.

ESSENTIAL POINT ▪ ▪ ▪

Genomics has led to a number of other related "omics" disciplines that are rapidly changing how modern biologists study DNA, RNA, and proteins and many aspects of cell function.

10 Years after the HGP: What Is Next?

In the ten years since completion of a draft sequence of the human genome, studies on the human genome continue at a rapid pace. For example, as the HGP was being completed, a group of about three dozen research teams around the world began the **Encyclopedia of DNA Elements (ENCODE) Project.**

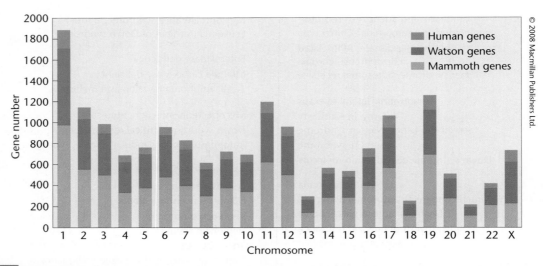

FIGURE 8   Plot showing the number of genes on each human chromosome (blue), the average fraction of protein-coding bases that align to Roche 454 reads from James D. Watson's genome (green), and the fraction of coding bases that align to one or more mammoth reads (orange), using predicted elephant genes that map to the human chromosome based on sequence similarity—approximately 50 percent for each autosome, but only 31 percent for the X chromosome because the mammoth used for this study was male.

The main goal of ENCODE is to use both experimental approaches and bioinformatics to identify and analyze functional elements (such as transcriptional start sites, promoters, and enhancers) that regulate expression of human genes. Prior to ENCODE approximately 532 promoters had been identified, but now in excess of 775 promoters have been identified in the human genome, with many other potential promoter sequences being analyzed.

As a result of the HGP, many other major theme areas for human genome research have emerged including a cancer genome project; analysis of the epigenome (including a Human Epigenome Project that is creating hundreds of maps of epigenetic changes in different cell and tissue types and evaluating potential roles of epigenetics in complex diseases); and characterization of SNPs (the International HapMap Project) and CNVs for their role in genome variation, disease, and pharmacogenomics applications. Here we consider two areas of human genome research that are extensions of the HGP: (1) the analysis of personal genomes, including haploid genomes and (2) the Human Microbiome Project.

Personalized Genome Projects and Personal Genomics

As we discussed earlier in this chapter, high-throughput sequencing and most recently next-generation sequencing technologies, capable of generating longer sequence reads at higher speeds with greater accuracy, have greatly reduced the cost of DNA sequencing, and expectations for continued cost reductions along with continued technological advances are high. These expectations have led several companies to propose personalized genome sequencing for individual people. In 2006, the X Prize Foundation announced the Archon X Prize for Genomics, an award of $10 million to the first private group that develops technology capable of sequencing 100 human genomes with a high degree of accuracy in 10 days for under $10,000 per genome. Other groups are working on sequencing a personalized genome for a mere $1000! Two programs funded by the National Institutes of Health are challenging scientists to develop sequencing technologies to complete a human genome for $1000 by 2014 .

Pursuit of the $1000 genome has become an indicator that DNA sequencing may eventually be affordable enough for individuals to consider acquiring a readout of their own genetic blueprint. The genome of James D. Watson, who together with Francis Crick discovered the structure of DNA, was the focus of "Project Jim" by the Connecticut company 454 Life Sciences, which wanted to sequence the genome of a high-profile person and decided that the co-discoverer of DNA structure and the first director of the U.S. Human Genome Project should be that person. This company used their next-generation approach for Project Jim, and

within two years it was announced that six-fold coverage of Watson's genome was complete at a rough cost of just under $1 million. James Watson was then presented with two DVDs containing his genome sequence.

Human genome pioneer J. Craig Venter had his genome completed by the J. Craig Venter Institute and deposited into GenBank in May 2007. George Church of Harvard and his colleagues have started a **Personal Genome Project (PGP)** and have recruited volunteers to provide DNA for individual genome sequencing on the understanding that the genome data will be made publicly available. Church's genome has been completed and been made available online. The concept of a personalized genome project raises the obvious question: would you have your genome sequenced for $10,000, $1000, or even for free?

Since the Watson and Venter genomes were completed, in 2008 the first complete genome sequence was provided for an individual "ancient" human, a Palaeo-Eskimo, obtained from ~4000-year-old permafrost-preserved hair. This work recovered about 78 percent of the diploid genome and revealed many interesting SNPs (of which about 7 percent have not been previously reported). As of early 2010, thirteen individual human genome sequences have been reported, including sequences for a Yoruba African, two individuals of northwest European origin, a Han Chinese individual, two persons from Korea, African Archbishop Desmond Tutu, and a family of four among several others.

Another particularly beneficial aspect of personal genome projects is the insight they are providing regarding genome variation. The HGP combined samples from different individuals to create a reference genome for a *haploid genome*. Personal genome projects sequence a diploid genome, and because of this such projects indicate that haploid genome comparisons may underestimate the extent of genome variation between individuals by five-fold or more. For example, when Venter's genome was analyzed, over 4 million variations were found between his maternal and paternal chromosomes alone. From what we are learning about personal genomes, genome variation between individuals may be closer to 0.5 percent than 0.1 percent, and in a 3 billion bp genome this is a significant difference in sequence variation. Integrating genome data from several complete individual genomes of persons from different ethnic groups will also be of great value in evolutionary genetics to address fundamental questions about human diversity, ancestry, and migration patterns.

The Human Microbiome Project

In 2008, the National Institutes of Health announced plans for the **Human Microbiome Project,** a $115 million, five-year project to complete the genomes of an estimated 600–1000 microorganisms, bacteria, viruses, and yeast that live on and inside humans. Microorganisms comprise ~1–2 percent of the human body, outnumbering human cells by about 10 to 1. Many microbes, such as *E. coli* in the digestive tract, have important roles in human health, and of course other microbes make us ill. The Human Microbiome Project has several major goals, including:

- Determining if individuals share a core human microbiome.

- Understanding whether changes in the microbiome can be correlated with changes in human health.

- Developing new methods, including bioinformatics tools, to support analysis of the microbiome.

- Addressing ethical, legal, and social implications raised by human microbiome research. Does this sound familiar? Recall that addressing ethical, legal, and social issues was a goal of the HGP.

The Human Microbiome Project is still very much in its infancy as a project, but it has already revealed that over 3.3 million human gut microbe genes characterized to date appear to be very similar among over 100 individuals. The saliva microbiome is also highly similar from individual to individual, and a Human Oral Microbiome database has been established from these studies, which seek to develop linkages between the oral microbiome and oral health.

No Genome Left Behind and the Genome 10K Plan

Without question, new sequencing technologies that have been developed as a result of the HGP are an important part of the transformational effect the HGP has had on modern biology. About ten years ago, a room full of sequencers and several million dollars were required to sequence the 97-Mb genome of *C. elegans*. As a sign of modern times in the world of genomics, in 2009 two sequencers and $500,000 produced a reasonably complete draft of the 750-Mb cod genome—in a month!

Modern sequencing technologies have prompted the question, "What would you do if you could sequence everything?" Partners around the world including genome scientists and museum curators have proposed sequencing 10,000 vertebrate genomes, the **Genome 10K** plan. Shortly after the HGP finished, the National Human Genome Research Institute (NHGRI) assembled a list of mammals and other vertebrates as priorities for genome sequencing in part because of their potential benefit for learning about the human genome through comparative genomics. Genome 10K will also provide insight into genome evolution and speciation. This ambitious plan proposes to assemble 10,000 genomes in five years—about one genome a day!

ESSENTIAL POINT ■   ■   ■

Since the Human Genome Project, human genome research has focused on individual human genomes (personalized genomics) and other efforts such as the Human Microbiome Project.

6 Comparative Genomics Analyzes and Compares Genomes from Different Organisms

As of 2010, the genomes of over 3800 prokaryotic and eukaryotic organisms—including many model organisms and a number of viruses—have been sequenced. This is quite extraordinary progress in a relatively short time span! Among these organisms are yeast (*Saccharomyces cerevisiae*)—the first eukaryotic genome to be sequenced to bacteria such as *E. coli*, the nematode roundworm (*Caenorhabditis elegans*), the thale cress plant (*Arabidopsis thaliana*), mice (*Mus musculus*), zebrafish (*Danio rerio*), and of course *Drosophila*. In the past few years, genomes for chimpanzees, dogs, chickens, sea urchins, honey bees, pufferfish, rice, and wheat have all been sequenced.

These studies have demonstrated not only significant differences in genome organization between prokaryotes and eukaryotes but also many similarities between genomes of nearly all species. In this section, we provide a basic overview of genome organization in prokaryotes and eukaryotes and discuss interesting aspects of genomes in selected organisms. Analysis of the growing number of genome sequences confirms that all living organisms are related and descended from a common ancestor. Similar gene sets are used in all organisms for basic cellular functions, such as DNA replication, transcription, and translation. These genetic relationships are the rationale for the use of model organisms to study inherited human disorders, the effects of the environment on genes, and interactions of genes in complex diseases, such as cardiovascular disease, diabetes, neurodegenerative conditions, and behavioral disorders.

Comparative genomics answers questions about genetics and other aspects of biology through the analysis of genomes from different organisms. It is a field with many research and practical applications, including gene discovery and the development of model organisms to study human diseases. It also incorporates the study of gene and genome evolution and the relationship between organisms and their environment. Comparative genomics can reveal genetic differences and similarities between organisms to provide insight into how those differences contribute to differences in phenotype, life cycle, or other attributes, and to ascertain the evolutionary history of those genetic differences.

Prokaryotic and Eukaryotic Genomes Display Common Structural and Functional Features and Important Differences

Since most prokaryotes have small genomes amenable to shotgun cloning and sequencing, many genome projects have focused on prokaryotes, and more than 900 additional projects to sequence prokaryotic genomes are now under way. Many of the prokaryotic genomes already sequenced are from organisms that cause human diseases, such as cholera, tuberculosis, and leprosy. Traditionally, the bacterial genome has been thought of as relatively small (less than 5 Mb) and contained within a single circular DNA molecule. *E. coli*, used as the prototypical bacterial model organism in genetics, has a genome with these characteristics. However, the flood of genomic information now available has challenged the validity of this viewpoint for bacteria in general (Table 2). Although most prokaryotic genomes are small, their sizes vary across a surprisingly wide range. In fact, there is some overlap in size between larger bacterial genomes (30 Mb in *Bacillus megaterium*) and smaller eukaryotic genomes (12.1 Mb in yeast). Gene number in bacterial genomes also demonstrates a wide range, from less than 500 to more than 5000 genes, a ten-fold difference.

In addition, although many bacteria have a single, circular chromosome, there is substantial variation in chromosome organization and number among bacterial species. An increasing number of genomes composed of linear DNA molecules are being identified, including the genome of *Borrelia burgdorferi*, the organism that causes Lyme disease.

We can make two generalizations about the organization of protein-coding genes in bacteria. First, gene density is very high, averaging about one gene per kilobase of DNA.

For example, the genome of *E. coli* strain K12, which, in 1997, was the second prokaryotic genome to be sequenced, is 4.6 Mb in size and it contains 4289 protein-coding genes in its single, circular chromosome. This close packing of genes in prokaryotic genomes means that a very high proportion of the DNA (approximately 85 to 90 percent) serves as coding DNA. Typically, only a small amount of a bacterial genome is noncoding DNA, often in the form of regulatory sequences or of transposable elements that can move from one place to another in the genome.

The second generalization we can make is that bacterial genomes contain operons (operons contain multiple genes functioning as a transcriptional unit whose protein products are part of a common biochemical pathway). In *E. coli*, 27 percent of all genes are contained in operons (almost 600 operons). In other bacterial genomes, the organization of genes into transcriptional units is challenging our ideas about the nature of operons. For example, in *Aquifex aeolicus*, one polygenic transcription unit contains six genes involved in several different cellular processes with no apparent common relationships: two genes for DNA recombination, one for lipid synthesis, one for nucleic acid synthesis, one for protein synthesis, and one that encodes a protein for cell motility. Other polygenic transcription units in this species also contain genes with widely different functions. This finding, combined with similar results from other genome projects, raises interesting questions about the consensus that operons encode products that control a single metabolic pathway in bacterial cells.

The basic features of eukaryotic genomes are similar in different species, although genome size in eukaryotes is highly variable (Table 3). Genome sizes range from about 10 Mb in fungi to over 100,000 Mb in some flowering plants (a ten thousand-fold range); the number of chromosomes per genome ranges from two into the hundreds (about a hundred-fold range), but the number of genes varies much less dramatically than either genome size or chromosome number.

Eukaryotic genomes have several features not found in prokaryotes:

- **Gene density.** In prokaryotes, gene density is close to 1 gene per kilobase. In eukaryotic genomes, there is a wide range of gene density. In yeast, there is about 1 gene/2 kb, in *Drosophila*, about 1 gene/13 kb, and in humans, gene density varies greatly from chromosome to chromosome. Human chromosome 22 has about 1 gene/64 kb, while chromosome 13 has 1 gene/155 kb of DNA.

- **Introns.** Most eukaryotic genes contain introns. There is wide variation among genomes in the number of introns they contain and also wide variation from gene to gene. The entire yeast genome has only 239 introns, whereas just a single gene in the human genome can contain more than

TABLE 2	Genome Size and Gene Number in Selected Prokaryotes		
		Genome Size (Mb)	Number of Genes
Archaea			
Methanosarcina berkeri		4.84	3680
Archaeoglobus fulgidis		2.17	2437
Methanococcus jannaschii		1.66	1783
Thermoplasma acidophilium		1.56	1509
Nanoarchaeum equitans		0.49	552
Eubacteria			
Pseudomonas aeruginosa		6.30	5570
Rhizobium radiobacter		4.67	5419
Escherichia coli		4.64	4289
Bacillus subtilis		4.21	4779
Haemophilus influenzae		1.83	1738
Aquifex aeolicus		1.55	1749
Rickettsia prowazekii		1.11	834
Mycoplasma pneumonia		0.82	680
Mycoplasma genitalium		0.58	483

TABLE 3   Comparison of Selected Genomes

Organism (Scientific Name)	Approximate Size of Genome (in million [megabase, Mb] or billion [gigabase, Gb] bases) (Date Completed)	Number of Genes	Approximate Percentage of Genes Shared with Humans
Bacterium (*Escherichia coli*)	4.1 Mb (1997)	4,403	not determined
Chicken (*Gallus gallus*)	1 Gb (2004)	~20,000–23,000	60%
Dog (*Canis familiaris*)	2.5 Gb (2003)	~18,400	75%
Chimpanzee (*Pan troglodytes*)	~3 Gb (2005)	~20,000–24,000	98%
Fruit fly (*Drosophila melanogaster*)	165 Mb (2000)	~13,600	50%
Human (*Homo sapiens*)	~2.9 Gb (2004)	~20,000	100%
Mouse (*Mus musculus*)	~2.5 Gb (2002)	~30,000	80%
Rat (*Rattus norvegicus*)	~2.75 Gb (2004)	~22,000	80%
Rhesus macaque (*Macaca mulatta*)	2.87 Gb (2007)	~20,000	93%
Rice (*Oryza sativa*)	389 Mb (2005)	~41,000	not determined
Roundworm (*Caenorhabditis elegans*)	97 Mb (1998)	19,099	40%
Sea urchin (*Strongylocentrotus purpuratus*)	814 Mb (2006)	~23,500	60%
Thale cress (plant) (*Arabidopsis thaliana*)	140 Mb (2000)	~27,500	not determined
Yeast (*Saccharomyces cerevisiae*)	12 Mb (1996)	~5,700	30%

Adapted from Palladino, M. A. *Understanding the Human Genome Project,* 2nd ed. Benjamin Cummings, 2006.

Note: Billion bp (gigabase, Gb).

100 introns. Regarding intron size, generally the size in eukaryotes is correlated with genome size. Smaller genomes have smaller average introns, and larger genomes have larger average intron sizes. But there are exceptions. For example, the genome of the pufferfish (*Fugu rubripes*) has relatively few introns.

- **Repetitive sequences.** The presence of introns and the existence of repetitive sequences are two major reasons for the wide range of genome sizes in eukaryotes. In some plants, such as maize, repetitive sequences are the dominant feature of the genome. The maize genome has about 2500 Mb of DNA, and more than two-thirds of that genome is composed of repetitive DNA. In humans, as discussed previously, about half of the genome is repetitive DNA.

ESSENTIAL POINT ■ ■ ■

Genomic analysis of model prokaryotes and eukaryotes has revealed similarities and important fundamental differences in genome size, gene number, and genome organization.

Comparative Genomics Provides Novel Information about the Genomes of Model Organisms and the Human Genome

As mentioned earlier, the Human Genome Project sequenced genomes from a number of model nonhuman organisms too, including *E. coli*, *Arabidopsis thaliana*, *Saccharomyces cerevisiae*, *Drosophila melanogaster*, the nematode roundworm *Caenorhabditis elegans*, and the mouse *Mus musculus*. Complete genome sequences of such organisms have been invaluable for comparative genomics studies of gene func-

tion in these organisms and in humans. As shown in Table 3, the number of genes humans share with other species is very high, ranging from about 30 percent of the genes in yeast to ~80 percent in mice and ~98 percent in chimpanzees. The human genome even contains around 100 genes that are also present in many bacteria. Comparative genomics has shown us that many mutated genes involved in human disease are also present in model organisms. For instance, approximately 60 percent of genes mutated in nearly 300 human diseases are also found in *Drosophila*. These include genes involved in prostate, colon, and pancreatic cancers; cardiovascular disease; cystic fibrosis; and several other conditions. Here we consider how comparative genomics studies of several model organisms (dogs, chimpanzees, Rhesus monkeys, and sea urchins) and the Neanderthal genome have revealed interesting elements of the human genome.

The Dog Genome

In 2005, the genome for "man's best friend" was completed, and it revealed that we share about 75 percent of our genes with dogs (*Canis familiaris*), providing a useful model with which to study our own genome. Dogs have a genome that is similar in size to the human genome: about 2.5 billion base pairs with an estimated 18,400 genes. The dog offers several advantages for studying heritable human diseases. Dogs share many genetic disorders with humans, including over 400 single-gene disorders, sex-chromosome aneuploidies, multifactorial diseases (such as epilepsy), behavioral conditions (such as obsessive-compulsive disorder), and genetic predispositions to cancer, blindness, heart disease, and deafness. The molecular causes of at least 60 percent of inherited diseases in dogs, such as point mutations and deletions, are similar or identical to those found in humans.

The Chimpanzee Genome

Although the chimpanzee (*Pan troglodytes*) genome was not part of the HGP, its nucleotide sequence was completed in 2004. Overall, the chimp and human genome sequences differ by less than 2 percent, and 98 percent of the genes are the same. Comparisons between these genomes offer some interesting insights into what makes some primates humans and others chimpanzees.

The speciation events that separated humans and chimpanzees occurred less than 6.3 million years ago (mya). Genomic analysis indicates that these species initially diverged but then exchanged genes again before separating completely. Differences in the time and place of gene expression also play a major role in differentiating the two primates. Using DNA microarrays (discussed in Section 8), researchers compared expression patterns of 202 genes in human and chimp cells from brain and liver. They found more species-specific differences in expression of brain genes than liver genes. To further examine these differences, Svante Pääbo and colleagues compared expression of 10,000 genes in human and chimpanzee brains and found that 10 percent of genes examined differ in expression in one or more regions of the brain. More importantly, these differences are associated with genes in regions of the human genome that have been duplicated subsequent to the divergence of chimps and humans. This finding indicates that genome evolution, speciation, and gene expression are interconnected. Further work on these segmental duplications and the genes they contain may identify genes that help make us human.

The Rhesus Monkey Genome

The Rhesus macaque monkey (*Macaca mulatta*), another primate, has served as one of the most important model organisms in biomedical research. Macaques have played central roles in our understanding of cardiovascular disease, aging, diabetes, cancer, depression, osteoporosis, and many other aspects of human health. They have been essential for research on AIDS vaccines and for the development of polio vaccines. The macaque's genome is the first monkey genome to have been sequenced.

A main reason geneticists are so excited about the completion of this sequencing project is that macaques provide a more distant evolutionary window that is ideally suited for comparing and analyzing human and chimpanzee genomes. As we discussed in the preceding section, humans and chimpanzees shared a common ancestor approximately 6 mya. But macaques split from the ape lineage that led to chimpanzees and humans about 25 mya. The macaque and human genome have thus diverged farther from one another, as evidenced

by the ~93 percent sequence identity between humans and macaques compared to the ~98 percent sequence identity shared by humans and chimpanzees.

The macaque genome was published in 2007, and it was no surprise to learn that it consists of 2.87 billion bp (similar to the size of the human genome) contained in 22 chromosomes (20 autosomes, an X, and a Y) with ~20,000 protein-coding genes. Although comparative analyses of this genome are ongoing, a number of interesting features have been revealed so far. As in humans, about 50 percent of the genome consists of repeat elements (transposons, LINEs, SINEs). Gene duplications and gene families are abundant, including cancer gene families found in humans.

A number of interesting surprises have also been observed. For instance, consider the genetic disorder phenylketonuria (PKU), an autosomal recessive inherited condition in which individuals cannot metabolize the amino acid phenylalanine due to mutation of the phenylalanine hydroxylase (*PAH*) gene. The histidine substitution encoded by a mutation in the *PAH* gene of humans with PKU appears as the wild-type amino acid in the protein from healthy macaques. Further analysis of the macaque genome and comparison to the human and chimpanzee genome will be invaluable for geneticists studying variations that played a role in primate evolution.

The Sea Urchin Genome

In 2006, researchers from the Sea Urchin Genome Sequencing Consortium completed the 814 million bp genome of the sea urchin *Strongylocentrotus purpuratus*. Sea urchins are shallow-water marine invertebrates that have served as important model organisms, particularly for developmental biologists. One reason is that the sea urchin is a nonchordate deuterostome, and humans, with their spinal cord, are chordate deuterostomes. Fossil records indicate that sea urchins appeared during the Early Cambrian period, around 520 mya.

Sea urchins have an estimated 23,500 genes, including representative genes for just about all major vertebrate gene families. Sequence alignment and homology searches demonstrate that the sea urchin contains many genes with important functions in humans, yet interestingly, important genes in flies and worms, such as certain cytochrome P-450 genes that play a role in the breakdown of toxic compounds, are missing from sea urchins. The sea urchin genome also has an abundance (~25 to 30 percent) of **pseudogenes**—nonfunctional relatives of protein-coding genes (we meet pseudogenes again in the next subsection). Sea urchins have a smaller average intron size than humans, supporting the general trend revealed by comparative genomics that intron size is correlated with overall genome size.

Another genome trend that urchins share with other eukaryotes is the presence of genes involved in innate immunity, the inborn defense mechanisms that provide broad-spectrum protection against many pathogens. Sea urchins have an extraordinarily rich number of genes providing innate immunity. For example, one very important category of innate immunity genes, the Toll-like receptors (*Tlr* genes), produce transmembrane proteins that are essential for pathogen recognition in nearly every cell type of vertebrates. Sea urchins have over 200 *Tlr* genes compared to 11 in humans. The abundance of these and other important innate immunity genes in sea urchins has led to categorizing these genes as the urchin "defensome." This characteristic of sea urchins may help explain how these organisms have adapted so well to the pathogen-loaded environments of seabeds.

Urchins have nearly 1000 genes for sensing light and odor, indicative of great sensory abilities. In this respect, their genome is more typical of vertebrates than invertebrates. A number of orthologs of human genes involved in hearing and balance are present in the sea urchin, as are many human-disease-associated orthologs, including protein kinases, GTPases, transcription factors, transporters, and low-density lipoprotein receptors. Sea urchins and humans share approximately 7000 orthologs. Further analysis of the urchin genome is expected to make important contributions to our understanding of evolutionary transitions between invertebrates and vertebrates.

The Neanderthal Genome and Modern Humans

In early 2009, a team of scientists led by Svante Pääbo at the Max Planck Institute for Evolutionary Anthropology in Germany and 454 Life Sciences reported completion of a rough draft of the Neanderthal (*Homo neanderthalensis*) genome encompassing more than 3 billion bp of Neanderthal DNA and about two-thirds of the genome.

Because Neanderthals are members of the human family, and closer relatives to humans than chimpanzees, the Neanderthal genome is expected to provide an unprecedented opportunity to use comparative genomics to advance our understanding of evolutionary relationships between modern humans and our predecessors. In particular, scientists are interested in identifying areas in the genome where humans have undergone rapid evolution since diverging from Neanderthals. Much of this analysis involves a comparative genomics approach to compare the Neanderthal genome to the human and chimpanzee genomes.

The human and Neanderthal genomes are 99 percent identical. Comparative genomics has identified 78 protein-coding sequences in humans that seem to have arisen since the divergence from Neanderthals and that may have helped modern humans adapt. Some of these sequences are involved in cognitive development and sperm motility. Of the many genes shared by these species, *FOXP2* is a gene that has been linked to speech and language ability. There are many genes

that influence speech, so this finding does not mean that Neanderthals spoke as we do. But because Neanderthals had the same modern human *FOXP2* gene, scientists have speculated that Neanderthals possessed linguistic abilities.

The realization that modern humans and Neanderthals lived in overlapping ranges as recently as 30,000 years ago has led to speculation about the interactions between modern humans and Neanderthals. Genome studies suggest that interbreeding took place between Neanderthals and modern humans an estimated 45,000 to 80,000 years ago in the eastern Mediterranean. In fact, the genome of non-African *H. sapiens* contains approximately 1–4 percent of sequence inherited from Neanderthals. These exciting studies, previously thought to be impossible, are having ramifications in many areas of human evolution, and it will be interesting indeed to follow the progress of this work.

Comparative Genomics Is Useful for Studying the Evolution and Function of Multigene Families

Comparative genomics has also proven to be valuable for identifying members of **multigene families**, groups of genes that share similar but not identical DNA sequences through duplication and descent from a single ancestral gene. Their gene products frequently have similar functions, and the genes are often, but not always, found at a single chromosomal locus. A group of related multigene families is called a **superfamily.** Sequence data from genome projects is providing evidence that multigene families are present in many, if not all, genomes. Here we examine the globin gene superfamily, whose members encode very similar but not identical polypeptide chains with closely related functions. Other well-characterized gene superfamilies include the histone, tubulin, actin, and immunoglobulin (antibody) gene superfamilies.

Recall that paralogs, which we defined in Section 3, are homologous genes present in the same single organism, believed to have evolved by gene duplication. The globin genes that encode the polypeptides in hemoglobin molecules are a paralogous multigene superfamily that arose by duplication and dispersal to occupy different chromosomal sites. One of the best-studied examples of gene family evolution is the **globin gene superfamily (Figure 9).**

In this family, an ancestral gene encoding an oxygen transport protein was duplicated about 800 mya, producing two sister genes, one of which evolved into the modern-day myoglobin gene. **Myoglobin** is an oxygen-carrying protein found in muscle. The other gene underwent further duplication and divergence about 500 mya and formed prototypes of the α-globin and β-globin genes. These genes encode proteins found in **hemoglobin,** the oxygen-carrying molecule in red blood cells. Additional duplications within these genes occurred within the last 200 million years. Events sub-

FIGURE 9 The evolutionary history of the globin gene superfamily. A duplication event in an ancestral gene gave rise to two lineages about 700 to 800 million years ago (mya). One line led to the myoglobin gene, which is located on chromosome 22 in humans; the other underwent a second duplication event about 500 mya, giving rise to the ancestors of the α-globin and β-globin gene subfamilies. Duplications beginning about 200 mya formed the β-globin gene subfamilies. In humans, the α-globin genes are located on chromosome 16, and the β-globin genes are on chromosome 11.

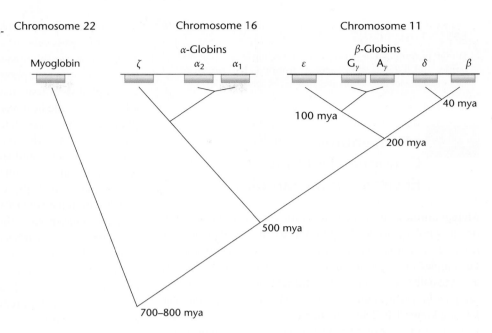

sequent to each duplication dispersed these gene subfamilies to different chromosomes, and in the human genome, each now resides on a separate chromosome. Adult hemoglobin is a tetramer containing two α- and two β-polypeptides . Each polypeptide incorporates a heme group that reversibly binds oxygen. The α-globin gene cluster on chromosome 16 and the β-globin gene cluster on chromosome 11 share nucleotide- and amino acid–sequence similarity (**Figure 10**), but the highest degree of sequence similarity is found within subfamilies.

As shown in Figure 9, the **α-globin** gene subfamily contains three genes: the ζ (zeta) gene, expressed only in early embryogenesis, and two copies of the α gene, expressed during the fetal (α_1) and adult stages (α_2). In addition, the cluster contains two pseudogenes (similar to ζ and α_1), which in this family are designated by the prefix ψ (psi) followed by the symbol of the gene they most resemble. Thus, the designation $\psi\alpha_1$ indicates a pseudogene of the fetal α_1 gene.

The human **β-globin** gene cluster contains five genes spaced over 60 kb of DNA. In this and the α-globin gene subfamily, the order of genes on the chromosome parallels their order of expression during development. Three of the five genes are expressed before birth. The ε (epsilon) gene is expressed only during embryogenesis, while the two nearly identical γ genes (G_γ and A_γ) are expressed only during fetal development. The polypeptide products of the two γ genes differ only by a single amino acid. The two remaining genes, δ and β, are expressed after birth and throughout life. A single pseudogene, $\psi\beta_1$, is present in this subfamily. All five functional genes in this cluster encode proteins with 146 amino acids and have two similar-sized introns at exactly the same positions. The second intron in the β-globin subfamily is significantly larger than its counterpart in the functional α-globin subfamily. These features reflect the evolutionary history of each subfamily and the events such as gene duplication, nucleotide substitution, and chromosome translocations that produced the present-day globin superfamily.

```
α-globin   V – L S P A D K T N V K A A W G K V G A H A G E Y G A E A L E R M F L S F P T T K T Y F P H F – D L S H
β-globin   V H L T P E E K S A V T A L W G K V – – N V D E V G G E A L G R L L V V Y P W T Q R F F E S F G D L S T

α-globin   – – – G S A Q V K G H G K K V A D A L T N A V A H V D D M P N A L S A L S D L H A H K L R V D P V N
β-globin   A V M G N P K V K A H G K K V L G A F S D G L A H L D N L K G T F A T L S E L H C D K L H V D P E N

α-globin   L L S H C L L V T L A A H L P A E F T P A V H A S L D K F L A S V S T V L T S K Y R 141 amino acids
β-globin   L L G N V L V C V L A H H F G K E F T P P V Q A A Y Q K V V A G V A N A L A H K Y H 146 amino acids
```

FIGURE 10 The amino acid sequences of the α- and β-globin proteins, depicted using the single-letter abbreviations for the amino acids. Shaded areas indicate identical amino acids. The two proteins are slightly different in length. α-globin contains 141 amino acids, while β-globin is 146 amino acids long. Gaps in the two sequences, representing areas that do not align, are indicated by horizontal dashes (—).

ESSENTIAL POINT ■ ■ ■

Studies in comparative genomics are revealing fascinating similarities and differences in genomes from different organisms, including the identification and analysis of gene families.

7   Metagenomics Applies Genomics Techniques to Environmental Samples

Metagenomics, also called **environmental genomics,** is the use of whole-genome shotgun approaches to sequence genomes from entire communities of microbes in environmental samples of water, air, and soil. Oceans, glaciers, deserts, and virtually every other environment on Earth are being sampled for metagenomics projects. Human genome pioneer J. Craig Venter left Celera in 2003 to form the J. Craig Venter Institute, and his group has played a central role in developing metagenomics as an emerging area of genomics research.

One of the institute's major initiatives has been a global expedition to sample marine and terrestrial microorganisms from around the world and to sequence their genomes. Through this project, called the *Sorcerer II* Global Ocean Sampling (GOS) Expedition, Venter and his researchers traveled the globe by yacht, in a sailing voyage described as a modern-day version of Charles Darwin's famous voyage on the *H.M.S. Beagle.*

A key benefit of metagenomics is its potential for teaching us more about millions of species of bacteria, of which only a few thousand have been well characterized. Many new viruses, particularly bacteriophages, are also identified through metagenomics studies of water samples. Metagenomics also has great potential for identifying genes with novel functions, some of which have potentially valuable applications in medicine and biotechnology.

The general method used in metagenomics to sequence genomes for all microbes in a given environment involves isolating DNA directly from an environmental sample without requiring cultures of the microbes or viruses. Such an approach is necessary because often it is difficult to replicate the complex array of growth conditions the microbes need to survive in culture. For the *Sorcerer II* GOS project, samples of water from different layers in the water column were passed through high-density filters of various sizes to capture the microbes. DNA was then isolated from the microbes and subjected to shotgun sequencing and genome assembly. High-throughput sequencers on board the yacht operated nearly around the clock. One of the earliest expeditions by this group sequenced bacterial genomes from the Sargasso Sea off Bermuda. This project yielded over 1.2 million novel DNA sequences from 1800 microbial species, including 148 previously unknown bacterial species, and

identified hundreds of photoreceptor genes. Many aquatic microorganisms rely on photoreceptors for capturing light energy to power photosynthesis. Scientists are interested in learning more about photoreceptors to help develop ways in which photosynthesis may be used to produce hydrogen as a fuel source. Medical researchers are also very interested in photoreceptors because, in humans and many other species, photoreceptors in the retina of the eye are key proteins that detect light energy and transduce electrical signals that the brain eventually interprets to create visual images.

By early 2007, the GOS database contained approximately 6 billion bp of DNA from more than 400 uncharacterized microbial species! These sequences included 7.7 million previously uncharacterized sequences, encoding more than 6 million different potential proteins. This is almost twice the total number of previously characterized proteins in all other known databases worldwide. **Figure 11(a)** shows the kingdom assignments for predicted protein sequences in publicly available databases worldwide, such as the NCBI-nonredundant protein database (NCBInr), which accesses GenBank, Ensembl, and other well-known databases. Eukaryotic sequences comprise the majority (63 percent) of predicted proteins in these databases. Reviewing the kingdom assignments of approximately 6 million predicted proteins in the GOS dataset shows that, in contrast, the largest majority (90.8 percent) of sequences in this database are from the bacterial kingdom [**Figure 11(b)**].

The GOS Expedition also examined protein families corresponding to the predicted proteins encoded by the genome sequences in the GOS database: 17,067 families were medium-sized (between 20 and 200 proteins) and large-sized (>200 proteins) clusters. These data demonstrate the value of the GOS Expedition and of metagenomics for identifying novel microbial genes and potential proteins.

ESSENTIAL POINT ■ ■ ■

Metagenomics, or environmental genomics, sequences genomes of microorganisms in environmental samples, often identifying new sequences that encoded proteins with novel functions.

8   Transcriptome Analysis Reveals Profiles of Expressed Genes in Cells and Tissues

Sequencing a genome is a major endeavor, and even once any genome has been sequenced and annotated, a formidable challenge still remains: that of understanding genome function by analyzing the genes it contains and the ways the genes expressed by the genome are regulated. **Transcriptome analysis,** also called **transcriptomics** or **global analysis of gene expression,** studies the expression of genes by a genome both

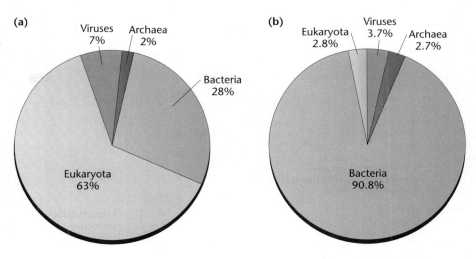

FIGURE 11 (a) Kingdom identifications for predicted proteins in NCBInr, NCBI Prokaryotic Genomes, the Institute for Genomics Research Gene Indices, and Ensembl databases. Notice that the publicly available databases of sequenced genomes and the predicted proteins they encode are dominated by eukaryotic sequences. (b) Kingdom identifications for novel predicted proteins in the Global Ocean Sampling (GOS) database. Bacterial sequences dominate this database, demonstrating the value of metagenomics for revealing new information about microbial genomes and microbial communities.

qualitatively—by identifying which genes are expressed and which genes are not expressed—and quantitatively—by measuring varying levels of expression for different genes.

As we know, all cells of an organism possess the same genome, but in any cell or tissue type, certain genes will be highly expressed, others expressed at low levels, and some not expressed at all. Transcriptome analysis provides gene-expression profiles that for the same genome may vary from cell to cell or from tissue type to tissue type. Transcriptome analysis provides insights into (1) normal patterns of gene expression that are important for understanding how a cell or tissue type differentiates during development, (2) how gene expression dictates and controls the physiology of differentiated cells, and (3) mechanisms of disease development that result from or cause gene-expression changes in cells.

A number of different techniques can be used for transcriptome analysis. **DNA microarray analysis** enables researchers to analyze all of a sample's expressed genes simultaneously. Most microarrays, also known as **gene chips,** consist of a glass microscope slide onto which single-stranded DNA molecules are attached, or "spotted," using a computer-controlled high-speed robotic arm called an arrayer. Arrayers are fitted with a number of tiny pins. Each pin is immersed in a small amount of solution containing millions of copies of a different single-stranded DNA molecule. For example, many microarrays are made with single-stranded sequences of complementary DNA (cDNA) or expressed sequenced tags (ESTs)—short fragments of cloned DNA from expressed genes. The arrayer fixes the DNA onto the slide at specific locations (points, or spots) that are recorded by a computer. A single microarray can have over 20,000 different spots of DNA, each containing a unique sequence for a different gene. Entire genomes are available on microarrays, including the human genome.

To prepare a microarray for use in transcriptome analysis, scientists typically begin by extracting mRNA from cells or tissues **(Figure 12)**. The mRNA is usually then reverse transcribed to synthesize cDNA tagged with fluorescently labeled nucleotides. The cDNA can be labeled in a number of ways, but most methods involve the use of fluorescent dyes. Typically, microarray studies often involve comparing gene expression in different cell or tissue samples. cDNA prepared from one tissue is usually labeled with one color dye, red for example, and cDNA from another tissue labeled with a different-colored dye, such as green. Labeled cDNAs are then denatured and incubated overnight with the microarray so that they will hybridize to spots on the microarray that contain complementary DNA sequences. Next, the microarray is washed, and then it is scanned by a laser that causes the cDNA hybridized to the microarray to fluoresce. The patterns of fluorescent spots reveal which genes are expressed in the tissue of interest, and the intensity of spot fluorescence indicates the relative level of expression. The brighter the spot, the more the particular mRNA is expressed in that tissue.

Microarrays are dramatically changing the way gene-expression patterns are analyzed. The biggest advantage of microarrays is that they enable thousands of genes to be studied simultaneously. As a result, however, they can generate an overwhelming amount of gene-expression data. In addition, even when properly controlled, microarrays often yield variable results. For example, one experiment under certain conditions may not always yield similar patterns of gene expression as another identical experiment. Some of these differences can be due to real differences in gene expression, but others can be the result of variability in chip preparation, cDNA synthesis, probe hybridization, or washing conditions, all of which must be carefully controlled to limit such variability. Commercially available microarrays can reduce the variability that can result when individual researchers make their own arrays.

Computerized microarray data analysis programs are essential for organizing gene-expression profile data from microarrays. In Section 6 we briefly discussed the sea urchin

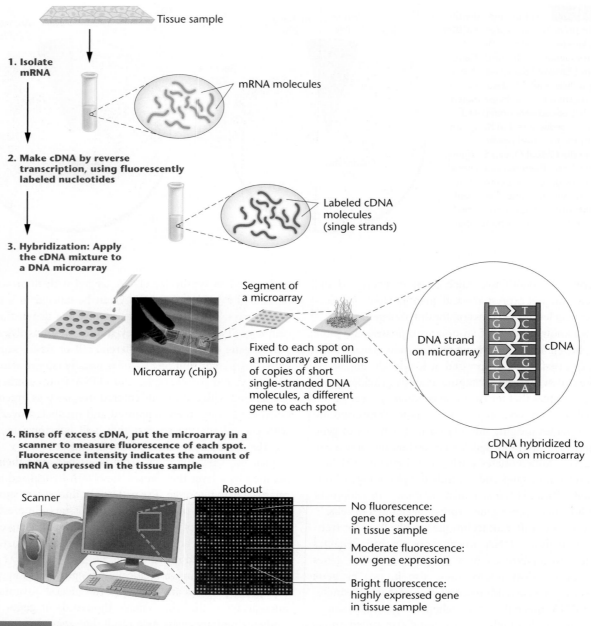

1. **Isolate mRNA**

mRNA molecules

2. **Make cDNA by reverse transcription, using fluorescently labeled nucleotides**

Labeled cDNA molecules (single strands)

3. **Hybridization: Apply the cDNA mixture to a DNA microarray**

Segment of a microarray

Microarray (chip)

Fixed to each spot on a microarray are millions of copies of short single-stranded DNA molecules, a different gene to each spot

DNA strand on microarray

cDNA

cDNA hybridized to DNA on microarray

4. **Rinse off excess cDNA, put the microarray in a scanner to measure fluorescence of each spot. Fluorescence intensity indicates the amount of mRNA expressed in the tissue sample**

Scanner

Readout

No fluorescence: gene not expressed in tissue sample

Moderate fluorescence: low gene expression

Bright fluorescence: highly expressed gene in tissue sample

FIGURE 12   Microarray analysis for analyzing gene-expression patterns in a tissue.

genome and the importance of the sea urchin as a key model organism. Sea urchins have a relatively simple body plan [Figure 13(a)]. They contain approximately 1500 cells with only a dozen cell types. Yet sea urchin development progresses through complex changes in gene expression that resemble vertebrate patterns of gene-expression changes during development. This is one reason that urchins are valuable model organisms for developmental biology.

Scientists at NASA's Ames Genome Research Facility in Moffett, California, used microarrays to carry out a transcriptome study on sea urchin gene expression during the first two days of development (to the mid-late gastrula stage, about 48 hours postfertilization). This work revealed that approximately 52 percent of all genes in the sea urchin are active during this period of development: 11,500 of the sea urchin's 23,500 known genes were expressed in the embryo. The functional categories of genes expressed in the embryo were diverse, including genes for about 70 percent of the nearly 300 transcription factors in the sea urchin genome, along with genes involved in cell signaling, immunity, fertilization, and metabolism [Figure 13(b)]. Incredibly, 51,000 RNAs of unknown function were also expressed. Studies are underway to explain the differences between gene number

FIGURE 13 (a) The sea urchin *Strongylocentrotus purpuratus*. (b) Transcriptome analysis of genes expressed in the sea urchin embryo. The *y*-axis of the histogram displays the percentage of annotated genes in different functional categories expressed in the embryo. The number at the top of each bar represents the total number of annotated genes in the corresponding functional category. Trans., transcription factors; Signal, signaling genes; Process, basic cellular processes, such as metabolism; Cytoskelet., cytoskeletal genes; Fertiliz., fertilization; and Biominer., biomineralization.

and transcripts expressed, although it is already known that many sea urchin genes are extensively processed through alternative splicing. Further analysis of the sea urchin genome will undoubtedly reveal interesting aspects of gene function during sea urchin development and advance our understanding of the genetics of embryonic development in both invertebrates and vertebrates.

Now that we have considered genomes and transcriptomes, we turn our attention to the ultimate end products of most genes, the proteins encoded by a genome.

ESSENTIAL POINT

DNA chips or microarrays are valuable for transcriptome analysis in studying expression patterns for thousands of genes simultaneously.

9 Proteomics Identifies and Analyzes the Protein Composition of Cells

As more genomes have been sequenced and studied, biologists in many different disciplines have focused increasingly on understanding the complex structures and functions of the proteins the genomes encode. This interest is not surprising given that in most of the genomes sequenced to date, many newly discovered genes and their putative proteins have no known function. Keep in mind, in the ensuing discussion, that although every cell in the body contains an equivalent set of genes, not all cells express the same genes and proteins. **Proteome** is a term that represents the complete set of proteins encoded by a genome, but it is also often used to mean the entire complement of proteins in a cell.

Proteomics—the identification, characterization, and quantitative analysis of the proteome of a cell, tissue, or organism—can be used to reconcile differences between the number of genes in a genome and the number of different proteins produced. But equally important, proteomics also provides information about a protein's structure and function; posttranslational modifications; protein–protein, protein–nucleic acid, and protein–metabolite interactions; cellular localization of proteins; protein stability and aspects of translational and posttranslational levels of gene-expression regulation; and relationships (shared domains, evolutionary history) to other proteins.

Proteomics is also of clinical interest because it allows comparison of proteins in normal and diseased tissues, which can lead to the identification of proteins as biomarkers for disease conditions. Proteomic analysis of mitochondrial proteins during aging, proteomic maps of atherosclerotic plaques from human coronary arteries, and protein profiles in saliva as a way to detect and diagnose diseases are examples of such work.

Reconciling the Number of Genes and the Number of Proteins Expressed by a Cell or Tissue

Genomics has revealed that the link between gene and gene product is often very complex. Genes can have multiple transcription start sites that produce several different types of transcripts. Alternative splicing and editing of pre-mRNA molecules can generate dozens of different proteins from a single gene. Remember the current estimate that approximately 95 percent of human genes produce more than one protein by alternative splicing. As a result, proteomes are substantially larger than genomes. For instance, the ~20,000 genes in the human genome encode ~100,000 proteins, although some es-

437

timates suggest that the human proteome may be as large as 150,000–200,000 proteins.

Proteomes undergo dynamic changes that are coordinated in part by regulation of gene-expression patterns—the transcriptome. However, a number of other factors affect the proteome profile of a cell, further complicating the analysis of protein function. For instance, many proteins are modified by co-translational or posttranslational events, such as cleavage of signal sequences that target a protein for an organelle pathway, propeptides, or initiator methionine residues; by linkage to carbohydrates and lipids; or by the addition of chemical groups through methylation, acetylation, and phosphorylation and other modifications. Over a hundred different mechanisms of posttranslational modification are known.

Well before a draft sequence of the human genome was available, scientists were already discussing the possibility of a "Human Proteome Project." One reason such a project never came to pass is that there is no single human proteome: different tissues produce different sets of proteins. But the idea of such a project led to the **Protein Structure Initiative (PSI)** by the National Institute of General Medical Sciences (NIGMS), a division of the National Institutes of Health, involving over a dozen research centers. Initiated in 2000, PSI is a multiphase project designed to analyze the three-dimensional structures of more than 4000 protein families. Proteins with interesting potential therapeutic properties are a top priority for the PSI, and to date, the structures of over 1000 proteins have been determined. Developing computation protein structural prediction methods, solving unique protein structures, disseminating PSI information, and focusing on the biological relevance of the work are major goals. There also are a number of other ongoing projects dedicated to identifying proteome profiles that correlate with diseases such as cancer and diabetes.

Proteomics Technologies: Two-Dimensional Gel Electrophoresis for Separating Proteins

With proteomics technologies, scientists have the ability to study thousands of proteins simultaneously, generating enormous amounts of data quickly and dramatically changing ways of analyzing the protein content of a cell. The early history of proteomics dates back to 1975 and the development of **two-dimensional gel electrophoresis (2DGE)** as a technique for separating hundreds to thousands of proteins with high resolution. In this technique, proteins isolated from cells or tissues of interest are loaded onto a polyacrylamide tube gel and first separated by **isoelectric focusing,** which causes proteins to migrate according to their electrical charge in a pH gradient. During isoelectric focusing, proteins migrate until they reach the location in the gel where their net charge is zero compared to the pH of the gel (**Figure 14**). Then in a second migration, perpendicular to the first, the proteins are separated by their molecular mass using **sodium dodecyl sulfate polyacrylamide gel electrophoresis (SDS-PAGE).** In

this step, the tube gel is rotated 90° and placed on top of an SDS polyacrylamide gel; an electrical current is applied to the gel to separate the proteins by mass.

Proteins in a 2D gel are visualized by staining with Coomassie blue, silver stain, or other dyes that reveal the separated proteins as a series of spots in the gel (Figure 14). It is not uncommon for a 2D gel loaded with a complex mixture of proteins to show several thousand spots in the gel, as in Figure 14, which displays the complex mixture of proteins in human platelets (thrombocytes). Particularly abundant protein spots in this gel have been labeled with the names of identified proteins. With thousands of different spots on the gel, how are the identities of the proteins ascertained?

In some cases, 2D gel patterns from experimental samples can be compared to gels run with reference standards containing known proteins with well-characterized migration patterns. Many reference gels for different biological samples such as human plasma are available, and computer software programs can be used to align and compare the spots from different gels. In the early days of 2DGE, proteins were often identified by cutting spots out of a gel and sequencing the amino acids the spots contained. Only relatively small sequences of amino acids can typically be generated this way; rarely can an entire polypeptide be sequenced using this technique. BLAST and similar programs can be used to search protein databases containing amino acid sequences of known proteins. However, because of alternative splicing or posttranslational modifications, peptide sequences may not always match easily with the final product, and the identity of the protein may have to be confirmed by another approach.

NOW SOLVE THIS

2 Annotation of a proteome attempts to relate each protein to a function in time and space. Traditionally, protein annotation depended on an amino acid sequence comparison between a query protein and a protein with known function. If the two proteins shared a considerable portion of their sequence, the query would be assumed to share the function of the annotated protein. Following is a representation of this method of protein annotation involving a query sequence and three different human proteins. Note that the query sequence aligns to common domains within the three other proteins. What argument might you present to suggest that the function of the query is not related to the function of the other three proteins?

——— Query amino acid sequence

Region of amino acid sequence match to query

■ **Hint** *This problem asks you to think about sequence similarities between four proteins and predict functional relationships. The key to its solution is to remember that although protein domains may have related functions, proteins can contain several different interacting domains that determine protein function.*

FIGURE 14 Two-dimensional gel electrophoresis (2DGE) is a useful method for separating proteins in a protein extract from cells or tissues that contains a complex mixture of proteins with different biochemical properties. The two-dimensional gel photo shows separations of human platelet proteins. Each spot represents a different polypeptide separated by molecular weight (*y*-axis) and isoelectric point, pH (*x*-axis). Known protein spots are labeled by name based on identification by comparison to a reference gel or by determination of protein sequence using mass spectrometry techniques. Notice that many spots on the gel are unlabeled, indicating proteins of unknown identity.

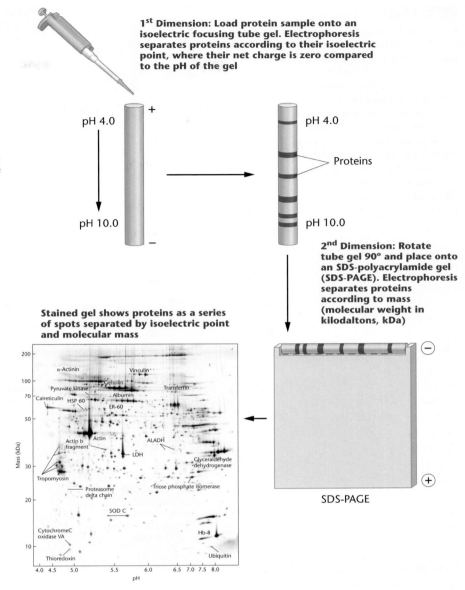

1st Dimension: Load protein sample onto an isoelectric focusing tube gel. Electrophoresis separates proteins according to their isoelectric point, where their net charge is zero compared to the pH of the gel

pH 4.0 +

pH 10.0 −

pH 4.0

Proteins

pH 10.0

2nd Dimension: Rotate tube gel 90° and place onto an SDS-polyacrylamide gel (SDS-PAGE). Electrophoresis separates proteins according to mass (molecular weight in kilodaltons, kDa)

Stained gel shows proteins as a series of spots separated by isoelectric point and molecular mass

SDS-PAGE

As you will learn in the next section, proteomics has incorporated other techniques to aid in protein identification, and one of these techniques is mass spectrometry.

Proteomics Technologies: Mass Spectrometry for Protein Identification

Mass spectrometry (MS) has been instrumental to the development of proteomics. Mass spectrometry techniques analyze ionized samples in gaseous form and measure the **mass-to-charge (m/z) ratio** of the different ions in a sample. Proteins analyzed by mass spectra generate m/z spectra that can be correlated with an m/z database containing known protein sequences to discover the protein's identity. Certain MS applications can provide peptide sequences directly from spectra. Some of the most valuable proteomics

applications of this technology are to identify an unknown protein or proteins in a complex mix of proteins, to sequence peptides, to identify posttranslational modifications of proteins, and to characterize multiprotein complexes.

One commonly used MS approach is **matrix-assisted laser desorption ionization (MALDI).** MALDI is ideally suited for identifying proteins and is widely used for proteomic analysis of tissue samples treated under different conditions. The proteins are first extracted from cells or tissues of interest and separated by 2DGE, after which MALDI (described below) is used to identify the proteins in the different spots. **Figure 15** shows an example in which two different sets of cells grown in culture are analyzed for protein differences. Just about any source providing a sufficient number of cells can be used: blood, whole tissues, and organs; tumor samples; microbes; and many other substances. Many proteins

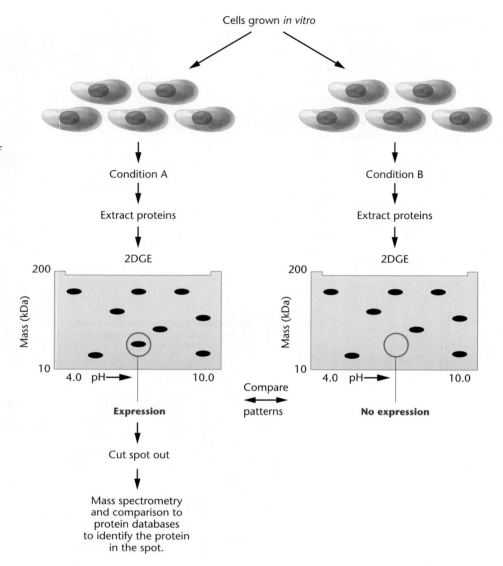

FIGURE 15 In a typical proteomic analysis, cells are exposed to different conditions (such as different growth conditions, drugs, or hormones). Then proteins are extracted from these cells and separated by 2DGE, and the resulting patterns of spots are compared for evidence of differential protein expression. Spots of interest are cut out from the gel, digested into peptide fragments, and analyzed by mass spectrometry to identify the protein they contain.

involved in cancer have been identified by the use of MALDI to compare protein profiles in normal tissue and tumor samples.

Protein spots are cut out of the 2D gel, and proteins are purified out of each gel spot. Computer–automated high-throughput instruments are available that can pick all of the spots out of a 2D gel. Isolated proteins are then enzymatically digested with a protease (a protein-digesting enzyme) such as trypsin to create a series of peptides. This proteolysis produces a complex mixture of peptides determined by the cleavage sites for the protease in the original protein. Each type of protein produces a characteristic set of peptide fragments, and these are identified by MALDI as follows.

In MALDI, peptides are mixed with a low molecular weight and ultraviolet (UV) light-absorbing acidic matrix material (such as dihydroxybenzoic acid) and then applied to a metal plate. A UV laser, often a nitrogen laser at a wavelength of 337 nm, is then fired at the sample. As the matrix absorbs energy from the laser, heat accumulating on the matrix vaporizes and ionizes the peptide fragments. Released ions are then analyzed for mass; MALDI displays the m/z ratio of each ionized peptide as a series of peaks representative of the molecular masses of peptides in the mixture and their relative abundance **(Figure 16)**. Because different proteins produce different sets of peptide fragments, MALDI produces a peptide "fingerprint" that is characteristic of the protein being analyzed.

Databases of MALDI-generated m/z spectra for different peptides can be analyzed to look for matches between m/z spectra of unknown samples and those of known proteins. One limitation of this approach is database quality. An unknown protein from a 2D gel can only be identified by MALDI if proteomics databases have a MALDI spectrum for that protein. But as is occurring with genomics databases, proteomics databases with thousands of well-characterized proteins from different organisms are rapidly developing.

Protein microarrays are also becoming valuable tools for proteomics research. These are designed around the same basic concept as microarrays (gene chips) and are often constructed with antibodies that specifically recognize and bind to different proteins. These microarrays are used, among other applications, for examining protein–protein interactions, for detecting protein markers for disease diagnosis, and for studying in biosensors designed to detect pathogenic microbes and potentially infectious bioweapons.

Identification of Collagen in *Tyrannosaurus rex* and *Mammut americanum* Fossils

Recently, a team of scientists reported results of mass spectrometry analysis of bone tissue from a *Tyrannosaurus rex* skeleton excavated from the Hell Creek Formation in eastern Montana and estimated to be 68 million years old. As mentioned earlier, DNA has been recovered from fossils, but the general assumption has been that proteins degrade in fossilized materials and cannot be recovered. This study demonstrated that fossilization does not fully destroy all proteins in well-preserved fossils under certain conditions. This research also demonstrates the power and sensitivity of MS as a proteomics tool.

In this work, medullary tissue was removed from the inside of the left and right femoral bones. Medullary tissue is porous, spongy bone that contains bone marrow cells, blood vessels, and nerves. *T. rex* proteins extracted from the tissue showed cross-reactivity with antibodies to chicken collagen and were digested by the collagen-specific protease collagenase. These results suggested that the *T. rex* protein samples contained collagen, a major matrix component of bone, ligaments, tendons, and skin. To definitively identify the presence of collagen, tryptic peptides from the *T. rex* samples were analyzed by liquid chromatography and MS (LC/MS). The m/z spectra for one of the *T. rex* peptides was identified from a database of m/z spectra as corresponding to collagen. Furthermore, the amino acid sequence of *T. rex* collagen peptide aligned with an isoform of chicken collagen, demonstrating sequence similarity. Such work has provided excellent experimental evidence to support the widely accepted theory that birds and dinosaurs are close relatives.

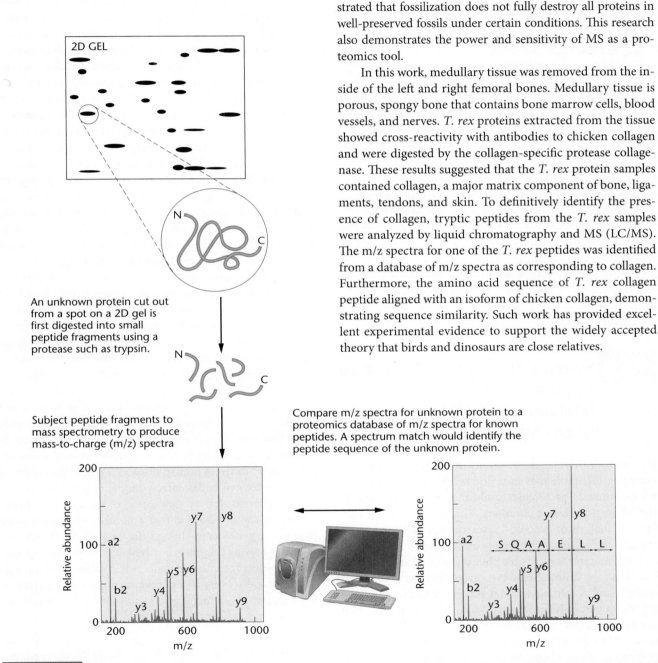

An unknown protein cut out from a spot on a 2D gel is first digested into small peptide fragments using a protease such as trypsin.

Subject peptide fragments to mass spectrometry to produce mass-to-charge (m/z) spectra

Compare m/z spectra for unknown protein to a proteomics database of m/z spectra for known peptides. A spectrum match would identify the peptide sequence of the unknown protein.

FIGURE 16 Mass spectrometry for identifying an unknown protein isolated from a 2D gel. The mass-to-charge spectrum (m/z) (determined, for example, by MALDI) for trypsin-digested peptides from the unknown protein can be compared to a proteomics database for a spectrum match to identify the unknown protein. The peptide in this example was revealed to have the amino acid sequence serine (S)-glutamine (Q)-alanine (A)-alanine (A)-glutamic acid (E)-leucine (L)-leucine (L), shown in single-letter amino acid code.

Similar results were obtained for 160,000- to 600,000-year-old mastodon (*Mammut americanum*) peptides that showed matches to collagen from extant species, including collagen isoforms from humans, chimps, dogs, cows, chickens, elephants, and mice.

ESSENTIAL POINT ■ ■ ■

Proteomics methods such as two-dimensional gel electrophoresis and mass spectrometry are valuable for analyzing proteomes—the protein content of a cell.

NOW SOLVE THIS

3 Because of its accessibility and biological significance, the proteome of human plasma has been intensively studied and used to provide biomarkers for such conditions as myocardial infarction (troponin) and congestive heart failure (B-type natriuretic peptide). Polanski and Anderson (Polanski, M., and Anderson, N. L., *Biomarker Insights*, 2: 1–48, 2006) have compiled a list of 1261 proteins, some occurring in plasma, that appear to be differentially expressed in human cancers. Of these 1261 proteins, only 9 have been recognized by the FDA as tumor-associated proteins. First, what advantage should there be in using plasma as a diagnostic screen for cancer? Second, what criteria should be used to validate that a cancerous state can be assessed through the plasma proteome?

■ **Hint** *This problem asks you to consider criteria that are valuable for using plasma proteomics as a diagnostic screen for cancer. The key to its solution is to consider proteomics data that you would want to evaluate to determine whether a particular protein is involved in cancer.*

10 Systems Biology Is an Integrated Approach to Studying Interactions of All Components of an Organism's Cells

We conclude this chapter by discussing **systems biology,** an emerging discipline that incorporates data from genomics, transcriptomics, proteomics, and other areas of biology, as well as engineering applications. Through the use of mutational analysis of genomes, researchers have been able to identify and characterize genes and proteins involved in creating mutants with visible phenotypes that share similar biochemical pathways. However, even extensive mutational analysis and screening will not provide a full understanding of complex cellular processes such as signal transduction pathways, metabolic pathways, and regulation of cell division, DNA replication, and gene expression. A more comprehensive, more integrated approach is needed.

In many ways, systems biology is interpreting genomic information in the context of the structure, function, and regulation of biological pathways. By studying relationships between all components in an organism, biologists are trying to build a "systems"-level understanding of how organisms function. Systems biologists typically combine recently acquired genomics and proteomics data with years of more traditional studies of gene and protein structure and function. Much of this data is retrieved from databases such as PubMed, GenBank, and other newly emerging genomics, transcriptomics, and proteomics resources.

Systems models are used to diagram interactions within a cell or an entire organism, such as protein–protein interactions, protein–nucleic acid interactions, and protein–metabolite interactions (e.g., enzyme-substrate binding). These models help systems biologists understand the components of interacting pathways and the interrelationships of molecules in an interacting pathway. In recent years, the term **interactome** has arisen to describe the interacting components of a cell. Systems biologists use several different types of models to diagram protein interaction pathways. One of the most common model types is a **network map**—a sketch showing interacting proteins, genes, and other molecules. These diagrams are essentially the equivalent of an electrical wiring diagram.

One disadvantage of network maps is that they are static diagrams that typically lack information about when and where each interaction occurs. Even so, they are a useful foundation for generating computational models that allow the running of simulations to determine how signaling events occur. For example, major groups of kinases, enzymes that phosphorylate other proteins to affect their activity, have been network mapped to show their interactions with each other. Because kinases play such important roles in the regulation of most critical cellular processes, such information about the "kinome" has been very valuable for companies developing drug treatments targeted to certain metabolic pathways.

Network maps are helping scientists model intricate potential interactions of molecules involved in normal and disease processes. **Figure 17** shows an example of a network map. This map depicts a human disease network model illustrating the complexity of interactions between genes involved in 22 different human diseases. Look at the cluster of turquoise-colored nodes corresponding to genes involved in several different cancers. One aspect of the map that should be immediately obvious is that a number of cancers share interacting genes even though the cancers affect different organs. Knowing the genes involved and the protein interaction networks for different cancers is a major breakthrough in the drug discovery and development process for informing scientists about target genes and proteins to consider for therapeutic purposes.

ESSENTIAL POINT ■ ■ ■

Systems biology approaches are designed to provide an integrated understanding of interactions between genes, proteins, and other molecules that govern complex biological processes.

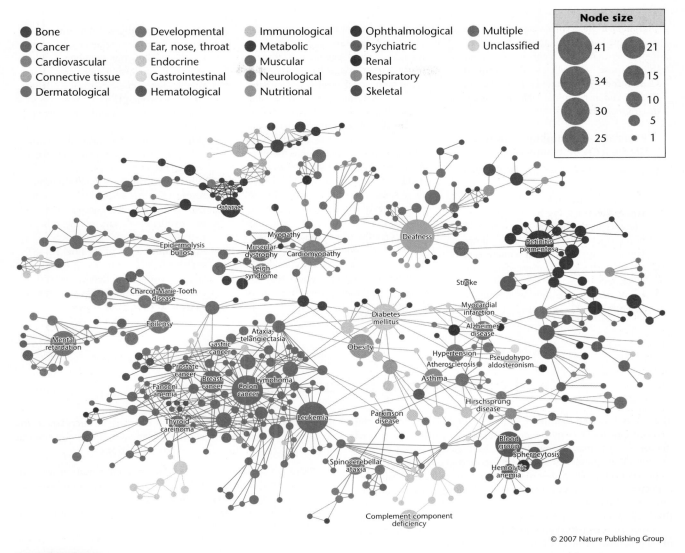

Node size

41	21
34	15
30	10
25	5
	1

- Bone
- Cancer
- Cardiovascular
- Connective tissue
- Dermatological
- Developmental
- Ear, nose, throat
- Endocrine
- Gastrointestinal
- Hematological
- Immunological
- Metabolic
- Muscular
- Neurological
- Nutritional
- Ophthalmological
- Psychiatric
- Renal
- Respiratory
- Skeletal
- Multiple
- Unclassified

FIGURE 17 A systems biology model of human disease gene interactions. The model shows nodes corresponding to 22 specific disorders colored by class. Node size is proportional to the number of genes contributing to the disorder.

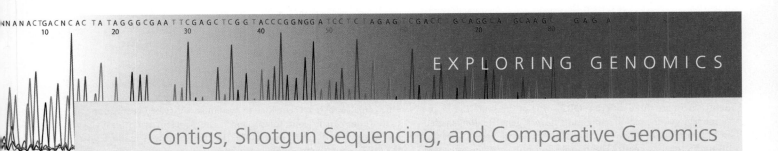

EXPLORING GENOMICS

Contigs, Shotgun Sequencing, and Comparative Genomics

MG Study Area: *Exploring Genomics*

In this chapter, we discussed how whole-genome shotgun sequencing methods can be used to assemble chromosome maps. Recall that in the technique of shotgun cloning, chromosomal DNA is digested with different restriction enzymes to create a series of overlapping DNA fragments called contiguous sequences, or "contigs." The contigs are then subjected to DNA sequencing, after which bioinformatics-based programs are used to arrange the contigs in their correct order on the basis of short overlapping sequences of nucleotides.

In this Exploring Genomics exercise you will carry out a simulation of contig alignment to help you to understand the underlying logic of this approach to creating sequence maps of a chromosome. For this purpose, you will use the **National Center for Biotechnology Information BLAST** site and apply a DNA alignment program called bl2seq.

■ Exercise I – Arranging Contigs to Create a Chromosome Map

1. Access BLAST from the NCBI Web site at http://www.ncbi.nlm.nih.gov/BLAST. Locate and select the "Align two sequences using BLAST (bl2seq)" category at the bottom of the BLAST homepage. The bl2seq feature allows you to compare two DNA sequences at a time to check for sequence similarity alignments.

2. Go to the Companion Web site for *Essentials of Genetics* and open the Exploring Genomics exercise for this chapter. Listed are eight contig sequences, called Sequences A through H, taken from an actual human chromosome sequence deposited in GenBank. For this exercise we have used short fragments; however, in reality contigs are usually several thousand base pairs long. To complete this exercise, copy and paste two sequences into the Align feature of BLAST and then run an alignment (by clicking on "Align"). Repeat these steps with other combinations of two sequences to determine which sequences overlap, and then use your findings to create a sequence map that places overlapping contigs in their proper order. Here are a few tips to consider:

■ Develop a strategy to be sure that you analyze alignments for all pairs of contigs.

■ Only consider alignment overlaps that show 100 percent sequence similarity.

3. On the basis of your alignment results, answer the following questions, referring to the sequences by their letter codes (A through H):

a. What is the correct order of overlapping contigs?

b. What is the length, measured in number of nucleotides, of each sequence overlap between contigs?

c. What is the total size of the chromosome segment that you assembled?

d. Did you find any contigs that do not overlap with any of the others? Explain.

4. Run a nucleotide-nucleotide BLAST search (BLASTn) on any of the overlapping contigs to determine which chromosome these contigs were taken from, and report your answer.

CASE STUDY » Bioprospecting in Darwin's wake

The Global Ocean Sampling (GOS) expedition followed the route of Charles Darwin's voyage to chart the genetic diversity of microbes in the marine environment. Metagenomics was used to catalog DNA sequences and their encoded proteins from thousands of previously undescribed organisms present in samples collected in diverse oceanic regions. Although many samples remain to be analyzed, the project has already identified more than 1.2 million new genes and thousands of species previously unknown to scientists, generated data on more than 1700 previously unknown families of proteins, and assembled information about more than 6 million specific proteins. Some, like those forming light-driven proton pumps and those that function in nitrogen fixation, may have immediate applications in technology and agriculture. For now, however, the emphasis is on better

understanding the distribution and diversity of microbes in the oceans. This massive project has raised several questions about bioprospecting in the waters of coastal nations.

1. Although sampling sites are selected in consultation with scientists in the host countries, who owns the organisms collected along the coast of these countries?

2. Who will own any processes or products developed from these genetic resources?

3. How can the findings from this metagenomics survey of the ocean be applied?

4. Is it surprising that so many previously undiscovered organisms are represented in the samples collected on this expedition?

INSIGHTS AND SOLUTIONS

1. One challenge in annotation is deciding how long a putative open reading frame (ORF) must be before it is accepted as a gene. Shown here are three different ORF scans of the same *E. coli* genome region—the region containing the *lacY* gene. Regions shaded in brown indicate ORFs. The top scan was set to accept ORFs of 50 nucleotides as genes. The middle and bottom scans accepted ORFs of 100 and 300 nucleotides as genes, respectively. How many potential genes are detected in each scan? The longest ORF covers 1254 bp; the next longest, 234 bp; and the shortest, 54 bp. How can we decide the actual number of genes in this region? In this type of an ORF scan, is it more likely that the number of genes in the genome will be overestimated or underestimated? Why?

Solution: Generally, one can examine conserved sequences in other organisms to indicate that an ORF is likely a coding region. One can also match a sequence to previously described sequences that are known to code for proteins. The problem is not easily solved—that is, deciding which ORF is actually a gene. The shorter the ORFs scan, the more likely the overestimate of genes because ORFs longer than 200 are less likely to occur by chance. For these scans, notice that the 50 bp scans produce the highest number of possible genes, whereas the 300 bp scan produces the lowest number (1) of possible genes.

2. Sequencing of the heterochromatic regions (repeat-rich sequences concentrated in centromeric and telomeric areas) of the *Drosophila*

genome indicates that within 20.7 Mb, there are 297 protein-coding genes (Bergman et al. 2002. *genomebiology3 (12)@genomebiology .com/2002/3/12/RESEARCH/0086*). Given that the euchromatic regions of the genome contain 13,379 protein-coding genes in 116.8 Mb, what general conclusion is apparent?

Solution: Gene density in euchromatic regions of the *Drosophila* genome is about one gene per 8730 base pairs, while gene density in heterochromatic regions is one gene per 70,000 bases (20.7 Mb/297). Clearly, a given region of heterochromatin is much less likely to contain a gene than the same-sized region in euchromatin.

100

Sequenced strand

Complementary strand

50

Sequenced strand

Complementary strand

300

Sequenced strand

Complementary strand

PROBLEMS AND DISCUSSION QUESTIONS

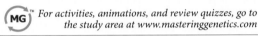

For activities, animations, and review quizzes, go to the study area at www.masteringgenetics.com

HOW DO WE KNOW?

1. In this chapter, we focused on the analysis of genomes, transcriptomes, and proteomes and considered important applications and findings from these endeavors. At the same time, we found many opportunities to consider the methods and reasoning by which much of this information was acquired. From the explanations given in the chapter, what answers would you propose to the following fundamental questions:

   (a) How do we know which contigs are part of the same chromosome?

   (b) How do we know if a genomic DNA sequence contains a protein-coding gene?

   (c) What evidence supports the concept that humans share substantial sequence similarities and gene functional similarities with model organisms?

   (d) How can proteomics identify differences between the number of protein-coding genes predicted for a genome and the number of proteins expressed by a genome?

   (e) What evidence indicates that gene families result from gene duplication events?

   (f) How have microarrays demonstrated that, although all cells of an organism have the same genome, some genes are expressed in almost all cells, whereas other genes show cell- and tissue-specific expression?

2. What is functional genomics? How does it differ from comparative genomics?

3. What, if any, features do bacterial genomes share with eukaryotic genomes?

4. What is bioinformatics, and why is this discipline essential for studying genomes? Provide two examples of bioinformatics applications.

5. List and describe three major goals of the Human Genome Project. Explain at least two findings about the Human Genome Project that surprised most geneticists.

6. How do high-throughput techniques such as computer-automated and next-generation sequencing and mass spectrometry facilitate research in genomics and proteomics? Explain.

7. BLAST searches and related applications are essential for analyzing gene and protein sequences. Define BLAST, describe basic features of this bioinformatics tool, and provide an example of information provided by a BLAST search.

8. Describe the human genome in terms of genome size, the percentage of the genome that codes for proteins, how much is composed of repetitive sequences, and how many genes it contains. Describe two other features of the human genome.

9. Compare the organization of bacterial genes to that of eukaryotic genes. What are some of the major differences?

10. The Human Genome Project has demonstrated that in humans of all races and nationalities approximately 99.9 percent of the sequence is the same, yet different individuals can be identified by DNA fingerprinting techniques. What is one primary variation in the human genome that can be used to distinguish different individuals? Briefly explain your answer.

11. Annotation involves identifying genes and gene-regulatory sequences in a genome. Describe characteristics of a genome that are hallmarks for identifying genes in an unknown sequence. What characteristics would you look for in a prokaryotic genome? A eukaryotic genome?

12. Through the Human Genome Project (HGP) a relatively accurate human genome sequence was published in 2003 from combined samples from different individuals. It serves as a reference for a haploid genome. Recently, genomes of a number of individuals have been sequenced under the auspices of the Personal Genome Project (PGP). How do results from the PGP differ from those of the HGP?

13. Describe the significance of the Genome 10K plan.

14. It can be said that modern biology is experiencing an "omics" revolution. What does this mean? Explain your answer.

15. Metagenomics studies generate very large amounts of sequence data. Provide examples of genetic insight that can be learned from metagenomics.

16. What are gene microarrays? How are microarrays used?

17. Annotation of the human genome sequence reveals a discrepancy between the number of protein-coding genes and the number of predicted proteins actually expressed by the genome. Proteomic analysis indicates that human cells are capable of synthesizing more than 100,000 different proteins and perhaps three times this number. What is the discrepancy, and how can it be reconciled?

18. Genomic sequencing has opened the door to numerous studies that help us understand the evolutionary forces shaping the genetic makeup of organisms. Using databases containing the sequences of 25 genomes, scientists (Kreil, D.P. and Ouzounis, C.A., *Nucl. Acids Res.* 29: 1608–1615, 2001) examined the relationship between GC content and global amino acid composition. They found that it is possible to identify thermophilic species on the basis of their amino acid composition alone, which suggests that evolution in a hot environment selects for a certain whole organism amino acid composition. In what way might evolution in extreme environments influence genome and amino acid composition? How might evolution in extreme environments influence the interpretation of genome sequence data?

19. The β-globin gene family consists of 60 kb of DNA, yet only 5 percent of the DNA encodes gene products. Account for as much of the remaining 95 percent of the DNA as you can.

20. M. Stoll and colleagues have compared candidate loci in humans and rats in search of loci in the human genome that are likely to contribute to the constellation of factors leading to hypertension. Through this research, they identified 26 chromosomal regions that they consider likely to contain hypertension genes. How can comparative genomics aid in the identification of genes responsible for such a complex human disease? The researchers state that comparisons of rat and human candidate loci to those in the mouse may help validate their studies. Why might this be so?

21. Comparisons between human and chimpanzee genomes indicate that a gene that may function as a wild type or normal gene in one primate may function as a disease-causing gene in another. For instance, the *PPARG* locus (regulator of adipocyte differentiation) is associated with type 2 diabetes in humans but functions as a wild-type gene in chimps. What factors might cause this apparent contradiction? Would you consider such apparent contradictions to be rare or common? What impact might such findings have on the use of comparative genomics to identify and design therapies for disease-causing genes in humans?

22. The discovery that *M. genitalium* has a genome of 0.58 Mb and only 470 protein-coding genes sparked interest in determining the minimum number of genes needed for a living cell. In the search for organisms with smaller and smaller genomes, a new species of Archaea, *Nanoarchaeum equitans*, was discovered in a high-temperature vent on the ocean floor. This prokaryote has one of the smallest cell sizes ever discovered, and its genome is only about 0.5 Mb. However, organisms such as *M. genitalium*, *N. equitans*, and other microbes with very small genomes are either parasites or symbionts. How does this affect the search for a minimum genome? Should the definition of the minimum genome size for a living cell be redefined?

SOLUTIONS TO SELECTED PROBLEMS AND DISCUSSION QUESTIONS

Answers to Now Solve This

1. **(a, b)** Most eukaryotic genes are organized in a particular manner and contain coding segments (exons), noncoding segments (introns), enhancers, promoters, untranslated regions (UTRs) and gene termination sequences. Protein-coding genes often contain one or more open-reading frames that are translated into an amino acid sequence of a protein. Such retions typically begin with an ATG (start codon) and end in some termination sequence (stop codons). It is through these types of landmarks that genes are typically identified.

   (c) Eight percent of 20,000 is 1600. Therefore, one would conclude that chromosome 1 is gene rich.

2. Since structural and chemical factors determine the function of a protein, it is likely to have several proteins share a considerable amino acid sequence identity, but not be functionally identical.

   Since the *in vivo* function of such a protein is determined by secondary and tertiary structures, as well as local surface chemistries in active or functional sites, the non-identical sequences may have considerable influence on function. Note that the query matches to different site positions within the target proteins. A number of other factors suggesting different functions include: associations with other molecules (cytoplasmic, membrane, or extracellular), chemical nature and position of binding domains, post-translational modification, signal sequences, etc.

3. Because blood is relatively easy to obtain in a pure state, its components can be analyzed without fear of tissue-site contamination. Second, blood is intimately exposed to virtually all cells of the body and may therefore carry chemical markers to certain abnormal cells it represents; theoretically, an ideal probe into the human body. However, when blood is removed from the body its proteome changes and those changes are dependent on a number of environmental factors. Thus, what might be a valid diagnostic for one condition, might not be so under others. In addition, the serum proteome is subject to change depending on the genetic, physiologic, and environmental state of the patient. Age and sex are additional variables that must be considered. Validation of a plasma proteome for a particular cancer would be strengthened by demonstrating that the stage of development of the cancer correlates with a commensurate change in the proteome in a relatively large, statistically-significant pool of patients. Second, the types of changes in the proteome should be reproducible and, at least until complexities are clarified, involve tumorigenic proteins. It would be helpful to have comparisons with archived samples of each individual at a disease-free time.

Solutions to Problems and Discussion Questions

6. High-throughput technologies allow comprehensive analyses of a number of labor-intensive tasks that would normally take days or weeks to be reduced to half-day activities. By shortening sequencing times for examples, numerous organisms can be sequenced to yield highly informative comparative sequences (comparative genomics). Applied to both genomics and proteomics, high-throughput technologies allow rapid analyses and deployment of genomic information.

8. The human genome is composed of over 3 billion nucleotides in which about 2 percent code for genes. Genes are unevenly distributed over chromosomes with clusters of gene-rich separated by gene-poor ones (deserts). Human genes tend to

be larger and contain more and larger introns than in invertebrates such as *Drosophila*. It is estimated that at least half of the genes generate products by alternative splicing. Hundreds of genes have been transferred from bacteria into vertebrates. Duplicated regions are common, which may facilitate chromosomal rearrangement. The human genome appears to contain approximately 20,000 protein-coding genes; however, there is still uncertainty as to the total number.

10. Because many repetitive regions of the genome are not directly involved in production of a phenotype, they tend to be isolated from selection and show considerable variation in redundancy. Length variation in such repeats is unique among individuals (except for identical twins) and, with various detection methods, provides the basis for DNA fingerprinting. Single nucleotide polymorphisms also occur frequently in the genome and can be used to distinguish individuals.

12. The PGP provides individual sequences of diploid genomes and results of such projects indicate that the HGP may underestimate genome variation by as much as five-fold. Genome variation between individuals may be 0.5 percent rather than the 0.1 percent estimated from the HGP. Since the PGP provides sequence information on individuals, fundamental questions about human diversity and evolution may be more answerable.

18. In general, one would expect certain factors (such as heat or salt) to favor evolution to increase protein stability: distribution of ionic interactions on the surface, density of hydrophobic residues and interactions, and number of hydrogen and disulfide bonds. By examining the codon table, a high GC ratio would favor amino acids Ala, Gly, Pro, Arg, and Trp and minimize the use of Ile, Phe, Lys, Asn, and Tyr. How codon bias influences actual protein stability is not yet understood. Most genomic sequences change by relatively gradual responses to mild selection over long periods of time. They strongly resemble patterns of common descent; that is, they are conserved. While the same can be said for organisms adapted to extreme environments, extraordinary physiological demands may dictate unexpected sequence bias.

20. Any time a DNA sequence is conserved in other species, it is likely that that sequence has an influence on similar phenotypes. The higher the number of species that conserve the sequence, the higher the likelihood of determining its function. Coupled with mutation analysis and physical mapping, comparative genomics provides a powerful method for linking DNA sequences with complex human diseases.

22. The issue here is whether the organism under consideration is independent and self-reproducing. It appears that the minimum number of genes for a free-living organism is in the range of 250–350. Symbionts can have much smaller genomes and exist with fewer genes because of materials supplied by the host cell. As long as one defines the life style (free-living or symbiont) of the organism in question, it is informative to consider how many genes are needed to accomplish the task of "living."

CREDITS

Credits are listed in order of appearance.

Photo

Text

Transgenic pigs generated by incorporating a viral vector carrying the jellyfish gene encoding green fluorescent protein into the pig genome. A nontransgenic pig is in the center of the photograph.

Applications and Ethics of Genetic Engineering and Biotechnology

CHAPTER CONCEPTS

- Recombinant DNA technology, genetic engineering, and biotechnology have revolutionized medicine and agriculture.

- Genetically modified plants and animals can serve as bioreactors to produce therapeutic proteins and other valuable protein products.

- Genetic modifications of plants have resulted in herbicide- and pest-resistant crops, and crops with improved nutritional value; similarly, transgenic animals are being created to produce therapeutic proteins and to protect animals from disease.

- A synthetic genome has been assembled and transplanted into a donor bacterial strain elevating interest in potential applications of synthetic biology.

- Applications of recombinant DNA technology and genomics have become essential for diagnosing genetic disorders, determining genotypes, and scanning the human genome to detect diseases.

- Genome-wide association studies scan for hundreds or thousands of genetic differences in an attempt to link genome variations to particular traits and diseases.

- Pharmacogenomics and rational drug design have led to customized medicines based primarily on a person's genotype.

- Gene therapy by transfer of cloned copies of functional alleles into target tissues is used to treat genetic disorders.

- Almost all applications of genetic engineering and biotechnology present unresolved ethical dilemmas that involve important moral, social, and legal issues.

Since the dawn of recombinant DNA technology in the 1970s, scientists have harnessed **genetic engineering** not only for biological research, but also for applications in medicine, agriculture, and biotechnology. Genetic engineering refers to the alteration of an organism's genome and typically involves the use of recombinant DNA technologies to add a gene or genes to a genome, but it can also involve gene removal. The ability to manipulate DNA *in vitro* and to introduce genes into living cells has allowed scientists to generate new varieties of plants, animals, and other organisms with specific gene traits, and to manufacture cheaper and more effective therapeutic products. These new varieties of organisms are called **genetically modified organisms,** or **GMOs.**

Biotechnology is the use of living organisms to create a product or a process that helps improve the quality of life for humans or other organisms. Biotechnology as a modern industry began in earnest shortly after recombinant DNA technology developed. However, biotechnology as a science dates back to ancient civilization where microbes were used to make many important products, including wine and beer, vinegar, breads, and cheeses. Modern biotechnology relies heavily on recombinant DNA technology, genetic engineering, and genomics applications, and these areas will be the focus of this chapter. Existing products and new developments that occur seemingly every day make the biotechnology industry one of the most rapidly developing branches of the workforce worldwide, encompassing nearly 5000 companies in 54 countries.

The development of the biotechnology industry and the rapid growth in the number of applications for DNA technologies have raised serious concerns about using our power to manipulate genes and to apply gene technologies. Genetic engineering and biotechnology have the potential to provide solutions to major problems globally and to significantly alter how humans deal with the natural world; hence, they raise ethical, social, and economic questions that are unprecedented in human experience. These complex issues cannot be fully explored in the context of an introductory genetics textbook.

This chapter will therefore present only a selection of applications that illustrate the power of genetic engineering and biotechnology and the complexity of the dilemmas they engender. We will begin by explaining how genetic engineering has modified agriculturally important plants and animals. We briefly describe how genetic engineering has affected the production of pharmaceutical products, and we examine the impact of genetic technologies on the diagnosis and treatment of human diseases, including gene therapy approaches. Finally, we explore some of the social, ethical, and legal implications of genetic engineering and biotechnology.

Genetically Engineered Organisms Synthesize a Wide Range of Biological and Pharmaceutical Products

The most successful and widespread application of recombinant DNA technology has been production by the biotechnology industry of recombinant proteins as **biopharmaceutical** products—particularly, therapeutic proteins to treat diseases. Prior to the recombinant DNA era, biopharmaceutical proteins such as insulin, clotting factors, or growth hormones were purified from tissues such as the pancreas, blood, or pituitary glands. Clearly, these sources were in limited supply, and the purification processes were expensive. In addition, products derived from these natural sources could be contaminated by disease agents such as viruses. Now that human genes encoding important therapeutic proteins can be cloned and expressed in a number of nonhuman host-cell types, we have more abundant, safer, and less expensive sources of biopharmaceuticals. **Biopharming** is a commonly used term to describe the production of valuable proteins in genetically modified (GM) animals and plants.

In this section, we outline several examples of therapeutic products that are produced by expression of cloned genes in transgenic host cells and organisms. It should not surprise you that cancers, arthritis, diabetes, heart disease, and infectious diseases such as AIDS are among the major diseases that biotechnology companies are targeting for treatment by recombinant therapeutic products. **Table 1** provides a short list of important recombinant products currently synthesized in transgenic bacteria, plants, yeast, and animals.

Insulin Production in Bacteria

Many therapeutic proteins have been produced by introducing human genes into bacteria. In most cases, the human gene is cloned into a plasmid, and the recombinant vector is introduced into the bacterial host. Large quantities of the transformed bacteria are grown, and the recombinant human protein is recovered and purified from bacterial extracts.

The first human gene product manufactured by recombinant DNA technology was human insulin, called Humulin, which was licensed for therapeutic use in 1982 by the **U.S. Food and Drug Administration (FDA),** the government agency responsible for regulating the safety of food and drug products and medical devices. In 1977, scientists isolated and cloned the gene for insulin and expressed it in bacterial cells. Previously, insulin was chemically extracted from the pancreas of cows and pigs obtained from slaughterhouses. **Insulin** is a protein hormone that regulates glucose metabolism. Individuals who cannot produce insulin have diabetes, a disease that, in its more severe form (type I), affects more than 2 million individuals in the United States. Although synthetic human insulin can now be produced by another process, a look at the original genetic engineering method is instructive, as it shows both the promise and the difficulty of applying recombinant DNA technology.

Clusters of cells embedded in the pancreas synthesize a precursor polypeptide known as preproinsulin. As this polypeptide is secreted from the cell, amino acids are cleaved from the end and the middle of the chain. These cleavages produce the mature insulin molecule, which contains two polypeptide chains (the A and B chains) joined by disulfide bonds. The A subunit contains 21 amino acids, and the B subunit contains 30.

In the original bioengineering process, synthetic genes that encode the A and B subunits were constructed by oligonucleotide synthesis (63 nucleotides for the A polypeptide and 90 nucleotides for the B polypeptide). Each synthetic oligonucleotide was inserted into a separate vector, adjacent to the *lacZ* gene encoding the bacterial form of the enzyme β-galactosidase. When transferred to a bacterial host, the *lacZ* gene and the adjacent synthetic oligonucleotide were transcribed and translated as a unit. The product is a **fusion protein (Figure 1)** that is purified from bacterial

TABLE 1 Examples of Genetically Engineered Biopharmaceutical Products Available or under Development

Gene Product	Condition Treated	Host Type
Erythropoitin	Anemia	*E. coli*; cultured mammalian cells
Interferons	Multiple sclerosis, cancer	*E. coli*; cultured mammalian cells
Tissue plasminogen activator tPA	Heart attack, stroke	Cultured mammalian cells
Human growth hormone	Dwarfism	Cultured mammalian cells
Monoclonal antibodies against vascular endothelial growth factor (VEGF)	Cancers	Cultured mammalian cells
Human clotting factor VIII	Hemophilia A	Transgenic sheep, pigs
C1 inhibitor	Hereditary angioedema	Transgenic rabbits
Recombinant human antithrombin	Hereditary antithrombin deficiency	Transgenic goats
Hepatitis B surface protein vaccine	Hepatitis B infections	Cultured yeast cells, bananas
Immunoglobulin IgG1 to HSV-2	Herpesvirus infections	Transgenic soybeans
Recombinant monoclonal antibodies	Passive immunization against rabies (also used in diagnosing rabies), cancer, rheumatoid arthritis	Transgenic tobacco, soybeans, cultured mammalian cells
Norwalk virus capsid protein	Norwalk virus infections	Potato (edible vaccine)
E. coli heat-labile enterotoxin	*E. coli* infections	Potato (edible vaccine)

extracts and treated with cyanogen bromide, which cleaves the insulin polypeptides from β-galactosidase. When the two fusion products were mixed, the two insulin subunits spontaneously united, forming an intact, active insulin molecule. Shortly after insulin became available, **growth hormone**—used to treat children who suffer from a form of dwarfism—was cloned. Soon, a wide variety of other medically important proteins became available. Over 200 recombinant products have now entered the market worldwide.

Transgenic Animal Hosts and Pharmaceutical Products

Although bacteria have been widely used to produce therapeutic proteins, they often cannot process and modify eukaryotic proteins for full biological activity, rendering them inactive. To increase the efficacy and yields, many biopharmaceuticals are now produced in eukaryotic hosts. As seen in Table 1, eukaryotic hosts may include cultured eukaryotic cells (plant or animal) or transgenic farm animals. A herd of goats or cows serve as very effective **bioreactors** or **biofactories**—living factories—that will continuously make milk containing the desired therapeutic protein that can then be isolated in a noninvasive way.

Yeast are also valuable hosts for expressing recombinant proteins. Even insect cells are valuable for this purpose, through the use of a gene delivery system (virus) called **baculovirus.** Recombinant baculovirus containing a gene of interest is used to infect insect cell lines, which then express the protein at high levels. Regardless of the host, therapeutic proteins may then be purified from the host cells—or when

transgenic farm animals are used, isolated from animal products such as milk.

An example of a biopharmaceutical product synthesized in transgenic animals is the human protein **α-antitrypsin.** A deficiency of the enzyme α-antitrypsin is associated with the heritable form of emphysema, a progressive and fatal respiratory disorder. To produce α-antitrypsin for use in treating this disease, the human gene was cloned into a vector at a site adjacent to a sheep promoter sequence that specifically activates transcription in mammary tissue. This fusion gene initially was microinjected into sheep zygotes fertilized *in vitro.* The fertilized zygotes were transferred to surrogate mothers. The resulting transgenic sheep developed normally and produced milk containing high concentrations of functional human α-antitrypsin. A small herd of lactating transgenic sheep can provide an abundant supply of this protein.

Using similar technology, in 2006, recombinant human **antithrombin,** an anticlotting protein, became the world's first drug extracted from the milk of farm animals to be approved for use in humans. Scientists introduced the human antithrombin gene into goats. By placing the gene adjacent to a promoter for beta casein, a common protein in milk, they were able to target antithrombin expression in the mammary gland, whereby the antithrombin protein is highly expressed in the milk. In one year, a single goat will produce the equivalent amount of antithrombin that in the past would have been isolated from ~90,000 blood collections.

(a)

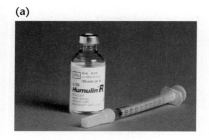

(b)

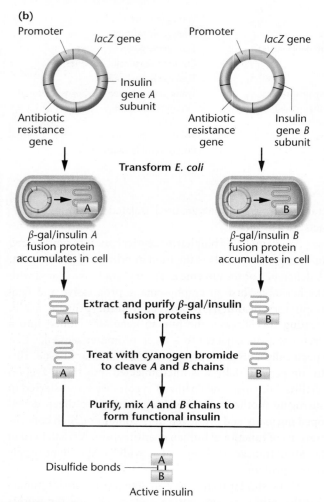

(a) Humulin, a recombinant form of human insulin, was the first therapeutic protein produced by recombinant DNA technology to be approved for use in humans. (b) To synthesize recombinant human insulin, synthetic oligonucleotides encoding the insulin A and B chains were inserted (in separate vectors) at the tail end of a cloned E. coli lacZ gene. The recombinant plasmids were transformed into E. coli host cells, where the β-gal/insulin fusion protein was synthesized and accumulated in the cells. Fusion proteins were then extracted from the host cells and purified. Insulin chains were released from β-galactosidase by treatment with cyanogen bromide. The insulin subunits were purified and mixed to produce a functional insulin molecule.

Recombinant DNA Approaches for Vaccine Production and Transgenic Plants with Edible Vaccines

One of the most promising applications of recombinant DNA technology for therapeutic purposes may be the production of vaccines, which stimulate the immune system to produce antibodies against disease-causing organisms and confer immunity against specific diseases. Traditionally, two types of vaccines have been used: **inactivated vaccines,** which are prepared from killed samples of the infectious virus or bacteria; and **attenuated vaccines,** which are live viruses or bacteria that can no longer reproduce. Inactivated vaccines include the vaccines for rabies and influenza; vaccines for tuberculosis, cholera, and chickenpox are examples of attenuated vaccines.

Genetic engineering is being used to produce a relatively new variety called a **subunit vaccine,** consisting of one or more surface proteins from the virus or bacterium but not the entire virus or bacterium. These act as antigens that stimulate the immune system to make antibodies against the organism from which they were derived. One of the first subunit vaccines was made against the **hepatitis B virus,** which causes liver damage and cancer. The gene that encodes the hepatitis B surface protein was cloned into a yeast expression vector, and the cloned gene was expressed in yeast host cells. The protein was then extracted and purified from the host cells and packaged for use as a vaccine.

In 2005, the FDA approved **Gardasil,** a subunit vaccine produced by the pharmaceutical company Merck and the first cancer vaccine to receive FDA approval. Gardasil targets four strains of **human papilloma virus (HPV)** that cause the majority of human cervical cancers. Approximately 70 percent of sexually active women will be infected by an HPV strain during their lifetime. Gardasil is designed to provide immune protection against HPV prior to infection.

Developing countries face serious difficulties in manufacturing, transporting, and storing vaccines that require refrigeration. To overcome these problems, scientists are attempting to develop vaccines that can be synthesized in edible food plants. 170–71 It warrants mentioning that **DNA-based vaccines** have been attempted for many years, and recently there has been renewed interest in using these vaccines to protect against viral pathogens. In this approach, DNA encoding proteins from a particular pathogen are inserted into a plasmid vector, which is then injected directly into an individual or delivered via a viral vector similar to the way certain viruses are used for gene therapy. The idea here is that pathogen proteins encoded by the delivered DNA would be produced and trigger an immune response that could provide protection should an immunized person be exposed to the pathogen in the future.

ESSENTIAL POINT ■   ■   ■

Recombinant DNA technology can be used to produce valuable biopharmaceutical protein products such as therapeutic proteins for treating disease.

NOW SOLVE THIS

1 DNA-based vaccines have not been particularly effective. What are some potential problems of this vaccine type, and what do you think may be the major limitation of DNA-based vaccines?

■ **Hint** *This problem asks you to consider reasons why DNA-based vaccines have generally not been effective. The key to its solution is to consider what kinds of molecules elicit an immune response and to think about how a DNA-based vaccine is intended to work.*

2 | Genetic Engineering of Plants Has Revolutionized Agriculture

For millennia, farmers have manipulated the genetic makeup of plants and animals to enhance food production. Until the advent of genetic engineering 30 years ago, these genetic manipulations were primarily restricted to **selective breeding**—the selection and breeding of naturally occurring or mutagen-induced variants. In the last 50 to 100 years, genetic improvement of crop plants through the traditional methods of artificial selection and genetic crosses has resulted in dramatic increases in productivity and nutritional enhancement. For example, maize yields have increased four-fold over the last 60 years, and more than half of this increase is due to genetic improvement by artificial selection and selective breeding (**Figure 2**). Modern maize has substantially larger ears and kernels than the predecessor crops, including hybrids from which it was bred.

Recombinant DNA technology provides powerful new tools for altering the genetic constitution of agriculturally important organisms. Scientists can now identify, isolate, and clone genes that confer desired traits and introduce these into organisms. As a result, it is possible to quickly confer insect resistance, herbicide resistance, or nutritional characteristics in farm plants and animals, a primary purpose of **agricultural biotechnology.** In this section, we primarily consider genetic manipulations to produce transgenic crop plants of agricultural value. In Section 3, we will discuss examples of genetic manipulations of agriculturally important animals.

Worldwide, as of 2010, over 150 million hectares have been planted with genetically engineered crops, particularly herbicide- and pest-resistant soybeans, corn, cotton, and canola; over 50 different transgenic crop varieties are available, including alfalfa, corn, rice, potatoes, tomatoes, tobacco, wheat,

Zea canina   Hybrid   *Zea mays*

FIGURE 2   Selective breeding is one of the oldest methods of genetic alteration of plants. Shown here is teosinte (*Zea canina*, left), a selectively bred hybrid (center), and modern corn (*Zea mays*).

and cranberries. These crops have been planted in both industrialized and developing countries by over 8 million farmers. Since 1996, there has been a 4000 percent increase in GM crop acreage worldwide. In the United States, 86 percent of the soybeans, 78 percent of the cotton, and 46 percent of the corn grown are genetically engineered to resist pests or herbicides.

Several of the main reasons for generating agriculturally valuable transgenic crops include improving growth characteristics and yield, increasing their nutritional value, and providing resistance against insect and viral pests, drought, and herbicides. In addition, many new GM crops that will soon be on the market will be designed for ethanol production and for making biodiesel fuel, providing sustainable sources of energy.

The first commercially available GM food was called the Flavr Savr tomato, which was designed to increase the shelf life of tomatoes by allowing them to stay ripe for several weeks without softening—a common problem for tomatoes. Scientists used antisense RNA technology to inhibit an enzyme called polygalacturonase, which digests pectin in the cell wall of tomatoes. Pectin digestion occurs naturally once tomatoes are picked from the plant, and this process is a main reason why tomatoes soften as they age. This GM approach was generally effective, but the attempts to remedy shipping problems that continued to cause bruising increased the cost of these tomatoes. Public skepticism about the safety of the first GM food was also in play, and so the Flavr Savr was eventually taken off the market.

453

Insights from plant genome sequencing projects will undoubtedly be the catalyst for analysis of genetic diversity in crop plants, identification of genes involved in crop domestication and breeding traits, and subsequent enhancement of a variety of desirable traits through genetic engineering. In the past several years, genome projects have been completed for many major food and industrial crops, including the three crops that account for most of the world's caloric intake: maize, rice, and wheat. We will now examine other, more successful examples of genetically engineered plants used in agriculture. Some of the ethical and social concerns associated with these practices will be examined in Section 7.

Transgenic Crops for Herbicide and Pest Resistance

Damage from weed infestation destroys about 10 percent of crops worldwide. In an attempt to combat this problem, farmers often apply herbicides to the soil to kill weeds prior to seeding a field crop. As the most efficient herbicides also kill crop plants, herbicide uses is limited. The creation of herbicide-resistant crops has opened the way to efficient weed control and increased yields of some major agricultural crops. At present, over 75 percent of soybeans and cotton in the United States are now resistant to the herbicide **glyphosate,** which is the active ingredient in Roundup, commonly sold to keep sidewalks and patios weed-free. Glyphosate is effective at very low concentrations, is not toxic to humans, and is rapidly degraded by soil microorganisms. It kills plants by inhibiting the action of a chloroplast enzyme called **EPSP synthase.** This enzyme is important in amino acid biosynthesis in both bacteria and plants. Without the ability to synthesize vital amino acids, plants wither and die.

The manipulations employed in creating herbicide-resistant plants is illustrated in **Figure 3** and described in the caption of that figure. Standard cloning techniques led to the insertion of the EPSP synthase gene into a specialized Ti plasmid, which is used to infect plant cells. These are then converted into plants expressing resistance. Transgenic techniques have also been used to make plants resistant to several other herbicides, to pathogens such as viruses, and also to insect pests.

Some of the most well-described and controversial GM crops are the so-called **Bt crops,** designed to be resistant to insects. The bacterium *Bacillus thuringiensis* (Bt) produces a protein that, when ingested by insects and larvae, will crystallize in the gut, killing pests such as corn-borer larvae that are responsible for millions of dollars of crop damage worldwide. Initially, applications of Bt involved spraying these bacteria on crops. But recombinant DNA technology has enabled scientists to produce Bt transgenic crops with built-in insecticide protection. The *cry* genes that encode the Bt crystalline protein have been effectively introduced into a

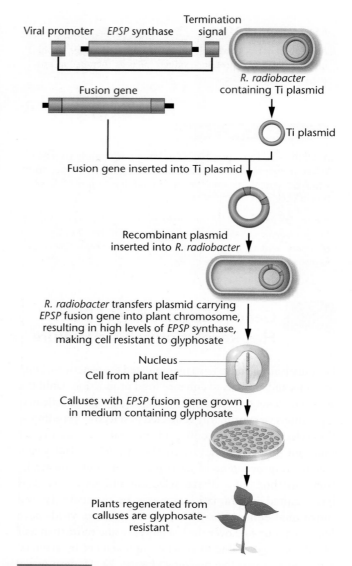

FIGURE 3 To create glyphosate-resistant transgenic plants, the EPSP synthase gene from bacteria is fused to a promoter such as the promoter from the cauliflower mosaic virus. This fusion gene is then ligated into a Ti-plasmid vector, and the recombinant vector is transformed into *R. radiobacter* host cells. *R. radiobacter* infection of cultured plant cells transfers the EPSP synthase fusion gene into a plant-cell chromosome. Cells that acquire the gene are able to synthesize large quantities of EPSP synthase, making them resistant to the herbicide glyphosate. Resistant cells are selected by growth in herbicide-containing medium. Plants regenerated from these cells are herbicide-resistant.

number of different crops, including corn, cotton, tomatoes, and tobacco.

Bt crops have been hailed as one of the greatest success stories of agricultural biotechnology, but they have also been one of the most controversial. Some studies had suggested a correlation between decreases in Monarch butterfly populations and ingestion of pollen from Bt corn (Monarchs do not feed on the corn itself). More recently, several long-term studies have demonstrated that exposure to Bt crops has no apparent effects on the Monarch; however, the possibility of danger to nontarget insect species must be considered whenever pest-resistant crops are used in the wild. In addition, concerns have been raised about the possibility that insect pests might develop resistance to Bt. Nonetheless, based on the success of Bt crops, many other transgenic crops are under development, including plants with increased tolerance to viral pests, drought, and salty soils.

Nutritional Enhancement of Crop Plants

Because crop plants can become deficient in some of the nutrients required in the human diet, biotechnology is being used to produce crops with enhanced nutritional value. One example is the production of **golden rice,** with enhanced levels of β-carotene, a precursor to vitamin A **(Figure 4)**. Vitamin A deficiency is prevalent in many areas of Asia and Africa, and more than 500,000 children a year become permanently blind as a result of this deficiency. Rice is a major staple food in these regions but does not contain vitamin A.

To create golden rice, scientists transferred into the rice genome, by recombinant DNA technology, three genes encoding enzymes required for the biosynthetic pathway leading to β-carotenoid synthesis. Two of these genes came from the daffodil and one from a bacterium. Although golden rice is currently available for planting, it produced only moderate levels of β-carotene. New varieties with higher levels of β-carotene production are in development. One of these strains, Golden Rice 2, incorporates the phytoene synthase gene from maize instead of daffodils and produces about 20 times more β-carotene than the original golden rice. Acceptance of genetically engineered varieties of rice, opposition and regulations on transgenic crops in some countries, and efficient distribution of Golden Rice remain challenges to its wider use. Nonetheless, encouraged by the effectiveness of Golden Rice 2,

researchers are working on developing rice with enhanced iron and protein content.

Many other varieties of nutritionally enhanced food crops have been, and are being, developed. These include plants with augmented levels of key fatty acids, antioxidants, and other vitamins and minerals. These efforts are directed at addressing nutrient deficiencies affecting more than 40 percent of the world's population. Other expected developments include decaffeinated teas and coffees, as well as crops enhanced for traits affecting taste, growth rates, yields, color, storage, ripening, and similar characteristics.

> **ESSENTIAL POINT** ◼ ◼ ◼
>
> Genetically modified (GM) plants, designed to improve crop yield and nutritional value and to increase resistance to herbicides, pests, and severe weather, are becoming more prevalent worldwide.

3 Transgenic Animals with Genetically Enhanced Characteristics Have the Potential to Serve Important Roles in Biotechnology

Although genetically engineered plants are major players in modern agriculture, commercial applications of transgenic animals are less widespread. Nonetheless, some high-profile examples of genetically engineered farm animals have aroused public interest and controversy.

The method of creating transgenic animals is conceptually rather simple, although there are species-specific challenges associated with creating transgenics. Many of the prevailing techniques used to make transgenics were developed in mice. One method is to isolate newly fertilized eggs from a female mouse (or female of the desired animal species) and to inject purified cloned DNA containing a vector and the transgene of interest into the nucleus of the egg **(Figure 5)**. In a relatively small percentage of transgenic eggs, the transgenic DNA becomes inserted into the egg cell genome by recombination due to the action of naturally occurring DNA recombination enzymes. Newer approaches involving stem cells are also popular for creating transgenic animals.

Injected eggs are then placed into the oviduct of a so-called pseudopregnant mouse, a mouse previously impregnated by mating with a male mouse. The pseudopregnant mouse offers a uterus that is receptive to implantation of the egg containing transgenic DNA. Once baby mice are born, researchers screen the transgenic mice by obtaining DNA from a sample of tail tissue, purifying the DNA, and performing PCR to verify that the transgene is present in the animal's

FIGURE 4 Golden rice, a strain genetically modified to produce β-carotene, a precursor to vitamin A. Many children in countries where rice is a dietary staple lose their eyesight because of diets deficient in vitamin A.

455

FIGURE 5 (a) Scientist microinjecting cloned DNA into a fertilized egg. The injections are performed by manipulating the egg and microinjection needle under a light microscope, seen in the background. The injection procedure is displayed on the screen in the foreground. The egg is held by a suction pipette (seen below the egg). (b) A transgenic mouse with its nontransgenic sibling. The mouse on the left is transgenic for a rat growth hormone gene, cloned downstream from a mouse metallothionein promoter. When the transgenic mouse was fed zinc, the metallothionein promoter induced the transcription of the growth hormone gene, stimulating the growth of the transgenic mouse.

(a)

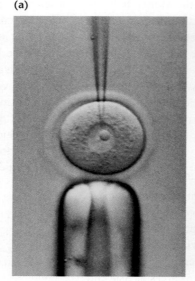

(b)

genome. As long as the integrated DNA is present in germ cells, the transgene will be inherited in all of the offspring generated with the transgenic mouse, but typically most F1 generation transgenic mice produced this way are not homozygous for the transgene. Sibling matings of F1 animals can then be used to generate homozygous transgenic animals.

Transgenic animals overexpressing certain genes, expressing human genes, and expressing mutant genes are among examples of transgenics that are valuable models for basic and applied research to understand gene function. Here we highlight several interesting examples of transgenic animals created with the purpose of producing a potentially commercially valuable biotechnology product.

Examples of Transgenic Animals

Oversize mice containing a human growth hormone transgene were some of the first transgenic animals created. Attempts to create farm animals containing transgenic growth hormone genes have not been particularly successful, probably because growth is a complex, multigene trait. One notable exception is the transgenic Atlantic salmon, bearing copies of a Chinook salmon growth hormone gene adjacent to a constitutive expressed (unregulated, constantly active) promoter. These salmon mature quickly, grow 400 to 600 percent faster than nontransgenic salmon, and appear to have no adverse health effects from the added gene **(Figure 6)**.

The major uses for transgenic farm animals are as bioreactors to produce useful pharmaceutical products, but a number of other interesting transgenic applications are under development. Several of these applications are designed to increase milk production or increase the nutritional value of milk. Significant research efforts are also being made to protect farm animals against common

pathogens that cause disease and animal loss (including potential bioweapons that could be used in a terrorist attack on food animals) and put the food supply at risk. For instance, controlling mastitis in cattle by creating transgenic cows has shown promise **(Figure 7)**. **Mastitis** is an infection of the mammary glands. It is the most costly disease affecting the dairy industry, leading to over $2 billion in losses in the United States. Mastitis can block milk ducts, reducing milk output, and can also contaminate the milk with pathogenic microbes. Infection by the bacterium *Staphylococcus aureus* is the most common cause of mastitis, and most cattle with mastitis typically do not respond well to conventional treatments with antibiotics. As a result, mastitis is a significant cause of herd reduction.

FIGURE 6 Transgenic Atlantic salmon (bottom) overexpressing a growth hormone (GH) gene display rapidly accelerated rates of growth compared to wild strains and nontransgenic domestic strains (top). GH salmon weigh an average of nearly 10 times more than nontransgenic strains.

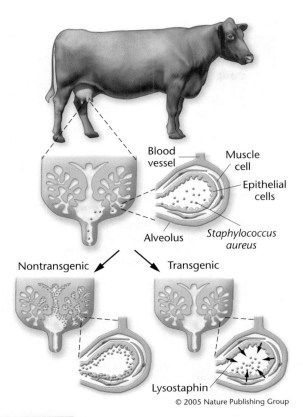

Blood vessel
Muscle cell
Epithelial cells
Alveolus
Staphylococcus aureus
Nontransgenic
Transgenic
Lysostaphin

© 2005 Nature Publishing Group

FIGURE 7 Transgenic cows for battling mastitis. The mammary glands of nontransgenic cows are highly susceptible to infection by the skin microbe *Staphylococcus aureus*. Transgenic cows express the lysostaphin transgene in milk, where it can kill *S. aureus* before they can multiply in sufficient numbers to cause inflammation and damage mammary tissue.

In an attempt to create cattle resistant to mastitis, transgenic cows were generated that possessed the lysostaphin gene from *Staphylococus simulana*. Lysostaphin is an enzyme that specifically cleaves components of the *S. aureus* cell wall.

Transgenic cows expressing this protein in milk produce a natural antibiotic that wards off *S. aureus* infections. These transgenic cows do not completely solve the mastitis problem because lysostaphin is not effective against other microbes such as *E. coli* and *S. uberis* that occasionally cause mastitis; moreover, there is also the potential that *S. aureus* may develop resistance to lysostaphin. Nonetheless, scientists are cautiously optimistic that transgenic approaches have a strong future for providing farm animals with a level of protection against major pathogens.

In one final and most interesting example, scientists have created the **GloFish,** a transgenic strain of zebrafish (*Danio rerio*) containing a red fluorescent protein gene from sea anemones. Marketed as the first GM pet in the United States, GloFish fluoresce bright pink when illuminated by ultraviolet light **(Figure 8)**. GM critics describe these fish as an abuse of genetic technology. However, GloFish may not be as frivolous a use of genetic engineering as some believe. A variation of this transgenic model, incorporating a heavy-metal-inducible promoter adjacent to the red fluorescent protein gene, has shown promise in a bioassay for heavy metal contamination of water. When these transgenic zebrafish are in water contaminated by mercury and other heavy metals, the promoter becomes activated, inducing transcription of the red fluorescent protein gene. In this way, zebrafish fluorescence can be used as a bioassay to measure heavy metal contamination and uptake by living organisms.

ESSENTIAL POINT ■ ■ ■

Transgenic animals with improved growth characteristics or desirable phenotypes are being genetically engineered for a number of different applications.

FIGURE 8 GloFish, marketed as the world's first GM-pet, are a controversial product of genetic engineering.

4 Synthetic Genomes and the Emergence of Synthetic Biology

Can scientists create artificial cells or designer organisms based on genes encoded by a **synthetic genome** constructed artificially? Before we discuss why a synthetic genome may be of value, let's examine a recent example of how an artificial genome was created and tested for its functionality. In 2008, scientists from the J. Craig Venter Institute (JCVI) reported a complete chemical synthesis of the 580-kb *Mycoplasma genitalium* genome, but they could not demonstrate the functionality of the synthetic genome they produced because it could not be transplanted into another bacterium. **Genome transplantation** is effectively the true test of the functionality of a synthetic genome. Recall that if a plasmid containing a gene of interest were introduced into bacteria, the expression of genes in the plasmid, whether they were antibiotic resistance genes or a human gene such as insulin, would demonstrate successful transformation of bacteria. Transplanting an entire synthetic genome with the expectation that genes in the synthetic genome would completely transform the phenotype of the cell is essential.

In 2010, JCVI scientists published the first report of a functional synthetic genome. In this approach they designed and had chemically synthesized more than one thousand 1080-bp segments called cassettes covering the entire 1.08-Mb *M. mycoides* genome **[Figure 9(a)]**. To assemble these segments correctly, the sequences had 80-bp sequences at each end that overlapped with their neighbor sequences. These sequences were cloned in *E. coli*. Then, using the yeast *Saccharomyces cerevisiae*, a homologous recombination approach was used to assemble the sequences into 11 separate 10-kb assemblies that were eventually combined to completely span the entire 1.08-Mb *M. mycoides* genome.

The entire assembled genome, called JCVI-syn1.0, was then transplanted into a close relative *M. capricolum* as recipient cells resulting in a new cell with the JCVI-syn1.0 genotype and phenotype of a new strain of *M. mycoides*. JDVI-syn1.0 was forced into the existing natural genome of *M. capricolum* (although many of the mechanistic details about genome transplantation are unclear). JCVI determined that the recipient cells were taken over to become JCVI-syn.10 *M. mycoides* in part because they were shown to express the *lacZ* gene which was incorporated into the synthetic genome. Selection for tetracycline resistance and a determination that recipient cells also made proteins characteristic of *M. mycoides* and not *M. capricolum* were also used to verify strain conversion **[Figure 9(b)]**.

(a)

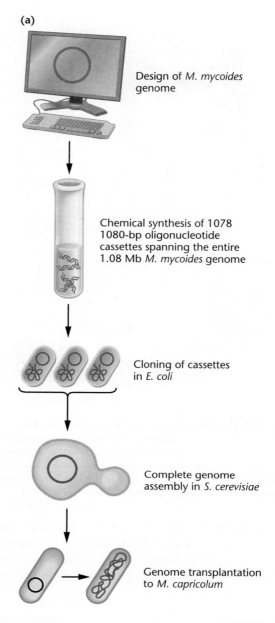

Design of *M. mycoides* genome

Chemical synthesis of 1078 1080-bp oligonucleotide cassettes spanning the entire 1.08 Mb *M. mycoides* genome

Cloning of cassettes in *E. coli*

Complete genome assembly in *S. cerevisiae*

Genome transplantation to *M. capricolum*

© 2010 Nature Publishing Group

(b)

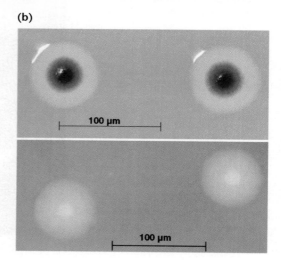

100 μm

100 μm

FIGURE 9   Building a synthetic version of the 1.08 Mb *Mycoplasma mycoides* genome JCVI-syn1.0. (a) Overview of the approach used to produce *M. mycoides* JCVI-syn1.0. (b) Images of *M. mycoides* JCVI-syn1.0 (top) and wild-type *M. mycoides* (bottom). Cells with the synthetic genome express the *lacZ* gene and thus produce colonies that are blue in color; wild-type cells do not express the *lacZ* gene and produce colonies that are white in color.

One particularly impressive accomplishment of these experiments was that the synthetic DNA was "naked" because it did not contain any proteins from *M. mycoides*. Therefore, it was capable of transcribing all of the appropriate genes and translating all of the protein products necessary for life as *M. mycoides*! This is not a trivial accomplishment. The synthetic genome effectively rebooted the *M. capricolum* recipient cells to change them from one form to another. When this work was announced, J. Craig Venter claimed: "This is equivalent to changing a Macintosh computer into a PC by inserting a new piece of PC software." This work did not create life from an inanimate object since it was based on converting one living strain into another. Also, the *Mycoplasma* strains used lack a cell wall typically found in other bacteria, which could be a barrier to genome transplantation in other bacterial species.

Many fundamental questions about synthetic genomes and genome transplantation need to be answered. But clearly these studies provided key "proof of concept" that synthetic genomes could be produced, assembled, and successfully transplanted to create a microbial strain encoded by a synthetic genome and bring scientists closer to producing novel synthetic genomes incorporating genes for specific traits of interest. Venter's recent work, a decade-long project that cost about $40 million and involved about 20 people, is being hailed as a defining moment in the emerging field of **synthetic biology.**

What are some of the potential applications of synthetic genomes and synthetic biology? JCVI claims that their ultimate goal is to create microorganisms that can be used to synthesize biofuels. Other possibilities exist such as creating synthetic microbes with genomes engineered to express gene products to degrade pollutants (bioremediation), the synthesis of new biopharmaceutical products, genetically programmed bacteria to help us heal, and the ability to make "prosthetic genomes." Work on synthetic genomes and synthetic biology has led to speculation of a future world in which new bacteria, and perhaps new animal and plant cells, can be designed and even programmed to be controlled as we want them to! In the future, could synthetic genomes be used to create life from inanimate components? Stay tuned!

ESSENTIAL POINT ■ ■ ■

Synthetic genomics and synthetic biology offer the potential for geneticists to create genetically engineered cells with novel characteristics that may have commercial value.

5 Genetic Engineering and Genomics Are Transforming Medical Diagnosis

Geneticists are applying knowledge about the human genome and the genetic basis of many diseases to a wide range of medical applications. Gene-based technologies have already had a major impact on the diagnosis of disease and are revolutionizing medical treatments and the development of specific and effective pharmaceuticals. In this section, we provide an overview of how gene-based technologies are being used to diagnose genetic diseases.

Using DNA-based tests, scientists can directly examine a patient's DNA for mutations associated with disease. Gene testing was one of the first successful applications of recombinant DNA technology, and currently more than 900 gene tests are in use. These tests usually detect DNA mutations associated with single-gene disorders that are inherited in a Mendelian fashion. Examples of such genetic tests are those that detect sickle-cell anemia, cystic fibrosis, Huntington disease, hemophilias, and muscular dystrophies. Other genetic tests have been developed for complex disorders such as breast and colon cancers. Gene tests are used to perform prenatal diagnosis of genetic diseases, to identify carriers, to predict the future development of disease in adults, to confirm the diagnosis of a disease detected by other methods, and to identify genetic diseases in embryos created by *in vitro* fertilization.

For genetic testing of adults, DNA from white blood cells is commonly used. Alternatively, many genetic tests can be carried out on cheek cells collected by swabbing the inside of the mouth, or by collecting hair samples. Some genetic testing can be carried out on gametes. For prenatal diagnosis, fetal cells are obtained by **amniocentesis** or **chorionic villus sampling. Figure 10** shows the procedure for amniocentesis, in which a small volume of the amniotic fluid surrounding the fetus is removed. Amniotic fluid contains fetal cells that can be used for karyotyping, genetic testing, and other procedures. For chorionic villus sampling, cells from the fetal portion of the placental wall (the chorionic villi) are sampled through a vacuum tube, and analyses can be carried out on this tissue.

Another approach called **fetal cell sorting** may eventually replace amniocentesis and chorionic villus sampling because it is noninvasive for the fetus. In pregnant women a small number of fetal cells are present in the maternal bloodstream. An instrument known as a fluorescence-activated cell sorter can be used to separate fetal cells from a maternal blood sample based on proteins expressed on the fetal cells but not the maternal cells. Captured fetal cells can then be subjected to genetic analysis, usually involving techniques that involve PCR (such as allele-specific oligonucleotide testing described later in this section).

Genetic Tests Based on Restriction Enzyme Analysis

A classic method of genetic testing is **restriction fragment length polymorphism (RFLP) analysis.** As we will discuss in the next section, PCR-based methods have largely replaced RFLP analysis; however, applications of this approach are still used occasionally, and for historical purposes it is also helpful to compare RFLP analysis to new approaches, which were rare prior to completion of the Human Genome Project.

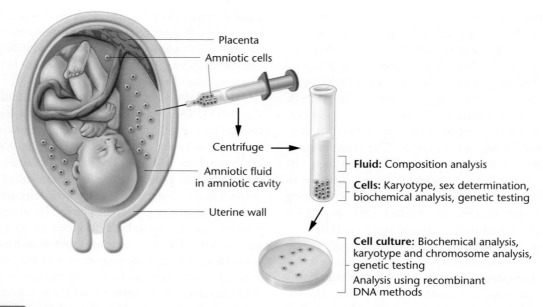

FIGURE 10 For amniocentesis, the position of the fetus is first determined by ultrasound, and then a needle is inserted through the abdominal and uterine walls to recover amniotic fluid and fetal cells for genetic or biochemical analysis.

To illustrate this method, we examine the prenatal diagnosis of **sickle-cell anemia.** This disease is an autosomal recessive condition common in people with family origins in West Africa, the Mediterranean basin, and parts of the Middle East and India. It is caused by a single amino acid substitution in the β-globin polypeptide, as a consequence of a single-nucleotide mutation in the β-globin gene. This mutation eliminates a cutting site in the β-globin gene for the restriction enzymes *Mst*II and *Cvn*I. As a result, the pattern of restriction fragments seen on Southern blots is altered. These differences are used to prenatally diagnose sickle-cell anemia and to establish the genotypes of parents and other family members who may be heterozygous carriers of this condition.

DNA is extracted from tissue samples and digested with *Mst*II. This enzyme cuts three times within a region of the normal β-globin gene, producing two small DNA fragments. In the mutant sickle-cell allele, the middle *Mst*II site is destroyed by the mutation, and one large restriction fragment is produced by *Mst*II digestion (**Figure 11**). The restriction-enzyme-digested DNA fragments are separated by gel electrophoresis, transferred to a nylon membrane, and visualized by Southern blot hybridization, using a probe from this region.

Figure 11 shows the results of RFLP analysis for sickle-cell anemia in one family. Both parents (I-1 and I-2) are heterozygous carriers of the mutation. *Mst*II digestion of both parents' DNA produces a large band (because of the mutant allele) and two smaller bands (from the normal allele). The first child (II-1) is homozygous normal because she shows only the two smaller bands. The second child (II-2) has sickle-cell anemia; he has only one large band and is homozygous for the mutant allele. The fetus (II-3) has a large band and two small bands and is therefore heterozygous for sickle-cell anemia. He or she will be unaffected but will be a carrier.

NOW SOLVE THIS

2 You are asked to assist with a prenatal genetic test for a couple, each of whom is found to be a carrier for a deletion in the β-globin gene that produces β-thalassemia when homozygous. The couple already has one child who is unaffected and is not a carrier. The woman is pregnant, and the couple wants to know the status of the fetus. You receive DNA samples obtained from the fetus by amniocentesis and from the rest of the family by extraction from white blood cells. Using a probe that binds to the mutant allele, you obtain the following blot. Is the fetus affected? What is its genotype for the β-globin gene?

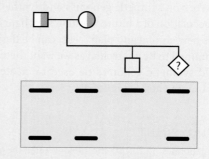

■ **Hint** *This problem is concerned with interpreting results of RFLP analysis of a fetus. The key to its solution is to remember that differences in the number and location of restriction sites create unique RFLPs that can be used to determine genotypes.*

Only about 5 to 10 percent of all point mutations can be detected by restriction enzyme analysis because most mutations occur in regions of the genome that do not contain restriction enzyme cutting sites. However, now that the HGP has been completed and many disease-associated mutations are known, geneticists can employ synthetic oligonucleotides to detect these mutations, as described next.

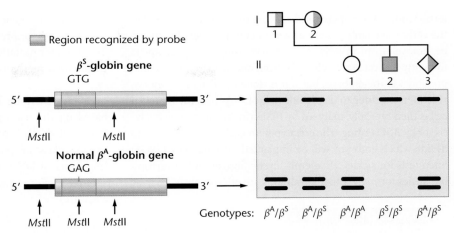

FIGURE 11 RFLP diagnosis of sickle-cell anemia. In the mutant β-globin allele (β^S), a point mutation (GAG $\rightarrow$ GTG) has destroyed a cutting site for the restriction enzyme *Mst*II, resulting in a single large fragment on a Southern blot. In the pedigree, the family has one unaffected homozygous normal daughter (II-1), an affected son (II-2), and an unaffected carrier fetus (II-3). The genotype of each family member can be read directly from the blot and is shown below each lane.

Genetic Tests Using Allele-Specific Oligonucleotides

Another method of genetic testing involves the use of synthetic DNA probes known as **allele-specific oligonucleotides (ASOs).** Scientists use these short, single-stranded fragments of DNA to identify alleles that differ by as little as a single nucleotide. In contrast to restriction enzyme analysis, which is limited to cases for which a mutation changes a restriction site, ASOs detect single-nucleotide changes (**single-nucleotide polymorphisms** or **SNPs**), including those that do not affect restriction enzyme cutting sites. As a result, this method offers increased resolution and wider application. Under proper conditions, an ASO will hybridize only with its complementary DNA sequence and not with other sequences, even those that vary by as little as a single nucleotide.

Genetic testing using ASOs and PCR analysis are now available to screen for many disorders, such as sickle-cell anemia. In the case of sickle-cell screening, DNA is extracted, and a region of the β-globin gene is amplified by PCR. A small amount of the amplified DNA is spotted onto strips of a DNA-binding membrane, and each strip is hybridized to an ASO synthesized to resemble the relevant sequence from either a normal or mutant β-globin gene (**Figure 12**). The ASO is tagged with a molecule that is either radioactive or fluorescent to allow for visualization of the ASO hybridized to DNA on the membrane. This rapid, inexpensive, and highly accurate technique is used to diagnose a wide range of genetic disorders caused by point mutations. Although highly effective, SNPs can affect probe binding leading to false positive or false negative results that may not reflect a genetic disorder, particularly if precise hybridization conditions are not used. Sometimes DNA sequencing is carried out on amplified gene segments to confirm identification of a mutation.

Because ASO testing makes use of PCR, small amounts of DNA can be analyzed. As a result, ASO testing is ideal for **preimplantation genetic diagnosis (PGD).** PGD is the genetic analysis of single cells from embryos created by *in vitro*

Region of β-globin gene amplified by PCR

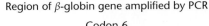

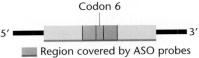

FIGURE 12 Allele-specific oligonucleotide (ASO) testing for the β-globin gene and sickle-cell anemia. The β-globin gene is amplified by PCR, using DNA extracted from white blood cells or cells obtained by amniocentesis. The amplified DNA is then denatured and spotted onto strips of DNA-binding membranes. Each strip is hybridized to a specific ASO and visualized on X-ray film after hybridization and exposure. (a) Results observed when the three possible genotypes are hybridized to an ASO from the normal β-globin allele: *AA*-homozygous individuals have normal hemoglobin that has two copies of the normal β-globin gene and will show heavy hybridization; *AS*-heterozygous individuals carry one normal β-globin allele and one mutant allele and will show weaker hybridization; *SS*-homozygous sickle-cell individuals carry no normal copy of the β-globin gene and will show no hybridization to the ASO probe for the normal β-globin allele. (b) Results observed when DNA for the three genotypes are hybridized to the probe for the sickle-cell β-globin allele: no hybridization by the *AA* genotype, weak hybridization by the heterozygote (*AS*), and strong hybridization by the homozygous sickle-cell genotype (*SS*).

461

fertilization. When sperm and eggs are mixed to create zygotes, the early-stage embryos are grown in culture. A single cell can be removed from an early-stage embryo using a vacuum pipette to gently aspirate one cell away from the embryo. This could possibly kill the embryo, but if it is done correctly the embryo will often continue to divide normally. DNA from the removed cell is then typically analyzed by FISH (for chromosome analysis) or by ASO testing. The genotypes for each cell can be used to decide which embryos will be implanted into the uterus. PGD often tests for sickle-cell anemia, cystic fibrosis, and dwarfism, but alleles for many other conditions are often analyzed.

NOW SOLVE THIS

3 The DNA sequence surrounding the site of the sickle-cell mutation in the β-globin gene, for normal and mutant genes, is as follows.

Each type of DNA is denatured into single strands and applied to a DNA-binding membrane. The membrane containing the two spots is hybridized to an ASO of the sequence

5'-GACTCCTGAGGAGAAGT-3'

Which spot, if either, will hybridize to this probe?

5'-GACTCCTGAGGAGAAGT-3'
3'-CTGAGGACTCCTCTTCA-5'
Normal DNA

5'-GACTCCTGTGGAGAAGT-3'
3'-CTGAGGACACCTCTTCA-5'
Sickle-cell DNA

■ **Hint** *This problem asks you to analyze the results of an ASO test. The key to its solution is to understand that ASO analysis is done under conditions that allow only identical nucleotide sequences to hybridize to the ASO on the membrane.*

Genetic Testing Using DNA Microarrays and Genome Scans

Both RFLP and ASO analyses are efficient methods of screening for gene mutations; however, they can only detect the presence of one or a few specific mutations whose identity and locations in the gene are known. There is also a need for genetic tests that detect complex mutation patterns or previously unknown mutations in genes associated with genetic diseases and cancers. For example, the gene that is responsible for cystic fibrosis, called the **cystic fibrosis transmembrane conductance regulator** (*CFTR*), contains 27 exons and encompasses 250 kilobases of genomic DNA. Of the 1000 known mutations of the *CFTR* gene, about half of these are point mutations, insertions, and deletions—and these are widely distributed throughout the gene. Moreover, additional *CFTR* mutations may yet be discovered. Similarly, over 500 different mutations are known to occur within the tumor

suppressor *p53* gene, and any of these mutations may be associated with, or predispose a patient to, a variety of cancers. To screen for mutations in these genes, comprehensive, high-throughput methods are required.

One emerging high-throughput screening technique is based on the use of **DNA microarrays.** DNA microarrays (also called DNA or gene chips) are small, solid supports, usually glass or polished quartz-based, on which known fragments of DNA are deposited in a precise pattern. Each spot on a microarray is called a **field** (sometimes also called a **feature**).

The DNA fragments that are deposited in a microarray field—called probes—are single-stranded and may be oligonucleotides synthesized *in vitro* or longer fragments of DNA created from cloning or PCR amplification. There are typically over a million identical molecules of DNA in each field. The numbers and types of DNA sequences on a microarray are dictated by the type of analysis that is required. For example, each field on a microarray might contain a DNA sequence derived from each member of a gene family, or sequence variants from one or several genes of interest, or a sequence derived from each gene in an organism's genome. Some microarrays use identical sequences as probes in each field for a particular gene; other microarrays use many different probes for the same gene.

What makes DNA microarrays so amazing is the immense amount of information that can be simultaneously generated from a single array. DNA microarrays the size of postage stamps (just over 1 cm square) can contain up to 500,000 different fields, each representing a different DNA sequence. Scientists are now using DNA microarrays in a wide range of applications, including the detection of mutations in genomic DNA and the detection of gene-expression patterns in diseased tissues.

Most human genes are available on a human genome microarray (**Figure 13**). DNA microarrays have been designed to scan for mutations in many disease-related genes, including the *p53* gene, which is mutated in a majority of human cancers, and the *BRCA1* gene, which, when mutated, predisposes women to breast cancer.

In addition to testing for mutations in single genes, DNA microarrays can contain probes that detect SNPs and copy number variations (CNVs) throughout the genome that may be involved in genetic disease conditions. CNVs are relatively large (~500 bp to several Mb) insertions or deletions, which together with SNPs comprise a majority of genome variations between individuals.

Genetic Analysis Using Gene-Expression Microarrays

Gene-expression microarrays are effective for analyzing gene-expression patterns in genetic diseases because the progression of a tissue from a healthy to a diseased state

FIGURE 13 A commercially available DNA microarray, called a GeneChip, marketed by Affymetrix, Inc. This microarray can be used to analyze expression for approximately 50,000 RNA transcripts. It contains 22 different probes for each transcript and allows scientists to simultaneously assess the expression levels of most of the genes in the human genome.

is almost always accompanied by changes in expression of hundreds to thousands of genes. These arrays provide a powerful tool for diagnosing genetic disorders and gene-expression changes. Expression microarrays may contain probes for only a few specific genes thought to be expressed differently in cell types or may contain probes representing each gene in the genome. Although microarray techniques provide novel information about gene expression, keep in mind that DNA microarrays do not directly provide us with information about protein levels in a cell or tissue. We often infer what predicted protein levels may be based on mRNA expression patterns, but this may not always be accurate.

In a typical expression microarray analysis, mRNA is isolated from two different cell or tissue types—for example, normal cells and cancer cells arising from the same cell type [Figure 14(a)]. The mRNA samples contain transcripts from each gene that are expressed in that cell type. Some genes are expressed more efficiently than others; therefore, each type of mRNA is present at a different level. The level of each mRNA can be used to develop a gene-expression profile that is characteristic of the cell type. Isolated mRNA molecules are converted into cDNA molecules, using reverse transcriptase. The cDNAs from the normal cells are tagged with fluorescent dye-labeled nucleotides (for example, green), and the cDNAs from the cancer cells are tagged with a different fluorescent dye-labeled nucleotide (for example, red).

The labeled cDNAs are mixed together and applied to a DNA microarray. The cDNA molecules bind to complementary single-stranded probes on the microarray but not to other probes. Keep in mind that each field or feature does not consist of just one probe, but rather they contain thousands of copies of the probe. After washing off the nonbinding cDNAs, scientists scan the microarray with a laser, and a

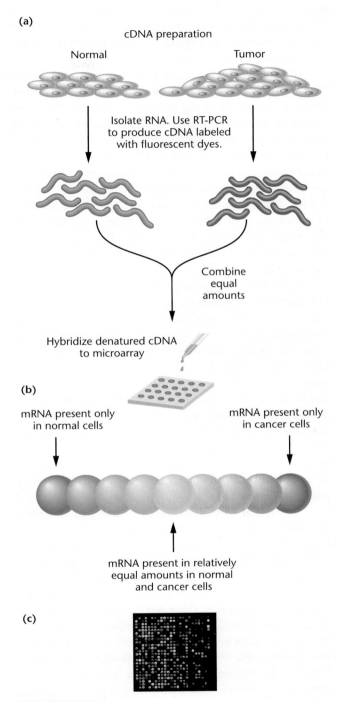

FIGURE 14 (a) Microarray procedure for analyzing gene expression in normal and cancer cells. (b) The method shown here is based on a two-channel microarray in which cDNA samples from the two different tissues are competing for binding to the same probe sets. Colors of dots on an expression microarray represent levels of gene expression. In this example, green dots represent genes expressed only in one cell type (e.g., normal cells), and red dots represent genes expressed only in another cell type (e.g., cancer cells). Intermediate colors represent different levels of expression of the same gene in the two cell types. (c) A small portion of a DNA microarray, showing different levels of hybridization to each field.

computer captures the fluorescent image pattern for analysis. The pattern of hybridization appears as a series of colored dots, with each dot corresponding to one field of the microarray [**Figure 14(b)**].

The color patterns revealed on the microarray fields [**Figure 14(c)**] provide a sensitive measure of the relative levels of each cDNA in the mixture. In the example shown here, if an mRNA is present only in normal cells, the probe representing the gene encoding that mRNA will appear as a green dot because only "green" cDNAs have hybridized to it. Similarly, if an mRNA is present only in the cancer cells, the microarray probe for that gene will appear as a red dot. If both samples contain the same cDNA, in the same relative amounts, both cDNAs will hybridize to the same field, which will appear yellow [Figure 14(b)]. Intermediate colors indicate that the cDNAs are present at different levels in the two samples.

Expression microarray profiling has revealed that certain cancers have distinct patterns of gene expression and that these patterns correlate with factors such as the cancer's stage, clinical course, or response to treatment. In one such experiment, scientists examined gene expression in both normal white blood cells and in cells from a white blood cell cancer known as **diffuse large B-cell lymphoma (DLBCL)**. About 40 percent of patients with DLBCL respond well to chemotherapy and have long survival times. The other 60 percent respond poorly to therapy and have short survival. The investigators assayed the expression profiles of 18,000 genes and discovered that there were two types of DLBCL, with almost inverse patterns of gene expression (**Figure 15**).

One type of DLBCL, called *GC B-like,* had an expression pattern dramatically different from that of a second type, called *activated B-like.* Patients with the activated B-like pattern of gene expression had much lower survival rates than patients with the GC B-like pattern. The researchers concluded that DLBCL is actually two different diseases with different outcomes. Once this type of analysis is introduced into routine clinical use, it may be possible to adjust therapies for each group of cancer patients and to identify new specific treatments based on gene-expression profiles. Similar gene-expression profiles have been generated for many other cancers, including breast, prostate, ovarian, and colon cancer. Gene-expression microarrays are providing tremendous insight into both substantial and subtle variations in genetic diseases.

Several companies are now promoting "nutrigenomics" services in which they claim to use genotyping and gene-expression microarrays to identify allele polymorphisms and gene-expression patterns for genes involved in nutrient metabolism. For example, polymorphisms in genes such as that for apolipoprotein A (*APOA1*), involved in lipid metabolism, and that for *MTHFR* (methylenetetrahydrofolate reductase), involved in metabolism of folic acid, have been implicated in

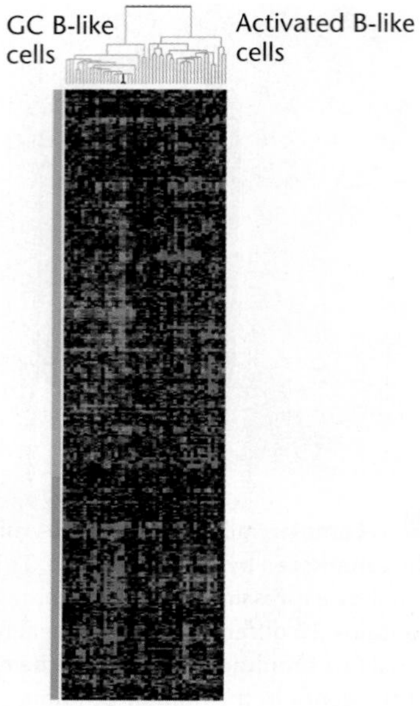

FIGURE 15 (a) Gene-expression analysis generated from expression DNA microarrays that analyzed 18,000 genes expressed in normal and cancerous lymphocytes. Each row represents a summary of the gene expression from one particular gene; each column represents data from one cancer patient's sample. The colors represent ratios of relative gene expression compared to normal control cells. Red represents expression greater than the mean level in controls, green represents expression lower than in the controls, and the intensity of the color represents the magnitude of difference from the mean. In this summary analysis, the cancer patients' samples are grouped by how closely their gene-expression profiles resemble each other. The cluster of cancer patients' samples marked with orange at the top of the figure are GC B-like DLBCL cells. The blue cluster contains samples from cancer patients within the activated B-like DLBCL group. Patients with activated B-like profiles have a much higher rate of death (16 in 21) than those with GC B-like profiles (6 in 19). Data such as these demonstrate the value of microarray analysis for diagnosing disease conditions.

cardiovascular disease. Nutrigenomics companies claim that microarray analysis of a patient's DNA sample for genes such as these and others enables them to judge whether a patient's allele variations or gene-expression profiles warrant dietary changes to potentially improve health and reduce the risk of diet-related diseases.

Application of Microarrays for Gene Expression and Genotype Analysis of Pathogens

Microarrays are also providing infectious disease researchers with powerful new tools for studying pathogens. Genotyping microarrays are being used to identify strains of emergent viruses, such as the virus that causes the highly contagious condition called Severe Acute Respiratory Syndrome (SARS) as well as the H5N1 avian influenza virus, the cause of bird flu.

Whole-genome transcriptome analysis of pathogens is being used to inform researchers about genes that are important for pathogen infection and replication. In this approach, bacteria, yeast, protists, or viral pathogens are used to infect host cells *in vitro*, and then expression microarrays are used to analyze pathogen gene-expression profiles. Patterns of gene activity during pathogen infection of host cells and replication are useful for identifying pathogens and understanding mechanisms of infection. Gene-expression profiling is a valuable approach for identifying important pathogen genes and the proteins they encode that may prove to be useful targets for subunit vaccine development or for drug treatment strategies to prevent or control infectious disease.

Similarly, researchers are evaluating host responses to pathogens **(Figure 16)**. This type of detection has been accelerated in part by the need to develop pathogen-detection strategies for military and civilian use both for detecting outbreaks of naturally emerging pathogens such as SARS and avian influenza and for potential detection of outbreaks such as anthrax (caused by the bacterium *Bacillus anthracis*) that could be the result of a bioterrorism event. Host-response gene-expression profiles are developed by exposing a host to a pathogen and then using expression microarrays to analyze host gene-expression patterns.

Figure 16 shows the different gene-expression profiles for mice following exposure to *Neisseria meningitidis*, the SARS virus, or *E. coli*. In this example, although there are several genes that are upregulated or downregulated by each pathogen, notice how each pathogen strongly induces different prominent clusters of genes that reveal a host gene-expression response to the pathogen and provide a signature of pathogen infection. Comparing such host gene-expression profiles following exposure to different pathogens provides researchers with a way to quickly diagnose and classify infectious diseases. In the future, scientists expect to develop databases of both pathogen and host response expression profile data that can be used to identify pathogens efficiently.

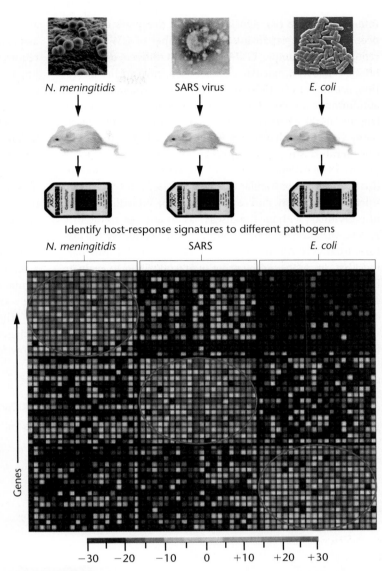

FIGURE 16 Gene-expression microarrays can reveal host-response signatures for pathogen identification. In this example, mice were infected with different pathogens: *Neisseria meningitidis,* the virus that causes Severe Acute Respiratory Syndrome (SARS), and *E. coli.* Mouse tissues were then used as the source of mRNA for gene-expression microarray analysis. Increased expression compared to uninfected control mice is shown in shades of yellow. Decreased expression compared to uninfected controls is indicated in shades of blue. Notice that each pathogen elicits a somewhat different response in terms of which major clusters of host genes are activated by pathogen infection (circles).

6 Genome-Wide Association Studies Identify Genome Variations That Contribute to Disease

Microarray-based genomic analysis has led geneticists to employ powerful new strategies called **genome-wide association studies (GWAS)** in their quest to identify genes that may

ESSENTIAL POINT

A variety of different molecular techniques, including restriction fragment length polymorphism analysis, allele-specific oligonucleotide tests, and DNA microarrays, can be used to identify genotypes associated with normal and diseased phenotypes.

influence disease risk. During the past five years there has been a dramatic expansion in the number of GWAS being reported. For example, GWAS for height differences, autism, obesity, diabetes, macular degeneration, myocardial infarction, arthritis, hypertension, several cancers, bipolar disease, autoimmune diseases, Crohn's disease, schizophrenia, amyotrophic lateral sclerosis, and multiple sclerosis are among the many GWAS that have been widely publicized in the scientific literature and popular press.

In a GWAS, the genomes of thousands of unrelated individuals with a particular disease are analyzed, typically by microarray analysis, and results are compared with genomes of individuals without the disease as an attempt to identify genetic variations that may confer risk of developing the disease. Many GWAS involve large-scale use of SNP microarrays that can probe on the order of 500,000 SNPs to evaluate results from different individuals. Other GWAS approaches can look for specific gene differences or evaluate CNVs or changes in the epigenome, such as methylation patterns in particular regions of a chromosome. By determining which SNPs, CNVs, or epigenome changes co-occur in individu-

als with the disease, scientists can calculate the disease risk associated with each variation. Analysis of GWAS results requires statistical analysis to predict the relative potential impact (association or risk) of a particular genetic variation on development of a disease phenotype.

Figure 17 shows a typical representation of one way that results from GWAS are commonly reported. Such representations are "scatterplots" that are used to display data with a large number of data points. The x-axis typically plots a particular position in the genome; in this case loci on each chromosome are plotted in a different color code. The y-axis plots results of a genotypic association test. Association can be calculated in several ways. Shown here is a negative log of p-values that shows loci determined to be significantly associated with a particular condition. The top line of this plot establishes a threshold value for significance. Marker sequences with significance levels exceeding 10^{-5}, corresponding to 5.0 on the y-axis, are likely disease-related sequences (Figure 17).

There are many questions and ethical concerns about patients involved in GWAS and their emotional responses to knowing about genetic risk data. For example,

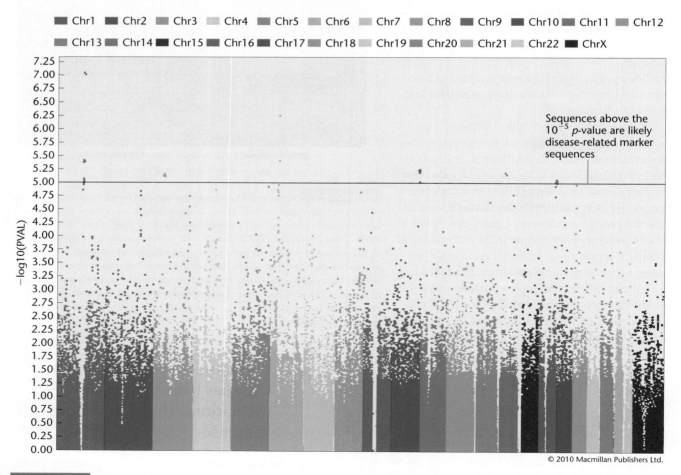

FIGURE 17   A GWAS study for type 2 diabetes revealed 386,371 genetic markers, clustered here by chromosome number. Markers above the black line appeared to be significantly associated with the disease.

- What does it mean if an individual has 3, 5, 9, or 30 risk alleles for a particular condition?

- How do we categorize rare, common, and low-frequency risk alleles to determine the overall risk for developing a disease?

- GWAS often reveal dozens of DNA variations, but many variations have only a modest effect on risk. How does one explain to a person that he or she has a gene variation that changes a risk difference for a particular disease from 12 to 16 percent over an individual's lifetime? What does this information mean?

- If the sum total of GWAS for a particular condition reveals about 50 percent of the risk alleles, what are the other missing elements of heritability that may contribute to developing a complex disease?

In some cases, risk data revealed by GWAS may help patients and physicians develop diet and exercise plans designed to minimize the potential for developing a particular disease. But the number of risk genes identified by most GWAS is showing us that, unlike single-gene disorders, complex genetic disease conditions involve a multitude of genetic factors contributing to the total risk for developing a condition. We need such information to make meaningful progress in disease diagnosis and treatment, which is ultimately a major purpose of what GWAS are all about.

ESSENTIAL POINT ■ ■ ■
Genome-wide association studies can reveal genetic variations linked with disease conditions within populations.

7 Gene Therapy Approaches for Treating Genetic Diseases

Genomic technologies are changing medical diagnosis and allowing scientists to manufacture abundant and effective therapeutic proteins. The examples already available today are a strong indication that in the near future, we will see even more transformative medical treatments based on genomics and advanced DNA-based technologies. Here we consider gene therapy approaches to treat genetic diseases.

Gene Therapy

Although drug treatments are often effective in controlling symptoms of genetic disorders, the ideal outcome of medical treatment is to cure these diseases. In an effort to cure genetic diseases, scientists are actively investigating **gene therapy gene therapy**—a therapeutic technique that aims to transfer normal genes into a patient's cells. In theory, the normal genes will be transcribed and translated into functional gene products, which, in turn, will bring about a normal phenotype.

One key to gene therapy is having a delivery system to transfer genes into a patient. In many gene therapy trials, scientists often used genetically modified retroviruses as vectors. An example is a vector based on a mouse virus called **Moloney murine leukemia virus (MLV).** Disabled forms of adeno-associated virus (AAV), which in its native form infects ~80–90 percent of the population during childhood, and nonviral methods are being used to transfer genes into cells. Retroviral vectors are created by removing a cluster of three genes from the virus and inserting a cloned human gene. After being packaged in a viral protein coat, the vector is used to infect cells. Once inside a cell, the virus cannot replicate because of the missing viral genes. In the cell, the recombinant virus with the inserted human gene moves to the nucleus, integrates into a site on a chromosome, and becomes part of the genome. If the inserted gene is expressed, it produces a normal gene product that may be able to correct the mutation carried by the affected individual.

Human gene therapy began in 1990 with the treatment of a young girl named Ashanti DeSilva [**Figure 18(a)**], who has a heritable disorder called **severe combined immunodeficiency (SCID)**. Individuals with SCID have no functional immune system and usually die from what would normally be minor infections. Ashanti has an autosomal form of SCID caused by a mutation in the gene encoding the enzyme **adenosine deaminase (ADA).** Her gene therapy began when clinicians isolated some of her white blood cells, called T cells [**Figure 18(b)**]. These cells, which are key components of the immune system, were mixed with a retroviral vector carrying an inserted copy of the normal *ADA* gene. The virus infected many of the T cells, and a normal copy of the *ADA* gene was inserted into the genome of some T cells. After being mixed with the vector, the T cells were grown in the laboratory and analyzed to make sure that the transferred *ADA* gene was expressed (Figure 18). Then a billion or so genetically altered T cells were injected into Ashanti's bloodstream. Some of these T cells migrated to her bone marrow and began dividing and producing daughter cells that also produce ADA. She now has ADA protein expression in 25 to 30 percent of her T cells, which is enough to allow her to lead a normal life.

To date, gene therapy has successfully restored the health of about 20 children affected by SCID. Although gene therapy was originally developed as a treatment for single-gene inherited diseases, the technique was quickly adapted for the treatment of acquired diseases such as cancer, neurodegenerative diseases, cardiovascular disease, and infectious diseases, such as HIV. There are nearly 1000 gene therapy trials actively underway in the United States alone. Over a 10-year period, from 1990 to 1999, more than 4000 people underwent gene therapy for a variety of genetic disorders.

FIGURE 18 (a) Ashanti DeSilva, the first person to be treated by gene therapy. (b) To treat SCID using gene therapy, a cloned human *ADA* gene is transferred into a viral vector, which is then used to infect white blood cells removed from the patient. The transferred *ADA* gene is incorporated into a chromosome and becomes active. After growth to enhance their numbers, the cells are inserted back into the patient, where they produce ADA, allowing the development of an immune response.

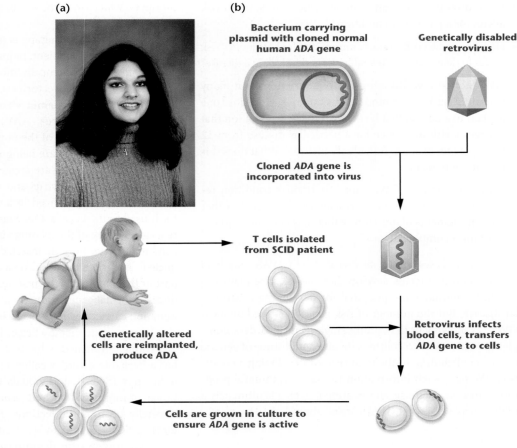

(a)

(b)

Bacterium carrying plasmid with cloned normal human *ADA* gene

Genetically disabled retrovirus

Cloned *ADA* gene is incorporated into virus

T cells isolated from SCID patient

Genetically altered cells are reimplanted, produce ADA

Retrovirus infects blood cells, transfers *ADA* gene to cells

Cells are grown in culture to ensure *ADA* gene is active

These trials often failed and thus led to a loss of confidence in gene therapy.

Hopes for gene therapy plummeted even further in September 1999 when teenager Jesse Gelsinger died while undergoing gene therapy to treat a liver disease condition. His death was triggered by a massive inflammatory response to the vector, a modified **adenovirus**, one of the viruses that cause colds and respiratory infections. Large numbers of adenovirus vectors bearing the *ornithine transcarbamylase* (*OTC*) gene were injected into his hepatic artery. The viral vectors were expected to lodge in his liver, enter the liver cells, and trigger the production of OTC protein. In turn, the OTC protein might correct his genetic defect and perhaps cure him of his liver disease. However, within hours of his first treatment, a massive immune reaction surged through Jesse's body. He developed a high fever, his lungs filled with fluid, multiple organs shut down, and he died four days later of acute respiratory failure.

In the aftermath of the tragedy, several government and scientific inquiries were conducted. Investigators learned that clinical trial scientists had not reported other adverse reactions to gene therapy and that some of the scientists were affiliated with private companies that could benefit financially from the trials. They found that serious side effects seen in animal studies were not explained to patients during informed-consent discussions, and that some clinical trials were proceeding too quickly in the face of data suggesting a need for caution. The U.S. Food and Drug Administration (FDA) scrutinized gene therapy trials across the country, halted a number of them, and shut down several gene therapy programs. Other research groups voluntarily suspended their gene therapy studies. Tighter restrictions on clinical trial protocols were imposed to correct some of the procedural problems that emerged from the Gelsinger case. Jesse's death had dealt a severe blow to the struggling field of gene therapy—a blow from which it was still reeling when a second tragedy hit.

The outlook for gene therapy brightened in 2000, when a group of French researchers reported the first large-scale success in gene therapy. Nine children with a fatal X-linked form of SCID developed functional immune systems after being treated with a retroviral vector carrying a normal gene. Published reports of the study were greeted with enthusiasm by the gene therapy community. But elation turned to despair in 2003, when it became clear that 2 of the 10 children who had been cured of X-SCID had developed leukemia as a direct result of their therapy, and one died as a result of the treatment. In two of the children, their cancer cells contained the retroviral vector, inserted near or into a gene called *LMO2*. This insertion activated the *LMO2* gene, causing uncontrolled white blood cell proliferation and development of leukemia. The FDA immediately halted 27 similar gene therapy clinical trials, and once again gene therapy underwent a profound reassessment. In 2005, a third child in the French X-SCID study developed leukemia, likely as a result of gene therapy.

Barriers to Gene Therapy as a Reliable Treatment

Up until the apparent success of the French X-SCID clinical trials, gene therapy had suffered not only from the scandals and scrutiny that emerged from Jesse Gelsinger's death but also from the skepticism of many scientists and the general public about the feasibility of this much-ballyhooed therapeutic technique. To date, no human gene therapy product has been approved for sale. Critics of gene therapy continue to berate research groups for undue haste, conflicts of interest, and sloppy clinical trial management, and for promising much but delivering little. Most problems associated with gene therapy have been traced to the vectors used to transfer therapeutic genes into cells.

These vectors, including MLV and adenovirus, have several serious drawbacks. First, integration of retroviral genomes (including the human therapeutic gene) into the host cell's genome occurs only if the host cells are replicating their DNA. In the body, only a small number of cells in any tissue are dividing and replicating their DNA. Second, most viral vectors are capable of causing an immune response in the patient, as happened in Jesse Gelsinger's case. Third, insertion of viral genomes into host chromosomes can activate or mutate an essential gene, as in the case of the three French patients. Viral integrase, the enzyme that allows for viral genome integration into the host genome, interacts with chromatin-associated proteins, often steering integration toward transcriptionally active genes. Unfortunately, it is not yet possible to reliably target insertion of therapeutic genes into specific locations in the genome, but targeted gene delivery is a major area of active research. Finally, there is a possibility that a fully infectious virus could be created if the vector were to recombine with another viral genome already present in the host cell.

To overcome these problems, new viral vectors and strategies for transferring genes into cells are being developed in an attempt to improve the action and safety of vectors. Researchers hope that the use of new gene delivery systems will circumvent the problems inherent in earlier vectors, as well as allow regulation of both insertion sites and the levels of gene product produced from the therapeutic genes.

In addition to the vector delivery issues addressed above, a number of other barriers must be overcome if gene therapy is to become a viable approach for reliably treating many genetic disorders. Issues include the following:

- What is the proper route for gene delivery in different kinds of disorders? For example, what is the best way to treat brain or muscle tissues? Tissue-specific gene delivery approaches are key.

- What percentage of cells in an organ or tissue need to express a therapeutic gene to alleviate the effects of a genetic disorder?

- What amount of a therapeutic gene product must be produced to provide lasting improvement of the condition, and how can sufficient production be ensured? Currently many gene therapy approaches provide only short-lived delivery of the therapeutic gene and its protein.

- Will it be possible to use gene therapy to treat diseases that involve multiple genes?

- Can expression of therapeutic genes be controlled in a patient?

Several well-publicized recent studies have been encouraging. For example, University of Pennsylvania and Children's Hospital of Philadelphia researchers reported beneficial results of treatments for Leber's congenital amaurosis (LCA), a degenerative disease of the retina that affects 1 in 50,000 to 1 in 100,000 infants each year and causes severe blindness. Young adult patients with defects in the *RPE65* gene were treated using AAV to deliver the normal gene. Similar human trial results were reported in studies in Italy and the United Kingdom, where human trials for LCA first began. While eyesight has improved in over two dozen patients, complete vision was not restored. Several months after a single treatment of the gene, the patients are still legally blind, but they can see more light, some of them can read lines of an eye chart, and two who had stumbled through an obstacle course were able to navigate through it.

Researchers at the University of Paris and Harvard Medical School have reported that after two years of gene therapy treatment for β-thalassemia, a blood disorder that involves a defect in the β-globin chain of hemoglobin which reduces the production of hemoglobin, a young man no longer needs transfusions and appears to be healthy. A modified, disabled HIV was used to carry a copy of the normal gene, although there have been reports of therapeutic gene integration near a growth factor gene called *HMGA2* resulting in activation of this gene, reminiscent of what occurred in the French X-SCID trials.

Gene Editing and Gene Silencing Approaches to Gene Therapy

Scientists are also working on gene replacement approaches that involve removing a defective gene from the genome. Recent work with enzymes called **zinc-finger nucleases** have shown promise in animal models and cultured cells. These enzymes can create site-specific cleavage in the genome and when coupled with certain integrases may lead to gene editing by cutting out defective sequences and introducing normal homologous sequences into the genome. Encouraging breakthroughs have taken place in this area using model organisms such as mice; however, this technology has not advanced sufficiently for use in humans. Attempts have been made to use **antisense oligonucleotides** in order to inhibit translation of mRNAs from defective genes, but this approach to gene therapy has generally not yet proven to be reliable.

The recent emergence of RNA interference as a powerful gene-silencing tool has reinvigorated gene therapy approaches by gene silencing. **RNA interference (RNAi)** is a form of gene-expression regulation in which **small interfering RNAs (siRNAs)** can block siRNA-bound mRNAs from being translated into protein or lead to degradation of siRNA-bound mRNAs so they cannot be translated into protein.

A main challenge to RNAi-based therapeutics so far has been *in vivo* delivery of double-stranded RNA or siRNA. RNAs degrade quickly in the body. It is also hard to get them to penetrate cells and to target the right tissue. For example, how does one deliver RNA-based therapies to cancer cells but not to noncancerous, healthy cells? Two common delivery approaches are to inject the siRNA directly or to deliver them via a plasmid vector that is taken in by cells and transcribed to make double-stranded RNA that can be cleaved by Dicer into siRNAs.

Several RNAi clinical trials to treat blindness are underway in the United States. One RNAi strategy to treat a form of blindness called macular degeneration targets a gene called *VEGF*. The VEGF protein promotes blood vessel growth. Overexpression of this gene, causing excessive production of blood vessels in the retina, leads to impaired vision and eventually blindness. Many expect that this disease will soon become the first condition to be treated by RNAi therapy. Other disease candidates for treatment by RNAi include several different cancers, diabetes, multiple sclerosis, and arthritis.

The question remains whether gene therapy can ever recover from past setbacks and fulfill its promise as a cure for genetic diseases. Perhaps we should view gene therapy as we have antibiotics, organ transplants, and manned space travel. There will be setbacks and even tragedies, but step by small step, we will move toward a technology that could—someday—provide cures for many severe genetic diseases.

ESSENTIAL POINT ■ ■ ■
Gene therapy involves delivering therapeutic genes or inhibiting expression of defective genes for treating genetic disorders.

NOW SOLVE THIS

4 Gene therapy for human genetic disorders involves transferring a copy of the normal human gene into a vector and using the vector to transfer the cloned human gene into target tissues. Presumably, the gene enters the target tissue and becomes active, and the gene product relieves the symptoms.

a. Why are disorders such as muscular dystrophy difficult to treat by gene therapy?
b. What are the potential problems of using retroviruses as vectors?
c. Should gene therapy involve germ-line tissue instead of somatic tissue? What are some of the potential ethical problems associated with the former approach?

■ **Hint** *This problem asks you to think about potential challenges associated with gene therapy for muscular dystrophy. The key to its solution is to consider the type of tissues affected in muscular dystrophy patients.*

8 Genetic Engineering, Genomics, and Biotechnology Create Ethical, Social, and Legal Questions

Geneticists use recombinant DNA and genomic technologies to identify genes, diagnose and treat genetic disorders, and produce commercial and pharmaceutical products among many other applications. However, these technologies raise important ethical, social, and legal issues that must be identified, debated, and resolved. Here we present a brief overview of some current ethical debates concerning the uses of genetic technologies.

Concerns about Genetically Modified Organisms and GM Foods

Most GM food products contain an introduced gene encoding a protein that confers a desired trait (for example, herbicide resistance or insect resistance). Much of the concern over GM plants centers on issues of consumer safety and environmental consequences. Are GM plants safe to eat? In general, if the proteins are not found to be toxic or allergenic and do not have other negative physiological effects, they are not considered to be a significant hazard to health. In Europe and Asia, labeling of food containing genetically modified ingredients is mandatory. But in the United States, such labeling is not required at the present time, and foods with less than 5 percent of their content from genetically modified organisms (GMOs) can be labeled as GMO-free. Certified organic foods, as designed by the USDA, must be GMO-free.

Environmental concerns generally have to do with any risks posed by releasing genetically modified organisms into the environment. Environmental risks include possible gene transfer by cross breeding with wild plants, toxicity, and invasiveness of the modified plant, resulting in loss of natural species (loss of biodiversity). Although laboratory and field studies suggest that cross-pollination and gene transfer can occur between some genetically engineered plants and wild relatives, there is little evidence that this has occurred in nature. If, for example, glyphosate resistance was transferred from cultivated plants such as canola into wild relatives, the herbicide-resistant weeds could make herbicide treatment ineffective. Biotechnology companies have engineered transgenic plants into sterile forms that are unable to transfer their genes into other plants. Built-in sterility was also designed to ensure that farmers could not produce their own seed from genetically modified crops, guaranteeing that biotechnology companies would have exclusive distribution of each year's crop. This, in itself, is an ethical issue, particularly in underdeveloped countries with limited resources to purchase genetically modified seeds.

Genetic Testing and Ethical Dilemmas

When the Human Genome Project was first discussed, scientists and the general public raised concerns about how genome information would be used and how the interests of both individuals and society can be protected. To address these concerns, the **Ethical, Legal, and Social Implications (ELSI) Program** was established in association with the Human Genome Project. The ELSI Program considers a range of issues, including the impact of genetic information on individuals, the privacy and confidentiality of genetic information, and implications for medical practice, genetic counseling, and reproductive decision making.

Many of the potential benefits and consequences of genetic testing are not always clear. For example,

- We have the technologies to test for genetic diseases for which there are no effective treatments. But *should* we test people for these disorders?

- With present genetic testing technologies, a negative result does not necessarily rule out future development of a disease; nor does a positive result always mean that an individual will get the disease. How can we effectively communicate the results of testing and the actual risks to those being tested?

- What information should people have before deciding to have a genome scan or a genetic test for a single disorder?

- How can we protect the information revealed by such tests?

- Since sharing of patient data through electronic medical records is a significant concern, what issues of consent need to be considered?

- How can we define and prevent genetic discrimination?

Earlier in this chapter we discussed preimplantation genetic diagnosis (PGD). As we learn more about genes involved in human traits, will other, nondisease-related genes

be screened for by PGD? Will couples be able to select embryos with certain genes encoding desirable traits for height, weight, intellect, and other physical or mental characteristics? What do you think of using genetic testing to purposely select for an embryo with a genetic disorder? Recently there have been several well-publicized cases of couples seeking to use prenatal diagnosis or PGD to select for embryos with dwarfism and deafness.

As identification of genetic traits becomes more routine in clinical settings, physicians will need to ensure genetic privacy for their patients. There are significant concerns about how genetic information could be used in negative ways by employers, insurance companies, governmental agencies, or the general public. In 2008, the **Genetic Information Nondiscrimination Act** was signed into law in the United States. This legislation is designed to prohibit the improper use of genetic information in health insurance and employment.

Direct-to-Consumer Genetic Testing

The past decade has seen dramatic developments in **direct-to-consumer (DTC) genetic tests**. A simple Web search will reveal many companies offering DTC genetic tests. There are approximately 1900 diseases for which such tests are now available (in 1993 there were about 100 such tests). Most DTC tests require that a person mail a saliva sample, hair sample, or cheek cell swab to the company. For a range of pricing options, DTC companies largely use SNP-based tests such as ASO tests to screen for different mutations. For example, in 2007, Myriad Genetics, Inc. began a major DTC marketing campaign of its tests for *BRCA1* and *BRCA2*. Mutations in these genes increase risk of developing breast and ovarian cancer. DTC testing companies report absolute risk, the probability that an individual will develop a disease, but how such risk results are calculated is highly variable and subject to certain assumptions.

Such tests are controversial for many reasons. For example, the test is purchased online by individual consumers and requires no involvement of a physician or other health-care professionals such as a nurse or genetic counselor to administer or to interpret results. There are significant questions about the quality, effectiveness, and accuracy of such products because currently the DTC industry is largely self-regulated. The FDA does not regulate DTC genetic tests. There is at present no comprehensive way for patients to make comparisons and evaluations about the range of tests available and their relative quality.

Most companies make it clear that they are not trying to diagnose or prevent disease, nor that they are offering health advice, so what is the purpose of the information that test results provide to the consumer?

Web sites and online programs from DTC companies provide information on what advice a person should pursue if positive results are obtained. But is this enough? If results are not understood, might negative tests not provide a false sense of security? Just because a woman is negative for *BRCA1* and *BRCA2* mutations *does not* mean that one cannot develop breast or ovarian cancer.

The National Institutes of Health plans to create the **Genetic Testing Registry (GTR)** designed to increase transparency by publicly sharing information about the utility of their tests, research for the general public, patients, health-care workers, genetic counselors, insurance companies, and others. The GTR is intended to allow individuals and families access to key resources to make better informed decisions about their health and genetic tests. But participation in the GTR by DTC companies has not been made mandatory yet, so will companies involved in genetic testing participate?

DNA and Gene Patents

Intellectual property (IP) rights are also being debated as an aspect of the ethical implications of genetic engineering, genomics, and biotechnology. Patents on intellectual property (isolated genes, new gene constructs, recombinant cell types, GMOs) can be potentially lucrative for the patent-holders but may also pose ethical and scientific problems. Why is protecting IP important for companies? Consider this issue. If a company is willing to spend millions or billions of dollars and several years doing research and development (R&D) to produce a valuable product, then shouldn't it be afforded a period of time to protect its discovery so that it can recover R&D costs and make a profit on its product?

Genes in their natural state cannot be patented. Consider the possibilities for a human gene that has been cloned and then patented by the scientists who did the cloning. The person or company holding the patent could require that anyone attempting to do research with the patented gene pay a licensing fee for its use. Should a diagnostic test or therapy result from the research, more fees and royalties may be demanded, and, as a result the costs of a genetic test may be too high for many patients to afford. But limiting or preventing the holding of patents for genes or genetic tools could reduce the incentive for pursuing the research that produces such genes and tools, especially for companies that need to profit from their research. Should scientists and companies be allowed to patent DNA sequences from naturally living organisms? And should there be a lower or an upper limit to the size of those sequences? For example, should patents be awarded for small pieces of genes, just because some individual or company wants to claim a stake in having cloned a piece of DNA first, even if no one knows whether the DNA sequence has a use? Can or should investigators be allowed to patent the entire genome of any organism they have sequenced?

To date the U.S. Patent and Trademark Office has granted patents for more than 35,000 genes or gene sequences, including an estimated 20 percent of human genes.

Incidentally the patenting of human genes has led some to use the term *patentome*! Some scientists are concerned that to award a patent for simply cloning a piece of DNA is awarding a patent for too little work. Given that computers do most of the routine work of genome sequencing, who should get the patent? What about individuals who figure out *what* to do with the gene? What if a gene sequence has a role in a disease for which a genetic therapy may be developed? Many scientists believe that it is more appropriate to patent novel technology and applications that make use of gene sequences than to patent the gene sequences themselves.

Congress is considering legislation that would ban the patenting of human genes and any sequences, functions, or correlations to naturally occurring products from a gene. The patenting of genetic tests is also under increased scrutiny in part because of concerns that a patented test can create monopolies in which patients cannot get a second opinion if only one company holds the rights to conduct a particular genetic test. Recent analysis has estimated that as many as 64 percent of patented tests for disease genes make it very difficult or impossible for other groups to propose a different way to test for the same disease.

In 2010, a landmark case brought by the American Civil Liberties Union against Myriad Genetics contended that Myriad could not patent the *BRCA1* and *BRCA2* sequences used to diagnose breast cancer. The U.S. Court of Appeals for the Federal Circuit ruled that isolated DNA sequences can be patented because these molecules do not exist in this same form in nature. This decision reversed a controversial decision by a District Court judge in 2010 who ruled Myriad's patents invalid on the basis that DNA in an isolated form is not fundamentally different from how it exists in the body.

Patents and Synthetic Biology

The J. Craig Venter Institute (JCVI) has filed two patent applications for what is being called "the world's first-ever human-made life form." The patents are intended to cover the minimal genome of *M. genitalium*, which JCVI believes are the genes essential for self-replication. One of these patent applications is designed to claim the rights to synthetically constructed organisms. Another U.S. patent recently issued to another group of researchers covers application of a minimal genome for *E. coli*, which has generated even more concern given its relative importance compared with *M. genitalium*. What do you think? Should it be possible to patent a minimal genome or a synthetic organism?

ESSENTIAL POINT ■ ■ ■

Applications of genetic engineering and biotechnology involve a wide range of ethical, legal, and social dilemmas with important scientific and societal implications.

GENETICS, TECHNOLOGY, AND SOCIETY

Personal Genome Projects and the Race for the $1000 Genome

The reference sequence created by the Human Genome Project was not derived from one person's DNA, but from a composite of DNA samples from numerous anonymous donors.

When the Human Genome Project was completed, the goal of routinely sequencing the genomes of individual humans seemed remote. However, the development of high-throughput sequencing technologies, capable of generating long sequence reads at high speeds with great accuracy, has reduced the cost of DNA sequencing dramatically. And now, the race is on to sequence a complete human genome for $1000!

The $1000 genome is coming closer to reality. In 2009, Stanford University professor Stephen Quake announced that he had sequenced his entire genome in a few weeks for under $50,000, using a new sequencing technology and a single sequencing machine. One year later, two California companies (Illumina and Life Technologies) offered sequencing instruments for sale that were capable of sequencing a human genome in one day for less than $6000. Then, in early 2012, Life Technologies announced that they would soon have the technology to decode an entire human genome in 24 hours at a cost approaching $1000.

The ability to provide low-cost and rapid genome sequencing will be a key step toward allowing genetic sequencing into routine medical diagnosis. However, an equally important feature of any medically useful $1000 genome is a high level of accuracy. In 2011, the X Prize Foundation set up a competition, worth $10 million, for any group able to sequence 100 human genomes in 30 days or less, at a cost of $1000 per genome sequence, and with an accuracy of no more than one error for every 1 million nucleotides.

As the $1000 genome approaches, scientists are making plans for even more ambitious projects. An international research consortium has initiated the "1000 Genomes Project," which aims to sequence the genomes of 1000 volunteers from various backgrounds including African, Asian, and European. The goal is to produce an extensive catalog of human DNA sequence variation. George Church of Harvard University and his colleagues have started the Personal Genome Project—a project that aims to sequence the genomes of 100,000 individuals. Volunteers for the Personal Genome Project must provide their DNA samples on the understanding that their genome data will be made publicly available. Church's genome has already been made available online. The goal of the project is to correlate genome sequences with phenotypic characteristics, from height and hair color to disease predisposition.

The race continues to lower the cost and time to accurately sequence an entire human genome. Ready or not, we are entering the age of personal genomics.

Your Turn

1. Would you have your genome sequenced, if the price was affordable? Why, or why not?

   *You can find a discussion of the pros and cons of knowing your own genome sequence in a series of articles under the heading of "My Genome. So What?" (*Nature* 456: 1, 2008).*

2. Would you make your genome sequence publicly available? How might such information be misused?

   The issue of genetic privacy is particularly important when considering making genomic sequences available to all. Read about the potential uses and misuses of genome sequences in: Taylor, P. 2008. When consent gets in the way. *Nature* 456: 32–33.

3. Private companies are now offering personal DNA sequencing along with interpretation. What services do they offer? Do you think that these services should be regulated, and if so, in what way?

   Investigate one such company, 23andMe, at http://www.23andMe.com. *You can read about the pros and cons of direct-to-consumer sequencing companies in* Ng, P.C. et al. 2010. An agenda for personalized medicine. *Nature* 461: 724–726.

CASE STUDY » A first for gene therapy

As discussed in this chapter, a child was born without a functional immune system and became the first person to undergo gene therapy. In an attempt to treat her autosomal recessive disorder, known as severe combined immunodeficiency (SCID), cloned copies of the gene encoding the missing enzyme (adenosine deaminase, or ADA) were inserted into some of her white blood cells, which were injected back into her bloodstream. Expression of the normal ADA allele led to the development of a functional immune system, allowing the child to lead a normal life. An understanding of this spectacular success depends on knowing several details of this process:

1. Is it important that the cloned gene becomes part of a chromosome when inserted into a cell?
2. Does the cloned gene replace the defective copy of the ADA gene?
3. Why were white blood cells chosen as targets for the transferred genes?
4. Would you expect that production of 50 percent of the normal levels of ADA would be enough to restore immune function?

INSIGHTS AND SOLUTIONS

1. Infection by HIV-1 (human immunodeficiency virus) weakens the immune system and results in the symptoms of AIDS (acquired immunodeficiency syndrome). Specifically, HIV infects and kills cells of the immune system that carry a cell-surface receptor known as CD4. An HIV surface protein known as gp120 binds to the CD4 receptor and allows the virus to enter the cell. The gene encoding the CD4 protein has been cloned. How might this clone be used along with recombinant DNA techniques to combat HIV infection?

Solution: Researchers hope that clones of the *CD4* gene can be used in the design of systems for the targeted delivery of drugs and toxins to combat the infection. For example, because infection depends on an interaction between the viral gp120 protein and the CD4 protein, the cloned *CD4* gene has been modified to produce a soluble form of the protein (sCD4) that, because of its solubility, would circulate freely in the body. The idea is that HIV might be

prevented from infecting cells if the gp120 protein of the virus first encounters and binds to extra molecules of the soluble form of the CD4 protein. Once bound to the extra molecules, the virus would be unable to bind to CD4 proteins on the surface of immune system cells. Studies in cell-culture systems indicate that the presence of sCD4 effectively prevents HIV infection of tissue culture cells. However, studies in HIV-positive humans have been somewhat disappointing, mainly because the strains of HIV used in the laboratory are different from those found in infected individuals. In another strategy, the *CD4* gene has been fused with genes encoding bacterial toxins. The resulting fusion protein contains CD4 regions that should bind to gp120 on the surface of HIV-infected cells and toxin regions that should then kill the infected cell. In tissue culture experiments, cells infected with HIV are killed by this fusion protein, whereas uninfected cells survive.

PROBLEMS AND DISCUSSION QUESTIONS

 For activities, animations, and review quizzes, go to the study area at www.masteringgenetics.com

HOW DO WE KNOW?

1. In this chapter, we focused on a number of interesting applications of genetic engineering, genomics, and biotechnology. At the same time, we found many opportunities to consider the methods and reasoning by which much of this information was acquired. From the explanations given in the chapter, what answers would you propose to the following fundamental questions:

   (a) How can we determine experimentally if a transgene has been successfully introduced into a genetically modified plant?

   (b) When creating *M. mycoides* JCVI-syn1.0, how did JCVI scientists prove that the synthetic genome transplanted into *M. capricolum* had been successfully expressed?

   (c) What is the molecular basis for why RFLP analysis for sickle-cell anemia is effective as a genetic test?

   (d) How can DNA microarray analysis reveal host-response signatures that are associated with different pathogens?

   (e) What basic determination do geneticists use to consider if genetic markers analyzed by GWAS are associated with a disease?

   (f) What are some of the technical reasons why gene therapy is difficult to carry out effectively?

2. What are some of the reasons why GM crops are controversial? Describe some of the primary concerns that have been raised about GM foods.

3. Should the United States require mandatory labeling of all foods that contain GMOs? Explain your answer.

4. Why are most recombinant human proteins produced in animal or plant hosts instead of bacterial host cells?

5. There are more than 1000 cloned farm animals in the United States. In the near future, milk from cloned cows and their offspring (born naturally) may be available in supermarkets. These cloned animals have not been transgenically modified, and they are no different than identical twins. Should milk from such animals and their natural-born offspring be labeled as coming from cloned cows or their descendants? Why?

6. One of the major causes of sickness, death, and economic loss in the cattle industry is *Mannheimia haemolytica*, which causes bovine pasteurellosis, or shipping fever. Noninvasive delivery of

a vaccine using transgenic plants expressing immunogens would reduce labor costs and trauma to livestock. An early step toward developing an edible vaccine is to determine whether an injected version of an antigen (usually a derivative of the pathogen) is capable of stimulating the development of antibodies in a test organism. The following table assesses the ability of a transgenic portion of a toxin (Lkt) of *M. haemolytica* to stimulate development of specific antibodies in rabbits.

   (a) What general conclusion can you draw from the data?

   (b) With regards to development of a usable edible vaccine, what work remains to be done?

Immunogen Injected	Antibody Production in Serum
Lkt50*—saline extract	+
Lkt50—column extract	+
Mock injection	−
Pre-injection	−

*Lkt50 is a smaller derivative of Lkt that lacks all hydrophobic regions.
+ indicates at least 50 percent neutralization of toxicity of Lkt; − indicates no neutralization activity.
Source: Modified from Lee et al. 2001. *Infect. and Immunity* 69: 5786–5793.

7. Describe how the team from the J. Craig Venter Institute created a synthetic genome. How did they demonstrate that the genome converted the recipient strain of bacteria into a different strain?

8. Suppose you develop a screening method for cystic fibrosis that allows you to identify the predominant mutation Δ508 and the next six most prevalent mutations. What must you consider before using this method to screen a population for this disorder?

9. Sequencing the human genome and the development of microarray technology promises to improve our understanding of normal and abnormal cell behavior. How are microarrays dramatically changing our understanding of complex diseases such as cancer?

10. A couple with European ancestry seeks genetic counseling before having children because of a history of cystic fibrosis (CF) in the husband's family. ASO testing for CF reveals that the husband

is heterozygous for the Δ508 mutation and that the wife is heterozygous for the R117 mutation. You are the couple's genetic counselor. When consulting with you, they express their conviction that they are not at risk for having an affected child because they each carry different mutations and cannot have a child who is homozygous for either mutation. What would you say to them?

11. When genome scanning technologies become widespread, medical records will contain the results of such testing. Who should have access to this information? Should employers, potential employers, or insurance companies be allowed to have this information? Would you favor or oppose having the government establish and maintain a central database containing the results of individuals' genome scans?

12. What limits the use of differences in restriction enzyme sites as a way of detecting point mutations in human genes?

13. What is the main purpose of genome-wide association studies (GWAS)? How can information from GWAS be used to inform scientists and physicians about genetic diseases?

14. Recombinant adenoviruses have been used in a number of preclinical studies to determine the efficacy of gene therapy for rheumatoid arthritis and osteoarthritis. Genes can be delivered by injection to the tissues that need them. Christopher Evans and colleagues (2001. *Arthritis Res.* 3: 142–146) estimated that approximately 20 percent of all human gene therapy trials have used adenoviruses for gene delivery. The death of a patient in 1999 after infusion of adenoviral vectors has caused concern. As you consider the use of viral vectors as therapy-delivery vehicles for human pathologies, what factors seem of paramount concern?

15. Provide examples of major barriers that need to be addressed if gene therapy is to become a safe and reliable treatment for genetic diseases.

16. The development of safe vectors for human gene therapy has been a goal since 1990. Among the problems associated with viral-based vectors is that many of the viruses (i.e., SV40) have transformation properties thought to be mediated by binding and inactivating gene products such as p53, retinoblastoma protein (pRB), and others. SV40-based vectors that are deficient in binding p53, pRB, and other proteins have been developed. Why would you specifically want to avoid inactivating p53, pRB, and related proteins?

17. Dominant mutations can be categorized according to whether they increase or decrease the overall activity of a gene or gene product. Although a loss-of-function mutation (a mutation that inactivates the gene product) is usually recessive, for some genes, one dose of the normal gene product, encoded by the normal allele, is not sufficient to produce a normal phenotype. In this case, a loss-of-function mutation in the gene will be dominant,

and the gene is said to be *haploinsufficient*. A second category of dominant mutation is the gain-of-function mutation, which results in a new activity or an increased activity or expression of a gene or gene product. The gene therapy technique currently used in clinical trials involves the "addition" to somatic cells of a normal copy of a gene. In other words, a normal copy of the gene is inserted into the genome of the mutant somatic cell, but the mutated copy of the gene is not removed or replaced. Will this strategy work for either of the two aforementioned types of dominant mutations?

18. In mice transfected with the rabbit β-globin gene, the rabbit gene is active in several tissues, including the spleen, brain, and kidney. In addition, some transfected mice suffer from thalassemia (a form of anemia) caused by an imbalance in the coordinate production of α- and β-globins. Which problems associated with gene therapy are illustrated by these findings?

19. The Genetic Testing Registry is intended to provide better information to patients, but companies involved in genetic testing are not required to participate. Should company participation be mandatory? Why or why not? Explain your answers.

20. Should the FDA regulate direct-to-consumer genetic tests, or should these tests be available as a "buyer beware" product?

21. An unapproved form of gene therapy—which raises great ethical concerns—is termed **enhancement gene therapy**, whereby people may be "enhanced" for some desired trait. Should genetic technology be used to enhance human potential? For example, should it be permissible to use gene therapy to increase height, enhance athletic ability, or extend intellectual potential?

22. In March 2010, Judge R. Sweet ruled to invalidate Myriad Genetics' patents on the *BRCA1* and *BRCA2* genes. Sweet wrote that since the genes are part of the natural world, they are not patentable. Myriad Genetics also holds patents on the development of a direct-to-consumer test for the *BRCA1* and *BRCA2* genes. Judge Sweet's ruling was subsequently reversed.

(a) Would you agree with Judge Sweet's ruling to invalidate the patenting of the *BRCA1* and *BRCA2* genes? If you were asked to judge the patenting of the direct-to-consumer test for the *BRCA1* and *BRCA2* genes, how would you rule?

(b) J. Craig Venter has filed a patent application for his "first-ever human made life form." This patent is designed to cover the genome of *M. genitalium*. Would your ruling for Venter's "organism" be different from Judge Sweet's ruling on patenting of the *BRCA1* and *BRCA2* genes?

23. Visit the website PooPrints.com and read about their DNA fingerprinting approach. What do you think of this application of genetic technology? Does this application present any ethical dilemmas for you?

SOLUTIONS TO SELECTED PROBLEMS AND DISCUSSION QUESTIONS

Answers to Now Solve This

1. In many cases, the natural immune system of the host mounts a destructive attack against the introduction of foreign macromolecules. Thus, any injected plasmid or virus might just be negated before it can be effective. In addition, regulating the output of the introduced DNA is difficult. In most cases studied to date, the stability of the injected material and its regulation have been shortcomings.

2. The child in question is a carrier of the deletion in the β-globin gene, just as the parents are carriers. Its genotype is therefore $\beta^A\beta^o$.

3. It will hybridize by base complementation to the normal DNA sequence.

4. **(a, b)** One of the main problems with gene therapy is delivery of the desired virus to the target tissue in an effective manner. Several of the problems involving the use of retroviral vectors are the following: (1) Integration into the host must be cell specific so as not to damage nontarget cells. (2) Retroviral integration into host cell genomes only occurs if the host cell is replicating. (3) Insertion of the viral genome might influence nontarget, but essential, genes. (4) Retroviral genomes have a low cloning capacity and cannot carry large inserted sequences as are many human genes. (5) There is a possibility that recombination with host viruses will produce an infectious virus that may do harm.

(c) The question posed here plays on the practical versus the ethical. It would certainly be more efficient (although perhaps more difficult technically) to engineer germ tissue, for once it is done in a family, the disease would be eliminated in that family. However, there are considerable ethical problems associated with germ-line therapy. It recalls previous attempts of the eugenics movements of past decades, which involved the use of selective breeding to purify the human stock. Some present-day biologists have said publically that germ-line gene therapy will *not* be conducted.

Solutions to Problems and Discussion Questions

2. There are concerns about proper testing of GM crops and foods for allergenicity, environmental impact, and the possibility of cross-pollination leading to the contamination of native species. If certain crops become the standard and under the control of a few manufacturers, it is likely that the world's supply of genetic variability might be reduced. Concern would increase if such crops routinely contained antibiotic-resistant genetic markers and genes conferring toxicity to pests. A broader concern is that the design and patenting of crops might allow domination of the world food supply by a few companies.

4. In general, bacteria do not process eukaryotic proteins in the same manner as eukaryotes. Transgenic eukaryotes are more likely to correctly process eukaryotic proteins, thus increasing the likelihood of their normal biological activity.

6. (a) Both the saline and column extracts of Lkt50 appear to be capable of inducing at least 50 percent neutralization of toxicity when injected into rabbits.

   (b) In order for a successful edible vaccine to be developed, numerous hurdles must be overcome. First, the immunogen must be stably incorporated into the host plant hereditary material and the host must express only that immunogen. During feeding, the immunogen must be transported across the intestinal wall unaltered, or altered in such a way as to stimulate the desired immune response. There must be guarantees that potentially harmful byproducts of transgenesis have not been produced. In other words, broad ecological and environmental issues must be addressed to prevent a transgenic plant from becoming an unintended vector for harm to the environment or any organisms feeding on the plant (directly or indirectly).

8. Even though you have developed a method for screening seven of the mutations described, it is possible that negative results can occur when the person carries the gene for CF. In other words, the specific probes (or allele-specific oligonucleotides) that have been developed will not necessarily be useful for screening all mutant alleles. In addition, the cost-effectiveness of such a screening proposal would need to be considered.

10. Since both mutations occur in the CF gene, children who possess both alleles will suffer from CF. With both parents heterozygous, each child born will have a 25 percent chance of developing CF.

12. Using restriction enzyme analysis to detect point mutations in humans is a tedious trial-and-error process. Given the size of the human genome in terms of base sequences and the relatively low number of unique restriction enzymes, the likelihood of matching a specific point mutation, separate from other normal sequence variations, to a desired gene is low.

14. As with all therapies, the cure must be less hazardous than the disease. In the case of viral-mediated gene therapy, the antigenicity of the virus must not interfere with the delivery system; such antigenicity can cause inflammation or more severe immunologic responses. The duration of desired gene expression at the diseased site is an issue. Short-period expression may require repeated exposure to the vehicle, which may present undesired responses. For some diseases, local gene therapy through inhalation or injection may produce fewer side effects than systemic exposure. Adenoviruses appear to be particularly useful for gene therapy because they can infect nondividing cells and they can accept relatively large amounts of additional DNA (30 kb or more).

16. p53 and pRB are tumor suppressor proteins and are required by the cell to effectively monitor the cell cycle. Reduction in their activity would diminish normal cell cycle controls and most likely lead to cancer. It would be especially important if such viral-vectors are intended to treat cancer where cell cycle control is likely already compromised.

18. The two major problems described here are common concerns related to genetic engineering. The first is the localization of the introduced DNA into the target tissue and target location in the genome. Inappropriate targeting may have serious consequences. In addition, it is often difficult to control the output of introduced DNA. Genetic regulation is complicated and subject to a number of factors including upstream and downstream signals as well as various posttranscriptional processing schemes. Artificial control of these factors will prove difficult.

20. Given the use to which genetic tests are put and their extreme personal nature, it would seem that FDA regulation is one way to decrease the distribution of misinformation that may be vital to individuals and families.

22. (a, b) Since a gene is a product of the natural world, it does not conform to section 101 of U.S. patent laws which govern patentable matter. Since both the direct-to-consumer test for the *BRCA1* and *BRCA2* genes and Venter's "first-ever human made life form" are original in their process or development, they should be patentable.

CREDITS

Credits are listed in order of appearance.

Photo

CO, The Roslin Institute; F-1 (a) SIU BIOMED COMM/Custom Medical Stock Photo; F-2, From Doebley, J. Plant Cell, 2005 Nov; 17(11): 2859-72. Courtesy of John Doebley/University of Wisconsin; F-4, Reprinted with permission from "New Genes Boost Rice Nutrients" by I. Potrykus and P. Beyer, Science, August 13, 1999, Vol. 285, pp. 994; F-05 (a) University of California Irvine University of California, Irvine Transgenic Mouse Facility; F-05 (b) R. L. Brinster, School of Veterinary Medicine, University of Pennsylvania; F-6, Garth Fletcher; F-8, GloFish; F-9 (b1, b2) Craig J. Venter; F-13,

Affymetrix, Inc.; F-14 © Cancer Genetics Branch/National Human Genome Research Institute/NIH; F-15, Reprinted with permission from Ash Alizadeh, *Nature Magazine* 2000, 403, pp. 503-511, figure 4 left panel. © 2000 Macmillan Magazines. F-16.1, Sebastian Kaultizki/Shutterstock.com; F-16.2, C. D. Humphrey and T. G. Ksiazek, CDC; F-16.3, Janice Haney Carr, CDC; F-16.4, C.D. Humphrey, T.G. Ksiazek, CDC; F-16.5, Affymetrix, Inc.; F-16.6, From Lorence, M. C. Application of High-Density Microarrays for Expression and Genotype Analysis in Infectious Disease, American Biotechnology Laboratory, pp. 10–12, January 2006. Figure, p. 10 Matthew Lorence, Ph.D., M.B.A.; F-18 (a), Van De Silva

Text

Developmental Genetics

From Chapter 20 of *Essentials of Genetics, Eighth edition,* William S. Klug, Michael R. Cummings, Charlotte A. Spencer, Michael A. Palladino. © 2013 by Pearson Education, Inc. All rights reserved.

This unusual four-winged *Drosophila* has developed an extra set of wings as a result of a mutation in a homeotic selector gene.

Developmental Genetics

CHAPTER CONCEPTS

- Gene expression during development is based on the differential transcription of selected genes.

- Animals use a small number of shared signaling systems and regulatory networks to construct a wide range of adult body forms from the zygote. These shared properties make it possible to use animal models to study human development.

- Differentiation is controlled by cascades of gene expression that are a consequence of events that specify and determine the developmental fate of cells.

- Plants independently evolved developmental regulatory mechanisms that parallel those of animals.

- In many organisms, cell–cell signaling systems program the developmental fate of adjacent and distant cells.

Over the last two decades, genetic analysis, molecular biology, and genomics have shown that, in spite of wide diversity in the size and shape of adult animals and plants, multicellular organisms share many genes, genetic pathways, and molecular signaling mechanisms that control developmental events leading from the zygote to the adult. At the cellular level, development is marked by three important events: **specification,** when the first cues confer spatially distinct identity, **determination,** the time when a specific developmental fate for a cell becomes fixed, and **differentiation,** the process by which a cell achieves its final adult form and function. Thanks to newly developed methods of analysis including microarrays, high-throughput sequencing, epigenetics, proteomics, and systems biology, we are beginning to understand how the action and interaction of genes and environmental factors control developmental processes in eukaryotes.

In this chapter, the primary emphasis will be on how genetic analysis has been used to study development. This field, called developmental genetics, laid the foundation for our understanding of developmental events at the molecular and cellular levels, which contribute to the continually changing phenotype of the newly formed organism.

1 Evolutionary Conservation of Developmental Mechanisms Can Be Studied Using Model Organisms

Genetic analysis of development across a wide range of organisms has shown that the size and shape of all animal bodies are controlled by a common set of genes and developmental mechanisms. For example, most of the differences in shape between zebras and zebrafish are the result of different patterns of expression in a single gene set, called the homeotic (abbreviated as *Hox*) genes, and not by expression of a host of species-specific genes. Genome-sequencing projects have confirmed that homeotic genes from a wide range of organisms have a common ancestry; this homology means that many aspects of normal human embryonic development and associated genetic disorders can be studied in model organisms such as *Drosophila,* where genetic methods including mutagenesis, genetic crosses, and large-scale experiments involving hundreds of offspring can be conducted .

Results from a relatively new field, called evo-devo, which combines evolutionary and developmental biology, have revealed that although many developmental mechanisms are common to all animals, evolution has generated several new and unique ways of transforming a zygote into an adult. These evolutionary changes result from several genetic mechanisms including mutation, gene duplication and evolutionary divergence, the assignment of new functions to old genes, and the recruitment of genes to new developmental pathways. However, the emphasis in this chapter will be on the similarities in genes and developmental mechanisms among species.

Analysis of Developmental Mechanisms

In the space of this chapter, we cannot survey all aspects of development, nor can we explore in detail how developmental mechanisms triggered by the fusion of sperm and egg were identified. Instead, we will focus on a number of general processes in development:

- how the adult body plan of animals is laid down in the embryo

- the program of gene expression that turns undifferentiated cells into differentiated cells

- the role of cell–cell communication in development

We will use three model systems—the fruit fly *Drosophila melanogaster,* the flowering plant *Arabidopsis thaliana,* and the nematode *Caenorhabditis elegans*—to illustrate these developmental processes and related topics. We will examine how patterns of differential gene expression lead to the progressive restriction of developmental options resulting in the formation of the adult body plan in *Drosophila* and *Arabidopsis.* We will then expand the discussion to consider how our knowledge of events in these organisms has contributed to our understanding of developmental defects in humans. Finally, we will consider the role of cell–cell communication in the development of adult structures in *C. elegans.*

2 Genetic Analysis of Embryonic Development in *Drosophila* Reveals How the Body Axis of Animals Is Specified

How does a given cell at a precise position in the embryo switch on or switch off specific genes at timed stages of development? This is a central question in developmental biology. To answer this question, we will examine the sequence of gene expression in the embryo of *Drosophila*. Although development in a fruit fly appears to have little in common with humans, recall that shared genes drive these steps in both species.

Overview of *Drosophila* Development

The life cycle of *Drosophila* is about 10 days long with a number of distinct phases: the embryo, three larval stages, the pupal stage, and the adult stage (**Figure 1**). Internally, the cytoplasm of the unfertilized egg is organized into a series of maternally constructed molecular gradients that play a key role in determining the developmental fates of nuclei located in specific regions of the embryo.

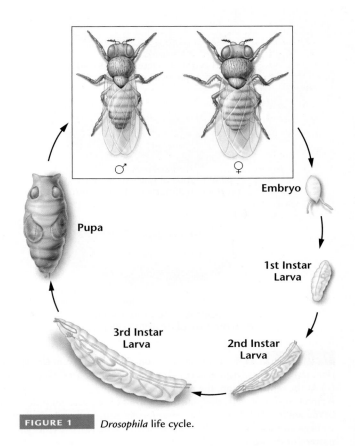

FIGURE 1 *Drosophila* life cycle.

Immediately after fertilization, the zygote nucleus undergoes a series of nuclear divisions without cytokinesis [**Figure 2(a)** and **(b)**]. The resulting cell, with multiple nuclei, is called a syncytium. At about the tenth division, nuclei move to the periphery of the egg, where the cytoplasm contains localized gradients of maternally derived mRNA transcripts and proteins [**Figure 2(c)**]. After several more divisions, the nuclei become enclosed in plasma membranes [**Figure 2(d)**], forming a cellular layer around the outside edge of the embryo. Interactions between the nuclei and the cytoplasmic components of these cells initiate and direct the pattern of embryonic gene expression.

Germ cells, which in the adult, are destined to undergo meiosis and produce gametes, form at the posterior pole of the embryo [Figure 2(c) and (d)]. Nuclei in other regions of the embryo normally form somatic cells. If nuclei from these regions are transplanted into the posterior cytoplasm, they will form germ cells and not somatic cells, confirming that the cytoplasm at the posterior pole of the embryo contains maternal components that direct nuclei to form germ cells.

Transcriptional programs activated by cytoplasmic components in somatic (non–germ-cell) nuclei form the embryo's anterior–posterior (front to back) and dorsal–ventral (upper to lower) axes of symmetry, leading to the formation of a segmented embryo [**Figure 2(e)**]. Under control of the *Hox* gene set (discussed in a later section), these segments will give rise to the differentiated structures of the adult fly [**Figure 2(f)**].

ESSENTIAL POINT ■ ■ ■

In *Drosophila*, both genetic and molecular studies have confirmed that the egg contains gradients of molecular information, which initiates a transcriptional cascade that specifies the body plan of the larva, pupa, and adult.

Genetic Analysis of Embryogenesis

Two different gene sets control embryonic development in *Drosophila*: maternal-effect genes and zygotic genes (**Figure 3**). Products of maternal-effect genes (mRNA and/or proteins) are placed in the developing egg by the "mother" fly.

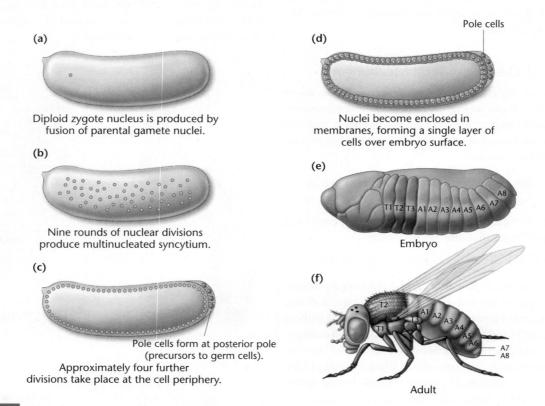

(a) Diploid zygote nucleus is produced by fusion of parental gamete nuclei.

(b) Nine rounds of nuclear divisions produce multinucleated syncytium.

(c) Pole cells form at posterior pole (precursors to germ cells). Approximately four further divisions take place at the cell periphery.

(d) Nuclei become enclosed in membranes, forming a single layer of cells over embryo surface.

Pole cells

(e) T1 T2 T3 A1 A2 A3 A4 A5 A6 A7 A8

Embryo

(f) T2 T3 T1 A1 A2 A3 A4 A5 A6 A7 A8

Adult

FIGURE 2 Early stages of embryonic development in *Drosophila*. (a) Fertilized egg with zygotic nucleus (2*n*), shortly after fertilization. (b) Nuclear divisions occur about every 10 minutes. Nine rounds of division produce a multinucleate cell, the syncytial blastoderm. (c) At the tenth division, the nuclei migrate to the periphery or cortex of the egg, and four additional rounds of nuclear division occur. A small cluster of cells, the pole cells, form at the posterior pole about 2.5 hours after fertilization. These cells will form the germ cells of the adult. (d) About 3 hours after fertilization, the nuclei become enclosed in membranes, forming a single layer of cells over the embryo surface, creating the cellular blastoderm. (e) The embryo at about 10 hours after fertilization. At this stage, the segmentation pattern of the body is clearly established. Behind the segments that will form the head, T1–T3 are thoracic segments, and A1–A8 are abdominal segments. (f) The adult fly showing the structures formed from each segment of the embryo.

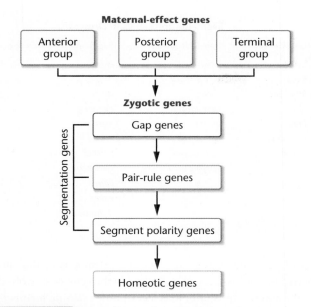

Maternal-effect genes

Anterior group | Posterior group | Terminal group

Zygotic genes

Segmentation genes:
- Gap genes
- Pair-rule genes
- Segment polarity genes

Homeotic genes

FIGURE 3 The hierarchy of genes involved in establishing the segmented body plan in *Drosophila*. Gene products from the maternal genes regulate the expression of the first three groups of zygotic genes (gap, pair-rule, and segment polarity, collectively called the segmentation genes), which in turn control expression of the homeotic genes.

Many of these products are distributed in a gradient or concentrated in specific regions of the egg cytoplasm. Female flies homozygous for deleterious recessive mutations of maternal-effect genes are sterile. None of their embryos receive wild-type maternal gene products, so all of the embryos develop abnormally and die. Maternal-effect genes encode transcription factors, receptors, and proteins that regulate gene expression. During development, these gene products activate or repress time- and location-specific programs of gene expression in the embryo.

Zygotic genes are those transcribed in the embryonic nuclei formed after fertilization. These products of the embryonic genome are differentially transcribed in specific regions of the embryo in response to the distribution of maternal-effect proteins. Recessive mutations in these genes can lead to embryonic lethality in homozygotes. In a cross between flies heterozygous for a recessive zygotic mutation, one-fourth of the embryos (the recessive homozygotes) fail to develop normally and die.

Much of our knowledge about the genes that regulate *Drosophila* development is based on the work of Ed Lewis, Christiane Nüsslein-Volhard, and Eric Wieschaus, who were awarded the 1995 Nobel Prize for Physiology or Medicine. Ed Lewis initially identified and studied one of these regulatory genes in the 1970s. In the late 1970s, Nüsslein-Volhard and Wieschaus devised a strategy to identify all the genes that control development in *Drosophila*. Their scheme required examining thousands of offspring of mutagenized flies, looking for recessive embryonic lethal mutations with defects in external structures. The parents were thus identified as heterozygous carriers of these mutations, which the researchers grouped into three classes: *gap, pair-rule,* and *segment polarity* genes. In 1980, on the basis of their observations, Nüsslein-Volhard and Wieschaus proposed a model in which embryonic development is initiated by gradients of maternal-effect gene products. The positional information laid down by these molecular gradients is interpreted by two sets of zygotic (embryonic) genes: (1) **segmentation genes** (gap, pair-rule, and segment polarity genes) and (2) **homeotic selector *(Hox)* genes.** Segmentation genes divide the embryo into a series of stripes or segments and define the number, size, and polarity of each segment. The homeotic genes specify the developmental fate of cells within each segment as well as the adult structures that will be formed from each segment (Figure 3).

Their model is shown in **Figure 4.** Most maternal-effect gene products placed in the egg during oogenesis are activated immediately after fertilization and help establish the anterior–posterior axis of the embryo by activating position-specific patterns of gene expression in the embryo's nuclei [Figure 4(a)]. Many maternal gene products encode

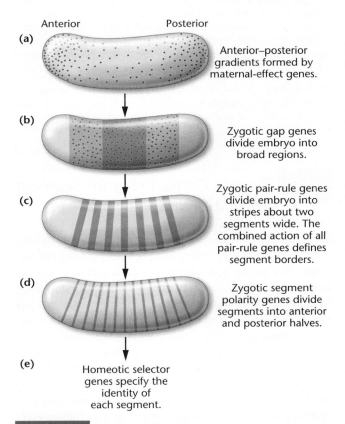

(a) Anterior / Posterior — Anterior–posterior gradients formed by maternal-effect genes.

(b) Zygotic gap genes divide embryo into broad regions.

(c) Zygotic pair-rule genes divide embryo into stripes about two segments wide. The combined action of all pair-rule genes defines segment borders.

(d) Zygotic segment polarity genes divide segments into anterior and posterior halves.

(e) Homeotic selector genes specify the identity of each segment.

FIGURE 4 (a) Progressive restriction of cell fate during development in *Drosophila*. Gradients of maternal proteins are established along the anterior-posterior axis of the embryo. (b), (c), and (d) Three groups of segmentation genes progressively define the body segments. (e) Individual segments are given identity by the homeotic genes.

transcription factors that activate gap genes, whose expression divides the embryo into a series of regions corresponding to the head, thorax, and abdomen of the adult [Figure 4(b)]. Gap genes encode other transcription factors that activate pair-rule genes, whose products divide the embryo into smaller regions about two segments wide [Figure 4(c)]. In turn, expression of the pair-rule genes activates the segment polarity genes, which divide each segment into anterior and posterior regions [Figure 4(d)]. The collective action of the maternal genes and the segmentation genes define the anterior–posterior axis, which is the field of action for the homeotic (*Hox*) genes [Figure 4(e)].

ESSENTIAL POINT ■   ■   ■

Maternal-effect gene products activate genes that lay down the anterior–posterior axis of the embryo and specify the location and number of segments, which in turn have their identity determined by homeotic selector genes.

NOW SOLVE THIS

1 Suppose you initiate a screen for maternal-effect mutations in *Drosophila* affecting external structures of the embryo and you identify more than 100 mutations that affect these structures. From their screenings, other researchers concluded that there are only about 40 maternal-effect genes. How do you reconcile these different results?

■ **Hint** *This problem involves an understanding of how mutants are identified when adult* Drosophila *have been exposed to mutagens. The key to its solution is an understanding of the differences between genes and alleles.*

3 | Zygotic Genes Program Segment Formation in *Drosophila*

To summarize, certain genes in the zygote's genome are activated or repressed according to a positional gradient of maternal gene products. Expression of three sets of segmentation genes divides the embryo into a series of segments along its anterior–posterior axis. These segmentation genes are normally transcribed in the developing embryo, and mutations of these genes have embryo-lethal phenotypes.

Over 20 segmentation genes (**Table 1**), have been identified, and they are classified on the basis of their mutant phenotypes: (1) mutations in gap genes delete a group of adjacent segments, causing gaps in the normal body plan of the embryo, (2) mutations in pair-rule genes affect every other segment and eliminate a specific part of each affected segment, and (3) mutations in segment polarity genes cause defects in portions of each segment.

In addition to these three sets of genes that determine the anterior–posterior axis of the developing embryo, another

TABLE 1	Segmentation Genes in *Drosophila*	
Gap Genes	**Pair-Rule Genes**	**Segment Polarity Genes**
Krüppel	*hairy*	*engrailed*
knirps	*even-skipped*	*wingless*
hunchback	*runt*	*cubitis*
giant	*fushi-tarazu*	*hedgehog*
tailless	*paired*	*fused*
buckebein	*odd-paired*	*armadillo*
caudal	*odd-skipped*	*patched*
	sloppy-paired	*gooseberry*
		paired
		naked
		disheveled

set of genes determines the dorsal–ventral axis of the embryo. Our discussion will be limited to the gene sets involved in the anterior–posterior axis. Let us now examine members of each group in greater detail.

Gap Genes

Transcription of **gap genes** in the embryo is controlled by maternal gene products laid down in gradients in the egg. Gap genes also cross-regulate each other to define the early stage of the body plan. Mutant alleles of these genes produce large gaps in the embryo's segmentation pattern. *Hunchback* mutants lose head and thorax structures, *Krüppel* mutants lose thoracic and abdominal structures, and *knirps* mutants lose most abdominal structures. Transcription of wild-type gap genes (which encode transcription factors) divides the embryo into a series of broad regions that become the head, thorax, and abdomen. Within these regions, different combinations of gene activity eventually specify both the type of segment that forms and the proper order of segments in the body of the larva, pupa, and adult. Expression domains of the gap genes in different parts of the embryo correlate roughly with the location of their mutant phenotypes: *hunchback* at the anterior, *Krüppel* in the middle (**Figure 5**), and *knirps* at the posterior. As mentioned earlier, gap genes encode transcription factors that bind to enhancer regions that control the expression of pair-rule genes.

Pair-Rule Genes

Pair-rule genes are expressed in a series of seven narrow bands or stripes of nuclei extending around the circumference of the embryo. The expression of this gene set does two things: first it establishes the boundaries of segments and then it programs the developmental fate of the cells within each segment by controlling expression of the segment polarity genes. Mutations in pair-rule genes eliminate segment-size sections of the embryo at every other segment. At least eight pair-rule genes act to divide the embryo into a series of

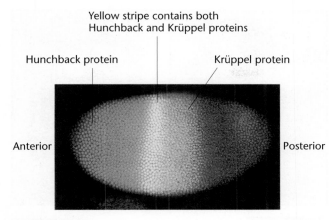

Yellow stripe contains both
Hunchback and Krüppel proteins

Hunchback protein

Krüppel protein

Anterior

Posterior

FIGURE 5 Expression of gap genes in a *Drosophila* embryo. The hunchback protein is shown in orange, and Krüppel is indicated in green. The yellow stripe is created when cells contain both hunchback and Krüppel proteins. Each dot in the embryo is a nucleus.

stripes. The transcription of the pair-rule genes is mediated by the action of maternal gene products and gap gene products. Initially, the boundaries of these stripes overlap, so that in each area of overlap, cells express a different combination of pair-rule genes (**Figure 6**). The resolution of boundaries in this segmentation pattern results from the interaction among the gene products of the pair-rule genes themselves (**Figure 7**).

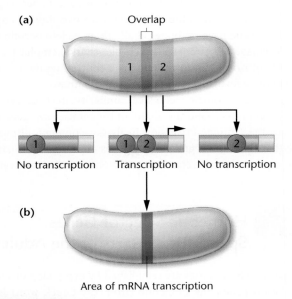

(a)

Overlap

1   2

No transcription   Transcription   No transcription

(b)

Area of mRNA transcription

FIGURE 6 New patterns of gene expression can be generated by overlapping regions containing two different gene products. (a) Transcription factors 1 and 2 are present in an overlapping region of expression. If both transcription factors must bind to the promoter of a target gene to trigger expression, the gene will be active only in cells containing both factors (most likely in the zone of overlap). (b) The expression of the target gene in the restricted region of the embryo.

(a)

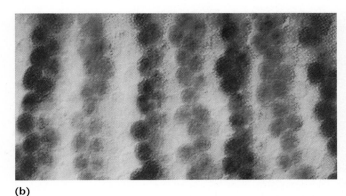

(b)

FIGURE 7 Stripe pattern of pair-rule gene expression in a *Drosophila* embryo. This embryo is stained to show patterns of expression of the genes *even-skipped* and *fushi-tarazu*; (a) low-power view and (b) high-power view of the same embryo.

Segment Polarity Genes

Expression of **segment polarity genes** is controlled by transcription factors encoded by pair-rule genes. Within each of the segments created by pair-rule genes, segment polarity genes become active in a single band of cells that extends around the embryo's circumference (**Figure 8**). This divides the embryo into 14 segments. The products of the segment polarity genes control the cellular identity within each of them and establish the anterior–posterior pattern (the polarity) within each segment.

Segmentation Genes in Mice and Humans

We have seen that segment formation in *Drosophila* depends on the action of three sets of segmentation genes. Are these genes found in humans and other mammals, and do they control aspects of embryonic development in these organisms? To answer this question, let's examine *runt*, one of the pair-rule genes in *Drosophila*. Later in development, it controls aspects of sex determination and formation of the nervous system. The gene encodes a protein that regulates transcription of its target genes. Runt contains a 128-amino-acid DNA-binding region (called the runt domain) that is

485

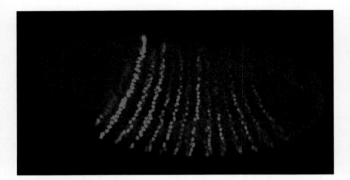

FIGURE 8 The 14 stripes of expression of the segment polarity gene *engrailed* in a *Drosophila* embryo.

highly conserved in *Drosophila*, mouse, and human proteins. In fact, *in vitro* experiments show that the *Drosophila* and mouse runt proteins are functionally interchangeable. In mice, *runt* is expressed early in development and controls formation of blood cells, bone, and the genital system. Although the target gene sets controlled by *runt* are different in *Drosophila* and the mouse, in both organisms, expression of *runt* specifies the fate of uncommitted cells in the embryo by regulating transcription of target genes.

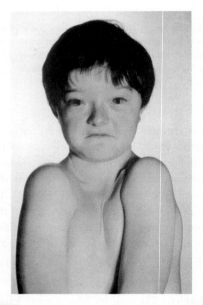

FIGURE 9 A boy affected with cleidocranial dysplasia (CCD). This disorder, inherited as an autosomal dominant trait, is caused by mutation in a human *runt* gene, *CBFA*. Affected heterozygotes have a number of skeletal defects, including a hole in the top of the skull where the infant fontanel fails to close, and collar bones that do not develop or form only small stumps. Because the collar bones do not form, CCD individuals can fold their shoulders across their chests. *Reprinted by permission from Macmillan Publishers Ltd.: Fig 1 on p. 244 from: British Dental Journal 195: 243–248 2003. Greenwood, M. and Meechan, J. G. "General medicine and surgery for dental practitioners." Copyright © Macmillan Magazines Limited.*

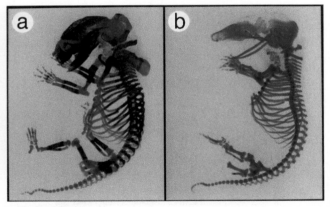

FIGURE 10 Bone formation in normal mice and mutants for the *runt* gene *Cbfa1*. (a) Normal mouse embryos at day 17.5 show cartilage (blue) and bone (brown). (b) The skeleton of a 17.5-day homozygous mutant embryo. Only cartilage has formed in the skeleton. There is complete absence of bone formation in the mutant mouse. Expression of a normal copy of the *Cbfa1* gene is essential for specifying the developmental fate of bone-forming osteoblasts.

In humans, mutations in *RUNX2*, a human homolog of *runt*, causes cleidocranial dysplasia (CCD), an autosomal dominantly inherited trait. Those affected with CCD have a hole in the top of their skull because bone does not form in the membranous gap known as the fontanel. Their collar bones (clavicles) do not develop, enabling affected individuals to fold their shoulders across their chest (**Figure 9**). Mice with one mutant copy of the *runt* homolog have a phenotype similar to that seen in humans; mice with two mutant copies of the gene have no bones at all. Their skeletons contain only cartilage (**Figure 10**), much like sharks, emphasizing the role of *runt* as an important gene controlling the initiation of bone formation in both mice and humans.

The *runt* domain sequence similarity in *Drosophila*, mice, and humans and the ability of the mouse *runt* gene to replace the *Drosophila* version in fly development all indicate that the same segmentation genes are found in organisms separated from a common ancestor by millions of years.

4 Homeotic Selector Genes Specify Body Parts of the Adult

As segment boundaries are established by expression of segmentation genes, the homeotic (from the Greek word for "same") genes are activated. Expression of homeotic selector genes determines which adult structures will be formed by each body segment. In *Drosophila*, this includes the antennae, mouth parts, legs, wings, thorax, and abdomen. Mutants of these genes are called **homeotic mutants** because one segment is transformed so that it forms the same structure as another segment. For example, the wild-type allele of *Antennapedia (Antp)* specifies formation of a leg on the second segment of the thorax. Dominant gain-of-function *Antp* mutations cause this gene to be expressed in the head *and* the thorax. The result is that mutant flies have a leg on their head instead of an antenna **(Figure 11)**.

Hox Genes in *Drosophila*

The *Drosophila* genome contains two clusters of homeotic selector genes (called *Hox* genes) on chromosome 3 that encode transcription factors **(Table 2)**. The *Antennapedia (ANT-C)* cluster contains five genes that specify structures in the head and first two segments of the thorax **[Figure 12(a)]**. The second cluster, the *bithorax (BX-C)* complex, contains three genes that specify structures in the posterior portion of the second thoracic segment, the entire third thoracic segment, and the abdominal segments **[Figure 12(b)]**.

TABLE 2	*Hox* Genes of *Drosophila*
Antennapedia Complex	**Bithorax Complex**
labial	*Ultrabithorax*
Antennapedia	*abdominal A*
Sex combs reduced	*Abdominal B*
Deformed	
proboscipedia	

(a)

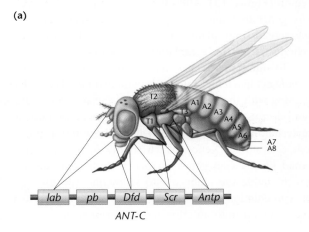

ANT-C

(b)

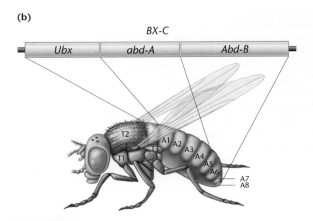

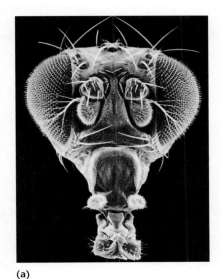

(a)

(b)

FIGURE 11 *Antennapedia (Antp)* mutation in *Drosophila*. (a) Head from wild-type *Drosophila*, showing the antenna and other head parts. (b) Head from an *Antp* mutant, showing the replacement of normal antenna structures with legs. This is caused by activation of the *Antp* gene in the head region.

FIGURE 12 Genes of the *Antennapedia* complex and the adult structures they specify. (a) In the *ANT-C* complex, the *labial* (*lab*) and *Deformed* (*Dfd*) genes control the formation of head segments. The *Sex comb reduced* (*Scr*) and *Antennapedia* (*Antp*) genes specify the identity of the first two thoracic segments, T1 and T2. The remaining gene in the complex, *proboscipedia* (*pb*), may not act during embryogenesis but may be required to maintain the differentiated state in adults. In mutants, the labial palps are transformed into legs. (b) In the *BX-C* complex, *Ultrabithorax* (*Ubx*) controls formation of structures in the posterior compartment of T2 and structures in T3. The two other genes, *abdominal A* (*abdA*) and *Abdominal B* (*AbdB*), specify the segmental identities of the eight abdominal segments (A1–A8).

Hox genes (listed in Table 2) from a wide range of species have two properties in common. First, each contains a highly conserved 180-bp nucleotide sequence known as a **homeobox.** (*Hox* is a contraction of homeobox.) The homeobox encodes a DNA-binding region of 60 amino acids known as a **homeodomain.** Second, in most species, expression of *Hox* genes is colinear with the anterior to posterior organization of the body. In other words, genes at one end of the cluster (the 3′ end) are expressed at the anterior end of the embryo, those in the middle are expressed in the middle of the embryo, and genes at the other end of a cluster (the 5′ end) are expressed at the embryo's posterior region **(Figure 13)**. Although first identified in *Drosophila, Hox* genes are found in the genomes of most eukaryotes with segmented body plans, including nematodes, sea urchins, zebrafish, frogs, mice, and humans **(Figure 14)**.

To summarize, genes that control development in *Drosophila* act in a temporally and spatially ordered cascade, beginning with the genes that establish the anterior–posterior (and dorsal–ventral) axis of the early embryo. Gradients of maternal mRNAs and proteins along the anterior–posterior axis activate gap genes, which subdivide the embryo into broad bands. Gap genes in turn activate pair-rule genes, which divide the embryo into segments. The final group of segmentation genes, the segment polarity genes, divides each segment into anterior and posterior regions arranged

(a) Expression domains of homeotic genes

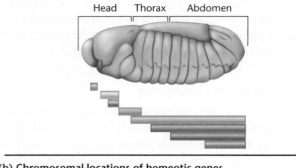

(b) Chromosomal locations of homeotic genes

FIGURE 13 The colinear relationship between the spatial pattern of expression and chromosomal locations of homeotic genes in *Drosophila*. (a) *Drosophila* embryo and the domains of homeotic gene expression in the embryonic epidermis and central nervous system. (b) Chromosomal location of homeotic selector genes. Note that the order of genes on the chromosome correlates with the sequential anterior borders of their expression domains.

FIGURE 14 Conservation of organization and patterns of expression in *Hox* genes. (Top) The structures formed in adult *Drosophila* are shown, with the colors corresponding to members of the *Hox* cluster that control their formation. (Bottom) The arrangement of the *Hox* genes in an early human embyro. As in *Drosophila*, genes at the (3′ end) of the cluster form anterior structures, and genes at the (5′ end) of the cluster form posterior structures.

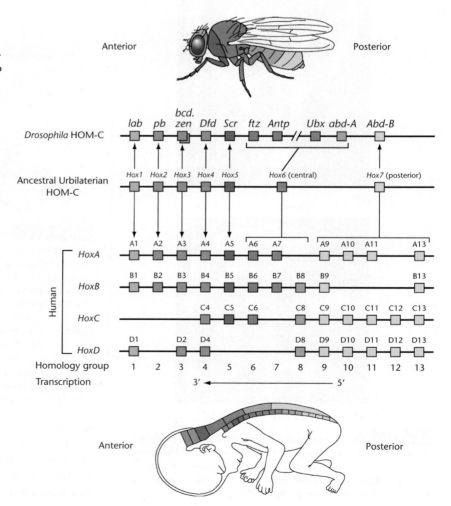

linearly along the anterior–posterior axis. The segments are then given identity by action of the *Hox* genes. Therefore, this progressive restriction of developmental potential of the *Drosophila* embryo's cells (all of which occurs during the first third of embryogenesis) involves a cascade of gene action, with regulatory proteins acting on transcription, translation, and signal transduction.

Hox Genes and Human Genetic Disorders

Although first described in *Drosophila, Hox* genes are found in the genomes of all animals where they play a fundamental role in shaping the body and its appendages. In vertebrates, the conservation of sequence, the order of genes in the *Hox* clusters, and their pattern of expression in vertebrates suggest that, as in *Drosophila*, these genes control development along the anterior–posterior and the formation of appendages. However, in vertebrates, there are four clusters of *Hox* genes: *HOXA, HOXB, HOXC,* and *HOXD* instead of a single cluster as in *Drosophila*. This means that in vertebrates, not just one, but a combination of 2 to 4 *Hox* genes is involved in forming specific structures. As a result, in vertebrates, mutations in individual *Hox* genes do not produce complete transformation as in *Drosophila*, where mutation of a single *Hox* gene can transform a haltere into a wing (see the photo at the beginning of this chapter). In spite of these differences, the role of *HOXD* genes in human development was confirmed by the discovery that several inherited limb malformations are caused by mutations in *HOXD* genes. For example, mutations in *HOXD13* cause synpolydactyly (SPD), a malformation characterized by extra fingers and toes, and abnormalities in bones of the hands and feet (**Figure 15**).

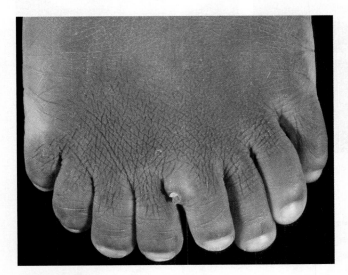

FIGURE 15 Mutations in posterior *Hox* genes (*HOXD13* in this case) in humans result in malformations of the limbs, shown here as extra toes. This condition is known as synpolydactyly. Mutations in *HOXD13* are also associated with abnormalities of the bones in the hands and feet.

NOW SOLVE THIS

2 In *Drosophila*, both *fushi tarazu* (*ftz*) and *engrailed* (*eng*) genes encode homeobox transcription factors and are capable of eliciting the expression of other genes. Both genes work at about the same time during development and in the same region to specify cell fate in body segments. To discover if *ftz* regulates the expression of *engrailed*; if *engrailed* regulates *ftz*; or if both are regulated by another gene, you perform a mutant analysis. In *ftz* embryos (*ftz/ftz*) engrailed protein is absent; in *engrailed* embryos (*eng/eng*) *ftz* expression is normal. What does this tell you about the regulation of these two genes—does the *engrailed* gene regulate *ftz*, or does the *ftz* gene regulate *engrailed*?

■ **Hint** *This problem involves an understanding of how genes are regulated at different stages of preadult development in* Drosophila. *The key to its solution lies in using the results of the mutant analysis to determine the timing of expression of the two genes being examined.*

5 Plants Have Evolved Developmental Regulatory Systems That Parallel Those of Animals

Plants and animals diverged from a common ancestor about 1.6 billion years ago, after the origin of eukaryotes and probably before the rise of multicellular organisms. Genetic analysis of mutants and genome sequencing in plants and animals indicate that basic mechanisms of developmental pattern formation evolved independently in animals and plants. We have already examined the genetic systems that control development and pattern formation in animals, using *Drosophila* as a model organism, and will now briefly examine these systems in plants.

In plants, pattern formation has been extensively studied using flower development in *Arabidopsis thaliana* (**Figure 16**), a small plant in the mustard family, as a model organism. A cluster of undifferentiated cells, called the *floral meristem*, gives rise to flowers (**Figure 17**). Each flower consists of four organs—sepals, petals, stamens, and carpels—that develop from concentric rings of cells within the meristem [**Figure 18(a)**]. Each organ develops from a different concentric ring, or whorl of cells.

Homeotic Genes in *Arabidopsis*

Three classes of floral homeotic genes control the development of these organs (**Table 3**). Acting alone, class A genes specify sepals; class A and class B genes expressed together specify petals. Acting together, class B and class C genes control stamen formation. Class C genes acting alone specify carpels. During flower development [**Figure 18(b)**], class A genes are active in whorls 1 and 2 (sepals and petals), class B

FIGURE 16 The flowering plant *Arabidopsis thaliana,* used as a model organism in plant genetics.

TABLE 3	Homeotic Selector Genes in *Arabidopsis**
Class A	*APETALA1 (AP1)*
	APETALA2 (AP2)
Class B	*APETALA3 (AP3)*
	PISTILLATA (P1)
Class C	*AGAMOUS (AG)*

*By convention, wild-type genes in *Arabidopsis* use capital letters.

genes are expressed in whorls 2 and 3 (petals and stamens), and class C genes are expressed in whorls 3 and 4 (stamens and carpels). The organ formed depends on the expression pattern of the three gene classes. In whorl 1, expression of class A genes alone causes sepals to form. Expression of class A *and* class B genes in whorl 2 leads to petal formation. Expression of class B and class C genes in whorl 3 leads to stamen formation. In whorl 4, expression of class C genes alone causes carpel formation.

As in *Drosophila,* mutations in homeotic genes cause organs to form in abnormal locations. For example, in *APETALA2* mutants (*ap2*), the order of organs is carpel, stamen, stamen, and carpel instead of the normal order,

(a)   **(b)**

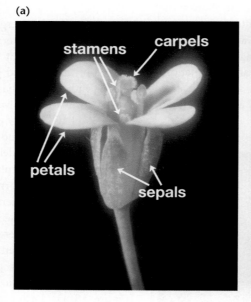

FIGURE 17 (a) Parts of the *Arabidopsis* flower. The floral organs are arranged concentrically. The sepals form the outermost ring, followed by petals and stamens, with carpels on the inside. (b) View of the flower from above.

FIGURE 18 Cell arrangement in the floral meristem. (a) The four concentric rings, or whorls, labeled 1–4, give rise to (b) arrangement of the sepals, petals, stamens, and carpels, respectively, in the mature flower.

(a)   **(b)**

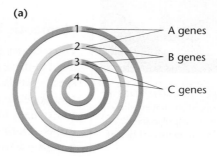
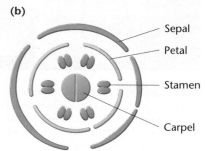

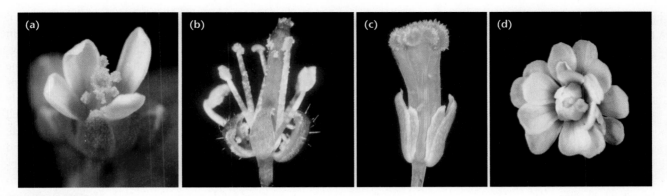

FIGURE 19 (a) Wild-type flowers of *Arabidopsis* have (from outside to inside) sepals, petals, stamens, and carpels. (b) A homeotic *APETALA2* mutant flower has carpels, stamens, stamens, and carpels. (c) *PISTILLATA* mutants have sepals, sepals, carpels, and carpels. (d) *AGAMOUS* mutants have petals and sepals at places where stamens and carpels should form.

sepal, petal, stamen, and carpel [**Figure 19(a) and (b)**]. In class B loss-of-function mutants (*ap3*, *pi*), petals become sepals, and stamens are transformed into carpels [**Figure 19(c)**], and the order of organs becomes sepal, sepal, carpel, carpel. Plants carrying a mutation for the Class 3 gene *AGAMOUS* will have petals in whorl 3 (instead of stamens) and sepals in whorl 4 (instead of carpels), and the order of organs will be sepal, petal, petal, and sepal [**Figure 19(d)**].

Evolutionary Divergence in Homeotic Genes

Drosophila and *Arabidopsis* use different sets of nonhomologous master regulatory genes to establish the body axis and specify the identity of structures along the axis. In *Drosophila,* this task is accomplished in part by the *Hox* genes, which encode a set of transcription factors sharing a homeobox domain. In *Arabidopsis,* the floral homeotic genes belong to a different family of transcription factors, called the **MADS-box proteins,** characterized by a common sequence of 58 amino acids with no similarity in amino acid sequence or protein structure with the *Hox* genes. Both gene sets encode transcription factors, both sets are master regulators of development expressed in a pattern of overlapping domains, and both specify identity of structures.

Reflecting their evolutionary origin from a common ancestor, the genomes of both *Drosophila* and *Arabidopsis* contain members of the homeobox and MADS-box genes, but these genes have been adapted for different uses in the plant and animal kingdoms. This indicates that developmental mechanisms evolved independently in each group.

In both plants and animals, the action of transcription factors depends on changes in chromatin structure that make genes available for expression. Mechanisms of transcription initiation are conserved in plants and animals, as is reflected in the homology of genes in *Drosophila* and *Arabidopsis* that maintain patterns of expression initiated by regulatory gene sets. Action of the floral homeotic genes is controlled

by a gene called *CURLY LEAF.* This gene shares significant homology with members of a *Drosophila* gene family called *Polycomb.* This family of regulatory genes controls expression of homeobox genes during development. Both *CURLY LEAF* and *Polycomb* encode proteins that alter chromatin conformation and shut off gene expression. Thus, although different genes are used to control development, both plants and animals use an evolutionarily conserved mechanism to regulate expression of these gene sets.

ESSENTIAL POINT ■ ■ ■

Flower formation in *Arabidopsis* is controlled by homeotic genes, but these gene sets are from a different gene family than the homeotic selector genes of *Drosophila* and other animals.

6 C. elegans Serves as a Model for Cell–Cell Interactions in Development

During development in multicellular organisms, cell–cell interactions influence the transcriptional programs and developmental fate of the interacting cells and surrounding cells. Cell–cell interaction is an important process in the embryonic development of most eukaryotic organisms, including *Drosophila,* mice, and humans.

Signaling Pathways in Development

In early development, animals use a number of signaling pathways to regulate development; after organ formation begins, additional pathways are added to those already in use. These newly activated pathways act both independently and in coordinated networks to generate specific transcriptional patterns. Signal networks establish anterior–posterior polarity and body axes, coordinate pattern formation, and direct the differentiation of tissues and organs. The signaling pathways

TABLE 4 — Signaling Pathways Used in Early Embryonic Development

Wnt Pathway
Dorsalization of body
Female reproductive development
Dorsal–ventral differences

TGF-β Pathway
Mesoderm induction
Left–right asymmetry
Bone development

Hedgehog Pathway
Notochord induction
Somitogenesis
Gut/visceral mesoderm

Receptor Tyrosine Kinase Pathway
Mesoderm maintenance

Notch Signaling Pathway
Blood cell development
Neurogenesis
Retina development

*Source: *Taken from Gerhart, J. 1999. 1998 Warkany lecture: Signaling pathways in development. *Teratology* 60: 226–239.

used in early development and some of the developmental processes they control are listed in **Table 4**. After an introduction to the components and interactions of one of these systems—the **Notch signaling pathway**—we will briefly examine its role in the development of the vulva in the nematode, *Caenorhabditis elegans*.

The Notch Signaling Pathway

The genes in the Notch pathway are named after the *Drosophila* mutants that were used to identify components of this signal transduction system (*Notch* mutants have an indentation or notch in their wings). The Notch signal system works through direct cell–cell contact to control the developmental

fate of interacting cells. The *Notch* gene (and the equivalent gene in other organisms) encodes a signal receptor protein embedded in the plasma membrane **(Figure 20)**. The signal is another membrane protein encoded by the *Delta* gene (and its equivalents). Because both the signal and receptor are membrane proteins, the Notch signal system works only between adjacent cells. When the Delta protein from one cell binds to the Notch receptor protein on a neighboring cell, the cytoplasmic tail of the Notch protein is cleaved off and binds to a cytoplasmic protein encoded by the *Su(H)* (suppressor of *Hairless*) gene. This protein complex moves into the nucleus and binds to transcriptional cofactors, activating transcription of a gene set that controls a specific developmental pathway (Figure 20).

One of the main roles of the Notch signal system is to specify different developmental fates for equivalent cells in a population. In its simplest form, this interaction involves two neighboring cells that are developmentally equivalent. We will explore the role of the Notch signaling system in development of the vulva in *C. elegans*, after a brief introduction to nematode embryogenesis.

Overview of *C. elegans* Development

The nematode *C. elegans* is widely used to study the genetic control of development. There are several advantages in using this organism: (1) its genetics are well known, (2) its genome has been sequenced, and (3) adults contain a small number of cells that follow a highly deterministic developmental program. Adult nematodes are about 1 mm long and develop from a fertilized egg in about two days **(Figure 21)**. The life cycle includes an embryonic stage (about 16 hours), four larval stages (L1 through L4), and the adult stage. Adults are of two sexes: XX self-fertilizing hermaphrodites that can make both eggs and sperm, and XO males. Self-fertilization of mutagen-treated hermaphrodites is used to develop homozygous stocks of mutant strains, and hundreds of such mutants have been generated, cataloged, and mapped.

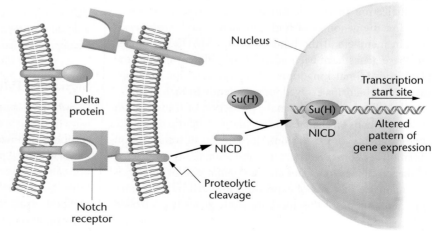

FIGURE 20 Components of the Notch signaling pathway in *Drosophila*. The cell carrying the Delta transmembrane protein is the sending cell; the cell carrying the transmembrane Notch protein receives the signal. Binding of Delta to Notch triggers a proteolytic-mediated activation of transcription. The fragment cleaved from the cytoplasmic side of the Notch protein, called the Notch intracellular domain (NICD), combines with the Su(H) protein and moves to the nucleus where it activates a program of gene transcription.

(a)

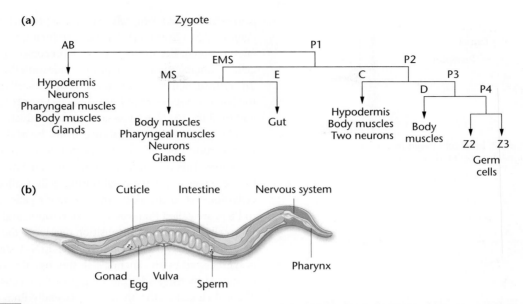

(b)

FIGURE 21 (a) A truncated cell lineage chart for *C. elegans,* showing early divisions and the tissues and organs formed from these lineages. Each vertical line represents a cell division, and horizontal lines connect the two cells produced. For example, the first division of the zygote creates two new cells, AB and P1. During embryogenesis, cell divisions will produce the 959 somatic cells of the adult hermaphrodite worm. (b) An adult *C. elegans* hermaphrodite. This nematode, about 1 mm in length, consists of 959 cells and is widely used as a model organism to study the genetic control of development.

Adult hermaphrodites have 959 somatic cells (and about 2000 germ cells). The lineage of each cell, from fertilized egg to adult has been mapped (Figure 21) and is invariant from individual to individual. Knowing the lineage of each cell, we can easily follow altered cell fates generated by mutations or by killing specific cells with laser microbeams or ultraviolet irradiation. In hermaphrodites, the developmental fate of cells in the reproductive system is determined by cell–cell interaction,

illustrating how gene expression and cell–cell interaction work together to specify developmental outcomes.

Genetic Analysis of Vulva Formation

Adult *C. elegans* hermaphrodites lay eggs through the vulva, an opening near the middle of the body (Figure 21). The vulva is formed in stages during larval development and involves several rounds of cell–cell interactions.

In *C. elegans,* interaction between two neighboring cells, Z1.ppp and Z4.aaa, determines which will become the gonadal anchor cell (from which the vulva forms) and which will become a precursor to the uterus **(Figure 22)**. The determination of which cell becomes which occurs during the second larval stage (L2) and is controlled by the Notch receptor gene, *lin-12.* In recessive *lin-12(0)* mutants (a loss-of-function mutant), no functional receptor protein is present, and both cells become anchor cells. The dominant mutation *lin-12(d)* (a gain-of-function mutation) causes both to become uterine precursors. Thus, expression of *lin-12* directs selection of the uterine pathway, because in the absence of the LIN-12 (Notch) receptor, both cells become anchor cells.

However, the situation is more complex than it first appears. Initially, the two neighboring cells are developmentally equivalent. Each synthesizes low levels of the Notch signal protein (encoded by the *lag-2* gene) *and* the Notch receptor protein. By chance, one cell ends up secreting more of the signal (LAG-2 or Delta protein) than the other cell. This

NOW SOLVE THIS

3 The identification and characterization of genes that control sex determination have been another focus of investigators working with *C. elegans.* As with *Drosophila,* sex in this organism is determined by the ratio of X chromosomes to sets of autosomes. A diploid wild-type male has one X chromosome, and a diploid wild-type hermaphrodite has two X chromosomes. Many different mutations have been identified that affect sex determination. Loss-of-function mutations in a gene called *her-1* cause an XO nematode to develop into a hermaphrodite and have no effect on XX development. (That is, XX nematodes are normal hermaphrodites.) In contrast, loss-of-function mutations in a gene called *tra-1* cause an XX nematode to develop into a male. Deduce the roles of these genes in wild-type sex determination from this information.

■ **Hint** *This problem involves an understanding of the mechanism of sex determination by the ratio of X chromosomes to sets of autosomes. The key to its solution is an understanding of the effect of loss-of-function mutations on expression of other genes or the action of other proteins.*

(a)

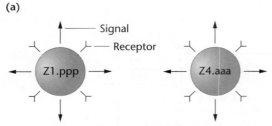

During L2, both cells begin secreting
signal for uterine differentiation

(b)

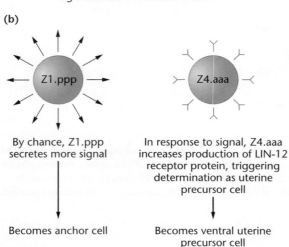

By chance, Z1.ppp
secretes more signal

In response to signal, Z4.aaa
increases production of LIN-12
receptor protein, triggering
determination as uterine
precursor cell

Becomes anchor cell

Becomes ventral uterine
precursor cell

FIGURE 22 Cell–cell interaction in anchor cell determination. (a) During L2, two neighboring cells begin the secretion of chemical signals for the induction of uterine differentiation. (b) By chance, cell Z1.ppp produces more of these signals, causing cell Z4.aaa to increase production of the receptor for signals. The action of increased signals causes Z4.aaa to become the ventral uterine precursor cell and allows Z1.ppp to become the anchor cell.

causes the neighboring cell to increase production of the receptor (LIN-12 protein). The cell producing more of the receptor protein becomes the uterine precursor, and the other cell, producing more signal protein, becomes the anchor cell. The critical factor in this first round of cell–cell interaction is the balance between the LAG-2 (Delta) signal gene product and the LIN-12 (Notch) receptor gene product.

Once the gonadal anchor cell has been determined, a second round of cell–cell interaction leads to formation of the vulva. This interaction involves the anchor cell (located in the gonad) and six neighboring cells (called precursor cells) located in the skin. The precursor cells, named P3.p to P8.p, are called Pn.p cells. The developmental fate of each Pn.p cell is specified by its position relative to the anchor cell.

During vulval development, the LIN-3 signal protein is synthesized by the anchor cell; this signal is received and processed by three adjacent Pn.p precursor cells (Pn.p 5-7). The cell closest to the anchor cell (usually Pn.p 6) becomes the primary vulval precursor cell, and the adjacent cells (Pn.p 5 and 7) become secondary precursor cells. A signal protein from the primary vulval precursor cell activates the *lin-12* receptor gene in the secondary cells, preventing them from becoming primary precursor cells. The other precursor cells (Pn.p 3, 4, and 8) receive no signal from the anchor cell and become skin cells.

ESSENTIAL POINT ■ ■ ■

In *C. elegans*, the well-studied pathway of cell lineage during embryonic development allows developmental biologists to study the cell–cell signaling required for organogenesis.

GENETICS, TECHNOLOGY, AND SOCIETY

Stem Cell Wars

Stem cell research is one of the most contentious and emotionally charged research areas to emerge since the beginning of recombinant DNA technology in the 1970s. Although stem cell research is the focus of presidential proclamations, media campaigns, and ethical debates, few people understand it sufficiently to evaluate its pros and cons.

Stem cells are primitive cells that replicate indefinitely and have the capacity to differentiate into cells with specialized functions, such as the cells of heart, brain, liver, and muscle tissue. All the cells that make up the approximately 200 distinct types of tissues in our bodies are descended from stem cells. Some types of stem cells are defined as *totipotent,* meaning that they have the ability to differentiate into any mature cell type in the adult body. Other types of stem cells are *pluripotent,* meaning that they are able to differentiate into any of a smaller number of mature cell types. In contrast, mature, fully differentiated cells do not replicate or undergo transformations into different cell types. Scientists are able to isolate pluripotent stem cells from tissues and grow the cells in culture dishes. When treated with growth factors or hormones, these pluripotent stem cells differentiate into cells that have the characteristics of neural, bone, kidney, liver, heart, or pancreatic cells.

The fact that pluripotent stem cells grow prolifically in culture and differentiate into more specialized cells has created great excitement. Some foresee a day when stem cells may be a cornucopia from which to harvest unlimited numbers of specialized cells to replace cells in damaged and diseased tissues. Hence, stem cells could be used to treat Parkinson disease, type 1 diabetes, chronic heart disease, Alzheimer disease, and spinal cord injuries. Some predict that stem cells will be genetically modified to eliminate transplant rejection or to deliver specific gene products, thereby correcting genetic defects or treating cancers. The excitement about stem cell therapies has been fueled by reports of dramatically successful experiments in animals. For example, mice with spinal cord injuries regained their mobility and bowel and bladder control after they were injected with human stem cells. Both proponents and critics of stem cell research agree that stem cell therapies could be revolutionary. Why, then, should stem cell research be so contentious?

The answer to that question lies in the source of the pluripotent stem cells. Until recently, all human pluripotent stem cell lines were derived from five-day-old embryonic blastocysts. Blastocysts at this stage consist of approximately 100 cells, most of which will develop into placental and supporting tissues for the early embryo. The inner cell mass of the blastocyst consists of about 30 pluripotent stem cells that can develop into all the embryo's tissues. *In vitro* fertilization clinics grow fertilized eggs to the five-day blastocyst stage prior to uterine transfer. *Embryonic stem cell (ESC)* lines are created by taking the inner cell mass out of five-day blastocysts and growing the cells in culture dishes.

The fact that early embryos are destroyed in the process of establishing human ESC lines disturbs people who believe that preimplantation embryos are persons with rights; however, it does not disturb people who believe that these embryos are too primitive to have the status of a human being. Both sides in the debate invoke fundamental questions of what constitutes a human being.

Recently, scientists have developed several types of pluripotent stem cells without using embryos. One of the most promising types—known as *induced pluripotent stem cells (iPSCs)*—uses adult somatic cells as the source of pluripotent stem cell lines. To prepare iPSCs, scientists isolate somatic cells (such as cells from skin) and introduce genes that encode transcription factors which in turn reprogram the adult cells into immortal pluripotent stem cells. Some research groups have introduced the transcription factor proteins directly into adult cells in order to reprogram these cells into iPSCs.

The development of iPS cell lines has generated renewed enthusiasm for stem cell research, as these cells bypass the ethical problems associated with the use of human embryos. In addition, they may become sources of patient-specific pluripotent stem cell lines that can be used for transplantation, without immune system rejection.

Will stem cell therapies be as miraculous as predicted? As stem cell research is progressing rapidly, we won't have long to wait to answer the question.

Your Turn

Take time, individually or in groups, to answer the following questions. Investigate the references and links to help you understand the technologies and controversies surrounding stem cell research.

1. What, in your opinion, are the scientific and ethical problems that still surround stem cell research? Are these problems solved by the new methods of creating pluripotent stem cells?

   You can find descriptions of some new methods of generating pluripotent stem cell lines, and the ethical issues that accompany these methods, in Kastenberg, Z. J. and Odorico, J. S. 2008. Alternative sources of pluripotency: science, ethics, and stem cells. *Transplantation* Rev. 22: 215–222.

2. What are the current stem cell research laws in your region, and how do these laws compare with national regulations?

   A starting point for information about stem cell research regulations can be found on the Stem Cell Information Web site of the National Institutes of Health (http://stemcells.nih.gov).

3. Do you oppose or support stem cell research? Why, or why not?

   An interesting online poll, along with arguments for and against stem cell research, is offered by the Public Broadcasting Corporation, at http://www.pbs.org/wgbh/nova/body/stem-cell-poll.html.

4. What, in your opinion, are the most significant developments in stem cell research in the last year?

   Some ideas to start your search are: the PubMed Web site (http://www.ncbi.nlm.nih.gov/sites/entrez? db=PubMed) *and the New York Times online Stem Cell page* (http://topics.nytimes.com/top/news/health/diseasesconditionsandhealthtopics/stemcells).

CASE STUDY » One foot or another

In humans the *HOXD* homeotic gene cluster plays a critical role in limb development. In one large family, 16 of 36 members expressed one of two dominantly inherited malformations of the feet known as rocker bottom foot (CVT) or claw foot (CMT). One individual had one foot with CVT and the other with CMT. Genomic analysis identified a single missense mutation in the *HOXD10* gene, resulting in a single amino acid substitution in the homeodomain of the encoded transcription factor. This region is crucial for making contact and binding to the target genes controlled by this protein. All family members with the foot malformations were heterozygotes; all unaffected members were homozygous for the normal allele.

1. Given that affected heterozygotes carry one normal allele of the *HOXD10* gene, how might a dominant mutation in a gene encoding a transcription factor lead to a developmental malformation?

2. How can two clinically different disorders result from the same mutation?

3. What might we learn about the control of developmental processes from an understanding of how this mutation works?

INSIGHTS AND SOLUTIONS

1. In the slime mold *Dictyostelium,* experimental evidence suggests that cyclic AMP (cAMP) plays a central role in the developmental program leading to spore formation. The genes encoding the cAMP cell-surface receptor have been cloned, and the amino acid sequence of the protein components is known. To form reproductive structures, free-living individual cells aggregate together and then differentiate into one of two cell types, prespore cells or prestalk cells. Aggregating cells secrete waves or oscillations of cAMP to foster the aggregation of cells and then continuously secrete cAMP to activate genes in the aggregated cells at later stages of development. It has been proposed that cAMP controls cell–cell interaction and gene expression. It is important to test this hypothesis by using several experimental techniques. What different approaches can you devise to test this hypothesis, and what specific experimental systems would you employ to test them?

Solution: Two of the most powerful forms of analysis in biology involve the use of biochemical analogs (or inhibitors) to block gene transcription or the action of gene products in a predictable way, and the use of mutations to alter genes and their products. These two approaches can be used to study the role of cAMP in the developmental program of *Dictyostelium*. First, compounds chemically related to cAMP, such as GTP and GDP, can be used to test whether they have any effect on the processes controlled by cAMP. In fact, both GTP and GDP lower the affinity of cell-surface receptors for cAMP, effectively blocking the action of cAMP.

Mutational analysis can be used to dissect components of the cAMP receptor system. One approach is to use transformation with wild-type genes to restore mutant function. Similarly, because the genes for the receptor proteins have been cloned, it is possible to construct mutants with known alterations in the component proteins and transform them into cells to assess their effects.

2. In the sea urchin, early development may occur even in the presence of actinomycin D, which inhibits RNA synthesis. However, if actinomycin D is present early in development but is removed a few hours later, all development stops. In fact, if actinomycin D is present only between the sixth and eleventh hours of development, events that normally occur at the fifteenth hour are arrested. What conclusions can be drawn concerning the role of gene transcription between hours 6 and 15?

Solution: Maternal mRNAs are present in the fertilized sea urchin egg. Thus, a considerable amount of development can take place without transcription of the embryo's genome. Because development past 15 hours is inhibited by prior treatment with actinomycin D, it appears that transcripts from the embryo's genome are required to initiate or maintain these events. This transcription must take place between the sixth and fifteenth hours of development.

3. If it were possible to introduce one of the homeotic genes from *Drosophila* into an *Arabidopsis* embryo homozygous for a homeotic flowering gene, would you expect any of the *Drosophila* genes to negate (rescue) the *Arabidopsis* mutant phenotype? Why or why not?

Solution: The *Drosophila* homeotic genes belong to the *Hox* gene family, whereas *Arabidopsis* homeotic genes belong to the MADS-box protein family. Both gene families are present in *Drosophila* and *Arabidopsis,* but they have evolved different functions in the animal and the plant kingdoms. As a result, it is unlikely that a transferred *Drosophila Hox* gene would rescue the phenotype of a MADS-box mutant, but only an actual experiment would confirm this.

PROBLEMS AND DISCUSSION QUESTIONS

 For activities, animations, and review quizzes, go to the study area at www.masteringgenetics.com

HOW DO WE KNOW?

1. In this chapter, we have focused on large-scale as well as the inter- and intracellular events that take place during embryogenesis and the formation of adult structures. In particular, we discussed how the adult body plan is laid down by a cascade of gene expression, and the role of cell-cell communication in development. Based on your knowledge of these topics, answer several fundamental questions:

   (a) How do we know how many genes control development in an organism like *Drosophila*?

   (b) What experimental evidence demonstrates that molecular gradients in the egg control development?

   (c) How did we discover that selector genes specify which adult structures will be formed by body segments?

   (d) How did we learn about the levels of gene regulation involved in vulval development in *C. elegans*?

2. Carefully distinguish between the terms *differentiation* and *determination.* Which phenomenon occurs initially during development?

3. Nuclei from almost any source may be injected into *Xenopus* oocytes. Studies have shown that these nuclei remain active in transcription and translation. How can such an experimental system be useful in developmental genetic studies?

4. Distinguish between the syncytial blastoderm stage and the cellular blastoderm stage in *Drosophila* embryogenesis.

5. (a) What are maternal-effect genes? (b) When are gene products from these genes made, and where are they located? (c) What aspects of development do maternal-effect genes control? (d) What is the phenotype of maternal-effect mutations?

6. (a) What are zygotic genes, and when are their gene products made? (b) What is the phenotype associated with zygotic gene mutations? (c) Does the maternal genotype contain zygotic genes?

7. List the main classes of zygotic genes. What is the function of each class of these genes?

8. Experiments have shown that any nuclei placed in the polar cytoplasm at the posterior pole of the *Drosophila* egg will differentiate into germ cells. If polar cytoplasm is transplanted into the anterior end of the egg just after fertilization, what will happen to nuclei that migrate into this cytoplasm at the anterior pole?

9. How can you determine whether a particular gene is being transcribed in different cell types?

10. You observe that a particular gene is being transcribed during development. How can you tell whether the expression of this gene is under transcriptional or translational control?

11. What are *Hox* genes? What properties do they have in common? Are all homeotic genes *Hox* genes?

12. The homeotic mutation *Antennapedia* causes mutant *Drosophila* to have legs in place of antennae and is a dominant gain-of-function mutation. What are the properties of such mutations? How does the *Antennapedia* gene change antennae into legs?

13. The *Drosophila* homeotic mutation *spineless aristapedia* (ss^a) results in the formation of a miniature tarsal structure (normally part of the leg) on the end of the antenna. What insight is provided by (ss^a) concerning the role of genes during determination?

14. Embryogenesis and oncogenesis (generation of cancer) share a number of features including cell proliferation, apoptosis, cell migration and invasion, formation of new blood vessels, and differential gene activity. Embryonic cells are relatively undifferentiated, and cancer cells appear to be undifferentiated or dedifferentiated. Homeotic gene expression directs early development, and mutant expression leads to loss of the differentiated state or an alternative cell identity. M. T. Lewis (2000. *Breast Can. Res.* 2: 158–169) suggested that breast cancer may be caused by the altered expression of homeotic genes. When he examined 11 such genes in cancers, 8 were underexpressed while 3 were overexpressed compared with controls. Given what you know about homeotic genes, could they be involved in oncogenesis?

15. Early development depends on the temporal and spatial interplay between maternally supplied material and mRNA and the onset of zygotic gene expression. Maternally encoded mRNAs must be produced, positioned, and degraded (Surdej and Jacobs-Lorena, 1998. *Mol. Cell Biol.* 18: 2892–2900). For example, transcription of the *bicoid* gene that determines anterior–posterior polarity in *Drosophila* is maternal. The mRNA is synthesized in the ovary by nurse cells and then transported to the oocyte, where it localizes to the anterior ends of oocytes. After egg deposition, *bicoid* mRNA is translated and unstable bicoid protein forms a decreasing concentration gradient from the anterior end of the embryo. At the start of gastrulation, *bicoid* mRNA has been degraded. Consider two models to explain the degradation of *bicoid* mRNA: (1) degradation may result from signals within the mRNA (intrinsic model), or (2) degradation may result from the mRNA's position within the egg (extrinsic model). Experimentally, how could one distinguish between these two models?

16. Formation of germ cells in *Drosophila* and many other embryos is dependent on their position in the embryo and their exposure to localized cytoplasmic determinants. Nuclei exposed to cytoplasm in the posterior end of *Drosophila* eggs (the pole plasm) form cells that develop into germ cells under the direction of maternally derived components. R. Amikura et al. (2001. *Proc. Nat. Acad. Sci. (USA)* 98: 9133–9138) consistently found mitochondria-type ribosomes outside mitochondria in the germ plasma of *Drosophila* embryos and postulated that they are intimately related to germ-cell specification. If you were studying this phenomenon, what would you want to know about the activity of these ribosomes?

17. One of the most interesting aspects of early development is the remodeling of the cell cycle from rapid cell divisions, apparently lacking G1 and G2 phases, to slower cell cycles with measurable G1 and G2 phases and checkpoints. During this remodeling, maternal mRNAs that specify cyclins are deadenylated, and zygotic genes are activated to produce cyclins. Audic et al. (2001. *Mol. and Cell. Biol.* 21: 1662–1671) suggest that deadenylation requires transcription of zygotic genes. Present a diagram that captures the significant features of these findings.

18. A number of genes that control expression of *Hox* genes in *Drosophila* have been identified. One of these homozygous mutants is *extra sex combs,* where some of the head and all of the thorax and abdominal segments develop as the last abdominal segment. In other words, all affected segments develop as posterior segments. What does this phenotype tell you about which set of *Hox* genes is controlled by the *extra sex combs* gene?

19. The *apterous* gene in *Drosophila* encodes a protein required for wing patterning and growth. It is also known to function in nerve development, fertility, and viability. When human and mouse genes whose protein products closely resemble *apterous* were used to generate transgenic *Drosophila* (Rincon-Limas et al. 1999. *Proc. Nat. Acad. Sci. [USA]* 96: 2165–2170), the *apterous* mutant phenotype was *rescued*. In addition, the whole-body expression patterns in the transgenic *Drosophila* were similar to normal *apterous*. (a) What is meant by the term *rescued* in this context? (b) What do these results indicate about the molecular nature of development?

20. In *Arabidopsis,* flower development is controlled by sets of homeotic genes. How many classes of these genes are there, and what structures are formed by their individual and combined expression?

21. The floral homeotic genes of *Arabidopsis* belong to the MADS-box gene family, while in *Drosophila,* homeotic genes belong to the homeobox gene family. In both *Arabidopsis* and *Drosophila,* members of the *Polycomb* gene family control expression of these divergent homeotic genes. How do *Polycomb* genes control expression of two very different sets of homeotic genes?

22. Based on the information in Problem 20-3 and the analysis of the phenotypes of single- and double-mutant strains, a model for sex determination in *C. elegans* has been generated. This model proposes that the *her-1* gene controls sex determination by establishing the level of activity of the *tra-1* gene, which in turn, controls the expression of genes involved in generating the various sexually dimorphic tissues. Given this information, (a) does the *her-1* gene product have a negative or a positive effect on the activity of the *tra-1* gene? (b) What would be the phenotype of a *tra-1, her-1* double mutant?

23. Dominguez et al. (2004) suggest that by studying genes that determine growth and tissue specification in the eye of *Drosophila,* much can be learned about human eye development.

    (a) What evidence suggests that genetic eye determinants in *Drosophila* are also found in humans? Include a discussion of orthologous genes in your answer.

    (b) What evidence indicates that the *eyeless* gene is part of a developmental network?

    (c) Are genetic networks likely to specify developmental processes in general? Explain fully and provide an example.

SOLUTIONS TO SELECTED PROBLEMS AND DISCUSSION QUESTIONS

Answers to Now Solve This

1. It is possible that your screen was more inclusive, that is, it identified more subtle alterations than the screen of others. You may have identified several different mutations (multiple alleles) in some of the same genes.

2. Because in *ftz/ftz* embryos, the engrailed product is absent and in *en/en* embryos *ftz* expression is normal, one can conclude that the *ftz* gene product regulates, either directly or indirectly, *en*. Because the *ftz* gene is expressed normally in *en/en* embryos, the product of the *engrailed* gene does not regulate expression of *ftz*.

3. Since *her-1⁻* mutations cause males to develop into hermaphrodites, and *tra-1⁻* mutations cause hermaphrodites to develop into males, one may hypothesize that the *her-1⁺* gene produces a product that suppresses hermaphrodite development, while the *tra-1⁺* gene product is needed for hermaphrodite development.

Solutions to Problems and Discussion Questions

4. The syncytial blasterm is formed as nuclei migrate to the egg's outer margin or cortex, where additional divisions take place. Plasma membranes organize around each of the nuclei at the cortex, thus creating the cellular blastoderm.

6. (a, b) Zygotic genes are activated or repressed depending on their response to maternal-effect gene products. Three subsets of zygotic genes divide the embryo into segments. These segmentation genes are normally transcribed in the developing embryo and their mutations have embryonic lethal phenotypes.

   (c) The maternal genotype contains zygotic genes and these are passed to the embryo as with any other gene.

8. Because the polar cytoplasm contains information to form germ cells, one would expect such a transplantation procedure to generate germ cells in the anterior region.

10. First, one may determine whether levels of hnRNA are consistent among various cell types of interest. This is often accomplished by either direct isolation of the RNA and assessment by northern blotting or by use of *in situ* hybridization.

If the hnRNA pools for a given gene are consistent in various cell types, then transcriptional control can be eliminated as a possibility. Support for translational control can be achieved directly by determining, in different cell types, the presence of a variety of mRNA species with common sequences. This can be accomplished only in cases where sufficient knowledge exists for specific mRNA trapping or labeling. Clues as to translational control *via* alternative splicing can sometimes be achieved by examining the amino acid sequence of proteins. Similarities in certain structural/functional motifs may indicate alternative RNA processing.

12. A dominant gain-of-function mutation is one that changes the specificity or expression pattern of a gene or gene product. The "gain-of-function" *Antp* mutation causes the wild type *Antennapedia* gene to be expressed in the eye-antenna disc and mutant flies have legs on the head in place of antenna.

14. There have been no homeotic transformations noted in mammary glands, so the typical expression of mutant homeotic genes in insects is not revealed in mammary tissue according to Lewis (2000). A substantial number of experiments will be needed to establish a functional link between homeotic gene mutation and cancer induction. Mutagenesis and transgenesis experiments are likely to be most productive in establishing a cause-effect relationship.

16. First, it would be interesting to know whether inhibitors of mitochondrial-ribosomal translation would interfere with germ cell formation. Second, one should know what types of mRNAs are being translated with these ribosomes.

18. Given the information in the problem, it is likely that this gene normally controls the expression of *BX-C* genes in all body segments. The wild type product of *esc* stored in the egg may be required to interpret the information correctly stored in the egg cortex.

22. If the *her-1⁺* product acts as a negative regulator, then when the gene is mutant, suppression over *tra-1⁺* is lost and hermaphroditism would be the result. This hypothesis fits the information provided. The double mutant should be male because even though there is no suppression from *her-1⁻*, there is no *tra-1⁺* product to support hermaphrodite development.

CREDITS

Credits are listed in order of appearance.

Photo

CO, Edward B. Lewis, California Institute of Technology; F-5, Jim Langeland, Stephen Paddock, and Sean Carroll, University of Wisconsin at Madison; F-7 (a, b) Peter A. Lawrence; F-8, Jim Langeland, Steve Paddock and Sean Carroll. HHMI, Dept. Molecular Biology, Univ. Wisconsin; F-9, British Dental Journal, MacMillan; F-10 (a, b) Elsevier Science Ltd.; F-11 (a, b) F. Rudolph Turner, Indiana University—Indiana Memorial Union; F-15, P. Barber/Custom Medical Stock Photo; F-16; Elliott M. Meyerowitz/California Institute of Technology; F-17 (a, b) Max-Planck-Institut fur Entwicklungsbiologie; F-19 (a, b, c, d) Dr. Elliot M. Meyerowitz, California Institute of Technology

Text

Source: PBS NOVA. Source: The New York Times

Quantitative Genetics and Multifactorial Traits

From Chapter 21 of *Essentials of Genetics, Eighth edition,* William S. Klug, Michael R. Cummings, Charlotte A. Spencer, Michael A. Palladino. © 2013 by Pearson Education, Inc. All rights reserved.

A field of pumpkins, where size is under the influence of quantitative inheritance.

Quantitative Genetics and Multifactorial Traits

CHAPTER CONCEPTS

- Quantitative inheritance results in a range of measurable phenotypes for a polygenic trait.

- Polygenic traits most often demonstrate continuous variation.

- Quantitative inheritance can be explained in Mendelian terms whereby certain alleles have an additive effect on the traits under study.

- The study of polygenic traits relies on statistical analysis.

- Heritability values estimate the genetic contribution to phenotypic variability under specific environmental conditions.

- Twin studies allow an estimation of heritability in humans.

- Quantitative trait loci (QTLs) can be mapped and identified.

Typically in traits such as human blood type, squash fruit shape, or fruit fly color, a genotype will produce a single identifiable phenotype, although phenomena such as variable penetrance and expressivity, pleiotropy, and epistasis can obscure the relationship between genotype and phenotype.

However, many traits are not as distinct and clear cut, including many that are of medical or agricultural impor-

tance. They show much more variation, often falling into a continuous range of multiple phenotypes. Most show what we call *continuous variation,* including, for example, height in humans, milk and meat production in cattle, and yield and seed protein content in various crops. Continuous variation across a range of phenotypes can be measured and described in quantitative terms, so this genetic phenomenon is known as **quantitative inheritance.** And because the varying phenotypes result from the input of genes at more than one, and often many, loci, the traits are said to be **polygenic** (literally "of many genes"). The genes involved are often referred to as **polygenes.**

To further complicate the link between the genotype and phenotype, the genotype generated at fertilization establishes a quantitative range within which a particular individual can fall. However, the final phenotype is often also influenced by environmental factors to which that individual is exposed. Human height, for example, is genetically influenced but is also affected by environmental factors such as nutrition. Quantitative (polygenic) traits whose phenotypes result from both gene action and environmental influences are termed **multifactorial,** or **complex traits**. Often these terms are used interchangeably. For consistency throughout the chapter, we will utilize the term *multifactorial* in our discussions.

In this chapter, we will examine examples of quantitative inheritance, multifactorial traits, and some of the statistical techniques used to study them. We will also consider how geneticists assess the relative importance of genetic versus environmental factors contributing to continuous phenotypic variation, and we will discuss approaches to identifying and mapping genes that influence quantitative traits.

1 Quantitative Traits Can Be Explained in Mendelian Terms

The question of whether continuous phenotypic variation could be explained in Mendelian terms caused considerable controversy in the early 1900s. Some scientists argued that, although Mendel's unit factors, or genes, explained patterns of discontinuous variation with discrete phenotypic classes, they could not account for the range of phenotypes seen in quantitative patterns of inheritance. However, geneticists William Bateson and G. Udny Yule, adhering to a Mendelian explanation, proposed the **multiple-factor** or **multiple-gene hypothesis,** in which many genes, each individually behaving in a Mendelian fashion, contribute to the phenotype in a *cumulative* or *quantitative* way.

The Multiple-Gene Hypothesis for Quantitative Inheritance

The **multiple-gene hypothesis** was initially based on a key set of experimental results published by Hermann Nilsson-Ehle in 1909. Nilsson-Ehle used grain color in wheat to test the concept that the cumulative effects of alleles at multiple loci produce the range of phenotypes seen in quantitative traits. In one set of experiments, wheat with red grain was crossed to wheat with white grain (**Figure 1**). The F_1 generation demonstrated an intermediate pink color, which at first sight suggested incomplete dominance of two alleles at a single locus. However, in the F_2 generation, Nilsson-Ehle did not observe the typical segregation of a monohybrid cross. Instead, approximately 15/16 of the plants showed some degree of red grain color, while 1/16 of the plants showed white grain color. Careful examination of the F_2 revealed that grain with color could be classified into four different shades of red. Because the F_2 ratio occurred in sixteenths, it appears that two genes,

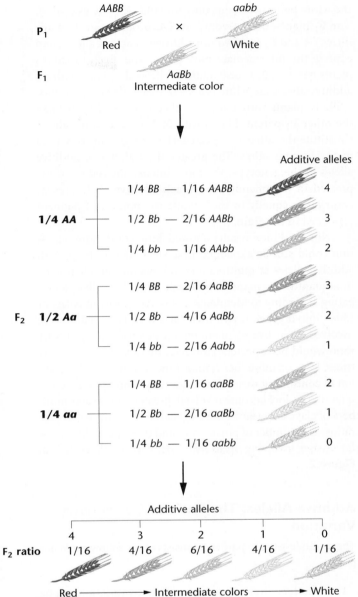

FIGURE 1   How the multiple-factor hypothesis accounts for the 1:4:6:4:1 phenotypic ratio of grain color when all alleles designated by an uppercase letter are additive and contribute an equal amount of pigment to the phenotype.

each with two alleles, control the phenotype and that they segregate independently from one another in a Mendelian fashion.

If each gene has one potential **additive allele** that contributes approximately equally to the red grain color and one potential **nonadditive allele** that fails to produce any red pigment, we can see how the multiple-factor hypothesis could account for the various grain color phenotypes. In the P_1 both parents are homozygous; the red parent contains only additive alleles (*AABB* in Figure 1), while the white parent contains only nonadditive alleles (*aabb*). The F$_1$ plants are heterozygous (*AaBb*), contain two additive (*A* and *B*) and two nonadditive (*a* and *b*) alleles, and express the intermediate pink phenotype. Each of the F$_2$ plants has 4, 3, 2, 1, or 0 additive alleles. F$_2$ plants with no additive alleles are white (*aabb*) like one of the P$_1$ parents, while F$_2$ plants with 4 additive alleles are red (*AABB*) like the other P$_1$ parent. Plants with 3, 2, or 1 additive alleles constitute the other three categories of red color observed in the F$_2$ generation. The greater the number of additive alleles in the genotype, the more intense the red color expressed in the phenotype, as each additive allele present contributes equally to the cumulative amount of pigment produced in the grain.

Nilsson-Ehle's results showed how continuous variation could still be explained in a Mendelian fashion, with additive alleles at multiple loci influencing the phenotype in a quantitative manner, but each individual allele segregating according to Mendelian rules. As we saw in Nilsson-Ehle's initial cross, if two loci, each with two alleles, were involved, then five F$_2$ phenotypic categories in a 1:4:6:4:1 ratio would be expected. However, there is no reason why three, four, or more loci cannot function in a similar fashion in controlling various quantitative phenotypes. As more quantitative loci become involved, greater and greater numbers of classes appear in the F$_2$ generation in more complex ratios. The number of phenotypes and the expected F$_2$ ratios for crosses involving up to five gene pairs are illustrated in **Figure 2**.

Additive Alleles: The Basis of Continuous Variation

The multiple-gene hypothesis consists of the following major points:

1. Phenotypic traits showing continuous variation can be quantified by measuring, weighing, counting, and so on.

2. Two or more gene loci, often scattered throughout the genome, account for the hereditary influence on the phenotype in an *additive way*. Because many genes may be involved, inheritance of this type is called *polygenic*.

3. Each gene locus may be occupied by either an *additive* allele, which contributes a constant amount to the phenotype, or a *nonadditive* allele, which does not contribute quantitatively to the phenotype.

4. The contribution to the phenotype of each additive allele, though often small, is approximately equal. While we now know this is not always true, we have made this assumption in the above discussion.

5. Together, the additive alleles contributing to a single quantitative character produce substantial phenotypic variation.

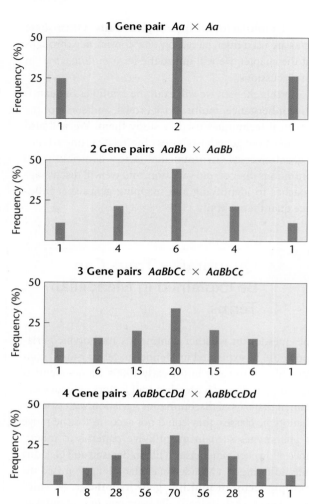

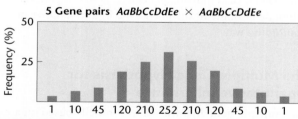

FIGURE 2 The genetic ratios (on the X-axis) resulting from crossing two heterozygotes when polygenic inheritance is in operation with 1 to 5 gene pairs. The histogram bars indicate the distinct F$_2$ phenotypic classes, ranging from one extreme (left end) to the other extreme (right end). Each phenotype results from a different number of additive alleles.

Calculating the Number of Polygenes

Various formulas have been developed for estimating the number of polygenes contributing to a quantitative trait. For example, if the ratio of F$_2$ individuals resembling *either* of the two extreme P$_1$ phenotypes can be determined, the number of polygenes (loci) involved (*n*) may be calculated as

$$1/4^n = \text{ratio of F}_2 \text{ individuals expressing either extreme phenotype}$$

In the example of the red and white wheat grain color summarized in Figure 1, 1/16 of the progeny are either red *or* white like the P_1 phenotypes. This ratio can be substituted on the right side of the equation to solve for n

$$\frac{1}{4^n} = \frac{1}{16}$$

$$\frac{1}{4^2} = \frac{1}{16}$$

$$n = 2$$

Table 1 lists the ratio and the number of F_2 phenotypic classes produced in crosses involving up to five gene pairs.

For low numbers of polygenes (n), it is sometimes easier to use the equation

$(2n + 1) =$ the number of distinct phenotypic
categories observed

For example, when there are two polygenes involved ($n = 2$), then $(2n + 1) = 5$ and each phenotype is the result of 4, 3, 2, 1, or 0 additive alleles. If $n = 3$, $2n + 1 = 7$ and each phenotype is the result of 6, 5, 4, 3, 2, 1, or 0 additive alleles. Thus, working backwards with this rule and knowing the number of phenotypes, we can calculate the number of polygenes controlling them.

It should be noted, however, that both of these simple methods for estimating the number of polygenes involved in a quantitative trait assume not only that all the relevant alleles contribute equally and additively, but also that phenotypic expression in the F_2 is not affected significantly by environmental factors. As we will see later, for many quantitative traits, these assumptions may not be true.

ESSENTIAL POINT ▪ ▪ ▪

Quantitative inheritance results in a range of phenotypes due to the action of additive alleles from two or more genes, as influenced by environmental factors.

NOW SOLVE THIS

1 A homozygous plant with 20-cm diameter flowers is crossed with a homozygous plant of the same species that has 40-cm diameter flowers. The F_1 plants all have flowers 30 cm in diameter. In the F_2 generation of 512 plants, 2 plants have flowers 20 cm in diameter, 2 plants have flowers 40 cm in diameter, and the remaining 508 plants have flowers of a range of sizes in between.

(a) Assuming that all alleles involved act additively, how many genes control flower size in this plant?

(b) What frequency distribution of flower diameter would you expect to see in the progeny of a backcross between an F_1 plant and the large-flowered parent?

▪ **Hint** *This problem provides F_1 and F_2 data for a cross involving a quantitative trait and asks you to calculate the number of genes controlling the trait. The key to its solution is to remember that unless you know the total number of distinct F_2 phenotypes involved, then the ratio (not the number) of parental phenotypes reappearing in the F_2 must be used in your determination of the number of genes involved.*

TABLE 1   **Determination of the Number of Polygenes (n) Involved in a Quantitative Trait**

n	Individuals Expressing Either Extreme Phenotype	Distinct Phenotypic Classes
1	1/4	3
2	1/16	5
3	1/64	7
4	1/256	9
5	1/1024	11

2   The Study of Polygenic Traits
Relies on Statistical Analysis

Before considering the approaches that geneticists use to dissect how much of the phenotypic variation observed in a population is due to genotypic differences among individuals and how much is due to environmental factors, we need to consider the basic statistical tools they use for the task. It is not usually feasible to measure expression of a polygenic trait in every individual in a population, so a random subset of individuals is usually selected for measurement to provide a *sample*. It is important to remember that the accuracy of the final results of the measurements depends on whether the sample is truly random and representative of the population from which it was drawn. Suppose, for example, that a student wants to determine the average height of the 100 students in his genetics class, and for his sample he measures the two students sitting next to him, both of whom happen to be centers on the college basketball team. It is unlikely that this sample will provide a good estimate of the average height of the class, for two reasons: First, it is too small; second, it is not a representative subset of the class (unless all 100 students are centers on the basketball team).

If the sample measured for expression of a quantitative trait is sufficiently large and also representative of the population from which it is drawn, we often find that the data form a **normal distribution;** that is, they produce a characteristic bell-shaped curve when plotted as a frequency histogram **(Figure 3)**. Several statistical concepts are useful in the analysis of traits that exhibit a normal distribution, including the mean, variance, standard deviation, standard error of the mean, covariance, and correlation coefficient.

The Mean

The mean provides information about where the central point lies along a range of measurements for a quantitative trait. **Figure 4** shows the distribution curves for two different sets of phenotypic measurements. Each of these sets of measurements clusters around a central value (as it happens, they both cluster around the same value). This clustering is called a *central tendency*, and the central point is the *mean*

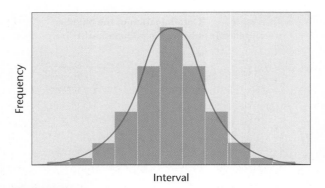

FIGURE 3 Normal frequency distribution, characterized by a bell-shaped curve.

Specifically, the **mean** ($\overline{X}$) is the arithmetic average of a set of measurements and is calculated as

$$\overline{X} = \frac{\Sigma X_i}{n}$$

where $\overline{X}$ is the mean, ΣX_i represents the sum of all individual values in the sample, and n is the number of individual values.

The mean provides a useful descriptive summary of the sample, but it tells us nothing about the range or spread of the data. As illustrated in Figure 4, a symmetrical distribution of values in the sample may, in one case, be clustered near the mean. Or a set of measurements may have the same mean but be distributed more widely around it. A second statistic, the variance, provides information about the spread of data around the mean.

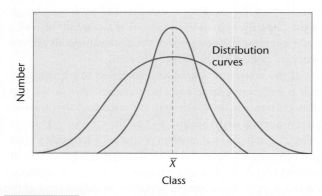

FIGURE 4 Two normal frequency distributions with the same mean but different amounts of variation.

Variance

The **variance** (s^2) for a sample is the average squared distance of all measurements from the mean. It is calculated as

$$s^2 = \frac{\Sigma (X_i - \overline{X})^2}{n - 1}$$

where the sum (Σ) of the squared differences between each measured value (X_i) and the mean ($\overline{X}$) is divided by one less than the total sample size ($n - 1$).

As Figure 4 shows, it is possible for two sets of sample measurements for a quantitative trait to have the same mean but a different distribution of values around it. This range will be reflected in different variances. Estimation of variance can be useful in determining the degree of genetic control of traits when the immediate environment also influences the phenotype.

Standard Deviation

Because the variance is a squared value, its unit of measurement is also squared (m^2, g^2, etc.). To express variation around the mean in the original units of measurement, we can use the square root of the variance, a term called the **standard deviation** (s)

$$s = \sqrt{s^2}$$

Table 2 shows the percentage of individual values within a normal distribution that fall within different multiples of the standard deviation. The values that fall within one standard deviation to either side of the mean represent 68 percent of all values in the sample. More than 95 percent of all values are found within two standard deviations to either side of the mean. This indicates that the standard deviation (s) can also be interpreted in the form of a probability. For example, a sample measurement picked at random has a 68 percent probability of falling within the range of one standard deviation.

TABLE 2	**Sample Inclusion for Various s Values**
Multiples of s	**Sample Included (%)**
$\overline{X} \pm 1s$	68.3
$\overline{X} \pm 1.96s$	95.0
$\overline{X} \pm 2s$	95.5
$\overline{X} \pm 3s$	99.7

Standard Error of the Mean

If multiple samples are taken from a population and measured for the same quantitative trait, we might find that their means vary. Theoretically, larger, truly random samples will represent the population more accurately, and their means will be closer to each other. To measure the accuracy of the sample mean we use the **standard error of the mean** ($S_{\overline{X}}$), calculated as

$$S_{\overline{X}} = \frac{s}{\sqrt{n}}$$

where s is the standard deviation and $\sqrt{n}$ is the square root of the sample size. Because the standard error of the mean is computed by dividing s by $\sqrt{n}$, it is always a smaller value than the standard deviation.

Covariance and Correlation Coefficient

Often geneticists working with quantitative traits find they have to consider two phenotypic characters simultaneously. For example, a poultry breeder might investigate the correlation between body weight and egg production in hens: Do heavier birds tend to lay more eggs? The covariance statistic measures how much variation is common to both quantitative traits. It is calculated by taking the deviations from the mean for each trait (just as we did for estimating variance) for each individual in the sample. This gives a pair of values for each individual. The two values are multiplied together, and the sum of all these individual products is then divided by one fewer than the number in the sample. Thus, the **covariance (cov$_{XY}$)** of two sets of trait measurements, X and Y, is calculated as

$$\text{cov}_{XY} = \frac{\Sigma\left[(X_i - \overline{X})(Y_i - \overline{Y})\right]}{n - 1}$$

The covariance can then be standardized as yet another statistic, the **correlation coefficient (r)**. The calculation is

$$r = \text{cov}_{XY}/S_X S_Y$$

where S_X is the standard deviation of the first set of quantitative measurements X, and S_Y is the standard deviation of the second set of quantitative measurements Y. Values for the correlation coefficient r can range from -1 to $+1$. Positive r values mean that an increase in measurement for one trait tends to be associated with an increase in measurement for the other, while negative r values mean that increases in one trait are associated with decreases in the other. Therefore, if heavier hens do tend to lay more eggs, a positive r value can be expected. A negative r value, on the other hand, suggests that greater egg production is more likely from less heavy birds. One important point to note about correlation coefficients is that even significant r values—close to $+1$ or -1—do not prove that a cause-and-effect relationship exists between two traits. Correlation analysis simply tells us the extent to which variation in one quantitative trait is associated with variation in another, not what causes that variation.

Analysis of a Quantitative Character

To apply these statistical concepts, let's consider a genetic experiment that crossed two different homozygous varieties of tomato. One of the tomato varieties produces fruit averaging 18 oz in weight, whereas fruit from the other averages 6 oz. The F_1 obtained by crossing these two varieties has fruit weights ranging from 10 to 14 oz. The F_2 population contains individuals that produce fruit ranging from 6 to 18 oz. The results characterizing both generations are shown in **Table 3**.

NOW SOLVE THIS

2 The following table shows measurements for fiber lengths and fleece weight in a small flock of eight sheep.

	Sheep Fiber Length (cm)	Fleece Weight (kg)
1	9.7	7.9
2	5.6	4.5
3	10.7	8.3
4	6.8	5.4
5	11.0	9.1
6	4.5	4.9
7	7.4	6.0
8	5.9	5.1

(a) What are the mean, variance, and standard deviation for each trait in this flock?
(b) What is the covariance of the two traits?
(c) What is the correlation coefficient for fiber length and fleece weight?
(d) Do you think greater fleece weight is correlated with an increase in fiber length? Why or why not?

■ **Hint** *This problem provides data for two quantitative traits and asks you to make numerous statistical calculations, ultimately determining if the traits are correlated. The key to its solution is that once the calculation of the correlation coefficient (r) is completed, you must interpret that value—whether it is positive or negative, and how close to zero it is.*

The mean value for the fruit weight in the F_1 generation can be calculated as

$$\overline{X} = \frac{\Sigma X_i}{n} = \frac{626}{52} = 12.04$$

| TABLE 3 | Distribution of F_1 and F_2 Progeny Derived from a Theoretical Cross Involving Tomatoes |

		Weight (oz)												
		6	7	8	9	10	11	12	13	14	15	16	17	18
Number of	F_1					4	14	16	12	6				
Individuals	F_2	1	1	2	0	9	13	17	14	7	4	3	0	1

The mean value for fruit weight in the F_2 generation is calculated as

$$\overline{X} = \frac{\Sigma X_i}{n} = \frac{872}{72} = 12.11$$

Although these mean values are similar, the frequency distributions in Table 3 show more variation in the F_2 generation. The range of variation can be quantified as the sample variance s^2, calculated the sum of the squared differences between each value and the mean, divided by one less than the total number of observations

$$s^2 = \frac{\Sigma (X_i - \overline{X})^2}{n - 1}$$

When the above calculation is made, the variance is found to be 1.29 for the F_1 generation and 4.27 for the F_2 generation. When converted to the standard deviation ($s = \sqrt{s^2}$), the values become 1.13 and 2.06, respectively. Therefore, the distribution of tomato weight in the F_1 generation can be described as 12.04 ± 1.13, and in the F_2 generation it can be described as 12.11 ± 2.06.

Assuming that both parental varieties are homozygous at the loci of interest and that the alleles controlling fruit weight act additively, we can estimate the number of loci involved in this trait. Since 1/72 of the F_2 offspring have a phenotype that overlaps one of the parental strains (72 total F_2 offspring; one weighs 6 oz, one weighs 18 oz; see Table 3), the use of the formula $1/4^n = 1/72$ indicates that n is between 3 and 4, providing evidence of the number of genes that control fruit weight in these tomato strains.

ESSENTIAL POINT ■   ■   ■

Numerous statistical methods are essential during the analysis of quantitative traits, including the mean, variance, standard deviation, standard error, covariance, and the correlation coefficient.

---

3 | Heritability Values Estimate the Genetic Contribution to Phenotypic Variability

The question most often asked by geneticists working with multifactorial traits and diseases is how much of the observed phenotypic variation in a population is due to genotypic differences among individuals and how much is due to environment. The term **heritability** is used to describe *the proportion of total phenotypic variation in a population that is due to genetic factors*. For a multifactorial trait in a given population, a high heritability estimate indicates that much of the variation can be attributed to genetic factors, with the environment having less impact on expression of the trait.

With a low heritability estimate, environmental factors are likely to have a greater impact on phenotypic variation within the population.

The concept of heritability is frequently misunderstood and misused. It should be emphasized that heritability indicates neither how much of a trait is genetically determined nor the extent to which an individual's phenotype is due to genotype. In recent years, such misinterpretations of heritability for human quantitative traits have led to controversy, notably in relation to measurements such as intelligence quotients, or IQs. Variation in heritability estimates for IQ among different racial groups led to incorrect suggestions that unalterable genetic factors control differences in intelligence levels among humans of different ancestries. Such suggestions misrepresented the meaning of heritability and ignored the contribution of genotype-by-environment interaction variance to phenotypic variation in a population. Moreover, heritability is not fixed for a trait. For example, a heritability estimate for egg production in a flock of chickens kept in individual cages might be high, indicating that differences in egg output among individual birds are largely due to genetic differences, as they all have very similar environments. For a different flock kept outdoors, heritability for egg production might be much lower, as variation among different birds may also reflect differences in their individual environments. Such differences could include how much food each bird manages to find and whether it competes successfully for a good roosting spot at night. Thus, a heritability estimate tells us the proportion of *phenotypic variation* that can be attributed to *genetic variation within a certain population in a particular environment*. If we measure heritability for the same trait among different populations in a range of environments, we frequently find that the calculated heritability values have large standard errors. This is an important point to remember when considering heritability estimates for traits in human populations. A mean heritability estimate of 0.65 for human height does not mean that your height is 65 percent due to your genes, but rather that in the populations sampled, on average, *65 percent of the overall variation in height could be explained by genotypic differences among individuals*.

With this subtle but important distinction in mind, we will now consider how geneticists divide the phenotypic variation observed in a population into genetic and environmental components. As we saw in the previous section, variation can be quantified as a sample variance: taking measurements of the trait in question from a representative sample of the population and determining the extent of the spread of those measurements around the sample mean. This gives us an estimate of the total **phenotypic variance** in the population (V_P). Heritability estimates are obtained by using different experimental and statistical techniques to partition V_P into **genotypic variance** (V_G) and **environmental variance** (V_E) components.

An important factor contributing to overall levels of phenotypic variation is the extent to which individual genotypes affect the phenotype differently depending on the environment. For example, wheat variety A may yield an average of 20 bushels an acre on poor soil, while variety B yields an average of 17 bushels. On good soil, variety A yields 22 bushels, while variety B averages 25 bushels an acre. There are differences in yield between the two genotypically distinct varieties, so variation in wheat yield has a genetic component. Both varieties yield more on good soil, so yield is also affected by environment. However, we also see that the two varieties do not respond to better soil conditions equally: The genotype of wheat variety B achieves a greater increase in yield on good soil than does variety A. Thus, we have differences in the interaction of genotype, with environment contributing to variation for yield in populations of wheat plants. This third component of phenotypic variation is **genotype-by-environment interaction variance ($V_{G \times E}$)** (**Figure 5**).

We can now summarize all the components of total phenotypic variance V_P using the following equation:

$$V_P = V_G + V_E + V_{G \times E}$$

(a)

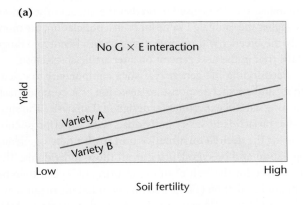

(b)

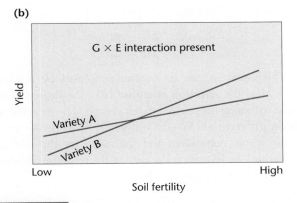

FIGURE 5   Differences in yield between two wheat varieties at different soil fertility levels. (a) No genotype-by-environment, or G × E, interaction: The varieties show genetic differences in yield but respond equally to increasing soil fertility. (b) G × E interaction present: Variety A outyields B at low soil fertility, but B yields more than A at high-fertility levels.

In other words, total phenotypic variance can be subdivided into genotypic variance, environmental variance, and genotype-by-environment interaction variance. When obtaining heritability estimates for a multifactorial trait, researchers often assume that the genotype-by-environment interaction variance is small enough that it can be ignored or combined with the environmental variance. However, it is worth remembering that this kind of approximation is another reason heritability values are *estimates* for a given population in a particular context, not a *fixed attribute* for a trait.

Animal and plant breeders use a range of experimental techniques to estimate heritabilities by partitioning measurements of phenotypic variance into genotypic and environmental components. One approach uses inbred strains containing genetically homogeneous individuals with highly homozygous genotypes. Experiments are then designed to test the effects of a range of environmental conditions on phenotypic variability. Variation *between* different inbred strains reared in a constant environment is due predominantly to genetic factors. Variation *among* members of the same inbred strain reared under different conditions is more likely to be due to environmental factors. Other approaches involve analysis of variance for a quantitative trait among offspring from different crosses, or comparing expression of a trait among offspring and parents reared in the same environment.

Broad-Sense Heritability

Broad-sense heritability (represented by the term H^2) measures the contribution of the genotypic variance to the total phenotypic variance. It is estimated as a proportion:

$$H^2 = \frac{V_G}{V_P}$$

Heritability values for a trait in a population range from 0.0 to 1.0. A value approaching 1.0 indicates that the environmental conditions have little impact on phenotypic variance, which is therefore largely due to genotypic differences among individuals in the population. Low values close to 0.0 indicate that environmental factors, not genotypic differences, are largely responsible for the observed phenotypic variation within the population studied. Few quantitative traits have very high or very low heritability estimates, suggesting that both genetics and environment play a part in the expression of most phenotypes for the trait.

The genotypic variance component V_G used in broad-sense heritability estimates includes all types of genetic variation in the population. It does not distinguish between quantitative trait loci with alleles acting additively as opposed to those with epistatic or dominance effects. Broad-sense heritability estimates also assume that the genotype-by-environment variance component is negligible. While broad-sense heritability estimates for a trait are of general genetic interest, these limitations mean this kind of heritability is not very useful in breeding programs. Animal or plant breeders wishing to develop improved strains of livestock or higher-yielding crop varieties need more precise heritability estimates for the traits they wish to manipulate in a population. Therefore, another type of estimate, narrow-sense heritability, has been devised that is of more practical use.

Narrow-Sense Heritability

Narrow-sense heritability (h^2) is the proportion of phenotypic variance due to additive genotypic variance alone. Genotypic variance can be divided into subcomponents representing the different modes of action of alleles at quantitative trait loci. As not all the genes involved in a quantitative trait affect the phenotype in the same way, this partitioning distinguishes among three different kinds of gene action contributing to genotypic variance. **Additive variance, V_A,** is the genotypic variance due to the additive action of alleles at quantitative trait loci. **Dominance variance, V_D,** is the deviation from the additive components that results when phenotypic expression in heterozygotes is not precisely intermediate between the two homozygotes. **Interactive variance, V_I,** is the deviation from the additive components that occurs when two or more loci behave epistatically. The amount of interactive variance is often negligible, and so this component is often excluded from calculations of total genotypic variance.

The partitioning of the total genotypic variance V_G is summarized in the equation

$$V_G = V_A + V_D + V_I$$

and a narrow-sense heritability estimate based only on that portion of the genotypic variance due to additive gene action becomes

$$h^2 = \frac{V_A}{V_P}$$

Omitting V_I and separating V_P into genotypic and environmental variance components, we obtain

$$h^2 = \frac{V_A}{V_E + V_A + V_D}$$

Heritability estimates are used in animal and plant breeding to indicate the potential response of a population to artificial selection for a quantitative trait. Narrow-sense heritability, h^2, provides a more accurate prediction of selection response than broad-sense heritability, H^2, and therefore h^2 is more widely used by breeders.

ESSENTIAL POINT ■ ■ ■

Heritability is an estimate of the relative contribution of genetic versus environmental factors to the range of phenotypic variation seen in a quantitative trait in a particular population and environment.

Artificial Selection

Artificial selection is the process of choosing specific individuals with preferred phenotypes from an initially heterogeneous population for future breeding purposes. Theoretically, if artificial selection based on the same trait preferences is repeated over multiple generations, a population can be developed containing a high frequency of individuals with the desired characteristics. If selection is for a simple trait controlled by just one or two genes subject to little environmental influence, generating the desired population of plants or animals is relatively fast and easy. However, many traits of economic importance in crops and livestock, such as grain yield in plants, weight gain or milk yield in cattle, and speed or stamina in horses, are polygenic and frequently multifactorial. Artificial selection for such traits is slower and more complex. Narrow-sense heritability estimates are valuable to the plant or animal breeder because, as we have just seen, they estimate the proportion of total phenotypic variance for the trait that is due to additive genetic variance. Quantitative trait alleles with additive impact are those most easily manipulated by the breeder. Alleles at quantitative trait loci that generate dominance effects or interact epistatically (and therefore contribute to V_D or V_I) are less responsive to artificial selection. Thus, narrow-sense heritability, h^2, can be used to predict the impact of selection. The higher the estimated value for h^2 in a population, the more likely the breeder will observe a change in phenotypic range for the trait in the next generation after artificial selection.

Partitioning the genetic variance components to calculate h^2 and predict response to selection is a complex task requiring careful experimental design and analysis. The simplest approach is to select individuals with superior phenotypes for the desired quantitative trait from a heterogeneous population and breed offspring from those individuals. The mean score for the trait of those offspring ($M2$) can then be compared to that of: (1) the original population's mean score (M) and (2) the selected individuals used as parents ($M1$). The relationship between these means and h^2 is

$$h^2 = \frac{M2 - M}{M1 - M}$$

This equation can be further simplified by defining $M2 - M$ as the **selection response (R)**—the degree of response to mating the selected parents—and $M1 - M$ as the **selection differential (S)**—the difference between the mean for the whole population and the mean for the selected population—so h^2 reflects the ratio of the response observed to the total response possible. Thus,

$$h^2 = \frac{R}{S}$$

A narrow-sense heritability value obtained in this way by selective breeding and measuring the response in the offspring is referred to as an estimate of **realized heritability.**

As an example of a realized heritability estimate, suppose that we measure the diameter of corn kernels in a population where the mean diameter M is 20 mm. From this population, we select a group with the smallest diameters, for which the mean $M1$ equals 10 mm. The selected plants are interbred, and the mean diameter $M2$ of the progeny kernels is 13 mm. We can calculate the realized heritability h^2 to estimate the potential for artificial selection on kernel size

$$h^2 = \frac{M2 - M}{M1 - M}$$

$$h^2 = \frac{13 - 20}{10 - 20}$$

$$= \frac{-7}{-10}$$

$$= 0.70$$

This value for narrow-sense heritability indicates that the selection potential for kernel size is relatively high.

The longest running artificial selection experiment known is still being conducted at the State Agricultural Laboratory in Illinois. Corn has been selected for both high and low oil content. After 76 generations, selection continues to result in increased oil content (**Figure 6**). With each cycle of successful selection, more of the corn plants accumulate a higher percentage of additive alleles involved in oil production. Consequently, the narrow-sense heritability h^2 of increased oil content in succeeding generations has declined (see parenthetical values at generations 9, 25, 52, and 76 in

TABLE 4	Estimates of Heritability for Traits in Different Organisms
Trait	**Heritability (h^2)**
Mice	
Tail length	60%
Body weight	37
Litter size	15
Chickens	
Body weight	50
Egg production	20
Egg hatchability	15
Cattle	
Birth weight	45
Milk yield	44
Conception rate	3

Figure 6) as artificial selection comes closer and closer to optimizing the genetic potential for oil production. Theoretically, the process will continue until all individuals in the population possess a uniform genotype that includes all the additive alleles responsible for high oil content. At that point, h^2 will be reduced to zero, and response to artificial selection will cease. The decrease in response to selection for low oil content shows that heritability for low oil content is approaching this point.

Table 4 lists narrow-sense heritability estimates expressed as percentage values for a variety of quantitative traits in different organisms. As you can see, these h^2 values vary, but heritability tends to be low for quantitative traits that are essential to an organism's survival. Remember, this does not indicate the absence of a genetic contribution to the observed phenotypes for such traits. Instead, the low h^2 values show that natural selection has already largely optimized the genetic component of these traits during evolution. Egg production, litter size, and conception rate are examples of how such physiological limitations on selection have already been reached. Traits that are less critical to survival, such as body weight, tail length, and wing length, have higher heritabilities because more genotypic variation for such traits is still present in the population. Remember also that any single heritability estimate can only provide information about one population in a specific environment. Therefore, narrow-sense heritability is a more valuable predictor of response to selection when estimates are calculated for many populations and environments and show the presence of a clear trend.

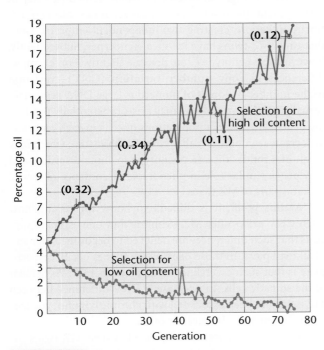

FIGURE 6 Response of corn selected for high and low oil content over 76 generations. The numbers in parentheses at generations 9, 25, 52, and 76 for the "high oil" line indicate the calculation of heritability at these points in the continuing experiment.

4 Twin Studies Allow an Estimation of Heritability in Humans

Twins are useful subjects for examining how much phenotypic variance for a human multifactorial trait is due to the genotype as opposed to the environment. In these studies, the underlying principle has been that **monozygotic (MZ),** or **identical, twins** are derived from a single zygote that divides mitotically and then spontaneously splits into two separate cells. Both cells give rise to a genotypically identical embryos. **Dizygotic (DZ),** or **fraternal, twins,** on the other hand, originate from two separate fertilization events and are only as genetically similar as any two siblings, with an average of 50 percent of their alleles in common. For a given trait, therefore, phenotypic differences between pairs of MZ twins will be equivalent to the environmental variance (V_E) (because the genotypic variance is zero). Phenotypic differences between DZ twins, however, display both environmental variance and approximately half the genotypic variance (V_G). Comparing the extent of phenotypic variance for the same trait in MZ and DZ sets of twins provides an estimate of broad-sense heritability for the trait.

Twins are said to be **concordant** for a given trait if both express it or neither expresses it. If one expresses the trait and the other does not, the pair is said to be **discordant**. Comparison of concordance values of MZ versus DZ twins reared together illustrates the potential value for heritability assessment. (See Now Solve This 3 on the next page, for example.)

Before any conclusions can be drawn from twin studies, the data must be examined carefully. For example, if concordance values approach 90 to 100 percent in MZ twins, we might be inclined to interpret that as a large genetic contribution to the phenotype of the trait. In some cases—for example, blood type and eye color—we know that this is indeed true. In the case of contracting measles, however, a high concordance value merely indicates that the trait is almost always induced by a factor in the environment—in this case, a virus.

It is more meaningful to compare the *difference* between the concordance values of MZ and DZ twins. If concordance values are significantly higher in MZ twins, we suspect a strong genetic component in the determination of the trait. In the case of measles, where concordance is high in both types of twins, the environment is assumed to be the major contributing factor. Such an analysis is useful because phenotypic characteristics that remain similar in different environments are likely to have a strong genetic component.

Twin Studies Have Several Limitations

Interesting as they are, human twin studies contain some unavoidable sources of error. For example, MZ twins are often treated more similarly by parents and teachers than are DZ twins, especially when the DZ siblings are of different sex. This circumstance may inflate the environmental variance for DZ twins. Another possible source of error is interactions between the genotype and the environment that produce variability in the phenotype. These interactions can increase the total phenotypic variance for DZ twins compared to MZ

twins raised in the same environment, influencing heritability calculations. Overall, heritability estimates for human traits based on twin studies should therefore be considered approximations and examined very carefully before any conclusions are drawn.

Although they must often be viewed with caution, classical twin studies, based on the assumption that MZ twins share the same genome, have been valuable for estimating heritability over a wide range of traits including multifactorial disorders such as cardiovascular disease, diabetes, and mental illness, for example. These disorders clearly have genetic components, and twin studies provide a foundation for studying interactions between genes and environmental factors. However, results from genomics research have challenged the view that MZ twins are truly identical and have forced a reevaluation of both the methodology and the results of twin studies. Such research has also opened the way to new approaches to the study of interactions between the genotype and environmental factors.

The most relevant genomic discoveries about twins include the following:

- By the time they are born, MZ twins do not necessarily have identical genomes.

- Gene-expression patterns in MZ twins change with age, leading to phenotypic differences.

We will address these points in order. First, MZ twins develop from a single fertilized egg, where sometime early in development, the resulting cell mass separates into two distinct populations creating two independent embryos. Until that time, MZ twins have identical genotypes. Subsequently, however, the genotypes can diverge slightly. For example, differences in **copy number variation (CNV)**—variation in the number of copies of numerous large DNA sequences (usually 1000 bp or more)—may arise, differentially producing genetically distinct populations of cells in each embryo . This creates a condition called *somatic mosaicism,* which may result in a milder disease phenotype in some disorders and may play a similar role in phenotypic discordance observed in some pairs of MZ twins.

At this point, it is difficult to know for certain how often CNV arises after MZ twinning, but one estimate suggests that such differences are believed to occur in 10 percent of all twin pairs. In those pairs where it does occur, one estimate is that such divergence takes place in 15 to 70 percent of the somatic cells. In one case, a CNV difference between MZ twins has been associated with chronic lymphocytic leukemia in one twin, but not the other.

The second genomic difference between MZ twins involves **epigenetics**—the chemical modification of their DNA and associated histones. An international study of epigenetic modifications in adult European MZ twins showed that MZ twin pairs are epigenetically identical at birth, but adult MZ twins show significant differences in the methylation patterns of both DNA and histones. Such epigenetic changes in turn affect patterns of gene expresson. The accumulation of epigenetic changes and gene-expression profiles may explain some of the observed phenotypic discordance and susceptibility to disease in adult MZ twins. For example, a difference in DNA methylation patterns is observed in MZ twins discordant for Beckwith-Wiedemann syndrome, a genetic disorder associated with developmental overgrowth of certain tissues and organs and an increased risk of cancer.

ESSENTIAL POINT ■ ■ ■

Twin studies, while having some limitations, are useful in assessing heritabilities for polygenic traits in humans.

NOW SOLVE THIS

3 The following table gives the percentage of twin pairs studied in which both twins expressed the same phenotype for a trait (concordance). Percentages listed are for concordance for each trait in monozygotic (MZ) and dizygotic (DZ) twins. Assuming that both twins in each pair were raised together in the same environment, what do you conclude about the relative importance of genetic versus environmental factors for each trait?

Trait	MZ %	DZ %
Blood types	100	66
Eye color	99	28
Mental retardation	97	37
Measles	95	87
Hair color	89	22
Handedness	79	77
Idiopathic epilepsy	72	15
Schizophrenia	69	10
Diabetes	65	18
Identical allergy	59	5
Cleft lip	42	5
Club foot	32	3
Mammary cancer	6	3

■ **Hint** *This problem asks you to evaluate the relative importance of genetic versus environmental contributions to specific traits by examining concordance values in MZ versus DZ twins. The key to its solution is to examine the difference in concordance values and to factor in what you have learned about the genetic differences between MZ and DZ twins.*

Progressive, age-related genomic modifications may be the result of MZ twins being exposed to different environmental factors, or from failure of epigenetic marking following DNA replication. These findings also indicate that

concordance studies in DZ twins must take into account genetic as well as epigenetic differences that contribute to discordance in these twin pairs.

The realization that *epigenetics* may play an important role in the development of phenotypes promises to make twin studies an especially valuable tool in dissecting the interactions among genes and the role of environmental factors in the production of phenotypes. Once the degree of epigenetic differences between MZ and DZ twin pairs has been defined, molecular studies on DNA and histone modification can link changes in gene expression with differences in the concordance rates between MZ and DZ twins.

5 | Quantitative Trait Loci Can Be Mapped

Environmental effects, interaction among segregating alleles, and the large number of genes that may contribute to a polygenic phenotype make it difficult to: (1) identify all polygenes that are involved; and (2) determine the effect of each gene on the phenotype. However, because many quantitative traits are of economic or medical relevance, it is often desirable to obtain this information. In such studies, a chromosome region known as a **quantitative trait locus (QTL)**[*] is identified as containing one or more genes contributing to a quantitative trait. When possible, the relevant gene or genes contained within a QTL are isolated and studied.

The modern approach used to find and map QTLs involves looking for associations between DNA markers and phenotypes. One way to do this is to begin with individuals from two lines created by artificial selection that are highly divergent for a phenotype (fruit weight, oil content, bristle number, etc.). For example, **Figure 7** illustrates a generic case of QTL mapping. Over many generations of artificial selection, two divergent lines become highly homozygous, which facilitates their use in QTL mapping. Individuals from each of the lines with divergent phenotypes [generation 25 in Figure 7(a)] are used as parents to create a generation whose members will be heterozygous at most of the loci contributing to the trait. Additional crosses, either among F_1 individuals or between the F_1 and the inbred parent lines, result in F_2 generations that carry different portions of the parental genomes [Figure 7(b)] with different QTL genotypes and associated phenotypes. This segregating is known as the **QTL mapping population.**

Researchers then measure phenotypic expression of the trait among individuals in the mapping population and identify genomic differences among individuals by using chromosome-specific DNA markers such as *restriction fragment*

---

[*]We utilize QTLs to designate the plural form, quantitative trait loci.

(a)

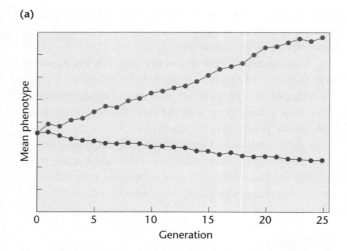

(b)

$P_1 \times P_1$

$F_1 \times F_1$

F_2

(c)

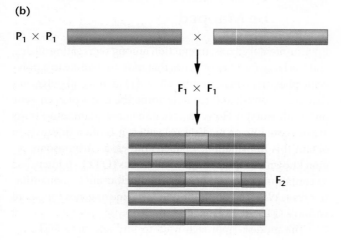

© 2001 Nature Publishing Group

FIGURE 7 (a) Individuals from highly divergent lines created by artificial selection are chosen from generation 25 as parents. (b) The thick bars represent the genomes of individuals selected from the divergent lines as parents. These individuals are crossed to produce an F_1 generation (not shown). An F_2 generation is produced by crossing members of the F_1. As a result of crossing over, individual members of the F_2 generation carry different portions of the P_1 genome, as shown by the colored segments of the thick bars. DNA markers and phenotypes in individuals of the F_2 generation are analyzed. (c) Statistical methods are used to determine the probability that a DNA marker is associated with a QTL that affects the phenotype. The results are plotted as the likelihood of association against chromosomal location. Units on genetic maps are measured in centimorgans (cM), determined by crossover frequencies. Peaks above the horizontal line represent significant results. The data shows five possible QTLs, with the most significant findings at about 10 cM and 60 cM.

genotypes at that marker locus will also differ in their phenotypic expression of the trait. When this occurs, the marker locus and the QTL are said to *cosegregate*. Consistent cosegregation establishes the presence of a QTL at or near the DNA marker along the chromosome—in other words, the marker and QTL are linked. When numerous QTLs for a given trait have been located, a genetic map is created, showing the probability that specific chromosomal regions are associated with the phenotype of interest [Figure 7(c)]. Further research using genomic techniques identifies genes in these regions that contribute to the phenotype.

QTL mapping has been extensively used in agriculture, including plants such as corn, rice, wheat, and tomatoes (**Table 5**), and livestock such as cattle, pigs, sheep, and chickens. For example, hundreds of QTLs have been located in the tomato, and its genome has been sequenced. Many chromosome regions responsible for quantitative traits such as fruit size, shape, soluble solid content, and acidity have been identified. These QTLs are distributed on all 12 chromosomes representing the haploid genome of this plant.

TABLE 5 QTLs for Quantitative Phenotypes

Organism	Quantitative Phenotype	QTLs Identified
Tomato	Soluble solids	7
	Fruit mass	13
	Fruit pH	9
	Growth	5
	Leaflet shape	9
	Height	9
Maize	Height	11
	Leaf length	7
	Grain yield	18
	Number of ears	10

Source: Used with permission of Annual Reviews of Genetics, from "Mapping Polygenes" by S. D. Tanksley, *Annual Review of Genetics*, Vol. 27:205–233, Table 1, December 1993. Permission conveyed through Copyright Clearance Center, Inc.

length polymorphisms (*RFLPs*), *microsatellites,* and *single-nucleotide polymorphisms* (*SNPs*) . Computer-based statistical analysis is used to search for linkage between the markers and a component of phenotypic variation associated with the trait. If a DNA marker (such as those markers described above) *is not* linked to a QTL, then the phenotypic mean score for the trait will not vary among individuals with different genotypes at that marker locus. However, if a DNA marker *is* linked to a QTL, then different

A research program conducted by Steven Tanksley and his colleagues at Cornell University has focused on mapping and characterizing quantitative traits in the tomato, including fruit shape and weight. We will describe one aspect of this research, which represents a model approach in the study of QTLs. While the cultivated tomato can weigh up to 1000 grams, fruit from the related wild species thought to be the ancestor of the modern tomato weighs only a few grams [Figure 8(a)]. QTL mapping has identified more than 28 QTLs related to this thousand-fold variation in fruit weight. More than ten years of work was required to localize, identify, and clone one of these QTLs, called *fw2.2* (on chromosome 2). Within this QTL, a specific gene, *ORFX,* has been identified, and alleles at this locus are responsible for about 30 percent of the variation in fruit weight.

The *ORFX* gene has been isolated, cloned, and transferred between plants, with interesting results. One allele of *ORFX* is present in all wild small-fruited varieties of tomatoes investigated, while another allele is present in all

domesticated large-fruited varieties. When a cloned *ORFX* gene from small-fruited varieties is transferred to a plant that normally produces large tomatoes, the transformed plant produces fruits that are greatly reduced in weight [(Figure 8(b)]. In the varieties studied by Tanksley's group, the reduction averaged 17 grams, a statistically significant phenotypic change caused by the action of a gene found within a single QTL.

Further analysis of *ORFX* revealed that this gene encodes a protein that negatively regulates cell division during fruit development. Differences in the time of gene expression and in the amount of transcript produced lead to small or large fruit. Higher levels of expression mediated by transferred *ORFX* alleles exert a negative control over cell division, resulting in smaller tomatoes.

Yet *ORFX* and other related genes cannot account for all the observed variation in tomato size. Analysis of another QTL, *fas* (located on chromosome 11), indicates that the development of extreme differences in fruit size resulting from artificial selection also involves an increase in the number of compartments in the mature fruit. The small, ancestral stocks produce fruit with two to four seed compartments, but the large-fruited present-day strains have eight or more compartments. Thus, the QTLs that affect fruit size in tomatoes work by controlling at least two developmental processes: cell division early in development and the determination of the number of ovarian compartments.

The discovery that QTLs can control levels of gene expression has led to new, molecular definitions of phenotypes associated with quantitative traits. For example, the phenotype investigated may be the amount of an RNA transcript produced by a gene (**expression QTLs,** or **eQTLs**), or the amount of protein produced (**protein QTLs,** or **pQTLs**). These molecular phenotypes are polygenically controlled in the same way as more conventional phenotypes, such as fruit weight. Gene expression, for example, is controlled by *cis* factors, including promoters, and by *trans*-acting transcription factors .

The use of these new methods of QTL analysis moves the field in a new direction by focusing on regulatory networks, protein–protein interactions, and systems biology. In plant biology, eQTLs are being used to study flowering time, pathogen resistance, and the influence of the environment on developmental events. These new methods will be useful not only in agriculture, but in other fields as well, including the dissection of multifactorial traits and diseases in humans such as *cleft lip*, *spina bifida*, *Type II diabetes*, and *coronary artery disease*.

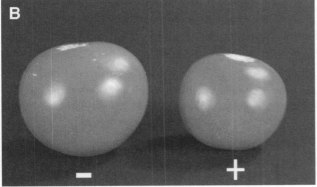

FIGURE 8 (a) The ancestral and modern-day tomato. (b) Phenotypic effect of the *fw2.2* transgene in the tomato. When the allele causing small fruit is transferred to a plant that normally produces large fruit, the fruit is reduced in size (+). A control fruit (−) that has not been transformed is shown for comparison.

ESSENTIAL POINT ■ ■ ■

Quantitative trait loci, or QTLs, may be identified and mapped using DNA markers.

GENETICS, TECHNOLOGY, AND SOCIETY

The Green Revolution Revisited: Genetic Research with Rice

Of the 7 billion people now living on Earth, over 800 million do not have enough to eat. That number is expected to grow by an additional 1 million people each year for the next several decades. How will we be able to feed the estimated 8 billion people on Earth by 2025?

The past gives us some reasons to be optimistic. In the 1950s and 1960s, in the face of looming population increases, plant scientists around the world set about to increase the production of crop plants, including the three most important grains—rice, wheat, and maize. These efforts became known as the *Green Revolution*. The approach was three-fold: (1) to increase the use of fertilizers, pesticides, and irrigation; (2) to bring more land under cultivation; and (3) to develop improved varieties of crop plants by intensive plant breeding.

The results were dramatic. Developing nations more than doubled their production of rice, wheat, and maize between 1961 and 1985. Nations that were facing widespread famine in the 1960s were able to feed themselves and became major exporters of grain.

The Green Revolution saved millions of people from starvation and improved the quality of life for millions more; however, its effects may be diminishing. The rate of increase in grain yields has slowed since the 1980s, due to slower growth in irrigation development and fertilizer use. If food production is to keep pace with the projected increase in the world's population, we will have to depend more and more on the genetic improvement of crop plants to provide higher yields. Is this possible, or are we approaching the theoretical limits of yield in important crop plants? Recent work with rice suggests that the answer to this question may be a resounding no.

About half of the Earth's population depends on rice for basic nourishment. The Green Revolution for rice began in 1960, aided by the establishment of the International Rice Research Institute (IRRI). One of their major developments was the breeding of a rice variety with improved disease resistance and higher yield. The IRRI research team crossed a Chinese rice variety (*Dee-geo-woo-gen*) and an Indonesian variety (*Peta*) to create a new cultivar known as IR8. IR8 produced a greater number of rice kernels per plant, when grown in the presence of fertilizers and irrigation. Under these cultural practices, IR8 plants were so top-heavy with grain that they tended to fall over—a trait called "lodging." To reduce lodging, IRRI breeders crossed IR8 with a dwarf native variety to create semi-dwarf lines. Due in part to the adoption of the semi-dwarf IR8 lines, the world production of rice doubled in 25 years.

Predictions suggest that a 40 percent increase in the annual rice harvest may be necessary to keep pace with anticipated population growth during the next 30 years. As land and water resources become more scarce and the prices of fertilizers and pesticides grow, more emphasis will be placed on creating new rice varieties that have even higher yields and greater disease resistance.

Geneticists are now using several strategies to develop more productive rice. Conventional hybridization and selection techniques, such as those used to create the IR8 strain, continue to produce varieties that contribute to a productivity increase of between 1 and 10 percent a year. Several quantitative trait loci (QTLs) from wild rice appear to contribute to increased yields, and scientists are attempting to introduce these traits into current dwarf varieties of domestic rice. Genomics and genetic engineering are also contributing to the new Green Revolution for rice. In 2002, the rice genome was the first cereal crop genome to be sequenced. Research is now concentrated on assigning a function to all the genes in the rice genome. Once identified and characterized, these genes may be transferred into crop plants, speeding the creation of rice varieties with desirable traits, such as disease resistance, tolerance to drought and salinity, and improved nutritional content.

Your Turn

Take time, individually or in groups, to answer the following questions. Investigate the references and links to help you understand some of the technologies and issues surrounding the new Green Revolution.

1. How do you think that genomics and genetic engineering will contribute to the development of more productive rice varieties?

   This topic and others are discussed in a series of articles entitled How to Feed a Hungry World, *which can be found in the July 29, 2010 issue of* Nature (Nature 466: 546–561).

2. A serious complication in our efforts to feed the Earth's population is global climate change. Given your assessment of current climate change predictions, what traits should we introduce into rice plants, and what methods can we use to do this?

   To learn more about the effects of global climate change on agriculture, read Battisti, D. S. and Naylor, R. L. 2009. Historical warnings of future food insecurity with unprecedented seasonal heat. *Science* 323: 240–244. *Also, search for "climate change" publications on the Web site of the International Rice Research Institute* (http://www.irri.org).

3. Despite its benefits, the Green Revolution has been the subject of controversy. What are the main criticisms of the Green Revolution, and how can we mitigate some of the Green Revolution's negative aspects?

   A discussion of this topic can be found in Bazuin, S., et al. 2011. Application of GM crops in sub-Saharan Africa: lessons learned from Green Revolution. *Biotechnol Adv* 29: 908–912.

CASE STUDY » A genetic flip of the coin

On July 11, 2008, twin sons were born to Stephan Gerth from Germany and Addo Gerth from Ghana. Stephan is very fair-skinned with blue eyes and straight hair; Addo is dark-skinned, with brown eyes and curly hair. The first born of the twins, Ryan, is fair-skinned, with blue eyes and straight hair; his brother, Leo, has light brown skin, brown eyes, and curly hair. Although the twins' hair texture and eye color were the same as those of one or the other parent, the twins had different skin colors, intermediate to that of their parents. Experts explained that the blending effect of skin color in the twins resulted from quantitative inheritance involving at least three different gene pairs, whereas hair texture and eye color are not quantitatively inherited. Using this

as an example of quantitative genetics, we can ask the following questions:

1. What approach is used in estimating how many gene pairs are involved in a quantitative trait? Why would this be extremely difficult in the case of skin color in humans?
2. Would either parent need to have mixed-race ancestry for the twins to be so different?
3. Would twins showing some parental traits (hair texture, eye color) but a blending of other traits (skin color in this case) seem to be a commonplace event, or are we looking at a "one in a million" event?

INSIGHTS AND SOLUTIONS

1. In a certain plant, height varies from 6 to 36 cm. When 6-cm and 36-cm plants were crossed, all F_1 plants were 21 cm. In the F_2 generation, a continuous range of heights was observed. Most were around 21 cm, and 3 of 200 were as short as the 6-cm P_1 parent.

(a) What mode of inheritance does this illustrate, and how many gene pairs are involved?

(b) How much does each additive allele contribute to height?

(c) List all genotypes that give rise to plants that are 31 cm.

Solution:

(a) Polygenic inheritance is illustrated when a trait is continuous and when alleles contribute additively to the phenotype. The $3/200$ ratio of F_2 plants is the key to determining the number of gene pairs. This reduces to a ratio of $1/66.7$, very close to $1/64$. Using the formula $1/4^n = 1/64$ (where $1/64$ is equal to the proportion of F_2 phenotypes as extreme as either P_1 parent), $n = 3$. Therefore, three gene pairs are involved.

(b) The variation between the two extreme phenotypes is

$$36 - 6 = 30 \text{ cm}$$

Because there are six potential additive alleles ($AABBCC$), each contributes

$$30/6 = 5 \text{ cm}$$

to the base height of 6 cm, which results when no additive alleles ($aabbcc$) are part of the genotype.

(c) All genotypes that include five additive alleles will be 31 cm (5 alleles × 5 cm/allele + 6 cm base height = 31 cm). Therefore, $AABBCc$, $AABbCC$, and $AaBBCC$ are the genotypes that will result in plants that are 31 cm.

2. A plant of unknown phenotype and genotype was testcrossed, with the following results

$$1/4 \ 11 \text{ cm}$$

$$2/4 \ 16 \text{ cm}$$

$$1/4 \ 21 \text{ cm}$$

An astute genetics student realized that the unknown plant could be only one phenotype but could be any of three genotypes. What were they?

Solution: When testcrossed (with $aabbcc$), the unknown plant must be able to contribute either one, two, or three additive alleles in its gametes in order to yield the three phenotypes in the off-spring. Since no 6-cm offspring are observed, the unknown plant never contributes all nonadditive alleles (abc). Only plants that are homozygous at one locus and heterozygous at the other two loci will meet these criteria. Therefore, the unknown parent can be any of three genotypes, all of which have a phenotype of 26 cm

$$AABbCc$$

$$AaBbCC$$

$$AaBBCc$$

For example, in the first genotype ($AABbCc$),

$$AABbCc \ \times \ aabbcc$$

$$\downarrow$$

$$1/4 \ AaBbCc \quad 21 \text{ cm}$$

$$1/4 \ AaBbcc \quad 16 \text{ cm}$$

$$1/4 \ AabbCc \quad 16 \text{ cm}$$

$$1/4 \ Aabbcc \quad 11 \text{ cm}$$

which is the ratio of phenotypes observed.

3. The mean and variance of corolla length in two highly inbred strains of *Nicotiana* and their progeny are shown in the following table. One parent (P_1) has a short corolla, and the other parent (P_2) has a long corolla. Calculate the broad-sense heritability (H^2) of corolla length in this plant.

Strain	Mean (mm)	Variance (mm)
P_1 short	40.47	3.12
P_2 long	93.75	3.87
F_1 ($P_1 \times P_2$)	63.90	4.74
F_2 ($F_1 \times F_1$)	68.72	47.70

Solution: The formula for estimating heritability is $H^2 = V_G/V_P$, where V_G and V_P are the genetic and phenotypic components of variation, respectively. The main issue in this problem is obtaining some estimate of two components of phenotypic variation: genetic and environmental factors. V_P is the combination of genetic and environmental variance. Because the two parental strains are true breeding, they are assumed to be homozygous, and the variance of 3.12 and 3.87 is considered to be the result of environmental influences. The average of these two values is 3.50. The F_1 is also genetically homogeneous and gives us an additional estimate of the impact of environmental factors. By averaging this value along with that of the parents,

$$\frac{4.74 + 3.50}{2} = 4.12$$

we obtain a relatively good idea of environmental impact on the phenotype. The phenotypic variance in the F_2 is the sum of the genetic (V_G) and environmental (V_E) components. We have estimated the environmental input as 4.12, so 47.70 minus 4.12 gives us an estimate of V_G of 43.58. Heritability then becomes 43.58/47.70, or 0.91. This value, when interpreted as a percentage, indicates that about 91 percent of the variation in corolla length is due to genetic influences.

PROBLEMS AND DISCUSSION QUESTIONS

 For activities, animations, and review quizzes, go to the study area at www.masteringgenetics.com

HOW DO WE KNOW?

1. In this chapter, we focused on a mode of inheritance referred to as quantitative genetics, as well as many of the statistical parameters utilized to study quantitative traits. Along the way, we found opportunities to consider the methods and reasoning by which geneticists acquired much of their understanding of quantitative genetics. From the explanations given in the chapter, what answers would you propose to the following fundamental questions:

   (a) How can we ascertain the number of polygenes involved in the inheritance of a quantitative trait?

   (b) What findings led geneticists to postulate the multiple-factor hypothesis that invoked the idea of additive alleles to explain inheritance patterns?

   (c) How do we assess environmental factors to determine if they impact the phenotype of a quantitatively inherited trait?

   (d) How do we know that monozygotic twins are not identical genotypically as adults?

2. What is the difference between continuous and discontinuous variation? Which of the two is most likely to be the result of polygenic inheritance?

3. Define the following: (a) polygenic, (b) additive alleles, (c) monozygotic and dizygotic twins, (d) heritability, and (e) QTL.

4. A dark-red strain and a white strain of wheat are crossed and produce an intermediate, medium-red F_1. When the F_1 plants are interbred, an F_2 generation is produced in a ratio of 1 dark-red: 4 medium-dark-red: 6 medium-red: 4 light-red: 1 white. Further crosses reveal that the dark-red and white F_2 plants are true breeding.

   (a) Based on the ratios in the F_2 population, how many genes are involved in the production of color?

   (b) How many additive alleles are needed to produce each possible phenotype?

   (c) Assign symbols to these alleles and list possible genotypes that give rise to the medium-red and light-red phenotypes.

   (d) Predict the outcome of the F_1 and F_2 generations in a cross between a true-breeding medium-red plant and a white plant.

5. Height in humans depends on the additive action of genes. Assume that this trait is controlled by the four loci *R*, *S*, *T*, and *U* and that environmental effects are negligible. Instead of additive versus nonadditive alleles, assume that additive and partially additive alleles exist. Additive alleles contribute two units, and partially additive alleles contribute one unit to height.

   (a) Can two individuals of moderate height produce offspring that are much taller or shorter than either parent? If so, how?

   (b) If an individual with the minimum height specified by these genes marries an individual of intermediate or moderate height, will any of their children be taller than the tall parent? Why or why not?

6. An inbred strain of plants has a mean height of 24 cm. A second strain of the same species from a different geographical region also has a mean height of 24 cm. When plants from the two strains are crossed together, the F_1 plants are the same height as the parent plants. However, the F_2 generation shows a wide range of heights; the majority are like the P_1 and F_1 plants, but approximately 4 of 1000 are only 12 cm high, and about 4 of 1000 are 36 cm high.

   (a) What mode of inheritance is occurring here?

   (b) How many gene pairs are involved?

   (c) How much does each gene contribute to plant height?

   (d) Indicate one possible set of genotypes for the original P_1 parents and the F_1 plants that could account for these results.

   (e) Indicate three possible genotypes that could account for F_2 plants that are 18 cm high and three that account for F_2 plants that are 33 cm high.

7. Erma and Harvey were a compatible barnyard pair, but a curious sight. Harvey's tail was only 6 cm long, while Erma's was 30 cm. Their F_1 piglet offspring all grew tails that were 18 cm. When inbred, an F_2 generation resulted in many piglets (Erma and Harvey's grandpigs), whose tails ranged in 4-cm intervals from 6 to 30 cm (6, 10, 14, 18, 22, 26, and 30). Most had 18-cm tails, while 1/64 had 6-cm tails and 1/64 had 30-cm tails.

   (a) Explain how these tail lengths were inherited by describing the mode of inheritance, indicating how many gene pairs were at work, and designating the genotypes of Harvey, Erma, and their 18-cm-tail offspring.

   (b) If one of the 18-cm F_1 pigs is mated with one of the 6-cm F_2 pigs, what phenotypic ratio would be predicted if many offspring resulted? Diagram the cross.

8. In the following table, average differences of height, weight, and fingerprint ridge count between monozygotic twins (reared together and apart), dizygotic twins, and nontwin siblings are compared:

Trait	MZ Reared Together	MZ Reared Apart	DZ Reared Together	Sibs Reared Together
Height (cm)	1.7	1.8	4.4	4.5
Weight (kg)	1.9	4.5	4.5	4.7
Ridge count	0.7	0.6	2.4	2.7

Variance	Vitamin A	Cholesterol
V_P	123.5	862.0
V_E	96.2	484.6
V_A	12.0	192.1
V_D	15.3	185.3

Based on the data in this table, which of these quantitative traits has the highest heritability values?

9. What kind of heritability estimates (broad sense or narrow sense) are obtained from human twin studies?

10. List as many human traits as you can that are likely to be under the control of a polygenic mode of inheritance.

11. Corn plants from a test plot are measured, and the distribution of heights at 10-cm intervals is recorded in the following table:

Height (cm)	Plants (no.)
100	20
110	60
120	90
130	130
140	180
150	120
160	70
170	50
180	40

Calculate (a) the mean height, (b) the variance, (c) the standard deviation, and (d) the standard error of the mean. Plot a rough graph of plant height against frequency. Do the values represent a normal distribution? Based on your calculations, how would you assess the variation within this population?

12. The following variances were calculated for two traits in a herd of hogs.

Trait	V_P	V_G	V_A
Back fat	30.6	12.2	8.44
Body length	52.4	26.4	11.70

(a) Calculate broad-sense (H^2) and narrow-sense (h^2) heritabilities for each trait in this herd.

(b) Which of the two traits will respond best to selection by a breeder? Why?

13. The mean and variance of plant height of two highly inbred strains (P_1 and P_2) and their progeny (F_1 and F_2) are shown here.

Strain	Mean (cm)	Variance
P_1	34.2	4.2
P_2	55.3	3.8
F_1	44.2	5.6
F_2	46.3	10.3

Calculate the broad-sense heritability (H^2) of plant height in this species.

14. A hypothetical study investigated the vitamin A content and the cholesterol content of eggs from a large population of chickens. The variances (V) were calculated, as shown below:

(a) Calculate the narrow-sense heritability (h^2) for both traits.

(b) Which trait, if either, is likely to respond to selection?

15. In a herd of dairy cows the narrow-sense heritability for milk protein content is 0.76, and for milk butterfat it is 0.82. The correlation coefficient between milk protein content and butterfat is 0.91. If the farmer selects for cows producing more butterfat in their milk, what will be the most likely effect on milk protein content in the next generation?

16. In an assessment of learning in *Drosophila*, flies were trained to avoid certain olfactory cues. In one population, a mean of 8.5 trials was required. A subgroup of this parental population that was trained most quickly (mean = 6.0) was interbred, and their progeny were examined. These flies demonstrated a mean training value of 7.5. Calculate realized heritability for olfactory learning in *Drosophila*.

17. Suppose you want to develop a population of *Drosophila* that would rapidly learn to avoid certain substances the flies could detect by smell. Based on the heritability estimate you obtained in Problem 16, do you think it would be worth doing this by artificial selection? Why or why not?

18. In a population of tomato plants, mean fruit weight is 60 g and (h^2) is 0.3. Predict the mean weight of the progeny if tomato plants whose fruit averaged 80 g were selected from the original population and interbred.

19. In a population of 100 inbred, genotypically identical rice plants, variance for grain yield is 4.67. What is the heritability for yield? Would you advise a rice breeder to improve yield in this strain of rice plants by selection?

20. A 3-inch plant was crossed with a 15-inch plant, and all F_1 plants were 9 inches. The F_2 plants exhibited a "normal distribution," with heights of 3, 4, 5, 6, 7, 8, 9, 10, 11, 12, 13, 14, and 15 inches.

(a) What ratio will constitute the "normal distribution" in the F_2?

(b) What will be the outcome if the F_1 plants are testcrossed with plants that are homozygous for all nonadditive alleles?

21. In a cross between a strain of large guinea pigs and a strain of small guinea pigs, the F_1 are phenotypically uniform, with an average size about intermediate between that of the two parental strains. Among 1014 F_2 individuals, 3 are about the same size as the small parental strain and 5 are about the same size as the large parental strain. How many gene pairs are involved in the inheritance of size in these strains of guinea pigs?

22. While most quantitative traits display continuous variation, there are others referred to as "threshold traits" that are distinguished by having a small number of discrete phenotypic classes. For example, Type II diabetes (adult-onset diabetes) is considered to be a polygenic trait, but demonstrates only two phenotypic classes: individuals who develop the disease and those who do not. By factoring in the concept of disease susceptibility, describe how a threshold trait such as Type II diabetes may be under the control of many polygenes, but express a limited number of phenotypes.

23. Many traits of economic or medical significance are determined by quantitative trait loci (QTLs) in which many genes, usually scattered throughout the genome, contribute to expression.

(a) What general procedures are used to identify such loci?

(b) What is meant by the term *cosegregate* in the context of QTL mapping? Why are markers such as RFLPs, SNPs, and microsatellites often used in QTL mapping?

SOLUTIONS TO SELECTED PROBLEMS AND DISCUSSION QUESTIONS

Answers to Now Solve This

1. (a) Since 1/256 of the F_2 plants are 20 cm and 1/256 are 40 cm, there must be 4 gene pairs involved in determining flower size.

   (b) Since there are nine size classes, one can conduct the following backcross: $AaBbCcDd \times AABBCCDD$. The frequency distribution in the backcross would be

$1/16 = 40$ cm	$4/16 = 32.5$ cm
$4/16 = 37.5$ cm	$1/16 = 30$ cm
$6/16 = 35$ cm	

2. (a) Taking the sum of the values and dividing by the number in the sample gives the following means

   mean sheep fiber length $= 7.7$ cm
   mean fleece weight $= 6.4$ kg

   The variance for each is

   variance sheep fiber length $= 6.097$
   variance fleece weight $= 3.12$

   The standard deviation is the square root of the variance

   sheep fiber length $= 2.469$
   fleece weight $= 1.766$

   (b, c) The covariance for the two traits is 30.36/7, or 4.34, while the correlation coefficient is $+0.998$.

   (d) There is a very high correlation between fleece weight and fiber length and it is likely that this correlation is not by chance. Even though correlation does not mean cause-and-effect, it would seem logical that as you increased fiber length, you would also increase fleece weight. It is probably safe to say that the increase in fleece weight is directly related to an increase in fiber length.

3. The role of genetics and the role of the environment can be studied by comparing the expression of traits in monozygotic and dizygotic twins. The higher concordance value for monozygotic twins as compared with the value for dizygotic twins indicates a significant genetic component for a given trait. Notice that for traits including blood type, eye color, and mental retardation, there is a fairly significant difference between MZ and DZ groups. However, for measles, the difference is not as significant, indicating a greater role of the environment. Hair color has a significant genetic component as do idiopathic epilepsy, schizophrenia, diabetes, allergies, cleft lip, and club foot. The genetic component to mammary cancer is present but minimal according to these data.

Solutions to Problems and Discussion Questions

4. (a) There are two alleles at each locus for a total of four alleles.

   (b, c) We can say that each gene (additive allele) provides an equal unit amount to the phenotype and the colors differ from each other in multiples of that unit amount. The number of additive alleles needed to produce each phenotype is given below

$1/16 =$ dark red	$= AABB$	
$4/16 =$ medium-dark red	$= 2AABb, 2AaBB$	
$6/16 =$ medium red	$= AAbb, 4AaBb, aaBB$	
$4/16 =$ light red	$= 2aaBb, 2Aabb$	
$1/16 =$ white	$= aabb$	

   (d) $F_1 =$ all light red
   $F_2 = $ 1/4 medium red   2/4 light red   1/4 white

6. (a, b) There are four gene pairs involved.

   (c) Since there is a difference of 24 cm between the extremes, 24 cm/8 $= 3$ cm for each increment (each of the additive alleles).

   (d) A typical F_1 cross that produces a "typical" F_2 distribution would be where all gene pairs are heterozygous ($AaBbCcDd$), independently assorting, and additive. There are many possible sets of parents that would give an F_1 of this type. The limitation is that each parent has genotypes that give a height of 24 cm as stated in the problem. Because the parents are inbred, it is expected that they are fully homozygous. An example is

   $$AABBccdd \times aabbCCDD$$

   (e) Since the $aabbccdd$ genotype gives a height of 12 cm and each uppercase allele adds 3 cm to the height, there are many possibilities for an 18 cm plant

   $$AAbbccdd, AaBbccdd, aaBbCcdd, etc.$$

   Any plant with seven uppercase letters will be 33 cm tall

   $$AABBCCDd, AABBCcDD, AABbCCDD, \text{ for example.}$$

8. For height, notice that average differences between MZ twins reared together (1.7 cm) and those MZ twins reared apart (1.8 cm) are similar (meaning little environmental influence) and considerably less than differences of DZ twins (4.4 cm) or sibs (4.5 cm) reared together. These data indicate that genetics plays a major role in determining height.

   However, for weight, notice that MZ twins reared together have a much smaller (1.9 kg) difference than MZ twins reared apart, indicating that the environment has a considerable impact on weight. By comparing the weight differences of MZ twins reared apart with DZ twins and sibs reared together one can conclude that the environment has almost as much an influence on weight as genetics.

   For ridge count, the differences between MZ twins reared together and those reared apart are small. For the data in the table, it would appear that ridge count and height have the highest heritability values.

10. Many traits, especially those we view as quantitative are likely to be determined by a polygenic mode with possible environmental influences. The following are some common examples: height, general body structure, skin color, and perhaps most common behavioral traits including intelligence.

12. (a) Using the following equations, H^2 and h^2 can be calculated as follows.

   For back fat: Broad-sense heritability $= H^2 = 12.2/30.6 = .398$

   Narrow-sense heritability $= h^2 = 8.44/30.6 = .276$

   For body length: Broad-sense heritability $= H^2 = 26.4/52.4 = .504$

   Narrow-sense heritability $= h^2 = 11.7/52.4 = .223$

   (b) Of the two traits, selection for back fat would produce more response.

14. (a) For vitamin A: $h_A^2 = V_A/V_P = V_A/(V_E + V_A + V_D) = 0.097$

   For cholesterol: $h_A^2 = 0.223$

(b) Cholesterol content should be influenced to a greater extent by selection.

16. $h^2 = (7.5 - 8.5/6.0 - 8.5) = 0.4$ (realized heritability)

18. $h^2 = 0.3 = (M_2 - 60/80 - 60)$ $M_2 = 66$ grams

20. **(a)** There are two ways to answer this section, a hard way and an easy way. The hard way would to take a big sheet of paper, make the cross ($AaBbCcDdEeFf \times AaBbCcDdEeFf$), collect the genotypes, and calculate the ratios.

This method would be very laborious and error-prone. The easy way would be to re-read the material on the binomial expansion and note the pattern preceding each expression. Notice that all numbers other than the 1's are equal to the sum of the two numbers directly above them. By enlarging the numbers to include six gene pairs, you can arrive at the thirteen classes and their frequencies

3" = 1	4" = 12	5" = 66
6" = 220	7" = 495	8" = 792
9" = 924	10" = 792	11" = 495
12" = 220	13" = 66	14" = 12
15" = 1		

To check your calculations, be certain that your frequencies total 4096. You will also notice an additional shortcut in that since the distribution is symmetrical, you need only calculate to the center and the remainder will be in the reverse order.

(b) To determine the outcome of a cross of the F_1 plants in the test cross, apply the formula that allows you to calculate any set of components: $n!/(s!t!)$ where n = total number of events (6), s = number of events of outcome a and t = number of events of outcome b. For example, to determine how many 6" plants would be recovered from the cross $AaBbCcDdEeFf \times aabbccddeeff$, we are really asking how many will have three additive alleles (uppercase) and three non-additive alleles (lowercase).

$$6!/(3!3!) = 20$$

Applying this formula throughout gives the following frequencies

3" = 1	4" = 6	5" = 15
6" = 20	7" = 15	8" = 6
9" = 1		

The total is 64. You can check your logic by considering that there should be only 1/64 with no additive alleles (3") and 1/64 with all additive alleles (9").

22. The level of blood sugar varies considerably from individual to individual, day to day, and hour to hour, and on a population level, it displays continuous variation. However the diagnosis of Type II diabetes is set by relatively fixed criteria. A fasting blood sugar level of 126 mg/dL or higher, repeated on different days, is diagnostic of diabetes. A casual (non-fasting) blood sugar level of 200 mg/dL or higher is suggestive of diabetes. In either case, while the level of blood sugar is influenced by a variety of factors (polygenic and environmental), the actual diagnosis of the disease leads one to be classified as diabetic or not diabetic. Since there are only two phenotypic classes (or three if one included the prediabetic state), diabetes is referred to as a threshold trait.

CREDITS

Population and Evolutionary Genetics

From Chapter 22 of *Essentials of Genetics, Eighth edition,* William S. Klug, Michael R. Cummings, Charlotte A. Spencer, Michael A. Palladino. ©2013 by Pearson Education, Inc. All rights reserved.

These ladybird beetles, from the Chiricahua Mountains in Arizona, show considerable phenotypic variation.

Population and Evolutionary Genetics

CHAPTER CONCEPTS

- Most populations and species harbor considerable genetic variation.

- This variation is reflected in the alleles distributed among populations of a species.

- The relationship between allele frequencies and genotype frequencies in an ideal population is described by the Hardy–Weinberg law.

- Selection, migration, and genetic drift can cause changes in allele frequency.

- Mutation creates new alleles in a population gene pool.

- Nonrandom mating changes population genotype frequency but not allele frequency.

- A reduction in gene flow between populations, accompanied by selection or genetic drift, can lead to reproductive isolation and speciation.

- Genetic differences between populations or species are used to reconstruct evolutionary history.

I n the mid-nineteenth century, Alfred Russel Wallace and Charles Darwin identified natural selection as the mechanism of evolution. In his book, *On the Origin of Species,* published in 1859, Darwin provided evidence that populations and species are not fixed, but change, or evolve, over time as a result of natural selection. However, Wallace and Darwin could not explain either the origin of the variations that provide the raw material for evolution or the mechanisms by which such variations are passed from parents to offspring. Gregor Mendel published his work on the inheritance of traits in 1866, but it received little notice at the time. The rediscovery of Mendel's work in 1900 began a 30-year effort to reconcile Mendel's concept of genes and alleles with the theory of evolution by natural selection. As twentieth-century biologists applied the principles of Mendelian genetics to populations, both the source of variation (mutation and recombination) and the mechanism of inheritance (segregation of alleles) were explained. We now view evolution as a consequence of changes in genetic material through mutation and changes in allele frequencies in populations over time. This union of population genetics with the theory of natural selection generated a new view of the evolutionary process, called *neo-Darwinism.*

In addition to natural selection, other forces including mutation, migration, and drift, individually and collectively, alter allele frequencies and bring about evolutionary divergence that eventually may result in **speciation,** the formation of new species. Speciation is facilitated by environmental diversity. If a population is spread over a geographic range encompassing a number of ecologically distinct subenvironments with different selection pressures, the populations occupying these areas may gradually adapt and become genetically distinct from one another. Genetically differentiated populations may remain in existence, become extinct, reunite with each other, or continue to diverge until they become reproductively isolated. Populations that are reproductively isolated are regarded as separate species. Genetic changes within populations can modify a species over time, transform it into another species, or cause it to split into two or more species.

Population geneticists investigate patterns of genetic variation within and among groups of interbreeding individuals. As changes in genetic structure form the basis for evolution of a population, population genetics has become an important subdiscipline of evolutionary biology. In this chapter, we examine the population genetics processes of **microevolution**—defined as evolutionary change within populations of a species—and then consider how molecular aspects of these processes can be extended to **macroevolution**—defined as evolutionary events leading to the emergence of new species and other taxonomic groups.

1 Genetic Variation Is Present in Most Populations and Species

A **population** is a group of individuals belonging to the same species that live in a defined geographic area and actually or potentially interbreed. In thinking about the human population, we can define it as everyone who lives in the United States, or in Sri Lanka, or we can specify a population as all the residents of a particular small town or village.

The genetic information carried by members of a population constitutes that population's **gene pool.** At first glance, it might seem that a population that is well-adapted to its environment must be highly homozygous because it is assumed that the most favorable allele at each locus is present at a high frequency. In addition, a look at most populations of plants and animals reveals many phenotypic similarities among individuals. However, a large body of evidence indicates that, in reality, most populations contain a high degree of heterozygosity. This built-in genetic diversity is not necessarily apparent in the phenotype; hence, detecting it is not a simple task. Nevertheless, the diversity within a population can be revealed by several methods.

Detecting Genetic Variation by Artificial Selection

One way to determine whether genetic variation exists in a population is to use artificial selection. If there is little or no variation present in the genes controlling a particular phenotype, selection will have little or no effect on the phenotype. If genetic variation is present, the phenotype will change over a few generations. A dramatic example of this test is the

domestic dog. The broad array of sizes, shapes, colors, and behaviors seen in different breeds of dogs all arose from the effects of selection on the genetic variation present in wolves, from which all domestic dogs are descended. Genetic and archaeological evidence indicates that the domestication of wolves took place at least 15,000 years ago and possibly much earlier. On a shorter time scale, laboratory selection experiments on the fruit fly *Drosophila melanogaster* can cause significant changes over a few generations in almost every phenotype imaginable, including size, shape, developmental rate, fecundity, and behavior.

Variations in Nucleotide Sequence

The most direct way to estimate genetic variation is to compare the nucleotide sequences of genes carried by individuals in a population. In one study, Martin Kreitman examined the *alcohol dehydrogenase* locus (*Adh*) in *Drosophila melanogaster* (Figure 1). This locus has two alleles, the *Adh-f* and the *Adh-s* alleles. The encoded proteins differ by only a single amino acid (thr versus lys at codon 192). To determine whether the amount of genetic variation detectable at the protein level (one amino acid difference) corresponds to the amount of variation at the nucleotide level, Kreitman cloned and sequenced *Adh* genes from individuals representing five natural populations of *Drosophila* (Figure 2).

The 11 cloned genes from these five populations contained a total of 43 nucleotide variations in the *Adh* sequence of 2721 base pairs. These variations are distributed throughout the gene: 14 in exon-coding regions, 18 in introns, and 11 in the untranslated flanking regions. Of the 14 variations in coding regions, only one leads to an amino acid replacement—the one in codon 192, producing the two alleles. The other 13 coding-region nucleotide substitutions do not lead to amino acid replacements and are silent variations in this gene.

Among the most intensively studied human genes is the locus encoding the cystic fibrosis transmembrane conductance regulator (CFTR). Recessive loss-of-function mutations in the *CFTR* locus cause **cystic fibrosis,** a disease that affects secretory glands and lungs, leading to susceptibility to bacterial infections. More than 1900 different mutations in

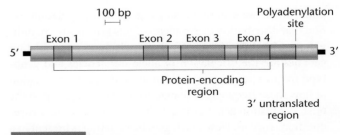

FIGURE 1 Organization of the *Adh* locus of *Drosophila melanogaster.*

	Exon 3	Intron 3	Exon 4
Consensus *Adh* sequence:	C C C C	G G A A T	C T C C A* C T A G
Strain			
Wa-S	T T • A	C A • T A	A C • • • • • • •
Fl1-S	T T • A	C A • T A	A C • • • • • • •
Ja-S	• • • •	• • • • •	• • • T • T • C A
Fl-F	• • • •	• • • • •	• • G T C T C C •
Ja-F	• • A •	• • G • •	• • G T C T C C •

© 1983 Macmillan Publishers Ltd.

FIGURE 2 DNA sequence variation in parts of the *Drosophila Adh* gene in a sample of the eleven laboratory strains derived from the five natural populations. The dots represent nucleotides that are the same as the consensus sequence; letters represent nucleotide polymorphisms. An A/C polymorphism (A*) in codon 192 creates the two *Adh* alleles (F and S). All other polymorphisms are silent or noncoding.

the *CFTR* gene have been identified. Among these are missense mutations, amino acid deletions, nonsense mutations, frameshifts, and splice defects.

Studies of other organisms, including the rat, the mouse, and the mustard plant *Arabidopsis thaliana,* have produced similar estimates of nucleotide diversity in various genes. These studies indicate that there is an enormous reservoir of genetic variability within most populations. At the DNA level most, and perhaps all, genes exhibit diversity from individual to individual. Alleles representing these variations are distributed among members of a population.

ESSENTIAL POINT ■ ■ ■

Genetic variation is widespread in most populations and provides a reservoir of alleles that serve as the basis for evolutionary changes within the population.

2 The Hardy–Weinberg Law Describes Allele Frequencies and Genotype Frequencies in Populations

Often when we examine a single gene in a population, we find that distribution of the alleles of this gene results in individuals with different genotypes. Key elements of population genetics are the calculation of allele frequencies and genotype frequencies in the population, and the determination of how these frequencies change from one generation to the next. Population geneticists use these calculations to answer questions such as: How much genetic variation is present in a population? Are genotypes randomly distributed in time and space, or do discernible patterns exist? What processes affect

the composition of a population's gene pool? Do these processes produce genetic divergence among populations that may lead to the formation of new species? Changes in allele frequencies in a population that do not result in reproductive isolation are examples of microevolution. In the following sections, we will discuss microevolutionary changes in population gene pools, and in later sections, we will consider macroevolution and the process of speciation.

The relationship between the relative proportions of alleles in the gene pool and the frequencies of different genotypes in the population was elegantly described in a mathematical model. This model, called the **Hardy–Weinberg law,** describes what happens to allele and genotype frequencies in an "ideal" population that is infinitely large and randomly mating, and that is not subject to any evolutionary forces such as mutation, migration, or selection.

The Hardy–Weinberg model uses Mendelian principles of segregation and simple probability to explain the relationship between allele and genotype frequencies in a population. We can demonstrate how this works by considering the example of a single autosomal gene with two alleles, A and $a,$ in a population where the frequency of A is 0.7 and the frequency of a is 0.3. Note that $0.7 + 0.3 = 1$, indicating that all the alleles of gene A present in the population are accounted for. This means that the probability that any female gamete will contain A is 0.7, and the probability that a male gamete will contain A is also 0.7. The probability that *both* gametes will contain A is $0.7 \times 0.7 = 0.49$. Thus we predict that in the offspring, the genotype AA will occur 49 percent of the time. The probability that a zygote will be formed from a female gamete carrying A and a male gamete carrying a is $0.7 \times 0.3 = 0.21$, and the probability of a female gamete carrying a being fertilized by a male gamete carrying A is $0.3 \times 0.7 = 0.21$ so the frequency of genotype Aa in the offspring is $0.21 + 0.21 = 0.42 = 42$ percent. Finally, the probability that a zygote will be formed from two gametes carrying a is $0.3 \times 0.3 = 0.09$, so the frequency of genotype aa is 9 percent. As a check on our calculations, note that $0.49 + 0.42 + 0.09 = 1.0$, confirming that we have accounted for all possible genotypic combinations in the zygotes. These calculations are summarized in **Figure 3**.

Now that we know the frequencies of genotypes in the next generation, what will be the frequency distribution of alleles in this new generation? Under the Hardy–Weinberg law, we assume that all genotypes have equal rates of survival and reproduction. This means that in the next generation, all genotypes contribute equally to the new gene pool. The AA individuals constitute 49 percent of the population, and we can predict that the gametes they produce will constitute 49 percent of the gene pool. These gametes all carry allele A. Similarly, Aa individuals constitute 42 percent of the population, so we predict that their gametes will constitute 42 percent of the new gene pool. Half (0.5) of these gametes will carry allele A. Thus, the frequency of allele A in the gene

Sperm

	fr(A) = 0.7	fr(a) = 0.3
fr(A) = 0.7	fr(AA) = 0.7 × 0.7 = 0.49	fr(Aa) = 0.7 × 0.3 = 0.21
fr(a) = 0.3	fr(aA) = 0.3 × 0.7 = 0.21	fr(aa) = 0.3 × 0.3 = 0.09

Eggs (label at left of lower rows)

FIGURE 3 Calculating genotype frequencies from allele frequencies. Gametes represent samples drawn from the gene pool to form the genotypes of the next generation. In this population, the frequency of the A allele is 0.7, and the frequency of the a allele is 0.3. The frequencies of the genotypes in the next generation are calculated as 0.49 for AA, 0.42 for Aa, and 0.09 for aa. Under the Hardy–Weinberg law, the frequencies of A and a remain constant from generation to generation.

pool is $0.49 + (0.5) 0.42 = 0.7$. The other half of the gametes produced by Aa individuals will carry allele a. The aa individuals constitute 9 percent of the population, so their gametes will constitute 9 percent of the new gene pool. All these gametes carry allele a. Thus, we can predict that the allele a in the new gene pool is $(0.5) 0.42 + 0.09 = 0.3$. As a check on our calculation, note that $0.7 + 0.3 = 1.0$, accounting for all of the gametes in the gene pool of the new generation.

For the general case, we use variables instead of numerical values for the allele frequencies in the Hardy–Weinberg model. Imagine a gene pool in which the frequency of allele A is p and the frequency of allele a is q, such that $p + q = 1$. If we randomly draw male and female gametes from the gene pool and pair them to make a zygote, the probability that both will carry allele A is $p \times p$. Thus, the frequency of genotype AA among the zygotes is p^2. The probability that the female gamete carries A and the male gamete carries a is $p \times q$, and the probability that the female gamete carries a and the male gamete carries A is $q \times p$. Thus, the frequency of genotype Aa among the zygotes is $2pq$. Finally, the probability that both gametes carry a is $q \times q$, making the frequency of genotype aa among the zygotes q^2. Therefore, the distribution of genotypes among the zygotes is

$$p^2 + 2pq + q^2 = 1$$

These calculations are summarized in **Figure 4**. They demonstrate the two main predictions of the Hardy–Weinberg model:

1. Allele frequencies in our population do not change from one generation to the next.

2. After one generation of random mating, genotype frequencies can be predicted from the allele frequencies.

In other words, this population does not change or evolve with respect to the locus being examined. Remember, however, the

Sperm

	fr(A) = p	fr(a) = q
fr(A) = p	fr(AA) = p^2	fr(Aa) = pq
fr(a) = q	fr(aA) = qp	fr(aa) = q^2

Eggs (label at left of lower rows)

FIGURE 4 The general description of allele and genotype frequencies under Hardy–Weinberg assumptions. The frequency of allele A is p, and the frequency of allele a is q. After mating, the three genotypes AA, Aa, and aa have the frequencies p^2, $2pq$, and q^2, respectively.

assumptions about the theoretical population described by the Hardy–Weinberg model:

1. Individuals of all genotypes have equal rates of survival and equal reproductive success—that is, there is no selection.

2. No new alleles are created or converted from one allele into another by mutation.

3. Individuals do not migrate into or out of the population.

4. The population is infinitely large, which in practical terms means that the population is large enough that sampling errors and other random effects are negligible.

5. Individuals in the population mate randomly.

These assumptions are what make the Hardy–Weinberg model so useful in population genetics research. By specifying the conditions under which the population does not evolve, the Hardy–Weinberg model can be used to identify the real-world forces that cause allele frequencies to change. Application of this model can also reveal "neutral genes" in a population gene pool—those not being operated on by the forces of evolution.

The Hardy–Weinberg model has three additional important consequences. First, it shows that dominant traits do not necessarily increase from one generation to the next. Second, it demonstrates that **genetic variability** can be maintained in a population since, once established in an ideal population, allele frequencies remain unchanged. Third, if we invoke Hardy–Weinberg assumptions, then knowing the frequency of just one genotype enables us to calculate the frequencies of all other genotypes at that locus. This is particularly useful in human genetics because we can calculate the frequency of heterozygous carriers for recessive genetic disorders even when all we know is the frequency of affected individuals.

1 The ability to taste the compound PTC is controlled by a dominant allele *T,* while individuals homozygous for the recessive allele *t* are unable to taste PTC. In a genetics class of 125 students, 88 can taste PTC and 37 cannot. Calculate the frequency of the *T* and *t* alleles in this population and the frequency of the genotypes.

■ **Hint** *This problem involves an understanding of how to use the Hardy–Weinberg law. The key to its solution is to determine which allele frequency (p or q) you must estimate first when homozygous dominant and heterozygous genotypes have the same phenotype.*

TABLE 1	CCR5 Genotypes and Phenotypes
Genotype	**Phenotype**
1/1	Susceptible to sexually transmitted strains of HIV-1
1/Δ32	Susceptible but may progress to AIDS slowly
Δ32/Δ32	Resistant to most sexually transmitted strains of HIV-1

3   The Hardy–Weinberg Law Can Be Applied to Human Populations

To show how allele frequencies are measured in a real population, let's consider a gene that influences an individual's susceptibility to infection by HIV-1, the virus responsible for AIDS (acquired immunodeficiency syndrome). A small number of individuals who make high-risk choices (such as unprotected sex with HIV-positive partners) remain uninfected. Some of these individuals are homozygous for a mutant allele of a gene called *CCR5*.

The *CCR5* gene **(Figure 5)** encodes a protein called the C-C chemokine receptor-5, often abbreviated CCR5. Chemokines are signaling molecules associated with the immune system. The CCR5 protein is also a receptor for strains of HIV-1, allowing it to gain entry into cells. The mutant allele of the *CCR5* gene contains a 32-bp deletion, making the encoded protein shorter and nonfunctional. The normal allele is called *CCR51* (also called *1*), and the mutant allele is called *CCR5-Δ32* (also called *Δ32*).

Δ32/Δ32 individuals are resistant to HIV-1 infection. Heterozygous *1/Δ32* individuals are susceptible to HIV-1 infection but progress more slowly to AIDS. **Table 1** summarizes the genotypes possible at the *CCR5* locus and the phenotypes associated with each.

The discovery of the *CCR5-Δ32* allele generates two important questions: Which human populations carry the *Δ32* allele, and how common is it? To address these questions, teams of researchers surveyed members of several

populations. Genotypes were determined by direct analysis of DNA **(Figure 6)**. In one population, 79 individuals had genotype *1/1*, 20 were (*1/Δ32*), and 1 was *Δ32/Δ32*. This population has 158 *1* alleles carried by the *1/1* individuals plus 20 *1* alleles carried by *1/Δ32* individuals, for a total of 178. The frequency of the *CCR51* allele in the sample population is thus 178/200 = 0.89 = 89 percent. Copies of the *CCR5-Δ32* allele were carried by 20 *1/Δ32* individuals, plus 2 carried by the *Δ32/Δ32* individual, for a total of 22. The frequency of the *CCR5-Δ32* allele is thus 22/200 = 0.11 = 11 percent. Notice that $p + q = 1$, confirming that we have accounted for all the alleles of this gene in the population. **Table 2** shows two methods for computing the frequencies of the *1* and *Δ32* alleles in the population surveyed.

Can we expect the *CCR5-Δ32* allele to increase in human populations because it offers resistance to infection by HIV? This specific question is difficult to answer directly, but as we have seen earlier in this chapter, when factors such as natural selection, mutation, migration, or genetic drift are present, the allele frequencies in a population may change from one generation to the next.

By determining allele frequencies over more than one generation, it is possible to determine whether the frequencies remain in equilibrium because the Hardy-Weinberg assumptions are operating, or whether allele frequencies or genotype frequencies are changing because one or more of these assumptions is not operating.

Testing for Hardy–Weinberg Equilibrium in a Population

One way to see whether one (or more) of the Hardy–Weinberg assumptions does not hold in a given population is to determine whether the population's genotypes are in equilibrium. To do this, we first determine the genotype

FIGURE 5   Organization of the *CCR5* gene in region 3p21.3. The gene contains 4 exons and 2 introns. The arrow shows the location of the 32-bp deletion in exon 4 that confers resistance to HIV-1 infection.

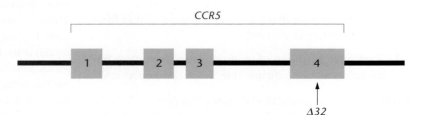

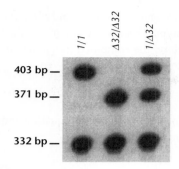

403 bp —
371 bp —
332 bp —

FIGURE 6 Allelic variation in the *CCR5* gene. Michel Samson and colleagues used PCR to amplify a part of the *CCR5* gene containing the site of the 32-bp deletion, cut the resulting DNA fragments with a restriction enzyme, and ran the fragments on an electrophoresis gel. Each lane reveals the genotype of a single individual. The *1* allele produces a 332-bp fragment and a 403-bp fragment; the Δ*32* allele produces a 332-bp fragment and a 371-bp fragment. Heterozygotes produce three bands.

frequencies either directly from the phenotypes (if heterozygotes are recognizable) or by analyzing proteins or DNA sequences. We then calculate the allele frequencies from the genotype frequencies, as demonstrated earlier. Finally, we use the allele frequencies in the parental generation to predict the offspring's genotype frequencies. According to the Hardy–Weinberg law, genotype frequencies are predicted to fit the $p^2 + 2pq + q^2 = 1$ relationship. If they do not, then one or more of the assumptions are invalid for the population in question.

To demonstrate, let's examine *CCR5* genotypes in a population. Our population includes 283 individuals; of these, 223 have genotype *1/1*; 57 have genotype *1/Δ32*; and 3 have genotype *Δ32/Δ32*. These numbers represent the following genotype frequencies: *1/1* = 223/283 = 0.788, *1/Δ32* = 57/283 = 0.201, and *Δ32/Δ32* = 3/283 = 0.011,

respectively. From the genotype frequencies, we can compute the *CCR51* allele frequency as 0.89 and the frequency of the *CCR5-Δ32* allele as 0.11. Once we know the allele frequencies, we can use the Hardy–Weinberg law to determine whether this population is in equilibrium. The allele frequencies predict the genotype frequencies in the next generation as follows

Expected frequency of genotype

$$1/1 = p^2 = (0.89)^2 = 0.792$$

Expected frequency of genotype

$$1/\Delta32 = 2pq = 2(0.89)(0.11) = 0.196$$

Expected frequency of genotype

$$\Delta32/\Delta32 = q^2 = (0.11)^2 = 0.012$$

These expected frequencies are nearly identical to the frequencies observed in the parental generation. Our test of this population has failed to provide evidence that Hardy–Weinberg assumptions are being violated. The conclusion can be confirmed by using the whole numbers utilized in calculating the genotype frequencies to perform a χ^2 analysis. In this case, neither the genotype frequencies nor the allele frequencies are changing in this population, meaning that the population is in equilibrium, and that selection, drift, migration, mutation, and nonrandom mating are not operating in this population.

ESSENTIAL POINT ■ ■ ■

Populations that are not in Hardy–Weinberg equilibrium may be undergoing changes in allele frequency owing to selection, drift, migration, or nonrandom mating.

TABLE 2 Methods of Determining Allele Frequencies from Data on Genotypes

| (a) Counting Alleles | Genotype | | | |
	1/1	*1/Δ32*	*Δ32/Δ32*	Total
Number of individuals	79	20	1	100
Number of *1* alleles	158	20	0	178
Number of Δ*32* alleles	0	20	2	22
Total number of alleles	158	40	2	200

Frequency of *CCR51* in sample: 178/200 = 0.89 = 89%
Frequency of *CCR5-Δ32* in sample: 22/200 = 0.11 = 11%

| (b) From Genotype Frequencies | Genotype | | | |
	1/1	*1/Δ32*	*Δ32/Δ32*	Total
Number of individuals	79	20	1	100
Genotype frequency	79/100 = 0.79	20/100 = 0.20	1/100 = 0.01	1.00

Frequency of *CCR51* in sample: 0.79 + (0.5)0.20 = 0.89 = 89%
Frequency of *CCR5-Δ32* in sample: (0.5)0.20 + 0.01 = 0.11 = 11%

Calculating Frequencies for Multiple Alleles in Populations

Although we have used one-gene two-allele systems as examples, many genes have several alleles, all of which can be found in a single population. The ABO blood group in humans is such an example. The locus *I* (isoagglutinin) has three alleles I^A, I^B, and i^O, yielding six possible genotypic combinations ($I^A I^A$, $I^B I^B$, $i^O i^O$, $I^A I^B$, $I^A i^O$, $I^B i^O$). Remember that in this case I^A and I^B are codominant alleles and that both of these are dominant to i^O. The result is that homozygous $I^A I^A$ and heterozygous $I^A i^O$ individuals are phenotypically identical, as are $I^B I^B$ and $I^B i^O$ individuals, so we can distinguish only four phenotypic combinations.

By adding another variable to the Hardy–Weinberg equation, we can calculate both the genotype and allele frequencies for the situation involving three alleles. Let p, q, and r represent the frequencies of alleles I^A, I^B, and i^O, respectively. Note that because there are three alleles

$$p + q + r = 1$$

Under Hardy–Weinberg assumptions, the frequencies of the genotypes are given by

$$(p + q + r)^2 = p^2 + q^2 + r^2 + 2pq + 2pr + 2qr = 1$$

If we know the frequencies of blood types for a population, we can then estimate the frequencies for the three alleles of the ABO system. For example, in one population sampled, the following blood-type frequencies are observed: A = 0.53, B = 0.133, O = 0.26. Because the i^O allele is recessive, the population's frequency of type O blood equals the proportion of the recessive genotype r^2. Thus,

$$r^2 = 0.26$$
$$r = \sqrt{0.26}$$
$$r = 0.51$$

Using r, we can calculate the allele frequencies for the I^A and I^B alleles. The I^A allele

is present in two genotypes, $I^A I^A$ and $I^A i^O$. The frequency of the $I^A I^A$ genotype is represented by p^2 and the $I^A i^O$ genotype by $2pr$. Therefore, the combined frequency of type A blood and type O blood is given by

$$p^2 + 2pr + r^2 = 0.53 + 0.26$$

If we factor the left side of the equation and take the sum of the terms on the right

$$(p + r)^2 = 0.79$$
$$p + r = \sqrt{0.79}$$
$$p = 0.89 - r$$
$$p = 0.89 - 0.51 = 0.38$$

Having calculated p and r, the frequencies of allele I^A and allele i^O, we can now calculate the frequency for the I^B allele

$$p + q + r = 1$$
$$q = 1 - p - r$$
$$= 1 - 0.38 - 0.51$$
$$= 0.11$$

The phenotypic and genotypic frequencies for this population are summarized in **Table 3**.

Calculating Heterozygote Frequency

A useful application of the Hardy–Weinberg law, especially in human genetics, allows us to estimate the frequency of heterozygotes in a population. The frequency of a recessive trait can usually be determined by counting individuals with the homozygous recessive phenotype in a sample of the population. With this information and the Hardy–Weinberg law, we can then calculate the allele and genotype frequencies for this gene.

Cystic fibrosis, an autosomal recessive trait, has an incidence of about $1/2500 = 0.0004$ in people of northern European ancestry. Individuals with cystic fibrosis are easily distinguished from the population at large by such symptoms as extra-salty sweat, excess amounts of thick mucus in the lungs, and susceptibility to bacterial infections. Because this is a recessive trait, individuals with cystic fibrosis must

| **TABLE 3** | Calculating Genotype Frequencies for Multiple Alleles in a Hardy–Weinberg Population Where the Frequency of Allele I^A = 0.38, Allele I^B = 0.11, and Allele i^O = 0.51 |

Genotype	Genotype Frequency	Phenotype	Phenotype Frequency
$I^A I^A$	$p^2 = (0.38)^2 = 0.14$	A	0.53
$I^A i^O$	$2pr = 2(0.38)(0.51) = 0.39$		
$I^B I^B$	$q^2 = (0.11)^2 = 0.01$	B	0.12
$I^B i^O$	$2qr = 2(0.11)(0.51) = 0.11$		
$I^A I^B$	$2pr = 2(0.38)(0.11) = 0.084$	AB	0.08
$i^O i^O$	$r^2 = (0.51)^2 = 0.26$	O	0.26

be homozygous. Their frequency in a population is represented by q^2 (provided that mating has been random in the previous generation). The frequency of the recessive allele is therefore

$$q = \sqrt{q^2} = \sqrt{0.0004} = 0.02$$

Because $p + q = 1$, then the frequency of p is

$$p = 1 - q = 1 - 0.02 = 0.98$$

In the Hardy–Weinberg equation, the frequency of heterozygotes is $2pq$. Thus,

$$2pq = 2(0.98)(0.02)$$
$$= 0.04 \text{ or 4 percent, or } 1/25$$

Thus, heterozygotes for cystic fibrosis are rather common (about $1/25$ individuals, or 4 percent of the population), even though the incidence of homozygous recessives is only $1/2500$, or 0.04 percent. Calculations such as these are estimates because the population may not meet all Hardy–Weinberg assumptions.

NOW SOLVE THIS

3 If the albino phenotype occurs in 1/10,000 individuals in a population at equilibrium and albinism is caused by an autosomal recessive allele *a*, calculate the frequency of: (a) the recessive mutant allele; (b) the normal dominant allele; (c) heterozygotes in the population; (d) mating between heterozygotes.

■ **Hint** *This problem involves an understanding of the method of calculating allele and genotype frequencies. The key to its solution is to first determine the frequency of the albinism allele in this population.*

4 — Natural Selection Is a Major Force Driving Allele Frequency Change

To understand evolution, we must understand the forces that transform the gene pools of populations and can lead to the formation of new species. Chief among the mechanisms transforming populations is **natural selection,** discovered independently by Darwin and by Alfred Russel Wallace. The Wallace–Darwin concept of natural selection can be summarized as follows:

1. Individuals of a species exhibit variations in phenotype—for example, differences in size, agility, coloration, defenses against enemies, ability to obtain food, courtship behaviors, and flowering times.

2. Many of these variations, even small and seemingly insignificant ones, are heritable and passed on to offspring.

3. Organisms tend to reproduce in an exponential fashion. More offspring are produced than can survive. This causes members of a species to engage in a struggle for survival, competing with other members of the community for scarce resources. Offspring also must avoid predators, and in sexually reproducing species, adults must compete for mates.

4. In the struggle for survival, individuals with particular phenotypes will be more successful than others, allowing the former to survive and reproduce at higher rates.

As a consequence of natural selection, populations and species change. The phenotypes that confer improved ability to survive and reproduce become more common, and the phenotypes that confer poor prospects for survival and reproduction may eventually disappear. Under certain conditions, populations that at one time could interbreed may lose that capability, thus segregating their adaptations into particular niches. If selection continues, it may result in the appearance of new species.

Detecting Natural Selection in Populations

Recall that measuring allele frequencies and genotype frequencies using the Hardy–Weinberg law is based on certain assumptions about an ideal population: large population size, lack of migration, presence of random mating, absence of selection and mutation, and equal survival rates of offspring.

However, if all genotypes do not have equal rates of survival or do not leave equal numbers of offspring, then allele frequencies may change from one generation to the next. To see why, let's imagine a population of 100 individuals in which the frequency of allele *A* is 0.5 and that of allele *a* is 0.5. Assuming the previous generation mated randomly, we find that the genotype frequencies in the present generation are $(0.5)^2 = 0.25$ for *AA*, $2(0.5)(0.5) = 0.5$ for *Aa*, and $(0.5)^2 = 0.25$ for *aa*. Because our population contains 100 individuals, we have 25 *AA* individuals, 50 *Aa* individuals, and 25 *aa* individuals. Now suppose that individuals with different genotypes have different rates of survival: All 25 *AA* individuals survive to reproduce, 90 percent or 45 of the *Aa* individuals survive to reproduce, and 80 percent or 20 of the *aa* individuals survive to reproduce. When the survivors reproduce, each contributes two gametes to the new gene pool, giving us $2(25) + 2(45) + 2(20) = 180$ gametes. What are the frequencies of the two alleles in the surviving population? We have 50 *A* gametes from *AA* individuals, plus 45 *A* gametes from *Aa* individuals, so the frequency of allele *A* is $(50 + 45)/180 = 0.53$. We have 45 *a* gametes from *Aa* individuals, plus 40 *a* gametes from *aa* individuals, so the frequency of allele *a* is $(45 + 40)/180 = 0.47$.

These differ from the frequencies we started with. The frequency of allele *A* has increased, whereas the frequency of allele *a* has declined. A difference among individuals in survival or reproduction rate (or both) is an example of **natural selection.** Natural selection is the principal force that shifts allele frequencies within large populations and is one of the most important factors in evolutionary change.

Fitness and Selection

Selection occurs whenever individuals with a particular genotype enjoy an advantage in survival or reproduction over other genotypes. However, selection may vary from less than 1 to 100 percent. In the previous hypothetical example, selection was strong. Weak selection might involve just a fraction of a percent difference in the survival rates of different genotypes. Advantages in survival and reproduction ultimately translate into increased genetic contribution to future generations. An individual organism's genetic contribution to future generations is called its **fitness.** Genotypes associated with high rates of reproductive success are said to have high fitness, whereas genotypes associated with low reproductive success are said to have low fitness.

Hardy–Weinberg analysis also allows us to examine fitness. By convention, population geneticists use the letter *w* to represent fitness. Thus, w_{AA} represents the relative fitness of genotype *AA*, w_{Aa} the relative fitness of genotype *Aa*, and w_{aa} the relative fitness of genotype *aa*. Assigning the values $w_{AA} = 1$, $w_{Aa} = 0.9$, and $w_{aa} = 0.8$ would mean, for example, that all *AA* individuals survive, 90 percent of the *Aa* individuals survive, and 80 percent of the *aa* individuals survive, as in the previous hypothetical case.

Let's consider selection against deleterious alleles. Fitness values, $w_{AA} = 1$, $w_{Aa} = 1$, and $w_{aa} = 0$ describe a situation in which *a* is a homozygous lethal allele. As homozygous recessive individuals die without leaving offspring, the frequency of allele *a* will decline. The decline in the frequency of allele *a* is described by the equation

$$q_g = \frac{q_0}{1 + gq_0}$$

where q_g is the frequency of allele *a* in generation *g*, q_o is the starting frequency of *a* (i.e., the frequency of *a* in generation zero), and *g* is the number of generations that have passed.

Figure 7 shows what happens to a lethal recessive allele with an initial frequency of 0.5. At first, because of the high percentage of *aa* genotypes, the frequency of allele *a* declines rapidly. The frequency of *a* is halved in only two generations. By the sixth generation, the frequency is halved again. By now, however, the majority of *a* alleles are carried by heterozygotes. Because *a* is recessive, these heterozygotes are not selected against. Consequently, as more time passes, the frequency of allele *a* declines ever more slowly. As long as heterozygotes continue to mate, it

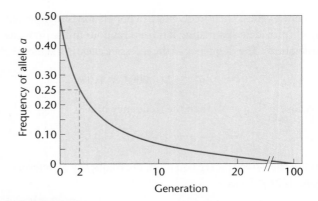

FIGURE 7   Change in the frequency of a lethal recessive allele, *a*. The frequency of *a* is halved in two generations and halved again by the sixth generation. Subsequent reductions occur slowly because the majority of *a* alleles are carried by heterozygotes.

is difficult for selection to completely eliminate a recessive allele from a population.

Figure 8 shows the outcome of different degrees of selection against a nonlethal recessive allele, *a*. In this case, the intensity of selection varies from strong (red curve) to weak (blue curve), as well as intermediate values (yellow, purple, and green curves). In each example, the frequency of the deleterious allele, *a*, starts at 0.99 and declines over time. However, the rate of decline depends heavily on the strength of selection. When selection is strong and only 90 percent of the heterozygotes and 80 percent of the *aa* homozygotes survive (red curve), the frequency of allele *a* drops from 0.99 to less than 0.01 in about 85 generations. However, when selection is weak, and 99.8 percent of the heterozygotes and 99.6 percent of the *aa* homozygotes survive (blue curve), it takes 1000 generations for the frequency of allele *a* to drop from 0.99 to 0.93. Two important conclusions can be drawn from this example. First, over thousands of generations, even weak selection can cause substantial changes in allele frequencies; because evolution generally occurs over a large number of generations, selection is a powerful force in

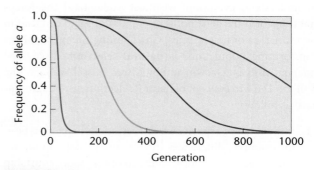

FIGURE 8   The effect of selection on allele frequency. The rate at which a deleterious allele is removed from a population depends heavily on the strength of selection.

evolutionary change. Second, for selection to produce rapid changes in allele frequencies, the differences in fitness among genotypes must be large.

The manner in which selection affects allele frequencies allows us to make some inferences about the *CCR5-Δ32* allele that we discussed earlier. Because individuals with genotype *Δ32/Δ32* are resistant to most sexually transmitted strains of HIV-1, while individuals with genotypes *1/1* and *1/Δ32* are susceptible, we might expect AIDS to act as a selective force causing the frequency of the *Δ32* allele to increase over time. Indeed, it probably will, but the increase in frequency is likely to be slow in human terms. In fact, it will take about 100 generations (about 2000 years) for the frequency of the *Δ32* allele to reach just 0.11. In other words, the frequency of the *Δ32* allele will probably not change much over the next few generations in most populations that currently harbor it.

There Are Several Types of Selection

The phenotype is the result of the combined influence of the individual's genotype at many different loci and the effects of the environment. Selection for traits can be classified as (1) directional, (2) stabilizing, or (3) disruptive.

In **directional selection (Figure 9)** phenotypes at one end of the spectrum present in the population become selected for or against, usually as a result of changes in the environment. A carefully documented example comes from research by Peter and Rosemary Grant and their colleagues, who study the medium ground finches (*Geospiza fortis*) of Daphne Major Island in the Galapagos Islands. The beak size of these birds varies enormously. In 1976, for example, some birds in the population had beaks less than 7 mm deep, while others had beaks more than 12 mm deep. In 1977, a severe drought killed some 80 percent of the finches. Big-beaked birds survived at higher rates than small-beaked birds because when food

became scarce, the big-beaked birds were able to eat a greater variety of seeds. When the drought ended in 1978, the offspring of the survivors inherited their parents' big beaks. Between 1976 and 1978, the beak depth of the average finch in the Daphne Major population increased by just over 0.5 mm, shifting the average beak size toward one phenotypic extreme.

Stabilizing selection, in contrast, tends to favor intermediate phenotypes, with those at both extremes being selected against. Over time, this will reduce the phenotypic variance in the population but without a significant shift in the mean. One of the clearest demonstrations of stabilizing selection is shown by a study of human birth weight and survival for 13,730 children born over an 11-year period. **Figure 10** shows the distribution of birth weight and the percentage of mortality at 4 weeks of age. Infant mortality increases on either side of the optimal birth weight of 7.5 pounds. Stabilizing selection acts to keep a population well adapted to its current environment.

Disruptive selection is selection against intermediates and for both phenotypic extremes. It can be viewed as the opposite of stabilizing selection because the intermediate types are selected against. This will result in a population with an increasingly bimodal distribution for the trait, as we can see in **Figure 11**. In one set of experiments using *Drosophila*, after several generations of disruptive artificial selection for bristle number, in which only flies with high- or low-bristle numbers were allowed to breed, most flies could be easily placed in a low- or high-bristle category. In natural populations, such a situation might exist for a population in a heterogeneous environment.

ESSENTIAL POINT ■ ■ ■

The rate of change under natural selection depends on initial allele frequencies, selection intensity, and the relative fitness of different genotypes.

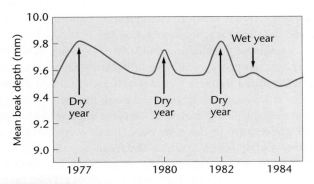

FIGURE 9 Beak size in finches during dry years increases because of strong selection. Between droughts, selection for large beak size is not as strong, and birds with smaller beak sizes survive and reproduce, increasing the number of birds with smaller beaks.

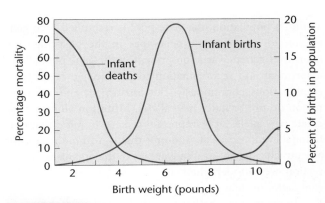

FIGURE 10 Relationship between birth weight and mortality in humans.

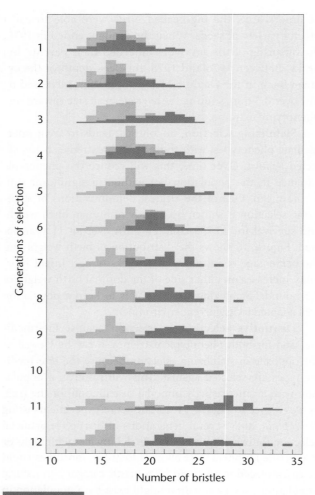

FIGURE 11 The effect of disruptive selection on bristle number in *Drosophila*. When individuals with the highest and lowest bristle number were selected, the population showed a nonoverlapping divergence in only 12 generations.

5 Mutation Creates New Alleles in a Gene Pool

Within a population, the gene pool is reshuffled each generation to produce new genotypes in the offspring. The enormous genetic variation present in the gene pool allows assortment and recombination to produce new genotypic combinations continuously. But assortment and recombination do not produce new alleles. **Mutation** alone acts to create new alleles. It is important to keep in mind that mutational events occur at random—that is, without regard for any possible benefit or disadvantage to the organism. In this section, we consider whether mutation, by itself, is a significant factor in changing allele frequencies.

To determine whether mutation is a significant force in changing allele frequencies, we measure the rate at which

they are produced. As most mutations are recessive, it is difficult to observe mutation rates directly in diploid organisms. Indirect methods use probability and statistics or large-scale screening programs to estimate mutation rates. For certain dominant mutations, however, a direct method of measurement can be used. To ensure accuracy, several conditions must be met:

1. The allele must produce a distinctive phenotype that can be distinguished from similar phenotypes produced by recessive alleles.

2. The trait must be fully expressed or completely penetrant so that mutant individuals can be identified.

3. An identical phenotype must never be produced by nongenetic agents such as drugs or chemicals.

Mutation rates can be stated as the number of new mutant alleles per given number of gametes. Suppose that for a given gene that undergoes mutation to a dominant allele, 2 out of 100,000 births exhibit a mutant phenotype. In these two cases, the parents are phenotypically normal. Because the zygotes that produced these births each carry two copies of the gene, we have actually surveyed 200,000 copies of the gene (or 200,000 gametes). If we assume that the affected births are each heterozygous, we have uncovered two mutant alleles out of 200,000. Thus, the mutation rate is 2/200,000 or 1/100,000, which in scientific notation is written as 1×10^{-5}. In humans, a dominant form of dwarfism known as **achondroplasia** fulfills the requirements for measuring mutation rates. Individuals with this skeletal disorder have an enlarged skull, short arms and legs, and can be diagnosed by X-ray examination at birth. In a survey of almost 250,000 births, the mutation rate (μ) for achondroplasia has been calculated as

$$\mu = 1.4 \times 10^{-5} \pm 0.5 \times 10^{-5}$$

Knowing the rate of mutation, we can estimate the extent to which mutation can cause allele frequencies to change from one generation to the next. We represent the normal allele as *d* and the allele for achondroplasia as *D*.

Imagine a population of 500,000 individuals in which everyone has genotype *dd*. The initial frequency of *d* is 1.0, and the initial frequency of *D* is 0. If each individual contributes two gametes to the gene pool, the gene pool will contain 1,000,000 gametes, all carrying allele *d*. Although the gametes are in the gene pool, 1.4 of every 100,000 *d* alleles mutate into a *D* allele. The frequency of allele *d* is now $(1,000,000 - 14)/1,000,000 = 0.999986$, and the frequency of allele *D* is $14/1,000,000 = 0.000014$. From these numbers, it will clearly be a long time before mutation, by itself, causes any appreciable change in the allele frequencies in this population. In other words, mutation generates new alleles but, by itself does not alter allele frequencies at an appreciable rate.

6   Migration and Gene Flow Can Alter Allele Frequencies

Occasionally, a species becomes divided into populations that are separated geographically. Various evolutionary forces, including selection, can establish different allele frequencies in such populations. **Migration** occurs when individuals move between the populations. Imagine a species in which a given locus has two alleles, A and a. There are two populations of this species, one on a mainland and one on an island. The frequency of A on the mainland is represented by p_m, and the frequency of A on the island is p_i. If there is migration from the mainland to the island, the frequency of A in the next generation on the island (p_i') is given by

$$p_i' = (1 - m)p_i + mp_m$$

where m represents migrants from the mainland to the island and migration is random with respect to genotype.

As an example of how migration might affect the frequency of A in the next generation on the island ($p_{i'}$), assume that $p_i = 0.4$ and $p_m = 0.6$ and that 10 percent of the parents of the next generation are migrants from the mainland ($m = 0.1$). In the next generation, the frequency of allele A on the island will therefore be

$$p_{i'} = [(1 - 0.1) \times 0.4] + (0.1 \times 0.6)$$
$$= 0.36 + 0.06$$
$$= 0.42$$

In this case, migration from the mainland has changed the frequency of A on the island from 0.40 to 0.42 in a single generation.

These calculations reveal that the change in allele frequency attributable to migration is proportional to the differences in allele frequency between the donor and recipient populations and to the rate of migration. If either m is large or p_m is very different from p_i, then a rather large change in the frequency of A can occur in a single generation. If migration is the only force acting to change the allele frequency on the island, then equilibrium will be attained only when $p_i = p_m$. These guidelines can often be used to estimate migration in cases where it is difficult to quantify. As m can have a wide range of values, the effect of migration can substantially alter allele frequencies in populations, as shown for the I^B allele of the ABO blood group in **Figure 12**.

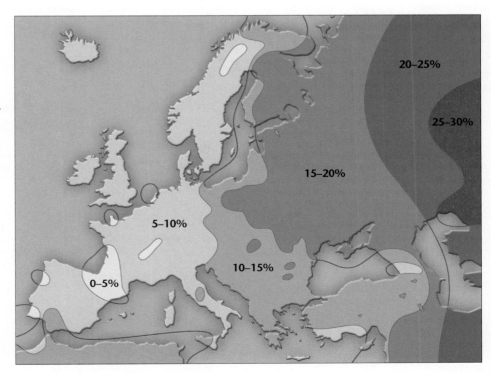

FIGURE 12 Migration as a force in evolution. The I^B allele of the *ABO* locus is present in a gradient from east to west. This allele shows the highest frequency in central Asia and the lowest in northeast Spain. The gradient parallels the waves of Mongol migration into Europe following the fall of the Roman Empire and is a genetic relic of human history.

7 Genetic Drift Causes Random Changes in Allele Frequency in Small Populations

In small populations, significant random fluctuations in allele frequencies are possible by chance alone, a situation known as **genetic drift.** The degree of fluctuation increases as the population size decreases. In addition to small population size, drift can arise through the **founder effect,** which occurs when a population originates from a small number of individuals. Although the population may later increase to a large size, the genes carried by all members are derived from those of the founders (assuming no mutation, migration, or selection, and the presence of random mating). Drift can also arise via a **genetic bottleneck.** Bottlenecks develop when a large population undergoes a drastic but temporary reduction in numbers. Even though the population recovers, its genetic diversity has been greatly reduced.

Founder Effects in Human Populations

Allele frequencies in certain human populations demonstrate the role of genetic drift in natural populations. Native Americans living in the southwestern United States have a high frequency of oculocutaneous albinism (OCA). In the Navajo, who live primarily in northeast Arizona, albinism occurs with a frequency of 1 in 1500–2000, compared with whites (1 in 36,000) and African-Americans (1 in 10,000). There are four different forms of OCA (OCA1–4), all with varying degrees of melanin deficiency in the skin, eyes, and hair. OCA2 is caused by mutations in the *P* gene, which encodes a plasma membrane protein. To investigate the genetic basis of albinism in the Navajo, researchers screened for mutations in the *P* gene. In their study, all Navajo with albinism were homozygous for a 122.5-kb deletion in the *P* gene, spanning exons 10–20 **(Figure 13)**. This deletion allele was not present in 34 individuals belonging to other Native American populations.

Using a set of PCR primers, researchers were able to identify homozygous affected individuals and heterozygous carriers **(Figure 14)** and surveyed 134 normally pigmented Navajo and 42 members of the Apache, a tribe closely related to the Navajo. Based on this sample, the heterozygote frequency in the Navajo is estimated to be 4.5 percent. No carriers were found in the Apache population that was studied.

The 122.5-kb deletion allele causing OCA2 was found only in the Navajo population and not in members of other Native American tribes in the southwestern United States, suggesting that the mutant allele is specific to the Navajo and may have arisen in a single individual who was one of a small number of founders of the Navajo population. Using other methods, workers originally estimated the age of the mutation to be between 400 and 11,000 years. To narrow this range, they relied on tribal history. Navajo oral tradition indicates that the Navajo and Apache became separate populations between 600 and 1000 years ago. Because the deletion is not found in the Apaches, it probably arose in the Navajo population after the tribes split. On this basis, the deletion is estimated to be 400 to 1000 years old and probably arose as a founder mutation.

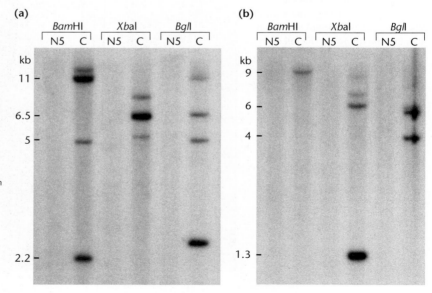

FIGURE 13 Genomic DNA digests from a Navajo affected with albinism (N5) and a normally pigmented individual (C). (a) Hybridization with a probe covering exons 11–15 of the *P* gene; there are no hybridizing fragments detected in N5. (b) Hybridization with a probe covering exons 15–20 of the *P* gene; there are no hybridizing fragments detected in N5. This confirms the presence of a deletion in affected individuals.

Courtesy of Murray Brilliant, "A 122.5 kilobase deletion of P gene underlies the high prevalence of oculocutaneous albinism type 2 in the Navajo population." From: American Journal Human Genetics 72:62–72, Figure 1, 65. Published by University of Chicago Press.

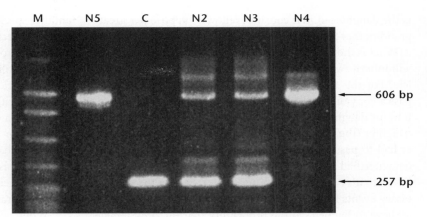

FIGURE 14 PCR screens of Navajo affected with albinism (N4 and N5) and the parents of N4 (N2 and N3). Affected individuals (N4 and N5), heterozygous carriers (N2 and N3), and a homozygous normal individual (C) each give a distinctive band pattern, allowing detection of heterozygous carriers in the population. Molecular size markers (M) are in the first lane.

Courtesy of Murray Brilliant, "A 122.5 kilobase deletion of P gene underlies the high prevalence of oculocutaneous albinism type 2 in the Navajo population." From: American Journal Human Genetics 72: 62–72, Figure 3, p. 67. Published by University of Chicago Press.

8 | Nonrandom Mating Changes Genotype Frequency but Not Allele Frequency

We have explored how violations of the first four assumptions of the Hardy–Weinberg law, in the form of selection, mutation, migration, and genetic drift, can cause allele frequencies to change. The fifth assumption is that the members of a population mate at random; in other words, any one genotype has an equal probability of mating with any other genotype in the population. Nonrandom mating can change the frequencies of genotypes in a population. Subsequent selection for or against certain genotypes has the potential to affect the overall frequencies of the alleles they contain, but it is important to note that nonrandom mating *does not itself directly change allele frequencies.*

Nonrandom mating can take one of several forms. In **positive assortative mating,** similar genotypes are more likely to mate than dissimilar ones. This often occurs in humans: A number of studies have indicated that many people are more attracted to individuals who physically resemble them (and are therefore more likely to be genetically similar as well). **Negative assortative mating** occurs when dissimilar genotypes are more likely to mate; some plant species have inbuilt pollen/stigma recognition systems that prevent fertilization between individuals with the same alleles at key loci. However, the form of nonrandom mating most commonly found to affect genotype frequencies in population genetics is **inbreeding.**

Inbreeding

Inbreeding occurs when mating individuals are more closely related than any two individuals drawn from the population at random; loosely defined, inbreeding is mating among relatives. For a given allele, inbreeding increases the proportion of homozygotes in the population. A completely inbred population will theoretically consist only of homozygous genotypes.

To describe the intensity of inbreeding in a population, geneticist Sewall Wright devised the **coefficient of inbreeding (F).** *F* quantifies the probability that the two alleles of a given gene in an individual are identical *because they are descended from the same single copy of the allele in an ancestor.* If $F = 1$, all individuals in the population are homozygous, and both alleles in every individual are derived from the same ancestral copy. If $F = 0$, no individual has two alleles derived from a common ancestral copy.

One method of estimating *F* for an individual is shown in **Figure 15.** The fourth-generation female (shaded pink)

FIGURE 15 Calculating the coefficient of inbreeding *(F)* for the offspring of a first-cousin marriage.

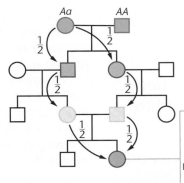

The chance that this female will inherit two copies of her great-grandmother's *a* allele is

$$F = \frac{1}{2} \times \frac{1}{2} \times \frac{1}{2} \times \frac{1}{2} \times \frac{1}{2} \times \frac{1}{2} = \frac{1}{64}$$

Because the female's two alleles could be identical by descent from any of four different alleles,

$$F = 4 \times \frac{1}{64} = \frac{1}{16}$$

is the daughter of first cousins (yellow). Suppose her great-grandmother (green) was a carrier of a recessive lethal allele, *a*. What is the probability that the fourth-generation female will inherit two copies of her great-grandmother's lethal allele? For this to happen, (1) the great-grandmother had to pass a copy of the allele to her son, (2) her son had to pass it to his daughter, and (3) his daughter had to pass it to her daughter (the pink female). Also, (4) the great-grandmother had to pass a copy of the allele to her daughter, (5) her daughter had to pass it to her son, and (6) her son had to pass it to his daughter (the pink female). Each of the six necessary events has an individual probability of 1/2, and they *all* have to happen, so the probability that the pink female will inherit two copies of her great-grandmother's lethal allele is $(1/2)^6 = 1/64$. To calculate an overall value of *F* for the pink female as a child of a first-cousin marriage, remember that she could also inherit two copies of any of the other three alleles present in her great-grandparents. Because any of four possibilities would give the pink female two alleles identical by descent from an ancestral copy,

$$F = 4 \times (1/64) = 1/16$$

ESSENTIAL POINT ■ ■ ■

Nonrandom mating in the form of inbreeding increases the frequency of homozygotes in the population and decreases the frequency of heterozygotes.

NOW SOLVE THIS

4 A prospective groom, who is normal, has a sister with cystic fibrosis (CF), an autosomal recessive disease. Their parents are normal. The brother plans to marry a woman who has no history of CF in her family. What is the probability that they will produce a CF child? They are both Caucasian, and the overall frequency of CF in the Caucasian population is 1/2500—that is, 1 affected child per 2500. (Assume the population meets the Hardy–Weinberg assumptions.)

■ **Hint** *This problem involves an understanding of how recessive alleles are transmitted and the probability of receiving a recessive allele from a heterozygous parent. The key to its solution is to first work out the probability that each parent carries the mutant allele.*

9 Reduced Gene Flow, Selection, and Genetic Drift Can Lead to Speciation

A **species** can be defined as a group of actually or potentially interbreeding organisms that is reproductively isolated in nature from all other such groups. In sexually reproducing organisms, speciation transforms the parental species into

another species, or divides a single species into two or more separate species. Changes in morphology or physiology and adaptation to ecological niches may also occur but are not necessary components of the speciation event.

Populations within a species may carry considerable genetic variation, present as differences in alleles or allele frequencies at a variety of loci. Genetic divergence of these populations can reflect the action of natural selection, genetic drift, or both. In an earlier section, we saw that the migration of individuals between populations tends to homogenize allele frequencies among populations. In other words, migration counteracts the tendency of populations to diverge.

When gene flow between populations is reduced or absent, the populations may diverge to the point that members of one population are no longer able to interbreed successfully with members of the other. When populations reach the point where they are reproductively isolated from one another, they have become different species, according to the biological species concept. The genetic changes that result in reproductive isolation between or among populations and that lead to the formation of new species or higher taxonomic groups represent an example of macroevolution. We will restrict our discussion to the macroevolutionary changes that lead to the formation of new species.

The biological barriers that prevent or reduce interbreeding between populations are called **reproductive isolating mechanisms.** These mechanisms may be ecological, behavioral, seasonal, mechanical, or physiological.

Prezygotic isolating mechanisms prevent individuals from mating in the first place. Individuals from different populations may not find each other at the right time, may not recognize each other as suitable mates, or may try to mate but find that they are unable to do so.

Postzygotic isolating mechanisms create reproductive isolation even when the members of two populations are willing and able to mate with each other. For example, genetic divergence may have reached the stage where the viability or fertility of hybrids is reduced. Hybrid zygotes may be formed, but all or most may be inviable. Alternatively, the hybrids may be viable, but be sterile or suffer from reduced fertility. Yet again, the hybrids themselves may be fertile, but their progeny may have lowered viability or fertility. In all these situations, hybrids are genetic dead-ends. These postzygotic mechanisms act at or beyond the level of the zygote and are generated by genetic divergence.

Postzygotic isolating mechanisms waste gametes and zygotes and lower the reproductive fitness of hybrid survivors. Selection will therefore favor the spread of alleles that reduce the formation of hybrids, leading to the development of prezygotic isolating mechanisms, which in turn prevent interbreeding and the formation of hybrid zygotes and offspring. In animal evolution, one of the most effective prezygotic mechanisms is behavioral isolation, involving courtship behavior.

Changes Leading to Speciation

Formation of the Isthmus of Panama about 3 million years ago created a land bridge connecting North and South America and separated the Caribbean Sea from the Pacific Ocean. After identifying seven Caribbean species of snapping shrimp **(Figure 16)** and seven similar Pacific species, researchers matched them in pairs. Members of each pair were closer to each other in structure and appearance than to any other species in its own ocean. Analysis of allele frequencies and mitochondrial DNA sequences confirmed that the members of each pair were one another's closest genetic relatives. In other words, the ancestors of each pair were members of a single species. When the isthmus closed, each of the seven ancestral species was divided into two separate populations, one in the Caribbean and the other in the Pacific. But after 3 million years of separation, were members of these populations different species?

Males and females were paired together, and successful matings between Caribbean–Pacific couples versus those of Caribbean–Caribbean or Pacific–Pacific pairs was calculated. In three of the seven species pairs, transoceanic couples refused to mate altogether. For the other four pairs, transoceanic couples were 33, 45, 67, and 86 percent as likely to mate with each other compared to the same-ocean pairs.

Of the transoceanic pairs that mated, only 1 percent produced viable offspring, while 60 percent of same-ocean pairs produced viable offspring. We can conclude that 3 million years of separation has resulted in complete or nearly complete speciation, involving strong pre- and postzygotic isolating mechanisms for all seven species pairs.

The Rate of Macroevolution and Speciation

How much time is required for speciation? As we saw in the example above, the time needed for genetic divergence and formation of new species can occur over a span of several million years. In fact, the average time for speciation ranges from 100,000 to 10,000,000 years. However, rapid speciation

over much shorter time spans has been reported in a number of cases, including fishes in East African lakes, marine salmon, palm trees on isolated islands, polyploid plants, and brown algae in the Baltic Sea.

In Nicaragua, Lake Apoyo was formed within the last 23,000 years in the crater of a volcano **(Figure 17)**. This small lake is home to two species of cichlid fish: the Midas cichlid, *Amphilophus citrinellus*, and the Arrow cichlid, *A. zalious*. The Midas is the most common cichlid in the region and is found in nearby lakes; the Arrow cichlid is found only in Lake Apoyo.

To establish the evolutionary origin of the Arrow cichlid, researchers used a variety of approaches, including phylogenetic, morphological, and ecological analyses. Sequence analysis of mitochondrial DNA established that the two species form a monophyletic group, indicating that there was first a population expansion of the Midas cichlid, followed by an ongoing expansion of the Arrowhead cichlid. Further analysis using a PCR-based method called amplified fragment length polymorphisms (AFLP) strengthened the conclusion that these two species are monophyletic and that *A. zalious* evolved from *A. citrinellus*. Members of the two species have distinctive morphologies **(Figure 18)**, including jaw specializations that reflect different food preferences, which were confirmed by analysis of stomach contents. In addition, the two species are reproductively isolated, a conclusion that was substantiated by laboratory experiments. Using a molecular clock calibrated for cichlid mtDNA, it is estimated that *A. zaliosus* evolved from *A. citrinellus* sometime within the last 10,000 years. This, and examples from other species, provide unambiguous evidence that depending on the strength of selection and that of other parameters of the Hardy-Weinberg law, species

FIGURE 16 A snapping shrimp (genus *Alpheus*).

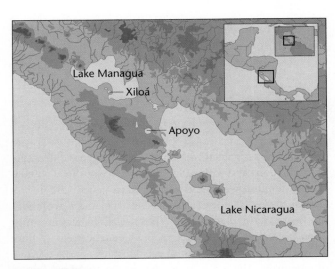

FIGURE 17 Lake Apoyo in Nicaragua occupies the crater of an inactive volcano. The lake formed about 23,000 years ago. Two species of cichlid fish in the lake share a close evolutionary relationship.

(a)

(b)

The two species of cichilds in Lake Apoyo exhibit distinctive morphologies: (a) *Amphilophus citrinellus,* (b) *Amphilophus zaliosus.*

formation can occur over a much shorter time scale than the usual range of 100,000–10,000,000 years.

---

10   Phylogeny Can Be Used to Analyze Evolutionary History

Speciation is associated with changes in the genetic structure of populations and with genetic divergence of those populations. Therefore, we should be able to use genetic differences among present-day species to reconstruct their evolutionary histories, or phylogenies. These relationships are most often presented in the form of phylogenetic trees (**Figure 19**), which show the ancestral relationships among a group of organisms. These groups can be species, or larger groups. In a phylogenetic tree, branches represent lineages over time. The length of a branch can be arbitrary, or it can be derived from a time scale, showing the length of time between spe-

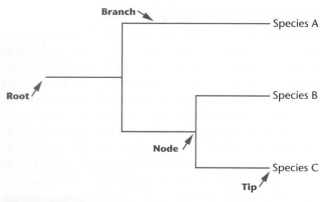

FIGURE 19   Elements of a phylogenetic tree showing the relationships among several species. The root represents a common ancestor to all species on the tree. Branches represent lineages through time. The points at which the branches separate are called nodes, and at the tips of the branches are the living or extinct species.

ciation events. Points at which lines diverge, called nodes, show when a species split into two or more species. Each node represents a common ancestor of the species diverging at that node. The tips of the branches represent species (or a larger group) alive today (or those that ended in extinction). Groups that consist of an ancestral species and all its descendants are called monophyletic groups. The root of a phylogenetic tree represents the oldest common ancestor to all the groups shown in the tree.

Constructing Phylogenetic Trees from Amino Acid Sequences

In an important early example of phylogenetic analysis, W. M. Fitch and E. Margoliash used data on the amino acid sequence of cytochrome c in a variety of organisms to construct a phylogenetic tree. **Cytochrome c** is a eukaryotic mitochondrial protein whose amino acid sequence has changed very slowly over time. For example, the amino acid sequence in humans and chimpanzees is identical, and humans and rhesus monkeys show only one amino acid difference. This high degree of similarity is remarkable considering that the fossil record indicates that the lines leading to humans and monkeys diverged from a common ancestor approximately 20 million years ago. Using these data, Margoliash postulated the idea of **genetic equidistance,** meaning that the differences in amino acid sequence between species and major groups of organisms is proportional to their evolutionary distance. For example, fish were early vertebrates, and the number of amino acid differences between individual species of fish and individual species of mammals should be similar. In fact, the differences between the sequence from tuna and horse, pig, rabbit, and human falls in the range of 17–21 amino acid differences, and the distance between yeast and vertebrates ranges from 43–48.

However, more than one nucleotide change may be required to generate an amino acid substitution. When the nucleotide changes necessary for all amino acid differences observed in a protein are totaled, the **minimal mutational distance** between the genes of any two species is established.

Fitch used data on the minimal mutational distances between the cytochrome c genes of 19 organisms to construct a phylogenetic tree showing their evolutionary history (**Figure 20**). The black dots on the tips of the branches represent existing species. Common ancestors are linked to them by green lines and red dots showing that ancestral species evolved and diverged to produce the modern species. The tree culminates in a single common ancestor for all the species on the tree, represented by the red dot on the extreme left. Numbers on the lines represent corrected minimal mutational differences.

Molecular Clocks Measure the Rate of Evolutionary Change

In many cases, we would like to estimate not only which members of a set of species are most closely related, but also when their common ancestors lived. The phylogenetic analysis of amino acid sequences and the concept of genetic equidistance led to the idea of constructing and using **molecular**

clocks to measure the rate of change in amino acid or nucleotide sequences as a way to infer evolutionary relationships and estimate time of divergence from a common ancestor.

To be useful, molecular clocks must be carefully calibrated. Molecular clocks can only measure changes in amino acids or nucleotides; they are linear over certain time scales, and times and dates must be added to the clock using independent evidence such as the fossil record. **Figure 21** shows a molecular clock showing divergence times for humans and other vertebrates based on the fossil record [Figure 21(a)], and molecular data [Figure 21(b)]. In both cases, changes in amino acid sequence and nucleotide sequence increase linearly with time.

Genomics and Molecular Evolution

The development of genome projects, bioinformatics, and the emergence of comparative genomics has revolutionized the field of molecular evolution. Using these methods, researchers are studying the origin, evolution, and function of gene duplications , the evolution of the human genome, and more recently, the role of natural selection in human adaptations. As we will see in the following section, comparative genomics can be used to provide a detailed picture of how our genome compares to those of our closest relatives and ancestors.

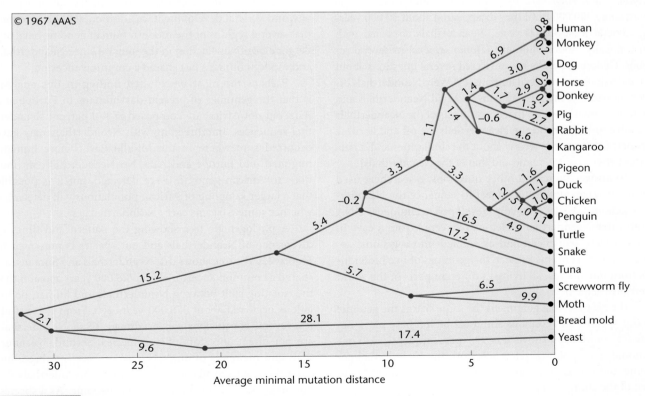

FIGURE 20 Phylogenetic tree constructed by comparing homologies in cytochrome c amino acid sequences.

Reprinted with permission from Fitch, W.M. and Margoliash, E. 1967. Construction of phylogenetic trees. Science 279: 279–284, Figure 2.

(a)

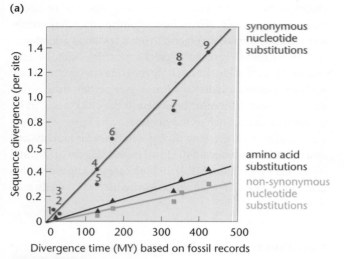

(b)

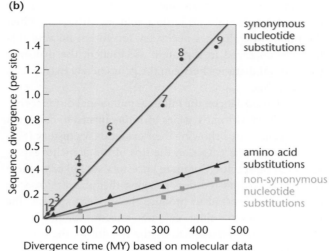

FIGURE 21 Relationship between the number of amino acid substitutions and the number of nucleotide substitutions for 4,198 nuclear genes from 10 vertebrate species. Humans versus (1) chimpanzee, (2) orangutan, (3) macaque, (4) mouse, (5) cow, (6) opossum, (7) chicken, (8) western clawed frog, and (9) zebrafish. In (a) the data are calculated by divergence times based on the fossil record, and in (b), based on molecular data. synonymous nucleotide substitutions do not result in the replacement of an amino acid; non-synonymous nucleotide substitutions result in a different amino acid in the protein.

Analysis of Genetic Divergence between Neanderthals and Modern Humans

Paleontological evidence indicates that the Neanderthals, *Homo neanderthalensis,* lived in Europe and western Asia from some 300,000 until they disappeared about 30,000 years ago. For at least 30,000 years, Neanderthals coexisted with anatomically modern humans (*Homo sapiens*) in several regions. Genome analysis has answered several questions about Neanderthals and modern humans: (1) Were Neanderthals direct ancestors of modern humans? (2) Did Neanderthals and *H. sapiens* interbreed, so that descendants of the Neanderthals are alive today? Or did the Neanderthals die off and become extinct? (3) What can we say about the similarities and differences between our genome and that of the Neanderthals?

To answer these and other questions, researchers used results from an ambitious project to sequence the Neanderthal genome. Using specially developed techniques, DNA from three Neanderthal skeletons recovered from a cave in Croatia were sequenced and assembled into a genome sequence. As part of this project, the genomes of five individuals from our species, all living in different parts of the world were also sequenced.

The Neanderthal genome is the same size as the genome of our species, and contains about 3.2 billion base pairs. Sequence comparison shows it is 99.7 percent identical to our genome. As a point of reference, both the Neanderthal genome and our genome are about 98.8 percent identical to that of the chimpanzee.

By comparing the Neanderthal genome sequence with the genomes of five present-day humans and with the chimpanzee genome, researchers were able to identify amino acid-coding differences in 78 genes that originated after the split between the Neanderthal and human lineages. These genes are involved in cognitive development, skin morphology, and skeletal development. In addition, there is evidence that several regions of the modern human genome have undergone positive selection in the interval since Neanderthals and modern humans last shared a common ancestor.

Perhaps the most unexpected finding in this analysis is that the genomes of present-day humans in Europe and Asia, but not Africa, are composed of 1–4 percent Neanderthal sequences. Interbreeding with Neanderthals may have occurred somewhere in the Middle East, before humans migrated into Europe and Asia. No Neanderthal contributions to African genomes were detected, but it is possible that a larger sampling of African populations will determine whether some Africans carry Neanderthal genes.

A phylogenetic tree showing the pattern and times of divergence of Neanderthals and our species is presented in **Figure 22**. The tree shows that Neanderthals and humans last shared a common ancestor about 706,000 years ago and that the isolating split between Neanderthals and human populations occurred about 370,000 years ago. From these studies, several conclusions can be drawn. First, Neanderthals are not direct ancestors of our species. Second, Neanderthals and members of our species may have interbred, but from the results available, it appears that Neanderthals did not make major contributions to our genome. As a species, Neanderthals are extinct, but some of their genes survive as part of our genome. Third, from what we know about the

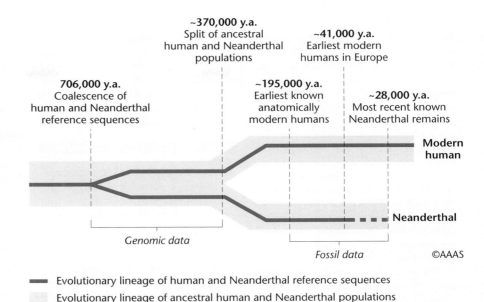

FIGURE 22 Estimated times of divergence of human and Neanderthal genomic sequences and, subsequently, of their populations, relative to landmark events in both human and Neanderthal evolution. These estimates are based on sequencing about 65,000 base pairs of Neanderthal DNA (y.a. = years ago). "Sequencing and analysis of Neanderthal genomic DNA," by J.P. Noonan et al., from SCIENCE, 314. Copyright © 2006 by AAAS. Reprinted with permission.

Neanderthal genome, we share most genes and other sequences with them.

More exciting answers to the question about the similarities and differences between our genome and that of Neanderthals will be derived from further analysis and will allow us to identify key differences that define our species and, in the process, revolutionize the field of human evolution.

GENETICS, TECHNOLOGY, AND SOCIETY

Tracking Our Genetic Footprints out of Africa

Based on the physical traits and distribution of hominid fossils, most paleoanthropologists agree that a large-brained, tool-using hominid they call *Homo erectus* appeared in east Africa about 2 million years ago. This species used simple stone tools and hunted, but did not fish, build houses, or follow ritual burial practices. About 1.7 million years ago, *H. erectus* spread into Eurasia and south Asia. Most scientists also agree that *H. erectus* likely developed into several hominid types, including Neanderthals (in Europe) and Denisovans (in Asia). These hominids were anatomically robust, with large, heavy skeletons and skulls. Neanderthals and other *H. erectus* groups disappeared 50,000 to 30,000 years ago—around the same time that anatomically modern humans (*H. sapiens*) appeared all over the world. It is at this point in our history—when ancient hominids gave way to anatomically modern humans—that controversy arises.

At present, two main hypotheses explain the origins of modern humans: the multiregional hypothesis and the out-of-Africa hypothesis. The multiregional hypothesis is based primarily on archaeological and fossil evidence. It proposes that *H. sapiens* developed gradually and simultaneously all over the world from existing *H. erectus* groups, including Neanderthals. Interbreeding between these groups eventually made *H. sapiens* a genetically homogeneous species. Natural selection then created the regional variants that we see today. In the multiregional view, our genetic makeup should include significant contributions from many *H. erectus* groups, including Neanderthals. In contrast, the out-of-Africa hypothesis, based primarily on genetic analyses of modern human populations, contends that *H. sapiens* evolved from the descendants of *H. erectus* in sub-Saharan Africa about 200,000 years ago. A small band of *H. sapiens* (probably fewer than 1000) then left Africa, expanded, and migrated into Europe and Asia around 60,000 years ago. By about 40,000 years ago, populations of *H. sapiens* reached Australia and later migrated into North America. In the out-of-Africa model, *H. sapiens* replaced all the preexisting *H. erectus* types, without interbreeding. In this way, *H. sapiens* became the only species in the genus by about 30,000 years ago.

Although the out-of-Africa hypothesis is still debated, most genetic evidence appears to support it. Humans all over the globe are remarkably similar genetically. DNA sequences from any two people chosen at random are 99.9 percent identical. More genetic identity exists between two persons chosen at random from a human population than between two chimpanzees chosen at random from a chimpanzee population. Interestingly, about 90 percent of the genetic differences that do exist occur between individuals rather than between populations.

This unusually high degree of genetic relatedness in all humans around the world supports the idea that our species arose recently from a small founding group of humans.

Studies of mitochondrial DNA sequences from current human populations reveal that the highest levels of genetic variation occur within African populations. Africans show twice the mitochondrial DNA sequence diversity of non-Africans. This implies that the earliest branches of *H. sapiens* diverged in Africa and had a longer time to accumulate mitochondrial DNA mutations, which are thought to accumulate at a constant rate over time.

DNA sequences from mitochondrial, Y-chromosome, and chromosome-21 markers support the idea that human roots are in east Africa and that the migration out of Africa occurred through Ethiopia, along the coast of the Arabian Peninsula, and outward to Eurasia and Southeast Asia. Recent data based on nuclear microsatellite variants and whole-genome single nucleotide polymorphism (SNP) analysis further support the notion that humans migrated out of Africa and dispersed throughout the world from a small founding population.

As with any explanation of human origins, the out-of-Africa hypothesis is actively debated. As methods to sequence DNA from ancient fossils improve, it may be possible to fill the gaps in the genetic pathway leading out of Africa and to resolve those age-old questions about our origins.

Your Turn

1. When the sequence of Neanderthal DNA was reported, it was suggested that the Neanderthal genome is so different from ours that they came from two separate species that diverged about 600,000 years ago. However, recent data suggest that there was some interbreeding between *H. sapiens* and *H. erectus* groups. Discuss evidence that supports, and refutes, the idea that Neanderthals and *H. sapiens* interbred in Europe and that Denisovans and *H. sapiens* interbred in Asia. How do these ideas affect the out-of-Africa hypothesis?
Start your investigations by reading Stoneking, M. and Krause, J. 2011. Learning about human population history from ancient and modern genomes. *Nat. Rev. Genet.* 12(9): 603–614.

2. If all people on Earth are very similar genetically, how did we come to have such a range of physical differences, which some describe as racial differences? How has modern genomics contributed to the debate about the validity and definition of the term *race*?
For an interesting discussion of race, human variation, and genomic studies, see Lewontin, R.C. 2006. Confusion about human races, on the *Social Sciences Research Center* website—raceandgenomics.ssrc.org/Lewontin. *A study of human population structure, based on microsatellite analysis, can be found at* Rosenberg, N.A. et al. 2002. Genetic structure of human populations. *Science* 298: 2381–2385.

3. Geneticists study mitochondrial and Y-chromosome DNA to determine the ancestry of modern humans. Why are these two types of DNA used in lineage studies? What is meant by the terms *mitochondrial Eve* and *Y-chromosome Adam*?
To read the original paper hypothesizing a mitochondrial Eve, see Cann, R.L. et al. 1987. Mitochondrial DNA and human evolution. *Nature* 325: 31–36. *For a discussion of Y-chromosome Adam, see* Gibbons, A. 1997. Y Chromosome shows that Adam was an African. *Science* 278: 804–805.

CASE STUDY » An unexpected outcome

A newborn screening program identified a baby with a rare autosomal recessive disorder called arginosuccinate aciduria (AGA), which causes high levels of ammonia to accumulate in the blood. Symptoms usually appear in the first week after birth and can progress to include severe liver damage, developmental delay, and mental retardation. AGA occurs with a frequency of about 1 in 70,000 births. There is no history of this disorder in either the father's or mother's family. The above case raises several questions:

1. Since it appears that the unaffected parents are heterozygotes, would it be considered unusual that there would be no family history of the disorder? How would they be counseled about risks to future children?
2. If the disorder is so rare, what is the frequency of heterozygous carriers in the population?
3. What are the chances that two heterozygotes will meet and have an affected child?

INSIGHTS AND SOLUTIONS

1. Tay–Sachs disease is caused by loss-of-function mutations in a gene on chromosome 15 that encodes a lysosomal enzyme. Tay–Sachs is inherited as an autosomal recessive condition. Among Ashkenazi Jews of Central European ancestry, about 1 in 3600 children is born with the disease. What fraction of the individuals in this population are carriers?

Solution: If we let p represent the frequency of the wild-type enzyme allele and q the total frequency of recessive loss-of-function alleles, and if we assume that the population is in Hardy–Weinberg equilibrium, then the frequencies of the genotypes are given by p^2 for homozygous normal, $2pq$ for carriers, and q^2 for individuals with Tay–Sachs. The frequency of Tay–Sachs alleles is thus

$$q = \sqrt{q^2} = \sqrt{\frac{1}{3600}} = 0.017$$

Since $p + q = 1$, we have

$$p = 1 - q = 1 - 0.017 = 0.983$$

Therefore, we can estimate that the frequency of carriers is

$$2pq = 2(0.983)(0.017) = 0.033 \text{ or about 1 in 30}$$

2. A single plant twice the size of others in the same population suddenly appears. Normally, plants of that species reproduce by self-fertilization and by cross-fertilization. Is this new giant plant simply a variant, or could it be a new species? How would you determine which it is?

Solution: One of the most widespread mechanisms of speciation in higher plants is polyploidy, the multiplication of entire sets of chromosomes. The result of polyploidy is usually a larger plant with larger flowers and seeds. There are two ways of testing the new variant to determine whether it is a new species. First, the giant plant should be crossed with a normal-sized plant to see whether the giant plant produces viable, fertile offspring. If it does not, then the two different types of plants would appear to be reproductively isolated. Second, the giant plant should be cytogenetically screened to examine its chromosome complement. If it has twice the number of its normal-sized neighbors, it is a tetraploid that may have arisen spontaneously. If the chromosome number differs by a factor of two and the new plant is reproductively isolated from its normal-sized neighbors, it is a new species.

PROBLEMS AND DISCUSSION QUESTIONS

 For activities, animations, and review quizzes, go to the study area at www.masteringgenetics.com

HOW DO WE KNOW?

1. Population geneticists study changes in the nature and amount of genetic variation in populations, the distribution of different genotypes, and how forces such as selection and drift act on genetic variation to bring about evolutionary change in populations and the formation of new species. From the explanation given in the chapter, what answers would you propose to the following fundamental questions?

   (a) How do we know how much genetic variation is in a population?

   (b) How do geneticists detect the presence of genetic variation as different alleles in a population?

   (c) How do we know whether the genetic structure of a population is static or dynamic?

   (d) How do we know when populations have diverged to the point that they form two different species?

   (e) How do we know the age of the last common ancestor shared by two species?

2. Are there nucleotide substitutions that will not be detected by electrophoretic studies of a gene's protein product?

3. Price et al. (1999. *J. Bacteriol.* 181: 2358–2362) conducted a genetic study of the toxin transport protein (PA) of *Bacillus anthracis,* the bacterium that causes anthrax in humans. Within the 2294-nucleotide gene in 26 strains they identified five point mutations—two missense and three synonyms—among different isolates. Necropsy samples from an anthrax outbreak in 1979 revealed a novel missense mutation and five unique nucleotide changes among ten victims. The authors concluded that these data indicate little or no horizontal transfer between different *B. anthracis* strains.

   (a) Which types of nucleotide changes (missense or synonyms) cause amino acid changes?

   (b) What is meant by horizontal transfer?

   (c) On what basis did the authors conclude that evidence of horizontal transfer is absent from their data?

4. The genetic difference between two *Drosophila* species, *D. heteroneura* and *D. sylvestris,* as measured by nucleotide diversity, is about 1.8 percent. The difference between chimpanzees (*P. troglodytes*) and humans (*H. sapiens*) is about the same, yet the latter species are classified in different genera. In your opinion, is this valid? Explain why.

5. The use of nucleotide sequence data to measure genetic variability is complicated by the fact that the genes of higher eukaryotes are complex in organization and contain 5′ and 3′ flanking regions as well as introns. Researchers have compared the nucleotide sequence of two cloned alleles of the *γ-globin* gene from a single individual and found a variation of 1 percent. Those differences include 13 substitutions of one nucleotide for another and 3 short DNA segments that have been inserted in one allele or deleted in the other. None of the changes takes place in the gene's exons (coding regions). Why do you think this is so, and should it change our concept of genetic variation?

6. Calculate the frequencies of the *AA, Aa,* and *aa* genotypes after one generation if the initial population consists of 0.2 *AA,* 0.6 *Aa,* and 0.2 *aa* genotypes and meets the requirements of the Hardy–Weinberg relationship. What genotype frequencies will occur after a second generation?

7. Consider rare disorders in a population caused by an autosomal recessive mutation. From the frequencies of the disorder in the population given, calculate the percentage of heterozygous carriers

   (a) 0.0064
   (b) 0.000081
   (c) 0.09
   (d) 0.01
   (e) 0.10

8. What must be assumed in order to validate the answers in Problem 7?

9. In a population where only the total number of individuals with the dominant phenotype is known, how can you calculate the percentage of carriers and homozygous recessives?

10. If 4 percent of a population in equilibrium expresses a recessive trait, what is the probability that the offspring of two individuals who do not express the trait will express it?

11. Consider a population in which the frequency of allele *A* is $p = 0.7$ and the frequency of allele *a* is $q = 0.3$, and where the alleles are codominant. What will be the allele frequencies after one generation if the following occurs?

    (a) $w_{AA} = 1, w_{Aa} = 0.9$, and $w_{aa} = 0.8$
    (b) $w_{AA} = 1, w_{Aa} = 0.95$, and $w_{aa} = 0.9$
    (c) $w_{AA} = 1, w_{Aa} = 0.99, w_{aa} = 0.98$
    (d) $w_{AA} = 0.8, w_{Aa} = 1, w_{aa} = 0.8$

12. If the initial allele frequencies are $p = 0.5$ and $q = 0.5$ and allele *a* is a lethal recessive, what will be the frequencies after 1, 5, 10, 25, 100, and 1000 generations?

13. Under what circumstances might a lethal dominant allele persist in a population?

14. Assume that a recessive autosomal disorder occurs in 1 of 10,000 individuals (0.0001) in the general population and that in this population about 2 percent (0.02) of the individuals are carriers for

the disorder. Estimate the probability of this disorder occurring in the offspring of a marriage between first cousins. Compare this probability to the population at large.

15. One of the first Mendelian traits identified in humans was a dominant condition known as *brachydactyly*. This gene causes an abnormal shortening of the fingers or toes (or both). At the time, some researchers thought that the dominant trait would spread until 75 percent of the population would be affected (because the phenotypic ratio of dominant to recessive is 3:1). Show that the reasoning was incorrect.

16. Describe how populations with substantial genetic differences can form. What is the role of natural selection?

17. Achondroplasia is a dominant trait that causes a characteristic form of dwarfism. In a survey of 50,000 births, five infants with achondroplasia were identified. Three of the affected infants had affected parents, while two had normal parents. Calculate the mutation rate for achondroplasia and express the rate as the number of mutant genes per given number of gametes.

18. A recent study examining the mutation rates of 5669 mammalian genes (17,208 sequences) indicates that, contrary to popular belief, mutation rates among lineages with vastly different generation lengths and physiological attributes are remarkably constant (Kumar, S., and Subramanian, S. 2002. *Proc. Natl. Acad. Sci. [USA]* 99: 803–808). The average rate is estimated at 12.2×10^{-9} per bp per year. What is the significance of this finding in terms of mammalian evolution?

19. A form of dwarfism known as Ellis–van Creveld syndrome was first discovered in the late 1930s, when Richard Ellis and Simon van Creveld shared a train compartment on the way to a pediatrics meeting. In the course of conversation, they discovered that they each had a patient with this syndrome. They published a description of the syndrome in 1940. Affected individuals have a short-limbed form of dwarfism and often have defects of the lips and teeth, and polydactyly (extra fingers). The largest pedigree for the condition was reported in an Old Order Amish population in eastern Pennsylvania by Victor McKusick and his colleagues (1964). In that community, about 5 per 1000 births are affected, and in the population of 8000, the observed frequency is 2 per 1000. All affected individuals have unaffected parents, and all affected cases can trace their ancestry to Samuel King and his wife, who arrived in the area in 1774. It is known that neither King nor his wife was affected with the disorder. There are no cases of the disorder in other Amish communities, such as those in Ohio or Indiana.

   (a) From the information provided, derive the most likely mode of inheritance of this disorder. Using the Hardy–Weinberg law, calculate the frequency of the mutant allele in the population and the frequency of heterozygotes, assuming Hardy–Weinberg conditions.

   (b) What is the most likely explanation for the high frequency of the disorder in the Pennsylvania Amish community and its absence in other Amish communities?

20. List the barriers that prevent interbreeding and give an example of each.

21. What are the two groups of reproductive isolating mechanisms? Which of these is regarded as more efficient, and why?

22. In a recent study of cichlid fish inhabiting Lake Victoria in Africa, Nagl et al. (1998. *Proc. Natl. Acad. Sci. [USA]* 95: 14,238–14,243) examined suspected neutral sequence polymorphisms in noncoding genomic loci in 12 species and their putative river-living ancestors. At all loci, the same polymorphism was found in nearly all of the tested species from Lake Victoria, both lacustrine and riverine. Different polymorphisms at these loci were found in cichlids at other African lakes.

   (a) Why would you suspect neutral sequences to be located in noncoding genomic regions?

   (b) What conclusions can be drawn from these polymorphism data in terms of cichlid ancestry in these lakes?

23. What genetic changes take place during speciation?

24. Some critics have warned that the use of gene therapy to correct genetic disorders will affect the course of human evolution. Evaluate this criticism in light of what you know about population genetics and evolution, distinguishing between somatic gene therapy and germ-line gene therapy.

25. Comparisons of Neanderthal mitochondrial DNA with that of modern humans indicate that they are not related to modern humans and did not contribute to our mitochondrial heritage. However, because Neanderthals and modern humans are separated by at least 25,000 years, this does not rule out some forms of interbreeding causing the modern European gene pool to be derived from both Neanderthals and early humans (called Cro-Magnons). To resolve this question, Caramelli et al. (2003. *Proc. Natl. Acad. Sci. [USA]* 100: 6593–6597) analyzed mitochondrial DNA sequences from 25,000-year-old Cro-Magnon remains and compared them to four Neanderthal specimens and a large dataset derived from modern humans. The results are shown in the graph.

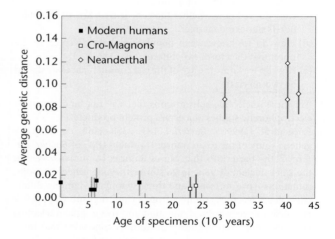

The *x*-axis represents the age of the specimens in thousands of years; the *y*-axis represents the average genetic distance. Modern humans are indicated by filled squares; Cro-Magnons, open squares; and Neanderthals, diamonds.

   (a) What can you conclude about the relationship between Cro-Magnons and modern Europeans? What about the relationship between Cro-Magnons and Neanderthals?

   (b) From these data, does it seem likely that Neanderthals made any mitochondrial DNA contributions to the Cro-Magnon gene pool or the modern European gene pool?

SOLUTIONS TO SELECTED PROBLEMS AND DISCUSSION QUESTIONS

Answers to Now Solve This

1. Because the alleles follow a dominant/recessive mode, one can use the equation $\sqrt{q^2}$ to calculate q, from which all other aspects of the answer depend. The frequency of aa types is determined by dividing the number of nontasters (37) by the total number of individuals (125).

$$q^2 = 37/125 = 0.296$$
$$q = 0.544$$
$$p = 1 - q$$
$$p = 0.456$$

The frequencies of the genotypes are determined by applying the formula $p^2 + 2pq + q^2$ as follows

Frequency of $AA = p^2 = (0.456)^2 = 0.208$ or 20.8%

Frequency of $Aa = 2pq = 2(0.456)(0.544) = 0.496$ or 49.6%

Frequency of $aa = q^2 = (0.544)^2 = 0.296$ or 29.6%

2. **(a)** For the CCR5 analysis, first determine p and q. Since one has the frequencies of all the genotypes, one can add 0.6 and 0.351/2 to provide p (= .7755); q will be 0.049 and $.351/2 = 0.2245$

The equilibrium values will be as follows

Frequency of $l/l = p^2 = (.7755)^2 = .6014$ or 60.14%

Frequency of $l/\Delta 32 = 2pq = 2(.7755)(.2245)$
$= .3482$ or 34.82%

Frequency of $\Delta 32/\Delta 32 = q^2 = (.2245)^2$
$= .0504$ or 5.04%

Comparing these equilibrium values with the observed values strongly suggests that the observed values are drawn from a population in Hardy-Weinberg equilibrium.

(b) For the AS (sickle-cell) analysis, first determine p and q. Since one has the frequencies of all the genotypes, one can add

.756 and .242/2 to provide p (= .877); q will be $1 - .877$ or .123

The equilibrium values will be as follows

Frequency of $AA = p^2 = (.877)^2 = .7691$ or 76.91%

Frequency of $AS = 2pq = 2(.877)(.123)$
$= .2157$ or 21.57%

Frequency of $SS = q^2 = (.123)^2 = .0151$ or 1.51%

Comparing these equilibrium values with the observed values suggests that the observed values may be drawn from a population that is not in equilibrium. Notice that there are more heterozygotes than predicted, and fewer SS types in the population. Since data are given in percentages, χ^2 values can not be computed.

3. Given that the recessive allele a is present in the homozygous state (q^2) at a frequency of 0.0001, the value of q is 0.01 and $p = 0.99$.

(a) q is 0.01.

(b) $p = 1 - q$ or 0.99

(c) $2pq = 2(0.01)(0.99) = 0.0198$ (or about 1/50)

(d) $2pq \times 2pq$
$= 0.0198 \times 0.0198 = 0.000392$ (or about 1/255)

4. The probability that the woman (with no family history of CF) is heterozygous is $2pq$ or $2(1/50)(49/50)$. The probability that the man is heterozygous is 2/3. The probability that a child with CF will be produced by two heterozygotes is 1/4. Therefore, the overall probability of the couple producing a CF child is $98/2500 \times 2/3 \times 1/4 = 0.00653$, or about 1/153.

Solutions to Problems and Discussion Questions

2. Because of degeneracy in the code, there are some nucleotide substitutions, especially in the third base, that do not change amino acids. In addition, if there is no change in the overall charge of the protein, it is likely that electrophoresis will not separate the variants. If a positively charged amino acid is replaced by an amino acid of like charge, then the overall charge on the protein is unchanged. The same may be said for other negatively charged and neutral amino acid substitutions.

4. There must be evidence that gene flow does not occur among the groups being called different species. Classifications above the species level (genus, family, etc.) are not based on such empirical data. Indeed, classification above the species level is somewhat arbitrary and based on traditions that extend far beyond DNA sequence information. In addition, recall that DNA sequence divergence is not always directly proportional to morphological, behavioral, or ecological divergence. While the genus classifications provided in this problem seem to be invalid, other factors, well beyond simple DNA sequence comparison, must be considered in classification practices.

6. Calculate p and q, then apply the equation $p^2 + 2pq + q^2$ to determine genotypic frequencies in the next generation.

$$p = \text{frequency of A} = 0.2 + 0.3 = 0.5$$
$$q = 1 - p = 0.5$$
Frequency of $AA = p^2 = 0.25$ or 25%
Frequency of $Aa = 2pq = 0.5$ or 50%
Frequency of $aa = q^2 = 0.25$ or 25%

The initial population was not in equilibrium; however, after one generation of mating under the Hardy–Weinberg conditions the population is in equilibrium and will continue to be so (and not change) until one or more of the Hardy–Weinberg conditions is not met. Note that *equilibrium* does not necessarily mean p and q equal 0.5.

8. In order for the Hardy–Weinberg equations to apply, the population must be in Hardy–Weinberg equilibrium.

10. Given that $q^2 = 0.04$, then $q = 0.2$, $2pq = 0.32$, and $p^2 = 0.64$. Of those not expressing the trait, only a mating between heterozygotes can produce an offspring that expresses the trait, and then only at a frequency of 1/4. The different types of matings possible (those without the trait) in the population, with their frequencies, are given below

$$AA \times AA = 0.64 \times 0.64 = 0.4096$$
$$AA \times Aa = 0.64 \times 0.32 = 0.2048$$
$$Aa \times AA = 0.64 \times 0.32 = 0.2048$$
$$Aa \times Aa = 0.32 \times 0.32 = 0.1024$$
$$Aa \times Aa = 0.32 \times 0.32 = 0.1024$$

Notice that of the matings of the individuals who do not express the trait, only the last two (about 20 percent) are capable of producing offspring with the trait. Therefore, one would arrive at a final likelihood of 1/4 × 20 percent or 5 percent of the offspring with the trait.

12. The general equation for responding to this question is

$$q_n = q_o/(1 + nq_o)$$

where n = the number of generations, q_o = the initial gene frequency, and q_n = the new gene frequency.

(a) $q_n = 0.33$ $p_n = 0.67$

(b) $q_n = 0.143$ $p_n = 0.857$

(c) $q_n = 0.083$ $p_n = 0.917$

(d) $q_n = 0.037$ $p_n = 0.963$

(e) $q_n = 0.0098$ $p_n = 0.9902$

(f) $q_n = 0.00099$ $p_n = 0.99901$

14. What one must do is predict the probability of one of the grandparents being heterozygous in this problem. Given the frequency of the disorder in the population as 1 in 10,000 individuals (0.0001), then $q^2 = 0.0001$, and $q = 0.01$. The frequency of heterozygosity is $2pq$, or approximately 0.02, as also stated in the problem. The probability for one of the grandparents to be heterozygous would therefore be $0.02 + 0.02$ or 0.04 or 1/25. (Note: If one considers the probability of both parents being carriers, 0.02×0.02, the answer differs slightly.) If one of the grandparents is a carrier, then the probability of the offspring from a first-cousin mating being homozygous for the recessive gene is 1/16. Multiplying the two probabilities together gives $1/16 \times 1/25 = 1/400$. Following the same analysis for the second-cousin mating gives $1/64 \times 1/25 = 1/1600$. Notice that the population at large has a frequency of homozygotes of 1/10,000; therefore, one can easily see how inbreeding increases the likelihood of homozygosity.

18. The approximate similarity of mutation rates among genes and lineages should provide more credible estimates of divergence times of species and allow for broader interpretations of sequence comparisons. It also provides for increased understanding of the mutational processes that govern evolution among mammalian genomes. For instance, if the rate of mutation is fairly constant among lineages or cells that have a more rapid turnover, it indicates that replication-related errors do not make a significant contribution to mutation rates.

22. (a) Since noncoding genomic regions are probably silent genetically, it is likely that they contribute little, if anything, to the phenotype. Selection acts on the phenotype; therefore, such noncoding regions are probably selectively neutral.

(b) These polymorphism data indicate that all the Lake Victoria area (lake and contributing rivers) cichlids are related by recent ancestry, whereas those from neighboring lakes are more distantly related. In addition, since Lake Victoria dried out about 14,000 years ago, it is likely that it was repopulated by a relatively small sample of cichlids.

24. Somatic gene therapy, like any therapy, allows some individuals to live more normal lives than those not receiving therapy. As such, the ability of such individuals to contribute to the gene pool increases the likelihood that less fit alleles will enter and be maintained in the gene pool. This is a normal consequence of therapy, genetic or not, and in the face of disease control and prevention, societies have generally accepted this consequence. Germ-line therapy could, if successful, lead to limited, isolated, and infrequent removal of an allele from a gene lineage. However, given the present state of the science, its impact on the course of human evolution will be diluted and negated by a host of other factors that afflict humankind.

CREDITS

Credits are listed in order of appearance.

Photo

CO, Simon Booth/Photo Researchers, Inc.; F-06, *Nature Magazine*; F-13 (a, b) Courtesy of Murray Brilliant, "A 122.5 kilobase deletion of P gene underlies the high prevalence of oculocutaneous albinism type 2 in the Navajo population": From: American Journal Human Genetics 72: 62-72, fig. 1 p. 65, fig2 p. 66. Published by University of Chicago Press; F-14, Courtesy of Murray Brilliant, "A 122.5 kilobase deletion of P gene underlies the high prevalence of oculocutaneous albinism type 2 in the Navajo population": From: American Journal Human Genetics 72: 62-72, fig. 3 p. 67. Published by University of Chicago Press; F-16, Smithsonian Institution (Reproductions); F-18 (a), Niels Poulsen/Alamy; F-18 (b) National Ciklid Society

Text

F-2, "Nucleotide polymorphism at the alcohol dehydrogenase locus of the Drosophila melanogaster," by M. Kreitman, from *Nature*, 304 (4). Copyright © 1983 by Nature Publishing Group. Reprinted with permission. **F-20**, "Construction of phylogenetic trees," by W.M. Fitch and E. Margoliash from *Science*, 279. Copyright © 1967 by AAAS. Reprinted with permission. **F-21 a and b**, "The Natural Theory in the Genomic Era," by M. Nie, Y. Suzuki, and M. Nozawa, from *Annual Review Genomics Human Genetics*, 11. Copyright © 2010 by Annual Reviews, Inc. Reprinted with permission. **F-22**, "Sequencing and analysis of Neanderthal genomic DNA," by J.P. Noonan et al., from *Science*, 314. Copyright © 2006 by AAAS. Reprinted with permission.

Epigenetics

The somatic cells of the human body contain 20,000 to 25,000 genes. In the more than 200 cell types present in the body, different cell-specific gene sets are transcribed, while the rest of the genome is transcriptionally inactive. During development, as embryonic cells gradually become specialized cells with adult phenotypes, programs of gene expression become more and more restricted. Until recently, it was thought that most regulation of gene expression is coordinated by *cis*-regulatory elements as well as DNA-binding proteins and transcription factors, and that this regulation can occur at any of the steps in gene expression . For example, transcriptional regulation, steps in mRNA processing, and other stages of posttranscriptional regulation control the amount of gene product synthesized from a DNA template. However, as we have learned more about genome organization and the regulation of gene expression, it is clear that classical regulatory mechanisms cannot fully explain how some phenotypes arise. For example, monozygotic twins have identical genotypes but often have different phenotypes. In addition, although one allele of each gene is inherited maternally and one is inherited paternally, in some cases, only the maternal or paternal allele is expressed, while the other is transcriptionally silent.

The newly emerging field of epigenetics is providing us with a basis for understanding how heritable changes other than those in DNA sequence can influence phenotypic variation (ST Figure 1). These advances greatly extend our understanding of the molecular basis of gene regulation and have application in wide-ranging areas including genetic disorders, cancer, and behavior.

An **epigenetic trait** is a stable, mitotically and meiotically heritable phenotype that results from changes in gene expression without alterations in the DNA sequence. **Epigenetics** is the study of the ways in which these changes alter cell- and tissue-specific patterns of gene expression. Epigenetic regulation of gene expression uses reversible modifications of DNA and chromatin structure to mediate the interaction of the genome with a variety of environmental factors and generates changes in the patterns of gene expression in response to these factors. The **epigenome** refers to

> The newly emerging field of epigenetics is providing us with a basis for understanding how heritable changes other than those in DNA sequence can influence phenotypic variation.

the epigenetic state of a cell. During its life span, an organism has one genome, but this genome can be modified in diverse cell types at different times to produce many epigenomes.

Current research efforts are focused on several aspects of epigenetics: how an epigenome arises in developing and differentiated cells and how these epigenomes are transmitted via mitosis and meiosis, making them heritable traits. In addition, because epigenetically controlled alterations to the genome are associated with common diseases such as cancer, diabetes, and asthma, efforts are also being directed toward developing drugs that can modify or reverse disease-associated epigenetic changes in cells.

Here we will focus on how epigenetics is associated with some heritable genetic disorders, cancer, and environment–genome interactions. Because epigenetic changes are potentially reversible, we will also examine how knowledge of molecular mechanisms of epigenetics is being used to develop drugs and treatments for human diseases.

Epigenetic Alterations to the Genome

Unlike the genome, which is identical in all cell types of an organism, the epigenome is cell-type specific and changes throughout the life cycle in response to environmental cues. Like the genome, the epigenome can be transmitted to daughter cells by mitosis and to future generations by meiosis. In the following sections, we will examine mechanisms of epigenetic changes and their role in imprinting, cancer, behavior, and environment–genome interactions, providing a snapshot of the many roles played by this recently discovered mechanism of gene regulation.

There are three major epigenetic mechanisms: (1) reversible modification of DNA by the addition or removal of methyl groups; (2) alteration of chromatin by the addition or removal of chemical groups to histone proteins; and (3) regulation of gene expression by small, noncoding RNA molecules.

BOX 1

The Beginning of Epigenetics

C.H. Waddington coined the term *epigenetics* in the 1940s to describe how environmental influences on developmental events can affect the phenotype of the adult. He showed that environmental alterations during development induced alternative phenotypes in organisms with identical genotypes. Using *Drosophila melanogaster,* Waddington found that wing vein patterns could be altered by administering heat shocks during pupal development. Offspring of flies with these environmentally induced changes showed the alternative phenotype without the need for continued environmental stimulus. He called this phenomenon "genetic assimilation." In other words, interactions between the environment and the genome during certain stages of development produced heritable phenotypic changes.

In the 1970s, Holliday and Pugh proposed that changes in the program of gene expression during development depends on the methylation of specific bases in DNA, and that altering methylation patterns affects the resulting phenotype. Waddington's pioneering work, the methylation model of Holliday and Pugh, and the discovery that expression of genes from both the maternal and paternal genomes is required for normal development, all helped set the stage for the birth of epigenetics and epigenomics as fields of scientific research.

Methylation

In mammals, DNA methylation takes place after replication and during cell differentiation. This process involves the addition of a methyl group ($-CH_3$) to cytosine, a reaction catalyzed by a family of enzymes called methyltransferases. Methylation takes place almost exclusively on cytosine bases located adjacent to a guanine base, a combination called a CpG dinucleotide. Many of these dinucleotides are clustered in regions called CpG islands, which are located in and near promoter sequences adjacent to genes (ST Figure 2). CpG islands and promoters adjacent to essential genes (housekeeping genes) and cell-specific genes are unmethylated, making these genes available for transcription. Genes with adjacent methylated CpG islands and methylated promoters are transcriptionally silenced. The methyl groups in CpG dinucleotides occupy the major groove of DNA and block the binding of transcription factors necessary to form transcription complexes.

As part of dosage compensation, one of the X chromosomes in mammalian females is inactivated by being converted to heterochromatin. These inactivated chromosomes have altered patterns of DNA methylation (). As mentioned above, CpG methylation in euchromatic regions also causes a parent-specific pattern of gene transcription.

Histone Modification and Chromatin Configuration

In addition to DNA methylation, histone modification is an important epigenetic mechanism of gene regulation. Recall that chromatin is a dynamic structure composed of DNA wound around a core of 8 histone proteins to form nucleosomes. The N-terminal region of each histone extends beyond the nucleosome, and the amino acids in these tails can be covalently modified in several ways, including the addition of acetyl, methyl, and phosphate groups (ST Figure 3). These modifications alter the structure of chromatin, making genes accessible or inaccessible for transcription. Histone acetylation, for example, makes genes on these modified nucleosomes available for transcription [ST Figure 4(a)]. This modification is reversible, and acetyl groups can be removed, changing the chromatin from an "open" to a "closed" configuration and silencing genes by making them unavailable for transcription [ST Figure 4(b)].

We are learning that specific combinations of histone modifications control the transcriptional status of a chromatin region. For example, whether or not lysine 9 on histone

ST FIGURE 1   The phenotype of an organism is the product of interactions between the genome and the epigenome. The genome is a constant from fertilization throughout life, but cells, tissues, and the organism develop different epigenomes as a result of epigenetic reprogramming of gene activity in response to environmental stimuli. These reprogramming events lead to phenotypic changes through the life cycle.

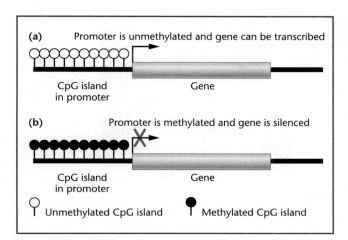

(a) Promoter is unmethylated and gene can be transcribed

CpG island in promoter — Gene

(b) Promoter is methylated and gene is silenced

CpG island in promoter — Gene

○ Unmethylated CpG island   ● Methylated CpG island

ST FIGURE 2 Methylation patterns of CpG dinucleotides in promoters control activity of the adjacent genes. CpG islands outside and within genes also have characteristic methylation patterns, contributing to the overall level of genome methylation.

H3 will be methylated is controlled by modifications made elsewhere on this protein. On one hand, if serine 10 is phosphorylated, methylation of lysine 9 is inhibited. On the other hand, if lysine 14 is deacetylated, methylation of lysine 9 is facilitated. A large number of histone modifications are possible, and the sum of the complex patterns and interactions of histone modifications that alter chromatin organization and gene expression is called the **histone code.** Combinations of these changes allow differentiated cells to carry out cell-specific patterns of gene transcription and to respond to external signals that modify these patterns without any changes in DNA sequence.

MicroRNAs

In addition to DNA methylation and histone modification, short, noncoding RNA molecules () called microRNAs (miRNAs) also participate in epigenetic regulation of gene expression. miRNAs are involved in controlling pattern formation in developing embryos and in timing of developmental events, as well as physiological processes such as cell signaling. Recent evidence shows that miRNAs also play roles in the development of cardiovascular disease and cancer. Over 1100 human miRNAs have been described, and their actions and interactions are emerging as important biological regulatory mechanisms.

miRNAs are transcribed as precursor molecules about 70–100 nucleotides long, containing a double-stranded stem loop. Processing removes the single-stranded regions, and the loops move to the cytoplasm where they are altered further. The resulting double-stranded RNA is incorporated into a protein complex where one RNA strand is removed and degraded, forming a mature RNA-Induced Silencing Com-

plex (RISC) containing the remaining single miRNA strand. RISCs act as posttranscriptional repressors of gene expression by binding to and destroying target mRNA molecules carrying sequences complementary to the RISC miRNA. Binding of target mRNAs that are partially complementary to the RISC miRNA causes these mRNAs to be uncapped and have their poly-A tails removed. This makes these target mRNAs less likely to be translated by ribosomes, resulting in downregulation of gene expression.

In addition to forming RISC complexes, miRNA can also associate with a different protein complex to form RNA-Induced Transcriptional Silencing (RITS) complexes. RITS complexes convert euchromatic chromosome regions into facultative heterochromatin, which silences the genes located within these newly created heterochromatic regions. Unlike the constitutive heterochromatin at telomeres and

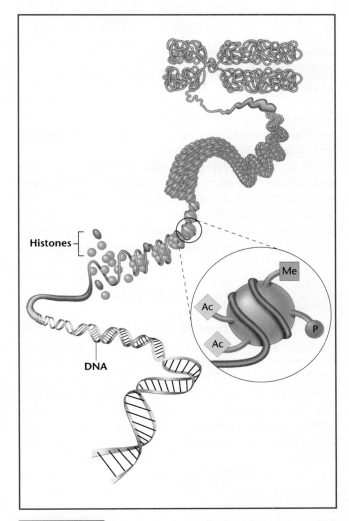

Histones

DNA

Me

Ac

Ac

P

ST FIGURE 3 Clusters of histones in nucleosomes have their N-terminal tails covalently modified in epigenetic modifications that alter patterns of gene expression. Ac = acetyl groups, Me = methyl groups, P = phosphate groups.

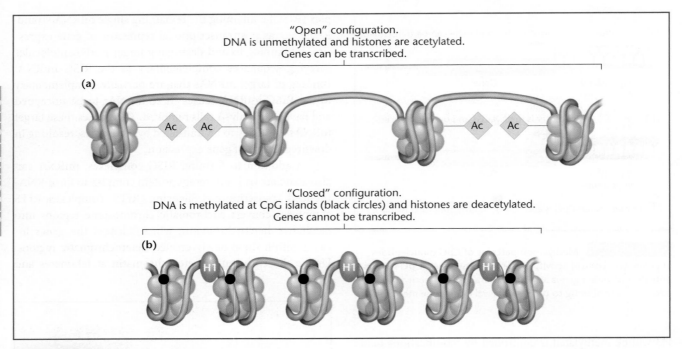

"Open" configuration.
DNA is unmethylated and histones are acetylated.
Genes can be transcribed.

(a)

"Closed" configuration.
DNA is methylated at CpG islands (black circles) and histones are deacetylated.
Genes cannot be transcribed.

(b)

ST FIGURE 4 Epigenetic modifications to the genome alter the spacing of nucleosomes and alter the availability of genes for transcription.

centromeres, the heterochromatic state in facultative heterochromatin is reversible, making genes in this region once again accessible for transcription.

In summary, epigenetic modifications alter chromatin structure by several mechanisms including DNA methylation, histone acetylation, and action of miRNAs, without changing the sequence of DNA. These epigenetic changes create an epigenome that in turn, can regulate normal development or generate physiological responses to environmental signals.

Epigenetics and Imprinting

Mammals inherit a maternal and a paternal copy of each autosomal gene, and usually both copies of these genes are expressed in the offspring. Imprinting is an epigenetically regulated process in which genes are expressed in a parent-of-origin pattern; that is, certain genes show expression of only the maternal allele or the paternal allele. Parent-specific patterns of allele expression are laid down in germ-line cells during gamete formation and are inherited via mitosis in somatic cells.

Differential methylation of CpG-rich regions and promoter sequences produce allele-specific imprinting and subsequent gene silencing. Once a gene has been imprinted, it remains transcriptionally silent during embryogenesis and development. Most imprinted genes direct aspects of growth during prenatal development. For example, in mice, genes on the maternal X chromosome are expressed in the placenta, while genes on the paternal X chromosome are silenced. Hav-

ing only one functional allele makes imprinted genes highly susceptible to the deleterious effects of mutations. Because imprinted genes are clustered at sites in the genome, mutation in one gene can have an impact on the function of adjacent imprinted genes, amplifying its impact on the phenotype. Mutations in imprinted genes can arise by changes in the DNA sequence or by dysfunctional epigenetic changes, called **epimutations**, both of which are heritable changes in the activity of a gene.

At fertilization, the mammalian embryo receives a maternal and a paternal set of chromosomes; the maternal chromosome set has female imprints, and the paternal set contains male imprints. When gamete formation begins in female germ cells, both the maternal and paternal chromosome sets have their imprints erased and are reprogrammed to a female imprint that is transmitted to the next generation through the egg (**ST Figure 5**). Similarly, in male germ cells, the paternal and maternal chromosome sets have their imprints erased and are reprogrammed by methylation to become a male imprinted set. Reprogramming occurs at two stages: in the parental germ cells and in the developing embryo just before implantation. In stage one, erasure by demethylation and reprogramming by remethylation lay down a male- or female-specific imprinting pattern in germ cells of the parent. In stage two, large-scale demethylation occurs in the embryo sometime before the 16-cell stage of development. After implantation, differential genomic remethylation in the embryo recalibrates which maternal alleles and which paternal alleles will be inactivated. It is important to remember that imprinted alleles remain transcriptionally

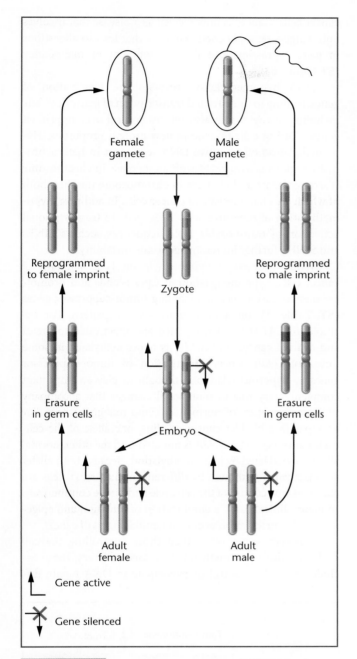

Female
gamete

Male
gamete

Reprogrammed
to female imprint

Zygote

Reprogrammed
to male imprint

Erasure
in germ cells

Embryo

Erasure
in germ cells

Adult
female

Adult
male

↑ Gene active

✕ Gene silenced

ST FIGURE 5 Imprinting patterns are reprogrammed each generation during gamete formation. A second round of epigenetic reprogramming occurs during early embryonic development.

remain to be discovered, the overall number of imprinting-related genetic disorders may be much higher.

In humans, most known imprinted genes encode growth factors or other growth-regulating genes. An autosomal dominant disorder of imprinting, Beckwith–Wiedemann syndrome (BWS), offers insight into how disruptions of epigenetically imprinted genes lead to an abnormal phenotype. BWS is a prenatal overgrowth disorder with abdominal wall defects, enlarged organs, large birth weight, and predisposition to cancer. BWS is not caused by mutation, nor is it associated with any chromosomal aberration. Instead it is a disorder of imprinting and is caused by abnormal methylation patterns and resulting altered patterns of gene expression.

Genes linked to BWS are located in a cluster of imprinted genes on the short arm of chromosome 11. All genes in this cluster regulate growth during prenatal development. Two genes in this cluster are *IGF2* (insulin growth factor 2) and *H19*. Normally, the paternal allele of *IGF2* is expressed, and the maternal allele is imprinted and silenced. In the case of *H19*, the maternal allele is expressed, and the paternal allele is imprinted and silenced.

In many individuals with BWS, the maternal *IGF2* allele is not imprinted. As a result, the maternal and paternal alleles are both transcriptionally active, resulting in the overgrowth of tissues that are characteristic of this disease.

The known number of imprinted genes represents only a small fraction (less than 1 percent) of the mammalian genome, but they play major roles in regulating growth during prenatal development. Because they act so early in life, external or internal factors that disturb the epigenetic pattern of imprinting or the expression of imprinted genes can have serious phenotypic consequences.

In farm animals, the introduction of *in vitro* fertilization (IVF) and embryo cloning led to a high incidence of birth defects and neonatal deaths. Many of these problems were linked to abnormal epigenetic alterations during embryonic growth and development. These findings in animals raised the possibility that the use of IVF and other assisted reproductive technologies (ART) in humans may cause problems with imprinted genes.

silent in all cells, while genes silenced by epigenetic methylation can be reactivated by external signals during or after differentiation.

Most human disorders associated with imprinting originate during fetal growth and development. Imprinting defects cause Prader–Willi syndrome, Angelman syndrome, Beckwith–Wiedemann syndrome, and several other diseases **(ST Table 1)**. However, given the number of candidate genes and the possibility that additional imprinted genes

ST TABLE 1	**Some Imprinting Disorders in Humans**
Disorder	**Locus**
Albright hereditary osteodystrophy	20q13
Angelman syndrome	15q11-q15
Beckwith–Wiedemann syndrome	11p15
Prader–Willi syndrome	15q11-q15
Silver–Russell syndrome	Chromosome 7
Uniparental disomy 14	Chromosome 14

In humans, several studies have shown that children born after IVF are at risk for low or very low birth weight, a condition that may result from abnormal imprinting. The use of IVF and other ART procedures has also been associated with a three- to six-fold increased risk of BWS. In one study, more than 90 percent of children born with BWS after ART had imprinting defects. Because imprinting disorders are uncommon (BWS occurs with a frequency of about 1 in 15,000 births following normal conception), large-scale and longitudinal studies will be needed to establish a causal relationship among imprinting abnormalities, growth disorders, and ART.

Epigenetics and Cancer

Epidemiological studies investigate the role of environmental factors in normal phenotypic variation and as risk factors for disease. For some complex diseases, there are strong links with environmental factors such as the association between smoking and lung cancer. The discovery that epigenetics mediates changing patterns of gene expression in response to environmental signals offers a new and potentially more direct approach to understanding the interactions between the genome and the environment in diseases such as cancer.

Following the discovery of cancer-associated genes, including those that promote (proto-oncogenes) or inhibit (tumor-suppressor genes) cell division, research into the genetic basis of cancer focused mainly on mutant alleles of genes involved in regulation of the cell cycle. Until recently, the conventional view has been that cancer is clonal in origin and begins in a single cell that has accumulated a suite of dominant and recessive mutations that allow it to escape control of the cell cycle. Subsequent mutations allow cells of the tumor to become metastatic, spreading the cancer to other locations in the body where new malignant tumors appear.

Converging lines of evidence are now clarifying the role of epigenetic changes in the initiation and maintenance of malignancy. These findings help researchers understand properties of cancer cells that are difficult to explain by the action of mutant alleles alone. Evidence for the role of epigenetic changes in cancer now challenges the conventional paradigm for the origin of cancer and establishes epigenomic changes as a major pathway for the formation and spread of malignant cells.

The relationship between epigenetics and cancer was first noted in the 1980s by Feinberg and Vogelstein who observed that colon cancer cells had much lower levels of methylation than normal cells derived from the same tissue. Subsequent research by many investigators showed that complex changes in DNA methylation patterns are associated with cancer. Global genomic hypomethylation is a property of all cancers examined to date. In addition, selective hypermethylation and gene silencing are also properties of

cancer cells. Cancer is now viewed as a disease that involves both epigenetic *and* genetic changes that lead to alterations in gene expression and the development of malignancy (ST Figure 6).

DNA hypomethylation reverses the inactivation of genes, leading to unrestricted transcription of many gene sets including oncogenes. It also relaxes control over imprinted genes, causing cells to acquire new growth properties. Hypomethylation of repetitive DNA sequences in heterochromatic regions is associated with an increase in chromosome rearrangements and changes in chromosome number, both of which are characteristic of cancer cells. In addition, hypomethylation of repetitive sequences leads to transcriptional activation of transposable DNA sequences such as LINEs and SINEs, further increasing genomic instability.

While widespread hypomethylation is a hallmark of cancer cells, hypermethylation at CpG islands and promoters silences certain genes, including tumor-suppressor genes (ST Table 2), often in a tumor-specific pattern. For example, *BRCA1* is hypermethylated and inactivated in breast and ovarian cancer, and *MLH1* is hypermethylated in some forms of colon cancer. Inactivation of tumor-suppressor genes by hypermethylation is thought to play an important complementary role to mutational changes that accompany the transformation of normal cells into malignant cells. For example, in a bladder cancer cell line, one allele of the cell-cycle control gene *CDKN2A* is mutated, and the other, normal allele is inactivated by hypermethylation. Because both alleles are inactivated (although by different mechanisms), cells are able to escape control of the cell cycle and divide continuously. In many clinical cases, a combination of mutation and epigenetic hypermethylation occurs in familial forms of cancer.

However, genes other than those controlling the cell cycle are also hypermethylated in some cancers; these include genes that control or participate in DNA repair, dif-

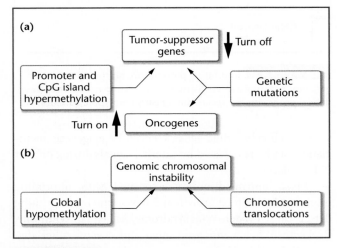

ST FIGURE 6 The development and maintenance of malignant growth in cancer involves gene mutations, hypomethylation, hypermethylation, overexpression of oncogenes, and the silencing of tumor-suppressor genes.

| | | **ST TABLE 2** | Some Cancer-Related Genes Inactivated by Hypermethylation in Human Cancers | |

Gene	Locus	Function	Related Cancers
BRCA1	17q21	DNA repair	Breast, ovarian
APC	5q21	Nucleocytoplasmic signaling	Colorectal, duodenal
MLH1	3p21	DNA repair	Colon, stomach
RB1	13q14	Cell-cycle control point	Retinoblastoma, osteosarcoma
AR	Xq11-12	Nuclear receptor for androgen; transcriptional activator	Prostate
ESR1	6q25	Nuclear receptor for estrogen; transcriptional activator	Breast, colorectal

ferentiation, apoptosis, and drug resistance. In fact, the majority of hypermethylated genes in cancer cells are not tumor-suppressor genes; this suggests that the pattern of hypermethylation may result from a widespread deregulation of the methylation process rather than a targeted event.

In addition to altered patterns of methylation, many cancer cells also have disrupted histone modification profiles. In some cases, mutations in the genes of the histone-modifying proteins such as histone acetyltransferase (HAT) and histone deacetylase (HDAC) are linked to the development of cancer. For example, individuals with Rubenstein–Taybi syndrome inherit a germ-line mutation that produces a dysfunctional HAT and have a greater than 300-fold increased risk of cancer. In other cases, HDAC complexes are selectively recruited to tumor-suppressor genes by mutated, oncogenic DNA binding proteins. Action of the HDAC complexes at these genes converts the chromatin to a closed configuration and inhibits transcription, causing the cell to lose control of the cell cycle.

Mechanisms that cause epigenetic changes in cancer cells are not well understood, partly because they take place very early in the conversion of a normal cell to a cancerous one and partly because by the time the cancer is detected, alterations in the methylation pattern have already occurred. The fact that such changes occur very early in the transformation process has led to the proposal that epigenetic changes leading to cancer may occur within adult stem cells in normal tissue. Three lines of evidence support this idea: (1) epigenetic mechanisms can replace mutations as a way of silencing individual tumor-suppressor genes or activating oncogenes; (2) global hypomethylation may cause genomic instability and the large-scale chromosomal changes that are a characteristic feature of cancer; and (3) epigenetic modifications can silence multiple genes, making them more effective in transforming normal cells into malignant cells than sequential mutations of single genes. A model of cancer based on epigenetic changes in colon stem cells as initial events in carcinogenesis followed by mutational events is shown in **ST Figure 7**.

In addition to changing ideas about the origins of cancer, the fact that epigenetic changes are potentially reversible makes it possible to develop new classes of drugs for chemotherapy. The focus of epigenetic therapy is the reactivation of genes silenced by methylation or histone modification, essentially reprogramming the pattern of gene expression in cancer cells. Several epigenetic drugs are in clinical trials, and one (decitabine, marketed as Vidaza) has been approved by the U.S. Food and Drug Administration for treatment of myelodysplastic syndrome, a precursor to leukemia, and for treatment of acute myeloid leukemia. This drug is an analog of cytidine and is incorporated into DNA during replication during the S phase of the cell cycle. Methylation enzymes (methyltransferases) bind irreversibly to decitabine, preventing methylation of DNA at many other sites, effectively reducing the amount of methylation in cancer cells. Other drugs that inhibit histone deacetylases (HDAC) are also being investigated for use in epigenetic therapy. Experiments

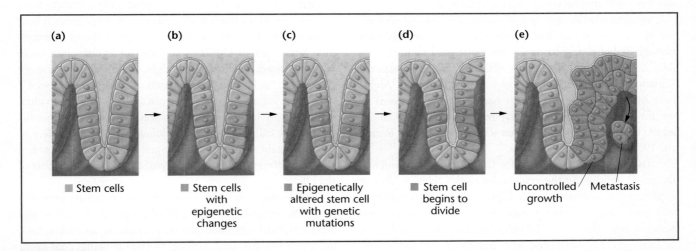

(a) ■ Stem cells

(b) ■ Stem cells with epigenetic changes

(c) ■ Epigenetically altered stem cell with genetic mutations

(d) ■ Stem cell begins to divide

(e) Uncontrolled growth / Metastasis

ST FIGURE 7 The epigenetic stem cell model proposes that both epigenetic changes and mutations are involved in the origins of cancer.

BOX 2

What More We Need to Know about Epigenetics and Cancer

The discovery that epigenetic changes may be as important as genetic changes in the origin, maintenance, and metastasis of cancers has opened new avenues of cancer research. Key discoveries about epigenetic mechanisms include the finding of tumor-specific deregulation of genes by altered DNA methylation profiles and

histone modifications, the discovery that epigenetic changes in histones or DNA methylation are interconnected, and the recognition that epigenetic changes can affect hundreds of genes in a single cancer cell. These advances were made in the span of a few years, and while it is clear that epigenetics plays a key role in cancer, many questions remain to be answered before we can draw conclusions about the relative contributions of genetics and epigenetics to the development of cancer. Some of these questions are as follows:

- Is global hypomethylation in cancer cells a cause or an effect of the malignant condition?

- Do these changes arise primarily in stem cells or in differentiated cells?
- Once methylation alterations begin, what triggers hypermethylation in cancer cells?
- Is hypermethylation a process that targets certain gene classes, or is it a random event?
- Can we develop drugs that target cancer cells and reverse tumor-specific epigenetic changes?
- Can we target specific genes for reactivation, while leaving others inactive?

---

with cancer cell lines indicate that inhibiting HDAC activity results in the re-expression of tumor-suppressor genes. The first HDAC inhibitor, vorinostat, (marketed as Zolinza), has recently been approved for the treatment of some forms of lymphoma. Further research into the mechanisms and locations of epigenetic genome modification in cancer cells will allow the design of more potent drugs to target epigenetic events as a form of cancer therapy.

Epigenetics and the Environment

Environmental agents including nutrition, chemicals, and physical factors, such as temperature, can alter gene expression by affecting the epigenetic state of the genome. In humans it is difficult to determine the relative contributions of environmental or learned behavior as factors in changing the epigenome, but there is indirect evidence that changes in nutrition and exposure to agents that affect the developing fetus can have detrimental effects during adulthood.

Women who were pregnant during the 1944–1945 famine in the Netherlands had children with increased risks for obesity, diabetes, and coronary heart disease. In addition, as adults, these individuals had significantly increased risks for schizophrenia and other neuropsychiatric disorders. Members of the F2 generation also had abnormal patterns of weight gain and growth. Similar results were found in the adult children of Chinese women who were pregnant during the 1959–1961 famine in China.

The most direct evidence for the role of environmental factors in modifying the epigenome comes from studies in experimental animals. A low-protein diet fed to pregnant rats results in permanent changes in the expression of several genes in both the F1 and F2 offspring. Increased expression of liver genes is associated with hypomethylation of their

respective promoter regions. Other evidence indicates that epigenetic changes triggered by this diet modification were gene-specific.

A dramatic example of how epigenome modifications affect the phenotype comes from the study of coat color in mice, where color is controlled by the dominant allele *Agouti* (*A*). In homozygous *AA* mice, the gene is active only during a specific time during hair development, resulting in a yellow band on an otherwise black hair shaft, resulting in the agouti phenotype. A nonlethal mutant allele (A^{vy}) causes yellow pigment formation along the entire hair shaft, producing yellow fur color. This allele is the result of the insertion of a transposable element near the transcription start site of the *Agouti* gene. A promoter element within the transposon is responsible for this change in gene expression. Researchers found that the degree of methylation in the transposon's promoter is related to the amount of yellow pigment deposited in the hair shaft and that the amount of methylation varies from individual to individual. The result is variation in coat color phenotypes even in genetically identical mice (**ST Figure 8**). In these mice, coat colors range from yellow (unmethylated promoter) to pseudoagouti (highly methylated promoter). In addition to a gradation in coat color, there is also a gradation in body weight. Yellow mice are more obese than the brown, pseudoagouti mice. Alleles such as A^{vy} that show variable expression from individual to individual in genetically identical strains caused by epigenetic modifications are called *metastable epialleles*. Metastable refers to the changeable nature of the epigenetic modifications, and epiallele refers to the heritability of the epigenetic status of the allele.

To evaluate the role of environmental factors in modifying the epigenome, the diet of pregnant A^{vy} mice was supplemented with methylation precursors, including folic acid, vitamin B_{12}, and choline. In the offspring, coat-color

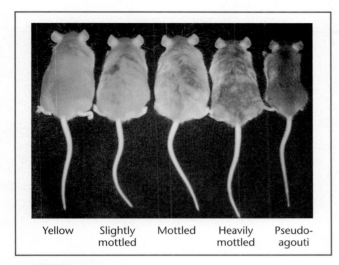

| Yellow | Slightly mottled | Mottled | Heavily mottled | Pseudo-agouti |

ST FIGURE 8 Variable expression of yellow phenotype in mice caused by diet-related epigenetic changes in the genome.

variation was reduced and shifted toward the pseudoagouti (highly methylated) phenotype. The shift in coat color was accompanied by increased methylation of the transposon's promoter. These findings have applications to epigenetic diseases in humans. For example, the risk of colorectal cancer is linked directly to folate dietary deficiency and activity differences in enzymes leading to the synthesis of methyl donors.

In addition to foods that mediate epigenetic changes in gene expression via methylation, it has recently been discovered that some foods such as rice, cabbage, wheat, and potatoes are the source of miRNAs circulating in the blood and serum of humans. Studies on one target gene showed that plant miRNAs downregulate expression of this gene and that this effect is reversible by treatment with anti-miRNAs. This preliminary but highly significant work shows that miRNAs can act across species and even across kingdoms to regulate expression of target genes, and suggests that environmental factors may play a major role in epigenetic regulation of gene expression.

Epigenome Projects

In conclusion, we will discuss several projects that are mapping the human epigenome. One of these is the NIH Roadmap Epigenomics Project. This undertaking is based on the idea that many aspects of health and susceptibility to human disease are related to epigenetic regulation or misregulation of gene activity. The program is focused on how epigenetic mechanisms controlling stem cell differentiation and organ formation generate biological responses to external and internal stimuli that result in disease. One program of the NIH Roadmap Project is the Human Epigenome Atlas, which collects and catalogs data on human epigenomes to serve as reference standards. The Atlas provides detailed information about epigenomic modifications at specific loci, in different cell types and physiological states, as well as genotypes. These data allow researchers to perform integrative and comparative analysis of epigenomic data across genomic regions or entire genomes.

Another project, called the Human Epigenome Project, is a multinational, public/private consortium organized to identify, map, and establish the functional significance of all DNA methylation patterns in the human genome. Analysis of these methylation patterns may show that genetic responses to environmental cues mediated by epigenetic changes are a pathway to disease.

Even though these projects are in the early stages of development, the information already available strongly suggests that we are on the threshold of a new era in genetics, one in which we can study the impact of environmental factors on the genome at the molecular level. The results of these projects may help explain how environmental settings in early life can affect predisposition to adulthood diseases.

SELECTED READINGS AND RESOURCES

Journal Articles and Reviews

Burdge, G.C., and Lillicrop, K.A. 2010. Nutrition, epigenetics, and developmental plasticity: Implications for understanding human disease. *Annu. Rev. Nutr.* 30: 315–339.

Crews, D. 2011. Epigenetic modifications of brain and behavior: Theory and practice. *Horm. Behav.* 59: 393-398.

Dulac, C. 2010. Brain function and chromatin plasticity. *Nature* 465: 728–735.

Gejman, P.V., Sanders, A.R., and Kendler, K.S. 2011. Genetics of schizophrenia: New findings and challenges. *Annu. Rev. Genomics Hum. Genet.* 12: 121–144.

Iacobuzio-Donahue, C.A. 2009. Epigenetic changes in cancer. *Ann. Rev. Pathol. Mech. Dis.* 4: 229–249.

Petronis, A. 2010. Epigenetics as a unifying principle in the aetiology of complex traits and diseases. *Nature* 465: 721–727.

Weaver, I.C., Cervoni, N., Chanpagne, F.A., D'Allesio, A.C., Sharma, S., Seckl, J.R., Dymov, S., Szyf, M., and Meany, M.J. 2004. Epigenetic programming by maternal behavior. *Nat. Neurosci.* 7: 847–854.

Zhang, L., Hou, D., Chen, X., Li, D., Zhu, L., Zhang, Y, Li, J., Bian, Z., et al. 2012. Exogenous plant MIR168a specifically targets mammalian LDLRAP1: Evidence of cross-kingdom regulation by microRNA. *Cell Res.* 22: 107–126.

Web Sites

National Institutes of Health Roadmap for Epigenomics. http://www.nihroadmap.nih.gov/epigenomics/initiatives.asp

Human Epigenome Project. www.epigenome.org

Human Epigenome Atlas. http://www.genboree.org/epigenomeatlas/index.rhtml

Computational Epigenetics Group. http://www.computational-epigenetics.de

CREDITS

Credits are listed in order of appearance.

Photo

CO, Paphrag. ST-8, Environmental Health Perspective

Text

Source: Bioinformatics Research Laboratory, Baylor College of Medicine.,
Source: Computational Epigenetics

Special Topics in Modern Genetics: DNA Forensics

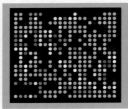
DNA Forensics

Genetics is dramatically affecting technologies in fields as diverse as agriculture, archaeology, medical diagnosis, and disease treatment. One of the areas that has been the most profoundly altered by modern genetics is forensic science. **Forensic science** (or *forensics*) uses technological and scientific approaches to answer questions about the facts of criminal or civil cases. Prior to 1986, forensic scientists had a limited array of tools with which to link evidence to specific individuals or suspects. These included some reliable methods such as blood typing and fingerprint analysis, but also many unreliable methods such as bite mark comparisons and hair microscopy.

Since the first forensic use of **DNA profiling** in 1986 **(Box 1)**, DNA forensics (also called **forensic DNA fingerprinting** or **DNA typing**) has become an important method for police to identify sources of biological materials. DNA profiles can now be obtained from saliva left on cigarette butts or postage stamps, pet hairs found at crime scenes, or bloodspots the size of pinheads. Even biological samples that are degraded by fire or time are yielding DNA profiles that help the legal system determine identity, innocence, or guilt. Investigators now scan large databases of stored DNA profiles in order to match profiles generated from crime scene evidence. DNA profiling has proven the innocence of hundreds of people who were convicted of serious crimes and even sentenced to death. Forensic scientists have used DNA profiling to identify victims of mass disasters such as the Asian Tsunami of 2004 and the September 11, 2001 terrorist attacks in New York. They have also used forensic DNA analysis to identify endangered species and animals trafficked in the illegal wildlife trade. The power of DNA forensic analysis has captured the public imagination, and DNA forensics is featured in several popular television series.

The applications of DNA profiling extend beyond forensic investigations. These include paternity and family relationship testing, identification of plant materials, verification of military casualties, and evolutionary studies.

It is important for all of us to understand the basics of forensic DNA analysis. As informed citizens, we need to monitor its uses and potential abuses. Although DNA profiling is well validated as a technique and is considered the gold standard of forensic identification, it is not without controversy and the need for legislative oversight.

In this Special Topics chapter, we will explore how DNA profiling works and how the results of profiles are interpreted. We will learn about DNA databases, the potential problems associated with DNA profiling, and the future of this powerful technology.

DNA Profiling Methods

VNTR-Based DNA Fingerprinting

The era of DNA-based human identification began in 1984, with Dr. Alec Jeffreys' publication on DNA loci known as **minisatellites,** or **variable number of tandem repeats (VNTRs).** VNTRs are located in noncoding regions of the genome and are made up of DNA sequences of between 15 and 100 bp long, with each unit repeated a number of times. The number of repeats found at each VNTR locus varies from person to person, and hence VNTRs can be from 1 to 20 kilobases (kb) in length, depending on the person. For example, the following VNTR is comprised of three tandem repeats of a 16-nucleotide sequence (highlighted in bold)

5′-GACTGCCTGCTAAGAT**GACTGCC
TGCTAAGAT**GACTGCCTGCTAAGAT-3′

VNTRs are useful for DNA profiling because there are as many as 30 different possible alleles (repeat lengths) at any VNTR in a population. This creates a large number of possible genotypes. For example, if one examined four different VNTR loci within a population, and each locus had 20 possible alleles, there would be more than 2 billion (4^{20}) possible genotypes in this four-locus profile.

To create a VNTR profile (also known as a DNA fingerprint), scientists extract DNA from a tissue sample and digest it with a restriction enzyme that cleaves on either side of the VNTR repeat region **(ST Figure 1)**. The digested DNA is separated by gel electrophoresis and subjected to Southern blot analysis . Briefly, separated DNA is transferred from the gel to a membrane and hybridized with a radioactive probe that rec-

> **Even biological samples degraded by fire or time are yielding DNA profiles that help determine identity, innocence, or guilt.**

The Pitchfork Case: The First Criminal Conviction Using DNA Profiling

In the mid-1980s, the bodies of two schoolgirls, Lynda Mann and Dawn Ashworth, were found in Leicestershire, England. Both girls had been raped, strangled, and their bodies left in the bushes. In the absence of useful clues, the police questioned a local mentally retarded porter named Richard Buckland who had a previous history of sexual offenses. During interrogation, Buckland confessed to the murder of Dawn Ashworth; however, police did not know whether he was also responsible for Lynda Mann's death. In 1986, in order to identify the second killer, the police asked Dr. Alec Jeffreys of the University of Leicester to try a new method of DNA analysis called DNA fingerprinting. Dr. Jeffreys had developed a method of analyzing DNA regions called *variable number of tandem repeats* (VNTRs), which vary in length between members of a population. Dr. Jeffreys's VNTR analysis revealed a match between the DNA profiles from semen samples obtained from both crime scenes, suggesting that the same person was responsible for both rapes. However, neither of the DNA profiles matched those from a blood sample taken from Richard Buckland. Having eliminated their only suspect, the police embarked on the first mass DNA dragnet in history by requesting blood samples from every adult male in the region. Although 4000 men offered samples, one did not. Colin Pitchfork, a bakery worker, paid a friend to give a blood sample in his place, using forged identity documents. Their plan was detected when their conversation was overheard at a local pub. The conversation was reported to police, who then arrested Pitchfork, obtained his blood sample, and sent it for analysis. His DNA profile matched the profiles from the semen samples left at both crime scenes. Pitchfork confessed to the murders, pleaded guilty, and was sentenced to life in prison. The Pitchfork Case was not only the first criminal case resolved by forensic DNA profiling, but also the first case in which DNA profiling led to the exoneration of an innocent person.

SPECIAL TOPIC

ognizes DNA sequences within the VNTR region. After exposing the membrane to X-ray film, the pattern of bands is measured, with larger VNTR repeat alleles remaining near the top of the gel and smaller VNTRs, which migrate more rapidly through the gel, being closer to the bottom. The pattern of bands is the same for a given individual, no matter what tissue is used as the source of the DNA. If enough VNTRs are analyzed, each person's DNA profile will be unique (except, of course, for identical twins) because of the huge number of possible VNTRs and alleles. In practice, scientists analyze about five or six loci to create a DNA profile.

A significant limitation of VNTR profiling is that it requires a relatively large sample of DNA (10,000 cells or about 50 μg of DNA)—more than is usually found at a typical crime scene. In addition, the DNA must be relatively intact (nondegraded). As a result, VNTR profiling has been used most frequently when large tissue samples are available—such as in paternity testing. Although VNTR profiling is still used in some cases, it has mostly been replaced by more sensitive methods, as described next.

Autosomal STR DNA Profiling

The development of the polymerase chain reaction (PCR) revolutionized DNA profiling. Using PCR-amplified DNA samples, scientists are able to generate DNA profiles from trace samples (e.g., the bulb of single hairs or a few cells from a bloodstain) and from samples that are old or degraded (such as a bone found in a field or an ancient Egyptian mummy).

The majority of human forensic DNA profiling is now done using commercial kits that amplify and analyze regions of the genome known as **microsatellites,** or **short tandem repeats (STRs).** STRs are similar to VNTRs, but the repeated motif is shorter—between two and nine base pairs, repeated from 7 to 40 times. For example, one locus known as D8S1179 is made up of the four base-pair sequence TCTA, repeated 7 to 20 times, depending on the allele. There are 19 possible alleles of the locus that are found within a population. Although hundreds of STR loci are present in the human genome, only a subset is used for DNA profiling. The FBI and other law enforcement agencies have selected 13 STR loci to be used as a core set for forensic analysis **(ST Figure 2)**.

Several commercially available kits are currently used for forensic DNA analysis of STR loci. The methods vary slightly but generally involve the following steps. As shown in **ST Figure 3**, each primer set is tagged by one of four fluorescent dyes—blue, green, yellow, or red. Each primer set is designed to amplify DNA fragments, the sizes of which vary depending on the number of repeats within the region amplified. For example, the primer sets that amplify the D19S433, vWA, TPOX, and D18S51 STR loci are all labeled with a yellow fluorescent tag. The sizes of the amplified DNA fragments produced allow scientists to differentiate between the yellow-labeled products. For example, the amplified products from the D19S433 locus range from about 100 to 150 bp in length, whereas those from the vWA locus range from about 150 to 200 bp, and so on.

After amplification, the DNA sample will contain a small amount of the original template DNA and a large

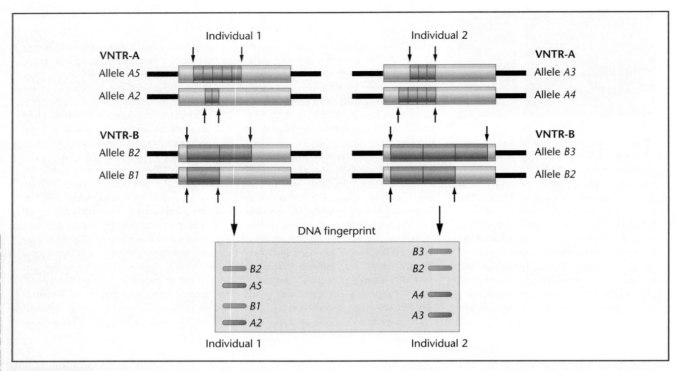

ST FIGURE 1 **DNA fingerprint at two VNTR loci for two individuals.** VNTR alleles at two loci (*A* and *B*) are shown for two different individuals. Arrows mark restriction-enzyme cutting sites that flank the VNTRs. Restriction-enzyme digestion produces a series of fragments that can be separated by gel electrophoresis and detected as bands on a Southern blot (bottom). The number of repeats at each locus is variable, so the overall pattern of bands is distinct for each individual. The DNA fingerprint profile shows that these individuals share one allele (B2).

amount of fluorescently labeled amplification products (**ST Figure 4**). The sizes of the amplified fragments are measured by **capillary electrophoresis.** This method uses thin glass tubes that are filled with a polyacrylamide gel material similar to that used in slab gel electrophoresis. The amplified DNA sample is loaded onto the top of the capillary tube and an electric current is passed through the tube. The negatively charged DNA fragments migrate through the gel toward the positive electrode, according to their sizes. Short fragments move more quickly through the gel, and larger ones more slowly. At the bottom of the tube, a laser detects each fluorescent fragment as it migrates through the tube. The data are analyzed by software that calculates both the sizes of the fragments and their quantities, and these are represented as peaks on a graph (**ST Figure 5**). Typically, automated capillary electrophoresis systems analyze as many as 16 samples at a time, and the analysis takes approximately 30 minutes.

After DNA profiling, the profile can be directly compared to a profile from another person, from crime scene evidence, or from other profiles stored in DNA profile databases (**ST Figure 6**). The STR profile genotype of an individual is expressed as the number of times the STR sequence is repeated. For example, in the profile shown in ST Figure 6, the person's profile would be expressed as shown in **ST Table 1**.

Scientists interpret STR profiles using statistics, probability, and population genetics, and these methods will be discussed in the section Interpreting DNA Profiles.

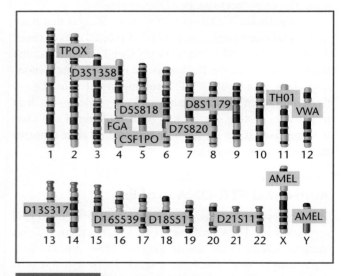

ST FIGURE 2 **Chromosomal positions of the 13 core STR loci used for forensic DNA profiling.** The *AMEL* (*Amelogenin*) locus is included with the 13 core loci and is used to determine the gender of the person providing the DNA sample. The *AMEL* locus on the X chromosome contains a 6-nucleotide deletion compared to that on the Y chromosome.

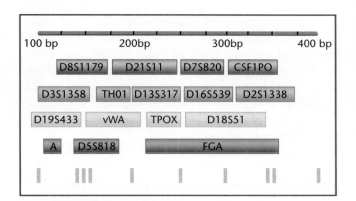

ST FIGURE 3 Relative size ranges and fluorescent dye labeling colors of 16 STR products generated by a commercially available DNA profiling kit. The DNA fragments shown in orange at the bottom of the diagram are DNA size markers. The *AMEL* locus is indicated as an *A*. (From Applied Biosystems AmpFLSTR Identifiler kit, http://www.appliedbiosystems.com)

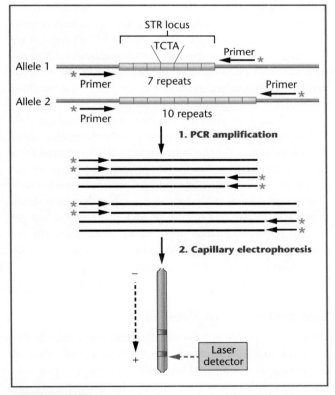

ST FIGURE 4 Steps in the PCR amplification and analysis of one STR locus (*D8S1179*). In this example, the person is heterozygous at the *D8S1179* locus: One allele has 7 repeats and one has 10 repeats. Primers are specific for sequences flanking the STR locus and are labeled with a blue fluorescent dye. The double-stranded DNA is denatured, the primers are annealed, and each allele is amplified by PCR in the presence of all four dNTPs and Taq DNA polymerase. After amplification, the labeled products are separated according to size by capillary electrophoresis, followed by fluorescence detection.

Y-Chromosome STR Profiling

In many forensic applications, it is important to differentiate the DNA profiles of two or more people in a mixed sample. For example, vaginal swabs from rape cases usually contain a mixture of female somatic cells and male sperm cells. In addition, some crime samples contain evidence material from a number of male suspects. In these types of cases, STR profiling of Y-chromosome DNA is useful. There are more than 200 STR loci on the Y chromosome that are useful for DNA profiling; however, fewer than 20 of these are used routinely for forensic analysis. PCR amplification of Y-chromosome STRs uses specific primers that do not amplify DNA on the X chromosome.

One limitation of Y-chromosome DNA profiling is that it cannot differentiate between the DNA from fathers and sons, or from male siblings. This is because the Y chromosome is directly inherited from the father to his sons, as a single unit. The Y chromosome does not undergo recombination, meaning that less genetic variability exists on the Y chromosome than on autosomal chromosomes. Therefore, all patrilineal relatives share the same Y-chromosome profile. Even two apparently unrelated males may share the same Y profile if they also shared a distant male ancestor.

Although these features of Y-chromosome profiles present limitations for some forensic applications, they are useful for identifying missing persons when a male relative's DNA is available for comparison. They also allow researchers to trace paternal lineages in genetic genealogy studies (**Box 2**).

Mitochondrial DNA Profiling

Another important addition to DNA profiling methods is **mitochondrial DNA (mtDNA)** analysis. Between 200 and 1700 mitochondria are present in each human somatic cell. Each mitochondrion contains one or more 16-kb circular DNA chromosomes. Mitochondria divide within cells and are distributed to daughter cells after cell division. Mitochrondria are passed from the human egg cell to the zygote during fertilization; however, as sperm cells contribute few if any mitochondria to the zygote, they do not contribute these organelles to the next generation. Therefore, all cells in an individual contain multiple copies of identical mitochondria inherited from the mother. Like Y-chromosome DNA, mtDNA undergoes little if any recombination and is inherited as a single unit.

Scientists create mtDNA profiles by amplifying regions of mtDNA that show variability between unrelated individuals and populations. Two commonly used regions are known as **hypervariable segment I and II (HVSI and HVSII) (ST Figure 7)**. After PCR amplification, the DNA sequence within these regions is determined by automated DNA sequencing. Scientists then compare the sequence with

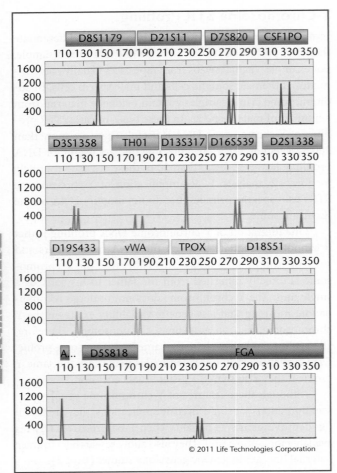

© 2011 Life Technologies Corporation

ST FIGURE 5 An electropherogram showing the results of a DNA profile analysis using the 16-locus STR profile kit shown in **ST Figure 3.** Heterozygous loci show up as double peaks and homozygous loci as single, higher peaks. The sizes of each allele can be calculated from the peak locations relative to the size axis shown at the top of each panel. The single peak for the *AMEL* (*A*) locus indicates that this DNA profile is that of a female individual, as described in the text.

sequences from other individuals or crime samples to determine whether or not they match.

The fact that mtDNA is present in high copy numbers in cells makes its analysis useful in cases where crime samples are small, old, or degraded. mtDNA profiling is particularly useful for identifying victims of mass murders or disasters, such as the Srebrenica massacre of 1995 and the World Trade Center attacks of 2001, where reference samples from relatives are available **(Box 3)**. The main disadvantage of mtDNA profiling is that it is not possible to differentiate between the mtDNA from maternal relatives or from siblings. Like Y-chromosome profiles, mtDNA profiles may be shared by two apparently unrelated individuals who also share a distant ancestor—in this case a maternal ancestor. Researchers use mtDNA profiles in scientific studies of genealogy, evolution, and human population migrations.

Mitochondrial DNA analyses have also been useful in wildlife forensics cases. Billions of dollars are generated throughout the world from the illegal wildlife trade. Often, the identification of the species or origin of plant and animal material is the key to successful prosecution of wildlife trafficking cases. A case of illegal smuggling of bird eggs in Australia, solved by mitochondrial sequence analysis, is presented in **Box 4**.

Single-Nucleotide Polymorphism Profiling

Single-nucleotide polymorphisms (SNPs) are single-nucleotide differences between two DNA molecules. They may be base-pair changes or small insertions or deletions **(ST Figure 8)**. SNPs occur randomly throughout the genome and on mtDNA, every 500 to 1000 nucleotides. This means that there are potentially millions of loci in the human genome that can be used for profiling. However, as SNPs usually have only two alleles, many SNPs (50 or more) must be used to create a DNA profile that can distinguish between two individuals as efficiently as STRs.

Scientists analyze SNPs by using specific primers to amplify the regions of interest. The amplified DNA regions are then analyzed by a number of different methods such as automated DNA sequencing or hybridization to immobilized probes on DNA microarrays that distinguish between DNA molecules with single-nucleotide differences.

Forensic SNP profiling has one major advantage over STR profiling. Because a SNP involves only one nucleotide of a DNA molecule, the theoretical size of DNA required for a PCR reaction is the size of the two primers and one more nucleotide (i.e., about 50 nucleotides). This feature makes SNP analysis suitable for analyzing DNA samples that are severely degraded. Despite this advantage, SNP profiling has not yet become routine in forensic applications. More frequently, researchers use SNP profiling of Y-chromosome and mtDNA loci for lineage and evolution studies.

Interpreting DNA Profiles

After a DNA profile is generated, its significance must be determined. In a typical forensic investigation, a profile derived from a suspect is compared to a profile from an evidence sample or to profiles already present in a DNA database. If the suspect's profile does not match that of the evidence profile or database entries, investigators can conclude that the suspect is not the source of the sample(s) that generated the other profile(s). However, if the suspect's profile matches the evidence profile or a database entry, the interpretation becomes more complicated. In this case, one could conclude that the two profiles either came from the same person—or they came from two different people who share the same DNA profile by chance. To determine the significance of any DNA profile

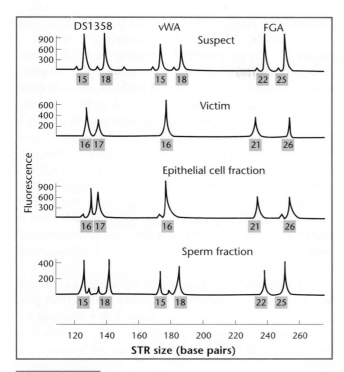

Electropherogram showing the STR profiles of four samples from a rape case. Three STR loci were examined from samples taken from a suspect, a victim, and two fractions from a vaginal swab taken from the victim. The x-axis shows the DNA size ladder, and the y-axis indicates relative fluorescence intensity. The number below each allele indicates the number of repeats in each allele, as measured against the DNA size ladder. Notice that the STR profile of the sperm sample taken from the victim matches that of the suspect.

ST TABLE 1 STR Profile Genotypes from the Four Profiles Shown in ST Figure 6

STR Locus	Profile Genotype from			
	Suspect	Victim	Epithelial Cells	Sperm Fraction
DS1358	15, 18	16, 17	16, 17	15, 18
vWA	15, 18	16, 16	16, 16	15, 18
FGA	22, 25	21, 26	21, 26	22, 25

match, it is necessary to estimate the probability that the two profiles are a random match.

The **profile probability,** or **random match probability,** method gives a numerical probability that a person chosen at random from a population would share the same DNA profile as the evidence or suspect profiles. The following example demonstrates how to arrive at a profile probability (ST Table 2).

The first locus examined in this DNA profile (D5S818) has two alleles: 11 and 13. Population studies show that the 11 allele of this locus appears at a frequency of 0.361 in this population and the 13 allele appears at a frequency of 0.141. In population genetics, the frequencies of two different alleles at a locus are given the designation p and q, following the Hardy–Weinberg law. We assume that the person having this DNA profile received the 11 and 13 alleles at random from each parent. Therefore, the probability that this person received allele 11 from the mother and allele 13 from the father is expressed as $p \times q = pq$. In addition, the probability that the person received allele 11 from the father and allele

BOX 2

Thomas Jefferson's DNA: Paternity and Beyond

For more than two centuries, historians debated whether U.S. President Thomas Jefferson had fathered one or more children with Sally Hemings, one of his slaves. In 1997, in an attempt to resolve the controversy, scientists analyzed the Y-chromosome DNA profiles from the Jefferson and Hemings male lineages. Because Jefferson did not have any surviving male-line descendants, the researchers tested the Y chromosome from five patrilineal descendants of Jefferson's paternal uncle. They also analyzed the DNA from a patrilineal descendant of Eston Hemings Jefferson, Sally Hemings' youngest son, as well as DNA from three patrilineal descendants of Jef-

ferson's sister's sons and five patrilineal descendants of Thomas Woodson, who claimed to be the first child of Sally Hemings. DNA profiles were generated from 11 Y-chromosome STR loci, one larger minisatellite (MSY1), and seven SNP-like loci that differ by one nucleotide. All five of the Jefferson-line males shared identical Y-chromosome profiles, except for one individual who had a mutation at one STR locus. The profiles from the Woodson family line did not match those from the Jefferson line, excluding Thomas Woodson as a son of Thomas Jefferson. Similarly, and not unexpectedly, Jefferson's nephews did not share his Y-chromosome profile. However, the Y-chromosome profile from Eston Hemings's male descendant was identical to profiles from the Jefferson line. Although these data support the idea that Jefferson fathered at least one child of Sally Hemings, it does not exclude the possibility that another male Jefferson (such as Thomas Jefferson's

brother, his brother's sons, or another male individual sharing the Jefferson Y chromosome) could have been Eston Hemings's father.

Another interesting finding to arise from the Jefferson DNA study was that Jefferson's Y-chromosome profile pattern (haplotype K2) is extremely rare in Europeans. It occurs predominantly in men from Africa and western Eurasia. It is also found in men of Jewish ancestry from North Africa, Egypt, and the Middle East. In 2007, geneticists discovered the K2 haplotype (and matches to Jefferson's Y-chromosome profile) in 2 out of 85 men with the surname Jefferson in Britain; hence, it is likely that Jefferson's family originally came from Britain, as he claimed. However, it also suggests that sometime in the past, a Y chromosome of African or Middle Eastern origin entered the Jefferson line. To add to the Jefferson speculations, some commentators have suggested that Jefferson was the "first Jewish president" of the United States.

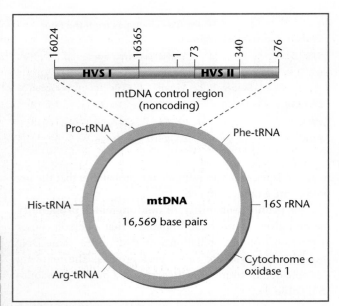

ST FIGURE 7 Human mtDNA molecule, showing the locations of HVS I and II regions relative to other mtDNA genes. The mtDNA control region is comprised of 1122 base pairs. Forensic DNA analysis involves sequencing the two HVS regions (610 base pairs total) and comparing sequences between individuals or samples. Sequence differences include short dinucleotide duplications as well as single base-pair changes.

13 from the mother is also *pq*. Hence, the total probability that this person would have the 11, 13 genotype at this locus, by chance, is 2*pq*. As we see from ST Table 2, 2*pq* is 0.102 or approximately 10 percent. It is obvious from this sample that using a DNA profile of only one locus would not be very informative, as about 10 percent of the population would also have the D5S818 11, 13 genotype.

The discrimination power of the DNA profile increases when we add more loci to the analysis. The next locus of this

person's DNA profile (TPOX) has two identical alleles—the 11 allele. Allele 11 appears at a frequency of 0.243 in this population. The probability of inheriting the 11 allele from each parent is $p \times p = p^2$. As we see in the table, the genotype frequency at this locus would be 0.059, which is about 6 percent of the population. If this DNA profile contained only the first two loci, we could calculate how frequently a person chosen at random from this population would have the genotype shown in the table, by multiplying the two genotype probabilities together. This would be $0.102 \times 0.059 = 0.006$. This analysis would mean that about 6 persons in 1000 (or 1 person in 166) would have this genotype. The method of multiplying all frequencies of genotypes at each locus is known as the **product rule.** It is the most frequently used method of DNA profile interpretation and is widely accepted in U.S. courts.

By multiplying all the genotype probabilities at the five loci, we arrive at the genotype frequency for this DNA profile: 9×10^{-7}. This means that approximately 9 people in every 10 million (or about 1 person in a million), chosen at random from this population, would share this 5-locus DNA profile.

The Uniqueness of DNA Profiles

As we increase the number of loci analyzed in a DNA profile, we obtain smaller probabilities of a random match. Theoretically, if a sufficient number of loci were analyzed, we could be *almost* certain that the DNA profile was unique. At the present time, law enforcement agencies in North America use a core set of 13 STR loci to generate DNA profiles. A hypothetical genotype comprised of the most common alleles of each STR locus in the core STR profile would be expected to occur only once in a population of 10 billion people. Hence, the frequency of this profile would be 1 in 10 billion.

The World Trade Center Attacks: Identifying Victims by DNA Profiling

Forensic DNA profiling has played an important role in identifying victims of the September 11, 2001 terrorist attacks in New York. The official death toll estimate was 2749. Only 293 of the victims were recovered as whole bodies; the remainder were represented by more than 20,000 pieces of bone and soft tissue. Most biological material was reduced to dust by the crush of collapsing buildings, the fires, and bacterial decay.

Very few of the victims could be identified by techniques such as fingerprint analysis or forensic dentistry. The major work of identification was carried out by DNA profiling, using the services of the Medical Examiner's Office and private biotechnology companies. To identify victims, the labs collected victims' personal items (such as combs, toothbrushes, and razor blades), as well as cheek swabs from relatives. Scientists then analyzed these personal items and the disaster

remains by STR profiling, attempting to match profiles. Because of the condition of the remains, only half yielded useful DNA profiles. In an attempt to identify more victims, scientists began using mtDNA profiling, SNP analysis, and a new technique involving mini-STRs. Mini-STRs are created by moving the PCR primers closer to the repeat regions of STR loci. In this way, shorter pieces of DNA can be successfully amplified by PCR methods. Using these methods, scientists identified several hundred more victims. Despite these refinements, more than 1000 victims have not been identified.

BOX 4
The Pascal Della Zuana Case: DNA Barcodes and Wildlife Forensics

On August 2, 2006, a freelance photographer named Pascal Della Zuana was stopped by customs officers at Australia's Sydney International Airport. While questioning him about his flight from Thailand to Australia, officers noticed that he was wearing an unusual white vest under his outer clothing. Inside the vest, they discovered 23 concealed bird eggs.

Due to Australia's strict quarantine regulations, the eggs had to be treated with radiation in order to sterilize them. Unable to hatch the eggs, authorities turned to DNA typing in an attempt to identify the origin and species of the eggs.

The eggs were sent to Dr. Rebecca Johnson at the DNA Laboratory of the Australian Museum for forensic identification. Dr. Johnson took a small sample from each egg and extracted the DNA. She used PCR methods to amplify an approximately 650-bp region of the mitochondrial genome, within the cytochrome c oxidase 1 gene. She then organized these sequences into a format known as a DNA barcode. In order to identify the species, Dr. Johnson compared each DNA barcode to barcode entries in a large DNA barcode database compiled at the University of Guelph in Canada. The database contains mitochondrial DNA barcode sequences from hundreds of universities and museums throughout the world, cataloging more than 70,000 different species.

The results of Dr. Johnson's barcode sequence comparisons were dramatic. Della Zuana's vest had concealed eggs of exotic bird species such as macaws, African grey and Eclectus parrots, as well as a rare threatened species, the Moluccan cockatoo.

On January 20, 2007, Pascal Della Zuana was found guilty of contravening the Convention on International Trade in Endangered Species (CITES), as well as three Australian Customs and Quarantine Acts. He was fined $10,000 and sentenced to two years in prison.

During the court case, it was learned that, if hatched, the birds would have fetched about $250,000 on the black market. The worldwide smuggling of wildlife and wildlife parts is thought to be worth as much as US$150 billion each year—surpassed only by drugs and arms in terms of illegal profit.

Although this would suggest that most DNA profiles generated by analysis of the 13 core STR loci would be unique on the planet, several situations can alter this interpretation. For example, identical twins share the same DNA, and their DNA profiles will be identical. Identical twins occur at a frequency of about 1 in 250 births. In addition, siblings can share one allele at any DNA locus in about 50 percent of cases and can share both alleles at a locus in about 25 percent of cases. Parents and children also share alleles, but are less likely than siblings to share both alleles at a locus. When DNA profiles come from two people who are closely related, the profile probabilities must be adjusted to take this into account. The allele frequencies and calculations that we describe here are based on assumptions that the population is large and has little relatedness or inbreeding. If a DNA profile is analyzed from a person in a small interrelated group, allele frequency tables and calculations may not apply.

The Prosecutor's Fallacy

It is sometimes stated, by both the legal profession and the public, that "the suspect must be guilty given that the chance of a random match to the crime scene sample is 1 in 10 billion—greater than the population of the planet." This type of statement is known as the **prosecutor's fallacy** because it equates guilt with a numerical probability derived from one piece of evidence, in the absence of other evidence. A match between a suspect's DNA profile and crime scene evidence does not necessarily prove guilt, for many reasons such as human error or contamination of samples, or even deliberate tampering. In addition, a DNA profile that does not match the evidence does not necessarily mean that the suspect is innocent. For example, a suspect's profile may not match that from a semen sample at a rape scene, but the suspect could still have been involved in the crime, perhaps by restraining the victim. For these and other reasons, DNA profiles must be interpreted in the context of all the evidence in a case. A more detailed description of problems with DNA profiles is given in the next section.

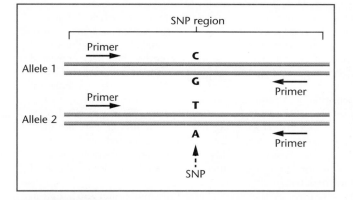

ST FIGURE 8 Example of a single-nucleotide polymorphism (SNP) from an individual who is heterozygous at the SNP locus. The arrows indicate the locations of PCR primers used to amplify the SNP region, prior to DNA sequence analysis. If this SNP locus only had two known alleles—the C and T alleles—there would be three possible genotypes in the population: CC, TT, and CT. The individual in this example has the CT genotype.

ST TABLE 2	A Profile Probability Calculation Based on Analysis of Five STR Loci		
STR Locus	**Alleles from Profile**	**Allele Frequency from Population Database***	**Genotype Frequency Calculation**
D5S818	11	0.361	$2pq = 2 \times 0.361 \times 0.141 = 0.102$
	13	0.141	
TPOX	11	0.243	$p^2 = 0.243 \times 0.243 = 0.059$
	11	0.243	
D8S1179	13	0.305	$2pq = 2 \times 0.305 \times 0.031 = 0.019$
	16	0.031	
CSF1PO	10	0.217	$p^2 = 0.217 \times 0.217 = 0.047$
	10	0.217	
D19S433	13	0.253	$2pq = 2 \times 0.253 \times 0.369 = 0.187$
	14	0.369	

Genotype frequency from this 5-locus profile = $0.102 \times 0.059 \times 0.019 \times 0.047 \times 0.187 = 0.0000009 = 9 \times 10^{-7}$

*A U.S. Caucasian population database (Butler, J.M., et al. 2003. *J. Forensic Sci.* 48: 908–911).

DNA Profile Databases

Many countries throughout the world maintain national DNA profile databases. The first of these databases was established in the UK in 1995 and now contains more than 5 million profiles—representing almost 10 percent of the population. In the UK, DNA samples can be taken from anyone arrested for an offense that could lead to a prison sentence. Although highly controversial, DNA profiles from suspects can be stored permanently in the national DNA database, even if the person is not convicted.

In the United States, both state and federal governments have DNA profile databases. The entire system of databases along with tools to analyze the data is known as the **Combined DNA Index System (CODIS)** and is maintained by the FBI. At the end of 2011, there were approximately 10 million DNA profiles stored within the CODIS system. The two main databases in CODIS are the **convicted offender database,** which contains DNA profiles from individuals convicted of certain crimes, and the **forensic database,** which contains profiles generated from crime scene evidence. In addition, some states have DNA profile databases containing profiles from suspects and from unidentified human remains and missing persons. Suspects who are not convicted can request that their profiles be removed from the databases.

BOX 5
The Kennedy Brewer Case: Two Bite-Mark Errors and One Hit

In 1992 in Mississippi, Kennedy Brewer was arrested and charged with the rape and murder of his girlfriend's 3-year-old daughter, Christine Jackson. Although a semen sample had been obtained from Christine's body, there was not sufficient DNA for profiling. Forensic scientists were also unable to identify the ABO blood group from the bloodstains left at the crime scene. The prosecution's only evidence came from a forensic bite-mark specialist who testified that the 19 "bite marks" found on Christine's body matched imprints made by Brewer's two top teeth. Even though the specialist had recently been discredited by the American Board of Forensic Odontology, and the defense's expert dentistry witness testified that the marks on Christine's body were actually postmortem insect bites, the court convicted Brewer of capital murder and sexual battery and sentenced him to death.

In 2001, more sensitive DNA profiling was conducted on the 1992 semen sample. The profile excluded Brewer as the donor of the semen sample. It also excluded two of Brewer's friends, and Y-chromosome profiles excluded Brewer's male relatives. Despite these test results, Brewer remained in prison for another five years, awaiting a new trial. In 2007, the Innocence Project took on Brewer's case and retested the DNA samples. The profiles matched those of another man, Justin Albert Johnson, a man with a history of sexual assaults who had been one of the original suspects in the case. Johnson subsequently confessed to Christine Jackson's murder, as well as to another rape and murder—that of a 3-year-old girl named Courtney Smith. Levon Brooks, the ex-boyfriend of Courtney's mother, had been convicted of murder in the Smith case, also based on bite-mark testimony by the same discredited expert witness.

On February 15, 2008, all charges against Kennedy Brewer were dropped, and he was exonerated of the crimes. Levon Brooks was subsequently exonerated of the Smith murder in March of 2008.

Since 1989, more than 250 people in the United States have been exonerated of serious crimes, based on DNA profile evidence. Seventeen of these people had served time on death row. In more than 100 of these exoneration cases, the true perpetrator has been identified, often through searches of DNA databases.

SPECIAL TOPIC

DNA profile databases have proven their value in many different situations. As of 2011, use of CODIS databases resulted in more than 100,000 profile matches that assisted criminal investigations and missing persons searches (Box 5). Despite the value of DNA profile databases, they remain a concern for many people who question the privacy and civil liberties of individuals versus the needs of the state.

Technical and Ethical Issues Surrounding DNA Profiling

Although DNA profiling is sensitive, accurate, and powerful, it is important to be aware of its limitations. One limitation is that most criminal cases have either no DNA evidence for analysis or DNA evidence that would not be informative to the case. In some cases, potentially valuable DNA evidence exists but remains unprocessed and backlogged. Another serious problem is that of human error. There are cases in which innocent people have been convicted of violent crimes based on DNA samples that had been inadvertently switched during processing. DNA evidence samples from crime scenes are often mixtures derived from any number of people present at the crime scene or even from people who were not present, but whose biological material (such as hair or saliva) was indirectly introduced to the site. Crime scene evidence is often degraded, yielding partial DNA profiles that are difficult to interpret.

One of the most disturbing problems with DNA profiling is its potential for deliberate tampering. DNA profile technologies are so sensitive that profiles can be generated from only a few cells—or even from fragments of synthetic DNA. There have been cases in which criminals have introduced biological material to crime scenes, in an attempt to affect forensic DNA profiles. It is also possible to manufacture artificial DNA fragments that match STR loci of a person's DNA profile. In 2010, a research paper[1] reported methods for synthesizing DNA of a known STR profile, mixing the DNA with body fluids, and depositing the sample on crime scene items. When subjected to routine forensic analysis, these artificial samples generated perfect STR profiles. In the future, it may be necessary to develop methods to detect the presence of synthetic or cloned DNA in crime scene samples. It has been suggested that such detections could be done, based on the fact that natural DNA contains epigenetic markers such as methylation.

Many of the ethical questions related to DNA profiling involve the collection and storage of biological samples and DNA profiles. Should police be able to collect DNA samples without a suspect's knowledge or consent? Who should have their DNA profiles stored on a database? Should law enforcement agencies reveal the identities of people whose DNA profiles partially match those of a suspect, on the chance that the two individuals are related? Should researchers have access to DNA databases for research purposes? DNA analyses can now predict eye color and provide age estimates. Soon it may be possible to predict a person's health, racial background, and other features of appearance from a small DNA sample. Should such information be admissible in court cases?

As DNA profiling becomes more sophisticated and prevalent, we should carefully consider both the technical and ethical questions that surround this powerful new technology.

SELECTED READINGS AND RESOURCES

Journal Articles

Brettell, T.A., et al., 2009. Forensic science. *Anal. Chem.* 81: 4695–4711.

Butler, J.M., et al., 2007. STRs vs. SNPs: Thoughts on the future of forensic DNA testing. *Forensic Sci Med Pathol* 3: 200–205.

Frumkin, D., et al., 2010. Authentication of forensic DNA samples. *Forensic Sci International* 4: 95–103.

Gill, P., Jeffreys, A.J., and Werrett, D.J. 2005. Forensic applications of DNA "fingerprints." *Nature* 318: 577–579.

Houck, M.M. 2006. CSI: Reality. *Scientific American* 295(1): 84–89.

Jobling, M.A., and Gill, P. 2004. Encoded evidence: DNA in forensic analysis. *Nature Reviews Genetics* 5: 739–750.

King, T.E., et al. 2007. Thomas Jefferson's Y chromosome belongs to a rare European lineage. *Am J Phys Anthropology* 132: 584–589.

Roewer, L. 2009. Y chromosome STR typing in crime casework. *Forensic Sci Med Pathol* 5(2): 77–84.

Whittall, H. 2008. The forensic use of DNA: Scientific success story, ethical minefield. *Biotechnol J* 3: 303–305.

Web Sites

Brenner, C.H., Forensic Mathematics of DNA Matching. http://dna-view.com/profile.htm

Butler, J.M., and Reeder, D.J. Short Tandem Repeat DNA Internet DataBase. http://www.cstl.nist.gov/div831/strbase

Wikipedia: DNA Profiling. http://en.wikipedia.org/wiki/DNA_profiling

DNA initiative: Advancing criminal justice through DNA technology. http://www.dna.gov

The Innocence Project. http://www.innocenceproject.org

Berson, S.B. Debating DNA collection, from Office of Justice Programs, National Institute of Justice Journal, November 2009. http://www.nij.gov/journals/264/debating-DNA.htm

CODIS-NDIS Statistics. Federal Bureau of Investigation Web site, http://www.fbi.gov/about-us/lab/forensic-science-communications/undermicroscope/table1.htm

---

[1]Frumkin, D., et al. 2010. Authentication of forensic DNA samples. *Forensic Sci Int Genetics* 4: 95–103.

CREDITS

Credits are listed in order of appearance.

Photo

CO, Paphrag.

Text

STT.2 and STF-2, Butler, J.M., et al. 2003. J. Forensic Sci. 48: 908–911. Copyright © 2003. Reprinted with permission. **STF-4**,

Figure kindly provided by John Butler (NIST). Copyright © 2011 bfgy Life Technologies Corporation. Reprinted with permission.

Special Topics in Modern Genetics:
Genomics and Personalized Medicine

Genomics and Personalized Medicine

Physicians have always practiced personalized medicine in order to make effective treatment decisions for their patients. Doctors take into account a patient's symptoms, family history, lifestyle, and data derived from many types of medical tests. However, within the last 20 years, personalized medicine has taken a new and potentially powerful direction based on genetics and genomics. Today, the phrase *personalized medicine* is used to describe the application of information from a patient's unique genetic profile in order to select effective treatments that have minimal side-effects and to detect disease susceptibility prior to development of the disease.

Despite the immense quantities of medical information and pharmaceuticals that are available, the diagnosis and treatment of human disease remains an imperfect process. It is sometimes difficult or impossible to accurately diagnose some conditions. In addition, some patients do not respond to treatments, while others may develop side-effects that can be annoying or even life-threatening. As much of the basis for disease susceptibility and the variation that patients exhibit toward drug treatments are genetically determined, progress in genetics, genomics, and molecular biology has the potential to significantly advance medical diagnosis and treatment.

The sequencing of the human genome, the cataloging of genetic sequence variants, and the linking of sequence variants with disease susceptibility form the basis of the newly emerging field of personalized medicine. By 2009, the number of diseases detectable by genetic tests rose to approximately 2000. In addition, a rapidly growing list of genetic tests helps physicians determine whether a patient will have an adverse drug reaction and whether a particular pharmaceutical will be effective for that patient.

Although much of the promise of personalized medicine remains in the future, significant progress is underway. As genome technologies advance and the cost of sequencing personal genomes declines, it is becoming easier to examine a patient's unique genomic profile in order to diagnose diseases and prescribe treatments. Proponents of personalized medicine foresee a future in which each person will have his or her genome sequence

determined at birth and will have the sequence stored in a digital form within a personal computerized medical file. Medical practitioners will use automated methods to scan the sequence information within these files for clues to disease susceptibility and reactions to drugs. In the near future, genomic profiling and personalized medicine will allow physicians to predict which diseases you will develop, which therapeutics will work for you, and which drug dosages are appropriate.

In this Special Topics chapter, we will outline the current uses of genetic and genomic-based personalized medicine in disease diagnosis and drug selection. In addition, we will outline the future directions for personalized medicine, as well as some ethical and technical challenges associated with it.

Personalized Medicine and Pharmacogenomics

At present, the most developed aspect of the new personalized medicine is in the field of pharmacogenomics. **Pharmacogenomics** is the study of how an individual's entire genetic make-up determines the body's response to drugs. The term *pharmacogenomics* is used interchangeably with *pharmacogenetics*, which refers to the study of how sequence variation within specific candidate genes affects an individual's drug responses.

In pharmacogenomics, scientists take into account many aspects of drug metabolism and how genetic traits affect these aspects. When a drug enters the body, it interacts with various proteins, including carriers, cell-surface receptors, transporters, and metabolizing enzymes. These proteins affect a drug's target site of action, absorption, pharmacological response, breakdown, and excretion. Because there are so many interactions that occur between a drug and proteins within the patient, many genes and many different genetic polymorphisms can affect a person's response to a drug.

In this subsection, we examine two ways in which genomics and personalized medicine are changing the field of pharmacogenomics: by optimizing drug therapies and by reducing adverse drug reactions.

> In the near future, personalized medicine will allow physicians to predict which diseases you will develop, which therapeutics will work for you, and which drug dosages are appropriate.

Optimizing Drug Therapies

When it comes to drug therapy, it is clear that "one size does not fit all." On average, a drug will be effective in only about 50 percent of patients who take it **(ST Figure 1)**. This situation means that physicians often must switch their patients from one drug to another until they find one that is effective. Not only does this waste time and resources, but also it may be dangerous to the patient who is exposed to a variety of different pharmaceuticals and who may not receive appropriate treatment in time to combat a progressive illness.

Pharmacogenomics increases the efficacy of drugs by targeting those drugs to subpopulations of patients who will benefit. One of the most common current applications of personalized pharmacogenomics is in the diagnosis and treatment of cancers. Large-scale sequencing studies show that each tumor is genetically unique, even though it may fall into a broad category based on cytological analysis or knowledge of its tissue origin. Given this genomic variability, it is important to understand each patient's mutation profile to select an appropriate treatment—particularly those newer treatments based on the molecular characteristics of tumors **(Box 1)**.

One of the first success stories in personalized medicine was that of the **HER-2** gene and the use of the drug **Herceptin®** in breast cancer. The human epidermal growth factor receptor 2 (*HER-2*) gene is located on chromosome 17 and codes for a transmembrane tyrosine kinase receptor protein called HER-2. These receptors are located within the cell membranes of normal breast epithelial cells and, when bound to an extracellular growth factor (ligand), send signals

Drug type		
Antidepressants (SSRIs)	38%	
Asthma drugs	40%	
Diabetes drugs	43%	
Arthritis drugs	50%	
Alzheimer's drugs	70%	
Cancer drugs	75%	

© 2011 Personalized Medicine Coalition

ST FIGURE 1 **Variations in patient response to drugs.** This figure gives a general summary of the percentages of patients for which a particular class of drugs is effective.

to the cell nucleus that result in the transcription of genes involved in cell growth and division.

In about 25 percent of invasive breast cancers, the *HER-2* gene is amplified and the protein is overexpressed on the cell surface. In some breast cancers, the *HER-2* gene is present in as many as 100 copies per cell. The presence of *HER-2* overexpression is associated with increased tumor invasiveness, metastasis, and cell proliferation, as well as a poorer patient prognosis.

SPECIAL TOPIC

BOX 1
The Story of Pfizer's Crizotinib

In 2007, Beverly Sotir was diagnosed with advanced non-small-cell lung cancer (NSCLC). Beverly, a 68-year-old grandmother and nonsmoker, received standard chemotherapy, but her cancer continued to proliferate. She was given six months to live. At this same time, an apparently unrelated scientific study was underway by the pharmaceutical company, Pfizer. Pfizer had developed a compound called crizotinib, which was designed to inhibit the activity of MET, a tyrosine kinase that is abnormal in a number of tumors. Although crizotinib also inhibited

another kinase called ALK (anaplastic lymphoma kinase), scientists did not consider it significant. After clinical trials for crizotinib began, an article was published[*] describing a chromosomal translocation found in a small number of NSCLCs. This translocation fused the *ALK* gene to another gene called *EML4,* leading to production of a fusion protein that stimulated cancer cell growth. Pfizer immediately changed its clinical trial to include NSCLC patients. Beverly's doctors at the Dana-Farber Cancer Institute in Boston tested her tumors, discovered that they contained the *ALK/ EML4* fusion gene, and enrolled Beverly in the trials. The results were dramatic. Within six months, Beverly's tumors shrank by more than 50 percent and some disappeared entirely. As of 2011, Beverly continued to do well.

Results of the clinical trials for crizotinib showed that tumors shrank or stabilized in 90 percent of the 82 patients whose tumors contained the *ALK* fusion gene. Those patients who responded well to treatment had positive responses for up to 15 months. Scientists report that the *ALK* fusion gene tends to occur most frequently in young NSCLC patients who have never smoked. Approximately 4 percent of patients with NSCLC have this translocation in their tumor cells. Although only a small percentage of people might benefit from crizotinib, this means that about 45,000 people a year, worldwide, may be eligible for this treatment. Crizotinib is now approved in the United States for treatment of NSCLCs.

[*]Choi, S.M., et al. 2007. Identification of the transforming EML4-ALK fusion gene in non-small-cell lung cancer. *Nature* 448: 561–566.

Using recombinant DNA technology, Genentech Corporation in California developed a monoclonal antibody known as trastuzumab (or Herceptin) that is designed to bind specifically to the extracellular region of the HER-2 receptor. When bound to the receptor, Herceptin appears to inhibit the signaling capability of HER-2 and may also flag the HER-2-expressing cell for destruction by the patient's immune system. In cancer cells that overexpress HER-2, Herceptin treatment causes cell-cycle arrest, and in some cases, death of the cancer cells.

Because Herceptin will only act on breast cancer cells that have amplified *HER-2* genes, it is important to know the HER-2 phenotype of each cancer. In addition, Herceptin has potentially serious side-effects. Hence, its use must be limited to those who could benefit from the treatment. A number of molecular assays have been developed to determine the gene and protein status of breast cancer cells. Two types of tests are used routinely to determine the amount of *HER-2* overexpression in cancer cells: immunohistochemistry (IHC) and fluorescence *in situ* hybridization (FISH). In IHC assays, an antibody that binds to the HER-2 protein is added to fixed tissue on a slide. The presence of bound antibody is then detected with a stain and observed under the microscope [**ST Figure 2(a)**]. The FISH assay assesses the number of HER-2 genes by comparing the fluorescence signal from a HER-2 probe with a control signal from another gene that is not amplified in the cancer cells [**ST Figure 2(b)**].

Herceptin has had a major effect on the treatment of HER-2 positive breast cancers. When Herceptin is used in combination with chemotherapy, there is a 25 to 50 percent increase in survival, compared with the use of chemotherapy alone. Herceptin is now one of the biggest selling biotechnology products in the world, generating more than $5 billion in annual sales.

There are now dozens of drugs whose prescription and use depend on the genetic status of the target cells. Approximately 10 percent of FDA-approved drugs have labels that include pharmacogenomic information (**ST Table 1**). For example, about 40 percent of colon cancer patients respond to the drugs **Erbitux®** (cetuximab) and **Vectibix®** (panitumab). These two drugs are monoclonal antibodies that bind to **epidermal growth factor receptors (EGFRs)** on the surface of cells and inhibit the EGFR signal transduction pathway. In order to work, cancer cells must express EGFR on their surfaces and must also have a wild-type *K-RAS* gene. The presence of EGFR can be assayed using a staining test and observation of cancer cells under a microscope. Mutations in the *K-RAS* gene can be detected using assays based on the polymerase chain reaction (PCR) method.

Another example of treatment decisions being informed by genetic tests is that of the **Oncotype DX®** Assay (Genomic Health Inc.). This assay analyzes the expression (amount of mRNA) from 21 genes in breast cancer samples, in order to help physicians select appropriate treatments and predict the course of the disease. These genes were chosen because their levels of gene expression correlate with breast cancer recurrence after initial treatment. Based on the mRNA expression levels revealed in the assay results, scientists calculate a "Recurrence Score," estimating the likelihood that the cancer will recur within a ten-year period. Those patients with a low-risk rating would likely not benefit by adding chemotherapy to their treatment regimens and so can be treated with hormones alone. Those with higher risk scores would likely benefit from more aggressive therapies.

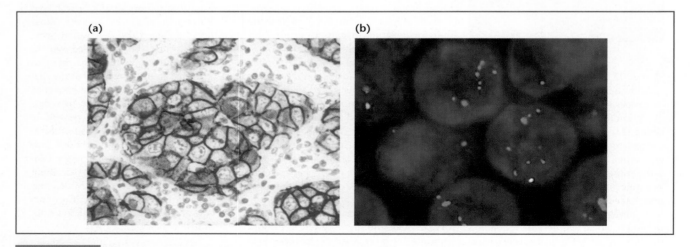

ST FIGURE 2   *HER-2* **gene and protein assays.** (a) Normal and breast cancer cells within a biopsy sample, stained by HER-2 immunohistochemistry. Cell nuclei are stained blue. Cancer cells that overexpress HER-2 protein stain brown at the cell membrane. (b) Cancer cells assayed for *HER-2* gene copy number by fluorescence *in situ* hybridization. Cancer cell nuclei appear blue under the fluorescence microscope and the *HER-2* gene DNA appears bright red. Chromosome 17 centromeres stain green. The degree of *HER-2* gene amplification is expressed as the ratio of red staining foci to green staining foci.

ST TABLE 1   Examples of Personalized Medicine Drugs and Diagnostics

Therapy	Gene Test	Description
Herceptin® (trastuzumab)	HER-2 amplification	Breast cancer test to accompany Herceptin use
Erbitux® (cetuximab)	EGFR expression, K-RAS mutations	Protein and mutation analysis prior to treatment
Gleevec® (imatinib)	BCR/ABL fusion	Gleevec used in treatment of Philadelphia chromosome+ chronic myelogenous leukemia
Gleevec® (imatinib)	C-KIT	Gleevec used in stomach cancers expressing mutated C-KIT
Tarceva® (erlotinib)	EGFR expression	Lung cancer for EGFR+ tumors
Drugs/surgery	MLH1, MSH2, MSH6	Gene mutations related to colon cancers
Hormone/chemotherapies	Oncotype DX® test	Selection of breast cancer patients for chemotherapy
Chemotherapies	Aviara Cancer TYPE ID®	Classifies 39 tumor types using gene-expression assays
Rituximab	PGx Predict® (FcRIIIa)	Detects CD-20 variants that predict response to rituximab in Non-Hodgkin's lymphoma

Reducing Adverse Drug Reactions

Every year, about 2 million people in the United States have serious side-effects from pharmaceutical drugs, and approximately 100,000 people die. The costs associated with these **adverse drug reactions (ADRs)** are estimated to be $136 billion annually. Although some ADRs result from drug misuse, others result from a patient's inherent physiological reactions to a drug.

Sequence variations in a large number of genes can affect drug responsiveness (**ST Table 2**). Of particular significance are the genes that encode the cytochrome P450 families of enzymes. These family members are encoded by 57 different genes. People with some cytochrome P450 gene variants metabolize and eliminate drugs slowly, which can lead to accumulations of the drug and overdose side-effects. In contrast, other people have variants that cause drugs to be eliminated quickly, leading to reduced effectiveness. An example of gene variants that affect drug responses is that of *CYP2D6* gene. This member of the cytochrome P450 fam-

ily encodes the debrisoquine hydroxylase enzyme, which is involved in the metabolism of approximately 25 percent of all pharmaceutical drugs, including diazepam, acetaminophen, clozapine, beta blockers, tamoxifen, and codeine. There are more than 70 variant alleles of this gene. Some mutations in this gene reduce the activity of the encoded enzyme, and others can increase it. Approximately 80 percent of people are homozygous or heterozygous for the wild-type *CYP2D6* gene and are known as extensive metabolizers (**ST Figure 3**). Approximately 10 to 15 percent of people are homozygous for alleles that decrease activity (poor metabolizers), and the remainder of the population have duplicated genes (ultra-rapid metabolizers). Poor metabolizers are at increased risk for ADRs, whereas ultra-rapid metabolizers may not receive sufficient dosages to have an effect on their conditions.

In 2005, the FDA approved a microarray gene test called the **AmpliChip® CYP450** assay (Roche Diagnostics) that detects 29 genetic variants of two cytochrome P450 genes—*CYP2D6* and *CYP2C19*. This test detects single-nucleotide

ST TABLE 2   Examples of Variant Gene Products that Affect Drug Responses

Gene Product	Variant Phenotype	Drugs Affected	Response
Acetyl transferase NAT2	Slow, rapid acetylators	Isoniazid, sulfamethazine, dapsone, paraminosalicylic acid, heterocyclic amines	Slow: toxic neuritis, lupus erythematosus, bladder cancer; Rapid: colorectal cancer
Thiopurine methyltransferase	Poor TPMT Methylators	6-mercaptopurine, 6-thioguanine, azathioprin	Bone marrow toxicity, liver damage
Catechol O-methyl transferase	High, low methylators	Levodopa, methyl dopa	Low or increased response
CYP2C19	Poor, extensive hydroxylators	Mephenytoin, hexobarbital, proguanil, etc.	Poor or increased toxicity, poor efficacy (proguanil)
β2 Adrenoceptor	Enhanced receptor downregulation	Albuterol, ventolin	Poor asthma control
5-HT2A serotonergic receptor	Multiple polymorphisms	Clozapine	Variable drug efficiencies
Multiple drug resistance transporter	Overexpression in cancer	Vinblastin, doxorubicin, paclitaxel, etc.	Drug resistance

Source: Adapted from Table 1 of Mancinelli, L., et al. 2000. Pharmacogenomics: The promise of personalized medicine. *AAPS PharmSci* 2(1): Article 4.

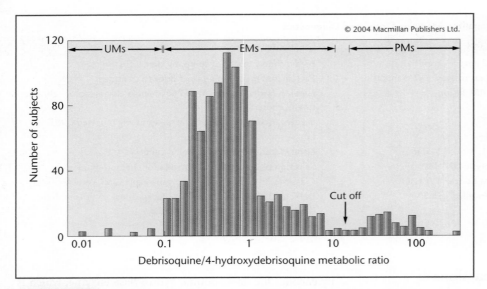

ST FIGURE 3 *CYP2D6* **pharmacogenetic profile in a Swedish population.** Individuals were tested for their ability to metabolize debrisoquine to 4-hydroxydebrisoquine, as an indication of the efficiency of debrisoquine hydroxylase enzyme activity. The population sample was divided into the categories UMs (ultra-rapid metabolizers), EMs (extensive metabolizers), and PMs (poor metabolizers). The "cut-off" label indicates the cut-off between extensive and poor metabolizers.

polymorphisms (SNPs) as well as gene duplications and deletions. The AmpliChip CYP450 assay is an example of a genotyping microarray. After scanning with an automated scanner, the data are analyzed by computer software, and the *CYP2D6/CYP2C19* genotype of the individual is generated.

Another example of pharmacogenomics in personalized medicine is that of the *CYP2C9* and *VKORC1* genes and the drug **warfarin.** Warfarin (also known as Coumadin) is an anticoagulant drug that is prescribed to prevent blood clots after surgery and to aid people with cardiovascular conditions who are prone to clots. Warfarin inhibits the vitamin K-dependent synthesis of several clotting factors. There is an approximately ten-fold variability between patients in the doses of warfarin that have a therapeutic response. In the past, physicians attempted to adjust the doses of warfarin through a trial-and-error process during the first year of treatment. If the dosage of warfarin is too high, the patient may experience serious hemorrhaging; if it is too low, the patient may develop life-threatening blood clots. It is estimated that 20 percent of patients are hospitalized during their first six months of treatment due to warfarin side-effects.

Variations in warfarin activity are affected by polymorphisms in several genes, particularly *CYP2C9* and *VKORC1.* Two single-nucleotide polymorphisms in *CYP2C9* lead to reduced elimination of warfarin and increased risk of hemorrhage. About 25 percent of Caucasians are heterozygous for one of these polymorphisms and 5 percent appear to be homozygous. About 5 percent of patients of Asian and African descent carry these variants. Patients who are heterozygous

or homozygous for some alleles of *CYP2C9* require a 10 to 90 percent lower dose of warfarin.

Recently, the FDA recommended the use of *CYP2C9* and *VKORC1* genetic tests to predict the likelihood that a patient may have an adverse reaction to warfarin. Several companies now offer tests to detect polymorphisms in these genes, using methods based on PCR amplification and allele-specific primers. It is estimated that the use of warfarin genetic tests could prevent 17,000 strokes and 85,000 serious hemorrhages per year. The savings in health care could reach $1.1 billion per year.

Pharmacogenomic tests and treatments, and the genetic information on which they are based, are rapidly advancing. A source of updated information on all aspects of pharmacogenomics can be found on the Pharmacogenomics Knowledge Base, which is described in **Box 2.**

Personalized Medicine and Disease Diagnosis

As of 2009, there were genetic tests for approximately 2000 different diseases. These tests are categorized according to their uses and can fall into one or more of the following groups: **Diagnostic tests** detect the presence or absence of gene variants linked to a suspected genetic disorder in a symptomatic patient; **predictive tests** detect a gene mutation in patients with a family history of having a known genetic disorder (for example, Huntington disease or BRCA-linked breast cancer); **carrier tests** help physicians identify patients who carry a gene mutation linked to a disorder that might be passed on to their offspring (for example Tay–Sachs or cystic fibrosis); **prenatal tests** detect potential genetic diseases in a fetus (for example, Down syndrome); and **preimplantation tests** are performed on early embryos in order to select embryos for implantation that do not carry a suspected disease. Most of these genetic tests detect the presence of known mutations in single genes that are linked to a disease **(ST Table 3).**

Although these genetic tests are extremely useful for detecting some future diseases and for guiding treatment, it is clear that most disorders are multifactorial and complex. It is likely that diseases such as diabetes, Alzheimer's, and heart

BOX 2

The Pharmacogenomics Knowledge Base (PharmGKB): Genes, Drugs, and Diseases on the Web

The Pharmacogenomics Knowledge Base (PharmGKB) is a publicly available Internet database and information source developed by Stanford University. It is funded by the National Institutes of Health (NIH) and forms part of the NIH Pharmacogenomics Research Network, a U.S. research consortium. The goal of PharmGKB is to provide researchers and the general public with information that will increase the understanding of how genetic variation contributes to an individual's reaction to drugs. On the PharmGKB Web site (see **ST Figure 4**), you may search for genes and more than 650 variants that affect drug reactions, information on a large number of drugs, diseases and their genetic links, pharmacogenomic pathways, gene tests, and relevant publications. Visit the PharmGKB Web site at http://www.pharmgkb.org.

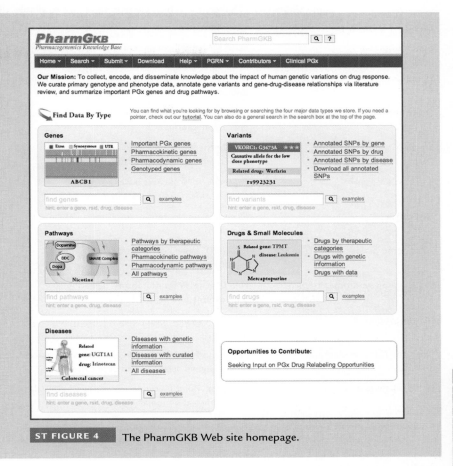

ST FIGURE 4 The PharmGKB Web site homepage.

disease are caused by interactions between many genes, as well as by factors contributed by epigenetic effects, lifestyle, and environment. These diseases tend to be chronic and have a significant burden on health-care systems.

Genome sequencing, SNP identification, and genome-wide association studies (GWAS) are beginning to reveal some of the DNA variants that may contribute to the risk of developing multifactorial diseases such as cancer, heart disease, and diabetes. For example, the Cancer Genome Atlas project is amassing data equivalent to 20,000 genome projects on normal and tumor DNA from patients with 20 different types of cancer. Such studies are revealing that cancers that were once classified broadly (such as "prostate cancer") are in fact many different diseases based on their genetic profiles. For example, in the past, blood cancers were categorized into two large groups: leukemias and lymphomas. Today, we know that each category can be broken down into more than 40 different types, based on gene mutation and expression characteristics. In addition, genome studies show that each individual tumor is unique genetically. This explosion in molecular information is now beginning to affect how these cancers are diagnosed and treated.

Although progress in genome science is accelerating, these studies have just begun to collect the vast amount of

genetic data required in order to make accurate predictions and treatment choices for many diseases. As whole-genome sequencing becomes faster and more economical, scientists predict that genomics and personal genome sequencing will become a significant part of personalized diagnosis and treatment by 2015.

But what will genome-based personalized medicine look like in 2015? And how will we use it to our benefit? Also, are there ethical and social questions that we need to address as we enter the age of the new personalized medicine? The next sections will address these questions.

Analyzing One Personal Genome

In 2010, the journal *Lancet* published a report illustrating the type of information that we can currently obtain from a personal genome sequence.[1] The patient was a healthy 40-year-old male who had a family history of arthritis, aortic aneurysm, coronary artery disease, and sudden cardiac death. The re-

---

[1]Ashley, E.A., et al. 2010. Clinical assessment incorporating a personal genome. *Lancet* 375: 1525–1535.

ST TABLE 3	Some Single-Gene Defects for Which Genetic Tests Are Available	
Disease	**Gene Mutation**	**Description**
Achondroplasia	FGFR3 gene. 99% of patients have a G to A point mutation at nucleotide 1138 (G380R substitution)	Abnormal bone growth
Hereditary breast/ovarian cancer	BRCA1 and BRCA2 genes. Deletions, duplications, and point mutations	Predisposition to breast, ovarian, prostate, and other cancers
Duchenne muscular dystrophy	DMD gene. Point mutations, deletions, insertions, splicing mutations	Early-onset progressive muscular weakness, heart disease
Fragile-X syndrome	FMR1 gene. Primarily expanded trinucleotide (CGG) repeats and loss of function	Mental retardation, developmental disorders
Friedrich's ataxia	FXN gene. 98% of cases have expanded trinucleotide (GAA) repeats in intron 1	Ataxia, muscle weakness, spasticity, heart and other organ dysfunctions
Hemophilia A	F8 gene. Point mutations, insertions, deletions, inversions	Factor VIII blood-clotting defects, bleeding
Huntington disease	HTT (HD) gene. Trinucleotide (CAG) repeat expansions	Midlife onset of progressive motor and cognitive disorders
Lesch-Nyhan syndrome	HPRT1 gene. Point mutations, deletions, duplications	Developmental, motor, and cognitive disorders
Marfan syndrome	FBN1 gene. Point mutations, splicing mutations, deletions	Connective tissue disorders affecting numerous organs
Polycystic kidney disease, dominant	PKD1 and PKD2 genes. Sequence variants, partial or whole-gene deletions and duplications	Cysts in kidney, liver, and other organs, vascular abnormalities
Sickle-cell disease	HBB gene. Point mutation leading to Glu to Val substitution at amino acid 6	Early-onset anemia

Source: GeneReviews at http://www.ncbi.nlm.nih.gov/bookshelf/br.fcgi?book=gene.

searchers obtained his genomic DNA sequence using a rapid single-molecule sequencing method (**Box 3**). By comparing the patient's sequence with other human genome sequences in databases, they discovered a total of 2.6 million SNPs and 752 copy number variations. The researchers then sorted through the genome sequence data to determine which of these variants might have an effect on phenotype. This was accomplished by searching known SNPs in several large databases, manually creating their own disease-associated SNP database, and calculating likelihood ratios for various disease risks. The analysis required the combined efforts of more than two dozen scientists and clinicians over a period of about a year, and information gleaned from more than a dozen sequence databases, new and existing sequence analysis tools, and hundreds of individually accessed research papers.

To determine how this patient may respond to pharmaceutical drugs, the researchers searched the PharmGKB database (see Box 2) for the presence of known variants within pharmacogenomically important genes. The patient was found to have 63 clinically relevant SNPs within genes associated with drug reactions. In addition, his genome contained six previously unknown SNPs that could alter amino acid sequences in drug-response genes. For example, the genome sequence revealed that this patient was heterozygous for a null mutation in the CYP2C19 gene. This muta-

tion could make him sensitive to a range of drugs, including those used to treat aspects of heart disease. He would also be more sensitive than normal to warfarin, based on SNPs within his VKORC1 and CYP4F2 genes. In contrast, the patient's sequence contained gene variants associated with good responses to statins; however, other gene variants suggested that he might require higher-than-normal statin dosages.

The search for mutations within genes that directly affect disease conditions revealed several potentially damaging variants. The patient was heterozygous for a SNP within the CFTR gene that would change a glycine to arginine at position 458. This mutation could lead to cystic fibrosis if it was passed on to a son or daughter who also inherited a defective CFTR gene from the other parent. Similarly, the patient was heterozygous for a recessive mutation in the hereditary haemochromatosis protein precursor gene (HFE), which is associated with the development of haemochromatosis, a serious condition leading to toxic accumulations of iron. Also, the patient was heterozygous for a recessive mutation in the solute carrier family 3 (SLC3A1) gene. This mutation is linked to cystinuria, an inherited disorder characterized by inadequate excretion of cysteine and development of kidney stones. The scientists discovered a heterozygous SNP within the parafibromin (CDC73) gene that would create a prematurely terminated protein. This gene is a tumor-suppressor

BOX 3

The Future of Personalized Genome Sequencing

The personal genome sequence described in this chapter was the first human genome to be sequenced using a method known as true single-molecule sequencing (tSMS™). Other methods such as the Sanger method and the second-generation high-throughput methods require cloning or PCR amplification of template DNA prior to sequencing. In contrast, the tSMS method directly sequences individual DNA strands with minimum processing. The sequencing of this genome took about a week, was performed with one machine and the services of three people, and cost $48,000. The sequence was that of Dr. Stephen Quake, a Stanford University professor who developed the technology and headed the research group.

Despite major advances in genome sequencing methods over the last decade, we still do not have rapid, cost-effective, whole-genome sequencing methods that are accurate enough for routine medical use. This may change in the near future, as many public and private companies are racing to develop such technologies. A partial list of these companies includes Helicos Biosciences, Illumina, Knome, 454 Life Sciences, Complete Genomics, and Pacific Biosciences.

In 2011, the X PRIZE Foundation announced its Archon Genomics X Prize. The competition winner will be awarded $10 million for sequencing 100 whole human genomes in 30 days or less at a cost of $1000 or less per genome. The error rate must be less than one error for every 1 million bases of DNA. The goal of the competition is to speed the development of rapid, accurate, and economical technologies for "medical grade" genomic sequencing.

Genetics, Technology, and Society essay 415

---

gene linked to the development of hyperparathyroidism and parathyroid tumors. The presence of this SNP increased the risk that the patient might develop these types of tumors, if any of the patient's cells experienced a loss-of-heterozygosity mutation in the other copy of the gene.

The analysis of this patient's genome sequence for the purpose of predicting future development of multifactorial disease was more challenging. Genome-wide association studies have revealed large numbers of sequence variants that are associated with complex diseases; however, each of these variants most often contributes only a small part of the susceptibility to disease. Because not all variants have been discovered or characterized, it is difficult to establish a numerical risk score for each of these diseases based on the presence of one or more SNPs. As an example, the researchers discovered SNPs within three genes (*TMEM43*, *DSP*, and *MYBPC3*) that may be associated with sudden cardiac death. However, the exact effects of two of these SNPs are still unclear, and the other SNP had not previously been described. The patient had five SNPs in genes associated with an increased risk of developing myocardial infarction and two SNPs associated with a lower risk. Among the SNPs associated with increased risk, a variant in the apolipoprotein A precursor (*LPA*) gene is associated with a five-fold increased plasma lipoprotein(a) concentration and a two-fold increased risk of coronary artery disease. By taking into consideration the simultaneous potential effects of many different SNPs, as well as the patient's own environmental and personal lifestyle factors, the researchers concluded that the patient's genetics contributed to a significantly increased risk for eight conditions (such as type 2 diabetes, obesity, and coronary artery disease) and a decreased risk for seven conditions (such as Alzheimer's disease). The patient was offered the services of clinical

geneticists, counselors, and clinical lab directors in order to help interpret the information generated from the genome sequence. Genetic counseling covered areas such as the psychological and reproductive implications of genetic disease risk, the possibilities of discrimination based on genetic test results, and the uncertainties in risk assessments.

Technical, Social, and Ethical Challenges

There are still many technical hurdles to overcome before personalized medicine will become a standard part of medical care. The technologies of genome sequencing and alignment, microarray analysis, and SNP detection need to be faster, more accurate, and cheaper. Scientists expect that these challenges will be overcome in the near future; however, personalized genome analysis needs to be used with caution until the technology becomes highly accurate and reliable. Even a low rate of error in genetic sequences or test results could lead to misdiagnoses and inappropriate treatments. Perhaps an even greater challenge lies in the ability of scientists to store and interpret the vast amount of emerging sequence data. Each personal genome generates the letter-equivalent of 200 large phone books, which must be stored in databases, mined for relevant sequence variants, and meaning assigned to each sequence variant. To undertake these kinds of analyses, scientists need to gather data from large-scale population genotyping studies that will link sequence variants to phenotype, disease, or drug responses. Experts suggest that such studies will take the coordinated efforts of public and private research teams and more than a decade

SPECIAL TOPIC

to complete. Scientists will also need to develop efficient automated systems and algorithms to deal with this massive amount of information. Moreover, these data analyses will have to consider that genetic variants contribute only partially to personal phenotype. Personalized medicine will also need to integrate information about environmental, personal lifestyle, and epigenetic factors.

Another technical challenge for personalized medicine is the development of automated health information technologies. Health-care providers will need to use electronic health records to store, retrieve, and analyze each patient's genomic profile, as well as to compare this information with constantly advancing knowledge about genes and disease. Currently, fewer than 10 percent of hospitals and physicians in the United States have access to these types of information technologies.

Personalized medicine has a number of societal implications. To make personalized medicine available to everyone, the costs of genetic tests, as well as the genetic counseling that accompanies them, must be reimbursed by insurance companies, even in cases where there is no prior disease or symptoms. Regulatory changes are required to ensure that genetic tests and genomic sequencing are accurate and that the data generated are reliably stored in databases that guarantee the patient's privacy. At the present time, less than 1 percent of genetic tests are regulated by agencies such as the FDA.

Personalized medicine also requires changes to medical education. In the future, physicians will be expected to use genomics information as part of their patient management. For this to be possible, medical schools will need to train future physicians to interpret and explain genetic data. In addition, more genetic counselors and genomics specialists will be required. These specialists will need to understand genomics and disease, as well as to manipulate bioinformatic data. As of 2010, there were only about 2500 genetic counselors and 1100 clinical geneticists in North America.

The ethical aspects of the new personalized medicine are also diverse and challenging. For example, it is sometimes argued that the costs involved in the development of genomics and personalized medicine are a misallocation of limited resources. Some argue that science should solve larger problems facing humanity, such as the distribution of food and clean water, before embarking on personalized medicine. Similarly, some critics argue that such highly specialized and expensive medical care will not be available to everyone and represents a worsening of economic inequality. There are also concerns about how we will protect the privacy of genome information that is contained in databases and private health-care records. In addition, there need to be effective ways to prevent discrimination in employment or insurance coverage, based on information derived from genomic analysis.

Most experts agree that we are at the beginning of a personalized medicine revolution. Information from genetics and genomics research is already increasing the effectiveness of drugs and enabling health-care providers to predict diseases prior to their occurrence. In the future, personalized medicine will touch almost every aspect of medical care. By addressing the upcoming challenges of the new personalized medicine, we can guide its use for the maximum benefit to the greatest number of people.

SELECTED READINGS AND RESOURCES

Journal Articles

Ashley, E.A., et al. 2010. Clinical assessment incorporating a personal genome. *Lancet* 375: 1525–1535.

Collins, F. 2010. Has the revolution arrived? *Nature* 464: 674–675.

Hamburg, M.A., and Collins, F.S. 2010. The path to personalized medicine. *New England Journal of Medicine* 363: 301–304.

Manolio, T.A., et al. 2009. Finding the missing heritability of complex diseases. *Nature* 461: 747–753.

Ormond, K.E., et al. 2010. Challenges in the clinical application of whole-genome sequencing. *Lancet* 375: 1749–1751.

Pushkarev, D., et al. 2009. Single-molecule sequencing of an individual human genome. *Nature Biotech* 27: 847–850.

Ross, J.S. 2009. The HER-2 receptor and breast cancer: Ten years of targeted anti-HER-2 therapy and personalized medicine. *The Oncologist* 14: 320–368.

Venter, J.C. 2010. Multiple personal genomes await. *Nature* 464: 676–677.

Weinshilboum, R., and Wang, L. 2004. Pharmacogenomics: Bench to bedside. *Nature Rev Drug Disc* 3: 739–748.

Web Sites

Personalized Medicine Coalition, 2011. The Case for Personalized Medicine. http://www.personalizedmedicinecoalition.org/sites/default/files/files/Case_for_PM_3rd_edition.pdf

University of Washington, 2010. GeneTests Web site: http://www.ncbi.nlm.nih.gov/sites/GeneTests/?db=GeneTests

U.S. Department of Energy Genome Program's Human Genome Project Information, 2008. Pharmacogenomics. http://www.ornl.gov/sci/techresources/Human_Genome/medicine/pharma.shtml

U.S. Food and Drug Administration, 2010. Table of Valid Genomic Biomarkers in the Context of Approved Drug Labels.

http://www.fda.gov/Drugs/ScienceResearch/ResearchAreas/Pharmacogenetics/ucm083378.htm

CREDITS

Credits are listed in order of appearance.

Photo

CO, Paphrag. ST-2 (a, b) Ren L. Ridolfi

Text

Source: Personalized Medicine Coalition. **STT.2**, With kind permission of Springer Science+Business media: *APPS Pharmsci* 2(1): Article 4, "Pharmacogenomics: The Promise of Personalized Medicince," 2000, Mancinelli, L., Table 1, Copyright 2000.

STF-1, Personalized Medicine Coalition. *The Case for Personalized Medicine*: 3ed, figure 2, Copyright © 2011. **STF-3**, "Pharmacogenomics: Bench to Bedside," by R. Weinshilboum and L. Wang, from *Nature Reviews Drug Discovery*, 3. Copyright © 2004 by Macmillan Publishers Ltd. Reprinted with permission.

Index

1
16S rRNA, 299, 302, 306, 564
18S rRNA, 299

3
30-nm fibers, 355
-35 Region, 283

4
454 Life Sciences, 427, 432, 577

5
5-bromouracil, 325, 339

A
A base, 197, 229, 292, 324, 341
a factor, 3, 7, 186, 189, 242, 255, 268, 510, 543
A protein, 1-3, 10-12, 20-21, 59, 123, 127-128, 138,
    207-208, 210-211, 272, 281, 311-312, 314,
    316, 321-322, 327, 329, 338, 341, 349, 359,
    362-363, 371, 376, 379, 383, 388, 396,
    418-421, 429, 437-440, 445-446, 450, 454,
    470, 485, 497, 513, 526, 539, 549
A site, 243, 302-304, 347, 409, 451, 467
ABO blood group, 83-84, 528, 533, 566
ABO blood type, 87, 89, 108
ABO blood types, 89, 107
ABO system, 83, 528
Abortion, 130, 139-140, 153-154
    spontaneous abortion, 140
Abscissa, 67
Absorption, 23, 214-215, 223-224, 228-229, 570
Absorption spectrum, 214-215
Accession number, 1, 226, 418
Acetaldehyde, 325
acetaminophen, 573
Acetyl group, 356
Achondroplasia, 71, 328, 338, 532, 544, 576
Acid, 1-14, 15, 20-22, 24, 28, 29, 81-82, 99, 104, 123,
    132, 143, 151, 155, 184, 186, 207-208,
    211-212, 214-218, 220, 223-224, 226,
    228-229, 232, 255-256, 258, 272-273,
    275-282, 285-286, 290, 292-294, 298-304,
    307, 309-314, 316-317, 321-325, 327, 338,
    341-342, 345-346, 351, 362, 367-369, 383,
    387, 393, 399, 404, 411, 413, 415, 418-419,
    423, 429, 431, 433, 437-438, 440-442, 446,
    454, 460, 464, 491, 495-496, 498, 523-524,
    538-540, 543, 545, 554, 573, 576
Acidity, 512
Acids, 3, 6-10, 13, 20-21, 28, 31, 84, 155, 184, 186,
    199, 207-209, 215-216, 220, 223, 225,
    228-229, 237, 255, 260, 262, 269, 272-282,
    286, 290-294, 298-299, 301-304, 306,
    308-312, 314, 316-317, 322, 338, 353-354,
    356, 370, 404, 433, 438, 446-447, 450,
    454-455, 488, 491, 539, 545, 548
    defined, 353
    pH of, 438
Acquired immunodeficiency, 7, 474, 526
Acquired immunodeficiency syndrome (AIDS), 7
Acridine, 1, 176
acridine dyes, 1
Acridine orange, 176
Actin, 21, 31, 313, 432, 439
Actinomycin, 291, 496
Action spectrum, 214-215
Activation, 6, 12, 21, 128, 251, 258, 265, 301, 340,
    344, 350, 352, 380, 388, 469, 487, 492, 552
Activator proteins, 356, 360
Activators, 344, 356, 359-362
Active site, 282, 288
    enzyme, 282, 288
ADA gene, 467-468, 473
Adaptation, 9, 319, 413, 536

evolutionary, 536
Adaptations, 529, 539
Adenine, 13, 20, 216-217, 220, 223, 227-229, 269,
    282-283, 289, 294, 302, 323-325, 328,
    339-340
    in DNA, 20, 216-217, 223, 227-228, 282, 289,
    324-325, 339-340
Adenine (A), 217, 228, 289
Adeno-associated virus (AAV), 467
Adenoma, 383
Adenosine, 31, 216, 218, 300, 350, 364, 467, 473
Adenosine deaminase, 467, 473
Adenosine deaminase (ADA), 467
Adenosine triphosphate (ATP), 31, 216, 218
Adenovirus, 425, 468-469
    genome, 469
Adenoviruses, 475-476
    size of, 476
Adenyl cyclase, 350, 370
Adherence, 62
Adhesion, 382-383, 423, 425, 437
Adipocytes, 420
ADP, 181, 183, 216, 357
Adsorption, 196, 201, 212
    in bacteriophages, 201
Adult stem cells, 249, 375, 553
Adults, 6, 23, 68, 100, 103-104, 118-119, 138, 387,
    459, 487, 492, 516, 529, 554
Aerobic cellular respiration, 103
Aerobic respiration, 324
Aflatoxin, 384
Africa, 411, 455, 460, 514, 540-542, 544, 563
African elephants, 426
Africans, 540, 542
    genomes of, 540
Agar, 186, 197-198, 209, 333, 395, 399
    blood, 209
    nutrient, 186, 197-198, 399
    properties of, 333
Agar plates, 395, 399
Agarose, 225, 392, 403, 405-406, 409
Agarose gel, 225, 392, 403, 405-406
Agarose gel electrophoresis, 403
Aging, 23, 247, 249, 431, 437
Agricultural biotechnology, 453-454
    plant, 453-454
Agricultural crops, 454
Agriculture, 15, 22, 24, 27, 203, 444, 449, 453-455,
    512-514, 558
    pest resistance, 454
    sustainable, 453
    transgenic plants, 203, 454
Agrobacterium, 202, 396, 413
Agrobacterium tumefaciens, 396
a-helix, 312
AIDS, 7, 148, 203, 216, 290, 431, 450, 474, 526, 531
    chimpanzees and, 431
    diagnosis, 450
    mortality, 531
    transmission, 7
    treatment, 148, 203, 290, 450
AIDS (acquired immunodeficiency syndrome), 474,
    526
Alanine, 275, 279, 292, 294, 300, 302, 309, 441
    genetic code, 275, 279, 292, 294
    structure, 292, 300, 309
    synthesis, 275, 279, 292, 294, 300, 302, 309
Albinism, 69-71, 74, 86-87, 307, 529, 534-535, 546
Albumin, 287, 439
Albuterol, 573
Alcohol, 384, 523, 546
Alcohol dehydrogenase, 523, 546
Alcohols, 217
Algae, 13, 31, 34-35, 116, 537
    brown, 537
    characteristics of, 13
    green, 34, 116

red, 31, 34, 116
    reproduction, 13, 34, 116
    unicellular, 13
Alkaptonuria, 71, 307
Alkylating agent, 325-327
alkylation, 325, 327, 341
Allele, 1, 4-6, 9, 11, 19, 28, 55, 59, 61, 65-66, 69-70,
    72-73, 77, 81-86, 88-92, 94-100, 104-108,
    110-111, 113, 125, 128, 131-132, 137, 155,
    160-161, 167, 169-172, 175, 180, 182,
    320-321, 329, 336, 340-341, 368, 379,
    381-383, 386-387, 404, 459-461, 464-465,
    473, 475-476, 486, 495, 501-502, 513, 515,
    518, 522-537, 542-546, 547, 550-552, 554,
    559-566
Alleles, 1, 3, 5-9, 11-13, 19, 28, 33, 43, 46, 55, 61, 63,
    72-75, 80-86, 88, 90, 92-93, 95, 97-98,
    100-101, 105-106, 108, 110, 112-113, 118,
    125-126, 131, 135, 149, 157, 159-161, 163,
    165-167, 169-175, 177, 179-180, 183, 247,
    268, 310, 319, 357, 380, 382-383, 386-387,
    449, 461-462, 467, 476, 484, 498, 500-503,
    506-511, 513, 515-519, 522-530, 532-533,
    535-536, 542-543, 545-546, 550-552, 554,
    558-560, 562-566, 573-574
    codominance, 3, 83, 86
    defined, 11, 19, 511, 523, 535-536
    dominant/recessive, 55, 80, 112, 545
    genetic variation and, 319
    homologous chromosomes and, 80
    incomplete dominance, 7, 82-83, 86, 106, 108, 501
    Punnett squares, 55
Allergies, 105, 518
Allergy, 511
Allolactose, 347
Allopatric speciation, 1
    process of, 1
Allopolyploids, 143
Alternation of generations, 43-44
Alternative RNA splicing, 363
Alternative splicing, 1, 128, 289, 293, 295, 362-364,
    367, 422-423, 437-438, 447, 498
Alzheimer's disease, 25, 267, 577
American Association for the Advancement of Science,
    317
Ames, Bruce, 1
Ames test, 1, 333-334, 340
Amines, 326, 573
Amino acid, 1-3, 5, 7-10, 12-14, 20-22, 28, 81-82, 104,
    123, 132, 143, 155, 186, 226, 272-273,
    275-282, 285-286, 290, 292-294, 298-304,
    307, 309-314, 316-317, 321-323, 327, 338,
    341-342, 345-346, 351, 362, 367-369, 387,
    411, 413, 418-419, 431, 433, 438, 441, 446,
    454, 460, 491, 495-496, 498, 523-524,
    538-540, 543, 545, 576
    amino group, 311-312, 323
    families, 5, 307, 387, 431, 438, 496
    nonpolar, 311
    R group, 311
    structure, 1-2, 7, 9-10, 12-14, 20, 155, 226, 292,
    298-300, 309-314, 316-317, 323, 338,
    346, 368, 418, 438, 491, 523, 538, 543
    synthesis, 2, 5, 9-10, 12-13, 20-22, 81, 104,
    272-273, 275-276, 278-282, 285, 290,
    292-294, 298, 300, 302, 304, 307, 309,
    317, 321, 342, 345-346, 351, 362,
    368-369, 496
Amino acids, 3, 6-8, 10, 20-21, 28, 155, 186, 199, 208,
    223, 237, 255, 260, 262, 272-282, 291-294,
    298-299, 301-304, 306, 308-312, 314,
    316-317, 322, 338, 353-354, 356, 433, 438,
    447, 450, 454, 488, 491, 539, 545, 548
    and transfer RNA (tRNA), 208, 223
    arrangement of, 3, 7, 20, 155, 312, 488
    biosynthesis of, 309, 353
    codons, 3, 21, 272-273, 275-276, 278-281,

291-294, 298, 302-304
essential, 8, 186, 199, 223, 237, 260, 262, 274, 276, 278, 280, 282, 298-299, 302, 306, 308, 310-312, 314, 316, 322, 353-354, 356, 491, 548
genetic code and, 272-282, 291-294
in protein structure, 311
in protein synthesis, 281, 302
list of, 450
structure, 6-7, 10, 20, 155, 208, 223, 255, 260, 262, 291-292, 298-299, 306, 309-312, 314, 316-317, 338, 353-354, 356, 438, 491, 548
Amino group, 228, 311-312, 323
Aminoacyl tRNA, 1, 301-302, 317
Ammonia, 542
Ammonium, 232
Amniocentesis, 1, 130, 139, 459-461
Amniotic fluid, 139, 459-460
AMP, 194, 351, 425, 496
Ampicillin, 194, 395, 397, 412
resistance, 194, 395, 397, 412
Ampicillin resistance, 395, 397, 412
ampicillin resistance gene, 395, 397, 412
amplified DNA, 376, 401, 461, 559-560, 562
ampR gene, 395
Amylase, 337
Amyotrophic lateral sclerosis (ALS), 23, 424
Anabolic pathway, 352
Anaerobe, 103
Anaphase, 2, 4, 32, 35-37, 39-41, 43, 47-48, 64, 127, 133, 139, 141-142, 148-149, 156-157, 235, 268
in meiosis, 39-40, 47-48, 139, 142, 148, 156
in mitosis, 39-40, 47
meiosis, 2, 4, 32, 35-37, 39-41, 43, 47-48, 64, 127, 133, 139, 142, 148-149, 156-157
mitosis, 2, 32, 35-37, 39-41, 43, 47-48, 141, 235
of meiosis, 2, 40-41, 43, 47-48, 133
of mitosis, 32, 35, 37, 39, 47-48
Anaphase I, 39-41, 43, 47-48, 133, 139, 157
Anaphase II, 40-41, 47
Androgen, 553
Androgens, 129
Anemia, 21, 23, 71, 309-310, 313, 316, 322, 332, 340, 424-425, 443, 451, 459-462, 474-475, 576
hemolytic, 425, 443
sickle-cell, 21, 23, 71, 309-310, 313, 316, 322, 424, 459-462, 576
types of, 332, 340, 462, 475
Aneuploidies, 430
Aneurysm, 93, 575
Angiogenesis, 138
Animal cells, 30-31, 35, 38, 42-43
Animal virus, 281
Animal viruses, 216, 223, 286
classes of, 223
Animals, 5, 16, 22-25, 27-28, 33, 38, 43, 56, 84, 86, 110, 117, 123, 135, 137, 143, 148, 215, 256, 313, 354, 364, 383-384, 449-451, 453, 455-457, 474, 480-481, 489, 491, 495, 508, 523, 551, 554, 558
research, 22, 24-25, 38, 84, 148, 215, 364, 449, 456, 495, 554
rise of, 489
transgenic animals, 22-23, 215, 449, 451, 455-457
Annealing, 10, 394, 401, 412
Annotation, 1, 418-420, 423, 438, 444-446
Antennae, 486, 497
Anterior end, 488, 496-497
Anterior-posterior axis, 5, 110, 483-484, 488-489
Anthranilic acid, 316
Anthrax, 203, 465, 543
vaccination, 203
Antibiotic, 7, 11, 111, 187, 194-195, 203, 227, 291, 316, 337, 394-395, 397, 452, 457-458, 476
natural, 457-458, 476
production, 195, 203, 227, 397, 452
protein synthesis, 111, 195
Antibiotic resistance, 7, 11, 187, 195, 203, 227, 337, 394, 397, 452, 458
genes, 7, 11, 187, 195, 203, 227, 337, 397, 458
plasmids and, 397
Antibiotics, 194-195, 203, 211, 316, 335, 384, 395-397, 412, 456, 470
and bacteria, 195
and viruses, 316
discovery of, 194, 335
resistance to, 194-195, 335, 395, 412

Antibodies, 2, 84, 202, 441, 451-452, 474, 572
diagram, 2
monoclonal, 451, 572
monoclonal antibodies, 451, 572
structure of, 2
Antibody, 14, 432, 474, 572
immunity, 432, 474
natural, 474
production, 474
anticancer drugs, 249
Anticodon, 1, 14, 276-277, 279, 281, 292, 299-301, 303, 306, 316-317
Anticodon loop, 300-301
Antigen, 83-84, 88, 110, 131, 202, 474
homologous, 83
Antigens, 31, 83-84, 88, 452
types of, 84, 452
Antioxidants, 455
antisense DNA, 290
Antisense nucleic acids, 290
RNA, 290
Antisense RNA, 1, 223, 290, 364-366, 453
Antisense RNA technology, 453
Antithrombin, 451
Aorta, 93
Apes, 411
Apoptosis, 1, 152, 340, 342, 373-374, 377, 379-381, 384, 387-388, 497, 553
in morphogenesis, 388
Appetite, 100, 415, 420
aquatic microorganisms, 434
Aquifex, 429
Aquifex aeolicus, 429
genome, 429
Arabidopsis, 25, 34, 256, 365, 428, 430, 481, 489-491, 496-497, 524
Arabidopsis thaliana, 25, 34, 428, 430, 481, 489-490, 524
Arber, Werner, 392
Archaea, 7, 273, 280, 354, 429, 435, 446
chromosomes, 7, 429, 446
cytoplasm, 273
DNA, 7, 273, 280, 429, 435, 446
evolution, 446
gene transfer, 7
genome of, 429, 446
metabolism, 7
protein synthesis, 273, 280, 429
replication, 7
RNA polymerase, 354
thermophilic, 446
transcription, 7, 273, 280, 354, 429
translation, 273, 280
viruses, 273, 280, 435
Archaeoglobus, 429
Arginine, 260, 275, 279-280, 292, 294, 322, 328, 576
genetic code, 275, 279-280, 292, 294
structure, 260, 292
synthesis, 275, 279-280, 292, 294
Aristotle, 16
Arms, 3, 8-9, 12, 32, 37, 120, 149-150, 176, 262, 268, 270, 356, 377, 532, 565
Arteries, 437
coronary, 437
structure and function, 437
Arterioles, 23
Arthritis, 175, 307, 450-451, 466, 470, 475, 571, 575
rheumatoid, 175, 307, 451, 475
Arthrobacter, 393
Arthrobacter luteus, 393
Artificial chromosomes, 396
Artificial selection, 1, 16, 453, 508-509, 511-513, 517, 523, 531
Asbestos, 384
Ascospore, 1
Ascospores, 1
Ascus, 1
Asexual reproduction, 34, 116
Asia, 455, 470, 533, 540-542
Asparagine, 275-276, 292-293, 295
genetic code, 275-276, 292-293, 295
structure, 292
synthesis, 275-276, 292-293
Aspergillus, 34
Aspergillus nidulans, 34
Assembly, 4, 9, 21, 31, 37, 99, 196, 212, 238, 244, 294, 311, 361, 379, 416-417, 434, 458
Assisted reproductive technologies, 551
Asthma, 16, 443, 547, 571, 573

treatment, 571, 573
Atherosclerosis, 443
Atom, 3, 13, 101, 216, 220-222, 311-313, 325
Atomic number, 7
Atomic structure, 218
Atoms, 7, 216, 220-221, 227, 262, 312, 326-327
isotopes, 7
structure, 7, 216, 220-221, 227, 262, 312
ATP, 31, 37, 196, 216-218, 239, 301-302, 350, 357, 388
synthesis of, 196, 239
use, 37, 388
Atrophy, 104-105, 249
Attachment, 7, 37, 212, 246, 317, 356, 378, 398
virus, 7
Attenuated vaccines, 452
Attenuation, 353-354, 368-369
Attenuator, 1, 352-353
Australia, 541, 562, 565
Autism, 148, 178, 466
Autoimmune diseases, 466
multiple sclerosis, 466
Automated DNA sequencing, 407-408, 415, 417, 561-562
Autopolyploids, 141, 143
Autoradiography, 1, 10, 12, 224-225, 234-236, 258, 264, 368, 400, 408
Autosomal genes, 94, 96-97, 108, 161, 170, 342
Autosome, 97, 108, 119, 128, 137, 161, 426
Autosomes, 1, 8, 11, 94, 107-108, 119-121, 123-124, 127-128, 131, 133, 137, 155, 161, 165, 182, 320, 421, 426, 431, 493
Auxotroph, 1, 186-187
auxotrophs, 11, 187-188, 333
Avery, Oswald, 19, 209
Avian influenza, 465
Axes of symmetry, 482
Axon, 363

B
B cells, 201
Bacillus, 211, 354, 393, 429, 454, 465, 543
B. anthracis, 543
Bacillus amyloliquefaciens, 393
Bacillus anthracis, 465, 543
Bacillus megaterium, 429
Bacillus subtilis, 211, 354, 429
genome, 354, 429
Bacillus thuringiensis, 454
Bacillus thuringiensis (Bt), 454
Backbones, 221
Bacteria, 1, 3, 7, 11, 13, 19, 22, 25, 30-32, 82, 99, 116, 185-205, 208-209, 212-215, 224, 227, 229, 234, 236, 238-240, 242-244, 251-252, 254-256, 260, 263, 269-270, 272, 274, 279-280, 282-285, 288, 292, 300, 302, 306, 313, 316, 321, 328-329, 333-335, 337, 341, 345, 347, 351, 354, 368, 392-397, 400, 402, 412-413, 417, 427-430, 434-435, 447, 450-452, 454, 458-459, 465, 474, 476
antibiotic-resistant, 203, 316, 476
antibiotics and, 396
cancer and, 25
cell wall of, 196
cholera, 202-203, 429, 452
chromosomes of, 202, 255, 263, 269
cloning of, 3, 458
conjugation in, 187
culture of, 186, 204, 251
cytoplasm, 3, 19, 30-31, 185, 189, 192, 194-198, 200, 202, 214, 256, 284-285, 306, 313
differential staining, 263
DNA replication in, 236, 238, 242, 244
domain, 354
eukaryotes compared, 272
gene expression in, 345, 354, 435
genomes of, 1, 334, 392, 417, 427-428, 430, 434
morphology, 197, 201, 209, 321
regulation of gene expression in, 354
reproduction of, 13, 185, 201, 213
size of, 3, 7, 199, 224, 252, 255-256, 269-270, 306, 394, 413, 430, 476
soil, 396, 434, 454
strains of, 7, 22, 188, 190-191, 193, 199, 201-204, 209, 242, 251, 316, 333-335, 396, 452, 474
structure of, 19, 30, 196, 208-209, 227, 229, 255, 269, 292, 368, 427
tetanus, 203

transcription in, 202, 272, 283-285, 451
transgenic bacteria, 450
translation in, 280, 302, 306
Bacterial artificial chromosomes, 396
Bacterial artificial chromosomes (BACs), 396
Bacterial cells, 3, 5, 10-11, 14, 22, 195-196, 198, 201, 210, 212-213, 232, 244, 274, 280, 345, 394-397, 399, 429, 450
bacterial chromosome, 5, 10, 31, 185, 188, 191-192, 195, 197-198, 200, 205, 236, 255, 345-347
Bacterial chromosomes, 247, 255-256, 269, 334-335, 413, 422
Bacterial conjugation, 5
bacterial diseases, 202-203
Bacterial genetics, 193
Bacterial genome, 429
bacterial species, 187, 194-195, 429, 434, 459
Bacterial transformation, 394
Bacterial viruses, 99, 196, 201, 213, 228
Bacteriophage, 1, 8, 10, 13-14, 185, 196-199, 201, 204-205, 211-212, 236, 240, 255-257, 269-270, 280, 291
    double-stranded DNA, 255, 270
    Escherichia coli, 211, 255-256
    MS2, 280, 291
    plaque assay, 196-198, 204-205
    RNA, 1, 8, 10, 13-14, 196, 211, 240, 255-256, 269, 280, 291
    single-stranded DNA, 10, 13, 255
    temperate, 8, 13, 198-199
    transposable, 14, 270
    virulent, 14, 198
bacteriophage lambda, 236, 255
Bacteriophages, 185-205, 209, 214, 216, 227, 236, 245, 255-256, 263, 272, 434
Baculovirus, 451
Balance, 59, 121, 128, 130, 133, 140, 328, 432, 494
Baldness, 12, 97, 112
Ball, 393
Barley, 320, 334
Barr body, 1, 4, 124-125, 132, 263
Basal, 31-32, 357-358, 360, 382
Basal body, 31
    eukaryotic cell, 31
Base, 1-5, 8-9, 12-14, 20, 59, 103, 127, 148, 175, 178, 188, 196-198, 216-224, 226, 228-229, 231, 236-237, 243-246, 251-252, 257-258, 261-263, 265-267, 269-276, 279, 282, 284-286, 288, 291-292, 294, 299-300, 302-303, 306-307, 317, 319-328, 330, 333-334, 339-341, 357-358, 387-388, 394, 398, 406, 408, 410, 417-419, 421, 423-424, 430, 444-445, 475-476, 515, 523, 540-541, 545, 548, 559, 562-564, 574-575
Base analog, 2, 339
Base pairing, 1, 3, 219, 221, 223, 229, 245, 288, 294, 300, 306, 324-325, 327, 394
    in DNA replication, 221, 223
base pairs, 4-5, 8, 13-14, 20, 59, 103, 127, 148, 178, 221-224, 226, 231, 236, 243-245, 251, 257-258, 261-263, 265-267, 269-270, 284, 286, 291, 300, 306, 320, 325, 334, 358, 394, 398, 417-418, 421, 424, 430, 444-445, 523, 540-541, 559, 563-564
Base sequence, 9, 14, 285, 326, 419
Base-pair substitution, 320
Base-pair substitutions, 333
Bases, 3, 5, 13, 20, 99, 152, 175, 207, 216-222, 226-229, 232-233, 239, 247, 267, 273, 279, 291, 294, 300-301, 304, 307, 320-324, 326-327, 341, 356, 369, 403, 408, 410, 415, 417-418, 420, 425-426, 430, 445, 548, 577
    complementary, 13, 20, 207, 222, 227, 229, 232, 239, 247, 273, 291, 294, 300, 323, 403, 445
    defined, 152, 267, 320
b-Carotene, 455
Beadle, George, 308
Beagle, 17, 434
Beaks, 531
Beans, 96
Bears, 105, 118, 131, 195, 215, 288, 316, 349, 426
beer, 250, 449
Bees, 428
Beetles, 522
Behavioral isolation, 536
Benign tumor, 374, 388
Benign tumors, 23, 177
b-galactosidase, 346-347, 349-350, 369, 395, 450-452

bicoid gene, 497
Biochemical pathways, 442
Biochemistry, 308, 314, 393
Biodiesel, 453
Biodiesel fuel, 453
Biodiversity, 471
Bioenergetics, 216
Bioethics, 130
Biofuels, 459
Bioinformatics, 2, 24, 45, 226, 314, 338, 403, 415-447, 539, 556
Biological molecules, 8
    proteins, 8
Biological species concept, 536
Biology, 3, 6, 13, 16-17, 22, 25-26, 30, 45, 53, 113, 126, 207, 209, 216, 221-222, 236, 252, 270, 290, 364, 371, 392-393, 401-402, 409-410, 413, 415, 428, 436, 442-443, 445, 447, 449, 458-459, 472, 480-481, 496, 498, 513, 523, 570
    genetics in, 17
Biomarkers, 437, 442, 578
Biomineralization, 437
Bioprospecting, 444
Biopsy, 125, 572
Bioreactors, 449, 451, 456
Bioremediation, 459
    overview of, 459
Biosensors, 441
biosynthesis, 9, 82, 309, 325, 340, 351, 353, 367, 369, 454
    amino acids, 309, 353, 454
    nucleotides, 9, 325, 353
Biosynthetic pathways, 345
Biotechnology, 2, 15, 22-28, 45, 178, 209, 268, 290, 364, 366, 385, 396, 400, 418, 434, 444, 449-477, 564, 572
    agricultural, 22, 27, 453-454
    analyzing gene expression, 463
    aquatic, 434
    DNA cloning, 22, 26
    DNA sequencing, 418, 461, 473
    experiments, 26, 209, 459, 474
    genetic engineering and, 449-477
    genetic testing and, 24, 471
    medical, 15, 22, 28, 209, 434, 450, 459, 467, 469, 471, 473, 475-476, 564
    microbial, 434, 459
    overview, 458-459, 470
    plant, 15, 25, 364, 396, 451, 453-454, 459, 471, 474, 476
    regulation of, 364, 366, 469
    tools, 45, 418, 453, 464, 472
    types of, 209, 268, 366, 396, 452, 462, 464, 475, 572
Biotechnology companies, 290, 450, 471, 564
    distribution of, 471
    production in, 450
Biotechnology industry, 22, 27, 396, 400, 450
Bioterrorism, 227, 410, 465
Biotin, 188
Bioweapons, 441, 456
Bird flu, 465
Birds, 120, 129, 441, 505-506, 531, 565
Birth defects, 150, 551
Bites, 566
Black bread mold, 34
Blades, 564
BLAST, 226, 314, 338, 366-367, 418-420, 438, 444-445
BLAST (Basic local alignment search tool), 418
Blastocyst, 4, 125, 495
Blastocysts, 495
Blindness, 71, 94-96, 99, 103, 107-108, 126, 430, 469-470
Blood, 9, 16, 21, 23, 31, 34, 70, 83-84, 86-89, 93, 99, 107-108, 110, 133, 139, 209, 214, 291, 309-310, 366, 374, 376-377, 382, 385, 387, 401, 411, 424, 426, 432, 439, 441, 443, 446, 450-451, 457, 459-461, 464, 467-470, 473, 486, 492, 497, 500, 510-511, 518-519, 528, 533, 542, 555, 558-559, 566, 574-576
    circulation, 21, 139
    components, 9, 21, 31, 310, 382, 387, 446, 457, 459, 467, 492, 497, 510, 519
    composition of, 16
    culture, 16, 209, 291, 439, 460, 468
    functions of, 387, 424
    pH of, 439

plasma, 31, 34, 70, 387, 446, 492, 497
pressure, 424, 426
types, 31, 83-84, 86-87, 89, 107-108, 110, 209, 214, 366, 374, 376-377, 382, 446, 450, 464, 510-511, 528, 542, 555, 575
vessels, 21, 93, 441, 470, 497
Blood cells, 21, 23, 31, 34, 83, 93, 214, 309-310, 366, 374, 376-377, 382, 432, 459-461, 464, 467-468, 473, 486
Blood pressure, 424, 426
Blood transfusions, 84, 291
Blood types, 83-84, 86-87, 89, 107-108, 511, 528
Blood typing, 558
Blood vessels, 21, 441, 470, 497
Bloom, 176-177
Blue-white selection, 397
Blunt ends, 393
Body axes, 491
Body plans, 488
Body temperature, 21, 99
    blood and, 99
Bond, 7, 9, 13, 217, 219-221, 223, 229, 241, 247, 252, 255, 260, 299, 303-304, 311-312, 323, 330, 406-407
    covalent, 9, 13, 252, 303, 311-312
    double, 9, 219, 221, 223, 229, 247, 252, 255, 323, 330, 406
    glycosidic, 323, 330
    peptide, 299, 303-304, 311
    triple, 219, 229
Bond angles, 219
Bonds, 4, 9-10, 20, 216, 219-223, 228-229, 232, 239, 245, 247, 282, 284, 300, 311-312, 323, 326, 329, 340, 394-395, 447, 450, 452
    hydrogen, 4, 20, 216, 219-223, 228-229, 232, 239, 245, 247, 284, 300, 312, 323, 340, 394, 447
    phosphodiester, 4, 9-10, 219, 282, 284, 326, 394-395
Bone, 123, 177, 250, 310, 377, 381, 387, 426, 441, 443, 467, 486, 492, 494, 559, 564, 573, 576
    marrow, 177, 250, 310, 377, 387, 441, 467, 573
Bone marrow, 177, 250, 310, 377, 387, 441, 467, 573
Bone marrow cells, 177, 441
Bones, 16, 93, 441, 486, 489
    development of, 489
Borrelia, 429
Borrelia burgdorferi, 429
    genome, 429
Bound ribosomes, 313
Boundaries, 156, 419, 484-486
Boveri, Theodor, 18, 63, 159
Bovine spongiform encephalopathy, 10
Brain, 21, 23, 25, 65, 94, 99, 103, 152, 293, 310, 313, 315, 322, 362-363, 375, 380-381, 398, 420, 431, 434, 469, 475, 494, 555
    components of, 21, 420
    development, 23, 25, 65, 99, 103, 322, 363, 375, 381, 398, 431, 469, 475, 494, 555
    tumors, 23, 152, 375, 380-381
Brains, 313, 431
Branch points, 59
BRCA, 385
BRCA1 and BRCA2 genes, 475-476, 576
Bread mold, 34, 102, 308, 539
Breast cancer, 23, 82, 337, 340, 342, 381-382, 384-388, 425, 443, 462, 472, 497, 571-574, 578
Breasts, 120, 385
Breathing, 23
Brown algae, 537
Brownian movement, 306
Bt crops, 454-455
Buffer, 400, 404-405
Buffers, 279
Bugs, 263
Burkitt's lymphoma, 374-375, 384

C
C3, 295
C5, 488
CAAT box, 2, 358, 419-420
cabbage, 555
Cadmium, 359
Caenorhabditis elegans, 25, 34, 117-118, 364, 428, 430, 481, 492
Calcitonin, 362
Calcium, 362, 395, 423
    ions, 395

Calcium ions, 395
Cambrian period, 431
Camels, 16
Cancer, 2, 4, 8, 23, 25, 35, 45, 82, 105, 138-139, 148,
    151-152, 249, 251-252, 265, 270, 290, 319,
    326-329, 332, 334, 337, 340-342, 344, 354,
    357, 368, 370, 373-389, 398, 406, 425, 427,
    430-431, 438, 440, 442-443, 446, 451-452,
    462-464, 467-468, 470-472, 474, 476,
    497-498, 511, 518, 547, 549, 551-555,
    571-576, 578
    and mutations, 386, 553
    bacteria and, 334
    brain tumors, 381
    breast, 23, 82, 152, 326, 332, 337, 340, 342, 375,
        380-382, 384-388, 425, 443, 462, 464,
        471-472, 497, 552-553, 571-574, 576,
        578
    cells, 2, 4, 8, 23, 25, 35, 45, 139, 151-152, 249,
        251, 265, 290, 326-327, 329, 332, 334,
        337, 340-341, 354, 357, 368, 370,
        373-388, 398, 406, 431, 438, 440, 442,
        446, 451-452, 463-464, 467-468, 470,
        474, 476, 497-498, 547, 549, 551-554,
        571-572
    cervical, 380, 384, 452
    chemotherapy, 386, 464, 553, 571-573
    colorectal, 376-377, 380, 553, 555, 573
    common types of, 328
    defined, 148, 152, 251, 374, 385, 511
    detection of, 340, 462
    development of, 4, 8, 25, 45, 105, 139, 326, 334,
        337, 373-379, 381-384, 386-387, 431,
        438, 446, 468, 471, 474, 511, 549,
        552-554, 576, 578
    early detection of, 340
    factors leading to, 446
    fiber and, 270
    gene therapy and, 468
    genetic testing for, 385
    hepatic, 468
    immune system and, 467, 474
    immunotherapy, 388
    liver, 152, 334, 377, 384, 431, 452, 468, 554, 573,
        576
    lung, 138, 148, 152, 340, 380-381, 552, 571, 573
    oncogenes, 370, 373, 379-380, 382-384, 386-388,
        552-553
    oncogenes in, 386
    ovarian, 332, 380, 385, 387, 464, 471-472,
        552-553, 576
    prostate, 152, 329, 342, 373, 375, 380, 384, 430,
        443, 464, 553, 575-576
    risk factors for, 383, 552
    skin, 23, 138, 341-342, 376-377, 518
    stem cell hypothesis, 375
    stomach, 152, 326, 553, 573
    telomerase in, 249, 265
    testicular, 151
    treatment of, 467, 553-554, 571-573
    treatments, 290, 373-374, 387-388, 442, 464, 467,
        471, 547, 571-572, 574
    vaccines, 431, 452
Cancer Genome Atlas, 575
Cancer Genome Atlas Project, 575
CAP binding site, 371
Capillaries, 21, 23
    blood, 21, 23
Capillary gel, 407-408
Capsule, 91, 137, 209, 227
Capsules, 210
Carbohydrate, 10, 84, 175
    structure, 10
Carbohydrates, 211, 382, 424, 438
    functions of, 424
Carbon, 8-9, 12, 101, 126, 186, 216, 227-228, 237,
    286, 308, 311, 350, 370
    nutrient, 186
Carbon (C), 311
Carbon source, 186, 350, 370
    culture medium, 186
Carboxyl group, 311-312
Carcinogen, 384
Carcinogenesis, 384, 553
Carcinogens, 334, 375-376, 384
    environmental factors, 384
    nitrosamines, 384
Carcinoma, 384, 443
Cardiac, 145, 363, 575, 577

muscle, 363
Cardiac muscle, 363
Cardiovascular, 25, 428, 430-431, 443, 464, 467, 510,
    549, 574
    disease, 25, 428, 430-431, 443, 464, 467, 510,
        549, 574
    system, 25, 467
Cardiovascular disease, 25, 428, 430-431, 464, 467,
    510, 549
Carotene, 455
Carotenoid, 308, 455
Carpels, 489-491
Carrier, 1-2, 4, 19, 22, 28, 69, 111, 150-152, 157, 324,
    423, 460-461, 475, 536, 546, 574, 576
Carriers, 2, 19, 23, 65, 69, 96, 105, 111, 144, 152, 310,
    385, 404, 459-460, 475, 483, 525, 534-535,
    542-543, 546, 570
Carroll, Sean, 498
Cartilage, 123, 486
Casein, 451
Catabolite repression, 350-351, 369
Catalyst, 454
Catalysts, 21, 314
Cataracts, 100, 424
Catechol, 573
Cation, 252
Cations, 220
Cats, 98, 108, 124-126
cattle, 10, 22, 34, 97, 107, 109-110, 113, 313,
    456-457, 474, 500, 508-509, 512
cauliflower mosaic virus, 454
Cavity, 460
CCR5 protein, 526
CD4 protein, 474
cDNA, 2, 5, 314-315, 368, 387, 398-399, 403, 411,
    435-436, 463-464
cDNA (complementary DNA), 2
cDNA libraries, 398, 403, 411
cDNA library, 398, 411
Cech, Thomas, 288
Celera Genomics, 422
Cell, 1-14, 16, 18, 21-24, 28, 29-45, 47-48, 62-63, 65,
    71, 81-82, 84, 98-101, 103, 111, 118, 121,
    125-126, 130-132, 141-142, 144, 146-148,
    151, 156-157, 185-186, 188-189, 191-200,
    204-205, 208-209, 211-212, 214-216, 223,
    227, 229, 231, 233-236, 238, 243-244,
    246-247, 249, 251-252, 254-256, 259-260,
    262-263, 268-269, 274-275, 277, 280-282,
    284, 290, 292, 295, 299-300, 307, 309-311,
    313-314, 316, 319-320, 322, 327, 329,
    331-333, 337, 339-342, 344-347, 349-351,
    353-356, 359-360, 362-366, 369-370,
    373-384, 386-388, 394-398, 403-405, 410,
    413, 415, 423-427, 429, 432, 435-438, 442,
    445-447, 450-455, 457-464, 467, 469,
    471-476, 480-483, 489-498, 510, 513, 528,
    545, 547-550, 552-555, 561, 563, 571-572,
    576
    characteristics, 13, 24, 47, 71, 121, 130, 194,
        208-209, 229, 238, 251, 255, 263, 320,
        349, 374, 377, 413, 429, 445, 453, 455,
        457, 459, 471, 473, 494, 510, 571
    chemical composition, 31, 208, 223
    origin, 2, 6, 14, 31, 40, 47, 100-101, 125, 141-142,
        146-147, 156, 191-192, 204, 235-236,
        243, 252, 263, 300, 319, 374-375, 395,
        427, 489, 491, 550, 552, 554, 563, 571
    primitive, 16, 31, 494-495
    structure, 1-2, 4, 6-7, 9-14, 16, 18, 24, 30-31, 33,
        35, 37-38, 125, 185, 188, 196, 204,
        208-209, 211-212, 214-216, 223, 227,
        229, 231, 238, 243, 246-247, 254-256,
        259-260, 262-263, 268-269, 284, 292,
        299-300, 309-311, 313-314, 316, 320,
        340, 346, 353-356, 359-360, 364-365,
        376, 386, 395-396, 403, 427, 437-438,
        442, 491, 497, 547-548, 550
Cell adhesion molecule, 425
Cell culture, 151, 460
cell cultures, 331
Cell cycle, 2, 5, 7, 28, 29, 31, 34-35, 38, 45, 47, 81,
    142, 208, 235, 243-244, 254, 260, 262-263,
    333, 341, 373, 377-381, 383-384, 386-388,
    476, 497, 552-553
    and meiosis, 2, 29, 31, 34-35, 38, 45, 47, 208, 254
    and mitosis, 34-35, 45, 388
    phases of, 2, 31, 35, 47, 260, 379, 381
    regulation of, 81, 262, 378, 384, 552

Cell death, 1, 99, 319, 332, 340, 342, 344, 364, 377,
    379-380
    programmed, 340, 342, 364, 377, 379-380
Cell differentiation, 34, 383, 548, 555
Cell division, 1-4, 7-9, 14, 18, 29-32, 34, 38, 43-44, 47,
    62, 100, 194-195, 231, 234, 247, 254, 260,
    263, 268, 345, 373-375, 378-379, 384, 388,
    442, 493, 513, 552, 561
    and meiosis, 2-3, 18, 29-32, 34, 38, 43-44, 47, 254
    and mitosis, 34, 388
    bacterial, 1, 3, 8-9, 14, 31, 194-195, 247, 260, 345
    cell cycle and, 38, 47, 378, 384, 552
    eukaryotic, 2, 7, 9, 14, 29-31, 34, 38, 194, 234,
        247, 254, 260, 263, 345
    evolution, 38, 62, 231, 247
    in meiosis, 7, 47
    in mitosis, 18, 31, 38, 47
    prokaryotic, 9, 30-31, 345
Cell division (cytokinesis), 373
Cell fate, 483, 489
Cell lines, 121, 251, 342, 451, 495, 554
    continuous, 251
    diploid, 121
    immortal, 495
Cell lysis, 212
Cell membrane, 31, 38, 196, 214, 380, 572
Cell migration, 382, 497
Cell motility, 429
Cell nucleus, 63, 260, 344, 378, 571
Cell plate, 36, 38
Cell reproduction, 29
Cell shape, 31
Cell signaling, 436, 480, 494, 549
Cell size, 141-142, 378
cell structure, 30
Cell theory, 16
Cell wall, 11, 30, 38, 196, 453, 457, 459
    bacteria, 11, 30, 196, 459
    eukaryotic, 11, 30, 38
    of bacteria, 30, 196
    prokaryotic, 30
    synthesis, 11, 196
Cells, 1-14, 17-18, 20-23, 25, 28, 29-38, 40, 42-43, 45,
    47-48, 62, 65, 70, 83, 93, 98, 100-101,
    103-104, 111, 117-122, 124-127, 132-133,
    139, 141-144, 146, 151-152, 176-177,
    185-196, 198-203, 208-215, 224, 227, 229,
    231-235, 238, 242-245, 247, 249-251,
    254-255, 257-258, 260, 265, 268, 274,
    279-281, 285, 288, 290, 293-294, 300,
    304-307, 309-310, 313-315, 320, 323-324,
    326-327, 329-337, 339-341, 345-349,
    352-357, 359-360, 362-370, 373-388, 392,
    394-403, 406, 410-411, 413, 415, 420,
    422-423, 428-429, 431-432, 434-442,
    445-446, 449-452, 454-461, 463-465,
    467-470, 473-476, 480-486, 489, 491-498,
    510, 526, 546, 547-554, 559, 561-563, 567,
    571-572, 577
    aging and, 249
    apoptosis, 1, 152, 340, 373-374, 377, 379-381,
        384, 387-388, 497, 553
    cleavage, 9-10, 12, 232, 260, 365, 394, 438, 440,
        470, 492
    cultured human cells, 370
    functions of, 212, 314, 386-387, 420, 423, 437
    morphogenesis, 388
    size of, 3, 7, 34, 100, 132, 152, 199, 224, 255, 257,
        268, 306, 370, 394, 413, 422, 431, 476,
        562
    structure and function, 306, 314-315, 359, 396,
        403, 437, 442
    types of, 47, 65, 120, 126-127, 143-144, 208-209,
        214, 244, 251, 255, 268, 285, 300, 314,
        324, 327, 330, 332-335, 340-341, 347,
        353, 355, 357, 365-369, 374, 380-381,
        383, 386, 396-398, 437, 442, 446, 452,
        464, 475, 494-495, 498, 510, 547, 561,
        572, 577
    typical animal cell, 30
Cellular differentiation, 366, 374
Cellular respiration, 103
    aerobic, 103
    electron transport, 103
Cellulose, 30
Central dogma, 2, 22, 28, 208, 292
Central dogma of molecular biology, 22
Central nervous system, 65, 488
Centrifugation, 4, 233, 251, 257, 264

Centrifuge, 211, 460
Centriole, 2, 30-31, 35, 47
Centrioles, 2, 31-32, 35-36
    in cell division, 2
Centromere, 1-2, 4, 7-9, 12-13, 32-33, 35, 37, 39-40,
    46-48, 63, 77, 116, 123, 138, 144, 147-149,
    154, 157, 176, 224, 265, 269, 396, 424
Centromeres, 2, 4, 11, 14, 36-37, 39-40, 47-48,
    148-149, 168, 183, 263-265, 268, 378, 422,
    550, 572
Centromeric DNA, 224, 265
Centrosome, 2, 31, 35
Centrosomes, 35
Cerebral cortex, 99
Cervical cancer, 380, 384
cervical cancers, 452
Cervix, 122, 152
chain, 3-4, 6-7, 10-13, 21, 81, 101, 208, 217, 219-220,
    237-238, 240, 252, 261, 280-284, 289, 291,
    298, 302-307, 309-313, 315-317, 342, 356,
    370, 388, 392, 400-402, 406-407, 410-411,
    416, 439, 450, 469, 559, 572
Chaperonins, 313
Characters, 4, 53-54, 57-59, 61, 87, 101, 106, 169,
    177, 322, 505
    multifactorial, 505
    traits and, 53, 505
Chargaff, Erwin, 208, 217
Charged tRNA, 1, 276, 302-304, 316-317
Chase, Martha, 212
chemical agents, 320, 334
    carcinogens, 334
    mutagenic, 320, 334
Chemical bonds, 20
    hydrogen, 20
Chemical equilibrium, 314
Chemical mutagens, 176, 324
Chemical reaction, 237, 314, 399
Chemical reactions, 31, 292, 307, 314
    of metabolism, 307
    synthesis, 31, 292, 307
Chemical signals, 31, 494
Chemicals, 143, 151, 319, 324-327, 333, 339, 384,
    387-388, 424, 532, 554
Chemistry, 26, 208, 216, 223, 247, 283, 340, 401
    atoms, 216
    elements, 283, 340
    molecules, 208, 216, 223, 247, 283, 340, 401
Chemokine, 526
Chemokine receptor, 526
chemokines, 526
Chemotherapy, 46, 386, 464, 553, 571-573
Chiasmata, 2, 39, 47, 159, 163, 166, 174-175, 259
Chickenpox, 452
chickens, 106, 120, 123, 132, 287, 428, 442, 506, 509,
    512, 517
Children, 23, 46, 72, 74, 83-84, 86, 88, 99, 105, 107,
    130-131, 138-139, 145, 153-156, 203, 250,
    310, 381, 387, 451, 455, 467-469, 474, 476,
    516, 531, 542, 552, 554, 563, 565
    development of, 88, 105, 139, 145, 381, 387,
        468-469, 474, 552, 554
Chimpanzees, 411, 423, 428, 430-432, 538, 541, 543
China, 130, 422, 554
Chlamydomonas, 34, 111, 257
Chloramphenicol, 194, 316
    resistance, 194
Chloride, 232-233, 340
chloride ion, 340
Chlorophyll, 101-102
Chloroplast, 101-102, 111, 257, 454
    DNA, 101-102, 111, 257, 454
    genome, 454
    structure, 257
Chloroplasts, 3, 5, 8, 31, 80, 101-102, 104, 214, 254,
    256-257, 269, 287-288, 313
Cholera, 202-203, 429, 452
    symptoms, 203
    treatment, 202-203, 452
    vaccine, 202-203, 452
Cholera enterotoxin, 202
Cholesterol, 23, 70, 517-519
    familial hypercholesterolemia, 23, 70
Choline, 554
Chorion, 139
Chorionic villi, 459
    sampling, 459
Chorionic villus sampling, 2, 139, 459
Chorionic villus sampling (CVS), 2, 139

Chromatid, 2, 5, 7, 12, 37, 39-41, 44, 46-47, 148,
    176-177, 182, 234-235, 251-252, 261-262,
    269-270, 332
Chromatids, 2-4, 9, 12-13, 32-33, 35, 37-41, 43-44,
    46-48, 141, 146, 148-149, 156, 159-160,
    165-166, 168, 170, 174-177, 180, 184,
    234-235, 251-252, 259, 262, 332-333
Chromatin, 2, 5, 9, 12, 30-31, 35-36, 38-39, 44-45, 47,
    124-125, 151, 243-244, 254, 258, 260-263,
    268-270, 284, 344-345, 355-358, 360-362,
    364-366, 370, 376-377, 469, 491, 547-550,
    553, 555
    animal cell, 30, 36
    cell division and, 44
    plant cell, 36
Chromatin fibers, 31, 35, 44-45, 254, 260, 262, 269
Chromatography, 441
Chromosome, 1-14, 18-19, 23, 26, 28, 29, 31-37, 39,
    41, 44-45, 47-48, 52, 63-65, 71, 75, 80-81,
    83, 86, 94-96, 98, 100, 107-108, 111,
    116-117, 119-133, 135-157, 159-184,
    185-186, 188-201, 203-205, 207-208, 224,
    226, 234-236, 239-240, 243, 245-247,
    251-252, 253-270, 281, 284, 300, 319-321,
    331-337, 344-347, 351, 356, 366, 369-370,
    374-379, 382-383, 386-388, 396-397,
    404-406, 412-413, 416-420, 422, 424-426,
    429-430, 433, 443-446, 454, 460, 462,
    466-468, 473, 487-488, 493, 511-513,
    542-543, 549-552, 560-563, 567, 571-573
    Archaea, 7, 429, 446
    artificial, 1, 7, 14, 320, 333, 396, 511-513, 567
    bacterial, 1, 3, 5, 8-11, 13-14, 31, 185-186, 188,
        190-201, 203-205, 236, 243, 247,
        255-257, 260, 269-270, 284, 300, 320,
        334-335, 345-347, 351, 369-370,
        396-397, 413, 422, 429, 445
    eukaryotic, 2, 5-7, 9, 11-14, 26, 29, 31, 34, 111,
        116, 174, 194, 201, 224, 234, 240, 243,
        245-247, 252, 254-258, 260, 262-267,
        269-270, 281, 284, 332, 334, 344-345,
        356, 370, 413, 419-420, 429, 445-446
    prokaryotic, 9-10, 12, 31, 243, 257, 284, 344-345,
        419, 429, 445
Chromosome theory of inheritance, 2, 18-19, 26, 28,
    95
Chromosome(s), 268
Chromosomes, 1-14, 15, 17-19, 26, 28, 29-41, 43-44,
    46-48, 52, 54, 62-64, 72, 74-75, 80-82,
    94-95, 100-101, 107, 115-134, 135-138,
    140-152, 154-157, 159-161, 163, 165, 168,
    175-178, 180, 182, 184, 188, 202, 205,
    207-208, 214, 216, 222-225, 231, 234,
    243-247, 249-250, 252, 254-256, 258-270,
    281, 300, 319, 326, 334-335, 337, 355-357,
    375-377, 383, 386-387, 394, 396, 404-405,
    413, 416, 418, 421-422, 424, 426-427, 429,
    431, 433, 446, 469, 493, 512, 543, 548, 550,
    561
    alterations of, 135, 137, 208
    alternation of, 43-44
    and sex determination, 131
    archaea, 7, 429, 446
    autosomes, 1, 8, 11, 94, 107, 119-121, 123-124,
        127-128, 131, 133, 137, 155, 161, 165,
        182, 421, 426, 431, 493
    bacterial, 1, 3, 5, 8-11, 13-14, 31, 82, 188, 202,
        205, 214, 243-244, 247, 255-256, 260,
        269-270, 300, 334-335, 394, 396, 413,
        422, 429
    bacterial artificial, 396
    cell division and, 14, 34, 44
    degradation of, 8, 355
    deletions, 1, 11, 143-145, 147-148, 152, 267, 326,
        337, 376, 405, 424, 431
    describing, 1, 8, 17, 30, 117, 142, 208, 258, 404
    diploid number of, 8, 18, 33, 36-38, 40-41, 46, 63,
        116, 141, 150, 155
    DNA and, 1, 3, 5, 8, 10, 126, 214, 216, 222-225,
        244, 255-256, 260, 264-265, 267, 281,
        356, 387, 394, 396, 418, 548
    eukaryotic, 2, 5-7, 9, 11-14, 26, 29-31, 34, 38, 116,
        202, 214, 224, 234, 243-247, 250, 252,
        254-256, 258, 260, 262-267, 269-270,
        281, 334, 355-357, 413, 429, 446
    genes in, 7, 13, 28, 82, 101, 124, 127, 159, 165,
        168, 177-178, 182, 202, 266-267, 337,
        356, 387, 404, 418, 421-422, 424, 429,
        431, 433, 446, 493, 512, 550

haploid number of, 5, 34, 43, 64, 127, 137, 161,
    184
homologous, 1-3, 6-7, 9, 12-13, 18, 29, 32-33,
    36-39, 41, 43, 46-48, 52, 63-64, 74,
    80-82, 122, 133, 142-143, 147-149, 155,
    157, 160-161, 163, 175-176, 178, 247,
    300, 319, 421
homologous chromosomes, 1-3, 6-7, 9, 13, 18, 29,
    32-33, 37, 39, 41, 46-48, 52, 63-64, 74,
    80-82, 142, 157, 163, 175-176
human, 1, 4-5, 7-9, 11, 15, 18, 26, 28, 33-35, 37,
    81-82, 95, 100, 116, 120-125, 130-132,
    135-138, 140, 148, 150-152, 155-156,
    175-178, 182, 184, 202, 214, 216, 224,
    231, 244, 249-250, 260, 262, 265-269,
    281, 319, 334, 337, 375, 383, 387, 396,
    404, 413, 418, 421-422, 424, 426-427,
    429, 431, 433, 446, 469, 561
    in cancer cells, 357, 376-377
    in gametes, 43, 63, 145, 175, 208
    in prokaryotic cells, 9, 355
    in somatic cells, 8, 125, 176, 245, 247, 258, 265,
        335, 550
    independent assortment of, 48, 64, 160
    karyotypes, 3, 120-121, 125, 132, 155, 263, 405
    karyotypes of, 120-121, 155
    karyotyping of, 135
    mapping of, 182
    maps of, 165, 178, 268, 427
    maternal, 5, 7-9, 12, 29, 33, 36, 39-41, 43, 46-48,
        63-64, 100-101, 107, 125, 131, 140, 152,
        155, 256, 375, 427, 548, 550
    number of, 3-8, 10-11, 13-14, 18-19, 28, 29, 32-34,
        36-41, 43, 46-47, 62-64, 75, 100,
        116-117, 119-121, 124-127, 130,
        132-133, 135-138, 140-141, 146, 148,
        150, 152, 154-156, 159-161, 165, 175,
        178, 184, 205, 214, 243-244, 249,
        259-260, 263-266, 268-269, 281, 300,
        319, 334, 356-357, 375-377, 383,
        386-387, 404, 413, 418, 421-422, 424,
        426, 429, 431, 446, 469, 512, 543, 561
    paternal, 7-8, 12, 29, 33, 36, 39-41, 43, 46-47,
        63-64, 100, 125, 131, 140, 152, 375, 427,
        550, 561
    prokaryotic, 9-10, 12, 30-31, 243-244, 355, 429
    recombinant, 3, 6, 11, 14, 15, 26, 148, 159-161,
        163, 165, 176, 202, 247, 394, 396,
        404-405, 413, 418
    sex, 1, 6-7, 10, 12, 14, 19, 29, 33, 47, 54, 63,
        80-81, 94, 107, 115-134, 136, 156, 182,
        188, 446, 493
    structure of, 2, 6, 14, 19, 26, 30, 184, 207-208, 216,
        223, 231, 255, 269, 421, 427, 543, 548
    translocations, 143, 149-150, 156, 326, 337, 376,
        387, 405, 433
    yeast artificial, 14, 396
Chronic inflammation, 384
Chronic myelogenous leukemia (CML), 376-377, 387
Chrysanthemums, 155
Chymotrypsin, 147
Cichlid fish, 537, 544
Cigarette smoking, 384
Cilia, 2, 31
    eukaryotic, 2, 31
Ciliate, 14, 265, 288
Ciliates, 281
Circular DNA, 8, 10, 31, 244, 254, 256, 269, 281, 429,
    561
Circulation, 5, 21, 132, 139
    fetal, 5, 132, 139
Circulatory system, 209
Cis configuration, 3
Cis-trans test, 3
Cistron, 3
Class, 1, 4, 14, 45, 112-113, 143, 147, 166-167, 171,
    190, 194, 223, 229, 298, 313, 322, 336, 347,
    352, 443, 489-491, 496, 503-504, 526, 571
    of humans, 322
Classes, 4, 74, 76, 88, 113, 167, 169, 171, 182-183,
    223, 264, 269, 311, 314, 378, 387, 483,
    489-490, 496-497, 501-503, 517-519,
    553-554
Classification, 171, 320-321, 447, 545
Clavicles, 486
Cleavage, 9-10, 12, 232, 260, 365, 393-394, 412, 438,
    440, 470, 492
Climate, 514
    global climate change, 514

Clinical trials, 23, 203, 249, 290, 373, 468-470, 475, 553, 571
clone, 2-3, 5, 14, 125-126, 202, 336-337, 387, 392, 395-396, 398-401, 403, 410, 413, 416, 422, 453, 474, 513
Cloned genes, 396, 450, 523
Clones, 2, 6, 12, 22, 24, 125-126, 197, 392, 395, 398-399, 404, 474
Cloning, 3, 10, 12, 14, 22, 26, 28, 105, 126, 185, 195, 334, 392-403, 407-408, 410-411, 413, 418, 423, 429, 454, 458, 462, 472, 475, 551, 577
   bacterial, 3, 10, 14, 22, 185, 195, 334, 394-397, 399, 410, 413, 429
   molecular, 3, 10, 22, 26, 28, 126, 185, 392-393, 401-403, 423
   shotgun, 12, 398, 429
   steps, 195, 401-403, 410-411
Cloning vector, 14, 407, 410
   artificial chromosome, 14
   hosts for, 410
   plasmid, 14
Cloning vectors, 393-394, 396, 411
Clotting factor VIII, 96, 451
Clotting factors, 450, 574
Coat protein, 280
Cod, 428
Codominance, 3, 83, 86, 107
Codominant alleles, 131, 528
Codon, 1, 3, 5, 7-9, 14, 20-21, 273, 275-281, 286, 291-292, 294, 298-300, 302-304, 306-307, 316-317, 321-322, 328, 348, 387-388, 419, 446-447, 461, 523-524
   start, 3, 7, 273, 280, 291, 302, 348, 419, 446
   stop, 3, 273, 279-280, 291, 304, 321, 328, 419, 446
Codon bias, 447
Codons, 3, 5, 9, 11-12, 14, 21, 272-273, 275-276, 278-281, 291-294, 298, 300, 302-304, 315, 321, 327, 341, 363, 419-420, 446
   codon recognition, 281
   genetic code and, 272-273, 275-276, 278-281, 291-294
   nonsense, 9, 280, 304, 321, 327, 341
   start, 3, 273, 280, 291, 302, 419, 446
   stop, 3, 273, 279-280, 291, 304, 321, 363, 419, 446
Cofactor, 288, 401
Cofactors, 354, 492
Coffee, 84, 141
Cohesive ends, 393-394
Col plasmid, 195
Colchicine, 3, 141-143, 156, 234
Cold, 141, 317
Colds, 468
Collagen, 21, 287, 313, 441-442
Collagenase, 441
Colon, 23, 25, 148, 152, 326, 329, 337, 342, 375-376, 382-384, 430, 443, 459, 464, 552-553, 572-573
   polyps, 23, 383
Colon cancer, 23, 25, 329, 342, 375, 382-383, 443, 464, 552, 572
Colony, 186, 209-211, 397, 400
Color blindness, 71, 95-96, 107-108, 126
   inheritance of, 71, 107
Color vision, 107, 133
Coloration, 101, 175, 529
   chromosome, 175
Colorectal cancer, 376, 555, 573
Combined DNA Index System (CODIS), 3, 566
Commercial applications, 455
Communication, 13, 148, 203, 481, 496
Communities, 434-435, 544
Community, 187, 408, 468, 529, 544
Comparative genomics, 415, 419, 428, 430-432, 434, 443, 445-447, 539
Competence, 3, 195
Competition, 422, 473, 577
Complement, 3, 7-8, 38-39, 63, 88, 93, 120, 127, 133, 137, 140-141, 149, 155-156, 167, 189, 232, 237-239, 245-246, 266, 272, 329, 355, 437, 443, 543
Complementary base pairing, 294, 394
Complementary base pairs, 223, 291, 394
Complementary DNA, 2, 215, 239, 314, 398-399, 413, 435, 461
Complementary DNA (cDNA), 314, 398, 435
Complementary RNA, 12-13, 20, 240, 282, 290-291
Complementation, 3, 11, 92-93, 110, 180, 332, 340, 342, 475
Complementation test, 3

Complete dominance, 82, 88
Compound, 3, 87, 124, 209, 315, 334, 340, 526, 571
Compounds, 1, 186, 315, 325, 333-334, 399, 431, 496
   elements and, 334
   inorganic, 186
Computer-automated DNA sequencing, 407-408, 415, 417
Concentration gradient, 497
Conclusions, 28, 62-63, 71, 75, 92, 105, 109, 123, 131-132, 167, 188, 210, 217-218, 223, 227, 238, 264, 268, 276, 293, 315, 336, 339, 369-370, 411, 496, 510, 530, 540, 544, 554
Congenital disorders, 16
Congestive heart failure, 442
Conidia, 308-309
Conjugation, 3, 5, 8, 11, 185, 187-195, 199, 204
   bacterial, 3, 5, 8, 11, 185, 187-188, 190-195, 199, 204
   in bacteria, 3, 11, 185, 187-195, 199, 204
Connective tissue, 21, 23, 93, 124, 287, 313, 331, 443, 576
   types of, 287
Consensus sequence, 3, 6, 10, 243, 283, 285, 358, 524
Consent, 16, 468, 471, 473, 567
Consortium, 422, 431, 473, 555, 575
Constitutive enzyme, 347
Consumers, 471
Contagious disease, 227
Contamination, 401, 413, 446, 457, 476, 565
   of water, 457
Contigs, 416-418, 424, 443-445
Continuous variation, 3, 62, 500, 502, 517, 519
Contractile proteins, 313
Contrast, 4-5, 28, 47, 59, 73, 80, 94, 96, 98, 103, 107, 110, 117, 119, 128, 131-133, 139, 146, 148, 155-156, 163, 180, 184, 186, 195-196, 198, 200, 208, 214, 227, 238, 245, 252, 254, 256, 264, 269, 272, 274, 279, 283, 285, 306, 309, 316, 320, 336, 340, 345, 350, 353-355, 357, 361-362, 367-368, 375, 377-378, 386, 434, 461, 493-494, 531, 541, 573, 576-577
Copolymer, 275-276, 278, 291-294
Copper, 117
Copy number, 3, 14, 148, 267, 395-396, 422-423, 462, 510, 572, 576
Core, 17, 196, 211-212, 215, 238-239, 241-242, 251, 261-262, 284-285, 292, 306, 356-358, 360-361, 428, 548, 559-560, 564-565
Corepressor, 352, 368, 371
Corn, 22, 34, 59, 117-118, 155, 170, 180, 319, 335, 384, 453-454, 509, 512, 517
   genetically engineered, 22, 453-454
   oil, 22, 509
Corn (Zea mays), 453
Coronary arteries, 437
Coronary artery disease, 175, 513, 575, 577
Correlation, 52, 63-64, 95, 100, 122, 159, 161, 173-174, 214, 223, 286, 454, 503, 505-506, 517-518
   in genetics, 52
Correlations, 127, 293, 472
Correns, Karl, 17, 101
Cortex, 99, 122, 482, 498
Cosmic rays, 325-326
Cosmid, 3
Cotyledons, 74, 76-77
Covalent bond, 13, 252, 303, 311-312
Covalent bonds, 9, 228
   described, 228
   DNA and, 228
   double, 9, 228
   single, 9, 228
cows, 132, 442, 450-451, 456-457, 474, 477, 517
   livestock, 474
Creutzfeldt-Jakob disease, 10, 313
Cri du chat, 144-145, 147, 157
Crick, Francis, 4, 14, 20, 26, 207, 216, 273, 279, 298, 306, 316, 427
Critical period, 129
Crocodiles, 129
Cro-Magnons, 544
Crop plants, 22, 28, 453-455, 514
Crops, 22, 27, 143, 203, 449, 453-455, 471, 474, 476, 500, 508, 514
   genetically altered, 22
Crossing over, 1-3, 7, 12, 14, 39-41, 43, 46-48, 133, 145-146, 148-149, 155-156, 159-161, 163, 165, 167, 171, 173, 175-177, 180-184, 186,

202, 205, 247, 316-317, 393, 512, 529, 536
   in meiosis, 7, 39-40, 46-48, 148, 156, 163, 176
   in recombination, 247
Cross-pollination, 471, 476
CRP, 369, 371
Crystallography, X-ray, 14, 301, 350
C-terminus, 312
Culture, 16, 35, 125, 151, 186, 190, 197-199, 201, 203-204, 209, 211, 249, 251, 291, 308, 323, 331, 340, 345, 370, 383, 395, 434, 439, 460, 462, 468, 474, 494-495
   continuous, 251
   pure, 186
Culture medium, 151, 186, 197, 199, 211, 345
   solid, 197
Cultures, 8, 130, 185-186, 193, 199, 210, 234, 308, 331, 366, 376, 434
   medium, 8, 186, 199, 234, 308, 434
   mixed, 199
   pure, 185-186, 193
Cuticle, 493
Cyclic adenosine monophosphate (cAMP), 350
Cyclic AMP, 351, 496
cyclic AMP (cAMP), 496
Cyclin, 82, 378-379, 381, 386, 388, 425
Cyclin-dependent kinases (cdks), 378
Cyclins, 38, 142, 378-380, 387, 497
Cysteine, 14, 277-278, 302, 576
   genetic code, 277-278
   structure, 14
   synthesis, 278, 302
Cystic fibrosis, 23, 25, 71, 175, 328, 340, 421, 424, 430, 459, 462, 474, 523, 528-529, 536, 574, 576
Cystic fibrosis transmembrane conductance regulator, 462, 523
Cystinuria, 307, 576
Cytidine, 218, 357, 553
Cytochrome, 431, 538-539, 564-565, 573
Cytochrome c, 538-539, 564-565
   cytochrome c oxidase, 564-565
Cytochromes, 102
Cytokinesis, 3, 10, 34, 36, 38, 41, 373, 482
   in mitosis, 38
Cytology, 3, 159
cytomegalovirus, 290
Cytoplasm, 2-3, 9-10, 19-20, 30-31, 34-35, 38, 43, 101-102, 125, 185, 189, 192, 194-198, 200, 202, 214, 256-257, 273, 281, 284-286, 290, 294, 306-307, 313, 331, 355, 360, 362, 364-365, 373, 378, 380, 387, 481-483, 496-498, 549
   eukaryotic cells, 9, 20, 30, 281, 285, 306, 313, 355, 360
   features of, 306, 355, 364, 497
   in eukaryotic cells, 9, 20, 285, 306, 313
   in prokaryotic cells, 9, 257, 355
   of bacteria, 30, 185, 195-198, 200
   of eukaryotes, 214
   viruses in, 197
Cytoplasmic determinants, 497
Cytosine, 13, 20, 101, 216-217, 220-222, 228-229, 323-325, 339-340, 356, 548
   in DNA, 20, 216-217, 221, 228, 324-325, 339-340, 356, 548
Cytosine (C), 222, 228
Cytoskeleton, 31, 382
   bacteria, 31
   eukaryotes, 31
   of prokaryotic cells, 31
Cytosol, 31
   in eukaryotes, 31

D

Dapsone, 573
Darkness, 104
Darwin, Charles, 17, 62, 434, 444, 522
Darwinism, 522
Data, 1-2, 5, 17, 24, 45, 52, 54, 56, 66-69, 71, 73-76, 81, 92, 94, 106-111, 118, 122, 124, 127, 130, 132, 163, 167, 169, 171-172, 177-179, 181-182, 191, 200-202, 212, 214, 216-221, 226, 228-229, 244, 252, 267-268, 270, 277, 292-293, 315-316, 334, 336, 340, 357, 367, 369-370, 374, 376, 385, 387, 403, 406-408, 411, 415, 418, 426-427, 432, 434-435, 438, 442, 444, 446, 464-468, 471, 473-474, 503-505, 510, 512, 517-518, 527-528, 538-546, 555, 560, 563, 566, 570, 574-578

dATP, 237, 252, 406-407
Daughter cells, 11, 28, 29, 31, 34-35, 38, 43, 47-48, 194, 200, 208, 215, 467, 547, 561
dCTP, 237, 252, 406-407
Deafness, 88, 96, 103, 105, 140, 424-425, 430, 443, 471
deamination, 323-324
Death, 1, 8, 13, 53-54, 72, 85-86, 94, 96, 99, 137-138, 156, 202, 209, 319, 332, 340, 342, 344, 364, 373, 377, 379-380, 425, 464, 468-469, 474-475, 558-559, 564, 566, 572, 575, 577
Deaths, 202, 342, 384, 531, 551
Defense mechanisms, 432
Defenses, 529
Deficiencies, 138, 150, 329, 455
Degenerate code, 3, 274, 294
Degenerative disease, 23, 469
Degradation, 8, 12, 196, 245, 274, 290, 292, 294, 307, 327, 355, 363-366, 470, 497
    mRNA, 8, 12, 274, 290, 292, 294, 307, 327, 355, 363-366, 497
    protein, 8, 12, 196, 274, 290, 292, 294, 307, 327, 355, 363-365, 470, 497
Dehydration, 202, 222, 311
Delbrück, Max, 5
Deletion, 2-3, 11, 86, 100, 127, 135, 144-145, 149, 155, 273, 292, 320-321, 323, 328, 336, 349, 369-370, 380, 386, 460, 475, 526-527, 534-535, 546, 560
Deletions, 1, 11, 143-145, 147-148, 152, 267, 291, 322-323, 325-326, 337, 341-342, 374, 376, 382, 388, 405, 424, 431, 462, 524, 562, 574, 576
Delivery, 451, 467, 469-470, 474-476
Dementia, 72, 103
Denaturation, 10, 223-224, 229, 250, 401
    DNA, 10, 223-224, 229, 250, 401
    proteins, 223, 229
Deoxyribonuclease, 4, 210-211, 227
Deoxyribonucleases, 4
Deoxyribonucleotide, 223, 282, 406-407
Deoxyribose, 4, 9, 20, 184, 211, 216-217, 220, 223, 228, 237, 252, 282, 323, 325, 330, 406
    and DNA, 217, 252, 325, 330
Depression, 7, 9, 431
Derived, 1, 3-5, 8-9, 14, 17, 22-23, 29, 31-33, 39, 43, 45-48, 53-54, 56, 84, 101, 104, 124-126, 139, 141-142, 151, 155, 167, 171, 174-175, 177, 179-180, 182, 185-186, 195, 197, 202-203, 208, 210, 215, 224-226, 228, 234, 236, 248-250, 252, 257-260, 269, 273, 277, 280, 282, 284, 288-289, 291, 303, 305-306, 308, 315, 331, 334, 365, 368, 373, 375, 380, 384, 393-394, 399, 404, 422, 450, 452, 462, 473, 482, 495, 497, 505, 510, 534-535, 538, 541, 544, 552, 562, 565, 567, 570, 578
Deserts, 422, 434, 446
Determination, 4, 12, 92-94, 115-134, 153, 169, 181, 276, 278, 412, 439, 460, 474, 480, 485, 493-494, 497, 503, 510, 513, 524
Development, 1, 4, 8-9, 13-14, 15-17, 22-23, 25-26, 30, 34, 42-43, 45, 65, 68, 85-88, 90-92, 99-100, 103-105, 108, 116, 118-129, 131-133, 137-140, 145, 155-156, 174, 186, 202-203, 216, 227, 234, 249-250, 258, 260, 269, 290, 295, 308, 320, 322, 326, 334-335, 337, 344, 354, 356, 359, 363, 366-367, 373-379, 381-384, 386-387, 392, 394, 398-399, 404-406, 408, 410, 415, 417-418, 424, 428, 431-433, 435-439, 442, 446, 450-451, 455-456, 459, 465-466, 468-469, 471-476, 480-483, 485-486, 488-489, 491-498, 510-511, 513-514, 536, 539-540, 547-555, 559, 570, 576-578
Developmental biology, 436, 481
Developmental plasticity, 555
dGTP, 237, 252, 406-407
Diabetes, 25, 100, 175, 366, 428, 431, 438, 446, 450, 466, 470, 477, 495, 510-511, 513, 517-519, 547, 554, 571, 574-575, 577
Diagnosis, 2, 5, 9-10, 16, 23-24, 72, 139, 154, 178, 315, 367, 373-374, 441, 450, 459-461, 467, 471, 473, 519, 558, 570-571, 574-575
Diagnostics, 573
Diamonds, 69, 544
Diarrhea, 202
Diazepam, 573
Dicer, 365, 470
Dictyostelium, 34, 496

Diet, 23, 384, 387, 424, 426, 455, 464, 467, 554-555
Diets, 455
Differential gene expression, 344, 481
    regulation of, 344
Differential stain, 176
Differential staining, 2, 263
Differentiation, 7, 12, 34, 42, 116-123, 128-129, 131-133, 294, 366, 374, 377, 380, 383, 446, 480, 491, 494, 496, 548, 551, 555
Digestion, 4, 10, 12, 14, 210, 260-261, 263, 268, 270, 346, 393-394, 403-404, 416, 453, 460, 560
    intracellular, 10
    of proteins, 260, 394
Digestive system, 202, 333
Dihybrid crosses, 58, 60-61, 67, 69, 106
Dilution, 186, 197-198, 204, 340
Dimer, 8, 13, 238, 241-242, 327, 329-331, 350, 357
Dimers, 4, 326, 329-330, 376
dimorphism, 116-117
    sexual, 116-117
Dinosaurs, 441
Dipeptide, 303-304, 311
Diploid, 1-2, 4, 6, 8, 14, 18, 28, 29, 32-34, 36-44, 46-47, 55, 63, 80-83, 116-118, 120-121, 126-128, 132, 135-137, 140-143, 150, 154-156, 161, 193, 200, 205, 214, 262, 269, 320, 339-341, 347, 349, 367, 369, 427, 447, 482, 493, 532
Diploid cells, 18, 117, 141, 214, 367
Diploid nucleus, 126
Diploid number, 8, 18, 32-33, 36-38, 40-41, 46, 63, 116, 120, 137, 141, 150, 155
Diploidy, 47, 121, 136
Directional selection, 4, 531
Disaccharide, 345-346, 395
disease, 6, 10, 13-14, 16, 23, 25, 27-28, 46, 65, 70-72, 82, 86, 96, 99-100, 103, 105, 110, 113, 124, 130, 138, 148, 152-154, 175-176, 178, 202-203, 209, 227, 267-268, 290, 293, 307, 309-310, 313, 315-316, 322-324, 327-328, 332, 337-338, 340, 366, 373-375, 381, 385-386, 388, 398, 410, 423-425, 427-432, 435, 437, 441-443, 446-447, 449-453, 456, 459-460, 462, 464-474, 476, 495, 510-511, 513-514, 517, 519, 523, 536, 542, 546, 549, 551-552, 554-555, 558, 570, 572, 574-578
    acute, 375, 425, 464-465, 468
    causes of, 323, 338, 373-374, 430, 474
    chronic, 23, 340, 425, 495, 510, 575
    contagious, 227, 464
    definition of, 446, 542
    diagnosis of, 23, 459-460, 519
    frequency of, 72, 176, 293, 338, 381, 536, 542, 546, 552
    fungi, 13, 25, 99, 429
    incidence, 105, 138, 375, 551
    infectious, 10, 25, 113, 313, 441, 450, 452, 464-465, 467, 469, 476
    manifestations of, 103, 175
    occurrence of, 366
    pathogens and, 465
    patterns of, 96, 307, 322, 431, 435, 447, 464-465, 511, 523, 549, 551-552, 554
    prevalence, 105, 153, 546
    prion, 10, 313
    sporadic, 103, 148
    spread of, 373, 536, 552
    syndromes, 124, 138, 332
Disease agents, 450
Disease-causing mutations, 293
Dispersal, 5, 432
Disruptive selection, 4, 531-532
Divergence, 5, 8, 11, 431-432, 481, 491, 510, 522, 524, 532, 536-541, 545-546
    morphological, 537, 545
Diversity, 6, 8, 10, 16, 21, 28, 33, 63-64, 208, 298, 311, 320, 357, 363, 422-423, 427, 444, 447, 454, 480, 522-524, 534, 542-543
Divisions, 7, 34, 39, 41-43, 45, 47, 126-127, 136, 141, 148, 186, 233, 247, 249, 320, 337, 375, 482, 493, 497-498
Dizygotic twins, 4, 516, 518
DNA, 1-14, 15, 17, 19-28, 29, 31-35, 38, 44-45, 47, 71, 80-82, 86, 100-104, 111, 116, 123-124, 126, 130, 139, 142, 145-148, 152, 174-177, 182, 185, 188-189, 191-192, 194-202, 207-229, 231-252, 253-270, 272-274, 280-287, 289-292, 294, 298, 300, 311-315, 319-342, 344-350, 353, 355-361, 363-366,

368-371, 373-381, 383-388, 391-413, 415-424, 426-437, 441-447, 449-450, 452-456, 459-465, 467, 469-470, 472-477, 485, 488, 494, 510-513, 524, 526-527, 534, 537, 540-546, 547-550, 552-555, 557-568, 572, 575-577
amplification of, 146, 386, 561, 577
amplifying, 10, 401, 413, 550, 561
bacterial chromosome, 5, 10, 31, 185, 188, 191-192, 195, 197-198, 200, 236, 255, 345-347
base pairs, 4-5, 8, 13-14, 20, 103, 148, 221-224, 226, 231, 236, 243-245, 251, 257-258, 261-263, 265-267, 269-270, 284, 286, 291, 300, 320, 325, 334, 358, 394, 398, 417-418, 421, 424, 430, 444-445, 540-541, 559, 563-564
base sequence, 9, 14, 285, 326, 419
bent, 262
blunt ends, 393
chloroplast, 101-102, 111, 257, 454
chromosomal, 1-5, 7, 9, 11-12, 14, 17, 31, 35, 71, 142, 145-148, 174, 182, 224, 256, 258-260, 264-265, 267, 269, 300, 320, 326, 334-336, 347, 370, 374, 376-377, 379-380, 388, 406, 432, 446-447, 488, 512, 552-553, 560
circular, 8, 10, 31, 185, 188, 191, 236, 239-240, 243-244, 254-257, 269-270, 281, 395, 429, 561
cohesive ends, 393-394
complementary, 1-2, 6, 11-13, 20, 194-195, 207, 215, 222-224, 227, 229, 232, 236, 239-241, 244-245, 247, 252, 257, 264-265, 273, 281-282, 284, 286, 290-291, 294, 298, 300, 314, 323, 329-330, 332-333, 364-366, 394, 398-406, 413, 435, 445, 461, 463, 549, 552
conjugation, 3, 5, 8, 11, 185, 188-189, 191-192, 194-195, 199
denaturation, 10, 223-224, 229, 250, 401
double helix, 3-4, 14, 20, 189, 195, 207, 216, 219, 221-223, 228, 231-232, 235, 239-240, 242, 247, 252, 259, 268, 281-282, 330
double-stranded, 1, 4, 6, 8, 11, 194, 222-223, 229, 239, 243, 245-247, 252, 255-257, 269-270, 290, 332, 339, 364-365, 368, 383-384, 393-394, 398-404, 406, 411, 413, 470, 549, 561
eukaryote, 228, 268, 369, 419
extrachromosomal, 5, 10, 194, 394
extranuclear, 5, 80-81, 101-102
function of, 6, 25-26, 81-82, 101, 175, 231, 237, 239, 242, 247, 250-251, 256, 258, 265, 268, 285, 298, 300, 311-313, 319-322, 334, 338, 341-342, 346, 369-370, 383, 403, 410, 415, 420, 432, 446, 550
gel electrophoresis, 14, 403-405, 407-408, 442, 460, 558, 560
hydrogen bonds, 4, 20, 219-223, 228-229, 232, 239, 245, 247, 284, 300, 312, 340
in eukaryotic genomes, 224, 254, 429
in prokaryotic genomes, 429
in transformation, 195, 214
insert, 22, 239, 322, 334-336, 342, 396-398, 400, 403, 410, 412
inverted repeats, 14, 335
inverted terminal repeats, 334
ionizing radiation, 11, 247, 326, 332, 380, 387
linear, 3, 9-11, 20-21, 28, 147-148, 194, 243-247, 249, 252, 255-256, 258-260, 265, 268, 270, 272-273, 280, 298, 300, 311-312, 326, 365, 395, 419, 429
major groove, 221-222, 548
melting, 8, 223-224, 229
methylation, 8, 101, 262, 265, 291, 328, 356-357, 368, 370, 373, 376-377, 511, 548-550, 552-555, 567
minor groove, 221, 262
mitochondrial, 101-103, 256-257, 280, 426, 437, 537, 542, 544, 561-562, 565
mobile, 11, 188, 266, 335-337
mutation and, 103, 323, 328, 338, 341, 475, 543, 552, 575
naked, 199, 214, 459
nicked, 189, 192, 247, 328
nuclear, 4, 7, 9, 12-13, 22, 31, 34-35, 38, 47, 101, 103-104, 124, 223, 244, 257, 281, 285,

294, 355-356, 368-369, 378-380, 388, 426, 447, 540, 542, 553
nucleotide bases in, 273
nucleotides in, 1, 3, 8, 11, 20, 217, 260, 284, 329, 357, 402, 407, 412, 446
of viruses, 197, 231, 255, 260, 269, 366, 383, 428
palindromic, 9, 11, 411, 413
patented, 28, 472
plasmid, 3, 5, 10-12, 14, 185, 189, 194-195, 248, 347, 368, 394-398, 400-401, 409, 411-413, 450, 452, 454, 470, 475
probes, 1, 4-5, 10-11, 116, 224, 264, 368, 388, 399, 404-405, 461-463, 476, 562
prokaryote, 242, 281, 369, 419, 446
protein synthesis and, 282
recombinant, 3, 6, 11, 14, 15, 22, 24-26, 148, 174, 176, 185, 192, 195, 201-202, 215, 229, 247-248, 280, 391-413, 415, 418, 423, 449-450, 452-455, 459-460, 467, 470, 472, 474-475, 494, 572
relaxed, 229, 239, 410
repair of, 8, 10, 238, 331-332, 342, 374, 378
repeated sequence, 245
replication of, 4, 32, 34, 208, 222, 228, 231-232, 234-236, 243, 245, 247, 250, 270, 272, 322, 394
restriction enzyme analysis, 459-461, 476
sequence determination, 12
size of, 3, 7, 34, 100, 102, 130, 152, 199, 223-224, 226, 252, 255-257, 268-270, 287, 370, 394, 404, 409, 413, 422, 430-431, 444, 462, 472, 476, 543, 558, 562
stem-loop structures, 353, 365
sticky ends, 147, 394-395
structure, 1-2, 4, 6-7, 9-14, 15, 19-20, 24-26, 31, 33, 35, 38, 124, 185, 188, 196, 202, 207-229, 231, 238, 241, 243, 245-248, 253-270, 284, 289, 291-292, 298, 300, 311-315, 320, 323, 325, 338, 340, 346, 353, 355-360, 364-365, 368, 371, 376, 386, 392, 395-396, 403, 406, 418, 421-422, 427, 437, 442, 537, 542-543, 547-548, 550
structure of, 2, 6, 14, 19-20, 26, 196, 207-209, 216, 219-220, 223, 227-229, 231, 255, 269, 291-292, 298, 311-312, 320, 323, 325, 340, 346, 353, 358, 368, 371, 421, 427, 542-543, 548
supercoiled, 251
synthesis of, 4, 11-13, 20-22, 81, 104, 196, 200, 208, 211, 215, 222, 227, 231-233, 236-237, 239-241, 243-244, 246, 250-252, 259, 265-266, 270, 274, 280, 282, 284-285, 290, 292, 298, 300, 321, 334, 344-345, 346, 368, 370-371, 407, 555
synthetic, 1, 12-13, 222, 274, 290-291, 320, 333, 365-366, 368, 384, 449-450, 452, 459-461, 472, 474, 477, 567
terminal repeats, 334, 336
unwinding, 12, 223, 239, 242, 244, 361
vaccines, 202, 431, 452-453
variable number of tandem repeats, 558
variations in, 4, 6, 258, 267-268, 314, 327, 424, 464
viral, 148, 196-198, 200-202, 212, 214-215, 227, 247, 255-256, 269, 290, 292, 320, 366, 380, 384, 393, 415, 423, 449, 452-455, 465, 467, 469, 474-476
DNA amplification, 376, 408
DNA chips, 437
DNA cloning, 22, 26, 185, 393-395, 398, 401, 408, 410, 423
DNA fingerprint, 558, 560
DNA fingerprinting, 4, 8, 14, 266, 445, 447, 475, 558-559
applications of, 558
in paternity testing, 559
DNA gyrase, 239, 242
DNA helicase, 177, 243
DNA libraries, 392, 398, 400-401, 410-411, 416
DNA ligase, 4, 241-243, 246, 251-252, 328-331, 378, 394-395, 399, 410
DNA methylation, 101, 328, 356, 368, 370, 373, 376-377, 511, 548-550, 552, 554-555
DNA microarrays, 431, 462-465, 562
DNA polymerase, 10-11, 215, 236-244, 246, 249-252, 282, 322-323, 328-332, 378, 398-399, 401-402, 406-407, 411, 413, 561
in eukaryotes, 215, 236, 242, 244, 246, 251-252,

330-332
in prokaryotes, 241, 243-244, 252, 331
Taq, 401, 561
DNA polymerase I, 236-238, 240-243, 250-252, 330-331, 399
DNA polymerase III, 238-244, 250-251, 328
DNA polymerases, 238-239, 241-244, 247, 252, 322
DNA probes, 461
DNA profiling, 4, 27, 558-561, 564, 566-567
DNA repair, 12, 25, 238, 244, 247, 319-342, 355, 357, 364, 370, 373, 376-377, 380, 384, 386, 388, 552-553
error-prone, 329, 333
mismatch repair, 328-329, 333, 340, 342, 376-377
role of methylation in, 357
DNA repair enzymes, 384
DNA replication, 2, 7-8, 10-13, 26, 47, 82, 145, 177, 195-196, 221, 223, 231-252, 256, 263, 320, 322-325, 327-329, 333, 339-342, 345, 357, 375, 378-379, 384, 386, 428, 442, 511
errors in, 10, 333
in eukaryotes, 2, 12-13, 177, 195, 223, 231, 234-236, 242, 244, 246-247, 251-252
in prokaryotes, 2, 12, 223, 231, 241, 243-244, 252, 345, 357, 428
replication fork, 7-8, 11, 231, 236, 238-242, 244, 252
semiconservative, 12, 231-236, 239, 245, 250-252, 324
DNA sequences, 2-6, 11-14, 123, 148, 174-175, 224-225, 227, 246, 249, 251, 254, 257, 263-267, 270, 283-286, 319, 327, 334, 344, 357, 375, 392, 398-399, 401, 403, 406, 408, 410, 416, 418, 420, 422, 432, 434-435, 444, 447, 462, 472, 510, 527, 537, 541-542, 544, 552, 558-559
determining, 6, 13-14, 123, 285, 447, 527
genes as, 432
noncoding, 13, 254, 257, 266-267, 327, 398, 420, 544, 558
DNA sequencing, 4, 7, 11-12, 225, 373, 375, 392, 395, 403, 406-408, 410-411, 415, 417-418, 427, 443, 461, 473, 561-562
automated, 407-408, 415, 417, 561-562
genomics, bioinformatics, and, 415, 417-418, 427, 443
machines, 407
sequence alignment, 417-418
DNA strands, 176, 224, 240, 245, 247-248, 258, 269, 325, 327, 332, 393, 401, 577
blunt ends, 393
complementary, 224, 240, 245, 247, 332, 401
DNA synthesis, 2, 5, 9-10, 34-35, 189, 231, 235-244, 246, 250-251, 329, 331-333, 340, 342, 378, 399, 407
DNA technology, 6, 11, 15, 22, 24-26, 215, 280, 391-413, 449-450, 452-455, 459, 494, 567, 572
DNA cloning, 22, 26, 393-395, 398, 401, 408, 410
recombinant, 6, 11, 15, 22, 24-26, 215, 280, 391-413, 449-450, 452-455, 459, 494, 572
DNA template strand, 273, 284, 286
DNA viruses, 383-384
replication, 384
DnaA, 82, 239, 243, 250
DnaB, 222, 239, 242-243
DNA-binding domains, 14, 314, 348
DnaC, 239, 242
dnaQ gene, 251
DNase, 4, 199, 211, 227
dogs, 104-105, 109, 113, 131, 153, 402, 415, 428, 430-431, 442, 523
Domains, 5, 7, 14, 261-262, 306, 314, 348, 354, 356, 418, 421, 423, 437-438, 446, 484, 488, 491
Dominant allele, 11, 59, 61, 70, 72, 82, 88, 90-92, 97, 107, 113, 155, 180, 526, 529, 532, 543, 554
Dominant alleles, 81, 92, 113, 183
Dominant traits, 57, 59, 66, 70-72, 75, 161, 182, 525
Dorsal-ventral axis, 484
Double fertilization, 117-118
Double helix, 3-4, 14, 20, 189, 195, 207, 216, 219, 221-223, 228, 231-232, 235, 239-240, 242, 247, 252, 259, 268, 281-282, 330
Double-stranded DNA, 11, 194, 229, 243, 245, 252, 255, 270, 383, 394, 398-399, 401-402, 406, 411, 413, 561
in viruses, 255
Double-stranded RNA, 290, 364-365, 384, 470, 549

Down syndrome, 137-140, 150-151, 153, 155-156, 425, 574
Drosophila, 5, 8-10, 12, 19, 24-25, 34, 47, 72, 74-75, 78, 82, 84, 92-95, 98-99, 104, 108-110, 117, 127-128, 131-133, 137, 146-147, 154-155, 161, 163-165, 167-168, 170, 173-176, 179-182, 243, 258, 260, 265-267, 289, 300, 308, 313, 320-321, 325, 327, 333, 336-337, 397, 399, 406, 412, 417, 422, 428-430, 444-445, 447, 480-493, 496-497, 517, 519, 523-524, 531-532, 543, 546, 548
genome, 5, 8, 10, 24-25, 146-147, 174-175, 266-267, 289, 300, 336-337, 399, 406, 417, 422, 428-430, 444-445, 447, 481, 483-484, 487, 489, 492, 496, 517, 548
Drought, 27, 453, 455, 514, 531
Drug resistance, 335, 553, 573
Drugs, 9, 25, 96, 148, 203, 227, 249, 290, 335, 341, 385-386, 440, 474, 532, 547, 553-554, 565, 570-573, 575-576, 578
antibiotics, 203, 335
dTTP, 237, 252, 406-407
Duchenne muscular dystrophy, 71, 96, 99, 116, 130, 287, 337, 341, 367, 576
Duplications, 143, 145-148, 150, 257, 267, 337, 405, 424, 431-433, 539, 564, 574, 576
Dwarfism, 338, 451, 462, 471, 532, 544
dyes, 1, 124, 405, 407-408, 435, 438, 463, 559
Dysentery, 195
Shigella, 195
Dysplasia, 123, 156, 486
Dystrophin, 96, 287, 337, 341, 367, 422

E
E site, 303-304, 413
Ear, 88, 117, 319, 443
Ears, 99, 140, 151, 307, 453, 512
disorders, 99, 140, 307
human, 99, 140, 151, 307
insect, 453
Earth, 31, 65, 207, 229, 272, 313, 324, 328, 422, 434, 514, 542
evolution, 65, 542
primitive, 31
Earthquakes, 202
Eating disorder, 100
Ecological niches, 536
EcoRI, 393-395, 398, 409, 412, 416-417
Ecosystem, 22
Ecosystems, 154
aquatic, 154
Edible vaccine, 202, 451, 474, 476
Edwards syndrome, 155
Effective population size, 4
Efficacy, 451, 475, 571, 573
Egg, 4, 16-18, 22, 25, 43-44, 54, 100, 103-104, 110, 124, 126, 133, 139, 155-156, 215, 322, 455-456, 481-484, 492-493, 496-498, 505-506, 509-510, 550, 561, 565
development, 4, 16-17, 22, 25, 43, 100, 103-104, 124, 126, 133, 139, 155-156, 322, 455-456, 481-483, 492-493, 496-498, 510, 550
Eggs, 6, 16, 34, 43, 116-118, 123, 127, 129, 132, 139, 175, 214, 280, 455, 462, 492-493, 495, 497, 505, 517, 525, 562, 565
amniotic, 139
amphibian, 280
fertilization and, 123
genetically engineered, 455
meiosis and, 43, 117
Electromagnetic radiation, 326
Electromagnetic spectrum, 326
Electron micrograph, 31, 44-45, 185, 194, 212, 231, 244, 248, 256-257, 259-260, 272, 287, 289, 373, 395
Electron microscope, 7, 18, 30, 243, 254-255, 286, 304-305
scanning, 18, 254
transmission, 7, 18, 30, 254
Electron microscopy, 44, 254, 258, 270, 289, 299, 306-307
scanning electron, 44
transmission electron, 254
Electron transport, 102-103
Electrons, 326
Electropherogram, 562-563
Electrophoresis, 1-2, 4, 9, 12, 14, 225, 228-229, 403-405, 407-408, 411, 413, 438-439, 442,

460, 527, 545, 558, 560-561
gel, 4, 14, 225, 228, 403-405, 407-408, 411, 438-439, 442, 460, 527, 558, 560
Electroporation, 4, 395
Element, 2, 9-10, 12-14, 188, 205, 285, 319, 334-337, 341-342, 353-354, 358-360, 364, 369, 371, 420, 554
Elements, 1, 6, 8, 11-12, 14, 59, 266, 269-270, 283, 285, 319, 334-337, 339-341, 344, 354, 357-362, 364, 369-371, 419-420, 422, 426-427, 429-431, 467, 524, 538, 547
  defined, 11, 266, 419
Elements in humans, 336
Elephants, 426, 442
Elimination, 4, 13, 153, 277, 413, 574
Elongation, 4, 237, 244, 251, 283-284, 298, 302-307, 317, 361
  in translation, 298, 302
  RNA transcript, 283-284
Elongation factors, 303
Embryo, 2, 5, 8-10, 12-13, 16, 25, 43-44, 104, 117-118, 121-123, 126, 129-130, 133, 140, 251, 382, 405, 436-437, 447, 462, 471, 481-486, 488-489, 495-498, 510, 550-551
  cloning, 10, 12, 126, 462, 551
  development of, 8-9, 25, 104, 118, 121-123, 126, 382, 471, 481, 489, 495
  growth of, 382
Embryonic, 4, 6-7, 25, 43, 86, 104, 118, 122-125, 127, 129, 140, 155-156, 249, 344, 356, 406, 437, 481-483, 485, 488, 491-492, 494-495, 497-498, 547, 551
  period, 43, 122, 129
  stem cells, 4, 7, 249, 494-495
Embryonic development, 25, 86, 104, 124-125, 127, 129, 140, 344, 406, 437, 481-483, 485, 491-492, 494, 551
  animal, 140, 491
Embryonic stem cells, 4, 7
  production of, 4
Emphysema, 23, 451
Encephalopathy, 10, 313
Endangered species, 558, 565
Endocrine gland, 23
Endometrium, 329, 342
Endonucleases, 9, 11, 247, 260, 270, 330
Endoplasmic reticulum, 30-31, 307, 313
  rough, 30-31, 307
  smooth, 30-31
Endoplasmic reticulum (ER), 31
  rough, 31
  smooth, 31
Endosperm, 117-118, 175, 335
Energy, 11, 21, 31, 37, 98, 103, 195, 216, 234, 239, 247, 282, 303, 305-306, 320, 324, 326, 346, 370, 416, 421, 434, 440, 453, 578
  activation, 21
  cellular, 21, 31, 103, 234, 247, 320, 324, 421
  forms of, 21, 282, 320
  free, 103, 326
  mitochondria and, 31
  proteins and, 21
Energy source, 346, 370
Energy sources, 320
Enhancers, 4, 284-285, 337, 357-359, 361-362, 368-370, 384, 418-420, 427, 446
enterococci, 316
Enterotoxin, 202-203, 451
Enterotoxins, 202
Entry, 1, 71, 123, 190-191, 195, 197, 302-303, 346, 526, 562
  virus, 1, 197, 526
Envelope, 30-31, 35, 38-39, 124
  nuclear, 30-31, 35, 38-39, 124
Environment, 7-8, 16-17, 27, 30, 80-81, 97-98, 100, 128, 294-295, 322, 325, 345, 350, 354, 367-370, 376, 378-379, 384, 413, 424, 428, 434, 444, 446, 470, 476, 504, 506-511, 513, 518, 523, 531, 547-548, 552, 554, 575
Environmental genomics, 434
Environmental issues, 476
Enzyme, 1-2, 4, 7, 9-14, 23, 37, 59, 65, 81-82, 84, 96, 99, 101, 103-104, 125-126, 129, 145, 153, 177, 192, 194, 196, 210, 213, 215, 231, 236-242, 244, 246-247, 249-251, 256, 263, 265-266, 272, 274, 282-285, 288, 290, 292, 294, 301-303, 308-310, 314-316, 328-330, 332, 334-336, 340, 345-347, 350, 365, 369, 371, 393-399, 401, 403-405, 409-411, 413,

416, 423, 425, 440, 442, 450-451, 453-454, 457, 459-461, 467, 469, 473, 475-476, 527, 542, 558, 560, 573-574
  active site, 282, 288
  constitutive, 345, 347, 369, 371
  DNA replication, 2, 7, 10-13, 82, 145, 177, 196, 231, 236-242, 244, 246-247, 249-251, 256, 263, 328-329, 340, 345, 442
  inactivation, 4, 14, 125-126
  induction, 12, 309
  isozyme, 7
  nomenclature, 9, 82, 283
  repression, 350, 369
  specificity, 84, 237, 284, 290, 336
  structure, 1-2, 4, 7, 9-14, 37, 125, 196, 210, 213, 215, 231, 238, 241, 246-247, 256, 263, 265-266, 284, 288, 292, 309-310, 314-316, 340, 346, 365, 371, 395-396, 403, 442, 542
Enzyme(s), 409
Enzymes, 4, 10-12, 21-22, 28, 38, 129, 175, 194-195, 231, 236, 238-239, 241-242, 244, 247-248, 252, 257, 265, 292, 301-302, 307-309, 312-316, 333-334, 337, 345-349, 351-353, 356, 364-370, 377, 382-384, 393-395, 398, 403-405, 409-411, 416, 424, 442-443, 455, 460, 469-470, 476, 548, 553, 555, 570, 573
  activity of, 129, 349, 364, 369, 404, 573
  adaptive, 10
  as biological catalysts, 21
  catalytic, 314
  components of, 11-12, 21, 316, 347-348, 369, 382, 384, 442
  constitutive, 345, 347-349, 352, 368-369
  defined, 11, 22, 313, 347, 353
  definition of, 38
  DNA repair enzymes, 384
  effects of temperature on, 129
  efficiency of, 307, 333
  evolution of, 337
  extracellular, 195, 377, 382
  in DNA replication, 12
  inducible, 345, 348, 351, 353, 368-369
  inhibitors of, 129, 382
  intracellular, 10-11
  kinases, 38, 244, 424, 442
  makeup of, 345
  pancreatic, 21, 292
  proteins as, 315-316
Enzyme(s)
  restriction, 409
Enzymes
  RNA polymerase, 10, 12, 292, 312, 347-348, 351, 353, 356, 365, 368, 370
  synthesis of, 4, 11-12, 21-22, 231, 236, 239, 241, 244, 252, 265, 292, 308-309, 334, 345-346, 368, 370, 555
  viral, 247, 292, 366, 384, 393, 455, 469, 476
Eosin, 84
Ephrussi, Boris, 102, 308
Epidemiological studies, 552
Epidermal growth factor, 148, 380, 571-572
Epidermis, 488
Epilepsy, 103, 105, 425, 430, 443, 511, 518
Episome, 198
Epistasis, 4, 88, 90, 106, 112-113, 500
Epithelial cells, 340, 383, 457, 563, 571
  epithelium, 383
Epstein-Barr virus, 384
Equational division, 4
Equilibrium, 6, 11, 140, 233, 264, 274, 314, 378, 526-529, 533, 542-543, 545
Error-prone repair, 333
Erythema, 331
Erythrocytes, 309-310, 313
Erythromycin, 111, 316
  inhibits protein synthesis, 111
  resistance, 111
  synthesis, 111
Escherichia, 12, 24, 31, 146, 211, 255-256, 394, 429-430
Escherichia coli, 12, 24, 31, 146, 211, 255-256, 394, 429-430
  bacteriophage, 211, 255-256
  blood, 31
  chromosome, 12, 31, 146, 255-256, 429-430
  culture medium, 211
  discovery of, 256, 394
  DNA, 12, 24, 31, 146, 211, 255-256, 394, 429-430

flagella, 31
genetically modified, 394
genome, 24, 146, 255, 429-430
  genome of, 255, 429-430
  nucleic acid probe, 12
  nucleoid, 31, 255
  number of genes, 429-430
  plasmids, 394
  replication, 12, 255-256, 394
  ribosomal subunits, 12
  RNA polymerase, 12
  size, 12, 255-256, 394, 429-430
  size of, 255-256, 394, 430
  transformation, 211, 394
  transformation of, 211
Escherichia coli (E. coli), 146
Esophagus, 152, 380
Essential nutrients, 8
  amino acids, 8
Ester, 217
  ester linkage, 217
Estradiol, 129
Estrogen, 362, 553
Estrogens, 129
Ethanol, 453
  biotechnology and, 453
Ethical issues, 130, 139, 495, 567
Ethics, 154, 449-477, 495
  of genetic testing, 459, 461, 471
  of recombinant DNA technology, 449-450, 452, 459
Ethidium bromide, 392, 403, 406
Euchromatin, 5, 123, 263, 445
Eugenics, 5, 153-154, 476
Eukaryote, 228, 268, 369, 419
  cell division, 268
  chromosomes, 268
  DNA, 228, 268, 369, 419
  gene expression, 268, 369
  genetics, 228, 268
  nucleus, 268
  RNA interference (RNAi), 369
  transcription, 369, 419
  viruses, 228
Eukaryotes, 2, 5, 8-9, 12-13, 18, 29, 31, 44, 101, 116, 159-184, 186, 190, 194-195, 201, 214-215, 223, 225, 227, 231, 234-236, 242-244, 246-247, 250-252, 255-256, 260, 263-265, 267-269, 272-273, 280-282, 284-286, 288-289, 291, 293-294, 299-300, 306-307, 313, 316, 330-332, 336, 344, 354-355, 357, 362, 369, 413, 415, 417, 428-431, 476, 480, 488-489, 543
  animals as, 354
  cell division in, 8
  distribution, 214, 234, 269, 476, 543
  DNA replication in, 2, 12, 236, 242, 244
  gene expression in, 12, 268, 289, 354, 488
  morphology, 201
  nutrition, 101
Eukaryotic cell, 31, 254, 260, 281, 299
Eukaryotic cells, 9, 20, 29-30, 243-244, 254-255, 258, 260, 281, 285, 306, 313, 332, 355-356, 360, 451
  animal cell, 30
  DNA replication in, 244
  initiation of transcription in, 285
Eukaryotic genes, 2, 6, 14, 280, 283-287, 355-357, 359, 368, 419-420, 429, 445-446
Eukaryotic genomes, 11, 224, 243, 254, 263, 266, 322, 334, 365, 429, 445
Eukaryotic ribosomes, 299
eukaryotic species, 356
Europe, 17, 130, 226, 411, 418, 470, 533, 540-542
Evidence, 2, 13, 16, 19, 26, 39, 53, 70, 95, 101, 116, 120, 123-125, 127, 131, 133, 139, 147, 152, 161, 176, 188, 207, 209-211, 213-215, 227-229, 232, 234, 236, 240-241, 249, 256, 264, 273, 282, 285-286, 298, 308-309, 324, 330, 340, 352, 356-357, 375, 419, 424, 432, 440-441, 445, 471, 496-497, 506, 522-523, 527, 537, 539-543, 549, 552-555, 558, 560-563, 565-567
Evolution, 5-6, 10, 17, 22, 24, 27, 38, 62, 65, 117, 135, 145, 147, 149, 155, 180, 187, 208-209, 231, 247, 251, 283, 287, 291, 295, 306, 319-320, 337, 341, 408, 426, 428, 431-432, 446-447, 481, 509, 522-523, 525, 529-530, 533, 536, 539, 541-542, 544, 546, 562
  Archaea, 446

chemical, 147, 208-209, 231, 319-320, 446
chordate, 431
definition of, 38, 446, 542
degenerative, 10
eukaryotes, 5, 180, 231, 247, 251, 291, 306, 428, 431
eukaryotic, 5-6, 38, 247, 283, 287, 291, 306, 428, 446
gene families, 431
human, 5, 22, 24, 65, 135, 155, 231, 251, 287, 291, 319, 337, 341, 408, 426, 428, 431-432, 446-447, 481, 509, 523, 525, 533, 539, 541-542, 544, 546, 562
mitochondria, 5, 287
mutations and, 337, 431
of animals, 481
of development, 446, 481
of enzymes, 231
of genes and genomes, 24
of introns, 10, 287
of mitosis, 208
of viruses, 231, 428
process, 5-6, 10, 38, 65, 117, 135, 180, 187, 208, 231, 247, 283, 306, 320, 408, 509, 522, 541
rRNA, 208, 306
viral, 247, 320
virulence, 209
Evolutionary biology, 523
evolutionary relationships, 418, 432, 539
Evolutionary time, 5
Excretion, 570, 576
Exercise, 45, 71, 178, 226, 267, 314, 338, 366, 403, 409, 426, 443-444, 467
Exon, 5, 10, 287-289, 337, 340, 362-363, 419-420, 423, 523-524, 526
Exons, 1, 286-289, 295, 340, 362-363, 419-420, 422-423, 446, 462, 526, 534, 543
Exonuclease, 238, 241, 244, 322, 328
Exonuclease proofreading, 322
exonucleases, 328
Experiments, 17-19, 26, 52-55, 57, 59, 62-63, 98, 124, 127, 163-165, 169, 173, 175, 177, 181, 187-188, 190, 193, 195, 199, 201, 203-204, 209-210, 212-214, 227, 250, 275, 277-279, 282, 299, 329, 348, 368-369, 387, 399, 403, 409, 459, 474, 481, 486, 495-496, 498, 501, 507, 523, 531, 537, 553
controlled, 17-18, 54, 459, 486, 495-496
designing, 409
natural, 17, 26, 53, 57, 62, 474, 523, 531
steps in, 403, 481
Expiration, 409
Expression vector, 5, 452
eukaryotic, 5
promoter, 5, 452
Expression vectors, 396
Extension, 7, 57, 224, 246, 262, 284, 315, 401, 407
Extreme environments, 446-447
Eye, 19, 23, 84, 87, 93-94, 98, 104, 108-109, 112, 132, 138, 146, 155, 161, 163, 167, 169, 180-183, 290, 308, 336, 366, 386, 409, 434, 469, 497-498, 510-511, 515, 518, 567
photoreceptors in, 434
structure, 19, 87, 155, 386, 497, 518
Eyes, 19, 84, 94-95, 98, 103-104, 109-110, 112, 127, 133, 138, 146-147, 154, 161-163, 167, 179-182, 381, 387, 409, 515, 534
color blindness, 95
structure of, 19

F

F factor, 5, 7, 12, 185, 188-195, 199, 203-205, 349, 371
F plasmid, 347
Factor B, 193
Factor VIII, 96, 337, 451, 576
Facts, 106, 203, 558
FAME, 90
Familial hypercholesterolemia, 23, 70, 328
Families, 5, 74, 84, 130, 140, 150, 153, 307, 336, 387, 431-432, 434, 438, 444-445, 472, 476, 496, 573
Family, 5, 8-9, 28, 37, 53, 65, 69, 72, 86, 103, 105, 111, 130, 148, 155-156, 265-267, 270, 307, 313, 324, 327, 338-339, 341, 374, 380, 385, 387-388, 421, 425, 427, 432-433, 446, 460-462, 474, 476, 489, 491, 495-497, 536, 542, 545, 548, 558, 563, 570, 573-576

of humans, 65, 542
Family histories, 105, 339
Family planning, 130
farm animals, 27, 451, 455-457, 474, 551
Fatty acids, 31, 84, 186, 455
Feathers, 97
Females, 1, 8, 10, 12, 14, 33, 43, 69-70, 72, 94-97, 108, 110-112, 116-117, 119-120, 122-133, 136, 146, 151, 161, 163-165, 167, 180-182, 263, 321, 413, 537, 548
Fertility, 5, 45-46, 127, 155, 185, 188, 194, 497, 507, 536
Fertility factor (F factor), 188
Fertilization, 1, 4, 6, 8, 10, 26, 33, 38-39, 43-44, 47, 54-57, 59, 61, 63, 65-66, 68, 74, 101, 103-104, 116-120, 123-124, 127, 130-131, 136-137, 139, 141-142, 148-150, 157, 161, 167, 169, 172, 436-437, 459, 462, 482-483, 492, 495-496, 500, 510, 535, 543, 548, 550-551, 561
completion of, 38, 47, 459
in vitro, 1, 6, 26, 130, 459, 462, 495, 551
process of, 1, 8, 33, 38-39, 43, 65, 74, 118, 123, 495
random, 4, 6, 8, 43, 55-56, 59, 61, 63, 66, 68, 116, 119-120, 131, 136, 535
Fertilizers, 514
Fetal circulation, 132
Fetal development, 122, 155, 433
Fetus, 1, 139-140, 153, 155, 338, 459-461, 554, 574
defined, 153
development of, 139, 459, 554
Fever, 94, 203, 468, 474
Fiber, 5, 37, 44, 47, 103, 196, 254, 260-263, 265, 268, 270, 284, 425, 505, 518
Fibers, 2-3, 31-32, 35-39, 44-45, 103, 196, 204, 212, 218-219, 234, 254, 259-260, 262, 269-270, 355, 378
Fibroblast, 125, 331
Fibroblasts, 124, 126, 177, 249, 331
Field trials, 22
Filament, 5
filters, 368, 434, 461
membrane, 461
Finches, 531
Fingerprint analysis, 558, 564
Fire, 364, 558
Fire, Andrew, 364
FISH, 5, 116, 120, 140, 154, 224, 326, 404-405, 411, 457, 462, 537-538, 541, 544, 572
genetically engineered, 457
transgenic, 457
FISH (fluorescent in situ hybridization), 224
Fishes, 129, 537
Fitness, 5, 9-10, 12, 81, 147, 149, 322, 530-531, 536
relative, 5, 12, 530-531
Fixation, 4, 444
nitrogen, 444
Flagella, 2, 31
bacterial, 31
control, 2, 31
eukaryotic, 2, 31
of prokaryotic cells, 31
staining of, 2
structure, 2, 31
synthesis, 2, 31
Flavr Savr tomato, 453
Flies, 19, 25, 59, 68, 73-74, 92-94, 98, 108, 127-128, 131, 137, 146, 161, 163-164, 169-170, 179-181, 258, 308, 336, 431, 483, 487, 498, 517, 531, 548
Flight, 322, 565
Flight muscles, 322
Flowers, 74, 82, 90-91, 98, 102, 143, 177, 489, 491, 503, 543
Flu, 465
Fluorescence, 5, 10, 116, 130, 176, 224, 405, 407-408, 435-436, 457, 459, 561, 563, 572
gel electrophoresis, 405, 407-408
Fluorescence in situ hybridization (FISH), 5, 116, 572
Fluorescence microscope, 572
Fluorescence microscopy, 10, 176
Fluorescent dyes, 407-408, 435, 463, 559
Fluorescent in situ hybridization (fiSH), 224, 404, 411
Folded proteins, 313
Folic acid, 151, 464, 554
Food, 22, 202-203, 333, 402, 450, 452-456, 468, 470, 476, 506, 514, 529, 531, 537, 553, 578
contamination, 476

genetically altered, 22, 468
pH, 476
safety, 450, 453, 470
safety of, 450, 453
selection, 450, 453, 514, 529, 531, 537
storage, 455
transgenic, 22, 202-203, 450, 452-456, 476
Food and Drug Administration (FDA), 450, 468
food production, 453, 514
Food products, 470
foods, 96, 470, 474, 476, 555
Forensic science, 5, 558, 567
formylmethionine, 280, 302, 307
Fossil, 11, 431, 538-541
eukaryotic, 11, 538
Fossil record, 11, 538-540
Fossilization, 441
Fossils, 401-402, 441, 541-542
human, 541-542
Founder effect, 5, 534
foxes, 108
FOXP2 gene, 432
Frameshift mutation, 5, 274, 294, 321, 386
Frameshift mutations, 273-274, 291, 321, 323, 325, 327, 333, 337, 340-341, 349
Franklin, Rosalind, 219, 222
Fraternal twins, 4, 132, 178
Free radicals, 103, 326, 384
Frogs, 109, 488
Fruits, 513
Functional genomics, 415, 420-421, 423, 445
Functional groups, 283
Fungal toxins, 325
Fungi, 1, 13, 25, 34-35, 43, 99, 182, 214, 256, 321, 329, 333, 354, 429
and cancer, 329
cells of, 34-35, 214
characteristics, 13, 321, 429
classification, 321
life cycles of, 43
morphology, 321
morphology of, 321
reproduction, 1, 13, 34, 43
unicellular, 13
Fusion protein, 368, 377, 450, 452, 474, 571
Fusion proteins, 452
fX174 bacteriophage, 255
genome, 255

G

G0 phase, 378, 381
G1 checkpoint, 5, 387
G1 phase, 5, 47, 142, 243, 378, 381
G2 phase, 333
gal operon, 334
Galactose, 23, 341-342, 345-346, 350, 395
Galactosidase, 96, 346-347, 349-350, 369-371, 395, 450-452
-galactosidase, 346, 370-371
Gall, Joe, 245, 264
Galls, 396
Gamete, 5-6, 9, 17-18, 29, 39-42, 44, 47, 52, 55-61, 63-64, 73-74, 80, 88, 100-101, 103, 119, 127, 131, 136-137, 139, 141-142, 148, 150-151, 155-157, 159-163, 165-167, 169, 175, 181, 320, 337, 339, 482, 524-525, 550-551
Gametes, 4-7, 12, 15, 17-18, 28, 29, 33, 38-39, 41-43, 52, 55-56, 59, 61, 63-65, 68, 72, 74, 76, 85, 90, 95, 100, 116-120, 123, 127-128, 131-132, 135-136, 139-142, 144-145, 148-151, 154-157, 159-163, 165-173, 175, 179, 181-182, 207-208, 214, 320-321, 339, 459, 482, 515, 524-525, 529, 532, 536, 544
algae, 116
and meiosis, 18, 29, 33, 38-39, 41-43, 63, 136, 208
chromosomes in, 7, 18, 29, 63-64, 120, 123, 131-132, 135-136, 148, 155, 159, 163
defined, 17, 148, 320, 536
variation in, 6, 12, 38, 41, 43, 100, 117, 135-136, 139-142, 144-145, 148-151, 154-157, 175, 524
Gametogenesis, 8, 10, 42, 122
and meiosis, 42
and oogenesis, 42
gamma rays, 326, 330
Gardasil, 452
Garrod, Archibald, 307
Gas, 327

Gastrointestinal tract, 310
Gastrula, 436
Gastrulation, 497
  in humans, 497
G-banding, 263
GC content, 412-413, 446
GC ratio, 447
Gel electrophoresis, 14, 403-405, 407-408, 438-439,
    442, 460, 558, 560
  DNA, 14, 403-405, 407-408, 442, 460, 558, 560
  in Southern blotting, 405
Gelsinger, Jesse, 468-469
Genbank, 226, 367, 418-419, 427, 434, 442, 444
Gender, 130, 560
Gene, 1-14, 16-28, 33, 45-46, 52, 55, 58-59, 61-63,
    65, 68, 70-75, 77, 80-101, 103-113, 116, 118,
    122-129, 131-133, 135, 137-138, 142,
    145-148, 152-153, 155, 157, 159-161, 163,
    166, 169-172, 175-183, 185, 187-188,
    190-192, 194-196, 201-205, 208, 215,
    221-223, 226, 229, 238-239, 242-243, 246,
    250-251, 256-259, 262-263, 266-270,
    272-273, 280-287, 289-291, 293, 295, 298,
    300, 308-310, 314-317, 319-342, 343-372,
    373, 375-388, 392, 394-406, 409-412,
    415-416, 418-426, 428-438, 441-446,
    449-452, 454-476, 480-498, 500-503, 508,
    510-511, 513, 515-519, 522-529, 532-536,
    539, 542-544, 546, 547-555, 565, 571-577
  definition, 269, 422, 446, 542
  eukaryotic, 2, 5-7, 9, 11-14, 20, 26, 111, 116, 194,
      201-202, 215, 243, 246, 250, 256-258,
      262-263, 266-267, 269-270, 280-281,
      283-287, 291, 314, 316, 322, 332, 334,
      344-345, 354-360, 362, 365, 368, 370,
      392, 399, 419-420, 428-429, 434-435,
      445-446, 451, 476, 491
  homologous, 1-3, 6-7, 9, 12-13, 18, 33, 46, 52, 63,
      74, 80-83, 122, 133, 142, 147-148, 155,
      157, 160-161, 163, 175-176, 178, 191,
      194-195, 283, 300, 319, 328-329,
      332-333, 420-421, 432, 458, 470
  housekeeping, 548
  information flow, 2, 208, 281
  open reading frame, 9, 425, 444
  overlapping, 3, 9, 204, 281, 353, 359-360, 398,
      416, 432, 443-444, 485, 491
Gene amplification, 5, 146, 380, 572
Gene annotation, 420
Gene cloning, 195, 398, 400, 418
Gene cluster, 155, 300, 345, 433, 495
Gene duplication, 5, 147, 432-433, 445, 481
Gene expression, 4, 8, 12, 14, 20-21, 26, 80, 82,
    96-98, 124-126, 128, 142, 221, 262, 268,
    272, 282, 289-290, 298, 314, 322, 335-337,
    340, 343-372, 378, 381, 392, 403-404, 406,
    415, 420, 424, 431, 434-436, 442, 463-464,
    476, 480-483, 485, 488, 491-493, 496-497,
    511, 513, 547-549, 551-555, 572
  analysis of, 12, 221, 268, 289, 336, 406, 415, 424,
      431, 434, 442, 464, 480-482, 493, 497,
      513, 555
  as transcription, 356, 359, 366
  control of, 82, 345-346, 350, 362-363, 369, 476,
      482, 492-493, 552-553
  DNA methylation and, 549
  eukaryotes, 8, 12, 268, 272, 282, 289, 336, 344,
      354-355, 357, 362, 369, 415, 431, 476,
      480, 488
  genetic code and, 272, 282, 289-290
  in bacteria, 82, 272, 282, 335, 345, 354, 368
  mutations and, 336-337, 431
  nucleic acids in, 290
  prokaryotes, 12, 272, 344-345, 349, 353-355, 357,
      359, 369, 415
  regulation, 8, 14, 26, 82, 262, 282, 322, 337,
      343-372, 403, 406, 420, 442, 476,
      496, 547-549, 552, 555
  regulation of, 8, 14, 262, 322, 343-372, 378, 403,
      420, 442, 547, 549, 552, 555
  silencing of, 552
  supercoiling, 12
  transgenic, 14, 476, 497
Gene families, 431, 434, 445, 496
Gene family, 5, 380, 421, 432, 446, 462, 491, 496-497
  evolution, 5, 432, 446
Gene flow, 5, 522, 533, 536
Gene function, 2, 11, 270, 322, 334, 366, 370,
    420-421, 424, 437, 456

  regulation of gene expression, 322, 366, 370, 420
Gene mapping, 182, 201
Gene microarrays, 446
Gene pool, 6, 10, 522-525, 529, 532, 544, 546
Gene pools, 524, 529
Gene regulation, 223, 262, 282, 344-345, 347,
    351-357, 362, 364-365, 368-369, 496,
    547-548
Gene silencing, 100, 290, 364-366, 469-470, 550, 552
  in plants, 364
Gene superfamily, 432-433
Gene testing, 459
Gene therapy, 16, 23, 26, 290, 449-450, 452, 467-470,
    473-476, 544, 546
  and cancer, 452
  for cystic fibrosis, 474
  in vivo, 470
  regulation of, 469
  safety of, 26, 450, 469
  targets for, 473
Gene transfer, 7, 14, 22, 27, 187-188, 191-192, 229,
    470-471
  horizontal, 7, 187
  vertical, 7, 14, 187
Genentech, 572
General transcription factors, 360-362, 370
Generalizations, 61, 320, 429
Generalized transduction, 200
Generation, 1-2, 4-8, 14, 15, 17-19, 24, 28, 29, 33, 38,
    41, 52-63, 69-70, 72-75, 82, 88, 90-92,
    95-96, 100, 106, 108-109, 113, 131-132,
    140, 149, 152, 161, 163, 167, 169-170, 181,
    207, 233-234, 250, 252, 320-321, 328, 379,
    408, 415, 417, 427, 445, 456, 497, 501-503,
    505-506, 508-509, 511-512, 515-517,
    524-527, 529-530, 532-533, 535-536,
    543-545, 550-551, 554, 561, 577
Genes, 1-14, 15-19, 22-26, 28, 29, 31, 33, 38, 45, 52,
    55-56, 62-65, 71, 73-75, 77, 80-82, 85-101,
    103-105, 107-108, 110, 113, 116-119,
    122-129, 132-133, 135, 137-138, 142-143,
    146-150, 152, 154-155, 157, 159-170,
    172-175, 177-184, 185-188, 190-196,
    199-205, 207-208, 215, 225-227, 242-244,
    247, 249, 251, 254, 256-258, 263-264,
    266-268, 270, 272, 280-281, 283-287,
    290-293, 295, 298-300, 307-310, 312-313,
    315-317, 319-320, 322, 325, 327-329,
    331-337, 339-342, 344-359, 362-371,
    373-388, 395-399, 401, 403-405, 410-412,
    415-416, 418-437, 442-447, 449-450,
    453-456, 458-459, 462-465, 467, 469-476,
    480-493, 495-498, 500-503, 506, 508,
    510-514, 516-517, 522-525, 528, 534,
    539-541, 543-544, 546, 547-555, 564,
    570-578
Gene(s), 131, 367
Genes
  artificial, 1, 7, 14, 16, 320, 333, 340, 396, 453, 458,
      476, 508, 511-513, 517, 523
  chemically synthesized, 13, 458
  cloning and, 185, 397, 399, 403, 429
  codominance, 3, 86, 107
  constitutive, 345, 347-349, 352, 368-369, 371, 387,
      456, 549
  differentiation and, 7, 496, 555
  evolution and, 428, 432
  expression of, 2, 4-6, 9-11, 13, 25, 55, 65, 74-75,
      80, 82, 88-90, 93, 96-101, 104, 108,
      123-128, 132, 142, 152, 208, 215, 257,
      299, 320, 322, 329, 335-337, 344,
      347-351, 353-357, 362, 364-368, 370,
      379-382, 384, 388, 396, 404, 427, 431,
      434, 450, 458, 463, 469-470, 473, 475,
      481, 483-486, 488-491, 493, 496-498,
      503, 506, 511-512, 548, 550-551,
      554-555
  function of, 6, 25-26, 65, 81-82, 87, 101, 175, 242,
      247, 251, 256, 258, 268, 285, 298, 300,
      312-313, 316, 319-320, 322, 334,
      341-342, 346, 369-370, 383, 403, 410,
      415, 420, 432, 446, 496, 539, 550
Gene(s)
  homolog, 131
Genes
  in plasmids, 195, 335, 397
  inducible, 7, 344-345, 348, 350-351, 353, 368-369,
      371
  linked, 1, 6, 9-10, 12, 86, 88, 94-96, 98-99, 101,

    105, 107-108, 110, 116, 119, 123-128,
    132, 137, 146, 152, 159-161, 163,
    165-167, 170, 173, 175, 177, 179-183,
    195-196, 200, 226, 247, 249, 256, 268,
    284, 308, 313, 316, 320, 322, 325, 327,
    337, 341, 346, 363, 385, 387, 432, 467,
    512, 539, 551, 553, 555, 574, 576-577
  mutation and, 103, 180, 328, 341, 475, 498, 522,
      543, 552, 575
  mutation rates, 320, 328, 336, 376, 381, 544, 546
  mutations of, 105, 462, 483-484, 553
  p53 gene, 380-381, 462
  prokaryotic, 9-10, 12, 31, 243-244, 257, 272,
      284-286, 292, 299, 307, 344-345,
      353-355, 358, 368, 419, 428-429, 435,
      445
Gene(s)
  regulatory, 367
Genes
  regulatory genes, 483, 491
  repressible, 11, 344-345, 351-353, 367-369, 371
  structural, 1, 4-5, 8-13, 31, 93, 143, 147, 207, 226,
      244, 263, 267, 312-313, 336, 345-353,
      355-356, 358, 367-371, 411, 415-416,
      418, 421, 423-424, 429, 446, 498
  structural genes, 4-5, 9-10, 345-353, 358, 367-371
  synthetic, 1, 12-13, 290-291, 320, 333, 365-366,
      368, 384, 449-450, 458-459, 472, 474
  transgenic organisms, 22, 25
Genetic code, 3, 6, 9, 20, 26, 257, 271-295, 298, 310,
    321, 340-341, 349, 413
  deciphering, 20, 274, 276
  degeneracy, 279, 294, 321
  degeneracy and, 279
  dictionary of, 278
  evolution of, 6
  mutations and, 321
  Mycoplasma, 280-281
  nucleotides and, 20
  universal, 272-273, 280
  universality of, 280
  wobble, 275, 279, 281, 292, 294
Genetic counseling, 139, 154-155, 471, 474, 577-578
Genetic crosses, 15, 53, 66, 102, 159, 453, 481
  dihybrid, 66
  monohybrid, 53, 66
Genetic disease, 338, 373-375, 385-386, 425, 462,
    467, 577
Genetic diseases, 8, 24, 105, 153, 319-320, 327-328,
    373-374, 459, 462, 464, 467, 470-471, 475,
    574
  screening for, 153, 462
Genetic disorders, 9, 23, 25, 71, 93, 100, 103, 148,
    250, 363, 402, 430, 449, 461, 463, 467,
    469-470, 481, 489, 525, 544, 547, 551
  alkaptonuria, 71
  multifactorial, 430
  testing for, 461
Genetic diversity, 6, 10, 16, 64, 320, 444, 454, 523,
    534
  viruses, 454
Genetic drift, 5-6, 10, 522, 526, 534-536
Genetic elements, 6, 11, 335
Genetic engineering, 6, 202, 394, 449-477, 514
  and evolution, 6
  antibiotics, 456, 470
  biotechnology and, 450, 453, 475
  methods, 453, 459-460, 462, 467, 474, 514
  monoclonal antibodies, 451
  of crop plants, 453, 455, 514
  of plants, 449, 453
  of transgenic animals, 455-456
  pharmaceutical products, 450-451, 456, 470
  therapeutic products, 449-450
  Ti plasmid, 454
  transgenic animals, 449, 451, 455-457
  genetic information, 1-8, 13-14, 15-17, 19-20, 26-28,
      29, 31, 33, 38, 41, 47, 63, 81, 99, 101,
      116, 120-122, 125-127, 132, 135, 140, 144,
      147, 149-150, 153, 178, 185-189, 194-195,
      199, 207-208, 215, 217, 222-225, 227, 229,
      247, 252, 254-256, 258, 270, 272-273,
      281-282, 284, 291, 298-300, 310-311, 313,
      344, 355, 368, 385, 422, 471, 523, 574
  transcription of, 1, 20, 81, 127, 208, 223, 284, 291,
      300, 368
  translation of, 6, 273, 291, 298
Genetic Information Nondiscrimination Act, 385, 471
Genetic map, 177, 181, 184, 190, 204, 512

bacterial, 190, 204
Genetic mapping, 165, 175, 190
  conjugation, 190
Genetic maps, 5, 175, 190, 512
Genetic marker, 12
Genetic markers, 6, 11, 466, 474, 476
genetic material, 1, 3, 11, 19, 23, 29-31, 33-34, 39, 43,
  63, 86, 98, 116, 135, 143-145, 148-150, 166,
  188, 195-196, 207-209, 211-216, 220-221,
  223, 226-229, 231, 254-256, 260, 262-263,
  269, 326, 355, 522
    genotype and, 98
    phenotype and, 135
Genetic profiles, 575
Genetic recombination, 3, 9, 26, 159, 180, 185-190,
    192, 194-196, 199, 201, 204, 231, 247-248
  by crossing over, 159
  conjugation, 3, 185, 187-190, 192, 194-195, 199,
    204
  in prokaryotes, 231
  plasmids, 185, 194-195
  transduction, 185, 187, 196, 199, 204
  transformation, 3, 185, 187, 195-196, 199, 204
Genetic screening, 153
Genetic testing, 1, 6, 23-24, 153, 385, 401, 404,
    459-462, 471-472, 475
  fetal, 1, 404, 459-460
  for breast cancer, 385
  for cystic fibrosis, 462
  newborn, 153
genetic transformation, 229
Genetic variation, 6, 9, 19, 38-39, 41, 43, 47, 64-65,
  116, 135, 147, 159, 175, 187, 195, 208, 319,
  466, 506-507, 522-524, 532, 536, 542-543,
  575
    and meiosis, 38-39, 41, 43, 47, 208
    between populations, 522, 536
    defined, 19, 523, 536
    in human populations, 506
    within populations, 319, 523
Genetically modified (GM) plants, 455
Genetically modified organism (GMO), 6
Genetically modified organisms (GMOs), 203, 470
genetically modified plants, 411, 413, 449
Genetics, 1, 3, 5-6, 8-9, 11, 14, 15-28, 29, 32-33, 45,
  51-78, 79-80, 86, 92, 99, 101, 103, 105-106,
  113, 115, 117, 130, 134, 135, 153, 155-156,
  159, 165, 177-178, 182, 184, 185, 193, 196,
  207-210, 213, 222-223, 225, 227-228, 231,
  247, 252, 253-254, 262, 268, 271-272,
  282-283, 286, 292, 297, 308-309, 313-314,
  319, 336, 338, 341, 343-344, 373-389,
  391-392, 398, 402, 409, 415, 421, 424,
  427-429, 437, 444, 449-450, 471-472,
  475-477, 479-498, 499-519, 521-546,
  547-556, 557-568, 569-579
  and correlation, 503, 505
  and pedigree analysis, 74
  bacterial, 1, 3, 5, 8-9, 11, 14, 22, 25, 99, 185, 193,
    196, 210, 213, 228, 247, 283, 429,
    449-450, 523, 528, 564
  central dogma of, 22, 28, 208, 292
  cytology and, 3, 159
  eukaryotic, 5-6, 9, 11, 14, 20, 26, 29, 247, 252,
    254, 262, 283, 286, 292, 313-314, 344,
    392, 428-429, 476, 491, 538
  gene therapy, 16, 23, 26, 449-450, 475-476, 544,
    546
  human, 1, 5, 8-9, 11, 15-16, 18, 21-26, 28, 33, 45,
    65, 69-71, 99, 103, 105, 130, 135, 153,
    155-156, 177-178, 182, 184, 227, 231,
    262, 268, 308, 313-314, 319, 336, 338,
    341, 373, 375, 380, 382-385, 387-388,
    398, 409, 415, 421, 424, 427-429, 437,
    444, 449-450, 471-472, 475-476,
    480-481, 486, 488-489, 495, 497, 500,
    506, 509-510, 517, 523, 525-526, 528,
    531, 533-535, 538-542, 544, 546, 547,
    549, 551, 553, 555, 558-559, 561-562,
    564-566, 570-571, 576-578
  law of segregation, 75
  microbial, 415
  molecular biology and, 113, 392
  nitrogen fixation, 444
  of speciation, 1, 524, 543
  polygenic inheritance, 502, 515-516
  prokaryotic, 9, 272, 286, 292, 344, 392, 428-429
  Punnett squares, 55
  recombinant DNA technology, 6, 11, 15, 22, 24-26,

391-392, 398, 402, 409, 449-450, 494,
    572
  reverse, 5, 11, 117, 156, 262, 388, 398, 519, 547,
    554
  transfer of genetic information, 14
  yeast, 1, 14, 24-25, 103, 182, 427-429, 450,
    538-539
genome, 1-8, 10-11, 13-14, 15-16, 22-26, 28, 33, 45,
  59, 71, 103, 105, 113, 122, 135-136,
  142-143, 145-150, 174-175, 178, 191, 202,
  204, 208, 215, 224, 231, 236, 244, 251,
  254-255, 262-264, 266-269, 280, 289,
  299-300, 319, 322-323, 329, 334-337,
  341-342, 354, 356, 359, 363, 365-367, 373,
  376, 381, 383-384, 387, 392-393, 396,
  398-400, 403-404, 406-408, 410-411,
  415-424, 426-438, 444-447, 449, 454-456,
  458-460, 462-463, 465-467, 469-476, 481,
  483-484, 487, 489, 492, 496, 502, 510, 512,
  514, 517, 539-542, 547-552, 554-555,
  558-559, 562, 565, 570, 575-578
  adenovirus, 469
  analysis, 1-3, 5-6, 10, 13, 15, 24, 59, 71, 103, 122,
    150, 174-175, 178, 191, 202, 204, 208,
    215, 224, 251, 254-255, 264, 266-268,
    280, 289, 300, 319, 323, 335-336, 396,
    400, 404, 406-407, 411, 415, 418,
    420-424, 426-428, 430-432, 434-438,
    445-447, 454, 459-460, 462-463,
    465-466, 472, 474, 476, 481, 489, 496,
    510, 512, 539-542, 555, 558-559, 562,
    565, 576-578
  annotating, 418, 420
  Archaeoglobus, 429
  assembly, 4, 244, 416-417, 434, 458
  chloroplast, 454
  core, 215, 251, 262, 356, 428, 548, 559, 565
  defined, 11, 22, 24, 148, 178, 251, 266-267, 363,
    419, 432
  eukaryotic, 2, 5-7, 11, 13-14, 26, 174, 202, 215,
    224, 244, 254-255, 262-264, 266-267,
    269, 280, 299, 322, 334, 354, 356, 359,
    365, 392, 399, 419-420, 428-429,
    434-435, 445-446, 476
  evolution, 5-6, 10, 22, 24, 135, 145, 147, 149, 208,
    231, 251, 319, 337, 341, 408, 426, 428,
    431-432, 446-447, 481, 539, 541-542,
    562
  HIV-1, 474
  mitochondrial, 103, 280, 426, 437, 542, 562, 565
  mycoplasma, 280, 429, 458-459
  Nanoarchaeum, 429, 446
  nuclear, 4, 7, 13, 22, 103, 244, 356, 426, 447, 540,
    542
  of viruses, 231, 255, 269, 366, 383, 428
  organelles, 3, 5, 8, 103
  pan, 430-431
  prokaryotic, 10, 244, 299, 354, 392, 419, 428-429,
    435, 445
  retrovirus, 7, 11, 14, 384
  RNA, 1-4, 6-8, 10-11, 13-14, 22, 146, 208, 215,
    224, 231, 244, 251, 255, 264, 266, 269,
    280, 289, 299-300, 337, 354, 356, 363,
    365-367, 384, 392, 398-399, 403-404,
    426, 463, 470, 496, 547, 549
  segmented, 483
  sequencing, 4, 6-7, 11, 24-25, 334, 336, 373, 392,
    396, 398-400, 403, 406-408, 410-411,
    415-418, 421-422, 426-429, 431, 434,
    438, 445-447, 454, 472-474, 481, 489,
    541, 562, 570, 575-578
  SV40, 475
  Thermoplasma, 429
  TMV, 215
  virus, 1, 7, 11, 22, 204, 215, 255, 280, 365-366,
    383-384, 387, 410, 415, 454, 465, 467,
    469, 474-476, 510
  yeast, 1, 14, 24-25, 103, 142, 269, 280, 300, 396,
    427-430, 458, 465, 539
Genome sequencing, 25, 334, 396, 400, 417-418,
    422, 427-428, 431, 454, 472-473, 489, 575,
    577-578
  applications of, 396, 418, 454, 472
  shotgun, 417, 422
  whole-genome, 400, 417-418, 422, 575, 577-578
Genomes, 1, 6, 8, 11-12, 24, 136, 142, 147-148, 153,
  185, 224, 243, 254, 263, 266, 322, 334,
  336-337, 365, 375, 383-384, 392, 398, 400,
  404, 408, 410, 415-417, 421-422, 424,

426-432, 434-435, 437, 442, 445-447,
  458-459, 466, 469, 473, 475, 488-489, 491,
  510-512, 540, 542, 546, 555, 570, 577-578
  complete, 136, 142, 243, 415, 427-428, 430, 437,
    458, 469, 473, 489, 577-578
DNA fingerprinting and, 8
DNA replication and, 243
  eukaryotic, 6, 11-12, 224, 243, 254, 263, 266, 322,
    334, 365, 392, 428-429, 434-435,
    445-446, 491
  evolution of, 6, 24, 337
  mapping, 6, 185, 447, 511-512, 555
  number of genes in, 437
  prokaryotic, 12, 243, 392, 428-429, 435, 445
  review of, 512
  sequencing, 6, 11-12, 24, 334, 336, 375, 392, 398,
    400, 408, 410, 415-417, 421-422,
    426-429, 431, 434, 445-447, 473, 489,
    546, 570, 577-578
  size of, 224, 404, 422, 430-431
  vertebrate, 428, 431, 540
  viral, 148, 384, 415, 469, 475
Genome-wide association studies, 6, 449, 465, 467,
    475, 577
Genomic analysis, 12, 397, 430-431, 465, 495, 578
Genomic DNA libraries, 398, 411
Genomic imprinting, 6, 100
Genomic libraries, 22, 24, 398
genomic library, 6, 397-398, 411
Genomics, 6, 15, 24, 26, 45, 71, 178, 226, 267, 314,
    338, 366, 385, 389, 392, 403, 408-409,
    415-447, 449-450, 459, 467, 470, 472-474,
    480, 510, 514, 539, 542, 546, 555, 569-579
  applications of, 403, 418-419, 439, 449, 459, 472,
    474, 571
  comparative, 415, 419, 428, 430-432, 434, 443,
    445-447, 539, 555
  definition of, 422, 446, 542
  environmental, 415, 434, 446, 470, 480, 510, 555,
    577-578
  ethical aspects of, 578
  functional, 6, 267, 314, 415, 420-421, 423, 427,
    429, 433, 436-438, 445-446, 449, 467,
    473, 555
  microbial, 415, 434-435, 459
  proteomics, 24, 314, 338, 415-447, 480
  viral, 366, 415, 423, 449, 467, 474
Genotype, 1, 3, 5-6, 9, 12-14, 19, 21-22, 28, 43,
  55-59, 61, 66, 68, 73-75, 81, 83-85, 88-91,
  96-98, 101, 104, 106, 108-110, 113, 121,
  131, 167, 170, 180-182, 187, 196, 201-202,
  347, 349, 368-369, 404, 413, 449, 458,
  460-461, 464, 475-476, 496, 498, 500, 502,
  506-507, 509-510, 515, 518, 522, 524-533,
  535, 543, 560, 563-566, 574
  designation, 56, 84, 202, 563
  homologous recombination, 458
Genotypes, 1, 5-6, 12-13, 55-56, 59, 61-62, 72-77, 82,
  85, 87-92, 98, 106-110, 112-113, 127, 161,
  167, 169-170, 181-182, 201-202, 204, 268,
  310, 348-349, 368, 449, 460-461, 465, 507,
  510-512, 515-516, 518-519, 524-532, 535,
  542-543, 545, 547-548, 555, 558, 563-565
  changes in, 12-13, 268, 368, 511, 524, 527,
    530-531, 543, 547-548, 555
  defined, 77, 511, 535
Genus, 141, 143, 537, 541, 545
Geographic range, 3, 13, 522
Germ cells, 122, 249, 320, 456, 482, 493, 496-498,
    550-551
  human, 122, 249, 456, 497, 551
German measles, 223
Germination, 13, 118
  seed, 118
Gestation, 122
Gibbons, 542
Gilbert, Walter, 349
Glands, 117, 450, 456-457, 493, 498, 523
  function, 456, 493, 498, 523
Gleevec, 377, 573
Global climate change, 514
-globin gene, 21, 286, 461, 475, 543
Glottis, 145
Glucocorticoids, 359
Glucose, 23, 96, 103, 125, 186, 345-346, 350-351,
    369-370, 395, 450
  as an energy source, 346, 370
  biosynthesis of, 351
  respiration, 103

synthesis of, 125, 345-346, 370
Glucose metabolism, 450
Glucose-6-phosphate, 96, 125
Glutamic acid, 21, 277, 310-311, 317, 441
Glutamine, 275-276, 280-281, 292-293, 295, 327, 441
   genetic code, 275-276, 280-281, 292-293, 295
   structure, 292
   synthesis, 275-276, 280-281, 292-293
Glyceraldehyde, 439
Glycerol, 316
   structure, 316
Glycerol phosphate, 316
Glycine, 275, 292, 294, 328, 576
   genetic code, 275, 292, 294
   structure, 292
   synthesis, 275, 292, 294
glycocalyx, 30-31
   eukaryotic cell, 31
Glycolysis, 103, 346
Glycomics, 424
Glycoprotein, 83, 382
Glycoproteins, 30
Glycosidic bond, 323, 330
Glyphosate resistance, 471
GM food products, 470
gold, 558
Golden Rice, 22, 455
Golgi body, 30
Gonads, 117, 122, 128, 132
   development of, 122
gp120, 474
Gradient, 3-4, 12, 233, 264, 438, 483-484, 497, 533
grains, 117, 137, 234-235, 258, 265, 514
granules, 30
   polysaccharide, 30
Granulocytes, 140
Graphs, 132, 268
   types of, 268
Grasshoppers, 119
Green fluorescent protein, 449
Griffith, Frederick, 209
Group I and group II introns, 288
Growth, 1-3, 5, 7-8, 10-11, 34, 42, 99-100, 102, 138, 140, 143, 148, 177, 186-188, 198-199, 290, 308-309, 315-316, 333, 340, 345, 351, 354, 360, 363-364, 370, 373-374, 376-384, 386-387, 397, 400, 434, 440, 450-451, 453-457, 468-470, 494, 497, 512, 514, 550-554, 571-572, 576
   control of, 2, 7, 34, 315, 345, 363, 377, 379, 384, 552-553
   foods, 470
   measurement, 3
   microbial, 434
   oxygen, 309, 384
   population, 2-3, 5, 8, 10, 138, 140, 387, 455, 514
   temperature effect, 99
Growth factor, 100, 138, 148, 315, 376, 380, 451, 469, 551, 571-572
Growth factors, 308, 377-378, 380-381, 384, 386, 494, 551
Growth hormone, 451, 456
   human, 451, 456
Growth hormone (GH), 456
Growth media, 351
Growth plate, 188
GTP, 216-217, 302-305, 317, 380, 386, 496
   in protein synthesis, 302
Guanine, 13, 20, 96, 99, 216-217, 220-221, 228-229, 323, 325-327, 339-341, 548
   in DNA, 20, 216-217, 221, 228, 325-326, 339-340, 548
Guanine (G), 228
Guanosine, 216, 218, 288, 380
Guthrie, Arlo, 86
Guthrie, Woody, 86
Gynecomastia, 120
GyrA, 243
Gyrase, 239, 242-243

H

Haemophilus, 256, 393, 417, 429
   H. influenzae, 417
Haemophilus influenzae, 256, 393, 417, 429
   genome, 393, 417, 429
Hair, 86, 89-90, 97, 313, 402, 411, 426-427, 459, 471, 473, 511, 515, 518, 534, 554, 558, 567
Half-life, 363-364
Hands, 120, 138, 367, 410, 489

Haploid, 1, 3-6, 8, 10-11, 13-14, 18, 22, 29, 32-34, 38-44, 47-48, 55, 63-64, 116-118, 127, 133, 135-137, 140-143, 146, 150, 155-157, 161, 184, 186, 214, 262-263, 266, 269, 300, 363, 422, 427, 445, 512
Haploid cells, 13, 43, 47
   eukaryotic, 13
Haploid number, 5, 8, 33-34, 39-41, 43, 63-64, 127, 137, 141, 161, 184
Haplotype, 6, 413, 563
HapMap, 6, 427
Hardy-Weinberg equilibrium, 526-528, 542, 545
Head, 65-66, 97, 104, 138, 196, 200, 202, 204, 211, 255-256, 269-270, 316, 482, 484, 487-488, 497-498
Heads, 65-66, 196, 199
Health, 15-16, 22, 25, 72, 105, 130, 184, 195, 203, 209, 223, 226, 249, 290-291, 342, 385, 389, 421, 426-428, 431, 438, 456, 464, 467, 470-472, 495, 555-556, 567, 572, 574-575, 578
Hearing, 105, 432
   disorders, 105
Heart, 16, 21, 70, 96, 105, 138, 225, 283, 310, 313, 344, 363, 366, 426, 430, 442, 450-451, 494-495, 554, 574-576
   circulation, 21
   congestive heart failure, 442
   fetal, 363
   insect, 451
   mammalian, 451
   structure and function, 442
   structure of, 310
Heart attack, 70, 451
Heart attacks, 70
Heart disease, 16, 105, 366, 430, 450, 495, 554, 575-576
Heart failure, 442
Heart rate, 426
Heat, 6, 16, 141, 209-211, 223, 228, 233, 263, 313, 370-371, 395, 400, 404-405, 413, 416, 425, 440, 447, 451, 514, 548
Heat-shock proteins, 313
Heavy metals, 335, 359-360, 457
helicase, 177, 242-243
Helicases, 239, 243, 332
Helix, 3-4, 6, 14, 20, 189, 194-195, 207, 216, 219-223, 228-229, 231-232, 235-236, 239-240, 242-243, 247, 250, 252, 259, 261-262, 268, 281-284, 298, 312-313, 330, 332, 340, 421
   a, 3-4, 6, 14, 20, 189, 194-195, 207, 216, 219-223, 228-229, 231-232, 235-236, 239-240, 242-243, 247, 250, 252, 259, 261-262, 268, 281-284, 298, 312-313, 330, 332, 340, 421
   double, 3-4, 6, 14, 20, 189, 194-195, 207, 216, 219, 221-223, 228-229, 231-232, 235, 239-240, 242-243, 247, 252, 259, 262, 268, 281-282, 330, 332, 421
Heme, 310, 433
Heme group, 433
Heme groups, 310
Hemoglobin, 21, 81, 93, 147, 155, 208, 280, 291, 305, 308-310, 312-313, 316-317, 432-433, 461, 469, 528
   fetal, 155, 291, 433
Hemolytic anemia, 425, 443
   autoimmune, 425
Hemophilia, 23, 71, 96, 110-111, 130, 137, 156, 322, 324, 337, 341, 451, 576
Hepatitis, 290, 384, 402, 451-452
Hepatitis B, 290, 384, 451-452
   and cancer, 452
Hepatitis B virus, 384, 452
Hepatitis C, 290, 384
Hepatitis C virus, 384
Herbicide, 22, 449, 453-454, 470-471
   resistance, 22, 453-454, 470-471
Herbicides, 22, 154, 333, 453-455
   and water pollution, 333
   resistance to, 22, 455
   RoundUp, 454
   transgenic, 22, 453-455
Herceptin, 571-573
Hereditary diseases, 23, 323
Heredity, 5, 14, 16-17, 28, 54-55, 101, 110, 113, 194, 207, 209, 213, 225, 227, 307-308
Heritability, 2, 6, 467, 500, 506-510, 515-519, 554, 578
Hermaphrodites, 118-119, 492-493, 498

Hermaphroditism, 498
Herpesvirus, 384, 451
Hershey, Alfred, 201, 212
Heterochromatin, 6, 13, 98, 123, 157, 263-264, 269-270, 366, 445, 548-550
Heterokaryon, 7, 331-332, 340
Heterozygous, 2, 6-7, 11, 55-61, 65, 69-70, 72-75, 78, 83, 85-86, 88-92, 95-98, 100, 104-105, 107-108, 110-112, 125-126, 132, 144, 146, 148-150, 154-156, 160-161, 166-167, 169-173, 175, 177, 179-183, 307, 310, 324, 337, 341, 382-383, 460-461, 475-476, 483, 502, 511, 515, 518, 525-526, 528, 532, 534-536, 542-543, 545-546, 561-562, 565, 573-574, 576
Heterozygous genotype, 96-97, 167
Hfr cell, 192-193
Hfr cells, 190-192
Hfr strain, 190-191, 204-205
HindIII, 393, 395, 411
Hippocrates, 16
Histidine, 1, 14, 152, 186, 199, 275-276, 278, 292-293, 295, 333, 353, 431
   genetic code, 275-276, 278, 292-293, 295
   structure, 1, 14, 292, 353
histocompatibility antigens, 31
Histone, 9, 126, 244, 260-262, 265, 269-270, 286, 307, 356-357, 373, 377, 422, 432, 511, 547-550, 553-554
Histone acetylation, 548, 550
Histones, 2, 244, 255, 260-262, 269, 287, 314, 355-356, 366, 510-511, 549-550, 554
HIV, 7, 148, 216, 290, 366, 402, 467, 469, 474, 526, 531
   resistance to, 526
   structure, 7, 216
HIV infection, 474
HIV-1, 290, 474, 526, 531
Hodgkin's disease, 46
Holliday junctions, 247
Holoenzyme, 238-239, 242, 251, 282-285
Homeotic genes, 481, 483, 488-491, 496-498
Hominid, 541
Hominids, 541
Homo, 34, 430, 432, 540-541
Homo erectus, 541
Homo neanderthalensis, 432, 540
Homo sapiens, 34, 430, 540
Homologies, 539
Homologous chromosome pair, 12
Homologous chromosomes, 1-3, 6-7, 9, 13, 18, 29, 32-33, 37, 39, 41, 46-48, 52, 63-64, 74, 80-82, 142, 157, 163, 175-176
   nondisjunction of, 48
Homologous genes, 178, 420-421, 432
Homologous pairs, 32-33, 36, 39, 43, 47, 63, 80, 155, 160
   described, 36, 39
   independent assortment of, 160
   segregation of, 63
Homologous recombination, 247-248, 319, 329, 332-333, 458
Homologs, 7, 12-14, 33, 39, 41, 43, 46-47, 52, 63-64, 93, 136, 139, 141-142, 145, 148-149, 159-160, 169-172, 175-176, 179, 181, 247-248, 258-259, 265, 268, 421, 529, 536
Homozygous, 1, 6-7, 13, 55-57, 59, 61, 69, 72-74, 78, 82, 84-86, 88-90, 92, 95-98, 104, 108, 110-113, 118, 128, 132, 146, 154-156, 161, 167, 170, 175, 177, 179-182, 310, 316, 321, 386, 456, 460-461, 475, 483, 486, 492, 495-497, 501, 503, 505-507, 511, 515-518, 523, 526, 528-530, 534-535, 542-543, 545-546, 554, 562, 573-574
Homozygous genotypes, 310, 507, 535
   defined, 535
Horizontal gene transfer, 7, 187
   transduction, 187
   transformation, 187
Hormone, 21, 96, 132, 359-360, 362, 380, 415, 418, 420, 425, 450-451, 456, 573
   interactions, 360, 415
   releasing, 425
   steroid, 96, 359-360
   types of, 96, 380
Hormones, 97, 122, 129, 133, 314, 359, 362, 378, 384, 440, 450, 494, 572
   embryonic, 122, 129, 494
   fetal, 122

in blood, 133
pituitary, 450
thyroid, 362
Horses, 16, 110, 113, 508
hospitals, 195, 203, 578
Host, 1, 10, 13-14, 22, 26, 185, 191-192, 194-198,
    200-201, 204-205, 212, 215, 227, 255-256,
    269, 337, 347, 369, 383-384, 392, 394-398,
    401, 410-411, 413, 444, 447, 450-452, 454,
    465, 469, 474-476, 481, 546
virus, 1, 22, 185, 196-197, 204, 215, 227, 255-256,
    383-384, 410, 451-452, 454, 465, 469,
    474-476
Host cell, 13, 185, 194-198, 200, 212, 215, 255-256,
    269, 347, 383-384, 394-397, 447, 469, 475
Host cells, 14, 22, 392, 394-398, 401, 410, 450-452,
    454, 465, 469, 474
Hosts, 185, 196, 201, 270, 410, 413, 451, 474
Hot springs, 401
Housekeeping genes, 548
Hox genes, 487-489, 491, 496-497
in plants, 489, 491
Human body, 251, 333-334, 375, 384, 428, 446, 547
elements in, 334
Human development, 480, 489
human disease, 25, 71, 105, 202, 268, 332, 337-338,
    424, 430, 442-443, 446-447, 555, 570
infectious, 25
Human evolution, 426, 432, 541-542, 544, 546
Human genetics, 9, 336, 424, 525, 528, 534-535, 546
Human genome, 4, 15-16, 24, 26, 71, 122, 148, 175,
    178, 231, 244, 251, 266-267, 269, 322, 334,
    336, 363, 365, 373, 387, 396, 398, 404,
    407-408, 415, 417-418, 420-424, 426-431,
    433-435, 437-438, 445-447, 449, 459,
    462-463, 471, 473-474, 476, 539-540, 555,
    559, 562, 570, 576-578
evolution of, 24, 539
sequencing of, 24, 408, 426, 570, 577-578
size of, 269, 404, 422, 430-431, 462, 476, 562
Human Genome Project, 4, 15, 24, 26, 71, 122, 175,
    178, 373, 396, 407-408, 415, 417, 421-422,
    424, 427-428, 430, 445, 459, 471, 473, 578
ethical aspects of, 578
history of, 26, 428
results of, 471
Human growth hormone, 451, 456
Human immunodeficiency virus (HIV), 7, 216
structure of, 216
Human Microbiome Project, 427-428
Human papillomavirus, 380
Human population, 83, 124, 140, 150, 152, 339, 388,
    523, 541-542, 562
Human Proteome Project, 438
Humans, 1, 6-7, 9-12, 16, 18, 25, 27, 33-34, 37-38, 43,
    59, 65, 69, 74-75, 82-83, 86-89, 94-95, 97,
    99-100, 103, 105, 107-108, 112, 116,
    120-124, 126-127, 130-133, 135-138, 144,
    147-150, 153, 174, 182, 203, 209, 247, 251,
    256-257, 265-268, 270, 280, 289, 300, 309,
    313, 320, 322, 324, 328-332, 334, 336-341,
    358, 363, 375, 383-384, 388, 402, 411-413,
    415, 417, 420-424, 428-434, 442, 445-446,
    449-452, 454, 470, 473-474, 476, 481,
    485-486, 488-489, 491, 495, 497, 500, 506,
    509, 511, 513, 515-516, 528, 531-532, 535,
    538-544, 551-552, 554-555
differences among, 506, 511, 538
Huntington disease, 14, 23, 25, 70-72, 86, 99-100,
    152, 313, 322-323, 328, 340, 459, 574, 576
Hybridization, 3-5, 9, 12, 14, 53, 59, 62, 116, 141, 143,
    146, 223-225, 228, 264-265, 286, 299, 331,
    366, 368, 399-401, 404-406, 411, 413,
    435-436, 460-461, 463-464, 498, 514, 534,
    562, 572
colony, 400
DNA-DNA, 399
genomic, 4, 9, 12, 401, 404, 411, 534, 572
in situ, 5, 116, 224, 264-265, 404-405, 411, 498,
    572
in Southern blotting, 405
microarrays, 4, 435-436, 463-464, 562
nucleic acid, 3, 9, 12, 223-224, 228, 399, 404
plant, 53, 59, 141, 143, 413, 514, 562
Hybrids, 10, 143, 156, 400, 404, 453, 536
sterility of, 143, 156
Hydrogen, 3-4, 13, 20, 207, 216, 219-223, 228-229,
    232, 239-240, 245, 247, 279, 284, 299-300,
    312, 317, 323-324, 340-341, 394, 406, 434,

447
formation, 3-4, 219, 247, 324
Hydrogen bond, 221, 223, 312
DNA, 221, 223, 312
protein, 223, 312
water, 221
Hydrogen bonds, 4, 20, 219-223, 228-229, 232, 239,
    245, 247, 284, 300, 312, 340
characteristics of, 228-229, 312
in transcription, 223
nucleic acid, 20, 220, 223, 228-229
water, 220-221
Hydrogen peroxide, 324
Hydrolase, 423
Hydrolases, 423
Hydrolysis, 37, 216, 228, 239, 303
enzymatic, 239
of ATP, 37, 216, 239
Hydrophilic, 221, 279, 311
Hydrophobic, 221, 279, 311, 447, 474
Hydroxyl group, 4, 92, 216, 229, 240, 406
hydroxyl radicals, 324
Hypercholesterolemia, 23, 70-71, 328
Hypersensitivity, 332
Hypertension, 175, 443, 446, 466
Hypodermis, 493
Hypotheses, 73-74, 78, 81, 132, 541
Hypothesis, 8-9, 13-14, 31, 66-68, 70, 73-74, 76-77,
    82, 84, 106, 108-109, 111, 120, 124-126,
    132, 140, 146, 155, 177, 180, 207-208, 215,
    218, 227, 259, 275, 279, 292, 294, 298, 306,
    308-310, 316, 375, 496, 498, 501-502, 516,
    541-542
modification of, 82, 84, 106, 108-109, 111, 292, 310

I
I gene, 347-348, 350
Identical DNA sequences, 416, 432
Identical twins, 8, 65, 178, 413, 447, 474, 559, 565
Imatinib, 573
Immune response, 203, 452-453, 468-469, 476
primary, 453
proteins in, 476
Immune system, 314, 337, 379, 452, 467, 473-475,
    495, 526, 572
disorders, 467, 495
evolution, 337
Immune systems, 249, 468
Immunity, 195, 202, 432, 436, 452, 474
acquired, 202, 474
active, 195, 436, 452
innate, 432
Immunization, 203, 451
active, 451
AIDS, 203
anthrax, 203
passive, 451
passive immunization, 451
recommended, 203
Immunodeficiency, 7, 216, 467, 473-474, 526
Immunogen, 474, 476
Immunoglobulins, 314
classes of, 314
Immunotherapy, 388
Implantation, 130, 382, 455, 550, 574
Imprinting, 6, 8, 100-101, 126, 366, 547, 550-552
In situ hybridization, 5, 116, 224, 404-405, 411, 498,
    572
In vitro fertilization, 26, 130, 459, 495, 551
In vitro fertilization (IVF), 551
Inactivated vaccines, 452
Inclusion, 504
Incomplete dominance, 7, 82-83, 86, 106-108, 111,
    501
Incubation period, 375
Independent assortment, 2, 7, 39, 48, 57-59, 61,
    63-66, 68, 77, 86, 92, 116, 149, 159-161
Independent assortment of chromosomes, 48
indole, 316
Induced mutations, 320, 324, 329, 339-340
Induced pluripotent stem cells, 495
Inducer, 7, 345, 347-348, 350, 370-371
Inducers, 346-347
Inducible operons, 350
Induction, 6, 12, 309, 320, 326-327, 351, 360, 492,
    494, 498
Infanticide, 130
Infants, 65, 130, 139-140, 145, 203, 310, 469, 544
development of, 139, 145, 469

Infection, 3, 99, 148, 185, 196-198, 200-205, 212-215,
    223, 227, 255, 269-270, 282, 290, 365-366,
    380, 382-384, 393, 425, 452, 454, 456-457,
    465, 474, 526
acute, 380, 384, 425, 465
chronic, 384, 425
T4, 196-197
virus, 185, 196-197, 204, 214-215, 223, 227, 255,
    290, 365-366, 383-384, 452, 454, 465,
    474, 526
infections, 290, 316, 366, 384, 451, 457, 467-468,
    523, 528
Infectious disease, 464-465, 476
emerging, 465
infectious diseases, 25, 450, 465, 467
acute, 465
patterns of, 465
Inflammation, 384, 457, 476
acute, 384
chronic, 384
tissue repair and, 384
Inflammatory bowel disease, 175, 366
inflammatory response, 468
Influenza, 366, 410, 425, 452, 465
avian, 465
bird flu, 465
vaccine, 452, 465
Influenza virus, 410, 465
Information flow, 2, 208, 228, 281
biological, 2
Ingestion, 203, 454
Inhalation, 476
Inheritance, 2-3, 5, 8-12, 14, 17-19, 24, 26, 28, 33,
    52-56, 61-63, 69-71, 75, 80-81, 83-92,
    94-97, 101-104, 106-113, 163, 165, 178,
    208, 215, 256, 307, 327, 500-503, 515-517,
    522, 544
chromosome theory of, 2, 18-19, 26, 28, 95
DNA and, 3, 5, 8, 10, 71, 215, 256
epigenetic, 8, 101
law of segregation, 75
nucleic acids and, 215
of genes, 2-3, 5, 8, 10, 12, 17-19, 24, 26, 33, 63,
    80, 91, 101, 108, 165, 178, 500, 503,
    516, 522
of hemophilia, 111
polygenic inheritance, 10, 502, 515-516
Punnett squares, 55
inherited disorders, 23, 71, 99, 104-105, 404
Inherited mutations, 382-383
Inhibition, 12, 201, 350, 365, 383
Inhibitors, 129, 363, 382, 496, 498
Initiation, 6-7, 9-12, 34, 37, 236, 238-240, 242-244,
    246, 251-252, 277-285, 294, 298, 302-307,
    315, 317, 344, 349, 353-354, 357-362, 366,
    369-371, 388, 419-420, 486, 491, 552
DNA replication, 7, 10-12, 236, 238-240, 242-244,
    246, 251-252, 357
transcription, 7, 10-12, 238, 246, 277-285, 294,
    298, 302, 305-306, 344, 349, 353-354,
    357-362, 366, 369-371, 388, 419-420,
    486, 491, 552
translation, 6, 9-10, 12, 277-281, 285, 298,
    302-307, 315, 317, 344, 349, 353, 362,
    366, 370, 420
Initiation complex, 303-304, 360-361
Initiation factors, 302-303, 317
Injection, 209-210, 364, 456, 474-476
Innate immunity, 432
Inner cell mass, 4, 495
Inner ear, 88
Innocence Project, 566-567
Inoculum, 186
dilution, 186
Inorganic salts, 186
Inosine, 279
Inosinic acid, 300
Insect, 22, 87, 131, 451, 453-455, 470, 566
pathogens, 454
insect bites, 566
Insect resistance, 22, 453, 470
Insecticide, 454
insects, 7, 22, 119-120, 258-259, 270, 454, 498
Insert DNA, 412
Insertion, 5-7, 20-21, 274-275, 281, 284, 320-321,
    323, 328, 334-337, 340-341, 386, 394, 397,
    454, 468-469, 475, 554
Insertion sequence (IS), 335
Insertions, 11, 267, 291, 322-323, 325, 328, 336-337,

341-342, 462, 562, 576
Insulin, 21, 100, 175, 215, 287, 292, 295, 315, 363, 401, 418-420, 450-452, 551
    as protein, 420
    genetically engineered, 450-451
    manufacturing, 452
    recombinant, 215, 401, 418, 450-452
    size of, 100, 287
Integrase, 469
Integration, 191-192, 469, 475
Intellectual property, 472
Interactions, 13, 24, 87, 92, 107, 178, 223, 260, 262, 265, 269, 283, 298, 305, 317, 347, 353, 360-361, 415, 424, 428, 432, 437, 441-443, 447, 482, 491-493, 510-511, 513, 547-549, 552, 570, 575
    plant and animal, 491
Interference, RNA, 12, 132, 290, 364, 369, 470
Interferon, 215, 287
Interferons, 451
Interphase, 2, 5-7, 34-36, 38-40, 44-45, 47, 124, 132, 141-142, 252, 254, 258, 260, 262-263, 344, 356, 370
Interrupted mating, 190-191
Intron, 7, 286-289, 337, 363, 419-420, 423, 430-431, 433, 524, 576
    self-splicing, 287
    yeast, 288, 430
Introns, 7, 10, 13, 256-257, 286-289, 320, 336, 340-341, 362, 365, 398, 419-420, 422, 429-430, 433, 446-447, 523, 526, 543
Invasiveness, 374, 470, 571
Inversion, 7, 9, 98, 144, 147-149, 154-156
Inversions, 1, 147-150, 155, 267, 337, 424, 576
Invertebrates, 13, 147, 422, 431-432, 437, 447
    deuterostomes, 431
Involuntary muscle, 100
Ion, 12, 340, 421, 423
Ion channel, 423
Ion channels, 421
Ionizing radiation, 11, 247, 326, 332, 380, 387
Ions, 14, 186, 354, 395, 439-440
Iron, 23, 117, 455, 576
Irradiation, 329, 493
Irrigation, 514
Isolation, 1, 10-11, 16, 105, 194, 210, 222, 224, 250, 309, 349, 352, 498, 522, 524, 536
Isoleucine, 277-278, 280-281, 292, 294
    genetic code, 277-278, 280-281, 292, 294
    structure, 292
    synthesis, 278, 280-281, 292, 294
Isomer, 347
Isomerase, 423, 439
Isoniazid, 573
Isotope, 11, 232-234
Isotopes, 7, 233, 399
    radioactive, 233, 399
Isozyme, 7
Italy, 469

J

Jacob, François, 190, 273, 282, 345
Japan, 195, 226, 411, 418, 422
Jaw, 537
Joints, 307, 310
Jumping genes, 334

K

Kanamycin, 194, 411
    resistance gene, 411
kanamycin resistance, 411
Karyotype, 7, 13, 18, 33, 120-123, 131, 138-141, 145, 155-156, 178, 374, 411, 460
Karyotypes, 3, 120-121, 125, 132, 155, 263, 405
    of human chromosomes, 120
Keratin, 313
Keto group, 323
Kidney, 23, 53, 96, 122, 140, 152, 376-377, 398, 475, 494, 576
    functions, 377, 494
    stones, 576
    structure, 376
Kidney stones, 576
Kidneys, 22-23, 310
    cancer, 23
    human, 22-23, 310
    structure, 310
    structure of, 310

Killing, 153, 326, 330, 454, 493
Kilobase, 424, 429, 534-535, 546
Kinases, 38, 243-244, 378, 387-388, 423-424, 432, 442
Kinetic energy, 98
Kinetochore, 7, 37, 265
Kinetochores, 3, 378
Kingdom, 15, 137, 434-435, 469, 555
    of humans, 555
Kingdoms, 117, 491, 496, 555
Klinefelter syndrome, 120-121, 125, 131-132, 136, 156
Klug, Aaron, 261, 300
Knockout, 5-7, 420
Knockout mice, 7, 420
Kornberg, Roger, 261
KpnI, 395
Kuru, 10

L

Labor, 401, 415, 446, 474
Lac operon, 322, 345-351, 353, 367-369, 371
lac promoter, 350-351, 371
lac repressor, 347, 349-351, 353, 369-371
lacI gene, 347, 369
Lactose, 186, 345-352, 368-371, 395
    metabolism, 345-348, 350-351, 368-369
lacZ, 346-347, 395, 397, 450, 452, 458
lacZ gene, 346, 395, 397, 450, 452, 458
Lagging strand, 4, 7, 9, 240-242, 244-247, 252
Lake, 537-538, 544, 546
    turnover, 546
Lake Victoria, 544, 546
Lakes, 537, 544, 546
Lambda phage, 24
Land, 514, 537
Landsteiner, Karl, 83
Language, 121, 145, 280, 432
Larva, 103-104, 481-482, 484
Larynx, 145
Law of segregation, 75
Lead, 16, 22, 25, 37, 65, 68-69, 88, 94, 101, 116, 120, 122, 129, 133, 135, 139, 144, 152, 156, 207, 223, 227, 245, 249, 310, 319-325, 327, 335, 340, 344, 354, 363, 370, 373-375, 377, 381, 383-385, 387-388, 437, 467, 470, 473, 476, 481, 483, 495, 513, 522-524, 529, 536, 546, 548, 551-552, 566, 573-574, 576-577
Leader sequence, 7, 12, 286, 353
Leading strand, 8, 240-242, 244-245, 252
Learning, 47, 56, 265, 268, 314, 385, 427-428, 434, 517, 542, 548
    social, 428, 542
Lederberg, Joshua, 199, 345-346
Legal issues, 130, 449, 470
    labeling, 470
Lens, 93, 425
Leprosy, 429
Leptin, 337, 415, 420-421
Lethal mutations, 322, 326, 483
Leucine, 8, 82, 188, 277-280, 291-292, 294, 309, 340, 353, 421, 425, 441
    genetic code, 277-280, 291-292, 294, 340
    structure, 188, 291-292, 309, 340, 353, 421
    synthesis, 278-280, 291-292, 294, 309, 340
Leucine (leu), 188, 277
Leucine zipper, 8, 421
Leukemia, 138, 376-377, 380-381, 384, 387-388, 425, 443, 467-469, 510, 553, 573
    viruses, 380, 384, 387, 468
Leukocytes, 177
Levan, 120
libraries, 22, 24, 45, 368, 392, 398-401, 403, 410-411, 416
    cDNA, 368, 398-399, 403, 411
    genomic, 22, 24, 45, 398, 401, 403, 411, 416
    plasmid, 368, 398, 400-401, 411
Licensing, 472
Life, 8, 11, 13, 16-17, 21-22, 24-25, 27-28, 32, 34, 43-44, 47, 53, 70, 72, 96-97, 99-100, 104, 116-118, 121, 123, 131, 137-138, 175, 184, 196-197, 201, 204, 209, 211-212, 231, 249-250, 270, 307, 313, 322, 363-365, 374, 378, 383, 385, 388, 427-428, 432-433, 447, 449, 453, 459, 467, 472-473, 475-476, 481, 492, 514, 547-548, 551, 555, 559, 562, 568, 570, 574, 577
    characteristics of, 13, 138, 175, 209
    molecules of, 364
    processes of, 32, 43-44, 116-117

Life cycle, 8, 11, 13, 25, 32, 43-44, 47, 99, 117-118, 131, 184, 196-197, 201, 204, 211-212, 322, 428, 481, 492, 547-548
    cellular, 47, 212, 428
    Saccharomyces cerevisiae, 25, 428
    temperate phage, 13
Life cycles, 24, 43, 116-117, 365
    of fungi, 43
Life expectancy, 138
Life processes, 231
Ligaments, 441
Ligands, 354
Ligase, 4, 189, 241-243, 246-248, 251-252, 288, 328-331, 378, 394-395, 399, 410, 423
Light, 5, 10, 13, 17, 30, 36, 84, 96, 101, 110, 176, 198, 214-215, 228-229, 238, 252, 254, 257-259, 263, 325-326, 329-330, 340, 370, 376, 380, 384, 392, 399, 403, 405, 408, 432, 434, 440, 444, 456-457, 469, 515-516, 518, 544
    characteristics of, 13, 228-229, 238
    ultraviolet (UV), 214, 325-326, 440
Light energy, 434
Light microscope, 30, 258-259, 263, 456
Light microscopy, 254, 258
Lily, 141
Limbs, 93, 338, 489
Linear DNA, 429
Linkage groups, 161, 195
Linked genes, 95-96, 124-125, 127, 159-161, 163, 165-167, 170, 175, 180-182, 195, 200
    mapping of, 182
Linker DNA, 261, 268
Linker sequences, 398
Lipid, 23, 82, 84, 99, 421, 429, 464
Lipids, 10, 211, 384, 438
    complex, 438
    metabolism, 384
    structure, 10, 211, 438
    synthesis, 10, 211
Lipoproteins, 70
Lists, 71, 84, 178, 226, 267, 327, 503, 509
Litter size, 509
Liver, 152, 282, 333-334, 377, 384, 431, 452, 468, 494, 542, 554, 573, 576
    hepatitis, 384, 452
Liver disease, 468
livestock, 22, 474, 508, 512
Lizards, 129, 132, 134, 140
Locus, 1, 5, 8-10, 14, 33, 63, 81-85, 88-90, 92, 94-95, 113, 127-128, 145, 172, 175, 178-181, 183-184, 205, 267-268, 339-340, 352, 371, 386, 425, 432, 446, 501-502, 511-513, 515, 518, 523, 525-526, 528, 533, 546, 551, 553, 558-566
Long bones, 93
Long terminal repeat (LTR), 8
Longevity, 154
Luciferase, 370
    bacterial, 370
Lumen, 313
Lung cancer, 138, 148, 152, 380, 552, 571, 573
    smoking and, 552
Lung disease, 340
Lungs, 21, 23, 310, 468, 523, 528
    diagram of, 23
    structure, 310, 523
Lupus, 573
Lupus erythematosus, 573
Luria, Salvador, 201
Lwoff, André, 345
Lyase, 423
Lyme disease, 429
Lymphatic system, 382
Lymphocytes, 464
Lymphoma, 374-375, 384, 425, 443, 464, 554, 571, 573
    human, 375, 384, 425, 443, 571
Lyon, Mary, 125
Lysine, 143, 260, 275-277, 279, 293, 295, 310-311, 317, 322, 548-549
    genetic code, 275-277, 279, 293, 295, 310
    structure, 260, 310-311, 317, 548
    synthesis, 275-276, 279, 293, 317
Lysis, 8, 197-199, 201, 204, 212, 304
Lysogeny, 8, 197, 204
Lysosome, 30
Lysosomes, 65
Lysozyme, 196, 213
    phage, 196, 213

T4, 196
Lytic cycle, 10, 200, 204

M

MacLeod, Colin, 19, 209
Macroevolution, 523-524, 536-537
Macromolecule, 4, 10, 13, 20
  amino acids, 10, 20
  chemical bonds, 20
  lipids, 10
  nucleic acids, 10, 13
  proteins, 20
Macromolecules, 313, 475
Macular degeneration, 290, 466, 470
Mad cow disease, 10, 313
MADS-box genes, 491
Magazines, 177, 184, 476, 486
Magic bullet, 290
Maize (corn), 59, 170
Males, 6, 10, 12, 33, 43, 69-70, 72, 94-97, 100, 108,
  110, 112, 116-129, 131-134, 136, 151, 161,
  163-165, 167, 180-182, 320-321, 492, 498,
  537, 561, 563
Malignant tumor, 374-375, 387
Malignant tumors, 8, 152, 374, 382, 386-388, 552
mammalian cells in culture, 323
Mammalian gene, 358
Mammals, 4, 7, 16, 85, 88, 100, 116-117, 119,
  123-124, 127, 131-133, 266, 269, 286, 288,
  313, 328, 412, 428, 485, 538, 548, 550
Mammary glands, 117, 456-457, 498
Mammograms, 342
Mammography, 340, 342
Mannheimia haemolytica, 474
Map units, 164, 174, 183-184, 190, 316
Mapping, 5-6, 159-184, 185-205, 346, 403, 409, 420,
  447, 501, 511-513, 517, 519, 555
  genetic, 5-6, 159, 161, 163, 165-166, 175, 177-178,
    180-182, 184, 185-205, 501, 511-512,
    555
Marfan syndrome, 71, 93-94, 328, 576
Marker genes, 14, 190, 395, 397
Mars, 117
Mass, 1, 3-4, 7, 124, 223, 225, 228-229, 299, 374,
  388, 438-442, 445, 495, 510, 512, 558-559,
  562
Mass spectrometry, 439-442, 445
Master regulatory genes, 491
Mastitis, 456-457, 477
Maternal alleles, 550
Maternal effect genes, 113
Maternal inheritance, 8, 103
Mating, 6, 9-10, 54, 73, 83, 86-87, 89, 102, 108-111,
  113, 118, 143, 184, 188, 190-191, 193, 203,
  321-322, 455, 508, 522, 524-525, 527, 529,
  534-536, 545-546
  fertilization and, 54
  human, 9, 83, 109-110, 184, 322, 455, 525,
    534-535, 546
  insect, 87, 455
  random, 6, 525, 529, 534-535
Mating behavior, 143, 322
Mating type, 102
  yeast, 102
Matrix, 382, 423, 439-441
Matter, 24, 101, 129, 229, 243, 311, 315, 388, 410,
  476, 559
Maturation, 42, 59, 96, 118, 146, 280, 286, 307
McCarty, Maclyn, 19, 209
McClintock, Barbara, 175, 335-336
Measles, 203, 223, 510-511, 518
  signs of, 203
Media, 188, 308, 351, 494, 579
  complex, 351
  enriched, 308
  specialized, 494
medicine, 15-17, 22-23, 26-28, 38, 45, 104, 203, 222,
  249, 295, 336, 364-366, 392, 424, 434, 449,
  473, 476, 483, 486, 556, 569-579
Medulla, 122
Megaspores, 44, 117-118
Meiosis, 2-4, 7-8, 11-13, 18, 26, 29-49, 52, 63-64, 74,
  94-95, 104, 116-118, 121-123, 127, 131-133,
  135-137, 139-140, 142, 144-145, 148-150,
  154-157, 159, 161, 163, 165, 176, 208, 247,
  254, 259, 265, 482, 529, 536, 547
  characteristics of, 13, 133
  errors in, 40
  fertilization and, 123

in animal cells, 42-43
  meiosis I and II, 47, 150
  nondisjunction during, 136-137, 139, 150, 155
  phases, 2, 31-32, 35, 41, 44, 47, 117, 127
  review of, 140
  stages of, 29-31, 40, 44, 47, 63, 482, 547
Meiosis I, 7, 29, 40, 42-43, 46-48, 131, 136, 139, 150,
  154, 156, 529, 536
Meiosis II, 40-43, 46-48, 136, 139, 156
Melanin, 208, 534
Melanoma, 23, 138, 382
Mello, Craig, 364
Membrane protein, 492, 534
Membrane proteins, 492
  membrane-bound ribosomes, 313
Membranes, 30, 65, 307, 314, 461, 482, 498, 571
Membranous organelles, 31
  chloroplasts, 31
  endoplasmic reticulum, 31
  mitochondria, 31
  nucleus, 31
Memory, 126
Mendel, Gregor, 17, 26, 52, 177, 522
Mendelian inheritance, 9, 71, 75, 80, 103, 178, 327
  exceptions to, 80
Mercury, 12, 194, 457
Meselson, Matthew, 232
Mesoderm, 492
Messenger RNA, 8, 10, 12, 20, 31, 208, 223, 272-273,
  282, 290, 292, 298, 346, 362
Messenger RNA (mRNA), 8, 31, 208, 223, 272-273,
  282, 290, 298
  classes, 223
  degradation of, 8, 290
  eukaryotic, 31
  in translation, 208, 298
  prokaryotic, 31, 272
  protein synthesis and, 282
  synthesis of, 208, 282, 290, 298
  transcription, 208, 223, 272-273, 282, 290, 298
  transcription of, 208, 223, 290
  translation, 208, 223, 273, 290, 298
Metabolic pathways, 81, 442
  engineering, 442
  regulation of, 81, 442
  types of, 442
Metabolism, 7, 13, 21, 23, 99, 270, 307, 313, 324,
  341, 345-348, 350-351, 355, 368-369, 384,
  421, 423-424, 426, 436-437, 450, 464, 570,
  573
  complementary, 13
  defined, 313, 347
  prokaryotic, 307, 345, 355, 368
Metabolite, 354, 437, 442
  secondary, 354
Metabolites, 11, 354
Metabolomics, 424
Metagenomics, 8, 415, 424, 434-435, 444, 446
Metal ions, 354
Metaphase, 3, 7, 32-33, 35-40, 44-48, 120, 151, 156,
  176, 224, 234-235, 251-252, 261-262, 268
  in meiosis, 7, 33, 39-40, 46-48, 156, 176
  in mitosis, 38-40, 46-47
  meiosis, 3, 7, 32-33, 35-40, 44-48, 156, 176
  mitosis, 3, 32-33, 35-40, 44-48, 176, 235
  of meiosis, 33, 38, 40, 44, 47-48
  of mitosis, 3, 32, 35, 37, 39, 44, 47-48, 176
Metaphase chromosomes, 7, 44, 151, 224, 252, 268
Metaphase I, 39-40, 46, 156, 252
Metaphase II, 40, 235
Metaphase plate, 36-40, 46-47
Metastasis, 8, 373-375, 378-379, 382-384, 398, 425,
  553-554, 571
Methanococcus, 429
Methanococcus jannaschii, 429
Methanosarcina, 429
Methionine, 7, 188, 199, 277, 279-281, 307, 438
  genetic code, 277, 279-281
  structure, 7, 188, 438
  synthesis, 279-281, 307
methionine (Met), 188
Methyl group, 229, 286, 325, 328, 548
Methyl transferase, 573
Methylation, 8, 101, 262, 265, 291, 328, 356-357, 368,
  370, 373, 376-377, 438, 466, 511, 548-555,
  567
  DNA, 8, 101, 262, 265, 291, 328, 356-357, 368,
    370, 373, 376-377, 511, 548-550,
    552-555, 567

Mexico, 202, 316
mice, 7, 85-86, 89, 110, 123, 125, 132, 138, 202-203,
  209-210, 215, 265, 280, 286, 329, 342, 387,
  399, 418-420, 428, 430, 442, 455-456, 465,
  470, 475, 485-486, 488, 491, 495, 509, 550,
  554-555
Microarray analysis, 435-436, 463-466, 474, 577
Microarrays, DNA, 431, 462-465, 562
Microbes, 25, 393, 428, 434, 439, 441, 444, 446, 449,
  456-457, 459
  definition of, 446
Microbial communities, 435
Microcephaly, 140
Microevolution, 523-524
Microfilaments, 31
Micrographs, 36, 211, 235, 255
Microinjection, 456
Microorganisms, 24, 186, 211, 236, 251, 345, 427-428,
  434, 454, 459
  agriculture, 24, 454
  beneficial, 427
MicroRNAs, 365, 369, 549
MicroRNAs (miRNAs), 365, 369, 549
Microsatellites, 174-175, 264, 512, 517, 559
Microscope, 7, 18, 30, 243, 254-255, 258-259, 263,
  286, 304-305, 435, 456, 572
  fluorescence, 435, 572
  light, 30, 254, 258-259, 263, 456
  resolution, 305
Microscopy, 10, 18, 44, 176, 254, 258, 270, 289, 299,
  306-307, 558
  electron, 18, 44, 254, 258, 270, 289, 299, 306-307
  light, 10, 176, 254, 258
  probe, 10, 306, 558
Microspores, 44, 117-118
Microtubule, 7, 37
Microtubules, 2, 30-32, 35, 37, 45, 313
  animal cell, 30
  of centrioles, 2, 35
  tubulin, 31-32, 45, 313
Middle lamella, 38
Migration, 4-6, 8, 32, 35, 37, 47, 126, 229, 247-248,
  299, 310, 382, 427, 438, 497, 522, 524,
  526-527, 529, 533-536, 542
Milk, 22, 97, 451, 456-457, 474, 500, 508-509, 517
Milkweed, 119
MinD, 81, 288, 302, 322, 345, 351, 364, 386, 437,
  463, 506, 532
Minerals, 455
miRNAs, 365-366, 369, 549-550, 555
Miscarriages, 124, 156
Mismatch repair, 8, 328-329, 333, 340, 342, 376-377
Missense mutation, 8, 321, 338, 387-388, 495, 543
Missense mutations, 327, 340, 524
Mitochondria, 3, 5, 8, 31, 80, 101, 103-104, 214, 244,
  254, 256-257, 269, 280-281, 287-288, 313,
  497, 561
  DNA, 3, 5, 8, 31, 80, 101, 103-104, 214, 244, 254,
    256-257, 269, 280-281, 287, 313, 561
  DNA in, 8, 31, 101, 214, 256, 269
  evolution, 5, 287
  features of, 497, 561
  genetic code, 3, 257, 280-281, 287-288
  genome, 3, 5, 8, 103, 244, 254, 269, 280
  in cellular respiration, 103
  proteins, 3, 5, 8, 31, 103, 214, 244, 254, 256-257,
    280-281, 287-288, 313
  ribosomes, 31, 281, 313, 497
  structure, 31, 214, 254, 256-257, 269, 288, 313,
    497
Mitochondrial DNA, 103, 256-257, 537, 542, 544,
  561-562, 565
Mitochondrion, 30, 47, 281, 561
Mitosis, 2-3, 8, 18, 29-49, 63, 118, 141, 144, 159,
  176-177, 180, 208, 235, 254, 260, 265, 378,
  386-388, 547, 550
  in animal cells, 42-43
  in plant cells, 38
  regulation of, 8, 378, 547
  stages of, 29-31, 40, 44, 47, 63, 547
  term, 37, 40
Model organisms, 8, 15, 24-25, 28, 117, 290, 325, 334,
  337, 421, 428, 430-431, 436, 445, 470,
  480-481
  Arabidopsis thaliana, 25, 428, 430, 481
  Caenorhabditis elegans, 25, 117, 428, 430, 481
  in developmental biology, 481
Models, 25, 203, 207, 219, 247, 251, 306, 366, 442,
  456, 470, 480, 497

atomic, 306
Mold, 34, 102, 308, 384, 496, 539
Molds, 32
Molecular biology, 6, 22, 113, 126, 221, 236, 252, 290, 364, 392-393, 401-402, 409, 480, 498, 570
central dogma of, 22
Molecular clock, 8, 537, 539
Molecular clocks, 539
Molecular genetics, 15, 20, 26, 28, 101, 185, 196, 208-210, 213, 222-223, 225, 227, 252, 272, 282-283, 286, 292, 344
Molecular mass, 438-439
molecular weight, 10, 223, 238, 240, 252, 282, 300, 311, 439-440
Molecule, 1-3, 6-14, 20-22, 28, 29, 31, 81, 84, 89, 92, 104, 146, 188, 191, 194, 198, 201, 207-208, 213-216, 218-219, 222-225, 227-228, 231-233, 238-239, 245, 247-248, 254-255, 257, 261, 266, 269, 272, 274-275, 280-285, 287, 289-292, 299-304, 306-308, 310-314, 316-317, 320-321, 323, 325, 333, 345, 347, 349-350, 352, 354, 362, 367-370, 376-377, 392, 394, 396, 398-399, 402, 406, 409-410, 417, 419, 423, 425, 429, 432, 435, 450-452, 461, 562, 564, 576-578
Molecules, 1-4, 6, 8-14, 21-22, 31, 38, 59, 84, 195-196, 208, 216, 218, 222-225, 228-229, 231-233, 236, 238-239, 244-245, 247-248, 254-256, 260, 269, 273-274, 277, 280-281, 283-285, 287-288, 290-292, 294, 298-303, 306, 309-311, 313-314, 317, 319, 326-327, 340, 344-345, 347-349, 353, 362-366, 368-372, 378-380, 382, 384, 386, 392-395, 397-399, 401-404, 408, 411, 413, 423, 429, 432, 435-437, 442, 446, 453, 462-463, 472, 474, 526, 547, 549, 562
covalent bonds, 9, 228
hydrogen bonds, 4, 222-223, 228-229, 232, 239, 245, 247, 284, 300, 340
hydrogen bonds and, 239
important biological, 549
inorganic, 8, 216, 274
macromolecules, 313
organic, 1, 283, 326
organic molecules, 326
polar, 4, 10, 311
Monkeys, 430, 538
Monoclonal antibodies, 451, 572
Monoclonal antibody, 572
Monod, Jacques, 273, 282, 345
Monohybrid crosses, 53-55, 57, 59, 74, 94
Monomer, 241
Monomers, 20
Monophyletic group, 8, 537
Monophyletic groups, 538
Monosaccharide, 346
Monosaccharides, 346, 395
Monozygotic twins, 8, 516, 518, 547
Moratorium, 410
Morphogenesis, 388
mechanisms of, 388
Morphology, 127-128, 137, 151, 197, 201, 209, 211, 259, 321, 536, 540
Mortality, 124, 139-140, 153, 249, 531
AIDS, 531
mortality rate, 124
Mortality rates, 153
Mosaicism, 154, 510
Mosquito, 34
Moths, 120
Motif, 6, 8, 358, 559
Motifs, 243, 358, 418, 421, 423, 498
motility, 124, 429, 432
eukaryotic, 429
prokaryotes, 429
prokaryotic, 429
Motor proteins, 37
Mountains, 522
Mouse, 8, 24, 34, 85, 90-91, 100, 110, 123, 127, 132, 138, 178, 210, 215, 264-266, 286-287, 320, 330, 334, 337, 418-421, 430, 446, 455-456, 465, 467, 476, 486, 497, 524, 540
Mouth, 138, 459, 486
Movement, 5, 14, 30, 32, 37, 39, 142, 148-149, 153, 189, 263, 306-307, 314, 334-335, 407
of bacteria, 30, 334-335
prokaryotic, 30, 306-307
MRI, 315
mRNA, 1, 6-10, 12-14, 20-21, 31, 208, 223, 226,

272-277, 279-282, 284-292, 294-295, 298-300, 302-307, 312, 315-317, 327, 337, 344-346, 348-349, 352-355, 362-366, 368, 370, 396, 398-399, 403-404, 419, 423, 435-437, 463-465, 482, 485, 497-498, 547, 549, 572
Mucosa, 124
Mucus, 23, 175, 528
Mullis, Kary, 401
Multicellular organism, 339
Multicellular organisms, 249, 265, 294, 377-378, 480, 489, 491
Multigene families, 432
Multiple alleles, 8, 83, 85-86, 106, 108, 498, 528
Multiple cloning site, 395, 397, 407
Multiple cloning sites, 395
Multiple drug resistance, 335, 573
Multiple resistance, 194
Multiple sclerosis, 451, 466, 470
Multiprotein complexes, 439
Multiregional hypothesis, 541
Mus musculus, 8, 24, 34, 419, 428, 430
Muscle, 21, 23, 96, 100, 103, 138, 202, 313, 363, 366-367, 398, 422-423, 432, 457, 469, 494, 576
Muscle cells, 366
animal, 366
Muscle cramps, 202
Muscle fibers, 103
Muscle tissue, 313, 494
Muscle tissues, 469
Muscle tone, 138
Muscles, 21, 23, 310, 322, 363, 493
Muscular dystrophy, 23, 71, 96, 99-100, 116, 130, 287, 337, 341, 363, 367, 422, 443, 470, 576
Mutagen, 9, 333, 340, 453, 492
Ames test, 333, 340
chemical, 9, 333
radiation, 340
Mutagenesis, 4, 12, 214, 322, 420, 481, 498
site-directed, 12
Mutagenicity, 333-334
Mutagens, 176, 324-325, 333-334, 340, 484
chemical, 176, 324-325, 333-334
frameshift, 325, 333, 340
radiation, 324, 340
Mutant, 1, 3, 5, 9-10, 12-13, 15, 19, 21, 23-25, 59, 65, 68, 70, 72, 75, 81-82, 84-86, 88, 92-95, 98-102, 104-105, 107, 110-111, 117-118, 125, 133, 147-148, 155, 161, 163, 165, 167, 169-170, 177, 180-182, 185-186, 188, 194-195, 201-202, 237-238, 241-242, 250-251, 268, 292-293, 308-309, 315, 317, 322, 333, 335-336, 339-342, 346-350, 352-353, 367, 369-371, 381-383, 386-388, 404, 422, 456, 460-462, 475-476, 484, 486-487, 489, 491-493, 496-498, 526, 529, 532, 534, 536, 544, 552, 554
isolation, 1, 10, 155, 194, 250, 309, 349, 352, 498, 536
kinds of, 5, 238, 268, 336
Mutant phenotypes, 10, 65, 147, 155, 202, 484
Mutants, 118, 133, 163-164, 187, 242, 308, 315, 331-332, 346-347, 442, 484, 486-487, 489-493, 496-497
Mutation, 2-14, 19, 21, 65, 70, 73, 75, 81-82, 84-86, 89, 92-94, 98-104, 108, 110-111, 123, 131-132, 144, 146, 148, 153-154, 176, 179-182, 186, 188, 195, 201, 208, 227, 237, 242, 250-251, 274, 279-281, 286, 292-294, 298, 304, 308-310, 313, 315-317, 319-342, 347-350, 352, 359, 367-370, 374-383, 385-388, 431, 447, 460-462, 467, 474-476, 480-481, 483, 486-487, 489, 491, 493, 495, 497-498, 522, 524-527, 529, 532, 534-535, 539, 543-544, 546, 550-553, 563, 571, 573-577
adaptive, 9-10
cis configuration, 3
complementation, 3, 11, 92-93, 110, 180, 332, 340, 342, 475
fitness, 5, 9-10, 12, 81, 322
hot spots, 7, 323
knockout, 5-7
molecular basis, 65, 86, 310, 316, 327, 474
rate, 6, 9, 103, 153, 188, 242, 251, 304, 320, 322, 324, 328-329, 334, 339-340, 359, 369, 375-376, 378, 532, 539, 544, 546, 577
replication errors, 319, 322

trans configuration, 13
mutation rate, 9, 103, 242, 320, 339, 376, 532, 544
Mutations, 3-5, 7-9, 11, 13-14, 19, 24-25, 38, 65, 81, 83, 85, 88, 92-94, 98-99, 101-103, 105, 108-110, 118, 135-157, 163-164, 179-182, 184, 186, 189, 194, 201, 214, 222, 242, 247, 250, 256, 265, 273-274, 279-280, 283, 291, 293, 298, 308, 310, 315-317, 319-329, 331-342, 346-349, 352, 358-359, 363, 366-367, 369-371, 373-388, 402, 406, 420, 423-424, 431, 459-462, 471-472, 474-476, 483-484, 486-487, 489-490, 493, 496-498, 523-524, 532, 534, 542-543, 550, 552-553, 572-574, 576
acquired, 7, 24, 155, 180, 291, 315, 376, 474
beneficial, 149, 153, 340-341
cancer-causing, 326, 334, 373, 375, 379, 388
causes of, 98, 140, 323, 338, 373-374, 474
definition of, 38, 542
disease-causing, 105, 293
DNA repair and, 247, 342, 377
effects of, 3, 25, 38, 93, 137, 145, 149, 320-322, 326-327, 329, 333, 335-336, 340, 358-359, 363, 370, 373, 424, 523, 550
frameshift, 5, 273-274, 291, 321, 323, 325, 327, 333, 337, 340-341, 349, 382, 386
frequency of, 3-4, 7-8, 139, 163-164, 180, 293, 308, 333, 338, 370, 381, 387, 524, 532, 534, 542-543, 552
genetic testing for, 385
gross, 376
Hox genes and, 489
identifying, 4, 19, 144, 319, 338, 385
in prokaryotes, 283, 331, 349, 359
inherited, 9, 11, 13, 65, 99, 102, 105, 108, 140, 145-146, 150-152, 156, 201, 256, 291, 308, 310, 315-316, 320, 322, 324, 338, 366-367, 373, 376, 381-383, 385-387, 424, 431, 459, 486, 489, 542, 550, 576
missense, 8, 321, 327-328, 338, 340, 387-388, 524, 543
nonsense, 9, 280, 321, 327-328, 341, 370, 524
point, 5, 19, 24-25, 38, 92-94, 99, 101, 137, 139-140, 143, 147-148, 150, 152-153, 179-181, 186, 189, 214, 222, 247, 256, 274, 280, 283, 308, 310, 321-324, 327, 333-337, 340-341, 348, 359, 363, 366, 374-377, 379-384, 386, 406, 424, 431, 459-462, 472, 475-476, 484, 524, 543, 553, 576
random, 4, 7-8, 11, 14, 136, 140, 155, 274, 323, 328-329, 340-341, 375, 532, 534
rate of, 9, 105, 143, 153, 189, 320, 324, 328-329, 334, 340, 358-359, 369, 375, 378, 532
repair of, 8, 331-332, 342, 374, 378
screening for, 153, 462
silent, 152, 321-322, 387-388, 523-524, 550
spontaneous, 9, 13, 140, 320, 322-324, 328-329, 333, 339
types, 4-5, 13, 65, 83, 108, 110, 135, 143-144, 152, 156-157, 164, 214, 320, 324, 327-328, 332-335, 338, 340-342, 347, 366-367, 369, 374, 376-377, 380-383, 386, 388, 462, 472, 475, 496, 498, 542-543, 572-573
types of, 65, 108, 110, 135, 143-144, 214, 324, 327-328, 332-335, 338, 340-342, 347, 366-367, 369, 374, 380-381, 383, 386, 462, 475, 498, 542-543, 572
Mycoplasma, 280-281, 429, 458-459
Mycoplasma genitalium, 429, 458
genome, 429, 458
genome of, 429, 458
Mycoplasma mycoides, 458
Myoglobin, 147, 312-313, 432-433
Myopia, 75
Myosin, 21, 313
Myotonia, 363

N
NADH, 103
NADH dehydrogenase, 103
Nails, 140, 250
Naked DNA, 199
Nanoarchaeum, 429, 446
genomics, 429, 446
Nanoarchaeum equitans, 429, 446
Nanotechnology, 408
Nathans, Daniel, 392

National Academy of Sciences, 45, 413
National Cancer Institute, 341, 389
National Center for Biotechnology Information (NCBI), 178, 268, 418
National Institutes of Health, 226, 389, 421, 427, 438, 472, 495, 555, 575
National Institutes of Health (NIH), 575
National Library of Medicine, 17, 28, 45
Natural selection, 9, 17, 62, 319, 509, 522, 526, 529-531, 536, 539, 541, 544
  and evolution, 17, 544
  Charles Darwin and, 17, 62
  genetic drift, 522, 526, 536
  theory of, 17, 62, 522
Nature, 12-13, 16, 19, 34, 44, 53, 57, 103, 111, 113, 129, 132, 139, 141-142, 153, 177, 184, 190, 200, 203, 207, 209, 216, 219-223, 232, 234, 256, 264-265, 273, 279-281, 283, 306-307, 309, 315, 349, 364-365, 369, 373, 375, 420, 429, 443, 446-447, 457-458, 471-473, 476-477, 497, 512, 514, 519, 536, 542-543, 546, 554-555, 567, 571, 578-579
Neanderthals, 426, 432, 540-542, 544
Nearsightedness, 75
Negative control, 347-348, 351-352, 367, 369, 513
Negative feedback, 364
Neisseria, 465
  N. meningitidis, 465
Neisseria meningitidis, 465
Nematodes, 488, 492-493
Neoplasia, 23, 382
Nerve, 23, 65, 88, 124, 322, 377, 497
Nerve cells, 124, 377
Nerve impulses, 88, 322
Nerve tissue, 23
Nerves, 23, 322, 441
Nervous system, 25, 65, 290, 362, 485, 488, 493
  disorders of, 25
Neurons, 65, 313, 493
Neurospora, 34, 102, 174, 308-309, 315, 320
Neurospora crassa, 34, 102, 308
Neutral mutations, 81, 322
Neutralization, 474, 476
Neutrons, 7, 11
Newborn, 153, 268, 542
Newborn screening, 542
Newborns, 16, 99, 110, 223, 307
Niches, 529, 536
Nick, 241, 247-248, 328, 330
Nickel, 65
Nirenberg, Marshall, 274
nitrocellulose filters, 368
Nitrogen, 8, 189, 216, 221, 227, 229, 232-233, 308, 440, 444
  cycle, 8, 233
Nitrogen fixation, 444
  genetics, 444
Nitrogen mustard, 189
Nitrogen source, 8, 232
Nitrogenous bases, 3, 207, 216-217, 219, 221, 227-229, 232-233, 300-301, 320, 323
  base pairing, 3, 219, 221, 229, 300
nitrosamines, 384
Nitrous acid, 324
Nobel Prizes, 26, 28
Nodes, 442-443, 538
Nomenclature, 9, 82, 181, 216, 283, 418
  scientific, 216
Noncoding DNA, 33, 254, 429
  transposable elements, 429
Noncoding RNA, 13, 365, 547, 549
Nondisjunction, 9, 40, 48, 121, 127, 131, 136-137, 139-140, 150, 155-156
  of homologous chromosomes, 48
Nonhomologous chromosomes, 11, 48, 64, 144, 149-150, 177
Nonsense mutation, 9, 280, 321
Nonsense mutations, 327, 524
Nonsister chromatids, 2-3, 39, 46-47, 148, 160, 165-166, 168, 176
Northern blot, 9, 404, 406, 409
Northern blotting, 404, 498
Nose, 99, 138, 307, 443
Notochord, 492
Nova, 495, 498
Nüsslein-Volhard, Christiane, 104, 483
N-terminus, 312, 317
Nuclear division, 7, 34, 482
  meiosis, 7, 34, 482

mitosis, 34
Nuclear DNA, 47, 103, 244, 257, 379, 388, 426
Nuclear envelope, 30-31, 35, 38-39, 124
  formation of, 31
Nuclear genome, 244, 447
Nuclear membrane, 40, 285, 355, 378
Nuclear pore, 30, 355
Nuclease, 9, 11, 260-261, 268, 270, 285, 331, 365
Nucleases, 195, 245, 307, 332, 469
Nucleic acid, 3, 6, 8-12, 20, 24, 29, 99, 184, 207-208, 211-212, 214-216, 220, 223-224, 226, 228-229, 255-256, 272, 280-281, 290, 317, 325, 383, 393, 399, 404, 415, 423, 429, 437, 442
  hydrogen bonds, 20, 220, 223, 228-229
  structure, 6, 9-12, 20, 24, 184, 207-208, 211-212, 214-216, 219-223, 223-224, 226, 228-229, 255-256, 317, 325, 437, 442
  synthesis, 9-12, 20, 208, 211, 215-216, 272, 280-281, 290, 317, 399, 429
Nucleic acid probe, 9, 12, 399
Nucleic acid probes, 224
Nucleic acids, 9-10, 13, 184, 207, 209, 215-216, 220, 223, 225, 228-229, 286, 290, 311, 317, 404
  components of, 216, 223
  digestion, 10, 404
  function of, 311
  structure of, 184, 207, 209, 216, 220, 223, 228-229, 311
  viral, 215, 290
Nuclein, 208
Nucleoid, 9, 31, 255
Nucleoids, 31
Nucleolus, 2, 9, 30-31, 35, 38-39, 47, 146, 285
Nucleoplasm, 12, 285
Nucleoside, 9, 216, 218, 236-237, 282, 325
nucleoside analog, 325
Nucleoside triphosphates, 237, 282
Nucleosides, 216, 218, 229
nucleosome, 9, 12, 244, 261-262, 268-269, 284, 356-357, 548
Nucleosomes, 243-244, 254, 260-262, 268-270, 355-358, 548-550
Nucleotide, 1-7, 9-13, 16, 20-22, 24-25, 100, 155, 175, 188, 195, 216-219, 225-228, 232, 236-241, 245-247, 252, 257, 264, 268, 273-275, 281, 283-286, 288, 291-292, 299-300, 310, 314, 316, 320-323, 328, 330-334, 340-341, 347, 349-350, 357-358, 367, 374, 376-377, 386, 392-393, 398-399, 401, 404, 406-408, 411-413, 415-416, 421-423, 431, 433, 444, 447, 460-463, 487, 523-524, 539-540, 542-543, 545-546, 558, 560, 562-563, 565, 573-574, 576
  function, 1-2, 5-6, 10-11, 13, 21, 24-25, 155, 175, 216, 226, 237-239, 241, 246-247, 257, 264, 268, 285, 300, 310, 314, 316, 320-322, 334, 340-341, 358, 374, 377, 406, 411, 415, 421-423, 444, 447, 487, 523, 539, 542, 576
  regulatory, 1, 3, 5, 7, 10-12, 284-285, 320-322, 334, 347, 350, 357, 367, 415-416, 422-423
  structure, 1-2, 4, 6-7, 9-13, 16, 20, 24-25, 155, 188, 216-219, 225-228, 238, 241, 245-247, 257, 264, 268, 284, 288, 291-292, 299-300, 310, 314, 316, 320, 323, 340, 357-358, 376, 386, 392, 406, 421-422, 523, 542-543
  synthesis, 2, 4-5, 9-13, 20-22, 195, 216, 227, 232, 236-241, 245-247, 252, 273-275, 281, 283-285, 291-292, 300, 321, 331-334, 340, 399, 407, 574
Nucleotide bases, 273, 341
  in DNA, 273
  in RNA, 273
nucleotide excision repair, 330-331, 333, 376-377
Nucleotides, 1, 3, 5-14, 20, 28, 152, 175, 208-209, 216-219, 223, 225, 228-229, 231-232, 236, 238-241, 243-245, 247, 251, 260, 273-274, 281-284, 286-288, 290, 292, 299-301, 303, 305-306, 316-317, 321-323, 325, 329-331, 334, 353, 357-359, 364, 368, 386, 393, 398, 401-402, 404-407, 412-413, 415, 419-422, 435-436, 443-444, 446, 450, 463, 473, 524, 539, 549, 562
  biosynthesis of, 353
  coding and noncoding, 322, 398
  in DNA replication, 12, 223

mutations and, 152, 321, 329
Nucleus, 2-3, 7, 9, 19-20, 22, 30-31, 34-35, 37-38, 40, 44-45, 62-63, 101, 103, 117-118, 125-126, 133, 141, 214, 257, 260-262, 268, 270, 273, 281, 284-285, 289, 306, 344, 355-356, 360, 363-365, 370, 376, 378, 380-381, 454-455, 467, 482, 485, 492, 571
  animal cell, 30, 40
  atomic, 7, 306
  cell, 2-3, 7, 9, 22, 30-31, 34-35, 37-38, 40, 44-45, 62-63, 101, 103, 118, 125-126, 141, 214, 260, 262, 268, 281, 284, 344, 355-356, 360, 363-365, 370, 376, 378, 380-381, 454-455, 467, 482, 492, 571
  division of, 34
  eukaryotic, 2, 7, 9, 20, 30-31, 34, 38, 214, 257, 260, 262, 270, 281, 284-285, 306, 344, 355-356, 360, 365, 370
  role of, 19, 117, 214, 260, 268, 306, 492
Nutrient, 186, 197-198, 399, 426, 455, 464
nutrient agar, 197-198, 399
Nutrient agar plates, 399
Nutrients, 8, 354, 455, 476
Nutrigenomics, 424, 426, 464
Nutrition, 98, 101, 426, 500, 554-555

O
ob gene, 420
Obesity, 100, 420, 443, 466, 554, 577
  and cancer, 554
  and genes, 420
  effects of, 577
Observations, 17, 31-32, 44, 52, 63, 81, 92, 94, 101, 106, 110, 123, 125, 127, 129, 131-132, 140, 147, 151, 163, 176, 188, 199, 204, 208-209, 211, 214-215, 227-229, 233, 238, 249-250, 259-261, 281-282, 291, 305-307, 335-336, 348, 356, 363, 383-384, 422, 483, 506
  of evolutionary change, 17
  scientific, 249
Ocean, 268, 434-435, 444, 446, 537
  open, 444
Oceans, 202, 434, 444
Offspring, 7-9, 14, 16-18, 22, 24, 29, 39, 41, 52, 54, 56-59, 61-62, 65-66, 68-71, 73-75, 80, 82, 86-92, 95-96, 102-108, 110-113, 116, 119, 123-124, 127-133, 135, 144, 148, 150-152, 154-157, 161, 163, 165, 167, 169-173, 175-176, 178-179, 181-184, 207-208, 307, 310, 319-322, 336, 341, 411, 413, 456, 474, 481, 483, 506-508, 515-516, 522, 524, 527, 529-532, 535-537, 543-546, 548, 550, 554, 574
  number of, 7-8, 14, 16, 18, 24, 29, 39, 41, 56, 61-62, 65-66, 69, 75, 92, 103, 108, 116, 119, 124, 127, 130, 132-133, 135, 148, 150, 152, 154-156, 161, 165, 167, 173, 175, 178, 184, 307, 319-321, 341, 413, 456, 474, 481, 506, 515-516, 522, 527, 530-532, 535, 537, 543-546
  survival of, 322
  survival rates of, 529-530
Oil, 22, 369, 371, 509, 511
Okazaki fragments, 4, 240-242, 244, 251-252
  eukaryotic, 240, 244, 252
Old age, 249
Olfaction, 424
Oligonucleotide synthesis, 450
OMIM (Online Mendelian Inheritance in Man), 178
Oncogene, 9, 11, 379-380, 383-384, 386, 423
Oncogenes, 11, 81, 370, 373, 379-380, 382-384, 386-388, 552-553
Oocytes, 10, 43, 47-48, 104, 122, 139, 146, 259, 270, 363, 496-497
  primary oocytes, 47, 122, 139
  secondary oocytes, 47-48
Oogenesis, 42-43, 47, 110, 118-119, 139, 155, 483, 529, 536
Oogonia, 122
Open reading frame (ORF), 9, 444
Operator, 9, 346-350, 352-353, 359, 367-371
Operator region, 9, 347, 350, 352-353, 367-370
Operon, 9, 322, 333-334, 342, 345-354, 367-371
Operons, 1, 344, 350, 353, 429
  inducible, 344, 350, 353
  repressible, 344, 353
Order, 1, 3, 20, 25, 27-28, 35, 49, 69-70, 77-78, 85, 113, 122, 134, 148, 157, 163, 165, 169-170, 172, 180-182, 184, 188, 190-191, 200,

204-205, 216, 226, 229, 239-242, 244, 247, 252, 262, 265, 270, 274, 287, 295, 311, 316-317, 320, 342, 345-347, 350, 353-355, 360, 363, 369, 372, 375, 379, 381-383, 386, 388-389, 403, 413, 415-417, 421, 424, 433, 443-444, 447, 466, 470, 476, 484, 488-491, 495, 498, 510, 515, 519, 543-546, 556, 558-559, 565, 568, 570, 572, 574-575, 577, 579

Orders, 169-170, 172, 191, 203, 312, 320
Ordinate, 67
Orf, 9, 444
Organ, 4, 7, 84, 87, 122, 133, 469-470, 489-491, 555, 576
Organ formation, 491, 555
Organ transplants, 470
Organelle, 2, 9, 12, 31, 101, 103, 110, 113, 244, 256-257, 370, 438
   genomes, 12
   photosynthetic, 257
Organelles, 3, 5, 8, 30-31, 81, 101, 103, 214, 256, 561
   eukaryotic, 5, 30-31, 214, 256
   membranous, 31
Organic chemistry, 283
Organic compounds, 1, 186
   structure, 1
Organic molecules, 326
Organisms, 1-3, 6-8, 11-13, 15-17, 21-22, 24-25, 28, 29-35, 38, 40, 43, 54, 59, 61-63, 65, 80, 82, 96, 99, 101, 116-117, 119-120, 129, 131, 133, 135-137, 141, 147-149, 154, 156, 159, 161, 165, 167, 174-175, 178, 182, 184, 185-186, 202-203, 207-208, 214, 216-217, 222, 224, 226-228, 231, 234-235, 240, 249, 251, 254, 256-258, 262, 264-265, 267, 272, 280, 283, 286, 289-290, 292-294, 298, 300, 308, 313-314, 316, 319-320, 325, 328-329, 331, 333-334, 337, 341, 344-345, 368-369, 377-378, 404, 410, 417-418, 421-424, 428-432, 434, 436, 440, 442, 444-447, 449-450, 452-453, 457-458, 470, 472, 476, 480-481, 485-486, 489, 491-492, 509, 524, 529, 532, 536, 538-539
   identification of, 120, 175, 286, 404, 421, 424, 446
   single-celled, 3, 34, 116
Organogenesis, 494
Organs, 42-43, 117-118, 122, 131-132, 175, 320, 377, 398, 439, 442, 468, 489-491, 493, 511, 551, 576
   animal, 42-43, 117, 131-132, 468, 491
   plant, 43, 117-118, 131, 175, 489-491
Orientation, 191, 204, 219, 222, 241, 359, 400
Origin, 2, 6, 14, 17, 25, 31, 40, 47, 100-101, 125, 136, 139, 141-143, 145-147, 150, 155-156, 168, 187, 190-192, 203-204, 235-236, 239, 242-243, 252, 263, 300, 319, 374-375, 395, 411, 427, 489, 491, 522, 537, 539, 550, 552, 554, 562-563, 565, 571
Origin of replication, 14, 235, 395
Origin recognition complex (ORC), 243
Origins of replication, 1, 243, 396
Ornithine, 468
Orthologous genes, 497
Orthologs, 421, 432
Oryza sativa, 137, 430
Osmotic pressure, 59
Osteoarthritis, 475
Osteoblasts, 486
Osteoporosis, 16, 431
Outbreak, 202, 543
Ovarian cancer, 332, 385, 387, 471-472, 552, 576
Ovaries, 43, 118, 120, 122-123, 129, 132, 385, 406
   human, 120, 122-123, 132, 385, 406
Ovary, 43, 122, 152, 329, 342, 375-376, 497
Overlapping genes, 281, 292
Oviduct, 455
Oviducts, 122
Ovulation, 43, 139
Ovules, 101, 148
Ovum, 33, 42-43, 54, 139, 141-142
Oxidase, 439, 564-565
Oxidation, 307
Oxidative phosphorylation, 103
Oxidizing agents, 330
Oxygen, 21, 220-221, 309, 313, 324, 384, 406, 411, 432-433
   forms of, 21
   in capillaries, 21

P

p arm, 32, 127, 145, 152, 266
P site, 303-304
p53, 364, 380-383, 388, 462, 475-476
p53 gene, 380-381, 462
Pace, Norman, 215
Pain, 21, 94, 367
Palate, 140
Palindrome, 9, 11, 393, 411
Pancreas, 152, 175, 375, 450
Panitumab, 572
Papaya, 22
Paper, 17, 26, 47, 163, 210, 216, 231, 249, 333, 349, 387, 404-405, 416, 519, 542
Papillomavirus, 380
Paralogs, 421, 432
Paralysis, 99, 220
Paramecium, 280-281
Parasites, 270, 446
   human, 446
Parental types, 164, 202
Parkinson's disease, 267
Parotid gland, 337
Passive immunization, 451
Pasteur, Louis, 17
Patau syndrome, 140, 155
Patents, 28, 472, 475
Paternal alleles, 550-551
Paternal chromosomes, 427
Paternity, 13, 27, 558-559, 563
Paternity testing, 559
Pathogen, 432, 452, 465, 474, 513
   attenuated, 452
   toxicity, 474
pathogenic bacteria, 402
Pathogenic microbes, 441, 456
Pathogenicity, 187
   altered, 187
Pathogens, 316, 432, 452, 454, 456-457, 464-465, 474
   bacterial, 316, 474
   identification of, 454
Pattern, 2, 4, 10-12, 14, 44, 48, 54-56, 70-71, 77, 89-90, 92, 94-95, 97, 101-103, 105-106, 108, 112-113, 125-126, 128-129, 132-133, 135, 146, 163, 190, 218-219, 228, 236, 244, 250, 258, 260, 263, 279, 281, 301, 308, 357, 367, 375, 378, 393, 399, 403-405, 411, 413, 460, 462, 464, 482, 484-485, 488-492, 498, 519, 535, 540, 548-553, 559-560, 563
Pattern formation, 489, 491, 549
Pauling, Linus, 219, 223, 310, 312
Pääbo, Svante, 431-432
PCR, 10-12, 392, 401-404, 409-413, 416, 455, 459, 461-463, 527, 534-535, 537, 559, 561-562, 564-565, 572, 574, 577
PCR (polymerase chain reaction), 11
   real-time PCR, 11
peanuts, 141, 384
Pectin, 453
Pedigree analysis, 69-71, 74, 310
Pedigrees, 69-71, 75, 96, 105, 107, 109, 111
Pentose, 9, 12, 216-217, 323
Peptide, 10, 299, 303-304, 311, 362, 438-442
Peptide bond, 299, 303-304, 311
Peptide bonds, 10, 311
Peptide hormones, 362
Peptides, 439-442
Peptidyl transferase, 303-304, 316-317
Perception, 125-126, 133
Peripheral nervous system, 362
Permafrost, 427
permease, 346
peroxide, 324
Personalized medicine, 473, 569-579
Pesticides, 333, 384, 514
   and water pollution, 333
   genetic, 333, 514
Petals, 489-491
Petri dish, 126, 186, 197, 397
Petri plates, 400
pH, 7, 65, 438-439, 476, 512
   food, 476
   gradient, 438
phage DNA, 196-197, 200, 212, 214, 256, 282
Phages, 5, 24, 185, 196-202, 204, 212-213, 215, 255, 269-270, 282
   virulent, 198

Phagocytic cells, 209, 379
Pharmaceutical products, 333, 450-451, 456, 470
   transgenic animals, 451, 456
Pharmaceuticals, 290, 459, 570-571
Pharmacogenomics, 9, 424, 427, 449, 570-571, 573-575, 578-579
Pharynx, 493
Phenotype, 3-7, 9, 11-14, 19-22, 28, 55-59, 61-62, 65-66, 69-70, 72-73, 75, 80-93, 95-99, 101, 103-104, 106-108, 112-113, 122-123, 129, 132-133, 135-137, 146-147, 150-151, 155-157, 167-170, 175, 179-182, 201, 242, 267, 310, 319-322, 331-332, 337-338, 340-342, 344, 369, 376-377, 379, 382, 388, 420, 428, 447, 458, 466-467, 475, 480, 486, 496-497, 500-504, 506-508, 510-513, 515-516, 518, 523, 526, 528-529, 531-532, 543, 546, 547-548, 550-551, 554-555, 572-573, 576-578
   designation, 56, 69, 84, 156, 321
Phenotypes, 3-5, 10, 14, 22, 25, 55-59, 61-62, 65, 69, 72-77, 80-82, 84, 87, 89, 91-93, 96-98, 100-101, 104, 106, 109-110, 112, 116, 118, 121, 124, 126, 132, 135, 137-138, 142, 145-147, 154-155, 161, 163-164, 167, 169-173, 175-176, 179-182, 194, 201-202, 309-310, 327, 338, 340, 423-424, 442, 447, 457, 465, 484, 497-498, 500-503, 507-509, 511-513, 515-517, 526-527, 529, 531-532, 547, 554
   changes in, 22, 147, 155, 423-424, 511, 527, 531, 547, 554
   defined, 22, 77, 511
   dominance and, 72-73, 182
   genes and, 3, 87, 91-92, 175, 181, 194, 202, 423-424, 442, 465, 484, 498, 511
   mutant, 3, 5, 10, 25, 59, 65, 72, 75, 81-82, 84, 92-93, 98, 100-101, 104, 110, 118, 147, 155, 161, 163, 167, 169-170, 180-182, 194, 201-202, 309, 340, 484, 497-498, 526, 529, 532, 554
   quantitative traits in, 509, 513
Phenotypic ratios, 62, 74, 80, 85, 91, 94, 108-110, 161, 173
Phenylalanine, 199, 275, 277-278, 287, 292, 309, 328, 353, 431
   genetic code, 275, 277-278, 287, 292
   structure, 292, 309, 353
   synthesis, 275, 278, 292, 309
Phenylalanine (Phe), 277
Phenylketonuria (PKU), 23, 431
Philadelphia chromosome, 9, 376-377, 387, 573
Phosphate, 9-10, 20, 96, 125, 216-217, 220-221, 223, 225, 228-229, 237, 255, 260, 288, 302, 316, 381, 439, 548-549
Phosphate group, 9, 216-217, 302
Phosphodiester bond, 9, 217, 219, 241, 406
Phosphodiester bonds, 4, 9-10, 282, 284, 326, 394-395
Phospholipids, 8, 31
   structure, 31
Phosphorus, 212
Phosphorylase, 274-275, 292
Phosphorylate, 38, 243-244, 442
Phosphorylation, 103, 262, 356, 359, 364, 381, 387-388, 438
   amino acid, 387, 438
   histone, 262, 356
   oxidative, 103
   oxidative phosphorylation, 103
Photons, 399
Photoreceptor, 434
Photoreceptors, 434
Photosynthesis, 31, 434
   bacterial, 31, 434
   prokaryotic, 31
Phylogenetic analysis, 538-539
Phylogenetic tree, 538-540
   eukaryotic, 538
Phylogenetic trees, 538-539, 546
   mammals, 538
Phylogenies, 538
   of mammals, 538
Phylogeny, 538
Physiology, 26, 38, 104, 222, 249, 336, 365, 392, 435, 483, 536
PI, 216, 274, 340, 491, 533
Picograms, 48, 214
Pigeons, 56

Pigmentation, 87-89, 103-104, 108, 125
Pigments, 308, 313
    respiratory, 313
pigs, 74, 85, 108, 449-451, 512, 516-517
Pili, 188
    conjugation, 188
    sex, 188
Pistils, 118
PKU, 23, 431
Placenta, 132-133, 139, 337, 460, 550
plant breeding, 177, 508, 514
Plant cells, 30-31, 38, 279, 396-397, 454, 459
plants, 5, 13, 16-18, 22, 24-28, 31-33, 35, 43, 53-61,
    69, 72-75, 77, 82, 84, 90-92, 97, 101-102,
    108, 117-118, 120, 135, 137, 140-143, 148,
    154, 175, 182, 202-203, 256-258, 308, 316,
    319, 329, 333, 335, 354, 364, 384, 396, 411,
    413, 429-430, 449-450, 452-455, 470-471,
    474, 480, 489, 491, 501-503, 507-509,
    512-519, 523, 537, 543
    chlorophyll in, 102
    diseases of, 25
    genetically engineered, 22, 202, 450, 453-455, 471
    green, 53, 57-61, 69, 74, 77, 90, 101-102, 108,
        257, 316, 449, 514
    photosynthesis, 31
    transgenic plants, 28, 203, 452, 454, 471, 474
Plaque, 10, 196-198, 201-202, 204-205, 368
    viral, 196-198, 201-202, 204-205
Plaque assay, 196-198, 204-205
Plaques, 197-198, 201-202, 204-205, 368, 437
Plasma, 4, 30-31, 34, 38, 70, 378, 387, 438, 442, 446,
    482, 492, 497-498, 534, 577
    components of, 31, 442, 482, 492
    proteins in, 378, 438
Plasma membrane, 4, 30-31, 34, 38, 378, 387, 492,
    534
    and hormones, 378
    functions, 31, 387
    structure, 4, 30-31, 38
Plasma membranes, 482, 498
Plasmid, 3, 5, 10-12, 14, 185, 189, 194-195, 248, 347,
    368, 394-398, 400-401, 409, 411-413, 450,
    452, 454, 458, 468, 470, 475
    cloning vector, 14
    copy number, 3, 14, 395-396
    engineered, 10, 395, 450, 454
    isolation, 10-11, 194
    replication, 5, 10-12, 14, 194-195, 248, 394-396
    single-stranded DNA, 10-11, 194, 400-401
    types, 5, 347, 368, 396-398, 413, 450, 452, 475
Plasmid DNA, 395, 397, 409, 411
Plasmids, 2, 5, 10-11, 185, 194-195, 334-335, 370,
    394-398, 410-412, 452
    as vectors, 395
    in bacterial conjugation, 5
    prokaryotic, 10
    recombinant, 11, 185, 195, 394-398, 410-412, 452
    Ti, 397
    Ti plasmids, 397
    transposons in, 335
Platelet-derived growth factor, 380
Platelets, 438
Platypus, 426
Pleiotropy, 10, 93-94, 500
Pluripotent, 4, 13, 494-495
Pluripotent stem cells, 494-495
Pneumonia, 209, 429
    bacterial, 429
    Haemophilus influenzae, 429
    Mycoplasma, 429
Point mutation, 10, 321, 323, 386, 461, 476, 576
    transition, 321
    transversion, 321
Point mutations, 322, 327, 340-341, 359, 380, 382,
    386, 431, 460-462, 475-476, 543, 576
    types of, 327, 340-341, 380, 386, 462, 475, 543
Polar bears, 426
Polar bodies, 10, 43, 47, 529, 536
Polar body, 10, 42-43, 48
    meiosis and, 43
Polar molecules, 4
Polarity, 12, 240, 329, 483-486, 488, 491, 497
Polio, 366, 431
Pollen grains, 117, 137
Pollen tubes, 117
Pollination, 54, 117-118, 471, 476
    seed plant, 117
Pollutants, 325

Pollution, 333
    water, 333
    water pollution, 333
Poly-A tail, 10, 286, 307, 317, 362, 364, 398-399, 419
Polyadenylation, 292, 337, 362, 419-420, 523
Polycistronic mRNA, 10, 284, 346, 349, 352-353
Polydactyly, 140, 544
Polygalacturonase, 453
Polygenic inheritance, 10, 502, 515-516
Polygenic traits, 500, 503, 511
Polylinker, 10
Polymerase, 1, 6, 10-13, 215, 236-244, 246, 249-252,
    272, 274, 282-285, 289, 292, 312, 322-323,
    328-332, 347-348, 350-351, 353-357,
    360-362, 365, 368, 370-372, 378, 388, 392,
    398-402, 406-407, 411, 413, 416, 559, 561,
    572
Polymerase chain reaction (PCR), 10, 392, 401-402,
    416, 559, 572
    applications, 392, 401-402
    applications of, 402
    DNA fingerprinting, 559
Polymerases, 10, 236, 238-239, 241-244, 247, 252,
    284-285, 289, 322, 355
Polymerization, 231, 238-241, 282, 284, 294, 298,
    306, 309, 328, 331
    DNA replication and, 231, 238-241
Polymers, 4, 20, 32, 311
Polymorphism, 6, 10-12, 387-388, 413, 459, 465, 524,
    542, 544, 546, 562, 565
Polymorphisms, 174-175, 387-388, 412, 421-423, 447,
    461, 464, 512, 524, 537, 544, 562, 570,
    573-574
Polynucleotide, 4, 6, 10, 12-14, 217, 219, 221-223,
    225, 231-232, 236, 238, 240, 274-275, 282,
    292
Polynucleotides, 217, 225, 229, 277-278
Polypeptide, 2-3, 5, 7-10, 12-14, 32, 81, 155, 194,
    202, 208, 226, 237-238, 274, 277-282,
    290-292, 298-300, 302-317, 321, 338, 418,
    421, 432-433, 438-439, 450, 460
Polypeptides, 8-9, 28, 155, 202, 274, 277-278,
    281-282, 291, 298, 304, 307, 311, 314, 338,
    432-433, 451
Polyploids, 140, 142
Polyploidy, 136, 140-142, 543
Polyps, 23, 383
Polyribosome, 10, 316-317
Polyribosomes, 304-305
Polysaccharide, 30, 209, 211, 227
    core, 211
    O, 227
    structure, 30, 209, 211, 227
    synthesis, 209, 211, 227
Polysaccharides, 30
Polysome, 10, 305
Population, 2-6, 8-10, 12-13, 16-17, 28, 33, 38, 70, 81,
    83, 85-86, 96-97, 101, 103, 122, 124, 130,
    138-140, 150, 152, 157, 192, 195, 208, 251,
    310, 321, 339, 341-342, 349, 385, 387-388,
    398, 401, 412-413, 455, 467, 474, 492,
    503-509, 511, 514, 516-517, 519, 521-546,
    558-560, 562-566, 573-574, 577
Population bottleneck, 10
Population genetics, 522-525, 535, 544, 560, 563
Population growth, 5, 514
    human, 5
Populations, 5-6, 11-13, 16-17, 80-81, 83, 122, 148,
    186, 203, 267, 319, 341, 384, 388, 413, 422,
    454, 467, 506-507, 509-510, 522-524,
    526-531, 533-534, 536-538, 540-544, 561
    genetic drift in, 534
Porphyrin, 93
Position, 7, 10, 21, 32, 53, 77, 98, 101, 103, 157, 191,
    216-217, 220, 229, 233, 241, 247, 263,
    267-269, 275, 279-281, 286, 292, 294, 298,
    300, 303, 310-311, 315, 317, 321, 323, 325,
    328, 341, 354, 357-359, 387-388, 406, 413,
    446, 460, 466, 481, 494, 497, 512, 576
Positional information, 483
Positive control, 345, 350-351, 369, 371
positive regulation, 350
Positive selection, 540
Posterior end, 497
Posttranslational modification, 10, 438
Posttranslational modifications, 344, 364, 437-439
Potatoes, 453, 555
poultry, 505
Predators, 529

Predictions, 66, 131, 347, 420, 423, 514, 525, 575
Pregnancy, 155, 338, 382
    human, 155, 338, 382
Preimplantation genetic diagnosis (PGD), 10, 461, 471
Pre-mRNA, 7, 10, 286, 289, 344, 355, 362-363, 437
Pressure, 12, 59, 188, 199, 424, 426
Pribnow box, 10, 283
Primary oocytes, 47, 122, 139
Primary spermatocytes, 47, 122
Primary structure, 310, 312, 346
    DNA, 312, 346
    protein, 312
    RNA, 312, 346
Primary transcript, 288-289, 355, 362
Primase, 240, 243, 246
Primate, 337, 431, 446
Primates, 1, 431
Primer, 10, 12, 238-240, 242-246, 251-252, 282, 284,
    395, 398-399, 401, 406-408, 412, 559, 561,
    565
Primer RNA, 240, 242, 244-246, 251
Primers, 11, 239-242, 244-245, 398-399, 401-403,
    409, 411-413, 534, 561-562, 564-565, 574
    nucleic acid, 11, 399
    RNA, 11, 239-242, 244-245, 398-399, 403, 409
Primordial germ cells, 122
Prion, 10, 313
Prions, 10, 313
    structure, 10, 313
Privacy, 16, 26, 471, 473, 567, 578
Probability, 3-4, 7, 9, 13, 46, 55, 58-59, 61-68, 70,
    73-74, 76, 78, 86-87, 108, 111, 124, 139,
    144, 155, 166, 169, 173, 183, 195-196, 200,
    274, 276, 294, 307, 325, 375, 387-388,
    412-413, 471, 504, 512, 524-525, 532,
    535-536, 543-546, 560, 563-566
    laws of, 61, 65-66, 196
Probes, 1, 4-5, 10-11, 116, 135, 224, 264, 368, 388,
    399, 404-405, 461-463, 476, 562
Problem solving, 72, 106, 227
Proboscis, 119, 133
Product, 1, 4-7, 9-11, 13-14, 58, 61, 65-66, 73, 81-82,
    85-86, 91-92, 99, 103-104, 110, 123-124,
    127-129, 132-133, 147, 152, 166, 194-196,
    202, 208, 215, 237, 242-243, 250, 265, 272,
    285-287, 291, 295, 298, 312-313, 315, 317,
    319-323, 327, 334, 337, 341-342, 344, 347,
    352, 354-355, 359, 362, 364, 366-367,
    369-371, 379-380, 383-384, 386, 388,
    401-403, 411, 413, 437-438, 449-451,
    456-457, 467, 469-470, 472, 475-476, 494,
    497-498, 543, 547-548, 564, 573
Production, 1, 3-5, 8, 29, 34, 42, 59, 64, 84, 97,
    99-100, 118-119, 127, 136, 143, 190, 195,
    199, 202-203, 212, 215, 227, 232, 240, 274,
    310, 314, 322, 341, 345, 347, 369, 371, 397,
    403, 447, 450, 452-453, 455-456, 468-470,
    473-475, 494, 500, 505-506, 509, 511, 514,
    516, 571
Product(s), 133
Products, 2, 4, 10, 13, 15, 21-23, 29, 38, 42-43, 48,
    65, 83, 85, 87, 91-92, 98-99, 101, 103-104,
    110, 113, 117-118, 123, 126-127, 133,
    146-149, 157, 180, 188, 194, 208, 215,
    242-243, 251-252, 256-257, 268, 272,
    280-281, 284, 289, 295, 298, 307, 312-313,
    315, 317, 319, 322, 324, 327-329, 331, 333,
    341-342, 344-346, 354, 362-366, 369, 373,
    376-384, 386-388, 392, 396, 401-402,
    407-408, 411-413, 423-424, 429, 432-433,
    437, 444, 446-447, 449-451, 453, 456, 459,
    467, 470-472, 475, 482-485, 495-498, 505,
    559-561, 572-573
Programmed cell death, 340, 342, 364, 377, 379-380
programmed cell death (apoptosis), 342
Proguanil, 573
Prokaryote, 242, 281, 369, 419, 446
    chromosome, 281, 369, 419, 446
    DNA, 242, 281, 369, 419, 446
    gene expression, 369
    mRNA, 281, 419
    ribosomes, 281
    transcription, 281, 369, 419
    translation, 281
Prokaryotes, 2, 12, 111, 185, 223, 231, 241, 243-244,
    250, 252, 264, 272-273, 283-286, 292, 299,
    302, 306-307, 330-331, 344-345, 349,
    353-355, 357, 359, 369, 413, 415, 428-430
    bacteria as, 185

first, 2, 223, 250, 252, 272, 283-286, 292, 299, 306-307, 330, 353, 355, 357, 369, 428-429
Prokaryotic cells, 9, 31, 257, 294, 307, 355
Prokaryotic genomes, 429, 435
  sizes, 429
Proline, 275-277, 293-295
  genetic code, 275-277, 293-295
  synthesis, 275-276, 293-294
Prometaphase, 29, 35-37
Promoter, 1-2, 5, 10, 53, 202, 283-285, 317, 336-337, 344, 346, 348, 350-353, 356-362, 366-372, 419-420, 427, 451-452, 454, 456-457, 485, 548-550, 552, 554-555
  -35 region, 283
  eukaryotic, 2, 5, 202, 283-285, 344, 356-360, 362, 368, 370, 419-420, 451
  expression vector, 5, 452
  in transcription, 358, 362, 367
  Pribnow box, 10, 283
  strong, 283, 350, 368, 419, 457, 552
Promoter sequence, 369, 451
Promoters, 4, 282-284, 327, 337, 353, 357-360, 362, 365-366, 368-370, 377, 384, 418, 420, 427, 446, 513, 548-549, 552
Proofreading, DNA, 320
Properties, 2, 7, 21, 202, 228, 238, 251, 279-280, 311, 313-314, 333, 374-375, 394, 412, 438-439, 475, 480, 487, 496-497, 552
Prophage, 10, 13, 198-199, 204
Prophages, 10, 199
Prophase, 2, 4, 13, 35-41, 43-44, 46-47, 64, 142, 159, 166, 175-176, 180, 259
  in meiosis, 39-40, 46-47, 142, 176, 259
  in mitosis, 38-40, 46-47
  meiosis, 2, 4, 13, 35-41, 43-44, 46-47, 64, 142, 159, 176, 259
  mitosis, 2, 35-41, 43-44, 46-47, 159, 176, 180
  of meiosis, 2, 13, 38, 40-41, 43-44, 47, 259
  of mitosis, 35, 37, 39, 44, 47, 176
Prophase I, 4, 39, 41, 43, 46-47, 175
Prophase II, 40-41
Prophylactic mastectomy, 385
Propidium iodide, 224
Prostate cancer, 373, 443, 575
Protease, 210-211, 440-441
Proteases, 227, 313, 387
Proteasome, 313, 439
Protein, 1-4, 6-14, 19-24, 32, 35, 37, 45, 59, 65, 71, 82, 93, 96, 104, 110-111, 123, 127-128, 138, 143, 146, 152, 178, 191, 194-197, 200, 202, 207-215, 219, 223, 226-227, 239, 243-244, 247, 250, 255, 260, 262, 267-269, 272-276, 278-282, 284-285, 287, 290-295, 298-299, 302-303, 305-307, 309, 311-317, 320-322, 327, 329, 333-334, 336-338, 340-341, 345, 347-356, 359-365, 367-371, 376-381, 383, 386-388, 392, 396, 401, 410-411, 415, 418-423, 425-426, 429, 431-432, 434, 437-442, 445-447, 449-455, 457, 459, 463, 467-470, 474-475, 485, 489, 491-497, 500, 513, 517, 523, 526, 534, 538-540, 543, 545, 549, 554, 571-573, 576
  catalytic, 227, 282, 299, 303, 306, 314
  denaturation, 10, 223, 250, 401
  domains, 7, 14, 262, 306, 314, 348, 354, 356, 418, 421, 423, 437-438, 446, 491
  genetically engineered, 22, 202, 450-451, 453-455, 457, 459
  hydrogen bonds, 4, 20, 219, 223, 239, 247, 284, 312, 340
  hypothetical, 9, 138, 178, 316, 340, 425, 517
  motility, 429, 432
  posttranslational modification, 10, 438
  primary structure, 312
  quaternary structure, 312
  secondary structure, 262, 284, 292, 312, 353-354, 368
  secretory, 523
  structural, 1, 4, 8-13, 21, 93, 143, 207, 223, 226, 244, 260, 262, 267, 312-314, 336, 345, 347-353, 355-356, 367-371, 411, 415, 418, 421, 423, 429, 438, 446
  structure, 1-2, 4, 6-7, 9-14, 19-20, 24, 35, 37, 196, 202, 207-215, 219, 223, 226-227, 243, 247, 255, 260, 262, 267-269, 284, 291-292, 298-299, 305-306, 309, 311-317, 320, 338, 340, 353-356, 359-360, 364-365, 368, 371, 376, 386,

392, 396, 418, 421-422, 437-438, 442, 491, 497, 523, 538, 543
synthesis, 2, 4, 9-13, 20-22, 35, 59, 104, 111, 146, 195-196, 200, 208-209, 211, 215, 227, 239, 243-244, 247, 250, 272-276, 278-282, 284-285, 290-294, 298, 302, 305, 307, 309, 315, 317, 321, 329, 333-334, 340, 345, 351, 354, 362-363, 368-371, 378, 429, 450, 455, 496
tertiary structure, 312
viral, 196-197, 200, 202, 212, 214-215, 227, 247, 255, 269, 290, 292, 320, 380, 415, 423, 449, 452-455, 467-469, 474-475
Protein A, 451
Protein folding, 312-313
  abnormal, 313
Protein kinase, 333, 376, 425
Protein kinases, 378, 423, 432
Protein microarrays, 441
Protein Structure Initiative, 438
Protein synthesis, 111, 146, 195-196, 272-274, 278-282, 294, 302, 305, 307, 317, 429
  genetic code and, 272-274, 278-282, 294
  overview, 272
  RNA and, 282
  transcription, 272-274, 278-282, 294, 302, 305, 429
  translation, 273-274, 278-281, 302, 305, 307, 317
Proteins, 2-3, 5-6, 8, 11-12, 14, 19-24, 26, 28, 31, 37-38, 65, 103, 126, 132, 194-195, 202, 204, 208-209, 211-212, 214-215, 223, 225, 227, 229, 231, 238-239, 242-245, 247-248, 251, 254-257, 260, 262-263, 265, 267-268, 272-273, 275, 280-282, 284-289, 291-293, 295, 297-317, 327-328, 331-333, 336, 338, 340, 344-347, 349, 353, 355-357, 359-365, 367-368, 370-371, 378-382, 384-388, 392, 394, 404, 415, 420-424, 426, 432-435, 437-442, 444-446, 449-453, 459, 465, 467, 469-470, 474-476, 482-483, 485-486, 488-489, 491-493, 495-496, 498, 523, 527, 547-548, 553, 570
  absorption of, 23, 229
  amino acids found in, 280
  apoptosis and, 340, 380
  biosynthesis of, 309, 353, 367
  denaturation, 223, 229
  digestion of, 260, 263
  domains of, 306, 488
  folding of, 2
  functions, 11-12, 14, 21, 24, 31, 212, 214, 227, 239, 242-243, 262-263, 268, 292-293, 295, 299, 313-314, 327, 344-345, 357, 362, 379, 381-382, 386-387, 415, 420-421, 423-424, 432, 434, 437-438, 446, 496
  functions of, 212, 292, 314, 386-387, 420-421, 423-424, 437
  genes coding for, 251, 267, 287, 299-300, 346
  heat-shock, 6, 313
  in viruses, 242, 254-256, 260
  matrix, 382, 423, 439-441
  membrane, 8, 12, 14, 31, 38, 214, 285, 307, 313, 355, 378, 380, 387, 404, 421, 446, 492
  primary structure of, 310, 312, 346
  prions, 313
  receptor proteins, 496
  regulatory, 3, 5, 11-12, 38, 243, 262, 282, 284-285, 289, 327, 336, 344-347, 353, 356-357, 359-360, 362, 364, 367-368, 370-371, 381-382, 387-388, 415, 420, 422-423, 445, 483, 489, 491, 547
  repressor, 11-12, 336, 346-347, 349, 353, 356, 360-361, 367-368, 370-371
  simple, 8, 214, 229, 242, 255, 265, 303, 370-371, 386, 420, 423, 523
  structure, 2, 6, 11-12, 14, 19-20, 24, 26, 31, 37-38, 202, 204, 208-209, 211-212, 214-215, 223, 225, 227, 229, 231, 238, 243, 245, 247-248, 254-257, 260, 262-263, 265, 267-268, 284, 288-289, 291-292, 298-300, 305-306, 309-317, 338, 340, 346, 353, 355-357, 359-360, 364-365, 368, 371, 386, 392, 421-422, 437-438, 442, 486, 491, 523, 547-548
  structure and function of, 26, 338
  structure of, 2, 6, 14, 19-20, 26, 208-209, 223, 227, 229, 231, 255, 291-292, 298, 305, 310-312, 340, 346, 353, 368, 371, 421, 548

synthesis of, 11-12, 20-22, 208, 211, 215, 227, 231, 239, 243-244, 251, 265, 280, 282, 284-285, 292-293, 298, 300, 308-309, 345-346, 362, 368, 370-371
transcription factors, 2, 284-285, 314, 356-357, 359-362, 370-371, 379-381, 388, 423, 432, 437, 483, 485, 489, 491, 495, 547-548
translation process, 302, 306
transport, 103, 195, 285, 313-314, 355, 365, 432
transporter, 423
water-soluble, 308
Proteolysis, 382, 440
Proteolytic enzymes, 364, 382
Proteome, 11, 363, 437-438, 442, 446
Proteomes, 437-438, 442, 445
Proteomics, 11, 24, 314, 338, 415-447, 480
Protists, 465
  infectious diseases, 465
Proton pumps, 444
Proto-oncogenes, 11, 81, 370, 373, 379-380, 382, 384, 386-387, 552
Protooncogenes, 383-384
Protoplast, 11
  formation, 11
Protoplasts, 214-215
Prototroph, 11, 186
Protozoans, 31, 34, 256, 258, 265, 288
PrP, 10, 313
PrPC, 313
PrPSc, 313
Pseudogenes, 133, 267, 341, 431, 433
Pseudomonas, 200, 429
Pseudomonas aeruginosa, 200, 429
  genome, 429
Puberty, 122, 139
Pufferfish, 428, 430
Pulse, 4, 94, 251, 395
Punnett square, 55-57, 59-62, 72, 74, 76-77, 90, 108
Punnett squares, 55
Pure cultures, 185-186, 193
Purine, 2, 9, 14, 99, 216-217, 220-221, 227-229, 285, 307, 321, 323, 325, 339-340, 342, 358
  synthesis, 2, 9, 216, 227, 285, 307, 321, 340, 342
Purines, 186, 216-217, 227-229, 302, 308-309, 323, 325-326, 354
Pyridoxine, 308
Pyrimidine, 2, 9, 12, 14, 216-217, 220-221, 223, 228-229, 321, 325-327, 329-330, 340, 342, 358
Pyrimidine dimer, 329
pyrimidine dimers, 326, 329-330
  effect of UV radiation, 326, 329
Pyrimidines, 186, 216-217, 227-228, 308-309, 323, 325-326
  biosynthesis of, 309
Pyruvate, 439
Pyruvate kinase, 439

Q
q arm, 32, 175
Quantitative data, 17
Quantitative genetics, 499, 501-519
Quantitative traits, 1, 3, 501, 503, 505-507, 509, 511-513, 516-517
Quarantine, 342, 565
Quaternary protein structure, 309
Quaternary structure, 310, 312
  proteins, 310, 312

R
R group, 311
R plasmid, 11, 194-195
R plasmids, 11, 194
Rabbit, 99, 210, 286-287, 305, 475, 538-539
  digestion, 210
Rabbits, 98, 179, 280, 286, 451, 474, 476
Rabies, 203, 451-452
  immunization against, 451
  symptoms, 203
  vaccine, 203, 451-452
  vaccines for, 203, 452
Race, 153, 216, 473, 515, 542
  genetic testing and, 153
Rad, 11
Radiation, 11-12, 143, 238, 247, 309, 319-320, 324, 326-327, 329, 331-332, 339-340, 375, 380, 384, 387, 565

ionizing, 11, 247, 326, 332, 380, 387
mutagenesis, 12
mutagenic, 238, 320, 326, 329, 340, 375, 384
mutagenic effects of, 340
prokaryotes and, 331
wavelengths, 326, 339
wavelengths of, 339
radicals, 103, 324, 326, 384
hydroxyl, 324
Radioactive decay, 1
Radioactive isotopes, 233, 399
Radioactive tracers, 258
Radioisotope, 234
Radioisotopes, 212
Random fertilization, 55-56, 59, 120
Random mating, 525, 529, 534
Range of species, 117, 487
ras gene, 380, 383, 386-387, 572
Ras protein, 380, 386-387
rats, 108, 399, 446, 554
Rays, 11, 14, 176, 218, 222, 308, 320, 325-327, 330,
338, 340, 342, 384, 388
Reading frame, 5, 9, 11, 281, 291-292, 294, 302, 315,
337, 341, 370, 425, 444
open, 9, 425, 444
reading frames, 274, 281, 314, 365, 419
real-time PCR, 11, 403
Reasoning, 47, 55, 72, 74, 131, 155, 165, 180, 182,
269, 291, 315, 410-411, 445, 474, 516, 544
recA gene, 194
RecA protein, 194, 247, 329
from E. coli, 194
Receptor, 31, 148, 195, 360, 362, 418-420, 423, 425,
474, 492-494, 496, 526, 553, 571-573, 578
hormone, 360, 362, 418, 420, 425, 573
olfactory, 425
Receptor proteins, 496
Receptor sites, 195
Receptors, 70, 133, 314, 378, 380, 423-424, 432, 483,
496, 570-572
hormone, 380
Recessive allele, 11, 69, 82, 88, 90, 95-96, 110, 113,
180, 526, 529-530, 536, 545
Recessive alleles, 105, 532, 536
defined, 536
Recessive traits, 57, 70-72, 96, 173, 181, 307
Recognition sequences, 393
recognition sites, 31, 316, 334, 394
Recombinant bacteria, 395
Recombinant chromosomes, 202
Recombinant DNA, 3, 6, 11, 14, 15, 22, 24-26, 174,
185, 195, 215, 229, 280, 391-413, 415, 418,
423, 449-450, 452-455, 459-460, 470, 474,
494, 572
Recombinant DNA technology, 6, 11, 15, 22, 24-26,
215, 280, 391-413, 449-450, 452-455, 459,
494, 572
applications of, 396, 402-403, 449, 452, 454-455,
459
development of, 22, 25-26, 392, 394, 399, 404,
408, 450, 459
diagram of, 395
plasmids in, 395
regulation of, 403
restriction enzymes in, 393, 395
techniques of, 408
vaccine produced by, 452
Recombinant plasmids, 395-398, 452
Recombinant proteins, 450-451
Recombinants, 161, 174, 182, 190, 201-205
Recombination, 3, 5, 7-9, 11, 13-14, 26, 139, 159,
163-166, 171-175, 177-178, 180, 185-192,
194-196, 199, 201-202, 204-205, 231-252,
319, 329, 332-333, 337, 345, 386, 429, 455,
458, 475, 522, 532, 561
detection, 11, 561
homologous, 3, 7, 9, 13, 163, 175, 178, 191,
194-195, 247-248, 319, 329, 332-333,
458
in transformation, 195
Recombination frequencies, 163
Recruitment, 481
Recycling, 368
Red blood cells, 21, 23, 31, 34, 83, 93, 214, 309-310,
432
size of, 34
Red-green color blindness, 95, 107-108, 126
Reduction, 5, 9-10, 18, 29, 98, 105, 119, 139, 148,
154, 173, 370-371, 456, 476, 513, 522, 534

Reductional division, 11
Refrigeration, 452
Regulation, 2, 8, 13-14, 26, 38, 81-82, 129, 193, 223,
262, 265, 282-284, 306, 322, 337, 343-372,
377-378, 380, 382, 384, 386, 403, 406, 410,
420, 423, 437-438, 442, 469-470, 475-476,
489, 496, 547-549, 552, 555
enzyme activity, 82
overview, 346, 369, 470
Regulators, 344, 358, 379, 381-382, 388, 491
Regulatory gene, 128, 322, 367, 491
Regulatory genes, 483, 491
Regulatory protein, 322
Regulatory proteins, 5, 284, 327, 353, 356-357, 359,
364, 381, 388, 489
repressors, 359
Regulatory RNA, 12
Relative abundance, 440
Relative fitness, 12, 530-531
Relaxation, 242
Release, 11, 216, 237, 302-305, 317, 361-362, 364,
375, 377
Release factors, 304, 317
Releasing, 7, 14, 212, 304, 311, 388, 425, 470
Repair, 5, 8, 10, 12, 25, 103, 238, 242, 244, 247-248,
265, 319-342, 345, 355, 357, 364, 370,
373-374, 376-380, 384, 386, 388, 552-553
DNA, 5, 8, 10, 12, 25, 103, 238, 242, 244, 247-248,
265, 319-342, 345, 355, 357, 364, 370,
373-374, 376-380, 384, 386, 388,
552-553
tissue, 320, 331, 336, 370, 377-378, 384, 552-553
Repair, DNA, 12, 25, 238, 244, 247, 319-342, 355,
357, 364, 370, 373, 376-377, 380, 384, 386,
388, 552-553
Repetitive DNA, 11-12, 224-225, 227, 254, 263-267,
269, 300, 336, 341, 422, 430, 552
Replicase, 215, 280
Replication, 1-2, 4-5, 7-8, 10-14, 26, 32, 34, 44, 47-48,
82, 145, 176-177, 191-196, 198, 200, 202,
208, 215, 221-223, 227-229, 231-252,
255-258, 262-263, 269-270, 272, 319-320,
322-329, 331, 333, 339-342, 345, 355, 357,
375, 378-379, 384, 386, 388, 394-396, 402,
428, 442, 465, 472, 511, 546, 548, 553
bidirectional, 2, 235-236, 239, 243, 251-252
DNA, 1-2, 4-5, 7-8, 10-14, 26, 32, 34, 44, 47, 82,
145, 176-177, 191-192, 194-196, 198,
200, 202, 208, 215, 221-223, 227-229,
231-252, 255-258, 262-263, 269-270,
272, 319-320, 322-329, 331, 333,
339-342, 345, 355, 357, 375, 378-379,
384, 386, 388, 394-396, 402, 428, 442,
465, 472, 511, 546, 548, 553
errors, 10, 228, 231, 239, 319, 322, 324, 326, 328,
333, 546
herpesvirus, 384
influenza virus, 465
initiation, 7, 10-12, 34, 236, 238-240, 242-244, 246,
251-252, 357, 388
lagging strand, 4, 7, 240-242, 244-247, 252
leading strand, 8, 240-242, 244-245, 252
origin of, 14, 47, 145, 192, 235-236, 239, 242, 319,
375, 395
plasmid, 5, 10-12, 14, 194-195, 248, 394-396
primers, 11, 239-242, 244-245, 402
proofreading, 10, 238-239, 241-242, 319-320, 322,
328, 333
retrovirus, 7, 11, 14, 384
semiconservative, 12, 222, 231-236, 239, 245,
250-252, 324
templates, 232, 237, 243, 250, 252, 333
termination, 236, 342
unidirectional, 194, 235-236, 251-252
unwinding of DNA, 244
viral, 196, 198, 200, 202, 215, 227, 247, 255-256,
269, 320, 384, 465
viral nucleic acid, 255
virus, 1, 7, 11, 196, 215, 223, 227, 229, 255-256,
384, 465
Replication fork, 7-8, 11, 231, 236, 238-242, 244, 252,
269
Replication origins, 243-244
Replisome, 11, 239
Repressible operon, 352
repression, 344, 348, 350-353, 364, 368-369, 377
Repressor, 7, 9, 11-12, 129, 336, 346-353, 356,
360-361, 367-371
lambda, 368

virus, 7, 11
Repressor protein, 9, 336, 347-348, 352, 360, 371
repressor proteins, 356, 361
Repressors, 344, 359-362, 549
Reproduction, 1, 8-10, 12-13, 16, 22, 29, 34, 38-39,
41, 43, 116, 119, 141, 185, 196-198, 201,
208, 212-213, 215, 290, 421, 524, 529-530
delayed, 22
eukaryotic, 9, 12-13, 29, 34, 38, 116, 201, 215
insect, 22
prokaryotic, 9-10, 12
rates of, 524, 529-530
viral, 196-198, 201, 212, 215, 290
Reproductive isolation, 11, 522, 524, 536
Reproductive organs, 42-43, 117-118, 122, 132
development of, 42, 118, 122
Reproductive success, 5, 142, 525, 530
Reproductive system, 493
Research, 4, 8-9, 13, 15, 17, 19, 22, 24-26, 38, 45,
52-53, 62, 70, 84, 95, 105, 109, 124, 127,
138, 148, 150, 152, 165, 178, 185, 195-196,
201-203, 208-210, 213, 215, 220, 225-228,
231, 236, 242, 244, 249, 262, 269, 272-273,
281, 306, 319, 333-334, 337, 345, 356-357,
364-366, 376, 380, 385, 387, 389, 392, 396,
401-403, 408-410, 415, 417-418, 421, 424,
427-428, 431, 434-436, 438, 441, 445-446,
449, 456, 468-469, 472-473, 476, 494-495,
510, 512-514, 525, 531, 542, 547-548, 552,
554, 556, 567, 575-578
Research and development (R&D), 472
Research funding, 203
resistance, 7, 11, 22, 111, 113, 187, 194-195, 203,
227, 334-335, 337, 363, 394-395, 397,
411-412, 425, 452-455, 457-458, 470-471,
513-514, 526, 553, 573
cross, 111, 113, 203, 470-471
development of, 22, 227, 334, 337, 394, 471,
513-514, 553
multiple, 187, 194, 203, 334-335, 395, 397, 411,
425, 470, 553, 573
Resistance genes, 11, 187, 203, 337, 397, 458
resistance transfer factor (RTF), 11
resistant bacteria, 203
Resolution, 222, 247, 262, 301, 305-306, 438, 461,
485
Respiration, 31, 103, 324
aerobic, 103, 324
cellular respiration, 103
Respiratory infections, 468
Respiratory pigments, 313
Responses, 295, 315, 340, 380, 388, 447, 465-466,
476, 550, 555, 570-571, 573, 576-577
Responsiveness, 374, 573
Restriction endonuclease, 11, 266
Restriction endonucleases, 9, 11
Restriction enzyme, 1, 4, 10, 12, 14, 393-398,
403-405, 409-411, 413, 416, 459-461,
475-476, 527, 558
recognition sequence, 398, 410-411
Restriction enzymes, 11, 22, 28, 175, 393-395, 398,
403-405, 409-411, 416, 443, 460, 476
action of, 410
cutting DNA with, 403
types of, 398, 404
Restriction fragment length polymorphism (RFLP), 11,
459
Restriction fragment length polymorphism (RFLP)
analysis, 459
restriction fragment length polymorphisms (RFLPs),
174
Restriction fragments, 395, 460
Restriction map, 403, 409
Restriction mapping, 403, 409
Restriction sites, 393-395, 397, 403, 409, 411, 413,
460
Reticulum, 30-31, 307, 313
Retina, 23, 25, 132, 434, 469-470, 492
function of, 25
Retinal, 25, 104-105, 126, 381
Retinitis, 23, 25, 443
Retinitis pigmentosa, 23, 25, 443
Retrotransposons, 264, 266, 269-270
Retrovirus, 7, 11, 14, 384, 468
gene therapy, 468
genes, 7, 11, 14, 384
genome, 7, 11, 14, 384
replication, 7, 11, 14, 384
structure, 7, 11, 14

Retroviruses, 215-216, 266, 270, 384, 467, 470
    as vectors, 467, 470
reverse genetics, 5, 11
Reverse transcriptase, 2, 11, 145, 215, 266, 398-399,
    403, 463
    telomerase, 145
Reverse transcription, 215, 246, 388, 403, 436
Reversion, 1, 11, 340
    Ames test, 1, 340
Rh factor, 11
Rhesus macaque (Macaca mulatta), 430
Rheumatoid arthritis, 175, 307, 451, 475
Rhizobium, 396-397, 429
Rhodopsin, 105
Ribonuclease, 210-211, 227
ribonucleic acid (RNA), 12
Ribonucleoproteins, 288
Ribonucleotide, 1, 273-275, 281-282, 284-286, 293,
    298, 366
Ribose, 9, 12, 20, 216-217, 220, 223, 228, 282, 286,
    300
    and RNA, 20, 216, 223, 228
    in nucleic acids, 9, 216
    in RNA, 12, 217, 223, 282
Ribosomal RNA, 12, 31, 146, 208, 223, 266, 289
Ribosomal RNA (rRNA), 12, 31, 208, 223
    16S, 12
    18S, 12
    translation, 12, 208, 223
Ribosome, 7, 9, 12, 14, 20-21, 26, 30, 47, 204, 208,
    223, 239, 273, 276-277, 280, 298-300,
    302-307, 311-312, 315-317, 346, 355
    eukaryotic, 7, 9, 12, 14, 20, 26, 30, 280, 299,
        306-307, 316, 355
    mitochondrial, 280
    prokaryotic, 9, 12, 30, 299, 305-307, 355
    reading frame, 9, 302, 315
    structure, 7, 9, 12, 14, 20, 26, 30, 204, 208, 223,
        298-300, 305-306, 311-312, 315-317,
        346, 355
    subunits, 7, 12, 20, 239, 277, 299, 302-307, 311,
        315, 317
    translation, 9, 12, 14, 20-21, 208, 223, 273,
        276-277, 280, 298-300, 302-307,
        311-312, 315-317, 346, 355
Ribosomes, 10, 12, 31, 146, 150, 205, 223, 272-274,
    276, 281-282, 284, 290, 298-300, 302,
    304-307, 313, 346, 353, 363, 366, 497-498,
    549
    eukaryotic, 12, 31, 281, 284, 299, 306-307, 313
    features of, 306, 497
    free, 274, 284, 304
    in bacteria, 205, 272, 282, 284, 300, 302, 306
    in protein synthesis, 281, 302
    in translation, 10, 298-299, 302
    membrane-bound, 313
    mitochondrial, 498
    polyribosomes, 304-305
    prokaryotic, 10, 12, 31, 272, 284, 299, 305-307,
        353
    protein synthesis and, 282
Riboswitch, 354
Riboswitches, 353-354, 368
Ribozyme, 303
Ribozymes, 288
Rice, 16, 22, 137, 428, 430, 453-455, 476, 512, 514,
    517, 555
Rickettsia, 429
    characteristics, 429
Rickettsia prowazekii, 429
    genome, 429
Ring structures, 216-217
Risk, 23-24, 93, 130, 138-139, 148, 152-153, 203,
    341, 376, 381, 383, 385, 387-388, 456, 464,
    466-467, 471, 475, 511, 526, 552-553, 555,
    572-575, 577
River, 544
Rivers, 202, 546
RNA, 1-4, 6-14, 20-22, 31, 127-128, 132, 146, 152,
    196, 207-208, 210-211, 215-218, 223-225,
    227-229, 231, 239-247, 249-251, 255-256,
    258-260, 264-266, 269, 272-278, 280-294,
    298-301, 303, 306-307, 312-314, 316, 337,
    345-351, 353-357, 360-372, 384, 392,
    398-399, 403-404, 409, 426, 453, 463, 470,
    496, 498, 513, 547, 549
    antisense, 1, 127, 223, 290, 364-366, 453, 470
    catalytic, 227, 246, 282, 288, 299, 303, 306, 314
    double-stranded, 1, 4, 6, 8, 11, 223, 229, 239, 243,

    245-247, 255-256, 269, 290, 364-365,
    368, 384, 398-399, 403-404, 470, 549
    function of, 6, 231, 239, 242, 247, 250-251, 256,
        258, 265, 278, 285, 298, 300, 303,
        312-313, 316, 346, 369-370, 403, 496
    in DNA replication, 12, 223
    in protein synthesis, 281
    in transcription, 223, 362, 367, 496
    in translation, 10, 208, 298-299
    messenger, 8, 10, 12, 20, 31, 208, 223, 272-273,
        282, 290, 292, 298, 346, 362
    noncoding, 13, 127, 256, 266, 291, 300, 362, 365,
        398, 547, 549
    nucleotide bases in, 273
    primary structure, 312, 346
    processing of, 12, 286
    regulatory, 1, 3, 7, 10-12, 127-128, 243, 282,
        284-285, 289, 337, 345-347, 350, 353,
        356-357, 360, 362, 364, 367-371, 513,
        547, 549
    ribosomal, 12, 31, 146, 208, 223, 266, 288-289,
        299-300, 303, 306-307, 316, 355, 404,
        498
    secondary structure, 284, 292, 300, 312, 353-354,
        368
    sequence determination, 12
    short interfering, 223, 290
    small, 2, 6, 12, 14, 21, 146, 223-224, 227, 229,
        240, 246, 255, 277, 280-281, 288, 290,
        299-300, 303, 313, 350, 354-355,
        364-366, 384, 463, 470, 513, 547
    small interfering, 290, 365, 470
    stable, 6, 11, 152, 224, 273, 292, 312, 314,
        363-364, 426, 547
    stem-loop structure, 365
    structure of, 2, 6, 14, 20, 196, 207-208, 216, 223,
        227-229, 231, 255, 269, 291-292, 298,
        312, 346, 353, 368, 371
    synthesis of, 4, 11-13, 20-22, 196, 208, 211, 215,
        227, 231, 239-241, 243-244, 246,
        250-251, 259, 265-266, 274, 276, 280,
        282, 284-285, 290, 292-293, 298, 300,
        345-346, 362, 368, 370-371
    synthetic, 1, 12-13, 274-275, 277-278, 290-291,
        365-366, 368, 384
    transfer, 3, 7-8, 11-14, 21-22, 31, 208, 223, 229,
        272, 281-282, 288, 298, 300-301, 307,
        316, 355, 384, 404, 470
    viral, 196, 215, 227, 247, 255-256, 269, 290, 292,
        366, 384, 453
RNA interference, 12, 132, 290, 364, 369, 470
RNA interference (RNAi), 12, 290, 364, 369, 470
RNA polymerase, 1, 6, 10, 12-13, 240, 243, 274,
    282-285, 289, 292, 312, 347-348, 350-351,
    353-357, 360-362, 365, 368, 370-372
    Archaea, 354
    core enzyme, 284, 292
    eukaryotic, 6, 12-13, 240, 243, 283-285, 292,
        354-357, 360, 362, 365, 368, 370
    in eukaryotes, 12-13, 284-285, 289, 354-355, 362
    in eukaryotic transcription, 360
    structure, 1, 6, 10, 12-13, 243, 284, 289, 292, 312,
        353-357, 360, 365, 368, 371
RNA polymerase II, 6, 285, 355-357, 360-362, 365,
    370-371
RNA polymerases, 284-285, 355
RNA primer, 240, 242, 244-246
    replication, 240, 242, 244-246
    reverse transcription, 246
    transcription of, 246
RNA primers, 239-242, 244-245
RNA probes, 264
RNA processing, 286, 362, 498
RNA replicase, 215, 280
RNA splicing, 128, 363, 367
RNA synthesis, 282-284, 291, 496
RNA transcript, 128, 283-284, 287, 513
RNA tumor viruses, 216
RNA virus, 229, 384
    double-stranded, 229, 384
    single-stranded, 229
RNA viruses, 384
RNA-induced silencing complex (RISC), 290, 365
RNase, 211, 227, 399
Roentgen, 11-12
Roots, 411, 542
Rotation, 221, 248
Rough colony, 210
Rough endoplasmic reticulum, 30

Rough ER, 307
rRNA, 9, 12, 31, 146, 150, 208, 223, 264, 266, 285,
    288-289, 299-300, 302-303, 306, 564
Rubella, 223, 229
Rubella virus, 223

S

S genes, 137, 177, 354
S phase, 34-35, 48, 157, 244, 252, 263, 378-379, 381,
    386-388, 553
Saccharomyces, 24-25, 34, 102, 142, 256-257, 265,
    428, 430, 458
    S. cerevisiae, 25, 458
Saccharomyces cerevisiae, 24-25, 34, 102, 142, 265,
    428, 430, 458
    gene expression, 142
    genome, 24-25, 142, 428, 430, 458
    life cycle, 25, 428
S-Adenosylmethionine, 8
Safety, 26, 410, 450, 453, 469-470
    of recombinant DNA technology, 410, 450
Salamanders, 334
Salinity, 514
Saliva, 428, 437, 471, 558, 567
Salmon, 456, 537
Salmonella, 199, 316, 333
Salmonella typhimurium, 199, 316, 333
Salt, 12, 220, 222, 233, 447
    molecules of, 233
Salts, 186, 202
Sample size, 66, 107, 170, 173, 184, 339, 504-505
Sampling error, 205
sanitation, 202-203
sarcoma, 384
Sargasso Sea, 434
SARS, 464-465
Scales, 539
scanning electron micrograph, 44-45, 259, 373
Schizophrenia, 425, 466, 511, 518, 554-555
Schleiden, Matthias, 16
Schwann, Theodor, 16
Science, 5, 13, 15-18, 26, 28, 45, 49, 53, 62, 105, 113,
    153, 203, 205, 226, 229, 252, 270, 295, 317,
    366, 385, 424, 447, 449, 476, 495, 498, 514,
    539, 541-542, 546, 558, 567, 575, 578-579
    and society, 15-16, 26
    defined, 17, 153, 385
    limits of, 229, 514
Scrapie, 10, 313
Screening, 153, 334, 339-340, 368, 388, 398-402,
    404, 411, 413, 416, 442, 461-462, 474, 476,
    532, 542
SDS-PAGE, 438-439
Sea anemones, 457
Sea urchins, 428, 430-432, 436, 488
Seals, 241, 247, 288, 328, 330, 394
Seawater, 119
Secondary oocytes, 47-48
Secondary protein structure, 313
Secondary sex characteristics, 121
Secondary spermatocytes, 42-43, 47-48
Secondary structure, 262, 284, 292, 300, 312,
    353-354, 368
    protein, 262, 284, 292, 312, 353-354, 368
    RNA, 284, 292, 300, 312, 353-354, 368
Seed coat, 177
Seed plants, 117
Seeding, 454
Seeds, 16, 57-59, 72, 74, 76-77, 101, 141, 155, 177,
    471, 531, 543
Segmentation, 5, 482-486, 488, 498
Seizures, 72, 100, 103
Selectable marker, 394-395, 397
Selection, 1, 3-4, 6, 9, 11-13, 16-17, 62, 105, 113, 130,
    186, 226, 319, 342, 395-397, 447, 450, 453,
    458, 493, 508-509, 511-514, 517-519,
    522-527, 529-537, 539-541, 543-544, 546,
    570, 573
    antibiotic, 11, 395, 397, 458
    artificial, 1, 16, 396, 453, 458, 508-509, 511-513,
        517, 523, 531
    negative, 4, 513-514, 535
Selective breeding, 5, 16, 453, 476, 508
Selfing, 54-55
Self-splicing, 287
Semen, 16, 23, 130, 401, 559, 565-566
Semiconservative replication, 12, 231-236, 245, 324
Semisolid medium, 186
Senescence, 249, 265

Sensitivity, 113, 148, 177, 325, 333, 376, 425, 441
Sensory, 87, 432, 437
    receptors, 432
Sepals, 489-491
Separase, 37
Sequence alignment, 338, 367, 417-418, 426, 431
serial dilution, 186, 197
Serine, 277-279, 292, 294, 387, 441, 549
    genetic code, 277-279, 292, 294
    structure, 292
    synthesis, 278-279, 292, 294
Serine (Ser), 277
serotypes, 209
Serum, 446, 474, 555
Severe acute respiratory syndrome (SARS), 464-465
severe combined immunodeficiency, 467, 473
Severe combined immunodeficiency (SCID), 467, 473
Sex, 1, 6-7, 10, 12, 14, 19, 29, 33, 47, 54, 63, 69,
    80-81, 94, 96-97, 107, 112, 115-134, 136,
    156, 181-182, 185, 188-189, 194, 430, 446,
    460, 485, 487, 493, 497, 510, 526
Sex chromosomes, 1, 14, 63, 80, 115-134, 182
    human, 1, 116, 120-125, 130-132, 182
Sex determination, 12, 94, 115-134, 460, 485, 493,
    497
Sex hormones, 97
Sex pilus, 185, 188-189, 194
Sex reversal, 118
Sexual dimorphism, 116-117
Sexual reproduction, 29, 38-39, 41, 43, 116, 119, 141
    animal, 38, 43
Shapes, 32, 91, 106, 229, 301, 325, 523
    cell, 32, 229
    enzyme, 301
    molecular, 523
Sharks, 259, 486
Sheep, 10, 22-23, 97, 313, 451, 505, 512, 518
Shigella, 195, 211
    toxin, 195
Shine-Dalgarno sequence, 12, 302, 307
Shock, 6, 141, 256-257, 313, 370-371, 395, 425
Short interfering RNA (siRNA), 223
Short tandem repeats (STRs), 559
Shotgun cloning, 12, 398, 429
Shotgun sequencing, 416-417, 422, 426, 434, 443
Shrimp, 537
Sibling species, 12
Sickle-cell anemia, 21, 23, 71, 309-310, 313, 316, 322,
    424, 459-462
    preimplantation genetic diagnosis (PGD), 461
Sickle-cell disease, 576
Sickle-cell trait, 310
Signal transduction, 373, 376, 378-380, 386, 442, 489,
    492, 572
Signal transduction pathways, 376, 378, 442
Silencing, 12, 100, 126, 290, 328, 364-366, 369, 382,
    469-470, 548-550, 552-553
    gene expression, 12, 126, 290, 364-366, 369,
        548-549, 552-553
    transcription, 12, 290, 328, 364-366, 369, 548-550,
        552-553
Silent mutation, 321, 387-388
Single-celled organisms, 3, 34
    first, 34
Single-nucleotide polymorphisms (SNPs), 421-423,
    562
Single-stranded RNA virus, 229
siRNA, 223, 290, 365-366, 470
Sister chromatids, 2-4, 9, 12, 32-33, 35, 37-41, 44,
    46-48, 141, 159, 176-177, 180, 184,
    234-235, 251-252, 259, 262, 332-333
    nondisjunction of, 48
Site-directed mutagenesis, 12
Size, 2-5, 7, 10, 12, 27, 33-34, 63, 66, 87, 98, 100,
    102, 105, 107, 120, 130, 132, 137-138,
    141-142, 151-152, 170, 173, 184, 188, 195,
    199, 223-226, 228-229, 252, 255-258,
    265-266, 268-270, 285, 287-288, 300, 303,
    305-306, 308-310, 339, 370, 378, 394, 396,
    403-405, 409, 411, 413, 415, 418, 422,
    429-431, 443-446, 462, 472, 476, 480, 483,
    500, 503-505, 509, 512-513, 517-518, 523,
    529, 531, 534-535, 540, 543, 558, 561-563,
    571
    of genomes, 415, 429, 445
    population, 2-5, 10, 12, 33, 130, 138, 152, 195,
        310, 339, 413, 503-505, 509, 517, 523,
        529, 531, 534-535, 540, 543, 558,
        562-563

prokaryote, 446
Skepticism, 101, 453, 469
Skin, 23, 34, 88, 96, 103, 120, 138, 156, 177, 208,
    250, 313, 331, 333, 341-342, 376-377, 401,
    441, 457, 494-495, 515, 518, 534, 540
    cancer, 23, 138, 341-342, 376-377, 518
    diagram of, 23, 156
    function of, 250, 313, 341-342
    human, 23, 34, 96, 103, 120, 138, 156, 177, 250,
        313, 331, 333, 341-342, 495, 534, 540
    lesions, 103, 331, 376
    mammalian, 333
    structure, 208, 313, 376, 518
    structure of, 208
Skin cancer, 341-342, 376
Skin color, 515, 518
Skin lesions, 331
Skull, 486, 532
Slime, 34, 496
Slime mold, 34, 496
    cellular, 496
SmaI, 395
Small interfering RNAs (siRNAs), 365, 470
Small intestine, 202-203
    enterotoxin, 202-203
Small nuclear RNA (snRNA), 12, 223
Smallpox, 227, 410
Smallpox virus, 227, 410
Smell, 517
Smith, Hamilton, 392
Smoking, 383-384, 387, 552
Smooth endoplasmic reticulum, 30
Snakes, 129
Snapdragons, 59, 82-83
Snapping shrimp, 537
Society, 15-16, 26, 62, 247, 413, 422, 471, 546, 577
Sodium, 426, 438
Soil, 396, 434, 454, 507
    layers, 434
Solute, 576
Solution, 46, 56, 61-62, 69, 72-73, 85-86, 92, 96,
    106-107, 119, 123, 126, 131, 137, 143, 149,
    154, 161, 167, 173, 179-180, 183, 191,
    196-197, 203-205, 214, 216, 222-223, 225,
    227, 234, 241, 246, 250, 256, 258, 262, 268,
    275, 278, 282, 291, 302, 310, 315, 322, 324,
    327, 333, 339, 349-351, 357, 362, 367-368,
    377, 381, 383, 386, 397, 399-400, 403-405,
    410, 435, 438, 442, 444-445, 453, 460, 462,
    470, 474, 484, 489, 493, 496, 503, 505, 511,
    515, 526, 528-529, 536, 542-543
    defined, 511, 536
Solutions, 4, 28, 48, 76-77, 111, 132-133, 156, 183,
    205, 227, 229, 252, 270, 294, 317, 341, 370,
    388, 412-413, 446, 450, 475-476, 498, 518,
    545
    saline, 476
Somatic cell, 7, 12, 47, 63, 125, 320, 331, 339, 373,
    475, 561
Somatic cells, 8, 32, 34, 47, 119, 121, 125, 127, 141,
    176, 214, 229, 245, 247, 249, 258, 265, 320,
    335, 337, 373-374, 381-382, 475, 482, 493,
    495, 510, 547, 550, 561
SOS response, 12
Sound, 16, 88, 153, 308, 428
Sound waves, 88
Southern blot, 4, 404, 406, 411, 460-461, 558, 560
Southern blotting, 1, 12, 404-405
Southern, Edwin, 12, 404
Soybean, 22
Soybeans, 22, 451, 453-454
Specialized transduction, 200, 204
Speciation, 1, 11, 13, 187, 428, 431, 522, 524,
    536-538, 543-544
    allopatric, 1
    macroevolution and, 524, 537
    sympatric, 13
Species, 1, 3-12, 14, 16-18, 22, 24-25, 27-28, 32-33,
    38, 42-43, 47, 63-65, 83, 94, 99, 116-117,
    119-120, 123, 127, 129, 131, 133, 135, 137,
    140-143, 147-149, 154-155, 159, 161, 178,
    185, 187, 194-195, 200, 202, 214, 226, 249,
    257-258, 263-266, 279, 281, 288, 300, 320,
    324, 326, 356, 367, 392, 396, 399, 403, 408,
    413, 417-418, 420-421, 423-424, 426,
    428-432, 434, 442, 444, 446-447, 455, 459,
    471, 476-477, 481, 487-488, 498, 503, 513,
    516-517, 522-524, 529, 533, 535-546, 555,
    558, 562, 565

defined, 11, 17, 22, 24, 148, 178, 266, 320, 432,
    523, 535-536
edge, 356
eukaryotic, 5-7, 9, 11-12, 14, 38, 116, 194, 202,
    214, 257-258, 263-266, 281, 356, 392,
    399, 413, 420, 428-429, 434, 446, 476,
    538
fusion of, 3, 12, 14, 18, 33, 481
homologous genes in, 421
human species, 5, 65
introduced, 5-6, 8, 22, 127, 154, 194, 202, 249,
    326, 420, 476
morphology and, 127
niches, 529, 536
origin of, 14, 17, 47, 141, 143, 147, 155, 187, 300,
    522, 537, 562
prokaryotic, 9-10, 12, 257, 392, 428-429
range of, 12, 24, 33, 116-117, 123, 320, 326, 392,
    403, 418, 429-430, 447, 459, 471, 481,
    487, 503, 516, 533, 538, 542
vertebrate, 25, 257, 428, 431, 540
viral, 99, 148, 200, 202, 214, 320, 423, 455, 476
Species concept, 536
Specific transcription factors, 366
Specificity, 84, 209, 221, 237, 284, 290, 336, 360, 370,
    498
Spectral karyotype, 13, 374, 411
Spectrophotometer, 229
Speech, 432
Sperm, 6, 16-18, 23, 33, 43-44, 46-47, 54, 100,
    117-120, 124, 127, 130, 133, 141-142, 157,
    175, 214, 229, 432, 462, 481, 492-493, 525,
    561, 563
    components of, 492
    conception and, 124
    development, 16-17, 23, 43, 100, 118-120, 124,
        127, 133, 432, 481, 492-493
    fertilization and, 54
    meiosis and, 43-44, 47, 117
    spermatogenesis and, 47
Sperm sorting, 130
Spermatids, 42-43, 47
Spermatocytes, 42-43, 47-48, 122, 259, 270
Spermatogenesis, 42-43, 45, 47, 119, 155
Spheroplast, 11
spheroplasts, 214
Spiegelman, Sol, 215
Spina bifida, 513
Spinal cord, 23, 431, 495
    development of, 431, 495
    diagram of, 23
Spinal cord injuries, 495
Spindle, 2-3, 31-32, 35-39, 141, 149, 234, 265, 378
Spleen, 475
Splice site, 289, 419-420
Spliceosome, 13, 288-289
Spliceosomes, 288
Splicing, 1, 10, 13, 128, 285-289, 293, 295, 327,
    336-337, 344, 355, 362-364, 367, 370, 392,
    410, 419, 422-423, 437-438, 447, 498, 576
Splicing of mRNA, 362
Splicing, RNA, 128, 363, 367
Spongy bone, 441
Spontaneous abortion, 140
Spontaneous generation, 17
Spontaneous mutation, 13, 188, 320, 322, 329, 339
Spontaneous mutations, 320, 322-324, 339
Spore, 1, 13, 29, 39-40, 47, 496
    slime mold, 496
Spores, 13, 29, 33, 38, 43, 308
    asexual, 308
    bacterial, 13
    meiosis and, 43
    reproductive, 43
    sexual, 29, 38, 43
Sporogenesis, 8
SpoT, 400, 435-436, 439-441, 462, 506
SRY gene, 123, 132
Stability, 11, 28, 221, 245, 265, 292, 294, 300, 327,
    344, 354, 361-364, 366, 376, 437, 447, 475
    population, 28
Stabilizing selection, 13, 531
Stahl, Franklin, 232
Stain, 130, 151, 176, 263, 438, 572
    Coomassie, 438
Staining, 1-2, 5-6, 124, 176, 263, 376, 403, 438, 572
    differential, 2, 176, 263
    special, 572
Stains, 124, 263

differential, 263
Stalk, 118
Stamens, 117-118, 489-491
Standard error, 503-506, 517
Staphylococcus, 393, 402, 456-457
    S. aureus, 457
Staphylococcus aureus, 393, 402, 456-457
    cell wall, 457
    growth of, 456
Starch, 59
Start codon, 3, 302, 348, 419, 446
Start point, 285
Statins, 576
Statistical tests, 9, 66
Stem cell, 375, 494-495, 553, 555
Stem cell lines, 495
Stem cells, 2, 4, 7, 249, 375, 455, 494-495, 553-554
    adult, 4, 249, 375, 494-495, 553-554
    research, 4, 249, 494-495, 554
    types of, 494-495
Stem-loop structure, 365
    DNA, 365
    mRNA, 365
Stems, 54-55, 63, 137, 244, 300, 413
Sterility, 121, 143, 154, 156, 336, 471
Sterilization, 153
    defined, 153
Steroid, 96, 359-360
Steroid hormone, 360
Steroid hormones, 359
Steroids, 129
Stigma, 117-118, 535
Stimuli, 198, 548, 555
Stimulus, 548
    environmental, 548
Stock, 131, 154, 157, 181-182, 413, 476, 498
Stomach, 152, 326, 537, 553, 573
Stop codon, 321, 328
Stop codons, 3, 291, 304, 363, 419, 446
Strains, 1, 4, 7-8, 11, 15, 22, 24, 53-54, 75, 86, 90, 92,
    102, 137, 142, 187-191, 193, 195, 199,
    201-204, 209, 242, 250-251, 308, 315-316,
    328, 333-336, 340, 369, 371, 396, 452,
    455-456, 459, 464, 474, 492, 497, 506-508,
    513, 515-517, 524, 526, 531, 543, 554
Streptococcus, 209
    characteristics, 209
Streptococcus pneumoniae, 209
    avirulent strains, 209
Streptomycin, 194, 316
    resistance, 194
Stress, 249, 364, 370, 380
Stroke, 16, 443, 451
Strokes, 574
Structural genes, 4-5, 9-10, 345-353, 358, 367-371
Structural proteins, 313, 423
Structure and function, 26, 269, 306, 314-315, 317,
    338, 359, 396, 403, 418, 437, 442
    molecular, 26, 269, 306, 315, 403
    of DNA, 26, 269, 396, 403, 418
Stylonychia, 280-281
Subcloning, 409
Substitutions, 7, 292, 294, 310, 316-317, 321, 333,
    374, 523, 540, 543, 545
Substrate, 4, 7, 81, 84, 237, 244, 282, 328, 344-345,
    442
substrate concentration, 7
Substrates, 265, 282, 345-346, 365
Subunit vaccine, 452, 465
Subunit vaccines, 452
Succession, 16
Sucrose, 59, 299
Sugar, 4, 9, 12, 20, 84, 216, 220-221, 223, 228-229,
    252, 282, 323, 330, 347, 424, 519
    biosynthesis, 9
    metabolism, 347, 424
    structure, 4, 9, 12, 20, 216, 220-221, 223, 228-229,
        323
Sugars, 84, 186, 216-217, 221, 227-228, 286, 351,
    354
    deoxyribose, 216-217, 228
    simple, 186
Sulfate, 438
Sulfonamide, 194
    resistance, 194
Sulfur, 212, 302, 346
Sunlight, 331
Suppressor mutation, 13
Surface, 10, 31, 83-84, 124, 131, 133, 186, 195-196,

201, 209, 234, 333, 378, 446-447, 451-452,
    474, 482, 496, 571-572
Surface area, 31
Surgery, 374, 385, 486, 573-574
Survival, 5, 17, 85, 138, 147, 149, 155-156, 187, 322,
    328, 337, 464, 509, 524-525, 529-531, 572
susceptibility, 6, 15, 82, 148, 152, 175, 374, 382-383,
    385, 477, 511, 517, 523, 526, 528, 555, 570,
    577
Swab, 471, 563
Sweden, 177
Swimming, 119
Symbionts, 446-447
Symmetry, 393, 482
    body, 482
Sympatric speciation, 13
Synapse, 7, 41, 46, 63, 94, 122-123, 139, 142, 149,
    176
Synapses, 132
Synapsis, 13, 39, 47-48, 133, 144-146, 148-150,
    154-155
Synaptic, 150, 155
syncytium, 482
syndrome, 7, 13, 23, 25, 71, 93-94, 96, 99-100,
    120-121, 123, 125, 127, 131-132, 136-140,
    144-145, 147, 150-153, 155-157, 176-177,
    268, 323, 327-328, 376, 381-382, 425, 443,
    464-465, 474, 511, 526, 544, 551, 553, 574,
    576
Syndromes, 116, 120-121, 124-125, 136, 138, 140,
    155, 249, 332, 382
Synergids, 117-118
Synthesis, 2, 4-5, 9-13, 20-22, 31, 34-35, 59, 81, 85,
    104, 111, 125, 129, 146, 189, 195-196, 200,
    208-209, 211, 215-216, 222, 227, 231-233,
    235-247, 250-252, 259, 265-266, 270,
    272-276, 278-285, 290-294, 298, 300, 302,
    304-305, 307-309, 315, 317, 321, 329,
    331-335, 340, 342, 344-346, 351, 354,
    362-363, 368-371, 378, 399, 407, 429, 435,
    450, 455, 458, 496, 555, 574
Synthetases, 301-302, 370
Synthetic biology, 13, 449, 458-459, 472
synthetic DNA, 222, 459, 461, 477, 567
Systems, 2, 13, 21, 28, 55, 82, 117, 129, 174, 216,
    220, 229, 243, 249, 280, 309, 314, 320, 328,
    330, 333, 341-342, 345, 396, 408, 415,
    442-443, 447, 468-469, 474, 480-481, 489,
    492, 496, 513, 528, 535, 560, 575, 578
Systems biology, 13, 415, 442-443, 447, 480, 513

T
T cells, 467-468
    development, 468
    development of, 468
T2 phage, 256
T4 phage, 196
Taq DNA polymerase, 561
Taq polymerase, 401
Target cells, 572
Target tissue, 470, 475-476
Taste, 455, 526
TATA box, 6, 13, 283, 285, 358, 360-361, 420
TATA boxes, 285, 360
Tatum, Edward, 187, 308
Taxa, 7
Taxon, 8
Tay-Sachs disease, 23, 65, 71, 82, 99, 322, 542
Technology, 6, 11, 15-16, 22, 24-26, 174, 202, 215,
    226, 237, 247, 259, 280, 290, 366, 391-413,
    415-416, 418, 427, 439, 444, 449-455, 457,
    459, 470, 472-475, 494, 498, 558, 567, 572,
    577
Teeth, 544, 566
    mammalian, 544
Telomerase, 13, 145, 223, 231, 246-247, 249-252,
    256, 265
Telomere, 13, 245-247, 249-250, 252, 256, 265
Telomeres, 14, 26, 145, 183, 223, 231, 245, 247,
    249-250, 252, 264-265, 337, 396, 422, 549
Telomeric DNA, 246-247, 249, 265
Telophase, 35-41, 43-44, 47
    in meiosis, 39-40, 47
    in mitosis, 38-40, 47
    meiosis, 35-41, 43-44, 47
    mitosis, 35-41, 43-44, 47
    of meiosis, 38, 40-41, 43-44, 47
    of mitosis, 35, 37, 39, 44, 47
Telophase I, 39-40, 43

Telophase II, 40-41
Temperate bacteriophage, 8
Temperate phages, 198-199
Temperature, 8, 13, 21, 98-99, 101, 116, 128-129, 132,
    134, 223-224, 229, 242, 250-251, 322, 401,
    411-413, 446, 554
    regulation, 8, 13, 129, 223, 322
Temperature range, 13
Temperature-sensitive mutant, 251
Template strand, 237-238, 240-241, 272-273,
    282-284, 286, 290, 323, 328, 407
Tendons, 441
Tension, 239, 242, 309
Teratology, 492
Termination, 3, 6, 9, 236, 278-281, 284, 298, 302,
    304-305, 307, 315-317, 321, 336-337, 342,
    349, 353-354, 363, 370, 406-407, 411,
    419-420, 446, 454, 496
    transcription, 278-281, 284, 298, 302, 305, 316,
        336-337, 349, 353-354, 370, 406,
        419-420, 496
    translation, 6, 9, 278-281, 298, 302, 304-305, 307,
        315-317, 321, 337, 349, 353, 363, 370,
        420, 496
Termination codons, 278, 280-281, 302, 304, 321
terminology, 55, 131, 136, 155-156, 177
Tertiary protein structure, 316
Tertiary structure, 312
    proteins, 312
Test, 1-3, 5, 9, 23, 66, 72, 74, 76, 78, 106, 108, 112,
    132, 153, 170, 180, 183, 202, 209, 215, 220,
    269, 276, 333-334, 340, 385, 387-388, 458,
    460, 462, 466, 471-472, 474-476, 496, 501,
    507, 517, 519, 523, 527, 566, 572-573, 577
Testcrosses, 74
Testes, 42, 118, 120, 122-123, 129, 132-133
Testis, 13, 122-123, 132-133
Testis-determining factor, 13, 123, 133
Testosterone, 129, 132
Tests, 5-6, 9, 27, 66, 68, 72, 105, 130, 153-155, 308,
    333, 338, 340, 367, 385, 387-388, 411, 424,
    426, 459, 461-462, 465, 471-472, 475-476,
    570, 572, 574-576, 578
Tetanus, 203
    control, 203
    symptoms, 203
    vaccine, 203
Tetracycline, 194, 316, 397, 412, 458
    production, 397
    structure, 316
    synthesis, 458
Tetracycline resistance, 397, 412, 458
Tetrads, 39, 41, 43, 47-48, 155, 165, 182, 247
    chromosomal, 48, 155, 165, 182
Tetrahymena, 245-247, 265, 280-281, 288
Tetraploids, 141
Thalassemia, 291, 340, 460, 469, 475
The Innocence Project, 566-567
Theories, 62
Theory, 2, 13, 16-19, 26, 28, 31, 62-63, 72, 95, 128,
    141, 145-146, 153, 159-161, 165, 209, 229,
    301, 345, 349, 366, 416, 441, 467, 522, 546,
    555
Thermal energy, 306
Thermocyclers, 401
Thermoplasma, 429
Thermus, 298, 306, 401
Thermus aquaticus, 401
Thermus thermophilus, 298, 306
Thiamine, 188, 308
Thiamine (thi), 188
Threatened species, 565
Threonine, 188, 275-280, 291-295, 353
    genetic code, 275-280, 291-295
    structure, 188, 291-292, 353
    synthesis, 275-276, 278-280, 291-294
Threonine (thr), 188
Threshold, 86, 128, 152, 466, 517, 519, 555
Throat, 443
Thrombocytes, 438
Thymidine, 176, 234-235, 251, 258, 327, 339
Thymine, 13, 20, 216-217, 220, 223, 228-229, 283,
    323, 325-327, 329-331, 341, 376
    in DNA, 20, 216-217, 223, 228, 325-326, 330-331
Thymine dimers, 329, 376
Thymine (T), 228
Thymus, 237
Ti plasmid, 397, 454
Time, 3, 5, 8, 10-11, 13, 16-17, 19, 24-28, 35, 43,

45-48, 53, 58, 63, 65, 68, 85-86, 99, 105,
110, 117, 120, 130-131, 133, 147, 153,
165-166, 183, 185, 190-191, 196-197, 203,
208, 212, 222, 224, 229, 233, 235-237, 242,
244, 249, 251-252, 260-262, 272-273, 275,
281, 284, 290-291, 303, 305-308, 328, 333,
344, 351, 358, 360, 368, 374, 376-379, 385,
388, 396, 398, 400-401, 403, 407-410, 413,
415-416, 422, 428, 431, 438, 444-447, 470,
472-474, 480, 483, 489, 495, 510, 513-514,
522-524, 529-532, 536-542, 544, 553-554,
558, 560, 564, 566, 571, 578
Tissue, 7, 11, 14, 21, 23, 93, 100, 117, 120, 122-124,
133, 152, 210, 215, 251-252, 287, 295,
313-314, 320, 331, 336, 356, 358-360, 363,
366-367, 370, 375, 377-378, 382, 384, 398,
404, 413, 424, 427, 435-437, 439-441, 443,
445-446, 451, 455, 457, 459-460, 462-463,
469-470, 474-476, 494, 497-498, 547,
552-553, 558-559, 564, 571-572, 576
Tissue plasminogen activator, 451
Tissue repair, 384
Tissues, 1, 4, 23, 34, 93, 99-100, 117, 122, 125, 202,
249-250, 258, 293, 295, 307, 309, 320, 326,
359, 362, 366-367, 373-374, 377, 382, 385,
398, 403, 409, 426, 434-435, 437-439,
449-450, 462-463, 465, 469-470, 475, 491,
493-495, 497, 511, 548, 551
    animal, 117, 202, 366, 470, 491
    defined, 320, 374, 385, 494, 511
    plant, 34, 117, 202, 491
    types of, 366-367, 374, 398, 437, 462, 475,
        494-495
Tm, 8, 223-224, 228-229, 369, 412
Tobacco, 34, 137, 215, 256, 325, 383-384, 451,
453-454
Tobacco mosaic virus (TMV), 215
Tobacco smoke, 325, 384
Tolerance, 455, 514
Tomatoes, 75, 453-454, 505, 512-513
Tongue, 138
Tools, 45, 105, 202, 274, 314-315, 334, 336, 338, 393,
404, 409, 418, 428, 441, 453, 464, 472, 503,
541, 558, 566, 576
Topoisomerases, 239
Tortoiseshell cats, 126
Totipotent, 13, 494
Touch, 43, 578
Toxicity, 94, 470, 474, 476, 573
    selective, 476
Toxin, 7, 195, 202, 474, 543
Toxins, 202, 325, 424, 474
    cholera, 202
    enterotoxins, 202
    fungal, 325
Trait, 2-6, 9-13, 17, 19, 33, 53-56, 61, 63, 69-71,
73-75, 78, 81-82, 88, 93-98, 101-103,
105-109, 111-112, 132, 137, 151, 155, 163,
169, 175, 201, 227, 268, 307, 310, 313, 319,
321, 324, 386, 456, 470, 475, 486, 500,
502-519, 528, 531-532, 543-545, 547
Traits, 1, 3, 5, 8-9, 15-19, 22, 28, 52-55, 57-59, 61-62,
65-66, 69-75, 80-82, 88, 95-96, 99, 101, 103,
135, 143, 153-154, 161, 163, 165, 167,
169-170, 173, 179-182, 203, 207, 211, 227,
307-308, 449, 453-455, 469-471, 499-519,
522, 525, 531, 541, 544, 547, 555, 570
    and pedigree analysis, 74
    inheritance of, 19, 52, 55, 61, 69, 71, 75, 80, 101,
        307, 502, 516-517, 522, 544
    polygenic, 500, 502-503, 508, 511, 515-519
    quantitative, 1, 3, 17, 54, 499-519
    recessive, 1, 9, 54-55, 57, 59, 61, 65-66, 69-75,
        80-82, 88, 95-96, 99, 154, 161, 167, 170,
        173, 179-182, 307, 525, 544
Trans configuration, 13
Transacetylase, 346
Transcription, 1-2, 4-5, 7, 10-13, 20, 28, 44, 81, 123,
127, 129, 133, 202, 208, 215, 223, 238, 246,
258-259, 269, 271-295, 298, 300, 302,
305-306, 314, 316, 322, 325, 327-328,
336-337, 344-362, 364-372, 377-381, 384,
386, 388, 403, 406, 419-420, 423, 428-429,
432, 436-437, 451, 456-457, 480, 483-489,
491-492, 495-497, 513, 547-550, 552-554,
571
    cell-type specific, 547
    control of, 2, 7, 81, 345-346, 350, 362, 369, 377,
        379, 384, 388, 492, 495, 552-553

direction, 4, 7, 20, 215, 238, 274-275, 282, 284,
    294, 300, 306, 328, 357, 367, 497, 513
elongation, 4, 283-284, 298, 302, 305-306, 361
enzymes required, 345
eukaryotic, 2, 5, 7, 11-13, 20, 202, 215, 246, 258,
    269, 280-281, 283-287, 291-292, 294,
    306, 314, 316, 322, 344-345, 354-360,
    362, 365, 368, 370, 419-420, 428-429,
    451, 491
in Archaea, 354
in bacteria, 11, 202, 208, 272, 279-280, 282-285,
    288, 292, 300, 302, 306, 316, 328, 345,
    354, 368, 429
in bacterial cells, 5, 274, 429
in eukaryotes, 2, 5, 12-13, 215, 223, 246, 272, 280,
    284-286, 288-289, 291, 294, 300, 306,
    354-355, 362, 369, 429, 480
in eukaryotic cells, 20, 285, 306
initiation, 7, 10-12, 238, 246, 277-285, 294, 298,
    302, 305-306, 344, 349, 353-354,
    357-362, 366, 369-371, 388, 419-420,
    486, 491, 552
initiation of, 10-12, 238, 246, 279, 281-285, 302,
    306, 353-354, 357, 361, 366, 369, 371,
    486
prokaryotic, 10, 12, 272, 284-286, 292, 294,
    305-306, 344-345, 353-355, 358, 368,
    419, 428-429
regulation, 2, 13, 81, 129, 223, 282-284, 306, 322,
    337, 344-362, 364-372, 377-378, 380,
    384, 386, 403, 406, 420, 423, 437, 489,
    496, 547-549, 552
regulation of, 13, 81, 129, 284, 306, 322, 344-362,
    364-372, 378, 384, 403, 420, 489, 547,
    549, 552
reverse, 2, 5, 11, 215, 238, 246, 328, 348, 388,
    403, 436, 547, 554
stages of, 44, 283, 305, 381, 489, 496, 547-548
termination, 278-281, 284, 298, 302, 305, 316,
    336-337, 349, 353-354, 370, 406,
    419-420, 496
termination of, 284, 298, 302, 305, 316, 336-337,
    349, 354, 370
translation and, 298, 300, 302, 305-306, 314, 316,
    355, 362, 364, 366
unit of, 2, 5, 11-12, 305
Transcription activators, 356, 361
Transcription factories, 356, 370
Transcription factors, 2, 133, 284-285, 314, 356-357,
    359-362, 366, 370-371, 379-381, 388, 423,
    432, 436-437, 483-485, 487, 489, 491, 495,
    513, 547-548
    general, 284, 356, 360-362, 370, 379, 388
    genes for, 359, 370, 432, 436
Transcription terminator, 354
Transcription units, 429
Transcriptional control, 355, 362, 498
    negative, 498
Transcriptional regulation, 369, 547
Transcriptome, 13, 415, 434-438, 447, 465
Transcriptomics, 424, 434, 442
Transducing phages, 200
Transduction, 1, 8, 14, 185, 187, 196, 199-200, 204,
    373, 376, 378-380, 386, 442, 489, 492, 572
    generalized, 200, 204
    specialized, 200, 204
Transfection, 213-215
Transfer of genetic material, 3, 188
Transfer RNA, 12, 14, 21, 208, 223, 298, 300-301,
    307, 316
Transfer RNA (tRNA), 14, 21, 208, 223, 298
    activation, 21
    cellular, 21, 208, 223
    in translation, 208, 298
    structure, 14, 208, 223, 298
    structure of, 14, 208, 223, 298
    translation, 14, 21, 208, 223, 298
Transferase, 84, 96, 99, 303-304, 316-317, 423, 573
Transferrin, 439
transformation, 3, 8, 10, 14, 22, 185, 187, 195-196,
    199-200, 202, 204-205, 209-211, 214, 227,
    229, 334, 336, 383, 394-397, 475, 489, 496,
    552-553
    in bacteria, 3, 185, 187, 195-196, 199-200, 202,
        204-205, 209, 214, 227, 229, 334, 396
    naturally occurring, 334, 394
Transgene, 455-457, 474, 513
Transgenes, 336
Transgenic animal, 451

Transgenic crops, 22, 453-455
    DNA technology and, 22
Transgenic organism, 14
Transgenic organisms, 22, 25
    animals, 22, 25
    plants, 22, 25
Transgenic plant, 202, 476
Transgenic plants, 28, 203, 452, 454, 471, 474
Transitions, 260, 269, 325, 363, 432
Translation, 5-6, 9-10, 12, 14, 20-21, 28, 103, 128,
    208, 223, 273-274, 276-281, 285, 289-292,
    295, 297-317, 321, 337, 341, 344, 346,
    348-349, 351, 353, 355, 362-366, 370, 420,
    428, 470, 489, 496, 498
    comparison of, 281, 299, 310-311, 349
    control of, 103, 315, 346, 362-363
    elongation, 298, 302-307, 317
    eukaryotic, 5-6, 9, 12, 14, 20, 280-281, 285,
        291-292, 299, 306-307, 313-314, 316,
        344, 355, 362, 365, 370, 420, 428
    genetic code and, 273-274, 276-281, 285, 289-292,
        295
    in bacterial cells, 5, 274
    in eukaryotes, 5, 9, 12, 223, 280, 285, 289, 291,
        299-300, 306-307, 355, 362
    in eukaryotic cells, 9, 20, 285, 306, 313
    initiation, 6, 9-10, 12, 277-281, 285, 298, 302-307,
        315, 317, 344, 349, 353, 362, 366, 370,
        420
    initiation of, 10, 12, 279, 281, 285, 302-303,
        306-307, 315, 353, 366
    mitochondrial proteins, 280
    prokaryotic, 9-10, 12, 285, 292, 299, 305-307, 344,
        353, 355, 428
    ribosomes in, 353
    site of, 5-6, 9, 12, 302, 314, 351
    stages of, 305, 489, 496
    steps, 20, 285, 301-304, 306, 315-316, 344, 362,
        365
    termination of, 6, 9, 298, 302, 305, 307, 316, 321,
        337, 349, 370
    transcription and, 5, 28, 208, 223, 306, 316, 496
Translation process, 302, 306
Translational control, 364, 496, 498
Translocations, 143, 149-150, 156, 326, 337, 374,
    376, 380, 387, 405, 433, 552
Transmembrane protein, 425, 492
Transmembrane proteins, 424, 432
Transmission, 5-10, 12, 14, 15, 17-18, 26, 29-30, 33,
    45, 47, 52, 54, 63, 66, 70, 72, 80, 86, 96,
    101, 103-104, 106, 116, 135, 152, 159, 161,
    177, 209, 231, 254, 319-320
transmission electron micrograph, 45, 231
transplant rejection, 495
Transplantation, 458-459, 495, 498
transplants, 470
Transport, 102-103, 195, 285, 313-314, 355, 365, 432,
    543
Transport protein, 432, 543
Transport proteins, 314
Transposable element, 14, 336, 554
Transposable elements, 59, 319, 334-337, 339, 422,
    429
Transposase, 334-336
transposition, 266, 270, 319-342
    discovery, 335, 341
    frequency of, 333, 338
    mechanism, 266, 270, 328, 331, 333
    replicative, 330
Transpositions, 337
Transposon, 337, 340-341, 554-555
Transposons, 334-337, 341-342, 365, 431
    as vectors, 336
    complex, 365
Travel, 470
trees, 537-539, 546
Triose phosphate, 439
tripeptide, 304, 311
Triphosphates, 216, 218, 236-237, 243, 252, 282, 401,
    406
Triplet code, 272-273, 275, 292, 294, 301
Triploidy, 14, 136
Trisomy 21, 137-138, 150-151, 157
tRNA, 1, 14, 20-21, 208, 223, 276-277, 279-281, 285,
    288, 292, 298-307, 315-317, 564
Tropomyosin, 439
Troponin, 442
Trout, 214
Trp operon, 352-354

Trypsin, 147, 263, 440-441
Tryptophan, 104, 199, 279-281, 292, 316, 345, 351-353, 368-370
  genetic code, 279-281, 292
  structure, 292, 316, 353, 368
  synthesis, 104, 279-281, 292, 345, 351, 368-370
Tryptophan (Trp), 351
Tryptophan (trp) operon, 351
Tuberculosis, 429, 452
  control, 429
  treatment, 452
Tubulin, 31-32, 45, 313, 425, 432
Tumor, 2, 4, 8, 14, 23, 34, 38, 152, 216, 249, 337, 370, 373-376, 379-384, 386-388, 397, 411, 413, 439-440, 442, 462-463, 476, 552-554, 571, 573, 575-576
tumor cells, 249, 382, 384, 571
Tumor suppressor genes, 370
Tumor viruses, 216
Tumors, 8, 23, 138, 152, 177, 320, 329, 342, 373-375, 377, 380-382, 386-388, 396, 413, 552, 571, 573, 577
Tumor-suppressor genes, 14, 373, 379-384, 386-388, 552-554
Tuna, 538-539
Turner syndrome, 120-121, 127, 131-132, 137, 140, 156
Turnover, 355, 546
Turtles, 129, 132, 134
Tutu, Desmond, 427
Twin studies, 3-4, 500, 509-511, 517
Twins, 3-4, 8, 65, 69, 132, 178, 413, 447, 474, 509-511, 515-516, 518, 547, 559, 565
Type 1 diabetes, 495
Type 2 diabetes, 175, 446, 466, 477, 577
Tyrannosaurus rex, 441
Tyrosine, 277-278, 291, 309, 311, 328, 380, 388, 492, 571
  codon, 277-278, 291, 328, 388
  genetic code, 277-278, 291
  structure, 291, 309, 311
  synthesis, 278, 291, 309

U

Ubiquitin, 313, 364, 439
Ultrasound, 130, 338, 460
Ultrastructure, 30
Ultraviolet light, 5, 10, 13, 176, 198, 228-229, 238, 325-326, 370, 376, 380, 384, 392, 403, 457
Ultraviolet radiation, 320
Ultraviolet (UV) light, 214, 325, 440
  mutagenic, 214, 325
Ultraviolet (UV) radiation, 326
United States, 15, 22-23, 28, 124, 138, 153, 287, 300, 310, 373, 385, 411, 421-422, 450, 453-454, 456-457, 467, 470-471, 474, 523, 534, 563, 566, 571, 573, 578
Units of measurement, 504
Unity, 128, 220
Universal code, 280
Unlinked genes, 195
  mapping, 195
  recombination of, 195
Untranslated regions (UTRs), 420, 446
Uracil, 12, 20, 216-217, 223, 228-229, 282, 294, 323-325, 330, 364
Ureter, 140
Uric acid, 99
Urine, 93, 307
  formation, 307
  formation of, 307
  proteins in, 307
U.S. Department of Energy, 578
U.S. Patent and Trademark Office, 472
Uterus, 122, 126, 130, 455, 462, 493
  cancer, 462

V

Vaccination, 203
Vaccine, 202-203, 451-453, 465, 474, 476
  anthrax, 203, 465
  DNA, 202, 452-453, 465, 474, 476
  edible, 202-203, 451-452, 474, 476
  genetically engineered, 202, 451, 453
  HPV, 452
  influenza, 452, 465
  production, 202-203, 452-453, 474
  recombinant, 202, 451-453, 474

  subunit, 202-203, 452, 465
  vector, 451-452, 476
Vaccines, 202-203, 431, 452-453
  against bacterial diseases, 203
  attenuated, 452
  chickenpox, 452
  DNA, 202, 431, 452-453
  DNA-based, 452-453
  edible, 202-203, 452
  first, 202-203, 431, 452-453
  hepatitis B, 452
  HPV, 452
  influenza, 452
  measles, 203
  oral, 203
  plant, 202, 453
  polio, 431
  production of, 202-203, 452
  rabies, 203, 452
  recombinant, 202, 452-453
  safety of, 453
  subunit, 202-203, 452
  tetanus, 203
  tuberculosis, 452
  types of, 452
Vagina, 122
Valine, 21, 278-280, 292, 294, 310-311, 317
  genetic code, 278-280, 292, 294, 310
  structure, 292, 310-311, 317
  synthesis, 21, 278-280, 292, 294, 317
Variable number tandem repeats, 8, 14, 266
Variables, 98, 229, 412, 446, 525
Variation, 3-4, 6, 9, 11-14, 17, 19, 35, 38-39, 41, 43, 47, 57, 62, 64-65, 81-82, 88, 92, 98, 100, 116-117, 128, 135-157, 159, 175, 186-187, 195, 208, 217, 223, 266-268, 294, 308, 311, 319-320, 340-341, 424, 427, 429, 445, 447, 457, 466-467, 473, 500-510, 512-513, 515-517, 519, 522-524, 527, 532, 536, 542-543, 547, 552, 554-555, 570, 575
Variegation, 14, 101
Variola, 227
variola virus, 227
Vector, 3, 5, 12, 14, 22, 368, 394-400, 406-407, 409-410, 449-452, 454-455, 467-470, 476
  cloning, 3, 12, 14, 22, 394-400, 407, 410, 454
  pathogen, 452
Vectors, 10, 22, 28, 290, 336, 392-398, 401, 410-411, 452, 467-470, 475-476
  bacteriophage, 10
  development of, 22, 392, 394, 468-469, 475
  DNA libraries and, 398
  expression, 10, 22, 290, 336, 392, 396, 452, 467, 469-470, 475-476
  in gene therapy, 468
  in recombinant DNA technology, 392
  Ti, 397
  types of, 396-398, 452, 475
Vegetative cells, 43
Vehicle, 476
Vein, 161, 308, 548
Veins, 161-162
Venter, Craig, 417, 422, 427, 434, 458-459, 472, 474-475
Vertebrates, 34, 129, 147, 208-209, 245, 256, 265, 269, 287, 314, 329, 363, 428, 432, 437, 447, 489, 538-539
  evolution of, 539
  mammals, 269, 428, 538
Vertical gene transfer, 7, 14, 187
Vessels, 21, 93, 441, 470, 497
Veterinary medicine, 476
Vibrio, 202
  V. cholerae, 202
Vibrio cholerae, 202
  chromosomes of, 202
Villi, 383, 459
Vinblastin, 573
Vinegar, 450
viral diseases, 203
Viral genome, 384, 469, 475
viral genomes, 415, 469
viral infections, 290, 366
viral multiplication, 215
Virulence, 209
  evolution, 209
Virulent phages, 198
Virus, 1, 7, 11, 22, 185, 196-197, 199, 204, 211, 214-216, 223, 227, 229, 255-256, 280-281,

290, 365-366, 383-384, 387, 410, 415, 451-452, 454, 464-465, 467-469, 474-476, 510, 526
Archaea, 7, 280
assembly, 196
attachment, 7
bacterial, 1, 11, 22, 185, 196-197, 199, 204, 211, 214-215, 255-256, 280, 410, 474
complex, 11, 185, 196, 215, 281, 290, 365-366, 387, 415, 467, 474
defective, 469
entry, 1, 197, 526
eukaryotic, 7, 11, 214-215, 255-256, 280-281, 365, 451, 476
genetic material, 1, 11, 196, 211, 214-216, 223, 227, 229, 255-256
genome, 1, 7, 11, 22, 204, 215, 255, 280, 365-366, 383-384, 387, 410, 415, 454, 465, 467, 469, 474-476, 510
helical, 227, 229
host, 1, 22, 185, 196-197, 204, 215, 227, 255-256, 383-384, 410, 451-452, 454, 465, 469, 474-476
icosahedral, 196
infection, 185, 196-197, 204, 214-215, 223, 227, 255, 290, 365-366, 383-384, 452, 454, 465, 474, 526
naked, 199, 214
polymorphic, 11, 387
proteins, 11, 22, 204, 211, 214-215, 223, 227, 229, 255-256, 280-281, 365, 384, 387, 415, 451-452, 465, 467, 469, 474-476
release, 11, 216
replication, 1, 7, 11, 196, 215, 223, 227, 229, 255-256, 384, 465
respiratory infections, 468
reverse transcriptase, 11, 215
RNA, 1, 7, 11, 22, 196, 211, 215-216, 223, 227, 229, 255-256, 280-281, 290, 365-366, 384
self-assembly, 196
size, 7, 199, 223, 229, 255-256, 415, 476
temperate, 199
transducing, 1
virulent, 227
Virus resistance, 22
  plants, 22
Viruses, 19, 22, 24, 29, 32, 99, 176, 185-186, 196-198, 201-202, 204-205, 207, 211-216, 223, 227-228, 231, 239-240, 242, 254-256, 260, 269, 273, 280-281, 286, 289-290, 316, 365-366, 373, 380, 383-384, 387, 393, 402, 427-428, 434-435, 450, 452, 454, 464, 468, 475
  animal, 202, 216, 223, 281, 286, 290, 366, 384, 468
  attachment of, 212
  beneficial, 427
  cancer and, 290
  cancer-causing, 373
  characteristics, 24, 207, 228, 255, 273
  characteristics of, 207, 228
  defined, 19, 22, 24
  definition of, 269
  discovery of, 242, 256
  DNA, 19, 22, 24, 29, 32, 176, 185, 196-198, 201-202, 207, 211-216, 223, 227-228, 231, 239-240, 242, 254-256, 260, 269, 273, 280-281, 286, 289-290, 365-366, 373, 380, 383-384, 387, 393, 402, 427-428, 434-435, 450, 452, 454, 464, 475
  emerging, 202, 434
  evolution of, 24
  genomes of, 427-428, 434
  helical, 227
  identification of, 286, 454
  in cancer, 373, 380, 383
  multiplication, 205, 214-215
  nomenclature, 216
  plant, 202, 216, 256, 316, 428, 454
  replication, 32, 176, 196, 198, 202, 215, 223, 227-228, 231, 239-240, 242, 255-256, 269, 384, 402, 428
  replication of, 32, 228, 231
  reproduction and, 197
  satellite, 269
  size of, 223, 255-256, 269
  sizes of, 269

structure, 19, 24, 185, 196, 202, 204, 207, 211-216, 223, 227-228, 231, 254-256, 260, 269, 289, 316, 365, 427
structure of, 19, 196, 207, 216, 223, 227-228, 231, 255, 269, 427
Visible light, 326, 340
Vision, 62, 94-95, 103, 107-108, 133, 469-470
  color blindness, 95, 107-108
  color vision, 107, 133
Vitamin, 308, 322, 354, 455, 517-518, 554, 574
Vitamin A, 455, 517-518
Vitamin B1, 308
vitamin B1 (thiamine), 308
Vitamin B12, 554
Vitamin B6, 308
vitamin B6 (pyridoxine), 308
Vitamin K, 574
Vitamins, 186, 308-309, 455
  antioxidants, 455
Volume, 34, 43, 130, 198, 255, 262, 268, 270, 459, 519
Vulva, 492-494

W
Walking, 2
Water, 34, 59, 105, 154, 202-203, 220-221, 308, 333, 431, 434, 457, 514, 578
  domestic, 514
  forms of, 34, 105
  hydrogen bonding, 221
  molecular structure, 220
  molecule, 308, 333, 578
  properties, 202, 333
  properties of, 333
  structure, 202, 220-221
Water pollution, 333
  chemicals, 333
water treatment, 203
Water-soluble vitamins, 308
Watson, James, 4, 20, 26, 207, 216, 421, 426-427
Wavelengths, 214-215, 229, 326, 339, 405, 408
  wavelengths of light, 408
Wavelengths of radiation, 339
Weight control, 421
Western blotting, 404
Wheat, 16, 34, 143, 428, 453-454, 501-502, 507, 512, 514, 516, 555
White blood cells, 374, 376-377, 382, 459-461, 464, 467-468, 473
White matter, 315
Wieschaus, Eric, 104, 483
Wild-type cells, 188, 369, 458
Wilkins, Maurice, 20, 26, 219
Wine, 449
Wings, 19, 68, 73-75, 77-78, 92-93, 132, 180-183, 322, 480, 486, 492
  seed, 73, 77
Wobble, 14, 275, 279, 281, 292, 294, 301, 306
Wolves, 16, 523
Wood, 196
Work, 17-20, 24, 26, 38, 47, 52, 59, 62-63, 67, 72-74, 81, 86-87, 94-95, 101, 104, 108, 127, 149, 163-165, 169, 172, 175-177, 185, 207, 209-211, 214-216, 219, 222, 234, 236, 241, 249-250, 259, 262-263, 274, 276, 306, 308-310, 315, 340, 362, 365, 369, 409, 417, 422-423, 426-427, 431-432, 436-438, 441, 453, 459, 469, 472, 474-475, 483, 489, 493, 513-514, 516, 522, 536, 548, 555, 564, 570, 572
Worms, 118-119, 431

X
X chromosome, 1-2, 4, 6, 8, 11, 14, 19, 33, 81, 94-96, 98, 107, 111, 116, 119-128, 131-133, 136-137, 146, 151-152, 156, 163-165, 167, 169, 173, 181, 263, 320-321, 356, 375, 426, 493, 550, 560-561
X chromosomes, 1, 8, 33, 94-95, 116, 119-121, 123-128, 131-133, 136, 163, 337, 357, 493, 548
X rays, 11, 176, 218, 222, 308, 320, 325-327, 338, 340, 342, 384, 388
  mutagenic effects of, 340
Xeroderma pigmentosum, 331-332, 340-342, 376
X-gal, 395-397
X-linked genes, 95-96, 124-125, 163, 165, 170, 181
X-linked traits, 70, 95-96

X-ray crystallography, 14, 301, 350
  of DNA, 14
X-rays, 14
XXX syndrome, 121

Y
Y chromosome, 11, 14, 33, 94-95, 107, 119-124, 127-128, 130-133, 136, 167, 263, 412-413, 422, 542, 560-561, 563, 567
Y chromosomes, 33, 116, 119-120, 122-123, 130, 132
Yeast, 1, 14, 24-25, 34, 38, 82, 102-103, 142, 182, 214, 243, 256, 265, 269, 280, 285, 288, 300, 307, 325, 328, 330, 332, 345, 396, 413, 427-430, 450-452, 458, 465, 538-539
  cloning vectors, 396
  genetics, 1, 14, 24-25, 103, 182, 427-429, 450, 538-539
  genome, 1, 14, 24-25, 103, 142, 269, 280, 300, 396, 427-430, 458, 465, 539
  genome of, 269, 427-430, 458
  life cycle, 25, 428
  mating type, 102
  nutritional, 1
  transfection, 214
  wild, 1, 14, 82, 102-103, 182, 458
Yeast artificial chromosomes, 396
Yeast cells, 243, 451
Yeasts, 2, 32
Yellowstone National Park, 401

Z
Zea mays, 34, 117-118, 131, 174-175, 453
Zebrafish, 15, 25, 34, 428, 457, 481, 488, 540
Zinc, 14, 359, 421, 456, 469
Zinc finger, 14
Zinder, Norton, 199
Zygote, 3, 8, 14, 33-34, 43-44, 59, 65, 101, 103, 117-118, 124, 126, 137, 141, 148, 150, 480-482, 484, 493, 510, 524-525, 536, 551, 561
  algae, 34
Zygotes, 12, 34, 47, 68, 127, 136, 451, 462, 524-525, 532, 536
  cleavage, 12